Meet the editor

Dr. Ali Gholamrezanezhad is a fellow of the European Board on Nuclear Medicine (FEBNM), with research experiences at Johns Hopkins University and University Klinikum Bonn (Germany). He is also a Senior Investigator (Research Institute for Nuclear Medicine, Tehran University of Medical Sciences), Editorial Board Member of Rare Tumors, Anatolian Journal of Cardiology, Journal of Thoracic Disease, Eastern Journal of Medicine and Reviewer of Clinical Neurology and Neurosurgery, Iranian Journal of Nuclear Medicine, Anatolian Journal of Cardiology, Plasma Chemistry and Plasma Processing, etc. Dr. Gholamrezanezhad has published 84 peer reviewed papers with experiences on stem cell imaging.

Contents

Preface

Recently, the field of stem cells and its biological and medical applications has attracted a remarkable attention, interest and accordingly funds. These cells are a source of great hope for clinical applications in regenerative medicine, for pharmaceutical industry and even for genetic studies. There are expanding world-wide efforts to discover more about stem cells and to understand how they transform into specialized human tissues. The topic is so expanding that every day the scientific community is bombarded with information on remarkable advances and achievements.

Although because of diversity and complexity of the topic and also limitations of open access publishing, the current book could not cover all aspects of stem cell research, I hope that the readers enjoy the book greatly and find it informative, practical and educational in the presented discussions. The book was broken to three parts: section one as an introduction, section two as an overview on clinical experiences of stem cell researches, and section three for the review on some of the recent advances and also perspectives in stem cell technology.

I would like to convey my respects and thanks to the authors for their efforts and contribution. Collaboration has always been important in academic environment. I am very happy to see that the book and chapters are the result of fruitful international collaboration involving more than 110 leading scientists and investigators from all over the world, from east to west. I hope the stem cell book will encourage continued close cooperation and collaboration among its authors.

Best regards,

Ali Gholamrezanezhad, MD, FEBNM
Research Institute for Nuclear Medicine,
Tehran University of Medical Sciences, Tehran, Iran
The Russell H. Morgan Department of Radiology and Radiological Science,
Johns Hopkins University, Baltimore, US

Part 1

Introduction

1

Stem Cells: General Features and Characteristics

Hongxiang Hui[1,2,5,6], Yongming Tang[2,4],
Min Hu[2,3] and Xiaoning Zhao[2,4]
[1]*Center for Metabolic Diseases, Southern Medical University (SMU), Guangzhou,*
[2]*Institute of Dongguan SMU Metabolic Science, Dongguan*
[3]*Regen Biotech Company, Beijing*
[4]*Cedars-Sinai Medical Center, Los Angeles, CA*
[5]*UCLA Center for Excellence in Pancreatic Diseases, Los Angeles, CA*
[6]*Department of Medicine, VA Greater Los Angeles Health Care System, Los Angeles, CA*
[1,2,3]*PR. China*
[4,5,6]*USA*

1. Introduction

Stem cells are a group of cells in our bodies, with capacity to self-renew and differentiate to various types of cells, thus to construct tissues and organs. In science, it is still a challenge to understand how a fertilized egg to develop germ layers and various types of cells, which further develop to multiple tissues and organs with different biological functions. In the battle to fight against diseases, stem cells present potencies to repair tissues by cell therapy and tissue regeneration. The study of stem cells turns to be a major frontier in 21 century biology and medicine.

There are many types of stem cells, differing in their degree of differentiation and ability to self-renewing. Gametes cells (eggs or sperms) are stem cells they will develop to a whole body with various tissues after fertilizing. Embryonic cells derived from the part of a human embryo or fetus, are stem cells also with full potential to differentiation. Adult stem cells are partially differentiated cells found among specialized (differentiated) cells in a tissue or organ. Based on current researches, adult stem cells appear to have a more restricted ability of producing different cell types and self-renewing compared with embryonic stem cells.

Cancer stem cells are a sub-group of cancer cells that respond the escaping of cancer chemotherapy and the relapse of tumors. This concept has a great impact on the strategy of cancer chemotherapy and anti-cancer drug design. The new understanding of stem cell has been applied to treat leukemia (induced differentiation) and bone/blood cancer (bone marrow transplants) for many years and has achieved great success.

In the medicine applications, the induced pluripotent stem cells (iPS) reveal a special significance, as they can be induced to derive from many adult tissues or organs by treatment of protein factors. Their features can be similar to the natural embryo stem cells. They provide the source for stem cells without an ethnic conflict.

2. Stem cells

Stem cells are certain biological cells found in all multicellular organisms. They are in small portion in body mass, but can divide through mitosis and differentiate into diverse specialized cell types and can self renew to produce more stem cells. Different types of stem cells vary in their degree of plasticity, or developmental versatility. Stem cells can be classified according to their plasticity and sources.

Classification		Characteristics
Sources/types	Embryonic stem cells	are pluripotent stem cells derived from the inner cell mass of the blastocyst, an early-stage embryo.
	Adult stem cells	Endodermal Origin: Pulmonary Epithelial SCs, Gastrointestinal Tract SCs, Pancreatic SCs, Hepatic Oval Cells, Mammary and Prostatic Gland SCs, Ovarian and Testicular SCs
		Mesodermal Origin: Hematopoietic SCs, Mesenchymal Stroma SCs, Mesenchymal SCs, mesenchymal precursor SCs, multipotent adult progenitor cells, bone marrow SCs, Fetal somatic SCs, Unrestricted Somatic SCs, Cardiac SCs, Satellite cells of muscle
		Ectodermal Origin : Neural SCs，Skin SCs，Ocular SCs
	Cancer stem cells	have been identified in almost all caner/tumor, such as Acute Myeloid leukemic SCs ($CD34^+/CD38^-$), Brain tumor SCs ($CD133^+$), Breast cancer SCs ($CD44^+/CD24^-$), Multiple Myeloma SCs ($CD138^+$), Colon cancer SCs ($CD133^+$), Liver cancer SCs ($CD133^+$), Pancreatic cancer SCs ($CD44^+/CD24^+$), Lung cancer SCs ($CD133^+$), Ovary cancer SCs ($CD44^+/CD117^+$), Prostate cancer SCs ($CD133^+/CD44^+$), Melanoma SCs ($CD4^+/CD25^+/FoxP3^+$), Gastric cancer SCs ($CD44^+$).
	Induced pluripotent stem cells	a type of pluripotent stem s artificially derived from a non-pluripotent cell, typically an adult somatic cell, by inducing a "forced" expression of specific genes.
Cell potency	Totipotent cells	Zygote, Spore, Morula; It has the potential to give rise to any and all human cells, such as brain, liver, blood or heart cells. It can even give rise to an entire functional organism.
	Pluripotent cells	Embryonic stem cell, Callus; They can give rise to all tissue types, but cannot give rise to an entire organism.
	Multipotent cells	Progenitor cell, such as hematopoietic stem cell and mesenchymal stem cell; They give rise to a limited range of cells within a tissue type.
	Unipotent cells	Precursor cell

Table 1. Classification of stem cells (SCs)

2.1 Embryonic stem cells

Human embryos consist of 50–150 cells when they reach the blastocyst stage, 4-5 days post fertilization. Embryonic stem cells (ES cells) are derived from the inner cell mass of the blastocyst. They present two distinctive properties: they are able to differentiate into all derivatives of three primary germ layers (pluripotency), and they are capable of propagating themselves indefinitely, under defined conditions (Ying & Chambers, 2003).

Dr. Evans first published a technique for culturing the mouse embryos in the uterus and derivation of ES cells from these embryos (Evans & Kaufman, 1981). Dr. Martin demonstrated that embryos could be cultured *in vitro* and ES cells could be derived from these embryos (Martin, 1981). In 1998, a research team led by James Thomson reported the success of isolating and growing human embryonic stem cells in cell culture (Thomason, et al., 2000).

The studies of gene expression in these SE cells have identified many proteins associated with the "stemness" phenotype and can serve as markers for ES cells. After several decades of investigates, a list of SE-specific markers has been established (The National Institutes of Health resource for stem cell research), such as 5T4, Nanog, ABCG2, Oct-3/4, Alkaline Phosphatase/ALPL, Oct-4A, E-Cadherin, Podocalyxin, CCR4, Rex-1/ZFP42, CD9, SCF R/c-kit, CD30/TNFRSF8, sFRP-2, CDX2, Smad2, Chorionic Gonadotropin, lpha Chain (alpha HCG), Smad2/3, Cripto, SOX2, DPPA4, SPARC/Osteonectin, DPPA5/ESG1, SSEA-1, ESGP, SSEA-3, FGF-4, SSEA-4, GCNF/NR6A1, STAT3, GDF-3, SUZ12, Integrin alpha 6/CD49f, TBX2, Integrin alpha 6 beta 4, TBX3, Integrin beta 1/CD29, TBX5, KLF5, TEX19, Lefty, THAP11, Lefty-1, TRA-1-60(R), Lefty-A, TROP-2, LIN-28, UTF1, LIN-41, ZIC3, c-Myc etc.

The potential to generate virtually any differentiated cell type from embryonic stem cells (ESCs) offers the possibility to establish new models of mammalian development and to create new sources of cells for regenerative medicine and genetic disease and toxicology tests in vitro (Aznar, et al., 2011). To realize this potential, it is essential to be able to control ESC differentiation and to direct the development of these cells along specific pathways. Current embryology has led to the identification of new multipotential progenitors for the hematopoietic, neural, and cardiovascular lineages and to the development of protocols for the efficient generation of a broad spectrum of cell types including hematopoietic cells, cardiomyocytes, oligodendrocytes, dopamine neurons, and immature pancreatic β cells (Murry & Keller, 2008). Today, the most challenges are to devise and optimize effective protocols to induce differentiation of the ES cells into functional adult cells, and to demonstrate the functional utility of these cells, both in vitro and in preclinical models of human disease. For example, effective protocols are expected not only to promote ES cells differentiation into hepatocytes, but also to induce hepatic functions such as albumin secretion, indocyanine green uptake and release, glycogen storage and p450 metabolism. Several recent protocols are efficient to produce high-purity (70%) hepatocytes in cultures, when these are transplanted into mice with acute liver injury, the human ES cells derived endoderm is capable to differentiate into hepatocytes and repopulated the damaged liver (Agarwal, et al., 2008). However, due to the difficulty in controlling of proliferation and differential potential, and the most controversial issue on ethical concerns, the applications of human ES cells are currently limited *in vitro* and in animal studies.

On January 23, 2009, Phase I clinical trials for transplantation of oligodendrocytes (a cell type of the brain and spinal cord) derived from human ES cells into spinal cord-injured individuals received approval from the U.S. Food and Drug Administration (FDA), marking it the world's first human ES cell human trial (CNN.com, 2009). The study leading to this scientific advancement was conducted by Hans Keirstead and his colleagues at the University of California, Irvine and supported by Geron Corporation of Menlo Park, CA. In October 2010 researchers enrolled and administered ESCs to the first patient at Shepherd Center in Atlanta (Vergano, 2010).

During the rapid development of medicine application of EC cells, safety is always a big concerning. The major concern is the risk of teratoma and other cancers as a side effect of ES cell applications, as their possibility to form tumors such as teratoma (Martin, 1981). The main strategy to enhance the safety of ESC for potential clinical use is to differentiate the ESC into specific cell types (e.g. neurons, muscle, liver cells) that have reduced or eliminated ability to cause tumors. Following differentiation, the cells are subjected to sorting by flow cytometry for further purification. While ESC are predicted to be inherently safer than iPS cells because they are not genetically modified with genes such as c-Myc that are linked to cancer. Nonetheless ESC express very high levels of the iPS inducing genes and these genes including Myc are essential for ESC self-renewal and pluripotency (Varlakhanova, et al., 2010), and potential strategies to improve safety by eliminating Myc expression are unlikely to preserve the cells' "stemness".

2.2 Embryonic germ stem cells

Embryonic germ (EG) cells are derived cells from primordial germline cells (PGCs) in early development. EG cells share many of the characteristics of human ES cells, but differ in significant ways. Human EG cells are derived from the primordial germ cells, which occur in a specific part of the embryo/fetus called the gonadal ridge, and which normally develop into mature gametes (eggs and sperm).

PGCs are mainly isolated from fetal tissue in a narrowed time window (Chapman, et al., 1999). These isolated cells are subsequently allowed to grow and divide in vitro. After one to three weeks in vitro, the human PGCs had formed dense, multilayered colonies of cells that resembled mouse ES or EG cells. Cells in these colonies expressed SSEA-1, SSEA-3, SSEA-4, TRA1-60, TRA-1-81, and alkaline phosphotase. A small, variable percentage (1 to 20 %) of the PGC-derived cell colonies spontaneously formed embryoid bodies. The growth medium for embryoid body cultures lacked LIF, bFGF, and forskolin (Roach, et al., 1993).

The range of cell types in the human PGC-derived embryoid bodies included derivatives of all three embryonic germ layers-endoderm, mesoderm, and ectoderm-based on the appearance of the cells and the surface markers they expressed. This result was interpreted to mean that the PGC-derived cells were pluripotent, however, it was not possible to demonstrate pluripotency in vivo by generating the formation of teratomas in mice (Shamblott, et al., 2001).

2.3 Fetal stem cells

Fetal stem cells are primitive cell types found in the organs of fetuses. Fetal stem cells are capable to differentiate into two types of stem cells: pluripotent stem cells and hematopoietic stem cells. Neural crest stem cells, fetal hematopoietic stem cells and pancreatic islet progenitors have been isolated in the fetuses (Beattie, et al., 1997). Fetal blood, placenta and umbilical cord are rich sources of fetal hematopoietic stem cells.

Human fetal stem cells have been used by many people including children and adults suffering from many of mankind's most devastating diseases (Sei, et al., 2009). Fetal neural stem cells found in the fetal brain were shown to differentiate into both neurons and glial cells (Villa, et al., 2000). Human fetal liver progenitor cells have shown enormous proliferation and differentiation capacity to generate mature hepatocytes after transplantation in immunodeficient animals (Soto-Guitierrez, et al., 2009). Suzuki et al. showed that a single cell in the c-Met^{+}CD49f$^{-\ low}$c-Kit^{-}CD45^{-}Ter119^{-} fraction from mid-gestational fetal liver has the capacity for self-renewal in vitro and for bipotential differentiation, indicating that this defined fraction contains hepatic stem cells (Suzuki, et al., 2002). Hepatic stem/progenitor cells can be enriched in mouse fetal hepatic cells based on several cell surface markers, including c-Met, Dlk, E-cadherin, and Liv2. Rat Dlk cells isolated from mid-gestational fetal liver exhibit characteristics expected for hepatic stem/progenitor cells. Thus, fetal liver cells may be suitable for overcoming the limitations in engraftment and to allow a functional correction of the disease phenotype (Khan, et al., 2010), as well as in use of artificial liver devices.

Hematopoietic cells are fetal stem cells in the umbilical cord after the birth of a baby. The only potential of these cells are to produce blood cells (Lee, et al., 2010). However, in current medicine practice, they are quite effective in treating blood diseases such as leukemia and anemia. It is a mature medical service today to store the frozen umbilical cord blood of a new born baby, and to use for leukemia, anemia and other predispositions if needed in future (Navarrete & Contreras, 2009).

The tissue rejection problems for fetal cell's application similar to those encountered in kidney and heart transplants may limit the usefulness of fetal stem cells. Further research to overcome this barrier is a hot topic in this field.

2.4 Bone Marrow (BM) stem cells

Adult BM mainly comprises two populations of precursor cells, *hematopoietic stem cells* (HSCs) and marrow stromal cells (MSCs) (Lagasse, et al., 2000). HSC and MSC are both multipotent stem cells. HSCs are present in circulating blood and umbilical cord blood (UCB) and are able to sustain production of all blood cells throughout life. MSCs can be isolated from several other tissues, including adipose tissue, placenta, amniotic fluid, UCB and fetal tissues are able to differentiate into osteocytes, adipocytes, chondrocytes, smooth muscle cells and haematopoietic supportive stroma (Herzog, et al., 2003; Yagi, et al., 2010).

Human HSCs have been defined with respect to staining for Lin, CD34, CD38, CD43, CD45RO, CD45RA, CD59, CD90, CD109, CD117, CD133, CD166, and HLA DR (human). In addition, metabolic markers/dyes such as rhodamine123 (which stains mitochondria), Hoechst33342 (which identifies MDR type drug efflux activity), Pyronin-Y (which stains RNA), and BAAA (indicative of Aldehyde dehydrogenase enzyme activity) have been described. The positive markers useful for MSC identification are CD106, CD105, CD73, CD29, CD44, and Sca-1 (Domen, et al., 2006).

Bone marrow transplantation (BMT) and peripheral blood stem cell transplantation (PBSCT) are the current clinical procedures to restore stem cells that have been destroyed by high doses of chemotherapy and/or radiation therapy. The isolation of a large number of potent HSC/MSC sets the basis of new methods for tissue regeneration and cell therapy (Körbling & Freireich, 2011). Nevertheless, the procedure of BM extraction is traumatic and the amount of material extracted is limited. Therefore, exploring new sources and isolation techniques for obtaining such cells is of great interest.

2.5 Adult stem cells

Adult stem cells are any stem cells taken from mature tissue. Because of the stage of development of these cells, they have limited potential compared to the stem cells derived from embryos and fetuses (Robinson, 2001). Most adult stem cells are lineage-restricted (multipotent) and are generally referred to by their tissue origin (mesenchymal stem cell, adipose-derived stem cell, endothelial stem cell, dental pulp stem cell, etc.) (Barrilleaux, et al., 2006; Gimble, et al., 2007). They play important roles on local tissue repair and regeneration.

The application of adult stem cells in research and therapy is not as controversial as embryonic stem cells, because the production of adult stem cells does not require the destruction of an embryo. Additionally, because in some instances adult stem cells can be obtained from the intended recipient (an autograft), the risk of tissue rejection is essentially non-existent in these situations. Consequently, more USA government funding is being provided for adult stem cell research (US Department of Health and Human Services, 2004).

2.6 Hepatic stem cells

Liver transplantation is the primary treatment for various end-stage hepatic diseases, but is hindered by the source of donor organs and by complications associated with tissue rejection and immunosuppression. Thus, the regenerative capabilities of adult hepatocytes, liver progenitors and stem cells are being studied with great interest.

Adult hepatocytes remain a low mitotic rate during periods of tissue homeostasis. However, extensive documents have been established of these mature hepatic cells to re-enter the cell cycle and to restore damaged parenchyma through both cell hypertrophy and hyperplasia following acute hepatic parenchymal loss when surgical resection or hepatotoxin. Under these circumstances, liver mass is restored primarily through the activation of hepatocytes (Fausto, et al., 2006), suggesting mature hepatocytes could serve their own physiologic precursors (Koniaris, et al., 2003). As evidence, the isolated adult hepatocytes have been showed suitable for the treatment of liver diseases in both animal and human livers. After transplantation of primary adult hepatocytes into Gunn rat, an animal model for UDP-glucuronosyl transferase (UGT1A1) deficiency (Crigler-Najjar syndrome type I), the high bilirubin level is markedly reduced (Matas, et al., 1976). This view is also supported by the current clinical practice of that the hepatocyte transplantation can cure or alleviate congenital metabolic diseases of the liver (Sokal, et al., 2003).

Liver oval cell, a blast-like cell and with the capability of self renewing and multipotent differentiation, is considered as the liver-specific stem cell. It can be identified only in the setting of chronic liver injury, when resident hepatocytes are unable to enter the cell cycle to restore liver mass. (Newsome, et al., 2004; Shafritz, et al., 2006). In multiple independent studies, these liver oval cells have been shown to present molecular markers of adult hepatocytes (albumin, cytokeratins 8 and 18), bile duct cells (cytokeratins 7 and 19, OV-6, A6), fetal hepatoblasts (AFP), and haematopoietic stem cells (Thy -1, Sca-1, c-kit). A recent study provides direct evidence that active Wnt/β-catenin signaling occurs preferentially during the transit amplifying of oval cell population and β-catenin clearly localizes to proliferating oval cells (Sekine, et al., 2007). Although it is not clear yet whether such a cell mass expanding in vitro is sufficient enough for clinical applications and its possible risk on carcinogenesis, oval cells isolated from the liver represent a promising source for cell-based therapy.

Human fetal liver progenitor cells have shown enormous proliferation and differentiation capacity to generate mature hepatocytes after transplantation in immunodeficient animals (Dan, et al., 2006). Hepatic stem/progenitor cells are enriched in mouse fetal hepatic cell fraction, identified with several cell surface markers including c-Met, Dlk, E-cadherin, and

Liv2. A single cell in the c-Met^{+}CD49f$^{-\ low}$c-Kit^{-}CD45^{-}Ter119^{-} fraction from mid-gestational fetal liver revealed the capacity of self-renewal in vitro and bipotential differentiation, indicating the containing of hepatic stem cells in this defined fraction, while the hepatic progenitor cells lack the capacity of self-renewal. As an in vitro cultivation protocol of fetal hepatic stem cells has been established, the fetal liver cells may be promised for the hepatic cell amount in engraftment and the functional correction of the disease phenotype (Khan, et al., 2010), which should be better over the artificial liver devices.
Extra hepatic stem cells have been demonstrated to be involved in liver regeneration too in mice and rats studies (Herzog, et al., 2003). For example, cells from multiple extra hepatic tissues (including BM, umbilical cord and umbilical cord blood (UCB), and amniotic fluid) may differentiate into hepatic cells with some or many hepatic features, and some of them have shown the ability of liver repopulation in vivo. Remarkable trans-differentiation of HSCs to hepatocyte-like cells has been described, mainly in animals with BM/HSC transplantations followed by induction of liver damage. Lagasse et al demonstrated that highly purified HSCs repopulated not only the haematopoietic system, but also the livers with hereditary tyrosinaemia, rescuing these animals from liver failure (Lagasse, et al., 2000). The published reports have suggested that MSCs may differentiate into hepatocyte-like cells both in vitro and in vivo. The cellular mechanism of trans-differentiation of MSCs to hepatocyte-like cells in vivo might be due to cell-fusion, while other reports suggested cell-autonomous trans-differentiation (Alvarez-Dolado, et al., 2003; Vassilopoulos & Russell, 2003).

2.7 Pancreatic stem cells

Pancreatic islet transplantation has demonstrated an efficient way to achieve the long-term insulin independence for the patients suffering from diabetes mellitus type 1. However, because of limited availability of islet tissue, new sources of insulin producing cells that are responsive to glucose are required. Development of pancreatic beta-cell lines from rodent or human origin has progressed slowly in recent years. To date, the best candidate sources for adult pancreatic stem or progenitor cells are: duct cells, exocrine tissue, nestin-positive islet-derived progenitor cells, neurogenin-3-positive cells, pancreas-derived multi-potent precursors; and mature β-cells.
The first report to describe in vitro generated insulin-producing islet-like clusters was based on the expansion of mouse pancreatic duct cells (Gupta, et al., 1999). Afterwards, Bonner-Weir et al (Bonner, 2000) generated the same type of insulin-producing islet-like clusters from cultivated islet buds developed from human pancreatic duct cells in vitro. Our previous study also provided evidence of that GLP-1 is able to induce pancreatic ductal cells with the expression of IDX-1 to differentiate into insulin producing cells (Hui H, 2001), and is able to stimulate glucose-derived de novo fatty acid synthesis and chain elongation during cell differentiation and insulin release (Bullota A , 2003). These data indicated pancreatic ductal cells are potential tissue source for insulin-producing islet cells. However, at this time, the expansion capacity of these cultivated cells is still limited, and protocols for in vitro amplification need further optimization for a sufficient number of fully differentiated cells to allow a successful transplantation.
A recent genetic lineage study (Dor, et al., 2004) claimed the replication success of pre-existing β-cells and that turned to be the dominant pathway for the formation of new β-cells in adult mice. Another similar study (Seaberg, et al., 2004) also showed a cloned isolation of multi-potential precursor cells from mouse adult pancreas called pancreas-derived multi-potent precursors. These precursor cells arise from single islet and duct cells.

The generation of insulin-producing cells from pancreatic exocrine tissue has recently been reported (Baeyens, et al., 2005). Both exocrine and endocrine pancreatic originate from a domain of the foregut endoderm, which expresses the pancreatic duodenal homeobox factor (Pdx-1) at early developmental stages. The inactivation of this gene leads to a non-pancreatic phenotype, demonstrating its major role in both exocrine and endocrine pancreatic development. In addition, signaling induced by soluble factors is a prerequisite to pancreatic lineage specification and triggers the emergence of pancreatic precursors expressing Pdx-1. Moreover, as Baeyens et al (ibid.) indicated, there were data suggesting the existence in vivo of acinar-islet transitional cells and the "spontaneous" trans-differentiation of acinar cells to insulin-expressing cells. Altogether, these may suggest that a population of acinar cells, in the presence of certain soluble factors, is competent to adopt an endocrine fate.
Some reports suggest that pancreatic precursor cells express nestin (Zulewski,2001), an intermediate filament protein that is a marker of neural stem cells. These nestin-positive islet-derived progenitor cells also express insulin, glucagon, and Pdx-1 as well as low with levels of insulin secretion. However, other studies suggest that nestin expression is not related to pancreatic precursor identity.
Recent data indicate that Ngn-3-positive cells are endocrine progenitors both in the adult pancreas and in the embryo and that Ngn-3 expression is not seen outside the islets (Gu, et al., 2002). Nevertheless, low levels of Ngn-3 expression within a population of duct cells are not excluded by these studies.
Pancreatic stem cells (PSCs) have the potential to differentiate into all three germ layers. Major markers present on the surface of PSCs include Oct-4, Nestin, and c-kit. DCAMKL-1 is a novel putative stem/progenitor marker, can be used to isolate normal pancreatic stem/progenitors, and potentially regenerate pancreatic tissues.

2.8 Eye stem cells

Human cornea is transparent and clear for vision. Unique to other human organs, there is no blood vessels to provide nutrition in corneas. It is the corneal stem cell existing in the nearby limbus ring, differentiate and move to the center of corneas to renew the transparent and clear cornea around every four months. Stem cells in human cornea play a unique and significant role to maintain the corneal function.
Human corneal stem cells locate on cornea limbus, which is between the colored and white part of the eye (where it joints the sclera). During homeostasis and following injury to the corneal epithelium, the limbal corneal stem cells divide to produce daughter transient amplifying cells that proliferate, migrate onto the central cornea and become terminally differentiated to replace the lost cells (Moore JE, 2002). When a stem cell divides, each new daughter cell has the potential to either remain a stem cell or become a differentiated corneal cell. The microenvironment within the corneal basement membrane is expected the primary factor responsible for the corneal terminal differentiation (Daniels JT, 2001). However, in the case of limbal stem cell deficiency, either due to injury or diseases, it is unable for the corneal ocular repairing and regeneration. In certain corneal disorder such as Keratoconus, some stem cell markers, such as CD34, p63, were reported significantly decreased from normal to keratoconus corneas (Daniels JT, 2001). It is speculated that many corneal disorders such as in keratoconus, anirdia and alkali burns are likely associated with the corneal stem cell deficiency.
Cornea transplantation is widely used to treat certain corneal diseases such as keratoconus. Due to the limited source of donated corneas, corneal stem cells are explored, instead of

corneal buttons. In a pioneering test on cornea damage patients, stem cells were taken from the biopsied limbus tissue, grew into healthy corneal tissue in a little over two weeks, and the healthy tissue was then grafted onto the damaged eye. In the study of 112 patients between 1998 and 2006, 77% of patients had a successful first or second graft. While the opaque cornea became clear again, the vision restored. As human cornea is the most tolerant organ to accept xenograft, the corneal stem cells might be among the first large scale produced stem cells for medical application.
Another frontier of stem cell applications in human eyes is the aged-related macular degeneration (AMD). Macular degeneration is a retinal degenerative disease which causes progressive loss of central vision. The risk of developing macular degeneration increases with age. This disease most often affects people over fifties, and is the most common cause of blindness in the elderly. The impact of AMD on patients includes, but not limits, vision impairment, difficulty with daily activities, increased risk of falls, more depression and emotional distress. It affects the quality of life for millions of elderly individuals worldwide (Pulido JS, 2006). It is not only a health challenge, but also a severe social problem across the world, no mater your ethnic group and gender.
The macula is the central portion of the retina responsible for perceiving fine visual detail. Light sensing cells in the macula, known as photoreceptors, convert light into electrical impulses and then transfer these impulses to the brain via the optic nerve. Central vision loss from macular degeneration occurs when photoreceptor cells in the macula degenerate.
During the stem cell treatment, macular patients are treated by implanting autologous (from selves) stem cells behind the eye via retrobulbar injection under local anesthesia. These re-injected stem cells have the potential to transform into multiple types of cells and are capable of regenerating damaged tissue. Stem cell treatment is so far the most promising approach to restore the vision from AMD among many strategies.

2.9 Cancer stem cell

Cancer stem cells theory is a finding on stem cell biology and an application of stem cell features on cancer studies. Cancer stem cells are those stem cells in tumor mass. They specifically are with the ability to give rise to all cell types found in a cancer sample. According to the hypothesis, the original tumor is developed and formed from these cancer stem cells by self-renewal and differentiation into multiple cell types. Cancer stem cell population consists of only a small potion of tumor mass (around 0.1-1% of total mass) and can be distinguished from the other cells in tumor mass by special cell surface antigens (such as $CD34^+$). Both stem cells and cancer stem cells share the characters of stemness, the capacity of differentiation, the multi-potential differentiation (Gupta PB, 2009). However, the unique character of cancer stem cells, different from normal stem cells, is the growth out of control. They, or their descendants, lost the behavior of "contact inhibition of growth", the most important character of a non-cancer cell.
During conventional cancer chemotherapies, the differentiated or differentiating cells are likely to be killed while the cancer stem cells, due to their stemness and inactivity, could remain untouched, therefore to escape from chemotherapies. It is believed they serve as "cancer seeds" and respond to the cancer relapse and metastasis by rising new tumors. Based on the concept of cancer stem cells, it is beneficial to include an induction of the cancer stem cell differentiation during chemotherapies (Perkel JM, 2010). This will be expected to increase the efficacy of chemotherapies and improve the survival rate of cancer patients.

2.9.1 Identify cancer stem cell in various types of cancers

The existence of cancer stem cells has been debated for many years until the first conclusive evidence was published in 1997 in Nature Medicine. Bonnet and Dick (Bonnet D, 1997) isolated a subpopulation of acute myeloid leukemic cells that express a specific surface antigen CD34, but lacks the antigen CD38. The authors established that the subpopulation, $CD34^+/CD38^-$, is capable of initiating tumors in NOD/SCID mice that is histologically similar to the donor. Later, Blair A et al reported a similar but slightly different cancer stem cell phenotype of $CD34^+/CD71^-/HLA^-/DR^-$ in acute myeloid leukemic cells (Takaishi S, 1998).

Evidence also comes from the rational of histology, the tissue structure of tumors. Many tumors are very heterogeneous and contain multiple types of cells. These multiple types of cells are believed to be developed from single cells (or a cluster of cells), rather than assembled by multiple cells. If the descendants of these multiple types of cells come from a same ascendant, this implies that the ancestor must have the capacity to generate multiple cell types. In other words, it possessed multi-differential potentials, the fundamental character of stem cells (Bonnet D, 1997).

Tumor type	Surface antigens	Year reported	Reference
Acute Myeloid leukemic	$CD34^+/CD38^-$	1997	Bonnet D, 1997
Brain tumor	$CD133^+$	2003	Singh SK, 2003
Breast cancer	$CD44^+/CD24^-$	2003	Al-Hajj M, 2003
Multiple Myeloma	$CD138^+$	2004	Matsui W, 2004
Colon cancer	$CD133^+$	2007	O'Brien CA, 2007
Liver cancer	$CD133^+$	2007	Ma S, 2007
Pancreatic cancer	$CD44^+/CD24^+$	2007	Li C, 2007
Lung cancer	$CD133^+$	2008	Eramo A, 2008
Ovary cancer	$CD44^+/CD117^+$	2008	Zhang S, 2008
Prostate cancer	$CD133^+/CD44^+$	2008	Maitland NJ, 2008
Melanoma	$CD4^+/CD25^+/FoxP3^+$	2008	Schatton T, 2008
Gastric cancer	$CD44^+$	2009	Takaishi S, 2009

Table 2. Reported cancer stem cell and their surface antigens

The existence of leukemic stem cells prompted further studies in this field. Cancer stem cells have been reported in more and more other cancer types. Followed the Acute Myeloid leukemic stem cells ($CD34^+/CD38^-$), cancer stem cells have also been identified in several solid human tumors respectively.
As cancer stem cells have been identified in various organ origin cancers, it is widely accepted that cancer stem cell is a general format and fundamental concept in all cancers (or tumors).

2.9.2 The origin of cancer stem cells

Where the cancer stem cell comes from? The origin of cancer stem cells is still a hot topic of discussion and argument. Several camps regarding the issue have formed within the scientific community, and it is likely that the correct answer is not limited in one, depending on the tumor types and their developments. Up to date, there is not yet an experimental model has been established to demonstrate a tumor formation in lab, as cancer stem cells are usually isolated from end-stage of tumors rather than the initial stage to tumors. Therefore, describing a cancer stem cell as the cell of origin is often an inaccurate claim, and as hypothesis.
As cancer stem cells share the features of stem cells and of cancer cells, it is not wonder that some researchers believe they are the results of cell mutants from developing stem cells, including progenitor cells, adult stem cells, and the most likely from stem cell niche populations during development. The rational behind is that these developing stem populations are mutated and then expand such that the mutation is shared by many of the descendants of the mutated stem cell. These daughter stem cells are then much easier to becoming tumors, and because of the large amount of cells, there is more chance of a mutation that can cause cancer (Wang ZY, 2000). Adult stem cells are with extremely long lifespan to accumulate mutants that drives cancer initiation. Thus, adult stem cells have also advantages on the logical backing of the theory of tumor formation.
It has also been proposed that the cancer stem cells are mutants from cancer cells after obtaining the stem cell-like features. De-differentiation is a reasonable hypothesis, which assumes these cells acquire stem cell like characteristics by reverse-differentiation from cancer cells. This is a potential alternative to any specific cell of origin, as it suggests that any cell might become a cancer stem cell.
The tumor hierarchy is another model to propose the origin of cancer stem cells. The main point of this model claims that a tumor is a heterogeneous population of mutant cells with various stages of stem cells. In this model, the tumor is made up of several types of stem cells, some stem cell lines will be more thrive than other cell lines, as they adapt to the specific environments. Within the tumor hierarchy model, it would be extremely difficult to pinpoint the cancer stem cell's origin. It is important to bear in mind that, due to the heterogeneous nature of cancers, it is possible that any individual cancer could come from an alternative origin.

2.9.3 The impact of cancer stem cell concept on cancer therapy

The concept of cancer stem cell has a great impact on the strategy of chemotherapy and cancer treatments. The classic view of cancer is that the tumor cell (and its progeny) arises from the progressive accumulation of mutations over time, giving it growth advantage over its neighbors. It also implies that all cells in a tumor have more or less an equivalent capacity to form another tumor - relapse or metastasis. Under the classic view of cancer, the anticancer drugs are designed to target rapid growth cells. However in CSC model, tumor cells have somehow been reprogrammed to be "stem-like", and thus grow slower than surrounding cells. It also implies that only CSCs have the ability to propagate new tumors. According to CSC model, the traditional therapies that target the bulk tumor are to some extent pointless, as the resulting shrinkage may look good on a CT scan, but the disease itself can still recur (Perkel JM, 2010).

Relapse and metastasis are major challenges in current cancer treatments. During the cancer chemotherapies, the cancer (or tumor) mass is initially shrink, but barely cleared up. After a while, they usually come back (relapse) with some new drug resistance features developed. It is believed the cancer stem cells serve as "cancer seeds" with stemness and inactivity features, which help them to escape from chemotherapy and survive from drug attack. They are responding to the cancer relapse. Based on this concept of CSC, it is beneficial to include an induction of the cancer stem cell differentiation before and during chemotherapies. This will be expected to increase the efficacy of chemotherapies and improve the survival rate of cancer patients. This induced differentiation strategy has achieved significant efficacy on blood cancer treatment, such as children's acute promyelocytic leukaemia (APL). A group of pioneer scientists in China used Arsenic and retinoic acid to induce children's APL and have achieved "a complete remission in 92 - 95% of patients with this disease" (Wang ZY, 2000). However in solid tumors, the differentiation inducers and chemotherapeutic agents are difficult to penetrate into the inside of solid tumors. How to improve this penetration is still a big challenge for pharmaceutical researchers.

3. Induced pluripotent stem cells

Induced pluripotent stem cells (Thomson, et al., 1998), commonly abbreviated as iPS cells or iPSCs are a type of pluripotent stem cell artificially derived from a non-pluripotent cell, typically an adult somatic cell, by inducing a "forced" expression of specific genes. Induced Pluripotent Stem Cells are similar to natural pluripotent stem cells, such as embryonic stem (ES) cells, in many respects, such as the expression of certain stem cell genes and proteins, chromatin methylation patterns, doubling time, embryoid body formation, teratoma formation, viable chimera formation, and potency and differentiability, but the full extent of their relation to natural pluripotent stem cells is still being assessed (Ying, et al., 2003).

iPSCs were first introduced in 2006 from mouse cells and in 2007 from human cells. This has been cited as an important advance in stem cell research, as it may allow researchers to obtain pluripotent stem cells, which are important in research and potentially have therapeutic uses, without the controversialuse of embryos. They also avoid the issue of graft-versus-host disease and immune rejection unlike embryonic stem cells because they are derived entirely from the patient.

Depending on the methods used, reprogramming of adult cells to obtain iPSCs may pose significant risks that could limit its use in humans. For example, if viruses are used to genomically alter the cells, the expression of cancer-causing genes or oncogenes may potentially be triggered. In February 2008, ground-breaking findings published in the journal Cell, scientists announced the discovery of a technique that could remove oncogenes after the induction of pluripotency, thereby increasing the potential use of iPS cells in human diseases (Evans & Kaufman, 1998). In April 2009, it was demonstrated that generation of iPS cells is possible without any genetic alteration of the adult cell: a repeated treatment of the cells with certain proteins channeled into the cells via poly-arginine anchors was sufficient to induce pluripotency (Martin, 1981). The acronym given for those iPSCs is piPSCs (protein-induced pluripotent stem cells).

4. References

Al-Hajj M, Wicha MS, Benito-Hernandez A, Morrison SJ, & Clarke MF. (2003). *Prospective identification of tumorigenic breast cancer cells.* PNAS 100 (7): 3983–8.

Alison MR, Poulsom R, & Jeffery R et al. (2000). *Hepatocytes from non-hepatic adult stem cells.* Nature. 406(6793):257.

Agarwal S, Holton KL, & Lanza REfficient. (2008). *Differentiation of functional hepatocytes from human embryonic stem cells.* Stem Cells. 2008 May;26(5):1117-27.

Alvarez-Dolado M, Pardal R, & Garcia-Verdugo JM, et al. (2003). *Fusion of bone-marrow-derived cells with Purkinje neurons, cardiomyocytes and hepatocytes.* Nature, 425(6961):968-73.

Aznar J, Sánchez JL. (2011).*Embryonic stem cells: are useful in clinic treatments?J Physiol Biochem.* 2011 Mar;67(1):141-4.

Baeyens L, De Breuck S, Lardon J, Mfopou JK, Rooman I, & Bouwens L. (2005).*In vitro generation of insulin-producing beta cells from adult exocrine pancreatic cells. Diabetologia* 48: 49–57, 2005.

Barrilleaux B, Phinney DG, Prockop DJ, & O'Connor KC. (2006).*Review: ex vivo engineering of living tissues with adult stem cells.* Tissue Eng 12 (11):3007-19, 2006.

Beattie GM, Otonkoski T, & Lopez AD, et al. (1997). *Functional beta-cell mass after transplantation of human fetal pancreatic cells: Differentiation or proliferation?* Diabetes 46: 244–248.

Blair A. Hogge DE. & Sutherland HJ,. (1998).*Most acute myeloid leukemia progenitor cells with long-term proliferative ability in vitro and in vivo have the phenotype CD34(+)/CD71(-)/HLA-DR-.* Blood. 92(11): 4325-35.

Bonner-Weir S, Taneja M, Weir GC, Tatarkiewicz K, Song KH, Sharma A, & O'Neil JJ. (2000). *In vitro cultivation of human islets from expanded ductal tissue. Proc Natl Acad Sci USA,* 97:7999–8004,.

Bonnet D, Dick JE. (1997). *Human acute myeloid leukemia is organized as a hierarchy that originates from a primitive hematopoietic cell.* Nature medicine 3 (7): 730–7.

Braun KM, Sandgren EP, (2000). *Cellular origin of regenerating parenchyma in a mouse model of severe hepatic injury.* Am J Pathol, 157:561–569,.

Brons IG et al. (2007). *Derivation of pluripotent epiblast stem cells from mammalian embryos.* Nature, 448:191-5.

Cantz T, Zuckerman DM, & Burda MR, et al. (2003). *Quantitative gene expression analysis reveals transition of fetal liver progenitor cells to mature hepatocytes after transplantation in uPA/RAG-2 mice.* Am J Pathol. 162(1):37-45.

Cantz T, Michael P, & Manns, et al. (2008). *Stem cells in liver regeneration and therapy.* Cell Tis Res. 331:271-82, 2008.

Chang HH, Hemberg M, & Barahona M, et al. (2008). *Transcriptome-wide noise controls lineage choice in mammalian progenitor cells.* Nature 453:4544-71, 2008.

Chapman, Audrey, Frankel, Mark, Garfinkel, & Michele. (1999). *Stem Cell Research and Applications: Monitoring the Frontiers of Biomedical Research,* "Index of Terms", Nov. 1999. Online: http://www.meta-library.net/stemcell/human1-body.html

Clarke MF, Dick JE, & Dirks PB, et al. (2006). *Cancer stem cells–perspectives on current status and future directions: AACR Workshop on cancer stem cells.* Cancer Res. 66:9339-44.

Dan YY, Riehle KJ, & Lazaro C, et al. (2006). *Isolation of multipotent progenitor cells from human fetal liver capable of differentiating into liver and mesenchymal lineages.* Proc Natl Aca Sci USA, 103:9912-7.

Daniels JT, Dart JKG, Turt SJ, & Khaw PT, (2001).*Corneal stem cells in review,* Wound repair and regeneration, 483-94.

De Vree JM, Ottenhoff, & Bosma PJ, et al. (2000). *Correction of liver disease by hepatocyte transplantation in a mouse model of progressive familial intrahepatic cholestasis.* Gastroenterology, 119:1720–30.

Domen, Amy Wagers, & Irving L. Weissman. (2006). *Bone Marrow (Hematopoietic) Stem Cells.* Regenerative Medicine 2006.

Dor Y, Brown J, Martinez OI, & Melton DA. (2004). *Adult pancreatic β-cells are formed by self-duplication rather than stem-cell differentiation. Nature* 429:41–6, 2004.

Dr. Yury Verlinsky, (2009). *Expert in reproductive technology,* Chicago Tribune, 1943-2009.

Eramo A, Lotti F, Sette G, Pilozzi E, Biffoni M, Di Virgilio A, Conticello C, Ruco L, Peschle C, & De Maria R. (2008). *Identification and expansion of the tumorigenic lung cancer stem cell population, Cell Death Differ.* 15(3):504-14, 2008.

Evans M, Kaufman M. (1981). *Establishment in culture of pluripotent cells from mouse embryos.* Nature 292:154–6, 1981.

Fausto N. (2004). *Liver regeneration and repair: hepatocytes, progenitor cells, and stem cells.* Hepatology 39:1477-87, 2004.

Fausto N, Campbell JS, & Riehle KJ. (2009). *Liver regeneration.* Hepatology 43 (2 suppl 1):S45–S53, 2006.

"FDA approves human embryonic stem cell study - CNN.com", January 23, 2009. http://www.cnn.com/2009/HEALTH/01/23/stem.cell/.

Fox IJ, Chowdhury JR, & Kaufman SS, et al. (1998).*Treatment of the Crigler–Najjar syndrome type I with hepatocyte transplantation.* N Engl J Med. 338:1422–6, 1998.

Gimble JM, Katz AJ, & Bunnell BA. (2007). *Adipose-derived stem cells for regenerative medicine.* Circ Res 100 (9):1249–60, 2007.

Grompe M, Laconi E, & Shafritz DA. (1999). *Principles of therapeutic liver repopulation.* Semin Liver Dis. 19:7–14, 1999.

Gupta S, Aragona E, & Vemuru RP, et al. (1999).*Permanent engraftment and function of hepatocytes delivered to the liver: implications for gene therapy and liver repopulation.* Hepatology 14:144–9.

Herzog EL, Chai L & Krause DS, (2003).*Plasticity of marrow-derived stem cells.* Blood. 102(10):3483-93.

Hu M, Kurobe M, & Jeong YJ, et al. (2007). *Wnt/ß-Catenin Signaling in Murine Hepatic Transit Amplifying Progenitor Cells.* Gastroetero, 133:1579–1591, 2007.

Jakubowski A, Ambrose C, & Parr M, et al. (2005). *TWEAK induces liver progenitor cell proliferation.* J Clin Invest. 115:2330-40.

Kakinuma S, Nakauchi H, & Watanabe M, (2009).*Hepatic stem/progenitor cells and stem-cell transplantation for the treatment of liver disease.* J Gastroenterol. 44(3):167-72.

Khan AA, Shaik MV, & Parveen N et al. (2010). *Human fetal liver-derived stem cell transplantation as supportive modality in the management of end-stage decompensated liver cirrhosis.* Cell Transplant. 19(4):409-18.

Koniaris LG, McKillop IH, & Schwartz SI et al. (2003). *Liver regeneration.* J Am Coll Surg. 197:634–659.

Körbling M, Freireich EJ. (2003).*25 years of peripheral blood stem cell transplantation.* Blood. 2011 Apr 1

Gimble JM, Katz AJ, & Bunnell BA, (2007). *Adipose-derived stem cells for regenerative medicine.* Circ Res 100 (9):1249–60, 2007.

Gu G, Dubauskaite J, & Melton DA. (2002). *Direct evidence for the pancreatic lineage: Ngn3+ cells are islet progenitors and are distinct from duct progenitors. Development.* 129:2447-57.

Gupta PB, Chaffer CL, & Weinberg RA. (2009). *Cancer stem cells: mirage or reality?.* Nat Med. 15 (9): 1010–2.

Lagasse E, Connors H, & Al-Dhalimy M. et al. (2000).*Purified hematopoietic stem cells can differentiate into hepatocytes in vivo.* Nat Med. 6(11):1229-34.

Li C, Heidt DG, Dalerba P, Burant CF, Zhang L, Adsay V, Wicha M, Clarke MF, & Simeone DM. (2007). *Identification of pancreatic cancer stem cells,* Cancer research 67 (3): 1030–7.

Maitland NJ, Collins AT. (2008). *Prostate cancer stem cells: a new target for therapy,* J. Clin. Oncol. 26 (17): 2862–70.

Martin G. (1981). *Isolation of a pluripotent cell line from early mouse embryos cultured in medium conditioned by teratocarcinoma stem cells.* Proc Natl Acad Sci USA 78 (12):7634–8.

Matas AJ, Sutherland DE, & Steffes MW et al. (1976). *Hepatocellular transplantation for metabolic deficiencies: decrease of plasma bilirubin in Gunn rats.* Science 192:892-4.

Matsui W, Huff CA, & Wang Q, et al. (2004). *Characterization of clonogenic multiple myeloma cells,* Blood 103 (6): 2332–6.

Ma S, Chan KW, Hu L, Lee TK, Wo JY, Ng IO, Zheng BJ, & Guan XY. (2007). *Identification and characterization of tumorigenic liver cancer stem/progenitor cells, Gastroenterology,* 132(7): 2542-56.

Michalopoulos GK, (2005). DeFrances M. Liver regeneration. *Adv Biochem Eng Biotechnol, 93:101–34*

Monga SP, Pediaditakis P, & Mule K, et al. (2001). Changes in WNT/beta-catenin pathway during regulated growth in rat liver regeneration. Hepatology, 33:1098-109.

Monga SP, Monga HK, & Tan X, et al. (2003).*Beta-catenin antisense studies in embryonic liver cultures: role in proliferation, apoptosis, and lineage specification.*Gastroenterology, 124:202-16.

Moore JE, McMullen CB, Mahon G, & Adamis AP, (2002). *The corneal epithelial stem cell.* DNA & Cell Biology, 21:443-51.

Murry C.E., Keller G. (2008). *Differentiation of Embryonic Stem Cells to Clinically Relevant Populations: Lessons from Embryonic Development.* Cell, 132 (4), pp. 661-680.

Muraca M, Gerunda G, & Neri D, et al. (2002). *Hepatocyte transplantation as a treatment for glycogen storage disease type 1a.* Lancet. 26;359(9303):317-8.

Muraca M, Galbiati G, & Vilei MT et al. (2006). *The future of stem cells in liver diseases.* Am Hepatol, 5(2):68-76.

Newsome PN, Hussain MA & Theise ND. (2004). *Hepatic oval cells: helping redefine a paradigm in stem cell biology.* Curr Top Dev Biol, 61:1-28.

Nelson WJ, Nusse R. (2004). *Convergence of Wnt, beta-catenin, and cadherin pathways.* Science 303:1483-7.

O'Brien CA, Pollett A, Gallinger S, & Dick JE. (2007). *A human colon cancer cell capable of initiating tumour growth in immunodeficient mice,* Nature 445 (7123): 106–10.

Overturf K, Al-Dhalimy M, & Finegold M, et al. (1999).*The repopulation potential of hepatocyte populations differing in size and prior mitotic expansion,* Am J Pathol, 155: 2135–43.

Paul S. Knoepfler. (2009). *Deconstructing stem cell tumorigenicity: a roadmap to safe regenerative medicine. Stem Cells,* Wiley on line library, 27(5):1050-6.

Perkel JM, (2010). *Rethinking the Classics, Bioscience Technology Online,* http://www.biosciencetechnology.com/Articles/2010/03

Petersen, B.E., Bowen, W.C., & Patrene, K.D., et al. (1999), *Bone marrow as a potential source of hepatic oval cells.* Science. 14;284(5417):1168-70.

Pulido JS, Winters JL, & Boyer D. (2006). *Preliminary analysis of the final multicenter investigation of rheopheresis for age related macular degeneration (AMD) trial (MIRA-1) results,* Trans Am Ophthalmol Soc, 104:221-31.

Quintana-Bustamante O, Alvarez-Barrientos A, & Kofman AV, et al. (2006). *Hematopoietic mobilization in mice increases the presence of bone marrow-derived hepatocytes via in vivo cell fusion.* Hepatology. 43(1):108-16.

Robinson BA, (2001).*Human Stem Cell Research. Ontario Consultants on Religious Tolerance.* Online: http://www.religioustolerance.org/res_stem1.htm

Roach, S., Cooper, S., Bennett, W. & Pera, M.F. (1993). *Cultured cell lines from human teratomas: windows into tumour growth and differentiation and early human development.* Eur. Urol. 23, 82-87.

Sandhu JS, Petkov PM, & Dabeva MD, et al. (2001). *Stem cell properties and repopulation of the rat liver by fetal liver epithelial progenitor cells.* Am J Pathol. 159(4):1323-34.

Schatton T, Murphy GF. et al. (2008). *Identification of cells initiating human melanomas.* Nature 451 (7176): 345–9, 2008.

Seaberg RM, Smukler SR,.et al. (2004). *Clonal identification of multipotent precursors from adult mouse pancreas that generate neural and pancreatic lineages. Nat Biotechnol.* 22:1115-24.

Sei Kakinuma, Hiromitsu Nakauchi. & Mamoru Watanabe. (1999). *Hepatic stem/progenitor cells and stem-cell transplantation for the treatment of liver disease.* Journal of Gastroenterology Volume 44, Number 3, 167-172.

Sekine S, Gutierrez PJ, & Yu-Ang LB, et al. (2007). *Liverspecific loss of beta-catenin results in delayed hepatocyte proliferation after partial hepatectomy.* Hepatology, 45:361-8, 2007.

Shafritz DA, Oertel M, & Menthena A, et al. (2006). *Liver stem cells and prospects for liver reconstitution by transplanted cells.* Hepatology 2006;43(2suppl1):S89–98, 2006.

Shamblott MJ, Axelman J, Littlefield JW, Blumenthal PD, Huggins GR, Cui Y, Cheng L, & Gearhart JD. (2001). *Human embryonic germ cell derivatives express a broad range of developmentally distinct markers and proliferate extensively in vitro.* Proc. Natl. Acad. Sci. USA, 98:113–118, 2001.

Singh SK, Clarke ID, Terasaki M, Bonn VE, Hawkins C, Squire J, & Dirks PB. (2003).*Identification of a cancer stem cell in human brain tumors,* Cancer research 63 (18): 5821–8.

Sokal EM, Smets F, Bourgois A, & Van Maldergem L, et al. (2003). *Hepatocyte transplantation in a 4-year-old girl with peroxisomal biogenesis disease: technique, safety, and metabolic follow-up.* Transplantation. 76(4):735-8.

Soto-Guitierrez A, Navarro-Alvarez N, & Yagi H, et al. (2009).*Stem cells for liver repopulation.* Curr Opin Organ Trans.14(6):667-73, 2009.

Suzuki A, Zheng YW, Kaneko S, Onodera M, Fukao K, & Nakauchi H, et al. (2002). *Clonal identification and characterization of selfrenewing pluripotent stem cells in the developing liver.* J Cell Biol 2002;156:173–84.

Takaishi S, Okumura T, Tu S, Wang SS, Shibata W, Vigneshwaran R, Gordon SA, Shimada Y, & Wang TC. (2009). *Identification of gastric cancer stem cells using the cell surface marker CD44,* Stem Cells, 27(5):1006-20.

Thomson J, Itskovitz-Eldor J, Shapiro S, Waknitz M, Swiergiel J, Marshall V, & Jones J. (1998). *Embryonic stem cell lines derived from human blastocysts.* Science 282: 1145-7.

Vassilopoulos G, Russell DW. (2003). *Cell fusion: an alternative to stem cell plasticity and its therapeutic implications.* Curr Opin Genet. 13(5):480-5.

Vergano, Dan. (2010). "*Embryonic stem cells used on patient for first time*". USA Today. http://www.usatoday.com/tech/science/2010-10-12-stemcells.

Villa A, Snyder EY, & Vescovi A, et al. (2000). *Establishment and properties of a growth factor dependent perpetual neural stem cell line from the human CNS.* Exp Neurol 161: 67–84.

Wang X, Foster M, & Al-Dhalimy M, et al. (2003). *The origin and liver repopulating capacity of murine oval cells.* Proc Natl Acad Sci USA 100(suppl 1):11881-8.

Wang ZY, Chen Z. (2000).*Differentiation and apoptosis induction therapy in acute promyelocytic leukaemia,* Lancet Oncol, 1:101-6.

Watt FM. Driskell RR. (2011). *The therapeutic potential of stem cells. Phil.* Trans. R. Soc. B 365:155-63.

Yagi H, Soto-Gutierrez A, & Kitagawa Y et al. (2010).*Bone marrow mesenchymal stromal cells attenuate organ injury induced by LPS and burn.* Cell Transplant. 19(6):823-30.

Ying; Nichols J, Chambers I, & Smith A. (2003). *BMP Induction of Id Proteins Suppresses Differentiation and Sustains Embryonic Stem Cell Self-Renewal in Collaboration with STAT3.* Cell 115 (3):281–292.

Yoshida Y, Tokusashi Y, & Lee GH, et al. (1996). *Intrahepatic transplantation of normal hepatocytes prevents Wilson's disease in Long-Evans cinnamon rats.* Gastroenterology, 111:1654-60, 1996.

Zhang S, Balch C, Chan MW, Lai HC, Matei D, Schilder JM, Yan PS, Huang TH, & Nephew KP. (2008). *Identification and characterization of ovarian cancer-initiating cells from primary human tumors,* Cancer research 68 (11): 4311-20.

Zulewski H, Abraham Ej Gerlach MJ, Daniel PB, Moritz W, Muller B, Vallejo M, Thomas MK, & Habener JF. (2001). *Multipotential nestin-positive stem cells isolated from adult pancreatic islets differentiated ex vivo into pancreatic endocrine, exocrine, and hepatic phenotypes. Diabetes.* 50:521-33.

2

Human Cord Blood-Derived Stem Cells in Transplantation and Regenerative Medicine

Cornel Iancu et al.*
"Iuliu Hatieganu" University of Medicine and Pharmacy Cluj-Napoca
Romania

1. Introduction

Regenerative medicine can be defined as the process of restoring diseased or damaged tissue by replacing it with biological substitutes and this issue represents one of the main concerns of medicine. Stem cells are likely to be used in the future for cardiac, neurologic, hepatic, orthopaedic and other applications. The stem cell populations have proved to be highly proliferative lately as isolated from a variety of embryonic, foetal and adult tissues. Their increased capacity to self-renew and to unlimitedly differentiate into derivates of all germ layers in vitro and in vivo have rendered embryonic stem cells (ESCs) the main concern of tissue engineering research and regenerative medicine (Hyslop et al., 2005). However these cell lines originate from the inner cell mass of human blastocysts. Thus, the relation of ESCs to the human blastocyst will always stir ethical, moral and emotional debate over their use in research. Moreover, besides these ethical and political issues, another drawback concerning their clinical application is their lack of accessibility, technique difficulties in purification and manipulation as well as the risk for teratoma development (Fan et al., 2011). Consequently, though ESC therapy has just come into attention and there still are unknowns, its success might result in it being followed by alternative stem cell therapies (Rogers & Casper, 2004).

Adult stem cells (ASCs) originate in a mature organism, including the brain, fat, skin, kidney, peripheral blood and bone marrow and they contribute to maintaining and repairing tissue that contain them. The adult-derived stem cells are also believed to have the ability to differentiate into tissues different from their tissue of origin (Krause et al., 2001; Jiang et al, 2002). Different from ESCs, ASCs can be easily harvested from various tissues, such as skin (Riekstina et al., 2008 as cited in Fan et al., 2011), bone marrow (BM) (Gastens et al., 2007 as cited in Fan et al., 2011) and adipose (Keiser et al., 2007 as cited in Fan et al., 2011), and might be employed in the clinical treatment of disorders of vulnerable vital organs. ASCs have the advantage that allows for small samples of tissues or even the patient's own cells to be used for implantation, avoiding problems of tissue rejection. Moreover, adult cells do not involve the typical ethical issues of embryonic research. In spite of these advantages, they are not the first choice and this is because of several reasons such

* Ioana Ilie, Lucian Mocan, Carmen Georgescu, Razvan Ilie, Ileana Duncea, Teodora Mocan, Dana Iancu, and Florin Zaharie

as their limited numbers, diminished growth and differential capacities age-dependent as well as invasive harvesting procedures (Roobrouck et al., 2008).
Compared to the extensive research on ESCs and ASCs, fetal stem cells have only recently come into attention (Fan et al., 2011). The transplant of fetal tissue has recently become a research therapy for chronic degenerative disease such as Parkinson's disease and insulin-dependent diabetes mellitus (DM). However, this method is widely opposed as fetal tissue originates from elective abortion (Rogers & Casper, 2004). However, the following two distinct sources have lately generated putative stem cells: the fetus proper (including fetal bone marrow (Chang et al., 2006 as cited in Fan et al., 2011), lung (Fan et al., 2005 as cited in Fan et al, 2011), spleen, liver (In't Anker et al., 2003 as cited in Fan et al., 2011), pancreas (Hu et al., 2003 as cited in Fan et al., 2011) and peripheral blood (Yu et al., 2004 as cited in Fan et al., 2011) and the supportive extra-embryonic structures (such as umbilical cord blood (Lu et al., 2005 as cited in Fan et al., 2011), umbilical cord (UC) (Fan et al. 2005 as cited in Fan et al., 2011), amniotic fluid (AF) (Mareschi et al., 2009 as cited in Fan et al., 2011), placenta (In't Anker et al., 2004 as cited in Fan et al., 2011) and amnion (Bilic et al., 2008 as cited in Fan et al., 2011), respectively.
New efficient stem cells sources would have to exhibit a similary potential to that of ESCs and to be highly proliferative or at least easy to harvest in large numbers. Out of these, the preferred ones are the BM-derived stem cells or UC-derived stem cells. There are two main cell types within the BM: mesenchymal, which are highly proliferative and show a propensity for multi-lineage differentiation (Rogers & Casper, 2004) and haematopoietic, also promising as multipotential stem cell (Kakinuma, 2003; Hao, 2003 as cited in Rogers & Casper, 2004). As mesenchymal cells and blood cells are likely to become potential alternatives to ESCs in some clinical situations, umbilical cord blood (UCB) and peripheral blood (PB) have started to be more and more regarded as sources of bone-marrow-like stem cells due to their easy harvesting compared to bone marrow (Rogers & Casper, 2004) Cord blood is already considered an important source for stem cells, mainly because, different from BM, there is no need for a perfect human leucocyte antigen (HLA) match, there is a lower risk of graft versus host disease (GVHD) with UCB transplantation than with BM transplantation and because UCB has more haematopoietic stem cells per volume than peripheral blood or bone marrow (Rogers & Casper, 2004; Barker & Wagner, 2003).
Consequently, the proliferation, the easy harvesting procedure, the advantages over embryonic and adults counterparts and no serious ethical issues are the main points that recommend the human cord blood (CB) as the most valuable solution against other stem cell sources, as detailed below.

2. The human umbilical cord: A source of stem cells

UCB has been widely considered an important stem cell source, because of its many pluses as compared with other stem cell sources. CB usually comes as a second choice after matched BM or PB but in the case of children suffering from leukaemia it represents the main stem cell source for transplant. (Stanevsky et al., 2009). Moreover, CB is also viewed as a primary stem cell source due to annual global human birth rate of more than 100 million a year. CB is easily and safely collected by CB banks and preserved as future therapeutic genetic material (Arien-Zakay et al., 2010). CB has been more and more considered as good alternative for embryonic stem cells lately (Figure 1 A), also because it has proved to contain populations of multipotent stem cells which are able to differentiate into a variety of cell types, including epithelial, endothelial, myotubes and neural [Harris & Rogers, 2007).

The human UC, a connecting tissue of extraembryonic origin, connects the fetus to the placenta. Developed during the fifth week of embryogenesis it reaches a final length of approximately 60–65 cm and a weight of about 40 g, and has a mean diameter of 1.5 cm in normal pregnancies. UC usually comprises two arteries and a vein, all immersed within a tissue that interconnects vessels- the so-called Wharton's jelly (WJ) and enclosed by a simple amniotic epithelium. WJ is a mucoid connective tissue abounding in proteoglycans and hyaluronic acid (HA), insulating and protecting umbilical vessels from torsion, compression, or bending. It is made of embedded stromal cells (Figure 1B). It thus assures a constant blood flow between fetus and placenta. (Fan et al., 2011; Malgieri et al., 2010). The blood remaining in the umbilical vein after birth is rich in hematopoetic stem and progenitor cells, and has been a successful alternative allogeneic donor source in the cure of a variety of pediatric genetic, hematologic, immunologic, and oncologic disorders (Broxmeyer et al., 1989; Gluckman et al., 1997; Han et al., 2003, Kim et al., 2002 as cited in Malgieri at al., 2010). Moreover, fresh cord blood is rich in non-hematopoietic stem cells, and also contains endothelial cells, mesenchymal stem cells (MSCs) and unrestricted somatic stem cells (USCC) (Kogler et al., 2004, 2006; Sensken et al., 2007; Greschat et al., 2008 as cited in Malgieri at al., 2010).

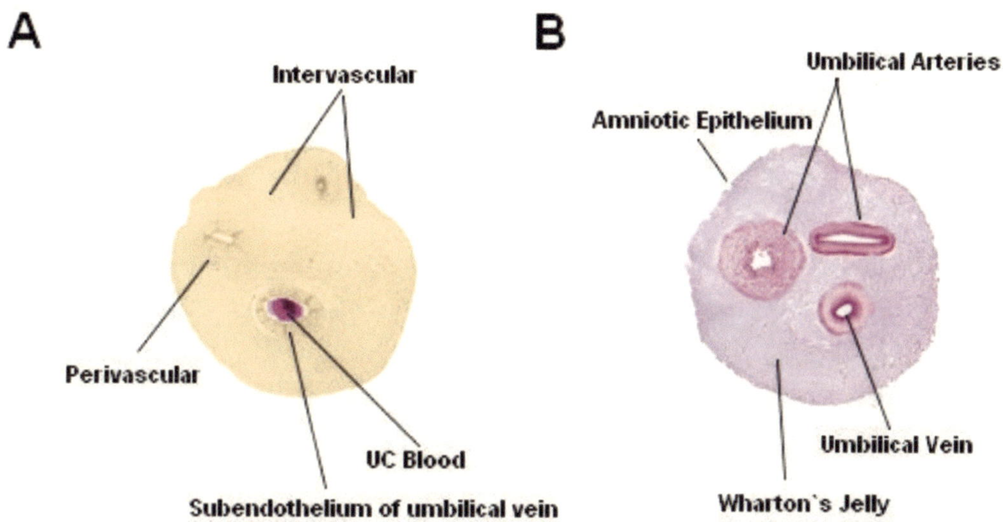

Fig. 1. Umbilical cord: cross section. (A)Localization of mesenchymal stromal cells. (B) Structure of umbilical cord wich contains two arteries and one vein surrounded by Wharton`s jelly and amniotic epithelium.

MSCs are refined as multipotent, undifferentiated cells that can self renew, regenerate mesenchymal tissues and blood cells and differentiate into several cell types such as chondrocyte, adipocyte, osteocyte, osteoblast, myocyte, cardiomyocyte, neuron-like cell as well as into insulin-producing cells (Jiang et al., 2002; Mareschi et al., 2006). There is not surprise, then, that they are the focus of regenerative medicine and tissue engineering (Arien-Zakay et al.,2010). Despite the fact that BM has been considered the main available source of MSCs (Pittenger et al., 1999), it is not always acceptable to use BM –derived cells because of the high degree of viral exposure, the significantly decreasing cell number and the proliferative/differentiation capacity along with age. Additionally, BM sampling involves a painfully invasive procedure. Therefore, it is of utmost clinical importance in so far as the easy accessibility and the reduced morbidity are concerned to try to identify alternative sources of MSCs (Malgieri et al., 2010). Anyway, Mitchell et al. (Mitchell et al., 2003) successfully isolated the matrix cells from porcine and human umbilical cord by explants' culture only in 2003 and, in it was only in the same year that Romanov et al. (Romanov et al., 2003) managed to isolate mesenchymal like cells from sub-endothelial layer of human umbilical cord vein. Consequently, it is only since 2003 that the umbilical cord-derived MSCs have started to be extensively researched (Fan et al., 2003). Human UC blood-derived mesenchymal stem cells (hUCB-MSCs) exhibit ability similar to that of BM-MSCs for multi-lineage differentiation (Gang et al., 2004).

Consequently, UCB has proved a legitimate source for haematopoietic stem cell transplantation. Moreover, the development of research and scientific understanding of hUCB-MSCs will show that they are a really fascinating source to be used in stem cell therapy.

3. Isolation of MSCs from the umbilical cord

In recent years, several investigators published protocols for isolating MSCs from the UC tissue. These protocols have been developed and adapted based on the region of the cord from where the cells are harvested. Schematic illustration of harvesting protocols is given in Figure 2.

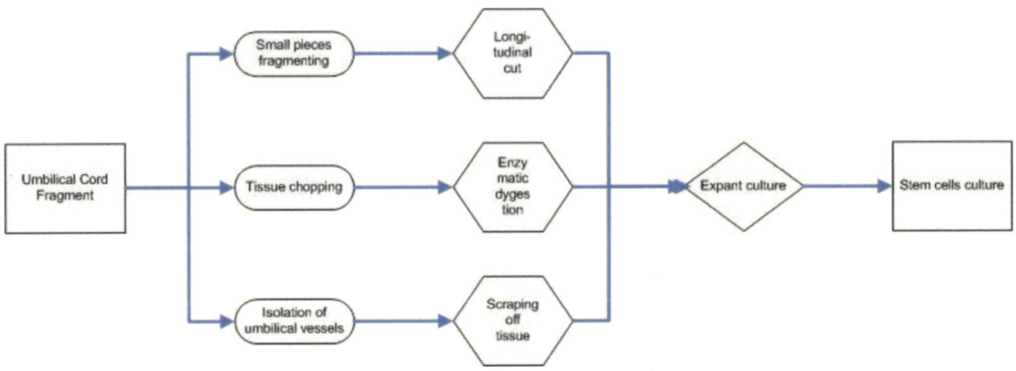

Fig. 2. Schematic illustration of various protocols used to isolate mesenchymal stem cells from the umbilical cord tissue.

The first step in the isolation procedure is represented by the removal of umbilical artery and vein followed by the splitting of the cord into smaller segments which are subsequently enzymatically digested (Weiss et al., 2006). Another techniques involve isolation methods without enzymatic digestions or removal of vessels (La Roca et al., 2009). Several methods were also described for the isolation of the cells from the perivascular tissue or the subendothelium of the umbilical vein (Covas et al., 2003).

Currently, the most used protocol is without enzymatic digestion in an explant culture approach without removal of umbilical vessels in order to isolate MSC-like cells from the whole UC tissue.

At the beginning, the blood from the umbilical vessels is removed by flushing phosphate buffered saline (PBS) through the artery and vein using a sterile syringe coupled to a small

tube. The next step is represented by the preservation of UC in a PBS medium enriched with 5g L^{-1} glucose, 50 ugmL^{-1} gentamicin, 2.5 ugmL^{-1} amphotericin B, 100 U.mL^{-1} penicillin, and 100 ugmL^{-1} streptomycin with the purpose of reducing the risk of contaminations. Then, the UC is serially cut first into 10-15 cm long segments and then into 0.5-0.8 cm^3 large pieces. During the isolation procedure, transfer medium is used to keep the cord and the minced pieces moist. Finally, the small pieces are transferred to 25 cm^3 flaks and incubated in a humidified atmosphere with 5% CO2 in a MEM supplemented with 15% of human serum and 50 ugm^{-1} gentamicin at 37°C which is changed every 48 hours. After 12 days, groups of adherent cells from single tissue pieces are observed. At this stage, the UC tissue is removed and the adherent cells are harvested by enzymatic treatment. The cell suspension is then centrifuged at 100 g for 10 min and the cells are resuspended in MEM supplemented with 10% human serum and 50 mgmL-1 gentamicin and subcultured at a density of 4000 cells cm~2. These culture conditions have demonstrated to support an optimal growth of the cells (Wobus et al., 2006).
Commonly, the isolated cells exhibited a high proliferation potential. For at least 20 population doublings, the cells can be expanded without loss of proliferative activity and viability (Figure 3). After approximately 50 population doublings the process of replicative senescence becomes common. At this point, UC-derived cells could be efficiently cryopreserved and revitalized in a cryo-medium containing 80% human serum, 10% culture medium, and 10 % DMSO.

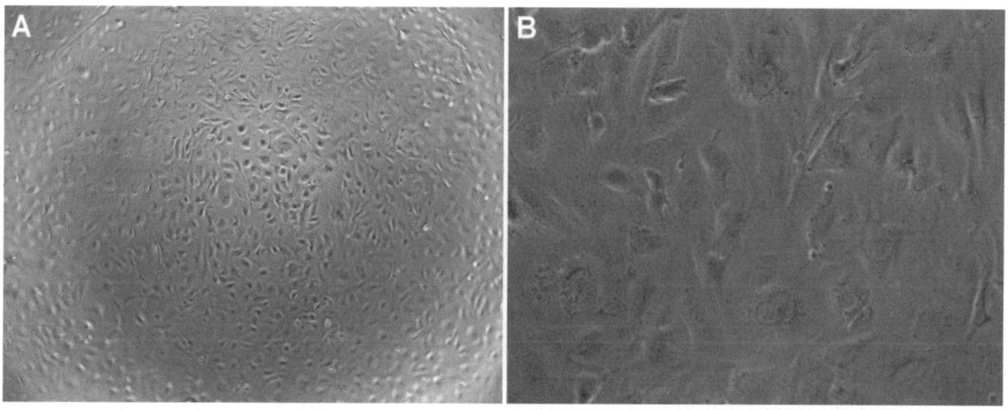

Fig. 3. Isolation of mesenchymal stem cells from umbilical cord: adherent growing monolayer of fibroblast-shaped cells after 21st days of culture (Magnification: A-10X, B-40X).

Depending on the isolation procedures, stem cells characteristics such as proliferation, differentiation potential and immunologic properties may be altered.

4. Characterization of UCSC

The evaluation of the nature of UCB HSC has demonstrated a higher proportion of primitive hematopoietic progenitors in UCB, with superior in vitro proliferative responses and in vivo engraftment capacity as compared to adult BM (Lewis et al., 2000; Wynter et al., 1999 as cited in Barker & Wagner, 2003) or PB (Barker & Wagner, 2003; Goldberg et al., 2007). Single UCB units can reconstitute entire lympho-hematopoietic systems in adults patients (Laughlin et al., 2001). In addition, UCB seems more tolerant of HLA mismatches, with less

GVHD. These unique biological features are further sustained by the fact that human UCB transplantation (UCBT) recipients show reliable myeloid and lymphoid hematopoietic reconstitution after myeloablative conditioning with greater than one log lesser nucleated cell (NC) dose than is used in BM transplantation (BMT).

Another unique characteristic of UCB is the reduced alloreactive response as compared with that seen with BMT, which is not fully explained. The median CD3+ cell dose of 8×106 kg^{-1} in UCB units makes it similar to a BM graft after modest T cell depletion (TCD). Such a T cell dose is fully capable of inducing significant GVHD, particularly in the setting of HLA mismatch. Therefore, reduced T cell number is insufficient to explain the relatively low incidence of severe GVHD with 1–2 antigen-mismatched UCBT. A more likely explanation is the functional differences reported with UCB lymphoid cells. These include higher levels of CD4+CD45RA+ cells, lower inflammatory cytokine production and responses, lower alloantigen and mitogen-specific T cell proliferation, a polyclonal T cell receptor repertoire, increased susceptibility to tolerance induction, and differences in NK cell and dendritic cell biology (Barker &Wagner, 2003; Canque et al., 2000).

In recent years, much excitement has been associated with studies that have suggested UCB does represent a source of multi-potent stem cells that could be used for generation of non-hematopoietic tissues. MSCs were isolated from the mononuclear cell fraction of CB using various criteria, including adherence to the surface of the culture medium, negative expression of hematopoietic markers such as CD34, CD133, CD45, CD14, and positive expression of mesenchymal markers such as CD90, CD105, CD73 (Flynn et al., 2007). The isolated UCB-MSCs were shown to differentiate into a wide range of cell types including bone, cartilage, cardiomyocytes, neural among others. Pre-clinical studies also revealed that MSCs from CB support ex vivo expansion and differentiation of haematopoietic stem/progenitor cells from CB. These MSCs improved the haematopoietic cells engrafted in non-obese diabetic/severe combined immune-deficient (NOD/SCID) mice (Huang et al., 2007). It is worth mentioning that there is no fully identified unique phenotype for MSCs derived from CB, as well as for MSCs from all other origins. This prevents reproducible isolation of MSC precursors with predictable potential for development, and their isolation and characterisation rely primarily on their ability to adhere to plastic and their ability to expand (Arien-Zakay et al., 2010). In the culture, the morphology of hUCB-MSCs demonstrates a typical MSCs immunophenothypic markers and fibroblastoid morphology. The absence of endothelial CD31 and leukocyte surface markers supports the involvement of UCB-MSCs as mesenchymal progenitors. Very important, as it was already mentioned, hUCB-MSCs are devoid of hematopoietic and endothelial markers such as CD14, CD28, CD31, CD33, CD34, CD45, CD56, CD133, HLA-DR, and for graft versus host disease CD80, CD86, CD40, which show that they are suitable for transplantation (Malgieri et al., 2010). The comparison between hUCB-MSCs and BM-MSCs showed that hUCB-MSCs share most of their immunophenotype with BM-MSCs, including a cluster of differentiating markers, neural markers and extracellular adhesion molecules as well as the cell cycle status, the adipogenic and osteogenic differentiation capacity and finally the cytokines and hematopoietic supportive function (Malgieri et al., 2010). Despite this, recent studies noted that there are still several differences between them. Firstly, the fibroblast colony-forming units (CFUF) frequency was significantly higher in UCB derived nucleated cells than in BM derived nucleated cells, which practically indicated a higher frequency of MSCs in the nucleated cells of UCB than in those of BM. Secondly, the proliferation analysis revealed that hUCB-MSCs have a faster population doubling time, as compared to BM-MSCs, that not change after 30 passages. UCB-MSCs also had a higher proliferative capacity of in comparison with BM-MSCs indicating that hUCB-MSCs may be a novel alternative source of human MSCs for clinical applications (Malgieri et al., 2010). Furthermore, Lu et al. (2006) noted the absence of HLA-DR expression on hUCB-MSCs and low expression of HLA-ABC, known as a hurdle for allogenic cell therapies, which may favour the use of hUCB-MSCs for allogenic cell therapy.

In a still growing set of experiments, another population of CB stroma cells was isolated, using the CD45 negative expression and the adherence abilities. These cells were termed unrestricted somatic stem cells (USSCs) (Kogler, 2004, 2006). They were reported to differentiate in vitro into osteoblasts, chondrocytes, adipocytes and neural progenitors as well as in vivo into bone, cartilage, haematopoietic cells and neural and heart tissue (Kim et al., 2005; Kogler et al., 2004). Furthermore, it was even suggested that because USSCs can easily be committed towards the mesenchymal cell lineage, they may represent an early mesodermal progenitor of MSC (Arien-Zakay et al. 2010).

5. *In vitro* differentiation potential of UCSC

5.1 Differential potential into osteogenic, chondrogenic and adipogenic lineages

HUCB-MSCs hold tremendous promise for tissue engineering and regenerative medicine as exposed to appropriate conditions and various factors they can differentiate in vitro along several cell lineages (Figure 4) of three germ layers such as the chondrocyte, osteoblast, adipocyte, myocyte, cardiomyocyte, endothelium, neuronal, astrocyte, oligodendrocyte pancreatic , hepatocyte lineages and others (as further detailed below) (Fan et al., 2011). Several researches reported that certain stimuli direct CB-derived MSCs and USSCs to differentiate into osteoblasts (Jager et al., 2007 as cited in Arien-Zakay et al., 2010; Kogler et al., 2006; Lee et al., 2004 as cited in Arien-Zakay et al., 2010). A recent study comparing the cell-mediated remodeling of three-dimensional collagen I/III gels during osteogenic differentiation of BM-MSCs and hUCB-MSCs showed that both cell types display all features needed for effective bone fracture healing (Schneider et al., 2010). On the contrary, comparing the the chondrogenic potential of hBM-MSCS and hUCB-MSCs it was noted that hUCB-MSC group had 3-times as much collagen as the hBM-MSC group, which supports the former may be a more desirable option for fibro-cartilage tissue engineering (Wang et al., 2009 as cited in Fan et al., 2011). Furthermore, the cell differentiation into chondrocytes was significantly augmented by bone-morphogenetic protein (BMP)-4, a mesodermal factor known to promote chondrogenesis. Additionally, the successful transformation of hUC stroma-derived stem cells into mature adipocytes supports their therapy for esthetic purposes (Karahuseyinoglu et al., 2008 as cited in Fan et al., 2011).

5.2 Differential potential into cardiomyocytes

Attempts to transdifferentiate CB-MSCs towards cardiomyocytes have yielded different results. 5-azacytidine, a chemical analogue of the cytosine nucleoside in the DNA and RNA helix, is currently employed to initiate myogenic differentiation. In spite of the fact that some researches successfully transformed Wharton's jelly derived hUC-MSCs into cardiomyocytes by 5-azacytidine or cardiomyocyte-conditioned medium and noted a slight spontaneous beating (Pereira et al., 2008; Wang et al., 2004 as cited in Fan et al., 2011), others failed to generate cardiomyocyte-like cells from hUC-MSCs, either spontaneously or after treatment with 5-azacytidine (Martin-Rendon et al., 2008). For instance, Roura et al. (2010)

failed to transdifferentiate CB-MSCs towards cardiomyocytes in vitro-the cells did not express cardiomyocyte-specific proteins or presence of calcium rhythmic oscillations or potential-dependent fluorescence emission after various protocols, including addition of chemicals (5-azacytidine and dimethyl sulphoxide), growth factors, Wnt signaling activators and direct contact with neonatal rat cardiomyocytes. Similarly, in other studies there wasn't' any evidence for neocardiomyocyte formation after systemic delivery of CB-mononuclear cells, or direct intramyocardial delivery of CB-133+ cardiac membrane cells, suggesting that there is a limitation in the potential for differentiation of unmodified CB-derived stem cells into cardiomyocytes (Murry et al., 2004). However, MSCs isolated from endothelial/subendothelial layers of the human umbilical cord veins have been proved to have the potential of transdifferentiating into cardiomyocyte-like cells with typical ultrastructure and sarcomers as well as expression of several cardiac-specific genes (Kadivar et al., 2006). Further studies are warranted to reduce the discrepancy between these research papers. However, they suggest that hUC-MSCs can be chemically transformed into cardiomyocytes and considered as a source of cells for cellular cardiomyoplasty.

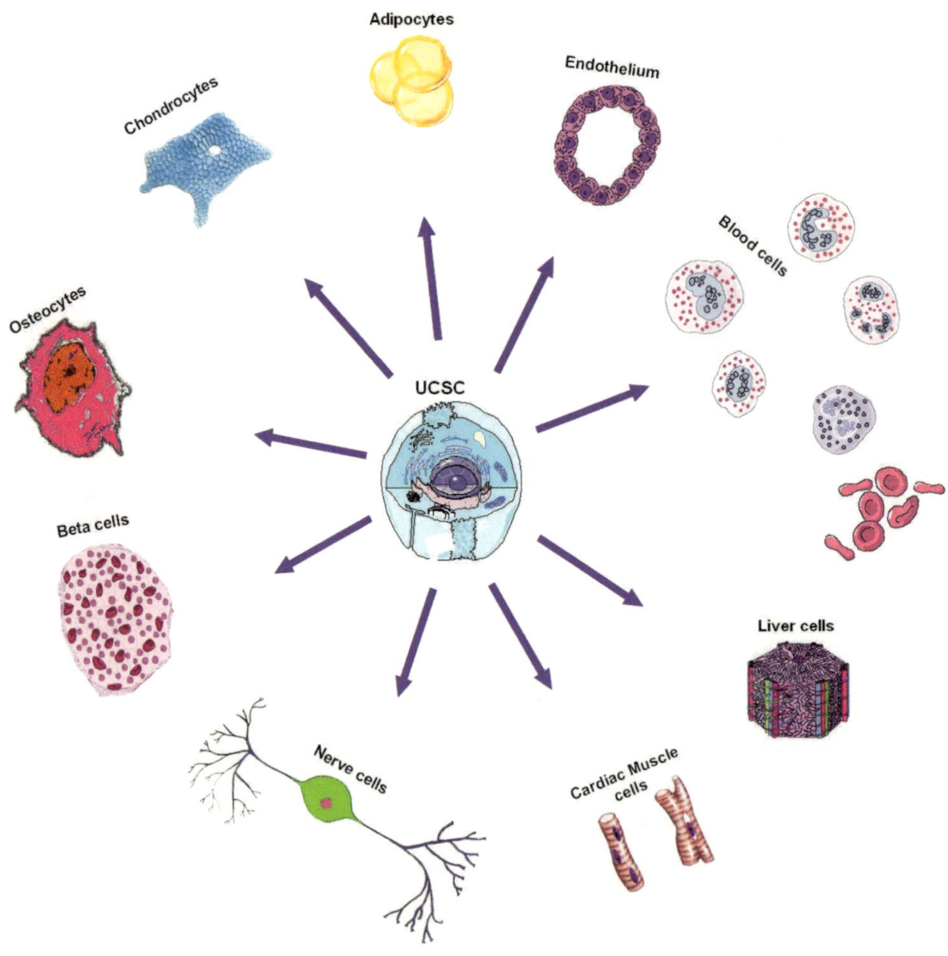

Fig. 4. Umbilical cord mesenchymal stem cell differentiation.

5.3 Differential potential into endothelial cells

Searching for sources with higher availability of endothelial progenitor cells, (Wu et al. 2007) cultured hUC-MSCs in an endothelial differentiation medium containing vascular endothelial growth factor (VEGF) and basic fibroblast growth factor (bFGF) hUC-MSCs differentiated into endothelial cells as demonstrated by acetylated low-density lipoprotein incorporation and expression of endothelial-specific proteins, such as platelet/endothelial cell adhesion molecule (PECAM) and cd34 (Wu et al., 2007). In comparison of endothelial differentiation of hUC-MSCs and BM-MSCs, some researches noted that hUC-MSCs had higher proliferate potential, higher expression of the endothelial-specific markers after induction and significantly higher total tubule length, diameter and area in angiogenesis assay than those of differentiated BM-MSCs (Chen et al., 2009).

Current studies have addressed the differentiation capacity of CD133 hematopoietic stem cell (HSC), a subset of UCB HSC. In particular, Peichev et al. (2000) have shown that cells isolated from CB expressing both CD34 and VEGFR2 (KDR) also expressed CD133. CD133 was not expressed on mature endothelial cells. The investigators further characterized these cells by illustrating their ability to migrate and differentiate into mature endothelial cells (Peichev et al., 2000). A subsequent study also demonstrated the differentiation of CB-derived CD34+ CD133+ cells into endothelial phenotype in vitro (Pesce et al., 2003).

5.4 Differential potential into neural cells

During the past decade, numerous in vitro studies have demonstrated the generation of neuronal cells from various CB-derived populations. For instance, embryonic stem (ES)-like cell population from CB were found to possess neuroglial progenitor morphology and primitive neuroglia cell markers (Mcguckin et al., 2004). Chen et al. (2005) detected positive expression for antigens typical of brain tissue in other adherent cell subpopulation from CB. CB-derived populations purified by antigen expression were also shown to differentiate into a neural phenotype. CB-CD133 positive cells (Jang et al., 2004 as cited in Arien-Zakay et al., 2010) or CD34− CD45− non-haematopoietic stem cells (Habich et al., 2006 as cited in Arien-Zakay et al., 2010) differentiated into neuronal and glial cells that expressed neuronal and astrocyte-specific markers. Moreover, the full range of neural differentiation ability of the CB-CD45+ multipotential stem cells was demonstrated by the achievement of positive phenotypic and functional markers for dopaminergic neurons, oligodendrocytes and astrocytes (Rogers et al., 2007).

For others, the differentiation potency of hUC-MSCs into neural lineage has attracted extensive attention. The pioneering study in the neuronal induction of hUC-MSCs was accomplished by Mitchell et al. (2004) using a relatively complex and multi-step neuronal induction procedure previously defined (Mitchell et al., 2004; Woodbury et al., 2000 as cited in Fan et al., 2011). In another experiment, hUC-MSCs began to express neuron-specific proteins and exhibit retraction of cell body, elaboration of processes, and clustering of cells after three days' induction with neuronal conditioned medium (NCM). Glutamate invoking inward current was found in the transformed cells between the 9th and 12th days, which suggested that the induced cells differentiated into mature neurons in the post mitosis phase at this stage (Fu et al., 2004). Using a three-step neural induction protocol 60% of the neuroglial cells transformed from hUC-MSCs were stained positive for microtubule-associated protein (MAP-2) and 32% stained positive for GFAP. Moreover, some of them expressed TuJ-1, synaptophysin and γ-aminobutyric acid (GABA). Expression of neuronal markers, such as neuron-specific nuclear protein (NeuN) and MAP2, by induced hUC-MSCs were also observed by other investigators. More specifically, other investigators observed that induction with protocol similar to Woodbury et al. (2000) resulted in a lower number of cells expressing markers for early neural progenitors (i.e., nestin), however, a greater number of cells expressed a mature neural marker for catecholaminergic cells - tyrosine hydroxylase (TH) (Ding et al. 2007; Kadam et al., 2009; Weiss et al., 2006 as cited in Fan et al., 2011). In addition, Fu et al. (2004) also succeeded in transforming hUC-MSCs into dopaminergic neurons in vitro through stepwise culture in NCM, sonic hedgehog, and FGF-8. These dopaminergic neurons were shown to express the rate-limiting catecholaminergic synthesizing enzyme, tyrosine hydroxylase and to release dopamine into the medium (Fu et al. 2006).

5.5 Differential potential into hepatocyte-like cells

Cirrhosis is a consequence of chronic hepatic injury characterized by replacement of liver tissue by fibrosis and scar tissue, with no effective therapy currently available. Recent developments in stem cell technology have struggled for identifying novel candidate sources of liver cells to be used for regenerative purpose. Instead of ESCs with ethical debates and safety concern, fetal or adult liver cells encumbered by organ availability, BM-MSCs plagued with decreased expansion and differentiation capacities in advanced ages (Roobrouck et al., 2008), Campard et al. (2008) demonstrated that in vitro expanded UC-MSCs constitutively expressed markers of hepatic lineage and genes of enzymes involved in hepatic metabolism after three steps of full term hepatogenic induction. However, absence of some hepatic markers in differentiating UC-MSCs, such as HepPar1 or HNF-4, implied that their differentiation did not reach the level of mature hepatocytes (Campard et al., 2008). A simple one-step induction protocol with hepatocytic growth factor (HGF) and fibroblast growth factor-4 (FGF-4) was also proved to be effective in transforming hUC-MSCs into hepatocyte-like cells expressing the hepatocyte-specific markers albumin (ALB), human α-fetoprotein (AFP) and cytokeratin 18 (CK-18) (Zhang et al., 2009 as cited in Fan et al.,2011). Further research showed that differentiated hUC-MSCs could store glycogen and uptake LDL (Zhang et al., 2009 as cited in Fan et al., 2011). Very recently, UCB derived HSC were also induced to hepatocyte like cells under a 2-step protocol with the combination of the same two growth factors-HGF and FGF-4. Hepatocyte like cells were observed at the end of the protocol (days 14). These differentiated cells were observed to show high expression of genes related to hepatocytes (tryptophan 2, 3-dioxygenase [TO], glucose 6-phosphate [G6P], CK 18, ALB and AFP (Sellamuthu et al., 2011). In addition, the differentiated functional hepatocyte-like cells may still retain their low immunogenicity in vitro (Zhao et al., 2009), which facilitate the allotransplantation to replace the diseased liver cells.

5.6 Differential potential into skeletal muscles

The first successful myogenic transformation of a subset of CD105(+)/CD31(-)/KDR(-) cells from WJ of UC into elongated, multinucleated cells expressing of Myf5 and MyoD in vitro was reported by Conconi et al.(2006).

5.7 Differential potential into Islet-like clusters

In the study by Wang et al. (2010) hMSCs from umbilical cord stroma were induced to differentiate into insulin-producing cells using differentiation medium. Pancreatic beta-cell development-related genes were expressed in the differentiated insulin-producing cells. Differentiated cells' C-peptide release in vitro increased after glucose challenge (Wang et al., 2010). Another recently published research also showed that WJ- derived hUC-MSCs may serve as a promising alternative cell source of transplantable islet-like clusters (Chao et al., 2008 as cited in Fan et al., 2011). Besides differentiating fresh human UCB into insulin-secreting cells, Phuc et al. (2011) showed that also MSCs derived from banked cord blood (cryopreserved for 1 year) can be differentiated into functional pancreatic islet-like cells in vitro. Having in mind their advantages including large potential donor pool, rapid availability, low risk of rejection and no risk of discomfort for the donor, hUC-MSCs might become an excellent candidate in β-cell replacement therapy of diabetes mellitus.

6. Immune properties of MSCs and in vivo applications

Besides their multi-lineage differentiation potential, UC-derived MSCs have been shown to have immune suppressive action on lymphocyte proliferation through alloantigen and mitogens such as phytohemagglutinin and to reduce the level of proinflammatory cytokines such as interferon-y (IFN-y) and tumor necrosis factor-a (TNF-a). These immune-privileged and immune-modulator properties, recommend them as ideal candidates for cell-based therapies. They fail to induce proliferation of allogeneic lymphocytes in vitro and do not induce an immune response when used in allogenic mismatched animal experimental models (Devine et al., 2003). Furthermore, they have regulatory effects on several cells of the immune system (e.g., T, B, dendritic, and natural killer cells) (DiNicola et al., 2002), prolong skin graft survival (Bartholomew et al., 2002) and have been used in clinical applications to reduce acute and chronic GVHD (Ringden et al., 2006). It has been demonstrated that UCB-MSCs can suppress not only the function of mature dendritic cells but also increase the portion of regulatory T cells (LeBlanc et al., 2004). This regulation of immune response by MSCs is mediated by soluble factors and cell to cell contact mechanisms. Co-culture experiments showed that UCMSCs did not induce any proliferation of resting immune cells and also suppressed the purified T cells or activated human peripheral blood mononuclear cells.

7. In vivo applications of UCSC

7.1 Ischemic vascular disease

7.1.1 Myocardial infarction

In spite of the fact that the treatment of cardiovascular disease and myocardial infarction has benefited over the past 15 years from advances in pharmacologic and intravascular intervention to include placement of drug-eluting stents and interventional re-vascularization, this condition is till associated with significant morbidity and mortality. Furthermore, the fact that there still aren't eligible donors for heart transplant candidates, combined with the limited ability of endogenous cardiomyocytes to divide and repair infarcted myocardium has encouraged the use of stem cell transplantation as adjuvant therapy for cardiovascular disease (Abbott & Giordano, 2003; Ramakrisham et al., 2003 as cited in Goldberg et al., 2007).

Both small clinical trials in humans and animal studies employing cellular therapies have produced data suggesting the efficacious nature of this modality lately. However, as far as patient-derived (adult marrow or mobilized peripheral blood) stem cells are concerned there are distinct disadvantages for their routine clinical application including the need to subject the patient to stem cell or large volume BM collection at a time of active cardiovascular disease and the needed time for cell culturing for some type of these cells (Goldberg et al., 2007). In addition, there is data suggesting that in patients either at high risk for future cardiovascular disease events or who have documented coronary disease, the body's endogenous stem cell supply and marrow response may be inadequate, as circulating endothelial progenitor cells isolated from patients with increased cardiovascular risk and compared to those isolated from healthy subjects are reduced in number and demonstrate increased degeneration in vitro (Goldberg et al., 2007; Hill et al., 2003).

Additional cellular sources such as embryonic and fetal cardiomyocytes have been studied as a means to improve the failing myocardium (Ramakrisham et al., 2003 as cited in Goldberg et al., 2007). However, besides ethical consideration and possible immune rejection, which are common to both lines, there are other several limitations such as the potential for malignant transformation and for malignant arrhythmias with respect to embryonic stem cells (Abbott & Giordano, 2003; Ramakrisham et al., 2003 as cited in Goldberg et al., 2007) as well as limited availability and sensitivity to ischemia of fetal cardiomyocytes (Etzion et al., 2001; Ramakrisham et al., 2003 as cited in Goldberg et al., 2007) that may preclude their clinical application, encouraging consequent search for other sources of stem cells.

Encouraging results were obtained increasingly from in vivo animal trials using CB stem cells to rejuvenate infracted myocardium, enhance healing and improve ventricular function. It was shown that the human CB mononuclear cells administered into the infracted border progressively improved the function of the left ventricle and reduced the size of infarction in rats with induced acute myocardial infarction (AMI) (Henning et al., 2004). Furthermore, Ma et al., (2005) who experimented with NOD/SCID mice with induced AMI, reported migration of CB mononuclear cells to the heart in 50% of the mice. The size of the infarct zone thus decreased and collagen deposition, as well as increased capillary density were detected, but no myogenesis. Further studies directed towards the ability of UCB CD34+ to improve cardiac function in a model of coronary ischemia were carried out by Hirata et al. (2005). In these experiments, AMI was induced in conditioned male Wistar rats and CB-derived CD34+ cells were injected into the peri-infarct zone. 4 weeks after transplantation, reduction of the left ventricular dilation and improvement of the cardiac function were reported. Moreover, intramyocardial CB cells survived, localized around the vessels of the ischaemic myocardium and capillary density was increased, but only about 1% of the injected cells were incorporated into the myocardium vessels of the rat, indicating that the angiogenic factors released by these cells account for the angiogenesis (Hirata et al., 2005). Similarly, significantly improved cardiac function, markedly increased capillary and arteriole density, and notably decreased apoptotic cells were reported by Wu et al. (2007), who investigated the therapeutic potential of hUC-MSCs in a rat myocardial infarction model. Moreover, it seems that some of transplanted cells managed to live in the infracted myocardium, gathered around arterioles and spread in capillary networks. Noteworthy is the fact that some of them exhibited cardiac troponin-T, von Willebrand factor, and smooth muscle actin, proving that the damaged myocardium regenerated by cardiomyocytic, endothelial, and smooth muscle differentiation of hUC-MSCs in the infracted myocardium (Wu et al., 2007). A specific subset of CB- CD34+ cells, the CD34+ KDR+ cell fraction was suggested as the factor for angiogenesis and improvement of cardiac function for both treatment with mononuclear cells and CD34+ cells (Botta et al., 2004). In a study by Botta et al. (2004), the myocardium of a NOD/SCID mouse was injected with the CD34+ /KDR+ double selected fraction, isolated from UCB, immediately after the ligation of the left anterior descending artery (LAD). Apoptosis was impaired in this cell population versus CD34+ /KDR- cells in extended culture (up to 22 days). The release of VEGF from the CD34+ /KDR+ cells is thought to provide an autocrine mechanism for survival. The left ventricular end-diastolic pressure decreased as well as the infarct size in the transplanted group for a period up 5 months after AMI induction. The treatment with this cell type also determined more angiogenesis compared with mononuclear cells or CD34+ (Botta et al., 2004). Moreover, immunostaining for human nuclei showed localization of human cells within cardiomyocytes. However, it was not clear whether the transplanted cells had fused, overlapped with or differentiated into cardiomyocytes (Botta et al., 2004; Goldberg et al., 2007). Similar results were reported by Leor et al. (2006), using another immature CB subset, CD133+. CB cells were infused into athymic nude rats with ligated coronary arteries and one month after infusion CB cells had migrated, colonized and survived in the infracted myocardium, being identified near the vessel walls and in the left ventricular cavity. Functionally, the left ventricular fractional shortening improved and the anterior wall thickness diminished (Leor et al., 2006). Ma et al. (2006) reported a comparison of CD133+ expressing cells from UCB and BM in an AMI model. Selected UCB CD133+ cells were injected into injured myocardium of NOD/SCID mice and 4 weeks later the analysis illustrated an approximate 30% absolute mortality reduction in UCB CD133+ -treated animals versus control. Further histological analysis focusing on capillary density produced a 25% increase in this number in cell treated animals versus controls (Ma et al., 2006). In a large animal study, Kim et al. (2005) used a porcine model with chronically occluded LAD and injected CB USSCs directly into the infracted tissue. The grafted cells were detected in the tissue 4 weeks later, the perfusion and wall motion of the infracted area improved whereas the scar thickness decreased (Kim et al., 2005). Overall, these studies suggest a positive effect of a variety of different UCB-derived cell types including cells expressing CD34 and CD133. There are significantly larger numbers of hemangioblasts in UCB which have the same efficacy as that of the BM stem cells. After cell transplantation the ejection fraction and hemodynamics showed cardiac function improvement. It seems that ex vivo long-term cultures do not to affect their ability to increase cardiac function post-infarction (Goldberg et al., 2007).

7.1.2 Buerger's disease

MSCs from CB were also proposed for the treatment of hind limb ischemia. In a pre-clinical study, ischemia was induced in the hind limb of athymic nude mice by femoral artery ligation and human UCB-derived MSCs were transplanted into the ischemic area. Up to 60% of the hind limbs were saved in the femoral artery-ligated animals and the human UCB-derived MSCs were detected by in situ hybridization in the arterial walls of the ischemic hind limb in the treated group (Kim et al., 2006). Buerger's disease, also known as thromboangiitis obliterans, is a nonatherosclerotic, inflammatory, vaso-occlusive disease, characterized as a panangiitis of medium and small blood vessels, including both arteries and adjacent veins, especially of the distal extremities (the feet and the hands). No curative medication or surgery for this disease has been discovered so far. The efficacy of CB-derived MSCs was verified in the treatment of limb ischemia and pain in a clinical case study of Buerger's disease (Kim et al., 2006). HLA-matched human UCB-derived MSCs were transplanted into four men with Buerger's disease who had already failed previous medical and surgical therapies. As a result, ischemic rest pain suddenly disappeared from their affected extremities, the necrotic skin lesions were healed, the number and size of digital capillaries increased and peripheral circulation improved (Kim et al., 2006), indicating that human UCB-derived MSC transplantation may be a new and useful therapeutic tool for Buerger's disease and similar ischemic diseases.

7.2 Metabolic diseases

7.2.1 Diabetes mellitus

In 2000 there were 150 million people in the world affected by diabetes mellitus (DM) and this number is estimated to double in 2025 (Zimmet et al., 2001 as cited in Fan et al., 2011). Moreover, in the US there are approximately, 15000 youth newly diagnosed with type 1 DM every year (Malgieri et al., 2010). Type 1 DM is the outcome of a cell-mediated autoimmune attack against pancreatic β-cells and the autoimmune response precedes the clinical diagnosis. When hyperglycaemia appears the autoimmune process is significantly advanced due to the fact that 70-80 % of the β-cell mass has been already damaged when the disease started (Couri & Voltarelli, 2008 as cited in Arien-Zakay et al., 2010). Therefore, besides tight control of blood glucose, self blood glucose monitoring and patient education, which have a significant role in the prevention of the development and retard the progression of chronic complication, β-cell mass preservation and/or increase are important targets in management of type 1 DM. Additionally, the replacement of a patient's islets is the only treatment of type 1 DM that achieves an insulin-independent, constant normoglycaemic state and avoids hypoglycaemic episodes (Bretzel et al., 1993 as cited in Fan et al., 2011). Recently, efforts have focused on the use of both autologous and allogenic stem cells as sources of new islets, and perhaps more surprisingly, as potential sources of safe and effective immunomodulation (Limbert et al., 2008). In vitro and in vivo studies have shown that hUCB-derived stem cells and CB-derived MSCs can differentiate into insulin-secreting β-cells (Reddi et al., 2010) expressing pancreatic β-cell markers and synthesizing and secreting functional islet proteins (Gao et al., 2008; Parech et al., 2009 as cited in Arien-Zakay et al., 2010). In the study by Wang et al. (2010) hMSCs from umbilical cord stroma were induced to differentiate into insulin-producing cells using differentiation medium. Pancreatic beta-cell development-related genes were expressed in the differentiated insulin-producing cells. Differentiated cells' C-peptide release in vitro increased after glucose challenge. Further, in vivo glucose tolerance tests showed that blood sugar levels decreased following cells' transplantation into nonobese diabetic (NOD) mice (Wang et al., 2010). Another recently published research also showed that WJ- derived hUC-MSCs may be a promising source of cells of transplantable islet-like clusters (Chao et al., 2008 as cited in Fan et al., 2011). Besides differentiating fresh human UCB into insulin-secreting cells, Phuc et al. (2011) showed that also MSCs derived from banked cord blood (cryopreserved for 1 year) can be differentiated into functional pancreatic islet-like cells in vitro. These results suggest that if human MSCs, especially MSCs from banked cord blood of diabetes patients themselves can be isolated, proliferated, differentiated into functional pancreatic islet-like cells, and transplanted back into them (autologous transplantation), their high-proliferation potency and rejection avoidance will provide one promising therapy for diabetes (Phuc et al., 2011). Recent reports evidenced also that WJ derived-MSCs possess immunomodulatory activities (acting on both innate and acquired immunity effectors) which should reduce the immunogenicity of transplanted cells, thus decreasing rejection. Moreover it has been proposed that MSC administration should be used to alleviate the autoimmune processes which lead to the destruction of beta cells (Anzalone et al., 2010). Besides reducing blood glucose levels and increasing survival in mouse models of type 1 and type 2 DM (Ende et al., 2004 as cited in Arien-Zakay et al., 2010), CB cell infusion was also proved to improve renal abnormalities and neuropathy caused by DM, suggesting a regenerative action in renal parenchyma and nerves (Ende et al. 2004; Naruse et al., 2005 as cited in Arien-Zakay et al., 2010). Autologous hUCB-MSCs are currently being evaluated in a clinical trial to treat type 1 DM in children (Haller et al., 2008). Moreover, autologous CB infusion in subjects with type 1 DM has been proved not only safe but also efficient by Haller et al. One-year follow-up suggest that autologous CB transfusion is safe and provides some slowing of the loss of endogenous insulin production in the treated children. However, the lack of significant adverse events associated with the study seems to have been the most important result. Nevertheless, we need broader randomized studies as well as 2-year post-infusion follow-up of this cohort to establish whether autologous CB-based approaches can be employed to decrease the endogenous insulin production in children with type 1 DM. This current study is still recruiting children with newly diagnosed type 1 DM, who also have their own CB deposited in a CB bank, for phase II trials. Consequently, the treatment of auto-immune type 1 and 2 DM may greatly benefit from the two positive effects of CB, namely its regenerative abilities and immunomodulatory function (Arien-Zakay et al., 2010).

7.2.2 Hurler syndrome

The mucopolysaccharidoses (MPSs) are inherited metabolic disorders (IMDs) caused by single-gene defects leading to progressive cellular accumulation of glycosaminoglycans (GAGs) and damage to multiple organs, including the central nervous, musculoskeletal or cardiorespiratory systems. Hurler syndrome, the most severe form, is the prototypical model. At present, while available for some conditions, exogenous enzyme replacement therapy is unable to correct cognitive and central nervous system disease because of its inability to cross the blood-brain barrier. In contrast, allogeneic HSC transplantation (HSCT) allows donor-derived, enzyme-producing cells to migrate to the brain and other organs to provide permanent enzyme therapy and thus help somatic organs, improve neurocognitive function and quality of life, and prolong survival, particularly when performed early in the course of the disease. HSCT may also mediate non-hematopoietic cell regeneration or repair. Bone marrow has been the graft source in the past. However, in the last 5 years many patients have been treated with unrelated donor (URD) UCBT, allowing rapid and increased access to transplantation with favorable outcomes (Prasad & Kurtzberg, 2010 a, 2010 b). Sustaining this, CB transplantations led to improved neurocognitive performance and decreased somatic features of Hurler's syndrome (Staba et al., 2004 as cited in Arien-Zakay et al., 2010) and therefore it might be considered as a preferred source of cells for treatment of this disease, as recently suggested after risk factor analysis of patients from the European Group for Blood and Marrow Transplantation (EBMT) (Boelens et al., 2007, 2009].

7.3 Heart valve replacement

Currently used replacements in cardiovascular surgery are made of foreign materials with well known disadvantages such as thrombo-embolic complications, infections, loss of functional and biological properties, and others. (Breymann et al., 2006). Tissue engineering of viable, autologous cardiovascular constructs with the potential to grow, repair, and remodel represents a promising new concept for cardiac surgery, especially for pediatric patients with congenital cardiac defects. Currently, vascular myofibroblast cells (VC) represent an established cell source for the construction of cardiovascular tissue. Cell isolation requires the invasive harvesting of venous or arterial vessel segments before scaffold seeding, a technique that may not be preferable, particularly in pediatric patients. (Kadner et al., 2002). It has been increasingly shown that, considering their excellent growth properties and tissue formation with biomechanical properties approaching native tissue in vitro, hUC cells seem to represent a promising alternative autologous cell source for cardiovascular tissue engineering, offering the additional benefits of using juvenile cells and avoiding the invasive harvesting of intact vascular structure (Breymann et al., 2006). Hoerstrup et al. (2002) successfully constructed living, autologous pulmonary artery conduits tissue with hUC cells (Hoerstrup et al., 2002 as cited in Fan et al., 2011), while another group published their successful attempt in tissue engineering of autologous human heart valves using cryopreserved vascular umbilical cord cells (Sodian et al., 2006). These results show that establishing autologous human cell banks for pediatric patients diagnosed intrauterinely with congenital defects and likely to need heart valve replacement in the early years of life is of utmost importance (Sodian et al., 2006). Thus, tissue engineering of autologous heart valves with the potential to grow and to remodel is likely to become a fructuous idea. Sustaining these observations, more recently, the same authors described the successful use of cryopreserved umbilical cord blood-derived CD133(+) cells as a single cell source for the tissue engineering of viable human heart valves (Sodian et al., 2006).

7.4 Neurological disorders

7.4.1 Stroke

Ischemic stroke with its immediate neurological injury due to interruption of blood flow to the brain, is a medical emergency of utmost severity: it can trigger severe functional defects in the brain, permanent neurological damage, complications and even death. If it is not quickly diagnosed and treated, it can turn the patient into a huge social and economic burden. The treatment of the stroke has been under constant research by several novel neuron-restorative approaches and it has become the main target for stem cell therapy. In this respect, a variety of stem cells are currently being studied in this research for best implant resource (Fan et al., 2011).

The ability and safety of CB and derived cell populations to protect against neurological deficits was proved in vivo in a variety of injuries models. The first evidence of a therapeutic effect of hUCB came from an experiment where rat was used for middle cerebral artery (MCA) occlusion to induce ischemia. Intravenous administration of hUCB resulted in improved behavioral deficits after stroke in rats (Sanberg et al., 2005). In another recently study the researchers examined the effects of hUCB-MSCs in canine thromboembolic brain model (Chung et al., 2009). Cerebral ischemia was induced through occlusion of the MCA by injecting thrombus emboli into 10 beagles. Infarct volume decreased 1 day after hUCB- MSCs were transplanted through the basilar artery in canine cerebral ischemia whereas infarct volume was increased in the control groups (Chung et al., 2009). Transplanted hUCB-MSCs were differentiated into neurons and astrocytes in and around endothelial cells and expressed neuroprotective factors, such as brain-derived neurotrophic factor (BDNF) and VEGF, at 4 wk transplantation. Additionally, Jeong et al. reported that transplantation of hUCB- MSCs into contralateral regions of injured rat brain at 7 day after injury resulted in significant behavioral improvement and that PKH26-labelled hUCB- MSCs differentiated into neural cells at the injured site 4 wk after transplantation (Jeong et al., 2006 as cited in Kim et al., 2010). Similarly, Chen et al. (2001) demonstrated in a medial carotid artery occlusion model of stroke that upon

CB intravenous administration, many of the physical and behavioral deficits associated with this disease were improved (Chen et al., 2001 as cited in Arien-Zakay et al., 2010). These observations were further supported by other groups, monitoring the dependency of the conferred beneficial effects in the CB cell dose (Vendrame et al., 2004 as cited in Arien-Zakay et al., 2010) and the timing of transplantation after injury (Newcomb et al., 2006 as cited in Arien-Zakay et al., 2010). As mechanisms, the observed beneficial effects afforded by CB therapies were suggested to include reduced inflammation (Vendrame et al., 2006 as cited in Arien-Zakay et al., 2010), apoptotic protection and a combination of trophic actions and nerve fibre reorganization (Xiao et al., 2005 as cited in Arien-Zakay et al., 2010). This observation is suggested to be mediated via the release of growth and repair factors triggered by anoxia (Newman et al., 2006 as cited in Arien-Zakay et al., 2010) as was explored by Arien-Zakay et al. (2009) in vitro, who showed that CB-derived progenitors confer protection towards an insulted neuron by a mechanism involving the release of antioxidants and neurotrophic and angiogenic factors (Arien-Zakay et al., 2009). In another experiment approximately $1*10^6$ clonally expanded hUCB-MSCs were transplanted into the cortex of MCA occlusion rat models. Consequently, significant improved neurological function as well as considerably increased cortical neuronal activity was observed. The transplanted hUCB-MSCs migrated towards the ischemic boundary zone and differentiated into glial, neuronal, doublecortin$^+$, CXCR4$^+$ and vascular endothelial cells. Moreover, hUCB-MSCs transplantation promoted the formation of new vessels to increase local cortical blood flow in the hemispheric and significantly increased expression of neurotrophic factors (Ding et al., 2006 as cited in Fan et al., 2011). It was hypothesized that modulation by stem cell-derived macrophage/microglial interactions and increased β1-integrin expression might enhance the angiogenic architecture as well as plasticity of the ischemic brain after the implantation of hUCB-MSCs (Ding et al., 2006 as cited in Fan et al., 2011). Koh et al. also noted improved neurobehavioral function, reduced infarct volume and increased nestin-positive endogenous stem cells in the hippocampus after transplantation of $6*10^5$ hUCB-MSCs into the damaged hemisphere of immunosuppressed ischemic stroke rats (Koh et al., 2008 as cited in Fan et al., 2011), but only few transplanted MSCs expressed detectable levels of neuron-specific markers. The authors believed that improvement in behavioral function might be connected to the neuroprotective effects of hUCB-MSCs which resulted in increase of endogenous neurogenesis and reduction of infarct volume rather than the formation of new networks between host neurons and the implanted hUCB-MSCs (Koh et al., 2008 as cited in Fan et al., 2011). In agreement with these results, other investigators also observed that transplanted hUCB-MSCs survived for at least 5 weeks in the ischemic brain, significantly reduced injury volume and neurological functional deficits of the subjected rats, widely incorporated into cerebral vasculature with partly differentiation into endothelail cells, and substantially increased vascular density in ipsilateral hemisphere of stroke. It seems that angiogenesis can thus mediate the mechanism for neurological functional recovery after hUCB-MSCs transplantation (Liao et al., 2009 as cited in Fan et al., 2011) and that transplantation of hUCB-MSCs can be useful in clinical trials for ischemia. However, there are serious issues involved, such as the route of transplantation, type of injected cells (hUCB vs hUCB-MSCs) and timing of transplanatation when using stem cell therapy with hUCB-MSCs.

7.4.2 Spinal cord injuries

The consequences of spinal cord injury (SCI), one of the major disabilities which clinical rehabilitation settings face, are: loss of neurons, degeneration of axons, formation of glial scar, and severe functional impairment (Hu et al., 2010). It also involves many negative factors in that the patients suffer from motor and sensory impairments as well as many other complications during their lifetimes. The restoration of damaged spinal cords has been under many clinical trials during the last two decades but the pharmacological therapies used in clinical settings are few and have limited effects on the regeneration, recovery speed, or retraining of the spinal cord (Hyun & Kim, 2010). Due to their ability to differentiate into neural cell lines replacing non functional tissue, stem cells have attracted the interest of researches. Efforts have been made to establish new synapses and provide a conducive environment. This involved grafting cells from autologous and fetal sources; including embryonic or adult stem cells, Schwann cells, genetically modified fibroblasts, bone stromal cells, and olfactory ensheathing cells and combinations/ variants thereof (Sobani et al., 2010) as well as umbilical cord blood cells (Zhao et al., 2004 as cited in Fan et al., 2011). The potential of transplanted CB and derived populations for the treatment of SCI is under investigation because of their advantages in comparison with other stem cell sources. It was reported that the result of the transplantation of both CB whole fraction and derived CD34 cells was the localization of these cells around the site of injury and improved functional recovery (Nishio et al., 2006; Saporta et al., 2003 as cited in Arien-Zakay et al., 2010). Furthermore, the transplantation of hCB in a spinal cord-injured female patient was associated with improved sensory perception and mobility, both morphologically and functionally and regeneration of the spinal cord at the injured site and some of the cauda equine below it, as shown by computed tomography (CT) and magnetic resonance imaging (MRI) (Kang et al., 2005 as cited in Arien-Zakay et al., 2010). Kuh et al. studied the effects of transplanted HUCBs ± BDNF on moderate degree spinal cord injured rats. The HUCBs transplanted group improved, more than the control group at every week after transplantation, and also, the BDNF enabled an improvement of the Basso, Beattie and Bresnahan (BBB) locomotion scores since the 1 week after its application. 8 weeks after transplantation, the HUCBs with BDNF transplanted group had more greatly improved BBB scores, than the other groups. Additionally, the transplanted HUCBs were differentiated into various neural cells. The HUCBs and BDNF each have individual positive effects on axonal regeneration. The HUCBs can differentiate into neural cells and induce motor function improvement in the cord injured rat models. Especially, the BDNF has effectiveness for neurological function improvement due to axonal regeneration in the early cord injury stage. The authors concluded that the HUCBs and BDNF have recovery effects of a moderate degree for cord injured rats (Kuh et al., 2005). Phase I/II clinical trials are in progress in the China Spinal Cord Injury Network (http://www.clinicaltrials.gov), aiming to determine the feasibility, safety, efficacy and optimal dose of CB mononuclear cell transplantation as a method to treat patients with chronic SCI (Arien-Zakay et al., 2010). Due to their high ability to differentiate into nerve like cells hUCB-MSCs hold great promise as tools for understanding development and as therapeutic agents for spinal cord injury. Transplantation of hUCB-MSCs into the injured spinal cord may have the following functions-: compensation for demyelination; removal of inhibition; promotion of axonal regeneration; direction of axons to appropriate targets and lost cells replacement (Malgieri et al., 2010). A promising research reported by Yang et al. (Yang et al., 2008 as cited in Fan et al., 2011) provided evidence that transplantation of WJ-derived hUC-MSCs was an effective strategy to promote the regeneration of corticospinal fibers and locomotor recovery after spinal cord transaction in the rat. The results were as follows: several weeks after transplantation of hUC-MSCs, significant improvements in locomotion and fewer astrocytes in the lesion site, activated microglia in rostral and caudal stumps of the lesion as well as increased numbers of regenerated axons in the corticospinal tract and neurofilament - positive fibers around the lesion site were observed as compared with the control group (Yang et al., 2008 as cited in Fan et al., 2011). Moreover, transplanted hUC-MSCs survived as long as 16 weeks, migrated from the implantation site for about 1.5 mm in the caudal direction of the rostrocaudal axis and produced large amounts of human neutrophil-activating protein-2, neurotrophin-3 (NT-3), bFGF and VEGF receptor 3 in the host spinal cord, which may help spinal cord repair (Yang et al., 2008 as cited in Fan et al., 2011). The regeneration of several corticospinal axons after transplantation of hUC-MSCs is likely to be sustained by the release of more cytokines or growth factors from the undifferentiated stem cells rather than differentiation of these cells into neuronal or glial cells. The authors also believe that the implanted hUC-MSCs are able to modulate the activities of microglia and reactive astrocytes (Yang et al., 2008 as cited in Fan et al., 2011). In tight connection with this, Zhang et al. (2009) noted that in the case of WJ cells-derived neurospheres transplanted in combination with BDNF into transected spinal cord rats, the number of survivor grafted cells was very small while BBB scores and axonal regeneration significantly increased and reduced cavitations were noticed. These results might be explained through the axonal regeneration and neuroprotective action activated by the grafted cells (Zhang et al., 2009 as cited in Fan et al., 2011). In another recent study, the treatment with hUC-MSCs proved to facilitate functional recovery of hindlimb locomotor function after traumatic spinal cord injury in rats, at 5 wks after transplantation. This recovery was accompanied by increased length of neurofilament-positive fibers and increased numbers of growth cone-like structures around the lesion site. Transplanted hUC-MSCs survived, migrated over short distances, and produced large amounts of glial cell line-derived neurotrophic factor and neurotrophin-3 in the host spinal cord. There were fewer reactive astrocytes in both the rostral and caudal stumps of the spinal cord in the hUC-MSCs group than in the control group (Hu et al., 2010). Mesenchymal stem cells can also differentiate into Schwann-like cells. Yan-Wu et al. (2011) induced hUC-MSCs in vitro into neurospheres constituted by neural stem-like cells, and further into cells bearing strong morphological, phenotypic and functional resemblances with Schwann-like cells. These HUMSC-derived Schwann-like cells, after grafting into the injured area of the rats' SCI, had a partial therapeutic effect in terms of improving the motor function. Furthermore, significant improvement in rats treated by Schwann-like cell grafting combined with NT-3 administration was demonstrated in the behavioral test as compared with that in animal models received the cell grafting only, suggesting that this combined administration may represent a new strategy of stem cell therapy for spinal cord injury (Yan-Wu et al., 2011). Taking into account these encouraging pre-clinical and clinical trials, reporting improvement of neurological deficits together with the increasing amount of data on the differentiation of CB-populations into a variety of neuronal phenotypes it may be assumed that CB will become an important factor for treatment of neurological illnesses.

7.4.3 Parkinson's disease

Several studies also investigated the CB therapeutic potential in neurodegenerative diseases, including Parkinson's disease (PD), Alzheimer's disease and amyotrophic lateral sclerosis. PD is a neurodegenerative disorder characterized by the progressive loss of striatal dopaminergic function. To date, a variety of stem cells have been explored to find a promising therapy for this unsatisfactorily treated disease in clinical settings (Yasuhara & Date, 2007 as cited in Fan et al., 2011). Fu et al. (2006) showed that in vitro hUC-derived

MSCs transformed into dopaminergic neurons expressing the rate-limiting catecholaminergic synthesizing enzyme, thyrosine hydroxylase and releasing dopamine into the medium (Fu et al., 2006). Transplantation of these dopaminergic neurons into the striatum of Parkinsonian rats could partially relieve the lesion-induced amphetamine-evoked rotation, indicating an amelioration of motor deficits. Similar results were demonstrated by Weiss et al. (2006), who also treated rat models affected by PD with hUC-MSCs. The transplantation of approximately 1,000 undifferentiated hUC-MSCs into striatum of hemiparkinsonian rats without immune suppression produced an amelioration in apomorphine-induced rotations in the pilot test without evidence of formation of brain tumors as well as a frank host immune rejection response. The hUC-MSCs produced significant amounts of glial cell line-derived neurotrophic factor (GDNF), one of the most potent trophic factors for dopaminergic neurons and FGF (Weiss et al., 2006 as cited in Arien-Zakay et al., 2010). More importantly, the TH-positive DA neurons in the substantia nigra (SN) and ventral tegmental area (VTA) showed a valid correlation between the number of cells and the number of apomorphine-induced rotations. Therefore, the behavioral recovery of PD model animals may contribute to rescue of the degenerating DA neurons in the SN and VTA (Weiss et al., 2006 as cited in Arien-Zakay et al., 2010). It appears that Li et al. (2010) are the first who described the induction of human umbilical vein mesenchymal stem cells (HUVMSCs) into dopaminergic-like cells. In their study, HUVMSCs were induced in vitro into neurospheres constituted by neural stem-like cells, and further into cells bearing strong morphological, phenotypic and functional resemblances with dopaminergic-like cells. These HUVMSC-derived dopaminergic-like cells, after grafting into the brain of a rat model of PD, showed a partial therapeutic effect in terms of the behavioral improvement. Moreover, nerve growth factor (NGF) administration significantly promoted the survival of the grafted cells in the host brain and enhanced the content of dopaminergic in the local brain tissue and behavioral test demonstrated a significant improvement of the motor function of the PD rats after dopaminergic-like cell grafting with NGF administration as compared with that of rats receiving the cell grafting only, which might indicate this combination as a good and new strategy of stem cell therapy for PD (Li et al., 2010).

7.4.4 Amyotrophic Lateral Sclerosis (ALS)

Stem-cell transplantation is an attractive strategy for neurological diseases and early successes in animal models of neurodegenerative disease generated optimism about restoring function or delaying degeneration in human beings (Silani et al., 2004). Considering the lack of effective drug treatments for amyotrophic lateral sclerosis (ALS), and compelling preclinical data, stem-cell research has highlighted this disease as a candidate for stem-cell treatment. In a mouse model of ALS, CB cells transplantation resulted with delayed disease progression and improved survival of diseased mice, while the transplanted cells survived 10-12 weeks after infusion entering regions of motor neuron degeneration in the brain and spinal cord (Garbuzova-Davis et al., 2003 as cited in Arien-Zakay et al., 2010), as monitored according to transplanted cell dose (Garbuzova-Davis et al., 2008 as cited in Arien-Zakay et al., 2010). ALS may also benefit from combined gene/stem-cell approaches. In the study by Rizvanov et al. (2011), mononuclear fraction of hUCBCs were transfected by electroporation with dual plasmid constructs, simultaneously expressing VEGF(165) and human fibroblast growth factor 2 (FGF(2)) (pBud-VEGF-FGF(2)). These genetically modified hUCBCs were injected retro-orbitally into presymptomatic ALS transgenic animal models ((G)93(A) mice).

The report results demonstrated that genetically naïve hUCBCs may differentiate into endothelial (CD34+) and microglial (iba1+) cells; however when over-expressing VEGF-FGF(2), hUCBCs transform into astrocytes (S100+). Autocrine regulation of VEGF and FGF(2) on hUCBCs, signal molecules from dying motor neurons in spinal cord, as well as self-differentiating potential may provide a unique microenvironment for the transformation of hUCBCs into astrocytes that eventually serve as a source of growth factors to increase the survive potential of surrounding cells in the diseased regions (Rizvanov et al., 2011).

7.4.5 Alzheimer's disease

Alzheimer's disease (AD) is a proeminent neurodegenerative disease, marked clinically by insidious dementia. The neuropathological hallmarks of AD include the presence of extracellular amyloid-beta peptide (Abeta) in the form of amyloid plaques in the brain parenchyma and neuronal loss. The mechanism associated with neuronal death by amyloid plaques is unclear but oxidative stress, glial activation and inflammatory changes have been implicated (Lee et al., 2010a, 2010b). The umbilical cord blood stem cells have been studied also in AD and they were shown to markedly diminish the beta-amyloid plaques and associated astrocytosis in an AD mouse model (Nikolic et al., 2008 as cited in Arien-Zakay et al., 2010). In addition, two very recent studies have demonstrated that hUCB-MSCs can be an efficient potential therapeutic agent in AD. Therefore, transplanted hUCB-MSC into amyloid precursor protein (APP) and presenilin1 (PS1) double-transgenic mice significantly improved spatial learning and memory decline and dramatically reduced amyloid-beta peptide (Abeta) deposition, beta-secretase 1 (BACE-1) levels and tau hyperphosphorylation. Interestingly, these effects were associated with reversal of disease-associated microglial neuroinflammation, as evidenced by decreased microglia-induced proinflammatory cytokines, elevated alternatively activated microglia, and increased anti-inflammatory cytokines. These results lead us to suggest that hUCB-MSC produced their sustained neuroprotective effect by inducing a feed-forward loop involving alternative activation of microglial neuroinflammation, thereby ameliorating disease pathophysiology and reversing the cognitive decline associated with Abeta deposition in AD mice (Lee et al., 2010 a). Moreover, the same authors examined in vitro the potential impact of hUCB-MSCs treatment on neuronal loss using a paradigm of cultured hippocampal neurons treated with Abeta, confirming that hUCB-MSCs co-culture reduced the hippocampal apoptosis induced by Abeta treatment. Furthermore, in an acute AD mouse model to directly test the efficacy of hUCB-MSCs treatment on AD-related cognitive and neuropathological outcomes, they shown that markers of glial activation, oxidative stress and apoptosis levels were decreased in AD mouse brain. Interestingly, hUCB-MSCs treated AD mice demonstrated cognitive rescue with restoration of learning/memory function (Lee et al., 2010 b).

7.4.6 Infantile cerebral paresis and other cerebral impairments

Traumatic brain injury is also a potential target for treatment using CB cells as it was first described by Lu et al. (2002) in a rat model. The transplanted cells migrated into the site of brain lesion and a decrease of neurological damage was documented (Lu et al. 2002 as cited in Arien-Zakay et al., 2010). Moreover, brain damage around birth may trigger lifelong neurodevelopmental deficits and cerebral palsy is estimated to affect 10 000 infants annually. Brain damage of the hypoxic-ischemic kind around birth is also treatable with CB cells. Using brain-damaged neonatal rats, Meyer et al. (2006) observed that the neurological effects of cerebral palsy were corrected- spastic paresis was largely alleviated and CB mononuclear cells were incorporated in the lesion brain area without obvious signs of transdifferentiation (Meyer et al., 2006 as cited in Arien-Zakay et al., 2010). The feasibility of CB collection, preparation and autologous infusion in babies born with signs of brain injury is now being tested in a pilot study conducted by Duke University (http://www.clinicaltrials.gov) (Arien-Zakay et al., 2010).

7.5 Hemoglobinopathies. Fanconi anemia

Allogeneic stem cell transplantation is the only curative option for patients with hereditary bone marrow failure syndromes. But because nearly two-thirds of patients requiring HSCT will not have a suitable related donor, the applicability of HSCT to larger numbers of patients has been augmented with the increasing availability of unrelated donors. Presently, alternative HSC sources include unrelated donor (URD) bone marrow (BM) or peripheral blood stem cells (PBSC) and unrelated donor umbilical cord blood (UCB). While URD BM and PBSC transplants have a proven track record of success, the search process takes 3-4 months, which is often longer than patients with high risk disease can wait. Despite nearly 13 million registered volunteer donors worldwide, nearly half of patients still do not have a closely HLA-matched donor. The applicability of HSCT markedly expanded with the introduction of UCBT, particularly for racial and ethnic minorities (Smith & Wagner, 2009). The first related donor UCBT was performed in 1988 (Gluckman et al., 1989 as cited in Barker & Wagner, 2003) and the first URD UCBT in 1993. Since those first reports, it has become clear that UCB is a safe and effective source of HSC for transplant. Moreover, the demonstration that hematopoiesis could be reliably reconstituted after myeloablative conditioning prompted the development of repositories of banked, HLA-typed UCB. With approximately 350,000 units banked worldwide (http://www.bmdw.org), the addition of UCB to the available stem cell sources makes it possible for nearly everyone who requires an HSCT to have a suitable donor available (Smith & Wagner, 2009). Sustaining this, a very recent multicenter, retrospective study, based on data reported to the Eurocord Registry about patients with hereditary bone marrow failure syndrome (other than Fanconi anemia) who underwent umbilical cord blood transplantation showed that related UCBT is associated with excellent outcomes while increasing cell dose and better HLA matching might provide better results in unrelated UCBT (Bizzeto et al., 2011).

Fanconi anemia (FA) is a rare autosomal recessive disease characterized by excessive chromosomal breakage, congenital abnormalities, progressive bone marrow failure and a predisposition to leukemia and epithelial malignancies (Smith AR & Wagner JE, 2009). The hematopoietic stem cell transplantation is the only curative treatment of this disease (Bielorai et al., 2004).

The first CB transplant was performed in 1988 in a patient with Fanconi anemia. The donor was his HLA-identical sister who was known by pre-natal diagnosis to be HLA identical and not affected by the Fanconi mutation. To date, there have been no formal comparisons between the alternative donor sources. As with hemoglobinopathies, URD transplant remains the gold standard until retrospective or prospective comparative trials can be performed. However, for patients who do not have an HLA-matched donor or who cannot wait the time it takes to complete a donor search, UCB is a reasonable alternative (Smith & Wagner, 2009).

More comparative studies are needed before definitive conclusions can be made in nonmalignant diseases, but the available data indicate UCB as a feasible alternative HSC source in most patient populations. In general, though, 8/8 HLA-matched BM remains the 'gold standard' for alternative donor HSCT, but UCB should be considered a reasonable option in those that do not have such a donor available and for those in whom the time to transplant is critical, such that waiting for an URD BM would not be in the best interest of the patient. Current practice has been to select the UCB unit that is 0-2 HLA antigen-mismatched with the patient with the highest cell dose. Further efforts focused on increasing the number, HLA diversity and quality of stored UCB units as well as addressing cell dose limitations using strategies, such as double UCB transplant and ex vivo expansion of a single unit are needed to continue to advance the field of UCBT (Barker & Wagner, 2003; Smith &Wagner, 2009).

7.6 Hematological malignancies

Hematological malignancy is the most common indication for allogeneic HSCT in both children and adults. The choice of HSC source depends on both patient and disease characteristics and the urgency of the transplant. Because of the rapid availability of units, UCB is a particularly attractive option (Smith & Wagner, 2009).

7.6.1 In children

Infant leukemia is a particularly challenging form of leukemia to treat and the decision of whether to treat with intensive chemotherapy or to proceed with URD transplant when a suitable related donor is not available is a difficult one.

On behalf of the Center for International Blood and Marrow Research (CIBMTR), Eapen et al. (2007) recently reported the outcomes of 785 children with acute leukemia comparing outcomes in recipients of UCB (n=503) and URD BM (n=282). The most notable finding was that UCB compared favorably to the 'gold standard' of 8/8 allele-matched unrelated BM. In fact, the 5-year leukemia-free survival (LFS) was similar after 8/8 matched unrelated BM (MUBM), mismatched unrelated bone marrow (MMUBM) and mismatched UCB (MMUCB) with higher survival in recipients of matched UCB (MUCB). The incidence of acute and chronic GVHD was similar between the groups. While treatment-related mortality (TRM) was higher after two-antigen MMUCB, a lower risk of relapse resulted in comparable survival outcomes for this cohort. This research was unique in that UCB was compared to the present day standard of allele-level HLA-matched BM donors. These findings support the use of HLA-matched or -mismatched UCB in children with high risk acute leukemia who need transplantation. (Eapen et al 2007 as cited in Smith &Wagner, 2009).

7.6.2 In adults

In contrast to the outcomes in children, HSCT in adults is typically associated with higher risks of GVHD, infections, delayed immune reconstitution and increased TRM, partly related to a higher likelihood of comorbidities at the time of transplant. In contrast to children, use of UCB in adults has been more restricted due to cell dose limitations. The safety and feasibility of UCBT in adults with hematological malignancies was first reported in 2004 (Laughlin et al., 2004; Rocha et al., 2004 as cited in Smith &Wagner, 2009).

The outcomes in 1240 adults (148 UCB, 243 MUBM, 111 MMUBM, 518 matched PBSC MPBSC and 210 mismatched PBSC MMPBSC) were examined and the results were published recently in an abstract on behalf of the CIBMTR (Eapen et al., 2008 as cited in Smith &Wagner, 2009). Contrary to the previous data (Laughlin et al., 2004; Rocha et al, 2004 as cited in Smith & Wagner, 2009), all unrelated donor BM and PBSC grafts were matched at allele level for HLA-A, -B, -C, and -DRB1. In this analysis, TRM was lower and LFS higher when MUBM and MPBSC were used as compared to the other sources, suggesting that these graft sources are preferred when available and time permits. However, partially HLA-matched UCB with an adequate cell dose (≥ 2.5 × 10^7 nucleated cells/kg) is a suitable alternative when an HLA-matched URD is not available or when the transplant is urgent (Eapen et al., 2008 as cited in Smith &Wagner, 2009).

Analyzing a more genetically heterogeneous group of patients, Kumar et al. (2008) also showed superior outcomes in UCB recipients relative to those transplanted with other sources of HSC. Patients receiving UCB had the lowest TRM and highest 3-year LFS (61% vs. 27%, 13% and 14% in the matched related donor, matched unrelated donor (MURD) and mismatched unrelated donor (MMURD) groups, respectively). Taken together, these results at least advocate for continued investigations into the use of UCB as an alternative stem cell source for the treatment of adults with hematological malignancy (Kumar et al., 2008 as cited in Smith & Wagner, 2009).

In summary, these retrospective studies suggest for children, the first line HSC source would be a 6/6 MUCB provided that the cell dose is adequate. The probability of finding a 6/6 MUCB, however, is low (~10%). However, results in recipients of 8/8 MUBM and 5/6 MMUCB and 4/6 MMUCB are similar, suggesting that any of these options are reasonable. In this case, the decision must be individualized and based on the urgency of the transplant and potential need for future donor lymphocyte infusion (DLI). However, it is clear that with UCB, TRM increases with each degree of HLA-mismatch, so higher cell doses are needed with more HLA disparity. In addition, PBSC or BM is often a more realistic HSC option in adults due to cell dose limitations. Therefore, in adults, cell dose limitations with UCB units give the advantage to HLA matched BM and PBSC. For all patients, if an 8/8 MUBM is not available, no one source stands out. UCB has the advantage of rapid availability while BM and PBSC have the advantage of availability of DLI (Smith & Wagner, 2009).

7.7 Liver diseases

Liver cell transplantation is a promising technique for the treatment of metabolic liver disease, liver fibrosis and other end stage liver diseases. In vivo investigations showed that in carbon tetrachloride (CCl_4)-induced cirrhosis in a rat model, hUCB-MSCs infusion demonstrated inhibition of TGF-β, collagen type I and α-smooth -muscle-actin (α-SMA) expression as well as expression of hepatocyte-specific markers human ALB and AFP in injured liver (Jung et al., 2009), which indicated they might become an alternative source for liver-directed cell therapies (Campard et al., 2008). Several weeks after transplantation into the lesion livers of CCl_4-induced liver fibrosis rats, significant reduction in liver fibrosis with lower levels of serum glutamic oxaloacetic transaminase, glutamic pyruvate transaminase, α-SMA and TGF-β in the liver was observed. Moreover, up-regulated expression of hepatic mesenchymal epithelial transition factor-phosphorylated type (Met-P) and hepatocyte growth factor was also found (Tsai et al., 2009 as cited in Fan et al., 2011). The engrafted hUCB-MSCs did not differentiate into hepatocytes expressing human ALB or AFP but secrete a variety of bioactive cytokines, including human cutaneous T cell-attracting chemokine, leukemia inhibitory factor, and prolactin, which may benefit the restoration of liver function and promotion of regeneration (Tsai et al., 2009 as cited in Fan et al., 2011). Similar data were noted by Yan et al. (2009). In similar CCl-4 -injured mouse hepatic injury, hUCB-MSCs were discovered to express tryptophan 2, 3-dioxygenase, human AFP, CK18, α-SMA after administration into the damaged livers. Transplanted hUCB-MSCs could reduce hepatocyte denaturation, inhibit hepatocyte apoptosis, diminish serum aminotransferases and facilitate hepatocyte proliferation (Yan et al., 2009 as cited in Fan et al., 2011).

Lin et al. (2010) tested the use of WJ stem cells (WJSCs) in treating chemically induced liver fibrosis via intraperitoneal injection of thioacetamide. They noticed that transplanted cells were distributed in the fibrotic area and around blood vessels, and hepatic recovery was accelerated. Serum prothrombin time significantly recovered, and serum albumin also improved at 21 days posttransplantation whereas collagen accumulation also decreased at 14 days. Thus, human WJSCs promoted recovery after chronic liver damage. Moreover, transplanted WJSCs produced albumin, hepatocyte growth factor (HGF), and metalloproteinase (MMP) after transplantation to chemically injured liver, indicating that WJSC may help to decrease liver collagen and thus may be useful for treating liver fibrosis (Lin et al., 2010). According to all these results, hUCB-MSCs could be useful in liver therapy. It should not be forgotten that liver contains endogenous abundant progenitor cells for recovery of liver damage. Therefore, it is currently difficult to determine which stem /progenitor cell populations are best for liver disease therapy (Kim et al., 2010).

7.8 Lung diseases

Progression of acute respiratory distress syndrome is documented by loss of lung tissue as a result of inflammation and fibrosis. The role of term human UC cells derived from WJ with a phenotype consistent with MSCs was tested in the treatment of a bleomycin-induced mouse model of lung injury. Injected MSCs were located in the lung 2 weeks later only in areas of inflammation and fibrosis but not in healthy lung tissue. The administration of MSCs decreased inflammation, inhibited the expression of TGF-beta, IFN-gamma, and the proinflammatory cytokines macrophage migratory inhibitory factor and TNF-α and reduced collagen concentration in the lung. MSCs also increased matrix metalloproteinase-2 levels and reduced their endogenous inhibitors, tissue inhibitors of matrix metalloproteinases, favoring a pro-degradative milieu following collagen deposition. Notably, injected human lung fibroblasts did not influence either collagen or matrix metalloproteinase levels in the lung. The results of this study suggest that MSCs participate in anti-fibrosis in lung injury and may increase lung repair if used to treat acute respiratory distress syndrome (Moodley et al., 2009). Others examined whether intratracheal or intraperitoneal transplantation of human UCB-derived MSCs can attenuate hyperoxia-induced lung injury in immunocompetent newborn rats. Wild-type rats were randomly exposed to 95% oxygen or air from birth and in the transplantation groups, a single dose of PKH26-labeled human UCB-derived MSCs was administered either intratracheally (2×10^6 cells) or intraperitoneally (5×10^5) cells) at postnatal day 5. The harvested lungs were examined after 10 days and two abilities of hUCB-MSCs, immune modulation and differentiation potential towards the respiratory epithelium were assessed. Despite one fourth dosage of MSCs, significantly more PKH26-labeled donor cells were recovered with intratracheal administration than with intraperitoneal administration both during normoxia and hyperoxia. The hyperoxia-induced impaired alveolarization and increased level of TNF-alpha and TGF-beta mRNA, alpha-SMA protein, and collagen were significantly alleviated only with intratracheal MSCs transplantation whereas the hyperoxia-induced increase in the number of TUNEL-positive cells, myeloperoixdase activity, and the level of IL-6 mRNA were significantly attenuated with both intratracheal and intraperitoneal MSCs transplantation. As MSCs differentiated into respiratory epithelium in vitro and a few PKH26-positive donor cells were colocalized with pro surfactant protein C in the damaged lungs, hUCB-MSCs could be employed in cell therapy via both anti-inflammation and regeneration in hypoxia induced lung injury, representing a possible candidate for the new therapeutic modality for the hyperoxia-induced neonatal lung diseases, such as clinical bronchopulmonary dysplasia (Chang et al., 2009; Malgieri et al., 2010).

7.9 Cartilage regeneration

Mature articular cartilage is vulnerable to injuries and disease processes that cause irreversible tissue damage because of its limited capacity for self-repair. UCB is a source of MSCs which can give rise to cells of different lineages, including cartilage or bone (Mara et al., 2010). In fact hUCB-MSCs have much higher chondrogenic differentiation potentials which might lead to regeneration of damaged cartilage. In addition to this chondrogenic differentiation potential of MSCs, recent advances in our understanding of the regeneration mechanism for cartilage defects have demonstrated that MSCs also show potent immunosuppression and anti-inflammatory effects (Di Nicola et al., 2002; Hao et al., 2009, Oh et al., 2008, Wang et al., 2009 as cited in Kim et al., 2010). These properties might be due in part to some specific secreted cytokines and growth factors, suggesting that chondrogenic differentiation and paracrine actions might be involved in replacement of damaged cartilage tissues and stimulation of the regeneration process (Kim et al., 2010). Wang et al. (2009) conducted for the first time a 6-week study comparing the differentiation potential of hBMSCs and hUCMSCs towards condrocytes in a three-dimensional (3D) scaffold. They concluded that hUCMSCs may be a desirable option for use as a mesenchymal cell source for fibrocartilage tissue engineering, based on abundant type I collagen and aggrecan production of hUCMSCs in a 3D matrix. Anyway, the authors stated that further investigation are warranted in order to find the signals that best promote also type II collagen production of hUCMSCs for hyaline cartilage engineering (Wang et al., 2009). Similar, a very recent study demonstrated that functionally, CB-MSC could be more readily induced to differentiate into chondrocytes than could BM-MSC and adipose tissue-MSC (AT-MSC). Moreover, CB-MSC showed immunosuppressive activity equal to that of BM-MSC and AT-MSC, indicating CB a practical source of MSC for cell therapy and regenerative medicine through the use of the well established CB banking system (Zhang et al., 2011). Another attempt to differentiate cells from UCB into chondrocytes with insulin-like growth factor 1 (IGF-1) and transforming growth factor-ss3 (TGF-ss3) showed that TGF-ss3 used in micromass culture is the best growth factor for promoting the proliferation and differentiation of mesenchymal cells from UCB during chondrogenesis. The western blot analysis revealed that after 3 weeks, the expression of type II collagen was greater in micromass culture with TGF-ss3 (Mara et al., 2010).

Two representative common diseases of cartilage degeneration include osteoarthritis and rheumatoid arthritis. Rheumatoid arthritis (RA) is a T-cell-mediated systemic autoimmune disease that primarily attacks synovial joints, leading to articular destruction and functional disability. The study by Liu et al. (2010) is the first report of the UC-MSCs use in the treatment of RA. The potential immunosuppressive effects of human UC-MSCs in RA were evaluated, too. Both the actions of UC-MSCs on the responses of fibroblast-like synoviocytes (FLSs) and T cells in RA patients and the possible molecular mechanism mediating this immunosuppressive effect of UC-MSCs and the therapeutic effects of systemic infusion of human UC-MSCs on collagen-induced arthritis (CIA) in a mouse model were explored. They demonstrated that UC-MSCs exerted a profound inhibitory effect on the proliferation, invasive behavior and inflammatory responses of FLSs, suppressed T cell activation in vitro and induced the generation of regular T cells (Tregs). More importantly, in mice, systemic infusion of UC-MSCs significantly reduced the severity of CIA, strongly suggesting that UC-MSCs might be a therapeutic strategy in RA. In addition, the immunosuppressive activitiy of UC-MSCs could be prolonged by the participation of Tregs (Liu et al., 2010). Despite growing evidence, the molecular mechanisms underlying cartilage repair and regeneration by hUCB-MSCs remain not settled. Therefore, extensive studies of hUCB-MSCs therapeutic mechanisms are required for an understanding of their regenerative potential and for efficient and safe clinical application. Supported by the clinical trial phase I/II results (NCT01041001), adult stem cell therapy using hUCB-MSCs for cartilage degenerative disease might be a promising alternative to previous treatments (Kim et al., 2010).

7.10 Gene therapy

Retroviral mediated gene transfer has been shown to be an efficient method for introducing genetic sequences into mammalian cells (Bernstein et al., 1985; Friedmann et al., 1989 as cited in Lu et al., 1996) and gene replacement therapy has been proposed for many single gene disorders (Lu et al., 1996). Adeno-associated viruses (AAV) can also be used as a vehicle for gene transfer. MSCs are amenable to be genetically modified, which makes them become a promising platform for cell and gene therapy and broadens their potential therapeutic applications in several fields, including improvement engraftment following stem cell transplantation and acceleration of hematopoietic reconstitution, treatment of severe graft-versus-host disease, utilization in targeting tumors and delivering anti-cancer molecules as well as cellular vehicle for protein-supplement gene therapy (Fritz & Jorgensen, 2008; Kumar et al., 2008 as cited in Fan et al., 2011). The strategy that uses therapeutic gene-transfected hUCMSCs as cellular vehicles for targeted biologic agent delivery has solved the problem of short half-life or excessive toxicity of biological agent(s) in vivo. Administration of hUC-MSCs expressing interferon-β with (Rachakatla et al., 2008 as cited in Fan et al., 2011) or without 5-fluorouracil (Rachakatla et al., 2007 as cited in Fan et al., 2011) was found to target to experimentally developed lung tumors and significantly reduce the tumor burden. Furthermore, the co-culture of a small numbe of IFN-beta gene-transfected hUCMSCs (IFN-beta-hUCMSCs) with the human bronchioloalveolar carcinoma cell lines H358 or SW1573 significantly inhibited growth of both types of carcinoma cell lines, whereas systemic administration of IFN-beta-hUCMSCs markedly attenuated growth of orthotopic H358 bronchioloalveolar carcinoma xenografts in SCID mice by increasing apoptosis. These results clearly indicate that IFN-beta-hUCMSCs caused cell death of bronchioloalveolar carcinoma cells through IFN-beta production, thereby attenuating tumor growth in vivo, suggesting that IFN-beta-hUCMSCs are a powerful anti-cancer cytotherapeutic tool for bronchioloalveolar carcinoma (Matsuzuka et al., 2010). HUC-MSCs modified to express BDNF were also found to improve neurological function and increase NSE-positive cells while decrease GFAP-positive cells and number of apoptosis cells after being delivered into the edge of lesion in athymic mice brain injury model (Zhang et al., 2009 as cited in Fan et al., 2011). In addition hUC-MSCs can be genetically modified to express biologically active human factor IX and serve as an efficient drug delivery vehicle for somatic gene therapy of hemophilia B (Chen et al., 2009 as cited in Fan et al., 2011). hUC-MSCs might have a migratory capacity toward glioma cells and therefore can also serve as a vector for gene therapy against brain tumors. Soluble factors including chemokines or growth factors expressed and released by glioma cells mediate the tropism of MSCs for gliomas. In the study by Park et al. (2011) the overexpression of the stromal cell-derived factor-1α (SDF-1α) receptor, chemokine receptor 4 (CXCR4), on hUCB-MSCs resulted in enhanced migratory capacity of MSCs toward gliomas. Furthermore, CXCR4 transfected hUCB-MSCs (hMSCs-CXCR4) showed a stronger migration capacity toward glioma cells in vitro compared with control MSCs, and also exhibited enhanced migration to glioma cells in an intracranial human malignant glioma xenograft model. These results indicate that SDF-1α/CXCR4 could be involved in recruitment of hUCB-MSCs to glioma cells and that overexpression of CXCR4 may be a useful tool for stem cell-based glioma therapy (Park et al., 2011). It has also been suggested that the genetic alteration of stem cells can function as a mechanism which could deliver specific gene products to parts of tissue ischemia to boost the repair process (Byun et al., 2001; Melo et al., 2004; Yi et al., 2006 as cited in Goldberg et al., 2007) and it was shown to be associated with restored physiological functions after AMI (Chen et al., 2005 as cited in Arien-Zakay et al., 2010). In preclinical studies, UCB CD34+ transfected with VEGF or Ang-1 have been shown to enhance the process of tissue repair in a murine LAD ligation model (Goldberg et al., 2007) and UCB MNCs modified to express VEGF have demonstrated their efficiency in a hindlimb model of chronic ischemia (Ikeda et al., 2004 as cited in Goldberg et al., 2007). Additionally, autologous CB CD 34+ transduced with human adenosine deaminase (ADA) gene were infused in children with ADA deficiency (Kohn et al., 1995 as cited in Lu et al., 1996).

8. Cord blood stem cells transplantation: Advantages and disadvantages

The UCB as a source of human stem cells has proved to be highly advantageous as compared to peripheral blood and bone marrow. UCB has more hematopoietic stem cells per volume than PB and BM (Rogers & Casper, 2004) and the immature hematopoietic stem cells and progenitor cells from cord blood seem to be of good quality (Lu et al., 1996). Stem cells remain in the placenta and umbilical cord after delivery and an average of 120 ml of blood can be collected without risking the health of either the mother or the baby (Rogers & Casper, 2004). Cord blood harvesting is thus simple, easy, painless and non-invasive for the donor, contrary to BM aspiration and consequently there are more potential donors than for BM (Malgieri et al., 2010). Moreover, the procedure does not involve ethical and technical issues (Wu et al., 2009 as cited in Malgieri et al., 2010). Different from BM transplants which require general anaesthesia and surgical transfer of the donor marrow to the recipient, UCB stem cell transplantation is a simple intravenous infusion of the hematopoietic stem cells, which find their way to the BM for engraftment (Rogers & Casper, 2004). Moreover, due to the fact that cord blood units are priory stored in public and private UCB banks (Malgieri et al., 2010) they are quickly available, an advantage which is crucial to high-risk patients with unstable disease (Barker et al., 2002 as cited in Barker & Wagner, 2003), whereas BM has to be collected from the donor just before transplantation and there is always the risk of last minute consent refusal (Malgieri et al., 2010). Successful transplantation require allogeneic grafts matching most, if not all, of the six to ten major HLA antigens between host and donor (Malgieri et al., 2010). The majority of the CB transplantations have been performed to treat patients with malignant and non-malignant diseases (Ballen et al., 2001 as cited in Rogers & Casper, 2004) and many of these originated from sibling donors with partial or complete HLA matching. The incidence of acute GVHD has been lower than the expected rate in a pediatric population who received transplants with BM. Interestingly, acute GVHD has not been related to the degree of HLA matching, thus proving that the use of cord blood itself is responsible for the lowered risk (Gluckman et al., 2001; Yu et al., 2001 as cited in Rogers & Casper, 2004). Consequently, the HLA type does not have to be a perfect match for allogeneic cord blood transplantation because the possibility that these cells induce immunological reactions is lower than in the case of BM cells (Malgieri et al., 2010). For a given HLA match, the risk of GVHD is lower with UCBT than with BMT, thus allowing UCBT with 1-2 antigen HLA-mismatched (Barker & Wagner, 2003). The UCB reduced graft-versus-host reactivity when compared with adult-derived marrow grafts can be explained through the relative immune immaturity of the newborn cells (MSCs and hematopoietic stem cells) (Malgieri et al., 2010). In this way, the donor poll could be extended and patients belonging to ethnic minorities could be better represented on BM donor registries. Moreover, UCB does not imply risk for the donor or donor attrition and has the advantage of lower cytomegalovirus (CMV) transmission.

The major disadvantage of UCB is the limited cell dose, which is of particular relevance to the application of UCB transplants to adults (Rogers & Casper, 2004). Moreover, while processing, testing and freezing the collected CB in a cord blood bank, there is usually a 10-20% loss from the initial blood volume and cell dose (Stanevsky et al., 2009). Moreover, there is an increased risk of contamination with any cell culture manipulation (Malgieri et al., 2010) and it appears that MSCs from UCB have lower success rate of isolation if compared with MSCs from BM (63% vs 100%) (Kern et al., 2006 as cited in Malgieri et al., 2010). The less experience, the present insufficient standardization of each steps between different banks, as well as inadequate storage policy, may lead to an even greater cell loss. Given that some of the CB transplants are performed with cell dose near the engraftment threshold, modest loss of potency of a product may have a major impact on clinical outcome. Indeed, delayed engraftment due to low cell dose represents one of the main disadvantages and the main restriction of cord blood transplants (Stanevski et al., 2009). However, several strategies- such as improved collection techniques and more efficient cell processing and cryopreservation methods are developing in order to overcome these obstacles and obtain successful engraftment of adults as well as children. Other limitations include the potential risk of genetic disease transmission and lack of donor recall for collection of stem cells or donor lymphocytes in the URD setting (Barker & Wagner, 2003).

9. Cord blood banking

Currently, there is a huge interest in establishing and developing cord blood banks. Worldwide, more than 400,000 cord blood grafts are available in more than 50 cord blood banks. The cord blood collections are managed by cord blood bank networks named Netcord (www.netcord.org), international registries for outcome data collection named Eurocord (www.eurocord.org) and CIBMTR www.cibmtr.org. Eurocord is an international registry developed by the European Blood and Marrow Transplant group (EBMT), which includes more than 190 transplant centers in 36 countries all performing cord blood transplants. Eurocord is collaborating with EBMT and Netcord banks to gather clinical information and monitor all patients transplanted in or outside Europe with Netcord units. The Netcord group was established in 1998 to provide good practice in umbilical cord blood storage, facilitate donor search, improve the quality of the grafts, standardize excellence criteria on an international scale and importantly, to establish procedures for bank accreditation in collaboration with FACT (Foundation on Accreditation in Cell Therapy). All the practical aspects of cord blood banking, such as mother informed consent, collection techniques, labelling and identification, infectious disease and genetic disease testing, HLA typing, methodology of cell processing, cryopreservation, transportation and release have been extensively published. All these aspects are detailed in the last version of the Netcord-FACT Standards (www.factwebsite.org).

Currently, there are increasing numbers of international exchanges of cord blood units. Profit entities such as private banks, store directed donations collected by obstetricians from babies born into families who intend to use the cord blood for the baby from whom it came (autologous donation) or for another family member in need of future transplantation therapy.

10. Future perspectives

The differentiation potential of UCMSC makes them the perfect candidate for a wide range of clinical applications. UC-derived MSCs showed to be multipotent and their differentiation abilities does not seem to be restricted to the mesodermal lineages, since the cells could be successfully induced to neurones, pancreatic and liver cells. Latest evidences suggest, that UC-MSC populations contain a subset of primitive cells. Therefore, the identification and characterization of these sub-populations in the future is of decisive importance. Another issue of great clinical importance is the question of whether the differentiation potential of the isolated populations is dependent on their location in the UC-tissue. Recently, MSCs markers were developed and may be helpful in this matter.

Additionally, first in vitro and in vivo animal studies showed immune-privileged and immune-modulatory properties of UC-derived MSCs. Due to immune-modulatory properties low levels of rejection were observed in all reports of in vitro transplantation experiments and good results in tissue engraftments were noted. Therefore further studies and clinical trials will clarify whether the in vivo properties of UC-derived MSCs after transplantation exhibits same effects.

One of the ambitious aims of regenerative medicine is the engineering of tissue in vitro. Few but very promising applications of UC-derived MSCs have been reported in this field. Biopolymer scaffolds, mechanical strain approaches, or 3D bioreactors for tissue generation, which were successfully applied with MSCs from other sources will be of great clinical interest in the future.

In the effort to overcome incurable diseases, it may only be a question of time until UC-derived MSCs will be successfully used for clinical and tissue engineering applications.

11. References

[1] Anzalone R, Lo Iacono M, Loria T, Di Stefano A, Giannuzzi P, Farina F & La Rocca G.(2010). Wharton's Jelly Mesenchymal Stem Cells as Candidates for Beta Cells Regeneration: Extending the Differentiative and Immunomodulatory Benefits of Adult Mesenchymal Stem Cells for the Treatment of Type 1 Diabetes. *Stem Cell Rev*.n.d.

[2] Arien-Zakay H, Lazarovici P & Nagler A. (2010). Tissue regeneration potential in human umbilical cord blood. Best Practice & Research Clinical Haematology, Vol. 23, No.2, pp. 291-303.

[3] Arien-Zakay H, Lecht S, Bercu MM, Tabakman R, Kohen R, Galski H, Nagler A & Lazarovici P.(2009). Neuroprotection by cord blood neural progenitors involves antioxidants, neurotrophic and angiogenic factors. *Exp Neurol.* Vol.216, No.1, pp.83-94.

[4] Barker J N., Wagner J E. (2003). Umbilical cord blood transplantation: current practice and future innovations. *Critical Reviews in Oncology/Hematology,* Vol.48, No.1, pp. 35-43.

[5] Bartholomew A, Sturgeon C, Siatskas M, Ferrer K, McIntosh K, Patil S, Hardy W, Devine S, Ucker D, Deans R, Moseley A & Hoffman R (2002) Mesenchymal stem cells suppress lymphocyte proliferation in vitro and prolong skin graft survival in vivo. *Exp Hematol.* Vol.30, pp.42-48.

[6] Bielorai B, Hughes MR, Auerbach AD, Nagler A, Loewenthal R, Rechavi G & Toren A.(2004). Successful umbilical cord blood transplantation for Fanconi anemia using preimplantation genetic diagnosis for HLA-matched donor. *Am J Hematol.* Vol.77, No.4, pp. 397-9.

[7] Bizzetto R, Bonfim C, Rocha V, Socié G, Locatelli F, Chan K, Ramirez O, Stein J, Nabhan S, Miranda E, Passweg J, de Souza CA & Gluckman E; Eurocord and SAA-WP from EBMT. (2011). Outcomes after related and unrelated umbilical cord blood transplantation for hereditary bone marrow failure syndromes other than Fanconi anemia. *Haematologica.* Vol.96, No.1, pp. 134-41.

[8] Boelens JJ, Rocha V, Aldenhoven M, Wynn R, O'Meara A, Michel G, Ionescu I, Parikh S, Prasad VK, Szabolcs P, Escolar M, Gluckman E, Cavazzana-Calvo M & Kurtzberg J; EUROCORD, Inborn error Working Party of EBMT and Duke University.(2009). Risk factor analysis of outcomes after unrelated cord blood transplantation in patients with hurler syndrome. *Biol Blood Marrow Transplant.* Vol.15, No.5, pp.618-25.

[9] Boelens JJ, Wynn RF, O'Meara A, Veys P, Bertrand Y, Souillet G, Wraith JE, Fischer A, Cavazzana-Calvo M, Sykora KW, Sedlacek P, Rovelli A, Uiterwaal CS & Wulffraat N.(2007). Outcomes of hematopoietic stem cell transplantation for Hurler's syndrome in Europe: a risk factor analysis for graft failure. *Bone Marrow Transplant.* Vol.40, No.3, pp. 225-33.

[10] Botta R, Gao E, Stassi G, Bonci D, Pelosi E, Zwas D, Patti M, Colonna L, Baiocchi M, Coppola S, Ma X, Condorelli G & Peschle C. (2004). Heart infarct in NOD-SCID mice: therapeutic vasculogenesis by transplantation of human CD34+ cells and low dose CD34+KDR+ cells. *FASEB J.* Vol.18, No.12, pp.1392-4.

[11] Breymann C, Schmidt D & Hoerstrup SP. (2006). Umbilical cord cells as a source of cardiovascular tissue engineering. *Stem Cell Rev.* Vol.2, No.2, pp.87-92.

[12] Campard D, Lysy PA, Najimi M & Sokal EM.(2008). Native umbilical cord matrix stem cells express hepatic markers and differentiate into hepatocyte-like cells. *Gastroenterology.* Vol. 134, No.3, pp.833-48

[13] Canque B, Camus S, Dalloul A, Kahn E, Yagello M, Dezutter-Dambuyant C, Schmitt D, Schmitt C & Gluckman JC.(2000). Characterization of dendritic cell differentiation pathways from cord blood CD34(+)CD7(+)CD45RA(+) hematopoietic progenitor cells. *Blood* .Vol. 96, No. 12, pp. 3748-56.

[14] Chang YS, Oh W, Choi SJ, Sung DK, Kim SY, Choi EY, Kang S, Jin HJ, Yang YS & Park WS. (2009). Human umbilical cord blood-derived mesenchymal stem cells

attenuate hyperoxia-induced lung injury in neonatal rats. *Cell Transplant.* Vol.18, No.8, pp. 869-86.

[15] Chen MY, Lie PC, Li ZL & Wei X. (2009). Endothelial differentiation of Wharton's jelly-derived mesenchymal stem cells in comparison with bone marrow-derived mesenchymal stem cells. *Exp Hematol.* Vol. 37, No.5, pp.629-40.

[16] Chen N, Hudson JE, Walczak P, Misiuta I, Garbuzova-Davis S, Jiang L, Sanchez-Ramos J, Sanberg PR, Zigova T & Willing AE.(2005). Human umbilical cord blood progenitors: the potential of these hematopoietic cells to become neural. *Stem Cells.* Vol. 23, No.10, pp. 1560-70.

[17] Chung DJ, Choi CB, Lee SH, Kang EH, Lee JH, Hwang SH, Han H, Lee JH, Choe BY, Lee SY & Kim HY. (2009). Intraarterially delivered human umbilical cord blood-derived mesenchymal stem cells in canine cerebral ischemia. *J Neurosci Res.*Vol.87, No.16, pp.3554-67.

[18] Conconi MT, Burra P, Di Liddo R, Calore C, Turetta M, Bellini S, Bo P, Nussdorfer GG & Parnigotto PP.(2006). CD105(+) cells from Wharton's jelly show in vitro and in vivo myogenic differentiative potential. *Int J Mol Med.* Vol.18, No.6, pp.1089-96.

[19] Covas DT, Siufi JL, Silva AR & Orellana MD. (2003). Isolation and culture of umbilical vein mesenchymal stem cells. *Braz J Med Biol Res*, Vol. 36, p.1179-1183.

[20] Devine SM, Cobbs C, Jennings M, Bartholomew A & Hoffman R (2003) Mesenchymal stem cells distribute to a wide range of tissues following systemic infusion into nonhuman primates. *Blood* Vol.101, pp.2999-3001.

[21] Di Nicola M, Carlo-Stella C, Magni M, Milanesi M, Longoni PD, Matteucci P, Grisanti S & Gianni AM. (2002). Human bone marrow stromal cells suppress T-lymphocyte proliferationinduced by cellular or nonspecific mitogenic stimuli. *Blood* Vol.99, pp.3838-3843.

[22] Fan CG, Zhang Q-j & Zhou J-r. (2011).Therapeutic Potentials of Mesenchymal Stem Cells Derived from Human Umbilical Cord. *Stem Cell Rev and Rep*, Vol 7, No 1, pp. 195-207

[23] Flynn A, Barry F, & O'Brien T. (2007). UC blood-derived mesenchymal stromal cells: an overview. *Cytotherapy.* Vol. 9, No. 8, pp.717-26.

[24] Fu YS, Cheng YC, Lin MY, Cheng H, Chu PM, Chou SC, Shih YH, Ko MH & Sung MS.(2006). Conversion of human umbilical cord mesenchymal stem cells in Wharton's jelly to dopaminergic neurons in vitro: potential therapeutic application for Parkinsonism. *Stem Cells.* Vol. 24, No.1, pp.115-24.

[25] Fu YS, Shih YT, Cheng YC & Min MY.(2004). Transformation of human umbilical mesenchymal cells into neurons in vitro. *J Biomed Sci.* Vol.11, No.5, pp.652-60.

[26] Gang EJ, Jeong JA, Hong SH, Hwang SH, Kim SW, Yang IH, Ahn C, Han H & Kim H.(2004). Skeletal myogenic differentiation of mesenchymal stem cells isolated from human umbilical cord blood. *Stem Cells*, Vol. 22, No.4, pp.617-624

[27] Goldberg JL, Laughlin MJ & Pompili VJ. (2007). Umbilical cord blood stem cells: implications for cardiovascular regenerative medicine. *J Mol Cell Cardiol*, Vol. 42, No. 5, pp. 912-920.

[28] Haller MJ, Viener HL, Wasserfall C, Brusko T, Atkinson MA & Schatz DA.(2008). Autologous umbilical cord blood infusion for type 1 diabetes. *Exp Hematol.* Vol. 36, No.6, pp.710-5.

[29] Harris D.T. & Rogers I. (2007).Umbilical cord blood: a unique source of pluripotent stem cells for regenerative medicine. *Curr Stem Cell Res Ther,* Vol. 2, pp. 301-309.

[30] Henning RJ, Abu-Ali H, Balis JU, Morgan MB, Willing AE & Sanberg PR.(2004). Human umbilical cord blood mononuclear cells for the treatment of acute myocardial infarction. *Cell Transplant.* Vol.13, No.7-8, pp.729-39.

[31] Hill JM, Zalos G, Halcox JP, Schenke WH, Waclawiw MA, Quyyumi AA & Finkel T.(2003). Circulating endothelial progenitor cells, vascular function, and cardiovascular risk. N Engl J *Med.* Vol.348, No.7, pp.593-600.

[32] Hirata Y, Sata M, Motomura N, Takanashi M, Suematsu Y, Ono M & Takamoto S.(2005). Human umbilical cord blood cells improve cardiac function after myocardial infarction. *Biochem Biophys Res Commun.* Vol.327, No.2, pp.609-14.

[33] Hu SL, Luo HS, Li JT, Xia YZ, Li L, Zhang LJ, Meng H, Cui GY, Chen Z, Wu N, Lin JK, Zhu G & Feng H. (2010). Functional recovery in acute traumatic spinal cord injury after transplantation of human umbilical cord mesenchymal stem cells. *Crit Care Med.* Vol.38, No. 11, pp.2181-9.

[34] Huang GP, Pan ZJ, Jia BB, Zheng Q, Xie CG, Gu JH, McNiece IK & Wang JF. (2007). Ex vivo expansion and transplantation of hematopoietic stem/progenitor cells supported by mesenchymal stem cells from human umbilical cord blood. *Cell Transplant.* Vol.16, No.6, pp.579-85.

[35] Hyslop, L. A., Armstrong, L., Stojkovic, M., & Lako, M. (2005). Human embryonic stem cells: biology and clinical implications. *Expert Reviews in Molecular Medicine, Vol.7,* No.19, pp. 1-21.

[36] Hyun JK & Kim HW.(2010). Clinical and experimental advances in regeneration of spinal cord injury. *J Tissue Eng.* 650857.

[37] Jiang Y, Jahagirdar BN, Reinhardt RL, Schwartz RE, Keene CD, Ortiz-Gonzalez XR, Reyes M, Lenvik T, Lund T, Blackstad M, Du J, Aldrich S, Lisberg A, Low WC, Largaespada DA & Verfaillie CM. (2002). Pluripotency of mesenchymal stem cells deived from adult marrow. *Nature,* Vol.418, pp. 41-49.

[38] Jung KH, Shin HP, Lee S, Lim YJ, Hwang SH, Han H, Park HK, Chung JH & Yim SV. (2009). Effect of human umbilical cord blood-derived mesenchymal stem cells in a cirrhotic rat model. *Liver Int.* Vol. 29, No.6, pp.898-909

[39] Kadivar M, Khatami S, Mortazavi Y, Shokrgozar MA, Taghikhani M Soleimani M.(2006). In vitro cardiomyogenic potential of human umbilical vein-derived mesenchymal stem cells. *Biochem Biophys Res Commun.* Vol. 340, No. 2, pp.639-47.

[40] Kadner A, Hoerstrup SP, Tracy J, Breymann C, Maurus CF, Melnitchouk S, Kadner G, Zund G & Turina M. (2002). Human umbilical cord cells: a new cell source for cardiovascular tissue engineering. *Ann Thorac Surg.* Vol.74, No. 4, pp.S1422-8.

[41] Kim BO, Tian H, Prasongsukarn K, Wu J, Angoulvant D, Wnendt S, Muhs A, Spitkovsky D & Li RK.(2005). Cell transplantation improves ventricular function after a myocardial infarction: a preclinical study of human unrestricted somatic stem cells in a porcine model. *Circulation.* Vol. 112, No.9 Suppl, pp.I96-104.

[42] Kim JY, Jeon HB, Yang YS, Oh W & Chang JW. (2010). Application of human umbilical cord blood-derived mesenchymal stem cells in disease models. *World J Stem Cells,* Vol.2, No.2, pp.34-38.

[43] Kim SW, Han H, Chae GT, Lee SH, Bo S, Yoon JH, Lee YS, Lee KS, Park HK & Kang KS. (2006). Successful stem cell therapy using umbilical cord blood-derived multipotent

stem cells for Buerger's disease and ischemic limb disease animal model. *Stem Cells*. Vol.24, No.6, pp.1620-6.

[44] Kögler G, Sensken S & Wernet P. (2006). Comparative generation and characterization of pluripotent unrestricted somatic stem cells with mesenchymal stem cells from human cord blood. *Exp Hematol*. Vol.34, No.11, pp.1589-95.

[45] Kögler G, Sensken S, Airey JA, Trapp T, Müschen M, Feldhahn N, Liedtke S, Sorg RV, Fischer J, Rosenbaum C, Greschat S, Knipper A, Bender J, Degistirici O, Gao J, Caplan AI, Colletti EJ, Almeida-Porada G, Müller HW, Zanjani E & Wernet P.(2004). A new human somatic stem cell from placental cord blood with intrinsic pluripotent differentiation potential.*J Exp Med*. Vol. 200, No.2, pp.123-35.

[46] Krause DS, Theise ND, Cottler MI et al. (2001). Multi-organ, multi-lineage engraftment by a single bone marrow-derived stem cell. *Cell*, Vol.105, No. 3, pp. 369-377.

[47] Kuh SU, Cho YE, Yoon DH, Kim KN & Ha Y. (2005). Functional recovery after human umbilical cord blood cells transplantation with brain-derived neutrophic factor into the spinal cord injured rat. *Acta Neurochir (Wien)*. Vol.147, No.9, pp. 985-92.

[48] La Rocca G, Anzalone R, Corrao S, Magno F, Loria T, Lo Iacono M, Di Stefano A, Giannuzzi P, Marasa L, Cappello F, Zummo G & Farina F. (2009). Isolation and characterization of Oct-4+/HLA-G+ mesenchymal stem cells from human umbilical cord matrix: differentiation potential and detection of new markers. *Histochem Cell Biol*, Vol. 131, pp. 267-282.

[49] Laughlin MJ, Barker J, Bambach B, Koc ON, Rizzieri DA, Wagner JE, Gerson SL, Lazarus HM, Cairo M, Stevens CE, Rubinstein P & Kurtzberg J.(2001). Hematopoietic engraftment and survival in adult recipients of umbilical-cord blood from unrelated donors. *N Engl J Med*, Vol. 344, No.24, pp.1815-22.

[50] Le Blanc K, Rasmusson I, Sundberg B, Gotherstrom C, Hassan M, Uzunel M & Ringden O (2004) Treatment of severe acute graft-versus-host disease with third party haploidentical mesenchymal stem cells. *Lancet*, Vol.363, pp.1439-1441.

[51] Lee HJ, Lee JK, Lee H, Carter JE, Chang JW, Oh W, Yang YS, Suh JG, Lee BH, Jin HK & Bae JS. (2010). Human umbilical cord blood-derived mesenchymal stem cells improve neuropathology and cognitive impairment in an Alzheimer's disease mouse model through modulation of neuroinflammation. *Neurobiol Aging*.n.d.

[52] Lee HJ, Lee JK, Lee H, Shin JW, Carter JE, Sakamoto T, Jin HK & Bae JS. (2010). The therapeutic potential of human umbilical cord blood-derived mesenchymal stem cells in Alzheimer's disease. *Neurosci Lett*. Vol.481, No.1, pp. 30-5.

[53] Leor J, Guetta E, Feinberg MS, Galski H, Bar I, Holbova R, Miller L, Zarin P, Castel D, Barbash IM & Nagler A.(2006). Human umbilical cord blood-derived CD133+ cells enhance function and repair of the infarcted myocardium. *Stem Cells*. Vol. 24, No.3, pp.772-80.

[54] Li M, Zhang SZ, Guo YW, Cai YQ, Yan ZJ, Zou Z, Jiang XD, Ke YQ, He XY, Jin ZL, Lu GH & Su DQ. (2010). Human umbilical vein-derived dopaminergic-like cell transplantation with nerve growth factor ameliorates motor dysfunction in a rat model of Parkinson's disease. *Neurochem Res*. Vol.35, No.10, pp.1522-9.

[55] Limbert C, Päth G, Jakob F & Seufert J.(2008). Beta-cell replacement and regeneration: Strategies of cell-based therapy for type 1 diabetes mellitus. *Diabetes Res Clin Pract*.Vol.79, No.3, pp. 389-99.

[56] Lin SZ, Chang YJ, Liu JW, Chang LF, Sun LY, Li YS, Luo GH, Liao CH, Chen PH, Chen TM, Lee RP, Yang KL, Harn HJ & Chiou TW. (2010). Transplantation of human Wharton's Jelly-derived stem cells alleviates chemically induced liver fibrosis in rats. *Cell Transplant.* Vol.19, No.11, pp. 1451-63.

[57] Liu Y, Mu R, Wang S, Long L, Liu X, Li R, Sun J, Guo J, Zhang X, Guo J, Yu P, Li C, Liu X, Huang Z, Wang D, Li H, Gu Z, Liu B & Li Z.(2010). Therapeutic potential of human umbilical cord mesenchymal stem cells in the treatment of rheumatoid arthritis. *Arthritis Res Ther.*Vol.12, No.6, pp. R210.

[58] Lu L, Shen RN & Broxmeyer HE. (1996). Stem cells from bone marrow, umbilical cord blood and peripheral blood for clinical application: current status and future application. *Crit Rev Oncol Hematol.* Vol.22, No.2, pp. 61-78.

[59] Lu LL, Liu YJ, Yang SG, Zhao QJ, Wang X, Gong W, Han ZB, Xu ZS, Lu YX, Liu D, Chen ZZ & Han ZC. (2006). Isolation and characterization of human umbilical cord mesenchymal stem cells with hematopoiesis-supportive function and other potentials. *Haematologica.* Vol. 91, No.8, pp. 1017-26.

[60] Ma N, Ladilov Y, Moebius JM, Ong L, Piechaczek C, Dávid A, Kaminski A, Choi YH, Li W, Egger D, Stamm C & Steinhoff G.(2006). Intramyocardial delivery of human CD133+ cells in a SCID mouse cryoinjury model: Bone marrow vs. cord blood-derived cells. *Cardiovasc Res.* Vol.71, No.1, pp.158-69.

[61] Ma N, Stamm C, Kaminski A, Li W, Kleine HD, Müller-Hilke B, Zhang L, Ladilov Y, Egger D & Steinhoff G. (2005). Human cord blood cells induce angiogenesis following myocardial infarction in NOD/scid-mice. *Cardiovasc Res.* Vol.66, No.1, pp.45-54.

[62] Malgieri A, Kantzari E, Patrizi MP, Gambardella S. (2010). Bone marrow and umbilical cord blood human mesenchymal stem cells: state of the art. *Int J Clin Exp Med,* Vol. 3, No. 4, pp.248-269.

[63] Mara CS, Duarte AS, Sartori A, Luzo AC, Saad ST & Coimbra IB. (2010). Regulation of chondrogenesis by transforming growth factor-beta 3 and insulin-like growth factor-1 from human mesenchymal umbilical cord blood cells. *J Rheumatol.* Vol. 37, No.7, pp. 1519-26.

[64] Mareschi K, Ferrero I, Rustichelli D, Aschero S, Gammaitoni L, Aglietta M, Madon E & Fagioli F. (2006). Expansion of mesenchymal stem cells isolated from pediatric and adult donor bone marrow. *J Cell Biochem.* Vol.97, No.4, pp.744-54.

[65] Martin-Rendon E, Sweeney D, Lu F, Girdlestone J, Navarrete C & Watt SM.(2008). 5-Azacytidine-treated human mesenchymal stem/progenitor cells derived from umbilical cord, cord blood and bone marrow do not generate cardiomyocytes in vitro at high frequencies. *Vox Sang.* Vol. 95, No.2, pp.137-48.

[66] Matsuzuka T, Rachakatla RS, Doi C, Maurya DK, Ohta N, Kawabata A, Pyle MM, Pickel L, Reischman J, Marini F, Troyer D & Tamura M.(2010). Human umbilical cord matrix-derived stem cells expressing interferon-beta gene significantly attenuate bronchioloalveolar carcinoma xenografts in SCID mice. *Lung Cancer.*Vol.70, No.1, pp. 28-36.

[67] Mitchell KE, Weiss ML, Mitchell BM, Martin P, Davis D, Morales L, Helwig B, Beerenstrauch M, Abou-Easa K, Hildreth T, Troyer D & Medicetty S.(2003). Matrix cells from Wharton's jelly form neurons and glia. *Stem Cells,* Vol. 21, No. 1, pp.50-60.

[68] Moodley Y, Atienza D, Manuelpillai U, Samuel CS, Tchongue J, Ilancheran S, Boyd R & Trounson A.(2009). Human umbilical cord mesenchymal stem cells reduce fibrosis of bleomycin-induced lung injury. *Am J Pathol.*Vol.175, No.1, pp. 303-13.

[69] Murry CE, Soonpaa MH, Reinecke H, Nakajima H, Nakajima HO, Rubart M, Pasumarthi KB, Virag JI, Bartelmez SH, Poppa V, Bradford G, Dowell JD, Williams DA & Field LJ.(2004). Haematopoietic stem cells do not transdifferentiate into cardiac myocytes in myocardial infarcts. *Nature.* Vol.428, No.6983, pp.664-8.

[70] Park SA, Ryu CH, Kim SM, Lim JY, Park SI, Jeong CH, Jun JA, Oh JH, Park SH, Oh W & Jeun SS. (2011). CXCR4-transfected human umbilical cord blood-derived mesenchymal stem cells exhibit enhanced migratory capacity toward gliomas. *Int J Oncol.* Vol.38, No.1, pp.97-103.

[71] Peichev M, Naiyer AJ, Pereira D, Zhu Z, Lane WJ, Williams M, Oz MC, Hicklin DJ, Witte L, Moore MA & Rafii S. (2000). Expression of VEGFR-2 and AC133 by circulating human CD34 (+) cells identifies a population of functional endothelial precursors. *Blood.* Vol. 95, No. 3, pp. 952-8

[72] Pesce M, Orlandi A, Iachininoto MG, Straino S, Torella AR, Rizzuti V, Pompilio G, Bonanno G, Scambia G & Capogrossi MC.(2003). Myoendothelial differentiation of human umbilical cord blood-derived stem cells in ischemic limb tissues. *Circ Res.* Vol. 93, No.5 pp.e51-62.

[73] Phuc PV, Nhung TH, Loan DT, Chung DC & Ngoc PK.(2011). Differentiating of banked human umbilical cord blood-derived mesenchymal stem cells into insulin-secreting cells. *In Vitro Cell Dev Biol Anim.* Vol.47, No.1, pp.54-63.

[74] Pittenger MF, Mackay AM, Beck SC, Jaiswal RK, Douglas R, Mosca JD, Moorman MA, Simonetti DW, Craig S & Marshak DR. (1999). Multilineage potential of mesenchymal cells. *Science*, Vol. 284, No. 5411, pp.143-147.

[75] Prasad VK & Kurtzberg J. (2010). Transplant outcomes in mucopolysaccharidoses. *Semin Hemato*l. Vol.47, No.1, pp.59-69.

[76] Prasad VK & Kurtzberg J.(2010). Cord blood and bone marrow transplantation in inherited metabolic diseases: scientific basis, current status and future directions. *Br J Haematol.* Vol. 148, No. 3, pp.356-72.

[77] Reddi AS, Kuppasani K & Ende N. (2010). Human umbilical cord blood as an emerging stem cell therapy for diabetes mellitus. *Curr Stem Cell Res The*r. Vol. 5, No.4, pp.356-61.

[78] Ringden O, Uzunel M, Rasmusson I, Remberger M, Sundberg B, Lonnies H, Marschall HU, Dlugosz A, Szakos A, Hassan Z, Omazic B, Aschan J, Barkholt L & Le Blanc K. (2006). Mesenchymal stem cells for treatment of therapy-resistant graft-versus-host disease. *Transplantation*, Vol.81, pp.1390-1397.

[79] Rizvanov AA, Guseva DS, Salafutdinov II, Kudryashova NV, Bashirov FV, Kiyasov AP, Yalvaç ME, Gazizov IM, Kaligin MS, Sahin F, Mukhamedyarov MA, Palotás A & Islamov RR. (2011). Genetically modified human umbilical cord blood cells expressing vascular endothelial growth factor and fibroblast growth factor 2 differentiate into glial cells after transplantation into amyotrophic lateral sclerosis transgenic mice. *Exp Biol Med (Maywood).* Vol.236, No.1, pp.91-8.

[80] Rogers I & Casper R F. (2004). Umbilical cord blood stem cells. *Best Practice & Research Clinical Obstetrics and Gynaecology*, Vol. 18, No. 6, pp. 893-908.

[81] Rogers I, Yamanaka N, Bielecki R, Wong CJ, Chua S, Yuen S & Casper RF. (2007). Identification and analysis of in vitro cultured CD45-positive cells capable of multi-lineage differentiation. *Exp Cell Res.* Vol. 313, No. 9, pp.1839-52.

[82] Romanov YA, Svintsitskaya VA & Smirnov VN. (2003). Searching for alternative sources of postnatal human mesenchymal stem cells: candidate MSC-like cells from umbilical cord.*Stem Cells,* Vol. 21, No. 1, pp. 105-110.

[83] Roobrouck VD, Ulloa-Montoya F & Verfaillie CM. (2008). Self-renewal and differentiation capacity of young and aged stem cells. *Experimental Cell Research,* Vol. 314, No.9, pp. 1937-1944.

[84] Roobrouck VD, Ulloa-Montoya F & Verfaillie CM.(2008). Self-renewal and differentiation capacity of young and aged stem cells. *Exp Cell Res.* Vol.314, No.9, pp.1937-44.

[85] Roura S, Farré J, Hove-Madsen L, Prat-Vidal C, Soler-Botija C, Gálvez-Montón C, Vilalta M & Bayes-Genis A. (2010). Exposure to cardiomyogenic stimuli fails to transdifferentiate human umbilical cord blood-derived mesenchymal stem cells. *Basic Res Cardiol.* Vol. 105, No. 3, pp. 419-30.

[86] Sanberg PR, Willing AE, Garbuzova-Davis S, Saporta S, Liu G, Sanberg CD, Bickford PC, Klasko SK & El-Badri NS. (2005). Umbilical cord blood-derived stem cells and brain repair. *Ann N Y Acad Sci.* Vol.1049, pp.67-83.

[87] Schneider RK, Puellen A, Kramann R, Raupach K, Bornemann J, Knuechel R, Pérez-Bouza A & Neuss S. (2010). The osteogenic differentiation of adult bone marrow and perinatal umbilical mesenchymal stem cells and matrix remodelling in three-dimensional collagen scaffolds. *Biomaterials.* Vol. 31, No.3, pp. 467-80.

[88] Sellamuthu S, Manikandan R, Thiagarajan R, Babu G, Dinesh D, Prabhu D & Arulvasu C.(2011). In vitro trans-differentiation of human umbilical cord derived hematopoietic stem cells into hepatocyte like cells using combination of growth factors for cell based therapy. *Cytotechnology.* n.d.

[89] Shaw PH, Haut PR, Olszewski M & Kletzel M.(1999). Hematopoietic stem-cell transplantation using unrelated cord-blood versus matched sibling marrow in pediatric bone marrow failure syndrome:one center's experience. *Pediatric Transplantation,* Vol. 3, No. 4, pp. 315-321.

[90] Silani V, Cova L, Corbo M, Ciammola A & Polli E. (2004). Stem-cell therapy for amyotrophic lateral sclerosis. *Lancet.* Vol.364, No.9429, pp.200-2.

[91] Smith AR & Wagner JE. (2009). Alternative Hematopoietic Stem Cell Sources for Transplantation: Place of Umbilical Cord Blood. Br J Haematol. Vol.147, No.2, pp. 246–261.

[92] Sobani ZA, Quadri SA & Enam SA.(2010). Stem cells for spinal cord regeneration: Current status. *Surg Neurol Int.* Vol.25, No.1, pp.93.

[93] Sodian R, Lueders C, Kraemer L, Kuebler W, Shakibaei M, Reichart B, Daebritz S & Hetzer R. (2006). Tissue engineering of autologous human heart valves using cryopreserved vascular umbilical cord cells. *Ann Thorac Surg.* Vol.81, No.6, 2207-16.

[94] Stanevski A, Goldstein G & Nagler A. (2009). Umbilical cord blood transplantation: Pros, cons and beyond. *Blood Reviews,* Vol. 23, pp. 199-204.

[95] Wang HS, Shyu JF, Shen WS, Hsu HC, Chi TC, Chen CP, Huang SW, Shyr YM, Tang KT & Chen TH.(2010). Transplantation of insulin producing cells derived from umbilical cord stromal mesenchymal stem cells to treat NOD mice. *Cell Transplant.*n.d.

[96] Wang L, Tran I, Seshareddy K, Weiss ML & Detamore MS. (2009). A comparison of human bone marrow-derived mesenchymal stem cells and human umbilical cord-derived mesenchymal stromal cells for cartilage tissue engineering. *Tissue Eng Part A*. Vol.15, No. 8, pp. 2259-66.

[97] Weiss ML, Medicetty S, Bledsoe AR, Rachakatla RS, Choi M, Merchav S, Luo Y, Rao MS, Velagaleti G & Troyer D. (2006). Human umbilical cord matrix stem cells: preliminary characterization and effect of transplantation in a rodent model of Parkinson's disease. *Stem Cells*, Vol. 24, pp. 781-792.

[98] Wobus, Anna M.; Boheler, Kenneth. Stem Cells (Eds.) 1st ed. 2006. 2nd printing, 2006, XII, 414 p. 53 illus., 18 in color., Softcover ISBN: 978-3-540-77854-7

[99] Wu KH, Zhou B, Lu SH, Feng B, Yang SG, Du WT, Gu DS, Han ZC & Liu YL. (2007). In vitro and in vivo differentiation of human umbilical cord derived stem cells into endothelial cells. *J Cell Biochem*. Vol. 100, No. 3, pp.608-16.

[100] Yan-Wu G, Yi-Quan K, Ming L, Ying-Qian C, Xiao-Dan J, Shi-Zhong Z, Wang-Ming Z & Chuan-Zhi D. (2011). Human Umbilical Cord-Derived Schwann-Like Cell Transplantation Combined with Neurotrophin-3 Administration in Dyskinesia of Rats with Spinal Cord Injury. *Neurochem Res*. n.d.

[101] Zhang X, Hirai M, Cantero S, Ciubotariu R, Dobrila L, Hirsh A, Igura K, Satoh H, Yokomi I, Nishimura T, Yamaguchi S, Yoshimura K, Rubinstein P & Takahashi TA. (2011). Isolation and characterization of mesenchymal stem cells from human umbilical cord blood: Reevaluation of critical factors for successful isolation and high ability to proliferate and differentiate to chondrocytes as compared to mesenchymal stem cells from bone marrow and adipose tissue. *J Cell Biochem*. Vol.112, No.4, pp. 1206-18.

[102] Zhao Q, Ren H, Li X, Chen Z, Zhang X, Gong W, Liu Y, Pang T & Han ZC.(2009). Differentiation of human umbilical cord mesenchymal stromal cells into low immunogenic hepatocyte-like cells. *Cytotherapy*. Vol.11, No.4, pp.414-26.

3

Mesenchymal Stem Cells: Immunology and Therapeutic Benefits

Najib El Haddad
Transplantation Research Center,
Brigham & Women's Hospital, Harvard Medical School
USA

1. Introduction

Bone marrow is a complex tissue containing hematopoietic cell progenitors and their progeny integrated within a connective-tissue network of mesenchymal-derived cells known as the stroma (1). The stroma cells, or Mesenchymal stem cells (MSCs), are multi-potent progenitor cells that constitute a minute proportion of the bone marrow, represented as a rare population of cells that makes up 0.001 to 0.01% of the total nucleated cells. They represent 10-fold less abundance than the haematopoietic stem cells (2), which contributes to the organization of the microenvironment supporting the differentiation of hematopoietic cells (3). MSC are present in tissues of young, as well as, adult individuals (4, 5), including the adipose tissue, umbilical cord blood, amniotic fluid and even peripheral blood (1, 6-8). MSCs were characterized over thirty years ago as fibroblast-like cells with adhesive properties in culture (9, 10). The term MSC has become the predominant term in the literature since the early 90s (11), after which their research field has grown rapidly due to the promising therapeutic potential, resulting in an increased frequency of clinical trials in the new millennium at an explosive rate.

As data accumulated, there was a need to establish a consensus on the proper definition of the MSCs. The International Society for Cellular Therapy has recommended the minimum criteria for defining multi-potent human MSCs (12, 13). The criteria included: (i) cells being adherent to plastic under standard culture conditions; (ii) MSC being positive for the expression of CD105, CD73 and CD90 and negative for expression of the haematopoietic cell surface markers CD34, CD45, CD11a, CD19 or CD79a, CD14 or CD11b and histocompatibility locus antigen (HLA)-DR; (iii) under a specific stimulus, MSC differentiate into osteocytes, adipocytes and chondrocytes *in vitro*. These criteria presented properties to purify MSC and to enable their expansion by several-fold *in-vitro,* without losing their differentiation capacity. When plated at low density, MSCs form small colonies, called colony-forming units of fibroblasts (CFU-f), and which correspond to the progenitors that can differentiate into one of the mesenchymal cell lineages (14, 15). It has been reported lately that MSCs are able to differentiate into both mesenchymal, as well as, non-mesenchymal cell lineages, such as adipocytes, osteoblasts, chondrocytes, tenocytes, skeletal myocytes, neurones and cells of the visceral mesoderm, both *in vitro* and *in vivo* (16, 17).

All cells have half-lives and their natural expiration must be matched by their replacement; MSCs, by proliferating and differentiating, can be the proposed source of these new replacement cells as characterized in their differentiation capacity. This replacement hypothesis mimics the known sequence of events involved in the turnover and maintenance of blood cells that are formed from haematopoietic stem cells (HSCs) (18). Unlike HSCs, MSCs can be culture-expanded *ex vivo* in up to 40 or 50 cell doublings without differentiation (19). A dramatic decrease in MSC per nucleated marrow cell can be observed when the results are grouped by decade, thus showing a 100-fold decrease from birth to old age. Being pericytes present in all vascularized tissues, the local availability of MSC decreases substantially as the vascular density decreases by one or two orders of magnitude with age (20). In recent years, the discovery of MSCs with properties similar, but not identical, to BM-MSCs has been demonstrated in the stromal fraction of the connective tissue from several organs, including adipose tissue, trabecular bone, derma, liver and muscle (21-24). It is important to note that the origin of MSCs might determine their fate and functional characteristics (25). Studies of human bone marrow have revealed that about one-third of the MSC clones are able to acquire phenotypes of pre-adipocytes, osteocytes and chondrocytes (16). This is in concordance with data showing that 30% of the clones from bone marrow have been found to exhibit a trilineage differentiation potential, whereas the remainder display a bi-lineage (osteo-chondro) or uni-lineage (osteo) potential (26). Moreover, MSC populations derived from adipose tissue and derma present a heterogeneous differentiation potential; indeed, only 1.4% of single cells obtained from adipose-derived adult stem cell (ADAS) populations were tri-potent, the others being bi-potent or unipotent (27).

2. Effect of Mesenchymal Stem Cells on Immune cells

Mesenchymal Stem Cells have been shown to possess immunomodulatory characteristics, as described through the inhibition of T-cell proliferation *in vitro* (28-30). These observations have triggered a huge interest in the immunomodulatory effects of MSCs. The *in vitro* studies have been complemented *in vivo,* where both confirmed the immunosuppressive effect of MSC. MSC activating stimuli *in vitro,* appears to include the secretion of cytokines and the interaction with other immune cells in the presence of proinflammatory cytokines (Fig 1) (31). Primarily, the *in vivo* effect has been originally shown in a baboon model, in which infusion of *ex vivo*-expanded matched donor or third-party MSCs delayed the time to rejection of histo-incompatible skin grafts (29). The delay indicated a potential role for MSC in the prevention and treatment of graft-versus-host disease (GVHD) in ASCT, in organ transplantation to prevent rejection, and in autoimmune disorders. Recently, MSCs were used to successfully treat a 9-year-old boy with severe treatment-resistant acute GVHD, further confirming the potent immunosuppressive effect in humans (32). The immunosuppressive potential has no immunologic restriction, whether the MSCs are autologous with the stimulatory or the responder lymphocytes or the MSCs are derived from a third party. The degree of MSC suppression is dose dependent, where high doses of MSC are inhibitory, whereas low doses enhance lymphocyte proliferation in MLCs (33). Broadly, MSC modulate cytokine production by the dendritic and T cell subsets DC/Th1 and DC/Th2 (34), block the antigen presenting cell (APC) maturation and activation (35), and increase the proportion of $CD4^{+}CD25^{+}$ regulatory cells in a mixed lymphocyte reaction (36).

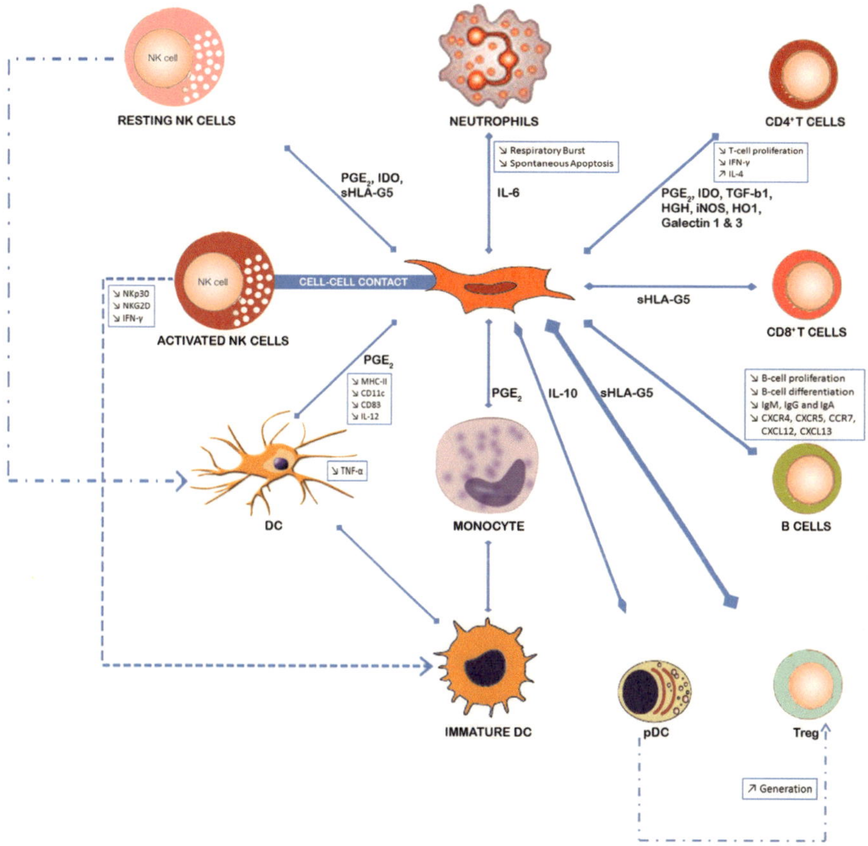

Fig. 1. **Potential mechanisms of the MSC interactions with immune cells**. Mesenchymal stem cells (MSCs) can inhibit both the proliferation and cytotoxicity of resting natural killer (NK) cells, as well as, their cytokine production by releasing prostaglandin E2 (PGE2), indoleamine 2,3-dioxygenase (IDO) and soluble HLA-G5 (sHLA-G5). Killing of MSCs by cytokine-activated NK cells involves the engagement of cell-surface receptors (Thick blue line) expressed by NK cells with its ligands expressed on MSCs. MSCs inhibit the differentiation of monocytes to immature myeloid dendritic cells (DCs), bias mature DCs to an immature DC state, inhibit tumour-necrosis factor (TNF) production by DCs and increase interleukin-10 (IL-10) production by plasmacytoid DCs (pDCs). MSC-derived PGE2 is involved in all of these effects. Immature DCs are susceptible to activated NK cell-mediated lysis. MSC Direct inhibition of $CD4^+$ T-cell function depends on their release of several soluble molecules, including PGE2, IDO, transforming growth factor-β1 (TGFβ1), hepatocyte growth factor (HGF), inducible nitric-oxide synthase (iNOS) and haem-oxygenase-1 (HO1). MSC inhibition of $CD8^+$ T-cell cytotoxicity and the differentiation of regulatory T cells mediated directly by MSCs are related to the release of sHLA-G5 by MSCs. In addition, the upregulation of IL-10 production by pDCs results in the increased generation of regulatory T cells through an indirect mechanism. MSC-driven inhibition of B-cell function seems to depend on soluble factors and cell–cell contact. Finally, MSCs dampen the respiratory burst and delay the spontaneous apoptosis of neutrophils by constitutively releasing IL-6.

3. Immunomodulatory effect of mesenchymal stem cells in innate immunity

Dendritic cells have the elementary role of antigen presentation to naïve T cells upon maturation, which in turn induce the proinflammatory cytokines. Immature DCs acquire the expression of co-stimulatory molecules and upregulate expression of MHC-I and II, as well as, other cell-surface markers (CD11c and CD83). Mesenchymal stem cells have profound effect on the development of DC, where in the presence of MSC, the percentage of DC with conventional phenotype is reduced, while that of plasmacytoid DC is increased, therefore biasing the immune system toward Th2 and away from Th1 responses in a PGE-2 dependent mechanism (37). Furthermore, MSCs inhibit the up-regulation of CD1a, CD40, CD80, CD86, and HLA-DR during DC differentiation and prevent an increase of CD40, CD86, and CD83 expression during DC maturation (38). When mature DCs are incubated with MSCs they have a decreased cell-surface expression of MHC class II molecules, CD11c, CD83 and co-stimulatory molecules, as well as, decreased interleukin-12 (Il-12) production, thereby impairing the antigen-presenting function of the DCs (Fig 1) (39, 40). MSCs can also decrease the pro-inflammatory potential of DCs by inhibiting their production of tumour-necrosis factor α (TNF-α) (40). Furthermore, plasmacytoid DCs (pDCs), which are specialized cells for the production of high levels of type-I IFN in response to microbial stimuli, upregulate production of the anti-inflammatory cytokine IL-10 after incubation with MSCs (34). These observations indicate a potent anti-inflammatory and immunoregulatory effect for MSC *in vitro* and potentially *in vivo.*

Natural killer (NK) cells are key effector cells of the innate immunity in anti-viral and anti-tumor immune responses through their Granzyme B mediated cytotoxicity and the production of pro-inflammatory cytokines (41). NK-mediated target cell lysis results from an antigen-ligand interaction realized by activating NK-cell receptors, and associated with reduced or absent MHC-I expression by the target cell (42). MSCs can inhibit the cytotoxic activity of resting NK cells by down-regulating expression of NKp30 and natural-killer group 2, member D (NKG2D), which are activating receptors involved in NK-cell activation and target-cell killing (Fig 1) (43). Resting NK cells proliferate and acquire strong cytotoxic activity when cultured with IL-2 or IL-15, but when incubated with MSC in the presence of these cytokines, resting NK-cell, as well as, pre-activated NK cell proliferation and IFN-γ production are almost completely abrogated (44, 45). It is worth noting that although the susceptibility of NK cells to MSC mediated inhibition is potent, the pre-activated NK cells showed more resistance to the immunosuppressive effect of MSC compared to resting NK cells (43). The susceptibility of human MSCs to NK-cell-mediated cytotoxicity depends on the low level of cell-surface expression of MHC class I molecules by MSCs and the expression of several ligands, that are recognized by activating NK-cell receptors. Autologous and allogeneic MSC were susceptible to lysis by NK cells (43), where NK cell-mediated lysis was inversely correlated with the expression of HLA class I on MSC (46). Incubation of MSCs with IFN-γ partially protected them from NK-cell-mediated cytotoxicity, through the up-regulation of expression of MHC-I molecules on MSCs (43). Taken together, a possible dynamic interaction between NK cells and MSC *in vivo* exists, where the latter partially inhibit activated MSC, without compromising their ability to kill MSC, reflecting on an interaction tightly regulated by IFN-γ concentration.

Neutrophils play a major role in innate immunity during the course of bacterial infections, where they are activated to kill foreign infectious agents and accordingly undergo a respiratory burst. MSCs have been shown to dampen the respiratory burst and to delay the spontaneous apoptosis of resting and activated neutrophils through an IL-6-dependent mechanism (47). MSC had no effect on neutrophil phagocytosis, expression of adhesion molecules, and chemotaxis in response to IL-8, f-MLP, or C5a (47). Stimulation with bacterial endotoxin induces chemokine receptor expression and mobility of MSCs, which secrete large amounts of inflammatory cytokines and recruit neutrophils in an IL-8 and MIF-dependent manner. Recruited and activated neutrophils showed a prolonged lifespan, an increased expression of inflammatory chemokines, and an enhanced responsiveness toward subsequent challenge with LPS, which suggest a role for MSCs in the early phases of pathogen challenge, when classical immune cells have not been recruited yet (48). Furthermore, MSC have shown the capability to mediate the preservation of resting neutrophils, a phenomenon that might be important in those anatomical sites, where large numbers of mature and functional neutrophils are stored, such as the bone marrow and lungs (49).

4. Immunomodulatory effect of mesenchymal stem cells in adaptive immunity

T-cells are major players of the adaptive immune response. After T-cell receptor (TCR) engagement, T cells proliferate and undergo numerous effector functions, including cytokine release and, in the case of $CD8^{+}$ T cells (CTL), cytotoxicity. Abundant reports have shown that T-cell proliferation stimulated with polyclonal mitogens, allogeneic cells or specific antigen is inhibited by MSCs (28, 29, 50-56). The observation that MSCs can reduce T cell proliferation *in vitro* is mirrored by the *in vivo* finding through infusions of *h*MSCs that control GVHD following bone marrow transplantation. Nevertheless, there is no demonstrable correlation between the measured effects of MSCs *in vitro* and their counter effect *in vivo* due to the lack of universality of methodology correlating the *in vitro* findings with the *in vivo* therapeutic potential.

MSC inhibition of T-cell proliferation is not MHC restricted, since it can be mediated by both autologous and allogeneic MSCs and depends on the arrest of T-cells in the G0/G1 phase of the cell cycle (55, 57). Thus, MSCs do not promote T-cell apoptosis, but instead maintain T cell survival upon subjection to overstimulation through the TCR and upon commitment to undergo CD95–CD95-ligand-dependent activation-induced cell death (57). MSC effects on T cell proliferation *in vitro* appear to have both contact-dependent and contact-independent components (58). Inhibition of T-cell proliferation by MSCs leads to decreased IFN-γ secretion *in vitro* and *in vivo* associated with increased IL-4 production by T helper 2 (TH2) cells (34, 59). Taken together, there is an implication for a shift from a pro-inflammatory state characterized by IFN-γ secretion to an anti-inflammatory state characterized by IL-4 secretion (Fig 1). An imperative role for effector T-cell is the MHC restricted killing of virally-infected or of allogeneic cells mediated via $CD8^{+}$ CTLs, and which is down-regulated by MSC (60).

Regulatory T cells (Tregs), a subpopulation of T cells, are vital to keep the immune system in check, help avoid immune-mediated pathology and contain unrestricted expansion of effector T-cell populations, which results in maintaining homeostasis and tolerance to self antigens. Tregs are currently identified by co-expression of CD4 and CD25, expression of the transcription factor FoxP3, production of regulatory cytokines IL-10 and TGF-β, and ability to suppress proliferation of activated $CD4^{+}CD25^{+}$ T cells in co-culture experiments. Differential expression of CD127 (α-chain of the IL-7 receptor) enable flow cytometry-based separation of human Tregs from $CD127^{+}$ non-regulatory T-cells (61). MSCs have been reported to induce the production of IL-10 by pDCs, which, in turn, trigger the generation of regulatory T cells (Fig 1) (34, 40). Furthermore, Tregs secrete TGF-β and when used *in vitro*, TGF-β in combination with IL-2 directs the differentiation of T-cells into Tregs, while Tregs suppress the proliferation of TCR-dependent proliferation of effector $CD25^{null}$ or $CD25^{low}$ T-cells in a non-autologous fashion. Also TGF-β alters angiogenesis following injury in experiments using MSC (62). In addition, after co-culture with antigen-specific T-cells, MSCs can directly induce the proliferation of regulatory T-cells through release of the immunomodulatory HLA-G isoform HLA-G5 (Fig 1) (63). Taken together, MSCs can modulate the intensity of an immune response by inhibiting antigen-specific T-cell proliferation and cytotoxicity and promoting the generation of regulatory T-cells.

Antibody producing B-cells constitute the second main cell type involved in adaptive immunity. Interactions between MSCs and B-cells have produced controversial results attributable to the inconsistent experimental conditions used (31, 55, 64). Initial reports on mice suggested that MSC exercise a dampening effect on the proliferation of B-cells (64), which is in concordance with most published works to date (31, 55, 64). Furthermore, MSCs can also inhibit B-cell differentiation and constitutive expression of chemokine receptors via the release of soluble factors and cell-cell contact mediated possibly by the Programmed Cell Death 1 (PD-1) and its ligand (31, 64). The addition of MSCs, at the beginning of a mixed lymphocyte reaction (MLC), considerably inhibited immunoglobulin production in standard MLC, irrespective of the MSC dose employed, which suggests that third-party MSC are able to suppress allo-specific antibody production, consequently, overcoming a positive cross-match in sensitized transplant recipients (65). However, other *in vitro* studies have shown that MSCs could support the survival, proliferation and differentiation to antibody-secreting cells of B-cells from normal individuals and from pediatric patients with systemic lupus erythematosus (66, 67). A major mechanism of B-cell suppression was *h*MSC production of soluble factors, as indicated by transwell experiments, where *h*MSCs inhibited B-cell differentiation shown as significant impairment of IgM, IgG, and IgA production. CXCR4, CXCR5, and CCR7 B-cell expression, as well as chemotaxis to CXCL12, the CXCR4 ligand, and CXCL13, the CXCR5 ligand, were significantly down-regulated by *h*MSCs, suggesting that these cells affect chemotactic properties of B-cells (Fig 1). B-cell costimulatory molecule expression and cytokine production were unaffected by *h*MSCs (64). Regardless of the controversial *in vitro* effects, B-cell response is mainly a T-cell dependent mechanism, and thus its outcome is significantly influenced by the MSC-mediated inhibition of T-cell functions. More recently, Corcione et al have shown that systemic administration of MSCs to mice affected by experimental autoimmune encephalomyelitis (EAE), a prototypical disease mediated by self-reactive T cells, results in striking disease amelioration mediated by the induction of peripheral tolerance (52). In addition, it has been shown that tolerance induction by MSCs may occur also through the inhibition of dendritic-cell maturation and function (34, 35), thus suggesting that activated T cells are not the only targets of MSCs.

Low concentrations of IFN-γ upregulate the expression of MHC-II molecules by MSCs, which indicates that they could act as antigen presenting cells (APCs) early in an immune response, when the level of IFN-γ are low (68, 69). However, this process of MHC-II expression by MSCs, along with the potential APC characteristics, was reversed as IFN-γ concentrations increased. These observations could suggest that MSCs function as conditional APC in the early phase of an immune response, while later switch into an immunosuppressive function (68). Since bone marrow might be a site for the induction of T-cell responses to blood-borne antigens (70), and since MSC are derived from the stromal progenitor cells that reside in the bone marrow, therefore, MSC express a yet unidentified role in the control of the immune response physiology of the bone marrow. Dendritic cells are the main APC for T-cell responses, and MSC-mediated suppression of DC maturation would prohibit efficient antigen presentation and thus, the clonal expansion of T-cells. Direct interactions of MSCs with T-cells *in vivo* could lead to the arrest of T-cell proliferation, inhibition of CTL-mediated cytotoxicity and generation of $CD4^{+}$ regulatory T-cells. As a consequence, impaired $CD4^{+}$ T-cell activation would translate into defective T-cell help for B-cell proliferation and differentiation to antibody-secreting cells.

The *h*MSCs express few to none of the B7-1/B7-2 (CD80/CD86) costimulatory-type molecules; this appears to contribute, at least in part, to their immune privilege characteristic. Mechanisms that lead to immune tolerance rely on interrelated pathways that involve complex cross talk and cross regulation of T-cells and APCs by one another. Both soluble mediators and modulation exerted via complex networks of cytokines and costimulatory molecules appear to play a role in MSC regulation of T cells, and these mechanisms invoke both contact-dependent and -independent pathways.

Although many of the studies use MSC-conditioned medium, both contact-dependent and -independent mechanisms are probably invoked in the therapeutic use of MSCs (20, 71). In addition to cytokines, the network of costimulatory molecules is hypothesized to play a prominent role in modulating tolerance and inflammation. MSCs down-regulate the expression of costimulatory molecules (30, 72, 73). The discovery of new functions for B7 family members, together with the identification of additional B7 and CD28 family members, is revealing new ways in which the B7:CD28 family may regulate T-cell activation and tolerance. Not only do CD80/86:CD28 interactions promote initial T-cell activation, they also regulate self-tolerance by supporting $CD4^{+}CD25^{+}$ Treg homeostasis (74-76). Cytotoxic T-lymphocyte antigen 4 (CTLA-4) can exert inhibitory effects in both B7-1/B7-2–dependent and –independent fashions. B7-1 and B7-2 can signal bi-directionally through engaging CD28 and CTLA-4 on T cells and by delivering signals into B7-expressing cells (77). The B7 family members—inducible co-stimulator (ICOS) ligand, PD-L1 (B7-H1), PD-L2 (B7-DC), B7-H3, and B7-H4 (B7x/B7-S1)—are expressed on professional APC cells, while B7-H4 and B7-H1 are expressed on *h*MSCs and on cells within non-lymphoid organs. These observations may provide a new means for regulating T-cell activation and tolerance in peripheral tissues (31, 71, 78). ICOS and PD-1 are expressed upon T-cell-induction, and they regulate previously activated T-cells (79). Both the ICOS:ICOSL pathway and the PD-1:PD-L1/PD-L2 pathway play a critical role in regulating T-cell activation and tolerance (79). There is consensus that both CTLA-4 and PD-1 inhibit T-cell and B-cell activation and may play a crucial role in peripheral tolerance (79, 80). Both CTLA-4 and PD-1 functions are associated with Rheumatoid Arthritis (RA) and other autoimmune diseases. PD-1 is over expressed on $CD4^{+}$ T cells in the synovial fluid of RA patients (81). Whether or not these costimulatory molecules are critical mediators of MSC-mediated immune suppression and/or tolerance *in vivo* is still under current investigation.

5. Mesenchymal Stem Cells escape the immune system *in vitro*

Studies have shown that MSCs escape the immune system, and this makes them a potential therapeutic tool for various transplantation procedures. MSCs express intermediate levels of HLA major histocompatibility complex (MHC) class I molecules

(16, 50, 82, 83), while they do not express HLA class II antigens of the cell surface. However, HLA class II is readily detectable by Western blot on whole-cell lysates of unstimulated adult MSCs, thus suggesting that MSCs contain intracellular deposits of HLA class II allo-antigens (83). Cell-surface expression can be induced by treatment of the cells with IFN-γ for 1 or 2 days. Unlike adult MSCs, the fetal liver- derived cells have no intracellular nor cell surface HLA class II expression (84), but incubation with IFN-γ initiated their intracellular expression followed by surface expression. Differentiation of MSCs into their mesodermal lineages of bone, cartilage, or adipose tissue, both in adult and fetal MSCs continued to express HLA-I, but not class II (84). Undifferentiated MSCs *in vitro* fail to elicit a proliferative response from allogeneic lymphocytes, thus suggesting that the cells are not inherently immunogenic (28, 30, 50). When pre-cultured with IFN-γ for full HLA class II expression, MSCs still escape recognition by allo-reactive T-cells, (83, 84) as is the case with MSCs differentiated adipocytes, osteoblasts, and chondrocytes. Limited *in vivo* data demonstrate the persistence of allogeneic MSCs into immunocompetent hosts after transplantation. In one patient treated with MSCs, DNA of donor MSC could not be detected in any organ at autopsy few weeks after the infusion, while in another patient receiving MSCs from two donors, the donor DNA from both donors was detected in lymph node and colon, the target organs of GVHD, within weeks after infusion (85). Data from our lab indicated that MSC were undetectable after two weeks in an allogeneic system (86). Therefore, the question of whether MSCs are recognized by an intact allogeneic immune system *in vivo* remains open, although the *in vitro* data support the theory that MSCs escape the immune system. MSCs do not express FAS ligand or costimulatory molecules, such as B7-1, B7-2, CD40, or CD40L (50). When costimulation is inadequate, T-cell proliferation can be induced by the addition of exogenous costimulation. However, MSCs differ from other cell types, and no T-cell proliferation can be observed when they are cultured with HLA-mismatched lymphocytes in the presence of a CD28-stimulating antibody (50). However, in agreement with the *in vitro* data, infusion or implantation of allogeneic and MHC-mismatched MSCs into baboons has been well tolerated (87-89). Unique immunologic properties of MSCs were also suggested by the fact that engraftment of human MSCs occurred after intra-uterine transplantation into sheep, even when the transplantation was performed after the fetuses became immunocompetent (90). MSC mainly fail to activate T-cells and show to be targets for $CD8^+$ T cell-cytotoxicity, althought controversial (60). Phyto-hemagglutinin (PHA) blasts, generated to react against a specific donor, will lyse chromium-labeled mononuclear cells from that individual but it will not lyse MSCs derived from the same donor. Furthermore, killer cell inhibitory receptor (KIR ligand)–mismatched natural killer cells do not lyse MSCs (60). Thus, MSCs, although incompatible at the MHC, tend to escape the immune system.

Although MSCs are transplantable across allogeneic barriers, a delayed type hypersensitivity reaction can lead to rejection in xenogenic models of human MSCs injected into immunocompetent rats (91). In this study, MSCs were identified in the heart muscle of severe compromised immune deficiency rats, in contrast to that of immunocompetent rats. In the latter group, peripheral blood lymphocytes proliferated after re-stimulation with human MSCs *in vitro*, thus suggesting cellular immunization. Such a proliferative response *in vitro* has not been detected in humans treated with intravenous (IV) infusion of allogeneic MSCs (Le Blanc and Ringdén, unpublished data, 2004).

6. Mechanisms of immunosuppression by Mesenchymal Stem Cells

Several studies have acknowledged the immunosuppressive activities of MSCs, but the underlying mechanisms are far from being fully characterized. The initial step in the interaction between MSCs and their target cells involves cell–cell contact mediated by adhesion molecules, in concordance with studies showing the dependence of T-cell proliferation on the engagement of PD-1 by its ligand (31). Several soluble immunosuppressive factors, either produced constitutively by MSCs or released following cross-talk with target cells have been reported, including nitric oxide and indoleamine 2,3-dioxygenase (IDO), which are only released by MSC after IFN-γ stimulation with target cells (92, 93), and thus not in a constitutive manner. IDO induces the depletion of tryptophan from the local environment, which is an essential amino acid for lymphocyte proliferation. MSC-derived IDO was reported as a requirement to inhibit the proliferation of IFN-γ-producing TH1 cells (92) and together with prostaglandin E2 (PGE-2) to block NK-cell activity (Fig 1) (44). In addition, IFN-γ, alone or in combination with TNF-α, IL-1α or IL-1β, stimulates the production of chemokines by mouse MSCs that attract T-cells and stimulate the production of inducible nitric-oxide synthase (iNOS), which in turn inhibits T-cell activation through the production of nitric oxide (56). It is worth noting that MSCs from IFN-γ receptor (IFN-γ-R1) deficient mice do not have immunosuppressive activity, which highlights the vital role of IFN-γ in the immunosuppressive function of MSC (56).

Additional soluble factors, such as transforming growth factor-β1 (TGF-β1), hepatocyte growth factor (HGF), IL-10, PGE-2, haem-oxygenase-1 (HO1), IL-6 and soluble HLA-G5, are constitutively produced by MSCs (28, 34, 51, 63, 94) or secreted in response to cytokines released by target cells upon interacting with MSC. TNF-α and IFN-γ have been shown to stimulate the production of PGE-2 by MSCs above constitutive level (34). Furthermore, IL-6 was shown to dampen the respiratory burst and to delay the apoptosis of human neutrophils by inducing phosphorylation of the transcription factor signal transducer and activator of transcription 3 (47), and to inhibit the differentiation of bone-marrow progenitor cells into DCs (95).

The failure to reverse suppression, when neutralizing antibodies against IL-10, TGF-β and IGF were added to MLR reactions does point to the possibility that MSC may secrete as yet uncharacterized immunosuppressive factors (93). Galectin-1 and Galectin-3, newly characterized lectins, are constitutively expressed and secreted by human bone marrow MSC. Inhibition of galectin-1 and galectin-3 gene expression with small interfering RNAs abrogated the suppressive effect of MSC on allogeneic T-cells (Fig 1) (96). The restoration of T-cell proliferation in the presence of β- lactose indicates that the carbohydrate-recognition domain of galectins is responsible for the immunosuppression of T-cells and highlights an extracellular mechanism of action for the MSC-secreted galectins. In this respect, the inhibition of T-cell proliferation could result from either direct effects of galectin-1 and galectin-3 on T cells and/or through a direct or an indirect on effect on dendritic cells (97).

HLA-G5 represents another important molecule involved in MSC mediated regulation of the immune response, where its production has been shown to suppress T-cell proliferation, as well as NK-cell and T-cell cytotoxicity, while promoting the generation of Tregs (63, 98). HLA-G protein expression is constitutive and the level is not modified upon stimulation by allogeneic lymphocytes in MSC/MLR. HLA-G5 is detected on MSCs by real-time reverse-phase polymerase chain reaction, immune-fluorescence, flow cytometry and enzyme-linked immunosorbent assay in the supernatant (99). Cell contact between MSCs and activated T-cells induces IL-10 production, which, in turn, stimulates the release of soluble HLA-G5 by MSCs (63). It is worth nothing that none of these molecules have an exclusive role and that MSC-mediated immune-regulation is a redundant system that is mediated by several molecules.

7. Mesenchymal Stem Cells in response to injury

One important characteristic of *h*MSCs is their ability to suppress inflammation resulting from injury, as well as, resulting from allogeneic solid organ transplants, and autoimmune disease. Consistent with *in vitro* studies, murine allogeneic MSCs are effective cellular therapy models in the treatment of murine models of human disease (52, 100-102). Several studies have documented the substantial clinical improvements observed in animal models, when MSC were systemically introduced as a therapy in mouse models of multiple sclerosis (102, 103), inflammatory bowel disease (104-106), infarct, stroke, and other neurologic diseases (107, 108), as well as diabetes (109). These findings strongly suggest that xenogeneic *h*MSCs are not immunologically recognized by various immunocompetent mouse models of disease and are able to home to sites of inflammation. However, the mechanisms behind the immunosuppressive actions at the site of inflammation and its association with the homing activity have not yet been completely elucidated.
Nitric Oxide (NO) mediate its effect partly through phosphorylation of Stat-5, which results in suppression of T- cell proliferation, partly through the inhibition of NO synthase or the inhibition of prostaglandin synthesis. This reveals the MSC-dependent effects on proliferation. Although indoleamine-2, 3-dioxygenase (IDO) has been hypothesized to be critical in mediating the effect of NO, neutralizing IDO by using a blocking antibody did not interfere with NO's suppressive effects (93, 110).
Within an *in vivo* setting, injury, inflammation, and/or foreign cells can lead to T-cell activation, which results in those T-cells producing proinflammatory cytokines including, but not limited to, TNF-α, IFN-γ, IL-1α, and IL-1β. Combinations of cytokines may also induce cell production of chemokines, some of which bind to CXCR3-R expressing cells (including T cells) that co-localize with MSCs. MSCs then produce NO, which inhibits Stat-5 phosphorylation, thereby leading to cell-cycle arrest (and thus halting T cell proliferation) (Fig 1) (110). In addition, iNOS appears to be important in mouse MSC *in vivo* effects. MSCs from mice that lack iNOS (or IFN-γR1) are unable to suppress GVHD. In contrast to mouse MSCs that use NO in mediating their immune-suppressive effect, *h*MSCs and MSCs from non-human primates appear to mediate their immune-suppressive effects via IDO (56). There is some controversy about whether the effect of IDO results from local depletion of tryptophan, or from the accumulation of tryptophan metabolites (which is suggested by data showing that use of a tryptophan antagonist, 1-methyl-L tryptophan, restored allo-reactivity that would otherwise have been suppressed by MSCs). In addition to its effect on the JAK-STAT pathway, NO may also influence mitogen activated protein kinase and nuclear factor κB, which would cause a reduction in the gene expression of proinflammatory cytokines.

8. Mesenchymal Stem Cell clinical applications

The clinical experience with and the safety of MSCs is of utmost interest for their wide therapeutic applications. The pioneering *in vivo* studies with MSC focused on the engraftment facilitation for the haematopoietic stem cells (111). Further work also focused on the regenerative functions of MSC in terms of functional repair of damaged tissues (112). Hypoimmunogenicity of MSC provided a critical advantage in their use for clinical and therapeutic purposes *in vitro* (50), followed by pre-clinical studies (29) and reaching the human clinical studies (32) with the use of allogeneic donors. Allogeneic MSC have proved to be an option with major advantages in clinical use, since the use of autologous MSC is hindered by the limited time frame for clonal expansion and the costly *in vitro* proliferation. However, some sub-acute conditions, such as autoimmune diseases, might allow the use of autologous MSCs and their culture *in vitro*. It is worth noting that some reports have recently challenged the belief that allogeneic MSCs are poorly immunogenic (113, 114), indicating that in some cases an autologous MSC source could be advantageous. Recent reports have shown that MSCs from patients with autoimmune disease have a normal capability to support hematopoiesis, (115) and to exercise immunomodulatory functions (116), and to show a normal phenotypic characteristics (117).

The perspective role of adult stem cells in degenerative disease conditions, where there is progressive tissue damage and an inability to repair, possibly due to the depletion of stem cell populations or functional alteration, has been considered. In cases of osteoarthritis, a disease of the joints where there is progressive and irreversible loss of cartilage characterized by changes in the underlying bone, Murphy et al showed that the proliferative capacity of the MSC was substantially reduced, and this was independent of the harvest site from patients with end-stage OA undergoing joint replacement surgery (118). In this study the marrow samples were harvested both from the site of surgery (either the hip or the knee) and also from the iliac crest. These effects were apparently disease-related, and not age-related. However, the data suggest that susceptibility to OA and perhaps other degenerative diseases may be due to the reduced mobilization or proliferation of stem cells. In addition, successfully recruited cells may have a limited capacity to differentiate, leading to defective tissue repair. Alternatively, the altered stem cell activity may be in response to the elevated levels of inflammatory cytokines seen in OA, which was confirmed by several other investigators (119, 120).

Similarly, the functional impairment of the anti-proliferative effect of MSCs derived from patients with aplastic anaemia (121) or multiple myeloma (122) might be resulting from an intrinsic abnormality in the microenvironment of the bone marrow, which is consistent with the possible use of autologous MSC for therapeutic purposes.

With the knowledge of the homing capacity of MSC and their capacity to engraft into the recipient's bone after systemic administration, MSCs have been utilized to treat children with severe osteogenesis imperfecta, leading to improved parameters of increased growth velocity and total body mineral content associated with fewer fractures (123). Systemic infusion of allogeneic MSCs also led to encouraging bone marrow recovery in patients with tumors following chemotherapy (123). The immunosuppressive effect of infused MSCs has been successfully shown in acute, severe graft-versus-host disease (GvHD) (32). The probable effect of MSC was the inhibition of donor T-cell reactivity to histocompatibility antigens of the recipient tissue. Currently, there is no successful therapy for steroid-refractory acute GVHD. The possible role of MSCs in this context is therefore of potential interest. Le Blanc et al reported a case of grade IV acute GVHD of the gut and liver in a patient who had undergone ASCT with cells from an unrelated female donor (32). The patient was unresponsive to all types of immunosuppression drugs. When the patient was infused with $2x\ 10^6$ MSCs per kilogram from his HLA-haploidentical mother, his GVHD

responded with a decline in bilirubin and normalization of stools. After the MSC infusion, DNA analysis of his bone marrow showed the presence of minimal residual disease (124). When immunosuppression was discontinued, the patient again developed severe acute GVHD, with its associated symptoms within a few weeks.
Modulation of host allo-reactivity led to accelerated bone-marrow recovery in patients co-transplanted with MSCs and haplo-identical HSCs (125). Clinical trials are being conducted on the immunomodulatory potential of MSCs in the treatment of Crohn's disease, with the potential for those cells to contribute to the regeneration of gastrointestinal epithelial cells (126).
As described previously, MSCs are characterized by their hypoimmunogenicity. In 2000, data from several research groups demonstrated long-term allo-MSC engraftment in a variety of non-cardiac tissues in the absence of immunosuppression (88, 90). On the basis of these observations, investigators began to look into the possibility of allo-MSCs engraftment into affected myocardium in rats, and later in swine, where allo-MSCs were found to readily engraft in necrotic myocardium and favorably alter ventricular function (2). The allo-MSC engraftment occurred without evidence of immunologic rejection or lymphocytic infiltration in the absence of assisted immunosuppressive therapy emphasizing some of the apparent advantages of these cells over other cell populations for cellular cardiomyoplasty. The immunologically privileged status of MSCs was also observed in xenogeneic setting, where Saito et al injected MSC intravenously from C57BL/6 mice into immunocompetent adult Lewis rats (127). When these animals were later subjected to MIs, murine MSCs could be identified in the region of necrosis, and these cells expressed muscle specific proteins not present before coronary ligation.

9. Animal models

Consistent with results from *in vitro* studies, murine allogeneic MSCs are effective in the treatment of murine models of human disease (52, 103, 128). Several studies have reported clinical improvements in mouse models of multiple sclerosis and amyotrophic lateral sclerosis, inflammatory bowel disease, stroke, diabetes, infarct and GVHD using I.V. injected xenogeneic *h*MSCs rather than allogeneic MSCs (108, 109). A major advantage in using *h*MSCs in mouse models of human disease is that the possibility of gathering mechanistic data through measuring biomarkers from body fluids or using noninvasive imaging technology, which may prove to be an advantage in clinical studies when applied on humans.
In experiments designed to study the trafficking of *h*MSCs, investigators used mouse models of severe erosive polyarthritis characterized by an altered transgene allele that results in chronic over-expression of TNF-α and which resemble human RA patients (60, 72). The motive behind utilizing these mice models was to investigate similarities in MSC homing with mouse models of chronic asthma and acute lung injury. Injected *h*MSC revealed a reduction in ankle arthritis parameters associated with decrease appendage related erythema, possibly due to the MSC localization to ankle joints as revealed by bioluminescence (129). Similar observations for inducing tolerance were made using adipose-derived MSC, where Treg were induced in RA PBMC and in mouse models of arthritis (36, 130). Furthermore, studies of rheumatoid arthritis T-cells showed a down-regulation of effector response using adipose-derived MSCs (131). Variations in this potential described by the capability of MSCs to down-regulate collagen-induced arthritis,

and in the ability to induce Tregs, depend on the source of MSC (mouse vs. human) and its characteristics (primary isolate of MSC line), which reflect on difference in function compared to primary expanded MSC (132). Other studies reported that in the collagen-induced model of arthritis, mice infused with MSCs have increased numbers of $CD4^+CD25^+$ cells that express FoxP3 and thus reveal a Treg phenotype (20). Recent data on collagen-induced arthritis model, where murine MSCs did not reveal therapeutic benefits against arthritis *in vivo*, but did show anti-proliferative effect *in vitro* suggest that there is no appropriate *in vitro* measures that can be accurately extrapolated into a potential therapeutic utility of MSCs *in vivo*, and that mouse MSCs show difference in functional characteristics to *h*MSC in terms of immunoregulatory capacity (133).

MSC's immunological properties appeared to have potential therapeutic advantages in other forms of autoimmune diseases, especially in type 1 diabetes. In NOD mouse model, several physiological defects that aim to maintain peripheral and central tolerance contribute to the development of autoimmune diabetes. These defects are summed up as a combination of immune cell dysfunction (including T-cell, NK cells, B-cells, and dendritic cells), associated with the presence of inflammatory cytokine milieu (134). MSCs possess specific immunomodulatory properties capable of halting autoimmunity through immunomodulation processes described in this chapter. The processes might be through a direct effect via the presentation of differential levels of negative costimulatory molecules and the secretion of regulatory cytokines that affect regulatory T-cells/autoreactive T-cells. Furthermore, MSCs could modulate the immunological dysregulation observed in antibody producing B-cells and cytotoxic NK cells. Dendritic cells have been shown to be defective in NOD mice characterized by higher levels of costimulation with a potential capability to shift to a TH-1 type of immune response.

In an experimental mouse model of diabetes induced by streptozotocin, it was observed that MSCs promoted the endogenous repair of pancreatic islets and renal glomeruli (109). Similarly, co-infusion of MSCs and bone-marrow cells inhibited the proliferation of β-cell-specific T-cells isolated from the pancreas of diabetic mice and restored insulin and glucose levels through the induction of recipient-derived pancreatic β-cell regeneration in the absence of trans-differentiation of MSCs (135). These studies show that the *in vivo* administration of MSCs is clinically efficacious through the modulation of pathogenic β- and T-cell responses and through potent bystander effects on the target tissue.

The timing of MSC infusion seems to be a critical parameter in their therapeutic efficacy. In the EAE mouse model of multiple sclerosis, MSC systematically injected at disease onset ameliorated myelin oligodendrocyte glycoprotein (MOG)-induced EAE and further decreased the infiltration T-cells, B-cells and macrophages into the central nervous system (CNS). Furthermore, T cells isolated from the lymph nodes of MSC-treated mice did not proliferate after *in vitro* re-challenge with MOG peptide, which is an indication of the induction of T-cell anergy (52). Systematic injection of MSCs was found to inhibit the *in vivo* production of pathogenic plp-specific antibodies and to suppress the encephalitogenic potential of plp-specific T cells in passive-transfer experiments. In this model, the MSCs migrated to the lymphoid organs, as well as, to the inflamed CNS, where they exercised a protective effect on the neuronal axons *in situ* (135, 136). In these studies, the therapeutic effect of MSCs depended on the release of anti-apoptotic, anti-inflammatory and trophic molecules, as occurred in the case of stroke in rats (137), and, possibly, on the recruitment of local progenitors and their subsequent induction to differentiate into neural cells (138). As trophic effect, the MSCs appeared to favor oligo-dendrogenisis by neural precursor cells (139).

Several other studies have provided insights into the effects of MSCs mediated by cytokines. In a model of acute renal failure, the administration of MSCs increased the recovery of renal function through the inhibition of production of proinflammatory cytokines, such as Il-1β, TNF and IFNγ, and through an anti-apoptotic effect on target cells (140). Along the same line, the anti-inflammatory activity of MSCs was revealed in a mouse model of lung fibrosis, where they inhibited the effects of IL-1α-producing T cells and TNF-producing macrophages through the release of IL-1 receptor antagonist (IL-1RA) (141). The release of trophic factors such as the WNT-associated molecule secreted frizzled-related protein 2 (SFRp2), which leads to the rescue of ischemic cardiomyocytes and the restoration of ventricular functions represent another important function for MSC (142).
With all the promising therapeutic potential of MSC, there seems to be a growing concern about their association with tumors. The immunoregulatory and anti-proliferative effects of MSCs led to several studies investigating the inhibitory effect of MSCs on tumor growth. Inhibition or, more frequently, stimulation of tumor-cell proliferation *in vitro* and/or tumor growth *in vivo* by MSCs has been reported (143-145). The heterogeneous nature of the MSC populations and the different experimental tumor models used, contribute to the effect of tumors on MSC in which the microenvironment generated by tumors influence the behavior of MSCs (146). Two main mechanisms are probably involved in the enhancement of tumor growth by MSCs. First, the cell-to-cell cross-talk between MSCs and tumor cells contribute to tumor progression, thus integrating within the tumor stroma (147), and second, the suppressive effects of MSCs on the immune system of tumor-bearing hosts might facilitate tumorigenesis, as shown for the inhibition of melanoma rejection, possibly mediated by regulatory $CD8^+$ T cells (144). Irrespective of the possible interactions between cancer cells, immune cells and MSCs, the potential risk of stimulating the growth cancer by MSCs must be considered.

10. Conclusion

As a whole, the data accumulated from preclinical and clinical data indicate that bone marrow-derived MSCs have, in addition to their therapeutic purposes in regenerative medicine, effects that can result from other characteristics, such as their anti-proliferative and anti-inflammatory properties. The immuno suppressive activity of MSCs provides means for inducing peripheral tolerance following systemic injection mediated through the inhibition of cell division, thereby preventing their responsiveness to antigenic triggers while maintaining them in a quiescent state. In addition, the clinical efficacy of MSCs in different experimental model seems to occur almost only during the acute phase of disease associated with limited trans-differentiation, which indicates that the therapeutic effectiveness of MSCs relies heavily on their ability to modify microenvironments. These modifications occur through the release of anti-inflammatory cytokines, and anti-apoptotic and trophic molecules that promote the repair and protection of damaged tissues, as well as, maintain the integrity of the immune cells.

11. References

[1] Zvaifler, N. J., L. Marinova-Mutafchieva, G. Adams, C. J. Edwards, J. Moss, J. A. Burger, and R. N. Maini. 2000. Mesenchymal precursor cells in the blood of normal individuals. *Arthritis Res* 2:477-488.

[2] Pittenger, M. F., and B. J. Martin. 2004. Mesenchymal stem cells and their potential as cardiac therapeutics. *Circ Res* 95:9-20.
[3] Sacchetti, B., A. Funari, S. Michienzi, S. Di Cesare, S. Piersanti, I. Saggio, E. Tagliafico, S. Ferrari, P. G. Robey, M. Riminucci, and P. Bianco. 2007. Self-renewing osteoprogenitors in bone marrow sinusoids can organize a hematopoietic microenvironment. *Cell* 131:324-336.
[4] Grove, J. E., E. Bruscia, and D. S. Krause. 2004. Plasticity of bone marrow-derived stem cells. *Stem Cells* 22:487-500.
[5] Prockop, D. J., C. A. Gregory, and J. L. Spees. 2003. One strategy for cell and gene therapy: harnessing the power of adult stem cells to repair tissues. *Proc Natl Acad Sci U S A* 100 Suppl 1:11917-11923.
[6] Gronthos, S., D. M. Franklin, H. A. Leddy, P. G. Robey, R. W. Storms, and J. M. Gimble. 2001. Surface protein characterization of human adipose tissue-derived stromal cells. *J Cell Physiol* 189:54-63.
[7] Erices, A., P. Conget, and J. J. Minguell. 2000. Mesenchymal progenitor cells in human umbilical cord blood. *Br J Haematol* 109:235-242.
[8] In 't Anker, P. S., S. A. Scherjon, C. Kleijburg-van der Keur, W. A. Noort, F. H. Claas, R. Willemze, W. E. Fibbe, and H. H. Kanhai. 2003. Amniotic fluid as a novel source of mesenchymal stem cells for therapeutic transplantation. *Blood* 102:1548-1549.
[9] Friedenstein, A. J., K. V. Petrakova, A. I. Kurolesova, and G. P. Frolova. 1968. Heterotopic of bone marrow. Analysis of precursor cells for osteogenic and hematopoietic tissues. *Transplantation* 6:230-247.
[10] Friedenstein, A. J. 1976. Precursor cells of mechanocytes. *Int Rev Cytol* 47:327-359.
[11] Caplan, A. I. 1991. Mesenchymal stem cells. *J Orthop Res* 9:641-650.
[12] Horwitz, E. M., K. Le Blanc, M. Dominici, I. Mueller, I. Slaper-Cortenbach, F. C. Marini, R. J. Deans, D. S. Krause, and A. Keating. 2005. Clarification of the nomenclature for MSC: The International Society for Cellular Therapy position statement. *Cytotherapy* 7:393-395.
[13] Dominici, M., K. Le Blanc, I. Mueller, I. Slaper-Cortenbach, F. Marini, D. Krause, R. Deans, A. Keating, D. Prockop, and E. Horwitz. 2006. Minimal criteria for defining multipotent mesenchymal stromal cells. The International Society for Cellular Therapy position statement. *Cytotherapy* 8:315-317.
[14] Prockop, D. J. 1997. Marrow stromal cells as stem cells for nonhematopoietic tissues. *Science* 276:71-74.
[15] Colter, D. C., I. Sekiya, and D. J. Prockop. 2001. Identification of a subpopulation of rapidly self-renewing and multipotential adult stem cells in colonies of human marrow stromal cells. *Proc Natl Acad Sci U S A* 98:7841-7845.
[16] Pittenger, M. F., A. M. Mackay, S. C. Beck, R. K. Jaiswal, R. Douglas, J. D. Mosca, M. A. Moorman, D. W. Simonetti, S. Craig, and D. R. Marshak. 1999. Multilineage potential of adult human mesenchymal stem cells. *Science* 284:143-147.
[17] Reyes, M., T. Lund, T. Lenvik, D. Aguiar, L. Koodie, and C. M. Verfaillie. 2001. Purification and ex vivo expansion of postnatal human marrow mesodermal progenitor cells. *Blood* 98:2615-2625.
[18] Thomas, E. D., and K. G. Blume. 1999. Historical markers in the development of allogeneic hematopoietic cell transplantation. *Biol Blood Marrow Transplant* 5:341-346.
[19] Welter, J. F., L. A. Solchaga, and K. J. Penick. 2007. Simplification of aggregate culture of human mesenchymal stem cells as a chondrogenic screening assay. *Biotechniques* 42:732, 734-737.

[20] da Silva Meirelles, L., A. I. Caplan, and N. B. Nardi. 2008. In search of the in vivo identity of mesenchymal stem cells. *Stem Cells* 26:2287-2299.

[21] Zuk, P. A., M. Zhu, H. Mizuno, J. Huang, J. W. Futrell, A. J. Katz, P. Benhaim, H. P. Lorenz, and M. H. Hedrick. 2001. Multilineage cells from human adipose tissue: implications for cell-based therapies. *Tissue Eng* 7:211-228.

[22] Noth, U., A. M. Osyczka, R. Tuli, N. J. Hickok, K. G. Danielson, and R. S. Tuan. 2002. Multilineage mesenchymal differentiation potential of human trabecular bone-derived cells. *J Orthop Res* 20:1060-1069.

[23] Lee, R. H., B. Kim, I. Choi, H. Kim, H. S. Choi, K. Suh, Y. C. Bae, and J. S. Jung. 2004. Characterization and expression analysis of mesenchymal stem cells from human bone marrow and adipose tissue. *Cell Physiol Biochem* 14:311-324.

[24] Toma, J. G., M. Akhavan, K. J. Fernandes, F. Barnabe-Heider, A. Sadikot, D. R. Kaplan, and F. D. Miller. 2001. Isolation of multipotent adult stem cells from the dermis of mammalian skin. *Nat Cell Biol* 3:778-784.

[25] Keyser, K. A., K. E. Beagles, and H. P. Kiem. 2007. Comparison of mesenchymal stem cells from different tissues to suppress T-cell activation. *Cell Transplant* 16:555-562.

[26] Muraglia, A., R. Cancedda, and R. Quarto. 2000. Clonal mesenchymal progenitors from human bone marrow differentiate in vitro according to a hierarchical model. *J Cell Sci* 113 (Pt 7):1161-1166.

[27] Zuk, P. A., M. Zhu, P. Ashjian, D. A. De Ugarte, J. I. Huang, H. Mizuno, Z. C. Alfonso, J. K. Fraser, P. Benhaim, and M. H. Hedrick. 2002. Human adipose tissue is a source of multipotent stem cells. *Mol Biol Cell* 13:4279-4295.

[28] Di Nicola, M., C. Carlo-Stella, M. Magni, M. Milanesi, P. D. Longoni, P. Matteucci, S. Grisanti, and A. M. Gianni. 2002. Human bone marrow stromal cells suppress T-lymphocyte proliferation induced by cellular or nonspecific mitogenic stimuli. *Blood* 99:3838-3843.

[29] Bartholomew, A., C. Sturgeon, M. Siatskas, K. Ferrer, K. McIntosh, S. Patil, W. Hardy, S. Devine, D. Ucker, R. Deans, A. Moseley, and R. Hoffman. 2002. Mesenchymal stem cells suppress lymphocyte proliferation in vitro and prolong skin graft survival in vivo. *Exp Hematol* 30:42-48.

[30] Le Blanc, K., L. Tammik, B. Sundberg, S. E. Haynesworth, and O. Ringden. 2003. Mesenchymal stem cells inhibit and stimulate mixed lymphocyte cultures and mitogenic responses independently of the major histocompatibility complex. *Scand J Immunol* 57:11-20.

[31] Augello, A., R. Tasso, S. M. Negrini, A. Amateis, F. Indiveri, R. Cancedda, and G. Pennesi. 2005. Bone marrow mesenchymal progenitor cells inhibit lymphocyte proliferation by activation of the programmed death 1 pathway. *Eur J Immunol* 35:1482-1490.

[32] Le Blanc, K., I. Rasmusson, B. Sundberg, C. Gotherstrom, M. Hassan, M. Uzunel, and O. Ringden. 2004. Treatment of severe acute graft-versus-host disease with third party haploidentical mesenchymal stem cells. *Lancet* 363:1439-1441.

[33] Maitra, B., E. Szekely, K. Gjini, M. J. Laughlin, J. Dennis, S. E. Haynesworth, and O. N. Koc. 2004. Human mesenchymal stem cells support unrelated donor hematopoietic stem cells and suppress T-cell activation. *Bone Marrow Transplant* 33:597-604.

[34] Aggarwal, S., and M. F. Pittenger. 2005. Human mesenchymal stem cells modulate allogeneic immune cell responses. *Blood* 105:1815-1822.

[35] Beyth, S., Z. Borovsky, D. Mevorach, M. Liebergall, Z. Gazit, H. Aslan, E. Galun, and J. Rachmilewitz. 2005. Human mesenchymal stem cells alter antigen-presenting cell maturation and induce T-cell unresponsiveness. *Blood* 105:2214-2219.

[36] Augello, A., R. Tasso, S. M. Negrini, R. Cancedda, and G. Pennesi. 2007. Cell therapy using allogeneic bone marrow mesenchymal stem cells prevents tissue damage in collagen-induced arthritis. *Arthritis Rheum* 56:1175-1186.
[37] Chen, L., W. Zhang, H. Yue, Q. Han, B. Chen, M. Shi, J. Li, B. Li, S. You, Y. Shi, and R. C. Zhao. 2007. Effects of human mesenchymal stem cells on the differentiation of dendritic cells from CD34+ cells. *Stem Cells Dev* 16:719-731.
[38] Zhang, W., W. Ge, C. Li, S. You, L. Liao, Q. Han, W. Deng, and R. C. Zhao. 2004. Effects of mesenchymal stem cells on differentiation, maturation, and function of human monocyte-derived dendritic cells. *Stem Cells Dev* 13:263-271.
[39] Ramasamy, R., H. Fazekasova, E. W. Lam, I. Soeiro, G. Lombardi, and F. Dazzi. 2007. Mesenchymal stem cells inhibit dendritic cell differentiation and function by preventing entry into the cell cycle. *Transplantation* 83:71-76.
[40] Maccario, R., M. Podesta, A. Moretta, A. Cometa, P. Comoli, D. Montagna, L. Daudt, A. Ibatici, G. Piaggio, S. Pozzi, F. Frassoni, and F. Locatelli. 2005. Interaction of human mesenchymal stem cells with cells involved in alloantigen-specific immune response favors the differentiation of CD4+ T-cell subsets expressing a regulatory/suppressive phenotype. *Haematologica* 90:516-525.
[41] Moretta, A. 2002. Natural killer cells and dendritic cells: rendezvous in abused tissues. *Nat Rev Immunol* 2:957-964.
[42] Moretta, A., C. Bottino, M. Vitale, D. Pende, R. Biassoni, M. C. Mingari, and L. Moretta. 1996. Receptors for HLA class-I molecules in human natural killer cells. *Annu Rev Immunol* 14:619-648.
[43] Spaggiari, G. M., A. Capobianco, S. Becchetti, M. C. Mingari, and L. Moretta. 2006. Mesenchymal stem cell-natural killer cell interactions: evidence that activated NK cells are capable of killing MSCs, whereas MSCs can inhibit IL-2-induced NK-cell proliferation. *Blood* 107:1484-1490.
[44] Spaggiari, G. M., A. Capobianco, H. Abdelrazik, F. Becchetti, M. C. Mingari, and L. Moretta. 2008. Mesenchymal stem cells inhibit natural killer-cell proliferation, cytotoxicity, and cytokine production: role of indoleamine 2,3-dioxygenase and prostaglandin E2. *Blood* 111:1327-1333.
[45] Poggi, A., C. Prevosto, A. M. Massaro, S. Negrini, S. Urbani, I. Pierri, R. Saccardi, M. Gobbi, and M. R. Zocchi. 2005. Interaction between human NK cells and bone marrow stromal cells induces NK cell triggering: role of NKp30 and NKG2D receptors. *J Immunol* 175:6352-6360.
[46] de Kuiper Jan, N. M. I. N. M. v. B. C. C. B. W. W. M. J. H. M. J. Human mesenchymal stem cells are susceptible to lysis by CD8 + T-cells and NK cells. *Cell Transplant*.
[47] Raffaghello, L., G. Bianchi, M. Bertolotto, F. Montecucco, A. Busca, F. Dallegri, L. Ottonello, and V. Pistoia. 2008. Human mesenchymal stem cells inhibit neutrophil apoptosis: a model for neutrophil preservation in the bone marrow niche. *Stem Cells* 26:151-162.
[48] Brandau, S., M. Jakob, H. Hemeda, K. Bruderek, S. Janeschik, F. Bootz, and S. Lang. Tissue-resident mesenchymal stem cells attract peripheral blood neutrophils and enhance their inflammatory activity in response to microbial challenge. *J Leukoc Biol* 88:1005-1015.
[49] Craddock, C. G., Jr., S. Perry, L. E. Ventzke, and J. S. Lawrence. 1960. Evaluation of marrow granulocytic reserves in normal and disease states. *Blood* 15:840-855.
[50] Tse, W. T., J. D. Pendleton, W. M. Beyer, M. C. Egalka, and E. C. Guinan. 2003. Suppression of allogeneic T-cell proliferation by human marrow stromal cells: implications in transplantation. *Transplantation* 75:389-397.

[51] Meisel, R., A. Zibert, M. Laryea, U. Gobel, W. Daubener, and D. Dilloo. 2004. Human bone marrow stromal cells inhibit allogeneic T-cell responses by indoleamine 2,3-dioxygenase-mediated tryptophan degradation. *Blood* 103:4619-4621.
[52] Zappia, E., S. Casazza, E. Pedemonte, F. Benvenuto, I. Bonanni, E. Gerdoni, D. Giunti, A. Ceravolo, F. Cazzanti, F. Frassoni, G. Mancardi, and A. Uccelli. 2005. Mesenchymal stem cells ameliorate experimental autoimmune encephalomyelitis inducing T-cell anergy. *Blood* 106:1755-1761.
[53] Rasmusson, I., O. Ringden, B. Sundberg, and K. Le Blanc. 2005. Mesenchymal stem cells inhibit lymphocyte proliferation by mitogens and alloantigens by different mechanisms. *Exp Cell Res* 305:33-41.
[54] Krampera, M., S. Glennie, J. Dyson, D. Scott, R. Laylor, E. Simpson, and F. Dazzi. 2003. Bone marrow mesenchymal stem cells inhibit the response of naive and memory antigen-specific T cells to their cognate peptide. *Blood* 101:3722-3729.
[55] Glennie, S., I. Soeiro, P. J. Dyson, E. W. Lam, and F. Dazzi. 2005. Bone marrow mesenchymal stem cells induce division arrest anergy of activated T cells. *Blood* 105:2821-2827.
[56] Ren, G., L. Zhang, X. Zhao, G. Xu, Y. Zhang, A. I. Roberts, R. C. Zhao, and Y. Shi. 2008. Mesenchymal stem cell-mediated immunosuppression occurs via concerted action of chemokines and nitric oxide. *Cell Stem Cell* 2:141-150.
[57] Benvenuto, F., S. Ferrari, E. Gerdoni, F. Gualandi, F. Frassoni, V. Pistoia, G. Mancardi, and A. Uccelli. 2007. Human mesenchymal stem cells promote survival of T cells in a quiescent state. *Stem Cells* 25:1753-1760.
[58] Groh, M. E., B. Maitra, E. Szekely, and O. N. Koc. 2005. Human mesenchymal stem cells require monocyte-mediated activation to suppress alloreactive T cells. *Exp Hematol* 33:928-934.
[59] Fadeel, B., A. Ahlin, J. I. Henter, S. Orrenius, and M. B. Hampton. 1998. Involvement of caspases in neutrophil apoptosis: regulation by reactive oxygen species. *Blood* 92:4808-4818.
[60] Rasmusson, I., O. Ringden, B. Sundberg, and K. Le Blanc. 2003. Mesenchymal stem cells inhibit the formation of cytotoxic T lymphocytes, but not activated cytotoxic T lymphocytes or natural killer cells. *Transplantation* 76:1208-1213.
[61] Figueroa-Tentori, D., S. Querol, I. A. Dodi, A. Madrigal, and R. Duggleby. 2008. High purity and yield of natural Tregs from cord blood using a single step selection method. *J Immunol Methods* 339:228-235.
[62] Oh, J. Y., M. K. Kim, M. S. Shin, H. J. Lee, J. H. Ko, W. R. Wee, and J. H. Lee. 2008. The anti-inflammatory and anti-angiogenic role of mesenchymal stem cells in corneal wound healing following chemical injury. *Stem Cells* 26:1047-1055.
[63] Selmani, Z., A. Naji, I. Zidi, B. Favier, E. Gaiffe, L. Obert, C. Borg, P. Saas, P. Tiberghien, N. Rouas-Freiss, E. D. Carosella, and F. Deschaseaux. 2008. Human leukocyte antigen-G5 secretion by human mesenchymal stem cells is required to suppress T lymphocyte and natural killer function and to induce CD4+CD25highFOXP3+ regulatory T cells. *Stem Cells* 26:212-222.
[64] Corcione, A., F. Benvenuto, E. Ferretti, D. Giunti, V. Cappiello, F. Cazzanti, M. Risso, F. Gualandi, G. L. Mancardi, V. Pistoia, and A. Uccelli. 2006. Human mesenchymal stem cells modulate B-cell functions. *Blood* 107:367-372.
[65] Comoli, P., F. Ginevri, R. Maccario, M. A. Avanzini, M. Marconi, A. Groff, A. Cometa, M. Cioni, L. Porretti, W. Barberi, F. Frassoni, and F. Locatelli. 2008. Human mesenchymal stem cells inhibit antibody production induced in vitro by allostimulation. *Nephrol Dial Transplant* 23:1196-1202.

[66] Traggiai, E., S. Volpi, F. Schena, M. Gattorno, F. Ferlito, L. Moretta, and A. Martini. 2008. Bone marrow-derived mesenchymal stem cells induce both polyclonal expansion and differentiation of B cells isolated from healthy donors and systemic lupus erythematosus patients. *Stem Cells* 26:562-569.
[67] Rasmusson, I., K. Le Blanc, B. Sundberg, and O. Ringden. 2007. Mesenchymal stem cells stimulate antibody secretion in human B cells. *Scand J Immunol* 65:336-343.
[68] Stagg, J., S. Pommey, N. Eliopoulos, and J. Galipeau. 2006. Interferon-gamma-stimulated marrow stromal cells: a new type of nonhematopoietic antigen-presenting cell. *Blood* 107:2570-2577.
[69] Chan, J. L., K. C. Tang, A. P. Patel, L. M. Bonilla, N. Pierobon, N. M. Ponzio, and P. Rameshwar. 2006. Antigen-presenting property of mesenchymal stem cells occurs during a narrow window at low levels of interferon-gamma. *Blood* 107:4817-4824.
[70] Feuerer, M., P. Beckhove, N. Garbi, Y. Mahnke, A. Limmer, M. Hommel, G. J. Hammerling, B. Kyewski, A. Hamann, V. Umansky, and V. Schirrmacher. 2003. Bone marrow as a priming site for T-cell responses to blood-borne antigen. *Nat Med* 9:1151-1157.
[71] Xue, Q., X. Y. Luan, Y. Z. Gu, H. Y. Wu, G. B. Zhang, G. H. Yu, H. T. Zhu, M. Wang, W. Dong, Y. J. Geng, and X. G. Zhang. The negative co-signaling molecule b7-h4 is expressed by human bone marrow-derived mesenchymal stem cells and mediates its T-cell modulatory activity. *Stem Cells Dev* 19:27-38.
[72] Majumdar, M. K., M. Keane-Moore, D. Buyaner, W. B. Hardy, M. A. Moorman, K. R. McIntosh, and J. D. Mosca. 2003. Characterization and functionality of cell surface molecules on human mesenchymal stem cells. *J Biomed Sci* 10:228-241.
[73] Fife, B. T., K. E. Pauken, T. N. Eagar, T. Obu, J. Wu, Q. Tang, M. Azuma, M. F. Krummel, and J. A. Bluestone. 2009. Interactions between PD-1 and PD-L1 promote tolerance by blocking the TCR-induced stop signal. *Nat Immunol* 10:1185-1192.
[74] Franceschini, D., M. Paroli, V. Francavilla, M. Videtta, S. Morrone, G. Labbadia, A. Cerino, M. U. Mondelli, and V. Barnaba. 2009. PD-L1 negatively regulates CD4+CD25+Foxp3+ Tregs by limiting STAT-5 phosphorylation in patients chronically infected with HCV. *J Clin Invest* 119:551-564.
[75] Fife, B. T., and J. A. Bluestone. 2008. Control of peripheral T-cell tolerance and autoimmunity via the CTLA-4 and PD-1 pathways. *Immunol Rev* 224:166-182.
[76] Salama, A. D., T. Chitnis, J. Imitola, M. J. Ansari, H. Akiba, F. Tushima, M. Azuma, H. Yagita, M. H. Sayegh, and S. J. Khoury. 2003. Critical role of the programmed death-1 (PD-1) pathway in regulation of experimental autoimmune encephalomyelitis. *J Exp Med* 198:71-78.
[77] Teft, W. A., M. G. Kirchhof, and J. Madrenas. 2006. A molecular perspective of CTLA-4 function. *Annu Rev Immunol* 24:65-97.
[78] Ni, X., Y. Q. Jia, W. T. Meng, L. Zhong, and Y. Zeng. 2009. [Expression of B7-H1 molecule on human bone marrow mesenchymal stem cells and its effects on T lymphocyte proliferation]. *Zhongguo Shi Yan Xue Ye Xue Za Zhi* 17:990-993.
[79] Greenwald, R. J., G. J. Freeman, and A. H. Sharpe. 2005. The B7 family revisited. *Annu Rev Immunol* 23:515-548.
[80] Martin-Orozco, N., Y. H. Wang, H. Yagita, and C. Dong. 2006. Cutting Edge: Programmed death (PD) ligand-1/PD-1 interaction is required for CD8+ T cell tolerance to tissue antigens. *J Immunol* 177:8291-8295.
[81] Hatachi, S., Y. Iwai, S. Kawano, S. Morinobu, M. Kobayashi, M. Koshiba, R. Saura, M. Kurosaka, T. Honjo, and S. Kumagai. 2003. CD4+ PD-1+ T cells accumulate as

unique anergic cells in rheumatoid arthritis synovial fluid. *J Rheumatol* 30:1410-1419.

[82] Deans, R. J., and A. B. Moseley. 2000. Mesenchymal stem cells: biology and potential clinical uses. *Exp Hematol* 28:875-884.

[83] Le Blanc, K., C. Tammik, K. Rosendahl, E. Zetterberg, and O. Ringden. 2003. HLA expression and immunologic properties of differentiated and undifferentiated mesenchymal stem cells. *Exp Hematol* 31:890-896.

[84] Gotherstrom, C., O. Ringden, C. Tammik, E. Zetterberg, M. Westgren, and K. Le Blanc. 2004. Immunologic properties of human fetal mesenchymal stem cells. *Am J Obstet Gynecol* 190:239-245.

[85] Ringden, O., M. Uzunel, I. Rasmusson, M. Remberger, B. Sundberg, H. Lonnies, H. U. Marschall, A. Dlugosz, A. Szakos, Z. Hassan, B. Omazic, J. Aschan, L. Barkholt, and K. Le Blanc. 2006. Mesenchymal stem cells for treatment of therapy-resistant graft-versus-host disease. *Transplantation* 81:1390-1397.

[86] El Haddad, N., D. Heathcote, R. Moore, S. Yang, J. Azzi, B. Mfarrej, M. Atkinson, M. H. Sayegh, J. S. Lee, P. G. Ashton-Rickardt, and R. Abdi. Mesenchymal stem cells express serine protease inhibitor to evade the host immune response. *Blood* 117:1176-1183.

[87] Devine, S. M., A. M. Bartholomew, N. Mahmud, M. Nelson, S. Patil, W. Hardy, C. Sturgeon, T. Hewett, T. Chung, W. Stock, D. Sher, S. Weissman, K. Ferrer, J. Mosca, R. Deans, A. Moseley, and R. Hoffman. 2001. Mesenchymal stem cells are capable of homing to the bone marrow of non-human primates following systemic infusion. *Exp Hematol* 29:244-255.

[88] Bartholomew, A., S. Patil, A. Mackay, M. Nelson, D. Buyaner, W. Hardy, J. Mosca, C. Sturgeon, M. Siatskas, N. Mahmud, K. Ferrer, R. Deans, A. Moseley, R. Hoffman, and S. M. Devine. 2001. Baboon mesenchymal stem cells can be genetically modified to secrete human erythropoietin in vivo. *Hum Gene Ther* 12:1527-1541.

[89] Devine, S. M., C. Cobbs, M. Jennings, A. Bartholomew, and R. Hoffman. 2003. Mesenchymal stem cells distribute to a wide range of tissues following systemic infusion into nonhuman primates. *Blood* 101:2999-3001.

[90] Liechty, K. W., T. C. MacKenzie, A. F. Shaaban, A. Radu, A. M. Moseley, R. Deans, D. R. Marshak, and A. W. Flake. 2000. Human mesenchymal stem cells engraft and demonstrate site-specific differentiation after in utero transplantation in sheep. *Nat Med* 6:1282-1286.

[91] Grinnemo, K. H., A. Mansson, G. Dellgren, D. Klingberg, E. Wardell, V. Drvota, C. Tammik, J. Holgersson, O. Ringden, C. Sylven, and K. Le Blanc. 2004. Xenoreactivity and engraftment of human mesenchymal stem cells transplanted into infarcted rat myocardium. *J Thorac Cardiovasc Surg* 127:1293-1300.

[92] Krampera, M., L. Cosmi, R. Angeli, A. Pasini, F. Liotta, A. Andreini, V. Santarlasci, B. Mazzinghi, G. Pizzolo, F. Vinante, P. Romagnani, E. Maggi, S. Romagnani, and F. Annunziato. 2006. Role for interferon-gamma in the immunomodulatory activity of human bone marrow mesenchymal stem cells. *Stem Cells* 24:386-398.

[93] Ryan, J. M., F. Barry, J. M. Murphy, and B. P. Mahon. 2007. Interferon-gamma does not break, but promotes the immunosuppressive capacity of adult human mesenchymal stem cells. *Clin Exp Immunol* 149:353-363.

[94] Chabannes, D., M. Hill, E. Merieau, J. Rossignol, R. Brion, J. P. Soulillou, I. Anegon, and M. C. Cuturi. 2007. A role for heme oxygenase-1 in the immunosuppressive effect of adult rat and human mesenchymal stem cells. *Blood* 110:3691-3694.

[95] Djouad, F., L. M. Charbonnier, C. Bouffi, P. Louis-Plence, C. Bony, F. Apparailly, C. Cantos, C. Jorgensen, and D. Noel. 2007. Mesenchymal stem cells inhibit the differentiation of dendritic cells through an interleukin-6-dependent mechanism. *Stem Cells* 25:2025-2032.

[96] Sioud, M. New insights into mesenchymal stromal cell-mediated T-cell suppression through galectins. *Scand J Immunol* 73:79-84.

[97] Ilarregui, J. M., D. O. Croci, G. A. Bianco, M. A. Toscano, M. Salatino, M. E. Vermeulen, J. R. Geffner, and G. A. Rabinovich. 2009. Tolerogenic signals delivered by dendritic cells to T cells through a galectin-1-driven immunoregulatory circuit involving interleukin 27 and interleukin 10. *Nat Immunol* 10:981-991.

[98] Morandi, F., L. Raffaghello, G. Bianchi, F. Meloni, A. Salis, E. Millo, S. Ferrone, V. Barnaba, and V. Pistoia. 2008. Immunogenicity of human mesenchymal stem cells in HLA-class I-restricted T-cell responses against viral or tumor-associated antigens. *Stem Cells* 26:1275-1287.

[99] Nasef, A., N. Mathieu, A. Chapel, J. Frick, S. Francois, C. Mazurier, A. Boutarfa, S. Bouchet, N. C. Gorin, D. Thierry, and L. Fouillard. 2007. Immunosuppressive effects of mesenchymal stem cells: involvement of HLA-G. *Transplantation* 84:231-237.

[100] Lanz, T. V., C. A. Opitz, P. P. Ho, A. Agrawal, C. Lutz, M. Weller, A. L. Mellor, L. Steinman, W. Wick, and M. Platten. Mouse mesenchymal stem cells suppress antigen-specific TH cell immunity independent of indoleamine 2,3-dioxygenase 1 (IDO1). *Stem Cells Dev* 19:657-668.

[101] Rafei, M., E. Birman, K. Forner, and J. Galipeau. 2009. Allogeneic mesenchymal stem cells for treatment of experimental autoimmune encephalomyelitis. *Mol Ther* 17:1799-1803.

[102] Bai, L., D. P. Lennon, V. Eaton, K. Maier, A. I. Caplan, S. D. Miller, and R. H. Miller. 2009. Human bone marrow-derived mesenchymal stem cells induce Th2-polarized immune response and promote endogenous repair in animal models of multiple sclerosis. *Glia* 57:1192-1203.

[103] Ding, Y., D. Xu, G. Feng, A. Bushell, R. J. Muschel, and K. J. Wood. 2009. Mesenchymal stem cells prevent the rejection of fully allogenic islet grafts by the immunosuppressive activity of matrix metalloproteinase-2 and -9. *Diabetes* 58:1797-1806.

[104] Gonzalez-Rey, E., P. Anderson, M. A. Gonzalez, L. Rico, D. Buscher, and M. Delgado. 2009. Human adult stem cells derived from adipose tissue protect against experimental colitis and sepsis. *Gut* 58:929-939.

[105] Gonzalez, M. A., E. Gonzalez-Rey, L. Rico, D. Buscher, and M. Delgado. 2009. Adipose-derived mesenchymal stem cells alleviate experimental colitis by inhibiting inflammatory and autoimmune responses. *Gastroenterology* 136:978-989.

[106] Tanaka, F., K. Tominaga, M. Ochi, T. Tanigawa, T. Watanabe, Y. Fujiwara, K. Ohta, N. Oshitani, K. Higuchi, and T. Arakawa. 2008. Exogenous administration of mesenchymal stem cells ameliorates dextran sulfate sodium-induced colitis via anti-inflammatory action in damaged tissue in rats. *Life Sci* 83:771-779.

[107] Kang, S. K., E. S. Jun, Y. C. Bae, and J. S. Jung. 2003. Interactions between human adipose stromal cells and mouse neural stem cells in vitro. *Brain Res Dev Brain Res* 145:141-149.

[108] Zhao, C. P., C. Zhang, S. N. Zhou, Y. M. Xie, Y. H. Wang, H. Huang, Y. C. Shang, W. Y. Li, C. Zhou, M. J. Yu, and S. W. Feng. 2007. Human mesenchymal stromal cells ameliorate the phenotype of SOD1-G93A ALS mice. *Cytotherapy* 9:414-426.

[109] Lee, R. H., M. J. Seo, R. L. Reger, J. L. Spees, A. A. Pulin, S. D. Olson, and D. J. Prockop. 2006. Multipotent stromal cells from human marrow home to and promote repair of pancreatic islets and renal glomeruli in diabetic NOD/scid mice. *Proc Natl Acad Sci U S A* 103:17438-17443.

[110] Sato, K., K. Ozaki, I. Oh, A. Meguro, K. Hatanaka, T. Nagai, K. Muroi, and K. Ozawa. 2007. Nitric oxide plays a critical role in suppression of T-cell proliferation by mesenchymal stem cells. *Blood* 109:228-234.

[111] Almeida-Porada, G., C. D. Porada, N. Tran, and E. D. Zanjani. 2000. Cotransplantation of human stromal cell progenitors into preimmune fetal sheep results in early appearance of human donor cells in circulation and boosts cell levels in bone marrow at later time points after transplantation. *Blood* 95:3620-3627.

[112] Orlic, D., J. Kajstura, S. Chimenti, I. Jakoniuk, S. M. Anderson, B. Li, J. Pickel, R. McKay, B. Nadal-Ginard, D. M. Bodine, A. Leri, and P. Anversa. 2001. Bone marrow cells regenerate infarcted myocardium. *Nature* 410:701-705.

[113] Eliopoulos, N., J. Stagg, L. Lejeune, S. Pommey, and J. Galipeau. 2005. Allogeneic marrow stromal cells are immune rejected by MHC class I- and class II-mismatched recipient mice. *Blood* 106:4057-4065.

[114] Nauta, A. J., G. Westerhuis, A. B. Kruisselbrink, E. G. Lurvink, R. Willemze, and W. E. Fibbe. 2006. Donor-derived mesenchymal stem cells are immunogenic in an allogeneic host and stimulate donor graft rejection in a nonmyeloablative setting. *Blood* 108:2114-2120.

[115] Papadaki, H. A., M. Tsagournisakis, V. Mastorodemos, C. Pontikoglou, A. Damianaki, K. Pyrovolaki, K. Stamatopoulos, A. Fassas, A. Plaitakis, and G. D. Eliopoulos. 2005. Normal bone marrow hematopoietic stem cell reserves and normal stromal cell function support the use of autologous stem cell transplantation in patients with multiple sclerosis. *Bone Marrow Transplant* 36:1053-1063.

[116] Bocelli-Tyndall, C., A. Barbero, C. Candrian, R. Ceredig, A. Tyndall, and I. Martin. 2006. Human articular chondrocytes suppress in vitro proliferation of anti-CD3 activated peripheral blood mononuclear cells. *J Cell Physiol* 209:732-734.

[117] Kastrinaki, M. C., P. Sidiropoulos, S. Roche, J. Ringe, S. Lehmann, H. Kritikos, V. M. Vlahava, B. Delorme, G. D. Eliopoulos, C. Jorgensen, P. Charbord, T. Haupl, D. T. Boumpas, and H. A. Papadaki. 2008. Functional, molecular and proteomic characterisation of bone marrow mesenchymal stem cells in rheumatoid arthritis. *Ann Rheum Dis* 67:741-749.

[118] Murphy, J. M., K. Dixon, S. Beck, D. Fabian, A. Feldman, and F. Barry. 2002. Reduced chondrogenic and adipogenic activity of mesenchymal stem cells from patients with advanced osteoarthritis. *Arthritis Rheum* 46:704-713.

[119] Fernandes, J. C., J. Martel-Pelletier, and J. P. Pelletier. 2002. The role of cytokines in osteoarthritis pathophysiology. *Biorheology* 39:237-246.

[120] Goldring, M. B. 2001. Anticytokine therapy for osteoarthritis. *Expert Opin Biol Ther* 1:817-829.

[121] Bacigalupo, A., M. Valle, M. Podesta, A. Pitto, E. Zocchi, A. De Flora, S. Pozzi, S. Luchetti, F. Frassoni, M. T. Van Lint, and G. Piaggio. 2005. T-cell suppression mediated by mesenchymal stem cells is deficient in patients with severe aplastic anemia. *Exp Hematol* 33:819-827.

[122] Arnulf, B., S. Lecourt, J. Soulier, B. Ternaux, M. N. Lacassagne, A. Crinquette, J. Dessoly, A. K. Sciaini, M. Benbunan, C. Chomienne, J. P. Fermand, J. P. Marolleau, and J. Larghero. 2007. Phenotypic and functional characterization of bone marrow

mesenchymal stem cells derived from patients with multiple myeloma. *Leukemia* 21:158-163.
[123] Horwitz, E. M., D. J. Prockop, L. A. Fitzpatrick, W. W. Koo, P. L. Gordon, M. Neel, M. Sussman, P. Orchard, J. C. Marx, R. E. Pyeritz, and M. K. Brenner. 1999. Transplantability and therapeutic effects of bone marrow-derived mesenchymal cells in children with osteogenesis imperfecta. *Nat Med* 5:309-313.
[124] Uzunel, M., J. Mattsson, M. Jaksch, M. Remberger, and O. Ringden. 2001. The significance of graft-versus-host disease and pretransplantation minimal residual disease status to outcome after allogeneic stem cell transplantation in patients with acute lymphoblastic leukemia. *Blood* 98:1982-1984.
[125] Ball, L. M., M. E. Bernardo, H. Roelofs, A. Lankester, A. Cometa, R. M. Egeler, F. Locatelli, and W. E. Fibbe. 2007. Cotransplantation of ex vivo expanded mesenchymal stem cells accelerates lymphocyte recovery and may reduce the risk of graft failure in haploidentical hematopoietic stem-cell transplantation. *Blood* 110:2764-2767.
[126] Okamoto, R., T. Yajima, M. Yamazaki, T. Kanai, M. Mukai, S. Okamoto, Y. Ikeda, T. Hibi, J. Inazawa, and M. Watanabe. 2002. Damaged epithelia regenerated by bone marrow-derived cells in the human gastrointestinal tract. *Nat Med* 8:1011-1017.
[127] Saito, T., J. Q. Kuang, B. Bittira, A. Al-Khaldi, and R. C. Chiu. 2002. Xenotransplant cardiac chimera: immune tolerance of adult stem cells. *Ann Thorac Surg* 74:19-24; discussion 24.
[128] Fiorina, P., M. Jurewicz, A. Augello, A. Vergani, S. Dada, S. La Rosa, M. Selig, J. Godwin, K. Law, C. Placidi, R. N. Smith, C. Capella, S. Rodig, C. N. Adra, M. Atkinson, M. H. Sayegh, and R. Abdi. 2009. Immunomodulatory function of bone marrow-derived mesenchymal stem cells in experimental autoimmune type 1 diabetes. *J Immunol* 183:993-1004.
[129] Singer, N. G., and A. I. Caplan. Mesenchymal stem cells: mechanisms of inflammation. *Annu Rev Pathol* 6:457-478.
[130] Gonzalez, M. A., E. Gonzalez-Rey, L. Rico, D. Buscher, and M. Delgado. 2009. Treatment of experimental arthritis by inducing immune tolerance with human adipose-derived mesenchymal stem cells. *Arthritis Rheum* 60:1006-1019.
[131] DelaRosa, O., E. Lombardo, A. Beraza, P. Mancheno-Corvo, C. Ramirez, R. Menta, L. Rico, E. Camarillo, L. Garcia, J. L. Abad, C. Trigueros, M. Delgado, and D. Buscher. 2009. Requirement of IFN-gamma-mediated indoleamine 2,3-dioxygenase expression in the modulation of lymphocyte proliferation by human adipose-derived stem cells. *Tissue Eng Part A* 15:2795-2806.
[132] Djouad, F., V. Fritz, F. Apparailly, P. Louis-Plence, C. Bony, J. Sany, C. Jorgensen, and D. Noel. 2005. Reversal of the immunosuppressive properties of mesenchymal stem cells by tumor necrosis factor alpha in collagen-induced arthritis. *Arthritis Rheum* 52:1595-1603.
[133] Schurgers, E., H. Kelchtermans, T. Mitera, L. Geboes, and P. Matthys. Discrepancy between the in vitro and in vivo effects of murine mesenchymal stem cells on T-cell proliferation and collagen-induced arthritis. *Arthritis Res Ther* 12:R31.
[134] Anderson, M. S., and J. A. Bluestone. 2005. The NOD mouse: a model of immune dysregulation. *Annu Rev Immunol* 23:447-485.
[135] Urban, V. S., J. Kiss, J. Kovacs, E. Gocza, V. Vas, E. Monostori, and F. Uher. 2008. Mesenchymal stem cells cooperate with bone marrow cells in therapy of diabetes. *Stem Cells* 26:244-253.

[136] Uccelli, A., V. Pistoia, and L. Moretta. 2007. Mesenchymal stem cells: a new strategy for immunosuppression? *Trends Immunol* 28:219-226.
[137] Li, Y., J. Chen, X. G. Chen, L. Wang, S. C. Gautam, Y. X. Xu, M. Katakowski, L. J. Zhang, M. Lu, N. Janakiraman, and M. Chopp. 2002. Human marrow stromal cell therapy for stroke in rat: neurotrophins and functional recovery. *Neurology* 59:514-523.
[138] Munoz, J. R., B. R. Stoutenger, A. P. Robinson, J. L. Spees, and D. J. Prockop. 2005. Human stem/progenitor cells from bone marrow promote neurogenesis of endogenous neural stem cells in the hippocampus of mice. *Proc Natl Acad Sci U S A* 102:18171-18176.
[139] Rivera, F. J., S. Couillard-Despres, X. Pedre, S. Ploetz, M. Caioni, C. Lois, U. Bogdahn, and L. Aigner. 2006. Mesenchymal stem cells instruct oligodendrogenic fate decision on adult neural stem cells. *Stem Cells* 24:2209-2219.
[140] Togel, F., Z. Hu, K. Weiss, J. Isaac, C. Lange, and C. Westenfelder. 2005. Administered mesenchymal stem cells protect against ischemic acute renal failure through differentiation-independent mechanisms. *Am J Physiol Renal Physiol* 289:F31-42.
[141] Ortiz, L. A., M. Dutreil, C. Fattman, A. C. Pandey, G. Torres, K. Go, and D. G. Phinney. 2007. Interleukin 1 receptor antagonist mediates the antiinflammatory and antifibrotic effect of mesenchymal stem cells during lung injury. *Proc Natl Acad Sci U S A* 104:11002-11007.
[142] Mirotsou, M., Z. Zhang, A. Deb, L. Zhang, M. Gnecchi, N. Noiseux, H. Mu, A. Pachori, and V. Dzau. 2007. Secreted frizzled related protein 2 (Sfrp2) is the key Akt-mesenchymal stem cell-released paracrine factor mediating myocardial survival and repair. *Proc Natl Acad Sci U S A* 104:1643-1648.
[143] Ramasamy, R., E. W. Lam, I. Soeiro, V. Tisato, D. Bonnet, and F. Dazzi. 2007. Mesenchymal stem cells inhibit proliferation and apoptosis of tumor cells: impact on in vivo tumor growth. *Leukemia* 21:304-310.
[144] Djouad, F., P. Plence, C. Bony, P. Tropel, F. Apparailly, J. Sany, D. Noel, and C. Jorgensen. 2003. Immunosuppressive effect of mesenchymal stem cells favors tumor growth in allogeneic animals. *Blood* 102:3837-3844.
[145] Ame-Thomas, P., H. Maby-El Hajjami, C. Monvoisin, R. Jean, D. Monnier, S. Caulet-Maugendre, T. Guillaudeux, T. Lamy, T. Fest, and K. Tarte. 2007. Human mesenchymal stem cells isolated from bone marrow and lymphoid organs support tumor B-cell growth: role of stromal cells in follicular lymphoma pathogenesis. *Blood* 109:693-702.
[146] Liotta, L. A., and E. C. Kohn. 2001. The microenvironment of the tumour-host interface. *Nature* 411:375-379.
[147] Karnoub, A. E., A. B. Dash, A. P. Vo, A. Sullivan, M. W. Brooks, G. W. Bell, A. L. Richardson, K. Polyak, R. Tubo, and R. A. Weinberg. 2007. Mesenchymal stem cells within tumour stroma promote breast cancer metastasis. *Nature* 449:557-563.

4

How do Mesenchymal Stem Cells Repair?

Patricia Semedo, Marina Burgos-Silva,
Cassiano Donizetti-Oliveira and Niels Olsen Saraiva Camara
Federal University of São Paulo (UNIFESP), University of São Paulo (USP)
Brazil

1. Introduction

Mesenchymal stem cells' field has flourished lately. It is undoubted that MSC have the ability to repair several damaged tissues. Multipotent mesenchymal stromal cells (MSC) can be isolated and efficiently expanded from almost every single body tissue and has the ability of self-renewal and differentiation into various mesodermal cell lineages. MSC can provide effective treatments for a wide range of diseases and posses several applications in regenerative medicine such as tissue repair and gene delivery. In fact, relevant data have been generated in animal models of human diseases and some clinical trials are even being started. However, little is known about how MSC can repair. It is suggested that the MSC has pleiotropic actions releasing pro-survival factors to cytoprotective microvesicles. In this chapter, we will review and discuss the main conjectures on the mechanisms of action of MSC in animal and human contexts.

2. Mesenchymal Stem Cells (MSC) biology: concept, terminology, sources, cell phenotype, gene profile and senescence

The so called "stem cell" called "stem cell" is the cell that by definition has two main properties: 1) Asymmetric division and 2) Differentiation capacity. Not all cells in our body follow these two properties.

In this sense, there are two main cells: embryonic stem cells and adult stem cells. For the purpose of this chapter, we will focus on adult stem cells.

The existence of adult stem cell is now well accepted. Tissue repair after injury and also the continuous homeostatic self renewal of some tissue consolidate this idea (1).

In 1970, Friendstein and collaborators described for the first time that bone marrow harbours not only hematopoetic stem cells (HSC), but also other type of cell that has clonogenic properties in vitro, characterizing these cells by their property to form fibroblast colonies in the primary passage in culture (CFU-F: colony-forming unit fibroblastic)(2). Later on, therapeutic approaches were approaches were achieved by using these cells in bone disorders (3, 4).

Years later, Caplan and co-workers named these cells as mesenchymal stem cells (MSC) (5, 6). In 1999, Pittenger et al demonstrated that these cells are multipotent stem cells with potential to differentiate into other cells from mesenchymal tissues (7).

Others sources of mesenchymal stem cells have been described since then. MSC are not only found at bone marrow. Adipose tissue has also been demonstrated as a potential niche (8). In addition, dental pulp (9), limbus (10), amniotic membrane (11) amniotic fluid (12), kidney

(13) and several other organs may harbor mesenchymal stem cell populations. In culture, all these cells posses a fibroblast-like phenotype and share similarities in their immunophenotype and differentiation assays.

Once it was seen that, several organs and tissue have MSC, the name needed to be standardized. In 2005, the International Society of Stem Cell Research (ISSCR) termed these cells as multipotent mesenchymal stromal cells for fibroblast-like plastic adherent cells isolated from any organ (14). And if this cell follows the minimal criteria of stem cell, so this cell can be called mesenchymal stem cell. There are three main criteria in order to determine the indentity of MSC: 1) must have adherence to plastic; 2) expression of CD105, CD73 and CD90, and lack of expression of CD45, CD34, CD14 or CD11b, CD79alpha or CD19 and HLA-DR surface molecules; 3) must have differentiation potential (15). More recently, Caplan has suggested, due to its properties; to again rename MSC to "Medicinal Signaling Cells" because of its main mechanism of action (16).

Despite this phenotype similarity, gene profile among all these cells may be different. A molecular signature of markers of stem lineage markers and genes regulating developmental and regenerative processes is more elucidative (17). Peroni et al, in a detailed work, demonstrated that MSC from bone marrow have an identical molecular profile when compared to MSC from adipose tissue. Nevertheless, a difference in some genes expression may happen with culture conditions (17). However, some authors suggest that the site of origin of MSC leads to the expression of specific genes pathways. A comparative study of MSC from bone marrow and MSC from umbilical cord from Panepucci et al showed that there are some gene that are more expressed in one cell than another. For instance, genes related to antimicrobial activity and osteogenesis are more expressed in MSC from bone marrow; and genes related to matrix remodeling and angiogenesis are more expressed in MSC from umbilical cord (18). Proteome analysis of MSC also demonstrated a full list of proteins expressed by MSC as described by Park et al (19).

In fact, MSC population is not spread all over the organ. To maintain its properties, stem cells are located at niches. Niche is an environment defined by extracellular matrix and others cells that secrete several factors to maintain stem cells under its undifferentiated state and quiescence, since adult stem cells are highly sensitive to external signals (20-22). Thus, if the niche is not well regulated, for instance in a pathologic state, stem cells may not react as expected: to repair and maintain homeostasis. Moreover, a problem in this delicate balance between stem cells and its niche may also lead to cancer (23).

The niche per se is very important; however stem cell stability is also necessary. DNA damage in melanocyte stem cells induce premature differentiation, leading to apoptosis and senescence (24). Regarding that cellular aging may be due to DNA alterations; this is probably an explanation for the inverse correlation between age and adult stem cell population and its functionality (25).

Despite all efforts, the potential niche for adult stem cell *in vivo* has not yet been defined. A perisvascular niche is one possible place (26, 27). Some authors support the idea that mesenchymal stem cells are located throughout the body as pericytes, since it is related to tissue homeostasis and blood vessels stabilization (27). In this sense, da Silva Meirelles et al showed the direct correlation between the quantity of blood vessel and stem cells (28).

3. MSC therapy and its ability to repair several diseases

As previously written, all tissues per se have the capacity of homeostasis maintenance. However, after an injury, this process is somehow disturbed by inflammation or by extracellular matrix disruption, not allowing proper stem cell action.

Thus, in a way to assist the repair process, several works have focused on the administration of exogenous stem cells. Surprisingly, the administration of adult stem cell in several experimental diseases have showed to improve in its clinical outcome and moreover amelioration of tissue architecture.

However, the mechanism of action that leads to this improvement is not well defined. Fusion of dying cells with MSC, differentiation of MSC to other cell types or paracrine action, by secretion of several bioactive factors are the main mechanism of action suggested. New mechanisms have also been suggested with the knowledge of microRNA. These topics will be better explained in the next sections.

3.1 Animals models of human diseases: kidney, heart, CNS, liver, lung and pancreas

The main experimental models treated with mesenchymal stem cells are summarized in the next table (Table I).

	Experimental Model	**Cell (amount/way/time)**	**Mechanism of Action suggested for better outcome of injury**	**Reference**
Kidney	Acute Kidney Injury	Rat MSC, $2x10^5$ cell/rat, e.v., 6h after ischemia-reperfusion injury	Immunomodulation	(29)
		Microvesicles from human MSC, 15 µg/animal, e.v., 3 days after glycerol administration	mRNA and microRNA carried by microvesicles	(30)
		MSC, $1x10^5$ cell/mouse, intra-arterially, 30 min after inschemia-reperfusion injury	Paracrine action - VEGF, HGF, IGF-1 secretion	(31)
	Chronic Kidney Injury (remnant model)	Rat MSC, $2x10^5$ cell/rat, e.v., 2nd week after 5/6 remnant model every other week until 8th week.	Immunomodulation and paracrine action	(32)
		Rat MSC, $1x10^5$ cell/rat, e.v., 1 week after 5/6 nephrotomy	MSC engraftment and paracrine action	(33)
	Chronic Kidney Injury (unilateral ureteral obstruction)	Human MSC, $1x10^6$ cell/rat, intra-arterially, after ureter obstruction	Immunomodulation	(34)
	Glomerulonephritis	Rat MSC, 2x106 cell/rat, intra-arterially, 2 days after Thy1.1 administration	Early results: paracrine effect Long-term results: maldifferentiation of MSC to adipocytes	(35)

	Experimental Model	Cell (amount/way/time)	Mechanism of Action suggested for better outcome of injury	Reference
Heart	Myocarcial infarct model	Swine MSC, 6x10^7 cell/swine, in situ at infracted area, two weeks after artery occlusion	Engraftment at myocardium	(36)
CNS	Spinal cord injury	Human MSC, 1.5x105 cell/rat, injury in situ, at time of surgery	Immunomodulation and engraftment	(37)
	Parkinson model	Rat MSC, 1.106 cell/rat, intranasally, 3 days post surgery.	Immunomodulation, anti-apoptotic effect and engraftment	(38)
Liver	Liver Fibrosis (CCl4)	Rat MSC, 3x106 cell/rat, e.v., 42 days after CCl4 administration	Paracrine and anti-apoptotic effect	(39)
Lung	Acute Lung Injury (LPS)	Mouse MSC, 1x10^5 cell/mouse, e.v., 2h after the first LPS inhalation	Immunomodulation	(40)
	Chronic Lung Injury (Bleomycin)	Mouse MSC, 2.5x10^5 cell/mouse, e.v., 7 days after bleomycin administration.	Engraftment, immunomodulation	(41)
Pancreas	Diabetes Autoimmune (NOD animal)	Mouse MSC, 1x10^5 cell/mouse, i.p. or e.v., NOD mice at 4 weeks old.	Immunomodulation	(42)

Table I. Some pre-clinical data on MSC therapy

3.2 Human trials with MSC

Pre-clinical studies have stimulated human clinical trials. Searching for "adult stem cell" at Clinicalstrials.gov, nearly 3300 studies were found. Within these studies, 1530 are recruiting patients. Some of these clinical trials are related in Table II.

These numbers are alarming. The real mechanism of action of MSC, as well as the long-term studies are not yet clarified. Despite it, others issues still hold doubts such as the amount of cells, site of injection, which cell is the best for each illness, which is the level of cell survival and engraftment, etc. These questions should be answered before clinical trials are started.

However, completed clinical trials have demonstrated efficient amelioration on clinical outcomes. MSC therapy has been focused on graft versus host diseases (GvHD) (43), limb ischemia (44), acute kidney injury (45), lung injury (46), myocardial infarction (47), etc. Safety and efficacy will be discussed next in this article. How MSC leads to this better clinical outcome, despite which disease is it, is described below.

Study	Condition	Status
Evaluation of Circulating Levels of Adult Stem Cells in the Peripheral Blood of Patients With Acute Decompensated Heart Failure and Following Stabilization, in Comparison With Healthy Volunteers NCT01027403	Heart Failure	Recruiting
Adult Stem Cell Therapy in Liver Insufficiency NCT00147043	Liver Cirrhosis	Completed
Safety Study of Adult Stem Cells to Treat Patients With Severe Leg Artery Disease NCT00913900	Critical Limb Ischemia	Recruiting
Long-term Safety and Efficacy of Adipose-derived Stem Cells to Treat Complex Perianal Fistulas in Patients Participating in the FATT-1 Randomized Controlled Trial NCT01020825	Complex Perianal Fistula	Recruiting
Evaluation of PROCHYMAL[tm] Adult Human Stem Cells for Treatment-resistant Moderate-to-severe Crohn's Disease NCT00482092	Crohn's Disease	Recruiting
Autologous Mesenchymal Stem Cells From Adipose Tissue in Patients With Secondary Progressive Multiple Sclerosis NCT01056471	Multiple Sclerosis	Recruiting
Mesenchymal Stem Cell Infusion as Prevention for Graft Rejection and Graft-Versus-Host Disease NCT00504803	Hematological Malignancies	Recruiting
Intravenous Stem Cells After Ischemic Stroke NCT00875654	Stroke	Recruiting
Induction Therapy With Autologous Mesenchymal Stem Cells for Kidney Allografts NCT00658073	Renal Transplant Rejection	Completed

Table II. Some clinical trials ongoing

4. MSC mechanisms of action

4.1 Differentiation

Before talking about differentiation it is important to consider the two categories into which stem cells can be classified (according to their developmental status): embryonic and adult (postnatal). Each represents a diverse differentiation potential status and a different potential application. Embryonic stem are pluripotent cells isolated from the inner cells mass of blastocysts, and are capable of giving rise to cells found in all three germ layers of the embryo. They are considered to have the greatest range of differentiation potential. On the other hand, adult stem cells have the capacity renew tissue after trauma, disease, or aging. These cells are found in different scales of quantity and potentiality (uni-, di-, tri-, and multi-potent), and their differentiation status is related to cell plasticity, where cells of one type give rise to cells of another type, however closely related (cell flexibility) (48). Mesechymal stem cells are adult stem cells with reduced differentiation capacity and higher plasticity capacity.

The differentiation and plasticity characteristics mentioned above exemplify the flexibility of MSC. It is well known that mesenchymal cells can shift from one differentiation pathway to another under modified external conditions, and can shift from quiescence to a proliferative state or that MSC differentiation can be reversed at least up to a certain stage (49, 50). MSCs have also been reported to differentiate into various epithelial cell types after systemic administration *in vivo*. Studies have shown that after bleomycin exposure, MSC lung engraftment was enhanced showing a small percentage of MSC localized to areas of lung injury. In addition, MSC also differentiated into type I pneumocytes (51) or assumed phenotypic characteristics of all major cell types in lung including fibroblasts, type I and type II epithelial cells, and myofibroblasts (52). MSCs have also been shown to differentiate into retinal pigment epithelial cells (53), skin epithelial cells (54) and tubular epithelial cells in kidney (55).

Different techniques have been used for MSC differentiation: use of biological and pharmacological reagents (56-59); mechanical cues (60-62) and external mechanical and electrical forces (63-66). Both mechanical and electrical stimulation have been applied separately and combined with soluble factors to facilitate MSC differentiation (67).

The criteria for differentiation need to be rigorously defined. It appears difficult to ascertain about differentiation process. Some markers are specific for certain cells; however they do not have functional relation. In addition these markers are not unique to characterize a specific cell. Delorme et al have found that MSC express cytoskeletal proteins usually expressed in neural stem cells (nestin), hepatocytes (cytokeratin-8 and -18), biliary cells (cytokeratin-19), and sarcomeric muscle (troponins, a-C-actin), without the expression of proneural or neuronal, prohepatocytic, or myogenic key transcription factors (68). Although cytoskeletal markers remain adequate indicators of a differentiation pathways, there are numerous exceptions to the rule, such as the expression of cytokeratin-18 in vascular smooth muscle cells in the synthetic phase and other similar misleading expression factors as reported by Montzka (69).

Some of the observed differentiations may also result from reprogramming. Dezawa and colleagues have shown that rodent and human bone marrow MSC can be reprogrammed into cells with skeletal muscle potential after specific treatment comprising first cytokines and then gene transfer of the notch intracellular domain (70).

Differentiation induction can be also marked by the decline of one factor at the expense of the other, like *runt-related transcription factor-2* (RUNX2) downregulation and *peroxisome*

proliferator-activated receptor-g (PPARγ) upregulation in the adipogenic condition, which remains to be tested at the single cell level (71).
It has been shown that MSC of various species may undergo myogenic differentiation (72, 73). In a mouse model of myocardial infarction it has been seen that MSCs are capable of engrafting in the site of injury and differentiating into cardiomyocyte-like cells expressing typical cardiomyocyte markers (74). Another study showed that in mice undergoing cisplatin-mediated acute renal failure, systemic injection of MSCs resulted in accumulation of MSC in the kidney and differentiation into tubular epithelial cells that exhibited the characteristic brush border of the proximal tubule cells (55). Besides, there have been a number of reports suggesting that MSC differentiate into epidermal keratinocytes and endothelial cells *in vivo* (75, 76). *In vivo,* adipose-derived MSC can migrate through the hemato-encephalic barrier after adhesion to the endothelium (77), and they can differentiate into neuroglial lineages after intraventricular in utero injection in rats (78). However, in humans there is no evidence concerning the neural regenerative potential of MSC.
MSC obtained from different tissue sources show some differences regarding differentiation potential and gene expression profiles. Thus, it has become clear that the microenviroment in which MSCs are transplanted, growth factors and local cellular interactions, play a pivotal role in determining both MSC biology (survival, proliferation, and specific differentiation) and eventually a clinical measurable improvement.

4.2 Fusion

Cell fusion is a process that has an important biological role in the development, physiology and disease of multicellular organisms. For example, we have the zygote formation and organogenesis of various tissues, such as placenta, bone and skeletal muscle. There are different types of cell fusion: homotypic and heterotypic cell fusion. Homotypic fusion happens between cells of the same type; like fusion between myoblasts for the formation and growth of multinuclear myofibres and multinucleated cells during chronic inflammatory conditions. In a heterotypic cell fusion, which describes the fusion between cells of different lineages, we have stem cells, where in specific adult stem cells can be used for clinical therapy by introducing a nuclei or functional genes in aged or degenerating cells (79). Recent reports have indicated that stem cells can fuse with differentiated cells in a range of tissues, including the brain, kidney, heart, lung and liver.
It's known that inflammation promotes migration and infiltration of bone marrow-derived stem cells to sites of tissue injury. Moreover, inflammation also increases the frequency of stem cell fusion. In the brain, chronic inflammation cen cause an increase in spontaneous fusion events through the increase of cytokine levels, by activating immune cells or by damaging the blood brain barrier leading to increased permeability (80, 81). Mechanistically, it is important to note that lipid bilayer membranes do not spontaneously fuse, and that fusion between membranes involves a highly intricate choreography of lipids and proteins. Alterations in the cell membrane that are likely to occur during chronic inflammation and the concomitant production of a variety of cytokines may predispose certain cells to fusion events.
Scolding et al hypothesized that endogenous factors associated with inflammation, such as tumour necrosis factor (TNF)-alpha and interferon (IFN)-gamma, may also directly activate stem cells and Purkinje cells to promote fusion. They demonstrated that fusion

between MSCs and cerebellar neurons can occur spontaneously in vitro, associated with a markedly higher incidence of the inflammatory mediators such as TNF-alpha and IFN-gamma. The same group, has shown that human bone marrow-derived MSC also exhibit this potentially reparative action in vivo, fusing with Purkinje cells in the rodent cerebellum. These function events are also increased in the neuroinflammatory environment of experimental autoimmune encephalomyelitis (EAE), with no apparent loss in Purkinje cell numbers (82).
To confirm this, previous reports have demonstrated fusion between rodent haematopoietic stem cells and rodent Purkinje cells, with increased levels occurring in the context of central nervous system (CNS) or systemic inflammation (80). In another CNS study, Bae et al demonstrated that bone marrow-derived MSC/Purkinje neuron fusion-like events develop into electrically active neurons with functional synaptic formation in the cerebellum of mice with neurodegeneration. Thus, MSCs may be able to integrate into the CNS and contribute to the essential properties of mature neurons (83).
Alvarez-Buylla et al used a simple method based on Cre/lox recombination to detect cell fusion events, demonstrated that bone-marrow-derived cells (BMDCs) fuse spontaneously with neural progenitors in vitro. Furthermore, bone marrow transplantation demonstrates that BMDCs fuse in vivo with hepatocytes in liver, Purkinje neurons in the brain and cardiac muscle in the heart, resulting in the formation of multinucleated cells, suggesting that genetic material derived from BMDCs contribute through cell fusion to the survival and function of these cells (84).
Adnan and colleagues have show that BMDCs can fuse with a long-lived progenitor or intestinal stem cell population of gamma-irradiated damaged intestinal epithelium and can also fuse with tumor epithelium. Fusion of BMDCs with progenitors or stem cells may play an important role in the regeneration of damaged tissue, and the observation that BMDCs can fuse with tumor epithelium is an important finding also, since the study of this event will enhance the understanding of the biology of tumorigenesis and may provide a novel strategy for the development of anticancer therapies (85).
Other clear beneficial effects of heterotypic cell fusion of BMDCs with other cells types have been demonstrated in a mouse model of a lethal liver disease in which the enzyme fumarylacetoacetate hydrolase is absent, and a large numbers of wild-type BMDCs fused with mutant liver cells corrected the metabolic deficiency and ameliorated the disease phenotype (86).
These studies demonstrate that stem cells are able to fuse with cells of different tissues. However, additional studies in animal models will be required to determine whether this fusion can be used in reparative cell therapy.

4.3 Paracrine factors

Although there are many ways through which stem cells may ameliorate injury, the main mechanism is considered is through paracrine and endocrine functions. Today, a wide range of cytokines and factors are known to be involved in the beneficial interaction between MSC and other cells.

4.3.1 Immunomodulation

Among all theories regarding the paracrine action for MSCs, by far the most explored and discussed by research is immunomodulation. Considering the inflammatory nature of most injuries, studies have indicated that the predominant role of MSCs in resolving tissue damage relies on toning down inflammation in specific sites of injury. While there are many factors suggested by literature, the main and most frequently pointed out factors and regulated immune cells are discussed ahead.

Monocytes/Macrophages

Since macrophages are one of the main cells responsible for cytokine production in injury sites, these cells are main focus points when immunomodulation is considered. Recent studies have indicated an important role in mesenchymal stem cell monocyte and macrophage immunomodulation. Maggini et al. amongst others have demonstrated that MSCs inhibit macrophage pro-inflammatory cytokine production, i.e. TNF-a, IL-6 and IFN-g (87-89). MSC also stimulate anti-inflammatory cytokine, IL-10 and IL-12p40 production (87, 88). This way, immune cell activation and the local inflammatory process is contained, reducing tissue damage. MSC also enhance apoptotic cell phagocytosis which is important in clearing injury sites. In addition, these cells inhibit monocyte derived dendritic cell differentiation (90). These effects seem to be due mainly to MSC-derived PGE2 secretion, which has direct effects on macrophage activation.

Dendritic Cells

Dendritic cells (DCs) have essential roles in antigen presentation and lymphocyte profile modulation. MSC are capable of inhibiting DC maturation through cell to cell contact and PDGE2 and IL-6 production, as seen through downregulated expression of CD1a, IL-12p70, MHC class II molecules, CD80 and CD86 (91, 92). As T cells depend importantly on antigen presenting cells to become activated, dendritc cell modulation is considered a possible mechanism for T lymphocyte tolerance induction. In addition, MSC treatment also impacts DC migration, as seen in reduced CCR7 expression and reduced migration in response to CCL19 (92). Also, mature type 1 DCs decrease TNF-a production and type 2 DCs increase IL-10 expression which may also induce Th2 profile lymphocytes (93).

T Lymphocytes

Research has shown that T lymphocyte modulation has an important role in the therapeutic character of MSC. For starters, in a disease setting, a considerable percentage of exogenous MSCs tend to accumulate in the spleen and lymph nodes, in particular around immune cells such as DCs, T and B lymphocytes, suggesting cell-specific interaction (94). MSCs are not considered to display important allogenic properties, being immune-privileged, due to low HLA -DR and costimulatory molecule expression and its basic immunomodulatory properties, although a full consensus does not exist towards the degree in which this happens (95, 96). *In vitro,* MSC inhibit phytohemagglutinin mitogen-induced and mixed lymphocyte reaction allo-antigen driven T cell proliferation (97-99). This is done by both cell-to-cell contact and humoral factors in specific, secreting IDO and Galectin-1. These stem cells also decrease TNF-a and IFN-g T cell production and raise IL-10 secretion, possibly mediated by IDO, PGE2 and B7-H amongst other molecules (100-102). MSCs also induce Th2 type lymphocyte and T regulatory cell differentiation through, HLA-G5 and other molecules (103-105).

B lymphocytes

Although many works have associated MSC and T lymphocyte regulation, few have indicated a direct role in B lymphocyte function, and many published studies are controversial. Yanfei et al have shown that B lymphocytes have decreased proliferation and antibody production when cultivated with MSC regulated allogenic DCs (106). In addition, other studies suggest that MSC directly inhibit B lymphocyte plasma cell differentiation through humoral factors (107, 108). On the other hand, other studies indicate that mesenchymal stem cells induce B-cell proliferation and differentiation into plasmocytes, when stimulated with a Toll-like receptor 9 agonist (109). These different results are probably due to the different stimuli used and its detailed effect on B cell outcome is still to be discovered.

NK / iNKT Cells

Little information is still known on MSC-mediated NK / invariant natural killer T Cells (iNKT) in comparison to other cell types. Studies demonstrate that MSCs inhibit NK and iNKT cell activation and IFN-g production by secretion of mediators as HLA-G5 and PGE2 respectively (110, 111). It is also important to mention that NK cells induce MSC lysis, probably due to low HLA class I molecule expression (112, 113). Also, important consideration needs to be given to MSC effects on the immune system in regards to tumor surveillance, due to NK cell inhibition (114).

4.3.2 Angiogenic factors

Angiogenic support provided by MSC can be considered one more supportive effect, since the re-establishment of blood supply is fundamental for recovery of damaged tissues. The pro-angiogenic effect of MSC have been demonstrated in several studies *in vitro* and *in vivo* (115-118).

It is known that MSCs express and secrete *Stromal cell-derived factors 1* (SDF-1), *vascular endothelial growth factor* (VEGF), and other cytokines important for angiogenesis (*Basic Fibroblast Growth Factor* (*bFGF*); *Matrix metalloproteinases* (MMPs). VEGF has been identified as a key component in the development of blood vessels, but VEGF alone may be insufficient to achieve functional and mature development of the vasculature. VEGF-induced vessels are often leaky and do not connect appropriately to the existing vasculature (119, 120). SDF-1 activity is essential for endothelial cell survival, vascular branching and pericyte recruitment (121). Interestingly, SDF-1a is not only a mobilization signal capable of recruiting CXCR4- positive progenitor cells into hypoxic tissues but also a retention signal for angiocompentent bone marrow-derived stem cells. It also recruits pericytes and smooth muscle cells to stabilize and mature newly formed blood vessels (122, 123). Thus, SDF-1 has been shown to augment neovascularization by the acceleration of *endothelial progenitor cell* (EPC) recruitment into ischemic foci (122, 124). In addition, VEGF is one of the powerful angiogenic cytokines that can also mobilize EPCs from bone marrow and inhibit EPC apoptosis (125). In the mouse ischemic hind limb model, VEGF-A-mediated angiogenesis partly depends on the activation of the SDF-1—CXCR4 pathway (124). Taken together, the chemokine SDF-1 likely plays a pivotal role in ASC-mediated angiogenesis (126).

Interestingly, studies have demonstrated that mesenchymal stem cells have the same angiogenesis property. Shintani et al demonstrated that implantation of autologous bone marrow mononuclear cells into ischemic skeletal muscles successfully augmented angiogenesis and collateral vessel formation in both animal studies and human trials (127-129). The same group found that implantation of ASC significantly augmented angiogenesis in a mouse model of hind limb ischemia, by release chemokines such as SDF-1 (126). Kim et al. demonstrated in a preliminary clinical trial that implantation of human cord blood-derived mesenchymal stem cells enhanced angiogenesis and collateral vessel formation in human cases with Buerger's disease (130). Kamihata and colleagues have reported that bone marrow mononuclear cells that survived engrafting can synthesize angiogenic factors such as VEGF, bFGF, and angiopoietin-1 to induce angiogenesis in the ischemic myocardium (128).

In summary, engrafting MSC in clinical therapy has a beneficial effect, which may be attributable to paracrine action. MSC secrete a number of angiogenic factors and stem cell homing factor that are able to re-establishment of blood supply, being this fundamental for the recovery of damaged tissues.

4.3.4 Anti-apoptotic factors

An important role in MSC-mediated protection is its inhibition of tissue apoptosis and augmentation of tissue turnover. Studies show that MSC are capable of inhibiting apoptosis in kidney, liver and brain injuries (30, 131-133). Recent evidence shows that this is done through increasing prosurvival factors as Akt expression in injured cells (55), BDNF and growth factors as IGF, VEGF and HGF expression which inhibit apoptosis and stimulate cell proliferation (134, 135). Together, this dynamic permits high cell turn-over, renewing damaged cells and decreasing excessive cell death, thus restoring normal tissue physiology.

4.3.5 Anti-oxidative factors

It is well known that MSC mediate tissue repair through paracrine mechanisms. Besides mediating directly in the inflammatory process, some studies have suggested that MSC also posses anti-oxidative characteristics. MSC have been observed to produce many anti-oxidative mediators such as IGF, PDGF, superoxide dismutase (SOD), HGF and IL-6 (136-139). MSC-conditioned media has also been seen to contain these anti-oxidative factors including G-CSF, GM-CSF and IL-12, as reviewed by Kim et al (140). In addition, evidence has supported that MSC and its conditioned media respectively decrease oxidative damage in culture when fibroblast cells are exposed to oxidative damage-inducing environments such as UVB (141) or tert-butyl hydroperoxide (tbOOH) (142).

In addition, research indicates that MSC posses not only considerable ability to support oxidative stress by producing anti-oxidative mediators, but may actually be stimulated by hipoxia-induced ROS (reactive oxigen species). During oxidative stress, MSC seem to have increased proliferation and migration mediated by the Akt and Erk pathways and platelet-derived growth factor receptor-β (PDGFR- β) phosphorylation (143, 144). This however is still controversial. Other studies indicate that oxidative stress correlates with MSC senescence (145, 146). These different results may be consequent to different ROS inductors, and consequently, different levels of ROS production and enviromental contributing factors envolved.

4.4 Microvesicles and miRNA

The paracrine action of MSC is now well accepted. In this sense, administration of conditioned medium of stem cell in an animal model of kidney injury has demonstrated to improve in clinical parameter outcomes correlated with decreased apoptosis and ameliorated histological parameters. Despite the bioactive molecules secreted by MSC, Bruno et al have showed that inside the conditioned medium also there are some microvesicules.

These microvesicles (MVs) are circular membranes fragments that shed from the cell surface membrane carrying protein and lipids from the membranes of the cells from which they

originate. Besides this, MVs may also carry mRNA and microRNA stablishing a communication between one cell and another (147). Shedding of microvesicles is a physiologic process, however some stress conditions lead to an increase in the number of MV shed (148).
Camussi's group has described that mesenchymal stem cells are able to shed microvesicles, and moreover, these MVs carry mRNA and microRNA leading to amelioration of acute and chronic experimental models of renal injury, accelerating regeneration of hepatectomized rats and activating endothelial cells (30, 149-151).

5. Long term aspects of MSC therapy: cancer, immunosuppression or maldifferentiation?

The majority of pre-clinical and clinical studies, which are going on, are incipient. They analyzed the results of treatment in few months or years, not allowing long-term analysis. In these early results, MSC therapy has not shown adverse effects, being designated safe by FDA (152). Thus it is difficult to tell if MSC treatment *per se* may lead to cancer. Nevertheless, several studies have showed that MSC after infusion do not engraft in any tissue, emphasizing the paracrine mechanism of action (152). However, if the treatment is continuous or the MSC administration is repeated, MSC can be found in the injured tissue (32). When MSC is engrafted or even when there is an in situ administration, MSC may maldifferentiate. In a model of glomerulonephritis, MSC maldifferentiate to adipose cells in the kidney (35). *In situ* administration of human autologous stem cell therapy in a patient with lupus nephritis has induced angiomyeloproliferative lesions at the sites of injection and hematuria (153).
All bioactive molecules secreted by MSC are prone to stimulate cancer cells to proliferate and migrate. Some works correlate MSC to suppression of tumor growth, others to supporting it. Several questions may be related to its discrepancy: animal host, timing of injection of MSC, differences in tumor models, etc (154). In addition, MSC may differentiate to tumor associated fibroblast (TAF), cells that support tumor growth (155). Thus, MSC therapy should exclude patients with cancer family history.
Also, the immunossupression level of MSC therapy cannot be controlled. Once MSC are injected endovenously, immunossupression is achieved. MSC therapy may compromise the host's defense against infections agents (156).

6. Final considerations

Many advances have been made in the field of mesenchymal stem cells. Nowadays much is known about what the MSC is and how is the mechanism of action. Currently, existing therapies administered exogenous MSC to treat several diseases. Perhaps, a deeper knowledge on endogenous stem cells may bring advances on cell therapy. To activate endogenous stem cells, via stimulation of the niche is an interesting therapeutic perspective. In addition, try to understand what happens to this endogenous stem cell in pathological state is one of the highlights to be discovered.
In the conventional therapy - administration of exogenous MSC - many issues are yet to be solved: amount of injected MSC, the best route of administration, the culture conditions which are not standardized with the use of materials of animal origin, which is the best source for these cells, etc. In conclusion, the field of MSC is very fruitful and still requires to be further studied.

7. References

[1] Gurtner GC, Callaghan MJ, Longaker MT. Progress and potential for regenerative medicine. Annu Rev Med. 2007;58:299-312.

[2] Friedenstein AJ, Chailakhjan RK, Lalykina KS. The development of fibroblast colonies in monolayer cultures of guinea-pig bone marrow and spleen cells. Cell Tissue Kinet. 1970 Oct;3(4):393-403.

[3] Friedenstein AJ, Latzinik NW, Grosheva AG, Gorskaya UF. Marrow microenvironment transfer by heterotopic transplantation of freshly isolated and cultured cells in porous sponges. Exp Hematol. 1982 Feb;10(2):217-27.

[4] Luria EA, Owen ME, Friedenstein AJ, Morris JF, Kuznetsow SA. Bone formation in organ cultures of bone marrow. Cell Tissue Res. 1987 May;248(2):449-54.

[5] Caplan AI. Mesenchymal stem cells. J Orthop Res. 1991 Sep;9(5):641-50.

[6] Caplan AI. The mesengenic process. Clin Plast Surg. 1994 Jul;21(3):429-35.

[7] Pittenger MF, Mackay AM, Beck SC, Jaiswal RK, Douglas R, Mosca JD, et al. Multilineage potential of adult human mesenchymal stem cells. Science. 1999 Apr 2;284(5411):143-7.

[8] Zuk PA, Zhu M, Ashjian P, De Ugarte DA, Huang JI, Mizuno H, et al. Human adipose tissue is a source of multipotent stem cells. Mol Biol Cell. 2002 Dec;13(12):4279-95.

[9] Gabbay JS, Heller JB, Mitchell SA, Zuk PA, Spoon DB, Wasson KL, et al. Osteogenic potentiation of human adipose-derived stem cells in a 3-dimensional matrix. Ann Plast Surg. 2006 Jul;57(1):89-93.

[10] Polisetty N, Fatima A, Madhira SL, Sangwan VS, Vemuganti GK. Mesenchymal cells from limbal stroma of human eye. Mol Vis. 2008;14:431-42.

[11] Diaz-Prado S, Muinos-Lopez E, Hermida-Gomez T, Rendal-Vazquez ME, Fuentes-Boquete I, de Toro FJ, et al. Multilineage differentiation potential of cells isolated from the human amniotic membrane. J Cell Biochem. Nov 1;111(4):846-57.

[12] Perin L, Sedrakyan S, Da Sacco S, De Filippo R. Characterization of human amniotic fluid stem cells and their pluripotential capability. Methods Cell Biol. 2008;86:85-99.

[13] Bussolati B, Bruno S, Grange C, Buttiglieri S, Deregibus MC, Cantino D, et al. Isolation of renal progenitor cells from adult human kidney. Am J Pathol. 2005 Feb;166(2):545-55.

[14] Horwitz EM, Le Blanc K, Dominici M, Mueller I, Slaper-Cortenbach I, Marini FC, et al. Clarification of the nomenclature for MSC: The International Society for Cellular Therapy position statement. Cytotherapy. 2005;7(5):393-5.

[15] Dominici M, Le Blanc K, Mueller I, Slaper-Cortenbach I, Marini F, Krause D, et al. Minimal criteria for defining multipotent mesenchymal stromal cells. The International Society for Cellular Therapy position statement. Cytotherapy. 2006;8(4):315-7.

[16] Caplan AI. What's in a name? Tissue Eng Part A. Aug;16(8):2415-7.

[17] Peroni D, Scambi I, Pasini A, Lisi V, Bifari F, Krampera M, et al. Stem molecular signature of adipose-derived stromal cells. Exp Cell Res. 2008 Feb 1;314(3):603-15.

[18] Panepucci RA, Siufi JL, Silva WA, Jr., Proto-Siquiera R, Neder L, Orellana M, et al. Comparison of gene expression of umbilical cord vein and bone marrow-derived mesenchymal stem cells. Stem Cells. 2004;22(7):1263-78.

[19] Park HW, Shin JS, Kim CW. Proteome of mesenchymal stem cells. Proteomics. 2007 Aug;7(16):2881-94.

[20] Ferraro F, Celso CL, Scadden D. Adult stem cels and their niches. Adv Exp Med Biol.695:155-68.

[21] Voog J, Jones DL. Stem cells and the niche: a dynamic duo. Cell Stem Cell. Feb 5;6(2):103-15.
[22] Walker MR, Patel KK, Stappenbeck TS. The stem cell niche. J Pathol. 2009 Jan;217(2):169-80.
[23] Li L, Neaves WB. Normal stem cells and cancer stem cells: the niche matters. Cancer Res. 2006 May 1;66(9):4553-7.
[24] Inomata K, Aoto T, Binh NT, Okamoto N, Tanimura S, Wakayama T, et al. Genotoxic stress abrogates renewal of melanocyte stem cells by triggering their differentiation. Cell. 2009 Jun 12;137(6):1088-99.
[25] Mimeault M, Batra SK. Aging of tissue-resident adult stem/progenitor cells and their pathological consequences. Panminerva Med. 2009 Jun;51(2):57-79.
[26] Feng J, Mantesso A, Sharpe PT. Perivascular cells as mesenchymal stem cells. Expert Opin Biol Ther. Oct;10(10):1441-51.
[27] da Silva Meirelles L, Caplan AI, Nardi NB. In search of the in vivo identity of mesenchymal stem cells. Stem Cells. 2008 Sep;26(9):2287-99.
[28] da Silva Meirelles L, Sand TT, Harman RJ, Lennon DP, Caplan AI. MSC frequency correlates with blood vessel density in equine adipose tissue. Tissue Eng Part A. 2009 Feb;15(2):221-9.
[29] Semedo P, Wang PM, Andreucci TH, Cenedeze MA, Teixeira VP, Reis MA, et al. Mesenchymal stem cells ameliorate tissue damages triggered by renal ischemia and reperfusion injury. Transplant Proc. 2007 Mar;39(2):421-3.
[30] Bruno S, Grange C, Deregibus MC, Calogero RA, Saviozzi S, Collino F, et al. Mesenchymal stem cell-derived microvesicles protect against acute tubular injury. J Am Soc Nephrol. 2009 May;20(5):1053-67.
[31] Togel F, Weiss K, Yang Y, Hu Z, Zhang P, Westenfelder C. Vasculotropic, paracrine actions of infused mesenchymal stem cells are important to the recovery from acute kidney injury. Am J Physiol Renal Physiol. 2007 May;292(5):F1626-35.
[32] Semedo P, Correa-Costa M, Antonio Cenedeze M, Maria Avancini Costa Malheiros D, Antonia dos Reis M, Shimizu MH, et al. Mesenchymal stem cells attenuate renal fibrosis through immune modulation and remodeling properties in a rat remnant kidney model. Stem Cells. 2009 Dec;27(12):3063-73.
[33] Choi S, Park M, Kim J, Hwang S, Park S, Lee Y. The role of mesenchymal stem cells in the functional improvement of chronic renal failure. Stem Cells Dev. 2009 Apr;18(3):521-9.
[34] Asanuma H, Vanderbrink BA, Campbell MT, Hile KL, Zhang H, Meldrum DR, et al. Arterially Delivered Mesenchymal Stem Cells Prevent Obstruction-Induced Renal Fibrosis. J Surg Res. Jul 8.
[35] Kunter U, Rong S, Boor P, Eitner F, Muller-Newen G, Djuric Z, et al. Mesenchymal stem cells prevent progressive experimental renal failure but maldifferentiate into glomerular adipocytes. J Am Soc Nephrol. 2007 Jun;18(6):1754-64.
[36] Shake JG, Gruber PJ, Baumgartner WA, Senechal G, Meyers J, Redmond JM, et al. Mesenchymal stem cell implantation in a swine myocardial infarct model: engraftment and functional effects. Ann Thorac Surg. 2002 Jun;73(6):1919-25; discussion 26.
[37] Samdani AF, Paul C, Betz RR, Fischer I, Neuhuber B. Transplantation of human marrow stromal cells and mono-nuclear bone marrow cells into the injured spinal cord: a comparative study. Spine (Phila Pa 1976). 2009 Nov 15;34(24):2605-12.

[38] Danielyan L, Schafer R, von Ameln-Mayerhofer A, Bernhard F, Verleysdonk S, Buadze M, et al. Therapeutic efficacy of intranasally delivered mesenchymal stem cells in a rat model of Parkinson disease. Rejuvenation Res. Feb;14(1):3-16.
[39] Abdel Aziz MT, Atta HM, Mahfouz S, Fouad HH, Roshdy NK, Ahmed HH, et al. Therapeutic potential of bone marrow-derived mesenchymal stem cells on experimental liver fibrosis. Clin Biochem. 2007 Aug;40(12):893-9.
[40] Xu J, Qu J, Cao L, Sai Y, Chen C, He L, et al. Mesenchymal stem cell-based angiopoietin-1 gene therapy for acute lung injury induced by lipopolysaccharide in mice. J Pathol. 2008 Mar;214(4):472-81.
[41] Ortiz LA, Gambelli F, McBride C, Gaupp D, Baddoo M, Kaminski N, et al. Mesenchymal stem cell engraftment in lung is enhanced in response to bleomycin exposure and ameliorates its fibrotic effects. Proc Natl Acad Sci U S A. 2003 Jul 8;100(14):8407-11.
[42] Madec AM, Mallone R, Afonso G, Abou Mrad E, Mesnier A, Eljaafari A, et al. Mesenchymal stem cells protect NOD mice from diabetes by inducing regulatory T cells. Diabetologia. 2009 Jul;52(7):1391-9.
[43] Le Blanc K, Frassoni F, Ball L, Locatelli F, Roelofs H, Lewis I, et al. Mesenchymal stem cells for treatment of steroid-resistant, severe, acute graft-versus-host disease: a phase II study. Lancet. 2008 May 10;371(9624):1579-86.
[44] Tateishi-Yuyama E, Matsubara H, Murohara T, Ikeda U, Shintani S, Masaki H, et al. Therapeutic angiogenesis for patients with limb ischaemia by autologous transplantation of bone-marrow cells: a pilot study and a randomised controlled trial. Lancet. 2002 Aug 10;360(9331):427-35.
[45] Togel FE, Westenfelder C. Mesenchymal stem cells: a new therapeutic tool for AKI. Nat Rev Nephrol. Mar;6(3):179-83.
[46] Matthay MA. Treatment of acute lung injury: clinical and experimental studies. Proc Am Thorac Soc. 2008 Apr 15;5(3):297-9.
[47] Yang Z, Zhang F, Ma W, Chen B, Zhou F, Xu Z, et al. A novel approach to transplanting bone marrow stem cells to repair human myocardial infarction: delivery via a noninfarct-relative artery. Cardiovasc Ther. Dec;28(6):380-5.
[48] Banas A, Yamamoto Y, Teratani T, Ochiya T. Stem cell plasticity: learning from hepatogenic differentiation strategies. Dev Dyn. 2007 Dec;236(12):3228-41.
[49] Loeffler M, Bratke T, Paulus U, Li YQ, Potten CS. Clonality and life cycles of intestinal crypts explained by a state dependent stochastic model of epithelial stem cell organization. J Theor Biol. 1997 May 7;186(1):41-54.
[50] Loeffler M, Roeder I. Tissue stem cells: definition, plasticity, heterogeneity, self-organization and models--a conceptual approach. Cells Tissues Organs. 2002;171(1):8-26.
[51] Kotton DN, Ma BY, Cardoso WV, Sanderson EA, Summer RS, Williams MC, et al. Bone marrow-derived cells as progenitors of lung alveolar epithelium. Development. 2001 Dec;128(24):5181-8.
[52] Rojas M, Xu J, Woods CR, Mora AL, Spears W, Roman J, et al. Bone marrow-derived mesenchymal stem cells in repair of the injured lung. Am J Respir Cell Mol Biol. 2005 Aug;33(2):145-52.
[53] Arnhold S, Absenger Y, Klein H, Addicks K, Schraermeyer U. Transplantation of bone marrow-derived mesenchymal stem cells rescue photoreceptor cells in the dystrophic retina of the rhodopsin knockout mouse. Graefes Arch Clin Exp Ophthalmol. 2007 Mar;245(3):414-22.

[54] Nakagawa H, Akita S, Fukui M, Fujii T, Akino K. Human mesenchymal stem cells successfully improve skin-substitute wound healing. Br J Dermatol. 2005 Jul;153(1):29-36.

[55] Morigi M, Imberti B, Zoja C, Corna D, Tomasoni S, Abbate M, et al. Mesenchymal stem cells are renotropic, helping to repair the kidney and improve function in acute renal failure. J Am Soc Nephrol. 2004 Jul;15(7):1794-804.

[56] Scintu F, Reali C, Pillai R, Badiali M, Sanna MA, Argiolu F, et al. Differentiation of human bone marrow stem cells into cells with a neural phenotype: diverse effects of two specific treatments. BMC Neurosci. 2006;7:14.

[57] Alhadlaq A, Elisseeff JH, Hong L, Williams CG, Caplan AI, Sharma B, et al. Adult stem cell driven genesis of human-shaped articular condyle. Ann Biomed Eng. 2004 Jul;32(7):911-23.

[58] Jorgensen NR, Henriksen Z, Sorensen OH, Civitelli R. Dexamethasone, BMP-2, and 1,25-dihydroxyvitamin D enhance a more differentiated osteoblast phenotype: validation of an in vitro model for human bone marrow-derived primary osteoblasts. Steroids. 2004 Apr;69(4):219-26.

[59] Sun S, Liu Y, Lipsky S, Cho M. Physical manipulation of calcium oscillations facilitates osteodifferentiation of human mesenchymal stem cells. FASEB J. 2007 May;21(7):1472-80.

[60] McBeath R, Pirone DM, Nelson CM, Bhadriraju K, Chen CS. Cell shape, cytoskeletal tension, and RhoA regulate stem cell lineage commitment. Dev Cell. 2004 Apr;6(4):483-95.

[61] Engler AJ, Sen S, Sweeney HL, Discher DE. Matrix elasticity directs stem cell lineage specification. Cell. 2006 Aug 25;126(4):677-89.

[62] Geiger B, Bershadsky A, Pankov R, Yamada KM. Transmembrane crosstalk between the extracellular matrix--cytoskeleton crosstalk. Nat Rev Mol Cell Biol. 2001 Nov;2(11):793-805.

[63] Altman GH, Horan RL, Martin I, Farhadi J, Stark PR, Volloch V, et al. Cell differentiation by mechanical stress. FASEB J. 2002 Feb;16(2):270-2.

[64] Yoshikawa T, Peel SA, Gladstone JR, Davies JE. Biochemical analysis of the response in rat bone marrow cell cultures to mechanical stimulation. Biomed Mater Eng. 1997;7(6):369-77.

[65] Sumanasinghe RD, Bernacki SH, Loboa EG. Osteogenic differentiation of human mesenchymal stem cells in collagen matrices: effect of uniaxial cyclic tensile strain on bone morphogenetic protein (BMP-2) mRNA expression. Tissue Eng. 2006 Dec;12(12):3459-65.

[66] Ward DF, Jr., Salasznyk RM, Klees RF, Backiel J, Agius P, Bennett K, et al. Mechanical strain enhances extracellular matrix-induced gene focusing and promotes osteogenic differentiation of human mesenchymal stem cells through an extracellular-related kinase-dependent pathway. Stem Cells Dev. 2007 Jun;16(3):467-80.

[67] Wu CC, Chao YC, Chen CN, Chien S, Chen YC, Chien CC, et al. Synergism of biochemical and mechanical stimuli in the differentiation of human placenta-derived multipotent cells into endothelial cells. J Biomech. 2008;41(4):813-21.

[68] Delorme B, Ringe J, Pontikoglou C, Gaillard J, Langonne A, Sensebe L, et al. Specific lineage-priming of bone marrow mesenchymal stem cells provides the molecular framework for their plasticity. Stem Cells. 2009 May;27(5):1142-51.

[69] Montzka K, Lassonczyk N, Tschoke B, Neuss S, Fuhrmann T, Franzen R, et al. Neural differentiation potential of human bone marrow-derived mesenchymal stromal cells: misleading marker gene expression. BMC Neurosci. 2009;10:16.
[70] Dezawa M, Ishikawa H, Itokazu Y, Yoshihara T, Hoshino M, Takeda S, et al. Bone marrow stromal cells generate muscle cells and repair muscle degeneration. Science. 2005 Jul 8;309(5732):314-7.
[71] Huang S. Reprogramming cell fates: reconciling rarity with robustness. Bioessays. 2009 May;31(5):546-60.
[72] Kotobuki N, Hirose M, Takakura Y, Ohgushi H. Cultured autologous human cells for hard tissue regeneration: preparation and characterization of mesenchymal stem cells from bone marrow. Artif Organs. 2004 Jan;28(1):33-9.
[73] Fouillard L, Chapel A, Bories D, Bouchet S, Costa JM, Rouard H, et al. Infusion of allogeneic-related HLA mismatched mesenchymal stem cells for the treatment of incomplete engraftment following autologous haematopoietic stem cell transplantation. Leukemia. 2007 Mar;21(3):568-70.
[74] Toma C, Pittenger MF, Cahill KS, Byrne BJ, Kessler PD. Human mesenchymal stem cells differentiate to a cardiomyocyte phenotype in the adult murine heart. Circulation. 2002 Jan 1;105(1):93-8.
[75] Wu Y, Chen L, Scott PG, Tredget EE. Mesenchymal stem cells enhance wound healing through differentiation and angiogenesis. Stem Cells. 2007 Oct;25(10):2648-59.
[76] Sasaki M, Abe R, Fujita Y, Ando S, Inokuma D, Shimizu H. Mesenchymal stem cells are recruited into wounded skin and contribute to wound repair by transdifferentiation into multiple skin cell type. J Immunol. 2008 Feb 15;180(4):2581-7.
[77] Anghileri E, Marconi S, Pignatelli A, Cifelli P, Galie M, Sbarbati A, et al. Neuronal differentiation potential of human adipose-derived mesenchymal stem cells. Stem Cells Dev. 2008 Oct;17(5):909-16.
[78] Munoz-Elias G, Woodbury D, Black IB. Marrow stromal cells, mitosis, and neuronal differentiation: stem cell and precursor functions. Stem Cells. 2003;21(4):437-48.
[79] Singec I, Snyder EY. Inflammation as a matchmaker: revisiting cell fusion. Nat Cell Biol. 2008 May;10(5):503-5.
[80] Johansson CB, Youssef S, Koleckar K, Holbrook C, Doyonnas R, Corbel SY, et al. Extensive fusion of haematopoietic cells with Purkinje neurons in response to chronic inflammation. Nat Cell Biol. 2008 May;10(5):575-83.
[81] Nygren JM, Liuba K, Breitbach M, Stott S, Thoren L, Roell W, et al. Myeloid and lymphoid contribution to non-haematopoietic lineages through irradiation-induced heterotypic cell fusion. Nat Cell Biol. 2008 May;10(5):584-92.
[82] Kemp K, Gordon D, Wraith DC, Mallam E, Hartfield E, Uney J, et al. Fusion between human mesenchymal stem cells and rodent cerebellar Purkinje cells. Neuropathol Appl Neurobiol. Feb;37(2):166-78.
[83] Bae JS, Han HS, Youn DH, Carter JE, Modo M, Schuchman EH, et al. Bone marrow-derived mesenchymal stem cells promote neuronal networks with functional synaptic transmission after transplantation into mice with neurodegeneration. Stem Cells. 2007 May;25(5):1307-16.
[84] Alvarez-Dolado M, Pardal R, Garcia-Verdugo JM, Fike JR, Lee HO, Pfeffer K, et al. Fusion of bone-marrow-derived cells with Purkinje neurons, cardiomyocytes and hepatocytes. Nature. 2003 Oct 30;425(6961):968-73.

[85] Rizvi AZ, Swain JR, Davies PS, Bailey AS, Decker AD, Willenbring H, et al. Bone marrow-derived cells fuse with normal and transformed intestinal stem cells. Proc Natl Acad Sci U S A. 2006 Apr 18;103(16):6321-5.

[86] Wang X, Willenbring H, Akkari Y, Torimaru Y, Foster M, Al-Dhalimy M, et al. Cell fusion is the principal source of bone-marrow-derived hepatocytes. Nature. 2003 Apr 24;422(6934):897-901.

[87] Maggini J, Mirkin G, Bognanni I, Holmberg J, Piazzon IM, Nepomnaschy I, et al. Mouse bone marrow-derived mesenchymal stromal cells turn activated macrophages into a regulatory-like profile. PLoS One.5(2):e9252.

[88] Nemeth K, Leelahavanichkul A, Yuen PS, Mayer B, Parmelee A, Doi K, et al. Bone marrow stromal cells attenuate sepsis via prostaglandin E(2)-dependent reprogramming of host macrophages to increase their interleukin-10 production. Nat Med. 2009 Jan;15(1):42-9.

[89] Zhang QZ, Su WR, Shi SH, Wilder-Smith P, Xiang AP, Wong A, et al. Human gingiva-derived mesenchymal stem cells elicit polarization of m2 macrophages and enhance cutaneous wound healing. Stem Cells. Oct;28(10):1856-68.

[90] Kronsteiner B, Peterbauer-Scherb A, Grillari-Voglauer R, Redl H, Gabriel C, van Griensven M, et al. Human mesenchymal stem cells and renal tubular epithelial cells differentially influence monocyte-derived dendritic cell differentiation and maturation. Cell Immunol.267(1):30-8.

[91] Djouad F, Charbonnier LM, Bouffi C, Louis-Plence P, Bony C, Apparailly F, et al. Mesenchymal stem cells inhibit the differentiation of dendritic cells through an interleukin-6-dependent mechanism. Stem Cells. 2007 Aug;25(8):2025-32.

[92] English K, Barry FP, Mahon BP. Murine mesenchymal stem cells suppress dendritic cell migration, maturation and antigen presentation. Immunol Lett. 2008 Jan 15;115(1):50-8.

[93] Aggarwal S, Pittenger MF. Human mesenchymal stem cells modulate allogeneic immune cell responses. Blood. 2005 Feb 15;105(4):1815-22.

[94] Lim JH, Kim JS, Yoon IH, Shin JS, Nam HY, Yang SH, et al. Immunomodulation of delayed-type hypersensitivity responses by mesenchymal stem cells is associated with bystander T cell apoptosis in the draining lymph node. J Immunol. Oct 1;185(7):4022-9.

[95] Fang L, Lange C, Engel M, Zander AR, Fehse B. Sensitive balance of suppressing and activating effects of mesenchymal stem cells on T-cell proliferation. Transplantation. 2006 Nov 27;82(10):1370-3.

[96] Deuse T, Stubbendorff M, Tang-Quan K, Phillips N, Kay MA, Eiermann T, et al. Immunogenicity and immunomodulatory properties of umbilical cord lining mesenchymal stem cells. Cell Transplant. Nov 5.

[97] Prasanna SJ, Gopalakrishnan D, Shankar SR, Vasandan AB. Pro-inflammatory cytokines, IFNgamma and TNFalpha, influence immune properties of human bone marrow and Wharton jelly mesenchymal stem cells differentially. PLoS One.5(2):e9016.

[98] Gieseke F, Bohringer J, Bussolari R, Dominici M, Handgretinger R, Muller I. Human multipotent mesenchymal stromal cells use galectin-1 to inhibit immune effector cells. Blood. Nov 11;116(19):3770-9.

[99] Meisel R, Zibert A, Laryea M, Gobel U, Daubener W, Dilloo D. Human bone marrow stromal cells inhibit allogeneic T-cell responses by indoleamine 2,3-dioxygenase-mediated tryptophan degradation. Blood. 2004 Jun 15;103(12):4619-21.

[100] Yoo KH, Jang IK, Lee MW, Kim HE, Yang MS, Eom Y, et al. Comparison of immunomodulatory properties of mesenchymal stem cells derived from adult human tissues. Cell Immunol. 2009;259(2):150-6.
[101] Tipnis S, Viswanathan C, Majumdar AS. Immunosuppressive properties of human umbilical cord-derived mesenchymal stem cells: role of B7-H1 and IDO. Immunol Cell Biol. Nov-Dec;88(8):795-806.
[102] Najar M, Raicevic G, Boufker HI, Fayyad Kazan H, De Bruyn C, Meuleman N, et al. Mesenchymal stromal cells use PGE2 to modulate activation and proliferation of lymphocyte subsets: Combined comparison of adipose tissue, Wharton's Jelly and bone marrow sources. Cell Immunol.264(2):171-9.
[103] Bouffi C, Bony C, Courties G, Jorgensen C, Noel D. IL-6-dependent PGE2 secretion by mesenchymal stem cells inhibits local inflammation in experimental arthritis. PLoS One.5(12):e14247.
[104] Augello A, Tasso R, Negrini SM, Cancedda R, Pennesi G. Cell therapy using allogeneic bone marrow mesenchymal stem cells prevents tissue damage in collagen-induced arthritis. Arthritis Rheum. 2007 Apr;56(4):1175-86.
[105] Selmani Z, Naji A, Zidi I, Favier B, Gaiffe E, Obert L, et al. Human leukocyte antigen-G5 secretion by human mesenchymal stem cells is required to suppress T lymphocyte and natural killer function and to induce CD4+CD25highFOXP3+ regulatory T cells. Stem Cells. 2008 Jan;26(1):212-22.
[106] Huang Y, Chen P, Zhang CB, Ko GJ, Ruiz M, Fiorina P, et al. Kidney-derived mesenchymal stromal cells modulate dendritic cell function to suppress alloimmune responses and delay allograft rejection. Transplantation. Dec 27;90(12):1307-11.
[107] Corcione A, Benvenuto F, Ferretti E, Giunti D, Cappiello V, Cazzanti F, et al. Human mesenchymal stem cells modulate B-cell functions. Blood. 2006 Jan 1;107(1):367-72.
[108] Asari S, Itakura S, Ferreri K, Liu CP, Kuroda Y, Kandeel F, et al. Mesenchymal stem cells suppress B-cell terminal differentiation. Exp Hematol. 2009 May;37(5):604-15.
[109] Traggiai E, Volpi S, Schena F, Gattorno M, Ferlito F, Moretta L, et al. Bone marrow-derived mesenchymal stem cells induce both polyclonal expansion and differentiation of B cells isolated from healthy donors and systemic lupus erythematosus patients. Stem Cells. 2008 Feb;26(2):562-9.
[110] Patel SA, Meyer JR, Greco SJ, Corcoran KE, Bryan M, Rameshwar P. Mesenchymal stem cells protect breast cancer cells through regulatory T cells: role of mesenchymal stem cell-derived TGF-beta. J Immunol. May 15;184(10):5885-94.
[111] Prigione I, Benvenuto F, Bocca P, Battistini L, Uccelli A, Pistoia V. Reciprocal interactions between human mesenchymal stem cells and gammadelta T cells or invariant natural killer T cells. Stem Cells. 2009 Mar;27(3):693-702.
[112] Jewett A, Arasteh A, Tseng HC, Behel A, Arasteh H, Yang W, et al. Strategies to rescue mesenchymal stem cells (MSCs) and dental pulp stem cells (DPSCs) from NK cell mediated cytotoxicity. PLoS One.5(3):e9874.
[113] Spaggiari GM, Capobianco A, Becchetti S, Mingari MC, Moretta L. Mesenchymal stem cell-natural killer cell interactions: evidence that activated NK cells are capable of killing MSCs, whereas MSCs can inhibit IL-2-induced NK-cell proliferation. Blood. 2006 Feb 15;107(4):1484-90.
[114] Li Y, Qu YH, Wu YF, Wang XP, Wei J, Huang WG, et al. Bone marrow mesenchymal stem cells reduce the antitumor activity of cytokine-induced killer/natural killer cells in K562 NOD/SCID mice. Ann Hematol. Jan 15.

[115] Kinnaird T, Stabile E, Burnett MS, Shou M, Lee CW, Barr S, et al. Local delivery of marrow-derived stromal cells augments collateral perfusion through paracrine mechanisms. Circulation. 2004 Mar 30;109(12):1543-9.

[116] Hung SC, Pochampally RR, Chen SC, Hsu SC, Prockop DJ. Angiogenic effects of human multipotent stromal cell conditioned medium activate the PI3K-Akt pathway in hypoxic endothelial cells to inhibit apoptosis, increase survival, and stimulate angiogenesis. Stem Cells. 2007 Sep;25(9):2363-70.

[117] Sorrell JM, Baber MA, Caplan AI. Influence of adult mesenchymal stem cells on in vitro vascular formation. Tissue Eng Part A. 2009 Jul;15(7):1751-61.

[118] Sanz L, Santos-Valle P, Alonso-Camino V, Salas C, Serrano A, Vicario JL, et al. Long-term in vivo imaging of human angiogenesis: critical role of bone marrow-derived mesenchymal stem cells for the generation of durable blood vessels. Microvasc Res. 2008 Apr;75(3):308-14.

[119] Zentilin L, Tafuro S, Zacchigna S, Arsic N, Pattarini L, Sinigaglia M, et al. Bone marrow mononuclear cells are recruited to the sites of VEGF-induced neovascularization but are not incorporated into the newly formed vessels. Blood. 2006 May 1;107(9):3546-54.

[120] Carmeliet P. VEGF gene therapy: stimulating angiogenesis or angioma-genesis? Nat Med. 2000 Oct;6(10):1102-3.

[121] Nagasawa T, Nakajima T, Tachibana K, Iizasa H, Bleul CC, Yoshie O, et al. Molecular cloning and characterization of a murine pre-B-cell growth-stimulating factor/stromal cell-derived factor 1 receptor, a murine homolog of the human immunodeficiency virus 1 entry coreceptor fusin. Proc Natl Acad Sci U S A. 1996 Dec 10;93(25):14726-9.

[122] Yamaguchi J, Kusano KF, Masuo O, Kawamoto A, Silver M, Murasawa S, et al. Stromal cell-derived factor-1 effects on ex vivo expanded endothelial progenitor cell recruitment for ischemic neovascularization. Circulation. 2003 Mar 11;107(9):1322-8.

[123] Grunewald M, Avraham I, Dor Y, Bachar-Lustig E, Itin A, Jung S, et al. VEGF-induced adult neovascularization: recruitment, retention, and role of accessory cells. Cell. 2006 Jan 13;124(1):175-89.

[124] Hiasa K, Ishibashi M, Ohtani K, Inoue S, Zhao Q, Kitamoto S, et al. Gene transfer of stromal cell-derived factor-1alpha enhances ischemic vasculogenesis and angiogenesis via vascular endothelial growth factor/endothelial nitric oxide synthase-related pathway: next-generation chemokine therapy for therapeutic neovascularization. Circulation. 2004 May 25;109(20):2454-61.

[125] Asahara T, Takahashi T, Masuda H, Kalka C, Chen D, Iwaguro H, et al. VEGF contributes to postnatal neovascularization by mobilizing bone marrow-derived endothelial progenitor cells. EMBO J. 1999 Jul 15;18(14):3964-72.

[126] Kondo K, Shintani S, Shibata R, Murakami H, Murakami R, Imaizumi M, et al. Implantation of adipose-derived regenerative cells enhances ischemia-induced angiogenesis. Arterioscler Thromb Vasc Biol. 2009 Jan;29(1):61-6.

[127] Shintani S, Murohara T, Ikeda H, Ueno T, Sasaki K, Duan J, et al. Augmentation of postnatal neovascularization with autologous bone marrow transplantation. Circulation. 2001 Feb 13;103(6):897-903.

[128] Kamihata H, Matsubara H, Nishiue T, Fujiyama S, Tsutsumi Y, Ozono R, et al. Implantation of bone marrow mononuclear cells into ischemic myocardium enhances collateral perfusion and regional function via side supply of angioblasts, angiogenic ligands, and cytokines. Circulation. 2001 Aug 28;104(9):1046-52.

[129] Kajiguchi M, Kondo T, Izawa H, Kobayashi M, Yamamoto K, Shintani S, et al. Safety and efficacy of autologous progenitor cell transplantation for therapeutic angiogenesis in patients with critical limb ischemia. Circ J. 2007 Feb;71(2):196-201.
[130] Kim SW, Han H, Chae GT, Lee SH, Bo S, Yoon JH, et al. Successful stem cell therapy using umbilical cord blood-derived multipotent stem cells for Buerger's disease and ischemic limb disease animal model. Stem Cells. 2006 Jun;24(6):1620-6.
[131] Zhang D, Jiang M, Miao D. Transplanted human amniotic membrane-derived mesenchymal stem cells ameliorate carbon tetrachloride-induced liver cirrhosis in mouse. PLoS One.6(2):e16789.
[132] Gaebel R, Furlani D, Sorg H, Polchow B, Frank J, Bieback K, et al. Cell origin of human mesenchymal stem cells determines a different healing performance in cardiac regeneration. PLoS One.6(2):e15652.
[133] Bi B, Schmitt R, Israilova M, Nishio H, Cantley LG. Stromal cells protect against acute tubular injury via an endocrine effect. J Am Soc Nephrol. 2007 Sep;18(9):2486-96.
[134] Imberti B, Morigi M, Tomasoni S, Rota C, Corna D, Longaretti L, et al. Insulin-like growth factor-1 sustains stem cell mediated renal repair. J Am Soc Nephrol. 2007 Nov;18(11):2921-8.
[135] Lu S, Lu C, Han Q, Li J, Du Z, Liao L, et al. Adipose-derived mesenchymal stem cells protect PC12 cells from glutamate excitotoxicity-induced apoptosis by upregulation of XIAP through PI3-K/Akt activation. Toxicology. Jan 11;279(1-3):189-95.
[136] Baregamian N, Song J, Jeschke MG, Evers BM, Chung DH. IGF-1 protects intestinal epithelial cells from oxidative stress-induced apoptosis. J Surg Res. 2006 Nov;136(1):31-7.
[137] Faiz M, Acarin L, Peluffo H, Villapol S, Castellano B, Gonzalez B. Antioxidant Cu/Zn SOD: expression in postnatal brain progenitor cells. Neurosci Lett. 2006 Jun 19;401(1-2):71-6.
[138] Hui L, Hong Y, Jingjing Z, Yuan H, Qi C, Nong Z. HGF suppresses high glucose-mediated oxidative stress in mesangial cells by activation of PKG and inhibition of PKA. Free Radic Biol Med. Aug 1;49(3):467-73.
[139] Kida H, Yoshida M, Hoshino S, Inoue K, Yano Y, Yanagita M, et al. Protective effect of IL-6 on alveolar epithelial cell death induced by hydrogen peroxide. Am J Physiol Lung Cell Mol Physiol. 2005 Feb;288(2):L342-9.
[140] Kim WS, Park BS, Sung JH. The wound-healing and antioxidant effects of adipose-derived stem cells. Expert Opin Biol Ther. 2009 Jul;9(7):879-87.
[141] Kim WS, Park BS, Park SH, Kim HK, Sung JH. Antiwrinkle effect of adipose-derived stem cell: activation of dermal fibroblast by secretory factors. J Dermatol Sci. 2009 Feb;53(2):96-102.
[142] Kim WS, Park BS, Kim HK, Park JS, Kim KJ, Choi JS, et al. Evidence supporting antioxidant action of adipose-derived stem cells: protection of human dermal fibroblasts from oxidative stress. J Dermatol Sci. 2008 Feb;49(2):133-42.
[143] Kim JH, Park SH, Park SG, Choi JS, Xia Y, Sung JH. The Pivotal Role of Reactive Oxygen Species Generation in the Hypoxia-Induced Stimulation of Adipose-Derived Stem Cells. Stem Cells Dev. Mar 2.
[144] Valle-Prieto A, Conget PA. Human mesenchymal stem cells efficiently manage oxidative stress. Stem Cells Dev. Dec;19(12):1885-93.
[145] Kasper G, Mao L, Geissler S, Draycheva A, Trippens J, Kuhnisch J, et al. Insights into mesenchymal stem cell aging: involvement of antioxidant defense and actin cytoskeleton. Stem Cells. 2009 Jun;27(6):1288-97.

[146] Ho JH, Chen YF, Ma WH, Tseng TC, Chen MH, Lee OK. Cell Contact Accelerates Replicative Senescence of Human Mesenchymal Stem Cells Independent of Telomere Shortening and p53 Activation: Roles of Ras and Oxidative Stress. Cell Transplant. Dec 22.

[147] Ratajczak J, Wysoczynski M, Hayek F, Janowska-Wieczorek A, Ratajczak MZ. Membrane-derived microvesicles: important and underappreciated mediators of cell-to-cell communication. Leukemia. 2006 Sep;20(9):1487-95.

[148] Hugel B, Martinez MC, Kunzelmann C, Freyssinet JM. Membrane microparticles: two sides of the coin. Physiology (Bethesda). 2005 Feb;20:22-7.

[149] Herrera MB, Fonsato V, Gatti S, Deregibus MC, Sordi A, Cantarella D, et al. Human liver stem cell-derived microvesicles accelerate hepatic regeneration in hepatectomized rats. J Cell Mol Med. Jun;14(6B):1605-18.

[150] Gatti S, Bruno S, Deregibus MC, Sordi A, Cantaluppi V, Tetta C, et al. Microvesicles derived from human adult mesenchymal stem cells protect against ischaemia-reperfusion-induced acute and chronic kidney injury. Nephrol Dial Transplant. Feb 15.

[151] Collino F, Deregibus MC, Bruno S, Sterpone L, Aghemo G, Viltono L, et al. Microvesicles derived from adult human bone marrow and tissue specific mesenchymal stem cells shuttle selected pattern of miRNAs. PLoS One.5(7):e11803.

[152] Parekkadan B, Milwid JM. Mesenchymal stem cells as therapeutics. Annu Rev Biomed Eng. Aug 15;12:87-117.

[153] Thirabanjasak D, Tantiwongse K, Thorner PS. Angiomyeloproliferative lesions following autologous stem cell therapy. J Am Soc Nephrol. Jul;21(7):1218-22.

[154] Klopp AH, Gupta A, Spaeth E, Andreeff M, Marini F, 3rd. Concise review: Dissecting a discrepancy in the literature: do mesenchymal stem cells support or suppress tumor growth? Stem Cells. Jan;29(1):11-9.

[155] Spaeth EL, Dembinski JL, Sasser AK, Watson K, Klopp A, Hall B, et al. Mesenchymal stem cell transition to tumor-associated fibroblasts contributes to fibrovascular network expansion and tumor progression. PLoS One. 2009;4(4):e4992.

[156] Sundin M, Orvell C, Rasmusson I, Sundberg B, Ringden O, Le Blanc K. Mesenchymal stem cells are susceptible to human herpesviruses, but viral DNA cannot be detected in the healthy seropositive individual. Bone Marrow Transplant. 2006 Jun;37(11):1051-9.

5

Immunogenicity and Immune-Modulating Properties of Human Stem Cells

Jan Nehlin, Adiba Isa and Torben Barington
Odense University Hospital & University of Southern Denmark
Denmark

1. Introduction

The future use of stem cell-based therapeutic applications in regenerative medicine is regarded as promising. In addition to autologous and allogeneic transplantation procedures, various innovative methods have been designed to generate patient-histocompatible stem cells from which lineage-specific cell progenies could be obtained (reviewed in Nehlin & Barington, 2009). Immunological aspects of the transplanted cells as well as the recipient need to be considered in order to predict the outcome of clinical cell therapies. Undifferentiated stem cells show initially a low degree of immunogenicity leading to weak immune responses when introduced into non-histocompatible hosts. In addition, stem cells possess immune-modulating properties that confer the capacity to withstand a cytotoxic response in a foreign host. The nature and significance of these strategies will be described in detail along this chapter.

Many valuable contributions dealing with immunogenicity and immunological tolerance have been possible by means of mouse embryonic and multipotent stem cells. However, this overview will explore in-depth the immunological features and clinical uses of two types of human stem cells, embryonic stem cells (Figure 1) and multipotent mesenchymal stem cells (Figure 2 & 3) that allow them to be considered in transplantation procedures.

2. Basic principles of antigen presentation, the adaptive immune response and immune histocompatibility

The highly polymorphic classical "Major Histocompatibility Complex" (MHC) class I protein family consists of extracellular, membrane spanning, alpha chains HLA-A, -B, and -C (Human leukocyte antigens) acting as ligands for T-lymphocyte receptors (TcR) expressed on T lymphocytes, the killer-immunoglobulin-like receptors (KIR) on Natural Killer (NK) cells and for certain members of the leukocyte immunoglobulin-like receptor (LILR / ILT / LIR) family. The non-classical MHC class I protein family include less polymorphic HLA members such as HLA-E, HLA-F, HLA-G, HLA-K and HLA-L (Li & Raghavan, 2010).

A major function of HLA molecules is presentation of intracellularly produced self and non-self peptides. During intracellular infection, pathogen-derived (non-self) peptides are presented by virtually all nucleated cells in the body to TcR of cytotoxic $CD8^+$ T lymphocytes leading to killing of infected cells. Endogenous (self) peptides are also presented, but usually T cells with specificities to self-peptides are eliminated in the thymus.

The introduction of allogeneic stem cell-derived tissues into an immunocompetent recipient is likely to result in T-cell-mediated rejection by either of two mechanisms. One is indirect allorecognition, where polymorphic donor-derived peptides are presented for recipient T cells, and the other is direct allorecognition, where polymorphic variants of donor MHC are recognized by recipient T cells. The latter is by far the strongest and because MHC is extremely polymorphic, this mechanism constitutes a major hurdle for allogeneic stem cell transplantation (Afzali et al., 2008; Gökmen et al. 2008; Hornick, 2006; Ingulli, 2010).

T-cell activation is dependent on binding of antigen-presenting MHC class I (signal 1) and non-antigen-specific co-stimulatory molecules CD40, CD80 and CD86 expressed by donor cells to their ligands on T cells (signal 2)(Jenkins, 1994). Presentation of peptides by MHC class I to TcR in the absence of co-stimulation induces anergy or apoptosis of T cells, abortive proliferation or tolerance-inducing immunoregulation (Ford & Larsen, 2009; Pearl et al., 2011).

Two types of T cells, cytotoxic and regulatory, have been implicated in immune responses in general and in relation to stem cell transplantation. Cytotoxic or cytolytic T cells (CTL or T_C) are activated T lymphocytes (usually CD8 positive) that can induce the killing of target cells, be it infected or allogeneic cells. CTL can induce apoptosis of targets cells by two independent mechanisms: release of cytotoxins such as perforin, granzymes and granulysin, or by binding of FasL (CD95L) to Fas (CD95) molecules expressed on the target cell (Brown, 2010).

Regulatory T cells (T_{reg}) comprise several specialized subsets of T cells that are able to control immune responses and promote and maintain immune tolerance in an antigen-specific way. Presence of T_{reg} has been associated with tolerance whereas their deficiency or defective function has been correlated with autoimmunity in many experimental models (Roncarolo et al., 2011).

The degree of histocompatibility at the HLA-locus can be determined by serologic or genomic tissue typing. In case of discrepancy for HLA class II antigens, the resulting incompatibility may be measured through Mixed Lymphocyte Reactions (MLR). Peripheral blood mononuclear cells (PBMC) from two different individuals (HLA mismatched allogeneic setting) are co-cultured for 4-6 days. One of the cell populations (stimulators) is irradiated to avoid its proliferation. The non-irradiated cells (responders) will start proliferating due to direct allorecognition, and this is measured through uptake of ^{3}H-(tritiated) thymidine. Suppression of T cell proliferation is often measured by adding unmodified or modified stem cells to the MLR (Aggarwal & Pittenger, 2005; Le Blanc et al., 2003a; Wonderlich et al., 2006).

MHC class II molecules and co-stimulatory molecules are primarily expressed on professional antigen-presenting cells (APC) such as B cells, monocytes and dendritic cells. MHC class II consists of three classes of HLA-II antigens: HLA-DP, -DQ and -DR presenting peptide antigens derived from extracellular proteins to CD4+ T helper cells to elicit an immune response. Because MHC class II expression is cell-type specific and mainly restricted to thymic epithelial cells and bone marrow-derived antigen presenting cells it is not expected to be a problem in many stem cell applications (Handunnetthi et al., 2010).

Natural killer (NK) cells participate in the innate immune response as they are capable of killing tumor cells and virally-infected cells. NK cells express a large number of cell surface receptors named Killer-cell Immunoglobulin-like Receptors (KIR) and NKG2 (CD94) that have classical and non-classical class I antigens as their ligands , respectively, and deliver either activating or inhibitory signals. The relative balance of these signals regulates NK cell activity (Orr & Lanier, 2010) and absence of HLA class-I expression therefore may lead to killing by NK cells. This is of relevance for stem cell therapy because some stem cells lack classical HLA class I expression all together and because more differentiated cells may fail to express alleles recognized by the NK cells of the recipient.

Organ transplantation is a routine therapeutic strategy for patients with end-stage organ failure (Newell, 2011) and bone marrow-derived hematopoietic stem cells are used to treat several hematopoietic malignancies and rare immunodeficiencies (Nehlin & Barington, 2009). Clinical experience through more than three decades with these treatments clearly demonstrates the relevance of both T cells and NK cells in transplantation. Allogeneic transplantation of organs often results in the development of either a) graft-versus-host-disease (GvHD) or b) immune rejection. GvHD is a condition where donor cells within a transplant launch an immune reaction against the recipient cells (Menendez et al., 2005; Shlomchik, 2007). The host can also build an acute and/or chronic rejection against antigens perceived as non-self by the recipient's immune system leading to destruction of the graft. Those antigens can belong to MHC, minor histocompatibility complex (mHC) or AB0 blood groups (Bradley et al., 2002; Charron et al. 2009; Feng et al., 2008; Shlomchik, 2007; Spencer et al., 2010).

3. Immunological tolerance at the fetal-maternal interface

A classical example of immune tolerance is found during pregnancy. The maternal immune system allows for the successful development of a semi-allograft consisting of a fetus and a placenta that express both maternal (self) and paternal (non-self) antigens. Cytotoxic adaptive immune responses are diminished, bypassed, or even abrogated (Leber et al., 2010). Immunological processes such as innate immunity remain intact to continue to provide host defense against infection and to interact with fetal tissues to promote successful placentation and pregnancy.

The trophoblasts are the cells that form the outer layer of the blastocyst, that develop into a large part of the placenta, and have a crucial role in the implantation of the developing embryo by interactions with the decidua, the lining of the maternal uterus containing endometrial stromal cells. The placenta and fetal trophoblast cells are directly exposed to maternal blood and tissues. The trophoblast layer protects the inner cell mass (ICM) from attack by cytotoxic components of the maternal immune system dedicated to destroying foreign tissues. The inner cell mass consists of a group of cells inside the primordial embryo that will eventually develop into defined fetal structures (Mor & Abrahams, 2009).

In humans, trophoblast cells differentiate from the trophoblast shell that surrounds the post-implantation embryo into two main lineages, villous trophoblast and extravillous trophoblast. Primary villous trophoblast cells do not express β2-microglobulin or any HLA class I or HLA class II molecules, whereas extravillous trophoblast cells express HLA-C, HLA-G and HLA-E, but not HLA-A, HLA-B or HLA-DR molecules in a normal pregnancy. Villous trophoblasts are in contact with the systemic immune system, whereas extravillous trophoblasts interact with the local mucosal immune cells. When extravillous trophoblasts were exposed to interferon-gamma (IFN-γ), a pro-inflammatory cytokine, they could not induce the expression of HLA-A, -B, -DR or up-regulate significantly HLA-G (Apps et al., 2009). A comprehensive overview of MHC expression at the fetal-maternal interface was recently reported (Tilburgs et al., 2010).

Human embryonic tissues possess a range of proteins and mechanisms that efficiently counteract and prevent maternal cytotoxic T cell attack and thereby provide protection and immune privilege to the fetus (Clark, 2005; Fändrich, 2002; Mor & Abrahams, 2009; Parhar et al. 1989; Petroff & Perchellet, 2010; Rebmann et al., 2010; Rizzo et al., 2011a; Verloes et al., 2011).

Knowledge of the immunological tolerance mechanisms taking place at the feto-maternal interface in mouse and rat models have contributed to extend such findings to human cells. Attention has focused on a group of pluripotent stem cells known as human embryonic stem cells (hESC), derived from the ICM of 6-8 days pre-implantation blastocysts obtained from *in vitro* fertilization procedures (Thomson et al., 1998), because they could potentially be used in cell therapy and regenerative medicine (Figure 1; Nehlin & Barington, 2009). However, the use of hESC may be limited by immunological incompatibility between the donor and the recipient as explained in the next sections.

4. Immunogenicity of human embryonic, multipotent and reprogrammed stem cells

4.1 General considerations

Ever since the development of the techniques that allowed researchers to establish blastocyst-derived hESC, there has been a great deal of interest in their potential use in regenerative medicine (Thomson et al., 1998). To satisfy clinical requirements, a number of matters need to be addressed including 1) the precise control of differentiation towards the tissue or cell-type of choice without remnant undifferentiated hESC; 2) safety issues pertaining the risk of transplanting undifferentiated hESC cells that could result in teratoma formation, the possible presence of genetic modifications as a result of *ex-vivo* culture, and immunogenicity concerns (see below); 3) the development of xenogeneic-free culture conditions and 4) potential ethical conflicts (Nehlin & Barington, 2009; section 6.1).

In-depth studies of the molecules and mechanisms responsible for the low grade of immunogenicity and allograft tolerance of cultured hESC in non-histocompatible recipients have been relatively few compared with bone marrow stem cell studies (section 4.3). The immune-privileged status of hESC resembles the tolerance properties exhibited by the developing embryo in the feto-maternal interface (section 3; English & Wood, 2011; Grinnemo et al., 2008a; Menendez et al., 2005). In comparison, multipotent stem cells such as bone marrow stromal cells still retain immune-privileged properties even after further differentiation (Le Blanc & Pittenger, 2005).

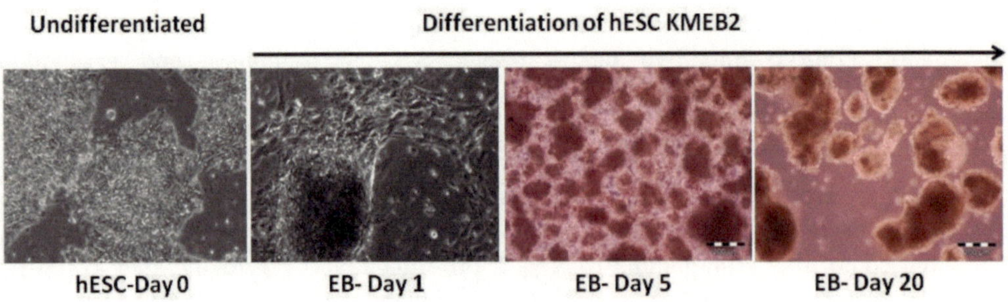

Fig. 1. Early differentiation of hESC

Phase contrast photographs of hESC KMEB2 (Harkness et al., 2010) growing on feeder-free Matrigel in an undifferentiated state (day 0), followed by differentiation towards embryoid body formation at days 1, 5 and 20 (left to right panels). Magnification: 100x

Traditionally, hESC have been isolated from the ICM of the blastocyst stage of the early developing embryo (Thomson et al., 1998). Also, in recent years, donor-specific hESC were obtained from single blastomeres at the 8-cell stage of embryogenesis or generated by somatic cell nuclear transfer (SCNT). Pluripotent stem cells with features resembling the hESC state have been generated by parthenogenetic activation of unfertilized oocytes and genetic reprogramming (Jopling et al. 2011; Nehlin & Barington, 2009; Yamanaka & Blau, 2010). The immunological aspects regarding hESC and hESC-like pluripotent stem cells are addressed below.

4.2 MHC class I expression in human embryonic stem cells, early stem cell progenitors and induced pluripotent stem cells

4.2.1 Expression studies

Immunogenicity concerns focus on the presentation of antigens and expression of MHC class I proteins, as well as expression of molecules that make hESC potential targets of cytotoxic responses by T and NK cells.

Almost all cellular studies dealing with HLA class I expression have relied on a single antibody (W6/32) targeting simultaneously HLA-A, -B, -C and cross-reacting with HLA-E and -F (Uchanska-Ziegler & Ziegler, 2007). Indirect immunofluorescence assays showed that human pre-implanted embryos expressed neither HLA class I, II antigens nor β2-microglobulin (Desoye et al., 1988). However, a later study using immunocytochemistry showed expression of HLA-G and MHC class I (Jurisicova et al., 1996). The status of expression of several immunogenicity markers on the surface of several hESC lines was later semi-quantified by flow cytometry: MHC class I proteins were expressed at very low levels, and increased moderately upon *in vitro* or *in vivo* differentiation. However, no cell surface staining of HLA-G or MHC class II was apparent in undifferentiated or differentiated hESC. Treatment of hESC *in vitro* with IFN-γ, a cytokine secreted during the course of an immune response, resulted in high-level expression of MHC class I (Drukker et al., 2002; Grinnemo et al., 2006). Another study showed that hESC express both MHC class I and β2-microglobulin at low levels on the cell surface regardless of using mouse feeder cells or feeder-free cultures. Upon IFN-γ stimulation, both MHC class I and β2-microglobulin were strongly up-regulated and this effect was considerably enhanced after cells had been pre-treated with retinoic acid, a differentiation stimulator (Draper et al., 2002).

The low rate of MHC class I expression was explained by the low expression of components of the antigen-processing machinery (APM) such as β2-microglobulin and tapasin, without detectable expression of TAP1, TAP2, LMP2, and LMP7 (Cabrera et al., 2007; Suarez-Alvarez et al., 2010).

Few studies exist where human MHC class I expression is followed during the differentiation process. Recently, an extensive study showed that hESC-derived embryoid bodies displayed significantly higher expression of HLA-B, HLA-E,-F and β2-microglobulin compared to the undifferentiated cells. Expression of NK cell receptor NKG2D ligands (MICA, MICB) was evident in all undifferentiated pluripotent stem cells lines analyzed, and their expression was maintained after differentiation. MHC expression was subject to epigenetic control in hESC. It was shown that methylation of histone H3K9me3 repressed the tapasin gene in undifferentiated cells whilst HLA-B and β2m acquired the histone

H3K4me3 modification during the differentiation to embryoid bodies. Absence of HLA-DR and HLA-G expression was regulated by DNA methylation (Suarez-Alvarez et al., 2010).
The expression of the HLA-A allele A*02 was found to be moderate in hemangioblasts or hematoendothelial precursors derived from hESC cell line H9, that exist transiently in early embryonic development, and express markers of immature hematopoietic and endothelial cells (CD31, CD34, VE-cadherin, Flt-1) and mature differentiated cells (CD45, CD33, CD146). However, HLA-A02* expression increased dramatically as cells were differentiated into endothelial or hematopoietic stem cells (Basak et al., 2009). Differential regulation of HLA-A,-B,-C alleles in multipotent stem cells has been revealed (Isa et al., 2010; section 4.3) and similar findings have been evaluated in embryonic and early hematopoietic precursors (Sabir et al., in preparation).
hESC can not only be obtained from the inner cell mass of day 5-6 developing embryos, or earlier, but can also be generated by SCNT, parthenogenesis, or by reprogramming adult somatic cells to generate inducible pluripotent stem cells (iPS cells; Jopling et al., 2011; Nehlin & Barington, 2009; Yamanaka & Blau, 2010). Parthenote-derived hESC show equivalent phenotypes to hESC in the undifferentiated state and can differentiate as demanded, but no data on the expression of molecules conferring immunogenicity is yet known (Harness et al., 2011), and is also unknown for SCNT-derived hESC.
Reprogrammed iPS cells resemble hESC in many ways, with their capacity of self-renewal, and pluripotency state. One of their main features is that they can be obtained from essentially any somatic, fully differentiated adult cells and converted to pseudo-hESC from which one can derive cells and tissues that are genetically compatible with the donor of the original cells. Such strategy, although seemingly expensive, could represent a favorable method to avoid any kind of immunological rejection. Thus, exponential interest in generating such cells has led to characterize the expression of various immunogenicity-associated molecules. Of particular interest, HLA-B, -C, -E, and β2-microglobulin mRNA levels were reduced in iPS cells compared to parental fibroblasts, whereas HLA-A, -G and MHC class II expression was absent. The mRNA levels of APM components TAP-1, TPN, LMP2 and RFX5 plummeted during the reprogramming process to iPS (Suarez-Alvarez et al., 2010).
However, a recent report showed that expression of MHC class I in iPS cells by flow cytometry is slightly higher in reprogrammed undifferentiated hESC than in ICM-derived hESC (Pearl et al., 2011). It is unclear if such difference would account for immunological discordance taking into account that iPS cells are expected to be histocompatible since the donor and recipient of iPS cells is the same individual. A drawback is represented by the genetic tools used to reprogram somatic cells to iPS cells, because they could raise immunogenicity concerns (Nehlin & Barington, 2009; Yamanaka & Blau, 2010).

4.2.2 Cellular and immunological studies

hESC underwent minimal killings when incubated with NK cells. NK cytotoxicity is mediated by engagement of NK lysis receptors, NKp30, NKp44, NKp46, and CD16. All of their ligands, except the NKp44 ligand, were absent on hESC and were not induced after IFN-γ treatment (Drukker et al., 2002).
Undifferentiated hESC were shown to possess immune-privileged characteristics and when transplanted into immune-competent mice, they did not elicit an immune response. Moreover, the inhibitory effect of hESC on alloreactive T cells was mediated by direct cell

membrane interactions rather than by secreted factors. Even slightly differentiated hESC-derived progenitors within cell aggregates known as embryoid bodies (Figure 1), did not induce proliferation of allogeneic T cells (Li et al., 2004). In contrast, a more recent study found that cellular extracts from hESC could indeed retain the immunoregulatory properties of intact cells e.g. inhibiting the function and maturation of monocyte-derived dendritic cells (Mohib et al., 2010).

In xenotransplantation and allotransplantation settings, when hESC were transplanted into various strains of immunocompetent mice and monitored during one month, the cells were totally eliminated (Drukker et al., 2006). In contrast, when a hESC-derived graft was transplanted into immune deficient mice lacking T, B or NK cells, it was found that T-cell deficient animals failed to reject the hESC-derived graft. The lack of NK cells or B cells did not interfere with vigorous hESC rejection, indicating that T cells play a pivotal role in hESC immune rejection. MHC class I molecules were expressed at low levels while MHC class II, and co-stimulatory proteins CD80 and CD86 were not expressed. The low immunostimulatory capacity of hESC was verified by transplanting undifferentiated or differentiated hESC into a mouse model, in which mice were pre-conditioned to carry PBMC from human origin. After one month, only a minute alloresponse was observed while control adult grafts were totally rejected. If MHC class I expression was induced, an increase in the alloresponse took place. These findings suggested that the use of immunosuppressants could be reduced in the case of hESC-derived transplants compared to solid organ transplantions (Drukker et al., 2006).

hESC-derived cells were found to be capable of long-term hematopoietic engraftment when transplanted into non-obese diabetic/severe combined immunodeficient (NOD/SCID) mice (Tian et al., 2006). hESC were shown to be immunologically inert when transplanted into the myocardium of immunocompetent mice and did not inhibit immune responses judged by an increase in lymphocyte infiltration with positive histological staining for CD11b, CD3, CD4 and CD8 during direct or indirect antigen presentation, and they were acutely rejected in a xenogeneic setting (Grinnemo et al., 2006), similar to previous findings (Drukker et al., 2006).

The earlier interpretations that hESC are immune privileged were contradicted by a new study whereby the fate of transplanted hESC were traced by bioluminescent imaging in immune-competent host mice. Graft infiltration by host immune cells occurred within 5 days, and already after 10 days, there was no evidence of hESC left. When hESC were transplanted into immune-deficient (NOD/SCID) host mice, the transplants expanded in number after only 10 days and teratoma formation was evident at 42 days. Rejection was demonstrated to be mediated predominantly by CD4+T cells and it was delayed when immunosuppressive therapies commonly used in the clinic were applied, such as tacrolimus and sirolimus (Swijnenburg et al., 2008).

Individual hESC may survive allotransplantation due to low immunogenicity compared to mature, adult cells, thereby escaping from NK cell- or T cell-mediated cytotoxicity. However, during differentiation MHC class I expression often increases and the presence of certain cytokines such as interferons may strongly induce its expression. In such cases, antigen disparities between donor hESC and recipient cells will lead to rejection, unless prevented (section 6.1). Alloreactive T cells are the major effectors of graft rejection and these cells usually prevent the generation of teratomas in immunocompetent animals injected with allogeneic ESC.

Recently, the immunogenicity of inducible pluripotent stem (iPS) cells was examined in the mouse model. In contrast to derivatives of mouse embryonic stem cells, abnormal gene expression in some cells differentiated from iPSC could induce T-cell-dependent immune responses in syngeneic recipients. Therefore, the immunogenicity of patient-specific iPS cells should be evaluated before any clinic application of these autologous cells into the patients (Zhao et al., 2011).

One of the factors used in reprogramming procedures towards generating iPS cells is OCT4, a transcription factor that plays a key role in the pluripotency program. Most healthy individuals harbor OCT4-specific $CD4^+$ memory T cells indicating a lack of immune tolerance to this antigen (Dhodapkar et al., 2010).

4.3 MHC class I expression in human multipotent mesenchymal stem cells

Human mesenchymal stem cells (hMSC) are multipotent stem cells with a fibroblast-like morphology and the capacity to self-renew *in vivo* that are distributed across virtually every tissue in the body. The "potency" of such cells is not restricted to skeletal tissues (bone, cartilage, fibrous tissue, fat, and myelosupportive stroma) but also includes non-skeletal mesodermal derivatives such as heart, endothelial cells and striated muscle. In contrast, osteogenic, stromal, or skeletal stem cells are multipotent $CD146^+$ cells found in the bone marrow stroma that can differentiate into bone, cartilage, fibrous tissue, adipose tissue, and myelosupportive stroma but not to skeletal muscle, other mesoderm-derived tissues, and non-mesodermally derived tissues (Figure 2; Bianco et al., 2010; Nombela-Arrieta et al., 2011).

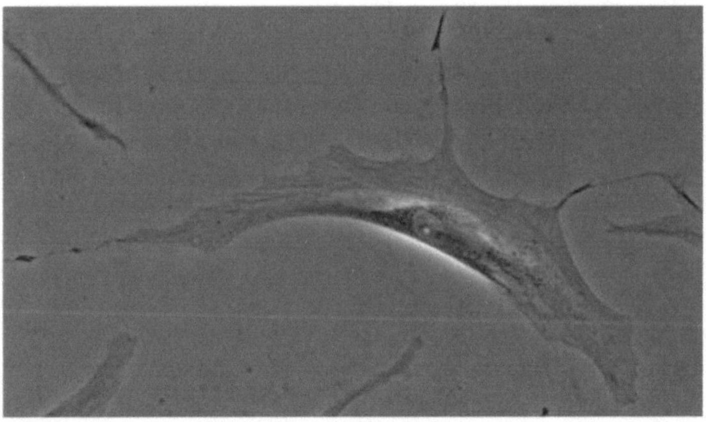

Fig. 2. Phase contrast photograph of a human bone marrow-derived stromal cell (hMSC). Magnification: 100x

Recently, CD146 expression was shown to differentiate between perivascular versus endosteal localization of non-hematopoietic bone marrow stem cell populations and this localization correlated with CD146 being expressed during normal oxygen conditions (normoxia) and absent during hypoxia (Tormin et al., 2011).

hMSC are an attractive source of stem cells for use in tissue engineering such as bone regeneration and cartilage repair, due to their differentiation capacity, their relative availability and their immune privilege properties (section 6.2; Niemeyer et al., 2006; Nombela-Arrieta et al., 2011).

An early study described the presence of various immunologically relevant markers such as neutral endopeptidase CD10, aminopeptidase CD13, neural cell adhesion molecule CD56, a typical NK cell marker, and MHC class I in hMSC isolated from fetal, mature, and geriatric individuals (Young et al. 1999). In another study, hMSC isolated from the bone marrow stroma were also shown to constitutively express MHC class I and even lymphocyte function-associated antigen (LFA)-3 antigens whereas MHC class II and intercellular adhesion molecule (ICAM)-1 antigens were only expressed upon IFN-γ treatment and CD80, CD86, or CD40 co-stimulatory molecules were not expressed at all. Moreover, hMSC failed to stimulate allogeneic PBMC or T-cell proliferation in mixed lymphocyte reactions (Tse et al., 2003).

In yet another study, undifferentiated hMSC were shown to express significant levels of MHC class I expression and no MHC class II expression, and when hMSC were differentiated into adipocytes, osteoblasts, and chondrocytes, they were shown to express lower levels of MHC class I, but still not MHC class II. Both undifferentiated hMSC and differentiated hMSC were not immunogenic as they did not stimulate allogeneic lymphocytes in co-culture experiments. Upon IFN-γ treatment, MHC class II expression increased dramatically in hMSC, but despite this, the inhibitory effect on lymphocyte alloreactivity persisted (Le Blanc et al., 2003b). Several terminally differentiated cell types like neurons, hepatocytes, skeletal and cardiac muscle cells failed to constitutively express HLA class I (Fleming et al., 1981).

hMSC can process and present HLA class I-restricted viral or tumor antigens to specific CTL with a limited efficiency, likely because of some defects in APM components such as lack of expression of LMP7, LMP10, and ERp57. However, they are protected from CTL-mediated lysis through a mechanism that is partly sHLA-G-dependent (see section 5.2.2.3; Morandi et al., 2008).

While hESC and the inner cell mass in blastocysts have been shown to express very low levels of HLA class I, lineage-committed stem cells like mesenchymal stem cells (MSC) have a much higher expression similar to that of lymphocytes. Constitutive expression of HLA class I is largely restricted to cells of the lymphoid organs, the epithelia and the lining of small vessels. HLA-A, -B- and -C are equally expressed in blood leukocytes and regulated primarily at the level of transcription through promoter elements that are conserved among the HLA genes. Using quantitative multicolour flow cytometry and allele-specific antibodies targeting classical MHC class I on muscle satellite cells, bone marrow stromal cells and adipose-derived stem cells, as well as in PBMC, we found high cell-surface expression of HLA-A whereas HLA-B and -C alleles were strongly down-regulated. IFN-γ stimulation of stem cells during 48-72h was required to induce full HLA-B protein expression. The major contributor to repression of HLA-B and -C in stem cells during basal, non-induced conditions may be a post-transcriptional mechanism leading to translational attenuation in stem cells. Since different HLA alleles have variable affinities for intracellularly-generated peptides, the lack of HLA-B and -C expression can influence antigen presentation and the resulting immune response (Isa et al., 2010).

5. Immune-modulating properties

5.1 Immune-modulating properties of human embryonic stem cells

As explained previously in section 3, embryonic tissues are endowed with powerful immune-protecting mechanisms. However, relatively little is known about molecules expressed by hESC of the ICM exerting immune tolerance functions, especially considering that such intrinsic hESC properties could perhaps be used in regenerative transplantation protocols.

hESC have limited antigen presentation capacity because of low MHC class I cell surface expression levels and the complete lack of MHC class II molecules and co-stimulatory molecules (Drukker et al., 2006; Grinnemo et al., 2006).
Among known immune tolerance molecules (section 5.2), ICM-derived hESC expressed the tolerogenic HLA-G molecule at the mRNA and protein levels, but it underwent down-regulation in ICM cells during blastocyst growth (Rizzo et al., 2011a; Verloes et al., 2011). Several immune-modulating proteins found in hMSC are also expressed in hESC (section 5.2; Nehlin et al., Isa et al.; Sabir et al., in preparation).

5.2 Immune-modulating properties of human mesenchymal stem cells

hMSC are capable of down-regulating allogeneic immune responses by a number of strategies that are illustrated in Figure 3. These strategies will be explained in detail below. Immunosuppression by MSC is species-specific, indicating that the valuable contributions from mouse studies in this area cannot necessarily be extrapolated to the human scenario (Ren et al. 2009). Thus, here we will explore the most relevant findings in the field of hMSC-mediated immunosuppression that have recently gained much attention.

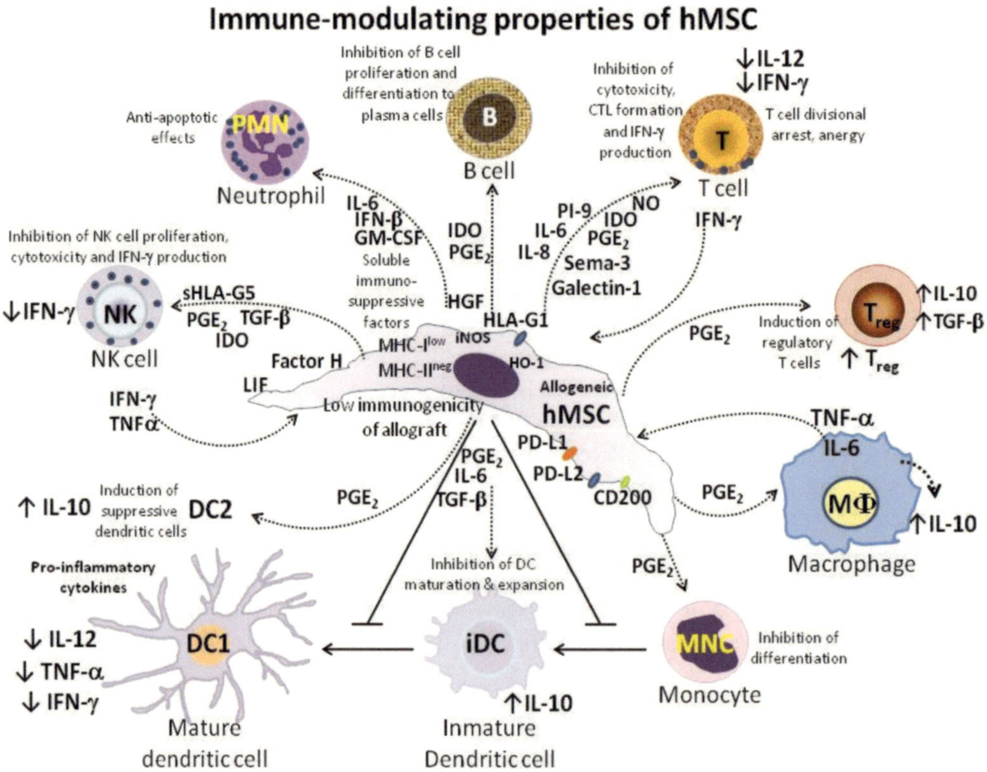

Fig. 3. Illustration depicting the main immune-modulating properties of human mesenchymal stem cells.

hMSC exhibit potent immune-modulating properties which could be useful in numerous clinical applications (section 6.2; Barry et al., 2005; English et al., 2010; Hoogduijn et al., 2010;

Le Blanc & Pittenger, 2005; Nauta & Fibbe, 2007; Rasmusson, 2006; Siegel et al., 2009; Sotiropoulou & Papamichail, et al., 2007; Trento & Dazzi, 2010). The immunomodulatory properties of MSC derived from adult human tissues including bone marrow (BM), adipose tissues (AT), umbilical cord blood (CB), and cord Wharton's jelly (WJ) were shown to be comparable (Yoo et al., 2009; Najar et al., 2010). CD4+ and CD8+ T-lymphocytes were equally targeted by hMSC of different origins, and the effects included prevention of lymphocyte activation as well as the suppression of T-cell proliferation regardless of the stimuli used to activate the lymphocytes (Najar et al., 2010).

Many different pathways mediating hMSC immunotolerance have been suggested such as suppression of T and B cell proliferation both by cell-mediated and soluble factors (Siegel et al., 2009; Uccelli et al., 2008).

Multiple cell-cell interactions and the secretion of soluble factors determine the grade of immunomodulatory capacity by hMSC. Adapted from Aggarwal & Pittenger 2005; Barry et al. 2005; Le Blanc & Ringden, 2007; Nasef et al., 2008; Nauta & Fibbe, 2007; Nemeth et al., 2009; Rasmusson, 2006; Uccelli et al. 2008, and many references in the text.

5.2.1 Origins of the hMSC immunosuppression: cell-cell interactions

Multiple interactions take place between hMSC and cells of the innate and adaptive immune system (Uccelli et al., 2008; Shi et al. 2011). The co-culture of hMSC with purified blood subpopulations in mixed lymphocyte reactions (MLR)(see section 3) has yielded valuable information as regards to hMSC-mediated immuno-modulatory mechanisms of action.

Several studies have explored the effect of hMSC on T cell populations. An early report indicated that hMSC were capable of inhibiting allogeneic T cell responses *in vitro* (Klyushnenkova et al., 1998). Autologous or allogeneic hMSC strongly suppressed T-lymphocyte proliferation, without induction of apoptosis, by both cellular as well as non-specific mitogenic stimuli and was likely due to the production of soluble factors (Di Nicola et al., 2002).

An in-depth analysis was later carried out by Rasmusson et al., 2003, where it was shown that hMSC inhibit T cells in the early activating phase of the allograft reaction, but not in the effector phase. When hMSC were added early in the MLR, they inhibited cytotoxicity, presumably by preventing the formation of active CTL. When MSC were added on day 3 to the 6-day MLR, little effect on cytotoxicity was observed, indicating that hMSC did not appear to inhibit activated CTL and even NK cells (Rasmusson et al., 2003).

Proliferation of allogeneic CD3+ T cell populations was suppressed by hMSC, in a dose-dependent, genetically unrestricted manner, regardless of being pre-treated or not with IFN-γ (Klyushnenkova et al., 2005). hMSC suppressed the proliferation of both CD4+ and CD8+ T lymphocytes, as well as of NK cells. The suppressive activity of hMSC was not only cell-contact dependent but required further the presence of IFN-γ produced by activated T cells and NK cells (Krampera et al., 2006; Pradier et al., 2011).

hMSC altered as well the cytokine secretion profile of naive and effector T cells: T helper 1 (Th1) and 2 (Th2), and NK cells to induce a more anti-inflammatory or tolerant phenotype, increased the proportion of T_{reg} and decreased the secretion of IFN-γ from NK cells induced by IL-2 (Aggarwal & Pittenger, 2005).

hMSC inhibited T-cell proliferation triggered either by allogeneic, mitogenic or antigen-specific stimuli. Interestingly, hMSC inhibited T-cell proliferation by inducing apoptosis once T cells were CD3-mitogenically activated, but had no effect on resting T cells (Plumas et al., 2005). hMSC were able to inhibit proliferation of not only resting thymocytes but also dividing thymocytes cultured in the absence of trophic factors. hMSC could support T cell survival in a quiescent G_0 state without inducing apoptosis, but T cells still regained their activation capacity once immunosuppression was not longer present. These results suggested that hMSC are endowed with the intrinsic capacity of promoting survival of T cells in a resting state. The protective effect of hMSC targets mainly the "death receptor" pathway of apoptosis, as suggested by the down-regulation of Fas receptor and Fas ligand on TCR activated T cells (Benvenuto et al., 2007).

hMSC were shown to target T-cell proliferation but not their effector function (cytotoxicity). This could be explained by inducing T cells to the G_0/G_1 cell cycle phase, in part through inhibition of cyclin D expression and up-regulation of $p27^{kip1}$ (Giuliani et al., 2011; Ramasamy et al., 2008; Siegel et al., 2009). According to a recent study, T cell inhibition by hMSC was not due to the soluble HLA-G5 isoform, but to the surface expression of HLA-G1, as shown by the need of cell-cell contact and by the use of neutralizing anti-HLA-G antibodies (Giuliani et al., 2011; section 5.2.2.3).

hMSC down-regulated alloantigen-induced lymphocyte expansion, especially that of CD8+T cells and of NK lymphocytes, decreased in a dose-dependent manner alloantigen-specific cytotoxic capacity mediated by either CTL or NK cells and favoured the differentiation of regulatory/suppressive $CD4^+$ T-cell subsets co-expressing CD25 and/or CTLA4. More effective suppressive activity on MLR-induced T-cell activation was observed when hMSC were used as third-party, rather than autologous, with respect to MLR-responder cells. These findings support the use of MSC to prevent immune complications related to both hematopoietic stem cell and solid organ transplantation and to the theory that hMSC are universal suppressors of immune reactivity (Maccario et al., 2005; Spaggiari et al. 2009).

Another report showed that the immunosuppressive effect of hMSC targets T cell proliferation of different subpopulations but their effector function or cytotoxicity was not affected in the presence of hMSC at different ratios (Ramasamy et al., 2008).

The significance of the hMSC:T or hMSC:NK co-culture ratios has been investigated in MLR. At low concentrations, hMSC supported rather than inhibited mitogen-stimulated PBMC. Higher concentrations of hMSC not only suppressed alloreactive effector cells, but greatly reduced lymphocyte proliferation caused by potent T-cell mitogens, in autologous or allogeneic PBMC, meaning that the responses were independent of MHC (Le Blanc et al., 2003a). Consistent with this early study, using purified $CD3^+$ T cells only, hMSC were shown to allow T-cell proliferation at a lower MSC:T-cell ratio (1:40) whereas an optimal inhibitory effect was shown when the target (hMSC): effector (T cell) ratio was 1:4 or 1:8 (Najar et al., 2009).

IL-15–stimulated NK cells from 4-day co-culture with hMSC were tested in cytotoxicity assays. When the hMSC:NK cell ratio was low (up to 1:10), hMSC altered the phenotype of NK cells and suppressed their proliferation, cytokine secretion, and cytotoxic capacity against T-cell specific peptide-HLA class I-complexes presented on cancer cells (Sotiropoulou et al., 2006). Some of these effects required cell-to-cell contact, whereas others were mediated by soluble factors (see below), suggesting the existence of diverse mechanisms for MSC-mediated NK cell suppression. On the other hand, hMSC from HLA-mismatched individuals are susceptible to lysis by activated NK cells (Selmani et al., 2009; Sotiropoulou et al., 2006).

Ex vivo-isolated human NK cells become activated upon interaction with bone marrow stromal cells, releasing high amounts of IFN-γ and TNF-α cytokines. These effects depend on the LFA1/ICAM1 interaction and the NKp30 receptor (Poggi et al., 2005). hMSC inhibited the surface expression of NKp30 and NKG2D activating NK receptors that are involved in NK-cell activation and target cell killing, and no cell surface expression of the NKp44 activating receptor (absent in resting NK cells and expressed upon cell activation) occurred in NK cells cultured with hMSC (Spaggiari et al., 2008).

The effects of hMSC on monocytes and dendritic cells were also examined. hMSC could suppress $CD14^+$ monocyte differentiation into dendritic cells (DC), the most potent antigen-presenting cells (APC)(Jiang et al., 2005), as well as maturation of APC (Beyth et al., 2005) and altered the cytokine secretion profile of dendritic cells (Aggarwal & Pittenger, 2005). Mature DC treated with hMSC decreased expression of antigen-presenting and co-stimulatory molecules, and down-regulated IL-12 secretion (Jiang et al., 2005). Similar results were shown where hMSC strongly inhibited the differentiation of alloantigen-induced monocytes to immature dendritic cells (DC1)(Maccario et al., 2005).

Allogeneic hMSC did not affect B lymphocyte proliferation during allo-stimulation with PBMC in mixed lymphocyte cultures at the the B-cell/hMSC ratio of 1:10 (Krampera et al., 2006). However, another study reported that hMSC inhibited B-cell proliferation by induction of cell cycle arrest at the G_0/G_1 phase. The differences were likely due to the cell ratios used, where maximum inhibition of B-cell proliferation was observed at the B-cell/hMSC ratio of 1:1, detected at a 1:2 ratio and non-measurable at 1:5 and 1:10 ratios. Also, hMSC inhibited B-cell differentiation because IgM, IgG, and IgA production was significantly impaired. CXCR4, CXCR5, and CCR7 B-cell expression, as well as chemotaxis to CXCL12, the CXCR4 ligand, and CXCL13, the CXCR5 ligand, were significantly down-regulated by hMSC, suggesting that these cells affect the chemotactic properties of B cells. B-cell co-stimulatory molecule expression and cytokine production were unaffected by hMSC (Corcione et al., 2006). hMSC were able to suppress allo-specific antibody production *in vitro*, and may therefore help overcome a positive cross-match in sensitized transplant recipients (Comoli et al., 2008).

hMSC also inhibited invariant Natural Killer T (iNKT) and $\gamma\delta$ T cell expansion from peripheral blood mononuclear cells. Such inhibition was neutralized by indomethacin, a non-steroid anti-inflammatory drug that inhibits the function of the prostaglandin E_2 molecule (see below, section 5.2.2). iNKT and $\gamma\delta$ T have protective and regulatory immune functions in common because they are involved in defense against infectious organisms, tumor rejection, autoimmune disease pathogenesis, and maintenance of transplant tolerance (Prigione et al., 2009).

hMSC have also been shown to influence human polymorphonuclear neutrophil (PMN) responses in co-culture experiments, exerting anti-apoptotic effects that sustained and amplified the functions of PMN in response to toll-like receptors TLR3- and TLR4-triggering, that may consequently contribute to inflammatory disorders. The biological effects exerted on PMN by TLR3-activated bone marrow-derived hMSC are mediated by the combined action of interleukin 6, interferon-β (IFN-β), and granulocyte macrophage colony-stimulating factor (GM-CSF), while those exerted by TLR4-activated BM-MSC mostly depend on GM-CSF (Cassatella et al., 2011). The key hMSC-derived soluble factor responsible for neutrophil protection from apoptosis was IL-6, (Raffaghello et al., 2008; see section 5.2.2.4).

Finally, an interesting study showed that the Stro-1-enriched population of hMSC isolated from the bone marrow elicits a significantly (~10 times) more profound dose-dependent inhibition of lymphocyte proliferation in mixed lymphocyte reactions than hMSC in general, suggesting its use in allogeneic transplantation (Nasef et al., 2009).

5.2.2 Origins of the hMSC immunosuppression: secreted factors

Even though the presence of hMSC in mixed lymphocyte cultures elicits stronger immunosuppressive effects than only hMSC-free culture supernatants, evidence suggests that secreted factors released to the media might account for a significant part of the hMSC-derived immunosuppression (Najar et al., 2009; Nasef et al., 2008a; Rasmusson et al. 2003). Soluble immunomodulatory factors such as indoleamine 2,3-dioxygenase (IDO), prostaglandin E2 (PGE_2), sHLA-G, transforming growth factor (TGF-β), interleukins IL-1β, IL-6, IL-8, IL-10, IL-11, hepatocyte growth factor (HGF), IFN-γ, stromal cell-derived factor-1 (SDF-1) are all secreted by hMSC (Di Nicola et al., 2002; Barry et al., 2005; Giuliani et al., 2011; Najar et al., 2009; Nasef et al., 2008a; Wu et al., 2010).

Secretion of pro-inflammatory and anti-inflammatory cytokines, chemokines, growth factors and prostaglandins by hMSC under resting and inflammatory conditions has been investigated (Hoogduijn et al., 2010; Wu et al., 2010). Exposure of hMSC to pro-inflammatory cytokines such as IFN-γ, tumour necrosis factor (TNF)-α and interleukin (IL)-6 enhances the immunosuppressive capacity of hMSC, suggesting that the use of pre-conditioning to pro-inflammatory conditions, improve hMSC properties for clinical immune therapy (Crop et al., 2010; Ryan et al., 2007).

Supernatants from hMSC exert suppression of T cell activity (Najar et al., 2009; Nasef et al., 2008a), in contrast to what has been found in hESC, where cell-cell contact is essential for immune tolerance (Li et al., 2004). The mechanisms by which some of these factors produced by hMSC act upon the immune system will be briefly explained below:

5.2.2.1 IDO

hMSC express the kynurenine pathway (KP) which is the central route that accounts for the degradation of the essential amino acid tryptophan to kynurenine and subsequently nicotinamide adenine dinucleotide (NAD+). KP has roles in antimicrobial activity, modulation of immune responses and in the creation of a tryptophan-depleted milieu that promotes immunosuppression (Croitoru-Lamoury et al. 2011; Munn et al. 1998). The KP pathway's first and rate limiting enzyme, indoleamine 2,3-dioxygenase (IDO), has two isoforms IDO1 and IDO2. hMSC-secreted IDO catabolizes tryptophan necessary for rapid T cell proliferation. IFN-γ induces the expression of IDO in hMSC, enabling them to inhibit T cell proliferation and modulate the function of major cell populations involved in both the innate and adaptive immune systems, including APC, NK cells, T- and B-cells (Djouad et al., 2007; Meisel et al., 2004; Uccelli et al., 2008). IDO inhibits the proliferation of activated T and NK cells (Krampera et al., 2006; Spaggiari et al., 2008). Also, IFN-γ-induced IDO activation in hMSC leads to their impaired proliferation and an alteration of their differentiation capacity (Croitoru-Lamoury et al., 2011).

When IDO inhibitors were used, however, it was shown that proliferation of PBMC was not restored in mixed lymphocyte cultures with hMSC. Insulin-like growth factor (IGF)–binding proteins were shown to contribute to the inhibitory mechanism of hMSC on PBMC proliferation, and this effect was independent of IFNγR1 signaling and IDO expression (Gieseke et al., 2007).

The mechanism of how MSC-mediated immunosuppression varies among different species has been investigated by Ren et al. (2009). Immunosuppression by hMSC is mediated by IDO whereas mouse MSC (mMSC) utilize nitric oxide under the same culture conditions. When the expression of IDO and inducible nitric oxide synthase (iNOS) were examined in hMSC and mMSC after stimulation with the pro-inflammatory cytokine IFN-γ in combination with TNFα, IL-1α, or IL-1β, it was shown that hMSC expressed extremely high levels of IDO, and very low levels of iNOS, whereas mMSC expressed abundant iNOS and very little IDO. Chemokines induced by IFN-γ and TNF-α also were released by MSC of mouse or human origin for optimal immunosuppression to attract immune cells to the vicinity, including T cells, which express the chemokine receptor CXCR3 (Ren et al., 2009).

Toll-like receptors (TLR) expressed on bone marrow-derived hMSC enhanced their immunosuppressive phenotype independent of IFN-γ, due to the production of immunosuppressive kynurenines by IDO1. Induction of IDO1 by TLR involved an autocrine IFN-β signaling loop, which was dependent on protein kinase R (PKR), but independent of IFN-γ (Opitz et al., 2009).

5.2.2.2 PGE_2

Prostaglandin E_2 (PGE_2) is, like all prostaglandins, a 20-carbon oxygenated lipid-signaling molecule, with pro-inflammatory functions, involved in producing swelling, redness and pain. Prostaglandin synthesis begins with the release of arachidonic acid from phospholipids by phospholipase A2, and arachidonic acid is then oxygenated by cyclooxygenase (COX) enzymes (COX1 and COX2) expressed by hMSC to form prostaglandin H2 (PGH2), from which PGE_2 and other prostaglandins are formed by specific enzymes. In the case of PGE_2, they are known as PGE_2 synthase enzymes: microsomal PGE_2 synthases 1 and 2 (mPGES1 and mPGES2) and cytosolic PGES (cPGES). PGE_2 plays a role in many immune functions, including the activation of B lymphocytes and the induction of T_{reg} cells. PGE_2 inhibition by indomethacin partially restored the proliferation of T cells in presence of MSC from human or murine origin (Aggarwal & Pittenger, 2005). In the clinic, the production of PGE_2 has been targeted with inhibitors of COX-2 function to treat a range of painful and inflammatory conditions.

PGE_2 can be produced by many cells and influence the function of a wide array of immune cells, including T cells, B cells, NK cells, macrophages and dendritic cells. Inhibition of PGE_2 synthesis by COX inhibitors restored to a great extent *in vitro* T cell proliferation, while blocking other known hMSC-secreted inhibitors did not have the same effect. hMSC inhibited activated T cells proliferation and pro-inflammatory cytokines production. Thus, PGE_2 appears to be a dominant secreted molecule involved in hMSC-induced suppression of an *in vitro* alloresponse (Yañez et al., 2010) and hMSC-mediated blocking of monocyte-derived DC maturation (Spaggiari et al., 2009). Also, PGE_2 and IDO represent key mediators of the hMSC-induced inhibition of NK cells (Spaggiari et al., 2008).

When hMSC were co-cultured with DC, high levels of PGE_2 were detected. PGE_2 blockade with indomethacin allowed maturation of plasmocytoid-DC but not myeloid-DC, and allowed T lymphocyte proliferation but did not restore pro-inflammatory cytokine secretion (Yañez et al., 2010).

hMSC reduced the expression of MHC class II, CD40, and CD86 co-stimulatory molecules on mature DC, which was responsible for a decrease in T-cell proliferation. The differentiation of bone marrow progenitors into DC was partly inhibited when cultured with conditioned supernatants from hMSC, and this effect was associated, at least in part with the secretion of IL-6 from hMSC. Suppression of T-lymphocyte activation was partially counteracted by anti-IL-6 but no enhanced effects were found by IL-6 and PGE_2 together suggesting that PGE_2 may act through the induction ofg IL-6 secretion (Djouad et al., 2007).

When mast cells (MC) are co-cultured with mMSC to allow cell-to-cell contact, mMSC suppressed mast cells degranulation, pro-inflammatory cytokine production, chemokinesis and chemotaxis. These inhibitory effects were dependent on up-regulation of COX2 in mMSC and were facilitated through the activation of EP4 receptors on MC. Whether a similar mechanism applies to hMSC remains to be investigated (Brown et al., 2011).

5.2.2.3 HLA-G and LIF

HLA-G is a non-classical MHC class I molecule, which is expressed in both membrane-bound and soluble isoforms. HLA-G expression is also claimed to be associated with embryo implantation, the protection of the allogeneic fetus from the maternal immune system, and placentation (section 3; Rebmann et al., 2010; Rizzo et al., 2011a).

HLA-G protein expression was found to be constitutive in hMSC and the level was not modified upon stimulation by allogenic lymphocytes in hMSC-mixed lymphocyte reaction assay (Nasef et al. 2007). Furthermore, hMSC secrete the soluble isoform HLA-G5 (sHLA-G5), which inhibits NK cell-mediated cytotoxicity and IFN-γ secretion and suppresses allogeneic T cell proliferation and expansion of CD4+CD25highFOXP3+ T_{reg} cells (Morandi et al., 2008; Selmani et al., 2009). A summary of HLA-G functions was recently presented (Menier et al., 2010). The HLA-G1 isoform, not the sHLA-G5 form, has recently been found to be crucial for the inhibition of T-cell proliferation (Giuliani et al., 2011).

Leukemia inhibitory factor (LIF) is a secreted glycoprotein cytokine that can inhibit the proliferation of myeloid leukemic cell lines and has several functions in hematopoietic expansion of bone marrow progenitors, pregnancy and in the humoral and cellular immune response. LIF and HLA-G expression in hMSC-mixed lymphocyte reactions is coordinated. When LIF was inhibited by a neutralizing antibody, HLA-G was not expressed (Nasef et al., 2008b).

5.2.2.4 Cytokines and chemokines

An early study showed that when hMSC were prevented from cell-cell contacts in transwell experiments, T-lymphocyte proliferation was significantly reduced. Soluble factors such as TGF-β1 and hepatocyte growth factor (HGF) were proposed as mediators of T-cell suppression (Di Nicola et al., 2002).

Among the secreted pro-inflammatory and anti-inflammatory cytokines, chemokines and prostaglandins characterized in hMSC supernatants under resting conditions is worth mentioning ICAM-1, IL-6, IL-8, CCL-2 and TIMP-2 (Wu et al. 2010). IFN-γ, or type II interferon, is a cytokine that is critical for innate and adaptive immunity against viral and intracellular bacterial infections and for tumor control. IFN-γ is not itself produced by hMSC, but originates from activated T and NK cells (Hoogduijn et al., 2010). IFN-γ is also known to play an important role in the induction of immune-modulatory molecules such as IDO (Krampera et al., 2006; see above). IFN-γ inhibited proliferation and altered human and mouse MSC neural, adipocytic and osteocytic differentiation via the activation of IDO (section 5.2.1; Croitoru-Lamoury et al., 2011).

Upon activation, CTL produce cytokines important for their effector functions such as IFN-γ and TNF-α. Incubation with peptide pulsed hMSC did not lead to any detectable induction of IFN-γ and TNF-α secretion by specific CTLs (Rasmusson et al., 2007).

TGF-β1 secreted by hMSC is an immunosuppressive factor capable of inhibiting NK-cell proliferation, cytotoxicity, and cytokine production and of downregulating the expression of activating receptors 2B4 and NKG2D in NK cells. PGE_2 and TGF-β1 had an additive inhibitory effect on NK-cell proliferation (Sotiropoulou et al., 2006).

5.2.2.5 Galectins and Sema-3A

Galectins are a family of β-galactoside binding proteins that bind not only to glycan structures expressed by host cells but can recognize β-galactoside carbohydrates on many pathogens. Galectins are considered as soluble pathogen recognition receptors. Within the immune system, galectins are expressed by virtually all immune cells, either constitutively or in an inducible fashion (Sioud et al., 2010; Sioud, 2011). Galectins-1 and -3 have been found to be main regulators of hMSC immunosuppressive function, and are constitutively expressed and secreted by human bone marrow MSC. Interfering RNAs abrogated their suppressive effect on allogeneic CD4+ and CD8+ T cells (Gieseke et al., 2010; Sioud et al., 2010; 2011). hMSC derived galectin-1 significantly modulated the release of cytokines involved in GvHD and autoimmunity, such as TNF-α, IFN-γ, IL-2 and IL-10. hMSC promote a shift from a pro-inflammatory Th1 toward a more anti-inflammatory Th2 T-cell response (Gieseke et al., 2010).

Galectin-1 is a homodimeric galactose-binding lectin with a single carbohydrate-recognition domain that binds to the neuropilin-1 receptor (NP-1) expressed on T cells. The main ligand of NP-1 is semaphorin-3A (Sema-3A) that arrests T cells in the G_0/G_1 phase. Galectin-1 and semaphorin-3A (Sema-3A) are two soluble factors highly expressed by hMSC capable to inhibit T-cell proliferation through neuropilin-1 (NP-1). Blocking the interaction to NP-1 abolished hMSC immunosuppression (Lepelletier et al., 2010).

5.2.2.6 PD-1 ligands

One of the well known co-stimulatory pathways is the programmed death (PD-1) pathway, which plays an important role in delivering inhibitory signals that regulate T cell activation, immune tolerance and immune-mediated tissue damage. PD-1 receptor expression is inducible on T cells, NK cells and activated monocytes. PD-1 interacts with the two ligands; PD-L1 (B7-H1; also called CD274) and PD-L2 (B7-DC also called CD273) which are transmembrane glycoproteins belonging to the B7 IgG superfamily (Keir et al. 2008; Petroff & Perchellet, 2010). The interaction leads to signalling via PD-1 receptor and deactivation of the immune cells such as T, B, NK, DC and macrophages, etc. While PD-L1 is widely expressed on low quantities in many cell types including trophoblasts, PD-L2 is more restricted to the myeloid cell types such as monocytes, DC and macrophages (Francisco et al., 2010; Petroff & Perchellet, 2010). hMSC express both PD-L1 and PD-L2 as well as several splice variants but their precise role in the induction of immune tolerance remains to be defined (Isa et al., in preparation). IFN-γ plays a critical role in triggering the immunosuppresion by mouse MSC through the up-regulation of PD-L1 (Sheng et al. 2008). Interestingly, the PD-L ligands may play a critical role in maintaining tolerance to the developing fetus (Petroff & Perchellet, 2010).

5.2.2.7 CD200

CD200 is a transmembrane glycoprotein involved in immune-modulation, such as graft rejection, autoimmune diseases, spontaneous fetal loss, inflammatory disorders and malignancy. The interaction of CD200 and CD200R on T cells results in inhibition of de-granulation and cytokine production which mediate immune regulation through a direct and/or indirect action on activated T-cells via DC (Gorczynski, 2005). CD200 is expressed in hMSC suggesting that it might exert immune suppressive effects on T cells (Delorme et al., 2008; Larsen et al., in preparation).

5.2.2.8 Factor H

hMSC constitutively secrete factor H, which potently inhibits complement activation, and its production is increased by pro-inflammatory cytokines, but suppressed by IDO and PGE_2 inhibitors. Factor H is the primary fluid phase complement regulator and it is mainly produced by hepatocytes in the liver. Complement is a pivotal part of the innate immunity whose primary roles are fighting infection and clearing out immune complexes. Excessive complement activation can lead to e.g. graft rejection (Tu et al. 2010).

5.2.2.9 Serpins

Serpins are a large family of proteins that control proteolytic cascades or have other cellular functions such as storage, hormone carrier proteins or tumor suppression. Serpins that inhibit the cytolytic enzyme granzyme B are expressed in the cytoplasm and nuclei of CTL and in cells of immunoprivileged sites, such as the placenta, testis, ovaries, and brain. In the mouse, one serpin known as serine protease inhibitor 6 (SPI6) is required to protect CTL from granzyme B-mediated death and facilitates the survival of virally infected cells and tumors. SPI6 helped mouse MSC to escape from host immune attack (El Haddad et al., 2011). The human orthologue of SPI6 is serpin B9 or PI-9/CAP-3, the only known intracellular inhibitor of granzyme B. PI-9 is expressed in many human cell types, but also in cancer cell lines where PI-9 is thought to protect them from granzyme B attack (Rousalova et al. 2010). The presence and potential role of PI-9 in hMSC remains to be investigated.

5.2.2.10 Nitric oxide

An important secreted factor that participates in suppression of T-cell proliferation is nitric oxide (NO) mediated by NO inhibition of transcription factor Stat5 phosphorylation. NO is produced by NO synthases (NOS), of which there are 3 subtypes: inducible NOS (iNOS), endothelial NOS, and neuronal NO. In the presence of a direct interaction between T cells and mouse MSC, there was a high level of NO production accompanied by a strong suppression of T-cell proliferation. The presence of T cells induced the expression of iNOS in mouse MSC (Sato et al., 2007). hMSC also express iNOS but at lower levels than in mouse MSC (Ren et al., 2009).

5.2.2.11 Heme-oxygenase 1

Heme oxygenases (HO) are rate-limiting intracellular enzymes that degrade heme to biliverdin, carbon monoxide, and free divalent iron. The stress inducible form, HO-1, has been described as an immunosuppressive molecule and mediator of the IL-10 anti-inflammatory cytokine. hMSC inhibited allogeneic PBMC proliferation by 60% in MLR experiments and HO-1 inhibition with SnPP completely abolished the immunosuppressive effect of hMSC (Chabannes et al., 2007). hMSC were shown to induce, in a HO-1-dependent fashion, formation of IL-10+ T_{reg} 1 and transforming growth factor-β+ Th3 T_{reg}-subsets in alloreactive and TCR-activated lymphocytes, and IL-10 production. HO-1 was down-regulated by soluble factors produced in the MLR and its functions were replaced by molecules such as COX-2. Two of HO-1 metabolic products, the heavy-chain ferritin and bilirubin, have been linked to T_{reg} activation and expansion (Mougiakakos et al., 2011).

6. Strategies leading to control of immunogenicity and immunotolerance of stem cells as means to overcome immune host responses in non-histocompatible transplantations

The first successful clinical organ transplantation was performed by Joseph E. Murray in 1954 between identical monozygotic and did not require immunosuppression (Murray et al., 2001). Successful transplantations between genetically diverse individuals require immunosuppression to suppress rejection by the recipient's immune system. In spite of improved immunosuppressants (see below, section 7), chronic rejection is still the leading cause of graft failure and happens in >50% of solid organ transplants within 10 years (Orlando et al., 2011a). During the first year after transplantation, the survival rates of the grafts are well above 90%, but the long-term survival of the graft is compromised (Li & Yang, 2009; Meier-Kriesche et al., 2004). Thus, immunological tolerance does not establish in practice.

Acute rejection usually manifests around the time of engraftment and the incidence ranges between 20 and 70%, depending on the extent of histocompatibility mismatches, the age of the recipient and the intensity of preparative regimens, while chronic rejection resembles autoimmune disorders and involves B-cell dysregulation (Busca, 2011).

Creation of custom-made bioengineered organs, where the cellular component is exquisitely autologous and have an internal vascular network will theoretically overcome the two major hurdles in transplantation, namely the shortage of organs and the toxicity deriving from lifelong immunosuppression. Advances in transplantation of custom-made organs have been described (Orlando et al., 2011b). The uses of hESC and hMSC in regenerative therapies will be described in section 6.1.

6.1 Uses of human embryonic stem cells in clinical practice

Adult stem cells have a limited lifespan and cannot be expanded endlessly to satisfy the numbers needed in clinical practice (Mason & Dunnill, 2009; Nehlin & Barington, 2009; Rayment & Williams, 2010). The earliest achievement regarding the isolation and culture of hESC from blastocysts (Thomson et al., 1998) suggested that their future use in regenerative medicine was possible.

However, despite their low immunogenicity, hESC are still immunogenic and immunosuppression or tolerance induction is needed for sustained engraftment in allogeneic transplantation protocols. Therefore, the focus of many laboratories has been to try to defeat potential immunological barriers against hESC. Immunogenicity concerns represents a challenge of future stem cell therapy approaches (Drukker, 2004; Drukker & Benvenisty, 2004; Chidgey et al., 2008; Charron et al., 2009; Fairchild et al., 2004; 2007).

Ideally, the use of self stem cells would be the most encouraging path but many technical issues need to be solved before it can be brought to clinical practice (Ahrlund-Richter et al., 2009; Bongso et al., 2008; Chidgey et al., 2008; Ginty et al., 2011, Nehlin & Barington, 2009; Yamanaka & Blau, 2010). Several strategies involved in sustaining antigen-specific tolerance through interplay between T_{reg} and DC could prolong acceptance of hESC-derived tissues with minimal use of immunosuppressants (Lui et al., 2009; Lui et al., 2010). However, given the promise of induced pluripotency (Yamanaka & Blau, 2010), stem cell transplantation tolerance protocols may well be displaced (Fairchild, 2009).

Clinical immunotolerance could be achieved by a) mixed hematopoietic chimaerism or b) co-stimulatory blockade. Mixed chimerism involves ablation of the host immune system, followed by its reconstitution with a mixture of host and donor T cell-depleted bone marrow. Reconstitution of the host immune system allows new emerging T cells to perceive the transplanted bone marrow cells as being "self", while donor-reactive T cells are eliminated in the thymus or differentiate into T_{reg}, leading to chimaeric hosts with both donor and self blood cells. The host needs no immunosuppression as it has become tolerant to the donor tissue (Ilstad & Sachs, 1984). The use of this approach has unfortunately been hampered by the toxic side effects inherent to conventional bone marrow transplantation protocols, the ablative regimen used and the use of immunosuppresants to avoid rejection and graft-versus-host disease (Pilat & Wekerle, 2010).

Approaches to facilitate immune tolerance by manipulation of transplantable hESC have been conceived previously such as development of a universal hESC line blocking HLA expression, somatic cell nuclear transfer (SCNT) and creation of a HLA genotyped hESC cell bank. Advantages and disadvantages have been discussed in detail (Boyd & Fairchild, 2010; Bradley et al., 2002; Cabrera et al. 2006; Chidgey et al. 2008; Drukker, 2008; Nehlin & Barington, 2009; O'Rourke et al. 2008).

A promising approach to achieve tolerance is to block co-stimulation during allogeneic transplantation. Co-stimulation blockade with anti-CD40L/anti-LFA-1 and CTLA4Ig blocking antibodies to induce tolerance to hESC transplanted into testicle, an immune-privileged environment, and heart was examined in immunocompetent and severe combined immunodeficient (SCID) mice. hESC injected into the testis of SCID mice and costimulation blockade treated C57BL/6 mice developed into teratoma in all animals, and were surrounded by CD4+CD25+Foxp3+ T cells, inducing tolerance to the grafts, while in the control treated mice, no surviving hESC were found. Thus, co-stimulation blockade induced tolerance to hESC in the immune-privileged environment of the testis (Grinnemo et al., 2008).

When hESC treated with co-stimulation blockade were transplanted into the hearts of SCID mice, hESC developed teratoma-like formations, whereas immune-competent mice exhibited loss of all hESC cells by 1-2 months due to lymphocytic infiltrates. However, if the co-stimulation blockade was repeated ~3 weeks after the initial transplantation, some surviving hESC-derived cells could be monitored 2 months later. Isolation of T_{reg} from intramyocardial transplanted recipients treated with co-stimulation blockade demonstrated specificity toward undifferentiated hESC and down-regulated naive T-cell activation toward hESC. hESC-specific T_{reg} developed to hESC transplanted to the heart. Thus, transplantation success in co-stimulation blockade treated mice was similar to that seen in SCID mice (Grinnemo et al. 2008b).

Recently, a successful co-stimulatory blockade protocol was created, by simultaneous blocking of CTLA-4, CD40L and IRF-1 using blocking antibodies during a short period of 6 days. Experimental mice were transplanted with e.g. transgenic human iPS or hESC, and their fate was examined using bio-imaging. The donor cells were able to survive and grow, and after 54 days, there was no evidence of rejection (Pearl et al., 2011). This finding suggested that allogeneic hESC transplantation with a brief co-stimulation blockade of leukocyte co-stimulatory pathways could be feasible. Unfortunately, the long-term effects of such procedures with respect to the risk of infections are not yet known.

The idea of creating a cell bank composed of donor HLA-typed hESC lines representing different haplotypes matching those of a large population that would help to reduce the risk of graft rejection and satisfy unmet clinical needs was envisioned (Taylor et al. 2005). However, a perfect 6/6 match between donor and recipient in terms of classical MHC class I

HLA-A,-B and -C alleles would not be enough to prevent rejection. Allograft survival is inversely correlated with the degree of mismatch between the donor's and the recipient's MHC antigens. The ideal transplant that carries lessened risk of rejection is the one where a perfect 12/12 match is ensured, whereby in addition to identical HLA-A,-B,-C between donor and recipient, also identical MHC class II proteins HLA-DR, -DP, and -DQ are sought (Loiseau et al. 2007). It appears unrealistic to build a stem cell bank of the required magnitude in order to match all HLA genotypes world wide. Also, even minor histocompatibility antigens may pose a risk (Bradley et al., 2002; Charron et al. 2009; Feng et al., 2008; Shlomchik, 2007; Spencer et al., 2010).

To assess if hMSC could be used to induce tolerance to co-transplanted hESC, ligation-induced myocardial infarction was performed in immunocompetent rats and 3 weeks later the hearts were injected with either hMSC, hESC or both. Co-transplantation of hESCs and hMSC provided better preservation of left ventricle function compared with single-cell treatment alone. The lack of clear evidence for an immunosuppressive or tolerogenic action of hMSC suggested that the benefits observed were mediated by synergistic trophic effects that enhanced repair of injured host tissue (Puymirat et al. 2009).

hESC could also be used in therapies to inhibit tumor progression. Supernatants from hESC have been shown to reduce the clonogenicity and tumorigenesis, as well as to increase apoptosis in aggressive cancer lines (Postovit et al., 2008).

The world's first clinical trial involving the use of hESC-derived cells is ongoing. A cell line known as GRNOPC1 contains hESC-derived oligodendrocyte progenitor cells that have demonstrated remyelinating and nerve growth stimulating properties leading to restoration of function in animal models of acute spinal cord injury. A phase I study has been initiated to assess safety and tolerability in a paralyzed patient with spinal cord injury. Each transplant recipient will be immune-suppressed from the time of injection with low-dose tacrolimus for 46 days, at which time the immune suppression will be tapered and withdrawn at 60 days (www.geron.com/GRNOPC1Trial/).

6.2 Uses of human mesenchymal stem cells in clinical practice

The first published human clinical study reporting positive results with hMSC was in a young boy with acute lymphoblastic leukemia in third remission that had received a transplant of blood stem cells from an HLA-A, HLA-B, HLA-DR1 identical, unrelated, female donor, but after 70 days, developed grade IV GvHD unresponsive to conventional therapy. Transplantation of his mother's hMSC exerted such a strong immunosuppressive effect *in vivo* that the patient made a remarkable recovery (Le Blanc et al., 2004).

After this successful case, the profound inmuno-modulatory and anti-inflammatory effects of hMSC have been exploited for clinical applications in a number of clinical trials to treat diseases such as hematological malignancies, autoimmune diseases such as Crohn's, type 1 diabetes, rheumatoid arthritis, systemic lupus erythematosus, neurodegenerative diseases such as Alzheimer's disease, Parkinson's disease, amyotrophic lateral sclerosis and multiple sclerosis, In addition, hMSC are included in prospective studies to prevent GvHD, treat refractory GvHD, cancer, stroke, acute myocardial infarction, critical limb ischemia, acute tubular necrosis, and use in plastic surgery, bone and cartilage tissue engineering, wound healing, dental pulp regeneration, heart transplantation, insulin-dependent diabetes mellitus, etc. (Abdi et al., 2008; Aggarwal & Pittenger, 2005; Dazzi et al., 2007; Doeppner & Hermann, 2010; Kan et al., 2007; Le Blanc & Ringden, 2007; Niemeyer et al., 2006; Sasportas et al., 2009; Sato et al., 2010; Singer & Caplan, 2011; Trento & Dazzi, 2010; Tögel &

Westenfelder, 2007; Vanikar et al., 2010; Wu et al., 2010). Several clinical trials where bone marrow hMSC have been used in the treatment of ischemic heart disease such as in clinical refractory angina, ischemic cardiomyopathy with left ventricular dysfunction, and end-stage heart failure have yielded promising results (Fuh & Brinton, 2009). Prochymal, a commercial hMSC line, underwent a successful clinical trial to treat myocardial infarction, giving insights into the preparation of hMSC (Hare et al., 2009). More recently, another study demonstrated that it was safe to treat patients with stable coronary artery disease, with autologous hMSC, showing significant improvement in left ventricular function, exercise capacity and clinical symptoms (Friis et al., 2011).

A summary of pre-clinical models and clinical trials where *in vitro* expanded hMSC were used and where the biological properties of hMSC were explained in detail have recently been reported (English et al., 2010; Garcia-Gomez et al., 2010; Parekkadan & Milwid, 2010; Salem & Thiemermann, 2010; Shi et al., 2011; Singer & Caplan, 2011). hMSC supported engraftment and survival of unrelated human donor hematopoietic stem cells infused into sublethally irradiated NOD-SCID mice (Maitra et al. 2004). The biological effects of MSC in mouse and rat pre-clinical models of disease have also been reported (Uccelli et al., 2008).

hMSC have the capacity to migrate (homing) and integrate into damaged tissues and provide immunomodulatory effects by paracrine (soluble factors) and/or cell-cell contact that is regulated by the inflammatory microenvironment. Local or systemic infusions are now being successfully used to co-transplant hMSC with other parenchymal cells, such as hepatocytes or islet cells, to enhance the engraftment and function of such cells in an immunoprotected fashion. These findings may extend future prospects of the clinical application of MSC into broader applications (Uccelli et al., 2008; Yagi et al. 2010). After *in vivo* administration, MSC induce peripheral tolerance and migrate to injured tissues, where they can inhibit the release of pro-inflammatory cytokines and promote the survival of damaged cells (Uccelli et al. 2008).

hMSC are easily isolated from bone marrow, fat and other tissues, and are readily propagated *in vitro*. Transplanted/injected MSC have been shown to migrate to a variety of organs and tissues, but they preferentially undergo homing to sites of inflammation and pathology for tissue remodeling and repair. Transplanted allogeneic MSC can be detected in recipients at extended time points, indicating a lack of immune recognition and clearance (Singer & Caplan, 2011). Because tumor microenvironments also appear to be a target of hMSC homing, there are various controversies surrounding these interactions regarding clinical outcomes (Kidd et al., 2008). hMSC have been used to secrete recombinant cytokine tumor necrosis factor apoptosis ligand (TRAIL) to induce apoptosis of glioma cells *in vivo* (Sasportas et al., 2009). Another study has shown that hMSC exhibit innate anti-tumor effects against human pancreatic carcinoma cells implantated in SCID mice and can serve as delivery vehicles for IFN-β for the treatment of pancreatic cancer (Kidd et al., 2010). Also of great concern is the potential tumorigenicity of hMSC. Although malignant transformation of primary hMSC has not been noted to date in clinical trials using hMSC, expansion *in vitro* for extended periods of time could confer the risk of chromosomal instability and malignant transformation as was reported for mMSC (Tolar et al., 2007).

The strong immunosuppressive activity of MSC has been exploited to attempt treating GvHD (Dazzi & Marelli-Berg, 2008). The clinical experience with hMSC for the treatment of GvHD is encouraging but incomplete. Results of clinical trials utilizing hMSC for the treatment of acute and chronic GvHD have been summarized (Kebriaei & Robinson, 2011). Longer follow-ups of current clinical trials are necessary to determine whether any long-term unwanted effects are associated with the use of hMSC to treat GvHD (Kebriaei & Robinson, 2011).

Clinical trials involving hMSC for the treatment of GvHD have been recently described (Sato et al. 2010). Infusion of hMSC expanded *in vitro*, irrespective of the donor, might be effective therapy for patients with steroid-resistant, acute GvHD (Le Blanc et al., 2008). hMSC appear capable of suppressing acute GvHD without increasing systemic infections (Sato et al., 2010). Donor hMSC significantly inhibited the proliferation of alloactivated recipient T cells before and after kidney transplantation suggesting that the application of hMSC in solid organ transplantation may facilitate graft acceptance and function (Crop et al., 2009).

Steroid-refractory GvHD may also be treated with hMSC infusions but hMSC have been almost impossible to detect after infusion when administered *in vivo*, and thus little is known regarding their migration, their mechanism of action, or their persistence (Paczesny et al., 2009). Advances in the immune reconstitution after hematopoietic stem cell transplantation (HSCT), a widely used method in cancer treatment have been reviewed elsewhere (Cavazzana-Calvo et al., 2009). One trial example involved co-transplantation of *ex vivo*-expanded donor hMSC with CD34+ cells from a relative in children with a hematological malignancy, leading to a reduced risk of graft failure in haploidentical hematopietic stem cell- transplant recipients (Ball et al., 2007). However, relapse of the underlying disease, GvHD, or severe opportunistic infections, account for the majority of deaths following HSCT. Approaches such as immune reconstitution, withdrawal of the immunosuppression, chemotherapy or novel drugs with or without donor lymphocyte infusions, and even second allogeneic stem cell transplantation are considered (Kröger, 2011; Seggewiss & Einsele, 2010). Cellular therapy including adoptive transfer of *ex vivo*-expanded immunomodulatory cells such as T_{reg} cells, NK/T_{reg} cells, donor-derived NK cells, and hMSC and adoptive transfer of allogeneic T cells specific for viral or tumor antigens appears promising to improve immune reconstitution after transplantation (Peters et al., 2009; Seggewiss & Einsele, 2010).

Immune responses against hMSC has also been studied in a kidney transplant setting. Donor hMSC and kidney recipient immune cells (PBMC) isolated at various time points after kidney transplantation were used in MLR assays. Donor hMSC significantly inhibited cytotoxic effector cells of the recipients isolated before transplantation. Allogeneic hMSC were susceptible to lysis by cytotoxic CD8+ T-cells and NK cells, while autologous hMSC were lysed by NK cells only. NK cell-mediated lysis was inversely correlated with the expression of HLA class I on MSC. PBMC isolated 3, 6 and 12 months after donor kidney transplantation showed increasing lysing ability against donor hMSC. Even 12 months after kidney transplantation, CD8+ T cell-mediated lysis of donor hMSC persisted, indicating that there was no evidence for desensitization against donor hMSC. Therefore, controlling the immunogenicity conferred by the HLA expression status, the survival over time of hMSC and avoidance of lysis by cytotoxic immune cells are important for the efficacy of MSC therapy in organ transplantation (Crop et al., 2009; Crop et al., 2011).

Treatment with allogeneic hMSC or the conditioned medium restored alveolar epithelial fluid transport and lung fluid balance in an *ex-vivo* perfused human lung preparation injured by *E. coli* endotoxin, and keratinocyte growth factor played a crucial role in this effect (Lee et al., 2009).

Also, the use of inducible T_{reg} or T regulatory type 1 cells are promising candidates for stem cell therapy because of their immunomodulatory activities such as ability to secrete suppressive cytokines and cell-to-cell contact-dependent killing of target myeloid cells mediated by granzyme B and perforin. Such properties are helpful not only to inhibit GvHD after allogeneic hematopoietic stem cell transplantation, but also in other transplantation settings, or to re-establish tolerance in autoimmune or allergic diseases (Peters et al., 2009; Roncarolo et al., 2011).

Although shown in a mouse sepsis model, it was found that injections of bone marrow mMSC protected cells from damage in affected vital organs and had reduced vascular permeability, one of the deadliest consequences of sepsis. The therapeutic effect was mediated by reprogramming of the macrophage after direct contact with MSC, resulting in macrophage-dependent production of IL-10 and lower expression of TNFα and IL-6. PGE_2 from MSC after activation of Toll-like receptor 4 by bacterial lipopolysaccharide (LPS) was responsible for the reprogramming (Nemeth et al., 2009). Also shown in a mouse model very recently, it was found that intramyocardial delivery of c-kit$^+$ bone marrow cells after myocardial infarction induces endogenous progenitor-derived cardiomyocyte renewal and improves ventricular function, an effect not observed with MSC (Loffredo et al., 2011).

Recently, methods have been developed to identify hMSC with the highest immunosuppressive capacity based on soluble HLA-G production (HLA-G5) in IL-10-treated bone marrow hMSC. A decreased positivity for CD90 is associated with loss of immunosuppressive capacity (Rizzo et al., 2011b).

Pre-clinical screening before allogeneic stem cell therapy is possible if isolated hMSC can undergo a cytotoxicity assay by means of mixed lymphocyte cultures and the subsequent measurement of their proliferation potential (Koppula et al., 2009).

hESC-derived hMSC demonstrated having immunosuppressive effects towards T and NK cells, similar to natural hMSC suggesting another origin from which hMSC can be obtained for therapy (Yen et al., 2009). As mentioned previously, the possibility of reprogramming self-cells into hESC from which one could develop the cell type and numbers needed for a given therapy, would be an ideal situation.

Many challenges need to be assessed in the future as regards the characterization and quantities of stem cells necessary for regenerative medicine which may vary by several orders of magnitude depending on the conditions to treat and their needs among the millions of people afflicted worldwide by a number of degenerative illnesses. Follow-up studies to understand the long-term *in vivo* effects of allogeneic transplantations are needed (Mason & Dunnill, 2009; Rayment & Williams, 2010). Also, optimization of the growth conditions to preserve hMSC immunomodulatory properties are merited (Samuelsson et al., 2009).

Although hMSC exhibit immune privileges as explained above, allogeneic hMSC infused intravenously into the host without immunosuppression or chemotherapeutic conditioning may still lead to adverse effects that require treatment. The purpose of deliberately induced immunosuppression in a host recipient is to prevent rejection during transplantation of non-histocompatible (allogeneic) cells, tissues or organs, and to treat GvHD. However, the use of immunosuppresants increases the vulnerability of the individual to opportunistic infections, nephrotoxicity, cancer and even accelerated aging (Li & Yang, 2009; Nehlin and Barington, 2009). Several families of immunosuppressants have been developed such as glucocorticoids, cytostatics, therapeutic monoclonal antibodies, and many others (Duncan & Wilkes, 2005). In the case of allogeneic hematopoietic stem cell transplants, even though response rates were reported to be more than 60%, long-term survival still remains sub-optimal, mainly due to the detrimental side effects of infectious complications, progressive GvHD and relapse due to the underlying malignancy (Busca, 2011).

7. Conclusion

hESC are pluripotent stem cells with low immunogenicity and immune-modulating properties conferred by molecules intrinsic to the cells themselves, but also found secreted in the microenvironment. Many studies remain to ascertain the role of the various components acting upon immune system cells in allogeneic settings. This is especially merited if off-shelf cell lines from hESC banks will be used in the future.

In-depth immunological characterization of iPS cells is required to envision their use in auto-transplantation protocols and clinical trials to reveal if any potential incompatibility might arise as a result of the reprogramming process. Reprogrammed cells (iPS) will offer great clinical advantages once potential hurdles are fully sorted out.

Multipotent stem cells such as hMSC are endowed with multiple immune-regulatory properties acting on different cells of the immune system that protect them from cytotoxic effects. The regulation and quantification of the hMSC-dependent immune-suppression properties both *in vitro* and *in vivo* in long-term transplantation studies await further analyses. The therapeutic uses of hMSC are becoming widespread in a number of clinical conditions, but its place in future medicine still needs to be clarified.

8. Acknowledgement

We wish to thank Linda Harkness, KMEB, University of Southern Denmark, for the generous contribution of hESC photographs. The authors wish to acknowledge the support from Odense University Hospital, the University of Southern Denmark and the Lundbeck Foundation.

9. References

Abdi, R.; Fiorina, P.; Adra, C.N.; Atkinson, M. & Sayegh, M.H. (2008). Immunomodulation by mesenchymal stem cells: a potential therapeutic strategy for type 1 diabetes. *Diabetes,* 57(7), pp. 1759-1767, ISSN 0012-1797

Afzali, B.; Lombardi, G. & Lechler, R.I. (2008). Pathways of major histocompatibility complex allorecognition. *Current Opinion in Organ Transplantation,* 13(4), pp. 438-444, ISSN 1087-2418

Aggarwal, S. & Pittenger, M.F. (2005). Human mesenchymal stem cells modulate allogeneic immune cell responses, *Blood,* 105(4), pp. 1815-1822, ISSN 0006-4971

Ahrlund-Richter, L.; De Luca, M.; Marshak, D.R.; Munsie, M.; Veiga, A. & Rao, M. (2009). Isolation and production of cells suitable for human therapy: challenges ahead. *Cell Stem Cell,* 4(1), pp. 20-26, ISSN 1934-5909

Apps, R.; Murphy, S.P.; Fernando, R.; Gardner, L.; Ahad, T. & Moffett, A. (2009). Human leucocyte antigen (HLA) expression of primary trophoblast cells and placental cell lines, determined using single antigen beads to characterize allotype specificities of anti-HLA antibodies. *Immunology,* 127(1), pp. 26-39, ISSN 0928-8244

Ball, L.M.; Bernardo, M.E.; Roelofs, H.; Lankester, A.; Cometa, A.; Egeler, R.M.; Locatelli, F. & Fibbe, W.E. (2007). Cotransplantation of ex vivo expanded mesenchymal stem cells accelerates lymphocyte recovery and may reduce the risk of graft failure in haploidentical hematopoietic stem-cell transplantation. *Blood,* 110(7), pp. 2764-2767, ISSN 0006-4971

Barry, F.P.; Murphy, J.M.; English, K. & Mahon, B.P. (2005). Immunogenicity of adult mesenchymal stem cells: lessons from the fetal allograft. *Stem Cells & Development,* 14(3), pp. 252-265, ISSN 1547-3287

Basak, G.W.; Yasukawa, S.; Alfaro, A.; Halligan, S.; Srivastava, A.S.; Min, W.P.; Minev, B. & Carrier, E. (2009). Human embryonic stem cells hemangioblast express HLA-antigens. *Journal of Translational Medicine* 7, pp. 27, ISSN 1479-5876

Benvenuto, F.; Ferrari, S.; Gerdoni, E.; Gualandi, F.; Frassoni, F.; Pistoia, V.; Mancardi, G. & Uccelli, A. (2007). Human mesenchymal stem cells promote survival of T cells in a quiescent state. *Stem Cells,* 25(7), pp. 1753-1760, ISSN 1549-4918

Beyth, S.; Borovsky, Z.; Mevorach, D.; Liebergall, M.; Gazit, Z.; Aslan, H.; Galun, E. & Rachmilewitz, J. (2005). Human mesenchymal stem cells alter antigen-presenting cell maturation and induce T-cell unresponsiveness. *Blood,* 105(5), pp. 2214-2219, ISSN 0006-4971

Bianco, P.; Robey, P.G.; Saggio, I. & Riminucci, M. (2010). "Mesenchymal" stem cells in human bone marrow (skeletal stem cells): a critical discussion of their nature, identity, and significance in incurable skeletal disease. *Human Gene Therapy,* 21(9), pp. 1057-1066, ISSN 1043-0342

Bongso, A.; Fong, C.Y. & Gauthaman, K. (2008). Taking stem cells to the clinic: Major challenges. *Journal of Cellular Biochemistry,* 105(6), pp. 1352-1360, ISSN 0730-2312

Boyd, A.S. & Fairchild, P.J. (2010). Approaches for immunological tolerance induction to stem cell-derived cell replacement therapies. *Expert Review of Clinical Immunology,* 6(3), pp. 435-448, ISSN 1744-666X

Bradley, J.A.; Bolton, E.M. & Pedersen, R.A. (2002). Stem cell medicine encounters the immune system. *Nature Reviews Immunology,* 2(11), pp. 859-871, ISSN 1474-1733

Brown, D.M. (2010). Cytolytic CD4 cells: direct mediators in infectious disease and malignancy. *Cellular Immunology,* 262, pp. 89-95, ISSN 0008-8749

Brown, J.M.; Nemeth, K.; Kushnir-Sukhov, N.M.; Metcalfe, D.D. & Mezey, E. (2011). Bone marrow stromal cells inhibit mast cell function via a COX2-dependent mechanism. *Clinical & Experimental Allergy,* 41(4), pp. 526-534, ISSN 0954-7894

Busca, A. (2011). The use of monoclonal antibodies for the treatment of graft-versus-host disease following allogeneic stem cell transplantation. *Expert Opinion on Biological Therapy,* 11(6), pp. 687-697, ISSN 1471-2598

Cabrera, C.M.; Nieto, A.; Cortes, J.L.; Montes, R.M.; Catalina, P.; Cobo, F.; Barroso-Del-Jesus, A. & Concha, A. (2007). The low rate of HLA class I molecules on the human embryonic stem cell line HS293 is associated with the APM components' expression level. *Cell Biology International,* 31(9), pp. 1072-1078, ISSN 1065-6995

Cabrera, C.M.; Cobo, F.; Nieto, A. & Concha, A. (2006). Strategies for preventing immunologic rejection of transplanted human embryonic stem cells. *Cytotherapy,* 8(5), pp. 517-518, ISSN 1465-3249

Cassatella, M.A.; Mosna, F.; Micheletti, A.; Lisi, V.; Tamassia, N.; Cont, C.; Calzetti, F.; Pelletier, M.; Pizzolo, G. & Krampera, M. (2011). Toll-like receptor-3-activated human mesenchymal stromal cells significantly prolong the survival and function of neutrophils. *Stem Cells,* 29(6), pp. 1001-1011, ISSN 1549-4918

Cavazzana-Calvo, M.; André-Schmutz, I.; Dal Cortivo, L.; Neven, B.; Hacein-Bey-Abina, S. & Fischer, A. (2009). Immune reconstitution after haematopoietic stem cell transplantation: obstacles and anticipated progress. *Current Opinion in Immunology,* 21(5), pp. 544-548, ISSN 0952-7915

Chabannes, D.; Hill, M.; Merieau, E.; Rossignol, J., Brion, R.; Soulillou, J.P.; Anegon, I. & Cuturi, M.C. (2007). A role for heme oxygenase-1 in the immunosuppressive effect of adult rat and human mesenchymal stem cells. *Blood,* 110(10), pp. 3691-3694, ISSN 0006-4971

Charron, D.; Suberbielle-Boissel, C. & Al-Daccak, R. (2009). Immunogenicity and allogenicity: a challenge of stem cell therapy. *Journal of Cardiovascular Translational Research,* 2(1), pp. 130-138, ISSN 1937-5387

Chidgey, A.P.; Layton, D.; Trounson, A. & Boyd, R.L. (2008). Tolerance strategies for stem-cell-based therapies. *Nature,* 453(7193), pp. 330-337, ISSN 0028-0836

Clark, D.A. (2005). Tolerance signaling molecules. In: Immunology of pregnancy. *Chemical Immunology & Allergy,* Karger, 89, pp.36-48 (ED. Market, U.R.), ISSN 1660-2242

Comoli, P.; Ginevri, F.; Maccario, R.; Avanzini, M.A.; Marconi, M.; Groff, A.; Cometa, A.; Cioni, M.; Porretti, L.; Barberi, W.; Frassoni, F. & Locatelli, F. (2008). Human mesenchymal stem cells inhibit antibody production induced in vitro by allostimulation. *Nephrology Dialysis Transplantation,* 23(4), pp. 1196-1202, ISSN 0931-0509

Corcione, A.; Benvenuto, F.; Ferretti, E.; Giunti, D.; Cappiello, V.; Cazzanti, F.; Risso, M.; Gualandi, F.; Mancardi, G.L.; Pistoia, V. & Uccelli ,A. (2006). Human mesenchymal stem cells modulate B-cell functions. *Blood,* 107(1), pp. 367-372, ISSN 0006-4571

Croitoru-Lamoury, J.; Lamoury, F.M.; Caristo, M.; Suzuki, K.; Walker, D.; Takikawa, O.; Taylor, R. & Brew, B.J. (2011). Interferon-γ regulates the proliferation and differentiation of mesenchymal stem cells via activation of indoleamine 2,3 dioxygenase (IDO). *PLoS One* 6(2),e14698, ISSN 1932-6203

Crop, M.J.; Baan, C.C.; Korevaar, S.S.; Ijzermans, J.N.; Alwayn, I.P.; Weimar, W.; Hoogduijn, M.J. (2009). Donor-derived mesenchymal stem cells suppress alloreactivity of kidney transplant patients. *Transplantation* 87(6), pp. 896-906, ISSN 0041-1337

Crop, M.J.; Baan, C.C.; Korevaar, S.S.; Ijzermans, J.N.; Pescatori, M.; Stubbs, A.P.; van Ijcken, W.F.; Dahlke, M.H.; Eggenhofer, E.; Weimar, W. & Hoogduijn, M.J. (2010). Inflammatory conditions affect gene expression and function of human adipose tissue-derived mesenchymal stem cells. *Clinical & Experimental Immunology,* 162(3), pp. 474-486, ISSN 0009-9104

Crop, M.J.; Korevaar, S.S.; de Kuiper, R.; Ijzermans, J.N., van Besouw, N.M.; Baan, C.C.; Weimar, W.; Della Bella, E. & Hoogduijn, M.J. (2011). Human mesenchymal stem cells are susceptible to lysis by CD8+ T-cells and NK cells. *Cell Transplantation* Mar 7. [Epub ahead of print], ISSN 0963-6897

Dazzi, F.; J. M. van Laar, A. Cope, and A. Tyndall. (2007). Cell therapy for autoimmune diseases. *Arthritis Research & Therapy,* 9(2), pp. 206, ISSN 1478-6354

Dazzi, F. & Marelli-Berg, F.M. (2008). Mesenchymal stem cells for graft-versus-host disease: close encounters with T cells. *European Journal of Immunology,* 38, pp. 1479-1482, ISSN 0014-2980

Delorme, B.; Ringe, J.; Gallay, N.; Le Vern, Y.; Kerboeuf, D.; Jorgensen, C.; Rosset, P.; Sensebé, L.; Layrolle, P.; Häupl, T. & Charbord, P. (2008). Specific plasma membrane protein phenotype of culture-amplified and native human bone marrow mesenchymal stem cells. *Blood,* 111(5), pp. 2631-2635, ISSN 0006-4971

Desoye, G.; Dohr, G.A.; Motter, W.; Winter, R.; Urdl, W.; Pusch, H.; Uchanska-Ziegler, B. & Ziegler, A. (1988). Lack of HLA class I and class II antigens on human preimplantation embryos. *Journal of Immunology,* 140(12), pp. 4157-4159, ISSN 0014-2980

Dhodapkar, K.M.; Feldman, D.; Matthews, P.; Radfar, S.; Pickering, R.; Turkula, S.; Zebroski, H. & Dhodapkar, M.V. (2010). Natural immunity to pluripotency antigen OCT4 in humans. *Proceedings of the National Academy of Sciences USA,* 107(19), pp. 8718-8723, ISSN 0027-8424

Di Nicola, M.; Carlo-Stella, C.; Magni, M.; Milanesi, M.; Longoni, P.D.; Matteucci, P.; Grisanti, S. & Gianni, A.M. (2002). Human bone marrow stromal cells suppress T-lymphocyte proliferation induced by cellular or nonspecific mitogenic stimuli. *Blood,* 99(10), pp. 3838-3843, ISSN 0006-4671

Djouad, F.; Charbonnier, L.M.; Bouffi, C.; Louis-Plence, P.; Bony, C.; Apparailly, F.; Cantos, C.; Jorgensen, C. & Noël, D. (2007). Mesenchymal stem cells inhibit the differentiation of dendritic cells through an interleukin-6-dependent mechanism. *Stem Cells,* 25(8), pp. 2025-2032, ISSN 1549-4918

Doeppner, T.R. & Hermann, D.M. (2010). Mesenchymal stem cells in the treatment of ischemic stroke: progress and possibilities. *Stem Cells & Cloning: Advances & Applications,* 3, pp. 157-163, ISSN 1178-6957

Draper, J.S.; Pigott, C.; Thomson, J.A. & Andrews, P.W. (2002). Surface antigens of human embryonic stem cells: changes upon differentiation in culture. *Journal of Anatomy,* 200(Pt 3), pp. 249-258, ISSN 0021-8782

Drukker, M.; Katz, G.; Urbach, A.; Schuldiner, M.; Markel, G.; Itskovitz-Eldor, J.; Reubinoff, B.; Mandelboim, O. & Benvenisty, N. (2002). Characterization of the expression of MHC proteins in human embryonic stem cells. *Proceedings of the National Academy of Sciences USA,* 99(15),9864-9869, ISSN 0027-8424

Drukker, M. (2004). Immunogenicity of human embryonic stem cells: can we achieve tolerance? *Springer Seminars in Immunopathology,* 26(1-2), pp. 201-213, ISSN 0172-6641

Drukker, M. & Benvenisty, N. (2004). The immunogenicity of human embryonic stem-derived cells. *Trends in Biotechnology,* 22(3), pp. 136-141, ISSN 0167-7799

Drukker, M.; Katchman, H.; Katz, G.; Even-Tov Friedman, S.; Shezen, E.; Hornstein, E.; Mandelboim, O.; Reisner, Y. & Benvenisty, N. (2006). Human embryonic stem cells and their differentiated derivatives are less susceptible to immune rejection than adult cells. *Stem Cells,* 24(2), pp. 221-229, ISSN 1549-4918

Drukker, M. (2008). Recent advancements towards the derivation of immune-compatible patient-specific human embryonic stem cell lines. *Seminars in Immunology,* 20(2), pp. 123-129, ISSN 1044-5323

Duncan, M.D. & Wilkes, D.S. (2005). Transplant-related immunosuppression: a review of immunosuppression and pulmonary infections. *Proceedings of the American Thoracic Society,* 2(5), pp. 449-455, ISSN 1546-3222

El Haddad, N.; Heathcote, D.; Moore, R.; Yang, S.; Azzi, J.; Mfarrej, B.; Atkinson, M.; Sayegh, M.H.; Lee, J.S.; Ashton-Rickardt, P.G. & Abdi, R. (2011). Mesenchymal stem cells express serine protease inhibitor to evade the host immune response. *Blood,* 117(4), pp. 1176-1183, ISSN 0006-4971

English, K.; French, A. & Wood, K.J. (2010). Mesenchymal stromal cells: facilitators of successful transplantation. *Cell Stem Cell,* 7(4), pp. 431-442, ISSN 1934-5909

English, K. & Wood, K.J. (2011). Immunogenicity of embryonic stem cell-derived progenitors after transplantation. *Current Opinion in Organ Transplantation,* 16, pp. 90–95, ISSN 1087-2418

Fairchild, P.J.; Cartland, S.; Nolan, K.F. & Waldmann, H. (2004). Embryonic stem cells and the challenge of transplantation tolerance. *Trends in Immunology,* 25(9), pp. 465-470, ISSN 1471-4906

Fairchild, P.J.; Robertson, N.J.; Minger, S.L. & Waldmann, H. (2007). Embryonic stem cells: protecting pluripotency from alloreactivity. *Current Opinion in Immunology,* 19(5), pp. 596-602, ISSN 0952-7915

Fairchild, P.J. (2009).Transplantation tolerance in an age of induced pluripotency. *Current Opinion in Organ Transplantation,* 14(4), pp. 321-325, ISSN 1087-2418

Feng, X.; Hui, K.M.; Younes, H.M. & Brickner, A.G. (2008). Targeting minor histocompatibility antigens in graft versus tumor or graft. *Trends in Immunology* 29(12), pp. 624-632, ISSN 1471-4906

Francisco, L. M.; Sage, P.T. & Sharpe, A.H. (2010). The PD-1 pathway in tolerance and autoimmunity. *Immunological Reviews,* 236, pp. 219-242, ISSN 1600-065X

Fleming, K.A.; McMichael, A.; Morton, J.A.; Woods, J. & McGee, J.O. (1981). Distribution of HLA class 1 antigens in normal human tissue and in mammary cancer. *Journal of Clinical Pathology,* 34, pp. 779-784, ISSN 0021-9746

Friis, T.; Haack-Sørensen, M.; Mathiasen, A.B.; Ripa, R.S.; Kristoffersen, U.S.; Jørgensen, E.; Hansen, L.; Bindslev, L.; Kjær, A.; Hesse, B.; Dickmeiss, E. & Kastrup, J. (2011), Mesenchymal stromal cell derived endothelial progenitor treatment in patients with refractory angina. *Scandinavian Cardiovascular Journal,* 45(3), pp. 161-168, ISSN 1401-7431

Ford, M.L. & Larsen, C.P. (2009). Translating costimulation blockade to the clinic: lessons learned from three pathways. *Immunological Reviews,* 229, pp. 294–306, ISSN 1600-065X

Fuh, E. & Brinton, T.J. (2009). Bone marrow stem cells for the treatment of ischemic heart disease: a clinical trial review. *Journal of Cardiovascular Translational Research,* 2(2), pp. 202-218, ISSN 1937-5387

Fändrich, F.; Dresske, B.; Bader, M. & Schulze, M. (2002). Embryonic stem cells share immune-privileged features relevant for tolerance induction. *Journal of Molecular Medicine,* 80(6), pp. 343-350, ISSN 0946-2716

García-Gómez, I.; Elvira, G.; Zapata, A.G.; Lamana, M.L.; Ramírez, M.; Castro, J.G.; Arranz, M.G.; Vicente, A.; Bueren, J. & García-Olmo, D. (2010). Mesenchymal stem cells: biological properties and clinical applications. *Expert Opinion on Biological Therapy,* 10(10), pp. 1453-1468, ISSN 1471-2598

Gieseke, F.; Schütt, B.; Viebahn, S.; Koscielniak, E.; Friedrich, W.; Handgretinger, R. & Müller, I. (2007). Human multipotent mesenchymal stromal cells inhibit proliferation of PBMCs independently of IFNgammaR1 signaling and IDO expression. *Blood,* 110(6), pp. 2197-2200, ISSN 0006-4971

Gieseke, F.; Böhringer, J.; Bussolari, R.; Dominici, M.; Handgretinger, R. & Müller, I. (2010). Human multipotent mesenchymal stromal cells use galectin-1 to inhibit immune effector cells. *Blood,* 116(19), pp. 3770-3779, ISSN 0006-4971

Ginty, P.J.; Rayment, E.A.; Hourd, P. & Williams, D.J. (2011). Regenerative medicine, resource and regulation: lessons learned from the remedi project. *Regenerative Medicine,* 6(2), pp. 241-253, ISSN 1746-0751

Giuliani, M.; Fleury, M.; Vernochet, A.; Ketroussi, F.; Clay, D.; Azzarone, B.; Lataillade, J.J.; & Durrbach, A. (2011). Long-lasting inhibitory effects of fetal liver mesenchymal stem cells on T-lymphocyte proliferation. *PLoS One,* 6(5),e19988, ISSN 1932-6203

Gorczynski, R. M. (2005). CD200 and its receptors as targets for immunoregulation. *Current Opinion in Investigational Drugs,* 6, pp. 483-488, ISSN 1472-4472

Grinnemo, K.H.; Kumagai-Braesch, M.; Månsson-Broberg, A.; Skottman, H.; Hao, X.; Siddiqui, A.; Andersson, A.; Strömberg, A.M.; Lahesmaa, R.; Hovatta, O.; Sylven, C.; Corbascio, M. & Dellgren, G. (2006). Human embryonic stem cells are immunogenic in allogeneic and xenogeneic settings. *Reproductive Biomedicine Online,* 13(5), pp. 712-724, ISSN 1472-6491

Grinnemo, K.H.; Sylvén, C.; Hovatta, O.; Dellgren, G. & Corbascio, M. (2008a). Immunogenicity of human embryonic stem cells. *Cell & Tissue Research,* 331(1), pp. 67-78, ISSN 0302-766X

Grinnemo, K.H.; Genead, R.;, Kumagai-Braesch, M.;, Andersson, A.; Danielsson, C.; Månsson-Broberg, A.; Dellgren, G.; Strömberg, A.M.; Ekberg, H.; Hovatta, O.; Sylvén, C.; & Corbascio, M. (2008b). Costimulation blockade induces tolerance to HESC transplanted to the testis and induces regulatory T-cells to HESC transplanted into the heart. *Stem Cells,* 26(7), pp. 1850-1857, ISSN 1549-4918

Gökmen, M.R.; Lombardi, G. & Lechler, R.I. (2008). The importance of the indirect pathway of allorecognition in clinical transplantation. *Current Opinion in Immunology,* 20(5), pp. 568-574, ISSN 0952-7915

Handunnetthi, L.; Ramagopalan, S.V.; Ebers, G.C. & Knight, J.C. (2010). Regulation of major histocompatibility complex class II gene expression, genetic variation and disease. *Genes & Immunity,* 11(2), pp. 99-112, ISSN 1466-4879

Hare, J.M.; Traverse, J.H.; Henry, T.D.; Dib, N.; Strumpf, R.K.; Schulman, S.P.; Gerstenblith, G.; DeMaria, A.N.; Denktas, A.E.; Gammon, R.S.; Hermiller, J.B. Jr.; Reisman, M.A.; Schaer, G.L. & Sherman, W. (2009). A randomized, double-blind, placebo-controlled, dose-escalation study of intravenous adult human mesenchymal stem cells (prochymal) after acute myocardial infarction. *Journal of the American College of Cardiology,* 54(24), pp. 2277-2286, ISSN 0735-1097

Harkness, L.; Rasmussen, I.A.; Erb, K. & Kassem, M. (2010) Derivation and characterisation of hESC lines from supernumerary embryos, experience from Odense, Denmark. *In Vitro Cellular & Developmental Biology-Animal,* 46(3-4), pp. 259-268, ISSN 1071-2690

Harness, J.V.; Turovets, N.A.; Seiler, M.J.; Nistor, G.; Altun, G.; Agapova, L.S.; Ferguson, D.; Laurent, L.C.; Loring, J.F. & Keirstead, H.S. (2011). Equivalence of conventionally-derived and parthenote-derived human embryonic stem cells. *PLoS One,* 6(1),e14499, ISSN 1932-6203

Hoogduijn, M.J.; Popp, F.; Verbeek, R.; Masoodi, M.; Nicolaou, A.; Baan, C. & Dahlke, M.H. (2010). The immunomodulatory properties of mesenchymal stem cells and their use for immunotherapy. *International Immunopharmacology,* 10(12), pp. 1496-1500, ISSN 1567-5769

Hornick, P. (2006). Direct and indirect allorecognition. *Methods in Molecular Biology,* 333, pp. 145-156, ISSN 1064-3745

Ildstad, S.T. & Sachs, D.H. (1984). Reconstitution with syngeneic plus allogeneic or xenogeneic bone marrow leads to specific acceptance of allografts or xenografts. *Nature,* 307(5947), pp. 168-170, ISSN 0028-0836

Ingulli, E. (2010) Mechanism of cellular rejection in transplantation. *Pediatric Nephrology,* 25(1), pp. 61-74, ISSN 0931-041X

Isa, A.; Nehlin, J.O.; Sabir, H.J.; Andersen, T.E.; Gaster, M.; Kassem, M. & Barington, T. (2010). Impaired cell surface expression of HLA-B antigens on mesenchymal stem cells and muscle cell progenitors. *PLoS One,* 5,e10900, ISSN 1932-6203

Jenkins, M.K. (1994). The ups and downs of T cell costimulation. *Immunity,* 1, pp. 443–446, ISSN 1074-7613

Jiang, X.X.; Zhang, Y.; Liu, B.; Zhang, S.X.; Wu, Y.; Yu, X.D. & Mao, N. (2005). Human mesenchymal stem cells inhibit differentiation and function of monocyte-derived dendritic cells. *Blood,* 105(10), pp. 4120-4126, ISSN 0006-4971

Jopling, C.; Boue, S. & Izpisua Belmonte, J.C. (2011). Dedifferentiation, transdifferentiation and reprogramming: three routes to regeneration. *Nature Reviews Molecular Cellular Biology,* 12(2), pp. 79-89, ISSN 1471-0072

Jurisicova, A.; Casper, R.F.; MacLusky, N.J.; Mills, G.B. & Librach, C.L. (1996). HLA-G expression during preimplantation human embryo development. *Proceedings of the National Academy of Sciences USA,* 93, pp. 161–165, ISSN 0027-842

Kan, I.; Melamed, E. & Offen, D. (2007). Autotransplantation of bone marrow-derived stem cells as a therapy for neurodegenerative diseases. *Handbook of Experimental Pharmacology,* 180, pp. 219-242, ISSN 0171-2004

Kebriaei, P. & Robinson, S. (2011). Treatment of graft-versus-host-disease with mesenchymal stromal cells. *Cytotherapy,* 13(3), pp. 262-268, ISSN 1465-3249

Keir, M.E.; Butte, M.J.; Freeman, G.J. & Sharpe, A.H. (2008). PD-1 and its ligands in tolerance and immunity. *Annual Review of Immunology,* 26, pp. 677-704, ISSN 0732-0582

Kidd, S.; Spaeth, E.; Klopp, A.; Andreeff, M.; Hall, B. & Marini, F.C. (2008). The (in) auspicious role of mesenchymal stromal cells in cancer: be it friend or foe. *Cytotherapy,* 10(7), pp. 657-667, ISSN 1465-3249

Kidd, S.; Caldwell, L.; Dietrich, M.; Samudio, I.; Spaeth, E.L.; Watson, K.; Shi, Y.; Abbruzzese, J.; Konopleva, M.; Andreeff, M. & Marini, F.C. (2010). Mesenchymal stromal cells alone or expressing interferon-beta suppress pancreatic tumors in vivo, an effect countered by anti-inflammatory treatment. *Cytotherapy,* 12(5), pp. 615-625, ISSN 1465-3249

Klyushnenkova, E.; Mosca, J.D. & McIntosh, K.R. (1998). Human mesenchymal stem cells suppress allogeneic T cell responses in vitro: implications forallogeneic transplantation [abstract]. *Blood,* 92(suppl 1, pt 1), 642a, ISSN 0006-4971

Klyushnenkova, E.; Mosca, J.D.; Zernetkina, V.; Majumdar, M.K.; Beggs, K.J.; Simonetti, D.W.; Deans, R.J. & McIntosh, K.R. (2005). T cell responses to allogeneic human mesenchymal stem cells: immunogenicity, tolerance, and suppression. *Journal of Biomedical Sciences,* 12(1), pp. 47-57, ISSN 1021- 7770

Koppula, P.R.; Chelluri, L.K.; Polisetti, N. & Vemuganti, G.K. (2009). Histocompatibility testing of cultivated human bone marrow stromal cells - a promising step towards pre-clinical screening for allogeneic stem cell therapy. *Cellular Immunology,* 259(1), pp. 61-65, ISSN 0008-8749

Krampera, M.; Cosmi, L.; Angeli, R.; Pasini, A.; Liotta, F.; Andreini, A.; Santarlasci, V.; Mazzinghi, B.; Pizzolo, G.; Vinante, F.; Romagnani, P.; Maggi, E.; Romagnani, S. & Annunziato, F. (2006). Role for interferon-gamma in the immunomodulatory activity of human bone marrow mesenchymal stem cells. *Stem Cells*, 24, pp. 386-398, ISSN 1549-4918

Kröger, N. (2011). Approaches to relapse after allogeneic stem cell transplantation. *Current Opinion in Oncology*, 23, pp. 203–208, ISSN 1040-8746

Le Blanc, K.; Tammik, L.; Sundberg, B.; Haynesworth, S.E. & Ringdén, O. (2003a). Mesenchymal stem cells inhibit and stimulate mixed lymphocyte cultures and mitogenic responses independently of the major histocompatibility complex. *Scandinavian Journal of Immunolology*, 57(1), pp. 11-20, ISSN 0300-9475

Le Blanc, K.; Tammik, C.; Rosendahl, K.; Zetterberg, E. & Ringdén, O. (2003b). HLA expression and immunologic properties of differentiated and undifferentiated mesenchymal stem cells. *Experimental Hematology*, 31(10), pp. 890-896, ISSN 0301-472X

Le Blanc, K.; Rasmusson, I.; Sundberg, B.; Götherström, C.; Hassan, M.; Uzunel, M. & Ringdén, O. (2004). Treatment of severe acute graft-versus-host disease with third party haploidentical mesenchymal stem cells. *Lancet*, 363(9419), pp. 1439-1441, ISSN 0140-6736

Le Blanc, K. & Pittenger, M.F. (2005). Mesenchymal stem cells: progress toward promise. *Cytotherapy*, 7(1), pp. 36-45, ISSN 1465-3249

Le Blanc, K. & Ringdén, O. (2007). Immunomodulation by mesenchymal stem cells and clinical experience. *Journal of Internal Medicine*, 262(5), pp. 509-525, ISSN 1365-2796

Le Blanc, K.; Frassoni, F.; Ball, L.; Locatelli, F.; Roelofs, H.; Lewis, I.; Lanino, E.; Sundberg, B.; Bernardo, M.E.; Remberger, M.; Dini, G.; Egeler, R.M.; Bacigalupo, A.; Fibbe, W.; Ringdén, O. & Developmental Committee of the European Group for Blood and Marrow Transplantation (2008). Mesenchymal stem cells for treatment of steroid-resistant, severe, acute graft-versus-host disease: a phase II study. *Lancet*, 371(9624), pp. 1579-1586, ISSN 0140-6736

Leber, A.; Teles, A. & Zenclussen, A.C. (2010). Regulatory T cells and their role in pregnancy. *American Journal of Reproductive Immunology*, 63(6), pp. 445-459, ISSN 1046-7408

Lee, J.W.; Fang, X.; Gupta, N.; Serikov, V. & Matthay, M.A. (2009). Allogeneic human mesenchymal stem cells for treatment of E. coli endotoxin-induced acute lung injury in the ex vivo perfused human lung. *Proceedings of the National Academy of Sciences USA*, 2106(38), pp. 16357-16362, ISSN 0027-8424

Lepelletier, Y.; Lecourt, S.; Renand, A.; Arnulf, B.; Vanneaux, V.; Fermand, J.P.; Menasché, P.; Domet, T.; Marolleau, J.P.; Hermine, O. & Larghero, J. (2010). Galectin-1 and semaphorin-3A are two soluble factors conferring T-cell immunosuppression to bone marrow mesenchymal stem cell. *Stem Cells & Development*, 19(7), pp. 1075-1079, ISSN 1547-3287

Li, L.; Baroja, M.L.; Majumdar, A.; Chadwick, K.; Rouleau, A.; Gallacher, L.; Ferber, I.; Lebkowski, J.; Martin, T.; Madrenas, J. & Bhatia, M. (2004). Human embryonic stem cells possess immune-privileged properties. *Stem Cells*, 22(4), pp. 448-456, ISSN 1549-4918

Li, C. & Yang, C.W. (2009). The pathogenesis and treatment of chronic allograft nephropathy. *Nature Reviews Nephrology,* 5, pp. 513–519, ISSN 1759-5061

Li, X.C. & Raghavan, M. (2010) Structure and function of major histocompatibility complex class I antigens. *Current Opinion in Organ Transplantation,* 15(4), pp. 499-504, ISSN 1087-2418

Loffredo, F.S.; Steinhauser, M.L.; Gannon, J. & Lee, R.T. (2011). Bone marrow-derived cell therapy stimulates endogenous cardiomyocyte progenitors and promotes cardiac repair. *Cell Stem Cell,* 8(4), pp. 389-398, ISSN 1934-5909

Loiseau, P.; Busson, M.; Balere, M.L.; Dormoy, A.; Bignon, J.D.; Gagne, K., Gebuhrer, L.; Dubois, V.; Jollet, I.; Bois, M.; Perrier, P.; Masson, D.; Moine, A.; Absi, L.; Reviron, D.; Lepage, V.; Tamouza, R.; Toubert, A.; Marry, E.; Chir, Z.; Jouet, J.P., Blaise, D.; Charron, D. & Raffoux, C. (2007). HLA Association with hematopoietic stem cell transplantation outcome: the number of mismatches at HLA-A, -B, -C, -DRB1, or -DQB1 is strongly associated with overall survival. *Biology of Blood & Marrow Transplantation,* 13(8), pp. 965-974, ISSN 1083-8791

Lui, K.O.; Waldmann, H. & Fairchild, P.J. (2009). Embryonic stem cells: overcoming the immunological barriers to cell replacement therapy. *Current Stem Cell Research & Therapy,* (1), pp. 70-80, ISSN 1574-888X

Lui, K.O.; Boyd, A.S.; Cobbold, S.P.; Waldmann, H. & Fairchild, P.J. (2010). A Role for Regulatory T Cells in Acceptance of Embryonic Stem Cell-Derived Tissues Transplanted Across an MHC Barrier. *Stem Cells,* 28(10), pp. 1905-1914, ISSN 1549-4918

Maccario, R.; Podestà, M.; Moretta, A.; Cometa, A.; Comoli, P.; Montagna, D.; Daudt, L.; Ibatici, A.; Piaggio, G.; Pozzi, S.; Frassoni, F. & Locatelli, F. (2005). Interaction of human mesenchymal stem cells with cells involved in alloantigen-specific immune response favors the differentiation of CD4+ T-cell subsets expressing a regulatory/suppressive phenotype. *Haematologica,* 90(4), pp. 516-525, ISSN 0390-6078

Maitra, B.; Szekely, E.; Gjini, K.; Laughlin, M.J.; Dennis, J.; Haynesworth, S.E. & Koç, O.N. (2004). Human mesenchymal stem cells support unrelated donor hematopoietic stem cells and suppress T-cell activation. *Bone Marrow Transplantation,* 33(6), pp. 597-604, ISSN 0268-3369

Mason, C. & Dunnill, P. (2009). Quantities of cells used for regenerative medicine and some implications for clinicians and bioprocessors. *Regenerative Medicine,* 4(2), pp. 153-157, ISSN 1746-0751

Meier-Kriesche, H.U.; Schold, J.D.; Srinivas, T.R. & Kaplan, B.(2004). Lack of improvement in renal allograft survival despite a marked decrease in acute rejection rates over the most recent era. *American Journal of Transplantation,* 4(3), pp. 378-383, ISSN 1600-6135

Meisel, R.; Zibert, A.; Laryea, M.; Göbel, U.; Däubener, W. & Dilloo, D. (2004). Human bone marrow stromal cells inhibit allogeneic T-cell responses by indoleamine 2,3-dioxygenase-mediated tryptophan degradation. *Blood,* 103(12), pp. 4619-4621, ISSN 0006-4971

Menendez, P.; Bueno, C.; Wang, L. & Bhatia, M. (2005). Human embryonic stem cells: potential tool for achieving immunotolerance? *Stem Cell Reviews,* 1(2), pp. 151-158, ISSN 1550-8943

Menier, C.; Rouas-Freiss, N.; Favier, B.; LeMaoult, J.; Moreau, P. & Carosella, E.D. (2010). Recent advances on the non-classical major histocompatibility complex class I HLA-G molecule. *Tissue Antigens,* 75(3), pp. 201-206, ISSN 0001-2815

Mohib, K.; Allan, D. & Wang, L. (2010). Human embryonic stem cell-extracts inhibit the differentiation and function of monocyte-derived dendritic cells. *Stem Cell Reviews,* 6(4), pp. 611-621, ISSN 1550-8943

Mor, G. & Abrahams, VM. (2009). The immunology of pregnancy, In: *Creasy and Resnik's Maternal-fetal medicine: Principles and practice,* 6th edition, Creasy, R.K.; Resnik, R.; Iams, J.D.; Lockwood, C.J. & Moore, T.R., Elsevier, pp. 87, ISBN-10 1416042245

Morandi, F.; Raffaghello, L.; Bianchi, G.; Meloni, F.; Salis, A.; Millo, E.; Ferrone, S.; Barnaba, V. & Pistoia, V. (2008). Immunogenicity of human mesenchymal stem cells in HLA-class I-restricted T-cell responses against viral or tumor-associated antigens. *Stem Cells,* 26(5), pp. 1275-1287, ISSN 1549-4918

Mougiakakos, D.; Jitschin, R.; Johansson, C.C.; Okita, R.; Kiessling, R. & Le Blanc, K. (2011). The impact of inflammatory licensing on heme oxygenase-1-mediated induction of regulatory T cells by human mesenchymal stem cells. *Blood,* 117(18), pp. 4826-4835, ISSN 0006-4971

Munn, D.H.; Zhou, M.; Attwood, J.T.; Bondarev, I.; Conway, S.J.; Marshall, B.; Brown, C. & Mellor, A.L. (1998). Prevention of allogeneic fetal rejection by tryptophan catabolism. *Science,* 281(5380), pp. 1191-1193, ISSN 0036-8075

Murray, J.E.; Merrill, J.P. & Harrison, J.H. (2001). Renal homotransplantation in identical twins. 1955. Journal of the *American Society of Nephrology,* 12(1), pp. 201-204, ISSN 1046-6673

Najar, M.; Rouas, R.; Raicevic, G.; Boufker, H.I.; Lewalle, P.; Meuleman, N.; Bron, D.; Toungouz, M.; Martiat, P. & Lagneaux, L. (2009). Mesenchymal stromal cells promote or suppress the proliferation of T lymphocytes from cord blood and peripheral blood: the importance of low cell ratio and role of interleukin-6. *Cytotherapy,* 11(5), pp. 570-583, ISSN 1465-3249

Najar, M.; Raicevic, G.; Boufker, H.I.; Fayyad Kazan, H.; De Bruyn, C.; Meuleman, N.; Bron, D.; Toungouz, M. & Lagneaux, L. (2010). Mesenchymal stromal cells use PGE_2 to modulate activation and proliferation of lymphocyte subsets: Combined comparison of adipose tissue, Wharton's Jelly and bone marrow sources. *Cellular Immunology,* 264(2), pp. 171-179, ISSN 0008-8749

Nasef, A.; Mathieu, N.; Chapel, A.; Frick, J.; François, S.; Mazurier, C.; Boutarfa, A.; Bouchet, S.; Gorin, N.C.; Thierry, D. & Fouillard, L. (2007). Immunosuppressive effects of mesenchymal stem cells: involvement of HLA-G. *Transplantation* 84(2), pp. 231-237, ISSN 0041-1337

Nasef, A.; Ashammakhi, N. & Fouillard, L. (2008a). Immunomodulatory effect of mesenchymal stromal cells: possible mechanisms. *Regenerative Medicine,* 3(4), pp. 531-546, ISSN 1746-0751

Nasef, A.; Mazurier, C.; Bouchet, S.; François, S.; Chapel, A.; Thierry, D.; Gorin, N.C. & Fouillard, L. (2008b). Leukemia inhibitory factor: Role in human mesenchymal stem cells mediated immunosuppression. *Cellular Immunology,* 253(1-2), pp. 16-22, ISSN 0008-8749

Nasef, A.; Zhang, Y.Z.; Mazurier, C.; Bouchet, S.; Bensidhoum, M.; Francois, S.; Gorin, N.C.; Lopez, M.; Thierry, D.; Fouillard, L. & Chapel, A. (2009). Selected Stro-1-enriched bone marrow stromal cells display a major suppressive effect on lymphocyte proliferation. *International Journal of Laboratory Hematology,* 31(1), pp. 9-19, ISSN 1751-5521

Nauta, A.J. & Fibbe, W.E. (2007). Immunomodulatory properties of mesenchymal stromal cells. *Blood,* 110(10), pp. 3499-506, ISSN 0006-4971

Nehlin, J.O. & Barington, T. (2009). Strategies for future histocompatible stem cell therapy. *Biogerontology,* 10(4), pp. 339-76, ISSN 1389-5729

Nemeth, K.; Leelahavanichkul, A.; Yuen, P.S.; Mayer, B.; Parmelee, A.; Doi, K.; Robey, P.G.; Leelahavanichkul, K.; Koller, B.H.; Brown, J.M.; Hu, X.; Jelinek, I.; Star, R.A. & Mezey, E. (2009). Bone marrow stromal cells attenuate sepsis via prostaglandin E(2)-dependent reprogramming of host macrophages to increase their interleukin-10 production. *Nature Medicine,* 15(1), pp. 42-49, ISSN 1078-8956

Newell, K.A. (2011). Clinical transplantation tolerance. *Seminars in Immunopathology,* 33(2), pp. 91-104, ISSN 1863-2297

Niemeyer, P.; Krause, U.; Kasten, P.; Kreuz, P.C.; Henle, P.; Südkam, N.P. & Mehlhorn, A. (2006). Mesenchymal stem cell-based HLA-independent cell therapy for tissue engineering of bone and cartilage. *Current Stem Cell Research & Therapy,* 1(1), pp. 21-27, ISSN 1574-888X

Nombela-Arrieta, C.; Ritz, J. & Silberstein, L.E. (2011). The elusive nature and function of mesenchymal stem cells. *Nature Reviews Molecular Cellular Biology,* 12(2), pp. 126-131, ISSN 1471-0072

Opitz, C.A.; Litzenburger, U.M.; Lutz, C.; Lanz, T.V.; Tritschler, I.; Köppel, A.; Tolosa, E.; Hoberg, M.; Anderl, J.; Aicher, W.K.; Weller, M.; Wick, W. & Platten, M. (2009). Toll-like receptor engagement enhances the immunosuppressive properties of human bone marrow-derived mesenchymal stem cells by inducing indoleamine-2,3-dioxygenase-1 via interferon-beta and protein kinase R. *Stem Cells,* 27(4), pp. 909-919, ISSN 1549-4918

Orlando, G.; Wood, K.J.; Stratta, R.J.; Yoo, J.J.; Atala, A. & Soker, S. (2011a). Regenerative Medicine and Organ Transplantation: Past, Present, and Future. *Transplantation,* 91(12), pp. 1310-1317, ISSN 0041-1337

Orlando, G.; Baptista, P.; Birchall, M.; De Coppi, P.; Farney, A.; Guimaraes-Souza, N.K.; Opara, E.; Rogers, J.; Seliktar, D.; Shapira-Schweitzer, K.; Stratta, R.J.; Atala, A.; Wood, K.J. & Soker, S. (2011b). Regenerative medicine as applied to solid organ transplantation: current status and future challenges. *Transplant International,* 24(3), pp. 223-232, ISSN 0934-0874

O'Rourke, P.P.; Abelman, M. & Heffernan, K.G. (2008). Centralized banks for human embryonic stem cells: a worthwhile challenge. *Cell Stem Cell,* 2(4), pp. 307-312, ISSN 1934-5909

Orr, M.T. & Lanier, L.L. (2010). Natural killer cell education and tolerance. *Cell,* 142, pp. 847-856, ISSN 0092-8674

Paczesny, S.; Choi, S.W. & Ferrara, J.L. (2009). Acute graft-versus-host disease: new treatment strategies. *Current Opininion in Hematology,* 16(6), pp. 427-436, ISSN 1065-6251

Parekkadan, B. & Milwid, J.M. (2010). Mesenchymal stem cells as therapeutics. *Annual Review of Biomedical Engineering,* 12, pp. 87-117, ISSN 1545-4274

Parhar, R.S.; Yagel, S. & Lala, P.K. (1989). PGE_2-mediated immunosuppression by first trimester human decidual cells blocks activation of maternal leukocytes in the decidua with potential anti-trophoblast activity. *Cellular Immunology,* 120(1), pp. 61-74, ISSN 0008-8749

Pearl, J.I.; Lee, A.S.; Leveson-Gower, D.B.; Sun, N.; Ghosh, Z.; Lan, F.; Ransohoff, J.; Negrin, R.S.; Davis, M.M. & Wu, J.C. (2011). Short-term immunosuppression promotes engraftment of embryonic and induced pluripotent stem cells. *Cell Stem Cell,* 8(3), pp. 309-317, ISSN 1934-5909

Peters, J.H.; Koenen, H.J., Hilbrands, L.B. & Joosten, I. (2009). Immunotherapy with regulatory T cells in transplantation. *Immunotherapy,* 1(5), pp. 855-871, ISSN 1750-743X

Petroff, M.G. & Perchellet, A. (2010). B7 family molecules as regulators of the maternal immune system in pregnancy. *American Journal of Reproductive Immunology,* 63(6), pp. 506-519, ISSN 8755-8920

Pilat, N. & Wekerle, T. (2010). Transplantation tolerance through mixed chimerism. *Nature Reviews Nephrology,* 6(10), pp. 594-605, ISSN 1759-5061

Plumas, J.; Chaperot, L.; Richard, M.J.; Molens, J.P.; Bensa, J.C. & Favrot, M.C. (2005). Mesenchymal stem cells induce apoptosis of activated T cells. *Leukemia,* 19(9), pp. 1597-1604, ISSN 0887-6924

Poggi, A.; Prevosto, C.; Massaro, A.-M.; Negrini, S.; Urbani, S.; Pierri, I.; Saccardi, R.; Gobbi, M. & Zocchi, M.R. (2005). Interaction between human NK cells and bone marrow stromal cells induces NK cell triggering: role of NKp30 and NKG2D receptors. *Journal of Immunology,* 175, pp. 6352-6360, ISSN 0014-2980

Postovit, L.M.; Margaryan, N.V.; Seftor, E.A.; Kirschmann, D.A.; Lipavsky, A.; Wheaton, W.W.L; Abbott, D.E.; Seftor, R.E. & Hendrix, M.J. (2008). Human embryonic stem cell microenvironment suppresses the tumorigenic phenotype of aggressive cancer cells. *Proceedings of the National Academy of Sciences USA,* 105(11), pp. 4329-4334, ISSN 0027-8424

Pradier, A.; Passweg, J.; Villard, J. & Kindler, V. (2011). Human bone marrow stromal cells and skin fibroblasts inhibit natural killer cell proliferation and cytotoxic activity. *Cell Transplantation,* 2010 Nov 5. [Epub ahead of print], ISSN 0041-1337

Prigione, I.; Benvenuto, F.; Bocca, P.; Battistini, L.; Uccelli, A. & Pistoia, V. (2009). Reciprocal interactions between human mesenchymal stem cells and gammadelta T cells or invariant natural killer T cells. *Stem Cells,* 27(3), pp. 693-702, ISSN 1549-4918

Puymirat, E.; Geha, R.; Tomescot, A.; Bellamy, V.; Larghero, J.; Trinquart, L.; Bruneval, P.; Desnos, M.; Hagège, A.; Pucéat, M. & Menasché, P. (2009). Can mesenchymal stem cells induce tolerance to cotransplanted human embryonic stem cells? *Molecular Therapy,* 17(1), pp. 176-182, ISSN 1525-0016

Raffaghello, L.; Bianchi, G.; Bertolotto, M.; Montecucco, F.; Busca, A.; Dallegri, F.; Ottonello, L. & Pistoia, V. (2008). Human mesenchymal stem cells inhibit neutrophil apoptosis: a model for neutrophil preservation in the bone marrow niche. *Stem Cells,* 26(1), pp. 151-162, ISSN 1549-4918

Ramasamy, R.; Tong, C.K.; Seow, H.F.; Vidyadaran, S. & Dazzi, F. (2008). The immunosuppressive effects of human bone marrow-derived mesenchymal stem cells target T cell proliferation but not its effector function. *Cellular Immunology,* 251(2), pp. 131-136, ISSN 0008-8749

Rasmusson, I.; Ringden, O.; Sundberg, B. & Le Blanc, K. (2003). Mesenchymal stem cells inhibit the formation of cytotoxic T lymphocytes, but not activated cytotoxic T lymphocytes or natural killer cells. *Transplantation,* 76, pp. 1208-1213, ISSN 0041-1337

Rasmusson, I. (2006). Immune modulation by mesenchymal stem cells. *Experimental Cell Research,* 312, pp. 2169-2179, ISSN 0014-4827

Rasmusson, I.; Uhlin, M.; Le Blanc, K. & Levitsky, V. (2007). Mesenchymal stem cells fail to trigger effector functions of cytotoxic T lymphocytes. *Journal of Leukocyte Biology,* 82(4), pp. 887-893, ISSN 0741-5400

Rayment, E.A. & Williams, D.J. (2010). Concise review: mind the gap: challenges in characterizing and quantifying cell- and tissue-based therapies for clinical translation. *Stem Cells,* 28(5), pp. 996-1004, ISSN 1549-4918

Rebmann, V.; Switala, M.; Eue, I. & Grosse-Wilde, H. (2010). Soluble HLA-G is an independent factor for the prediction of pregnancy outcome after ART: a German multi-centre study. *Human Reproduction,* 25(7), pp. 1691–1698, ISSN 0268-1161

Ren, G.; Su, J. ; Zhang, L.; Zhao, X.; Ling, W.; L'huillie, A.; Zhang, J.; Lu, Y.; Roberts, A.I.; Ji, W.; Zhang, H.; Rabson, A.B. & Shi, Y. (2009). Species variation in the mechanisms of mesenchymal stem cell-mediated immunosuppression *Stem Cells,* 27(8), pp. 1954-1962, ISSN 1549-4918

Rizzo, R.; Vercammen, M.; van de Velde, H.; Horn, P.A. & Rebmann, V. (2011a). The importance of HLA-G expression in embryos, trophoblast cells, and embryonic stem cells. *Cellular & Molecular Life Sciences,* 68(3), pp. 341-352, ISSN 1420-682X

Rizzo, R.; Lanzoni, G.; Stignani, M.; Campioni, D.; Alviano, F.; Ricci, F.; Tazzari, P.L.; Melchiorri, L.; Scalinci, S.Z.; Cuneo, A.; Bonsi, L.; Lanza, F.; Bagnara, G.P. & Baricordi, O.R. (2011b). A simple method for identifying bone marrow mesenchymal stromal cells with a high immunosuppressive potential. *Cytotherapy,* 13(5), pp. 523-527, ISSN 1465-3249

Roncarolo, M.G.; Gregori, S.; Lucarelli, B.; Ciceri, F. & Bacchetta, R. (2011). Clinical tolerance in allogeneic hematopoietic stem cell transplantation. *Immunological Reviews,* 241(1), pp. 145-163, ISSN 1600-065X

Rousalova, I.; Krepela, E.; Prochazka, J.; Cermak, J. & Benkova, K. (2010). Expression of proteinase inhibitor-9/serpinB9 in non-small cell lung carcinoma cells and tissues. *International Journal of Oncology,* 36(1), pp. 275-283, ISSN 1019-6439

Ryan, J.M.; Barry, F.; Murphy, J.M. & Mahon, B.P. (2007). Interferon-gamma does not break, but promotes the immunosuppressive capacity of adult human mesenchymal stem cells. *Clinical & Experimental Immunology,* 149(2), pp. 353-363, ISSN 0009-9104

Salem, H.K. & Thiemermann, C. (2010). Mesenchymal stromal cells: current understanding and clinical status. *Stem Cells* 28(3), pp. 585-596, ISSN 1549-4918

Samuelsson, H.; Ringdén, O.; Lönnies, H. & Le Blanc, K. (2009). Optimizing in vitro conditions for immunomodulation and expansion of mesenchymal stromal cells. *Cytotherapy* 11(2), pp. 129-136, ISSN 1465-3249

Sasportas, L.S.; Kasmieh, R.; Wakimoto, H.; Hingtgen, S.; van de Water, J.A.; Mohapatra, G.; Figueiredo, J.L.; Martuza, R.L.; Weissleder, R. & Shah, K. (2009). Assessment of therapeutic efficacy and fate of engineered human mesenchymal stem cells for cancer therapy. *Proceedings of the National Academy of Sciences USA,* 106(12), pp. 4822-4827, ISSN 0027-8424

Sato, K.; Ozaki, K.; Oh, I.; Meguro, A.; Hatanaka, K.; Nagai, T., Muroi, K. & Ozawa, K. (2007). Nitric oxide plays a critical role in suppression of T-cell proliferation by mesenchymal stem cells. *Blood,* 109(1), pp. 228-234, ISSN 0006-4971

Sato, K.; Ozaki, K.; Mori, M.; Muroi, K. & Ozawa, K. (2010). Mesenchymal stromal cells for graft-versus-host disease: basic aspects and clinical outcomes. *Journal of Clinical Experimental Hematopathology* 50(2), pp. 79-89, ISSN 1346-4280

Shlomchik, W.D. (2007). Graft-versus-host-disease. *Nature Reviews Immunology,* 7(5), pp. 340-352, ISSN 1474-1733

Seggewiss, R. & Einsele, H. (2010). Immune reconstitution after allogeneic transplantation and expanding options for immunomodulation: an update. *Blood,* 115(19), pp. 3861-3868, ISSN 0006-4971

Selmani, Z.; Naji, A.; Gaiffe, E.; Obert, L.; Tiberghien, P.; Rouas-Freiss, N.; Carosella, E.D. & Deschaseaux, F. (2009). HLA-G is a crucial immunosuppressive molecule secreted by adult human mesenchymal stem cells. *Transplantation,* 87(9 Suppl), pp. S62-66, ISSN 0041-1337

Sheng, H.; Wang, Y.; Jin, Y.; Zhang, Q.; Zhang, Y.; Wang, L.; Shen, B.; Yin, S.; Liu, W.; Cui, L. & Li, N. (2008). A critical role of IFNgamma in priming MSC-mediated suppression of T cell proliferation through up-regulation of B7-H1. *Cell Research,* 18, pp. 846-857, ISSN 1001-0602

Shi, M.; Liu, Z.W. & Wang, F.S. (2011). Immunomodulatory properties and therapeutic application of mesenchymal stem cells. *Clinical & Experimental Immunology* 164(1), pp. 1-8, ISSN 0009-9104

Siegel, G.; Schafer, R. & Dazzi., F. (2009). The immunosuppressive properties of mesenchymal stem cells. *Transplantation,* 87, pp. S45-S49, ISSN 0041-1337

Singer, N.G. & Caplan, A.I. (2011) Mesenchymal stem cells: mechanisms of inflammation. *Annual Review of Pathology,* 6, pp. 457-478, ISSN 1553-4006

Sioud, M.; Mobergslien, A.; Boudabous, A. & Fløisand, Y. (2010). Evidence for the involvement of galectin-3 in mesenchymal stem cell suppression of allogeneic T-cell proliferation. *Scandinavian Journal of Immunology,* 71(4), pp. 267-274, ISSN 0300-9475

Sioud, M.; Mobergslien, A.; Boudabous, A. & Fløisand, Y. (2011). Mesenchymal stem cell-mediated T cell suppression occurs through secreted galectins. *International Journal of Oncology,* 38(2), pp. 385-390, ISSN 1019-6439

Sioud, M. (2011). New insights into mesenchymal stromal cell-mediated T-cell suppression through galectins. *Scandinavian Journal of Immunology,* 73(2), pp. 79-84, ISSN 0300-9475

Sotiropoulou, P.A.; Perez, S.A.; Gritzapis, A.D.; Baxevanis, C.N. & Papamichail, M. (2006). Interactions between human mesenchymal stem cells and natural killer cells. *Stem Cells,* 24(1), pp. 74-85, ISSN 1549-4918

Sotiropoulou, P.A. & Papamichail, M. (2007), Immune properties of mesenchymal stem cells. *Methods in Molecular Biology,* 407, pp. 225-243, ISSN 1064-3745

Spaggiari, G.M.; Capobianco, A.; Abdelrazik, H.; Becchetti, F.; Mingari, M.C. & Moretta, L. (2008). Mesenchymal stem cells inhibit natural killer-cell proliferation, cytotoxicity, and cytokine production: role of indoleamine 2,3-dioxygenase and prostaglandin E2. *Blood,* 111(3), pp. 1327-1333, ISSN 0006-4971

Spaggiari, G.M.; Abdelrazik, H.; Becchetti, F. & Moretta, L. (2009). MSCs inhibit monocyte-derived DC maturation and function by selectively interfering with the generation of immature DCs: central role of MSC-derived prostaglandin E2. *Blood,* 113(26), pp. 6576-6583, ISSN 0006-4971

Spencer, C.T.; Gilchuk, P.; Dragovic, S.M. & Joyce, S. (2010). Minor histocompatibility antigens: presentation principles, recognition logic and the potential for a healing hand. *Current Opinion in Organ Transplantation,* 15(4), pp. 512-525, ISSN 1087-2418

Suárez-Alvarez, B.; Rodriguez, R.M.; Calvanese, V.; Blanco-Gelaz, M.A.; Suhr, S.T.; Ortega, F.; Otero, J.; Cibelli, J.B.; Moore, H.; Fraga, M.F. & López-Larrea, C. (2010). Epigenetic mechanisms regulate MHC and antigen processing molecules in human embryonic and induced pluripotent stem cells. *PLoS One,* 5(4),e10192, ISSN 1932-6203

Swijnenburg, R.J.; Schrepfer, S.; Govaert, J.A.; Cao, F.; Ransohoff, K.; Sheikh, A.Y.; Haddad, M.; Connolly, A.J.; Davis, M.M.; Robbins, R.C. & Wu, J.C. (2008). Immunosuppressive therapy mitigates immunological rejection of human embryonic stem cell xenografts. *Proceedings of the National Academy of Sciences USA,* 105(35), pp. 12991-12996, ISSN 0027-8424

Taylor, C.J.; Bolton, E.M.; Pocock, S.; Sharples, L.D.; Pedersen, R.A. & Bradley, J.A. (2005). Banking on human embryonic stem cells: estimating the number of donor cell lines needed for HLA matching. *Lancet,* 366(9502), pp. 2019-2025, ISSN 0140-6736

Tian, X.; Woll, P.S.; Morris, J.K.; Linehan, J.L. & Kaufman, D.S. (2006). Hematopoietic engraftment of human embryonic stem cell-derived cells is regulated by recipient innate immunity. *Stem Cells,* 24(5), pp. 1370-1380, ISSN 1549-4918

Thomson, J.A.; Itskovitz-Eldor, J.; Shapiro, S.S.; Waknitz, M.A.; Swiergiel, J.J.; Marshall, V.S. & Jones, J.M. (1998). Embryonic stem cell lines derived from human blastocysts. *Science,* 282, pp. 1145–1147, ISSN 0036-8075

Tilburgs, T.; Scherjon, S.A. & Claas, F.H. (2010). Major histocompatibility complex (MHC)-mediated immune regulation of decidual leukocytes at the fetal-maternal interface. *Journal of Reproductive Immunology,* 85(1), pp. 58-62, ISSN 0165-0378

Tolar, J.; Nauta, A.J.; Osborn, M.J.; Panoskaltsis Mortari, A.; McElmurry, R.T.; Bell, S.; Xia, L.; Zhou, N.; Riddle, M.; Schroeder, T.M.; Westendorf, J.J.; McIvor, R.S.; Hogendoorn, P.C.; Szuhai, K.; Oseth, L.; Hirsch, B.; Yant, S.R.; Kay, M.A.; Peister, A.; Prockop, D.J.; Fibbe, W.E. & Blazar, B.R. (2007). Sarcoma derived from cultured mesenchymal stem cells. *Stem Cells,* 25, pp. 371–379, ISSN 1549-4918

Tormin, A,; Li, O.; Brune, J.C.; Walsh, S.; Schütz, B.; Ehinger, M.; Ditzel, N.; Kassem, M. & Scheding, S. (2011). CD146 expression on primary nonhematopoietic bone marrow stem cells is correlated with in situ localization. *Blood,* 117(19), pp. 5067-5077, ISSN 0006-4971

Trento, C. & Dazzi, F. (2010). Mesenchymal stem cells and innate tolerance: biology and clinical applications. *Swiss Medical Weekly,* 140,w13121, ISSN 1424-7860

Tse, W.T.; Pendleton, J.D.; Beyer, W.M.; Egalka, M.C. & Guinan, E.C. (2003). Suppression of allogeneic T-cell proliferation by human marrow stromal cells: implications in transplantation. *Transplantation,* 75(3), pp. 389-397, ISSN 0041-1337

Tu, Z. ; Li, Q. ; Bu, H. & Lin, F. (2010). Mesenchymal stem cells inhibit complement activation by secreting factor H. *Stem Cells & Development,* 19(11), pp. 1803-1809, ISSN 1547-3287

Tögel, F. & Westenfelder, C. (2007). Adult bone marrow-derived stem cells for organ regeneration and repair. *Developmental Dynamics,* 236(12), pp. 3321-3331, ISSN 1097-0177

Uccelli, A.; Moretta, L. & Pistoia, V. (2008) Mesenchymal stem cells in health and disease. *Nature Reviews of Immunology,* 8(9), pp. 726-736, ISSN 1474-1733

Uchanska-Ziegler, B. & Ziegler, A. (2007). On the reactivity of monoclonal antibodies specific for different forms of HLA class I molecules. *Rheumatology (Oxford),* 46(3), pp. 555-556, ISSN 1462-0332

Vanikar, A.V.; Dave, S.D.; Thakkar, U.G. & Trivedi, H.L. (2010). Cotransplantation of adipose tissue-derived insulin-secreting mesenchymal stem cells and hematopoietic stem cells: a novel therapy for insulin-dependent diabetes mellitus. *Stem Cells International,* 2010, pp. 582382, ISSN 1687-9678

Verloes, A.; Van de Velde, H.; LeMaoult, J.; Mateizel, I.; Cauffman, G.; Horn, P.A.; Carosella, E.D.; Devroey, P.; De Waele, M.; Rebmann, V. & Vercammen, M. (2011). HLA-G expression in human embryonic stem cells and preimplantation embryos. *Journal of Immunology,* 186(4), pp. 2663-2671, ISSN 0014-2980

Wonderlich, J.; Shearer, G.; Livingstone, A. & Brooks, A. (2006). assays for T cell function. Induction and measurement of cytotoxic T lymphocyte activity. Current Protocols in Immunology, 3.11.1-3.11.23, ISSN 1934-3671

Wu, Y.; Zhao, R.C. & Tredget, E.E. (2010). Concise review: bone marrow-derived stem/progenitor cells in cutaneous repair and regeneration. *Stem Cells,* 28(5), pp. 905-915, ISSN 1549-4918

Yagi, H.; Soto-Gutierrez, A.; Parekkadan, B.; Kitagawa, Y.; Tompkins, R.G.; Kobayashi, N. & Yarmush, M.L. (2010). Mesenchymal stem cells: Mechanisms of immunomodulation and homing. *Cell Transplantation,* 19(6), pp. 667-679, ISSN 0963-6897

Yamanaka, S. & Blau, H. (2010). Nuclear reprogramming to a pluripotent state by three approaches. *Nature,* 465, pp. 704-712, ISSN 0028-0836

Yañez, R.; Oviedo, A.; Aldea, M.; Bueren, J.A. & Lamana, M.L. (2010). Prostaglandin E2 plays a key role in the immunosuppressive properties of adipose and bone marrow tissue-derived mesenchymal stromal cells. *Experimental Cell Research,* 316(19), pp. 3109-3123, ISSN 0014-4827

Yen, B.L.; Chang, C.J.; Liu, K.J.; Chen, Y.C.; Hu, H.I.; Bai, C.H. & Yen, M.L. (2009). Brief report--human embryonic stem cell-derived mesenchymal progenitors possess strong immunosuppressive effects toward natural killer cells as well as T lymphocytes. *Stem Cells* 27(2), pp. 451-456, ISSN 1549-4918

Yoo, K.H.; Jang, I.K.; Lee, M.W.; Kim, H.E.; Yang, M.S.; Eom, Y.; Lee, J.E.; Kim, Y.J.; Yang, S.K.; Jung, H.L.; Sung, K.W.; Kim, C.W. & Koo, H.H. (2009). Comparison of immunomodulatory properties of mesenchymal stem cells derived from adult human tissues. *Cellular Immunology,* 259(2), pp. 150-156, ISSN 0008-8749

Young, H.E.; Steele, T.A.; Bray, R.A.; Detmer, K.; Blake, L.W.; Lucas, P.W. & Black, A.C. Jr. (1999). Human pluripotent and progenitor cells display cell surface cluster differentiation markers CD10, CD13, CD56, and MHC class-I. *Proceedings of the Society for Experimental Biology & Medicine,* 221(1), pp. 63-71, ISSN 0037-9727

Zhao, T. ; Zhang, Z.N.; Rong, Z. & Xu, Y. (2011). Immunogenicity of induced pluripotent stem cells. *Nature,* 474(7350), pp. 212-215, ISSN 0028-0836

6

Stem Cell Culture Collection - Promising Strategy for Animal Genetic Resource Preservation

Weijun Guan, Xiangchen Li, Dapeng Jin,
Xiaohong He, Yabin Pu, Qianjun Zhao, Taofeng Lu,
Chunyu Bai, Shen Wu, Xiaohua Su and Yuehui Ma
Institute of Animal Sciences, Chinese Academy of Agricultural Sciences, Beijing 100193
PR China

1. Introduction

With the continuous increase of world population, intensified industrial activities, and aggravating environmental pollution, biodiversity is severely endangered to an unprecedented extent. Animal resources, a basis of agriculture and the whole society closely related to living and production, supply human beings with meat, eggs, milk, furs, medicinal materials, products for athletic and ornamental purposes, etc. In most developed countries, massive feeding is restricted within a limited number of high yield breeds or crossbreeds for an intensified operating system of animal husbandry, virtually reducing the variety of local animal breeds. In the meanwhile, despite the existence of enormous animal genetic resources, the lack of efficient preservation strategies and blind introduction of exotic breeds for hybridization have significantly compromised the diversity. As a result, only a few high-yield breeds and hybrids are made more widespread, and gradually supersede indigenous breeds, therefore leading to a shrunken genetic resource pool, progressive narrowing of genetic variation and subsequent crisis of genetic treasures. Nowadays, the livestock and poultry breeds are disappearing at the speed of 1 to 2 per week, so it's definitely far-reaching to explore an efficient and reasonable preservation method for the development of animal husbandry, utilization of animal resources and ecological balance.

As evolution has it, livestock and poultry breeds, the best narration of human labour, diet, religion and customs, is culturally a tangible carrier of civilization vicissitudes (Zhang, 2003). Those animal breeds with precious genome, physiological characteristics, disease resistance, adaptability, and so forth, serve as ideal research models. Moreover, the animal biodiversity provides abundant original materials for the thremmatologists, create infinite selection possibilities, reduce the risks and challenge of animal husbandry, and enhance its interior tenacity and exterior opportunities, thereby enabling people to handle environmental and marketing changes, and invigorating its long term development.

China has the most abundant genetic resources of livestock and poultry, featuring balanced breed range and characteristic distinction. Some genetic and phenotypic properties par excellence, such as adaptability, hardiness, fecundity, etc., are essentially the outcome of thousands-of-year interaction between natural environment and artificial breeding.

According to the statistics from Food and Agriculture Organization (FAO), among the 3019 breeds of livestock and poultry all around the world, one third is located in Asia, of which china accounts for a half.

A survey of genetic resources and the assessment of "Chinese Committee of Livestock and Poultry Evaluation" in 2001 reported that the animal genetic resources of China involve chicken (*Gallus gallus*), duck (*Anseriformes Anatidae*), goose (*Anser cygnoides orientalis*), sandpiper (*Scolopacidae*), cattle (*Bos taurus*), sheep (*Ovis aries*), goat (*Capra hircus*), pig (*Suidae*), ferret-polecat (*Mustela Pulourius Furot*), raccoon dog (*Nyctereutes procyonoides*), horse (*Equus caballus*), red deer (*Cervus elaphus*), sika deer (*Cervus nippon*) bactrian camel (*Camelus bactrianus*), etc., which, altogether amount to some 20 species, 576 breeds (426 indigenous breeds, 73 fostered ones and 77 introduced ones) (Liu et al., 2004). However, in these years, the genetic resources are shrinking considerably due to the brunt of the massive introduction of high yield breeds. Data in the 70's and 80's suggested that within China, 10 indigenous breeds are disappearing, 8 are on the edge of extinction, and 20 are decreasing rapidly. Furthermore, a recent research suggested that more than 50% of local breed populations are extremely reduced, and that a large number of the rest are severely endangered (Ma et al., 2001).

In containing the huge loss of animal genetic resources, the preservation procedure has become a very concern of more and more researchers. The most important is to protect existential genetic materials from adulterating and extinction in a comprehensive and proper manner, which virtually means to preserve available genetic resources as integrated as possible, no matter whether there are application potentials from current perspective. For the preservation of population genome, there are optional forms, e.g. individuals, organs, semen, embryos, cell strains, genomic libraries, and cDNA libraries. It's noteworthy that the above-mentioned methods all have their defects, so appropriate strategies should be devised to fit in with specific species. A new preservation protocol using stem cells, is both novel and complementary to the existing multi-approach tactics, and thus will become a major technique in a long term. With the strenuous efforts scientists have ever made, preservation media of semen, embryos, cell strains, genomic libraries and natural reserves are primarily shaped. In contrast, preservation via stem cells is still lacking, which apparently has its distinguished advantages.

Stem cells can be categorized into embryonic stem cells (ESCs) and adult stem cells (ASCs) by origins. ESCs derived from inner cell mass have totipotency and continuous self-renewal ability, and therefore are widely believed as the stem cells with the most therapeutic and research values. ASCs are ubiquitous in almost every organ of adult animals, to maintain the structural and functional homeostasis. The applications of ESCs in clinical therapies are open to doubt, mainly for ethnic reasons, propelling people to resort to the more applicable ASCs. Emerging evidence on the plasticity of ASCs and the presence of multipotent stem cells in adult tissues deepened the comprehension of their developmental repertoire.

For the preservation of animal genetic resources, stem cells, by virtue of their potent self-renewal ability, can provide a large amount of serviceable cells with relatively small volume. Meanwhile, the plasticity of stem cells confers them more advantages in applications, for instance, in nuclear transfer. Stem cell cryopreservation is not only an efficient and safe strategy for the maintenance of animal genetic resources, but also promising to show scientific values in other fields of research.

This chapter will introduce the preservation of animal genetic resources in terms of animal cells and its applications by detailed experimental description.

2. Isolation, *in vitro* culture and identification of stem cell lines

Preservation of animal genetic resources in terms of stem cells is essentially to store as many purified cells as possible, which impose strict criteria on the *in vitro* culture and identification of stem cells with various origins. Therefore, it plays a crucial role in the entire technical system to fulfil efficient and high quality culture and purification.

2.1 *In vitro* culture

2.1.1 Sampling

Stem cells are widely distributed in a variety of tissues and organs, thus it is of great importance to pinpoint and dissect the parts where the most stem cells populated. For instance, ESCs is located precisely in inner cell mass, while adult neural stem cells mainly reside in subventricular zone and hippocampal dentate gyrus. In addition, sterile operation and quick isolation are items of very concern as well, e.g. proper sterilization during sampling, hypothermal transportation, etc.

2.1.2 Isolation

Stem cells are unhomogenously distributed *in vivo,* which requires a lot to isolate as pure stem cells as possible using various available methods. Strategies for stem cell isolation of common types are as listed in Table 1 (except specially indicated otherwise, all the methods are primarily for avian species):

2.1.3 *In vitro* cell culture

Primary culture is the first step of cells into *ex vivo* environment. The time of this phase is contingent upon the adaptability of different stem cells, which, in turn, is just the basis for further purification. When primary cells grow to a certain density, leading to contact inhibition, along comes subculture process. During all this progress, most stem cells have got used to new circumstances, and go on rapid proliferation, reentering proliferation - inhibition -proliferation cycles.

Because of the distinctive characteristics, each type of stem cells needs specific factors in its niche, e.g. epidermal growth factor (EGF), basic fibroblast growth factor (bFGF), glutamine, etc. In addition, feeder layer is necessary for successful culture of PGCs.

2.1.4 Cryopreservation

The purpose of *in vitro* stem cell culture is to conserve animal genetic resources, for which cyropreservation is optimal at cell level. Repetitive tests confirmed that cryonics of stem cells don't differ from those of somatic cells. However, for the preciousness of stem cells, serum concentration is elevated to an appropriate extent, which is to say, the cryogenic media are composed of 50% basic media, 40% serum and 10% cryoprotectant (DMSO), fundamentally capable of ensuring the viability of resuscitated cells.

2.2 Identification

Stem cells preserved as genetic resources should be subjected to evaluation of at least two major respects. One is assessment of general biological characteristics. The other is about stem cell properties in terms of specific markers, self renewal and plasticity.

Cell type	Isolation methods	Protocols
Bone marrow mesenchymal stem cells (MSCs)	Total blood adherent method	Bone marrow was suspended into 10 ml serum-free L-DMEM medium containing 100 IU/ml penicillin and 100 μg/ml streptomycin using syringe, and then was pipetted into cell suspension with 4# needle gently. The cell suspension was centrifuged at 1000 rpm for 10 min, the top fat impurities were removed. The bottom cells were harvested and washed twice using serum-free L-DMEM medium, resuspended with complete medium, and subsequently plated into a culture flask with 10 ml complete medium.
	Density gradient centrifugation method	Bone marrow single cell suspension was prepared as above, and gently added into a 10 ml centrifuge tube with isovolumic, 1.073 g/ml percoll solution underneath. Then the nebulous white ring on the interface of percoll and cell suspension was pipetted out and washed twice using L-DMEM and centrifuged for 5 min at 1 000 rpm. After counting, these cells were plated into flasks at $2 \times 10^5/cm^2$, and cultured at 37 °C, 5% CO_2.
Primordial germ cells (PGCs)	Trypsinization method	PGCs were retrieved from the embryonic gonads incubated at 38 °C and 60% humidity for 5.5 days. After rinsing 3 times with PBS to remove residual yolk, gonadal tissues were collected carefully with sharp tweezers under a microsurgery microscope, and then dissociated in 0.25% trypsin-0.02% EDTA at room temperature (RT) for 5 min. After inactivation of the trypsin-EDTA with DMEM containing 15% FBS, the cells were harvested by centrifugation (Zhang, 2003). These cells were plated into flasks at $2 \times 10^5/cm^2$, and cultured at 37 °C, 5% CO_2.
Adipose derived stem cells (ADSCs)	Collagenase method	Adipose tissues were separated from subcutaneous tissues of abdomen and inguinal fat pads of 1-day newborns. All the operation steps were conducted under aseptic condition. The tissues were washed 3 times with PBS containing 100 IU/mL penicillin/streptomycin to remove connective tissue membrane and capillaries. The tissues were chopped into small pieces, and digested with 0.1% (m/v) type I collagenase at 37 °C for 1 hr. Enzymatic digestion was then neutralized with DMEM (Gibco, USA) supplemented with 10% (v/v) FBS (Biochrom, Germany). The suspension was filtered with 74-μm-mesh sieve, and centrifuged at 300g for 10 min. Then the pellet was resuspended with complete medium containing DMEM/F-12 Ham's (Gibco), 10% (v/v) FBS (Biochrom), 10 ng/ml bFGF (Peprotech, USA), 2 mM L-glutamine, 1% B-27 (m/v) (Gibco) and 100 IU/mL penicillin/streptomycin. The cell suspension was plated and incubated at 37 °C with 5% CO_2.

Cell type	Isolation methods	Protocols
Skeletal muscle satellite cells (SCs)	Collagenase method	Skeletal muscles were isolated from embryos and chopped into pieces using ophthalmic scissors. The comminuted tissues were disaggregated by combinatorial digestion with 0.1% collagenase I for 30 min and 0.25% trypsin for 1 h. Then add DMEM medium containing 20% FBS to terminate reaction. The cell suspension was centrifuged at 1,500 rpm for 8 min with the supernatant discarded, whereafter the cells were resuspended with complete medium(DMEM/F12 +20% FBS+ 2.5 ng/ml bFGF) and plated into flasks. Cells were cultured in 5% CO_2 incubator at 37 °C for 2h, and then plate the cell suspension to petri dishes, to continue culturing at 37 °C, in 5% CO_2 (Qu et al., 1998).
Neural stem cells (NSCs)	Mechanical isolation	Embryonic brains were isolated and rinsed 3 times and then placed in precooled normal saline water. The dorsal ventricular ridges of the brain were isolated, rinsed, and transferred to complete neural stem cell media 1:1 DMEM/F12 (Gibco, Carlsbad, CA), 2% B27 supplement (Gibco), 20 ng/mL of EGF and bFGF (PeproTech, Rocky Hill, NJ), 100 IU/ml penicillin/streptomycin, cleaved into 1.0 mm^3 pieces, and pipetted repeatedly to prepare a homogeneous monoblast suspension, which was subsequently filtered through 400- and 800-mesh sieves in order. The entire operation was performed under a low temperature to protect the cortex tissues. The cells were plated in flasks at a concentration of 2×10^5 cells/mL and were cultured in a humidified incubator with 5% CO_2 at 37 °C.

Table 1. Isolation methods of several types of stem cells

General biological characteristics include hereditary stability (karyotyping), growth dynamics (growth curve), microbial detection, cross-contamination detection, viabilty before and after cryopreservation and the expression of exogenous genes. As for stem cell nature, specific markers are detected via immunofluorescence and immunochemistry, RT-PCR assay, Western blotting, etc. Self-renewal is evaluated using clonogenic assay. To verify the plasticity, the stem cells are induced for multi-lineage differentiation, which are then identified functionally. The fundamental principles will be introduced in detail in the following paragraphs.

2.2.1 Growth dynamics

Following Bai's method (2010), the stem cells were plated in 24-well plates at a concentration of 1×10^4 cells/well and cultured for 9 days (Bai et al., 2010). The cell concentration was counted using hematometer and then recorded from 3 wells per day until the plateau phase was reached. The growth curve was plotted and the population doubling time (PDT) was calculated accordingly. The formula is as follows:

$$PDT=(t-t_0)\ lg2/\ (lgN_t-lgN_0)$$

t_0: the initiating time of culture; t: the end time of culture; N_0: the cell numbers of initiating culture; N_t: the cell number of end culture.

2.2.2 Microbial detection

Detection of bacteria and fungi:

The cells were cultured in DMEM containing 10% fetal bovine serum without antibiotics and tested for the presence of microbes 3 days after subculture according to the method of Doyle et al. (1990).

Mycoplasma detection:

The cells were cultured in medium free of antibiotics for at least one week and then fixed and stained with Hoechst 33258 according to Masover (1998) and Freshney's method (2000). Results of DNA staining were confirmed by ELISA using the ELISA Mycoplasma Detection kit (Roche, Lewes, East Sussex, UK.), which can identify the four most common Mycoplasma species: *M .arginini, M. hyorhinis, A .laidlawii,* and *M. orale.*

Virus detection:

Routine examination for cytopathogenic effects using phase-contrast microscopy was performed according to Hay's haemadsorption protocol (Hay, 1992).

2.2.3 Cryopreservation and resuscitation

Cells were cultured in fresh medium 24 h prior to cryopreservation to ensure sufficient nutrition and optimal cellular condition. The monoplast suspension was prepared by dissociating cells in 0.25% (m/v) Trypsin. The suspension was centrifuged at 1000rpm for 8 min and the supernatant was discarded. Then, the cells were resuspended at a density of approximately 4×10^6/mL in freezing media of 10% dimethyl sulfoxide (DMSO), 40% FBS and 50% DMEM, and then subpackaged in cryovials which labeled the species, breeding, gender, date and serial numbers. The vials were placed at 4°C for 20-30 min to enable the DMSO to reach equilibrium, and then placed in liquid nitrogen for long term storage (Ren et al., 2002). For resuscitation, they were placed in prewarmed water bath at 42 °C. As soon as it was nearly thawed, the pellet and suspension were transferred into a sterile tube containing DMEM and centrifuged at 1000 rpm for 10 min to remove DMSO. The cells were then resuspended in fresh DMEM and plated onto petri dishes, and cultured in 5% CO_2, 37 °C. Medium should be refreshed after 24 h (Ren et al., 2002; Freshney, 2000).

2.2.4 Karyotyping

Metaphase spreads were prepared from cells at exponential phase following treatment with 0.1 μg/mL colcemid (Gibco/BRL). The cells were treated with a hypotonic solution (KCl/HEPES/EDTA) and harvested according to standard dissociation procedures. Slides of fixed cells were Giemsa banded to identify individual metaphase chromosomes. Representative chromosome sets were photographed and analyzed. The percent of diploid was counted from 100 cells. Karyotypes were processed following the protocol described in the Reading Conference report (Ford et al., 1980).
These chromosomal parameters were calculated using the formulas:

Arm ratio =long arm length (q) vs short arm length (p)

Centromere index =short arm length vs chromosomal length

Relative length=single chromosomal length vs (total autosome lengths + X-chromosome length)

2.2.5 Expression of exogenous genes

According to the method described by Tsuchiya et al. (2002), the same quantity of the fluorescent protein vectors pEGFP-N3, pDsRed-N1 and pEYFP-N1 were transfected into the stem cells with Lipofectamine™ 2000 transfection reagent (Invitrogen Corp, Carlsbad, CA). The plasmid DNA (μg) to Lipofectamine 2000 (μl) ratio was 1:3. After 8 h, the cells were removed from non-serum medium and transferred to serum containing medium. Cell morphology was observed, and the cells were dyed with Trypan Blue to estimate the viability. The cells were observed after being transfected for 24 h, 48 h and 72 h, respectively, to estimate the transfection efficiency. Cell morphology was observed by confocal microscopy (Nikon TE-2000-E, Japan), and a comparative analysis of expression was made according to the intensity of the different fluorescent proteins in the cell nuclei and cytoplasm. For each individual experiment, images were captured from 10 visual fields, and confocal microscopy was used to measure the total and positive cell counts in each field to determine the transfection efficiency. The mean values were accordingly calculated. Multiple comparisons of the test data were made to analyze the statistical differences (Tsuchiya et al., 2002).

2.2.6 Identification of characteristic markers

Stem cells with various origins possess different markers, providing a major approach to identify their lineages (Table 2). Prior to cryopreservation, it is extremely necessary to find molecular evidence for their identity. These markers are generally detected by three commonly used assays, i.e. immunofluorescence, immunochemistry and RT-PCR.

Cell types	Markers
BMSCs	CD44, ICAM-1, SSEA-4
PGCs	SSEA-1, SSEA-4, TRA-1-60, TRA-1-81
ADSCs	CD29, CD44 , CD71, CD73
SMSCs	Pax7, Desmin, Myod
nscs	Nestin

Table 2. Characteristic markers of some kinds of stem cells

Immunofluorescence

Surface markers of different passages of the stem cells were detected by immunofluorescence. Stem cells were fixed in 4% (m/v) paraformaldehyde (in PBS) for 15-20 min, and then permeabilized for 20 min with methanol containing 0.1% Triton X-100 and 0.3% hydrogen peroxide (H_2O_2) to eliminate endogenetic hydrogen peroxidise. Incubated in goat serum working solution for 30 min to block nonspecific binding, the cells were then incubated with primary antibodies at 4 °C overnight, followed by incubation with secondary antibodies conjugated with FITC. For negative control, 0.01 mol/L PBS was used to replace primary antibodies. Fluorescence images were observed using confocal microscope (Nikon TE-2000-E, Japan). Ten non-overlapped visual fields (x100) were photographed randomly from stem cells of different passages, then the percentage of positive cells to total count of stem cells was calculated and the results were formulated as mean±SD, and subjected to variance analysis using SPSS 10.0 software.

Genes	Primer Sequences	Tm (°C)	Cycle No.	Size (bp)
CD29	F 5' GAACGGACAGATATGCAACGG 3' R 5' TAGAACCAGCAGTCACCAACG 3'	60	30	300
CD44	F 5' CATCGTTGCTGCCCTCCT 3' R 5' ACCGCTACACTCCACTCTTCAT 3'	58	30	290
CD71	F 5' CCCAGGCTTCCCTTCGT 3' R 5' GGGCTCCAATCACAACATAC 3'	56	30	305
CD73	F 5' AGTGCAAACATTAAGGGAAAA 3' R 5' CCTCCAATAACAACATCCACTCCT 3'	58	30	310
Collage type I	F 5' AAGGATGGTCGCAATG 3' R 5' GGTGGCTAAGTCTGAGGT 3'	48.5	30	310
Osteopontin	F 5' CAGAACAGCCGGACTTTC 3' R 5' CTTGCTCGCCTTCACCAC 3'	51	30	227
PPARγ	F 5' CTGTCTGCGATGGATGAT 3' R 5' AATAGGGAGGAGAAGGAG 3'	47.3	30	199
Lipoproteinlipase (LPL)	F 5' AGTGAAGTCAGGCGAAAC 3' R 5' ACAAGGCACCACGATT 3'	48.7	30	477
Desmin	F 5' GGGCTTTCTCCTACCTGC 3' R 5' GCTTCCTTGCCATCCTGT 3'	57	30	240
MyoD1	F 5' GCTACTACACGGAATCACCA 3' R 5' GGGCTCCACTGTCACTCA 3'	57	30	198
GAPDH	F 5' TAAAGGCGAGATGGTGAAAG 3' R 5' ACGCTCCTGGAAGATAGTGAT 3'	53	30	244

Table 3. Primers for RT-PCR assay

RT-PCR assay

RNA was extracted from cells of different passages using Trizol reagent (Invitrogen, USA). Template cDNA was prepared with reverse transcription system (Takara, China) and then amplified by PCR using specific primers listed in Table 3. The PCR products were visualized by 2% (m/v) agarose gel electrophoresis.

2.2.7 Clonogenic assay

Stem cells of different passages were plated in 24-well microplates at the density of 1×10^4 cells per well, cultured for 7 d, and then counted for the numbers of colony-forming units (CFU) to calculate colony-forming rate, which is formulated as CFU number/ plating cell number ×100%.

2.2.8 Induced differentiation and identification

Induced differentiation

Stem cells are characterized by the potentials to escape cell cycle and to differentiate into terminal cells upon exposure to inducing media. Therefore, their plasticity is one important aspect which is a constitutional factor to evaluate for the sake of genetic preservation.

Osteogenic differentiation

The stem cells of 80% confluence were divided into two groups. The induction group was incubated in osteogenic media (β-sodium glycerophosphate, dexamethasone, vitamin C) containing osteoblasts. The control group were incubated in the same inducing medium without osteoblasts. Culture medium was changed every 3 days. Two weeks later, alkaline phosphatase levels were measured by the Gomori Ca-Co method. Three weeks later, Alizarin Red staining was used to detect calcium nodules. Four weeks later, Von Kossa's method and tetracycline fluorescence labeling of calcium were used to determine calcium nodules (Li et al., 2009).

Adipogenic differentiation

Stem cells were plated and divided into 2 groups as above mentioned. When the cells grew to 50%-60% confluence, the induced group was incubated in adipogenic medium supplemented with dexamethasone (Sigma), isobutyl-methylxanthine (IBMX; Sigma), and insulin (Sigma), while the control group was still cultured in complete medium. After 3 weeks, the two groups were stained with Oil Red O to assess intracellular lipid accumulation. The RNA from the two groups was extracted for further RT-PCR assay.

Neurogenic differentiation

The preparation of stem cells was the same as above mentioned. Stem cells in the induction group were induced with medium containing 20% fetal bovine serum and β -mercaptoethanol (BME, Sigma, USA) for 24 h, washed thrice with PBS, and then induced with serum-free medium containing dimethyl sulphoxide (DMSO, Sigma) and butylated hydroxyanisole (BHA, Sigma). Stem cells in the control group were incubated with normal culture medium. The neurogenic differentiation was then detected using immunofluorescence and observed under confocal microscope (Nikon TE-2000-E, Japan). Ten non-overlapped visual fields (×100) were randomized from induced cells, followed by the same data processing as previously mentioned.

Cardiomyogenic differentiation

Cells were plated and divided into 2 groups as above mentioned. The induced group was incubated in serum-free cardiomyogenic medium containing 5-Azacytidine (5-aza; Sigma) for 24 h, and then the medium was replaced with normal culture medium. After 28 days, the cells were harvested and the RNA from the two groups was extracted for further RT-PCR assays.

2.3 Case study

The Animal Population Culture Collection of China (APCCC) has been making efforts to preserve animal genetic resources in terms of stem cells, which involve bone marrow mesenchymal stem cells (MSCs), primordial germ cells (PGCs), adipose derived stem cells (ADSCs), skeletal muscle satellite cells (SCs), neural stem cells (NSCs), etc. Now they will be exemplified one by one.

2.3.1 Evaluation of general biological indices

The stem cells with different origins display fusiform or round shapes and swirl-like or sphere-like patterns, and most of them have plump cytoplasm, one of the indications of good vitality (Fig. 1).

The growth curve of stem cells typically display typical "S" shapes, which are composed of latency phase, exponential growth phase and stationary phase, based on which PDT is calculated as a reflection of proliferative activity (Fig. 2). It's also worth mentioning that there are slight differences among different passages.

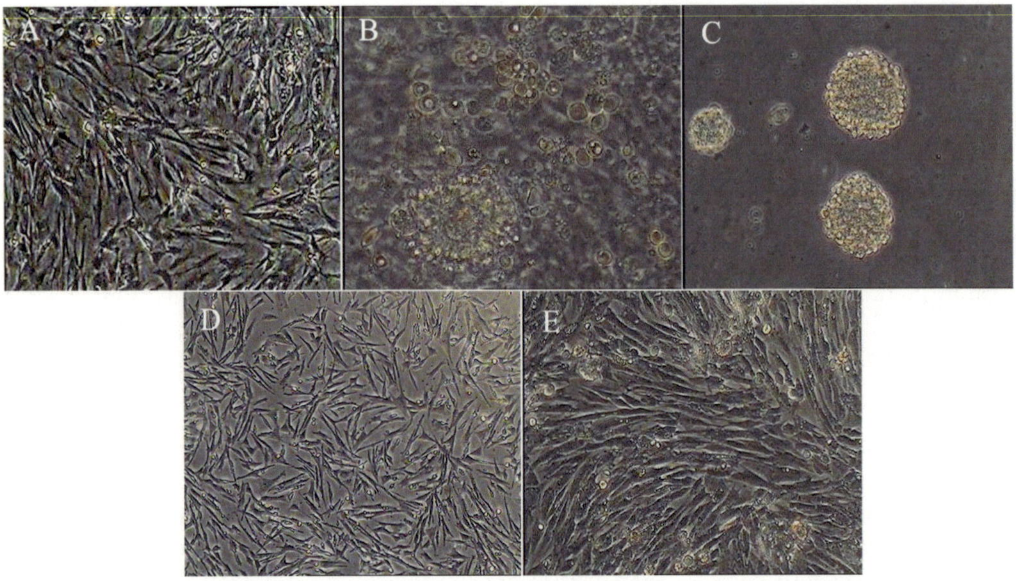

Fig. 1. Morphology of (A) duck bone marrow MSCs; (B) chicken PGCs; (C) duck NSPCs; (D) chicken ADSCs; (E) chicken skeletal muscle SCs.

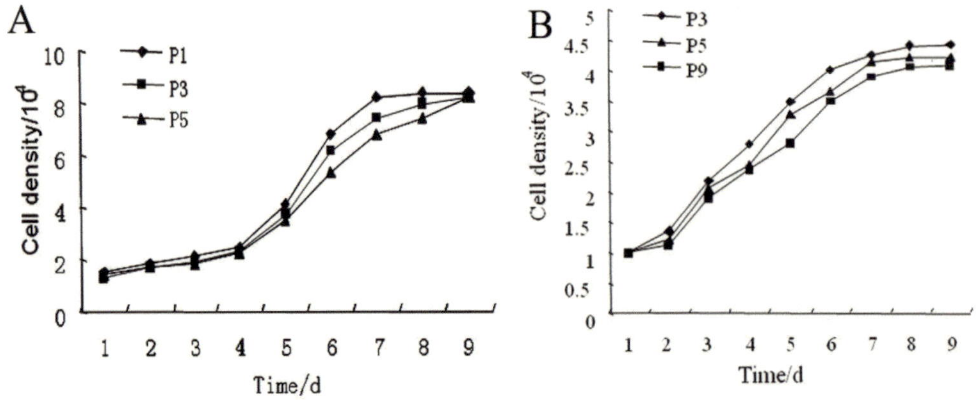

Fig. 2. Growth curves of (A) duck bone marrow MSCs of passages 1, 3 and 5; and (B) chicken ADSCs of passages 3, 5 and 9. The growth curve of the different passages of duck MSCs and chicken ADSCs display typical "S" shapes, which are composed of latency phase, exponential growth phase and stationary phase.

In a sharp contrast with infections by bacteria, fungi and yeasts (Fig. 3 B, C and D), characterized by turbidity, colony or hypha which can be observed by unaided eyes, the mycoplasma contamination (Fig. 3 F), usually undistinguishable, is only accompanied with slightly slower growth and increased cell fragmentation. As a result, Hoechst 33258 staining or molecular assays are required further. Therefore, all the stem cells are subjected to microbial detection prior to cryopreservation to ensure they are free of contamination (Fig.3 A and E).

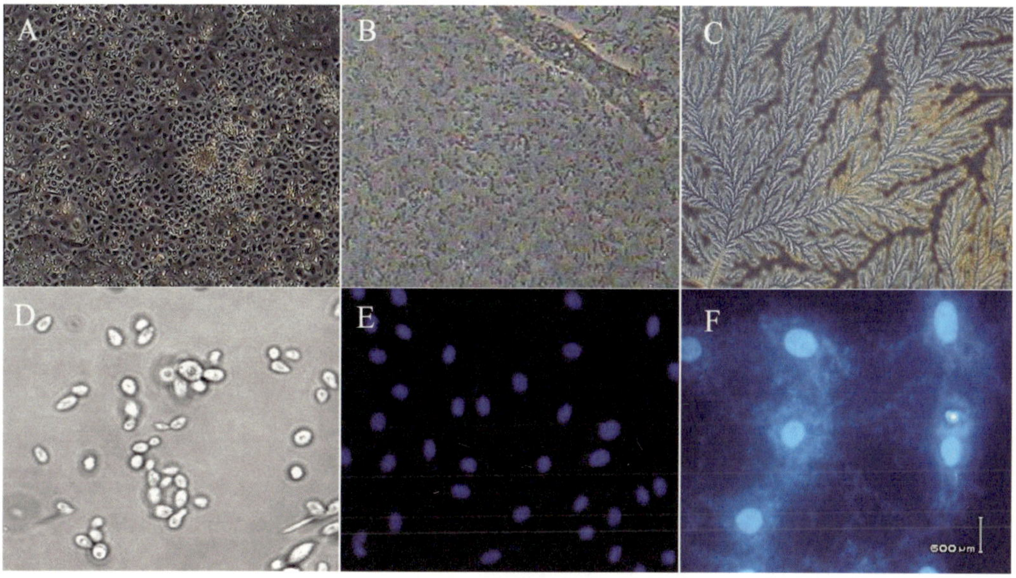

Fig. 3. Microbial detection of duck bone marrow MSCs. (A) normal bone marrow MSCs (40×); positive control infected by (B) bacteria (200×) and (C) fungi (200×); and (D) yeasts (200×); (E) bone marrow MSCs, mycoplasma negative (200×); (F)positive control infected by mycoplasma (400×).

Cells possess a characteristic chromosome number, shape and structure, which remain very stable in normal cells (Fig. 4). Therefore, karyotype analysis is a major method for distinguishing normal cells from mutants. The percentage of diploid cells tends to decrease with increasing passage number. However, the fact that the diploid proportion is normally higher than 90% warrants the hereditary stability.

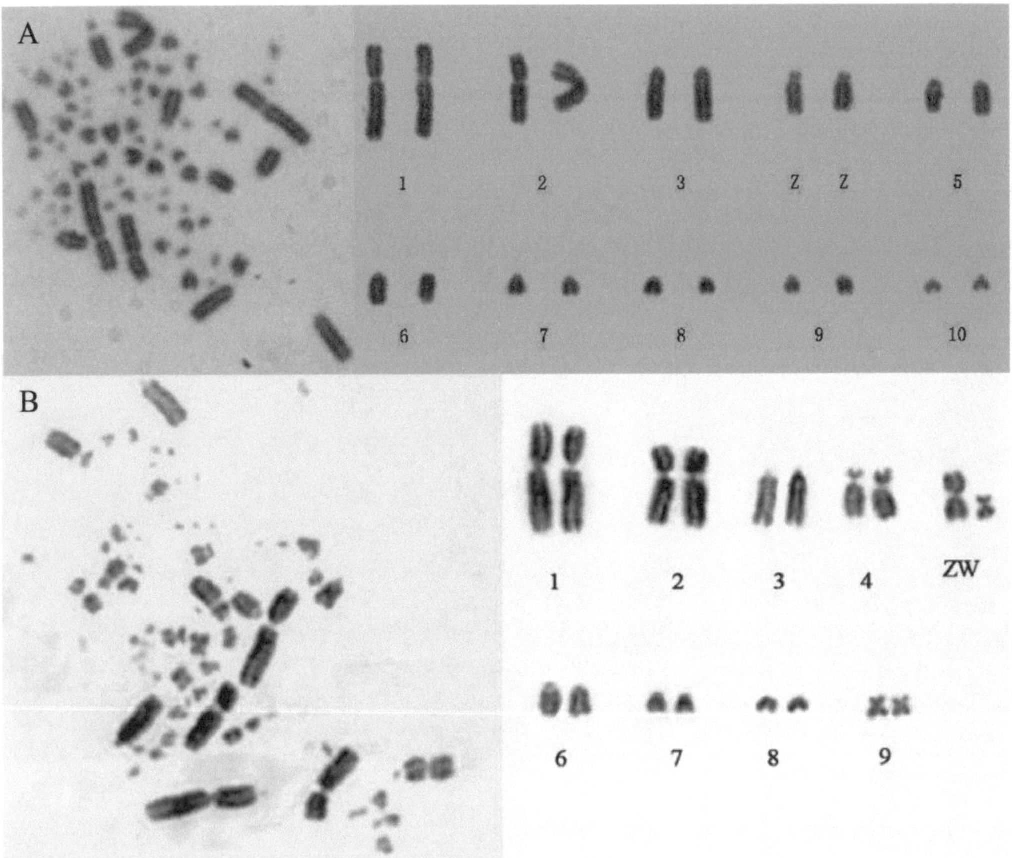

Fig. 4 Representative spreads at metaphase (left) and karyotypes (right) of (A) duck bone marrow MSCs, ZZ type (♂); (B) chicken PGCs, ZW type (♀). The chromosomal number of duck bone marrow MSCs is 78, consisting of 10 pairs of macrochromosomes and 29 pairs of microchromosomes, while that of chicken PGCs (2n=78) is composed of 9 pairs of macrochromosomes and about 30 pairs of microchromosomes.

Fluorescent genes, in light of their stable expression and species-independent efficiency, have long been used as markers to monitor the function and distribution target proteins in live cells and organisms (Heim, 1995). The expression levels of EGFP, EYFP, and DsRed1 are usually maximal at 48 h (Fig. 5). In addition, different fluorescent protein genes may have different transfection efficiency for the same cell line. As for most types of stem cells preserved, the transfection efficiencies of the yellow (pEYFP-N1) and red (pDsRed1-N1) fluorescent protein genes are significantly lower than those of the green fluorescent protein gene (pEGFP-N3).

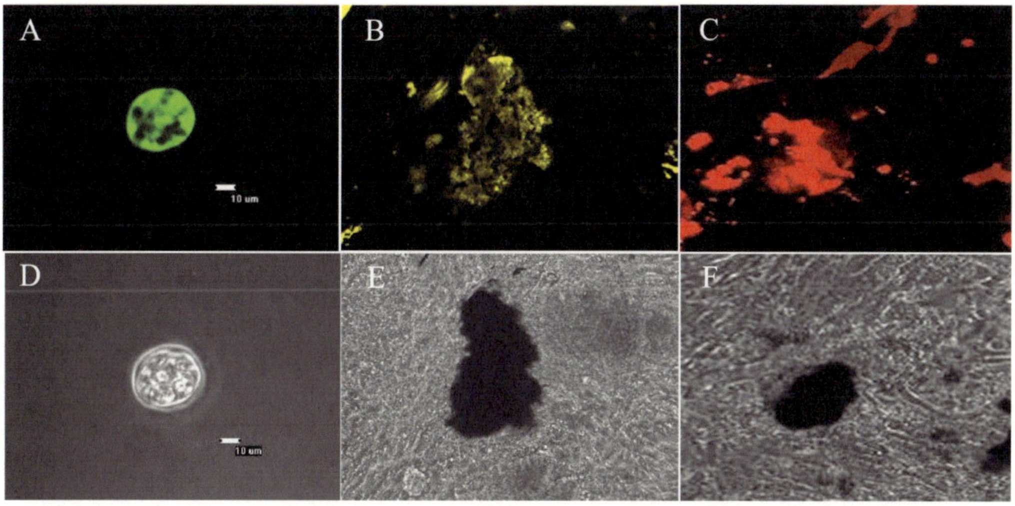

Fig. 5. The expression of (A, D) pEGFP-N3; (B, E) pEYFP-N1; and (C, F) pDsRed-N1 at 48 h in chicken PGCs. The expression of the three types of exogenous fluorescent genes is optimal at 48 h, and the transfected cells exhibit no obvious difference in morphology and proliferation compared with controls. The pEGFP-N3, pEYFP-N1, and pDsRed-N1 refer to plasmids encoding the green, yellow and red fluorescent genes, respectively. Scale bars: 10 µm in A and D, 80 µm in B, C, E and F.

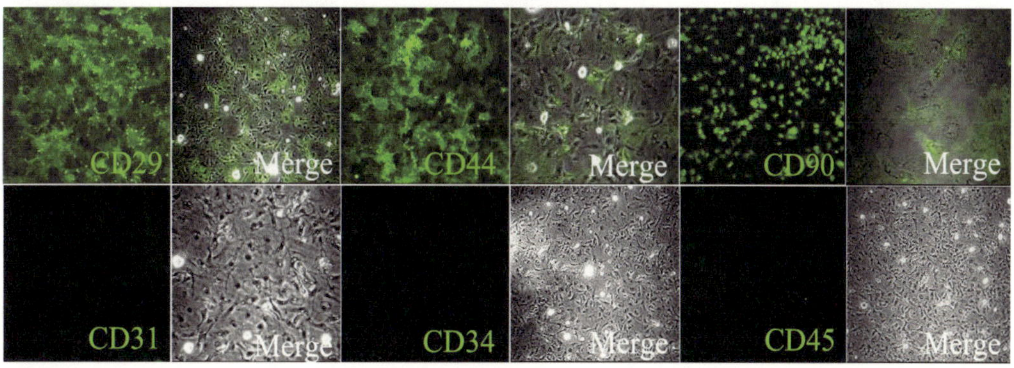

Fig. 6. Surface marker expression of chicken bone marrow MSCs of passage 5 (100×). Chicken bone marrow MSCs express numerous surface markers including CD29, CD44 and CD90, but no hematopietic markers such as CD31, CD34 and CD45.

2.3.2 Stem cell characteristics

Both the characteristic markers and the ability of multi-lineage differentiation are detected in this section, which are indicative of stem cell nature.

Identification of characteristic markers

The specific surface markers of stem cells were detected via immunofluorescence and RT-PCR assay. The results of immunofluorescence staining and RT-PCR assay are as shown in Figs. 6- 9.

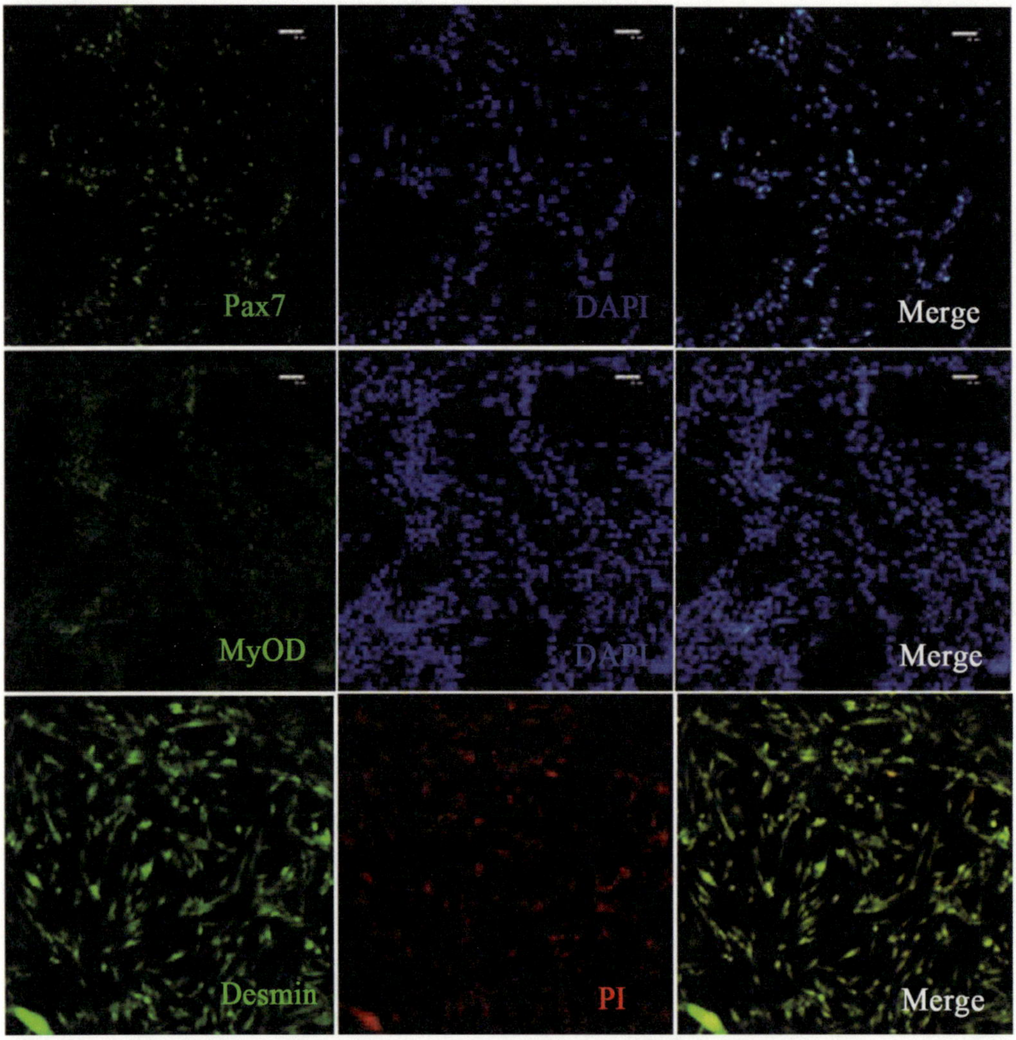

Fig. 7. Identification of chicken skeletal muscle satellite cells. Chicken skeletal muscle satellite cells express Pax7 and MyoD in nucleus, and Desmin in cytoplasm (100×).

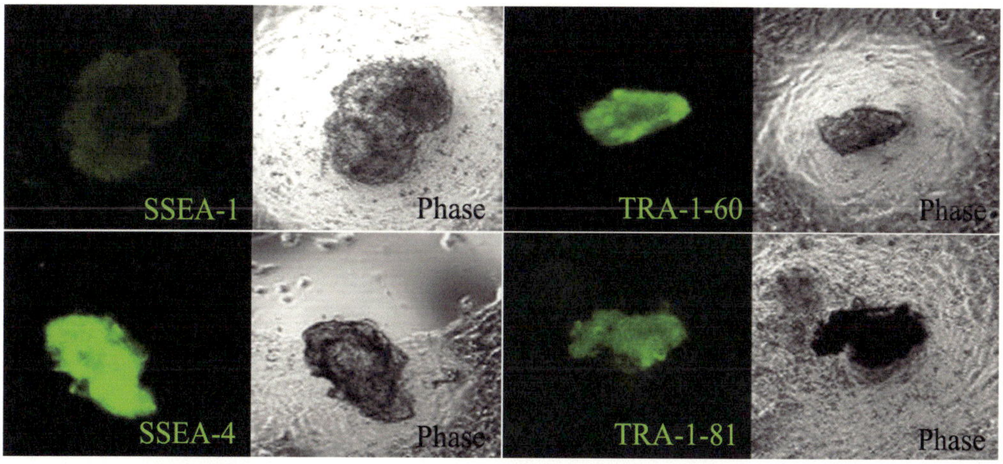

Fig. 8. Identification of colony formation in chicken PGCs of passage 3 with a set of antibodies recognizing specific cell surface antigens. Chicken PGCs express SSEA-1, SSEA-4, TRA-1-60 and TRA-1-81.

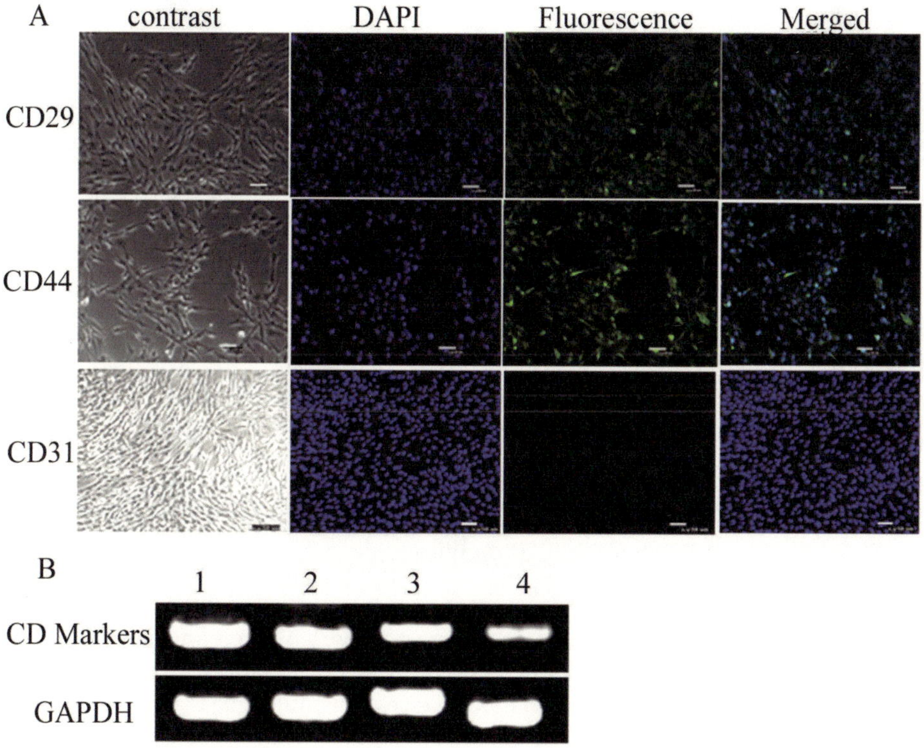

Fig. 9. Surface markers of chicken ADSCs. (a) Immunofluorescence showed that CD29 and CD44 are positively expressed, while CD31 detection is negative. Scale bars=50 μm. (b) RT-PCR analysis shows that the ADSCs express CD29, CD44, CD71 and CD73. In Panel (b), the lanes are in accordance to CD29, CD44, CD71 and CD73 from left to right. GAPDH in the lower picture serves as internal control.

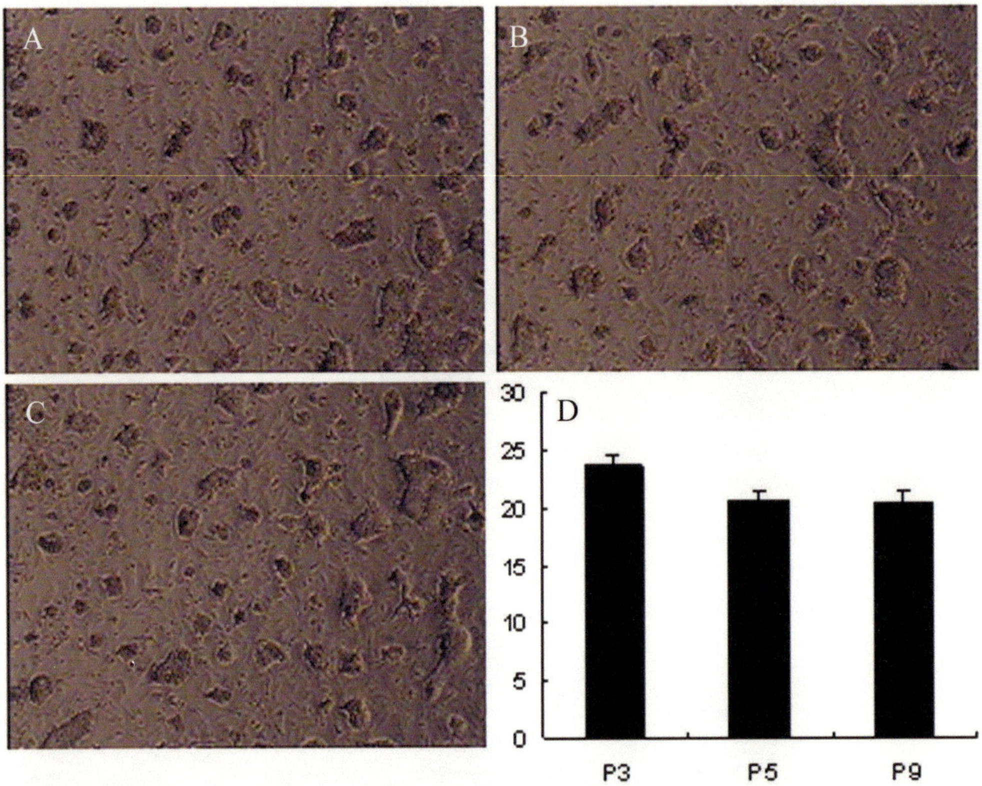

Fig. 10. Colony-forming assay. Colony-forming units of P3, P5 and P9 ADSCs were counted, which indicated that colony-forming rates decreased but didn't disappear with increasing passage number. (A), (B) and (C) are colonies of P3, P5 and P9 respectively, (D) is the bar chart of colony-forming rates for chicken ADSCs of different passages.

Clonogenic assay

The self-renewal of the stem cells was evaluated via clonogenic assay. Colony formation was observed 4 days after plating under the microscope. The colony-forming rates of chicken ADSCs were 23.61±0.14%, 20.54±0.31%, 20.37±0.46% for passages 3, 5 and 9 respectively, demonstrating their self-renewal ability (Fig. 10).

Differentiation detection

The multi-potency of stem cells is one of the most important prerequisites for autologous cell therapy. Therefore, different types of stem cells, e.g. bone marrow MSCs, ADSCs, PGCs, etc., were subjected to induced differentiation to assess the multi-lineage potentials.

Osteogenic differentiation

After incubation in osteogenic medium for about 15 days, morphological changes of the stem cells were obviously observed. The cells changed to tri-dimensional firstly, and then aggregated and formed mineralized nodules with increasing incubation time. Furthermore, the nodules were identified calcium positive. (Figs. 11-12)

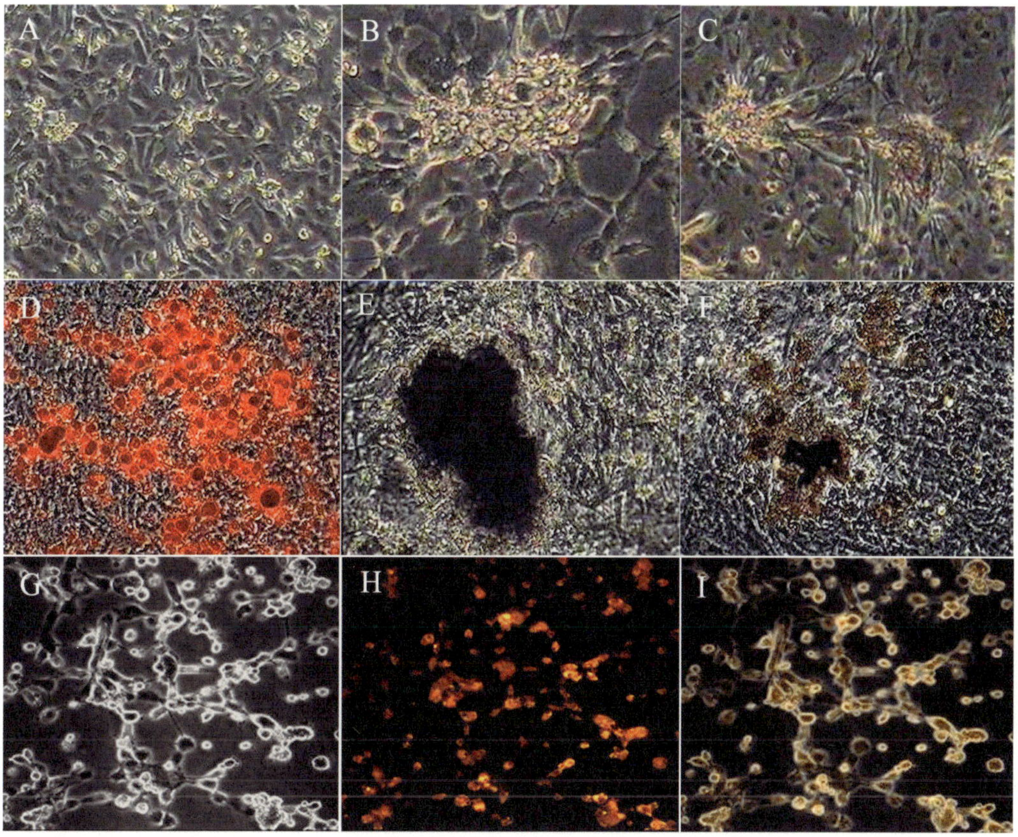

Fig. 11. Identification of osteogenic differentiation of chicken bone marrow MSCs of passage 5. (A) induction group at day 6 (100×); (B) osteogenesis at day 9 (100×); (C) osteogenesis at day 14; (D) alizarin red staining (100×) and (E) AKP identification with the Gomori Ca-Co's method two weeks after osteogenic induction (100×); (F) Von Kossa staining positive four weeks after osteogenic induction, indicative of calcium deposition (100×); (G) phase, (H) fluorescence, and (I) merge of tetracycline labelling at day 21 post osteogenic induction.

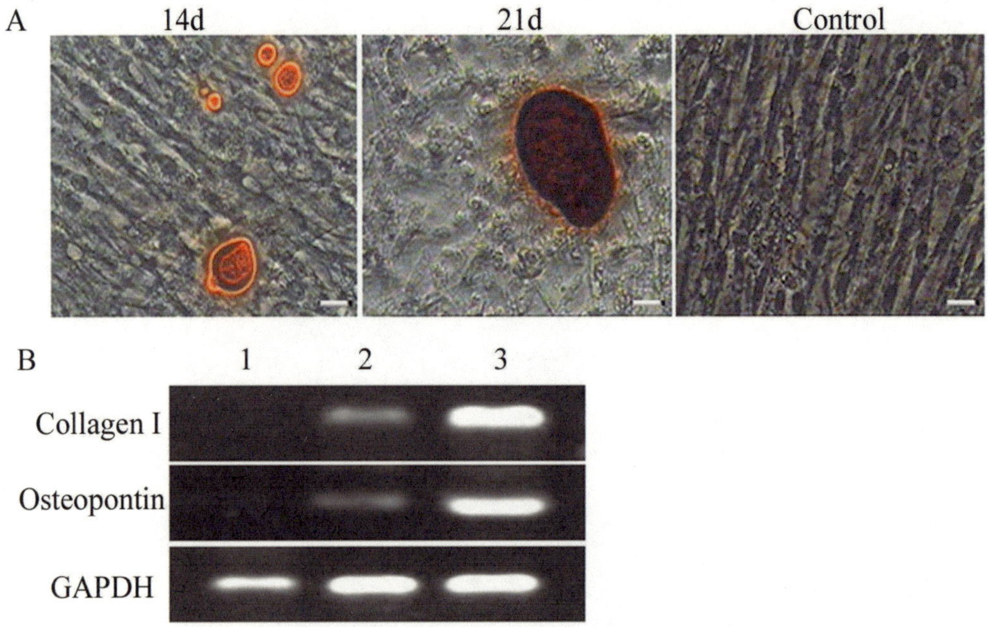

Fig. 12. Osteogenic differentiation of chicken ADSCs. (a) After incubation in osteogenic medium for 14 days, the cells metamorphosed from fusiform to tridimensional shapes, and Alizarin Red staining was positive. The nodules became more and larger with prolonged inducing time. About 21 days later, the nodules were obviously observed following Alizarin Red staining. Cells cultured in complete medium were not influenced in morphology or stained by Alizarin Red. Scale bars=25 μm. (b) RT-PCR assay revealed the expression of osteoblast specific genes, including collage type I and osteopontin in the induced group after incubation for 14 days (Lane 2) and for 21 days (Lane 3); while these genes were not expressed in control (Lane 1).

Adipogenic differentiation

Adipogenic differentiation of the stem cells can be evidenced by positive Oil Red O staining (Jing et al., 2007). After incubation in adipogenic medium for 3 weeks, the stem cells changed their morphology, and there were many lipid droplets in the cells. The number of droplets increased in a time dependent manner and tiny lipid droplets aggregated to form larger ones. In the control, cells cultured in complete medium all through the culture process were not stained by Oil Red O (Figs. 13-14).

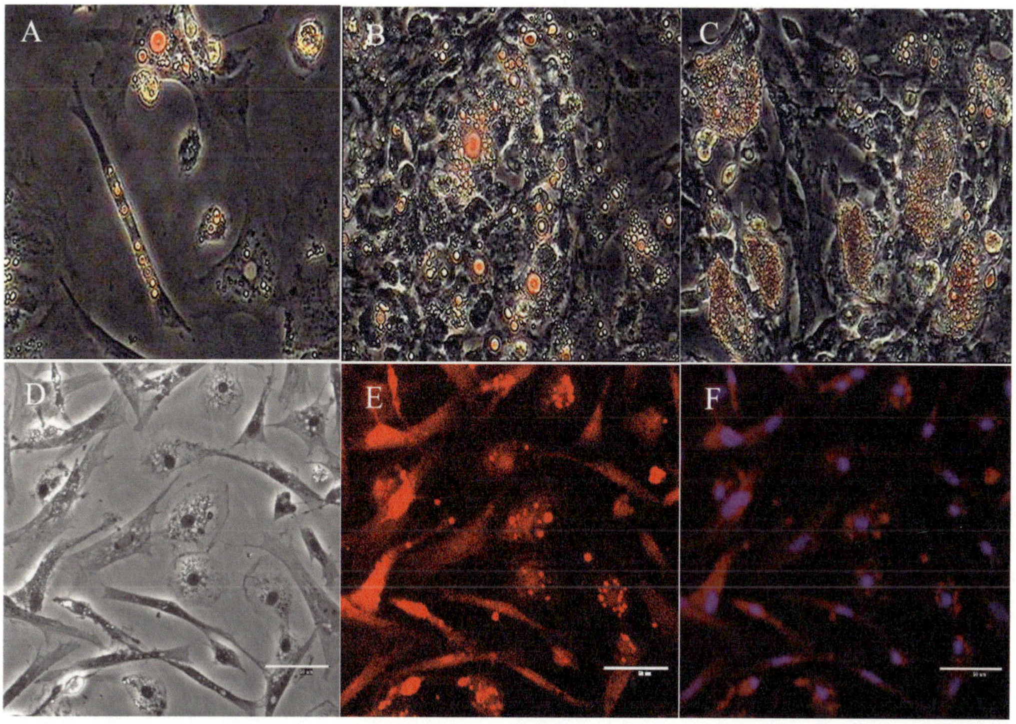

Fig. 13. Oil red-O staining and immunofluorescence of chicken bone marrow MSCs after adipogenic induction. Cells stained with Oil red-O post induction at (A) day 5; (B) day 7; and (C) day 21; the induced chicken bone marrow MSCs are FABP positive as revealed by immunofluorescence; (D) phase; (E) FABP+ (F) merge of FABP and DAPI.

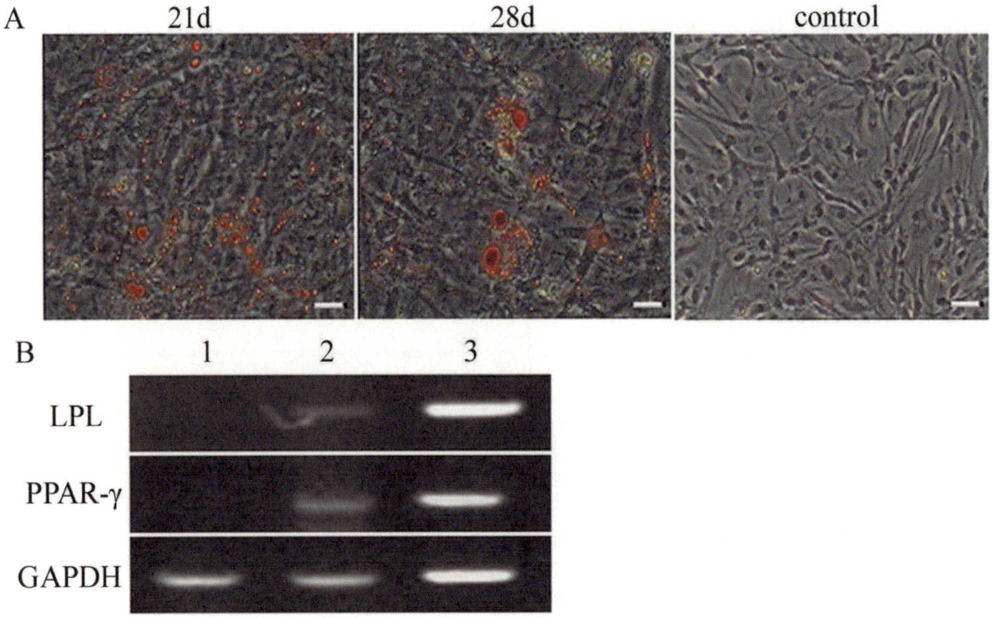

Fig. 14. Adipogenic differentiation of the ADSCs. (a) After 3-week induction, ADSCs metamorphosed from fibroblast-like to oblate and formed many lipid droplets in cells. Along with the prolongation of inducing time, droplets increased and aggregated to form larger ones gradually. As for negative control, cells cultured in complete medium all through the culture process didn't change in morphology and wasn't stained by Oil Red O. Scale bars=25 μm. (b) the expression of adipocyte specific genes, including LPL and PPAR-γ, were detected using RT-PCR assay in induced group at day 21 (Lane 2) and day 28 (Lane 3), while these genes were not expressed in control (Lane 1).

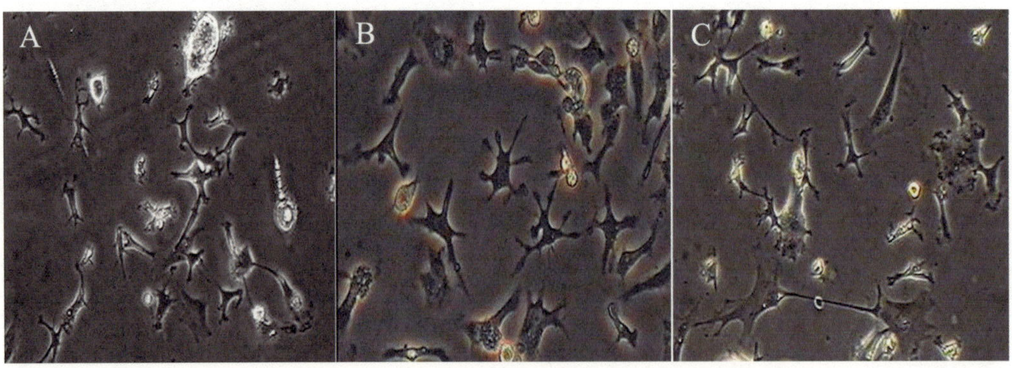

Fig. 15. Neural differentiation of chicken bone marrow MSCs (100×) treated for (A) 1 h; (B) 3 h; and (C) 5 h.

Neurogenic differentiation

Once the induction began, cell bodies of chicken bone marrow MSCs further contracted and became round, triangular or cone-shaped with multi-polar processes. Processes continued to ramify, displaying many branches, and growing cone-like dendrites. Some cells underwent a long process with evident varicosities, similar to the long axon of Golgi I neuron (Fig. 15). The expression of neural markers including Nestin, NSE and GFAP then became positive (Fig. 16).

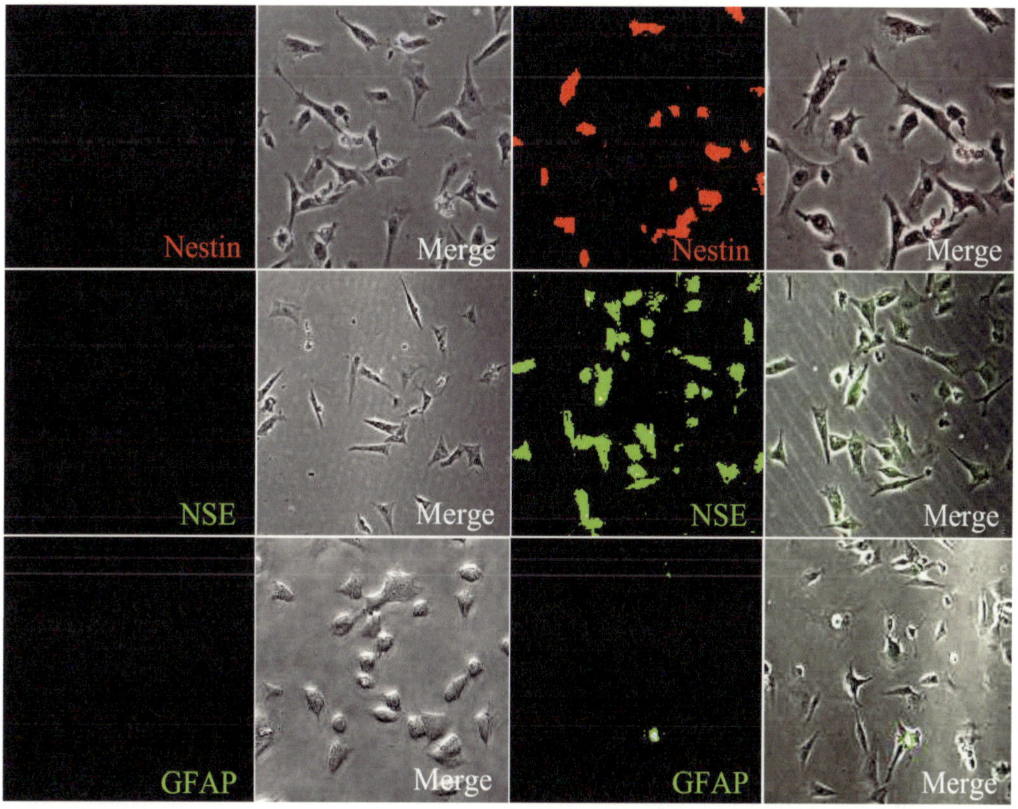

Fig. 16. Immunofluorescence detection of Nestin, NSE and GFAP expression in chicken bone marrow MSCs of passage 3 post 6 h neurogenic induction.

Cardiomyogenic differentiation

After incubation in cardiomyogenic medium, the cells polymerized to form myotubules, and the myotubules increased and fused to form fascicle as time passed by. Autopulse of the myotubes was observed after about 21 days. There were no obvious morphological changes in the control group (Fig. 17).

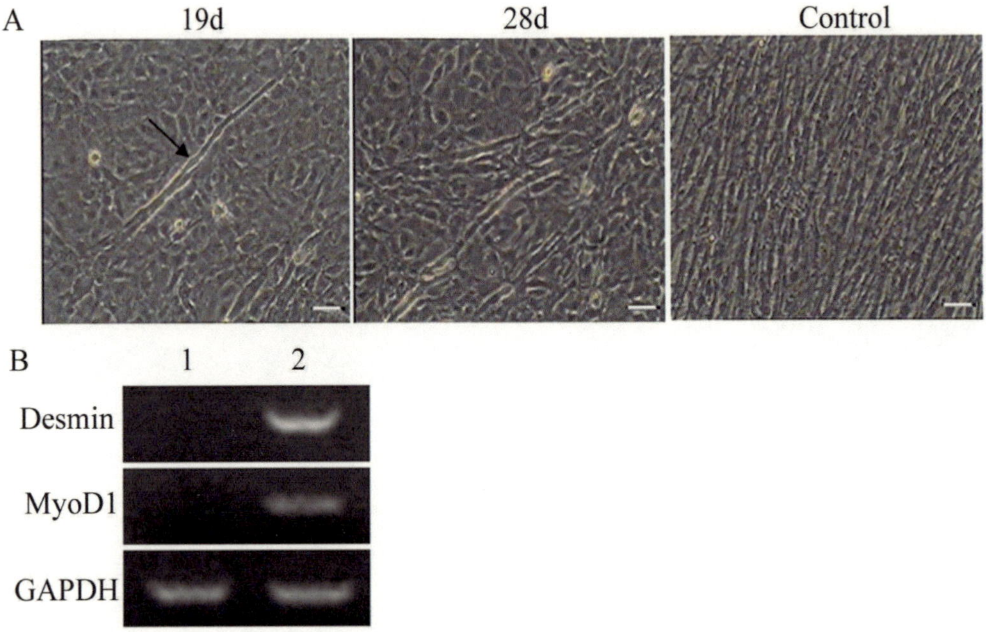

Fig. 17. Cardiomyogenic differentiation of chicken ADSCs. (A) The cells polymerized to form myotubules after culture in cardiomyogenic medium for 19 days (arrow). Around day 28, the myotubules increased and fused to form fascicles. There was no obvious metamorphosis in control. Scale bar=50 μm. (B) Myocyte specific genes, Desmin and MyoD1, were detected via RT-PCR assay after incubation in cardiomyogenic medium for 28 days (Lane 2), while these genes were not detected in control (Lane 1).

3. Conclusion

Animal genetic resources are encountering a challenging moment, for which reason scientists are making every effort to store the genetic materials in a long term, so that they can be explored completely and appropriately, however valuable they may seem from current point of view.

Researchers have been making every effort to preserve and to exploit animal genetic resources. At present, preservation in terms of individual animals, semen, embryos, genomic libraries and cDNA libraries are all alternative methods. However, myriads of practical problems exist on the following grounds: i) endangered species and breeds are incredibly diversified, making it unlikely for individual preservation; ii) some key techniques remain flawed, ruling out semen and embryos as an option; iii) confined by limited self-proliferation potential, mere genome DNA or organ preservation is insufficient for long term utilization; iv) despite of the proliferative properties of genomic libraries and cDNA libraries, they are not the basic unit of life activities, moreover, their cellular function can only be embodied by transgenic techniques.

Stem cells, due to their self-renewal ability and multi-lineage differentiation potentiality, have attracted extensive attention in medicine and biological sciences. Preservation of stem cells not only inherits the merits of somatic cell preservation, but also obtains extra advantages owing to their intrinsic characteristics.

It's worth mentioning that stem cells are far from being thoroughly investigated, especially for the *ex vivo* culture system. The development relies very much on daring and creative attempt, as well as repetitive testing. Fortunately, in Animal Population Culture Collection of China (APCCC), a series of stem cell lines from animal embryos and adults, such as primordial germ cells, bone marrow mesenchymal stem cells, neural stem cells, cardiac progenitor cells, endothelium progenitor cells, adipose stem cells and umbilical cord mesenchymal stem cells have been established and cyropreserved, which serve for the preservation of animal genetic resources after identification for their biological characteristics and multi-lineage differentiation potentials. Comprehensive assays, including morphology, microbial contamination, isozyme testing, karyotype, growth dynamics and surface antigen detection, are performed. After serial passage cell growth curves still display typical "S" types; microbial tests are all negative; isozyme patterns maintain specificity; and cells possess sound chromosome genetic stability. Exogenous genes are introduced into stem cells and get stable expression, testifying the cell lines have good performance at gene expression level. The identification results show that the established stem cell lines have stable and normal biological properties, meeting all the cell line quality control standards enacted by American Type Culture Collection (ATCC). Furthermore, the preserved cells retain good stem characteristics. Molecular markers are detected for the proof of their identity. Multi-germ-layer differentiation potential are tested, including the differentiation to ectoderm cells represented by neural cells, to mesoderm cells, e.g. the osteoblasts, adipocytes and cardiac muscle cells, and to endoderm cells, mainly epithelial cells.

In this sense, the APCCC has established a set of technical system suitable for the preservation of animal genetic resources in terms of stem cells, which is promising in generalizing to all kinds of animal species, therefore effectively protecting genetic treasures shaped in millions of years.

4. Acknowledgment

This research was supported by the "863" National Major Research Program (2006AA10Z198, 2007AA10Z170), National Key Technology R&D Program (2006BAD13B08) and National Scientific Foundation of China (30671539).

5. Abbreviations

ADSCs	Adipose derived stem cells
AKP	Alkaline phosphatase
APCCC	Animal Population Culture Collection of China
ATCC	American Type Culture Collection
ESCs	Embryonic stem cells
ASCs	Adult stem cells

MSCs Mesenchymal stem cells
NSCs Neural stem cells
NSPCs Neural stem and progenitor cells
PDT Population doubling time
PGCs Primordial germ cells
SCs Satellite cells

6. References

Bai C.Y., Li C.Y., Jin D.P., Guo Y., Guan W.J., Ma Y.H. & Zhao Q.J. (2010). Establishment and characterization of a fibroblast line from Landrace. Artif Cell Blood Sub, Vol. 38. No. 3. Pp. (129-135).

Doyle A., Hay R., Kirsop. & B.E., (1990). Animal cells, living resources for biotechnology. Cambridge University Press, pp(81-100). Cambridge, U.K.

Ford C. E., Pollock D. L. & Gustavsson I. (1980). Proceedings of the First International Conference for the Standardization of Banded Karyotypes of Domestic Animals. Hereditas. Vol. 92. No. 1. pp. (145-162).

Freshney R.I., 2000. Culture of animal cells: A manual of basic technique. 4th ed. Wiley-Liss, Hoboken, NJ. Pp. (149-175). ISBN: 9780471453291. New York.

Hay, R. I., (1992). Cell line preservation and characterization. Animal cell culture: a practical approach. 2nd edn (Ed. R. I. Freshney). Oxford University Press, pp. (104-135). Oxford.

Heim R, Cubitt AB & Tsien RY. (1995). Improved green fluorescence. Nature, Vol. 373. No. 6516. pp. (663–664).

Jing W., Lin Y., Wu L., Li X., Nie X., Liu L., Tang W., Zheng X. & Tian W., (2007). Ectopic adipogenesis of preconditioned adipose-derived stromal cells in an alginate system. Cell Tissue Res, Vol. 330. No. 3. pp. (567-572).

Li L.F., Bai X.J., Gong X.L., Liu H.K., Chen L.N., Guan W.J. & Ma Y.H. (2009). Differentiation potential of bone marrow mesenchymal stem cells in duck. J Genet Genomics, Vol. 36. No. 3. Pp. (133-140).

Liu C.S., Wang Z.G. & Li N. (2004). Status and strategy of protection of endangered animal genetic resourse in China. China Animal Husbandry Bulletin, Vol. 21, No. 017, pp. (28-31).

Ma Y.H. & Wu C.X. (2001). Evaluation of Threatening Degree of the Animal Genetic Resources in China. Acta Ecologae Animalis Domastici, Vol. 22, No. 2, pp. (8-13).

Masover G. K. & Becker, F. A., (1998). Detection of mycoplasmas in cell cultures by cultural methods. Methods in molecular biology. Vol. 104: Mycoplasma protocols. Humana Press nc. pp. 207-15, 217-26. Totawa NJ.

Qu Z., Balkir L., Van Deutekom J.C., Robbins P.D., Pruchnic R. & Huard J. (1998). Development of approaches to improve cell survival in myoblast transfer therapy. J Cell Biol, Vol. 142, No. 5, pp.(1257-1267).

Ren ,F. L., Li ,Y., Zhang, Y., (2002). In vitro cultivation and freezing of bovine skin fibroblast cells. Scalper Magazine. Vol. 28. pp. (8-10).

Tsuchiya R, Yoshiki F, Kudo Y, Morita M.l (2002). Cell type-selective expression of green fluorescent protein and the calcium indicating protein, yellow cameleon, in rat cortical primary cultures. Brain Research. Vol. 956. No. 2. pp. (221-229).

Zhang L. (2003). Animal Breeding, China Radio & TV University Press. BeiJing.

7

The Role of *SOX2* in Maintaining Pluripotency and Differentiation of Human Embryonic Stem Cells

Keiko Adachi[1], Hirofumi Suemori[2], Norio Nakatsuji[1,3] and Eihachiro Kawase[4]

[1]*Department of Development and Differentiation, Institute for Frontier Medical Sciences, Kyoto University, Kyoto*

[2]*Laboratory of Embryonic Stem Cell Research, Stem cell Research Center, Institute for Frontier Medical Sciences, Kyoto University, Kyoto*

[3]*Institute for Integrated Cell-Material Sciences, Kyoto University, Kyoto*

[4]*Laboratory of Cell Processing, Stem Cell Research Center, Institute for Frontier Medical Sciences, Kyoto University, Kyoto*

Japan

1. Introduction

Embryonic stem cells (ESCs) are pluripotent and self-renewing cells that are derived from the inner cell mass (ICM) of the developing blastocysts (Evans and Kaufman, 1981; Martin, 1981). ESCs have the ability to maintain self-renewal and to differentiate into all types of cells. With respect to primates, ESCs were first established from the rhesus monkey (*Macaca mulatta*) (Thomson et al., 1995) and subsequently from humans (Thomson et al., 1998). Human ESCs are of particular interest because of their potential application to regenerative medicine and drug discovery. In addition, human ESCs may provide insights into human embryo development in culture. Therefore, identification of the molecular mechanisms that govern human ESC self-renewal, differentiation and proliferation is of considerable interest.

A small number of genes, the so-called "core transcription factors", is thought to have a central role in the control of the stem cell state in concert with other genes including other transcription factors. The ESC state is largely governed by three core transcription factors, *OCT4*, *SOX2*, and *NANOG*. In the mouse, Oct4 and Nanog have a distinctive expression pattern in ESCs and during embryonic development, and genetic and molecular analyses have shown that both genes have essential roles for maintaining the stable pluripotent state (Chambers et al., 2003; Mitsui et al., 2003; Nichols et al., 1998; Niwa et al., 2000). By contrast, Sox2 is widely expressed during embryonic development and occurs not only in the inner cell mass and epiblast but also in neural tissues, extra embryonic ectoderm, gut endoderm, esophagus and trachea (Avilion et al., 2003; Williamson et al., 2006; Wood and Episkopou, 1999). Indeed, *SOX2* is required for development during the peri-implantation period, and is required for trophoblast formation and neural development (Avilion et al., 2003; Kelberman et al., 2006; Taranova et al., 2006). Oct4 acts as a heterodimer with Sox2 in mouse

ESCs; Sox2 assigns a position for key regulators for the maintenance of mouse ESCs with undifferentiated state (Ambrosetti et al., 2000; Avilion et al., 2003; Masui et al., 2007).
Mouse ESCs have often been used as a model system for human ESCs. However, various differences are known to exist between the ESCs of the two species, including the molecular mechanisms for self-renewal. For example, the LIF/STAT3 pathway is involved in mouse ESC self-renewal (Niwa et al., 1998), but is dispensable in human ESCs (Daheron et al., 2004; Humphrey et al., 2004). Moreover, BMP4 signaling is required for the maintenance of self-renewal in mouse ESCs grown in serum-free medium, and acts by inhibiting neural differentiation (Ying et al., 2003). However, the addition of BMP4 to human ESC culture promotes primitive endoderm or trophectoderm differentiation (Pera et al., 2004; Xu et al., 2002). These inter-species differences have become a highly contentious issue following the establishment of a new type of pluripotent cell line called EpiSCs (Brons et al., 2007; Tesar et al., 2007). EpiSCs were first established by explanting late epiblast from mouse embryos; these explanted cells expressed the core transcription factors Oct4, Sox2 and Nanog. EpiSCs can differentiate into three germ layers but they are inefficient in generating chimeras, suggesting that they have a more limited developmental potential than ESCs. Interestingly, mouse EpiSCs and human ESCs have similar growth requirements and gene expression patterns, and both types of cells are distinguishable from mouse ESCs. It is unclear whether human ESCs are really a counterpart of mouse EpiSCs, since EpiSCs have not been extensively studied to date.
Disregarding the issue of EpiSCs, mouse and human ESCs are believed to have similar molecular mechanisms for maintenance of the undifferentiated state. Chromatin immunoprecipitation (ChIP) assays combined with genome-wide location methodologies showed that in both human and mouse ESCs, *OCT4*, *SOX2* and *NANOG* have common target sites within the regulatory regions of many genes. This is true of active, highly expressed genes, such as *OCT4*, *SOX2* and *NANOG* themselves, and inactive genes, such as developmental regulators that maintain the pluripotent state (Boyer et al., 2005; Loh et al., 2006). Downregulation of *OCT4* in mouse and human ESCs induces trophectoderm differentiation (Babaie et al., 2007; Hay et al., 2004; Matin et al., 2004; Niwa et al., 2000; Zaehres et al., 2005). By contrast, overexpression of *OCT4* induces ESCs to differentiate into endoderm or mesoderm cells (Niwa et al., 2000; Rodriguez et al., 2007). Reduction in expression of *NANOG* in mouse ESCs induces endoderm differentiation (Chambers et al., 2003; Mitsui et al., 2003), while in human ESCs, it induces both endoderm and trophectoderm differentiation (Hyslop et al., 2005; Zaehres et al., 2005). Overexpression of *NANOG* promotes stabilization of an undifferentiated state in ESCs. Mouse ESCs can maintain an undifferentiated state in the absence of LIF (Chambers et al., 2003; Chambers et al., 2007; Mitsui et al., 2003). The human ESCs allows feeder-free growth or growth without conditioned medium from feeder cells (Darr et al., 2006). In mouse ESCs, repression of *Sox2* expression induces trophectoderm differentiation (Masui et al., 2007), whereas overexpression induces non-specific lineage differentiation, neuronal differentiation or massive cell death (Kopp et al., 2008; Mitsui et al., 2003; Zhao et al., 2004). Furthermore, *Sox2*-deficient mice are defective in the maintenance of ICM/epiblast and trophoblast development (Avilion et al., 2003), leading to the conclusion that *SOX2* may have an important role in trophoblast development as well as in ESC maintenance. However, the role of *SOX2* in human ESCs is not fully understood.
Reprogramming of somatic cells into induced pluripotent stem (iPS) cells provides another approach to investigating the nature of the undifferentiated state in ESCs. Surprisingly, the overexpression of only four transcription factors, Oct4, Sox2, Klf4 and c-Myc, can convert

somatic fibroblasts to pluripotent cells that can contribute to the germline in chimeric mice, similarly to ESCs (Okita et al., 2007; Takahashi and Yamanaka, 2006; Wernig et al., 2007). Reprogramming can be induced not only by Oct4, Sox2, Klf4 and c-Myc but also by use of combinations including Nanog, Lin28 and other genes. Generally, overexpression of Sox2 is required unless the somatic cells already have endogenous Sox2 expression (Hanna et al., 2010; Stadtfeld and Hochedlinger, 2010). Moreover, the level of *SOX2* expression has a direct effect on the rate of direct reprogramming of somatic cells to iPSCs (Yamaguchi et al., 2011). These results provide further support for the assumption that SOX2 has a definitive role in the maintenance of ESCs.

In this study, we investigated the role of *SOX2* in human ESCs by manipulating the level of *SOX2* expression. We show that depletion or overexpression of *SOX2* in human ESCs induced trophectoderm differentiation. We further showed that overexpression of *SOX2* during human ESC differentiation promoted neural and glandular epithelium development.

2. The role of *SOX2* in maintaining pluripotency and differentiation of human ESCs

2.1 The role of SOX2 in maintaining pluripotency of human ESCs

2.1.1 Depletion of *SOX2* in human ESCs induces trophectodermal and some endodermal differentiation

To investigate the role of *SOX2* in human ESCs, we used a small interfering RNA (siRNA) to knockdown gene expression. We first optimized the protocol by examining the effect of a previously described siRNA on OCT4 expression in the KhES1 cell line (Dharmacon ID D-019591-05) (Babaie et al., 2007) (a detailed protocol is described in Adachi et al., 2010). This siRNA reduced OCT4 expression in our protocol as determined by quantitative real-time PCR (Q-PCR) (Figure 1A) and Western blotting (40-, 44-kDa signal, Figure 1B), indicating that this protocol was appropriate for the study. Next, we analyzed the levels of *SOX2* expression in KhES1 cells transfected with a range of siRNAs, since selection of appropriate siRNA sequences was crucial to the study. Overall, we found that D-011778-01 siRNA (Dharmacon) was the most effective of the tested siRNAs. A Q-PCR analysis indicated that D-011778-01 siRNA reduced *SOX2* mRNA expression to 25% of the control level (Fig. 1A). Western blotting confirmed a reduction in the level of SOX2 protein in the cells (34-kDa signal, Figure 1B). SOX2 repression in human ESCs caused changes to cell morphology from 48 hours post-transfection. Cells with enlarged nuclei and a flattened morphology were present after 72 hours. These morphological changes were accompanied by a reduction in the SOX2 protein level as determined by immunohistochemistry (Fig. 1C). In addition, the cells showed downregulated expression of ESC-specific surface markers, anti-stage specific embryonic antigen (SSEA)-4, anti-tumor rejection antigens (TRA)-1-60 and TRA-1-81, and also upregulated expression of the differentiation marker SSEA-1 (Fig. 1D, and data not shown). These results indicate that *SOX2* expression may be required for maintenance of human ESCs in a pluripotent state.

As mentioned earlier, three core transcription genes, *OCT4*, *SOX2* and *NANOG*, are thought to form an interconnected autoregulatory loop in human ESCs (Boyer et al., 2005; Loh et al., 2006). In agreement with previous reports, *OCT4* knockdown in human ESCs caused a decrease in both *NANOG* and *SOX2* levels to 10-30% of the control level, indicating that *OCT4* may regulate expression of these core transcription factors. In the KhES1 cells with knockdown of SOX2, the *OCT4* transcript level was slightly decreased, while the *NANOG* level decreased to 60% of the control (Fig. 1A). When we examined another human ESC line,

HES-3, we found that D-011778-01 siRNA reduced *SOX2* expression to 28% of the control. In this cell line, the *OCT4* and *NANOG* transcript levels decreased to 60% and 55% of the control, respectively. These results indicate that *SOX2* may play a role in regulating expression of *NANOG* and, perhaps, *OCT4*. Future studies using *SOX2*-deficient human ESCs should provide conclusive evidence on whether *SOX2* regulates the expression of *OCT4* and *NANOG*.

Our data indicate that *SOX2* expression is required to maintain human ESCs in an undifferentiated state, although the exact differentiation status of the cells was uncertain. Immunohistochemical analysis showed that the majority of SOX2 downregulated cells (98.5%, n = 200) were positive for the trophectodermal marker cytokeratin 8, and a small population of cells with downregulation of SOX2 were positive for the endodermal marker GATA6 (0.8%, n = 1,000) (Fig. 1E). We could not detect any increase in other linage markers including the ectodermal marker PAX6 and the mesodermal marker brachyury (T) homolog (K.A. and E.K., unpublished observation). Q-PCR showed that *SOX2* knockdown in human ESCs resulted in upregulation of the trophectodermal markers *EOMES*, *BMP4* and *HAND1* at 72 hours after transfection. Interestingly, *CDX2* was not significantly upregulated (Fig. 1F). Upregulation of trophectodermal markers, except CDX2, was also present at 48 hours after transfection. At this time point, GATA3 was upregulated (Fig 1G). GATA3 is a gene expressed in the trophectoderm; *Gata3* directly regulates *Cdx2* transcription via a conserved GATA motif at the intron 1 region of the *Cdx2* locus in the mouse (Home et al., 2009). Thus, *CDX2* shows an exceptional expression pattern compared to other early trophectodermal markers, possibly indicating that in these cells some other gene may be involved in its regulation. We also found that expression of the endodermal markers *GATA6* and *FOXA2* was upregulated in cells with downregulation of SOX2 (Fig. 1F). Furthermore, expression of ectodermal and mesodermal markers in *SOX2* siRNA transfected cells was not significantly altered relative to controls. However, we observed that the ectodermal marker *OTX1* and the mesodermal marker *VIM* were upregulated. Thus, the results from assays of transcripts were consistent with the immunohistological observations. Taken together with similar results from the HES-3 cell line, we conclude that *SOX2* expression maintains ESC pluripotency by suppressing trophectodermal and, to a lesser extent, endodermal differentiation.

It was previously shown that reduction in *SOX2* expression in human ESCs resulted in the loss of the undifferentiated state with an associated increase in the expression of trophectoderm lineage markers including CDX2 (Fong et al., 2008). However, in this study, human ESCs were cultured with mouse MEF feeder cells; consequently, it was not possible to exclude the possibility that the feeder cells also influenced the undifferentiated or differentiated state of the ESCs. In this study, we used a feeder-free culture system and were thus able to address directly the effect of *SOX2* reduction on human ESCs. Our results without feeder cells showed that *CDX2* was not upregulated in cells with suppressed expression of SOX2; this finding is consistent with previous results from ES cells of the Sox2 null mouse (Masui et al., 2007). Furthermore, our results were the first to show that a reduction in SOX2 expression in human ESCs specifically caused differentiation of endoderm as well as trophectoderm. A lack of *SOX2* has been found to induce differentiation into trophectoderm or extraembryonic endoderm cells from the ICM/epiblast lineage in mouse embryos (Avilion et al., 2003), consistent with our results in human ESCs. Thus, we believe that feeder cells may bias against the intrinsic lineage differentiation in human ESCs. Our analysis of human ESCs in a feeder cell-free culture system identified new roles for *SOX2*. Clearly, this system could be also be used to investigate the role of other genes in human ESCs.

Figure 1A-C

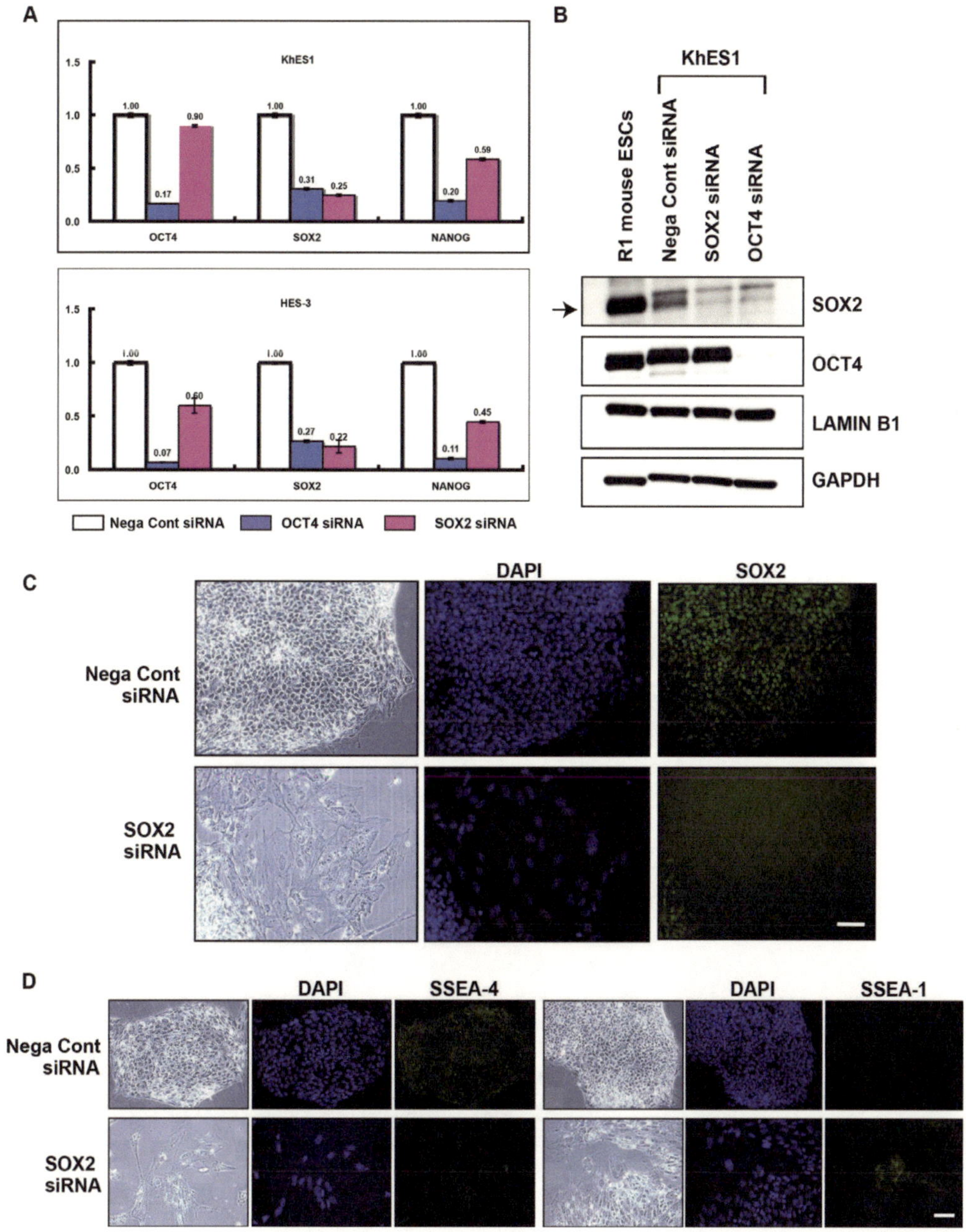

Figure 1E-G

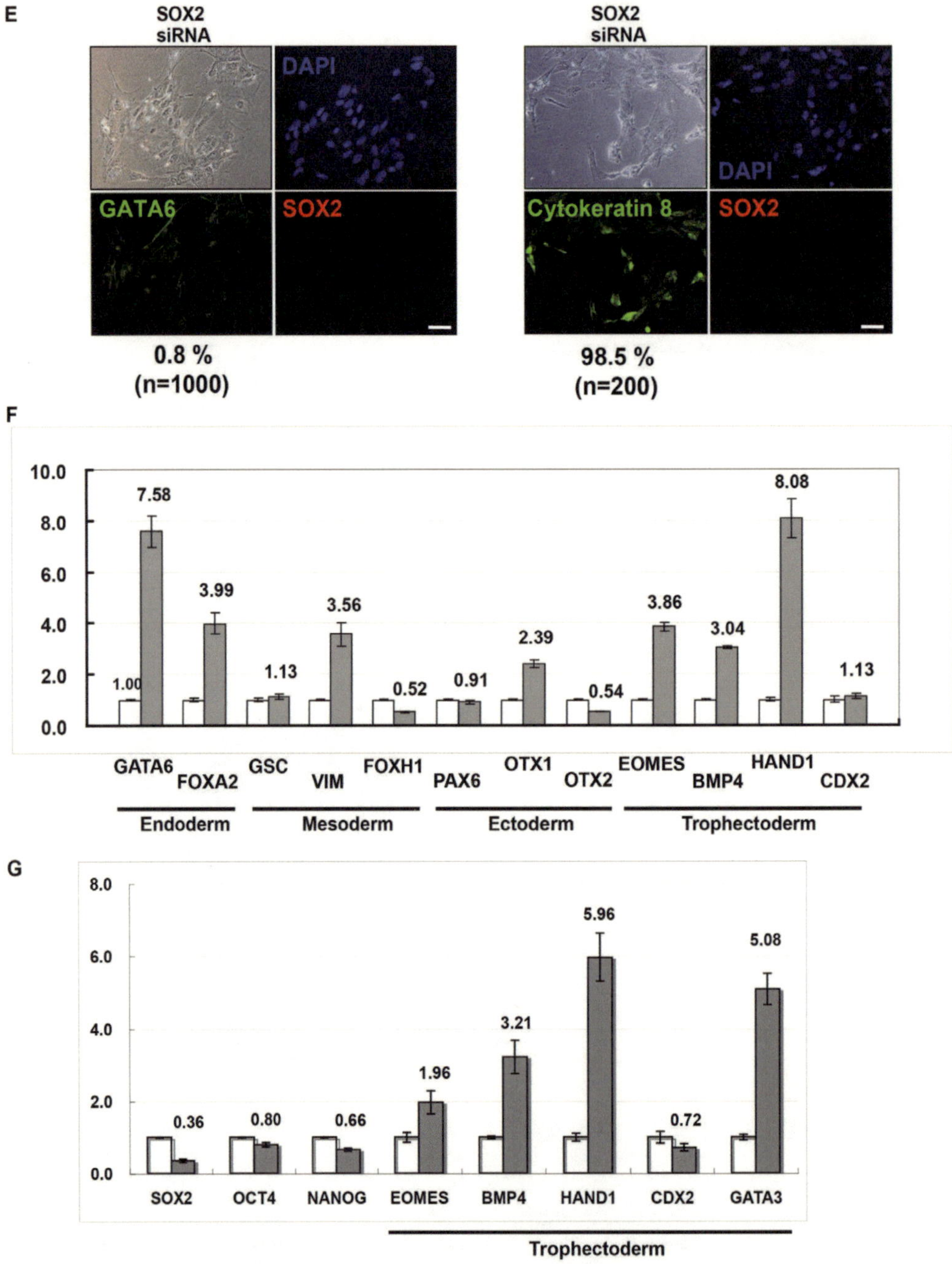

Fig. 1. Depletion of *SOX2* in human ESCs induces trophectodermal and some endodermal differentiation.

Transfection with *SOX2* siRNA reduced *SOX2* mRNA levels to approximately 25% of the negative control (Nega Control siRNA). *OCT4* and *NANOG* mRNA levels also decreased. (B) Western blotting analysis of SOX2 protein levels at 72 hours after siRNA transfection. Both mouse and human SOX2 protein was detected as a 34 kDa band (indicated by arrow). Decreased levels of SOX2 protein was present in human ESCs after transfection with SOX2 or OCT4 siRNAs. Lamin B1 and GAPDH was used as loading controls. (C) Morphology of KhES1 cells 72 hours after transfection with either Nega Control siRNA or *SOX2* siRNA. The level of SOX2 protein was confirmed by immunohistochemistry. (D) *SOX2* knockdown cells showed downregulation of the ESC-specific surface marker SSEA-4 and upregulation of the differentiation marker SSEA-1. (E) Immunohistochemical detection of the trophectodermal marker cytokeratin 8 (98.5%, n = 200) and endodermal marker GATA6 (0.8%, n = 1,000) in *SOX2*-downregulated cells. (F, G) *SOX2* knockdown in human ESCs resulted in upregulation of the trophectodermal markers *EOMES, BMP4* and *HAND1* (F, G) and endodermal markers *GATA6* and *FOXA2* (F). The Q-PCR data (A, D) represent the means (white bar for Nega control siRNA, and gray bar for SOX2 siRNA) ± SEM relative to the negative control (=1.0). The assays were normalized to *GAPDH*. The transfection experiments were carried out three times in triplicate (n = 9) for KhES1 cells (A, E) and twice (n = 6) for HES-3 cells (A). Statistical significance of the results was assessed using the Student's *t-test*. Scale bars = 100 mm (C, D, E). Analyses were performed at 48 hours (G) or 72 hours (others) after transfection. Abbreviations: *GSC*, goosecoid homeobox. This figure is modified from (Adachi et al., 2010).

2.1.2 Overexpression of *SOX2* in human ESCs induces trophectodermal differentiation

We used a lipofection method to induce transient *SOX2* overexpression in human ESCs and analyzed the effects with an RT-PCR analysis (Adachi et al., 2010). We monitored *SOX2* expression for 72 hours after transfection. *SOX2* expression levels peaked at 24 hours post-transfection and, thereafter, the levels decreased. An immunohistochemical analysis indicated that about 20% of the human ESCs overexpressed the *SOX2* protein. When compared with wild type or mock transfected cells, *SOX2* overexpressing cells showed significant upregulation of the early trophectodermal markers *EOMES* or *CDX2*, but not of later trophectodermal markers, such as *BMP4* or the glycoprotein hormones alpha chain. By contrast, the expression patterns were not changed in the endodermal markers *GATA6* and alpha-fetoprotein (*AFP*), the mesodermal markers goosecoid homeobox (*GSC*) and brachyury (*T*) homolog or the ectodermal markers *PAX6, OTX1* and *OTX2*. Interestingly, we did not detect significant downregulation of the pluripotency markers *OCT4* and *NANOG*. Thus, transient overexpression of *SOX2* in human ESCs may induce expression of early trophectodermal genes, but this was not sufficient for re-specification of the cells into the trophectoderm lineage.

In order to achieve constitutive overexpression of SOX2 in human ESCs, we used the Tet-On/Off system of monkey ESCs and developed this for human ESCs (Adachi et al., 2006; Adachi et al., 2010). We isolated two independent cell lines (#14 and #27) that showed doxycycline (Dox)-induced *SOX2* expression in a dose-dependent manner. Human ESCs, in general, are sensitive to chemicals and we found that high levels of Dox (more than 2 mg/ml) had some detrimental effects on the survival of wild-type human ESCs (data not shown). Therefore, we used 1 mg/ml Dox in our experiments. Human ES clones #14 and #27 were morphologically indistinguishable from parental KhES1 ESCs when cultured under standard human ESC culture conditions. The addition of 1 mg/ml Dox to the medium resulted in a two- to four-fold increase in *SOX2* expression (as determined by Q-PCR) on day 5. From day 3 of culture, 30-50% of the human ESC colonies displayed a flattened morphology in their middle. Moreover, *SOX2* overexpression in human ESCs induced upregulation of the trophectodermal markers *EOMES* and *CDX2*, but not of other lineage markers as determined by semi-quantitative RT-PCR and Q-PCR analysis (data not shown). After day 5, immunohistochemical staining showed that *SOX2* overexpression in human ESCs resulted in trophectoderm differentiation. Exposure to Dox increased *SOX2* expression (Figure 2A); however, the levels of expression of OCT4 and NANOG, which are essential for maintenance of the undifferentiated state, decreased in these cells (Figure 2B-C). Furthermore, SOX2 overexpression in human ESCs upregulated the expression of a differentiation cell surface marker SSEA-1 (Figure 2D). This result further supports the notion that *SOX2* overexpression induces differentiation in human ESCs. We found that there was a substantial increase in the expression of the trophectoderm marker cytokeratin 8 (Figure 2E), but there were no significant increases in expression of the endoderm marker GATA6, the ectoderm marker PAX6 or the mesoderm marker brachyury (data not shown). Finally, we found that CDX2 expression could be detected in cells with the morphological appearance of trophectoderm, consistent with our Q-PCR results. Following a further 2 days induction (day 7), we found that the number of cells expressing CGα increase more than 10-fold (Fig. 2G). Interestingly, immunohistochemical staining did not detect increases in other trophectodermal, placental lactogen positive cells among *SOX2*-overexpressing cells (data not shown).

Taken together, our observations indicate that overexpression of *SOX2* in human ESCs resulted in trophectodermal differentiation accompanied by increased *CDX2* expression. On the basis of the decrease in expression of OCT4 and NANOG, the increase in expression of trophectoderm markers, and the absence of a significant increase in expression of other lineage markers, we concluded that overexpression of *SOX2* in human ESCs causes trophectoderm differentiation.

Figure 2A-C

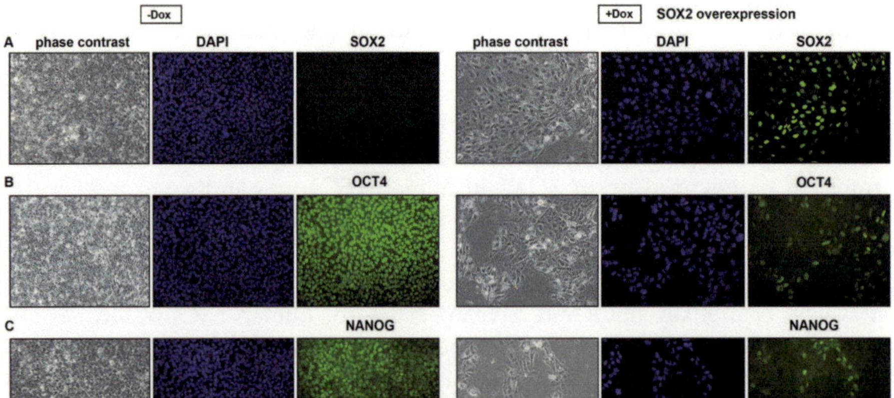

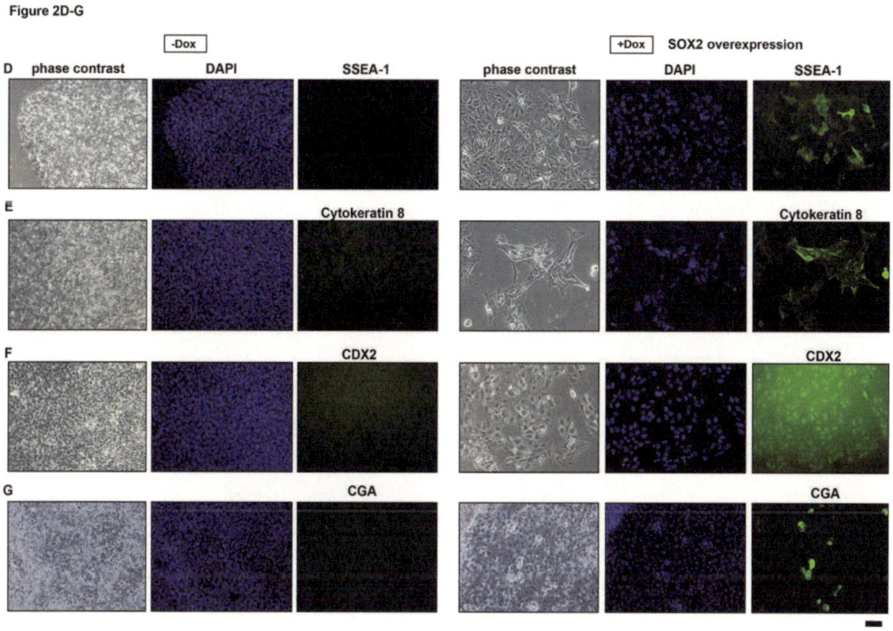

Fig. 2. Overexpression of *SOX2* in human ESCs induces trophectodermal differentiation. Immunohistochemical staining at day 5 (A-F) or day 7 (G) after induction of *SOX2* overexpression. (A) DOX induces increased SOX2 expression compared to control human ESCs. (B, C) Overexpression of SOX2 repressed expression of the core transcription factors OCT4 (B) and NANOG (C) in human ESCs. (D) SOX2 overexpression in human ESCs resulted in upregulation of the differentiation marker SSEA-1. (E-G) Immunohistochemical staining of trophectoderm for cytokeratin 8, CDX2 and CGA. (E) Increased cytokeratin 8 expression was induced by SOX2 overexpression compared to control cells. (F) CDX2 was only detected in human ESCs overexpressing *SOX2*. (G) A few CGA positive cells were present in control samples, but a more than 10 fold increase in positive cells with stronger fluorescence was present after *SOX2* overexpression. Scale bars = 100 mm. Abbreviations: Dox, doxycycline; -Dox, without Dox in the medium; +Dox, with Dox in the medium; *CGA*, glycoprotein hormones alpha chain. This figure is adapted from (Adachi et al., 2010).

2.1.3 Expansion of CDX2 positive cells from *SOX2*-overexpressing human ESCs

Our investigation demonstrated that overexpression of *SOX2* induced trophectoderm differentiation, which was accompanied by an increase in *CDX2* expression. *Cdx2* is essential for the maintenance of trophoblast stem (TS) cells in the mouse (Niwa et al., 2005). Therefore, we speculated whether overexpression of *SOX2* in human ESCs could induce TS-like cells and/or expansion of trophoblast cells under mouse TS cell culture conditions. At day 4, human ESCs that had been induced by Dox to overexpress *SOX2* started to change morphologically, to resemble trophoblast-like cells prior to flattening. The cells continued to proliferate, while retaining immunohistochemically positive staining for CDX2 (Figure 3A, B). On rare occasions, CDX2 positive cells were found in wild type or non-Dox induced human ESCs (Figure 3B).

CDX2 positive cells were detected for 3 to 4 weeks after Dox induction. Most of the CDX2 positive cells formed a glandular epithelium-like structure on the feeder cells (Figure 3C I).

Cells that were morphologically similar to mouse TS-like cells surrounded these structures, but CDX2 expression was weak or undetectable by immunohistochemistry in the majority of the cells (data not shown). A minority of the cells was CDX2 positive, and may represent candidate human TS cells (Figure 3C II). The numbers of other trophoblast cells that were positive for CGα immunohistochemical staining (data not shown) also greatly increased. Thus, under mouse TS cell culture conditions, overexpression of *SOX2* in human ESCs resulted in an increase in the rate of differentiation of trophoblast cells. However, the majority of the trophoblast cells lost CDX2 expression, thereby preventing derivation of human TS cell lines. SOX2 overexpression was therefore not sufficient for maintenance of CDX2 expression in the TS-like cells. Instead, these culture conditions induced an increase in

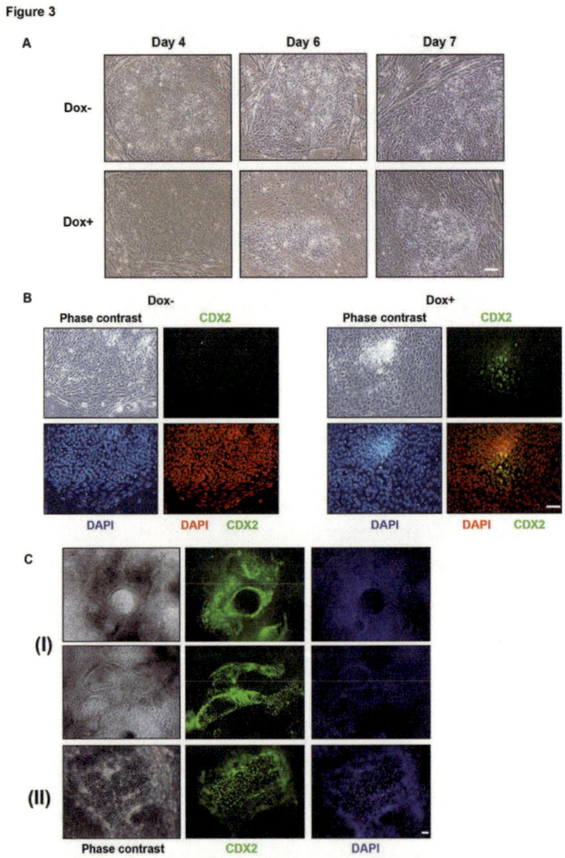

Fig. 3. Expansion of trophoblast and glandular epithelium cells from human ESCs by *SOX2* overexpression using the TS cell culture medium.
(A) Initial morphological changes in trophoblast-like cells differentiating from human ESCs. Trophoblast-like cells were detectable at day 4 after induction of *SOX2* overexpression. (B,C) Immunohistochemical staining for the trophectoderm marker *CDX2*. (B) *CDX2* was only detected in human ESCs induced to overexpress *SOX2*. No *CDX2* positive cells were present in parental cell lines or double transfected cell lines without Dox treatment. (C) *CDX2* positive cells were detected one month after Dox induction. (I) Glandular epithelium-like cells, (II) putative TS-like cells. Scale bars = 100 mm.

the number of CDX2 positive putative glandular epithelium-like cells, which showed more rapid proliferation than putative TS cells. Glandular differentiation did not occur only in response to *SOX2* overexpression in human ESCs, suggesting that initiation of glandular epithelium differentiation is not related to SOX2 overexpression.

2.2 Increased *SOX2* expression in human ESCs induces differentiation of neural and glandular epithelium

The teratoma formation assay has often been used to investigate the developmental potential of ESCs. We, therefore, used this assay to determine whether *SOX2* overexpression in human ESCs enabled formation of trophoblastic tumors or caused them to develop into other tissues. To examine the effect of *SOX2* overexpression during human ESC differentiation, we injected the double-transfected ES clones (lines #14 and #27) subcutaneously into the backs of SCID mice and monitored their differentiation with or without Dox treatment. Under this system, SOX2 should be an effective inducer of differentiation not only for human ESCs but also their differentiated descendants (Figure 4A) Parental human ESC lines formed teratomas comprised of all three germ layers in the presence or absence of Dox treatment (Suemori et al., 2006; data not shown). Similarly, double-transfected ES clones formed teratomas comprised of all three germ layers in the absence of Dox treatment (Figure 4B, data not shown). However, after Dox treatment (i.e. *SOX2* overexpression conditions), the double-transfected ES clones did not form teratomas like those observed in the absence of Dox. In this case, the teratomas were mainly comprised of neural epithelium and glandular epithelium (Figure 4C). We were unable to investigate how glandular epithelium differentiated from human ESCs. Glandular epithelium can differentiate from extraembryonic trophoblastic tissue and also the embryo proper, such as from gut or lung tissues, where SOX2 is expressed during development. Our results from use of the mouse TS cell culture protocol indicated that glandular epithelium could differentiate from human ESCs with or without SOX2 overexpression, and may suggest that SOX2 increases expansion of cell numbers rather than inducing initial differentiation.

By contrast to culture of mouse ESCs, we had to remove differentiated colonies to enable continual culture of human ESCs. As described earlier, we did not detect any cells at day 5 that were positive for the neuronal marker PAX6 among SOX2-overexpressing human ESCs. Two days later, PAX6-positive cells could be detected but only in the piled-up, embryoid body-like differentiated colonies formed by *SOX2*-overexpressing cells (K.A. and E.K., unpublished observation). We believe that SOX2 overexpression promoted expansion of PAX6-positive neural cells or neural progenitors rather than neuronal differentiation from human ESCs. In combination with our cell culture data, these observations indicated that overexpression of *SOX2* enhanced specific-lineage development of neural and glandular epithelium in differentiating human ESCs.

2.3 A putative model of SOX2 function in human ESCs

We have shown that the regulated expression of *SOX2* plays an important role in the maintenance of pluripotency in human ESCs. Decrease or increase of the level of *SOX2* expression resulted in human ESCs differentiating into the trophectodermal lineage. However, our results also suggest that trophectodermal differentiation may occur through different pathways. Although trophectoderm genes were upregulated in human ESCs with either repression or overexpression of SOX2, as determined by mRNA and protein studies, *CDX2* expression was exceptional and seemed to alter in concert with the level of expression of SOX2. Overexpression of either *CDX2* or *EOMES* induces differentiation of mouse ESCs into the trophectoderm lineage (Niwa et al., 2005). *OCT4* and *SOX2* are known to interact physically and co-occupy many sites in ES cells, including those on the EOMES gene (Boyer et al., 2005). In our study, *EOMES* was significantly upregulated upon *SOX2* repression in human ESCs, whereas *CDX2* was not, suggesting that induction of EOMES by *SOX2* repression in human ESCs may be a key step in trophectodermal differentiation. In contrast, overexpression of *SOX2* in human ESCs and activation of *CDX2* expression may be key for trophectodermal differentiation. In this study, we found that SOX2 overexpression in human ESCs resulted in decreased expression of OCT4 and NANOG, leading to activation of trophectodermal genes. Indeed, OCT4 repression induces CDX2 and EOMES expression in human ESCs (Babaie et al., 2007; K.A. and E.K., unpublished data). It will be interesting to determine how SOX2 overexpression reduces stem cell related genes while, at the same

Figure 4

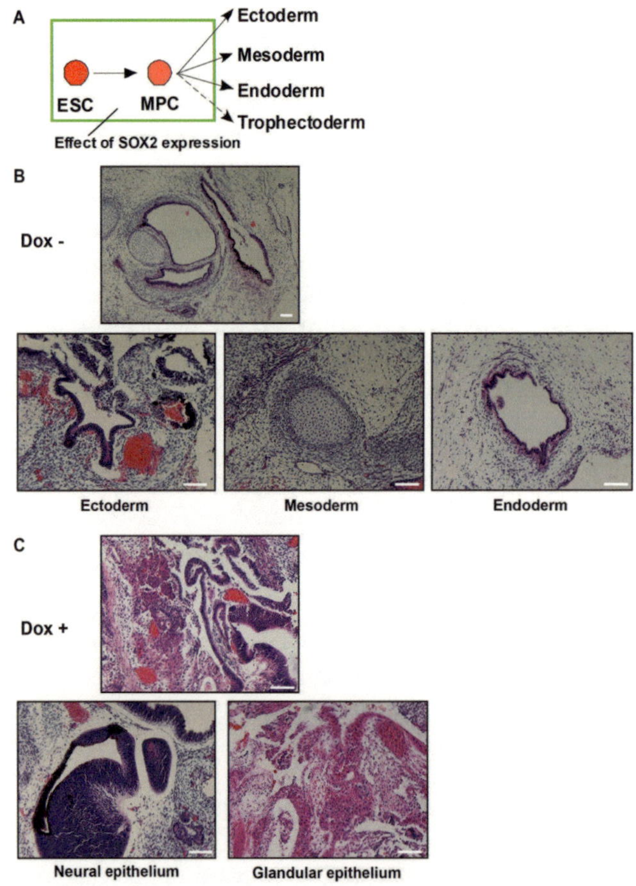

Fig. 4. Induction of teratomas by transplantation of human ESCs overexpressing *SOX2*. (A) SOX2 overexpression was effective for both human ES cells and their descendants. (B) In the absence of Dox, human ESCs generated teratomas composed of all three germ layers. (C) *SOX2* overexpressing ESCs induced by Dox in drinking water produced teratomas composed mainly of neuronal and glandular epithelium-like tissue. Scale bars = 100 mm.

time, increasing trophectodermal lineage genes. Elucidation of this process may provide novel insights into the gene regulatory networks for human ESC maintenance and differentiation.

3. Conclusion

In summary, the regulation of *SOX2* expression is essential for maintaining the pluripotent state of human ESCs. Both *SOX2* downregulation and upregulation caused trophectodermal differentiation. *SOX2* overexpression also promoted neural and glandular epithelium tissues during human ESC differentiation as a late or secondary effect. Our results demonstrate that in human ESCs, *SOX2* plays an important role in the maintenance of pluripotency and, possibly, trophoblast, neural and glandular epithelium development.

4. Acknowledgments

We thank Miss Mari Hamao for animal assistance, Dr. Tatsuaki Tsuruyama for valuable comments on teratoma development, and the members of the laboratory of Prof. N. Nakatsuji for discussions and support. This study was also supported by the New Energy and Industrial Technology Development Organization (NEDO) and the Japan Society for the Promotion of Science.

5. References

Adachi, K., Kawase, E., Yasuchika, K., Sumi, T., Nakatsuji, N., and Suemori, H. (2006). Establishment of the gene-inducible system in primate embryonic stem cell lines. *Stem Cells*, Vol.24, No.11, (November 2006), pp. 2566-2572, ISSN 1066-5099.

Adachi, K., Suemori, H., Yasuda, S.Y., Nakatsuji, N., and Kawase, E. (2010). Role of SOX2 in maintaining pluripotency of human embryonic stem cells. *Genes Cells*, Vol.15, No.5, (May 2010) pp. 455-470, ISSN 1365-2443.

Ambrosetti, D.C., Scholer, H.R., Dailey, L., and Basilico, C. (2000). Modulation of the activity of multiple transcriptional activation domains by the DNA binding domains mediates the synergistic action of Sox2 and Oct-3 on the fibroblast growth factor-4 enhancer. *J. Biol. Chem.*, Vol.275, No.30, (July 2000), pp. 23387-23397, ISSN 0021-9258.

Avilion, A.A., Nicolis, S.K., Pevny, L.H., Perez, L., Vivian, N., and Lovell-Badge, R. (2003). Multipotent cell lineages in early mouse development depend on SOX2 function. Genes Dev., Vol.17, No.1, (January 2003), pp. 126-140, ISSN 0890-9369.

Babaie, Y., Herwig, R., Greber, B., Brink, T.C., Wruck, W., Groth, D., Lehrach, H., Burdon, T., and Adjaye, J. (2007). Analysis of Oct4-dependent transcriptional networks regulating self-renewal and pluripotency in human embryonic stem cells. *Stem Cells*, Vol.25, No.2, (February 2007), pp. 500-510, ISSN 1066-5099.

Boyer, L.A., Lee, T.I., Cole, M.F., Johnstone, S.E., Levine, S.S., Zucker, J.P., Guenther, M.G., Kumar, R.M., Murray, H.L., Jenner, R.G., *et al.* (2005). Core transcriptional regulatory circuitry in human embryonic stem cells. *Cell*, Vol.122, No.6, (September 2005), pp. 947-956, ISSN 0092-8674.

Brons, I.G., Smithers, L.E., Trotter, M.W., Rugg-Gunn, P., Sun, B., Chuva de Sousa Lopes, S.M., Howlett, S.K., Clarkson, A., Ahrlund-Richter, L., Pedersen, R.A., *et al.* (2007).

Derivation of pluripotent epiblast stem cells from mammalian embryos. *Nature*, Vol.448, No. 7150, (July 2007), pp. 191-195, ISSN 1476-4687.

Chambers, I., Colby, D., Robertson, M., Nichols, J., Lee, S., Tweedie, S., and Smith, A. (2003). Functional expression cloning of Nanog, a pluripotency sustaining factor in embryonic stem cells. *Cell*, Vol.113, No.5, (May 2003), pp. 643-655, ISSN 0092-8674.

Chambers, I., Silva, J., Colby, D., Nichols, J., Nijmeijer, B., Robertson, M., Vrana, J., Jones, K., Grotewold, L., and Smith, A. (2007). Nanog safeguards pluripotency and mediates germline development. *Nature*, Vol.450, No.7173, (December 2007), pp. 1230-1234, ISSN 1476-4687.

Daheron, L., Opitz, S.L., Zaehres, H., Lensch, W.M., Andrews, P.W., Itskovitz-Eldor, J., and Daley, G.Q. (2004). LIF/STAT3 signaling fails to maintain self-renewal of human embryonic stem cells. *Stem Cells*, Vo.l22, No.5, (May 2004), pp. 770-778, ISSN 1066-5099.

Darr, H., Mayshar, Y., and Benvenisty, N. (2006). Overexpression of NANOG in human ES cells enables feeder-free growth while inducing primitive ectoderm features. *Development, Vol.* 133, No. 6, (March 2006), pp. 1193-1201, ISSN 0950-1991.

Evans, M.J., and Kaufman, M.H. (1981). Establishment in culture of pluripotential cells from mouse embryos. *Nature*, Vol.292, No.5819, (July 1981), pp. 154-156, ISSN 0028-0836.

Fong, H., Hohenstein, K.A., and Donovan, P.J. (2008). Regulation of self-renewal and pluripotency by Sox2 in human embryonic stem cells. *Stem Cells*, Vol.26, No.8 (August 2008), pp. 1931-1938, ISSN 1066-5099.

Hanna, J.H., Saha, K., and Jaenisch, R. (2010). Pluripotency and cellular reprogramming: facts, hypotheses, unresolved issues. *Cell, Vol.143*, No.4 (November 2010), pp. 508-525, ISSN 0092-8674.

Hay, D.C., Sutherland, L., Clark, J., and Burdon, T. (2004). Oct-4 knockdown induces similar patterns of endoderm and trophoblast differentiation markers in human and mouse embryonic stem cells. *Stem Cells*, Vol.22, No.2 (February 2004), pp. 225-235, ISSN 1066-5099..

Home, P., Ray, S., Dutta, D., Bronshteyn, I., Larson, M., and Paul, S. (2009). GATA3 is selectively expressed in the trophectoderm of peri-implantation embryo and directly regulates Cdx2 gene expression. *J. Biol. Chem.* Vol.284, No. 42, (October 2009), pp. 28729-28737, ISSN 0021-9258.

Humphrey, R.K., Beattie, G.M., Lopez, A.D., Bucay, N., King, C.C., Firpo, M.T., Rose-John, S., and Hayek, A. (2004). Maintenance of pluripotency in human embryonic stem cells is STAT3 independent. *Stem Cells*, Vol.22 (July 2004), pp. 522-530, ISSN 1066-5099, .

Hyslop, L., Stojkovic, M., Armstrong, L., Walter, T., Stojkovic, P., Przyborski, S., Herbert, M., Murdoch, A., Strachan, T., and Lako, M. (2005). Downregulation of NANOG induces differentiation of human embryonic stem cells to extraembryonic lineages. Stem Cells, Vol.23, No. 8 (September 2005), pp. 1035-1043, ISSN 1066-5099.

Kelberman, D., Rizzoti, K., Avilion, A., Bitner-Glindzicz, M., Cianfarani, S., Collins, J., Chong, W.K., Kirk, J.M., Achermann, J.C., Ross, R., *et al.* (2006). Mutations within Sox2/SOX2 are associated with abnormalities in the hypothalamo-pituitary-gonadal axis in mice and humans. *J Clin. Invest.* , Vol.116, No.9 (September 2006), pp. 2442-2455, ISSN 0021-9738.

Kopp, J.L., Ormsbee, B.D., Desler, M., and Rizzino, A. (2008). Small increases in the level of sox2 trigger the differentiation of mouse embryonic stem cells. *Stem Cells*, Vol.26, No.4 (April 2008), pp. 903-911, ISSN 1066-5099.

Loh, Y.H., Wu, Q., Chew, J.L., Vega, V.B., Zhang, W., Chen, X., Bourque, G., George, J., Leong, B., Liu, J., *et al.* (2006). The Oct4 and Nanog transcription network regulates pluripotency in mouse embryonic stem cells. *Nat Genet.* Vol.38, No.4, (April 2006), pp. 431-440, ISSN 1061-4036.

Martin, G.R. (1981). Isolation of a pluripotent cell line from early mouse embryos cultured in medium conditioned by teratocarcinoma stem cells. *Proc. Natl. Acad. Sci. U.S.A.*, Vol.78, No.12, (December 1981), pp. 7634-7638, ISSN 0027-8424.

Masui, S., Nakatake, Y., Toyooka, Y., Shimosato, D., Yagi, R., Takahashi, K., Okochi, H., Okuda, A., Matoba, R., Sharov, A.A., *et al.* (2007). Pluripotency governed by Sox2 via regulation of Oct3/4 expression in mouse embryonic stem cells. *Nat. Cell. Biol.*, Vol.9, No.6, (June 2007), pp. 625-635, ISSN 1465-7392.

Matin, M.M., Walsh, J.R., Gokhale, P.J., Draper, J.S., Bahrami, A.R., Morton, I., Moore, H.D., and Andrews, P.W. (2004). Specific knockdown of Oct4 and beta2-microglobulin expression by RNA interference in human embryonic stem cells and embryonic carcinoma cells. *Stem Cells*, Vol.22, No. 5, (September 2004), pp. 659-668, ISSN 1066-5099.

Mitsui, K., Tokuzawa, Y., Itoh, H., Segawa, K., Murakami, M., Takahashi, K., Maruyama, M., Maeda, M., and Yamanaka, S. (2003). The homeoprotein Nanog is required for maintenance of pluripotency in mouse epiblast and ES cells. *Cell*, Vol.113, No. (5), (May 2003), pp. 631-642, ISSN 0092-8674.

Nichols, J., Zevnik, B., Anastassiadis, K., Niwa, H., Klewe-Nebenius, D., Chambers, I., Scholer, H., and Smith, A. (1998). Formation of pluripotent stem cells in the mammalian embryo depends on the POU transcription factor Oct4. *Cell*, Vol.95, No. 3, (July 1998), pp. 379-391, ISSN 0092-8674.

Niwa, H., Burdon, T., Chambers, I., and Smith, A. (1998). Self-renewal of pluripotent embryonic stem cells is mediated via activation of STAT3. *Genes Dev.*, Vol.12, No. 13, (July 1998), pp. 2048-2060, ISSN 0890-9369.

Niwa, H., Miyazaki, J., and Smith, A.G. (2000). Quantitative expression of Oct-3/4 defines differentiation, dedifferentiation or self-renewal of ES cells. Nat Genet. Vol.24, No.4 (April 2000), pp. 372-376, ISSN 12061-4036.

Niwa, H., Toyooka, Y., Shimosato, D., Strumpf, D., Takahashi, K., Yagi, R., and Rossant, J. (2005). Interaction between Oct3/4 and Cdx2 determines trophectoderm differentiation. *Cell*, Vol.123, No.5, (December 2005), pp. 917-929, ISSN 0092-8674.

Okita, K., Ichisaka, T., and Yamanaka, S. (2007). Generation of germline-competent induced pluripotent stem cells. *Nature*, Vol.448, No. 7151, (June 2007), pp. 313-317, ISSN 0028-0836.

Pera, M.F., Andrade, J., Houssami, S., Reubinoff, B., Trounson, A., Stanley, E.G., Ward-van Oostwaard, D., and Mummery, C. (2004). Regulation of human embryonic stem cell differentiation by BMP-2 and its antagonist noggin. *J. Cell Sci.* , Vol.117, No.7, (March 2004), pp.1269-1280, ISSN 0021-9533.

Rodriguez, R.T., Velkey, J.M., Lutzko, C., Seerke, R., Kohn, D.B., O'Shea, K.S., and Firpo, M.T. (2007). Manipulation of OCT4 levels in human embryonic stem cells results in induction of differential cell types. *Exp. Biol. Med. (Maywood)*, Vol.232, No.10, (November 2007), pp. 1368-1380, ISSN 1535-3702.

Stadtfeld, M., and Hochedlinger, K. (2010). Induced pluripotency: history, mechanisms, and applications. *Genes Dev.*, Vol. 24, No. 20, (October 2010), pp. 2239-2263, ISSN 0890-9369.

Suemori, H., Yasuchika, K., Hasegawa, K., Fujioka, T., Tsuneyoshi, N., and Nakatsuji, N. (2006). Efficient establishment of human embryonic stem cell lines and long-term maintenance with stable karyotype by enzymatic bulk passage. *Biochem. Biophys. Res. Commun.*, Vol.345, No.3, (July 2006), pp. 926-932, ISSN 0006-291X.

Takahashi, K., and Yamanaka, S. (2006). Induction of pluripotent stem cells from mouse embryonic and adult fibroblast cultures by defined factors. *Cell*, Vol.126, No.4, (August 2006), pp. 663-676, ISSN 0092-8674.

Taranova, O.V., Magness, S.T., Fagan, B.M., Wu, Y., Surzenko, N., Hutton, S.R., and Pevny, L.H. (2006). SOX2 is a dose-dependent regulator of retinal neural progenitor competence. *Genes Dev.*, Vol.20, No.9, (May 2006), pp. 1187-1202, ISSN 0890-9369 .

Tesar, P.J., Chenoweth, J.G., Brook, F.A., Davies, T.J., Evans, E.P., Mack, D.L., Gardner, R.L., and McKay, R.D. (2007). New cell lines from mouse epiblast share defining features with human embryonic stem cells. *Nature*, Vol.448, No.7150, (July 2007), pp. 196-199, ISSN 1476-4687.

Thomson, J.A., Itskovitz-Eldor, J., Shapiro, S.S., Waknitz, M.A., Swiergiel, J.J., Marshall, V.S., and Jones, J.M. (1998). Embryonic stem cell lines derived from human blastocysts. *Science*, Vol.282, No.5391 (November 1998), pp. 1145-1147, ISSN 0036-8075.

Thomson, J.A., Kalishman, J., Golos, T.G., Durning, M., Harris, C.P., Becker, R.A., and Hearn, J.P. (1995). Isolation of a primate embryonic stem cell line. *Proc. Natl. Acad. Sci. U.S.A.*, Vol.92, No.17, (August 1995), pp. 7844-7848, ISSN 0027-8424.

Wernig, M., Meissner, A., Foreman, R., Brambrink, T., Ku, M., Hochedlinger, K., Bernstein, B.E., and Jaenisch, R. (2007). In vitro reprogramming of fibroblasts into a pluripotent ES-cell-like state. *Nature*, Vol.*448*, No.7151, (July 2007), pp. 318-324, ISSN 0028-0836.

Williamson, K.A., Hever, A.M., Rainger, J., Rogers, R.C., Magee, A., Fiedler, Z., Keng, W.T., Sharkey, F.H., McGill, N., Hill, C.J., *et al.* (2006). Mutations in SOX2 cause anophthalmia-esophageal-genital (AEG) syndrome. *Hum. Mol. Genet.*, Vol.15, No.9, (May 2006), pp. 1413-1422, ISSN 0963-6906.

Wood, H.B., and Episkopou, V. (1999). Comparative expression of the mouse Sox1, Sox2 and Sox3 genes from pre-gastrulation to early somite stages. *Mech. Dev.*, Vol.86, No.1-2, (August 1999), pp. 197-201, ISSN 0925-4773.

Xu, R.H., Chen, X., Li, D.S., Li, R., Addicks, G.C., Glennon, C., Zwaka, T.P., and Thomson, J.A. (2002). BMP4 initiates human embryonic stem cell differentiation to trophoblast. *Nat. Biotechnol.*, Vol.20, No.12, (December 2002), pp. 1261-1264, ISSN 1087-0156.

Yamaguchi, S., Hirano, K., Nagata, S., and Tada, T. (2011). Sox2 expression effects on direct reprogramming efficiency as determined by alternative somatic cell fate. *Stem Cell Res.*, Vol.*6*, No.2, (March 2011), pp. 177-186, ISSN 1876-7753.

Ying, Q.L., Nichols, J., Chambers, I., and Smith, A. (2003). BMP induction of Id proteins suppresses differentiation and sustains embryonic stem cell self-renewal in collaboration with STAT3. Cell, Vol.115, No.3, (October 2003pp. 281-292, ISSN 0092-8674.

Zaehres, H., Lensch, M.W., Daheron, L., Stewart, S.A., Itskovitz-Eldor, J., and Daley, G.Q. (2005). High-efficiency RNA interference in human embryonic stem cells. *Stem Cells*, Vol.*23*, No.3, pp. 299-305, ISSN 1066-5099.

Zhao, S., Nichols, J., Smith, A.G., and Li, M. (2004). SoxB transcription factors specify neuroectodermal lineage choice in ES cells. Mol. Cell Neurosci., Vol.27, No.3, (November 2004), pp. 332-342, ISSN 1044-7431.

8

The Role of MicroRNAs in Regulating Cancer Stem Cells

Musaffe Tuna and Christopher I. Amos
Department of Epidemiology,
The University of Texas MD Anderson Cancer Center
Houston, Texas
USA

1. Introduction

Stem cells are a rare population of cells that have the ability to self-renew (to replenish the stem cell pool) and to differentiate (to produce daughter cells that will perform the physiological functions of tissues and organs). Although stem cells exist in different tissues, organs, and developmental stages. However, stem cells differ to some degree with regard to their developmental potency; life span, and notably their potential for self-renewal and proliferation capacity.

Stem cell self-renewal and differentiation is regulated by signaling pathways, transcription factors, and micro RNAs (miRNAs). Some key transcription factors directly regulate the expression of miRNAs in stem cells. Meanwhile, miRNAs target key transcription factors and either repress or induce their expression in stem cells to regulate self-renewal and differentiation. Thereby, the miRNA regulatory network and the signaling pathways cross-talk to each other to orchestrate stem cell maintenance and cell fate decision. Dysregulation of core signaling pathways, transcription factors and miRNAs associated with normal stem cells can lead to carcinogenesis. Thus, understanding the regulation of normal stem cell is crucial for understanding the molecular mechanisms underline carcinogenesis.

In this chapter, we review the characteristics and functions of miRNAs and cancer stem cells (CSCs), focusing on the roles of miRNAs in regulating CSCs. First, we provide an introduction to stem cells and CSCs. Then, we describe the signaling pathways that regulate stem cell self-renewal and differentiation. In particular, we review the Wnt/β-catenin, Hedgehog (HH), and Notch pathways. Next, we discuss the epithelial-mesenchymal transition (EMT), CSCs, and miRNAs that play roles in regulating stem cells. Finally, we summarize the current status and discuss future perspectives.

2. Stem cells

Depending on their differentiation potentials, human stem cells can be classified into totipotent, pluripotent, and multipotent (http://stemcells.nih.gov/info/scireport). **Totipotent** cells have the potential to form any of the differentiated cells in a living organism from a single cell. Thus, these cells have ability to form extraembryonic membranes and tissues; the embryo itself, and all postembryonic tissues and organs. At the very early stage of embryo development, each cell in the blastomere is totipotent. **Pluripotent** cells can differentiate to form tissues of any of the three germ layers: ectoderm, endoderm, or mesoderm. However, a pluripotent cell cannot form an entire living organism. Embryonic stem cells (ESCs) are pluripotent stem cells derived from the inner cell mass of the human blastocyst. ESCs can differentiate into specialized cells, and have an unlimited capacity for self-renewal. **Multipotent** cells—adult stem cells—have a differentiation ability that is limited to a specific tissue- or organ. Tissue-specific adult stem cells are responsible for organogenesis; tissue maturation, repair and regeneration, and maintenance; and balancing the cellular turnover. To fulfill these responsibilities; first, an adult stem cell is an undifferentiated cell that is found in a differentiated tissue and has the capacity to become specialized to yield all of the cell types of the tissue from which it originated; second, a stem cell has capacity to self-renewal (Spradling et al., 2001). They can undergo two kinds of cell division: symmetric and asymmetric. In symmetric division, a stem cell divides into two identical daughter cells, which are both identical to the originating stem cell. This type of division is crucial for expanding the stem cell pool, most likely in very early embryonic development. In contrast, in asymmetric division, a stem cell divides into one daughter progenitor cell (also known as a precursor cell), which eventually differentiates into a mature cell, and one new stem cell, which is identical to the originating stem cell. This process maintains stem cell number, and this feature also distinguishes the stem cell self-renewal from other proliferative processes. Normal adult stem cell can divide asymmetrically to maintain the population of stem cells and differentiated cells. The processes that regulate the balance between asymmetric and symmetric division of stem cells are unclear.

A progenitor cell is a partially specialized cell that can divide and yield two specialized cells. Progenitor cells can be distinguished from adult stem cells as follows. When a stem cell divides, at least one of the two new cells is always identical to the originating stem cell and can replicate itself. In contrast, when a progenitor cell divides, it gives rise to two progenitor cells or two specialized cells, neither of which can replicate itself. Progenitor cells can replace cells that are damaged or dead, thereby maintaining the integrity and functions of a tissue or an organ such as the liver or the brain. Examples of stem and progenitor cells:

- Hematopoietic stem cells (adult stem cells) from the bone marrow that give rise to erythrocytes, lymphocytes, plateles, monocytes, and granulocytes.
- Mesenchymal stem cells (MSCs) are a subset of nonhematopoietic multipotent stem cells (adult stem cells) that are found primarily within the bone marrow and give rise to stromal cells; within the adipose tissue that give rise to adipocytes (Bieback et al., 2008; Digirolamo et al., 1999). MSCs have also been isolated from the umblical cord (fetal stem cells). Mesenchymal stem cells can self-renew and are defined as cells that differentiate into a variety of mesenchyme-derived cell types: fibroblasts, chondrocytes, osteoblasts, myoblasts, and neural stem cells; the latter cells have the potential to differentiate into neurons, astrocytes, and oligodendrocytes (Barry and Murphy, 2004; Halleux et al., 2001).
- Epithelial stem cells (progenitor cells) that give rise to the various types of skin cells.
- Muscle stem cells that give rise to differentiated muscle tissue.
- Intestinal stem cells.

On the other hand, accumulating data show that different stem cells have distinct potential to proliferate, and some adult stem cells from one tissue are capable of differentiating into the specialized cell types of another tissue (Herzog et al., 2003; Krause, 2002a). This phenomenon is referred to as stem cell plasticity. For example, under specific experimental conditions, adult stem cells from bone marrow can differentiate into cells that resemble neurons (Herzog et al., 2003; Krause, 2002a). Growing evidence indicates that, given the right environment (environmental niche), some adult stem cells are capable of being genetically reprogrammed to differentiate into tissues other than the ones from which they originated.

Regardless of division type, stem cell self-renewal is especially important in tissues with high self-renewal capacity, such as the intestinal cells and bone marrow, and also in tissue repair after injury. Adult tissues that undergo turnover throughout life are maintained via a very small portion of cells—adult stem cells that live through the entire life span of an organism. These stem cells can maintain homeostasis even in mitotically inactive adult tissues, such as the brain (Bartlett, 1982; Ricci-Vitiani et al., 2008). Even though stem cells have an extensive capacity for self-renewal, in fact they remain quiescent most of the time and may undergo a limited number of self-renewing divisions in adult life (Cheshier et al., 1999). This may be because, despite their proliferative capacity, stem cells often arrest at a G_0-like cell cycle phase or checkpoint (Cheshier et al., 1999). In addition, the differentiation and self-renewal rates differ depending on the stem cell type (Ahn and Joyner, 2005; Hu et al., 2004).

Adult stem cells are not easy to characterize. To date, adult stem cells have been characterize *in vitro* by using their differentiations patterns and cell surface markers. Stem cells have been identified in bone marrow, blood, the cornea and retina, the brain, skeletal muscle, dental pulp, liver, skin, the intestinal tract, pancreas, ovary, breast, lung, prostate and head and neck (http://stemcells.nih.gov/info/scireport). Thus, stem cells have been found in tissues that develop from all three embryonic germ layers.

3. Signaling pathways in stem cells

In both pluripotent and multipotent cells, self-renewal and cell fate decision are regulated by a complex set of factors and pathways. Each process: self-renewal and differentiation requires unique molecular programs specific to each pluripotent or multipotent cell. For example, in ESCs, self-renewal requires that the unique molecular program of the pluripotent state be maintained, whereas to differentiate into various lineages, ESCs must shift to alternative molecular programs that inhibit self-renewal and promote differentiation (Marson et al., 2008a). Understanding how cells switch between self-renewal and differentiation, and discovering which factors or signaling pathways control which daughter cell of an adult stem cell remains a stem cell and which undergoes differentiation, is crucial to understand the mechanism of tumorigenesis.

Several "stemness" factors are required to ensure appropriate ESC behavior (pluripotency). A core network of factors, including transcription factors and RNA binding proteins (Oct4, Sox2, Nanog, Klf4, c-Myc, Tcf3, and Lin28), is involved in the circuits that regulate ESC pluripotency. (Marson et al., 2008a). Some of these regulatory factors are tissue or cancer specific; for example, Oct4 is expressed only in the inner cell mass of the embryo and not in the trophectoderm. Some of these key regulators of ESC identity, such as Oct4, Sox2, and Nanog are expressed only in specific human cancer types (Gidekel et al., 2003; Rodriguez-Pinilla et al., 2007; Santagata et al., 2007). Thus, regulatory networks can determine classes of stem cells, such as ESCs, neural stem cells, or breast stem cells or other tissue specific stem cells (Muller et al., 2008). Sox2 and Nanog can also reprogram differentiated human cells into ESC-like induced pluripotent stem cells (Park et al., 2008; Wernig et al., 2007).

Pluripotency and the unlimited potential for self-renewal are the characteristics that distinguish ESCs from adult (tissue-specific) stem cells, which have more limited self-renewal and tissue-specific differentiation potential. The common feature of stem cells (ESCs and adult stem cells) is self-renewal. Not surprisingly, the stem cell niche and signaling pathways such as Wnt/β-catenin, Notch, Hedgehog TGF-β, and Bmi-1 are involved in the regulation of normal self-renewal programs, the balance between self-renewal and differentiation (Dontu et al., 2004; Reya and Clevers, 2005; Schofield, 1978; Song et al., 2007; Taipale et al., 2002). Accumulating evidence indicates that networks that balance proto-oncogenes (promoting self-renewal) and tumor suppressors, which act as gatekeepers (limiting self-renewal) and caretakers (maintaining self-renewal) is also involved in tissue stem cell self-renewal programs (He et al., 2009). For example, the p53, PTEN, and INK4A pathways are involved in stem cell self-renewal (Armesilla-Diaz et al., 2009b; Cicalese et al., 2009; Lowe and Sherr, 2003; Nagao et al., 2008; Zheng et al., 2008a) (Table 1). Therefore, it is not surprising that these transcription factors (*PTEN, TP53* and *INK4A*) are deleted or mutated in multiple CSCs.

Signaling pathway	Type of stem or progenitor cell	References
Wnt/β-catenin	Hematopoietic stem cells Epidermal stem cells Gastrointestinal stem cells Neural stem cells Embryonic stem cells Dental pulp stem cells	(Luis et al., 2009; Reya et al., 2003) (Zhu and Watt, 1999) (Brittan and Wright, 2002; He et al., 2004) (Kalani et al., 2008) (Melchior et al., 2008; Tam et al., 2008) (Scheller et al., 2008)
Notch	Gastrointestinal progenitor cells Mammary stem/progenitor cells Liver stem cells Muscle progenitor cells Hematopoietic stem cells	(Fre et al., 2005) (Bouras et al., 2008; Dontu et al., 2004) (Zong et al., 2009) (Buas and Kadesch, 2010; Conboy et al., 2003; Conboy and Rando, 2002) (Varnum-Finney et al., 2000)
Hedgehog	Hematopoietic stem cells Neural stem cells Mammary stem cells	(Bhardwaj et al., 2001) (Palma et al., 2005; Wechsler-Reya and Scott, 1999) (Liu et al., 2006)
Bmi-1	Mammary stem cells Hematopoietic stem cells	(Liu et al., 2006) (Park et al., 2003)
PTEN	Neural stem cells	(Groszer et al., 2006; Groszer et al., 2001; Nagao et al., 2008; Zheng et al., 2008a)
p53	Mammary stem cells Neural stem cells	(Cicalese et al., 2009) (Armesilla-Diaz et al., 2009a; Zheng et al., 2008a)

Table 1. Signaling pathways involved in stem cell self-renewal.

Moreover, signaling pathways cross-talk or interact with each other to regulate stem cell behavior. For example, hypoxia-inducible factor-1α and Notch signaling interact to regulate medulloblastoma precursor cell proliferation and differentiation (Pistollato et al., 2010). The Notch and EGFR pathways interact with each other to regulate the number of neural stem cells (NSCs) (Aguirre et al., 2010). Key pathways including Wnt, HH, Notch, and Bmi-1 and transcription factors including *TP53* and *PTEN* are involved in the development of various organs during embryogenesis and in the regulation of self-renewal and differentiation in both normal adult stem cells (Molofsky et al., 2004), and CSCs such in normal adult SCs and CSCs in glioblastoma (Zheng et al., 2008b). Dysregulation of these core pathways (e.g. Wnt, HH, Notch) and transcription factors (*TP53* and *PTEN*), which associated with normal stem cells is also plays a role in the cancer development (Zheng et al., 2008b). Here, we focus on the Wnt/β-catenin, HH, and Notch pathways.

3.1 Wnt/β-catenin pathway

Two kinds of Wnt signaling pathways exist: the canonical Wnt pathway, in which Wnt ligands signal through the stabilization of β-catenin, and the noncanonical Wnt pathway, which is β-catenin-independent. The canonical Wnt pathway is activated when Wnt ligands bind to cell surface receptors composed of a member of the Frizzled protein family and one of the co-receptors LRP5 and LRP6 and hyperphosphorylate the Dishevelled (Dsh) protein, thereby activating it. Activation of Dsh prevents the phosphorylation of β-catenin and inhibits the formation of β-catenin destruction complex (glycogen synthase kinase 3β [GSK-3β], adenomatous polyposis coli [APC], casein kinase 1α [CK1α] and Axin) which leads to the stabilization of hypophosphorylated β-catenin and, thereby, to its translocation to the nucleus where it interacts with transcription factors (T cell factor/lymphoid enhancer factor [TCF/LEF]). Thus the β-catenin/TCF/LEF complex activates the transcription of target genes. In the absence of Wnt ligands, β-catenin destruction complex hyperphosporylates β-catenin, thereby this complex (hyperphosphorylated β-catenin, APC, Axin, GSK-3β, and CK1α) is thus a target for ubiquitination and degradation by the proteasome (Schweizer and Varmus, 2003).

The noncanonical Wnt signaling pathway requires Frizzled receptors and the proteoglycan co-receptor Knypek. In this pathway, Dsh localizes in the cell membrane and activates Rho through Daam1. Dsh induce cellular response by stimulating calcium flux and activating the calcium-sensitive kinases protein kinase C and calmodulin-dependent protein kinase II (Veeman et al., 2003).

Wnt signaling has been studied intensively in embryonic development. The response of cells to the Wnt pathway is tissue-dependent. Wnt signaling is involved in many key developmental processes, such as cell survival, proliferation, inhibition of apoptosis, stem cell maintenance, differentiation, and cell fate decision, and in the development of a variety of organ systems, including the cardiovascular system, central nervous system, kidney, and lung (Ille and Sommer, 2005; Peifer, 2000; Vainio et al., 1999a; Vainio et al., 1999b; Wodarz and Nusse, 1998). For example, the canonical Wnt pathway plays a crucial role in the development of intestinal tissue by regulating the self-renewal, migration and proliferation of intestinal stem and progenitor cells, and tissue self-renewal in hair follicles and bone growth plates (Clevers, 2006). The Wnt pathway also interacts with other pathways to regulate stem cell processes. For example, bone morphogenetic protein (BMP) inhibits Wnt signaling to negatively regulate stem cell proliferation (He et al., 2004), BMP signaling thereby represses de novo crypt formation and polyp growth, and mutations in BMP pathway genes lead to formation of crypts and generation of benign polyps (Reya and Clevers, 2005).

In addition to biologic and developmental process, Wnt signaling is also involved in genetic processes. For example, APC has been shown to be involved in regulating mitotic spindle assembly, orientation of chromosomes during mitotic division, and chromosome segregation (Kaplan et al., 2001). Abnormalities in the orientation of chromosomes during mitotic division may contribute to numeric chromosomal aberrations in cancer cells (Peifer, 2000).

3.2 Hedgehog pathway

Three HH ligands have been identified—Sonic HH, Desert HH, and Indian HH (Cohen, 2003). In the presence of ligands, these ligands bind to the transmembrane receptor Patched 1 (Ptch 1), which inhibits Smothened (Smo). The binding of HH ligands relieves Smo inhibition, leading to activation of the Gli transcription factors Gli1 and Gli2 (activator) and Gli3 (suppressor). Activated Gli accumulates in the nucleus and controls the transcription of HH target genes. In the absence of HH ligands, Ptch1 inhibits Smo, and cannot activate Gli (Pasca di Magliano and Hebrok, 2003). The HH signaling pathway regulates cell proliferation through Cyclin D1 and FoxM1, apoptosis through Bcl-2, EMT through Snail and E-cadherin, and self-renewal through Bmi-1 (Kasper et al., 2009). Bmi-1 has been shown to be a key regulator of the self-renewal of NSCs and both normal and leukemic stem cells (Lessard and Sauvageau, 2003; Molofsky et al., 2003; Park et al., 2003). HH signaling has been shown to play a critical role in the development of many systems, including the limb, brain, spinal cord, thalamus, and teeth. The HH pathway is also important in cell proliferation, differentiation, and stem cell maintenance during embryogenesis (Ma et al., 2002), and in the self-renewal and maintenance of NSCs (Ahn and Joyner, 2005; Lai et al., 2003; Palma et al., 2005), mammary stem cells (Liu et al., 2006).

Either aberration of genes in the HH signaling pathway or aberrant activation of HH signaling results in tumorigenesis. For example, germline mutations in Patch, which functions as a tumor suppressor has been found in basal cell carcinomas and Gorlin syndrome (Cohen, 2003). Activation of HH signaling is implicated in small cell lung cancer, digestive tract tumor, pancreatic carcinoma, breast cancer and prostate cancer (Karhadkar et al., 2004; Olsen et al., 2004).

3.3 Notch pathway

Members of the Notch gene family encode transmembrane receptors that are crucial for cell fate decision. Four Notch receptors (Notch1, Notch2, Notch3 and Notch4) and five ligands (Jagged-1 [JAG1] and JAG2, three Delta-like [DLL1, DLL2, and DLL4]) have been found. These receptors and ligands are expressed in different combinations in most cell types (Mumm and Kopan, 2000). After ligand binding, Notch receptors are activated via cleavages of ADAM metallopeptidase domain 17 (ADAM17) and presenilin-1, which result in the release and translocation of the Notch intracellular domain (NICD) to the nucleus and the activation of HES (Hes/E(spl) family) and HEY (Hesr/Hey family) families through interaction of NICD with sequence-binding protein (Mumm and Kopan, 2000). Notch signaling is crucial for arterial-venous differentiation, for self-renewal and differentiation in hematopoietic stem cells (Krause, 2002b), maintenance of the mammary stem cell population (Bouras et al., 2008), for adult neurogenesis (Androutsellis-Theotokis et al., 2006), and for the activity of myogenic muscle stem and progenitor cells (Buas and Kadesch, 2010; Conboy et al., 2003; Conboy and Rando, 2002). Notch signaling is involved in the self-renewal process mostly in rapidly renewing tissues, such as the hematopoietic system (Mercher et al., 2008; Wu et al., 2007), the intestine, skin, highly proliferative ESCs, and the intestine, in which the epithelium is renewed every 4-5 days (Dontu et al., 2004). Notch and Wnt signaling cooperate to regulate self- renewal and cell fate in the adult intestine (Chiba, 2006; Fre et al., 2005; Wang and Hou, 2010), and inhibition of Notch/γ-secretase induces proliferation in intestinal crypt cells and the formation of polyps (van Es et al., 2005). Adult epidermal stem cells reside in the epidermal basal layer and in the bulge region of the hair follicle (Ambler and Maatta, 2009). In addition to differentiation and self-renewal, the Notch pathway is also involved in other developmental processes, including EMT, proliferation, apoptosis, and cell adhesion during embryogenesis (Zong et al., 2009).

EMT was originally defined as a cellular reorganization process that is essential for embryonic development. EMT results in a loss of cell to cell adhesive properties, a loss of cell polarity, and a gain of the invasive and migratory features of mesenchymal cells (Thiery et al., 2009). During embryogenesis, EMT leads progenitor/precursor cells to migrate to distant sites within the embryo to form new tissues (Shook and Keller, 2003). The EMT process is reversible. EMT also occurs during tumorigenesis; the process is similar to EMT during the embryogenesis, but instead of forming new tissue, it allows some CSCs to become metastatic while keeping the features of the original tissue. It is not surprising that, the same, or similar, core signaling pathways (Wnt, HH and Notch) that regulate stem cell self-renewal are also involve regulation of EMT together as are niche factors (Mani et al., 2008; Vincan and Barker, 2008; Yang and Weinberg, 2008).

The dysregulation of signaling pathways by mutations and/or by genomic and epigenetic aberrations, which are involved in the regulation of stem cell function as well as in EMTs during embryonic development may play a crucial role in the development of cancer. CSCs and normal stem cells use many of the same signaling pathways, such as Wnt, HH, and Notch, but the difference is CSC use dysregulated way of these signaling pathways (Takebe et al., 2010). Upto now, aberrant Notch signaling has been shown in multiple human cancers including hepatocellular carcinoma, hepatoblastoma, colorectal cancer, acute myeloid leukemia, chronic myeloid leukemia, multiple myeloma, gastric cancer, and Wilms' tumor which also shows dysregulation of Wnt signaling (de La Coste et al., 1998; Kim et al., 2009; Koesters et al., 1999; Martin et al., 2010 ; Reya and Clevers, 2005) (Table 2).

4. Cancer stem cells

Normal stem cells and CSCs share several important properties, including the ability to self-renew. The signaling pathways and transcription factors that are involved in the self-renewal of normal stem cells have all been implicated in the development of cancers, but in CSCs the pathways are dysregulated and the factors are aberrantly expressed. CSCs can be distinguished from normal stem cells by the following.

1. CSCs have the capacity for self-renewal like normal stem cells, but CSCs have a different self-renewal rate from normal stem cells.
2. CSCs have the capacity to differentiate into cells of the specific tissue, but aberrantly (Singh et al., 2003).
3. CSCs have the ability to develop tumor when transplanted into the proper environment.
4. CSCs have the capacity for tumor metastasis.

5. CSCs have the abilty to repopulate the tumor, causing relapse, and can become resistant to different therapeutic agents.
6. CSCs are identified by characteristic cell surface markers.

Aberrant activation of an individual signaling pathway or cross-talk between pathways may result in tissue-specific carcinogenesis (Sun et al., 2010). Thus, an understanding of the pathways that govern the self-renewal and cell fate decisions of normal stem cells, and how these pathways are dysregulated and which of them are dysregulated during carcinogenesis, is of utmost importance. In many cases, self-renewal regulators have surprisingly similar functions in CSCs and normal stem cells (Tables 1 and 2). For example,

Signaling pathway	**Type of cancer**	**References**
Wnt/ β-catenin	Liver	(Ma et al., 2007)
	Breast	(Korkaya et al., 2009)
	Chronic myeloid leukemia	(Zhao et al., 2007)
	Acute myeloid leukemia	(Wang et al., 2010)
	Colon	(Polakis, 2000; Vermeulen et al.)
	Prostate	(Bisson and Prowse, 2009; Shahi et al., 2011)
	Intestine	(Fre et al., 2009)
	Skin	(Chan et al., 1999)
Notch	Liver	(Ma et al., 2007)
	Colon	(Sikandar et al., 2010)
	Breast	(Bouras et al., 2008; Dontu et al., 2004)
	Intestine	(Fre et al., 2009)
	Prostate	(Shahi et al., 2011)
	T-cell leukemia	(Aster et al., 2010)
Hedgehog	Liver	(Ma et al., 2007)
	Breast	(Liu et al., 2006)
	Pancreatic	(Li et al., 2007)
	Glioblastoma	(Ingham, 2008)
	Chronic myeloid leukemia	(Dierks et al., 2008; Zhao et al., 2009)
	Colon	(Varnat et al., 2009)
	Multiple myeloma	(Peacock et al., 2007)
	Medulloblastoma	(Berman et al., 2002)
	Basal cell carcinoma	(Gailani and Bale, 1999)
Bmi-1	Breast	(Liu et al., 2006)
	Head and neck squamous cell cancer	(Prince et al., 2007)
	Acute myeloid leukemia	(Lessard and Sauvageau, 2003)
PTEN	Breast	(Korkaya et al., 2009)
	Glioblastoma	(Zheng et al., 2008b)

Table 2. Signaling pathways that are involved in stem cell self-renewal and are dysregulated in cancer stem cells

the proto-oncogene Bmi-1 is required to maintain both the proliferative potential of leukemic stem cells (Lessard and Sauvageau, 2003) and the self-renewal potential of normal hematopoietic stem cells, mammary stem cells, and NSCs (Liu et al., 2006; Molofsky et al., 2003; Park et al., 2003). Similarly, *PTEN* and *TP53* are required for differentiation and to maintain self-renewal not only in normal NSCs but also in neoplastic stem cells of glioblastoma (Zheng et al., 2008a; Zheng et al., 2008b). Notch signaling is also required to maintain self-renewal in normal and glioma stem cells (Hu et al., 2011), and HH signaling is required not only for normal NSC maintenance but also for brain tumor cell proliferation (Balordi and Fishell, 2007).

Whereas some key transcription factors share some of their target genes and participate in autologous feedback loops to control one another's transcription, others directly regulate self-renewal. On the other hand, in addition to key transcription factors and RNA-binding proteins that regulate self-renewal, miRNAs are also involved in this complex regulatory network.

5. miRNAs

Small noncoding RNAs, which include miRNAs, are a new class of gene that do not code mRNA or protein but are post-transcriptional regulators of gene expression. This regulation generally occurs by binding of a small (~22-nucleotide-long) mature miRNA to mRNA via direct canonical base-pairing between nucleotides 2–8 at the 5′ end of the miRNA (the seed region) and the 3′ untranslated region (UTR) of the target mRNA (its complementary seed-match sequence). Mature single-stranded miRNA is unwound by the helicase activity of Dicer and the RNA-induced silencing complex, resulting in the inhibition of translation, destabilization, and localization of target mRNA. miRNAs are not only post-transcriptional regulators of target genes but also play roles in establishing epigenetic programs (Filipowicz et al., 2008; Stefani and Slack, 2008). miRNAs are not translated into protein, rather, their function is to regulate gene expression by binding to other RNAs, particularly mRNA (Bartel, 2004) (Table 3).

The first miRNAs were discovered in *Caenorhabditis elegans* when mutations in *lin-4* (Lee et al., 1993) and *let-7* (Reinhart et al., 2000) were found to result in defective stem cell maturation (Bartel, 2004). Since then, the miRNA field has been explored extensively and miRNAs have been found to be key regulators of many gene expression networks. In humans, thousands of miRNAs regulate thousands of mRNAs, and each miRNA targets and regulates hundreds of mRNAs to either induce their degradation or prevent their translation. Accumulating data have shown that miRNAs are involved in almost every biological process, and therefore dysregulation of miRNAs is involved in many human diseases, most notably cancer (Esquela-Kerscher and Slack, 2006; Yu et al., 2007) (Table 4).

miRNAs play crucial roles as regulators of stem cell function, differentiation, and embryonic development (Filipowicz et al., 2008; Stefani and Slack, 2008), as well as act as oncogenes and tumor suppressor genes (Garzon et al., 2006). Recent discoveries have revealed that a complex regulatory network of miRNAs, transcription factors, and signaling pathways orchestrate cell-renewal and differentiation (Ferretti et al., 2008; Kato et al., 2009; Kennell et al., 2008; Marson et al., 2008b). The switch from pluripotent to lineage-specific cells is characterized by suppression of pluripotency by activation of expression of lineage-specific genes and repression of self-renewal genes in ESCs, and miRNAs are involved in the regulation of genetic programs. For example, miR-145 promotes the switch from the

Target	Effect (positive or negative)	miRNA	Biological process	References
I. Wnt signaling				
β-catenin	–	*miR-200a*	Meningioma	(Saydam et al., 2009)
APC	–	*miR-135a, miR-135b*	Colorectal cancer	(Nagel et al., 2008)
Wnt1		*miR-34a, miR-21*	Dendritic cell differentiation	(Hashimi et al., 2009)
II. Hedgehog signaling				
Smoothened (Smo)	+	*miR-324-5p* *miR-125b* *miR-326*	Neural stem cell proliferation, medulloblastoma	(Ferretti et al., 2008)
Gli1	+ & –	*miR-324-5p*	Neural stem cell proliferation, medulloblastoma	(Ferretti et al., 2008)
Dkk1 Kremen2 SFRP2	+	*miR-29*	Osteoblast differentiation	(Kapinas et al., 2010)
III. Receptor tyrosine kinase signaling				
NRAS, KRAS	–	*let-7*	Cancer stem cell differentiation, tumor formation	(Johnson et al., 2005)
IV. Notch signaling				
HES1	–	*miR-159b-5p*	Medulloblastoma	(Garzia et al., 2009)
JAG1	–	*miR-34a, miR-21*	Dendritic cell differentiation	(Hashimi et al., 2009)
JAG1	–	*miR-200*	Pancreatic adenocarcinoma Basal-type breast cancer	(Brabletz et al., 2011)
Notch1 JAG1	–	*miR-34a*	Cervical carcinoma	(Pang et al., 2010)
LATS	+	*miR-372, miR-373*	Testicular germ cell tumor	(Voorhoeve et al., 2006)
V. p53 signaling				
TP53	+	*miR-125b*	Apoptosis in the brain	(Le et al., 2009)

Target	Effect (positive or negative)	miRNA	Biological process	References
VI. PTEN signaling				
PTEN Bim Prkaa1 PP2A	-	*miR-19*	T-cell acute lymphoblastic leukemia	(Mavrakis et al., 2010)
PTEN	-	*miR-21*	Hepatocellular cancer	(Meng et al., 2007)

Table 3. MicroRNAs that regulate signaling pathways that determine properties of cancer stem cells

pluripotent state to lineage-specific differentiation by supressing pluripotency factors (e.g., Klf4, Sox2, and Oct4) (Xu et al., 2009). Similarly, the switch from multipotent to lineage specific cells is marked by inhibition of self-renewal and proliferation and induction of cell fate decision. For example, miR-124 promotes neuronal differentiation by downregulating Sox9 in adult neural stem cells (Cheng et al., 2009). miRNAs that are involved in stem cell self-renewal and differentiation and thus regulate cell type specification and differentiation are summarized in Table 3.

Recent reports indicate that miRNAs are central players in stem cell biology (Gangaraju and Lin, 2009), and may have a crucial role in future stem cell therapies. Each type of cell has a distinct miRNA signature. For example, Suh and colleagues reported the first miRNA signature in human ESCs and grouped those miRNAs into four classes; (1) miRNAs found to be specific to ESCs (miR-154, miR-200c, miR-368, miR-371, miR-372, and miR-373); (2) miRNAs found in both ESCs and their malignant counterpart, embryonal carcinoma cells (miR-302a, miR-302b, miR-302c, miR-302d, and miR-367); (3) miRNAs found to be rare in ESCs but abundant in HeLa and STO cells (let-7a, , miR-21, miR-29, miR-29b, miR-301, and miR-374); and (4) miRNAs found to be expressed in most of the cell lines tested (miR-16, miR-17-5p, miR-19b, miR-26a, miR-92, miR-103, miR-130a, and miR-222) (Suh et al., 2004).

miRNA	Type of cell	Biological process	References
let-7	Breast cancer stem cells	Self-renewal	(Yu et al., 2007)
let-7	Breast cancer stem cells	Differentiation	(Yu et al., 2007)
let-7a-1 *let-7d* *let-7-f-1*	ESCs	Pluripotency	(Navarro et al., 2009)
let-7a-2 *let-7a-3,* *let-7b*	ESCs	Pluripotency	(Navarro et al., 2009)
miR-92a	ESC	Self-renewal and differentiation	(Sengupta et al., 2009)
miR-124	Adult neuronal stem cell	Differentiation	(Cheng et al., 2009)

miRNA	Type of cell	Biological process	References
miR-200 *miR-205*	ESCs	Epithelial–mesenchymal transition	(Bracken et al., 2008; Gregory et al., 2008)
miR-150	B cells	Differentiation	(Xiao et al., 2007)
miR-1	Myoblasts	Differentiation	(Chen et al., 2006)
miR-430 *miR-427* *miR-302*	ESCs	Repress formation of ectoderm progenitor cells	(Ivey and Srivastava, 2010)
miR-109 *miR-24*	ESCs	Repress formation of endoderm progenitor cells	(Ivey and Srivastava, 2010)
miR-122 *miR-192*	ESCs	Promote formation of endoderm progenitor cells	(Ivey and Srivastava, 2010)
miR-17-92 *miR-15a* *miR-16-1* *miR-21*	ESCs	Self-renewal	(Navarro et al., 2009)
miR-199a	Mesoderm progenitor cells	Repress differentiation into chondrocytes	(Ivey and Srivastava, 2010)
miR-296 *miR-2861*	Mesoderm progenitor cells	Promote differentiation into osteoblasts	(Ivey and Srivastava, 2010)
miR-214 *miR-206* *miR-1* *miR-26a*	Mesoderm progenitor cells	Promote differentiation into skeletal muscle cells	(Ivey and Srivastava, 2010; Chen et al., 2006)
miR-133 *miR-221* *miR-222*	Mesoderm progenitor cells	Repress differentiation into skeletal muscle cells	(Ivey and Srivastava, 2010)
miR-1	Mesoderm progenitor cells	Promote differentiate into cardiac muscle cells	(Ivey and Srivastava, 2010)
miR-133	Mesoderm progenitor cells	Repress differentiation into cardiac muscle cells	(Ivey and Srivastava, 2010)
miR-145	Neural crest stem cells	Promote differentiation into smooth muscle cells	(Ivey and Srivastava, 2010)
miR-203	Ectoderm progenitor cells	Promote differentiation into keratinocytes	(Ivey and Srivastava, 2010)
miR-9 *miR-124a*	Neural stem cells	Promote differentiation into glial cells and neurons	(Ivey and Srivastava, 2010)

miRNA	Type of cell	Biological process	References
miR-223 *miR-181*	Hematopoietic progenitor cells	Promote differentiation into lymphoid progenitor cells	(Ivey and Srivastava, 2010)
miR-223	Hematopoietic progenitor cells	Promote differentiation into myeloid progenitor cells	(Ivey and Srivastava, 2010)
miR-146 *miR-128a* *miR-181a*	Hematopoietic progenitor cells	Repress differentiation into lymphoid progenitor cells	(Ivey and Srivastava, 2010)
miR-128a *miR-181a* *miR-155* *miR-24a* *miR-17*	Hematopoietic progenitor cells	Repress differentiation into myeloid progenitor cells	(Ivey and Srivastava, 2010)
miR-150	Lymphoid progenitor cells	Promote differentiation into T cells	(Ivey and Srivastava, 2010)
miR-223	Myeloid progenitor cells	Repress differentiation into granulocytes	(Ivey and Srivastava, 2010)
miR-17-5p *miR-20a* *miR-106a*	Myeloid progenitor cells	Repress differentiation into monocytes	(Ivey and Srivastava, 2010)
miR-150 *miR-155* *miR-221* *miR-222*	Myeloid progenitor cells	Repress differentiation into red blood cells	(Ivey and Srivastava, 2010)
miR-451 *miR-16*	Myeloid progenitor cells	Promote differentiation into red blood cells	(Ivey and Srivastava, 2010)
miR-355	Mesenchymal stem cells	Repress proliferation and migration	(Ivey and Srivastava, 2010)
miR-92a	ESC	Repress G_1–S transition	(Sengupta et al., 2009)
miR-372 *miR-195*	ESC ESC	Repress G_1–S transition Repress G_2–M transition	(Qi et al., 2009)

Table 4. miRNAs involved in self-renewal and differentiation processes in normal stem cells and cancer stem cells.

Nanog, Oct4, and Sox2 have been found to be key regulators of ESC pluripotency. miR-134, miR-296, and miR-470 have been shown to modulate ESC pluripotency by regulating Nanog, Oct4, and Sox2, which are key regulators of ESC pluripotency (Tay et al., 2008). Recent studies have identified two groups of miRNAs: markers of pluripotency, which are expressed in the undifferentiated state (miR-200c, miR-371, miR-372, miR-302a, miR-320d, miR-373, miR-302c, miR-21, miR-222, miR-296, miR-494, and miR-367) and miRNAs that regulate the differentiation of cells into one of the different lineages (miR-17, miR-92, and miR-93, which are overexpressed in differentiated cells; and miR-154, miR-29a, miR-143, miR-29c, and let-7a, which are underexpressed in differentiated cells) (Lakshmipathy et al.,

2007). miR-302d and miR-372 target the transcription factors TRPS1 and KLF13 and the RNA binding protein MBNL2 to regulate ESC self-renewal (Li et al., 2009).
miRNAs of the let-7 family (let-7a-1, let-7a-2, let-7a-3, let-7b, let-7c, let-7d, let-7e, let-7f-1, let-7f-2, let-7g, let-7i and miR-98) are key regulators of self-renewal and proliferation and act as tumor suppressors. Numerous genes that promote the G_1/S or G_2/M transition, such as CDK6, CDC25A, and CCND2, are direct targets of let-7. Let-7 also negatively regulates oncogenes such as NRAS, KRAS, HMGA2, and c-Myc, and pluripotency-regulating genes such as Lin28 (Chivukula and Mendell, 2008). Let-7 modulates self-renewal by targeting HRAS and differentiation by targeting HMGA2 in breast cancer cells (Yu et al., 2007). Expression of the let-7 family of miRNAs has been found to be downregulated both in embryonic lung tissue and in lung tumors (Navarro et al., 2009), colon cancer (Akao et al., 2006), and breast cancer (Iorio et al., 2005). Moreover, let-7 has been shown to be downregulated in ESCs and high during differentiation, in which LIN28 expression is high in ESC, but decreases during differentiation (Marson et al., 2008b). Let-7 and LIN28 form a tight feedback loop that is fundamental for stem cell self-renewal and differentiation (Gunaratne, 2009; Martinez and Gregory, 2010). miR-150 regulates differentiation by targeting c-Myb in B-cells (Xiao et al., 2007), while miR-1 regulates differentiation by targeting Mef2c in myoblasts (Chen et al., 2006). miRNAs regulate self-renewal in ESC by controlling the G1-S and G2-M transition. For example, miR-92a is a negative regulator of G1-S transition by targeting CDKN1C (Sengupta et al., 2009). miR-372 targets CDKN1A to negatively regulate G1-S transition, while miR-195 negatively regulates G2-M transition by targeting WEE1 in ESCs (Qi et al., 2009). miR-125b, miR-504, miR-25 and miR-30d directly target and negatively regulate TP53 (Kumar et al., 2011).
The following miRNAs have been found to be key regulators of EMT: miR-200a, miR-200b, miR-200c, miR-141, and miR-429 (Gregory et al., 2008). The miR-200 family regulates EMT by targeting different genes. For example, miR-200b, miR-141 and miR-205 target ZEB2 (Gregory et al., 2008), miR-141 and miR-155 targets TGF-β2 (Bracken et al., 2008; Burk et al., 2008), miR-200a targets ZEB2 and CTNNB1 (Xia et al., 2010) to regulate EMT. In addition, miR-335 has been found to regulate differentiation, proliferation, and migration in mesenchymal stem cells (Tome et al., 2011).

6. Conclusion

In the past decade, tremendous progress has been made in discovering molecular mechanisms (signaling pathways, transcription factors and miRNAs) that regulate stem cell self-renewal and differentiation, but many questions remain to be answered. For example, which factors and signaling pathways determine which daughter cell of an adult stem cell remains a stem cell and which undergoes differentiation. How do cells decide whether to self-renew? How do cells decide whether to migrate to develop organs during embryogenesis, and how do cells decide when that specific organogenesis process is complete? How do cells decide to stop proliferating? Which regulatory factors are involved in normal cell differentiation, and which factors are aberrantly expressed in cancer?
New discoveries will add to our understanding of the balance between self-renewal and differentiation in normal stem cells and, therefore, provide new insights into development and progression of cancer, which may lead to the development of more effective molecular cancer therapies. Most current cancer therapeutic agents aim to kill cancer cells. These

therapeutic agents kill cancer cells as well as normal cells, but do not kill CSCs. A more effective approach to the treatment of cancer may be to use therapeutic agents that block self-renewal and that induce cell to complete differentiation instead of killing cells. Since miRNAs are key regulators in self-renewal and differentiation, thereby miRNAs can be used as potential therapeutic agents or targets.

7. Acknowledgments

This research is supported in part by the National Institutes of Health through MD Anderson's Cancer Center Support Grant CA016672, U19CA148127, and CA133996.

8. References

Aguirre, A., Rubio, M. E., and Gallo, V. (2010). Notch and EGFR pathway interaction regulates neural stem cell number and self-renewal. *Nature 467*, 323-327.

Ahn, S., and Joyner, A. L. (2005). In vivo analysis of quiescent adult neural stem cells responding to Sonic hedgehog. *Nature 437*, 894-897.

Akao, Y., Nakagawa, Y., and Naoe, T. (2006). let-7 microRNA functions as a potential growth suppressor in human colon cancer cells. *Biol Pharm Bull 29*, 903-906.

Ambler, C. A., and Maatta, A. (2009). Epidermal stem cells: location, potential and contribution to cancer. *J Pathol 217*, 206-216.

Androutsellis-Theotokis, A., Leker, R. R., Soldner, F., Hoeppner, D. J., Ravin, R., Poser, S. W., Rueger, M. A., Bae, S. K., Kittappa, R., and McKay, R. D. (2006). Notch signalling regulates stem cell numbers in vitro and in vivo. *Nature 442*, 823-826.

Armesilla-Diaz, A., Bragado, P., Del Valle, I., Cuevas, E., Lazaro, I., Martin, C., Cigudosa, J. C., and Silva, A. (2009a). p53 regulates the self-renewal and differentiation of neural precursors. *Neuroscience 158*, 1378-1389.

Armesilla-Diaz, A., Elvira, G., and Silva, A. (2009b). p53 regulates the proliferation, differentiation and spontaneous transformation of mesenchymal stem cells. *Exp Cell Res 315*, 3598-3610.

Aster, J. C., Blacklow, S. C., and Pear, W. S. (2010). Notch signalling in T-cell lymphoblastic leukaemia/lymphoma and other haematological malignancies. *J Pathol 223*, 262-273.

Balordi, F., and Fishell, G. (2007). Hedgehog signaling in the subventricular zone is required for both the maintenance of stem cells and the migration of newborn neurons. *J Neurosci 27*, 5936-5947.

Barry, F. P., and Murphy, J. M. (2004). Mesenchymal stem cells: clinical applications and biological characterization. *Int J Biochem Cell Biol 36*, 568-584.

Bartel, D. P. (2004). MicroRNAs: genomics, biogenesis, mechanism, and function. Cell *116*, 281-297.

Bartlett, P. F. (1982). Pluripotential hemopoietic stem cells in adult mouse brain. *Proc Natl Acad Sci U S A 79*, 2722-2725.

Berman, D. M., Karhadkar, S. S., Hallahan, A. R., Pritchard, J. I., Eberhart, C. G., Watkins, D. N., Chen, J. K., Cooper, M. K., Taipale, J., Olson, J. M., and Beachy, P. A. (2002).

Medulloblastoma growth inhibition by hedgehog pathway blockade. *Science 297*, 1559-1561.

Bhardwaj, G., Murdoch, B., Wu, D., Baker, D. P., Williams, K. P., Chadwick, K., Ling, L. E., Karanu, F. N., and Bhatia, M. (2001). Sonic hedgehog induces the proliferation of primitive human hematopoietic cells via BMP regulation. *Nat Immunol 2*, 172-180.

Bieback, K., Kern, S., Kocaomer, A., Ferlik, K., and Bugert, P. (2008). Comparing mesenchymal stromal cells from different human tissues: bone marrow, adipose tissue and umbilical cord blood. *Biomed Mater Eng 18*, S71-76.

Bisson, I., and Prowse, D. M. (2009). WNT signaling regulates self-renewal and differentiation of prostate cancer cells with stem cell characteristics. *Cell Res 19*, 683-697.

Bouras, T., Pal, B., Vaillant, F., Harburg, G., Asselin-Labat, M. L., Oakes, S. R., Lindeman, G. J., and Visvader, J. E. (2008). Notch signaling regulates mammary stem cell function and luminal cell-fate commitment. *Cell Stem Cell 3*, 429-441.

Brabletz, S., Bajdak, K., Meidhof, S., Burk, U., Niedermann, G., Firat, E., Wellner, U., Dimmler, A., Faller, G., Schubert, J., and Brabletz, T. (2011). The ZEB1/miR-200 feedback loop controls Notch signalling in cancer cells. *Embo J 30*, 770-782.

Bracken, C. P., Gregory, P. A., Kolesnikoff, N., Bert, A. G., Wang, J., Shannon, M. F., and Goodall, G. J. (2008). A double-negative feedback loop between ZEB1-SIP1 and the microRNA-200 family regulates epithelial-mesenchymal transition. *Cancer Res 68*, 7846-7854.

Brittan, M., and Wright, N. A. (2002). Gastrointestinal stem cells. J Pathol *197*, 492-509.

Buas, M. F., and Kadesch, T. (2010). Regulation of skeletal myogenesis by Notch. *Exp Cell Res 316*, 3028-3033.

Burk, U., Schubert, J., Wellner, U., Schmalhofer, O., Vincan, E., Spaderna, S., and Brabletz, T. (2008). A reciprocal repression between ZEB1 and members of the miR-200 family promotes EMT and invasion in cancer cells. *EMBO Rep 9*, 582-589.

Chan, E. F., Gat, U., McNiff, J. M., and Fuchs, E. (1999). A common human skin tumour is caused by activating mutations in beta-catenin. *Nat Genet 21*, 410-413.

Chen, J. F., Mandel, E. M., Thomson, J. M., Wu, Q., Callis, T. E., Hammond, S. M., Conlon, F. L., and Wang, D. Z. (2006). The role of microRNA-1 and microRNA-133 in skeletal muscle proliferation and differentiation. *Nat Genet 38*, 228-233.

Cheng, L. C., Pastrana, E., Tavazoie, M., and Doetsch, F. (2009). miR-124 regulates adult neurogenesis in the subventricular zone stem cell niche. *Nat Neurosci 12*, 399-408.

Cheshier, S. H., Morrison, S. J., Liao, X., and Weissman, I. L. (1999). In vivo proliferation and cell cycle kinetics of long-term self-renewing hematopoietic stem cells. *Proc Natl Acad Sci U S A 96*, 3120-3125.

Chiba, S. (2006). Notch signaling in stem cell systems. *Stem Cells 24*, 2437-2447.

Chivukula, R. R., and Mendell, J. T. (2008). Circular reasoning: microRNAs and cell-cycle control. *Trends Biochem Sci 33*, 474-481.

Cicalese, A., Bonizzi, G., Pasi, C. E., Faretta, M., Ronzoni, S., Giulini, B., Brisken, C., Minucci, S., Di Fiore, P. P., and Pelicci, P. G. (2009). The tumor suppressor p53 regulates polarity of self-renewing divisions in mammary stem cells. *Cell 138*, 1083-1095.

Clevers, H. (2006). Wnt/beta-catenin signaling in development and disease. *Cell 127*, 469-480.

Cohen, M. M., Jr. (2003). The hedgehog signaling network. *Am J Med Genet* A *123A*, 5-28.

Conboy, I. M., Conboy, M. J., Smythe, G. M., and Rando, T. A. (2003). Notch-mediated restoration of regenerative potential to aged muscle. *Science 302*, 1575-1577.

Conboy, I. M., and Rando, T. A. (2002). The regulation of Notch signaling controls satellite cell activation and cell fate determination in postnatal myogenesis. *Dev Cell 3*, 397-409.

de La Coste, A., Romagnolo, B., Billuart, P., Renard, C. A., Buendia, M. A., Soubrane, O., Fabre, M., Chelly, J., Beldjord, C., Kahn, A., and Perret, C. (1998). Somatic mutations of the beta-catenin gene are frequent in mouse and human hepatocellular carcinomas. *Proc Natl Acad Sci U S A 95*, 8847-8851.

Dierks, C., Beigi, R., Guo, G. R., Zirlik, K., Stegert, M. R., Manley, P., Trussell, C., Schmitt-Graeff, A., Landwerlin, K., Veelken, H., and Warmuth, M. (2008). Expansion of Bcr-Abl-positive leukemic stem cells is dependent on Hedgehog pathway activation. *Cancer Cell 14*, 238-249.

Digirolamo, C. M., Stokes, D., Colter, D., Phinney, D. G., Class, R., and Prockop, D. J. (1999). Propagation and senescence of human marrow stromal cells in culture: a simple colony-forming assay identifies samples with the greatest potential to propagate and differentiate. *Br J Haematol 107*, 275-281.

Dontu, G., Jackson, K. W., McNicholas, E., Kawamura, M. J., Abdallah, W. M., and Wicha, M. S. (2004). Role of Notch signaling in cell-fate determination of human mammary stem/progenitor cells. *Breast Cancer Res 6*, R605-615.

Esquela-Kerscher, A., and Slack, F. J. (2006). Oncomirs - microRNAs with a role in cancer. *Nat Rev Cancer 6*, 259-269.

Ferretti, E., De Smaele, E., Miele, E., Laneve, P., Po, A., Pelloni, M., Paganelli, A., Di Marcotullio, L., Caffarelli, E., Screpanti, I., *et al.* (2008). Concerted microRNA control of Hedgehog signalling in cerebellar neuronal progenitor and tumour cells. *Embo J 27*, 2616-2627.

Filipowicz, W., Bhattacharyya, S. N., and Sonenberg, N. (2008). Mechanisms of post-transcriptional regulation by microRNAs: are the answers in sight? *Nat Rev Genet 9*, 102-114.

Fre, S., Huyghe, M., Mourikis, P., Robine, S., Louvard, D., and Artavanis-Tsakonas, S. (2005). Notch signals control the fate of immature progenitor cells in the intestine. *Nature 435*, 964-968.

Fre, S., Pallavi, S. K., Huyghe, M., Lae, M., Janssen, K. P., Robine, S., Artavanis-Tsakonas, S., and Louvard, D. (2009). Notch and Wnt signals cooperatively control cell proliferation and tumorigenesis in the intestine. *Proc Natl Acad Sci U S A 106*, 6309-6314.

Gailani, M. R., and Bale, A. E. (1999). Acquired and inherited basal cell carcinomas and the patched gene. Adv Dermatol *14*, 261-283; discussion 284.

Gangaraju, V. K., and Lin, H. (2009). MicroRNAs: key regulators of stem cells. *Nat Rev Mol Cell Biol 10*, 116-125.

Garzia, L., Andolfo, I., Cusanelli, E., Marino, N., Petrosino, G., De Martino, D., Esposito, V., Galeone, A., Navas, L., Esposito, S., *et al.* (2009). MicroRNA-199b-5p impairs cancer stem cells through negative regulation of HES1 in medulloblastoma. *PLoS One 4*, e4998.

Garzon, R., Fabbri, M., Cimmino, A., Calin, G. A., and Croce, C. M. (2006). MicroRNA expression and function in cancer. *Trends Mol Med 12*, 580-587.

Gidekel, S., Pizov, G., Bergman, Y., and Pikarsky, E. (2003). Oct-3/4 is a dose-dependent oncogenic fate determinant. *Cancer Cell 4*, 361-370.

Gregory, P. A., Bert, A. G., Paterson, E. L., Barry, S. C., Tsykin, A., Farshid, G., Vadas, M. A., Khew-Goodall, Y., and Goodall, G. J. (2008). The miR-200 family and miR-205 regulate epithelial to mesenchymal transition by targeting ZEB1 and SIP1. *Nat Cell Biol 10*, 593-601.

Groszer, M., Erickson, R., Scripture-Adams, D. D., Dougherty, J. D., Le Belle, J., Zack, J. A., Geschwind, D. H., Liu, X., Kornblum, H. I., and Wu, H. (2006). PTEN negatively regulates neural stem cell self-renewal by modulating G0-G1 cell cycle entry. *Proc Natl Acad Sci U S A 103*, 111-116.

Groszer, M., Erickson, R., Scripture-Adams, D. D., Lesche, R., Trumpp, A., Zack, J. A., Kornblum, H. I., Liu, X., and Wu, H. (2001). Negative regulation of neural stem/progenitor cell proliferation by the Pten tumor suppressor gene in vivo. *Science 294*, 2186-2189.

Gunaratne, P. H. (2009). Embryonic stem cell microRNAs: defining factors in induced pluripotent (iPS) and cancer (CSC) stem cells? *Curr Stem Cell Res Ther 4*, 168-177.

Halleux, C., Sottile, V., Gasser, J. A., and Seuwen, K. (2001). Multi-lineage potential of human mesenchymal stem cells following clonal expansion. *J Musculoskelet Neuronal Interact 2*, 71-76.

Hashimi, S. T., Fulcher, J. A., Chang, M. H., Gov, L., Wang, S., and Lee, B. (2009). MicroRNA profiling identifies miR-34a and miR-21 and their target genes JAG1 and WNT1 in the coordinate regulation of dendritic cell differentiation. *Blood 114*, 404-414.

He, S., Nakada, D., and Morrison, S. J. (2009). Mechanisms of stem cell self-renewal. *Annu Rev Cell Dev Biol 25*, 377-406.

He, X. C., Zhang, J., Tong, W. G., Tawfik, O., Ross, J., Scoville, D. H., Tian, Q., Zeng, X., He, X., Wiedemann, L. M., *et al.* (2004). BMP signaling inhibits intestinal stem cell self-renewal through suppression of Wnt-beta-catenin signaling. *Nat Genet 36*, 1117-1121.

Herzog, E. L., Chai, L., and Krause, D. S. (2003). Plasticity of marrow-derived stem cells. *Blood 102*, 3483-3493.

Hu, A. B., Cai, J. Y., Zheng, Q. C., He, X. Q., Shan, Y., Pan, Y. L., Zeng, G. C., Hong, A., Dai, Y., and Li, L. S. (2004). High-ratio differentiation of embryonic stem cells into hepatocytes in vitro. *Liver Int 24*, 237-245.

Hu, Y. Y., Zheng, M. H., Cheng, G., Li, L., Liang, L., Gao, F., Wei, Y. N., Fu, L. A., and Han, H. (2011). Notch signaling contributes to the maintenance of both normal neural stem cells and patient-derived glioma stem cells. *BMC Cancer 11*, 82.

Ille, F., and Sommer, L. (2005). Wnt signaling: multiple functions in neural development. *Cell Mol Life Sci 62*, 1100-1108.

Ingham, P. W. (2008). Hedgehog signalling. *Curr Biol 18*, R238-241.

Iorio, M. V., Ferracin, M., Liu, C. G., Veronese, A., Spizzo, R., Sabbioni, S., Magri, E., Pedriali, M., Fabbri, M., Campiglio, M., *et al.* (2005). MicroRNA gene expression deregulation in human breast cancer. *Cancer Res 65*, 7065-7070.

Ivey, K. N., and Srivastava, D. (2010). MicroRNAs as regulators of differentiation and cell fate decisions. *Cell Stem Cell 7*, 36-41.

Johnson, S. M., Grosshans, H., Shingara, J., Byrom, M., Jarvis, R., Cheng, A., Labourier, E., Reinert, K. L., Brown, D., and Slack, F. J. (2005). RAS is regulated by the let-7 microRNA family. *Cell 120*, 635-647.

Kalani, M. Y., Cheshier, S. H., Cord, B. J., Bababeygy, S. R., Vogel, H., Weissman, I. L., Palmer, T. D., and Nusse, R. (2008). Wnt-mediated self-renewal of neural stem/progenitor cells. *Proc Natl Acad Sci U S A 105*, 16970-16975.

Kapinas, K., Kessler, C., Ricks, T., Gronowicz, G., and Delany, A. M. (2010). miR-29 modulates Wnt signaling in human osteoblasts through a positive feedback loop. *J Biol Chem 285*, 25221-25231.

Kaplan, K. B., Burds, A. A., Swedlow, J. R., Bekir, S. S., Sorger, P. K., and Nathke, I. S. (2001). A role for the Adenomatous Polyposis Coli protein in chromosome segregation. *Nat Cell Biol 3*, 429-432.

Karhadkar, S. S., Bova, G. S., Abdallah, N., Dhara, S., Gardner, D., Maitra, A., Isaacs, J. T., Berman, D. M., and Beachy, P. A. (2004). Hedgehog signalling in prostate regeneration, neoplasia and metastasis. *Nature 431*, 707-712.

Kasper, M., Jaks, V., Fiaschi, M., and Toftgard, R. (2009). Hedgehog signalling in breast cancer. *Carcinogenesis 30*, 903-911.

Kato, M., Putta, S., Wang, M., Yuan, H., Lanting, L., Nair, I., Gunn, A., Nakagawa, Y., Shimano, H., Todorov, I., *et al.* (2009). TGF-beta activates Akt kinase through a microRNA-dependent amplifying circuit targeting PTEN. *Nat Cell Biol 11*, 881-889.

Kennell, J. A., Gerin, I., MacDougald, O. A., and Cadigan, K. M. (2008). The microRNA miR-8 is a conserved negative regulator of Wnt signaling. Proc Natl Acad Sci U S A *105*, 15417-15422.

Kim, M. S., Kim, S. S., Ahn, C. H., Yoo, N. J., and Lee, S. H. (2009). Frameshift mutations of Wnt pathway genes AXIN2 and TCF7L2 in gastric carcinomas with high microsatellite instability. *Hum Pathol 40*, 58-64.

Koesters, R., Ridder, R., Kopp-Schneider, A., Betts, D., Adams, V., Niggli, F., Briner, J., and von Knebel Doeberitz, M. (1999). Mutational activation of the beta-catenin proto-oncogene is a common event in the development of Wilms' tumors. *Cancer Res 59*, 3880-3882.

Korkaya, H., Paulson, A., Charafe-Jauffret, E., Ginestier, C., Brown, M., Dutcher, J., Clouthier, S. G., and Wicha, M. S. (2009). Regulation of mammary stem/progenitor cells by PTEN/Akt/beta-catenin signaling. *PLoS Biol 7*, e1000121.

Krause, D. S. (2002a). Plasticity of marrow-derived stem cells. Gene Ther *9*, 754-758.

Krause, D. S. (2002b). Regulation of hematopoietic stem cell fate. *Oncogene 21*, 3262-3269.

Kumar, M., Lu, Z., Takwi, A. A., Chen, W., Callander, N. S., Ramos, K. S., Young, K. H., and Li, Y. (2011). Negative regulation of the tumor suppressor p53 gene by microRNAs. *Oncogene 30*, 843-853.

Lai, K., Kaspar, B. K., Gage, F. H., and Schaffer, D. V. (2003). Sonic hedgehog regulates adult neural progenitor proliferation in vitro and in vivo. *Nat Neurosci 6*, 21-27.

Lakshmipathy, U., Love, B., Goff, L. A., Jornsten, R., Graichen, R., Hart, R. P., and Chesnut, J. D. (2007). MicroRNA expression pattern of undifferentiated and differentiated human embryonic stem cells. *Stem Cells Dev 16*, 1003-1016.

Le, M. T., Teh, C., Shyh-Chang, N., Xie, H., Zhou, B., Korzh, V., Lodish, H. F., and Lim, B. (2009). MicroRNA-125b is a novel negative regulator of p53. *Genes Dev 23*, 862-876.

Lee, R. C., Feinbaum, R. L., and Ambros, V. (1993). The C. elegans heterochronic gene lin-4 encodes small RNAs with antisense complementarity to lin-14. *Cell 75*, 843-854.

Lessard, J., and Sauvageau, G. (2003). Bmi-1 determines the proliferative capacity of normal and leukaemic stem cells. *Nature 423*, 255-260.

Li, C., Heidt, D. G., Dalerba, P., Burant, C. F., Zhang, L., Adsay, V., Wicha, M., Clarke, M. F., and Simeone, D. M. (2007). Identification of pancreatic cancer stem cells. *Cancer Res 67*, 1030-1037.

Li, S. S., Yu, S. L., Kao, L. P., Tsai, Z. Y., Singh, S., Chen, B. Z., Ho, B. C., Liu, Y. H., and Yang, P. C. (2009). Target identification of microRNAs expressed highly in human embryonic stem cells. *J Cell Biochem 106*, 1020-1030.

Liu, S., Dontu, G., Mantle, I. D., Patel, S., Ahn, N. S., Jackson, K. W., Suri, P., and Wicha, M. S. (2006). Hedgehog signaling and Bmi-1 regulate self-renewal of normal and malignant human mammary stem cells. *Cancer Res 66*, 6063-6071.

Lowe, S. W., and Sherr, C. J. (2003). Tumor suppression by Ink4a-Arf: progress and puzzles. *Curr Opin Genet Dev 13*, 77-83.

Luis, T. C., Weerkamp, F., Naber, B. A., Baert, M. R., de Haas, E. F., Nikolic, T., Heuvelmans, S., De Krijger, R. R., van Dongen, J. J., and Staal, F. J. (2009). Wnt3a deficiency irreversibly impairs hematopoietic stem cell self-renewal and leads to defects in progenitor cell differentiation. *Blood 113*, 546-554.

Ma, S., Chan, K. W., Hu, L., Lee, T. K., Wo, J. Y., Ng, I. O., Zheng, B. J., and Guan, X. Y. (2007). Identification and characterization of tumorigenic liver cancer stem/progenitor cells. *Gastroenterology 132*, 2542-2556.

Ma, Y., Erkner, A., Gong, R., Yao, S., Taipale, J., Basler, K., and Beachy, P. A. (2002). Hedgehog-mediated patterning of the mammalian embryo requires transporter-like function of dispatched. *Cell 111*, 63-75.

Mani, S. A., Guo, W., Liao, M. J., Eaton, E. N., Ayyanan, A., Zhou, A. Y., Brooks, M., Reinhard, F., Zhang, C. C., Shipitsin, M., *et al.* (2008). The epithelial-mesenchymal transition generates cells with properties of stem cells. *Cell 133*, 704-715.

Marson, A., Foreman, R., Chevalier, B., Bilodeau, S., Kahn, M., Young, R. A., and Jaenisch, R. (2008a). Wnt signaling promotes reprogramming of somatic cells to pluripotency. *Cell Stem Cell 3*, 132-135.

Marson, A., Levine, S. S., Cole, M. F., Frampton, G. M., Brambrink, T., Johnstone, S., Guenther, M. G., Johnston, W. K., Wernig, M., Newman, J., *et al.* (2008b). Connecting microRNA genes to the core transcriptional regulatory circuitry of embryonic stem cells. *Cell 134*, 521-533.

Martin, V., Valencia, A., Agirre, X., Cervera, J., San Jose-Eneriz, E., Vilas-Zornoza, A., Rodriguez-Otero, P., Sanz, M. A., Herrera, C., Torres, A., *et al.* (2010). Epigenetic regulation of the non-canonical Wnt pathway in acute myeloid leukemia. *Cancer Sci 101*, 425-432.

Martinez, N. J., and Gregory, R. I. (2010). MicroRNA gene regulatory pathways in the establishment and maintenance of ESC identity. *Cell Stem Cell 7*, 31-35.

Mavrakis, K. J., Wolfe, A. L., Oricchio, E., Palomero, T., de Keersmaecker, K., McJunkin, K., Zuber, J., James, T., Khan, A. A., Leslie, C. S., *et al.* (2010). Genome-wide RNA-mediated interference screen identifies miR-19 targets in Notch-induced T-cell acute lymphoblastic leukaemia. *Nat Cell Biol 12*, 372-379.

Melchior, K., Weiss, J., Zaehres, H., Kim, Y. M., Lutzko, C., Roosta, N., Hescheler, J., and Muschen, M. (2008). The WNT receptor FZD7 contributes to self-renewal signaling of human embryonic stem cells. *Biol Chem 389*, 897-903.

Meng, F., Henson, R., Wehbe-Janek, H., Ghoshal, K., Jacob, S. T., and Patel, T. (2007). MicroRNA-21 regulates expression of the PTEN tumor suppressor gene in human hepatocellular cancer. *Gastroenterology 133*, 647-658.

Mercher, T., Cornejo, M. G., Sears, C., Kindler, T., Moore, S. A., Maillard, I., Pear, W. S., Aster, J. C., and Gilliland, D. G. (2008). Notch signaling specifies megakaryocyte development from hematopoietic stem cells. *Cell Stem Cell 3*, 314-326.

Molofsky, A. V., Pardal, R., Iwashita, T., Park, I. K., Clarke, M. F., and Morrison, S. J. (2003). Bmi-1 dependence distinguishes neural stem cell self-renewal from progenitor proliferation. *Nature 425*, 962-967.

Molofsky, A. V., Pardal, R., and Morrison, S. J. (2004). Diverse mechanisms regulate stem cell self-renewal. *Curr Opin Cell Biol 16*, 700-707.

Muller, F. J., Laurent, L. C., Kostka, D., Ulitsky, I., Williams, R., Lu, C., Park, I. H., Rao, M. S., Shamir, R., Schwartz, P. H., *et al.* (2008). Regulatory networks define phenotypic classes of human stem cell lines. *Nature 455*, 401-405.

Mumm, J. S., and Kopan, R. (2000). Notch signaling: from the outside in. Dev Biol *228*, 151-165.

Nagao, M., Campbell, K., Burns, K., Kuan, C. Y., Trumpp, A., and Nakafuku, M. (2008). Coordinated control of self-renewal and differentiation of neural stem cells by Myc and the p19ARF-p53 pathway. *J Cell Biol 183*, 1243-1257.

Nagel, R., le Sage, C., Diosdado, B., van der Waal, M., Oude Vrielink, J. A., Bolijn, A., Meijer, G. A., and Agami, R. (2008). Regulation of the adenomatous polyposis coli gene by the miR-135 family in colorectal cancer. *Cancer Res 68*, 5795-5802.

Navarro, A., Marrades, R. M., Vinolas, N., Quera, A., Agusti, C., Huerta, A., Ramirez, J., Torres, A., and Monzo, M. (2009). MicroRNAs expressed during lung cancer development are expressed in human pseudoglandular lung embryogenesis. *Oncology 76*, 162-169.

Olsen, C. L., Hsu, P. P., Glienke, J., Rubanyi, G. M., and Brooks, A. R. (2004). Hedgehog-interacting protein is highly expressed in endothelial cells but down-regulated during angiogenesis and in several human tumors. *BMC Cancer 4*, 43.

Palma, V., Lim, D. A., Dahmane, N., Sanchez, P., Brionne, T. C., Herzberg, C. D., Gitton, Y., Carleton, A., Alvarez-Buylla, A., and Ruiz i Altaba, A. (2005). Sonic hedgehog controls stem cell behavior in the postnatal and adult brain. *Development 132*, 335-344.

Pang, R. T., Leung, C. O., Ye, T. M., Liu, W., Chiu, P. C., Lam, K. K., Lee, K. F., and Yeung, W. S. (2010). MicroRNA-34a suppresses invasion through downregulation of Notch1 and Jagged1 in cervical carcinoma and choriocarcinoma cells. *Carcinogenesis 31*, 1037-1044.

Park, I. H., Zhao, R., West, J. A., Yabuuchi, A., Huo, H., Ince, T. A., Lerou, P. H., Lensch, M. W., and Daley, G. Q. (2008). Reprogramming of human somatic cells to pluripotency with defined factors. *Nature 451*, 141-146.

Park, I. K., Qian, D., Kiel, M., Becker, M. W., Pihalja, M., Weissman, I. L., Morrison, S. J., and Clarke, M. F. (2003). Bmi-1 is required for maintenance of adult self-renewing haematopoietic stem cells. *Nature 423*, 302-305.

Pasca di Magliano, M., and Hebrok, M. (2003). Hedgehog signalling in cancer formation and maintenance. *Nat Rev Cancer 3*, 903-911.

Peacock, C. D., Wang, Q., Gesell, G. S., Corcoran-Schwartz, I. M., Jones, E., Kim, J., Devereux, W. L., Rhodes, J. T., Huff, C. A., Beachy, P. A., *et al.* (2007). Hedgehog signaling maintains a tumor stem cell compartment in multiple myeloma. *Proc Natl Acad Sci U S A 104*, 4048-4053.

Peifer, M. (2000). Cell biology. Travel bulletin--traffic jams cause tumors. Science *289*, 67-69.

Pistollato, F., Rampazzo, E., Persano, L., Abbadi, S., Frasson, C., Denaro, L., D'Avella, D., Panchision, D. M., Della Puppa, A., Scienza, R., and Basso, G. (2010) Interaction of hypoxia-inducible factor-1alpha and Notch signaling regulates medulloblastoma precursor proliferation and fate. *Stem Cells 28*, 1918-1929.

Polakis, P. (2000). Wnt signaling and cancer. *Genes Dev 14*, 1837-1851.

Prince, M. E., Sivanandan, R., Kaczorowski, A., Wolf, G. T., Kaplan, M. J., Dalerba, P., Weissman, I. L., Clarke, M. F., and Ailles, L. E. (2007). Identification of a subpopulation of cells with cancer stem cell properties in head and neck squamous cell carcinoma. *Proc Natl Acad Sci U S A 104*, 973-978.

Qi, J., Yu, J. Y., Shcherbata, H. R., Mathieu, J., Wang, A. J., Seal, S., Zhou, W., Stadler, B. M., Bourgin, D., Wang, L., *et al.* (2009). microRNAs regulate human embryonic stem cell division. *Cell Cycle 8*, 3729-3741.

Reinhart, B. J., Slack, F. J., Basson, M., Pasquinelli, A. E., Bettinger, J. C., Rougvie, A. E., Horvitz, H. R., and Ruvkun, G. (2000). The 21-nucleotide let-7 RNA regulates developmental timing in Caenorhabditis elegans. *Nature 403*, 901-906.

Reya, T., and Clevers, H. (2005). Wnt signalling in stem cells and cancer. *Nature 434*, 843-850.

Reya, T., Duncan, A. W., Ailles, L., Domen, J., Scherer, D. C., Willert, K., Hintz, L., Nusse, R., and Weissman, I. L. (2003). A role for Wnt signalling in self-renewal of haematopoietic stem cells. *Nature 423*, 409-414.

Ricci-Vitiani, L., Pallini, R., Larocca, L. M., Lombardi, D. G., Signore, M., Pierconti, F., Petrucci, G., Montano, N., Maira, G., and De Maria, R. (2008). Mesenchymal differentiation of glioblastoma stem cells. *Cell Death Differ 15*, 1491-1498.

Rodriguez-Pinilla, S. M., Sarrio, D., Moreno-Bueno, G., Rodriguez-Gil, Y., Martinez, M. A., Hernandez, L., Hardisson, D., Reis-Filho, J. S., and Palacios, J. (2007). Sox2: a possible driver of the basal-like phenotype in sporadic breast cancer. *Mod Pathol 20*, 474-481.

Santagata, S., Ligon, K. L., and Hornick, J. L. (2007). Embryonic stem cell transcription factor signatures in the diagnosis of primary and metastatic germ cell tumors. *Am J Surg Pathol 31*, 836-845.

Saydam, O., Shen, Y., Wurdinger, T., Senol, O., Boke, E., James, M. F., Tannous, B. A., Stemmer-Rachamimov, A. O., Yi, M., Stephens, R. M., *et al.* (2009). Downregulated microRNA-200a in meningiomas promotes tumor growth by reducing E-cadherin and activating the Wnt/beta-catenin signaling pathway. *Mol Cell Biol 29*, 5923-5940.

Scheller, E. L., Chang, J., and Wang, C. Y. (2008). Wnt/beta-catenin inhibits dental pulp stem cell differentiation. *J Dent Res 87*, 126-130.

Schofield, R. (1978). The relationship between the spleen colony-forming cell and the haemopoietic stem cell. *Blood Cells 4*, 7-25.

Schweizer, L., and Varmus, H. (2003). Wnt/Wingless signaling through beta-catenin requires the function of both LRP/Arrow and frizzled classes of receptors. *BMC Cell Biol 4*, 4.

Sengupta, S., Nie, J., Wagner, R. J., Yang, C., Stewart, R., and Thomson, J. A. (2009). MicroRNA 92b controls the G1/S checkpoint gene p57 in human embryonic stem cells. *Stem Cells 27*, 1524-1528.

Shahi, P., Seethammagari, M. R., Valdez, J. M., Xin, L., and Spencer, D. M. (2011). Wnt and Notch Pathways have Interrelated Opposing Roles on Prostate Progenitor Cell Proliferation and Differentiation. *Stem Cells.*

Shook, D., and Keller, R. (2003). Mechanisms, mechanics and function of epithelial-mesenchymal transitions in early development. *Mech Dev 120*, 1351-1383.

Sikandar, S. S., Pate, K. T., Anderson, S., Dizon, D., Edwards, R. A., Waterman, M. L., and Lipkin, S. M. (2010). NOTCH signaling is required for formation and self-renewal of tumor-initiating cells and for repression of secretory cell differentiation in colon cancer. *Cancer Res 70*, 1469-1478.

Singh, S. K., Clarke, I. D., Terasaki, M., Bonn, V. E., Hawkins, C., Squire, J., and Dirks, P. B. (2003). Identification of a cancer stem cell in human brain tumors. *Cancer Res 63*, 5821-5828.

Song, S., Song, S., Zhang, H., Cuevas, J., and Sanchez-Ramos, J. (2007). Comparison of neuron-like cells derived from bone marrow stem cells to those differentiated from adult brain neural stem cells. *Stem Cells Dev 16*, 747-756.

Spradling, A., Drummond-Barbosa, D., and Kai, T. (2001). Stem cells find their niche. *Nature 414*, 98-104.

Stefani, G., and Slack, F. J. (2008). Small non-coding RNAs in animal development. *Nat Rev Mol Cell Biol 9*, 219-230.

Suh, M. R., Lee, Y., Kim, J. Y., Kim, S. K., Moon, S. H., Lee, J. Y., Cha, K. Y., Chung, H. M., Yoon, H. S., Moon, S. Y., *et al.* (2004). Human embryonic stem cells express a unique set of microRNAs. *Dev Biol 270*, 488-498.

Sun, L., Tian, Z., and Wang, J. (2010). A direct cross-talk between interferon-gamma and sonic hedgehog signaling that leads to the proliferation of neuronal precursor cells. *Brain Behav Immun 24*, 220-228.

Taipale, J., Cooper, M. K., Maiti, T., and Beachy, P. A. (2002). Patched acts catalytically to suppress the activity of Smoothened. *Nature 418*, 892-897.

Takebe, N., Harris, P. J., Warren, R. Q., and Ivy, S. P. (2010). Targeting cancer stem cells by inhibiting Wnt, Notch, and Hedgehog pathways. *Nat Rev Clin Oncol 8*, 97-106.

Tam, W. L., Lim, C. Y., Han, J., Zhang, J., Ang, Y. S., Ng, H. H., Yang, H., and Lim, B. (2008). T-cell factor 3 regulates embryonic stem cell pluripotency and self-renewal by the transcriptional control of multiple lineage pathways. *Stem Cells 26*, 2019-2031.

Tay, Y., Zhang, J., Thomson, A. M., Lim, B., and Rigoutsos, I. (2008). MicroRNAs to Nanog, Oct4 and Sox2 coding regions modulate embryonic stem cell differentiation. *Nature 455*, 1124-1128.

Thiery, J. P., Acloque, H., Huang, R. Y., and Nieto, M. A. (2009). Epithelial-mesenchymal transitions in development and disease. *Cell 139*, 871-890.

Tome, M., Lopez-Romero, P., Albo, C., Sepulveda, J. C., Fernandez-Gutierrez, B., Dopazo, A., Bernad, A., and Gonzalez, M. A. (2011). miR-335 orchestrates cell proliferation, migration and differentiation in human mesenchymal stem cells. *Cell Death Differ.*

Vainio, S., Heikkila, M., Kispert, A., Chin, N., and McMahon, A. P. (1999a). Female development in mammals is regulated by Wnt-4 signalling. *Nature 397*, 405-409.

Vainio, S. J., Itaranta, P. V., Perasaari, J. P., and Uusitalo, M. S. (1999b). Wnts as kidney tubule inducing factors. *Int J Dev Biol 43*, 419-423.

van Es, J. H., van Gijn, M. E., Riccio, O., van den Born, M., Vooijs, M., Begthel, H., Cozijnsen, M., Robine, S., Winton, D. J., Radtke, F., and Clevers, H. (2005). Notch/gamma-secretase inhibition turns proliferative cells in intestinal crypts and adenomas into goblet cells. *Nature 435*, 959-963.

Varnat, F., Duquet, A., Malerba, M., Zbinden, M., Mas, C., Gervaz, P., and Ruiz i Altaba, A. (2009). Human colon cancer epithelial cells harbour active HEDGEHOG-GLI signalling that is essential for tumour growth, recurrence, metastasis and stem cell survival and expansion. *EMBO Mol Med 1*, 338-351.

Varnum-Finney, B., Xu, L., Brashem-Stein, C., Nourigat, C., Flowers, D., Bakkour, S., Pear, W. S., and Bernstein, I. D. (2000). Pluripotent, cytokine-dependent, hematopoietic stem cells are immortalized by constitutive Notch1 signaling. *Nat Med 6*, 1278-1281.

Veeman, M. T., Axelrod, J. D., and Moon, R. T. (2003). A second canon. Functions and mechanisms of beta-catenin-independent Wnt signaling. *Dev Cell 5*, 367-377.

Vermeulen, L., De Sousa, E. M. F., van der Heijden, M., Cameron, K., de Jong, J. H., Borovski, T., Tuynman, J. B., Todaro, M., Merz, C., Rodermond, H., *et al.* (2010). Wnt activity defines colon cancer stem cells and is regulated by the microenvironment. *Nat Cell Biol 12*, 468-476.

Vincan, E., and Barker, N. (2008). The upstream components of the Wnt signalling pathway in the dynamic EMT and MET associated with colorectal cancer progression. *Clin Exp Metastasis 25*, 657-663.

Voorhoeve, P. M., le Sage, C., Schrier, M., Gillis, A. J., Stoop, H., Nagel, R., Liu, Y. P., van Duijse, J., Drost, J., Griekspoor, A., *et al.* (2006). A genetic screen implicates miRNA-372 and miRNA-373 as oncogenes in testicular germ cell tumors. *Cell 124*, 1169-1181.

Wang, P., and Hou, S. X. (2010). Regulation of intestinal stem cells in mammals and Drosophila. *J Cell Physiol 222*, 33-37.

Wang, Y., Krivtsov, A. V., Sinha, A. U., North, T. E., Goessling, W., Feng, Z., Zon, L. I., and Armstrong, S. A. (2010). The Wnt/beta-catenin pathway is required for the development of leukemia stem cells in AML. *Science 327*, 1650-1653.

Wechsler-Reya, R. J., and Scott, M. P. (1999). Control of neuronal precursor proliferation in the cerebellum by Sonic Hedgehog. *Neuron 22*, 103-114.

Wernig, M., Meissner, A., Foreman, R., Brambrink, T., Ku, M., Hochedlinger, K., Bernstein, B. E., and Jaenisch, R. (2007). In vitro reprogramming of fibroblasts into a pluripotent ES-cell-like state. *Nature 448*, 318-324.

Wodarz, A., and Nusse, R. (1998). Mechanisms of Wnt signaling in development. *Annu Rev Cell Dev Biol 14*, 59-88.

Wu, M., Kwon, H. Y., Rattis, F., Blum, J., Zhao, C., Ashkenazi, R., Jackson, T. L., Gaiano, N., Oliver, T., and Reya, T. (2007). Imaging hematopoietic precursor division in real time. *Cell Stem Cell 1*, 541-554.

Xia, H., Ng, S. S., Jiang, S., Cheung, W. K., Sze, J., Bian, X. W., Kung, H. F., and Lin, M. C. (2010). miR-200a-mediated downregulation of ZEB2 and CTNNB1 differentially inhibits nasopharyngeal carcinoma cell growth, migration and invasion. *Biochem Biophys Res Commun 391*, 535-541.

Xiao, C., Calado, D. P., Galler, G., Thai, T. H., Patterson, H. C., Wang, J., Rajewsky, N., Bender, T. P., and Rajewsky, K. (2007). MiR-150 controls B cell differentiation by targeting the transcription factor c-Myb. *Cell 131*, 146-159.

Xu, N., Papagiannakopoulos, T., Pan, G., Thomson, J. A., and Kosik, K. S. (2009). MicroRNA-145 regulates OCT4, SOX2, and KLF4 and represses pluripotency in human embryonic stem cells. *Cell 137*, 647-658.

Yang, J., and Weinberg, R. A. (2008). Epithelial-mesenchymal transition: at the crossroads of development and tumor metastasis. *Dev Cell 14*, 818-829.

Yu, F., Yao, H., Zhu, P., Zhang, X., Pan, Q., Gong, C., Huang, Y., Hu, X., Su, F., Lieberman, J., and Song, E. (2007). let-7 regulates self renewal and tumorigenicity of breast cancer cells. *Cell 131*, 1109-1123.

Zhao, C., Blum, J., Chen, A., Kwon, H. Y., Jung, S. H., Cook, J. M., Lagoo, A., and Reya, T. (2007). Loss of beta-catenin impairs the renewal of normal and CML stem cells in vivo. *Cancer Cell 12*, 528-541.

Zhao, C., Chen, A., Jamieson, C. H., Fereshteh, M., Abrahamsson, A., Blum, J., Kwon, H. Y., Kim, J., Chute, J. P., Rizzieri, D., *et al.* (2009). Hedgehog signalling is essential for maintenance of cancer stem cells in myeloid leukaemia. *Nature 458*, 776-779.

Zheng, H., Ying, H., Yan, H., Kimmelman, A. C., Hiller, D. J., Chen, A. J., Perry, S. R., Tonon, G., Chu, G. C., Ding, Z., *et al.* (2008a). p53 and Pten control neural and glioma stem/progenitor cell renewal and differentiation. *Nature 455*, 1129-1133.

Zheng, H., Ying, H., Yan, H., Kimmelman, A. C., Hiller, D. J., Chen, A. J., Perry, S. R., Tonon, G., Chu, G. C., Ding, Z., *et al.* (2008b). Pten and p53 converge on c-Myc to control differentiation, self-renewal, and transformation of normal and neoplastic stem cells in glioblastoma. *Cold Spring Harb Symp Quant Biol 73*, 427-437.

Zhu, A. J., and Watt, F. M. (1999). beta-catenin signalling modulates proliferative potential of human epidermal keratinocytes independently of intercellular adhesion. *Development 126*, 2285-2298.

Zong, Y., Panikkar, A., Xu, J., Antoniou, A., Raynaud, P., Lemaigre, F., and Stanger, B. Z. (2009). Notch signaling controls liver development by regulating biliary differentiation. *Development 136*, 1727-1739.

Part 2

Stem Cells in Clinical Experiences

9

Are We There Yet? A Story About Cardiac Stem Cells

Shizuka Uchida, Piera De Gaspari and Thomas Braun
Max-Planck-Institute for Heart and Lung Research
Germany

1. Introduction

The heart is the first organ which becomes functional during embryonic development. During lifetime, it cannot rest or stop working without sacrificing the whole organism. Nobody would question that the heart is one of the most vital organs in the body. Severe diseases of the heart compromise the quality of life and often lead to premature death. Cardiovascular diseases are currently the number one killer around the world, and it seems likely that this will not change in the near future. It is therefore not a surprise that huge efforts are made to treat cardiac disease and more recently to repair damaged hearts. However, up to now, there is no causative and definitive cure for cardiovascular diseases although major therapeutic improvements were made. Yet, most therapeutic approaches tackle symptoms and not causes, which is in part due to the complex etiology of cardiovascular diseases (e.g. environmental, genetic, life style, physiological, etc. (Kullo & Cooper, 2010)).

During the last halve of the century, the view predominated that the heart is essentially a post-mitotic organ, overthrowing previous assumptions, which reasoned that the heart is able to regenerate. The latter theory persisted from 1850 to the first quarter of the 20th century and was then replaced by the belief that remodelling events of the heart are solely based on growth rather than on proliferation of cardiomyocytes (Carvalho & de Carvalho, 2010; Karsner et al., 1925). More recently, the view that cardiomyocytes cannot renew has been put into question. Piero Anversa's group published a series of articles claiming that ongoing cell death of cardiomyocytes in the heart requires extensive replacement of cardiomyocytes. It is not surprising that this hypothesis stirred a lot of controversy (Anversa et al., 2007). One should not forget, however, that the concept of cardiomyocyte renewal is not completely new. A very low but nevertheless detectable level of cardiomyocyte cell cycle activity was reported in rodent studies almost 50 years ago (Soonpaa & Field, 1998).

One of the major challenges in the field has been the detection of newly formed cardiomyocytes and/or the visualization of cell division events. This problem has been, at least partially, overcome by utilizing clever labelling schemes, which rely on the intentional or non-intentional labelling of the genetic material of cardiomyocytes. A study by Jonas Frisén and his colleagues (Bergmann et al., 2009) took advantage of the nuclear fall-out that was generated during nuclear bomb testing until 1963. The carbon-14 (C14) isotope that was released into the atmosphere was incorporated into plants and animals or diffused into the oceans to enter the food cycle and eventually "labelled" human beings. Using the C14 concentration in cardiomyocyte genomic DNA, Frisén's group calculated the age of cardiomyocytes. The authors estimated that human cardiomyocytes renew at a rate of 1% per year at the age of 20, which declines to 0.45% at the age of 75. According to their calculation, fewer than 50% of cardiomyocytes are renewed during a normal life span of a human being. In 2010, Anversa's group reported a much higher rate of turnover using materials from cancer patients who received infusion of the radiosensitizer iododeoxyuridine (IdU) for therapeutic purposes. IdU is rapidly incorporated into cycling cells thereby setting up a pulse-and-chase "experiment" (similar to bromodeoxyuridine (BrdU)-labelling (DuFrain et al., 1984)). According to the IdU study, the turnover rate of cardiomyocytes is 22% per year (20% and 13% for fibroblasts and endothelial cells, respectively). The lifespan of cardiomyocytes was calculated to be ~4.5 years resulting in the claim that cardiomyocytes renew several times during a normal life span (Kajstura et al, 2010a). To further support their hypothesis, Anversa's group published another report (Kajstura et al, 2010b), which demonstrated that the ageing human heart of women is more adoptive to cardiomyocyte loss than that of men (Olivetti et al., 1995). By counting cardiomyocytes and c-kit+ cardiac stem cells (CSCs) undergoing apoptosis, cellular senescence and proliferation, the authors concluded that from 20 to 100 years of age, the entire cardiomyocyte compartment of women is replaced 15 times, whereas that of men is renewed 11 times. These numbers, which are provocatively high, are awaiting independent confirmation from other labs.

At the first glance, results of the above mentioned 3 reports in human seem to contradict findings in mice published in 2007. By utilizing transgenic mice to trace the fate of adult cardiomyocytes based on alpha-MHC-driven GFP expression, Hsieh et al. (Hsieh et al., 2007) observed cardiomyocyte turnover by stem or precursor cells after myocardial infarction or pressure overload. However, during normal ageing up to one year (which is equivalent to a 34-year-old human (Holaska, 2008)), the authors did not record a significant turnover of cardiomyocytes. Yet, a careful analysis of the published data reveals that labelling based on tamoxifen injections is not 100% (82.7 ± 1.7% in the above study), and there is a 1-2 percent fluctuation of labelled cells, which explain differences to the results by Bergmann et al. (Bergmann et al., 2009).

Although the above mentioned studies in human cardiomyocyte turnover are intriguing and provide impressive results, some problems remain. Human cardiomyocytes initiate cell cycle activity in response to mechanical stress (e.g. hypertrophy) without nuclear or cell division (karyo- and cyto-kinesis, respectively) (Adler & Friedburg, 1986; Soonpaa & Field, 1998), which might lead to an increase of the number of labelled cardiomyocytes. Surprisingly, there is no consensus in the literature regarding the number of nuclei in human cardiomyocytes. While murine cardiomyocytes are >90% binucleated (Liu et al., 2010; Soonpaa et al., 1996), it has been reported that 74% of human cardiomyocytes are mononucleated, compared to 25.5% for bi-, 0.4% for tri- and 0.1% for tetranucleated cells (Olivetti et al., 1996). The calculations of Anversa's group in recent publications relied mostly on those numbers (Kajstura et al, 2010a, 2010b). On the other hand, Frisén's group argued that the bulk of human cardiomyocytes is not mononucleated. According to their calculations, the majority of cardiomyocyte nuclei from the human adult left ventricle has more than two complete sets of chromosomes, (i.e. 33.5, 55.8 and 10.7% are di-, tetra- and octa-ploid, respectively (Bergmann et al., 2011)). It should not be too difficult to resolve this discrepancy in the future.

Given the findings discussed above, it seems safe to state that cardiomyocytes undergo a certain degree of renewal during the lifetime of a mammalian organism. The cellular source of cardiomyocyte renewal, however, remains an open question. In principal, it is possible that new cardiomyocytes are generated from already existing cardiomyocytes, a principle that dominates embryonic heart development. Alternatively, different types of cardiac stem cells might generate new cardiomyocytes and other cell types following an endogenous program or inductive stimuli. To address these possibilities, it is required to utilize proper animal models since the required experimental manipulation is not applicable to human beings. Therefore, we will first survey the origin of cells in the heart and then pay special attention to putative cardiac stem cells (CSC) in the murine heart.

2. Cell types in heart and their lineages

Researchers from other fields often assume that the heart is mostly made up from cardiomyocytes, which is not true. The mouse heart is composed of ~56% cardiomyocytes, 27% fibroblasts, 10% vascular smooth muscle cells and 7% endothelial cells (Banerjee et al., 2007). Most likely, several other cell types are hidden within these principal groups awaiting further characterization. Interestingly, the distribution of cell types in the heart varies between different rodent species. Rat hearts contain only 26.4% cardiomyocytes, a number that is similar to human beings. The major cell type in rat and human hearts is the fibroblast (62.6%) (Banerjee et al., 2007; Nag, 1980). If we look at the composition of cell types in the heart, rats appear to be rather similar to humans (Rubart & Field, 2006). Moreover, there are several other reasons which added to the popularity of the rat as a cardiovascular model system compared to mice (Aitman et al., 2008). However, advanced genetic manipulation of rats was not possible so far (Cui et al., 2011; Tong et al., 2010) although the advent of zinc finger nucleases (Geurts et al., 2009) and rat ES cells (Buehr et al., 2008; Li et al., 2008) might change that picture in the future. Given that the composition of cell types differs greatly between human and mice, it might not be possible to translate all findings in mice directly to humans. Nevertheless, the advanced state of mouse genetics greatly facilitates analysis of CSCs.

2.1 Cardiomyocytes

Cardiomyocytes constitute the functionally most relevant part of the myocardium, which generates the necessary force enabling the heart to act as a pump. Although the view about cardiomyocyte turnover might have changed over the years, several studies carefully documented birth and growth of cardiomyocytes. For example, Loren Field's group (Soonpaa et al., 1996) reported that DNA synthesis in cardiomyocytes occurs in two distinct phases during the murine development. The first phase is associated with cardiomoycyte proliferation, which occurs during fetal life. The second phase follows after the cession of reduplication of cardiomyocytes (transition from the first phase takes place before day 10 after birth). The feature of this phase is binucleation of cardiomyocytes due to a round of genomic duplication and karyokinesis without cytokinesis.

Currently, the following four sources of cardiomyocyte renewal are considered (Parmacek & Epstein, 2009):

1. Adult cardiomyocytes reentering the cell cycle and divide
2. Bone-marrow-derived cardiac stem/progenitor cells
3. Cells derived from the embryonic epicardium
4. Cardiac stem/progenitor cells

Since a lack of sufficient number of cardiomyocytes can cause many forms of congenital and adult cardiovascular diseases, intensive research has been conducted to find a set of genes/proteins that might drive adult cardiomyocytes into cell cycle to regenerate cardiomyocytes (Rubart & Field, 2006). However, one needs to keep in mind that such approaches bear some inherent problems and that it is absolutely required to prevent uncontrolled proliferation. Since primary cardiac tumours are very rare (Devbhandari et al., 2007), there must be an intrinsic, biological block, which prevents division of cardiomyocytes soon after birth. Studies by several laboratories revealed that it is rather difficult to overcome this block and to achieve controlled proliferation of adult cardiomyocytes (Ebelt et al., 2005, 2006, 2008a, 2008b). There is still hope, however, to utilize the remaining, low-level potential of cardiomyocytes to proliferate for therapeutic purposes.
The second acclaimed source of cardiomyocyte renewal is bone-marrow-derived mesenchymal stem cells (BMSCs). Transdifferentiation of such cells has been fiercely debated over the years, which is well beyond the scope of this chapter. Interested readers might consult the excellent articles (Alaiti et al., 2010; Phinney & Prockop, 2007; Psaltis et al., 2008), which cover this topic. The majority view in the field sees BMSCs as beneficial to treat patients suffering from acute myocardial infarction and ischemic heart failure (Chugh et al., 2009). However, transdifferentiation of transplanted or injected BMSCs (fresh or cultured) are not very likely to contribute to the success of these therapies; instead, BMSCs secrete growth factors and cytokines that might enhance survival of surviving cardiomyocytes and stimulate endogenous repair mechanisms via activation of resident CSCs and other stem cells (Wen et al., 2010). Regarding the issue of transdifferentiation, an interesting study was conducted by Nern et al. (Nern et al., 2009). In this study, the authors employed a lineage tracing system based on the haematopoietic-specific promoter vav to monitor cell fusion events under physiological conditions to challenge transplantation studies using BMSCs. In the case of the heart, the authors found only a single LacZ-positive (the reporter gene from Rosa26 LacZ allele) cardiomyocyte in four hearts of non-irradiated healthy trasgenic mice (vav-iCre/Rosa26 LacZ). Therefore, it is unlikely that haematopoietic cells contribute to renewal of cardiomyocytes. Of course, one might argue that the labelled cells are different from BMSCs but current evidence does not support the view that BMSCs act as major players for the cardiomyocyte renewal.
The third potential source of cardiomyocyte renewal is so-called "epicardially derived mesenchymal cells (EPDCs)" (Morabito et al., 2001). EPDCs are derived from a subpopulation of epicardial cells and were shown to differentiate into cardiac vessels, cardiomyocytes and connective tissue of heart (Limana et al., 2011). Some authors name these cells "cardiac stem cells" (Wessel & Pérez-Pomares, 2004) although several other cell populations, which reside within the myocardium have also been dubbed this way. A more comprehensive description of different population of putative cardiac stem cells is given below in Section 3.

2.2 Smooth muscle cells

Smooth muscle cells are highly plastic and might toggle between a contractile and synthetic phenotype in response to extracellular cues unlike their striated muscle cousins (cardiac and skeletal muscle cells) (Owens, 1995). In the heart, the tone of vascular smooth muscle cells regulates the diameter of blood vessels, blood pressure and blood flow (Rzucidlo et al., 2007), which are integral to the function of the heart. Dysregulation of smooth muscle cells promote various diseases, including atherosclerosis, which results in over 55% of all deaths in Western civilization (Owens et al., 2004). Therefore, significant efforts have been made to elucidate pathological mechanisms affecting vascular smooth muscle cells. Vascular progenitor cells (Kumar & Caplice, 2010) have been claimed to represent a potential source for the renewal of damaged or diseased smooth muscle cells. Since the major source of such progenitor cells are from vascular walls, which consists of endothelial cells, they will be discussed in the next subsection.

2.3 Endothelial cells

In arteries and arterioles, endothelial cells are closely associated with vascular smooth muscle cells while capillaries are devoid of smooth muscle cells (Ergün, 2011). Due to their importance, endothelial cells have been studied intensively and numerous excellent reviews are available. Here, we discuss briefly potential progenitor cells of the endothelium known as "vascular progenitor cells" or "endothelial progenitor cells (EPCs)".

The first report about EPCs in 1997 (Asahara et al., 1997) gave rise to numerous other studies. Most researchers relied on surface markers to identify EPCs, sort them and put them in culture to differentiate them in vitro to test their plasticity. Such studies resulted in lab-to-lab, equipment-to-equipment and reagent-to-reagent (e.g. antibodies) differences in the results, which lead to discussions of contaminations by other cell types (Prokopi et al., 2009). However, the therapeutic potential of EPCs appears relatively high. Numerous clinical trials have been conducted, which mostly reported beneficial effects (Kumar & Caplice, 2010).

The most confusing issue about EPCs is the lack of defined marker and niches (Ergün, 2011; Psaltis et al., 2010). This lead to the claim that EPCs might belong to the haematopoietic lineage and share its niche and origin (Richardson & Yoder, 2011). The contribution of EPCs to neo-angiogenesis is controversial. Some reports describe a very strong contribution to the endothelium of different vessels whereas other studies demonstrated only a moderate if any contribution. A contribution of EPCs to parenchymal cells of the heart appears unlikely. Lineage tracing experiments using the haematopoietic-specific promoter vav did not indicate a contribution to cardiomyocytes (Nern et al., 2009) although it is possible that some EPCs are not derived from the haematopoietic lineage.

Endothelial and smooth muscle cells are relatively plastic, which might allow them to acquire different fates after de-differentiation and re-differentiation including fibroblasts (Stintzing et al., 2009; Zeisberg et al., 2007). All these considerations make it difficult to judge the contributions of such EPCs to the renewal of myocardial cell types without a lineage tracing study using a clearly defined marker for EPCs, which is currently non-existing.

2.4 Cardiac fibroblasts

Cardiac fibroblasts comprise a rather heterogenous group of cells. Due to the absence of a single pathognomonic marker for fibroblasts, virtually all interstitial cells, which are not associated with vessels and which are not cardiomyocytes, are considered fibroblasts. The unusual phenotypic plasticity of fibroblasts does also raise the question whether they represent a single mature cell type or also comprise different types of precursor cells (Eyden, 2004). This point becomes clearer when considering the morphology of primary BMSCs in culture, which are often described to have "fibroblast-like" morphology (Friedenstein et al., 1976), an attribute that is also used to describe adult stem cells in culture (Rios & Williams,1990). CSCs, for example, show a very similar morphology, which makes it difficult to distinguish them from fibroblasts without additional molecular markers (Messina et al., 2004). By definition, CSCs should give rise to all the lineages in heart, including cardiac fibroblasts. It probably will remain a lasting challenge for some time to distinguish CSCs from fibroblasts and from intermediary cell types that have been generated from CSCs but not yet acquired a classical fibroblastic phenotype.

3. Current status of cardiac stem cells

When one speaks about CSCs, currently there are 6 schools (Table 1).

Type of CSCs	Other Markers	References
Sca1+ CSCs	CD34-, CD45-, FLK1-, c-kit+/-, GATA4+, NKX2-5+/-, MEF2C+	(Oh et al., 2003; Forte et al., 2008; Matsuura et al., 2004; Rosenblatt-Velin et al., 2005; Tateishi et al., 2007; Wang et al., 2006; Wu et al., 2006)
c-kit+ CSCs	CD34-, CD45-, Sca1+, GATA4+, NKX2-5+, MEF2C+	(Bearzi et al., 2007; Beltrami et al., 2003; Dawn et al., 2005; Linke et al., 2005; Miyamoto et al., 2010; Tillmanns et al., 2008; Urbanek et al., 2003)
Isl-1+ CSCs	CD31-, Sca1-, c-kit-, GATA4+, NKX2-5+	(Laugwitz et al., 2005; Moretti et al., 2006)
Side population (SP) cells	CD34+, CD45+, ABCG2+, Sca1+, c-kit+, NKX2-5-, GATA4-	(Martin et al., 2004; Liang et al., 2010; Oyama et al., 2007; Pfister et al., 2005)
Cardiospheres	CD34+, CD45+, ABCG2+, Sca1+, c-kit+, NKX2-5-, GATA4-	(Messina et al., 2004; Andersen et al., 2009; Cheng et al., 2010; Davis et al., 2010; Smith et al., 2007; Tateishi et al., 2007)
Cardiac mesangioblasts (EPCs)	CD31+, CD34+, CD44+, CD45-, Sca1+, c-kit+	(Barbuti et al., 2010; Galvez et al., 2008; Gálvez et al., 2009)

Table 1. List of resident CSCs

Due to the importance of heart as a vital organ, several attempts have been made to characterize CSCs using various techniques. Numerous review articles were published covering the field of CSCs (Anversa et al., 2006; Bollini et al., 2011; Di Nardo et al., 2010; Musunuru et al., 2010; Tateishi et al., 2008). Considering the acclaimed types of CSCs, it seems that heart, once considered to be a post-mitotic organ, harbours the highest number of adult stem cells in the body. At present, it is not clear whether CSCs can be called "true" adult stem cells with a "true" multipotency (Ellison et al., 2010; Stamm et al., 2009). The concept of multipotent adult stem cells and progenitor cells (sometime classified as "transient-amplifying cells") (Bryder et al., 2006) is derived from the haematopoietic system and might not necessarily apply to CSCs.

As described in previous chapters, postnatal cardiomyocytes are likely to be renewed throughout the life of a mammal. The cellular sources of postnatally emerging cardiomyocytes are probably CSCs although definitive evidence is missing. Similarly, CSCs appear to be situated within the heart and not in the bone marrow or other locations outside the heart although such a possibility has not been unequivocally ruled out. Recent findings of circulating EPCs and "very small embryonic-like stem cells (VSELs)" (Ratajczak et al., 2009) renewed the debate about the origin of CSCs although the impact of such studies seems limited. Virtually all studies devoted to "circulating" ("non-resident") cells are based on ex vivo experiments. In such experiments, putative stem cells are isolated using antibodies against surface markers followed by transplantation into "hosts". This procedure is similar to bone marrow transplantation experiments, which sometimes results in fusion of transplanted cells to host cells or uptake of marker proteins (e.g. GFP) by host cells (Nern et al., 2009). In contrast, no clear evidence exists that intercellular fusions occur under physiological conditions in the heart involving, for example, cardiomyocytes and endothelial cells. In fact, classical genetic lineage tracing experiments and injection of genetically labelled cells into mouse blastocyts argue against widespread cellular fusion events (Schulze et al., 2005).

3.1 Lineage tracing

The "gold standard" to define the origin of a particular set of cells is a "lineage tracing" experiment using a defined and established stem cell marker gene. Long-term lineage tracing experiments rely on model organisms using transgenic technology. To understand the theoretical basis of lineage tracing, one needs to distinguish between transient and permanent cell labelling. Transient labelling is based on knock-in of a reporter gene (e.g. GFP) into a gene of interest to visualize the current activity of a specific gene. This has been successfully applied to observe the contributions of c-kit+ cells to the revascularizing infarct regions of the myocardial infarcted heart by utilizing c-kit(BAC)-EGFP mice (Tallini et al., 2009). Labelled cells in the infarct regions were not cardiomyocytes but endothelial and smooth muscle cells. Under normal physiological condition, the authors observed an increasing number of labelled cells in the heart up to postnatal day 2. Thereafter, the number declined, and the labelled cells were rarely observed in an adult heart. The authors concluded from these results that "c-kit expression in cardiomyocytes in the adult heart after injury does not identify cardiac myogenesis." (Tallini et al., 2009). As one can easily imagine, this is due to the limitation of transient labelling, since labelled cells will loose the label upon differentiation.

In contrast, a permanent labelling strategy, which directly targets the genetic material, can bypass problems that arise from changes in the transcriptional program. This can be done by utilizing mouse strains that carry a conditionally active Cre-recombinase and a Cre-dependent reporter gene (Smedley et al., 2010). Some researchers use inducible systems (e.g. tamoxifen-inducible Mer-Cre-Mer system (Petrich et al., 2003)) or double transgenic mice (Cre-reporter) and triple transgenic mice based on tet-Cre system (Tang et al., 2008).

3.2 A potential lineage tracing study to monitor contribution of CSCs to the myocardium

A permanent labelling strategy to trace the fates of adult stem cells might utilize a tetracycline transactivator (tTA) placed under the control of stem cell marker gene as shown in Fig. 1. Preferably, such a system should be based on a "tet-off" design, in which the system is activated until doxycycline (DOX) is added (Gossen & Bujard, 1992; Urlinger et al., 2000). The opposite system is called "tet-on" (rtTA is used), which is claimed to allow a tight control (Stary et al., 2010). However, when tet-on system is used, DOX must be administered in the drinking water of mice, which can be rather expensive. In addition, DOX inhibits angiogenesis in mouse (Cox et al., 2010; Fainaru et al., 2008). In our lab, the DOX-treated murine hearts (1mg/ml in the drinking water) tend to be ~30% smaller compared to the age-matched untreated hearts (data not shown). Furthermore, DOX might also have effects on the attenuation of cardiac hypertrophy through the inhibition of matrix metalloproteases (Errami et al., 2008), which makes it difficult to challenge such mice by transverse aortic constriction (TAC) to observe the contribution of marked adult stem cells to regeneration and remodelling of hypertrophied hearts. Recent studies shows that DOX can suppresses doxorubicin-induced oxidative stress and cellular apoptosis in the murine heart (Lai et al., 2010), which is another cardiovascular disease model commonly used in the field of cardiovascular medicine.

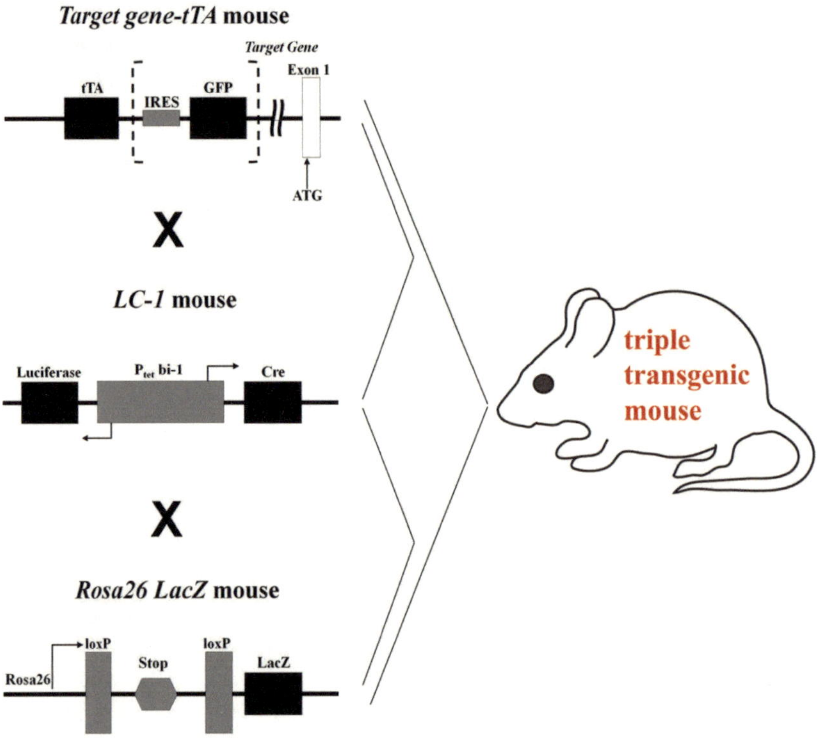

Fig. 1. Triple transgenic mouse model.

To monitor the current expression of a stem cell marker gene, one might insert an additional reporter gene (e.g. GFP) into the genomic locus using an internal ribosomal entry site (IRES) element (Attal et al., 1999). Yet, fluorescent reporter gene will limit the number of fluorescent signals that can be used for further studies to identify the fates of adult stem cells. Furthermore, autofluorescence of the heart caused by lipofuscin, which is breakdown product of old red blood cells (Van de Lest et al., 1995) might obstruct detection of GFP signals, which might lead to misinterpretation of experimental results (Laflamme & Murry, 2005). Therefore, the usage of such reporter genes must be considered carefully.

When such a tTA mouse is created, it should be crossed with a mouse containing a tTA/rtTA responsive element (TRE), such as "Ptet-1" (Baron & Bujard, 2000). In the example shown in Fig. 1, a mouse line called "LC-1", which ubiquitously expresses Cre-recombinase and luciferase gene in all tissues in an adult mouse upon induction, is shown (Schönig et al., 2002). As shown by Schönig et al. (Schönig et al., 2002), this mouse line was used in the triple transgenic mouse system based on tet-Cre crossed to a reporter line as we propose here. Alternatively, Tet-O-Cre transgenic mice, which expresses Cre-recombinase in a TRE-dependent manner, might be used (Hsu et al., 2010; Le et al., 2008; Radomska et al., 2002; Tang et al., 2008). The offspring of this mating are called "tet-Cre mice".

Upon successful creation of tet-Cre mice, these mice should be crossed to a reporter line allowing permanent labelling of cells, which have express tTA. Labelling has to be permanent (i.e. based on genomic recombination) to identify derivatives of adult stem cells even after differentiation into mature cells (e.g. cardiomyocyte). Various reporter lines are available to achieve this goal. For example, the Rosa26-LacZ mouse (Soriano, 1999), which expresses β-galactosidase from the Rosa26 locus; Z/AP mice (Lobe et al., 1999), which express β-galactosidase before and human placental alkaline phosphatase (AP) after Cre-mediated recombination. Another option is the Z/EG mouse (Novak et al., 2000), which is similar to the Z/AP mouse, but expresses enhanced GFP instead of AP after Cre-recombinase mediated recombination. Progeny of labelled cells can be easily identified using various cell type specific markers in combination with the reporter system that has been chosen.

Recent studies indicate that transgenes used to label distinct cell types are subject to gene silencing probably due to methylation, which will reduce the efficiency of Cre-mediated labelling (Long & Rossi, 2009). Therefore, it seems prudent to employ at least two different reporter lines. To allow clonal analysis of stem cell derivatives, future approaches will take advantage of reporter mice based on the "*Brainbow*" system. Up to 166 different colours can be generated in these mice by random, alternative use of variant loxP-sites creating combinations of different fluorescent proteins (OFP, M-RFP, M-YFP and M-CFP) (Livet et al., 2007). The original *Brainbow* mice are based on the Thy1 promoter, which restricted its usage to a certain tissues (primarily neuronal cells). Recently, Hans Clevers's group modified the *Brainbow* system by integrating the *Brainbow* cassette into the Rosa26 locus. The resulting "R26R-*Confetti*" mice allow ubiquitous expression in numerous cell types (Snippert et al., 2010). R26R-*Confetti* mice allow random labelling of single adult stem cell with a unique colour. When labelled cells proliferate, the label will be inherited to daughter cells, and it will be possible to distinguish the progeny from that of other individual stem cells. When this strategy is employed to the heart, it should be possible to characterize the identity of adult stem cells, which gives rise to all three lineages of heart (namely, cardiomyocytes, endothelial cells and smooth muscle cells (Moretti et al., 2006)). Further profiling of such cells might facilitate identification of additional markers and effectors.

The tracing system might be combined with different pathological conditions to explore effects of pathogenic stimuli on stem cells. For this purpose, we recently introduce a model to induce right ventricular hypertrophy through pulmonary artery clipping (PAC), which avoids detrimental right ventricular pressure overload, and thus allows long-term survival of operated mice (Kreymborg et al., 2010). Given that the origins of right ventricle and outflow tract are from the secondary heart field (Waldo et al, 2001; Verzi et al., 2005), our PAC model should primarily affect the regions of the secondary heart field so that the behaviour of descendents of Isl1-positive CSCs (Cai et al., 2003; Laugwitz et al., 2005) can be monitored. It is also possible to combine the tracing system with conditional knockout mice. Cre-mediated recombination in such quadruple transgenic mice will allow conditional inactivation of the gene of interest precisely in stem cell-derived, labelled cells.

3.2 Adverse effects of stem cell therapy

Stimulation of cardiac regeneration is one of the major goals for cardiovascular stem cell research. The use of ES cells poises several safety concerns spurred by potential oncogenicity and immunogenicity. Moreover, the generation of ES cells from human preimplantation embryos has raised several ethical issues. The main problem with other cell types (e.g. BMSCs, skeletal myoblasts) is their limited ability to acquire a fully functional cardiomyocyte state (Haider et al, 2010). Beneficial effects from such cells might be limited to the secretion of cytokines, growth factors and other signaling molecules to provide paracrine and/or trophic effects. Both clinical and basic science studies have provided evidences that some types of cells are able to cause cardiac arrhythmias in patients, although other studies revealed no harmful side-effects . In order to avoid mistakes that tainted gene therapy attempts, it is utmost importance to obtain an optimal control on all types of injected cells. In principal, it seems much safer (and more efficient) to manipulate resident CSCs to differentiate cell types in needs. Such activation of CSCs has been attempted in various studies through injection of growth factors (e.g. Hepatocyte Growth Factor and Insulin-Like Growth Factor-1 (Bocchi et al., 2011) avoiding potential side effects such as arrhythmias). However, it is clear that stimulated signalling pathways needs to be shut off once CSCs have differentiated. The generation of mouse reporter strains to monitor in vivo activation of CSCs would be very helpful. The tracing system that we proposed in the previous subsection should help in this respect

4. Conclusion

Recent findings have challenged the classical view of the heart as a post-mitotic organ. While replacement of cardiomyocytes and other specialized cells of the heart might be taken as granted, much more needs to be learned about the origin of cells that are instrumental for this process. A thorough analysis of resident CSCs in the heart but also of mature cells to contribute to the tissue rejuvenation is of outmost interest. Careful lineage tracing experiments are instrumental to achieve this goal. A potential approach for such an analysis was described in this chapter, which might yield definitive answers about the contribution of different CSCs. Lastly, we would like to answer the question that we asked in the title of this chapter: "Are we there yet?" Our answer is: "No, not yet. But soon, we might be closer."

5. Acknowledgment

We thank the members of our laboratory (Katharina Jenniches, Pascal Gellert, Mizue Teranishi and David John) for their help and support. We also thank Dr. Petra Uchida and Dr. Stefan Momma for their valuable advice and comments on this book chapter. This work was supported by a start-up-grant of the Excellence Cluster Cardio-Pulmonary System (ECCPS) (to SU), by fellowships of the International Max Planck Research School for Heart and Lung Research (IMPRS-HLR) (to PD-G), and by the Max-Planck-Society, the DFG (Br1416), the EU Commisson (MYORES network of excellence), the Kerckhoff-Foundation and the Excellence Initiative "Cardiopulmonary System" (to TB).

6. References

Adler, C.P. & Friedburg, H. (1986). Myocardial DNA content, ploidy level and cell number in geriatric hearts: post-mortem examinations of human myocardium in old age. *Journal of Molecular and Cellular Cardiology*, Vol.18, No.1, (January 1986), pp. 39-53.

Aitman, T.J., Critser, J.K., Cuppen, E., Dominiczak, A., Fernandez-Suarez, X.M., Flint, J., Gauguier, D., Geurts, A.M., Gould, M., Harris, P.C., Holmdahl, R., Hubner, N.,

Izsvák, Z., Jacob, H.J., Kuramoto, T., Kwitek, A.E., Marrone, A., Mashimo, T., Moreno, C., Mullins, J., Mullins, L., Olsson, T., Pravenec, M., Riley, L., Saar, K., Serikawa, T., Shull, J.D., Szpirer, C., Twigger, S.N., Voigt, B. & Worley, K. (2008). Progress and prospects in rat genetics: a community view. *Nature Genetics*, Vol.40, No.5 (May 2008), pp. 516-22.

Alaiti, M.A., Ishikawa, M. & Costa, M.A. (2010). Bone marrow and circulating stem/progenitor cells for regenerative cardiovascular therapy. *Translational Research*, Vol.156, No.3 (September 2010), pp. 112-29.

Andersen, D.C., Andersen, P., Schneider, M., Jensen, H.B. & Sheikh, S.P. (2009). Murine "cardiospheres" are not a source of stem cells with cardiomyogenic potential. *Stem Cells*, Vol.27, No.7 (July 2009), pp. 1571-81.

Anversa, P., Kajstura, J., Leri, A. & Bolli, R. (2006). Life and death of cardiac stem cells: a paradigm shift in cardiac biology. *Circulation*, Vol. 113, No.11 (March 2006), pp. 1451-63.

Anversa, P., Leri, A., Rota, M., Hosoda, T., Bearzi, C., Urbanek, K., Kajstura, J. & Bolli, R. (2007). Concise review: stem cells, myocardial regeneration, and methodological artifacts. *Stem Cells*, Vol.25, No.3 (March 2007), pp. 589-601.

Asahara, T., Murohara, T., Sullivan, A., Silver, M., van der Zee, R., Li, T., Witzenbichler, B., Schatteman, G. & Isner, J.M. (1997). Isolation of putative progenitor endothelial cells for angiogenesis. *Science*, Vol.275, No.5302 (February 1997), pp. 964-7

Attal, J., Theron, M.C., Puissant, C. & Houdebine, L.M. (1999). Effect of intercistronic length on internal ribosome entry site (IRES) efficiency in bicistronic mRNA. *Gene Expression*, Vol.8, No.5-6, pp. 299-309.

Banerjee, I., Fuseler, J.W., Price, R.L., Borg, T.K. & Baudino, T.A. (2007). Determination of cell types and numbers during cardiac development in the neonatal and adult rat and mouse. *American Journal of Physiology - Heart and Circulatory Physiology*, Vol.293, No.3 (September 2007), pp. H1883-91.

Barbuti, A., Galvez, B.G., Crespi, A., Scavone, A., Baruscotti, M., Brioschi, C., Cossu, G. & DiFrancesco, D. (2010). Mesoangioblasts from ventricular vessels can differentiate in vitro into cardiac myocytes with sinoatrial-like properties. *Journal of Molecular and Cellular Cardiology*, Vol.48, No.2 (February 2010), pp. 415-23.

Baron, U. & Bujard, H. (2000). Tet repressor based systems for regulated gene expression in eukaryotic cells: principles and advances. *Methods in Enzymology*, Vol.327, pp. 401-421.

Bearzi, C., Rota, M., Hosoda, T., Tillmanns, J., Nascimbene, A., De Angelis, A., Yasuzawa-Amano, S., Trofimova, I., Siggins, R.W., Lecapitaine, N., Cascapera, S., Beltrami, A.P., D'Alessandro, D.A., Zias, E., Quaini, F., Urbanek, K., Michler, R.E., Bolli, R., Kajstura, J., Leri, A. & Anversa, P. (2007). Human cardiac stem cells. *Proceedings of the National Academy of Sciences*, Vol.104, No.35 (August 2007), pp. 14068-73.

Beltrami, A.P., Barlucchi, L., Torella, D., Baker, M., Limana, F., Chimenti, S., Kasahara, H., Rota, M., Musso, E., Urbanek, K., Leri, A., Kajstura, J., Nadal-Ginard, B. & Anversa, P. (2003). Adult cardiac stem cells are multipotent and support myocardial regeneration. *Cell*, Vol.114, No.6 (September 2003), pp. 763-76.

Bergmann, O., Bhardwaj, R.D., Bernard, S., Zdunek, S., Barnabé-Heider, F., Walsh, S., Zupicich, J., Alkass, K., Buchholz, B.A., Druid, H., Jovinge, S. & Frisén, J. (2009).

Evidence for cardiomyocyte renewal in humans. *Science*, Vol.324, No.5923 (April 2009), pp. 98-102.

Bergmann, O., Zdunek, S., Alkass, K., Druid, H., Bernard, S. & Frisén, J. (2011). Identification of cardiomyocyte nuclei and assessment of ploidy for the analysis of cell turnover. *Experimenal Cell Research*, Vol.317, No.2 (January 2011), pp. 188-94.

Bocchi, L., Savi, M., Graiani, G., Rossi, S., Agnetti, A., Stillitano, F., Lagrasta, C., Baruffi, S., Berni, R., Frati, C., Vassalle, M., Squarcia, U., Cerbai, E., Macchi, E., Stilli, D., Quaini, F., & Musso, E. (2011) Growth factor-induced mobilization of cardiac progenitor cells reduces the risk of arrhythmias, in a rat model of chronic myocardial infarction. *PLoS One*, Vol.6, No.3(March 2011), pp. e17750.

Bollini, S., Smart, N. & Riley, P.R. (2011). Resident cardiac progenitor cells: At the heart of regeneration. *Journal of Molecular and Cellular Cardiology*, Vol.50, No.2 (February 2011), pp. 296-303.

Bryder, D., Rossim, D.J. & Weissman, I.L. (2006). Hematopoietic stem cells: the paradigmatic tissue-specific stem cell. *American Journal of Pathology*, Vol.169, No.2 (August 2006), pp. 338-46.

Buehr, M., Meek, S., Blair, K., Yang, J., Ure, J., Silva, J., McLay, R., Hall, J., Ying, Q.L. & Smith,

A. (2008). Capture of authentic embryonic stem cells from rat blastocysts. *Cell*, Vol.135, No.7 (December 2008), pp. 1287-98.

Cai, C.L., Liang, X., Shi, Y., Chu, P.H., Pfaff, S.L., Chen, J. & Evans, S. (2003). Isl1 identifies a cardiac progenitor population that proliferates prior to differentiation and contributes a majority of cells to the heart. *Developmental Cell*, Vol.5, No.6 (December 2003), pp. 877-89.

Carvalho, A.B. & de Carvalho, A.C. (2010). Heart regeneration: Past, present and future. *World Journal of Cardiology*, Vol.2, No.5 (May 2010), pp. 107-11.

Cheng, K., Li, T.S., Malliaras, K., Davis, D.R., Zhang, Y. & Marbán, E. (2010). Magnetic targeting enhances engraftment and functional benefit of iron-labeled cardiosphere-derived cells in myocardial infarction. *Circulation Research*, Vol.106, No.10 (May 2010), pp. 1570-81.

Chugh, A.R,. Zuba-Surma, E.K. & Dawn, B. (2009). Bone marrow-derived mesenchymal stems cells and cardiac repair. *Minerva Cardioangiologica*, Vol.57, No.2 (April 2009), pp. 185-202.

Cox, C.A., Amaral, J., Salloum, R., Guedez, L., Reid, T.W., Jaworski, C., John-Aryankalayil, M., Freedman, K.A., Campos, M.M., Martinez, A., Becerra, S.P. & Carper, D.A. (2010). Doxycycline's effect on ocular angiogenesis: an in vivo analysis. *Ophthalmology*, Vol.117, No.9 (September 2010), pp. 1782-91.

Cui, X., Ji, D., Fisher, D.A., Wu, Y., Briner, D.M. & Weinstein, E.J. (2011). Targeted integration in rat and mouse embryos with zinc-finger nucleases. *Nature Biotechnology*, Vol.29, No.1 (January 2011), pp. 64-7.

Davis, D.R., Kizana, E., Terrovitis, J., Barth, A.S., Zhang, Y., Smith, R.R., Miake, J. & Marbán, E. (2010). Isolation and expansion of functionally-competent cardiac progenitor cells directly from heart biopsies. *Journal of Molecular and Cellular Cardiology*, Vol.49, No.2 (August 2010), pp. 312-21.

Dawn, B., Stein, A.B., Urbanek, K., Rota, M., Whang, B., Rastaldo, R., Torella, D., Tang, X.L., Rezazadeh, A., Kajstura, J., Leri, A., Hunt, G., Varma, J., Prabhu, S.D., Anversa, P.

& Bolli, R. (2005). Cardiac stem cells delivered intravascularly traverse the vessel barrier, regenerate infarcted myocardium, and improve cardiac function. *Proceedings of the National Academy of Sciences*, Vol.102, No.10 (March 2005), pp. 3766-71.

Devbhandari, M.P., Meraj, S., Jones, M.T., Kadir, I. & Bridgewater, B. (2007). Primary cardiac sarcoma: reports of two cases and a review of current literature. *Journal of Cardiothoracic Surgery*, Vol.2 (July 2007), pp. 34.

Di Nardo, P., Forte, G., Ahluwalia, A. & Minieri, M. (2010). Cardiac progenitor cells: potency and control. *Journal of Cellular Physiology*, Vol.224, No.3 (September 2010), pp. 590-600.

DuFrain, R.J., McFee, A.F., Linkous, S., Jennings, C.J. & Lowe, K.W. (1984). In vivo SCE analysis using bromodeoxyuridine, iododeoxyuridine, and chlorodeoxyuridine. *Mutation Research*, Vol.139, No.2 (February 1984), pp. 57-60.

Ebelt, H., Hufnagel, N., Neuhaus, P., Neuhaus, H., Gajawada, P., Simm, A., Müller-Werdan, U., Werdan, K. & Braun, T. (2005). Divergent siblings: E2F2 and E2F4 but not E2F1 and E2F3 induce DNA synthesis in cardiomyocytes without activation of apoptosis. *Circulation Research*, Vol.96, No.5 (March 2005), pp. 509-17.

Ebelt, H., Liu, Z., Müller-Werdan, U., Werdan, K. & Braun, T. (2006). Making omelets without breaking eggs: E2F-mediated induction of cardiomyoycte cell proliferation without stimulation of apoptosis. *Cell Cycle*, Vol.5, No.21 (November 2006), pp. 2436-9.

Ebelt, H., Zhang, Y., Köhler, K., Xu, J., Gajawada, P., Boettger, T., Hollemann, T., Müller-Werdan, U., Werdan, K. & Braun, T. (2008). Directed expression of dominant-negative p73 enables proliferation of cardiomyocytes in mice. *Journal of Molecular and Cellular Cardiology*, Vol.45, No.3 (September 2008), pp. 411-9.

Ebelt, H., Zhang, Y., Kampke, A., Xu, J., Schlitt, A., Buerke, M., Müller-Werdan, U., Werdan, K. & Braun, T. (2008). E2F2 expression induces proliferation of terminally differentiated cardiomyocytes in vivo. Cardiovascular Research, Vol.80, No.2 (November 2008), pp. 219-26.

Ellison, G.M., Galuppo, V., Vicinanza, C., Aquila, I., Waring, C.D., Leone, A., Indolfi, C. & Torella, D. (2010). Cardiac stem and progenitor cell identification: different markers for the same cell? *Frontier in Bioscience (Scholor Edition)*, Vol.2 (January 2010), pp. 641-52.

Ergün, S., Tilki, D. & Klein, D. (2011). Vascular Wall as a Reservoir for Different Types of Stem and Progenitor Cells. *Antioxidants & Redox Signaling*, [Epub ahead of print] (January 2011).

Errami, M., Galindo, C.L., Tassa, A.T., Dimaio, J.M., Hill, J.A. & Garner, H.R. (2008). Doxycycline attenuates isoproterenol- and transverse aortic banding-induced cardiac hypertrophy in mice. *Journal of Pharmacology and Experimental Therapeutics*, Vol.324, No.3 (March 2008), pp. 1196-203.

Eyden, B. (2004). Fibroblast phenotype plasticity: relevance for understanding heterogeneity in "fibroblastic" tumors. *Ultrastructural Pathology*, Vol.28, No.5-6 (September-December 2004), pp. 307-19.

Fainaru, O., Adini, I., Benny, O., Bazinet, L., Pravda, E., D'Amato, R. & Folkman, J. (2008). Doxycycline induces membrane expression of VE-cadherin on endothelial cells and prevents vascular hyperpermeability. *FASEB Journal*, Vol.22, No.10 (October 2008), pp. 3728-35.

Forte, G., Carotenuto, F., Pagliari, F., Pagliari, S., Cossa, P., Fiaccavento, R., Ahluwalia, A., Vozzi, G., Vinci, B., Serafino, A., Rinaldi, A., Traversa, E., Carosella, L., Minieri, M. & Di Nardo, P. (2008). Criticality of the biological and physical stimuli array inducing resident cardiac stem cell determination. *Stem Cells*, Vol.26, No.8 (August 2008), pp. 2093-103.

Friedenstein, A.J., Gorskaja, J.F. & Kulagina, N.N. (1976). Fibroblast precursors in normal and irradiated mouse hematopoietic organs. *Experimental Hematology*, Vol.4, No.5 (September 1976), pp. 267-74.

Galvez, B.G., Sampaolesi, M., Barbuti, A., Crespi, A., Covarello, D., Brunelli, S., Dellavalle, A., Crippa, S., Balconi, G., Cuccovillo, I., Molla, F., Staszewsky, L., Latini, R., Difrancesco, D. & Cossu, G. (2008). Cardiac mesoangioblasts are committed, self-renewable progenitors, associated with small vessels of juvenile mouse ventricle. *Cell Death and Differetiation*, Vol.15, No.9 (September 2008), pp. 1417-28.

Gálvez, B.G., Covarello, D., Tolorenzi, R., Brunelli, S., Dellavalle, A., Crippa, S., Mohammed, S.A., Scialla, L., Cuccovillo, I., Molla, F., Staszewsky, L., Maisano, F., Sampaolesi, M., Latini, R. & Cossu, G. (2009). Human cardiac mesoangioblasts isolated from hypertrophic cardiomyopathies are greatly reduced in proliferation and differentiation potency. *Cardiovascular Research*, Vol.83, No.4 (September 2009), pp. 707-16.

Geurts, A.M., Cost, G.J., Freyvert, Y., Zeitler, B., Miller, J.C., Choi, V.M., Jenkins, S.S., Wood, A., Cui, X., Meng, X., Vincent, A., Lam, S., Michalkiewicz, M., Schilling, R., Foeckler, J., Kalloway, S., Weiler, H., Ménoret, S., Anegon, I., Davis, G.D., Zhang, L., Rebar, E.J., Gregory, P.D., Urnov, F.D., Jacob, H.J. & Buelow, R. (2009). Knockout rats via embryo microinjection of zinc-finger nucleases. *Science*, Vol.325, No.5939 (July 2009), pp. 433

Gossen, M. & Bujard, H. (1992). Tight control of gene expression in mammalian cells by tetracycline-responsive promoters. *Proceedings of the National Academy of Sciences*, Vol.89, No.12 (June 1992), pp. 5547-51.

Haider, K.H., Buccini, S., Ahmed, R.P., & Ashraf, M. (2010) De novo myocardial regeneration: advances and pitfalls. *Antioxidants & Redox Signaling*, Vol.13, No.12 (December 2010), pp. 1867-77.

Holaska, J.M. (2008). Emerin and the nuclear lamina in muscle and cardiac disease. *Circulation Research*, Vol.103, No.1 (July 2008), pp. 16-23.

Hsieh, P.C., Segers, V.F., Davis, M.E., MacGillivray, C., Gannon, J., Molkentin, J.D., Robbins, J. & Lee, R.T. (2007). Evidence from a genetic fate-mapping study that stem cells refresh adult mammalian cardiomyocytes after injury. *Nature Medicine*, Vol.13, No.8 (August 2007), pp. 970-4.

Hsu, W., Mirando, A.J. & Yu, H.M. (2010). Manipulating gene activity in Wnt1-expressing precursors of neural epithelial and neural crest cells. *Developmental Dynamics*, Vol.239, No.1 (January 2010), pp. 338-45.

Kajstura, J., Urbanek, K., Perl, S., Hosoda, T., Zheng, H., Ogórek, B., Ferreira-Martins, J., Goichberg, P., Rondon-Clavo, C., Sanada, F., D'Amario, D., Rota, M., Del Monte, F., Orlic, D., Tisdale, J., Leri, A. & Anversa, P. (2010). Cardiomyogenesis in the adult human heart. *Circulation Research*, Vol.107, No.2 (July 2010), pp. 305-15.

Kajstura, J., Gurusamy, N., Ogórek, B., Goichberg, P., Clavo-Rondon, C., Hosoda, T., D'Amario, D., Bardelli, S., Beltrami, A.P., Cesselli, D., Bussani, R., Del Monte, F., Quaini, F., Rota, M., Beltrami, C.A., Buchholz, B.A., Leri, A. & Anversa, P. (2010). Myocyte Turnover in the Aging Human Heart. *Circulation Research*, Vol.107, No.11 (November 2010), pp. 1374-86.

Karsner, H.T., Saphir, O. & Todd, T.W. (1925). The State of the Cardiac Muscle in Hypertrophy and Atrophy. *American Journal of Pathology*, Vol.1, No.4 (July 1925), pp. 351-372.1.

Kreymborg, K., Uchida, S., Gellert, P., Schneider, A., Boettger, T., Voswinckel, R., Wietelmann, A., Szibor, M., Weissmann, N., Ghofrani, A.H., Schermuly, R., Schranz, D., Seeger, W. & Braun, T. (2010). Identification of right heart-enriched genes in a murine model of chronic outflow tract obstruction. *Journal of Molecular and Cellular Cardiology*, Vol.49, No.4 (October 2010), pp. 598-605.

Kullo, I.J. & Cooper, L.T. (2010). Early identification of cardiovascular risk using genomics and proteomics. *Nature Review Cardiology*, Vol.7, No.6 (June 2010), pp. 309-17.

Kumar, A.H. & Caplice, N.M. (2010). Clinical potential of adult vascular progenitor cells. *Arteriosclerosis, Thrombosis, and Vascular Biology*, Vol.30, No.6 (June 2010), pp. 1080-7.

Laflamme, M.A. & Murry, C.E. (2005). Regenerating the heart. *Nature Biotechnology*, Vol.23, No.7 (July 2005), pp. 845-56.

Lai, H.C., Yeh, Y.C., Ting, C.T., Lee, W.L., Lee, H.W., Wang, L.C., Wang, K.Y., Lai, H.C., Wu, A. & Liu, T.J. (2010). Doxycycline suppresses doxorubicin-induced oxidative stress and cellular apoptosis in mouse hearts. *European Journal of Pharmacology*, Vol.644, No.1-3(October 2010), pp. 176-87.

Laugwitz, K.L., Moretti, A., Lam, J., Gruber, P., Chen, Y., Woodard, S., Lin, L.Z., Cai, C.L., Lu, M.M., Reth, M., Platoshyn, O., Yuan, J.X., Evans, S. & Chien, K.R. (2005). Postnatal isl1+ cardioblasts enter fully differentiated cardiomyocyte lineages. *Nature*, Vol.433, No.7026 (February 2005), pp. 647-53.

Le, Y.Z., Zheng, W., Rao, P.C., Zheng, L., Anderson, R.E., Esumi, N., Zack, D.J. & Zhu, M. (2008). Inducible expression of cre recombinase in the retinal pigmented epithelium. *Investigative Ophthalmology & Visual Science*, Vol.49, No.3 (March 2008), pp. 1248-53.

Li, P., Tong, C., Mehrian-Shai, R., Jia, L., Wu, N., Yan, Y., Maxson, R.E., Schulze, E.N., Song, H., Hsieh, C.L., Pera, M.F. & Ying, Q.L. (2008). Germline competent embryonic stem cells derived from rat blastocysts. *Cell*, Vol.135, No.7 (December 2008), pp. 1299-310.

Liang, S.X., Tan, T.Y., Gaudry, L. & Chong, B. (2010). Differentiation and migration of Sca1+/CD31- cardiac side population cells in a murine myocardial ischemic model. *International Journal of Cardiology*, Vol.138, No.1 (January 2010), pp. 40-9.

Limana, F., Capogrossi, M.C. & Germani, A. (2011). The epicardium in cardiac repair: from the stem cell view. *Pharmacology & Therapeutics*, Vol.129, No.1 (January 2011), pp. 82-96.

Linke, A., Müller, P., Nurzynska, D., Casarsa, C., Torella, D., Nascimbene, A., Castaldo, C., Cascapera, S., Böhm, M., Quaini, F., Urbanek, K., Leri, A., Hintze, T.H., Kajstura, J. & Anversa, P. (2005). Stem cells in the dog heart are self-renewing, clonogenic, and multipotent and regenerate infarcted myocardium, improving cardiac function.

Proceedings of the National Academy of Sciences, Vol.102, No.25 (June 2005), pp. 8966-71.

Liu, Z., Yue, S., Chen, X., Kubin, T. & Braun, T. (2010). Regulation of cardiomyocyte polyploidy and multinucleation by CyclinG1. *Circulation Research*, Vol.106, No.9 (May 2010), pp. 1498-506.

Livet, J., Weissman, T.A., Kang, H., Draft, R.W., Lu, J., Bennis, R.A., Sanes, J.R. & Lichtman, J.W. (2007). Transgenic strategies for combinatorial expression of fluorescent proteins in the nervous system. *Nature*, Vol.450, No.7166 (November 2007), pp. 56-62.

Lobe, C.G., Koop, K.E., Kreppner, W., Lomeli, H., Gertsenstein, M. & Nagy, A. (1999). Z/AP, a double reporter for cre-mediated recombination. *Developmental Biology*, Vol.208, No.2 (April 1999), pp. 281-92.

Long, M.A. & Rossi, F.M. (2009). Silencing inhibits Cre-mediated recombination of the Z/AP and Z/EG reporters in adult cells. *PLoS One*, Vol.4, No.5 (May 2009), pp. e5435.

Martin, C.M., Meeson, A.P., Robertson, S.M., Hawke, T.J., Richardson, J.A., Bates, S., Goetsch, S.C., Gallardo, T.D. & Garry, D.J. (2004). Persistent expression of the ATP-binding cassette transporter, Abcg2, identifies cardiac SP cells in the developing and adult heart. *Developmental Biology*, Vol.265, No.1 (January 2004), pp. 262-75.

Matsuura, K., Nagai, T., Nishigaki, N., Oyama, T., Nishi, J., Wada, H., Sano, M., Toko, H., Akazawa, H., Sato, T., Nakaya, H., Kasanuki, H. & Komuro, I. (2004). Adult cardiac Sca-1-positive cells differentiate into beating cardiomyocytes. *Journal of Biological Chemistry*, Vol.279, No.12 (March 2004), pp. 11384-91.

Messina, E., De Angelis, L., Frati, G., Morrone, S., Chimenti, S., Fiordaliso, F., Salio, M., Battaglia, M., Latronico, M.V., Coletta, M., Vivarelli, E., Frati, L., Cossu, G. & Giacomello, A. (2004). Isolation and expansion of adult cardiac stem cells from human and murine heart. *Circulation Research*, Vol.95, No.9 (October 2004), pp. 911-21.

Miyamoto, S., Kawaguchi, N., Ellison, G.M., Matsuoka, R., Shin'oka, T. & Kurosawa, H. (2010). Characterization of long-term cultured c-kit+ cardiac stem cells derived from adult rat hearts. *Stem Cells and Development*, Vol.19, No.1 (January 2010), pp. 105-16.

Morabito, C.J., Dettman, R.W., Kattan, J., Collier, J.M. & Bristow, J. (2001). Positive and negative regulation of epicardial-mesenchymal transformation during avian heart development. *Developmental Biology*, Vol.234, No.1 (June 2001), pp. 204-15.

Moretti, A., Caron, L., Nakano, A., Lam, J.T., Bernshausen, A., Chen, Y., Qyang, Y., Bu, L., Sasaki, M., Martin-Puig, S., Sun, Y., Evans, S.M., Laugwitz, K.L. & Chien, K.R. (2006). Multipotent embryonic isl1+ progenitor cells lead to cardiac, smooth muscle, and endothelial cell diversification. *Cell*, Vol.127, No.6 (December 2006), pp. 1151-65.

Musunuru, K., Domian, I.J. & Chien, K.R. (2010). Stem cell models of cardiac development and disease. *Annual Review of Cell and Developmental Biology*, Vol.26 (November 2010), pp. 667-87.

Nag, A.C. (1980). Study of non-muscle cells of the adult mammalian heart: a fine structural analysis and distribution. *Cytobios*, Vol.28, No.109, pp. 41-61.

Nern, C., Wolff, I., Macas, J., von Randow, J., Scharenberg, C., Priller, J. & Momma, S. (2009). Fusion of hematopoietic cells with Purkinje neurons does not lead to stable heterokaryon formation under noninvasive conditions. *Journal of Neuroscience*, Vol.29, No.12 (March 2009), pp. 3799-807.

Novak, A., Guo, C., Yang, W., Nagy, A. & Lobe, C.G. (2000). Z/EG, a double reporter mouse line that expresses enhanced green fluorescent protein upon Cre-mediated excision. *Genesis*, Vol.28, No.3-4 (November-December 2000), pp. 147-55.

Oh, H., Bradfute, S.B., Gallardo, T.D., Nakamura, T., Gaussin, V., Mishina, Y., Pocius, J., Michael, L.H., Behringer, R.R., Garry, D.J., Entman, M.L. & Schneider, M.D. (2003). Cardiac progenitor cells from adult myocardium: homing, differentiation, and fusion after infarction. *Proceedings of the National Academy of Sciences*, Vol.100, No.21 (October 2003), pp. 12313-8.

Olivetti, G., Giordano, G., Corradi, D., Melissari, M., Lagrasta, C., Gambert, S.R. & Anversa, P. (1995). Gender differences and aging: effects on the human heart. *Journal of the American College of Cardiology*, Vol.26, No.4 (October 1995), pp. 1068-79.

Olivetti, G., Cigola, E., Maestri, R., Corradi, D., Lagrasta, C., Gambert, S.R. & Anversa, P. (1996). Aging, cardiac hypertrophy and ischemic cardiomyopathy do not affect the proportion of mononucleated and multinucleated myocytes in the human heart. *Journal of Molecular and Cellular Cardiology*, Vol.28, No.7 (July 1996), pp. 1463-77.

Owens, G.K., Kumar, M.S. & Wamhoff, B.R. (2004). Molecular regulation of vascular smooth muscle cell differentiation in development and disease. *Physiological Reviews*, Vol.84, No.3 (July 2004), pp. 767-801.

Owens, G.K. (1995). Regulation of differentiation of vascular smooth muscle cells. *Physiological Reviews*, Vol.75, No.3 (July 1995), pp. 487-517.

Oyama, T., Nagai, T., Wada, H., Naito, A.T., Matsuura, K., Iwanaga, K., Takahashi, T., Goto, M., Mikami, Y., Yasuda, N., Akazawa, H., Uezumi, A., Takeda, S. & Komuro, I. (2007). Cardiac side population cells have a potential to migrate and differentiate into cardiomyocytes in vitro and in vivo. *Journal of Cell Biology*, Vol.176, No.3 (January 2007), pp. 329-41.

Parmacek, M.S. & Epstein, J.A. (2009). Cardiomyocyte renewal. *New England Journal of Medicine*, Vol.361, No.1 (July 2009), pp. 86-8.

Petrich, B.G., Molkentin, J.D. & Wang, Y. (2003). Temporal activation of c-Jun N-terminal kinase in adult transgenic heart via cre-loxP-mediated DNA recombination. *FASEB Journal*, Vol.17, No.6 (April 2003), pp. 749-51.

Pfister, O., Mouquet, F., Jain, M., Summer, R., Helmes, M., Fine, A., Colucci, W.S. & Liao, R. (2005). CD31- but Not CD31+ cardiac side population cells exhibit functional cardiomyogenic differentiation. *Circulation Research*, Vol.97, No.1 (July 2007), pp. 52-61.

Phinney, D.G. & Prockop, D.J. (2007). Concise review: mesenchymal stem/multipotent stromal cells: the state of transdifferentiation and modes of tissue repair--current views. *Stem Cells*, Vol.25, No.11 (November 2007), pp. 2896-902.

Prokopi, M., Pula, G., Mayr, U., Devue, C., Gallagher, J., Xiao, Q., Boulanger, C.M., Westwood, N., Urbich, C., Willeit, J., Steiner, M., Breuss, J., Xu, Q., Kiechl, S. & Mayr, M. (2009). Proteomic analysis reveals presence of platelet microparticles in endothelial progenitor cell cultures. *Blood*, Vol.114, No.3 (July 2009), pp. 723-32.

Psaltis, P.J., Zannettino, A.C., Worthley, S.G. & Gronthos, S. (2008). Concise review: mesenchymal stromal cells: potential for cardiovascular repair. *Stem Cells*, Vol.26, No.9 (September 2008), pp. 2201-10.

Psaltis, P.J., Harbuzariu, A., Delacroix, S., Holroyd, E.W. & Simari, R.D. (2010). Resident Vascular Progenitor Cells-Diverse Origins, Phenotype, and Function. *Journal of Cardiovascular Translational Research*, Vol.4, No.2 (April 2011), pp. 161-76.

Radomska, H.S., Gonzalez, D.A., Okuno, Y., Iwasaki, H., Nagy, A., Akashi, K., Tenen, D.G. & Huettner, C.S. (2002). Transgenic targeting with regulatory elements of the human CD34 gene. *Blood*, Vol.100, No.13 (December 2002), pp. 4410-9.

Ratajczak, M.Z., Kucia, M., Ratajczak, J. & Zuba-Surma, E.K. (2009). A multi-instrumental approach to identify and purify very small embryonic like stem cells (VSELs) from adult tissues. *Micron*, Vol.40, No.3 (April 2009), pp. 386-93.

Richardson, M.R. & Yoder, M.C. (2011). Endothelial progenitor cells: Quo Vadis? *Journal of Molecular and Cellular Cardiology*, Vol.50, No.2 (February 2011), pp. 266-72.

Rios, M. & Williams, D.A. (1990). Systematic analysis of the ability of stromal cell lines derived from different murine adult tissues to support maintenance of hematopoietic stem cells in vitro. *Journal of Cellular Physiology*, Vol.145, No.3 (December 1990), pp. 434-43.

Rosenblatt-Velin, N., Lepore, M.G., Cartoni, C., Beermann, F. & Pedrazzini, T. (2005). FGF-2 controls the differentiation of resident cardiac precursors into functional cardiomyocytes. *Journal Clinical Investigation*, Vol.115, No.7 (July 2005), pp. 1724-33.

Rubart, M. & Field, L.J. (2006). Cardiac regeneration: repopulating the heart. *Annual Review of Physiology*, Vol.68, pp. 29-49.

Rzucidlo, E.M., Martin, K.A. & Powell, R.J. (2007). Regulation of vascular smooth muscle cell differentiation. *Journal of Vascular Surgery*, Vol.45, Suppl.A (June 2007), pp. A25-32.

Schönig, K., Schwenk, F., Rajewsky, K. & Bujard, H. (2002). Stringent doxycycline dependent control of CRE recombinase in vivo. *Nucleic Acids Research*, Vol.30, No.23 (December 2002), pp. e134.

Schulze, M., Belema-Bedada, F., Technau, A. & Braun, T. (2005). Mesenchymal stem cells are recruited to striated muscle by NFAT/IL-4-mediated cell fusion. *Genes & Development*, Vol.19, No.15 (August 2005), pp. 1787-98.

Smedley, D., Salimova, E. & Rosenthal, N. (2010). Cre recombinase resources for conditional mouse mutagenesis. *Methods*, Vol.53, No.4 (April 2011), pp. 411-6.

Smith, R.R., Barile, L., Cho, H.C., Leppo, M.K., Hare, J.M., Messina, E., Giacomello, A., Abraham, M.R. & Marbán, E. (2007). Regenerative potential of cardiosphere-derived cells expanded from percutaneous endomyocardial biopsy specimens. *Circulation*, Vol.115, No.7 (February 2007), pp. 896-908.

Snippert, H.J., van der Flier, L.G., Sato, T., van Es, J.H., van den Born, M., Kroon-Veenboer, C., Barker, N., Klein, A.M., van Rheenen, J., Simons, B.D. & Clevers, H. (2010). Intestinal crypt homeostasis results from neutral competition between symmetrically dividing Lgr5 stem cells. *Cell*, Vol.143, No.1 (October 2010), pp. 134-44.

Soonpaa, M.H. & Field, L.J. (1998). Survey of studies examining mammalian cardiomyocyte DNA synthesis. *Circulation Research*, Vol.83, No.1 (July 1998), pp. 15-26.

Soonpaa, M.H., Kim, K.K., Pajak, L., Franklin, M. & Field, L.J. (1996). Cardiomyocyte DNA synthesis and binucleation during murine development. *American Journal of Physiology*, Vol.271, No.5 Pt 2 (November 1996), pp. H2183-9.

Soriano, P. (1999). Generalized lacZ expression with the ROSA26 Cre reporter strain. *Nature Genetics*, Vol.21, No.1 (January 1999), pp. 70-1.

Stamm, C., Choi, Y.H., Nasseri, B. & Hetzer, R. (2009). A heart full of stem cells: the spectrum of myocardial progenitor cells in the postnatal heart. *Therapeutic Advances in Cardiovascular Disease*, Vol.3, No.3 (2009), pp. 215-29.

Stary, E., Gaupp, R., Lechner, S., Leibig, M., Tichy, E., Kolb, M. & Bertram, R. (2010). New architectures for Tet-on and Tet-off regulation in Staphylococcus aureus. *Applied and Environmental Microbiology*, Vol.76, No.3 (February 2010), pp. 680-7.

Stintzing, S., Ocker, M., Hartner, A., Amann, K., Barbera, L. & Neureiter, D. (2009). Differentiation patterning of vascular smooth muscle cells (VSMC) in atherosclerosis. *Virchows Archiv*, Vol.455, No.2 (August 2009), pp. 171-85.

Tallini, Y.N., Greene, K.S., Craven, M., Spealman, A., Breitbach, M., Smith, J., Fisher, P.J., Steffey, M., Hesse, M., Doran, R.M., Woods, A., Singh, B., Yen, A., Fleischmann, B.K. & Kotlikoff, M.I. (2009). c-kit expression identifies cardiovascular precursors in the neonatal heart. *Proceedings of the National Academy of Sciences*, Vol.106, No.6 (February 2009), pp. 1808-13.

Tang, W., Zeve, D., Suh, J.M., Bosnakovski, D., Kyba, M., Hammer, R.E., Tallquist, M.D. & Graff, J.M. (2008). White fat progenitor cells reside in the adipose vasculature. *Science*, Vol.322, No.5901 (October 2008), pp. 583-6.

Tateishi, K., Ashihara, E., Takehara, N., Nomura, T., Honsho, S., Nakagami, T., Morikawa, S., Takahashi, T., Ueyama, T., Matsubara, H. & Oh, H. (2007). Clonally amplified cardiac stem cells are regulated by Sca-1 signaling for efficient cardiovascular regeneration. *Journal of Cell Science*, Vol.120, No.Pt 10 (May 2007), pp. 1791-800.

Tateishi, K., Takehara, N., Matsubara, H. & Oh, H. (2008). Stemming heart failure with cardiac- or reprogrammed-stem cells. *Journal of Cellular and Molecular Medicine*, Vol.12, No.6A (December 2008), pp. 2217-32.

Tillmanns, J., Rota, M., Hosoda, T., Misao, Y., Esposito, G., Gonzalez, A., Vitale, S., Parolin, C., Yasuzawa-Amano, S., Muraski, J., De Angelis, A., Lecapitaine, N., Siggins, R.W., Loredo, M., Bearzi, C., Bolli, R., Urbanek, K., Leri, A., Kajstura, J. & Anversa, P. (2008). Formation of large coronary arteries by cardiac progenitor cells. *Proceedings of the National Academy of Sciences*, Vol.105, No.5 (February 2008), pp. 1668-73.

Tong, C., Li, P., Wu, N.L., Yan, Y. & Ying, Q.L. (2010). Production of p53 gene knockout rats by homologous recombination in embryonic stem cells. *Nature*, Vol.467, No.7312 (September 2010), pp. 211-3.

Urbanek, K., Quaini, F., Tasca, G., Torella, D., Castaldo, C., Nadal-Ginard, B., Leri, A., Kajstura, J., Quaini, E. & Anversa, P. (2003). Intense myocyte formation from cardiac stem cells in human cardiac hypertrophy. *Proceedings of the National Academy of Sciences*, Vol.100, No.18 (September 2003), pp. 10440-5.

Urlinger, S., Baron, U., Thellmann, M., Hasan, M.T., Bujard, H. & Hillen, W. (2000). Exploring the sequence space for tetracycline-dependent transcriptional activators: novel mutations yield expanded range and sensitivity. *Proceedings of the National Academy of Sciences*, Vol.97, No.14 (July 2000), pp. 7963-8.

Van de Lest, C.H., Versteeg, E.M., Veerkamp, J.H. & Van Kuppevelt, T.H. (1995). Elimination of autofluorescence in immunofluorescence microscopy with digital image processing. *Journal of Histochemistry and Cytochemistry*, Vol.43, No.7 (1995), pp. 727-30.

Verzi, M.P., McCulley, D.J., De Val, S., Dodou, E. & Black, B.L. (2005). The right ventricle, outflow tract, and ventricular septum comprise a restricted expression domain within the secondary/anterior heart field. *Developmental Biology*, Vol.287, No.1 (November 2005), pp. 134-45.

Waldo, K.L., Kumiski, D.H., Wallis, K.T., Stadt, H.A., Hutson, M.R., Platt, D.H. & Kirby, M.L. (2001). Conotruncal myocardium arises from a secondary heart field. *Development*, Vol.128, No.16 (August 2001), pp. 3179-88.

Wang, X., Hu, Q., Nakamura, Y., Lee, J., Zhang, G., From, A.H. & Zhang, J. (2006) The role of the sca-1+/CD31- cardiac progenitor cell population in postinfarction left ventricular remodeling. *Stem Cells*, Vol.24, No.7 (July 2006), pp. 1779-88.

Wen, Z., Zheng, S., Zhou, C., Wang, J. & Wang, T. (2010). Repair mechanisms of bone marrow mesenchymal stem cells in myocardial infarction. *Journal of Cellular and Molecular Medicine*, (December 2010) [Epub ahead of print].

Wessels, A. & Pérez-Pomares, J.M. (2004). The epicardium and epicardially derived cells (EPDCs) as cardiac stem cells. *Anatomical Record. Part A, Discoveries in Molecular, Cellular, and Evolutionary Biology*, Vol.276, No.1 (January 2004), pp. 43-57.

Wu, S.M., Fujiwara, Y., Cibulsky, S.M., Clapham, D.E., Lien, C.L., Schultheiss, T.M. & Orkin, S.H. (2006). Developmental origin of a bipotential myocardial and smooth muscle cell precursor in the mammalian heart. *Cell*, Vol.127, No.6 (December 2006), pp. 1137-50.

Zeisberg, E.M., Tarnavski, O., Zeisberg, M., Dorfman, A.L., McMullen, J.R., Gustafsson, E., Chandraker, A., Yuan, X., Pu, W.T., Roberts, A.B., Neilson, E.G., Sayegh, M.H., Izumo, S. & Kalluri, R. (2007). Endothelial-to-mesenchymal transition contributes to cardiac fibrosis. *Nature Medicine*, Vol.13, No.8 (August 2007), pp. 952-61.

10

Stem Cell Therapy in Myocardial Infarction Clinical Point of View and the Results of the REANIMA Study (REgenerAtion of Myocardium with boNe Marrow Mononuclear Cells in MyocArdial Infarction)

Slobodan Obradovic, Bela Balint and Zoran Trifunovic
Clinic of Emergency Medicine, Institute of Transfusiology,
Clinic of Cardiovascular Surgery, Military Medical Academy, Belgrade
Serbia

1. Introduction

The incidence of heart failure (HF) after acute myocardial infarction (AMI) is around 10-40% during the hospital stay depending on its definition (Weir & McMurray, 2006; Cleland & Torabi, 2005). Also, another 10-20% of patients will develop heart failure symptoms during the next few months and years (Torabi et al., 2008). The mortality of patients with heart failure symptoms after AMI is very high and it reaches up to 50% in 5 years (Weir & McMurray, 2006; Fox et al, 2006). The left ventricle dilatation occurs in even 30% in patients reperfused successfully with primary angioplasty during six months follow-up (Bolognese el al, 2002) and the occurrence of dilatation is more pronounced in patients with lower baseline left ventricle ejection fraction (LVEF). The incidence of HF after AMI has increased, and mortality decreased over time with the better reperfusion therapy (Velagaleti et al, 2008). According to these facts, it is extremely important to develop therapeutic modalities in order to prevent the remodeling of myocardium after infarction. The adult stem cell therapy is a relatively new and promising method of an infarcted heart healing and HF prevention.

In the last two decades three important discoveries regarding different regenerative steps of damaged myocardium promoted the completely new era in the treatment of ischemic heart disease. First of all, several adult multipotent and pluripotent stem cells from different tissues may trans-differentiate in certain circumstances to cardiomyocytes or other needed cells, such as endothelial cells (Körbling M & Estrov Z, 2003; Müller et al, 2005). However, in vivo, this mechanism of heart regeneration seems to be negligible (Wagers et al, 2002; Murry et al, 2004), at least for the acute injury. The second is the fact that a significant number of cardiac cells are in the proliferative state in the areas of myocardium adjunction to infarction (Beltrami et al, 2001). The first source of these regenerative cells is very probably resident cardiac stem cells which are in the quiescent state out of injury, but in the time of infarction they proliferate and differentiate to cardiomyocytes, smooth muscle cells and endothelial cells (Bollini et al, 2011). And the third important discovery is that in the time of infarction, myocardial ischemia initiates the eruption of cytokines, growth-factors and chemokines from the injured myocardium which promote mobilization of stem cells from other niches and their homing into the damaged myocardium (Frangogiannis, 2008). The most likely function of these cells in the ischemic myocardium are various paracrine effects which enable survival of severely damaged cardiomyocytes, promote differentiation and the proliferation of cardiac stem cells and participate in the creation of new blood vessels which all halted myocardial remodeling and the development of heart failure (Mirotsou et al, 2011).

The knowledge of these processes is very important because the regenerative therapy depends on artificial augmentation of some steps in order to make regenerative process more efficient. The most important steps are shown in figure one. Ischemic injury induces the hypoxia-inducible factor-alpha which in turns stimulates the expression of several growth factors and chemokines in the infracted heart (Dong et al, 2010). Those cytokines, especially stromal derived factor-1, interleukin-8 and vascular-endothelial growth factor promote mobilization of local and remote stem cells and enable engraftment of them into the damaged tissue (Figure 1).

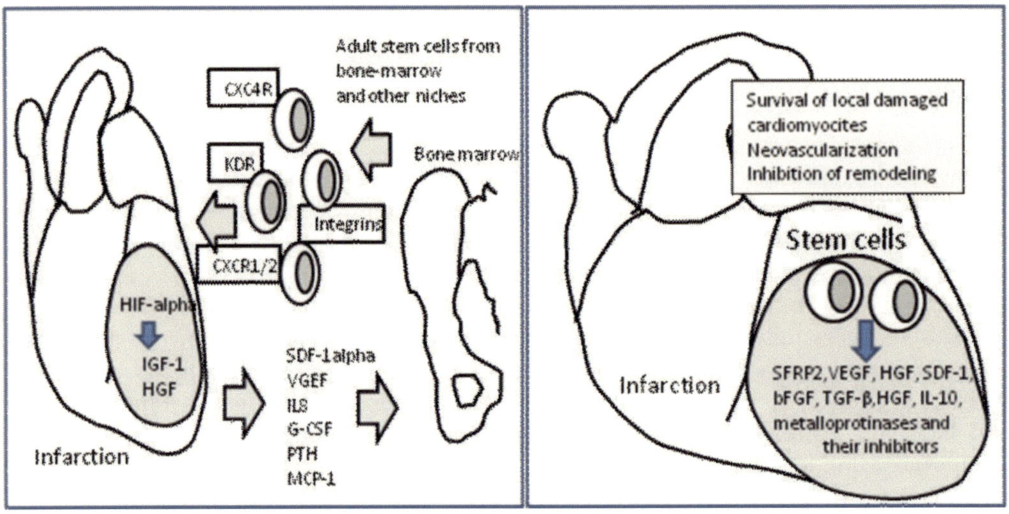

Fig. 1. Mobilization of stem cells by the cytokine and chemokine storm after myocardial infarction and potential paracrine effect of stem cells in the infracted heart and beneficial effect on cardiomyocytes survival, promotion of angiogenesis and inhibition of remodeling HIF-α - hypoxia inducible factor-alpha, IGF-1 - insulin-like growth factor - 1, HGF - hepatocyte growth factor, SDF-1alpha - stromal cell-derived factor 1 alpha, VEGF - vascular endothelial growth factor, G-CSF- granulocyte colony-stimulating factor, IL8 - interleukin 8, PTH - parathyroid hormone, MCP-1 - monocyte chemoattractant protein-1, KDR - receptor for VGEF, CXC4R - receptor for SDF-1, CXCR1/2 receptors for other chemokines, SFRP2 - signaling protein important for cardiomyocyte survival.

Chemokine receptors (CXC-R1 and CXC4R), growth receptors (VGFR) and several selectins and integrins on stem cells are important for the successful homing of these cells in the ischemic myocardium (Chavakis et al, 2008). Expression of matrix metalloproteinases such as MMP-2, 9 and cathepsin by stem cells represent the final step of their transmigration into the damaged tissue (Cheng et al, 2007; Huang et al, 2009). Several growth factors up-regulated by ischemia (insulin growth factor-1, hepatocyte growth factor, fibroblast growth factor) enable the survival of these cells in the hostile environment (Frangogiannis, 2008). Paracrine effects of stem cells promote local cardiomyocytes survival, neovascularization, attenuate the remodeling and improve cardiac function. Among several niches of stem cell residency, myocardium itself, bone marrow and adipose tissue are probably the most important reservoir of this regenerative capacity. The advance age, large necrosis and enhanced inflammatory reaction decrease the stem cell mobilization after infarction (Turan et al, 2007).

2. Important clinical trials on stem cell therapy in acute myocardial infarction

Several clinical studies investigated the usage of bone marrow derived cells for the treatment of AMI. The most of them used autologous bone marrow derived mononuclear (MNC) cell suspensions with intracoronary delivery through the inflated balloon placed on the spot of previous stent placement (Abdel-Latif et al, 2007; Tongers et al, 2011). The pioneering study of Strauer (Strauer et al, 2002), on 20 AMI was not randomized, but had the well matched control group that showed improved left ventricular systolic function and perfusion in the short and long-term follow-up. After that study several randomized studies were published with the conflicting results (Table 1). Transplantation of Progenitor Cells and Regeneration Enhancement in Acute Myocardial Infarction (TOPCARE study) compared bone marrow derived MNC and circulating progenitor cells (CPS) given intracoronary but without the control group (Schachinger et al, 2004). Both systolic function and viability improved in the similar way after 4 months follow-up. In the study of Chen et al (Chen et al, 2004) intracoronary injections of mesenchymal stem cells were used for the first time in humans, and with the sophisticated methodology they demonstrated that this method was safe, feasible and that it significantly improved global and regional left ventricle function. Interestingly, there was no trial with the use of MSC intracoronary after Chen's study. In BOO transfer to enhance ST-elevation infarct regeneration (BOOST) trial (Schafer et al, 2006) with magnetic resonance imaging (MRI) of left ventricle ejection fraction (LVEF) and volumes for follow-up, single dose of intracoronary bone marrow cell provided the accelerate improvement of systolic function (after 6 months) with the late catch-up of the control group (after 18 months). In the study of Janssens et al, intracoronary transfer of bone marrow MNC was done 24 hours after primary percutaneous coronary intervention (PCI) and did improve only regional, but not the global left ventricle systolic function after 4 months by the MRI imaging (Jansenss et al, 2006). Reinfusion of Enriched Progenitor Cells and Infarct Remodeling in Acute Myocardial Infarction (REPAIR-AMI) trial (Schachinger et al, 2006a, 2006b) is the largest randomized trial that examined the intracoronary transfer of bone marrow derived MNC and it brought interesting results. For the first time one of the inclusion criteria for the participation in the study was the baseline LVEF measured at the time of primary PCI. The significant improvement of LVEF was detected in the cell therapy group compared to controls and it was more pronounced in patients with the baseline LVEF less than median (48.9%) and in those in whom cell transfer was performed later than the 4-post infarction day. The most important result of this trial was that the combined end point death and recurrence of myocardial infarction and rehospitalization for heart failure, was significantly reduced in the BMC group after two years follow-up (Assmus et al, 2010).

Method of SC delivery	Number of patients and type of cells N	Timing (d)	Bone marrow volume, method of cell preparation and the number of cells	Criteria for patient selection	The basic result 4-6 months after STEMI
I.C. short FU					
Strauer et al.	10 BMMNC/10 C	5-9	40 ml, Ficoll, 2.8±2.2x10^7 MNC	First STEMI, pPCI	ECHO, LVA, PET - EF, volumes, perfusion ↑
TOPCARE	29 BM-MNC/30 CPC	4-6	50 ml, Ficoll 5±3x10^6 CD34+/16±12x10^6CPC	First STEMI, pPCI	ECHO, RVA, MRI – EF, volumes, perfusion↑
Chen	34 BM-MSC/35 C	8/16 Harv/ deli	60 ml, MSC culture 8-10x10^9 MSC	First STEMI, pPCI	ECHO, PET – EF, volumes, perfusion↑
BOOST	30 BMC/30 C	5-7	120 ml, gelatin-polysuccinate, 9.5±6.3x10^6 CD34+	First STEMI, pPCI	MRI - EF↑ at six but not at 18 months
Jansens	33 BMMNC/34 C	1	130 ml, Ficoll, 2.8±1.7x10^6 CD34+	First STEMI, pPCI	ECHO, MRI - EF↔, regional function↑
REPEAR-AMI	101 BMMNC/103 C	3-6	50 ml, Ficoll, 3.6±3.6x10^6 CD34+	First STEMI, pPCI, EF≤45%	LVA - EF↑ Comp hard end point↓
ASTAMI	50 BMMNC/47 C	4-7	50 ml, Lymphoprep, 0.7x10^6 CD34+	First STEMI, pPCI on LAD	ECHO-EF, SPECT, MRI – EF and volumes↔
Meluzin	22 HD-BMMNC/22 LD-BMMNC/22 C	5-9	NS, Histopaque-buffy-coat, HD-10^8 MNC, LD-10^7 MNC	First STEMI pPCI	ECHO, gSPECT - ↑EF, ↓volumes, HD better
REGENT	80 NS-BMMNC/80 CD34+/CXC4R+BM Cells/40 C*	3-12	100-120 ml-selected cell group and 50-70 ml-unselected group, Ficoll/selection 1.8x10^8 cells/1.9x10^6 CD34+CXCR4+	First STEMI, LAD-IRA, EF≤40%	MRI – EF and volumes↔, EF and volumes↑ in pts with EF<37% (median)
FINCELL	40 BMMNC/40 C	2-6 (after PES stent)	80 ml, Ficoll, 2.6±1.6x10^6 CD34+	First STEMI, Fibrinolysis	ECHO, LVA - EF↑ IVUS - MLA↔
HEBE	69 BMMNC/66 PBMNC/ 65 C	3-8	60 ml BM, 150-200 ml PB, Lymphoprep, 4.0(2.1-6.5)x10^6 CD34+/0.3(0.2-0.4)x10^6 CD34+	First STEMI, pPCI	MRI - EF, IS and regional function↔
I.C.long FU					
BALANCE	62 BMMNC/62 C	5-10	80-120 ml, Ficoll, 6.1±3.9x10^7 BMC	First STEMI, pPCI	LVA, dECHO - EF↑, arrhythmias↓, mortality↓

Method of SC delivery	Number of patients and type of cells N	Timing (d)	Bone marrow volume, method of cell preparation and the number of cells	Criteria for patient selection	The basic result 4-6 months after STEMI
CAO	41 BMMNC/46 C	7	40 ml, Lymphoprep,5x10^{8}MNC (1.8±0.6%CD34+)	First STEMI, pPCI on LAD	gSPECT - EF↑, viability↔
BOOST 5y	27 BMMNC/26 C	5-7	120 ml, gelatin-polysuccinae, 9.5±6.3x10^6 CD34+	First STEMI, pPCI	MRI - EF and volumes↔
Repeated I.C.					
Yao	12 S-i.c.BMMNC transfer/15 R-i.c.BMMNC - 3 months/12 C	7 d and 90 d	90 ml, Ficoll, 1.9-2.1x10^8 BMC in both groups and in repeat infusion	First STEMI, EF 20-39%	MRI EF↑ highest in repeat cell group
I.V.					
Hare	39 alloMSC (0.5 vs 1.6 vs 5.0x10^6/kg) /21 C	1-10	Single unrelated donor no HLA matched	First STEMI, pPCI	ECHO-EF antMI↑, MRI-EF↑
Endocardial					
MYSTAR	30 EG/30 LG	3-6 w vs. 3-4 m	300 ml, COBE-vol. depl. EG: 3.6x10^6 CD34+i.m. +23.2.4x10^{6}CD34+i.c. LG: 3.0.3x10^6 CD34+i.m. +22.5x10^{6}CD34+ i.c.	First STEMI, pPCI, 30-45%EF	SPECT-EF↑ in both groups, no difference between groups

SC- stem cells, I.C.- intracoronary, I.V. intravenously, FU- follow-up, BMC-Bone marrow cells, BM-MNC - Bone marrow mononuclear cells, CPC-circulating progenitor cells, Bone marrow mesenchimal stem cells, HD-MMNC - higher dose of BMMNC-10^8, LD-BMMNC-lower dose BMMNC-10^7, PBMNC-peripheral blood mononuclear cells, C-controls, PES- paclitaxel eluting stent, STEMI- ST elevation myocardial infarction, pPCI- primary percutaneous coronary intervention, antMI - anterior myocardial infarction, LAD- left anterior descending, IA- infarction related artery, EG - early group, LG - late group, dECHO- dobutamine echocardiography, gSPECT- gated single-photon emission computed tomography, PET- positron emission tomography, MRI- magnetic resonance imaging, LVA- left ventricle angiography, EMM- electro-mechanical-mapping, IVUS- intravascular ultrasound, MLA-minimal lumen area, EF- ejection fraction.

Table 1. Important clinical trials of stem cell therapy in acute myocardial infarction.

Autologous Stem-Cell Transplantation in Acute Myocardial infarction trial (ASTAMI) also used some inclusion criteria for attention to recruit more severe seek patients (Lunde et al, 2006). The inclusion criterion in this study, among the presence of the first STEMI, was the finding on coronarography with the culprit lesion on the proximal part of the left anterior descending artery (LAD). However, more than 25% of patients in both groups (cell group and control) had the TIMI-2/3 flow before the primary percutaneous coronary intervention (PCI) and the baseline mean LVEF measured by three methods (echocardiography, single photon computed tomography-SPECT and MRI) was greater than 40%, which means that this group did not represent the anterior STEMI realistically. This study showed no effects of cell therapy on global LVEF. The other probably important pitfall of this study was the late baseline MRI imaging, after 3 weeks of stem cell infusion which could have missed some early action of stem cells. Different protocols of bone marrow mononuclear cell preparation (for instance - Lymphoprep gradient media in ASTAMI and Ficoll in REPAIR-AMI) among the studies might be the reason for these discrepant results, but there are certain controversies about that issue (Seeger et al, 2007; Yeo et al, 2009). Meluzin et al, addressed the question of "cell dosage" for the intracoronary infusion after STEMI in their study (Meluzin et al, 2006). Although some other studies did not found such relationship (TOPCARE, REGENT), improvement of regional LV function was "cell-dose" dependant in this study. Regeneration by Intracoronary Infusion of Selected Population of Stem Cell in Acute Myocardial Infarction (REGENT) trial (Tendera et al, 2009) is important for two reasons. The first is the patients' selection, with the enrollment of patients with more severe LVEF impairment (LVEF≤40%) and the second is the immunomagnetic selection of bone marrow MNC for CD34+/CXC4R+ cells which represents the "selection" arm in this study. Unfortunately MRI follow-up was paired in only 59% of patients. Again, patients with baseline LVEF less than median had the significant improvement of LVEF after 6 months in both cell groups (selected and non-selected). However, the median baseline LVEF value in this study was 37%, meaning that a half of patients have had the baseline LVEF between 37-40%, probably indicating the recruitment bias in this study. The FIN study of autologous bone marrow-derived stem CELLs in acute myocardial infarction (FINCELL) for the first time used intracoronary stem cell therapy a few days after successful thrombolysis (Huikuri et al, 2008). The intracoronary injections of bone marrow MNC were given immediately after percutaneous coronary intervention which was performed on the already opened infarct related artery. Intracoronary injections of stem cells in these patients were feasible and associated with the improvement of LVEF after 6-months. Meticulous assessment of arrhythmogenic potential of stem cells was done in this study using three non-invasive methods (Holter monitoring, microvolt T wave alternans and Signal-averaged electrocardiogram) having proved that intracoronary bone marrow cell therapy did not seriously aggravate arrhythmias. Intravascular ultrasound imaging performed in this study confirmed that cell therapy did not cause restenosis. The HEBE trial (Hirsch et al, 2010) investigated the influence of bone marrow compared to peripheral blood derived MNC intracoronary and controls to global and regional LV function measured by MRI. This relatively large trial resulted in neutral influence of cell therapy on LV performance after 6 months. The relatively short ischemia time in this trial may explain the equal and significant recovery of LVEF in all three arms of this trial. Besides, the baseline LVEF was above the 40% (median=43.4%) pointing that the majority of patients in this study had good prognosis and no additional benefit of stem cell therapy should be expected. Indeed, there was a trend toward better results of stem cell therapy according to percent of the regional segment improvement in patients with baseline LVEF bellow the median value. The French study (Roncalli et al, 2010) was concentrated to the scintigraphy analysis of viability after intracoronary infusion of bone marrow derived MNC. Patients with more severe infarction (LVEF≤45%) were enrolled in this study. Bone marrow cells slightly improved viability in cell therapy group. This study also emphasized the negative impact of smoking on the improvement of viability during time.

Only three trials published their long-term results of intracoronary bone marrow derived cell therapy in the acute phase of STEMI. Strauer's group, in their non-randomized, but well controlled study had showed that the benefit on intracoronary bone marrow derived MNC infusion after infarction for the myocardial performance sustained after 5 years and that even decreased the abnormal heart rate variability, late potentials and ectopic beats (Yousef et al, 2009). And the most important, mortality of BMC-treated patients was significantly reduced in comparison with the control group. The long-term study of Chinese group (Cao et al, 2009), also indicated the persistent improvement of LVEF (over 4 years) in AMI patients treated with intracoronary bone marrow MNC compared to controls, but interestingly without significant improvement on viability. In BOOST trial (Meyer et al, 2009) patients with more transmural extension of infarction appeared to benefit from BMC transfer throughout the five years.

Most likely, single intracoronary cell infusion cannot bring enough stem cells into the infarction area for the sustained beneficial effect on the myocardial function. There is probably the saturation level of stem cell delivery in such short period of time which precludes their significant influence on myocardial regeneration in patients with very large myocardial necrosis. Yao's group, in their relatively small study suggested that repeated intracoronary stem cell therapy, after 3-7 days from STEMI and again after 3 months may have an additional advantage in comparison to single early stem cell treatment (Yao et al, 2009).

The extraordinary trial comes from the Hare's group, who for the first time used intravenous allogeneic mesenchymal stem cells infusion from the healthy unrelated bone marrow donor in patients with STEMI (Hare et al, 2009). Mesenchymal stem cells lack major histocompatibility complex and costimulatory cell-surface antigens which enable their allogeneic transfer and secret various anti-inflammatory cytokines promoting healing. They are also rich in the homing properties which allow intravenous application. This study performed detailed safety assessment including pulmonary function and computed tomography of chest abdomen and pelvis in the follow-up. Mesencymal stem cell therapy demonstrated reduced ventricular tachycardia, better pulmonary function and increase of LVEF in patients with anterior infarction compared to controls.

Two trials examine the safety, feasibility and efficacy of trans-endocardial route of bone marrow derived MNC delivery using electromechanical mapping as the guidance (NOGA system) after AMI. MYSTAR trial (Gyöngyösi et al, 2009) compared early (3 weeks after AMI) and late (3 months after AMI) combined trans-endocardial and intracoronary bone marrow derived MNC. In both arms cell therapy achieved small but significant improvement of LVEF measured by g-SPECT. This study used a large number of CD34+ cells, and the majority of cells were given intracoronary. Unfortunately this study had no arms with intracoronary and trans-endocardial route of delivery separately and we do not know if the combined route of stem cell delivery has any synergistic effect. Krause et al, published their small, uncontrolled study with early trans-endocardial delivery of bone marrow MNC in AMI, and they proved its safety with the significant improvement of LVEF after six months (Krause et al, 2009).

Several studies (Table 2) investigated the usage of granulocyte growth factor (G-CSF) for induction of longer and increased mobilization of stem cells during the first days of AMI (Valgimigli et al, 2008). The application of G-CSF for several days achieved the 10-30 times, increased of CD34+ cells number in peripheral blood (Ince et al, 2005; Valgimigli et al, 2005; Zohlnhöfer et al, 2006; Engelman et al, 2006; Ripa et al, 2006; Takano et al, 2007; Leone et al, 2007). When we analyzed the results of these studies it seemed that very early start of G-CSF after STEMI (during the first day) and its application in patients with lower LVEF (lower than 40%) had a positive effect on systolic function (Ince et al, 2005; Takano et al, 2007; Leone et al, 2007). However, G-CSF had some potential prothrombotic and pro-inflammatory effects (Le Blanc et al, 1999; Falanga et al, 1999) which could be deleterious for patients with AMI, but it was not seen in the current published trials. Parathyroid hormone or its analogs may be an alternative drug for stem cell mobilization in this setting (Huber et al, 2010).

Study	The number of patients	Time GCSF	Duration of G-CSF therapy and dosage	Patient selection	Results of the study
FIRSTLINE-AMI	25GCSF/10 C	1.5 h-pPCI	6d, 10 μg/kg/d s.c.	1st-AIM, pPCI	ECHO-EF I WMSI↑, PET ↑
STEMMI	39 GCSF/39C	2 d	6d, 10 μg/kg/d s.c.	1st AIM, pPCI	MRI wall thick, EF↔,
G-CSF-STEMI	23 GCSF/21C	2 d	5 d, 10 μg/kg/d sc	1st AIM, pPCI	MRI-EF, vol. and reg. function↔, perfusion↑
REVIVAL-2	58 GCSF/56C	5 d	5d, 10 μg/kg/d s.c.	1st AIM-lysis, PCI 5d	SPECT IS ↔, MRI-EF↔
REGENERA	14 GCSF/27C	≥5 d	5d,10 μg/kg/d s.c.	1st ant AIM EF<50%	ECHO-EF, vol. and WMSI↑
TAKANO	22 GCSF/18C	1 d	5d,2.5 μg/kg/d s.c.	1st ant AIM pPCI	gSPECT-EF, vol. IS↑

Table 2. Important clinical trials used mobilization of stem cells to treat acute myocardial infarction.GCSF- Granulocyte colony-stimulating factor, AIM - ST elevation myocardial infarction, pPCI- primary percutaneous coronary intervention, dECHO- dobutamine echocardiography, EF- ejection fraction, WMSI- wall motion score index, EDV- end-diastolic volume, gSPECT- gated single-photon emission computed tomography, PET- positron emission tomography, MRI- magnetic resonance imaging

3. Important clinical trials on stem cell therapy in chronic myocardial infarction

The chronic myocardial infarction (CMI) represents a completely different environment for the stem cell therapy. The precise definition of chronic is not established, but it seems that it would be accepted that the chronic MI may be old at least 1-2 months after the necrotic event. Highly dynamic inflammatory reaction with cellular and cytokine storm is finished and slow fibrotic process replaces it (Frangogiannis, 2008). The abundance of chemokines, growth factors, adhesion molecules and other biologically active substances in the acute inflammatory phase of infarction not longer exist. Some parts of myocardium adjacent to infarction core due to long time of ischemia and because of partly damaged structure after the index event are alive but not capable for fully function. Those areas need revitalization with stem cells, but the question is whether the same cells are needed for the chronic IM as for the acute MI, and whether the same route of delivery would be equally efficient? Very interested human pilot study of tracking the labeled circulating progenitor cells (CPC) with indium oxine (^{111} In-oxine) after intracoronary injections in patients with acute (<15 days), intermediate phase (15 days-1 year) and a late chronic stage of MI (>1 year), demonstrated that amount of progenitor cells retained in the myocardium decreased progressively over the time (Schächinger et al, 2008) alludes the answer on the second question. Human trials comparing bone marrow derived MNC and peripheral blood progenitor cells (PBPC) exist at least for AMI patients with inconclusive and contradictory results on their regenerative capacity (Schächinger et al, 2004; Hirsch et al, 2010). However, those cells are very similar but the only difference is that bone marrow MNC cells have more primitive cell subpopulation then PBPC which are more commitment to endothelial lineage. The comparison of mesenchymal stem cells and hematopoietic CD34+ cells in animal model of myocardial infarction showed that mesenchymal stem cells were more potent for the healing of the heart (Arminan et al, 2010).

Method of SC delivery Study	Number of patients In groups	Timing of SC therapy after MI	Bone marrow volume, the number of cells	Selection of the patients	The main results of the study
I.C.					
TOPCA RE-CHD	28 BMMNC/24 CPC/23 C	>3 m	50 ml BM, 270 m PB, Ficoll, 2.0±1x10⁶MNC/ 22±11x10⁶CPC	Patent IRA	LVA, MRI, PET - EF↑, regional function↑ - BMMNC
STAR	191 BMMNC/200 C	8.5±3.2 y	80-120 ml, Ficoll, 6.6x10⁷BMC	Patent IRA by PCI, EF≤35%	LVA, EF and regional function↑, exercise capacity↑, Mortality↓
MAGIC-DES	25 BMMNC-AMI/25 AMI-C/ 16 BMMNC-CMI/16 CMI-C	≤14 d, 2±3 d-AIM/>1 4 d-CMI ≈2 y	GCSF s.c.10μg/kg 3d, 4 d COBE-BCT, 1.4x10⁹ Leu, CD34-9.2±10.4%	Patent IRA	MRI-EF ↑ in AMI CPC group, in CMI ↔
I.C. vs. I.M.					
Ang et al	21 BMMNC IC/ 21 BMMNC IM/ 20 C	>6 w	80 ml, Lymphoprep, 1.4x10⁵ CD34+ I.M./2.4x10⁵ CD34+ I.C.	Graftable infarct area	d-ECHO, MRI - EF and regional function↔ in all groups
Epicardial					
Patel	10 BMMNC/ 10 c	NS	550 ml, Ficoll, immuno-magnetic sel. 22x10⁶CD34+	Graftable infract area, EF≤35%	ECHO, gSPECT - EF↑
Mocini	18 BMMNC/18 C	Recent MI >4 w and < 6 m	50 ml, seeded with HES, centrifugation, 3.7x10⁹CD34+ after CABG during arrest	Graftable infract area, LVEF≥35%	MRI - EF↑ and WMSI↓

Method of SC delivery Study	Number of patients In groups	Timing of SC therapy after MI	Bone marrow volume, the number of cells	Selection of the patients	The main results of the study
Hendrik x	10 BMMNC/10 C	217±162 d	40 ml, Lymphoprep, 60.2x10^6 BMC, CD34% 1.4±1.0	Graftable infract area	MRI - EF↔,, SPECT - viability ↔
Stamm	20 BMMNC/20 C	7-9 w	90-250 ml, immune sel. 133+/CD34+ 6.0x10^6	Graftable infract area	ECHO EF↑, SPECT-viability↑
Zhao	18 BMMNC/18 C	18-21±17 m	30 ml, Ficoll, 6.6x10^8 BMMNC	Graftable infract area, EF<40%	ECHO - EF↑, volumes↓, regional function ↑, SPECT↑
MAGIC	30 HDMy/33 LDMy/34 C	>4 w	10 g of tigh muscle, 3 w of culturing, HDMy-800x10^6, LDMy-400x10^6	Graftable 15%≥EF ≤35%	ECHO - EF↔, ESV↓ in HDMy group
Endocardial					
Perin	14 BMMNC/7 C	>3 m	50 ml, Ficoll, 5.7±6.1x10^{4}CD34+	Ineligible for revasc. EF<40%	LVA - EF↑, EMM - viability↑
Pokusha lov	55 BMMNC/54 C	>12 m, 9±8 y	NS, Ficoll, 1.0±0.6x10^{6}CD34+	Ineligible for revasc.EF< 45%	ECHO - EF↑, SPECT - viability↑, functional status↑, 6 min WT↑
SEISMIC	26 My/14 C	8 y IQR 4-12	10 g of tigh muscle, 2-3 w of culturing, My-100-400x10^6	Ischemic HF	RNV-MUGA - EF↔, functional status trend↑
Ramshor st	25 BMMNC/25 C	NS	80 ml, Ficoll, 40x10^{6}MNC	Ineligible for revasc. EF<40%	SPECT↑, MRI - EF↑

I.C.- intracoronary, I.M. intramyocardial- during CABG, BMC-Bone marrow cells, BM-MNC – Bone marrow mononuclear cells, CPC-circulating progenitor cells, EPC- endothelial progenitor cells, My-Myoblasts, HD- high dose, LD- low dose, Bone marrow mesenchymal stem cells, PB-peripheral blood, C-controls, IQR- interquartile range, CABG- coronary artery bypass grafting, revac.- revascularization, HF- heart failure, dECHO- dobutamine echocardiography, gSPECT- gated single-photon emission computed tomography, PET- positron emission tomography, MRI- magnetic resonance imaging, RNV-radionuclide ventriculography, 6 min WT- six minutes walking test, LVA- left ventricle angiography, EMM- electro-mechanical-mapping, IVUS- intravascular ultrasound, MLA- minimal lumen area, EF-ejection fraction.

Table 3. Important clinical trials of stem cell therapy for CMI.

Clinical trials of stem cell therapy in CMI (Table 3) are smaller and not so well conducted as trials of stem cell therapy in AMI (Sanz-Ruiz et al, 2010; Donndorf et al, 2011). According to coronary status we can divide patients with CMI in two groups, the first one eligible for revascularization of the infracted area and the second with no option of revascularization. We believe that it is very important to perform as complete as possible revascularization before the stem cell therapy and not to proceed to sophisticated stem cell trial in ischemic cardiomyopathy without knowing the coronary status of enrolled patients (C-CURE, NCT00810238). Again, the different modes of stem cells and methods of delivery might be necessary in those two groups. Based on some animal models (Hou et al, 2005) and on the logical assumption the direct intramyocardial (trans-epicardial in patients who need surgical revascularization and trans-endocardial in patients who have no option of revascularization) route of stem cell delivery might be a preferred option.

Transplantation of Progenitor Cells and Recovery of LV Function in Patients with Chronic Myocardial Infarction (TOPCARE-CHD) was the first randomized, cross-over study examining the role of intracoronary bone marrow stem cell therapy for CMI (Assmus et al, 2006). The transplantation of bone marrow derived MNC was associated with the modest but significant improvement of six-months LVEF (ΔLVEF=4.8% measured by MRI) and regional myocardial function. The improvement of the functional status assessed by the NYHA classification was also significant in the BMMNC group. The second large, not randomized but well controlled study was Stem cell Transplantation in 191 patients with chronic heart failure - STAR-heart study (Strauer et al, 2010).

The intracoronary injections of BMMNC had sustained (3 months - 5 years) beneficial effect on LV global and regional function, increased exercise capacity, improved functional capacity and reduced mortality compared to controls. Myocardial Regeneration and Angiogenesis in Myocardial Infarction study (MAGIC-DES) compared the influence of intracoronary injections of G-CSF mobilized PBPC on LV performance between patients in AMI and CMI previously treated with drug-eluting stents (Kang et al, 2006). Only patients with AMI had improvement of LVEF after 6 months. The study of Ang, compared intra coronary (through the graft) and intramyocardial injection of bone marrow derived MNC and controls during CABG (Ang et al, 2006). Stress echocardiography did not reveal any improvement in viability in the akinetic segment. MRI follow-up was done in only one third of patients in this study.

The first small study of application bone marrow derived MNC into the myocardium was the study of Hamano (Hamano et al, 2001). They injected bone marrow MNC into the non-graftable area during the coronary artery bypass grafting (CABG) and found that the procedure was feasible, safe and that induced improvement of myocardial perfusion assessed by SPECT in 3/5 patients. The detailed description of patients was not presented.

Patel conducted the first randomized trial with intramyocardial injections of enriched suspension of CD34+ cells during the off-pump CABG in patients with severe ischemic cardiomyopathy (Patel et al, 2005). Intramyocardial bone marrow derived MNC transplantation with off-pump CABG led to significant improvement of 6 months LVEF and functional status compared to patients treated with surgery only.

In the randomized trial of Mocini injections of bone marrow MNC into the peri-infarcted and infracted region (only patients with recent infarction were included) after the CABG during the cardiac arrest was compared to CABG alone (Mocini et al, 2006). The patient cohorts had moderate LV systolic dysfunction (inclusion criteria was baseline EF>35%).

There was no increase of serious arrhythmias. Transplanted patients had significant improvement of EF and WMSI measured by MRI after 6 months compared to the controls.
The relatively small randomized study of Hendrikx demonstrated only improvement of regional, but not the global systolic function by 6-months MRI follow-up with the bone marrow MNC myocardial injections after CABG (Hendikx et al, 2006). Interestingly the number of CD34+ cells injected was significantly higher in the responder group what implied the possible importance of cell dosing.
The randomized study of Stamm, investigated the influence of more premature CD133+ cell myocardial injections after CABG on myocardial function and perfusion (Stamm et al, 2007). The significant improvement of EF and myocardial viability was detected after 6 months in the cell therapy group. Subgroup analysis showed that patients with the lower EF had the greater benefit for selected stem cell therapy. The injection of selected more premature cells was safe.
The study of Zhao corroborated with the previous investigations, and verified the benefit of intramyocardial injections of MNC during CABG in patients with severe impaired EF post-infarction on global and regional myocardial function and perfusion (Zhao et al, 2008).
Very interesting non-randomized, case control study comes from Thailand's group, who used thoracoscopic delivery of in-vitro expanded endothelial progenitors (EPC) isolated from the peripheral blood into the peri-infarction area (Arom et al, 2008). The subset of patients received combined EPC therapy with off-pump CABG. They enrolled patients with very severe ischemic heart disease and low basal EF (26±7%). EPC transplantation improved significantly LVEF even combined or not with CABG. This study is important because it gives a possible solution for very ill patients with chronic ischemic cardiomyopathy and with no option for revascularization. The procedure is minimally invasive, safe and might help.
The clinical application of stem cell therapy had started with intra-myocardial injection of myoblasts. Menasche reported the first successful case on myoblast implantation during CABG and significant improvement of EF throughout 6 months (Menasche et al, 2001). Seven years later definitive results of Myoblast Autologous Grafting in Ischemic Cardiomyopathy (MAGIC) trial were published (Menasce et al, 2008). The study had three arms, high and low-dose myoblast groups and a placebo group. Myoblasts were obtained from thigh biopsy and in vitro cultivation for three weeks. All patients received implantable cardioverter-defibrilator at the time of tight biopsy. Myoblasts were injected neighboring the akinetic segments. The modest increase of EF after 6 months was noticed in all groups equally. Nevertheless, patients who received high number of myoblasts had a significant decrease of end-systolic volume.
Transmyocardial route of stem cell delivery guided with electro-mechanic mapping (NOGA system) represents an alternative option for the treatment of patients ineligible for conventional revascularization. Perin's group conducted the pioneering, non-randomized but controlled study of trans-endocardial bone-narrow MNC injections using the electromechanical mapping (NOGA system) to guide cell injections into the viable but not mechanically functional myocardium (Perin et al, 2003). Patients treated with cell therapy had a greater increase of EF measured by RNA, reduction of reversible defect on SPECT and improved functional status after 2 and 4 months follow-up.
Four relatively larger randomized studies with trans-endocardial application of bone marrow derived MNC or myoblasts have been recently published. Pokushalov's group (Pokushalov et al, 2010) randomized patients with end-staged ischemic cardiomyopathy with chronic MI were assigned to trans-endocardial injections of bone marrow MNC and

the control group. Cell therapy provided the significant improvement in functional status, angina score, myocardial perfusion and global EF in comparison to control arm. Mortality also significantly decreased in cell therapy group (10.9% compared to 38.9%, p<0.001). The extremely valuable study comes from Ramshort group. They examined a role of trans-endocardial injections of bone marrow derived MNC into the electrically alive but functionally inactive myocardium in patients with severe, refractory angina and without additional option of revascularization (Ramshort et al, 2009). More than half number of patients had had previous MI in both cell therapy group and control group. Stress-induced ischemia was reduced after 3 months and slight improvement of LVEF was demonstrated with cell therapy. This therapy also significantly improved the clinical status of patients.

Other two studies (SEISMIC and CAuSMIC) implanted cultured autologous myoblasts vie NOGA guiding system in patients with severe ischemic heart disease, previous infarction and chronic heart failure symptoms (Dib et al, 2009; Duckers et al, 2011). A high percentage of patients in both studies had previously ICD implanted. There was favorable safety with no difference between groups in arrhythmias and deaths. In both studies there was a functional improvement in myoblast groups, but SEISMIC study did not show any EF increase, and CAUSMIC sustained reduction of LV diameters.

Two pilot trials with adipose derived stem cells (ADSC), one with intracoronary injections of ADSC in patients with STEMI (A Randomized Clinical Trial of Adipose-Derived Stem cells in the Treatment of Patients With ST-Elevation myocardial Infarction - The APOLLO Trial) and one with intra-myocardial injections of ADPC in patients with severe ischemic heart disease and illegible for revascularization (adipose-derived stem & Regenerative Cells In the Treatment of Patients With Non revascularizable ischemic Myocardium - The PRECISE Trial) showed feasibility, safety and initial promising results.

4. Our experience

Forty two patients enrolled in the REANIMA study **(Regeneration of myocardium with bone marrow mononuclear cells in myocardial infarction)** underwent the autologous, bone marrow derived stem cell therapy for myocardial infarction in our Institution (Military Medical Academy, Belgrade) in the period from February 2004 to September 2010. The Local Ethical Committee approved the study and the informed written consent was obtained from all participants. All patients reperfused successfully with primary percutaneous coronary intervention or by thrombolytic therapy (accelerated protocol with Actilyse, Boehringer-Ingelheim) between 2-12 hours from the pain onset.

Three groups were formed. Group I received intracoronary injection of bone marrow derived MNC on 6-12 day after MI; group II received intracoronary injection of bone marrow derived MNC in the chronic phase of infarction; and group III received bone marrow derived MNC intramyocardially during the CABG. The inclusion criteria for the first group were the presence of the first MI, age under 70 years, opened infarct related artery on the 5th day of infarction, LVEF≤40% on the 5th day, and the clinically stable patient. The inclusion criteria for the II group were age under 70, MI at least 2 months before stem cell therapy, clinically stable patient, and LVEF≤40%. Finally, the inclusion criteria for the III group were age under 70, indication for CABG, the graftable infarction related area, LVEF≤45%, and the clinically stable patient. The common exclusion criteria were the presence of the important comorbidities (systemic or cardiac).

In AMI group baseline echocardiography assessment was performed between 4-7 days. The LVEF was determined according to the Simpson's rule, wall motion score index at rest and end-diastolic and end-systolic volume indices were measured. Examination was repeated in the sixth month after MI.

Single-photon emission tomography (SPECT) with Technetium 99^{m}-sestamibi was done between 4-7 day and in the 6th month. The infarction size (IS) of left ventricle (LV) was quantified by the commercial software (AutoQuant software, Cedars-Sinai QPS/QGS component of AutoQuant) as an area of LV (in percentage) with the uptake of tracer less than 50% of the maximal value.

Twenty-four hour ECG Holter was done in all patients in the second month from cell therapy.

The harvest of bone marrow was done in the morning of cell therapy. For intracoronary MNC delivery, between 250-350 ml of bone marrow was harvested under the general anesthesia with the multiple aspirations from the posterior iliac crests. After harvesting, cell suspension was filtered twice and processed with the COBE SPECTRA to reduce the number of red cells and platelets. The total final cell suspension volume was 150 ml, and the total cell number was between $10\text{-}50 \times 10^{9}/\mu l$. MNC represented 25-40% of these cells, and CD34+ cells were between 1.5-2.0% of it. Cell suspension was given through the diagnostic catheter deeply positioned in the LM. Boluses of 20 ml were given during 1 minute with 2 minutes pauses apart from the injections. Transient ST segment elevation was noticed in every patient. A slight increase of troponin was detected in one patient in CMI and one in AMI group after the procedure with minimally prolonged chest pain.

The bone marrow harvest (150 ml) for intramyocardial cell transfer was done under the general anesthesia immediately before the CABG. Cells were processed manually and after several filtration and centrifugation steps total volume of 15-20 ml was prepared. Preparation of cells was done during the operation, and cell injections of 20-30x0.3 ml per injection were performed after the end of operation during the cardiac arrest in the myocardial area adjacent to necrotic core. The mean number of intramyocardial injected CD34+ cells was $2.2 \pm 1.1 \times 10^{6}$ cells. Time from the bone marrow harvest to MNC application was 3-4 hours in all three groups.

Charcteristics	Intracoronary BMMNC in AMI (N=19)	Intracoronary BMMNC in CMI (N=9)	Intramyocard.BMMNC CMI-CABG (N=14)	P
Age - y±SD	50±11	50±12	54±11	NS
Gender - n (%)				
Female	3 (15.8)	2 (22.2)	0 (0.0)	NS
Risk factors				
Diabetes – n (%)	2 (10.5)	1 (11.1)	3 (21.4)	NS
Hypertension – n (%)	8 (42.1)	3 (33.3)	7 (50.0)	NS
Active smoking – n (%)	13 (68.4)	5 (55.6)	5 (35.7)	NS
Hypercholesterolemia – n (%)	12 (63.2)	5 (55.6)	8 (57.1)	NS
Infarct related artery – n (%)				
LAD	18 (94.7)	9 (100.0)	11 (78.6)	NS
LCX	1 (5.3)	-	1 (7.1)	
RCA	-	-	2 (14.3)	

Table 4. Baseline demographic data of study patients. BMMNC- bone marrow mononuclear cells, AMI- acute myocardial infarction, CMI- chronic myocardial infarction, CABG- coronary artery bypass grafting, LAD- left anterior descending, Left circumflex artery, RCA- right coronary artery.

Baseline demographic characteristics of patients (Table 4) were similar throughout groups. Patients with CMI treated with intracoronary injections of bone marrow MNC had lower LVEF, larger end-diastolic and end-systolic volumes indices and larger infarction size at baseline and after 6 months.
Left ventricle EF significantly increased in patients with intracoronary injections of bone marrow MNC after AMI (ΔLVEF=5.5±6.6%) and in patients treated with intramyocardial injections of bone marrow MNC (ΔLVEF=5.0± 4.2) and there was no change of LVEF in patients with intracoronary injections of bone marrow MNC in CMI (Figure 2). The infarction size was significantly reduced in patients with intracoronary injections of bone marrow MNC after AMI (ΔIS=6.2±5.0%) and in patients treated with intramyocardial injections of bone marrow MNC (ΔIS=4.9± 4.3) and there was no change of infarction size in patients with intracoronary injections of bone marrow MNC in CMI (Figure 2).

End-points	Intracoronary BMMNC in AMI N=19	Intracoronary BMMNCin CMI N=9	Intramyocardial BMMNC in CMI after CABG N=14	P value between 3 groups
Baseline LVEF (%)	33.1±4.1	30.8±4.4	35.3±3.9	0.05
6-m LVEF (%)	38.6±8.3	29.9±6.53	40.3±5.4	0.01
ΔLVEF %	5.5±6.6	-0.9±2.7	5.0±4.2	0.01
	P =0.002	P=0.354	P=0.001	
Baseline EDVCI ml/m^2	68.2±11.3	90.8±29.3	70.3±22.5	0.02
6-m EDVCI ml/m^2	72.5±12.8	94.2±35.1	70.7±15.3	0.02
ΔEDVCI ml/m^2	-4.4±10.1	-3.5±12.4	-0.4±13.2	0.63
	P=0.080	P=0.428	P=0.920	
Baseline ESVCI ml/m^2	44.1±9.9	63.4±23.7	48.4±15.3	0.01
6-m ESVCI ml/m^2	44.5±11.0	65.3±30.3	42.1±10.9	0.01
ΔESVCI ml/m^2	-0.3±7.8	-1.9±9.6	6.3±11.0	0.07
	P=0.852	P=0.575	P=0.050	
Baseline IS (%)	30.3±8.5	37.9±9.1	28.9±4.1	0.19
6-m IS (%)	25.3±11.0	37.4±8.4	22.7±5.2	0.01
ΔIS	4.9±4.3	0.4±1.4	6.2±5.0	0.02
	P<0.001	P=0.377	P<0.001	

Table 5. Left ventricle ejection fraction (LVEF) and infarction size (IS) at baseline and after 6 months.

Although improved LVEF, intracoronary bone marrow MNC transfer in patients with AMI did not block remodeling of the left ventricle. There was a trend toward significant increase of LV end-diastolic volume index in those patients (Table 5). On the other side, patients treated with intramyocardial bone marrow MNC injections with CABG had a positive effect on end-systolic volume index which significantly decreased after 6 months. In patients with CMI, there were no significant changes of either LVEF, or volume indices, or IS after six months (Table 5).
After six months of follow up, there were no deaths in any group (Table 6). Other important clinical event is showed in the table 6. We did not observe any significant arrhythmias on 24

hours ECG Holter during the follow-up of six months. Patients with CABG and cell therapy were the most stable. Also, functional NYHA class in six months was lower in CABG plus cell therapy treated patients compared to other two groups.

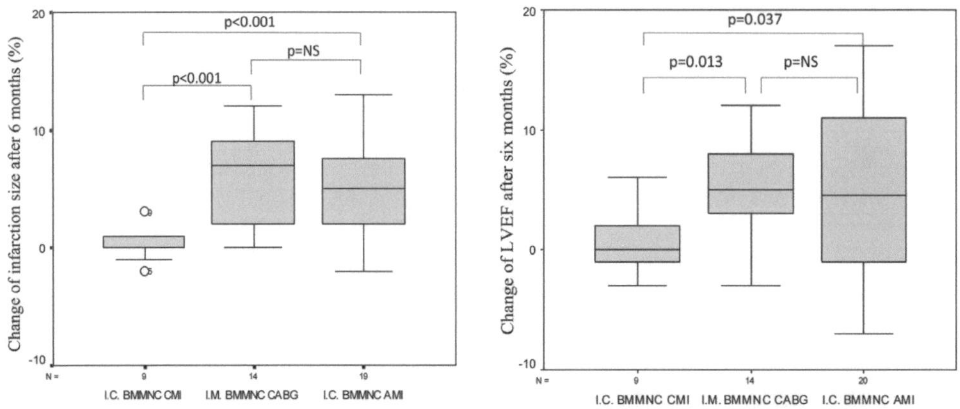

Fig. 2. Changes of LVEF and IS between 6-months and baseline measurements across the three groups. I.C. BMMNC-CMI- Intracoronary bone marrow mononuclear cells in chronic myocardial infarction; I.M.BMMNC-CABG- intramyocardial bone marrow mononuclear cells in chronic myocardial infarction after coronary artery bypass grafting; I.C.BMMNC AIM- intracoronary bone marrow mononuclear cells in acute myocardial infarction

Major adverse cardiac events	**Intracoronary BMMNC in AMI (N=19)**	**Intracoronary BMMNC in CMI (N=19)**	**Intramyocardial BMMNC in CMI after CABG (N=14)**
Revascularization n (%)	4 (21.1)	1 (11.1)	-
Heart failure - n (%)	3 (15.8)	4 (44.4)	1 (7.1)
NYHA class 6 months	1.58	1.14	1.89

Table 6. Major cardiac adverse events after 6 month follow-up.BMMNC- bone marrow mononuclear cells, AMI - acute myocardial infarction, CMI - chronic myocardial infarction, CABG - coronary artery bypass grafting

Our study is small and non-randomized, but nevertheless, suggests two important conclusions. The first is that bone marrow derived, native stem cells showed the improvement of the left ventricle function and a decrease of infarction size in both patients with AMI and CMI, and the second, direct intramyocardial delivery of bone marrow derived MNC is probably more efficient than intracoronary route of administration in patients with CMI. In our previous study (Obradovic et al, 2009a, 2009b) we compared function of LV and reduction of infarction size in patients with acute myocardial infarction treated with intracoronary bone marrow cell injections to well-matched control group and showed trend of improvement of LVEF and significant reduction in infarction size in cell therapy group. The improvement of LVEF by 5% in our trial of AMI patients is in accordance with the results of REPAIR AMI trial (Schachinger et al, 2006a, 2006b) and the result of meta-analysis of intracoronary bone marrow derived stem cell transplantation in AMI patients (Abdel-Latif et al, 2007).

The outcome of stem cell therapy depends on different factors. The proper selection of patients, timing and methodology of stem cell therapy is crucial for improvement. In AMI we have a reasonable assumption that patients with lower LVEF had increased benefit of bone marrow derived stem cell therapy. However, among larger trials with intracoronary bone marrow derived stem cell therapy for AMI, only REGENT trial (Tendera et al, 2009) have had the entrance criteria of LVEF<40%, but it has not been stated when and how LVEF was measured, because it is not equal if it is measured on admission, or 2-3 days after the reperfusion therapy, and it is difficult to explain how the median of LVEF in this study was 37% with the such entrance criteria for LVEF. This implies some recruitment bias. The entrance echocardiogram in our study was performed on the 4-5 days after AMI to avoid myocardial stunning which is very pronounced in the first few days of AMI, and we suggest that entrance LV performance should be measured on the 3rd-4th day after AMI and not on admission or within the first 48 hours.

But is there a lower border of infarction damage when the stem cell therapy has no benefit? In our study (Obradovic et al, 2009) we showed that patients with too large myocardial infarction (measured by the perfusion defect on SPECT and by the maximum serum lactate dehydrogenase activity during the acute phase of STEMI) have no benefit of single, intracoronary stem cell therapy. Those patients might need repeated stem cell injections like in Chinese study (Yao et al, 2009) with the repeated intracoronary bone marrow cell transplantation three months after the AMI with the similar cohort of patients as ours.

It seems that intracoronary bone marrow stem cell therapy in early days of stem cell therapy also has no effect (Zhang et al, 2009), because the stem cells are injected in a very hostile, inflammatory, ischemic environment full of toxins. But, there are no randomized trials comparing stem cell therapy, for instance between 1-5 days to 6-12 days after infarction. Like in the most studies with intracoronary transplantation of stem cells in AMI, we injected cells intracoronary in the second week of infarction, not too soon from the initial event and not too late from it, to be in the burst of reparation process.

However, MYSTAR trial (Gyöngyösi et al, 2009) demonstrated that stem cell therapy after 3 weeks and 3 months had resulted in similar benefit on LV function. Having in mind that finding, our experience and previous reports we can only conclude that we do not know the proper timing for stem cell therapy after AMI.

What kind of cells we need in AMI, and do we need another cell type or types for CMI? In our study we only use the filtration of bone marrow and concentration of their mononuclear cells. We suppose that different kind of cells and their interplay is important for the successful cell therapy in AMI. The immune selection of bone marrow stem cells without some in-vitro manipulation of cells is probably unnecessary, especially in AMI patients. What do we gain and what do we lose with this procedure? The same number of cells with certain phenotype would be given with, or without selection, and a selection procedure would for sure prolong the timing from the bone marrow harvest to its application and further damage. The preparation of cells is important, however, at least for patients with AMI it is more important to give appropriate number of viable and functioning cells and the duration of bone marrow processing should be short. In the REGENT trial, immune-selection of CD-34+/CXC4R+ cells did not bring any advantage compared to un-selected bone marrow mononuclear cells.

Again there is no clinical randomized trial comparing different methods of stem cell processing. Do we need mesenchymal stem cells for AMI or CMI patients? Very well conducted study (Chen et al, 2004), with successful intracoronary implantation of mesenchymal stem cells in patients with AMI is almost neglected and those results are not challenged.

The way of cell delivery is also a matter of controversy. For intracoronary delivery almost all studies have used the same method (Strauer et al, 2002) nevertheless, the animal model suggests that the injections of cells through the inflated balloon currently applied in clinical studies are not necessary for cell deposit (Tossios P, et al, 2008). So, our study also has showed that non-selective injections of bone marrow MNC into the left coronary artery proved to be efficient in improving the LVEF and diminishing the infarction size. There is no human trial addresses for that issue. There are also numerous tips and tricks for stem cell delivery that might be important. Strauer used albumin-microaggregates to ensure prolonged passage of stem cells through the infracted microcirculation, and dobutamine infusion (Strauer et al, 2010) to increase the demand of oxygen in myocardium and probably to enhance engratment of stem cells with such treatment. However, does the freshly infracted myocardium need such an ischemic push? We have noticed very clear ischemic changes on electrocardiography monitoring in every patient during the delivery of cell suspension.

Intracoronary way of cell delivery is probably more suitable for the AMI patients because it enables homogenous spread of stem cells throughout the infracted microcirculation full of chemoattractants. On the other hand, a direct intramyocardial injection of stem cells in patients with CMI seems to be preferred mode of cell delivery. Some animal model and pilot human trial confirm this assumption (Hou et al, 2005; Schächinger et al, 2008). Our results have shown benefit of bone marrow derived stem cells given into the myocardium during CABG improving LVEF and myocardial perfusion which is in accordance with other studies of bone marrow derived stem cell therapy with CABG (Donndorf et al, 2011). On the other hand, our results have not shown any benefit of intracoronary transplantation of bone marrow stem cells in patients with CMI. There are only 2 published studies with intracoronary transplantation of bone marrow derived mononuclear cells in CMI and the both of them demonstrated improvement of LV performance after the procedure (Assmus et al, 2006; Strauer et al, 2009). The way of trans-balloon application of stem cells was used in both studies and on the contrary we used non-selective intracoronary implantation of stem cells. This distinction might have the different outcome between our and the mentioned studies and underlines the importance of ischemic preconditioning in this cohort of patients.

Finaly, and probably the most important aspect of stem cell therapy is a clinical benefit. REPEAR-AMI (Schächinger et al, 2010), studies of Strauer's group in AMI (Yousef et al, 2009) and CMI (Strauer et al, 2010) and the largest study with endocardial implantation of bone marrow derived stem cells in CMI (Pokushalov et al, 2010) have showed clear clinical benefit with hard end points during the relatively long period of follow-up. In our study, we have not a sufficient number of patients to show the difference of major adverse cardiac events in several groups of our patients. Nevertheless, there were no deaths during the 6 months follow-up, and the number of patients with restenosis and symptomatic heart failure was low.

When we take into account the benefit of stem cell therapy in the treatment of myocardial infarction one scenario is possible. Stem cells do not improve significantly global or even regional myocardial infarction after MI but do stabilize myocardium on the molecular level with the long-term clinically important benefits through yet unknown mechanisms.

As you can easily realize, there are too many confounding, important factors. It is impossible to randomize all the possibilities. Logic is important but it does not mean that it is always right. Clinical trials in stem cell therapy are being done too fast, and many trials did not meet the entrance criteria of sample size for the right statistical power. The European Task Force for stem cell therapy in cardiovascular diseases does not recommend the stem cell therapy in wider clinical practice and recommends large, placebo controlled trials (Bartunek et al, 2006). However, do we know enough to create the proper, large clinical trial for stem cell therapy? We believe that centrally coordinated, well-organized, small, always multicentric, pilot trials that address the various issues of stem cell therapy must precede the creation of a large randomized trial.

5. References

Abdel-Latif A., Bolli R., Tleych IM., Montori V., Perin EC., Hornung CA & Zuba-Surma E. (2007). Adult bone marrow-derived cells for cardiac repair. *Archives of Internal Medicine*, Vol.167, pp. 989-997

Ang K., Chin D., Leyva F., Foley P., Kubal C., Chalil S. et al. (2008). Randomized, controlled trial of intramuscular or intracoronary injection of autologous bone marrow cells into scarred myocardium during CABG versus CABG alone. *Nature clinical practice*, Vol.5, No.10, pp. 663-670

Arminan A., Gandia C., Garcia-Verdugo M., Lledo E., Trigueros C., Ruiz-Sauri A., et al. (2010). Mesenchimal stem cells provide better results than hematopoietic precursors for the treatment of myocardial infarction. *Journal of the American College of Cardiology*, Vol.55, pp. 2244-53

Arom KV., Ruengsakulrach P. & Jotisakulratana V. (2008). Efficacy of myocardial injection of angiogenic cell precursors for ischemic Cardiomyopathy. Innovations, Vol.3, No.1, pp. 38-45

Assmus B., Honold J., Schachinger V., Britten M., Fischer-Rasokat U., Lehmann R. et all. (2006). Transcoronary Transplantation of Progenitor Cells after Myocardial Infarction. *N Engl J Med*, Vol.355, pp. 1222-32

Assmus B., Rolf A., Erbs S., Elsässer A., Haberbosch W., Hambrecht R., et al. (2010). Clinical outcome 2 years after intracoronary administration of bone marrow-derived progenitor cells in acute myocardial infarction. *Circulation Heart Failure*, Vol.3, No.1, pp. 89-96

Bartunek J., Dimmeler S., Drexler H., Fernandez-Aviles F., Galinanes M., Janssens S., et al. The consensus of the task force of the European Society of Cardiology concerning the clinical investigation of the use of autologous adult stem cells for repair of the heart. *European Heart Journal*, Vol.27, pp. 1338-1340

Beltrami AP., Urbanek K., Kajstura J., Yan S-M., Finato N., Bussani R., et al. (2001). Evidence that human cardiac myocytes divide after myocardial infarction. *New England Journal of Medicine*, Vol.344, pp. 1750-1757

Bollini S., Smart N., Riley PR. (2011). Resident cardiac progenitor cells: at the heart of regeneration. *Journal of Molecular and Cellular Cardiology*, Vol.50, pp. 296-303.

Bolognese L., Neskovic AN., Parodi GM., Antoniucci D. (2002). Left ventricular remodeling after primary coronary angioplasty: patterns of left ventricular dilatation and long-term prognostic implications. *Circulation*, Vol.106, pp. 2351-2357

Cao F., Sun D., Li C., Narsinh K., Zhao L., Li X., et al. (2009). Long-term myocardial functional improvement after autologous bone marrow mononuclear cells transplantation in patients with ST-elevation myocardial infarction: 4 years follow-up. *European Heart Journal*, Vol.30, pp. 1986-1994

Chavakis E., Urbich C, Dimmler S. (2008). Homing and engraftment of progenitor cells: a prerequisiste for cell therapy. *Journal of Molecular and Cellular Cardiology*, Vol.45, pp. 514-522.

Chen S-L., Fang W-w., Ye F., Liu Y-H., Qian J., Shan S-j, et al. (2004). Effect on left ventricular function of intracoronary transplantation of autologous bone marrow mesenchymal stem cell in patients with acute myocardial infarction. *American Journal of Cardiology*, Vol.94, pp. 92-95

Cheng XW., Kuzuya M., Nakamura K., Maeda K., Tsuzuki M., Kim W., et al. (2007). Mechanisms underlying the impairment of ischemia-induced neovascularization in matrix metalloproteinase 2-deficient mice. *Circulation Research*, Vol.100, pp. 904-913

Cleland JGF., Torabi A., Khan NK. (2005). Epidemiology and management of heart failure and left ventricular systolic dysfunction in the aftermath a myocardial infarction. *Heart*, Vol.91, Suppl II, pp. 7-13

Dib N., Dinsmore J., Lababidi Z., White B., Moravec S., Campbell A. et al. (2009). One- year follow-up of feasibility and safety of the first U.S., randomized, controlled study using 3-dimensional guided catheter-based delivery of autologous skeletal myoblasts for ischemic cardiomyopathy (CAuSMIC Study). *Journal od the American College of Cardiology: Cardiovascular Interventions*, Vol.2, No.1, pp.9-16

Dong F., Khalil M., Kiedrovski M., O'Connor C., Petrovic E., Yhou X, Penn MS. (2010). Critical role for leukocyte hypoxia inducible factor-1α expression in post-myocardial infarction left ventricle remodeling. *Circulation Research*, Vol.106, pp. 601-610

Donndorf P., Kundt G., Kaminski A., Yerebaken C., Liebold A., Steinhoff G & Glass A. (2011). Intramyocardial bone marrow stem cell transplantation during coronary artery bypass surgery: a meta analysis. *Journal of Thoracic Cardiovascular Surgery, in press.*

Duckers H., Houtgraaf., Hehrlein C., Schofer J., Waltenberger J., Gershlick A., et al. Final results of a phase IIa, randomized, open-label trial to evaluate the percutaneous intramyocardial transplantation of autologous skeletal myoblasts in congestive heart failure patients: the SEISMIC trial. *Eurointervention*, Vol.6, pp. 805-812

Engelmann MG., Theiss HD., Henning-Theiss C., Huber A., Wintersperger BJ., Werle-Ruedinger A-E., et al. (2006). Autologous bone marrow stem cell mobilization induced by granulocyte colony-stimulating factor after subacute ST-segement elevation myocardial infarction. *Journal of the American College of Cardiology*, Vol.48, No.8, pp. 1712-1721

Erbs S., Linke A., Adams V., Lenk K., Thiele H., Diederich K. et all. (2005). Transplantation of Blood-Derived Progenitor Cells After Recanalisation of Chronic Coronary Artery Occlusion First Randomised and placebo -Controlled Study. *Circulation Research*, Vol.97, pp. 756-762

Falanga A., Marchetti M., Evangelista V., Manarini S., Oldani E., Giovanelli S., et al. (1999). Neutrophil activation and Hemostatic changes in healthy donors receiving granulocyte colony-stimulating factor. *Blood*, Vol.93, No.8, pp. 2506-2514

Fox KAA., Dabbous OH., Goldberg RJ., Pieper KS., Eagle KA., Van deWerf F., et al. (2006). Prediction of risk of death and myocardial infarction in six months after presentation with acute coronary syndrome: prospective multinational observational study (GRACE). *British Medical Journal*, Vol.333, pp. 1-6

Frangogiannis NG. (2008). The immune system and cardiac repair. *Pharmacology Research*, Vol.58, No.2, pp. 88-111

Gyöngyösi M., Lang I., Dettke I., Beran G., Graf S., Sochor H., et al. (2009). Combined delivery approach of bone marrow mononuclear stem cells early after myocardial infarction: the MYSTAR prospective, randomized study. *Nature*, Vol.6, No.1, pp. 70-80

Hamano K., Nishida M., Hirata K., Mikamo A., Tao-Sheng li, Harada M. et al. (2001). Local implantation of autologous bone marrow cells for therapeutic angiogenesis in patient with ishemic heart disease. *Japanese Circulation Journal*, Vol.65, pp. 845-847

Hare JM., Traverse JH., Henry TD., Dib N., Strumpf RK., Schulman SP., et al. (2009). A randomized, doble-blind, placebo-controlled, dose-escalation study of intravenous adult human mesenchimal stem cells (Prochymal) after acute myocardial infarction. *Journal of American College of Cardiology*, Vol.54, pp.2277-2286

Hendrikx M., Hensen K., Clijsters C., Jongen H., Koninckx R., Bijnens E. et al. (2006). Recovery of regional but not global contractile function by the direct intramyocardial autologous bone marrow transplantation: results from a randomized controlled clinical trial. *Circulation*, Vol.114, pp. I-101-I-107

Hirsch A., Nijveldt R., Vleuten P.A., Tijssen J., Giessen W., Tio R. et al. (2010). Intracoronary infusion of mononuclear cells from bone marrow or peripheral blood compared with standard therapy in patients after acute myocardial infarction treated by primary percutaneous coronary intervention: results of the randomized controlled HEBE trial. *Eur Heart J*, doi:10.1093/eurheart/ehq449

Hou D., Youssef EA., Brinton TJ., Zhang P., Rogers P., Price ET., et al. (2005). Intracoronary, and interstitial retrograde coronary venous delivery. *Circulation*, Vol.112, suppl I, pp. I-150-I-156

Huang P-H, Chen Y-H, Wang C-H, Chen J-S, Tsai H-Y, Lin F-Y, et al. (2009). Matrix metalloproteinase-9 is essential for ischemia-induced neovascularization by modulating bone marrow derived endothelial progenitor cells. *Arteriosclerosis, Thrombosis, and Vascular Biology*, Vol.29, pp. 1179-1184

Huber BC., Fischer R., Brunner S., Groebner M., Rischpler C., Segeth A., et al. (2010). Comparison of Parathyroid hormone and G-CSF treatment after myocardial infarction on perfusion and stem cell homing. *American Journal of Physiology - Heart Circulatory Physiology*, Vol.298, pp. H1466-H1471c

Huikuri H., Kervinen K., Niemela M., Ylitalo K., Saily M., Koistinen P. et al. (2008). Effect of intracoronary injection of mononuclear bone maroww cells on left ventricular function, arrhythmia risk profile, and restenosis after trombolytic therapy of acute myocardial infarction. *Eur Heart J*, Vol.29, pp. 2723-2732

Ince H., Petzsch M., Kleine D., Schmidt H., Rehderhs T., Köber T., et al. (2005). Preservation from left ventricular remodeling by front-integrated revascularization and stem cell liberation in evolving acute myocardial infarction by use of granulocyte-colony-stimulating factor (FIRSTLINE-AMI). *Circulation*, Vol.112, pp. 3097-3106

Janssens S., Dubois C., Bogaert J., Theunissen K., Deroose C. et al. (2006). Autologous bone marrow-derived stem-cell transfer in patient with ST-segment elevation myocardial infarction: double-blind, randomized controlled trial. *Lancet*, Vol.367, pp. 113-21

Kang H-J., Lee H-Y., Na S-H., Chang S-A., Park K-W, Kim H-K. et al. (2006). Differential effect of intracoronary infusion of mobilized peripheral blood stem cells by granulocyte colony-stimulating factor on left ventricular function and remodeling in patients with acute myocardial infarction versus old myocardial infarction: The MAGIC Cell-3-DES Randomized, Controlled Trial. (2006). *Circulation*, Vol.144, pp. I-145-I-151

Körbling M. & Estrov Z. Adult stem cells for tissue repair – a new therapeutic Concept? *New England Journal of Medicine*, Vol.349, No.6, pp. 570-582

Krause K., Jaquet K., Schneider C., Haupt S., Lioznov MV., Otte K-M., et al. (2009). Percutaneous intramyocardial stem cell injection in patients with acute myocardial infarction: first-in-men study. *Heart*, Vol.95, pp. 1145-52

Laan A.M., Hirsch A., Nijveldt R., Vleuten P.A., Giessen W.J., Doevendans P.A. et al. (2008). Bone marrow cell therapy after acute myocardial infarction: the HEBE trial in perspective, first results. *Neth Heart J*, Vol.16, No.12, pp.436-9

Le Blanc R., Roy J., Demers C., Vu L & Cantin G. (1999). A prospective study of G-CSF effects on hemostasis in allogeneic blood stem cell donors. *Bone Marrow Transplantation*, Vol.23, pp. 991–996

Leone AM., Galiuto L., Garramone B., Rutella S., Giannico MB., Brugaletta S., et al. (2007). Usefulness of granulocyte colony-stimulating factor in patients with large anterior wall acute myocardial infarction to prevent ventricular remodeling (The Rigenera Study). *American Journal of Cardiology*, Vol.100, pp. 397-403

Lunde K., Solheim S., Aakhus S., Arnesen H., Abdelnoor M., Egeland T. et al. (2006). Intracoronary Injection of Bone Marrow Cells in Acute Myocardial Infarction. *N Eng J Med*, Vol.355, No.12, pp. 1199-1209

Meluzin J., Mayer J., Groch L., Janousek S., Hornacek I., Hlinomaz O. et al. (2006). Autologous transplantation of bone marrow cells in patient with acute myocardial infarction: The effect of the dose of transplanted cells on myocardial function. *Am Heart J*, Vol.152, pp. 975.e9-15

Menasce P., Hagage AA., Scorsin M., Pouzet B., Desnos M., Duboc D., et al. (2001). Myoblast transplantation for heart failure. *Lancet*, Vol. 357, pp. 279-80

Menasche P., Alfieri O., Janssens S., McKenna W., Reichenspurner H., Trinquart L. et all. (2008). The myoblast autologous grafting in ischaemic cardiomyopathy (MAGIC) trial- first randomized placebo-controlled study of myoblast transplantation. *Circulation*, Vol.117, pp. 1189-1200

Meyer GP., Wollert KC., Lotz J., Pirr J., Rager U., Lippolt P., et al. (2009). Intracoronary bone marrow cell transfer after myocardial infarction: 5-year follow-up from the randomized-controlled BOOST trial. *European Heart Journal*, Vol.30, pp. 2978-2984

Mirotsou M., Jayawardena TM., Schmeckpeper J., Gnecchi M., Dzau VJ. (2011). Paracrine mechanisms of stem cell reparative and regenerative actions in the heart. *Journal of Molecular and Cellular Cardiology*, Vol.50, pp: 280-289

Mocini D., Staibano M., Mele L., Giannantoni P., Menichella G., Colivicchi F. et al. (2006). Autologous bone marrow mononuclear cell transplantation in patients undergoing coronary bypass grafting. *American Heart Journal*, Vol.151, No1, pp. 192-7

Müller P., Beltrami AP., Cesselli D., Pfeiffer P., Kazakov A., Böhm M. (2005). Myocardial regeneration by endogenous adult progenitor cells. *Journal of Molecular and Cellular Cardiology*, Vol.39, pp. 377-387

Murry CE., Soonpaa MH., Reinecke H., Nakajima H., Nakajima HO., Rubart M., et al. (2004). Haematopoietic stem cells do not transdifferentiate into cardial myocytes in myocardial infarcts. *Nature*, Vol.428, pp. 664-668

Obradovic S., Balint B., Romanovic R., Trifunovic Z., Rusovic S., Baskot B., et al. (2009). Influence of intracoronary injections of bone marrow derived mononuclear cells on large myocardial infarction outcome: quantum of initial necrosis is the key. *Vojnosanitetski Pregled* Vol.66, No.12, pp 998-1004

Obradovic S., Balint B., Trifunovic Z., Rusovic S., Baskot B., Ostojic G., et al. (2009). Non-selective intracoronary injections of bone marrow derived mononuclear cells in subacute phase of large myocardial infarction reduce the infarction size without significant improvement of left ventricle ejection fraction. *Bulletein of Transfusion Medicine*, Vol.55, No.1-2, pp. 84-88

Patel A., Geffner L., Vina R., Saslavsy J., Urschel H., Kormos R.&Benetti F. (2005). Surgical treatment for congestive heart failure with autologous adult stem cell transplantation: A prospective randomized study. *Journal of Thoracic Cardiovascular Surgery*, Vol.130, pp. 1631-1638

Perin EC., Dohmann HFR., Borojevic R., Silva SA., Sousa ALS., Mesquita CT., et al. (2003). Transendocardial, autologous bone marrow cell transplantation for severe, chronic ischemic heart failure. *Circulation*, Vol.107, pp. 2294-2302

Pokushalov E., Romanov A., Chernyavsky A., Larionov P., Terekhov I., Artyomenko S., et al. (2010). Efficiency of intramyocardial injections of autologous bone marrow mononuclear cells in patients with ischemic heart failure: a randomized study. *Journal of Cardiovascular Translational Research*, Vol.3, pp.160-168

Ramshorst J., Bax J., Beeres S., SchneiderP., Roes S., Stokkel M. et all. (2009). Intramyocardial Bone Marrow Cells Injection for Crhonic Myocardial Ischemia. *JAMA*, Vol.301, No.19, pp. 1997-2004

Ripa RS., Jørgensen E., Wang Y., Thune JJ., Nilsson JC., Søndergaard L., et al. (2006). Stem cell mobilization induced by subcutaneous granulocyte-colony stimulating factor to improve cardiac regeneration after acute ST-elevation myocardial infarction: result of double-blind, randmonized, placebo-controlled stem cells in myocardial infarction (STEMMI) trial. *Circulation*, Vol.113, pp. 1983-1992

Roncalli J., Mouquet F., Piot C., Trochu J-N., Le Corvoisier P., Neuder Y., et al. (2010). Intracoronary autologous mononucleated bone marrow cell infusion for acute myocardial infarction: results of the randomized multicenter BONAMI trial. *European Heart Journal*, doi:10.1093/eurheartj/ehq455

Sanz-Ruiz R., Ibanes G., Arranz AV., Santos MEF, Fernandez PLS., Fernandez-Aviles F. (2010). Phases I-III clinical trials using adult stem cells. *Stem Cells International, Article* ID 579142, pp.1-12, doi:10.4061/2010/579142

Schächinger V., Assmus B., Britten M., Honold J., Lehmann R., Teupe C. et al. (2004). Transplantation of Progenitor Cells and Regeneration Enhancement in Acute Myiocardial Infarction: Finale One-Year Results of the TOPCARE-AMI Trial. JACC, Vol.44, No.8. pp. 1690-9

Schächinger V., Erbs S., Elsasser A., Elsasser A., Haberbosch W., Hambrecht R., Holchermann H. et al. (2006). Intracoronary Bone Marrow-Derived Progenitor Cells in Acute Myocardial Infarction. *N Eng J Med*, Vol.355, pp. 1210-21

Schächinger V., Erbs S., Elsasser A., Haberbosch W., Hambrecht R., Holchermann H. et al. (2006). Improved clinical outcome after intracoronary administration of bone marrow/derived progenitor cells in acute myocardial infarction: final 1-year results of the REPAIR-AMI trial. *Eur Heart J*, Vol.27, pp. 2775-2783

Schächinger V., Aicher A., Döbert N., Röver R., Diener J., Fichtlscherer., et al. (2008). Pilot trial on determinants of progenitor cell recruitment to the infracted human myocardium. Circulation, Vol.118, pp. 1425-1432

Schafer A., Meyer G., Fuchs M., Klein G., Kaplan M. et al. (2006). Impact of intracoronary bone marrow cell transfer on diastolic function in patient after acute myocardial infarction: results from the BOOST trial. *Eur Heart J*, Vol.27, pp. 929-935

Seeger FH., Tonn T., Krzossok N., Zeiher AM., Dimmeler S. (2007). Cell isolation procedures matter: a comparison of different isolation protocols of bone marrow mononuclear cells used foe cell therapy in patients with acute myocardial infarction. *European Heart Journal*, Vol.28,pp. 766-772

Stamm C., Kleine H., Choi Y., Dunkelmann S., Lauffs J., Lorenzen B. et al. (2007). Intramyocardial delivery of CD133+ bone marrow cells and coronary artery bypass grafting for chronic ischemic heart disease: Safety and efficasy studies. *Journal of Thoracic Cardiovascular Surgery*, Vol.133, pp. 717-725

Strauer BE., Brehm M., Zeus T., Köstering M., Hernandez A., Sorg RV, et al. (2002). Repair of infracted myocardium by autologous intracoronary mononuclear bone marrow cell transplantation in humans. *Circulation*, Vol106, pp. 1913-1918

Strauer BE., Yousef M., Schannwell CM. (2010). The long-term effects of intracoronary stem cell transplantation in 191 patients with chronic heart failure: the STAR-heart study. *European Journal of Heart Failure*, Vol.12, No.7, pp. 721-729

Takano H., Hasegawa H., Kuwubara Y., Nakayama T., Matsuno K., Miyazaki Y., et al. (2007). Feasibility and safety of granulocyte colony-stimulating factor treatment in patients with acute myocardial infarction. *International Journal of Cardiology*, Vol.122, pp. 41-47

Tendera M., Wojakowski W., Ruzytto W., Chojnowska L., Kepka C., Tracz W. Et al. (2009). Intracoronary infusion of bone marrow-derived selected CD34+CXCR4+ cells and non-selected mononuclear cells in patient with acute STEMI and reduced left ventricular ejection fraction: results of randomized, multicentre Myocardial Regeneration by Intracoronary Infusion of Selected Population of Stem Cells in

Acute Myocardial Infarction (REGENT) Trial. *European Heart Hournal*, Vol.30, No.11, pp. 1313-21

Tongers J., Losordo DW., Landmesser U. (2011). Stem and progenitor cell-based therapy in ischaemic heart disease: promise, uncertainties, and challenges. *European Heart Journal* doi:10.1093/eurheart/ehr018

Torabi A., Cleland JGF., Khan NK., Loh PH., Clark AL., Alamgir F., et al. (2008).The timing of development subsequent clinical course of heart failure after myocardial infarction. *European Heart Journal*, Vol.29, pp. 859-870

Tossios P., Krausgrill B., Schmidt M., Fischer T., Halbach M., Fries JWU., et al. (2009). Role of balloon occlusion for mononuclear bone marrow deposition after intracoronary injection in pigs with reperfused Myocardial infarction. *European Heart Journal*, Vol.29, pp. 1911-1921

Turan RG., Brehm M., Koestering M., Tobias Z., Bartsch T., Steiner S., et al. (2007). Factors influencing spontaneous mobilization of CD34+ and CD133+ progenitor cells after myocardial infarction. European Journal of Clinical Investigation, Vol.37, No.11, pp. 842-851

Valgimigli M., Rigolin GM., Cittani C., Malagutti P., Curello S., Percoco G., et al. (2005). Use of granulocyte-colony stimulating factor during acute myocardial infarction to enhance bone marrow stem cell mobilization in humans: clinical and angiographic safety profile. *European Heart Journal*, Vol.26, pp. 1838-1845

Valgimigli M., Biondi-Zoccai GGL., Malagutti P., Leone AM., Abbate A. (2008). Autologous bone marrow stem cell mobilization induced by granulocyte colony-stimulating factor after myocardial infarction. *European Heart Journal Supplements*. Vol.10, Supplement K, pp. K27-K34

Velagaleti RS., Pencina MJ., Murabito JM., Wang TJ., Parikh NI, D'Agostino RB, et al. (2008). Long-term trends in incidence of heart failure after myocardial infarction. *Circulation*, Vol.118, pp. 2057-2062

Wagers AJ., Sherwood RI., Cristensen JL., Weissman IL. (2002). Little evidence for developmental plasticity of adult hematopoietic stem cells. Science Vol.297, pp. 2256-2259

Weir, R. & McMurray. (2006). Epidemiology of heart failure and left ventricle dysfunction after acute myocardial infarction. *Current Heart Failure Reports*, Vol.3, pp.175-180

Yao K., Huang R., Sun A., Qian J., Liu X., Ge L., et al. (2009). Repeated autologous bone marrow mononuclear cell therapy in patients with large myocardial infarction. *European Journal of Heart Failure*, Vol.11, pp. 691-698.

Yeo C., Saunders N., Locca D., Flett A., Preston M., Brookman P., et al. (2009). Ficoll-Paque vs. Lymphoprep: a comparative study of two density gradient media for therapeutic bone marrow mononuclear cell preparations. *Regenerative Medicine*, Vol.4, pp. 689-696

Yousef M., Schannwell CM., Köstering M., Zeus T., Brehm M., Strauer BE. (2009). The BALANCE study: clinical benefit and long-term outcome after intracoronary autologous bone marrow cell transplantation in patients with acute myocardial infarction. *Journal of American College of Cardiology*, Vol.53, pp. 2262-2269

Zhang S., Sun A., Xu D., Yao K., Huang Z., Jin H., et al. Impact of timing and safety of intracoronary autologous bone marrow stem cells transplantation in acute myocardial infarction: pooled subgroup analysis of randomized controlled trials. *Clinical Cardiology*, Vol.32, No.8, pp. 458-66

Zhao Q., Sun Y., Xia L., Chen A. & Wang Z. (2008). Randomized study of bone marrow cell transplantation in patients with coronary surgery. *Ann Thorac Surg*, Vol.86, pp-1833-1840

Zohlnhöfer D., Ott I., Mehilli J., Schömig K., Michalk F., Ibrahim T., et al. (2006). Stem cell mobilization by granulocyte colony-stimulating factor in patients with myocardial infarction. A randomized controlled trial. The Journal of American Medical Association, Vol.295, No.9, pp. 1003-101

11

Randomized Clinical Trials in Stem Cell Therapy for the Heart - Old and New Types of Cells for Cardiovascular Repair

Ricardo Sanz-Ruiz, Adolfo Villa Arranz,
Enrique Gutiérrez Ibañes, María Eugenia Fernández Santos,
Pedro Luis Sánchez Fernández and Francisco Fernández-Avilés
Cardiology Department, Hospital General Universitario Gregorio Marañón, Madrid, Spain

1. Introduction

Recent advances in reperfusion strategies have dramatically reduced early mortality after acute myocardial infarction (AMI), but as a result there is a higher incidence of heart failure among survivors. Optimal medical therapy and device implantation can improve the prognosis and the quality of life of these patients. Nevertheless, mortality and rehospitalization rates are still high and entail an overwhelming cost.
The field of cardiac cell therapy has emerged as a new alternative in this situation, and has made rapid progress. Its final goal is to repair the damaged myocardium and to restore cardiac function. Nevertheless, this goal is hindered by the massive loss of cardiomyocytes after an AMI (in the order of 1 billion cells) and because ischemic myocardium loses cellular and extracellular signals which guide stem cells to the cardiac lineage or to the secretion of paracrine factors (Wollert & Drexler, 2010).
Studies evaluating this new approach during the last 15 years have overall succeeded to a greater or lesser extent, and evidence available so far is encouraging. Phase I and II randomized clinical trials (RCT) indicate that cell therapy is a safe treatment which can improve cardiac function after AMI and in the chronic phase of coronary artery disease (CAD). Trial results are not uniform, however, probably due 1) to a lack of standardization of cell isolation and delivery procedures, 2) to the absence of a universally accepted nomenclature, and 3) to the large number of stem cell types under investigation in different clinical settings. Nevertheless, these inconsistencies can be avoided or reduced if classical scientific methodology is followed. Although considered a relatively new field of research, stem cell experimentation must invariably walk on the path of the scientific method. Since Aristotle´s time, scientific method has been used as a way to ask and answer scientific questions by making observations and doing experiments. It includes a series of steps, i.e. 1) asking a question, 2) doing background research, 3) constructing a hypothesis, 4) testing the hypothesis by doing an experiment, 5) analyzing the data and drawing a conclusion, and 6) communicating the results.

In the case of stem cell therapy, RCT started questioning if there was a possibility of repairing the heart after different types of tissular damage. Background evidence has already demonstrated that this possibility exists through stem cell administration in several preclinical models of cardiomyopathy. Thus, the key points in the design of present and future RCT in humans are 1) to formulate an adequate hypothesis, 2) to select the ideal population, cell type and delivery method and 3) to develop a correct and precise protocol. These decisions must be made in the light of previous evidence and with a translational mentality, in which experimental/preclinical data should help to design new RCT. Inversely, results of human studies should transfer new questions and hypothesis to the laboratory/bench side.

2. Types of stem cells in cardiovascular research. Preclinical background

There are several types of stem cells which have been used with the aim of repairing damaged myocardium. Broadly speaking, stem cells can be subdivided into two large groups: embryonic and adult.

2.1 Embryonic stem cells

Embryonic stem cells (ESC) are present in the earliest stage of embryonic development - the blastocyst (this is the stage before the embryo is implanted into the uterine wall, one week after fertilization). ESC can divide indefinitely in vitro and are pluripotent: they are capable of generating any terminally differentiated cell in the human body that is derived from any one of the three embryonic germ layers (Perin & Silva 2006).

Classic experimental studies have demonstrated cardiomyocyte obtention from ESC, with the same structural and functional properties as cardiomyocytes present in the cardiac muscle (Kehat et al., 2001), and showing even successful electromechanical coupling with host myocardium. After their transplantation into the infarcted myocardium, these cells can engraft and survive in the myocardial tissue network, providing an improvement in ventricular function in several small animal preclinical models (Klug et al., 1996; Min et al., 2002). On the other hand, in chronic heart failure models, ESC have been proved to differentiate into new cardiomyocytes and also into endothelial and smooth muscle cells forming new blood vessels (Yamashita et al., 2000).

Nevertheless, despite the enormous evidence available regarding ESC isolation, growth, differentiation and transplantation in animal models, there are several issues that have restricted their clinical applications in human RCT in most countries: ethical and legal considerations, limited numbers of cell lines, limited resources, the difficulty of obtaining autologous ESC, the need for immunosuppression in case of allogenic cells and the potential risk for tumorigenesis (Thomson et al., 1998).

2.2 Adult stem cells

Adult stem cells are intrinsic to specific tissues of the postnatal organism into which they are committed to differentiate. Theoretically, adult stem cells are capable of self-renewal, yielding mature differentiated cells which are: 1) integrated into a particular tissue and 2) capable of performing the specialized function of that tissue. Adult-tissue specific stem cells are present in organs with self-renewal capacity, including the liver, pancreas, skeletal muscle, skin and bone marrow. These are the most investigated cells in regenerative medicine and include different cellular types:

2.2.1 Skeletal myoblasts

Skeletal myoblasts are satellite cells that remain in a quiescent stage between the basal lamina and the sarcolemma on the periphery of mature skeletal muscle fibers. They can be obtained from muscular biopsies and easily expanded in vitro (Murry et al., 1996; Taylor et al., 1998). This kind of stem cells has received considerable attention in the setting of chronic ischemic ventricular dysfunction, because these cells: 1) do not need a specific microenvironment to differentiate, 2) can be expanded without problems in undifferentiated stages, and 3) are highly resistant to ischemia and can multiply after injury. These cells are programmed to differentiate into myogenic lineages and, after their transplantation into scarred myocardium, they become myotubes and myocytes with the characteristics and function of skeletal muscle (Leobon et al., 2003; Pagani et al., 2003). In other words, transdifferentiation into cardiomyocytes, which was their first hoped mechanism of action, is an exceptional event rarely seen at the graft-host interface. Therefore, their benefitial effect is thought to be mediated through myocyte contraction, paracrine effects and by an increase in the infarcted wall resistance and stiffness, thus limiting ventricular dilatation and adverse remodeling. Safety concerns include basically the risk of ventricular arrhythmias observed in the first clinical trials, due to the absence of electromechanical coupling of the myoblasts with another cells or with host myocardium. As a consequence, in most of these protocols patients were treated at the same time with prophylactic implantable cardiac defibrillators and/or chronic amiodarone. Administration of skeletal myoblasts after certain culture conditions (with autologous human serum) could avoid this risk (Herreros et al., 2003).

2.2.2 Bone marrow-derived cells

The bone marrow is one of the most investigated sources of adult stem cells. It is a complex organ with a specific geometric organization and an intrincate system of cell-to-cell interaction and signaling, and it contains several cell subpopulations: 1) differentiated cells like stromal supporting mesenchymal cells, vascular cells, adipocytes, osteoblasts and osteocytes, and 2) progenitor cells. The most important subpopulation is the bone marrow mononuclear cell (BMMC) fraction, that includes mesenchymal stem cells (MSC), hematopoietic progenitor cells (HPC) and endothelial progenitor cells (EPC) (Shizuru et al., 2005; Suva et al., 2004). It also contains more committed cell lineages, such as natural killers, T and B lymphocytes. However, after the finding that MSC and EPC represent, respectively, only 0.01% and 1-2% of the total amount of cells included in BMMC fraction, culture techniques have been developed to select a specific type of cell and to expand it to obtain solutions with higher numbers of cells.

BMMC can be easily harvested, isolated, in vitro expanded and administered to the patient with several delivery devices. The plasticity of these cells has been demonstrated in classic preclinical studies, showing even transdifferentiation into mature cells from different germinal layers.

It has been observed that BMMC contribute to the formation of new cardiomyocytes and endothelial cells in ischemic areas when injected after an AMI in animal models. In treated animals, an improvement of hemodynamic parameters was demonstrated (Jackson et al., 2001; Orlic et al., 2001a).

Regarding HPC, no real specific surface antigen has been described to identify these cells. The antigens CD34, CD133 and CD117 and the surface markers c-kit, Sca-1 and Thy-1 have been used to sort this subtype of BMMC. C-kit+ cells without hematopoietic markers (lin-) have been administered after AMI, with a noticeable increase in the number of cardiomyocytes in the infarcted area, improvement of the ventricular function and survival benefits (Orlic et al., 2001b). However, other researchers have proved differentiation of these HPC into hematopoietic lineages and with the same results of reversing adverse remodeling and preventing post-infarction ventricular dysfunction (Balsam et al., 2004). Finally, improvements of myocardial perfusion, capillary density/collateral vessel formation and left ventricular ejection fraction have also been observed in other studies after HPC injection in AMI models, with a significant reduction of the infarct size (Kamihata et al., 2001).

EPC were first defined as cells positive for both hematopoietic stem cell markers such as CD34 and endothelial marker proteins such as vascular endothelial growth factor receptor 2 (VEGFR-2) or CD133. CD133 (also known as prominin or AC133) is expressed on HPC and on early stages of EPC, but is absent on mature EPC and monocyte cells. CD133+ cells promote angiogenesis in ischemic tissues, differentiating into adult endothelial cells and secreting angiogenic factors. The main drawback of CD133+ cells is their low number in the BMMC fraction (they represent less than 1% of the total population) and the impossibility of in vitro expansion. In fact, cytokine treatment with granulocyte colony-stimulating factor (G-CSF) is needed in order to mobilize a sufficient number of cells from the bone marrow (Virmani et al., 2006).

EPC have similar phenotypic and functional characteristics to those shown by fetal angioblasts. Indeed, they arise from a common hemangioblast precursor in the adult bone marrow (Virmani et al., 2006). Endothelial cell lineage markers used to sort these cells include CD34, Flk-1, VE-cadherin, platelet-endothelial cell adhesion molecule 1 (PECAM-1), von Willebrand factor (vWF) and E-selectin. Conversely to what happens with other stem cells, EPC have been proved to transdifferentiate into cardiomyocytes and smooth muscle cells in vivo. Besides, they increase capillary density after an AMI by means of angiogenesis and arteriogenesis mechanisms, reduce collagen deposits and cardiomyocyte apoptosis, and improve cardiac function (Kocher et al., 2001). Again, the ability to expand these cells is limited by their scarcity in peripheral blood. Furthermore, functional impairment of EPC has been observed in the elderly and in several pathologic conditions (i.e., diabetes mellitus). Recently, specific subpopulations of EPC have been described: CD14+/CD34- cells, which have shown very high plasticity (Yoon et al., 2005), and CD14+/CD34+ cells, which have been demonstrated to induce a paracrine response by releasing angiogenic growth factors.

MSC are one of the most promising types of stem cells for cardiac repair. They have been isolated from the bone marrow stroma (although they can also be found around blood vessels, in muscle, skin and adipose tissue) and exhibit unique characteristics, such as a multipotent differentiaton potential capacity and the lack of surface markers. Indeed, they are CD34-, CD45- and CD133- cells that show an immunophenotype positive for adhesion proteins like CD29, CD44, CD71, CD90, CD105, CD106, CD117, CD120a, CD124, SH2, SH3 and SH4 (Pittenger & Martin, 2004). This is the reason why MSC can be isolated after density centrifugation by means of their ability to adhere to culture plates. Purified human MSC have been shown to migrate and differentiate to a cardiomyocyte phenotype and to endothelial cells in both healthy and infarcted myocardium (Toma et al., 2002). In the former MSC express cardiac surface markers, and in the latter they improve wall motion and prevent the adverse remodeling process. One advantage of MSC is that they are considered to be immuneprivileged. Thus, allogenic MSC have been successfully transplanted into murine hearts without the need for immunosuppression (Zimmet & Hare, 2005).

2.2.3 Adipose-derived stem cells

Recently, it has been shown that, besides committed adipogenic, endothelial and pluripotent vascular progenitor cells, the adipose tissue stroma contains multipotent adipose-derived stem cells (ADSC) (Sanz-Ruiz et al., 2009). These self-renewal cells possess high potentiality, and are capable of differentiating into myogenic, neural and cardiomyocytic lineages, in this last case showing even spontaneous beating. The phenotype of ADSC is similar to that of MSC (i.e., ADSC also express adhesion molecules in their surface) (Sanz-Ruiz et al., 2008). The similarities of both cell types in terms of potentiality and the fact that adipose tissue can be easily obtained in large amounts by means of a simple liposuction procedure, have pointed at the human adipose tissue as a novel promising alternative source of stem cells for cardiovascular repair that has shown encouraging results in the preclinical field (Fraser et al., 2004). In fact, in our center two first-in-man phase II randomized trials have been conducted with ADSC: with intracoronary administration after an AMI (APOLLO trial) and with transendocardial injection in "no-option" patients with advanced CAD (PRECISE trial).

2.2.4 Resident cardiac stem cells

A newly discovered population of resident cardiac stem cells (CSC) has been recently identified in the adult heart, which contributes to myocardial regeneration in animal models of AMI (Beltrami et al., 2003; Oh et al., 2003; Urbanek et al., 2003; Urbanek et al., 2005). These cells express surface markers such as c-kit, Sca-1 and Abcg2. However, the phenotypic characterization of this lineage is not definitely clear. Indeed, currently all CSC types are thought to be distinct from each other. In the adult human heart, these clusters of CSC have shown their capacity to differentiate into cardiomyocytes, to fuse with pre-existing cardiomyocytes and to differentiate also into smooth muscle and endothelial cells, providing both myocardial and vascular regeneration. A novel cardiac cell type has been identified using the transcription factor islet-1 (isl1). These isl1+ cells have been proposed even for pacemaker cell and conduction system repair (Laugwitz et al., 2005). Finally, CSC have been isolated and expanded for the first time from human myocardial biopsies, offering another new source of stem cells in cardiovascular regenerative medicine (Messina et al., 2004).

2.2.5 Cord-blood derived stem cells

The umbilical cord blood contains a high proportion of hematopoietic and mesenchymal stem cells, and in higher numbers than in peripheral blood or in the bone marrow. Also known as "somatic non-restricted stem cells", these cells are negative for c-kit, CD34 and CD45, and can differentiate into cardiomyocytes. They have been administered after an AMI in animal models, improving ventricular perfusion and contractility, and reducing the infarct size (Ma et al., 2005). Nevertheless, cord-blood stem cells have not been used in clinical trials so far.

2.3 Induced or "embryonic-like" stem cells

At the end of 2007, two groups reported for the first time the generation of pluripotent stem cell lines derived from adult human cells (Takahashi et al., 2007; Yu et al., 2007). Differentiated adult somatic human cells were successfully reprogrammed into a pluripotent state by transduction of four defined transcription factors. The resulting induced-pluripotent stem cells (iPS) were proved to have the same morphological characteristics, surface markers, proliferative capacity and potentiality as those known in ESC. In the first study (Takahashi et al., 2007), the factors Oct3/4, Sox2, Klf4 and c-Myc were transduced in human dermal fibroblasts, and in the second one (Yu et al., 2007), Oct4, Sox2, NANOG and LIN28 were induced in the same type of adult cells. These iPS can differentiate into any kind of cell from any of the three embryonic germ layers, but two safety concerns have arisen: 1) the risk of mutations after viral transfection, and 2) the risk of teratoma formation. In other words, although these studies have meant a revolution in the field of regenerative medicine, we still have a long way to cover before their application in human pathology.

3. Host tissue and cell related issues. Understanding the ischemic myocardium

The two main determiners of cardiovascular repair are stem cells and injured myocardial tissue in which these cells are delivered. Both play the central role that will establish the efficacy of the treatment, and knowledge of the molecular/cellular changes and interactions between them is crucial when designing new RCT.

After AMI, if blood flow is not restored quickly, cell death and myocardial necrosis are definitive. This activates a complement cascade with free radical and cytokine generation that recruit leukocytes and initiate the inflammatory response. Inflammation, while potentially detrimental to surviving cardiomyocytes, is necessary to clear away the debris (clearance of necrotic cells) and orchestrate downstream healing events. Chronic inflammatory cells such as macrophages and mast cells secrete cytokines and growth factors, which in turn activate fibroblasts to proliferate and synthesize collagen, a major component of the scar that replaces cardiomyocyte loss. Neovascularization is also stimulated by the release of growth factors from the inflammatory cells. Scar remodeling may continue for months to years, depending on the extent of the initial ischemic event (Lindsey et al., 2003).

Left ventricle (LV) remodeling, defined as post-AMI changes in wall structure, chamber geometry and pump function, is mainly caused by changes in extracellular matrix (ECM). Cardiac ECM not only supports and aligns cardiomyocytes, thereby preserving a fundamental mechanical relationship by which sarcomeric shortening is translated to muscle force contraction, but also has signaling functions. Indeed, ECM is a storage depot for growth factors, hormones and cytokines, and uses integrins to communicate with cells (Lindsey et al., 2003). All these functions are lost after myocardial ischemia due to the release from inflammatory and endogenous cells of matrix metalloproteinases (MMP) and cytokines. MMP degrade ECM, disengage integrins and stimulate reparative fibrosis. Cytokines like tumor necrosis factor α (TNF-α) and interleukins like IL-1 and IL-6 induce MMP synthesis and are related to the development of LV dysfunction, pulmonary edema, endothelial dysfunction and cardiomyocyte apoptosis (Dewald et al., 2004).

These cellular and signaling processes that constitute the proliferative phase of infarct healing in the myocardium influence and determine the fate of implanted stem cells. Ischemic myocardium constitutes an inflammatory hostile environment for stem cells, which is devoid of nutrients and oxygen and lacks survival signals from the ECM and cell-to-cell interactions. Indeed, only a small fraction of them survive in such adverse conditions. Nevertheless, some studies have shown that certain implanted stem cells may improve or counteract this situation. Intramyocardial transplantation of EPC after AMI induces

significant and sustained increase in angiogenic, antiapoptotic and chemoattractant factors, that are up-regulated in both transplanted and host cells (i.e., vascular endothelial growth factor-A [VEGF-A], fibroblast growth factor-2 [FGF-2], angiopoietin-1 [Ang-1], angiopoietin-2 [Ang-2], placenta growth factor [PIGF], hepatocyte growth factor [HGF], insulin-like growth factor-1 [IGF-1], platelet-derived growth factor-B [PDGF-B] and stromal cell-derived factor-1 [SDF-1]) (Cho et al., 2007). These humoral factors provide an additional favorable milicu for neovascularization and repair or regeneration of ischemic myocardium. Furthermore, there is a cross-talk between the heart and the bone marrow mediated by humoral effects that may improve this therapeutic effect: it has been proved that EPC transplantation further mobilizes endogenous BMMC into peripheral circulation, recruiting them into the ischemic myocardium (Cho et al., 2007).
Having these considerations in mind, new lines of research are being developed to improve cell survival rates in the ischemic myocardium, between them (Wollert & Drexler 2010):

1. Preconditioning of the myocardium to retain a higher number of cells: low-energy shock waves, ultrasound-mediated destruction of microbubbles in the coronary circulation and extracorporeal shock wave treatment have proved to increase retention of EPC, BMMC and MSC.
2. Activation or increase of chemotactic factors to attract cells to the damaged area: high mobility group box-1 (HMGB-1), SDF-1 or its receptor CXCR4, β_2 integrin and endothelial nitric oxide synthase can be activated to increase the rate of homing of different types of stem cells (i.e., progenitor blood cells, EPC).

Regarding stem cells administered to the myocardium, their functional activity is determined by age and cardiovascular risk factors. As a consequence, future phase II-III RCT will explore cell enhancement strategies intended to increase their therapeutic potential. Several strategies are currently under investigation (Wollert & Drexler 2010):

1. Pretreatment of the patients with drugs to stimulate cell potenciality: statins, rosiglitazone and nitric oxide synthase enhancer AVE9488 can improve the migratory, invasive and neovascularization capacity of EPC.
2. Strategies to prolong cell survival: between them, the use of a combination of growth factors to stimulate the expression of cardiomyocyte genes in MSC, the use of heat shock to increase the resistance of cells to external stressors and the pretreatment of ESC-derived cardiomyocytes with heat shock and a cocktail of survival factors, are being explored.
3. Genetic modification of the cells prior to administration: overexpression of antiapoptotic genes or genetic manipulation to maintain cell´s functionality (i.e., capacity to secrete paracrine mediators, to connect with host myocardium or to differentiate into specialized cardiac cell types) can be achieved through genetic cell engineering.
4. Combined injection of cells and biomaterials: BMMC encapsulation within scaffolds (epicardial patches) or peptide nanofibers represents another strategy that needs further investigation.

4. Patient selection and delivery methods

Patients with larger AMI or with severely depressed baseline left ventricular ejection fraction (LVEF) and stroke volumes, or those with transmural extent of the infarct seem to benefit the most after BMMC treatment (Wollert & Drexler 2010). Conversely, patients with

microvascular obstruction may not respond to intracoronary infusion of cells. Therefore, patient selection before conducting a RCT must take into account the pathophysiologic basis of the disease and baseline characteristics of the patients. For instance, it is well known that age, cardiovascular risk factors and previous heart failure have a negative impact on the potentiality and functional capacity of most types of stem cells in cardiovascular research.
On the other hand, exploration of new delivery methods is mandatory, due to the low rate of cell retention, engraftment and survival in the myocardium with the present routes of administration. New devices include transcoronary arterial injection into the perivascular space, improvements in transendocardial injection needle design and the fusion of different imaging techniques for a more precise delivery (i.e., X-ray/MRI suites used in conjunction with electroanatomic maps of the LV).

5. Mechanisms of action

Nowadays, it is believed that stem cell therapy could lead to successful cardiac regeneration or repair by any or a combination of three main general mechanisms (figure 1): 1) differentiation of the administered cells into all of the cellular constituents of the heart (i.e., cardiomyogenesis and vasculogenesis processes), or, less probably, fusion of the administered cells with those, 2) release of factors capable of paracrine signaling from the administered cells and 3) stimulation of endogenous repair by injected cells, through stem cell cardiac niches activation (Mazhari & Hare, 2007).

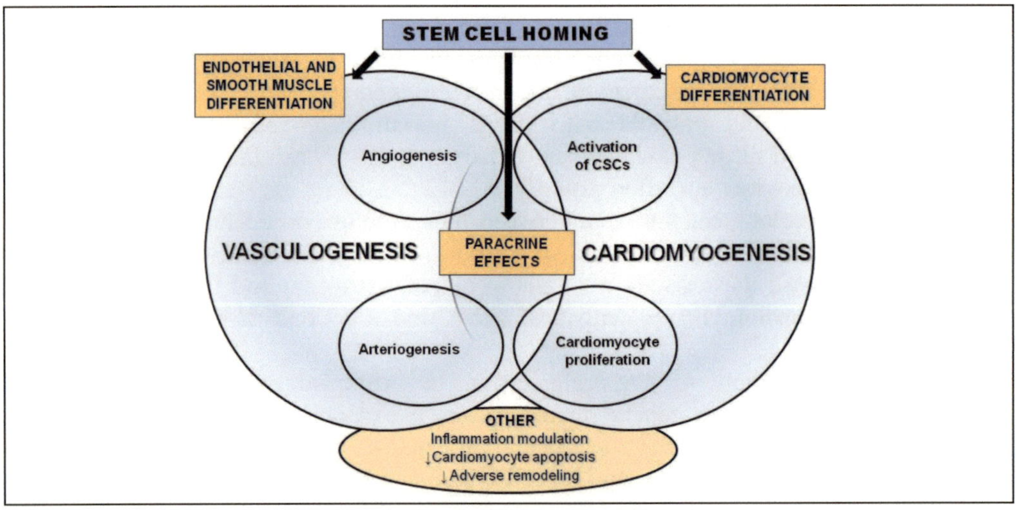

Fig. 1. Proposed mechanisms of stem cell function after homing into the damaged heart. Note that differentiation processes and paracrine effects activate a cascade of events that interact actively to create new blood vessels and cardiomyocytes, with the final objective of functional cardiac repair. CSCs: cardiac stem cells.

5.1 Cardiomyogenesis and vasculogenesis.

While in the classic studies of the beginning of the decade (trans)differentiation of BMMC into cardiomyocytes, smooth muscle cells and endothelial cells was postulated as the main mechanism that might explain the cardiac recovery resulting from stem cell therapy, this phenomenon has been demonstrated in low proportions in more recent studies. Regarding cellular fusion of administered cells with host myocardial ones, to date there is little evidence to support this hypothesis.

5.2 Paracrine actions

Given that differentiation debate is still ongoing and that the number of newly generated cardiomyocytes and blood vessels is too low to explain significant functional improvements, the paracrine hypothesis is now considered the most plausible. According to this idea, the functional benefits of stem cells might be related to secretion of soluble factors that, acting in a paracrine fashion, protect the heart, attenuate pathological LV remodeling, induce neovascularization and promote regeneration (Gnecchi et al., 2008). BMMC and MSC have been extensively studied and proved to produce and secrete a broad variety of cytokines, chemokines and growth factors, between them VEGF, FGF, HGF, IGF, adrenomedullin, thymosin β4 (TB4), SDF-1, PDGF and angiopoietin. These paracrine mediators are expressed/released in a temporal and spatial manner exerting different effects depending on the microenvironment after injury. In addition, these released factors may have autocrine actions on the biology of stem cells themselves (Gnecchi et al., 2008).

The paracrine factors may influence adjacent cells and exert their actions via several mechanisms, including:

1. Myocardial protection: MSC and BMMC in an ischemic environment release cytoprotective molecules that increase cardiomyocyte survival (VEGF, FGF, HGF, IGF-1, TB4, SDF-1, PDGF and IL-1).
2. Neovascularization: BMMC, MSC and EPC can give rise to vascular structures. The molecular processes leading to angiogenesis and arteriogenesis involve mediators such as nitric oxide, VEGF, SDF-1, FGF, HGF and angiopoietin.
3. Cardiac remodeling: paracrine factors released by transplanted stem cells may alter the ECM (i.e., inhibiting cardiac fibroblast proliferation and types I and III collagen synthesis), resulting in more favorable post-AMI remodeling and strengthening of the infarct scar. Stem cells (MSC) may also produce molecules that limit local inflammation, thus reducing the remodeling process.
4. Cardiac contractility and metabolism: it has been demonstrated that stem cell therapy limits infarct size and prevents LV dysfunction. On the other hand, MSC and BMMC secrete inotropic factors (i.e., IGF-1) that positively modulate cell contractility, and these cells attenuate the profound bioenergetic abnormalities found in the border zone of myocardial infarction.
5. Cardiac regeneration: differentiation and cell fusion with native cardiomyocytes occur in very low rates after stem cell administration. Therefore, it is now believed that exogenous stem cell transplantation may activate resident CSC and/or stimulate cardiomyocyte replication via paracrine action. Factors secreted by BMMC, MSC and EPC, including HGF, SDF-1, VEGF and IGF-1, enhance proliferation, mobilization, differentiation, survival and function of CSC or even restoration of cardiac stem cell niches.

5.3 Endogenous repair

As we have seen, clonogenic and self-replicating endogenous CSC have been isolated and cultured from human hearts. These CSC - located in cardiac stem cell niches - have the capacity to differentiate into endothelial cells, smooth muscle myocytes and cardiomyocytes. Though insufficient for a complete repair of the myocardium after any kind of insult, these cells can be activated by extracardiac delivered cells. Thus, administered allogeneic MSC participate in maintaining stem cell niches, and through cell-to-cell interactions - apart from paracrine effects - may not only restore lost cellular constituents (differentiation) but also these niches with an ongoing and regulated self-replicating capacity (Mazhari & Hare, 2007).

6. Clinical scenarios in stem cell therapy. Evidence available

Stem cell therapy has accumulated growing evidence in different pathophysiological conditions in small and large animal models, but human research has been almost limited to CAD. In this chapter we will focus on the largest and most relevant randomized placebo-controlled clinical trials in humans (tables 1 through 4).

Natural history of CAD can be divided into acute (AMI) and chronic phases (chronic ischemic heart disease). In the latter stem cell therapy has been investigated in the subset of 1) ischemic heart failure (ventricular dysfunction) and 2) chronic myocardial ischemia (refractory angina).

In patients where restoration of contractile function is the clinical goal - such as those with end-stage ischemic heart failure or those early post-infarction - delivering cells with contractile potential may be of high priority. Under these conditions, naturally myogenic cells (i.e., skeletal myoblasts, cardiomyocytes or any progenitor cell driven down a muscle lineage) appear to be the first choice. However, on the one hand, formation of new myocardial mass has only been strictly established for ESC, and is a process that has been achieved in very few trials and in small percentages with adult stem cells. And on the other hand, most of the studies after AMI have used BMMC as an easily accessible source of adult stem and progenitor cells.

In conditions where chronic ischemia prevails, the angiogenic potential of the cells seems a more reasonable approach. In this case, BMMC, EPC, vascular progenitor cells or blood-derived multipotent adult progenitor cells and MSC may be better choices than myogenic precursors.

6.1 Stem cell therapy after acute myocardial infarction

Several trials have evaluated stem cell therapy after AMI, some with positive results and some with neutral ones. All of them used the intracoronary route, once the patency of the infarct-related artery was restored, and most of them with the mononuclear fraction of the bone marrow (table 1). Four main RCT have been published with positive findings so far. In the BOOST trial (Wollert et al., 2004), BMMC were proved to improve left ventricular contractility in the infarct border zone and global LVEF by 6%. However, only patients with larger infarcts showed maintained benefits in terms of LVEF at long follow-up (18 months). In the REPAIR-AMI trial (Schachinger et al., 2006), infusion of BMMC promoted an increase in LVEF of 2.8% at 12 months. The FINCELL trial (Huikuri et al., 2008) reported an improvement of 5% in LVEF after BMMC delivery. Finally, in the REGENT trial (Tendera et al., 2009), patients treated with BMMC and with $CXCR4^+/CD34^+$ BMMC showed an increase of 3% in LVEF which was not observed in the control group, but these differences were not significant between treated and control patients at 6-months follow-up. This trial was limited by imbalances in baseline LVEF and by incomplete follow-up.

Trial (year)	n	Cell type	Cell count	Days after AMI	Primary endpoint (follow-up)	Comments
Chen (2004)	69	MSC	9 x 10^9	18	Improved LVEF at 6m	LVEF by echocardiography
BOOST (2004)	60	BMMC	2 x 10^9	6±1	Improved LVEF at 6m	Effect diminished after 18/61m
REPAIR-AMI (2006)	187	BMMC	2 x 10^8	3-6	Improved LVEF at 4m	LVEF by ventriculography
Janssens (2006)	66	BMMC	2 x 10^8	1	No change LVEF at 4m	Improved regional contractility and reduction in infarct size
ASTAMI (2006)	97	BMMC	7 x 10^7	6±1	No change LVEF at 6m	LVEF ↑8% by SPECT,↑1% by MRI
TCT-STAMI (2006)	20	BMMC	4 x 10^7	1	Improved LVEF at 6m	LVEF by echocardiography
FINCELL (2008)	77	BMMC	4 x 10^8	3	Improved LVEF at 6m	LVEF by ventriculography
Meluzin (2006)	66	BMMC	1 x 10^7 (low d) 1 x 10^8 (high d)	7	Improved LVEF at 3m in high dose group	LVEF by SPECT
Penicka (2007)	27	BMMC	3 x 10^9	9	No change LVEF at 4m	LVEF by echocardiography
HEBE (2008)	189	BMMC vs PBC	-	3-8	No changes in global or regional LV function	Final results pending
REGENT (2009)	117	BMMC (unselected, $CD34^+$/$CXCR4^+$)	2 x 10^8 (unsel), 2 x 10^6 ($CD34^+$)	3-12	Improved LVEF with both cell types	LVEF by MRI (in 117/200 patients)

MSC: mesenchymal stem cells (bone marrow origin); BMMC: bone marrow mononuclear cells; PBC: peripheral blood cells; LVEF: left ventricular ejection fraction; LV: left ventricle; SPECT: single-photon emission computed tomography; MRI: magnetic resonance imaging.

Table 1. Randomized clinical trials with stem cells in patients with acute myocardial infarction (intracoronary delivery).

On the other hand, three RCT resulted in neutral findings. Janssens et al. (Janssens et al., 2006) reported no changes in LVEF after BMMC infusion, but a reduction in the infarct volume and an improvement in regional contractility in the greatest transmural infarct cases were observed in treated patients. In the ASTAMI trial (Lunde et al., 2006) no significant effects on LVEF, LV volumes, or infarct size were observed after BMMC administration. The smaller number of cells and differences in the cell isolation protocol were invocated to explain these findings. Finally, in the HEBE trial (van der Laan et al., 2008), presented at the AHA Scientific Sessions in 2008, no changes in global or regional LV systolic function were reported after BMMC and mononuclear cells isolated from peripheral blood injection.

So far, no safety concerns after BMMC intracoronary infusion have emerged. The risk of a higher rate of instent restenosis was not confirmed in the FINCELL trial (Huikuri et al., 2008) and in two recent meta-analyses (Lipinski et al., 2007; Martin-Redon et al., 2008). Moreover, none of the trials reported an increased incidence of malignant arrhythmias with BMMC (Wollert & Drexler, 2010).

Two trials have used MSC after AMI. The study by Chen et al. (Chen et al., 2004) demonstrated an improvement in LVEF and perfusion with intracoronary infusion of these cells, but these results have not been duplicated. Hare et al. (Hare et al., 2009) intravenously administered allogeneic MSC after an AMI with no higher rate of MACE and some benefits in terms of LVEF.

New types of cells are also being explored, like ADSC (figure 2). No evidence is available to date, but the first-in-man RCT with intracoronary administration of freshly isolated ADSC after AMI (the APOLLO trial) has been recently completed.

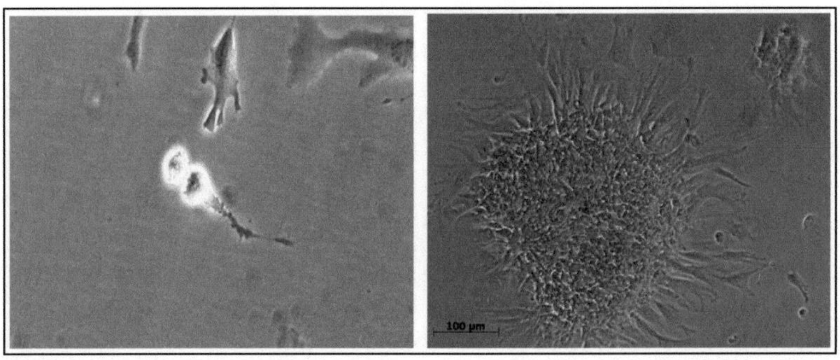

Fig. 2. Autologous adipose-derived mesenchymal stem cells during mitosis (left pannel) and growing in colonies in the 6th day of culture (magnification x 10, right pannel). These cells were expanded from the adipose tissue stroma-vascular fraction under good manufacturing practice (GMP) conditions in our Cell Production Unit (Hospital Gregorio Marañón, Madrid).

Another approach for stem cell therapy after AMI is cell mobilization from the bone marrow with the administration of G-CSF. Several RCT have been published, but results have been somehow less encouraging (table 2). Only three trials have reported positive results. In the FIRSTLINE-AMI trial (Ince et al., 2005), the RIGENERA study (Leone et al., 2007) and in the study by Takano et al. (Takano et al., 2007), significant improvements in LVEF were observed. The rest of the trials showed negative findings.

Finally, the MAGIC trials used a combination of G-CSF and intracoronary injection of peripheral blood progenitor cells. In the first trial no differences in LVEF were noted, and an increase in instent restenosis rate was observed (G-CSF administration before bare-metal stent implantation) (Kang et al., 2004). Then the investigators changed the design and used drug-eluting stents. In the MAGIC 3-DES trial, positive results in terms of LVEF were found after mobilization and intracoronary injection of isolated cells (Kang et al., 2006).

Trial (year)	n	Dose	Timing after AMI	Follow-up	Comments
Valgimigli (2005)	20	5 µg/kg x 4 d	1 d	No change LVEF at 6m	LVEF by SPECT
FIRSTLINE -AMI (2005)	50	10 µg/kg x 6 d	90 min	Improved LVEF at 4m	LVEF by echocardiography
REVIVAL-2 (2006)	114	10 µg/kg x 5 d	5 d	No change LVEF at 5m	LVEF by MRI
STEMMI (2006)	78	10 µg/kg x 6 d	28 h	No change LVEF at 6m	LVEF by echocardiography and MRI
G-CSF-STEMI (2006)	44	10 µg/kg x 5 d	35 h	No change LVEF at 3m	LVEF by MRI
Ellis (2006)	18	5µg/kgx5d (low d), 10µg/kgx5d (high d)	< 30 h	Improved LVEF at 30d	LVEF by echocardiography
RIGENERA (2007)	41	10 µg/kg x 5 d	5 d	Improved LVEF at 6m	LVEF by echocardiography
Takano (2007)	40	2,5 µg/kg x 5 d	1 d	Improved LVEF at 6m	LVEF by SPECT
MAGIC (2004)*	27	10 µg/kg x 4 d; PBC: 1 x 10^9	1 d	No change LVEF at 6m	LVEF by SPECT
MAGIC 3-DES (2006)*	50	10 µg/kg x 3 d; PBC: 2 x 10^9	1 d	Improved LVEF at 6m	LVEF by MRI

Table 2. Randomized clinical trials with granulocyte colony-stimulating factor in patients with acute myocardial infarction (subcutaneous). *: MAGIC trials used a combination of indirect mobilization (G-CSF) and direct intracoronary injection of peripheral blood cells (PBC); LVEF: left ventricular ejection fraction; SPECT: single-photon emission computed tomography; MRI: magnetic resonance imaging.

6.2 Stem cell therapy for chronic ischemic heart disease

6.2.1 Ischemic heart failure

Skeletal myoblasts and BMMC have been used in heart failure patients (table 3). The MAGIC trial (Menasche et al., 2008), with transepicardial injection of skeletal myoblasts during coronary artery bypass graft surgery, reported no changes in global or regional contractility. However, a reduction in LV end-diastolic and end-systolic volumes was observed in the high-dose group. Moreover, a trend towards a higher incidence of ventricular arrhythmias was noted. Dib et al.(Dib et al., 2005) reported an improvement in LVEF and viability after myoblasts transendocardial injection, in contradiction with the SEISMIC trial (presented by Serruys at the 2008 ACC meeting) which showed no benefit of the same procedure at 6 months.

In the TOPCARE-CHD trial (Assmus et al., 2006), BMMC intracoronary delivery into the coronary artery supplying the most dyskinetic LV area showed an increase in LVEF of 2.9%, whereas progenitor circulating cells infusion and controls did not show any positive change. No major adverse cardiac events (MACE) were reported in this trial. Another trial has been recently published with BMMC by the group of Strauer (the STAR-heart study) (Strauer et al., 2010). Although it lacks of a randomized design, this study represents the largest trial with BMMC in patients with severe LV dysfunction. Intracoronary transplantation improved haemodynamics at rest (including LVEF), exercise capacity, LV contractility and geometry (decrease in LV end-diastolic and end-systolic volumes, decrease in infarct size), and a remarkable reduction in mortality at 5-years follow-up was observed.

Trial	n	Cell type	Delivery	Timing	Primary endpoint	Comments
MAGIC	97	SM	epi	>4 weeks	No change LVEF	Reduction in LVEDV/LVESV
Dib	23	SM	endo	>10 years	Improved LVEF and viability	-
SEISMIC	47	SM	endo	chronic	No change LVEF	
TOPCARE-CHD	58	BMMC vs CPC	ic	81±72 months	Improved LVEF w/BMMC	-
STAR	391	BMMC vs controls	ic	8.5±3.2 years	Improved LVEF, LVEDV/LVESV, exercise capacity	No randomized design. Reduction of infarct size and mortality

Table 3. Randomized clinical trials in patients with chronic ischemic heart failure. SM: skeletal myoblasts; BMMC: bone marrow mononuclear cells; CPC: circulating progenitor cells; epi: transepicardial; endo: transendocardial; ic: intracoronary; LVEF: left ventricular ejection fraction; LVEDV: left ventricular end-diastolic volume; LVESV: left ventricular end-systolic volume.

6.2.2 Chronic myocardial ischemia

Patients with advanced CAD and no further options of revascularization ("no-option" patients) have also been studied in stem cell therapy trials (table 4). Three RCT have been completed using the transendocardial route after electromechanical mapping of the LV, with BMMC or blood-derived progenitor cells. Losordo et al. (Losordo et al., 2007) studied peripheral $CD34^+$ cells isolated after G-CSF injections. Angina frequency and exercise time were improved, but no clear effects on myocardial perfusion were observed. In the PROTECT-CAD trial (Tse et al., 2007), BMMC injections improved NYHA functional class, exercise time, LVEF, wall thickening and stress-induced perfusion defects. Finally, Van Ramshorst et al. (van Ramshorst et al., 2009) reported better LVEF, myocardial perfusion, angina functional class, exercise capacity and quality of life after BMMC administration.

ADSC have also been studied in this type of patients. The PRECISE trial is a prospective, double blind, RCT that has randomised 27 patients with end-stage CAD not amenable for revascularization and with moderate-severe LV dysfunction to receive freshly isolated ADSC or placebo in a 3:1 ratio. The cells were delivered via transendocardial injections after

LV electromechanical mapping with the NOGA XP™ delivery system (BDS, Cordis Corporation, Johnson and Johnson) (figure 3), and results are still waiting for publication.

Trial	n	Cell type	Delivery	Timing	Primary endpoint	Comments
Losordo	24	$CD34^+$	endo	chronic	Improved angina parameters	No clear perfusion benefit
PROTECT-CAD	28	BMMC	endo	chronic	Improved angina parameters	Improved LVEF and perfusion
Van Ramshorst	50	BMMC	endo	chronic	Improved angina parameters	Improved LVEF and perfusion

Table 4. Randomized clinical trials in patients with chronic myocardial ischemia. BMMC: bone marrow mononuclear cells; endo: transendocardial; LVEF: left ventricular ejection fraction.

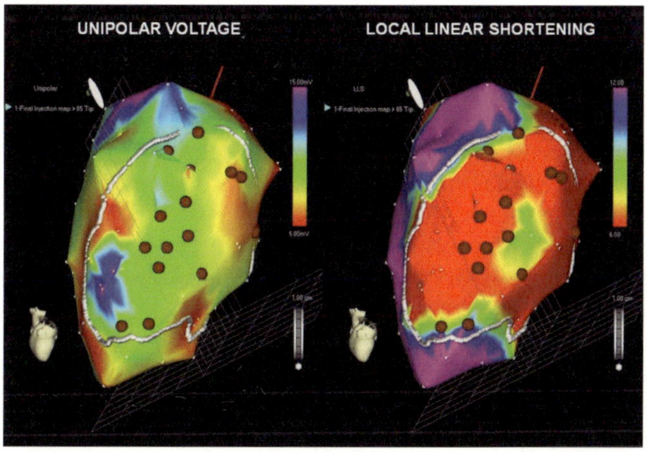

Fig. 3. Electromechanical mapping of the left ventricle with the NOGA XP™ System (BDS, Cordis Corporation, Johnson and Johnson) from a patient enrolled in the PRECISE trial in our centre. Myocardial areas with low contractility (red means most akinetic areas) and impaired endocardial voltage (red means necrotic myocardium, green hibernating tissue) are identified as viable and targeted for cell injection (brown dots).

7. Safety concerns of stem cell therapy

As a novel therapy, stem cell therapy has raised several safety concerns, which are listed below:

1. Arrhythmias: malignant arrhythmias can be induced 1) because transplanted cells do not engraft in a physiologic electromechanical way, forming functionally isolated clusters of cells, 2) because of intrinsic arrhythmogenicity of cells, 3) because reinervation of the myocardium, 4) due to local damage or edema after cell injection or 5) because of the use of heterologous proteins in cell cultures. Ventricular arrhythmias

have been reported specifically with skeletal myoblasts (Menasche et al., 2003; Siminiak et al., 2004; Dib N 2005) and with CD133+ cells (Bartunek et al., 2005), but without compromising the overall safety of this therapy.
2. Restenosis: the only study that reported a higher incidence of restenosis used G-CSF four days before stent implantation (Kang et al., 2004), which was considered to be a non-adequate methodology. This risk has not been observed in further trials.
3. Extracardiac engraftment: administered stem cells are located in the heart in low proportions, whereas kidneys, lungs and spleen receive the most of the cells. Local intramyocardial injections achieve higher retention rates in the heart. Nevertheless, clinical consequences of extracardiac engraftment are so far unknown or even benefitial.
4. Accelerated atherogenesis: some studies have shown a high proportion of de novo lesions in non-treated coronary arteries (Bartunek et al., 2005), but this issue needs further evaluation.
5. Microvascular obstruction: MSC have been proved to cause microinfarctions in canine models, but these findings have not been reproduced in human trials (Chen et al., 2004).
6. Oncogenicity: plutipotent and ESC have the risk of inducing cardiac tumors, but this risk is reduced if cells are in culture for no more than 6 or 8 weeks.

8. The consensus of the task force of the European society of cardiology for future trials

The Task Force of the European Society of Cardiology on stem cells and repair of the heart was created in 2006 to investigate and regulate the role of progenitor/stem cell therapy in the treatment of cardiovascular disease. It was almost four years ago that this group of experts and opinion leaders stated the type of studies needed (Bartunek et al., 2006):

1. Further large, double-blind, controlled RCT for the use of autologous BMMC in the treatment of AMI. The patient population should be all those presenting within 12 h of AMI and treated with immediate revascularization, be it primary angioplasty or fibrinolysis.
2. Double-blind, controlled RCT for the use of autologous BMMC in the treatment of AMI in those patients presenting late (>12 h) or who fail to respond to therapy (candidates for 'rescue' angioplasty). Although these groups may represent a small proportion of all patients with AMI, their prognosis remains poor.
3. Double-blind, controlled RCT for the use of autologous BMMC or skeletal myoblasts in the treatment of ischemic heart failure. At some stage, the role of autologous stem/progenitor cells in the treatment of cardiomyopathies (in particular, dilated cardiomyopathy) will need to be examined.
4. A series of well-designed small studies to address safety or mechanism to test specific hypotheses (i.e., studies with labelled cells or to investigate paracrine or autocrine mechanisms). Such hypotheses would have arisen from basic science experiments.
5. Studies to confirm the risk/benefit ratio of the use of cytokines alone (i.e., G-CSF) or in conjunction with stem/progenitor cell therapy.

This Task Force also underlined the necessity for studies with hard clinical endpoints, MACE, subjective benefit and economic gain (Bartunek et al., 2006). Another key point is standardization, both in outcome measures and in the processing of cells (better achieved in specialized centers following GMP routines), in order to derive meaningful comparisons. Since these trials will need to include thousands of patients, they should be multicentre and ideally pan-European. On the other hand, the Task Force stated that small uncontrolled trials with BMMC should be avoided, as they are unlikely to add anything new to the field.

9. Next directions in stem cell therapy research.

Finally, next directions of cardiac cell therapy include:

1. The study of the array of bioactive molecules that are secreted by stem cells, which have been demonstrated to induce neovascularization, modulate inflammation, fibrogenesis, cardiac metabolism and contractility, increase cardiomyocyte proliferation and activate resident stem cells. The exhaustive analysis of this "secretomes" of BMMC, MSC or EPC would lead to a better understating of the mechanisms of action of the cells and to a hypothetical protein-based therapy (off-the-shelf, noninvasive, systemic and repetitive administration).
2. The use of different sources of pluripotent stem cells, like ESC, spermatogonial stem cells and oocytes. A new era has been initiated with the possibility of reprogramming adult cells (skin fibroblasts) to a pluripotent state by retroviral transduction (iPS) (Takahashi et al., 2007; Yu et al., 2007). New retroviral vectors and even nonviral vectors have been developed to reduce the risk of mutagenesis, and genetic modification of cells with suicide genes have been proposed to reduce the risk of tumor formation.
3. The creation of bioartificial hearts after a process of decelularization with detergents, obtention of the underlying extracellular matrix (cardiac architecture) and stem cell repopulation (Ott et al., 2008). The "acellular" heart can then be reseeded with cardiac stem cells or EPC, showing contractile activity in animal models. This new approach of tissue bioengineering has opened a fascinating era in cardiovascular medicine.

10. Conclusion

Although mixed results have emerged from the first stem cell therapy RCT in cardiovascular medicine, the overall data suggest that these procedures are feasible and safe in both acute and chronic scenarios of ischemic heart disease. After phase I-II RCT, it is clear that BMMC transfer after AMI has the potential to improve the recovery of LV systolic function beyond what can be achieved by current interventional and medical therapies. In chronic ischemic heart disease, skeletal myoblasts and BMMC have proved to improve myocardial perfusion and contractile performance.

New types of cells (including ADSC and iPS), improvements in delivery and imaging methods, strategies to enhance cell potentiality or to improve the myocardial pro-inflammatory microenvironment, and the creation of bioartificial hearts are the main new directions of research in the near future. Finally, large-scale, phase III, double-blind, controlled RCT performed under rigorous safety standards are being initiated to prove unequivocal clinical benefits, including improved survival. These trials will definitively establish the effectiveness of stem cell therapy in improving clinical outcomes, confirming the real potential of cardiac regenerative therapy.

11. Acknowledgment

We thank the Spanish Cooperative Cardiovascular Disease Research Network (RECAVA, Red Temática de Investigación Cooperativa en Enfermedades Cardiovasculares) and the Spanish Cell Therapy Research Network (TERCEL, Red Temática de Terapia Celular from the Instituto de Salud Carlos III, ISCIII) for their sponsorship. We appreciate the efforts of the members of the Task Force of the European Society. We would also like to thank the research nurses (María José Lorenzo, Alexandra Vázquez and Lucía Fernández) and Ana Fernández-Baza for her work, enthusiasm and invaluable help.

12. References

Assmus, B., Honold, J., Schächinger, V., Britten, M.B., Fischer-Rasokat, U., Lehmann, R., Teupe, C., Pistorius, K., Martin, H., Abolmaali, N.D., Tonn, T., Dimmeler, S., & Zeiher, A.M. (2006). Transcoronary transplantation of progenitor cells after myocardial infarction. *The New England Journal of Medicine,* Vol.355, No.12, (September 2006), pp. 1222-1232, ISSN 1533-4406

Balsam, L.B., Wagers, A.J., Christensen, J.L., Kofidis, T., Weissman, I.L., & Robbins, R.C. (2004). Haematopoietic stem cells adopt mature haematopoietic fates in ischaemic myocardium. *Nature,* Vol.428, No.6983, (April 2004), pp. 668-673, ISSN 1476-4687

Bartunek, J., Vanderheyden, M., Vandekerckhove, B., Mansour, S., De Bruyne, B., De Bondt, P., Van Haute, I., Lootens, N., Heyndrickx, G., & Wijns, W. (2005). Intracoronary injection of CD133-positive enriched bone marrow progenitor cells promotes cardiac recovery after recent myocardial infarction: feasibility and safety. *Circulation,* Vol.112, No.9 suppl, (August 2005), pp. I178-183, ISSN 1524-4539

Bartunek, J., Dimmeler, S., Drexler, H., Fernández-Avilés, F., Galinanes, M., Janssens, S., Martin, J., Mathur, A., Menasche, P., Priori, S., Strauer, B., Tendera, M., Wijns, W., & Zeiher, A.; task force of the European Society of Cardiology. (2006). The consensus of the task force of the European Society of Cardiology concerning the clinical investigation of the use of autologous adult stem cells for repair of the heart. *European Heart Journal,* Vol.27, No.11, (June 2006), pp. 1338-1340, ISSN 0195-668X

Beltrami, A.P., Barlucchi, L., Torella, D., Baker, M., Limana, F., Chimenti, S., Kasahara, H., Rota, M., Musso, E., Urbanek, K., Leri, A., Kajstura, J., Nadal-Ginard, B., & Anversa, P. (2003). Adult cardiac stem cells are multipotent and support myocardial regeneration. *Cell,* Vol.114, No.6, (September 2003), pp. 763-776, ISSN 0092-867

Chen, S.L., Fang, W.W., Ye, F., Liu, Y.H., Qian, J., Shan, S.J., Zhang, J.J., Chunhua, R.Z., Liao, L.M., Lin, S., & Sun, J.P. (2004). Effect on left ventricular function of intracoronary transplantation of autologous bone marrow mesenchymal stem cell in patients with acute myocardial infarction. *The American journal of cardiology,* Vol.94, No.1, (July 2004), pp. 92-95, ISSN 0002-9149

Cho, H.J., Lee, N., Lee, J.Y., Choi, Y.J, Ii, M., Wecker, A., Jeong, J.O., Curry, C., Qin, G., & Yoon, Y.S. (2007). Role of host tissues for sustained humoral effects after endothelial progenitor cell transplantation into the ischemic heart. *The Journal of Experimental Medicine,* Vol.204, No.13, (December 2007), pp. 3257-3269, ISSN 1540-9538

Dewald, O., Ren, G., Duerr, G.D., Zoerlein, M., Klemm, C., Gersch, C., Tincey, S., Michael, L.H., Entman, M.L., & Frangogiannis, N.G. (2004). Of mice and dogs: species-specific differences in the inflammatory response following myocardial infarction. *The American Journal of Pathology,* Vol.164, No.2, (February 2004), pp. 665-677, ISSN 0002-9440

Dib, N., Michler, R.E., Pagani, F.D., Wright, S., Kereiakes, D.J., Lengerich, R., Binkley, P., Buchele, D., Anand, I., Swingen, C., Di Carli, M.F., Thomas, J.D., Jaber, W.A., Opie, S.R., Campbell, A., McCarthy, P., Yeager, M., Dilsizian, V., Griffith, B.P., Korn, R.,

Kreuger, S.K., Ghazoul, M., MacLellan, W.R., Fonarow, G., Eisen, H.J., Dinsmore, J., Diethrich, E. (2005). Safety and feasibility of autologous myoblast transplantation in patients with ischemic cardiomyopathy: four-year follow-up. *Circulation,* Vol.112, No.12, (September 2005), pp. 1748-1755, ISSN 1524-4539

Fraser, J.K., Schreiber, R.E., Zuk, P.A., & Hedrick, M.H. (2004) Adult stem cell therapy for the heart. *International journal of biochemistry & cell biology,* Vol.36, No.4, (April 2004), pp. 658-666, ISSN 1357-2725

Gnecchi, M., Zhang, Z., Ni, A., & Dzau, V.J. (2008). Paracrine mechanisms in adult stem cell signaling and therapy. *Circulation research,* Vol.103, No.11, (November 2008), pp. 1204-1219, ISSN 1524-4571

Hare, J.M., Traverse, J.H., Henry, T.D., Dib, N., Strumpf, R.K., Schulman, S.P., Gerstenblith, G., DeMaria, A.N., Denktas, A.E., Gammon, R.S., Hermiller, J.B. Jr., Reisman, M.A., Schaer, G.L., & Sherman, W. (2009). A randomized, double-blind, placebo-controlled, dose-escalation study of intravenous adult human mesenchymal stem cells (prochymal) after acute myocardial infarction. *Journal of the American College of Cardiology,* Vol.54, No.24, (December 2009), pp. 2277-2286, ISSN 1558-3597

Herreros, J., Prosper, F., Perez, A., Gavira, J.J., Garcia-Velloso, M.J., Barba, J., Sanchez, P.L., Canizo, C., Rabago, G., Marti-Climent, J.M., Hernandez, M., Lopez-Holgado, N., Gonzalez-Santos, J.M., Martin-Luengo, C., & Alegria, E. (2003). Autologous intramyocardial injection of cultured skeletal muscle-derived stem cells in patients with non-acute myocardial infarction. *European Heart Journal,* Vol.24, No.22, (November 2003), pp. 2012-2020, ISSN 0195-668X

Huikuri, H.V., Kervinen, K., Niemelä, M., Ylitalo, K., Säily, M., Koistinen, P., Savolainen, E.R., Ukkonen, H., Pietilä, M., Airaksinen, J.K., Knuuti, J., & Mäkikallio, T.H.; FINCELL Investigators. (2008). Effects of intracoronary injection of mononuclear bone marrow cells on left ventricular function, arrhythmia risk profile, and restenosis after thrombolytic therapy of acute myocardial infarction. *European Heart Journal,* Vol.29, No.22, (November 2008), pp. 2723-2732, ISSN 1522-9645

Ince, H., Petzsch, M., Kleine, H.D., Schmidt, H., Rehders, T., Körber, T., Schümichen, C., Freund, M., & Nienaber, C.A. (2005). Preservation from left ventricular remodeling by front-integrated revascularization and stem cell liberation in evolving acute myocardial infarction by use of granulocyte-colony-stimulating factor (FIRSTLINE-AMI). *Circulation,* Vol.112, No.20, (November 2005), pp. 3097-3106, ISSN 1524-4539

Jackson, K.A., Majka, S.M., Wang, H., Pocius, J., Hartley, C.J., Majesky, M.W., Entman, M.L., Michael, L.H., Hirschi, K.K., & Goodell, M.A. (2001). Regeneration of ischemic cardiac muscle and vascular endothelium by adult stem cells. *Journal of Clinical Investigation,* Vol.107, No.11, (June 2001), pp. 1395-1402, ISSN 0021-9738

Janssens, S., Dubois, C., Bogaert, J., Theunissen, K., Deroose, C., Desmet, W., Kalantzi, M., Herbots, L., Sinnaeve, P., Dens, J., Maertens, J., Rademakers, F., Dymarkowski, S., Gheysens, O., Van Cleemput, J., Bormans, G., Nuyts, J., Belmans, A., Mortelmans, L., Boogaerts, M., & Van de Werf, F. (2006). Autologous bone marrow-derived stem-cell transfer in patients with ST-segment elevation myocardial infarction: double-blind, randomised controlled trial. *Lancet,* Vol.367, No.9505, (January 2006), pp. 113-121, ISSN 1474-547X

Kamihata, H., Matsubara, H., Nishiue, T., Fujiyama, S., Tsutsumi, Y., Ozono, R., Masaki, H., Mori, Y., Iba, O., Tateishi, E., Kosaki, A., Shintani, S., Murohara, T., Imaizumi, T., & Iwasaka, T. (2001). Implantation of bone marrow mononuclear cells into ischemic myocardium enhances collateral perfusion and regional function via side supply of angioblasts, angiogenic ligands, and cytokines. *Circulation*, Vol.104, No.9, (August 2001), pp. 1046-1052, ISSN 1524-4539

Kang, H.J., Kim, H.S., Zhang, S.Y., Park, K.W., Cho, H.J., Koo, B.K., Kim, Y.J., Soo Lee, D., Sohn, D.W., Han, K.S., Oh, B.H., Lee, M.M., & Park, Y.B. (2004). Effects of intracoronary infusion of peripheral blood stem-cells mobilised with granulocyte-colony stimulating factor on left ventricular systolic function and restenosis after coronary stenting in myocardial infarction: the MAGIC cell randomised clinical trial. *Lancet*, Vol.363, No.9411, (March 2004), pp. 751-756, ISSN 1474-547X

Kang, H.J., Lee, H.Y., Na, S.H., Chang, S.A., Park, K.W., Kim, H.K., Kim, S.Y., Chang, H.J., Lee, W., Kang, W.J., Koo, B.K., Kim, Y.J., Lee, D.S., Sohn, D.W., Han, K.S., Oh, B.H., Park, Y.B., & Kim, H.S. (2006). Differential effect of intracoronary infusion of mobilized peripheral blood stem cells by granulocyte colony-stimulating factor on left ventricular function and remodeling in patients with acute myocardial infarction versus old myocardial infarction: the MAGIC Cell-3-DES randomized, controlled trial. *Circulation*, Vol.114, No.1 Suppl, (July 2006), pp. I145-I151, ISSN 1524-4539

Kehat, I., Kenyagin-Karsenti, D., Snir, M., Segev, H., Amit, M., Gepstein, A., Livne, E., Binah, O., Itskovitz-Eldor, J., & Gepstein, L. (2001). Human embryonic stem cells can differentiate into myocytes with structural and functional properties of cardiomyocytes. *Journal of Clinical Investigation*, Vol.108, No.3, (August 2001), pp. 407-414, ISSN 0021-9738

Klug, M.G., Soonpaa, M.H., Koh, G.Y., & Field, L.J. (1996). Genetically selected cardiomyocytes from differentiating embronic stem cells form stable intracardiac grafts. *Journal of Clinical Investigation*, Vol.98, No.1, (July 1996), pp. 216-224, ISSN 0021-9738

Kocher, A.A., Schuster, M.D., Szabolcs, M.J., Takuma, S., Burkhoff, D., Wang, J., Homma, S., Edwards, N.M., & Itescu, S. (2001). Neovascularization of ischemic myocardium by human bone-marrow-derived angioblasts prevents cardiomyocyte apoptosis, reduces remodeling and improves cardiac function. *Nature Medicine*, Vol.7, No.4, (April 2001), pp. 430-436, ISSN 1078-8956

Laugwitz, K.L., Moretti, A., Lam, J., Gruber, P., Chen, Y., Woodard, S., Lin, L.Z., Cai, C.L., Lu, M.M., Reth, M., Platoshyn, O., Yuan, J.X., Evans, S., & Chien, K.R. (2005). Postnatal isl1+ cardioblasts enter fully differentiated cardiomyocyte lineages. *Nature*, Vol.433, No.7026, (February 2005), pp. 647-653, ISSN 1476-4687

Leobon, B., Garcin, I., Menasche, P., Vilquin, J.T., Audinat, E., & Charpak, S. (2003). Myoblasts transplanted into rat infarcted myocardium are functionally isolated from their host. *Proceedings of the National Academy of Sciences USA*, Vol.100, No.13, (June 2003), pp. 7808-7811, ISSN 0027-8424

Leone, A.M., Galiuto, L., Garramone, B., Rutella, S., Giannico, M.B., Brugaletta, S., Perfetti, M., Liuzzo, G., Porto, I., Burzotta, F., Niccoli, G., Biasucci, L.M., Leone, G., Rebuzzi, A.G., & Crea, F. (2007). Usefulness of granulocyte colony-stimulating factor in patients with a large anterior wall acute myocardial infarction to prevent left ventricular remodeling (the rigenera study). *The American journal of cardiology*, Vol.100, No.3, (August 2007), pp. 397-403, ISSN 0002-9149

Lindsey, M.L., Mann, D.L., Entman, M.L., & Spinale, F.G. (2003). Extracellular matrix remodeling following myocardial injury. *Annals of Medicine*, Vol.35, No.5, (2003), pp. 316-326, ISSN 0785-3890

Lipinski, M.J., Biondi-Zoccai, G.G., Abbate, A., Khianey, R., Sheiban, I., Bartunek, J., Vanderheyden, M., Kim, H.S., Kang, H.J., Strauer, B.E., & Vetrovec, G.W. (2007).

Impact of intracoronary cell therapy on left ventricular function in the setting of acute myocardial infarction: a collaborative systematic review and meta-analysis of controlled clinical trials. *Journal of the American College of Cardiology*, Vol.50, No.18, (October 2007), pp.1761-1767, ISSN 1558-3597

Losordo, D.W., Schatz, R.A., White, C.J., Udelson, J.E., Veereshwarayya, V., Durgin, M., Poh, K.K., Weinstein, R., Kearney, M., Chaudhry, M., Burg, A., Eaton, L., Heyd, L., Thorne, T., Shturman, L., Hoffmeister, P., Story, K., Zak, V., Dowling, D., Traverse, J.H., Olson, R.E., Flanagan, J., Sodano, D., Murayama, T., Kawamoto, A., Kusano, K.F., Wollins, J., Welt, F., Shah, P., Soukas, P., Asahara, T., & Henry, T.D. (2007). Intramyocardial transplantation of autologous CD34+ stem cells for intractable angina: a phase I/IIa double-blind, randomized controlled trial. *Circulation*, Vol.115, No.25, (June 2007), pp. 3165-3172, ISSN 1524-4539

Lunde, K., Solheim, S., Aakhus, S., Arnesen, H., Abdelnoor, M., Egeland, T., Endresen, K., Ilebekk, A., Mangschau, A., Fjeld, J.G., Smith, H.J., Taraldsrud, E., Grøgaard, H.K., Bjørnerheim, R., Brekke, M., Müller, C., Hopp, E., Ragnarsson, A., Brinchmann, J.E., & Forfang, K. (2006). Intracoronary injection of mononuclear bone marrow cells in acute myocardial infarction. *The New England journal of medicine*, Vol.355, No.12, (September 2006), pp. 1199-1209, ISSN 1533-4406

Ma, N., Stamm, C., Kaminski, A., Li, W., Kleine, H.D., Muller-Hilke, B., Zhang, L., Ladilov, Y., Egger, D., & Steinhoff, G. (2005). Human cord blood cells induce angiogenesis following myocardial infarction in NOD/scid-mice. *Cardiovascular Research*, Vol.66, No.1, (April 2005), pp. 45-54, ISSN 0008-6363

Martin-Rendon, E., Brunskill, S.J., Hyde, C.J., Stanworth, S.J., Mathur, A., & Watt, S.M. (2008). Autologous bone marrow stem cells to treat acute myocardial infarction: a systematic review. *European heart journal*, Vol.29, No.15, (August 2008), pp. 1807-1818, ISSN 1522-9645

Mazhari, R., & Hare, J.M. (2007). Mechanisms of action of mesenchymal stem cells in cardiac repair: potential influences on the cardiac stem cell niche. *Nature Clinical Practice Cardiovascular Medicine*, Vol.4, No.Suppl 1, (2007), pp. S21-26, ISSN 1743-4300

Menasché, P., Hagège, A.A., Vilquin, J.T., Desnos, M., Abergel, E., Pouzet, B., Bel, A., Sarateanu, S., Scorsin, M., Schwartz, K., Bruneval, P., Benbunan, M., Marolleau, J.P., & Duboc, D. (2003). Autologous skeletal myoblast transplantation for severe postinfarction left ventricular dysfunction. *Journal of the American College of Cardiology*, Vol.41, No.7, (April 2003), pp.1078 –1083, ISSN 0735-1097

Menasché, P., Alfieri, O., Janssens, S., McKenna, W., Reichenspurner, H., Trinquart, L., Vilquin, J.T., Marolleau, J.P., Seymour, B., Larghero, J., Lake, S., Chatellier, G., Solomon, S., Desnos, M., & Hagège, A.A. (2008). The Myoblast Autologous Grafting in Ischemic Cardiomyopathy (MAGIC) trial: first randomized placebo-controlled study of myoblast transplantation. *Circulation*, Vol.117, No.9, (March 2008), pp. 1189-1200, ISSN 1524-4539

Messina, E., De Angelis, L., Frati, G., Morrone, S., Chimenti, S., Fiordaliso, F., Salio, M., Battaglia, M., Latronico, M.V., Coletta, M., Vivarelli, E., Frati, L., Cossu, G., & Giacomello, A. (2004). Isolation and expansion of adult cardiac stem cells from human and murine heart. *Circulation Research*, Vol.95, No.9, (October 2004), pp. 911-921, ISSN 1524-4571

Min, J.Y., Yang, Y., Converso, K.L., Liu, L., Huang, Q., Morgan, J.P., & Xiao, Y.F. (2002). Transplantation of embryonic stem cells improves cardiac function in postinfarcted rats. *Journal of Applied Physiology*, Vol.92, No.1, (January 2002), pp. 288-296, ISSN 8750-7587

Murry, C.E., Wiseman, R.W., Schwartz, S.M., & Hauschka, S.D. (1996). Skeletal myoblast transplantation for repair of myocardial necrosis. *Journal of Clinical Investigation,* Vol.98, No.11, (December 1996), pp. 2512-2523, ISSN 0021-9738

Oh, H., Bradfute, S.B., Gallardo, T.D., Nakamura, T., Gaussin, V., Mishina, Y., Pocius, J., Michael, L.H., Behringer, R.R., Garry, D.J., Entman, M.L., & Schneider, M.D. (2003). Cardiac progenitor cells from adult myocardium: homing, differentiation, and fusion after infarction. *Proceedings of the National Academy of Sciences USA,* Vol.100, No.21, (October 2003), pp. 12313-12318, ISSN 0027-842

Orlic, D., Kajstura, J., Chimenti, S., Jakoniuk, I., Anderson, S.M., Li, B., Pickel, J., McKay, R., Nadal-Ginard, B., Bodine, D.M., Leri, A., & Anversa, P. (2001a). Bone marrow cells regenerate infarcted myocardium. *Nature,* Vol.410, No.6829, (April 2001), pp. 701-705, ISSN 0028-0836

Orlic, D., Kajstura, J., Chimenti, S., Limana, F., Jakoniuk, I., Quaini, F., Nadal-Ginard, B., Bodine, D.M., Leri, A., & Anversa, P. (2001b). Mobilized bone marrow cells repair the infarcted heart, improving function and survival., *Proceedings of the National Academy of Sciences USA,* Vol.98, No.18, (August 2001), pp. 10344-10349, ISSN 0027-8424

Ott, H.C., Matthiesen, T.S., Goh, S.K., Black, L.D., Kren, S.M., Netoff, T.I., & Taylor, D.A. (2008). Perfusion-decellularized matrix: using nature's platform to engineer a bioartificial heart. *Nature Medicine,* Vol.14, No.2, (February 2008), pp. 213-221, ISSN 1546-170X

Pagani, F.D., DerSimonian, H., Zawadzka, A., Wetzel, K., Edge, A.S., Jacoby, D.B., Dinsmore, J.H., Wright, S., Aretz, T.H., Eisen, H.J., & Aaronson, K.D. (2003). Autologous skeletal myoblasts transplanted to ischemia-damaged myocardium in humans. Histological analysis of cell survival and differentiation. *Journal of the American College of Cardiology,* Vol.41, No.5, (March 2003), pp. 879-888, ISSN 0735-1097

Perin, E.C. & Silva, G.V. (2006). What are stem cells and what do they do?, In: *An essential guide to cardiac cell therapy,* E.C. Perin, G.V. Silva, J.T. Willerson (Ed), 3-12, Informa Healthcare, ISBN 978-1-84184-471-8, Houston, United States of America

Pittenger, M.F., & Martin, B.J. (2004) Mesenchymal stem cells and their potential as cardiac therapeutics. *Circulation Research,* Vol.95, No.1, (July 2004), pp. 9-20, ISSN 1524-4571

Sanz-Ruiz, R., Fernández-Santos, M.E., Domínguez-Muñoa, M., Ludwig Martín, I., Parma, R., Sánchez, P.L., & Fernández-Avilés, F. (2008). Adipose tissue-derived stem cells: the friendly side of a classic cardiovascular foe. *Journal of Cardiovascular Translational Research,* Vol.1, No.1, (March 2008), pp.55-63, ISSN 1937-5395

Sanz-Ruiz, R., Fernández-Santos, M.E., Domínguez-Muñoa, M., Parma, R., Villa, A., Fernández, L., Sánchez, P.L., & Fernández-Avilés, F. (2009). Early translation of adipose derived cell therapy for cardiovascular disease. *Cell transplantation,* Vol.18, No.3, (March 2009), pp.245-254, ISSN 0963-6897

Schächinger, V., Erbs, S., Elsässer, A., Haberbosch, W., Hambrecht, R., Hölschermann, H., Yu, J., Corti, R., Mathey, D.G., Hamm, C.W., Süselbeck, T., Assmus, B., Tonn, T., Dimmeler, S., & Zeiher, A.M.; REPAIR-AMI Investigators. (2006). Intracoronary bone marrow-derived progenitor cells in acute myocardial infarction. *The New England Journal of Medicine,* Vol.355, No.12, (September 2006), pp. 1210-1221, ISSN 1533-4406

Shizuru, J.A., Negrin, R.S., & Weissman, I.L. (2005). Hematopoietic stem and progenitor cells: clinical and preclinical regeneration of the hematolymphoid system. *Annual review of medicine,* Vol.56, (February 2005), pp. 509-538, ISSN 0066-4219

Siminiak, T., Kalawski, R., Fiszer, D., Jerzykowska, O., Rzeźniczak, J., Rozwadowska, N., & Kurpisz, M. (2004). Autologous skeletal myoblast transplantation for the treatment of postinfarction myocardial injury: phase I clinical study with 12 months of follow-up. *American Heart Journal*, Vol.148, No.3, (September 2004), pp. 531-537, ISSN 1097-6744

Strauer, B.E., Yousef, M., & Schannwell, C.M. (2010). The acute and long-term effects of intracoronary stem cell transplantation in 191 patients with chronic heart failure: the STAR-heart study. *European Journal of Heart Failure*, Vol.12, No.7, (July 2010), pp. 721-729, ISSN 1879-0844

Suva, D., Garavaglia, G., Menetrey, J., Chapuis, B., Hoffmeyer, P., Bernheim, L., & Kindler, V. (2004). Non-hematopoietic human bone marrow contains long-lasting, pluripotential mesenchymal stem cells. *Journal of cellular physiology*, Vol.198, No.1, (January 2004), pp. 110-118, ISSN 0021-9541

Takahashi, K., Tanabe, K., Ohnuki, M., Narita, M., Ichisaka, T., Tomoda, K., & Yamanaka, S. (2007). Induction of pluripotent stem cells from adult human fibroblasts by defined factors. *Cell*, Vol.131, No.5, (November 2007), pp. 861-872, ISSN 0092-8674

Takano, H., Hasegawa, H., Kuwabara, Y., Nakayama, T., Matsuno, K., Miyazaki, Y., Yamamoto, M., Fujimoto, Y., Okada, H., Okubo, S., Fujita, M., Shindo, S., Kobayashi, Y., Komiyama, N., Takekoshi, N., Imai, K., Himi, T., Ishibashi, I., & Komuro, I. (2007). Feasibility and safety of granulocyte colony-stimulating factor treatment in patients with acute myocardial infarction. *International journal of cardiology*, Vol.122, No.1, (October 2007), pp. 41-47, ISSN 1874-1754

Taylor, D.A., Atkins, B.Z., Hungspreugs, P., Jones, T.R., Reedy, M.C., Hutcheson, K.A., Glower, D.D., & Kraus, W.E. (1998). Regenerating functional myocardium: improved performance after skeletal myoblast transplantation. *Nature Medicine*, Vol.4, No.8, (August 1998), pp. 929-933, ISSN 1078-8956

Tendera, M., Wojakowski, W., Ruzyłło, W., Chojnowska, L., Kepka, C., Tracz, W., Musiałek, P., Piwowarska, W., Nessler, J., Buszman, P., Grajek, S., Breborowicz, P., Majka, M., Ratajczak, M.Z.; REGENT Investigators. (2009). Intracoronary infusion of bone marrow-derived selected CD34+CXCR4+ cells and non-selected mononuclear cells in patients with acute STEMI and reduced left ventricular ejection fraction: results of randomized, multicentre Myocardial Regeneration by Intracoronary Infusion of Selected Population of Stem Cells in Acute Myocardial Infarction (REGENT) Trial. *European Heart Journal*, Vol.30, No.11, (June 2009), pp. 1313-1321, ISSN 1522-9645

Thomson, J.A., Itskovitz-Eldor, J., Shapiro, S.S., Waknitz, M.A., Swiergiel, J.J., Marshall, V.S., & Jones, J.M. (1998). Embryonic stem cell lines derived from human blastocysts. *Science*, Vol.282, No5391, (November 1998), pp. 1145-1147, ISSN 0036-8075

Toma, C., Pittenger, M.F., Cahill, K.S., Byrne, B.J., & Kessler, P.D. (2002). Human mesenchymal stem cells differentiate to a cardiomyocyte phenotype in the adult murine heart. *Circulation*, Vol.105, No.1, (January 2002), pp. 933-938, ISSN 1524-4539

Tse, H.F., Thambar, S., Kwong, Y.L., Rowlings, P., Bellamy, G., McCrohon, J., Thomas, P., Bastian, B., Chan, J.K., Lo, G., Ho, C.L., Chan, W.S., Kwong, R.Y., Parker, A., Hauser, T.H., Chan, J., Fong, D.Y., & Lau CP. (2007). Prospective randomized trial of direct endomyocardial implantation of bone marrow cells for treatment of severe coronary artery diseases (PROTECT-CAD trial). *European heart journal*, Vol.28, No.24, (December 2007), pp. 2998-3005, ISSN 0195-668X

Urbanek, K., Quaini, F., Tasca, G., Torella, D., Castaldo, C., Nadal-Ginard, B., Leri, A., Kajstura, J., Quaini, E., & Anversa, P. (2003). Intense myocyte formation from cardiac stem cells in human cardiac hypertrophy. *Proceedings of the National Academy of Sciences USA*, Vol.100, No.18, (September 2003), pp. 10440-10445, ISSN 0027-8424

Urbanek, K., Torella, D., Sheikh, F., De Angelis, A., Nurzynska, D., Silvestri, F., Beltrami, C.A., Bussani, R., Beltrami, A.P., Quaini, F., Bolli, R., Leri, A., Kajstura, J., & Anversa, P. (2005). Myocardial regeneration by activation of multipotent cardiac stem cells in ischemic heart failure. *Proceedings of the National Academy of Sciences USA*, Vol.102, No.24, (June 2005), pp. 8692-8697, ISSN 0027-8424

van der Laan, A., Hirsch, A., Nijveldt, R., van der Vleuten, P.A., van der Giessen, W.J., Doevendans, P.A., Waltenberger, J., Ten Berg, J.M., Aengevaeren, W.R., Zwaginga, J.J., Biemond, B.J., van Rossum, A.C., Tijssen, J.G., Zijlstra, F., & Piek, J.J. (2008). Bone marrow cell therapy after acute myocardial infarction: the HEBE trial in perspective, first results. *Netherlands Heart Journal,* Vol.16, No.12, (December 2008), pp. 436-439, ISSN 1568-5888

van Ramshorst, J., Bax, J.J., Beeres, S.L., Dibbets-Schneider, P., Roes, S.D., Stokkel, M.P., de Roos, A., Fibbe, W.E., Zwaginga, J.J., Boersma, E., Schalij, M.J., & Atsma, D.E. (2009). Intramyocardial bone marrow cell injection for chronic myocardial ischemia: a randomized controlled trial. *JAMA*, Vol.301, No.19, (May 2009), pp. 1997-2004, ISSN 1538-3598

Virmani, R., Kolodgie, F.D., & Ladich, E. (2006). Mechanistic insights into cardiac stem cell therapy from a histopathologic perspective, In: *An essential guide to cardiac cell therapy,* E.C. Perin, G.V. Silva, J.T. Willerson (Ed), 43-54, Informa Healthcare, ISBN 978-1-84184-471-8, Houston, United States of America

Wollert, K.C., Meyer, G.P., Lotz, J., Ringes-Lichtenberg, S., Lippolt, P., Breidenbach, C., Fichtner, S., Korte, T., Hornig, B., Messinger, D., Arseniev, L., Hertenstein, B., Ganser, A., & Drexler, H. (2004). Intracoronary autologous bone-marrow cell transfer after myocardial infarction: the BOOST randomised controlled clinical trial. *Lancet,* Vol.364, No.9429, (July 2004), pp.141-148, ISSN 1474-547X

Wollert, K.C., & Drexler, H. (2010). Cell therapy for the treatment of coronary heart disease: a critical appraisal. *Nature Reviews Cardiology,* Vol.7, No.4, (April 2010), pp. 204-215, ISSN 1759-5010

Yamashita, J., Itoh, H., Hirashima, M., Ogawa, M., Nishikawa, S., Yurugi, T., Naito, M., Nakao, K., & Nishikawa, S. (2000). Flk1-positive cells derived from embryonic stem cells serve as vascular progenitors. *Nature*, Vol.408, No.6808, (November 2000), pp. 92-96, ISSN 0028-0836

Yoon, Y.S., Wecker, A., Heyd, L., Park, J.S., Tkebuchava, T., Kusano, K., Hanley, A., Scadova, H., Qin, G., Cha, D.H., Johnson, K.L., Aikawa, R., Asahara, T., & Losordo, D.W. (2005). Clonally expanded novel multipotent stem cells from human bone marrow regenerate myocardium after myocardial infarction. *Journal of Clinical Investigation,* Vol.115, No.2, (February 2005), pp. 326-338, ISSN 0021-9738

Yu, J., Vodyanik, M.A., Smuga-Otto, K., Antosiewicz-Bourget, J., Frane, J.L., Tian, S., Nie, J., Jonsdottir, G.A., Ruotti, V., Stewart, R., Slukvin, I.I., & Thomson, J.A. (2007). Induced pluripotent stem cell lines derived from human somatic cells. *Science,* Vol.318, No.5858, (December 2007), pp. 1917-1920, ISSN 1095-9203

Zimmet, J.M., & Hare, J.M. (2005). Emerging role for bone marrow derived mesenchymal stem cells in myocardial regenerative therapy. *Basic Research in Cardiology*, Vol.100, No.6, (November 2005), pp. 471-481, ISSN 0300-8428

12

Autologous Stem Cells for Cardiac Repair: New Insights on Clinical Trial Safety and Best Cell Source

Nicolas Noiseux[1,2], Louis Mathieu Stevens[1,2], Jessica Forcillo[1] and Samer Mansour[2,3]
[1]*Cardiac Surgery Department, Centre Hospitalier de l'Université de Montréal (CHUM),*
[2]*CHUM Research Center (CRCHUM),*
[3]*Cardiology Department, CHUM,*
Canada

1. Introduction

Acute myocardial infarction (AMI) is a leading cause of death worldwide. (Wollert and Drexler, 2010) Despite improvements in survival rate after AMI with recent medical advances (Jackson, et al., 2001), the reduced heart function attributed to irreversible loss of viable cardiomyocytes remains a major clinical problem (Timmermans, et al., 2003). Typical MI results in large-scale loss of cardiac muscle, often a billion or more myocytes. (Laflamme and Murry, 2005) The loss of myocardial cell mass and the inability of remaining cells to adequately compensate through hypertrophic or hyperplastic responses result in the development of heart failure due to ventricular dysfunction (Jain, et al., 2005). Endogenous repair mechanisms of the adult human heart are not sufficient for meaningful tissue regeneration, so muscle lost is replaced by non-contractile fibrotic scar tissue, initiating progressive heart failure (Laflamme, et al., 2007). Heart failure is the number one cause of hospitalization in US citizens over 65 of age and has a significant economic impact (Laflamme and Murry, 2005). Despite recent advances in treatment options, mortality remains unacceptably high (Ertl, et al., 1993, Pfeffer and Braunwald, 1990, Towbin and Bowles, 2002) with >50% of all patients succumbing within 5 years of the initial diagnosis of heart failure (Jain, et al., 2005). Improved medical and surgical treatments for patients after AMI have led to the decrease in early mortality, but as a result there of a higher incidence of heart feailure. (Wollert and Drexler, 2010) In some patients with advanced disease, cardiac transplantation may be an option, but due to organ shortage, its practical use is limited to end-stage heart failure.

In the current clinical setting, early reperfusion by percutaneous coronary intervention (PCI) including intra-coronary stenting is the treatment of choice in the situation of acute MI (Topol, 2003). In patients with diffused coronary artery disease associated with severe ischemic cardiomyopathy, coronary artery bypass grafting (CABG) remains the optimal way of coronary revascularization, and is associated with improved survival and reduction of angina (Detre, et al., 1996, Pell, et al., 2004). Although surgical and catheter-based revascularization of ischemic myocardium can treat angina, reduce the risk of MI, and improve cardiac function, the viability of severely ischemic myocardium, necrotic myocardium, or both cannot be restored. (Patel, et al., 2005) Current pharmacologic and interventional strategies are unsuccessful to regenerate dead myocardium and failed to address the clinical challenge caused by the early loss of cardiomyocytes (Charwat, et al., 2008). Moreover, none of our current therapies addresses the underlying cause of the remodeling process, namely the damage to the cardiomyocytes and the vasculature in the ischemic tissue (Wollert, 2008). Therefore, new cardiovascular therapies must be elaborated to promote myocardial repair and regeneration, this "Holy Grail" represents a major challenge in the treatment of ischemic cardiovascular diseases.

2. Stem cells for myocardial repair

One approach to counteract the effects of myocardial dysfunction could be replacement of damaged contractile cells by healthy myocytes or progenitor cells with the potential of becoming new functional cardiomyocytes (Evans, et al., 2007). There is growing evidence suggesting that heart muscle has the ability to regenerate through activation of cardiac resident stem cells (CSC), or through recruitment of progenitor stem cells population from other tissues, such as the bone marrow (BM) (Charwat, et al., 2008). Cellular transplantation is a potential approach to improve healing of the ischemic heart, to repopulate the injured myocardium (Dowell, et al., 2003), to treat heart failure (Dowell, et al., 2003, Raeburn, et al., 2002) and restore cardiac function (Hassink, et al., 2003, Orlic, et al., 2002). Experimental studies have shown that adult bone marrow stem cells (BMSC) are capable of differentiation (Blau, et al., 2001, Goodell, et al., 2001, Jackson, et al., 2002, Krause, et al., 2001), regeneration of infarcted myocardium, induction of myogenesis, as well as promotion of angiogenesis, ultimately leading to a better cardiac contractile performance (Dowell, et al., 2003, Kocher, et al., 2001, Orlic, et al., 2001, 2003, Orlic, et al., 2001, Orlic, et al., 2001, Shake, et al., 2002, Strauer, et al., 2002, Tomita, et al., 1999, Tomita, et al., 2002). A variety of embryonic and adult-derived cell types have been investigated for their capacity to mediate cardiac and vascular repair (Psaltis, et al., 2008). Nowadays, several cell candidates have been recently investigated for the treatment of ischemic cardiomyopathy such as fetal (Li, et al., 1997) and neonatal (Watanabe, et al., 1998) cardiomyocytes, embryonic stem cells (Min, et al., 2002, Min, et al., 2003), cardiac resident stem cells, skeletal myoblasts (Leor, et al., 1996, Menasche, et al., 2001, Menasche, et al., 2003, Murry, et al., 1996, Taylor, et al., 1998) and endothelial progenitor cells (Kawamoto and Losordo, 2008, Kawamoto, et al., 2003, Kocher, et al., 2001). The bone marrow is known to be an excellent reservoir for many adult stem cells and BMSC have been used to treat hematologic disorders for decades (Liao, et al., 2007). Bone marrow, which is easily accessible, is presently the most frequent source of cells used for clinical cardiac repair (Dimmeler, et al., 2005, Kocher, et al., 2007). Bone marrow-derived cells contain a complex assortment of progenitor cells, including hematopoietic stem cells (HSC), side population (SP) cells, mesenchymal or stromal, stem cells (MSC), and multipotent adult progenitor cells (MAPC), a subset of MSC (Barbash, et al., 2003, Jackson, et al., 2001, Perin, et al., 2003, Strauer, et al., 2002, Toma, et al., 2002). Other progenitor/stem cell populations investigated include (Kocher, et al., 2007): fat tissue-derived multipotent stem cells (Planat-Benard, et al., 2004), multipotential cells from bone marrow (Jiang, et al., 2002), skeletal muscle, somatic stem cells from placental cord blood (Kogler, et al., 2004), amniotic fluid-derived stem cells (De Coppi, et al., 2007), circulating endothelial progenitor cells (EPC), and cardiac resident progenitor cells (CPC) are already predisposed to adopt a cardiomyocyte

phenotype (Beltrami, et al., 2003, Oh, et al., 2003). Positive results from animal studies have prompted several clinical trials to ascertain the safety, feasibility and efficacy of cell therapy. Despite the fact that the exact mechanisms underpinning stem cell therapy have not been yet elucidated and are still intensely debated, cell therapy has already been introduced into the clinical setting (Dowell, et al., 2003, Weisel, et al., 2004, Wollert and Drexler, 2005).
Cell transplantation treatments are faced with many technical and practical issues, and are limited to autologous cells, which require bone marrow aspirations, muscle biopsies or blood sampling with cells sorting and/or culture. Moreover, they are unable to deliver large cell numbers that would survive the peri-transplantation period (Reinecke, et al., 1999, Shake, et al., 2002, Wollert and Drexler, 2005, Zhang, et al., 2001). Over the past several years, a surge of experimental data from pre-clinical and clinical studies has emerged providing both a proof of concept and therapeutic promises for post-MI cardiac repair and regeneration using these cell therapies (Jain, et al., 2005). Despite all the excitement in stem cell research resulting from initial experimental data and preliminary clinical trials, many crucial questions regarding stem cells therapy still remain to be answered. The most important ones being what are the exact mechanisms underlying their beneficial effects, and which cell type is the most appropriate for clinical application? A major obstacle to the identification of the optimal cell therapy is that the fate of the implanted cells and the nature of their beneficial effects are ill defined (Evans, et al., 2007). Moreover, a clear characterization of the cellular effects of stem cell transplantation is critical to avoid potentially adverse consequences and to improve the outcome (Evans, et al., 2007). A better understanding is fundamental for the development of new therapeutics, and to optimize stem cell applications in the treatment of ischemic cardiovascular diseases.

3. Various cell candidates for cardiac repair

A myriad of cell types have been tested experimentally, each of them being usually credited by its advocates of a high "regeneration" potential. This has led to a flurry of clinical trials entailing the use of skeletal myoblasts or bone marrow-derived cells either unfractionated or enriched in progenitor subpopulations. (Menasche, 2009) There is currently uncertainty as to which of the stem cell population is most potent in stimulating angiogenesis and cardiac repair. Theoretically, many stem and progenitor cell populations are potential candidates to be used for cardiac repair and to treat ischemic cardiomyopathy. Each cell type possesses its own profile of advantages, limitations and practicability issues (Wollert, 2008). It, appears that not "one cell fits all" but that the selection of the cell type should be tailored to the primary clinical indication. (Menasche, 2009) Studies comparing the regenerative capacity of distinct cell populations are scarce. (Wollert, 2008) As replacement therapy in humans, the estimates are that 10^8-10^9 cells might be needed to replace those lost after a moderate size AMI. Accordingly, an important issue in cell therapy is scalability and the ability to deliver a large amount of cells into the ischemic myocardium. It does not make sense to develop an "ideal" cell in a culture dish, if we remain unable to deliver it appropriately and to keep it alive, at least for a while, which requires to improve on the delivery techniques and to provide cells along with the vascular and extracellular matrix type of support necessary for their survival and patterning. (Menasche, 2009) Another issue which must be considered is autologous versus non-autologous source of donor cells. (Evans, et al., 2007) Autologous have the advantage of avoiding inflammatory response and immune rejection. However, to be useful for broad clinical application, cells must be readily available in sufficient number or easily to harvest and isolate. Herein are some characteristics of the most commonly used adult stem cells for cardiac repair.

3.1 Primary remuscularization by myogenic cells

When contemplating how cell therapy might benefit the recently infarcted heart, perhaps the most obvious strategy is remuscularization, that is, the replacement of the necrotic myocardium with viable new muscle cells. (Laflamme, et al., 2007) Strong proof-of-principle data to support this approach were provided by studies in which committed cardiomyocytes from fetal (Li, et al., 1997) and neonatal (Watanabe, et al., 1998) sources were transplanted into murine or rat model of ischemic myocardium. The implanted myocytes formed stable intracardiac grafts and resulted in improvement in cardiac function. (Leor, et al., 1996, Li, et al., 1996) As expected for terminally differentiated cardiomyocytes, the implanted cells retained their contractile phenotype and expressed the necessary elements for intercellular electrical communication such as gap junction proteins. (Laflamme, et al., 2007) Another myogenic cell type, skeletal muscle satellite cells commonly referred to as skeletal myoblast has been extensively studied in animals and clinical trials. (Menasche, 2008, Menasche, et al., 2001, Menasche, et al., 2003, Murry, et al., 1996, Taylor, et al., 1998) Skeletal myoblasts have undergone extensive preclinical testing that has consistently demonstrated their ability to preserve postinfarct left ventricular function and to limit remodelling. (Menasche, 2008) Indeed, skeletal myoblasts, the progenitor cell that mediate normal regeneration of skeletal striated muscle, have a number of attractive properties for cardiac cell therapies, including relative resistance to ischemia, potentially autologous donor source from readily accessible muscle biopsies, paracrine effects by secretions of growth factors, and the capacity for tremendous in vitro expansion. (Dowell, et al., 2003, Laflamme, et al., 2007, Menasche, 2008, 2009) The primary disadvantage of skeletal myoblasts is simply that they differentiate into the wrong form of striated muscle, i.e. skeletal rather than cardiac, with the caveat of severe ventricular arrhythmias requiring the implantation of automatic implantable cardioverter-defibrillator (AICD) devices. (Laflamme, et al., 2007, Menasche, 2011, Menasche, et al., 2003)

3.2 Hematopoietic stem cells (HSC)

Commonly referred as bone marrow stem cells (BMSC), HSC constitute a small cell population of the bone marrow, perhaps as few as 1:10,000 bone marrow mononuclear cells. (Anversa, et al., 2004, Harrison, et al., 1988) HSC can be enriched using various technologies to high purity on the basis of surface markers (lineage$^-$, c-kit$^+$, Sca-1$^+$, CD38^{high}, CD34, or the more immature marker protein CD133) (Dimmeler and Zeiher, 2004, Orlic, et al., 2002). These cells represent the prototypic adult stem cell population. The ability of HSC to reconstitute the hematopoietic system of a myeloablated host led to the first clinical application of adult stem cells more than three decades ago. (Liao, et al., 2007) HSC can self-renew and were shown to trans-differentiate into multiple lineages, including endothelial cells (Nygren, et al., 2004, Orlic, et al., 2002, Urbich and Dimmeler, 2004), and can contribute to the regeneration of a variety of non-hematopoietic lineages in multiple organs, including myocardium (Jackson, et al., 2001, Orlic, 2003, 2004). The interest in stem cells for cardiac regeneration started in 2001 with the observation of Orlic et al who injected Lin$^-$ ckit$^+$ bone marrow-derived HSC that were able to repair acute MI in mice by transdifferentiating into cardiomyocytes and vascular cells (Orlic, et al., 2001). To date, the HSC appear to be the most versatile stem cells across all lineages, since adult somatic HSC may share a similar developmental plasticity commonly seen in embryonic stem cells (Anversa, et al., 2004). The use of hematopoietic progenitors stem cells appears to uniformly induce neovascularization that seems to be a prerequisite of the successful functional repair. HSC can be used to repair infarcted hearts by regenerating new cardiomyocytes and vascular endothelium in response to ischemic injury (Jackson, et al., 2001, Kocher, et al., 2001, Nygren, et al., 2004, Orlic, et al., 2001).

3.3 Circulating stem cells

Endothelial progenitor cells (EPCs) are a heterogeneous group of endothelial cell (EC) precursors originating from the hematopoietic compartment of the bone marrow. (Asahara, et al., 1999) EPCs share many surface markers (hematopoietic markers CD133, CD34, vascular endothelial growth factor receptor-2 (Flk-1)) and biological properties with hematopoietic stem cell (HSC). (Asahara, et al., 1999, Gehling, et al., 2000, Shi, et al., 1998) EPCs are isolated from the bone marrow or from the peripheral circulation and expanded in vitro. (Liao, et al., 2007) After a few days in culture in endothelial cell medium supplemented with growth factors, EPC phenotype can be confirmed by direct fluorescent staining with 1,1′-dioctadecyl-3-3-3,3′-tetramethylindocarbocyanine (Dil)-labeled acetylated low-density lipoprotein (DiLDL) and fluorescein isocyanate (FITC)-labeled Ulex europaeus agglutinin-I (lectin). Dual stained cells positive for both DiLDL and lectin are considered EPC, see Fig. 1. EPCs are subdivided into "early", and "late" also referred as outgrowth of EC (OECs). (Hur, et al., 2004, Smadja, et al., 2007, Yoon, et al., 2005) EPCs have the homing capacity to sites of active angiogenesis, where they participate in the repair of various tissues, including the heart. (Asahara, et al., 1999, Crosby, et al., 2000, Fujiyama, et al., 2003, Vasa, et al., 2001, Yoon, et al., 2005) Recent studies suggested that circulating EPC can home to site of tissue ischemia and injury, express endothelial antigens and play a significant role in new blood vessels formation and re-endothelialization (Asahara, et al., 1999, Crosby, et al., 2000, Fujiyama, et al., 2003, Vasa, et al., 2001). Normal adults have a small amount of circulating EPC in the peripheral blood (Choi, et al., 2004). In response to cytokine stimulation, ischemic insult, drugs, and under the influence of other pathological conditions, these cells are mobilized from the bone marrow (Asahara, et al., 1999, Choi, et al., 2004, Dimmeler, et al., 2001). For example, numbers and angiogenic functions of EPC have been reported to be reduced in patients with risk factors for coronary artery disease (Vasa, et al., 2001), [10], chronic renal failure (Choi, et al., 2004), older age (Scheubel, et al., 2003) and type II diabetics (Tepper, et al., 2002). In contrast, limb ischemia (Asahara, et al., 1999), acute myocardial infarction (Shintani, et al., 2001) and HMG-CoA reductase inhibitors (statins) (Dimmeler, et al., 2001) were associated with increased EPC in the circulation and better angiogenic potential. EPC accelerate re-endothelialization and improve vascular healing[14]. A reduction of their numbers and impaired function have been correlated with inflammatory states, adverse vascular events and death (George, et al., 2003, Lambiase, et al., 2004, Ruel, et al., 2005). The exact mechanism of EPCs-mediated neovascularization is unknown, and possibly the EPCs differentiate into EC, or by paracrine mechanisms via growth factors secretion. (Smadja, et al., 2007) There are limits of the EPCs clinical application. (Choi, et al., 2004, Kawamoto and Asahara, 2007) Amount of available EPCs is restricted, so large blood sample is required for cell preparation.

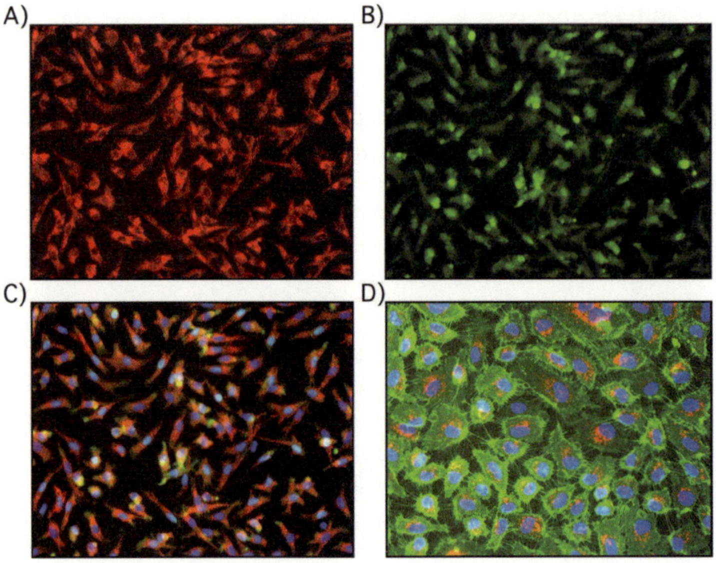

Fig. 1. Characterization of cultured EPC by direct fluorescent staining. A) Early EPC after a few days in culture, adherent cells are stained with (Dil)-labeled acetylated low-density lipoprotein (DiLDL, red) and B) Lectin staining is performed by incubation with fluorescein isocyanate (FITC, green)-labeled Ulex europaeus agglutinin-I. C) Nuclei are stained with Hoechst (blue) and typical merge image demonstrates dual stained cells positive for both DiLDL and lectin are considered EPC (Dimmeler, et al., 2001) (10X original magnification). After staining, samples are viewed with an inverted fluorescent microscope. D) After many days in culture some late EPC can adopt new phenotype from small spindle-like to large, round and flat cells characteristic of mature endothelial cells (10X original magnification). Dr Noiseux, unpublished data.

3.4 Mesenchymal Stem Cells (MSC)

Adult MSC are a population of stromal cells isolated from bone marrow-derived stem cells (BMSC) (Barbash, et al., 2003, Strauer, et al., 2002, Toma, et al., 2002). MSC have been isolated from many species including humans, and are present at a concentration several fold lower than their hematopoietic counterparts. MSC are present at a concentration several fold lower than their hematopoietic counterparts, representing approximatively 0.001-0.01% of the total nucleated marrow cell population. (Jain, et al., 2005, Meirelles Lda and Nardi, 2003) MSC are separated from other cells in culture by their preferential attachment to plastic surfaces (Colter, et al., 2000, Colter, et al., 2001, Meirelles Lda and Nardi, 2003). In their undifferentiated state, MSC do not express hematopoietic or endothelial cell surface markers (Annabi, et al., 2003, Colter, et al., 2000, Mangi, et al., 2003, Meirelles Lda and Nardi, 2003, Minguell, et al., 2001, Peister, et al., 2004) such as CD14, CD31, CD34, CD45, but may express CD-117 (c-kit) and the majority express CD29, CD44, CD90, and Sca-1 (Annabi, et al., 2003, Colter, et al., 2000, Mangi, et al., 2003, Meirelles Lda and Nardi, 2003, Minguell, et al., 2001, Peister, et al., 2004, Pittenger and Martin, 2004). MSC are identified by their specific antigens (SH2, SH3, SH4, STRO-1) and adhesion molecules (ALCAM, CD44). (Jain, et al., 2005) MSC are easily expandable in culture without losing their differentiation potential (Meirelles Lda and Nardi, 2003), and they constitute an unlimited pool of transplantable cells, Fig. 2. MSC fulfilled the stem cells criteria including self-renewing cell division and potential for differentiation (Verfaillie, et al., 2002). MSC are multipotent (Grove, et al., 2004), and can differentiate into multiple lineages (Grove, et al., 2004, Mackay, et al., 1998, Meirelles Lda and Nardi, 2003, Pittenger, et al., 1999, Song and Tuan, 2004), including fibroblasts, osteoblasts (Liu, et al., 2001), chondroblasts and adipocytes. Differentiation of MSC to cardiomyocyte-like cells has been observed in vitro under specific conditions (Makino, et al., 1999, Rangappa, et al., 2003, Tomita, et al., 1999), and in vivo after injection into the myocardium (Dai, et al., 2005, Saito, et al., 2002, Shake, et al., 2002, Toma, et al., 2002, Tomita, et al., 1999, Wang, et al., 2000).

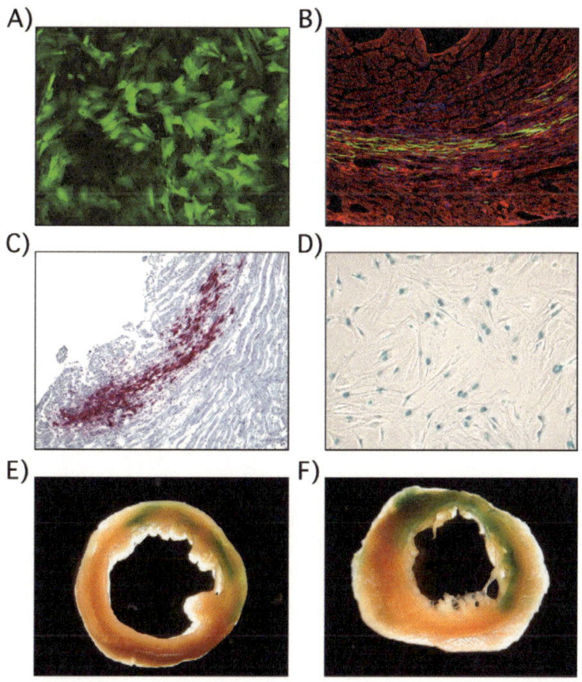

Fig. 2. A) MSC were retrovirally transduced to express GFP (Noiseux, et al., 2006), image under inverted fluorescent microscope. B) MSC were injected into ischemic heart in a mouse model of acute MI. GFP immunostaining (green) was used for tracking the implanted cells at 3 days post-MI, and cardiomyocytes were labeled in red by immunostaining for cardiac alpha-sarcomeric actin, and nuclei are identified by Hoechst staining (blue). C) MSC were stained in red by immonostaining for GFP 3 days after MI, demonstrating massive engraftment in the infarcted area. D) MSC were retrovirally transduced to express LacZ gene (nuclear localizating sequence), image under light microscope after X-gal staining demonstrating blue nuclei. E) Thick section of rat infarcted heart after X-gal staining demonstrating engraftment of MSC expressing LacZ (bleu) in the infarct border zone at the level of the papillary muscles. F) Same as in E, but in the apical segment of the left ventricle. 10X original magnification in A and B, 20X original magnification in B and C. Dr Noiseux, unpublished data.

MSC are known to secrete a wide spectrum of biologically active factors and angiogenic/arteriogenic cytokines (Heil, et al., 2004, Kamihata, et al., 2001, Kinnaird, et al.,

2004, Kinnaird, et al., 2004, Kinnaird, et al., 2004) that can be found in the culture-conditioned medium (MSC-CM) (Heil, et al., 2004, Kamihata, et al., 2001, Kinnaird, et al., 2004, Kinnaird, et al., 2004, Kinnaird, et al., 2004), see Fig. 3. Under hypoxic culture conditions, the expression of several factors is significantly up-regulated (Gnecchi, et al., 2005, Gnecchi, et al., 2006, Kinnaird, et al., 2004, Noiseux N, 2004, Noiseux N, 2004). MSC implantation into the injured heart has been associated with improvement of cardiac performance and repair in animal studies and clinical trials. Because myocardial infarction leads to permanent loss of tissue with subsequent impaired function, the reports highlighting the capacity of MSC to differentiate into new cardiomyocytes have prompted new perspectives in the treatment of cardiovascular diseases using stem cell transplantation. Among adult stem cells, MSC possess unique properties that make them eligible for convenient and highly effective cell therapy (Jain, et al., 2005). MSC are particularly suitable for cellular therapy because of their multipotency, low immunogenicity, amenability to ex vivo expansion, and genetic modification. However, a recent study by Vulliet et al. raised concerns with the use of MSC, and their findings showed acute myocardial ischemia and sub-acute myocardial micro-infarction following intra-coronary injection of MSC into ischemic dog hearts due to the formation of small capillary emboli (Vulliet, et al., 2004).

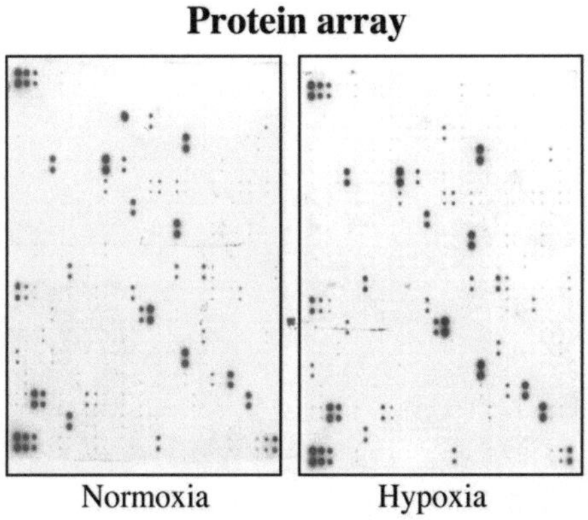

Fig. 3. Human MSC express genes encoding several cytokines and growth factors that are released in the MSC-conditioned medium. Hypoxic culture condition alters the production of several factors. Protein arrays from RayBiotech inc. allow the detection of several proteins simultaneously. Dr Noiseux, unpublished data.

3.5 Cardiac resident stem cells

Recently identified by Hierlihy and colleagues in the adult heart, these cardiac stem cell-like populations were identified based on their specific ability to efflux Hoechst dye (Hierlihy, et al., 2002). These progenitor cells, expressing c-kit, are capable of differentiation into cardiomyocytes and/or vascular tissue. They are self-renewing, clonogenic, and multipotent (Bearzi, et al., 2007). Recent studies have suggested that the heart has an inherent ability to replace its parenchymal cells continuously by these resident stem cells (Braun and Martire, 2007). These observations suggest the capacity of adult myocardium to maintain physiological homeostasis, at least partially, through resident cardiac stem cells (Liao, et al., 2007). The origin of these cardiac stem cells is unclear. These cells could either constitute the remaining endogeneous cardiac stem cells dormant in the myocardium or alternatively, cells that have homed to the heart from another organ (maybe the bone marrow) or in response from injury (Lyngbaek, et al., 2007). These cells can be clonally expanded in vitro from an endomyocardial biopsy and be used for cardiac repair.

4. Insights on mechanisms of action and fate of the implanted cells

Over the last several years, there has been a surge of data detailing the use of cell-based therapies for post-MI treatment. Stem cells implantation has been shown to prevent deleterious remodeling and improve recovery of infarcted myocardium by cytoprotective effects, myogenesis and angiogenesis (Dai, et al., 2005, Davani, et al., 2003, Dowell, et al., 2003, Kocher, et al., 2001, Mangi, et al., 2003, Min, et al., 2002, Orlic, et al., 2002, Orlic, et al., 2001, 2003, Orlic, et al., 2001, Orlic, et al., 2001, Pittenger and Martin, 2004, Shake, et al., 2002, Strauer, et al., 2002, Tomita, et al., 1999, Tomita, et al., 2002, Wang, et al., 2001, Wollert and Drexler, 2005). The ultimate goal for cell therapy is the stem cell engraftment into the damaged heart and, ultimately, differentiation into new functional cardiomyocytes, vascular smooth muscle and endothelial cells. Although differentiation of stem cells into cardiomyocytes has been observed, there is much debate over the frequency of this phenomenon. Recent work has questioned the ability of stem cells to generate new cardiomyocytes. (Laflamme, et al., 2007, Murry, et al., 2004, Noiseux, et al., 2006) It remains unclear whether the beneficial effect results as a direct consequence of the transplanted cells participating and integrating in a functional syncytium with the host myocardium, or alternatively if the cells could benefit cardiac function without directly contributing to contraction (Dowell, et al., 2003). This indirect effect may be attributed to secretion of biologically active factors (Nguyen, et al., 2010) that could enhance the angiogenic process (Dowell, et al., 2003, Timmermans, et al., 2003), protect cardiomyocytes from apoptosis and induce proliferation, improve heart function with inotropic properties, or recruit resident cardiac stem cells (Wollert and Drexler, 2005). Implanted stem cells may also fuse with the native dysfunctional myocytes to augment function. (Noiseux, et al., 2006, Nygren, et al., 2004, Ying, et al., 2002) Indeed, mechanistic underpinnings of stem cell therapy appear to be far more complex that formerly anticipated, such indirect effects are referred to paracrine action, Fig. 4. Interestingly, while only few groups have observed differentiation of BMSC into cardiomyocytes, most groups have reported a beneficial effect on post-MI remodeling and cardiac functional recovery (Liao, et al., 2007). As such, these data are certainly encouraging, given the improvement in objective measures such as infarct size and contractility. However, they are also disappointing, since they fail to demonstrate cardiac differentiation.

The mechanisms by which stem cells can repair damaged myocardium or lead to improvement in cardiac function are still largely unknown. However, the foremost apparent and proposed pathways are: a) direct or indirect improvement in neovascularization, i.e., vasculogenesis, angiogenesis and arteriogenesis; and b) differentiation/fusion into cardiomyocytes and formation of regenerated myocardial tissue (Kocher, et al., 2007). To which extent these different mechanisms of action are involved may critically depend on the cell type and the clinical setting, such as acute versus chronic ischemic injury (Kocher, et al., 2007). For example, in patients presenting acute MI, stem cell transplantation is thought to significantly improve post-MI ventricular remodeling and function through enhanced neovascularization and reduced cardiomyocytes apoptosis, irrespective of long-term engraftment and trans-differentiation. On the other hand, these

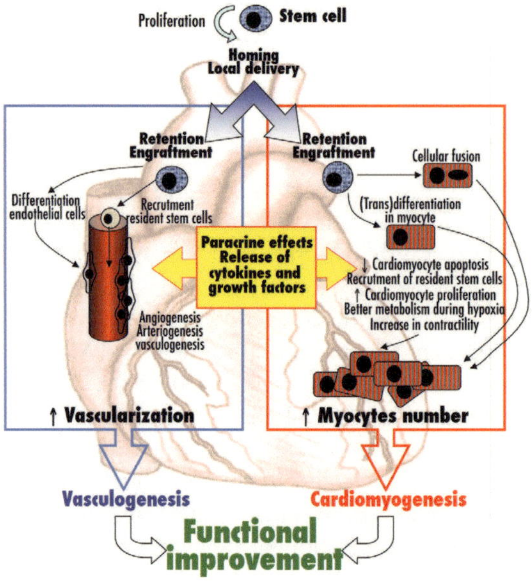

Fig. 4. Working hypothesis for mechanism of stem cell transplantation for the treatment of ischemic cardiomyopathy. Stem and progenitor cell transplantation can have a favorable impact on tissue perfusion and contractile performance by promoting vascularization and myocytes formation. Improved vascularization may facilitate beneficial effects in the myocyte compartment. Adapted from (Wollert and Drexler, 2005).

later mechanisms may have little or no benefit in patients with long established scars, apart from the functional rescue of hibernating cardiomyocytes. Virtually all currently available mechanistic insights have come from pre-clinical studies in animal models, whereas the human experience is rather limited (Laflamme, et al., 2007).

4.1 Stem cells (Trans)differentiation

The foremost purpose of cell-based therapies remains the regeneration of lost cardiac cells via stem cells (trans)differentiation into new contractile cardiomyocytes. (Liao, et al., 2007) Crucial proof-of-principle data in support of this contention were provided by early animal studies with transplantation of fetal and neonatal cardiomyocytes following acute MI (Laflamme, et al., 2007). Implanted myocytes formed stable intra-cardiac grafts and resulted in improved left ventricular function (Jia, et al., 1997, Li, et al., 1996, Li, et al., 1997, Sakai, et al., 1999, Scorsin, et al., 2000). Skeletal muscle satellite cells (referred commonly as myoblasts) have also been exhaustively examined as candidate cells for cardiac repair. Unlike native cardiomyocytes which form an electrical syncytium via gap junction coupling, implanted myoblasts are isolated and do not contract in synchrony with the host myocardium (Leobon, et al., 2003). Embryonic stem cells (ESC) are unambiguously cardiogenic cell type, capable of differentiating into cells with ventricular, atrial and pacemaker/conduction system phenotypes (He, et al., 2003, Kehat, et al., 2001). Cardiomyocytes derived from ESC are able to form stable intra-cardiac implant, and provide functional benefit in murine experimental infarct models (Hodgson, et al., 2004, Singla, et al., 2006).

Orlic et al were among the first to demonstrate in a mice model of AMI that bone marrow could provide an extra-cardiac source of progenitor cells with the ability to differentiate into cardiomyocytes and restore cardiac function (Orlic, et al., 2001, Orlic, et al., 2001, Orlic, et al., 2001). Studies in several species demonstrated that BMSC are stem cells found in various mesenchymal tissues, and precursors of peripheral tissue such as the heart muscle (Strauer and Kornowski, 2003). Cardiogenic potential has also been demonstrated for a surprisingly high number of adult progenitor cell types, including HSC, MSC, circulating EPC, resident cardiac progenitor cells, adipose-derived stem cells and others (Badorff, et al., 2003, Beltrami, et al., 2003, Laflamme, et al., 2007, Oh, et al., 2003, Planat-Benard, et al., 2004, Toma, et al., 2002). When BMSC are transplanted into an ischemic heart, they can express the cardiac specific markers troponin I and cardiac myosin, suggesting transformation into functional cardiomyocytes (Grigoropoulos and Mathur, 2006, Kajstura, et al., 2005, Yoon, et al., 2008).

Using genetic markers (e.g., Y chromosome) and/or labeled fluorescent dyes (e.g., Green fluorescent protein GFP), many groups reported the transdifferentiation of bone marrow-derived HSC into cardiomyocytes following implantation into ischemic heart (Jackson, et al., 2001, Liao, et al., 2007, Orlic, et al., 2001, Orlic, et al., 2001). However, these early observations have been called into question by others, who failed to identify similar cardiomyocytes derived from implanted stem cells (Balsam, et al., 2004, Murry, et al., 2004). Further experimental studies addressing the capacity of transplanted BMSC to differentiate into the cardiomyogenic lineage yielded conflicting results (Anversa, et al., 2007, Dimmeler, et al., 2008, Murry, et al., 2004).

4.2 Cellular fusion

Despite the initially observation that implanted HSC could generate new cardiomyocytes in the infarcted heart, several groups have subsequently revisited this hypothesis of stem cells differentiation (Laflamme and Murry, 2005, Laflamme, et al., 2007, Murry, et al., 2004, Noiseux, et al., 2006). Using state-of-the-art techniques and genetic markers, it has been suggested that rarely cardiomyocytes would display lineage markers of the transplanted stem cells, but that myocardial cells were derived from the fusion of implanted stem cells with damaged host cardiomyocytes (Alvarez-Dolado, et al., 2003, Laflamme, et al., 2007, Murry, et al., 2004, Nygren, et al., 2004). Fusion of transplanted stem cells with resident cardiomyocytes has been offered as an alternative for the previous claims of transdifferentiation (Alvarez-Dolado, et al., 2003, Balsam, et al., 2004, Murry, et al., 2004, Noiseux, et al., 2006, Nygren, et al., 2004, Wollert, 2008). Fused implanted cells may adopt phenotype of recipient cells, and without detailed genetic analysis, might be falsely interpreted as differentiation (Terada, et al., 2002). Fusion of BMSC with purkinje neurons, hepatocytes and cardiomyocytes has been reported by Alvarez-Dolado et al in 2003 (Alvarez-Dolado, et al., 2003). Using different genetic markers to track the fate of transplanted cells, fusion with recipient cardiomyocytes has been also reported for peripheral blood $CD34^+$ cells (Zhang, et al., 2004), hematopoietic stem cells (Murry, et al., 2004, Nygren, et al., 2004), cardiac progenitor cells (Oh, et al., 2003), and skeletal muscle cells (Reinecke, et al., 2004).

It is possible that many examples of putative stem cell plasticity reported in the literature are actually due to transplanted cells fusing with a different tissue type, followed by reprogramming of the donor and recipient cell genome (Vassilopoulos and Russell, 2003).

To address the issue of cellular fusion, it is imperative to use a method that allows the tracking of the implanted cells, but also a genetic marker that can be triggered exclusively by fusion event. A well-validated technique suitable for this purpose relies on the use of Cre/LoxP system. Using a model of AMI in transgenic mice, we previously demonstrated that transplanted MSC could fuse with recipient cardiomyocytes in the infarcted area (Noiseux, et al., 2006). MSC from wild type C57BL/6 mice were retro-virally transduced to express GFP as reporter gene and Cre recombinase. These MSC were transplanted into infarcted hearts of histocompatible R26R mice (similar C57BL/6 genetic background). In these transgenic mice, a loxP-flanked stop sequence is present 5' of the LacZ expression cassette to prevent transcriptional read-through until selective excision by Cre mediated recombination from implanted MSC (Soriano, 1999). Thus, the LacZ gene is expressed exclusively after a donor MSC expressing Cre fuses with a recipient cell from the R26R mice. Consequently, X-gal staining is used to detect cellular fusion events (Alvarez-Dolado, et al., 2003, Nygren, et al., 2004) As early as 3 days following MSC injection into the ischemic heart, we observed cellular fusion with individual blue cells having typical cardiomyocyte morphology. Interestingly, the majority of the fusion events with $LacZ^{+}$ cells were detected within the infarct border zone, in areas with viable cardiomyocytes.

4.3 Indirect and paracrine effects

Despite the lack of significant myocardial regeneration by implanted stem cells through differentiation into new cardiomyocytes, numerous studies reported that cardiac function can be improved by cell therapy, suggesting further evidence for a non-myogenic pathway of cardiac repair (Laflamme, et al., 2007). Regardless of the mechanism responsible for the beneficial effects of stem cell therapy on post-MI, cardiac function improvement remains disproportionate to the degree of cardiomyogenic differentiation. (Liao, et al., 2007, Noiseux, et al., 2006) Indeed, the variable observations relating to cell transdifferentiation and fusion, as well as transient cell retention and survival in the ischemic heart (Noiseux, et al., 2006), have prompted a rethinking of the mechanisms that account for the functional benefits of cell therapy in cardiac repair (Psaltis, et al., 2008). It has been proposed that stem cell transplantation may improve myocardial function recovery predominantly by facilitating endogenous repair mechanisms, rather than through direct regeneration of lost cardiomyocytes and vascular cells. Thus, it is possible that implanted cells could secrete bioactive factors which may stimulate angiogenesis, suppress apoptosis of cardiomyocytes, increase efficiency of cardiomyocyte metabolism, improve inotropy of survival myocytes, modulate interstitial matrix composition and remodeling, and maybe even recruit cardiac resident stem cells (Gnecchi, et al., 2006, Kamihata, et al., 2001, Kinnaird, et al., 2004, Noiseux, et al., 2006, Wollert, 2008). The fact that it is possible to improve the heart's function without remuscularizing the infarct is already known from numerous pharmacological studies, e.g. with beta-adrenergic blocker or angiotensin-converting enzyme inhibitors and this has resulted in a search for other potential mechanisms of action (Laflamme, et al., 2007).

MSC are known to secrete a wide spectrum of biologically active factors, (Heil, et al., 2004, Kamihata, et al., 2001, Kinnaird, et al., 2004, Kinnaird, et al., 2004, Kinnaird, et al., 2004) and under hypoxic culture condition and serum starvation, conditions that mimic myocardial infarct, the expression of several factors is significantly up-regulated (Gnecchi, et al., 2005, Gnecchi, et al., 2006, Kinnaird, et al., 2004, Noiseux N, 2004, Noiseux N, 2004). These growth factors and cytokines are released in the culture medium (MSC-conditioned medium, MSC-CM), and can be concentrated and use therapeutically. We previously reported that transplantation of MSC into ischemic myocardium improved cardiac function recovery and repair as early as 72 hours, which is too early to be explained solely by myocardial regeneration from MSC (Gnecchi, et al., 2005, Gnecchi, et al., 2006, Noiseux, et al., 2006). Transplanted cells were found within the infarcted myocardium (detection of MSC by GFP immunoreactivity). Early massive engraftment was observed at 3 days, but the number of implanted MSC decreased significantly over time, and by 28 days post-MI very few cells remained. Since such an early recovery cannot be explained entirely by de novo myogenesis, we proposed that MSC could achieve cardioprotection by indirect effects, through paracrine mediators, rather than cardiac regeneration. Furthermore, to investigate how MSC could achieve protection of the ischemic myocardium by paracrine mediators and indirect effects, we demonstrated that MSC-CM reduced hypoxia-induced apoptosis, and also triggered vigorous spontaneous contraction of isolated hypoxic adult rat cardiomyocytes in vitro, suggesting the presence of anti-apoptotic and inotropic factors (Gnecchi, et al., 2006). Moreover, when injected directly into infarcted rat hearts, the MSC-CM limited infarct size as early as 72 hours and improved ventricular function (Gnecchi, et al., 2005, Gnecchi, et al., 2006) at levels comparable to those observed following MSC transplantation. These remarkable data strongly support the concept that the effects observed after intramyocardial injection of MSC into infarcted hearts are to a great extent attributable to paracrine protection and indirect effects on ischemic myocardium.

Recently, we investigated the effects of MSC-secreted growth factors on extent of early recovery from MI in a large animal pre-clinical model (Nguyen, et al., 2010). Swine subjected to acute MI by temporary balloon occlusion of the left anterior descending coronary artery using percutaneous techniques received intra-coronary injection of either concentrated MSC-derived growth factors or control medium. Treatment with MSC-derived factors significantly reduced cardiac tropinin-T elevation, and improved echocardiographic parameters, including fractional area shortening, stroke volume, and cardiac output. Quantitative evaluation of fibrosis by Verhoff staining revealed a reduction of the fibrotic area in the infracted zone at 7 days. Similarly, Masson's trichrome staining revealed reduced myocardial damage as demonstrated by areas of relatively preserved myocardium in the infracted area. TUNEL assay demonstrated less cardiomyocyte apoptosis. Protein array detected the presence of angiogenic (VEGF, endothelin, epiregulin), anti-apoptic (Galectin-3, Smad-5, sRFP-1), and anti-remodeling factors (TIMP). RT-PCR confirmed the expression of these factors. In summary, a single intra-coronary injection of concentrated biologically active factors secreted by MSC could achieve early protection of ischemic myocardium, improve cardiac repair, and partially preserve contractility. MSC-derived growth factors injection (rather than MSC themselves) should be evaluated as a novel therapy to treat ischemic heart disease, many practical and technical issues of cell therapy.

4.4 Healing response, inflammation and tissue perfusion

Following myocardial infarct, inflammation and cytokine production regulate myocytes survival/apoptosis, and trigger additional cellular responses, contributing to the healing and remodeling of the injured tissue, which may ultimately influence clinical outcome (Nian, et al., 2004). The immune system plays a crucial role during the acute phase of MI and this constitutes a most appropriate timing where any immunomodulatory effects following stem cell implantation could exert their effects (Laflamme, et al., 2007). An adequate microenvironment supporting nutrient delivery and waste removal is necessary to sustain survival, growth and possibly differentiation of the transplanted cells (Timmermans, et al., 2003). Therefore, concurrent revascularization must keep pace with cell repopulation of the infarcted tissue (Luttun and Carmeliet, 2003). Transplanted cells may act in an indirect supportive role, optimizing the milieu for host vasculature and cells to respond to ischemia and healing. (Gnecchi, et al., 2005, Kinnaird, et al., 2004, Kinnaird, et al., 2004, Noiseux N, 2004) Normal growth and ultimate stem cell fate depend on engraftment in an appropriate "niche". Thus, it appears that the fate of stem cells is determined by the environment in which they are engrafted rather than by an intrinsically programmed fate, but the mechanisms explaining how the local milieu influences stem cell differentiation are yet to be characterized (Strauer and Kornowski, 2003). Importantly, an angiogenic response has been observed in many studies following cell transplantation (Dowell, et al., 2003).

Much attention has been focused on the direct consequence of cell-based therapy by the renewal of lost parenchymal cells, but restoration of the extra-cellular matrix and vascular supply are also important issues since both components are essential for structural and functional support of the surviving cardiomyocytes (Timmermans, et al., 2003) in the ischemic myocardium. It is also important to consider other potential mechanisms by which transplanted cells could benefit cardiac function without directly contributing to systolic contraction. Blood vessel formation in the heart proceeds mainly through 2 mechanisms: angiogenesis and arteriogenesis. Arteriogenesis is by far the most efficient adaptative mechanism for survival of ischemic organs because of its ability to conduct relatively large blood flow (Buschmann and Schaper, 1999). Angiogenesis and arteriogenesis are complex processes sharing several common mechanisms of action, growth factors and cytokines dependency (Buschmann and Schaper, 1999). Many of these cytokines act not only in a co-ordinated time- and concentration-dependent manner, but one cytokine may potentiate (or inhibit) the effect of another (Kinnaird, et al., 2004), and the complexity of the process of collateral formation has led to speculate that multiple factor strategies could be used therapeutically to modulate vessel formation (Kinnaird, et al., 2003). The recruitment of monocytes that differentiate into macrophages and produce abundant angiogenic growth factors such as VEGF, NO, monocyte chemoatractant protein-1 (MCP-1) and other cytokines, is also essential and ultimately leads to endothelial and smooth muscle cell proliferation, migration, vessel remodeling and extra-cellular matrix synthesis.

The observation that BMSC contain a subpopulation of endothelial progenitor cells with the potential for angiogenesis suggests that these cells may contribute to cardiac repair by enhancing the local blood supply in ischemic myocardium (Grigoropoulos and Mathur, 2006). Some reports have suggested that MSC implantation into ischemic tissue promote collaterals development through paracrine mechanisms, but not through direct cell incorporation into growing vasculature (Dowell, et al., 2003, Fuchs, et al., 2001, Heil, et al., 2004, Hirata, et al., 2003, Kamihata, et al., 2001, Kinnaird, et al., 2004, Kinnaird, et al., 2004, Kinnaird, et al., 2003, Kobayashi, et al., 2000, Li, et al., 2002, Mangi, et al., 2003, Strauer, et al., 2002, Ziegelhoeffer, et al., 2004). MSC express genes encoding a broad spectrum of cytokines with angiogenic properties including VEGF, HGF, FGF, MCP-1, PGF, IL-1 and IL-6, SDF-1, MMP-9 (Kinnaird, et al., 2004, Kinnaird, et al., 2004, Timmermans, et al., 2003). These cytokines can be found in the media of cultured cells, and have all individually been shown to have positive effects on experimental blood flow recovery (Buschmann, et al., 2003, Heil, et al., 2004, Kinnaird, et al., 2004, Kinnaird, et al., 2004, Kinnaird, et al., 2004, Timmermans, et al., 2003). Moreover, media collected from MSC cultures (MSC-derived conditioned medium, MSC-CM) promote in vitro proliferation and migration of endothelial cells (EC) and vascular smooth muscle cells (VSMC), and enhance in vivo collateral blood flow recovery when injected into ischemic hindlimb (Kinnaird, et al., 2004, Kinnaird, et al., 2004, Timmermans, et al., 2003). It is possible that neovascularization by cell therapy is leading to enhanced blood supply in the peri-infarct region and thereby promoting salvage of stunned, hibernating, or otherwise susceptible cardiomyocytes (Laflamme, et al., 2007). Controversy persists regarding the exact mechanisms through which cell transplantation may enhance repair and collateral development in ischemic tissue (Kinnaird, et al., 2004).

5. Lessons from early clinical trials: feasibility, efficacy and safety

Over the past few years, cell therapy has emerged as a potential new treatment of a variety of cardiac diseases, including AMI, refractory angina, and chronic heart failure. (Menasche, 2009) A variety of cell types have been tested experimentally, each of them being usually recognized by its beneficial "regeneration" potential. This has led to a flurry of clinical trials entailing the use of skeletal myoblasts or bone marrow-derived cells either unfractionated or enriched in progenitor subpopulations. Following acute MI, the observation that mobilization of BMSC occurs naturally to heal the myocardium led several groups to investigate their potential for cardiac repair (Shintani, et al., 2001). In patients with recent MI, Strauer et al demonstrated that IC administration of unselected BMSC was safe and associated with improved cardiac performance (Strauer, et al., 2002). In the same way, Zeiher et al reported that IC administration of BM progenitors after MI improved cardiac performance and enhanced the myocardial perfusion of the infarcted myocardium up to 1 year of follow-up (Assmus, et al., 2002, Britten, et al., 2003, Schachinger, et al., 2004). The landmark multi-center placebo-controlled REPAIR-AMI trial showed also a higher increase in the LVEF in the cell treated group compared to controls (Schachinger, et al., 2006), this effect was sustained up to 2 years and was associated with a significant reduction in the MACE rate (Assmus, et al., 2010). Recently, the FINCELL trial showed also that following acute MI, BMSC therapy significantly improved the LVEF recovery in treated patients as compared to controls (Huikuri, et al., 2008). Interestingly the randomized "BOOST" trial showed an improvement in the cardiac function following IC injection of autologous BMSC without adverse effect. Furthermore, this beneficial effect was sustained at 6 months (Wollert, et al., 2004) and 1 year of follow-up (Schaefer, et al., 2006), but not at 18 months15 raising questions about the potential transient effect of heterogeneous cells on cardiac repair (Meyer, et al., 2006).

Although many trials demonstrated a slight benefit on cardiac functional recovery, recent studies have yielded disparate results (Janssens, et al., 2006, Lunde, et al., 2006, Tendera, et al., 2009). In the randomized ASTAMI trial, the use of IC unselected BMSC following acute MI failed to demonstrate at 6 months any improvement in LVEF (Lunde, et al., 2006). Similarly, Janssens et al reported no significant difference in overall LVEF at 4 months, but showed a decreased infarct size and better regional function in cell-treated patients (Janssens, et al., 2006). Moreover, the REGENT acute MI trial failed to show any significant improvement in the LVEF, LVEDV and LVESV at 6 months in patients treated with unselected mononuclear cells or selected CD34+ CXCR4+ cells as compared to controls. Finally, Lunde et al (Lunde, et al., 2008) failed to show any improvement in the LV function in patients with anterior MI treated with BMSC.

In the most recently published SCAMI (Wohrle, et al., 2010) (Stem Cell therapy in patients with Acute Myocardial Infarction) study which used serial MRI for assessing results in 42 patients, of whom 29 were allocated to the treated arm), there was no either evidence for a positive effect of intracoronarily infused stem cells treatment versus placebo with regard to LVEF, volumes or infarct size. In this study, the centrifugation technique used for collecting mononuclear cells was similar to that used in the REPAIR-MI trial and a greater number of cells (381×10^6 versus 240×10^6 in REPAIR-MI) was injected at was has been identified as the optimal time point, i.e., at a median of 6.1 days after infarction, a noticeable difference between these two conflicting trials being the longer interval between symptom onset and revascularization in the SCAMI patients (14.3 hours versus 4.5 hours in REPAIRMI). However, a salient feature of the SCAMI protocol has been the rigorousness of the blinding since the control preparation consisted of autologous erythrocytes, which made the placebo syringes indistinguishable from those of the treated group), the design of the SCAMI trial strongly validates its conclusions.

Put altogether, these data clearly show that the potential benefit of bone marrow-derived stem cell therapy shortly after AMI still remains conflicting and the major lesson drawn from this first wave of clinical studies is therefore that there is a real need for a large, adequately powered trial incorporating some of the key findings of the previous studies regarding cell preparation, dosing and timing of delivery and focusing on clinically relevant "hard" end points such as mortality, re-infarction and heart failure. Four recent meta-analyses of BMC in the setting of acute MI incorporating 5, 10, 13, and 18 trials, helped to position results of individual trials into perspective (Abdel-Latif, et al., 2007, Hristov, et al., 2006, Lipinski, et al., 2007, Martin-Rendon, et al., 2008). Overall, results of these randomized placebo-controlled trials and cohort studies are promising in that they demonstrate feasibility, safety, and a slight but positive improved LVEF of 3.66%, a reduced infarct scar size of -5.49% and a reduced LVESV volume of -4.80 ml (Abdel-Latif, et al., 2007). Thus, available evidence suggests that BMSC transplantation is associated with modest improvement in physiologic and anatomic parameters in patients with both acute MI and chronic ischemic cardiomyopathy, but above and beyond conventional therapy (Abdel-Latif, et al., 2007) and could leads to significant and longstanding reduction in the mortality of patients after acute MI as demonstrated in the Balance study (Martin-Rendon, et al., 2008). Therefore, therapy with stem cells appears to be safe, but well-designed double-blinded and randomized studies are clearly needed to confirm promising findings from early studies.

To date, fewer randomized trials of IC trans-catheter transplantation of BMSC have been performed in the setting of chronic coronary artery disease and chronic heart failure (Assmus, et al., 2007, Erbs, et al., 2005, Tse, et al., 2007). Nonetheless, the results are similar to those in patients with acute MI, showing an improvement in LV function, perfusion, and relief of angina pectoris (Losordo, et al., 2007, van Ramshorst, et al., 2009). Assmus et al. reported a registry of 121 consecutive patients with chronic ischemic heart disease treated with intracoronary infusion of BMC (Assmus, et al., 2007). Importantly, they demonstrated that infusion of high number of colony forming cells is associated with a significantly lower mortality during further follow-up (Assmus, et al., 2007). Recently, Strauer et al. (Strauer, et al., 2010) reported in patients with chronic ischemic cardiomyopathy treated with intracoronary BMSC therapy a significant improvement in the LVEF with a significant decrease in long-term mortality as compared to the control group over 3 months to 5 years of follow-up. These 2 studies were not blinded and randomized, and either lacked or only had a matched cases control groups.

5.1 What about safety?

As often in medicine, the hype generated by the early uncontrolled and small-sized clinical trials has been dampened by the marginally successful outcomes of the subsequent, more rigorously conducted randomized trials. (Menasche, 2009) Although they may have failed or succeed to achieve their primary functional end points, these trials have been positive in the sense that they have allowed to identify a major key issue regarding the safety and feasibility of this therapy. Cardiac cell therapy is overall safe in surgical studies where cells have been injected with hand-held syringes directly in the myocardium, no bleeding complications have been reported. Catheter-based endocardial injections have been equally safe without an increased risk of re-infarction, stent thrombosis or in-stent restenosis (Zhang, et al., 2009). Although the use of the intramyocardial injection of myoblasts was associated with the occurrence of sustained ventricular arrhythmias (Menasche, 2008), none of the trials testing the BMNC has reported an increased incidence of malignant arrhythmias (Menasche, 2011). Likewise, there has been no report of cell-derived tumor formation in the myocardium or elsewhere. This is clinically relevant as the longest follow-up periods now span almost 10 years (Menasche, 2009, Zhang, et al., 2009).

5.2 What about used endpoints in the previous trials?

So far, ejection fraction (EF) has usually been the gold standard for assessing outcomes in the first generation of clinical trials, regardless of the method on which its calculation was based (echocardiography, angiography, radionuclide imaging or magnetic resonance imaging which is likely the most reliable but may not be always possible because of a previously implanted ICD). There is mounting evidence that changes in EF may not be the most suited criteria for assessing the effects of cell therapy. Quantitative assessment of regional systolic function, such as echocardiographic strain rate, could be more sensitive than measuring global LVEF for the evaluation of cell therapy after AMI. (Wollert and Drexler, 2010) In trials exploring a new cell type in a limited number of patients, measurements in regional geometry and function with state of-the-art imaging modalities may be more useful than global assessments for establishing the proof of concept and providing mechanistic cues (Herbots, et al., 2009).

5.3 Selected vs. unselected cells for myocardial transfer?

There is currently uncertainty as to which of the stem cell population is most potent in stimulating angiogenesis and cardiac repair. Nevertheless, the use of hematopoietic progenitor stem cells appears to uniformly induce neovascularization that seems to be a prerequisite of the successful functional repair. The hematopoietic stem cells are characterized by the presence of the surface marker CD34. In addition, CD133 has been identified as a marker that is present on the stem cells that co-express not only CD34 but also other markers such as c-kit. It is hypothesized that use of $CD133^+$ cells may involve larger and more primitive group of stem cells than selection based only on the use of $CD34^+$ marker. The use of non-homogenous or un-selected stem cells may contribute to the

regeneration of necrotic myocardium and blood vessels, but does not allow the characterization of optimal cell type for cardiac repair. Moreover, different cell types may compete for the engraftment in the injured myocardium (Rosenzweig, 2006). The group of Drexler showed that selected BMSC displayed a 7-fold higher retention in the infarcted myocardium, when compared to unfractionated and unselected BMSC (Hofmann, et al., 2005). Furthermore, Stamm (Stamm, et al., 2003) used CD133+ cells for intramyocardial injections performed during an open-chest procedure, and observed an improvement in the tissue perfusion and LVEF during follow-up. Experimental studies demonstrated that selected, well-defined hematopoeitic stem cells contribute to cardiac repair of the acutely infarcted myocardium by inducing neovascularization, inhibition of apoptosis and cardiomyogenesis. Indeed, the hematopoietic CD133+ cells possess high engraftment, multipotent and angiogenic capacity, and appear to be valuable for cardiac repair in experimental myocardial infarction (Bhatia, 2001, Kuci, et al., 2003, Quirici, et al., 2001). Menasche showed in the situation of post-MI scars, transplantation of $CD133^+$-derived BMSC improves cardiac function, and benefit was similar to that afforded by myogenic cells. (Agbulut, et al., 2004)
Bartunek et al. tested in a phase I study the feasibility, safety and functional effects of intracoronary administration of selected $CD133^+$ BMSC in patients with recent MI. (Bartunek, et al., 2005) They noted a significant increase of 7% in the LVEF in the treated group compared to 3% in controls at 4 months of follow-up. Importantly, they showed in cell-treated patients that improvement of the LV function was paralleled with increased myocardial perfusion and viability (Bartunek, et al., 2005). However, they noted in the treated patients an increase of coronary events with greater in-stent proliferation and larger luminal loss in the non-stented segments of the infarct-related artery that resulted in a significant decrease in pressure-derived FFR (Fractional Flow Reserve) (Mansour, et al., 2006) mainly in early compared to late intracoronary CD133+ cells administration (Vanderheyden, et al., 2007). Yet, these data were obtained from retrospective analysis; they lack randomized controls and systematic use of intravascular ultrasound imaging to track changes in the vascular wall.

5.4 The COMPARE-AMI trial

Our group initiated the first Canadian clinical trial evaluating the intracoronary injection of autologous highly selected $CD133^+$ bone marrow-derived stem cells in patients presenting acute MI treated by percutaneous intervention and intra-coronary stent implantation: the COMPARE-AMI trial (Mansour, et al., 2010, Mansour, et al., 2009). This is a randomized, double blind, placebo controlled study. We are investigating the safety, feasibility and efficacy, and the change in the coronary atherosclerotic burden progression in the treated artery, in addition to the change in LVEF measured by MRI. For more details on the study, please visit http://www.anzctr.org.au, study#ACTRN12609001045202. We published a preliminary safety analysis on the first twenty patients that were successfully randomized and treated in the COMPAREAMI trial. The mean age was 52.2±8.9 years with a predominance of males (90%); culprit lesion was located on the left anterior descending artery in 90%, and peak troponin and CKMB were 10.5±8.3 Ug/L and 341±260 U/L, respectively suggesting large infarct. (Mansour, et al., 2011) To date, there is no protocol-related complication to report such as death, MI, stroke, or sustained ventricular arrhythmia. Re-PCI was necessary at 4 months of follow-up in three patients to treat bare-metal stent restenosis. These patients were asymptomatic; however, silent ischemia was documented in the target territory. Baseline fractional flow reserve (FFR) was significantly lower in the stented culprit artery compared to the non-culprit artery at baseline: 0.88±0.05 vs 0.96±0.04, $P<0.001$. However, at 4 months of follow-up (n=20), no significant difference was found in the delta FFR compared to baseline in the culprit vs. non-culprit artery (−3.7%±5.4 vs −1.1%±4.6, respectively, P=0.148), suggesting no acceleration of the atherosclerosis by the treatment. Finally, at 4 months of follow-up, MRI assessment of the LVEF (n=18 patients) showed a significant improvement compared to baseline with LVEF 51.1%±2.5 vs 41.2%±1.1, respectively (P<0.001;). This improvement was sustained at 12-month followup (52.3% ± 2.0, P < .001 versus baseline, n = 13). Fig. 5. illustrates the procedure and typical results obtained in patients enrolled in our research protocol.

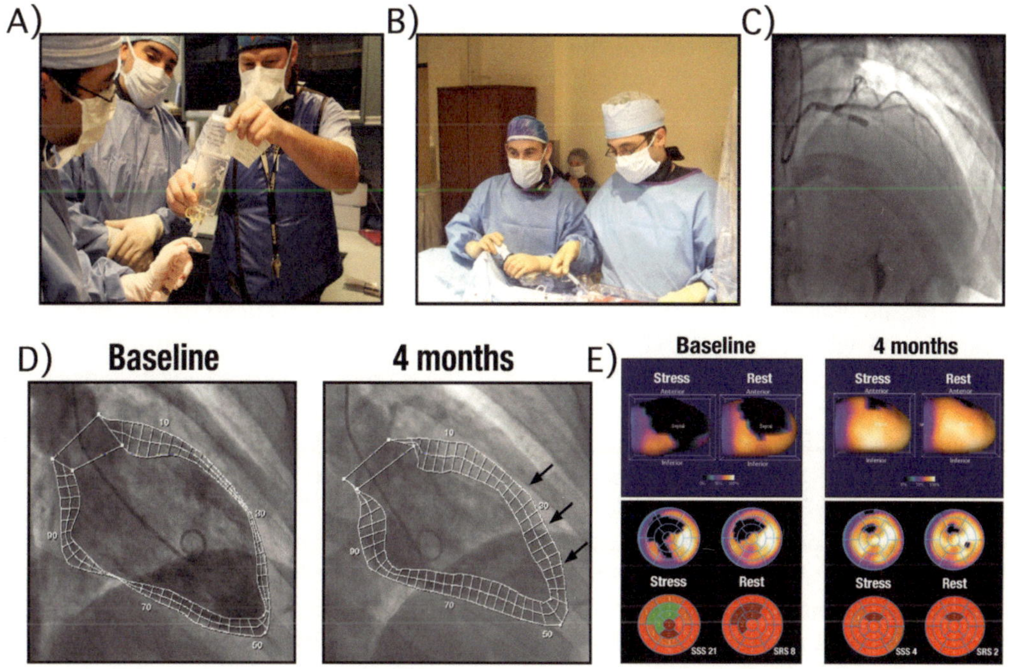

Fig. 5. A) Selected CD133+ bone marrow-derived cells or placebo is injected intracoronary in patients participating to the COMPARE-AMI trial. B) Using a small catheter under fluoroscopic guidance, the cells or placebo are injected in the stented coronary artery. C) During injection, a balloon is occluding the coronary artery proximally. D) A representative left heart ventriculography showing a near complete recovery of cardiac function in a 54-year-old man after his randomization in COMPARE-AMI trial. E) Technetium-99 m Sestamibi single-photon emission computed tomography (MIBI SPECT) scintigraphy looking at myocardial perfusion. The same patient as in D. A, B and C from Dr Noiseux and Dr Mansour, unpublished data. D and E, adapted from (Mansour, et al., 2010).

In conclusion, the findings of the present studies are as follows: i) BMC therapy after MI provide a faster improvement in systolic cardiac function, including LVEF, LVESV and LVEDV, compared with controls. These improvements were sustained after at least six months; ii) Statistically and clinically significant benefits were observed in the regional cardiac anatomy, but these did not provide a physiological benefit; iii) In the baseline-impaired LVEF subgroup, LVEF improved after BMC therapy compared with the control treatment; iv) BMC therapy was safe, but a reduction in cardiovascular events was not observed; v) Subgroup analyses suggested that cell infusion after AMI had a positive effect on LVEF. Cell infusion within 4-15 days with a higher number of CD34+ cells may have a beneficial impact on LVEF; and vi) Selected BMC are safe and may have a beneficial effect on the healing of the infarcted myocardium.

Recently, we initiated a second study in our stem cell therapy program, looking at a different patient population presenting chronic ischemic cardiomyopathy associated with reduced LV function. IMPlantation of Autologous CD133$^+$ sTem cells in patients undergoing coronary artery bypass grafting (CABG) surgery or IMPACT-CABG trial is a randomized, prospective, double blind, placebo-controlled phase II clinical trial designed a to assess the safety, feasibility and functional benefit of intramyocardial injection of autologous CD133$^+$ BMSC as compared to placebo at the time of CABG surgery. The first 5 patients were treated in an open label fashion and received 10 millions autologous CD133+ cells with CABG. For more details, please visit http://clinicaltrials.gov, study #NCT01033617. To date, 7 patients were enrolled, and no protocol related complications were observed. Our preliminary work suggests the safety of CD133$^+$ autologous cells for cardiac repair, and possibly beneficial effects on cardiac function are observed.

5.5 IMPACT-CABG trial: case report, presentation of the first treated patient

A 59 year-old male with angina at rest (CCS class 4) and congestive heart failure symptoms: dyspnea on slight exertion (NYHA functional class III) and edema of the lower extremities was referred for coronary artery bypass graft (CABG) surgery. He was an active smoker and known for hypertension, type II diabetes since 20 years treated with insulin and a slight chronic renal insufficiency (creatinine 134 umol/L, estimated glomerular filtration rate of 47 ml/min/1.73m^2). His coronary angiogram showed a left dominance system with 90% stenosis on the left anterior descending coronary artery (LAD), 80% on the first diagonal branch and 80% on the posterior descending artery (PDA). In addition, he was known to have a severe left ventricular (LV) dysfunction as assessed by left ventriculography (30%) and echocardiography (LVEF 35-40%). He consented to participate in the IMPACT-CABG study, a phase II clinical trial testing the safety and feasibility of selected CD133$^+$ bone marrow stem cells. Pre-operative stress echocardiography and magnetic resonance imaging (MRI) depicted on the left ventricle necrosis of the apical segment with aneurysm, hypokinesia of mid and basal regions the antero-septal and antero-lateral segments, and akinesia of the infero-apical segments. On the morning of the surgery, the patient underwent bone marrow (BM) aspiration from the iliac crest under local anesthesia. Stem cells were prepared in the cell therapy laboratory and CD133$^+$ cells were purified using the CliniMACS® CD133 Reagent System from Miltenyi Biotech Inc®. On the evening of the same day, the patient underwent CABG surgery and received the left internal thoracic artery on the LAD and a saphenous vein graft on the PDA. Immediately following distal anastomoses, autologous CD133$^+$ cells (10 millions cells, 15 injection sites) were injected directly into the myocardium using a 26g needle in the anterior and lateral wall of the left ventricle (Figure 6). The aortic cross-clamp time and the total cardiopulmonary bypass (CPB) time were 29 minutes and 45 minutes respectively. The peri-operative course was uneventful without any in-hospital complication related to neither the research protocol nor the surgery. The patient was discharged from the hospital after 7 days. At 6 months follow-up, the patient symptoms improved to NHYA class I and LVEF was increased to 60% assessed by echocardiography (Table 1). Regional motion also improved: contractility of the apical region enhanced significantly, and the left antero-septal segments were only slightly hypokinetic. The MRI study demonstrated a spectacular improvement of the perfusion in all territories with disappearance of the ischemia in the antero-apical and inferior territories and with mild

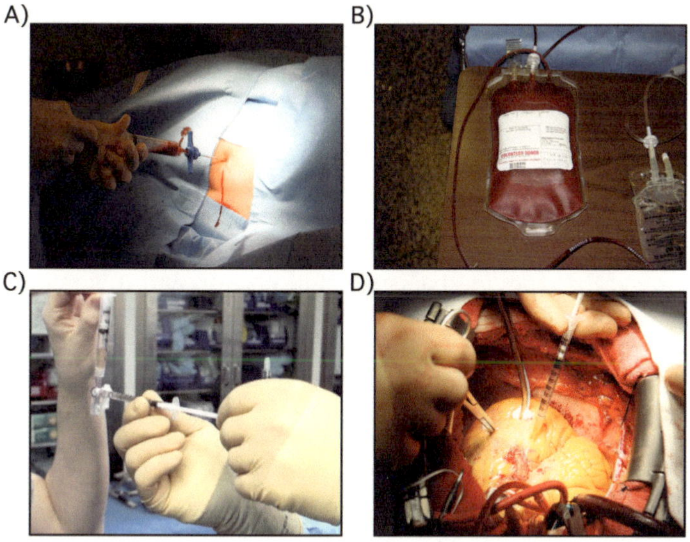

Fig. 6. A) Bone marrow aspiration in the iliac crest in the IMPACT-CABG trial protocol. B) Bone marrow is transferred into a blood collection bag with heparin. C) After CD133+ stem cells isolation and selection using the CliniMACS system from Miltenyi Biotech© cells are transferred into 1.0 ml syringe for injection. D) Intra-operative injection of the $CD133^+$ stem cells into the infarcted area and infarct border zone. Dr Noiseux, unpublished data.

	Baseline	6 months post-$CD133^+$
Echocardiography:		
LVEF bi plan %	41	60
WMS	37	22
WMSI	2.3	1.4
MRI:		
LVEDV ml (ml/m^2)	179 (107)	178 (106)
LVESV ml (ml/m^2)	111 (66)	94 (56)
LVEF %	38	48
LV mass gr (gr/m^2)	118 (70)	140 (83)
Stroke volume ml	68	84

Table 1. Echocardiography and magnetic resonance imaging results at baseline and at 6 months post $CD133^+$ injections. LVEF: left ventricular ejection fraction; WMS: wall motion score; WMSI: wall motion score index (normal=1); LVEDV: left ventricular end diastolic volume; LVESV: left ventricular end systolic volume; MRI: magnetic resonance imaging.

ischemia in the antero and inferoseptal basal segments. Moreover, the LV dilatation was reduced, with smaller volume and an increased myocardial mass. No arrhythmia was detected by 24 hrs Holter monitoring.

We believe $CD133^+$ stem cells to be amongst the most potent cells for myocardial repair. This work represents the first Canadian experience with $CD133^+$ stem cells for the treatment of chronic ischemic cardiomyopathy. The remarkable and encouraging results from the fist patient support the continuation of IMPACT-CABG trial, and by randomization between CABG combined to stem cell versus CABG alone, this trial will prove the safety of the procedure and possibly the beneficial effects of the cellular therapy. This novel therapy may become an important therapeutical adjunct to conventional treatment for coronary artery diseases.

6. Pitfalls and important issues on cell therapy

Although great enthusiasm was created by the possibility of reconstituting the damaged heart by cell therapy, the exact mechanism is still unclear, and it is possible that findings supporting myocardial regeneration by stem cells differentiation possibly result form technical artifacts. Controversy exists surrounding the ability of BMSC to undergo transdifferentiation, as some techniques that have been used to demonstrate this phenomenon have been questioned (Grigoropoulos and Mathur, 2006). Current failure to label the donor cells adequately and to follow them in vivo makes it very difficult to distinguish them from background tissue and could lead to misinterpretation (Hassink, et al., 2003). Indeed, the use of GFP reporter gene is attractive because it is compatible with a variety of imaging techniques, but dead and dying cardiomyocytes have an autofluorescent spectrum that partially overlaps with that of GFP. (Burdon, et al., 2011) After injury, autofluorescence increases due to accumulated lipofuscin, blood-derived pigments, and other intrinsic fluors such as flavins and reduced nicotinamide adenine dinucleotide (NADH) (Laflamme and Murry, 2005). Evidence for regeneration includes colocalization of GFP fluorescence from donor cells, with immunostaining for cardiomyocytes markers, including sarcomeric actin. Thus, it is possible to misidentify a GFP positive cardiomyocyte as the result from a donor GFP positive cell.

An important issue limiting cell therapy is the extensive cell death following transplantation into the ischemic heart. The survival of implanted cells is often limited in the infarcted tissue (Koc and Gerson, 2003, Zhang, et al., 2001) and cell death is worsened by the hostile environment caused by the reduced blood flow, hypoxemia, inflammation and scarring in the ischemic myocardium. Regardless of cell type, multiple studies suggest that more than 90% of cells successfully delivered to the heart will die within one week (Laflamme and Murry, 2005, Noiseux, et al., 2006, Zhang, et al., 2001). Most cells die within hours of transplantation because of interplay of ischemia, inflammation, and apoptosis. (Menasche, 2009, Rosenzweig, 2006) Recent clinical trials yielded inconsistent data in cardiac function reporting mixed results. (Janssens, et al., 2006, Lunde, et al., 2006, Schachinger, et al., 2006) These conflicting data have re-ignited interest in the unresolved questions regarding the biology of candidate cells, and how to improve these cells for clinical therapy. Our group investigated the over-expression of anti-apoptotic proteins such as Akt, to improve the cell survival, but also the reparative effects of MSC therapy in an animal model of acute MI (Gnecchi, et al., 2005, Gnecchi, et al., 2006, Mangi, et al., 2003, Noiseux, et al., 2006). We genetically engineered MSC to over-express the pro-survival gene Akt1 (Datta, et al., 1999,

Franke, et al., 2003, Franke, et al., 1997) (Akt-MSC). Akt over-expression resulted in better protective and anti-apoptotic effects against ischemia in vitro, but also in vivo as shown by the reduction of the proportion of apoptotic cells following implantation into rat infarcted myocardium and increased cell retention. Moreover, intra-cardiac injection of $5x10^6$ Akt-MSC inhibited the process of cardiac remodeling and resulted in normalization of cardiac function at 2 weeks, to a level indistinguishable from sham-operated animals (Mangi, et al., 2003). Although improvements in cardiac function were also observed in other studies, complete recovery of LV function occurred only following transplantation of MSC over-expressing Akt, and these results represent a new threshold in cardiac regeneration using cell therapy (Koc and Gerson, 2003). This cell-based therapy, combined with a gene-therapy approach, has the potential to address multiple issues in cell availability and scalability.

A clear understanding of events stimulated by and consequent of transplant of various cell types is critical to avoid potentially adverse consequences. (Evans, et al., 2007) In particular, appropriate electrical and mechanical coupling in the myocardium is essential for optimal cardiac function and avoiding lethal arrhythmias such as observed with skeletal myoblasts. An additional issue of possible concern may be tumorigenesis. (Evans, et al., 2007) Recent experimental study demonstrated that engraftment of undifferentiated embryonic stem cells into the myocardium could result in tumor formation. (Cai, et al., 2007)

The identification of the appropriate route for cell administration to the damaged heart is an essential prerequisite for successful tissue repair (Strauer and Kornowski, 2003). The goal of any cell delivery strategy is to transplant sufficient number of cells into the myocardial region of interest and to achieve maximum retention of cells within that area (Kocher, et al., 2007). Furthermore, the success of cell delivery is determined by the local milieu since it will influence short-term cell survival, cell properties in regard to cell adhesion, transmigration through the vascular wall, and tissue invasion. High cell concentrations within the area of interest and prevention of homing of transplanted cells into other organs are required, therefore targeted and regional administration of cells are preferred (Strauer and Kornowski, 2003). Cell homing, adhesion, transmigration through the vascular wall, and tissue invasion involves many complex steps.

The most frequently used routes of cell delivery for ischemic cardiomyopathy are percutaneous intra-coronary injection, percutaneous endomyocardial using 3D-guiding systems, and direct intra-myocardial injection during open chest cardiac surgery procedure. Intra-coronary sinus or intravenous systemic deliveries have also been described, as well as stem cell mobilization following G-CSF treatment as described in the FIRSTLINE-AMI trial (Ince, et al., 2005)[5]. Intra-coronary infusion requires migration through the vessel wall into the ischemic heart tissue, which is helped by damaged and permeabilized endothelium as found after acute myocardial infarct. Cells like bone marrow-derived and blood-derived progenitor cells are known to extravasate and migrate efficiently into ischemic areas, whereas skeletal myoblasts do not (Kocher, et al., 2007). The intravenous route is the easiest, but the main disadvantage is that approximately only 3% of normal cardiac output will flow through the left ventricle via the coronary arteries, and is limited because of transpulmonary first-pass captation and sequestration (Strauer and Kornowski, 2003).

The intra-coronary route has been used safely and effectively to achieve selective administration and higher first-pass delivery to the heart than systemic therapy (Psaltis, et al., 2008). However, there have been reports of micro-circulation obstruction following intra-coronary infusion of satellite cells and MSC, resulting in embolic myocardial damage and sub-acute microinfarction (Vulliet, et al., 2004). Unlike intra-vascular infusion, direct intramyocardial injection targets specific regions of myocardium without relying on the up-regulation of inflammatory signals to assist transvascular cell migration and tissue invasion (Psaltis, et al., 2008). Pre-clinical and clinical results suggest that direct MSC injection may result in less non-cardiac cell entrapment than intra-coronary or systemic intravenous infusion, along with better retention culminating in greater benefit for cardiac function recovery (Freyman, et al., 2006, Heldman and Hare, 2008, Perin, et al., 2008). An intra-myocardial injection delivery approach appears specially well-suited for transplantation of larger and adherent cells (e.g. MSC) and particularly relevant to chronic myocardial diseases such as chronic ischemic cardiomyopathy or dilated failing heart (Psaltis, et al., 2008). After more than a decade since the beginning of cell therapy, no one has identified the technique for optimal administration of stem cell into the heart. As a result, more than 90% of the cells delivered to the heart through a needle are lost to the circulation or leak out of the injection site (Laflamme and Murry, 2005). Indeed, mechanical leakage and washout may account for a major proportion of cell loss after cell implantation (Kocher, et al., 2007). Experiments by Teng et al with microspheres revealed a retention rate of only 11% following direct intra-myocardial injection in the beating porcine heart versus 67% in the non-beating heart illustrate the complexity of the approach (Teng, et al., 2006).

7. Conclusion

The initial hypothesis underpinning cell therapy was that new muscle cells would be generated in the injured myocardium, and this would restore cardiac function through systolic force generation (Laflamme, et al., 2007). Although the prospect of cardiac tissue regeneration provided an initial stimulus for cell-based therapies, subsequent experimental work has questioned the ability of stem cells to effectively regenerate cardiomyocytes (Murry, et al., 2004, Noiseux, et al., 2006). Unfortunately, hardly any clinical studies demonstrated convincing evidence for electrical or mechanical activation of engrafted cells within the infarct. There is currently uncertainty as to the optimal stem cell population to use clinically for cardiac repair. It appears that not "one cell fits all" but that the selection of the cell type should be tailored to the primary clinical indication and expected outcome (Menasche, 2009). Despite recent significant progress, answer to basic questions such as the best cell type has not been addressed so far. Different cell type may compete for the engraftment in the injured myocardium. The fact that numerous cell types and preparations can enhance myocardial repair, when none of these cells can even generate new cardiomyocytes, suggests that multiple mechanisms may be implicated. Presently, the field of stem cell therapy in regenerative medicine has extremely limited insight about how any candidate cell or treatment may work (Laflamme, et al., 2007). Indeed, the mechanisms responsible for the improvement in cardiac function have not yet been elucidated, and clinical studies have suggested that only a small fraction of implanted cells survived in the heart (Rosenzweig, 2006). This lack of mechanistic understanding has not prevented rapid clinical translation of cellular therapy, and numerous trials have been recently initiated. Recent randomized studies of cell therapy represent a milestone in this rapidly developing field, while serving as a cogent reminder that many important clinical and fundamental questions have yet to be addressed in carefully designed human studies (Rosenzweig, 2006). While stem cell therapy trials in cardiovascular disease are promptly progressing from bench to bedside, more extensive and vigorous basic science and clinical research are clearly needed to thoroughly investigate the therapeutic merits and potential adverse effects of stem cell transplantation (Jain, et al., 2005). Over the past few years, several promising results have been reported, but many hurdles remain before cell therapy can actually be commonly applied to treat patients with damaged hearts (Smits, et al., 2005). With a better understanding of adult stem cell biology and of the underpinning mechanisms involved in myocardial repair, we may eventually harness the therapeutic potential of these unique cells to better fulfill the beneficial promise of regenerative medicine in future years to come. Nevertheless, it appears that the exact mechanism underlying stem cell therapy is far more complex that previously anticipated, but so far cardiac cell therapy is feasible and overall safe in the clinical arena.

8. References

Abdel-Latif, A., Bolli, R., Tleyjeh, I. M., Montori, V. M., Perin, E. C., Hornung, C. A., Zuba-Surma, E. K., Al-Mallah, M. and Dawn, B. (2007). Adult bone marrow-derived cells for cardiac repair: a systematic review and meta-analysis. *Arch Intern Med*, 167. 10, pp.(989-97), 0003-9926 (Print).

Agbulut, O., Vandervelde, S., Al Attar, N., Larghero, J., Ghostine, S., Leobon, B., Robidel, E., Borsani, P., Le Lorc'h, M., Bissery, A., Chomienne, C., Bruneval, P., Marolleau, J. P., Vilquin, J. T., Hagege, A., Samuel, J. L. and Menasche, P. (2004). Comparison of human skeletal myoblasts and bone marrow-derived CD133+ progenitors for the repair of infarcted myocardium. *J Am Coll Cardiol*, 44. 2, pp.(458-63).

Alvarez-Dolado, M., Pardal, R., Garcia-Verdugo, J. M., Fike, J. R., Lee, H. O., Pfeffer, K., Lois, C., Morrison, S. J. and Alvarez-Buylla, A. (2003). Fusion of bone-marrow-derived cells with Purkinje neurons, cardiomyocytes and hepatocytes. *Nature*, 425. 6961, pp.(968-73. Epub 2003 Oct 12.).

Annabi, B., Lee, Y. T., Turcotte, S., Naud, E., Desrosiers, R. R., Champagne, M., Eliopoulos, N., Galipeau, J. and Beliveau, R. (2003). Hypoxia promotes murine bone-marrow-derived stromal cell migration and tube formation. *Stem Cells*, 21. 3, pp.(337-47).

Anversa, P., Leri, A., Rota, M., Hosoda, T., Bearzi, C., Urbanek, K., Kajstura, J. and Bolli, R. (2007). Concise review: stem cells, myocardial regeneration, and methodological artifacts. *Stem Cells*, 25. 3, pp.(589-601), 1066-5099 (Print).

Anversa, P., Sussman, M. A. and Bolli, R. (2004). Molecular genetic advances in cardiovascular medicine: focus on the myocyte. *Circulation*, 109. 23, pp.(2832-8).

Asahara, T., Masuda, H., Takahashi, T., Kalka, C., Pastore, C., Silver, M., Kearne, M., Magner, M. and Isner, J. M. (1999). Bone marrow origin of endothelial progenitor cells responsible for postnatal vasculogenesis in physiological and pathological neovascularization. *Circ Res*, 85. 3, pp.(221-8).

Assmus, B., Fischer-Rasokat, U., Honold, J., Seeger, F. H., Fichtlscherer, S., Tonn, T., Seifried, E., Schachinger, V., Dimmeler, S. and Zeiher, A. M. (2007). Transcoronary transplantation of functionally competent BMCs is associated with a decrease in natriuretic peptide serum levels and improved survival of patients with chronic postinfarction heart failure: results of the TOPCARE-CHD Registry. *Circ Res*, 100. 8, pp.(1234-41), 1524-4571 (Electronic).

Assmus, B., Rolf, A., Erbs, S., Elsasser, A., Haberbosch, W., Hambrecht, R., Tillmanns, H., Yu, J., Corti, R., Mathey, D. G., Hamm, C. W., Suselbeck, T., Tonn, T., Dimmeler, S., Dill, T., Zeiher, A. M. and Schachinger, V. (2010). Clinical outcome 2 years after intracoronary administration of bone marrow-derived progenitor cells in acute myocardial infarction. *Circ Heart Fail*, 3. 1, pp.(89-96), 1941-3297 (Electronic), 1941-3289 (Linking).

Assmus, B., Schachinger, V., Teupe, C., Britten, M., Lehmann, R., Dobert, N., Grunwald, F., Aicher, A., Urbich, C., Martin, H., Hoelzer, D., Dimmeler, S. and Zeiher, A. M. (2002). Transplantation of Progenitor Cells and Regeneration Enhancement in Acute Myocardial Infarction (TOPCARE-AMI). *Circulation*, 106. 24, pp.(3009-17).

Badorff, C., Brandes, R. P., Popp, R., Rupp, S., Urbich, C., Aicher, A., Fleming, I., Busse, R., Zeiher, A. M. and Dimmeler, S. (2003). Transdifferentiation of blood-derived human adult endothelial progenitor cells into functionally active cardiomyocytes. *Circulation*, 107. 7, pp.(1024-32), 1524-4539 (Electronic).

Balsam, L. B., Wagers, A. J., Christensen, J. L., Kofidis, T., Weissman, I. L. and Robbins, R. C. (2004). Haematopoietic stem cells adopt mature haematopoietic fates in ischaemic myocardium. *Nature*, 428. 6983, pp.(668-73).

Barbash, I. M., Chouraqui, P., Baron, J., Feinberg, M. S., Etzion, S., Tessone, A., Miller, L., Guetta, E., Zipori, D., Kedes, L. H., Kloner, R. A. and Leor, J. (2003). Systemic delivery of bone marrow-derived mesenchymal stem cells to the infarcted myocardium: feasibility, cell migration, and body distribution. *Circulation*, 108. 7, pp.(863-8).

Bartunek, J., Vanderheyden, M., Vandekerckhove, B., Mansour, S., De Bruyne, B., De Bondt, P., Van Haute, I., Lootens, N., Heyndrickx, G. and Wijns, W. (2005). Intracoronary injection of CD133-positive enriched bone marrow progenitor cells promotes cardiac recovery after recent myocardial infarction: feasibility and safety. *Circulation*, 112. 9 Suppl, pp.(I178-83), 1524-4539 (Electronic).

Bearzi, C., Rota, M., Hosoda, T., Tillmanns, J., Nascimbene, A., De Angelis, A., Yasuzawa-Amano, S., Trofimova, I., Siggins, R. W., Lecapitaine, N., Cascapera, S., Beltrami, A. P., D'Alessandro, D. A., Zias, E., Quaini, F., Urbanek, K., Michler, R. E., Bolli, R., Kajstura, J., Leri, A. and Anversa, P. (2007). Human cardiac stem cells. *Proc Natl Acad Sci U S A*, 104. 35, pp.(14068-73), 0027-8424 (Print).

Beltrami, A. P., Barlucchi, L., Torella, D., Baker, M., Limana, F., Chimenti, S., Kasahara, H., Rota, M., Musso, E., Urbanek, K., Leri, A., Kajstura, J., Nadal-Ginard, B. and Anversa, P. (2003). Adult cardiac stem cells are multipotent and support myocardial regeneration. *Cell*, 114. 6, pp.(763-76).

Bhatia, M. (2001). AC133 expression in human stem cells. *Leukemia*, 15. 11, pp.(1685-8),

Blau, H. M., Brazelton, T. R. and Weimann, J. M. (2001). The evolving concept of a stem cell: entity or function? *Cell*, 105. 7, pp.(829-41).

Braun, T. and Martire, A. (2007). Cardiac stem cells: paradigm shift or broken promise? A view from developmental biology. *Trends Biotechnol*, 25. 10, pp.(441-7), 0167-7799 (Print).

Britten, M. B., Abolmaali, N. D., Assmus, B., Lehmann, R., Honold, J., Schmitt, J., Vogl, T. J., Martin, H., Schachinger, V., Dimmeler, S. and Zeiher, A. M. (2003). Infarct remodeling after intracoronary progenitor cell treatment in patients with acute myocardial infarction (TOPCARE-AMI): mechanistic insights from serial contrast-enhanced magnetic resonance imaging. *Circulation*, 108. 18, pp.(2212-8).

Burdon, T., Paul, A., Noiseux, N., Prakash, s. and Shum-Tim, D. (2011). Bone marrow stem cell derived paracrine factors for regenerative medicine: current perspectives and therapeutic potential. *Bone Marrow Research*, Vol 2011. Open acces.

Buschmann, I., Heil, M., Jost, M. and Schaper, W. (2003). Influence of inflammatory cytokines on arteriogenesis. *Microcirculation*, 10. 3-4, pp.(371-9).

Buschmann, I. and Schaper, W. (1999). Arteriogenesis Versus Angiogenesis: Two Mechanisms of Vessel Growth. *News Physiol Sci*, 14. pp.(121-125).

Cai, J., Yi, F. F., Yang, X. C., Lin, G. S., Jiang, H., Wang, T. and Xia, Z. (2007). Transplantation of embryonic stem cell-derived cardiomyocytes improves cardiac function in infarcted rat hearts. *Cytotherapy*, 9. 3, pp.(283-91), 1465-3249 (Print), 1465-3249 (Linking).

Charwat, S., Gyongyosi, M., Lang, I., Graf, S., Beran, G., Hemetsberger, R., Nyolczas, N., Sochor, H. and Glogar, D. (2008). Role of adult bone marrow stem cells in the repair of ischemic myocardium: current state of the art. *Exp Hematol*, 36. 6, pp.(672-80), 0301-472X (Print).

Choi, J. H., Hur, J., Yoon, C. H., Kim, J. H., Lee, C. S., Youn, S. W., Oh, I. Y., Skurk, C., Murohara, T., Park, Y. B., Walsh, K. and Kim, H. S. (2004). Augmentation of therapeutic angiogenesis using genetically modified human endothelial progenitor cells with altered glycogen synthase kinase-3beta activity. *J Biol Chem*, 279. 47, pp.(49430-8), 0021-9258 (Print), 0021-9258 (Linking).

Choi, J. H., Kim, K. L., Huh, W., Kim, B., Byun, J., Suh, W., Sung, J., Jeon, E. S., Oh, H. Y. and Kim, D. K. (2004). Decreased number and impaired angiogenic function of endothelial progenitor cells in patients with chronic renal failure. *Arterioscler Thromb Vasc Biol*, 24. 7, pp.(1246-52).

Colter, D. C., Class, R., DiGirolamo, C. M. and Prockop, D. J. (2000). Rapid expansion of recycling stem cells in cultures of plastic-adherent cells from human bone marrow. *Proc Natl Acad Sci U S A*, 97. 7, pp.(3213-8).

Colter, D. C., Sekiya, I. and Prockop, D. J. (2001). Identification of a subpopulation of rapidly self-renewing and multipotential adult stem cells in colonies of human marrow stromal cells. *Proc Natl Acad Sci U S A*, 98. 14, pp.(7841-5).

Crosby, J. R., Kaminski, W. E., Schatteman, G., Martin, P. J., Raines, E. W., Seifert, R. A. and Bowen-Pope, D. F. (2000). Endothelial cells of hematopoietic origin make a significant contribution to adult blood vessel formation. *Circ Res*, 87. 9, pp.(728-30).

Dai, W., Hale, S. L., Martin, B. J., Kuang, J. Q., Dow, J. S., Wold, L. E. and Kloner, R. A. (2005). Allogeneic mesenchymal stem cell transplantation in postinfarcted rat myocardium: short- and long-term effects. *Circulation*, 112. 2, pp.(214-23).

Datta, S. R., Brunet, A. and Greenberg, M. E. (1999). Cellular survival: a play in three Akts. *Genes Dev*, 13. 22, pp.(2905-27).

Davani, S., Marandin, A., Mersin, N., Royer, B., Kantelip, B., Herve, P., Etievent, J. P. and Kantelip, J. P. (2003). Mesenchymal progenitor cells differentiate into an endothelial phenotype, enhance vascular density, and improve heart function in a rat cellular cardiomyoplasty model. *Circulation*, 108 Suppl 1. pp.(II253-8).

De Coppi, P., Bartsch, G., Jr., Siddiqui, M. M., Xu, T., Santos, C. C., Perin, L., Mostoslavsky, G., Serre, A. C., Snyder, E. Y., Yoo, J. J., Furth, M. E., Soker, S. and Atala, A. (2007). Isolation of amniotic stem cell lines with potential for therapy. *Nat Biotechnol*, 25. 1, pp.(100-6), 1087-0156 (Print).

Detre, K. M., Rosen, A. D., Bost, J. E., Cooper, M. E., Sutton-Tyrrell, K., Holubkov, R., Shemin, R. J. and Frye, R. L. (1996). Contemporary practice of coronary revascularization in U.S. hospitals and hospitals participating in the bypass angioplasty revascularization investigation (BARI). *J Am Coll Cardiol*, 28. 3, pp.(609-15), 0735-1097 (Print).

Dimmeler, S., Aicher, A., Vasa, M., Mildner-Rihm, C., Adler, K., Tiemann, M., Rutten, H., Fichtlscherer, S., Martin, H. and Zeiher, A. M. (2001). HMG-CoA reductase inhibitors (statins) increase endothelial progenitor cells via the PI 3-kinase/Akt pathway. *J Clin Invest*, 108. 3, pp.(391-7).

Dimmeler, S., Burchfield, J. and Zeiher, A. M. (2008). Cell-based therapy of myocardial infarction. *Arterioscler Thromb Vasc Biol*, 28. 2, pp.(208-16), 1524-4636 (Electronic).

Dimmeler, S. and Zeiher, A. M. (2004). Wanted! The best cell for cardiac regeneration. *J Am Coll Cardiol*, 44. 2, pp.(464-6).

Dimmeler, S., Zeiher, A. M. and Schneider, M. D. (2005). Unchain my heart: the scientific foundations of cardiac repair. *J Clin Invest*, 115. 3, pp.(572-83).

Dowell, J. D., Rubart, M., Pasumarthi, K. B., Soonpaa, M. H. and Field, L. J. (2003). Myocyte and myogenic stem cell transplantation in the heart. *Cardiovasc Res*, 58. 2, pp.(336-50).

Erbs, S., Linke, A., Adams, V., Lenk, K., Thiele, H., Diederich, K. W., Emmrich, F., Kluge, R., Kendziorra, K., Sabri, O., Schuler, G. and Hambrecht, R. (2005). Transplantation of blood-derived progenitor cells after recanalization of chronic coronary artery occlusion: first randomized and placebo-controlled study. *Circ Res*, 97. 8, pp.(756-62), 1524-4571 (Electronic).

Ertl, G., Gaudron, P. and Hu, K. (1993). Ventricular remodeling after myocardial infarction. Experimental and clinical studies. *Basic Res Cardiol*, 88 Suppl 1. pp.(125-37).

Evans, S. M., Mummery, C. and Doevendans, P. A. (2007). Progenitor cells for cardiac repair. *Semin Cell Dev Biol*, 18. 1, pp.(153-60), 1084-9521 (Print).

Franke, T. F., Hornik, C. P., Segev, L., Shostak, G. A. and Sugimoto, C. (2003). PI3K/Akt and apoptosis: size matters. *Oncogene*, 22. 56, pp.(8983-98).

Franke, T. F., Kaplan, D. R. and Cantley, L. C. (1997). PI3K: downstream AKTion blocks apoptosis. *Cell*, 88. 4, pp.(435-7).

Freyman, T., Polin, G., Osman, H., Crary, J., Lu, M., Cheng, L., Palasis, M. and Wilensky, R. L. (2006). A quantitative, randomized study evaluating three methods of mesenchymal stem cell delivery following myocardial infarction. *Eur Heart J*, 27. 9, pp.(1114-22), 0195-668X (Print).

Fuchs, S., Baffour, R., Zhou, Y. F., Shou, M., Pierre, A., Tio, F. O., Weissman, N. J., Leon, M. B., Epstein, S. E. and Kornowski, R. (2001). Transendocardial delivery of autologous bone marrow enhances collateral perfusion and regional function in pigs with chronic experimental myocardial ischemia. *J Am Coll Cardiol*, 37. 6, pp.(1726-32).

Fujiyama, S., Amano, K., Uehira, K., Yoshida, M., Nishiwaki, Y., Nozawa, Y., Jin, D., Takai, S., Miyazaki, M., Egashira, K., Imada, T., Iwasaka, T. and Matsubara, H. (2003). Bone marrow monocyte lineage cells adhere on injured endothelium in a monocyte chemoattractant protein-1-dependent manner and accelerate reendothelialization as endothelial progenitor cells. *Circ Res*, 93. 10, pp.(980-9).

Gehling, U. M., Ergun, S., Schumacher, U., Wagener, C., Pantel, K., Otte, M., Schuch, G., Schafhausen, P., Mende, T., Kilic, N., Kluge, K., Schafer, B., Hossfeld, D. K. and Fiedler, W. (2000). In vitro differentiation of endothelial cells from AC133-positive progenitor cells. *Blood*, 95. 10, pp.(3106-12).

George, J., Herz, I., Goldstein, E., Abashidze, S., Deutch, V., Finkelstein, A., Michowitz, Y., Miller, H. and Keren, G. (2003). Number and adhesive properties of circulating endothelial progenitor cells in patients with in-stent restenosis. *Arterioscler Thromb Vasc Biol*, 23. 12, pp.(e57-60).

Gnecchi, M., He, H., Liang, O. D., Melo, L. G., Morello, F., Mu, H., Noiseux, N., Zhang, L., Pratt, R. E., Ingwall, J. S. and Dzau, V. J. (2005). Paracrine action accounts for marked protection of ischemic heart by Akt-modified mesenchymal stem cells. *Nat Med*, 11. 4, pp.(367-8).

Gnecchi, M., He, H., Noiseux, N., Liang, O. D., Zhang, L., Morello, F., Mu, H., Melo, L. G., Pratt, R. E., Ingwall, J. S. and Dzau, V. J. (2006). Evidence supporting paracrine hypothesis for Akt-modified mesenchymal stem cell-mediated cardiac protection and functional improvement. *Faseb J*, 20. 6, pp.(661-9).

Goodell, M. A., Jackson, K. A., Majka, S. M., Mi, T., Wang, H., Pocius, J., Hartley, C. J., Majesky, M. W., Entman, M. L., Michael, L. H. and Hirschi, K. K. (2001). Stem cell plasticity in muscle and bone marrow. *Ann N Y Acad Sci*, 938. pp.(208-18; discussion 218-20).

Grigoropoulos, N. F. and Mathur, A. (2006). Stem cells in cardiac repair. *Curr Opin Pharmacol*, 6. 2, pp.(169-75), 1471-4892 (Print).

Grove, J. E., Bruscia, E. and Krause, D. S. (2004). Plasticity of bone marrow-derived stem cells. *Stem Cells*, 22. 4, pp.(487-500).

Harrison, D. E., Astle, C. M. and Lerner, C. (1988). Number and continuous proliferative pattern of transplanted primitive immunohematopoietic stem cells. *Proc Natl Acad Sci U S A*, 85. 3, pp.(822-6).

Hassink, R. J., Brutel de la Riviere, A., Mummery, C. L. and Doevendans, P. A. (2003). Transplantation of cells for cardiac repair. *J Am Coll Cardiol*, 41. 5, pp.(711-7).

He, J. Q., Ma, Y., Lee, Y., Thomson, J. A. and Kamp, T. J. (2003). Human embryonic stem cells develop into multiple types of cardiac myocytes: action potential characterization. *Circ Res*, 93. 1, pp.(32-9), 1524-4571 (Electronic).

Heil, M., Ziegelhoeffer, T., Mees, B. and Schaper, W. (2004). A different outlook on the role of bone marrow stem cells in vascular growth: bone marrow delivers software not hardware. *Circ Res*, 94. 5, pp.(573-4).

Heldman, A. W. and Hare, J. M. (2008). Cell therapy for myocardial infarction: Special delivery. *J Mol Cell Cardiol*, 44. 3, pp.(473-6), 1095-8584 (Electronic).

Herbots, L., D'Hooge, J., Eroglu, E., Thijs, D., Ganame, J., Claus, P., Dubois, C., Theunissen, K., Bogaert, J., Dens, J., Kalantzi, M., Dymarkowski, S., Bijnens, B., Belmans, A., Boogaerts, M., Sutherland, G., Van de Werf, F., Rademakers, F. and Janssens, S. (2009). Improved regional function after autologous bone marrow-derived stem cell transfer in patients with acute myocardial infarction: a randomized, double-blind strain rate imaging study. *Eur Heart J*, 30. 6, pp.(662-70), 1522-9645 (Electronic), 0195-668X (Linking).

Hierlihy, A. M., Seale, P., Lobe, C. G., Rudnicki, M. A. and Megeney, L. A. (2002). The post-natal heart contains a myocardial stem cell population. *FEBS Lett*, 530. 1-3, pp.(239-43), 0014-5793 (Print).

Hirata, K., Li, T. S., Nishida, M., Ito, H., Matsuzaki, M., Kasaoka, S. and Hamano, K. (2003). Autologous bone marrow cell implantation as therapeutic angiogenesis for ischemic hindlimb in diabetic rat model. *Am J Physiol Heart Circ Physiol*, 284. 1, pp.(H66-70).

Hodgson, D. M., Behfar, A., Zingman, L. V., Kane, G. C., Perez-Terzic, C., Alekseev, A. E., Puceat, M. and Terzic, A. (2004). Stable benefit of embryonic stem cell therapy in myocardial infarction. *Am J Physiol Heart Circ Physiol*, 287. 2, pp.(H471-9), 0363-6135 (Print).

Hofmann, M., Wollert, K. C., Meyer, G. P., Menke, A., Arseniev, L., Hertenstein, B., Ganser, A., Knapp, W. H. and Drexler, H. (2005). Monitoring of bone marrow cell homing into the infarcted human myocardium. *Circulation*, 111. 17, pp.(2198-202).

Hristov, M., Heussen, N., Schober, A. and Weber, C. (2006). Intracoronary infusion of autologous bone marrow cells and left ventricular function after acute myocardial infarction: a meta-analysis. *J Cell Mol Med*, 10. 3, pp.(727-33), 1582-1838 (Print).

Huikuri, H. V., Kervinen, K., Niemela, M., Ylitalo, K., Saily, M., Koistinen, P., Savolainen, E. R., Ukkonen, H., Pietila, M., Airaksinen, J. K., Knuuti, J. and Makikallio, T. H. (2008). Effects of intracoronary injection of mononuclear bone marrow cells on left ventricular function, arrhythmia risk profile, and restenosis after thrombolytic therapy of acute myocardial infarction. *Eur Heart J*, 29. 22, pp.(2723-32), 1522-9645 (Electronic) 0195-668X (Linking).

Hur, J., Yoon, C. H., Kim, H. S., Choi, J. H., Kang, H. J., Hwang, K. K., Oh, B. H., Lee, M. M. and Park, Y. B. (2004). Characterization of two types of endothelial progenitor cells and their different contributions to neovasculogenesis. *Arterioscler Thromb Vasc Biol*, 24. 2, pp.(288-93), 1524-4636 (Electronic) 1079-5642 (Linking).

Ince, H., Petzsch, M., Kleine, H. D., Eckard, H., Rehders, T., Burska, D., Kische, S., Freund, M. and Nienaber, C. A. (2005). Prevention of left ventricular remodeling with granulocyte colony-stimulating factor after acute myocardial infarction: final 1-year results of the Front-Integrated Revascularization and Stem Cell Liberation in Evolving Acute Myocardial Infarction by Granulocyte Colony-Stimulating Factor (FIRSTLINE-AMI) Trial. *Circulation*, 112. 9 Suppl, pp.(I73-80), 1524-4539 (Electronic).

Jackson, K. A., Majka, S. M., Wang, H., Pocius, J., Hartley, C. J., Majesky, M. W., Entman, M. L., Michael, L. H., Hirschi, K. K. and Goodell, M. A. (2001). Regeneration of ischemic cardiac muscle and vascular endothelium by adult stem cells. *J Clin Invest*, 107. 11, pp.(1395-402).

Jackson, K. A., Majka, S. M., Wulf, G. G. and Goodell, M. A. (2002). Stem cells: a minireview. *J Cell Biochem Suppl*, 38. pp.(1-6).

Jain, M., Pfister, O., Hajjar, R. J. and Liao, R. (2005). Mesenchymal stem cells in the infarcted heart. *Coron Artery Dis*, 16. 2, pp.(93-7).

Janssens, S., Dubois, C., Bogaert, J., Theunissen, K., Deroose, C., Desmet, W., Kalantzi, M., Herbots, L., Sinnaeve, P., Dens, J., Maertens, J., Rademakers, F., Dymarkowski, S., Gheysens, O., Van Cleemput, J., Bormans, G., Nuyts, J., Belmans, A., Mortelmans, L., Boogaerts, M. and Van de Werf, F. (2006). Autologous bone marrow-derived stem-cell transfer in patients with ST-segment elevation myocardial infarction: double-blind, randomised controlled trial. *Lancet*, 367. 9505, pp.(113-21), 1474-547X (Electronic).

Jia, Z. Q., Mickle, D. A., Weisel, R. D., Mohabeer, M. K., Merante, F., Rao, V., Li, G. and Li, R. K. (1997). Transplanted cardiomyocytes survive in scar tissue and improve heart function. *Transplant Proc*, 29. 4, pp.(2093-4).

Jiang, Y., Jahagirdar, B. N., Reinhardt, R. L., Schwartz, R. E., Keene, C. D., Ortiz-Gonzalez, X. R., Reyes, M., Lenvik, T., Lund, T., Blackstad, M., Du, J., Aldrich, S., Lisberg, A., Low, W. C., Largaespada, D. A. and Verfaillie, C. M. (2002). Pluripotency of mesenchymal stem cells derived from adult marrow. *Nature*, 418. 6893, pp.(41-9).

Kajstura, J., Rota, M., Whang, B., Cascapera, S., Hosoda, T., Bearzi, C., Nurzynska, D., Kasahara, H., Zias, E., Bonafe, M., Nadal-Ginard, B., Torella, D., Nascimbene, A., Quaini, F., Urbanek, K., Leri, A. and Anversa, P. (2005). Bone marrow cells differentiate in cardiac cell lineages after infarction independently of cell fusion. *Circ Res*, 96. 1, pp.(127-37. Epub 2004 Nov 29.).

Kamihata, H., Matsubara, H., Nishiue, T., Fujiyama, S., Tsutsumi, Y., Ozono, R., Masaki, H., Mori, Y., Iba, O., Tateishi, E., Kosaki, A., Shintani, S., Murohara, T., Imaizumi, T. and Iwasaka, T. (2001). Implantation of bone marrow mononuclear cells into ischemic myocardium enhances collateral perfusion and regional function via side supply of angioblasts, angiogenic ligands, and cytokines. *Circulation*, 104. 9, pp.(1046-52).

Kawamoto, A. and Asahara, T. (2007). Role of progenitor endothelial cells in cardiovascular disease and upcoming therapies. *Catheter Cardiovasc Interv*, 70. 4, pp.(477-84), 1522-1946 (Print).

Kawamoto, A. and Losordo, D. W. (2008). Endothelial progenitor cells for cardiovascular regeneration. *Trends Cardiovasc Med*, 18. 1, pp.(33-7), 1050-1738 (Print).

Kawamoto, A., Tkebuchava, T., Yamaguchi, J., Nishimura, H., Yoon, Y. S., Milliken, C., Uchida, S., Masuo, O., Iwaguro, H., Ma, H., Hanley, A., Silver, M., Kearney, M., Losordo, D. W., Isner, J. M. and Asahara, T. (2003). Intramyocardial transplantation of autologous endothelial progenitor cells for therapeutic neovascularization of myocardial ischemia. *Circulation*, 107. 3, pp.(461-8), 1524-4539 (Electronic).

Kehat, I., Kenyagin-Karsenti, D., Snir, M., Segev, H., Amit, M., Gepstein, A., Livne, E., Binah, O., Itskovitz-Eldor, J. and Gepstein, L. (2001). Human embryonic stem cells can differentiate into myocytes with structural and functional properties of cardiomyocytes. *J Clin Invest*, 108. 3, pp.(407-14), 0021-9738 (Print).

Kinnaird, T., Stabile, E., Burnett, M. S. and Epstein, S. E. (2004). Bone marrow-derived cells for enhancing collateral development: mechanisms, animal data, and initial clinical experiences. *Circ Res*, 95. 4, pp.(354-63).

Kinnaird, T., Stabile, E., Burnett, M. S., Lee, C. W., Barr, S., Fuchs, S. and Epstein, S. E. (2004). Marrow-derived stromal cells express genes encoding a broad spectrum of arteriogenic cytokines and promote in vitro and in vivo arteriogenesis through paracrine mechanisms. *Circ Res*, 94. 5, pp.(678-85).

Kinnaird, T., Stabile, E., Burnett, M. S., Shou, M., Lee, C. W., Barr, S., Fuchs, S. and Epstein, S. E. (2004). Local delivery of marrow-derived stromal cells augments collateral perfusion through paracrine mechanisms. *Circulation*, 109. 12, pp.(1543-9).

Kinnaird, T., Stabile, E., Epstein, S. E. and Fuchs, S. (2003). Current perspectives in therapeutic myocardial angiogenesis. *J Interv Cardiol*, 16. 4, pp.(289-97).

Kobayashi, T., Hamano, K., Li, T. S., Katoh, T., Kobayashi, S., Matsuzaki, M. and Esato, K. (2000). Enhancement of angiogenesis by the implantation of self bone marrow cells in a rat ischemic heart model. *J Surg Res*, 89. 2, pp.(189-95).

Koc, O. N. and Gerson, S. L. (2003). Akt helps stem cells heal the heart. *Nat Med*, 9. 9, pp.(1109-10),

Kocher, A. A., Schlechta, B., Gasparovicova, A., Wolner, E., Bonaros, N. and Laufer, G. (2007). Stem cells and cardiac regeneration. *Transpl Int*, 20. 9, pp.(731-46), 0934-0874 (Print).

Kocher, A. A., Schuster, M. D., Szabolcs, M. J., Takuma, S., Burkhoff, D., Wang, J., Homma, S., Edwards, N. M. and Itescu, S. (2001). Neovascularization of ischemic myocardium by human bone-marrow-derived angioblasts prevents cardiomyocyte apoptosis, reduces remodeling and improves cardiac function. *Nat Med*, 7. 4, pp.(430-6).

Kogler, G., Sensken, S., Airey, J. A., Trapp, T., Muschen, M., Feldhahn, N., Liedtke, S., Sorg, R. V., Fischer, J., Rosenbaum, C., Greschat, S., Knipper, A., Bender, J., Degistirici, O., Gao, J., Caplan, A. I., Colletti, E. J., Almeida-Porada, G., Muller, H. W., Zanjani, E. and Wernet, P. (2004). A new human somatic stem cell from placental cord blood with intrinsic pluripotent differentiation potential. *J Exp Med*, 200. 2, pp.(123-35), 0022-1007 (Print).

Krause, D. S., Theise, N. D., Collector, M. I., Henegariu, O., Hwang, S., Gardner, R., Neutzel, S. and Sharkis, S. J. (2001). Multi-organ, multi-lineage engraftment by a single bone marrow-derived stem cell. *Cell*, 105. 3, pp.(369-77).

Kuci, S., Wessels, J. T., Buhring, H. J., Schilbach, K., Schumm, M., Seitz, G., Loffler, J., Bader, P., Schlegel, P. G., Niethammer, D. and Handgretinger, R. (2003). Identification of a novel class of human adherent CD34- stem cells that give rise to SCID-repopulating cells. *Blood*, 101. 3, pp.(869-76).

Laflamme, M. A. and Murry, C. E. (2005). Regenerating the heart. *Nat Biotechnol*, 23. 7, pp.(845-56), 1087-0156 (Print) 1087-0156 (Linking).

Laflamme, M. A., Zbinden, S., Epstein, S. E. and Murry, C. E. (2007). Cell-based therapy for myocardial ischemia and infarction: pathophysiological mechanisms. *Annu Rev Pathol*, 2. pp.(307-39), 1553-4006 (Print).

Lambiase, P. D., Edwards, R. J., Anthopoulos, P., Rahman, S., Meng, Y. G., Bucknall, C. A., Redwood, S. R., Pearson, J. D. and Marber, M. S. (2004). Circulating humoral factors and endothelial progenitor cells in patients with differing coronary collateral support. *Circulation*, 109. 24, pp.(2986-92).

Leobon, B., Garcin, I., Menasche, P., Vilquin, J. T., Audinat, E. and Charpak, S. (2003). Myoblasts transplanted into rat infarcted myocardium are functionally isolated from their host. *Proc Natl Acad Sci U S A*, 100. 13, pp.(7808-11), 0027-8424 (Print).

Leor, J., Patterson, M., Quinones, M. J., Kedes, L. H. and Kloner, R. A. (1996). Transplantation of fetal myocardial tissue into the infarcted myocardium of rat. A potential method for repair of infarcted myocardium? *Circulation*, 94. 9 Suppl, pp.(II332-6).

Li, R. K., Jia, Z. Q., Weisel, R. D., Mickle, D. A., Zhang, J., Mohabeer, M. K., Rao, V. and Ivanov, J. (1996). Cardiomyocyte transplantation improves heart function. *Ann Thorac Surg*, 62. 3, pp.(654-60; discussion 660-1), 0003-4975 (Print).

Li, R. K., Mickle, D. A., Weisel, R. D., Mohabeer, M. K., Zhang, J., Rao, V., Li, G., Merante, F. and Jia, Z. Q. (1997). Natural history of fetal rat cardiomyocytes transplanted into adult rat myocardial scar tissue. *Circulation*, 96. 9 Suppl, pp.(II-179-86; discussion 186-7).

Li, T. S., Hamano, K., Suzuki, K., Ito, H., Zempo, N. and Matsuzaki, M. (2002). Improved angiogenic potency by implantation of ex vivo hypoxia prestimulated bone marrow cells in rats. *Am J Physiol Heart Circ Physiol*, 283. 2, pp.(H468-73).

Liao, R., Pfister, O., Jain, M. and Mouquet, F. (2007). The bone marrow--cardiac axis of myocardial regeneration. *Prog Cardiovasc Dis*, 50. 1, pp.(18-30), 0033-0620 (Print).

Lipinski, M. J., Biondi-Zoccai, G. G., Abbate, A., Khianey, R., Sheiban, I., Bartunek, J., Vanderheyden, M., Kim, H. S., Kang, H. J., Strauer, B. E. and Vetrovec, G. W. (2007). Impact of intracoronary cell therapy on left ventricular function in the setting of acute myocardial infarction: a collaborative systematic review and meta-analysis of controlled clinical trials. *J Am Coll Cardiol*, 50. 18, pp.(1761-7), 1558-3597 (Electronic).

Liu, P., Kalajzic, I., Stover, M. L., Rowe, D. W. and Lichtler, A. C. (2001). Human bone marrow stromal cells are efficiently transduced by vesicular stomatitis virus-pseudotyped retrovectors without affecting subsequent osteoblastic differentiation. *Bone*, 29. 4, pp.(331-5).

Losordo, D. W., Schatz, R. A., White, C. J., Udelson, J. E., Veereshwarayya, V., Durgin, M., Poh, K. K., Weinstein, R., Kearney, M., Chaudhry, M., Burg, A., Eaton, L., Heyd, L., Thorne, T., Shturman, L., Hoffmeister, P., Story, K., Zak, V., Dowling, D., Traverse, J. H., Olson, R. E., Flanagan, J., Sodano, D., Murayama, T., Kawamoto, A., Kusano, K. F., Wollins, J., Welt, F., Shah, P., Soukas, P., Asahara, T. and Henry, T. D. (2007). Intramyocardial transplantation of autologous CD34+ stem cells for intractable angina: a phase I/IIa double-blind, randomized controlled trial. *Circulation*, 115. 25, pp.(3165-72), 1524-4539 (Electronic).

Lunde, K., Solheim, S., Aakhus, S., Arnesen, H., Abdelnoor, M., Egeland, T., Endresen, K., Ilebekk, A., Mangschau, A., Fjeld, J. G., Smith, H. J., Taraldsrud, E., Grogaard, H. K., Bjornerheim, R., Brekke, M., Muller, C., Hopp, E., Ragnarsson, A., Brinchmann, J. E. and Forfang, K. (2006). Intracoronary injection of mononuclear bone marrow cells in acute myocardial infarction. *N Engl J Med*, 355. 12, pp.(1199-209), 1533-4406 (Electronic).

Lunde, K., Solheim, S., Forfang, K., Arnesen, H., Brinch, L., Bjornerheim, R., Ragnarsson, A., Egeland, T., Endresen, K., Ilebekk, A., Mangschau, A. and Aakhus, S. (2008). Anterior myocardial infarction with acute percutaneous coronary intervention and intracoronary injection of autologous mononuclear bone marrow cells: safety, clinical outcome, and serial changes in left ventricular function during 12-months' follow-up. *J Am Coll Cardiol*, 51. 6, pp.(674-6), 1558-3597 (Electronic).

Luttun, A. and Carmeliet, P. (2003). De novo vasculogenesis in the heart. *Cardiovasc Res*, 58. 2, pp.(378-89).

Lyngbaek, S., Schneider, M., Hansen, J. L. and Sheikh, S. P. (2007). Cardiac regeneration by resident stem and progenitor cells in the adult heart. *Basic Res Cardiol*, 102. 2, pp.(101-14), 0300-8428 (Print).

Mackay, A. M., Beck, S. C., Murphy, J. M., Barry, F. P., Chichester, C. O. and Pittenger, M. F. (1998). Chondrogenic differentiation of cultured human mesenchymal stem cells from marrow. *Tissue Eng*, 4. 4, pp.(415-28).

Makino, S., Fukuda, K., Miyoshi, S., Konishi, F., Kodama, H., Pan, J., Sano, M., Takahashi, T., Hori, S., Abe, H., Hata, J., Umezawa, A. and Ogawa, S. (1999). Cardiomyocytes can be generated from marrow stromal cells in vitro. *J Clin Invest*, 103. 5, pp.(697-705).

Mangi, A. A., Noiseux, N., Kong, D., He, H., Rezvani, M., Ingwall, J. S. and Dzau, V. J. (2003). Mesenchymal stem cells modified with Akt prevent remodeling and restore performance of infarcted hearts. *Nat Med*, 9. 9, pp.(1195-201. Epub 2003 Aug 10.).

Mansour, S., Roy, D. C., Bouchard, V., Nguyen, B. K., Stevens, L. M., Gobeil, F., Rivard, A., Leclerc, G., Reeves, F. and Noiseux, N. (2010). COMPARE-AMI trial: comparison of intracoronary injection of CD133+ bone marrow stem cells to placebo in patients after acute myocardial infarction and left ventricular dysfunction: study rationale and design. *J Cardiovasc Transl Res*, 3. 2, pp.(153-9), 1937-5395 (Electronic).

Mansour, S., Roy, D. C., Bouchard, V., Nguyen, B. K., Stevens, L. M., Gobeil, J. F., Rivard, A., Leclerc, G. and Noiseux, N. (2010). COMPARE-AMI Trial: Comparison of Intracoronary Injection of CD133+ Bone Marrow Stem Cells to Placebo in Patients After Acute Myocardial Infarction and Left Ventricular Dysfunction: Study Rationale and Design. *J. of Cardiovasc. Trans. Res.*, 3. 2, pp.(153-159).

Mansour, S., Roy, D. C., Bouchard, V., Stevens, L. M., Gobeil, J. F., Rivard, A., Leclerc, G. and Noiseux, N. (2011). One-Year Safety Analysis of the COMPARE-AMI Trial: Comparison of Intracoronary Injection of CD133+ Bone Marrow Stem Cells to Placebo in Patients after Acute Myocardial Infarction and Left Ventricular Dysfunction. *Bone Marrow Research*, 2011. pp.(1-6),

Mansour, S., Roy, D. C., Lemieux, B., Ouellet, C., Stevens, L. M. and Noiseux, N. (2009). Stem cell therapy for the broken heart: mini-organ transplantation. *Transplant Proc*, 41. 8, pp.(3353-7), 1873-2623 (Electronic).

Mansour, S., Vanderheyden, M., De Bruyne, B., Vandekerckhove, B., Delrue, L., Van Haute, I., Heyndrickx, G., Carlier, S., Rodriguez-Granillo, G., Wijns, W. and Bartunek, J. (2006). Intracoronary delivery of hematopoietic bone marrow stem cells and luminal loss of the infarct-related artery in patients with recent myocardial infarction. *J Am Coll Cardiol*, 47. 8, pp.(1727-30), 1558-3597 (Electronic).

Martin-Rendon, E., Brunskill, S. J., Hyde, C. J., Stanworth, S. J., Mathur, A. and Watt, S. M. (2008). Autologous bone marrow stem cells to treat acute myocardial infarction: a systematic review. *Eur Heart J*, 29. 15, pp.(1807-18), 1522-9645 (Electronic).

Meirelles Lda, S. and Nardi, N. B. (2003). Murine marrow-derived mesenchymal stem cell: isolation, in vitro expansion, and characterization. *Br J Haematol*, 123. 4, pp.(702-11),

Menasche, P. (2008). Skeletal myoblasts and cardiac repair. *J Mol Cell Cardiol*, 45. 4, pp.(545-53), 1095-8584 (Electronic).

Menasche, P. (2009). Cell-based therapy for heart disease: a clinically oriented perspective. *Mol Ther*, 17. 5, pp.(758-66), 1525-0024 (Electronic).

Menasche, P. (2011). Cardiac cell therapy: lessons from clinical trials. *J Mol Cell Cardiol*, 50. 2, pp.(258-65), 1095-8584 (Electronic) 0022-2828 (Linking).

Menasche, P., Hagege, A. A., Scorsin, M., Pouzet, B., Desnos, M., Duboc, D., Schwartz, K., Vilquin, J. T. and Marolleau, J. P. (2001). Myoblast transplantation for heart failure. *Lancet*, 357. 9252, pp.(279-80).

Menasche, P., Hagege, A. A., Vilquin, J. T., Desnos, M., Abergel, E., Pouzet, B., Bel, A., Sarateanu, S., Scorsin, M., Schwartz, K., Bruneval, P., Benbunan, M., Marolleau, J. P. and Duboc, D. (2003). Autologous skeletal myoblast transplantation for severe postinfarction left ventricular dysfunction. *J Am Coll Cardiol*, 41. 7, pp.(1078-83).

Meyer, G. P., Wollert, K. C., Lotz, J., Steffens, J., Lippolt, P., Fichtner, S., Hecker, H., Schaefer, A., Arseniev, L., Hertenstein, B., Ganser, A. and Drexler, H. (2006). Intracoronary bone marrow cell transfer after myocardial infarction: eighteen months' follow-up data from the randomized, controlled BOOST (BOne marrOw transfer to enhance ST-elevation infarct regeneration) trial. *Circulation*, 113. 10, pp.(1287-94), 1524-4539 (Electronic).

Min, J. Y., Sullivan, M. F., Yang, Y., Zhang, J. P., Converso, K. L., Morgan, J. P. and Xiao, Y. F. (2002). Significant improvement of heart function by cotransplantation of human mesenchymal stem cells and fetal cardiomyocytes in postinfarcted pigs. *Ann Thorac Surg*, 74. 5, pp.(1568-75).

Min, J. Y., Yang, Y., Converso, K. L., Liu, L., Huang, Q., Morgan, J. P. and Xiao, Y. F. (2002). Transplantation of embryonic stem cells improves cardiac function in postinfarcted rats. *J Appl Physiol*, 92. 1, pp.(288-96).

Min, J. Y., Yang, Y., Sullivan, M. F., Ke, Q., Converso, K. L., Chen, Y., Morgan, J. P. and Xiao, Y. F. (2003). Long-term improvement of cardiac function in rats after infarction by transplantation of embryonic stem cells. *J Thorac Cardiovasc Surg*, 125. 2, pp.(361-9).

Minguell, J. J., Erices, A. and Conget, P. (2001). Mesenchymal stem cells. *Exp Biol Med (Maywood)*, 226. 6, pp.(507-20).

Murry, C. E., Soonpaa, M. H., Reinecke, H., Nakajima, H., Nakajima, H. O., Rubart, M., Pasumarthi, K. B., Virag, J. I., Bartelmez, S. H., Poppa, V., Bradford, G., Dowell, J. D., Williams, D. A. and Field, L. J. (2004). Haematopoietic stem cells do not transdifferentiate into cardiac myocytes in myocardial infarcts. *Nature*, 428. 6983, pp.(664-8).

Murry, C. E., Soonpaa, M. H., Reinecke, H., Nakajima, H., Nakajima, H. O., Rubart, M., Pasumarthi, K. B., Virag, J. I., Bartelmez, S. H., Poppa, V., Bradford, G., Dowell, J. D., Williams, D. A., Field, L. J., Shi, D., Torok-Storb, B. and Minami, E. (2004). Haematopoietic stem cells do not transdifferentiate into cardiac myocytes in myocardial infarcts Myogenic fusion of human bone marrow stromal cells, but not hematopoietic cells Evidence for fusion between cardiac and skeletal muscle cells. *Nature*, 428. 6983, pp.(664-8. Epub 2004 Mar 21.).

Murry, C. E., Wiseman, R. W., Schwartz, S. M. and Hauschka, S. D. (1996). Skeletal myoblast transplantation for repair of myocardial necrosis. *J Clin Invest*, 98. 11, pp.(2512-23).

Nguyen, B. K., Maltais, S., Perrault, L. P., Tanguay, J. F., Tardif, J. C., Stevens, L. M., Borie, M., Harel, F., Mansour, S. and Noiseux, N. (2010). Improved function and myocardial repair of infarcted heart by intracoronary injection of mesenchymal stem cell-derived growth factors. *J Cardiovasc Transl Res*, 3. 5, pp.(547-58), 1937-5395 (Electronic).

Nian, M., Lee, P., Khaper, N. and Liu, P. (2004). Inflammatory cytokines and postmyocardial infarction remodeling. *Circ Res*, 94. 12, pp.(1543-53).

Noiseux N, G. M., Mirotsou M, Pratt R, Dzau VJ. (2004). Adult Mesenchymal Stem Cells Genetically Engineered to Overexpress Akt Exhibit Improved Resistance to

Apoptosis, Glucose Metabolism, and Angiogenic Potential. *Can J Cardiol*, 20. pp.((Suppl D):97D).

Noiseux, N., Gnecchi, M., Lopez-Ilasaca, M., Zhang, L., Solomon, S. D., Deb, A., Dzau, V. J. and Pratt, R. E. (2006). Mesenchymal stem cells overexpressing Akt dramatically repair infarcted myocardium and improve cardiac function despite infrequent cellular fusion or differentiation. *Mol Ther*, 14. 6, pp.(840-50), 1525-0016 (Print).

Noiseux N, L.-I. M., Gnecchi M, Zang L, Solomon SD, Dzau VJ, Pratt RE. (2004). Transplantation of Bone Marrow-Derived Mesenchymal Stem Cells Expressing Akt Into Infracted Murine Heart Produces Dramatic Improvement in Cardiac Function Despite Infrequent Cellular Fusion. *Circulation*, 110. (17), pp.((supplement III): III-68).

Nygren, J. M., Jovinge, S., Breitbach, M., Sawen, P., Roll, W., Hescheler, J., Taneera, J., Fleischmann, B. K. and Jacobsen, S. E. (2004). Bone marrow-derived hematopoietic cells generate cardiomyocytes at a low frequency through cell fusion, but not transdifferentiation. *Nat Med*, 10. 5, pp.(494-501).

Oh, H., Bradfute, S. B., Gallardo, T. D., Nakamura, T., Gaussin, V., Mishina, Y., Pocius, J., Michael, L. H., Behringer, R. R., Garry, D. J., Entman, M. L. and Schneider, M. D. (2003). Cardiac progenitor cells from adult myocardium: homing, differentiation, and fusion after infarction. *Proc Natl Acad Sci U S A*, 100. 21, pp.(12313-8).

Orlic, D. (2003). Adult bone marrow stem cells regenerate myocardium in ischemic heart disease. *Ann N Y Acad Sci*, 996. pp.(152-7).

Orlic, D. (2004). The strength of plasticity: stem cells for cardiac repair. *Int J Cardiol*, 95 Suppl 1. pp.(S16-9).

Orlic, D., Hill, J. M. and Arai, A. E. (2002). Stem cells for myocardial regeneration. *Circ Res*, 91. 12, pp.(1092-102).

Orlic, D., Kajstura, J., Chimenti, S., Bodine, D. M., Leri, A. and Anversa, P. (2001). Transplanted adult bone marrow cells repair myocardial infarcts in mice. *Ann N Y Acad Sci*, 938. pp.(221-9; discussion 229-30).

Orlic, D., Kajstura, J., Chimenti, S., Bodine, D. M., Leri, A. and Anversa, P. (2003). Bone marrow stem cells regenerate infarcted myocardium. *Pediatr Transplant*, 7 Suppl 3. pp.(86-8).

Orlic, D., Kajstura, J., Chimenti, S., Jakoniuk, I., Anderson, S. M., Li, B., Pickel, J., McKay, R., Nadal-Ginard, B., Bodine, D. M., Leri, A. and Anversa, P. (2001). Bone marrow cells regenerate infarcted myocardium. *Nature*, 410. 6829, pp.(701-5).

Orlic, D., Kajstura, J., Chimenti, S., Limana, F., Jakoniuk, I., Quaini, F., Nadal-Ginard, B., Bodine, D. M., Leri, A. and Anversa, P. (2001). Mobilized bone marrow cells repair the infarcted heart, improving function and survival. *Proc Natl Acad Sci U S A*, 98. 18, pp.(10344-9).

Patel, A. N., Geffner, L., Vina, R. F., Saslavsky, J., Urschel, H. C., Jr., Kormos, R. and Benetti, F. (2005). Surgical treatment for congestive heart failure with autologous adult stem cell transplantation: a prospective randomized study. *J Thorac Cardiovasc Surg*, 130. 6, pp.(1631-8), 1097-685X (Electronic).

Peister, A., Mellad, J. A., Larson, B. L., Hall, B. M., Gibson, L. F. and Prockop, D. J. (2004). Adult stem cells from bone marrow (MSCs) isolated from different strains of inbred mice vary in surface epitopes, rates of proliferation, and differentiation potential. *Blood*, 103. 5, pp.(1662-8).

Pell, J. P., Pell, A. C., Jeffrey, R. R., Jennings, K., Oldroyd, K., Eteiba, H., Hogg, K. J., Murday, A., Faichney, A., Colquhoun, I., Berg, G., Starkey, I. R., Flapan, A. and Mankad, P. (2004). Comparison of survival following coronary artery bypass grafting vs. percutaneous coronary intervention in diabetic and non-diabetic patients: retrospective cohort study of 6320 procedures. *Diabet Med*, 21. 7, pp.(790-2), 0742-3071 (Print).

Perin, E. C., Geng, Y. J. and Willerson, J. T. (2003). Adult stem cell therapy in perspective. *Circulation*, 107. 7, pp.(935-8).

Perin, E. C., Silva, G. V., Assad, J. A., Vela, D., Buja, L. M., Sousa, A. L., Litovsky, S., Lin, J., Vaughn, W. K., Coulter, S., Fernandes, M. R. and Willerson, J. T. (2008). Comparison of intracoronary and transendocardial delivery of allogeneic mesenchymal cells in a canine model of acute myocardial infarction. *J Mol Cell Cardiol*, 44. 3, pp.(486-95), 1095-8584 (Electronic).

Pfeffer, M. A. and Braunwald, E. (1990). Ventricular remodeling after myocardial infarction. Experimental observations and clinical implications. *Circulation*, 81. 4, pp.(1161-72).

Pittenger, M. F., Mackay, A. M., Beck, S. C., Jaiswal, R. K., Douglas, R., Mosca, J. D., Moorman, M. A., Simonetti, D. W., Craig, S. and Marshak, D. R. (1999). Multilineage potential of adult human mesenchymal stem cells. *Science*, 284. 5411, pp.(143-7).

Pittenger, M. F. and Martin, B. J. (2004). Mesenchymal stem cells and their potential as cardiac therapeutics. *Circ Res*, 95. 1, pp.(9-20).

Planat-Benard, V., Menard, C., Andre, M., Puceat, M., Perez, A., Garcia-Verdugo, J. M., Penicaud, L. and Casteilla, L. (2004). Spontaneous cardiomyocyte differentiation from adipose tissue stroma cells. *Circ Res*, 94. 2, pp.(223-9), 1524-4571 (Electronic).

Psaltis, P. J., Zannettino, A. C., Worthley, S. G. and Gronthos, S. (2008). Concise review: mesenchymal stromal cells: potential for cardiovascular repair. *Stem Cells*, 26. 9, pp.(2201-10), 1549-4918 (Electronic).

Quirici, N., Soligo, D., Caneva, L., Servida, F., Bossolasco, P. and Deliliers, G. L. (2001). Differentiation and expansion of endothelial cells from human bone marrow CD133(+) cells. *Br J Haematol*, 115. 1, pp.(186-94).

Raeburn, C. D., Zimmerman, M. A., Arya, J., Banerjee, A. and Harken, A. H. (2002). Stem cells and myocardial repair. *J Am Coll Surg*, 195. 5, pp.(686-93).

Rangappa, S., Fen, C., Lee, E. H., Bongso, A. and Wei, E. K. (2003). Transformation of adult mesenchymal stem cells isolated from the fatty tissue into cardiomyocytes. *Ann Thorac Surg*, 75. 3, pp.(775-9).

Reinecke, H., Minami, E., Poppa, V. and Murry, C. E. (2004). Evidence for fusion between cardiac and skeletal muscle cells. *Circ Res*, 94. 6, pp.(e56-60. Epub 2004 Mar 4.).

Reinecke, H., Zhang, M., Bartosek, T. and Murry, C. E. (1999). Survival, integration, and differentiation of cardiomyocyte grafts: a study in normal and injured rat hearts. *Circulation*, 100. 2, pp.(193-202).

Rosenzweig, A. (2006). Cardiac cell therapy--mixed results from mixed cells. *N Engl J Med*, 355. 12, pp.(1274-7), 1533-4406 (Electronic).

Ruel, M., Suuronen, E. J., Song, J., Kapila, V., Gunning, D., Waghray, G., Rubens, F. D. and Mesana, T. G. (2005). Effects of off-pump versus on-pump coronary artery bypass grafting on function and viability of circulating endothelial progenitor cells. *J Thorac Cardiovasc Surg*, 130. 3, pp.(633-9).

Saito, T., Kuang, J. Q., Bittira, B., Al-Khaldi, A. and Chiu, R. C. (2002). Xenotransplant cardiac chimera: immune tolerance of adult stem cells. *Ann Thorac Surg*, 74. 1, pp.(19-24; discussion 24).

Sakai, T., Li, R. K., Weisel, R. D., Mickle, D. A., Jia, Z. Q., Tomita, S., Kim, E. J. and Yau, T. M. (1999). Fetal cell transplantation: a comparison of three cell types. *J Thorac Cardiovasc Surg*, 118. 4, pp.(715-24), 0022-5223 (Print).

Schachinger, V., Assmus, B., Britten, M. B., Honold, J., Lehmann, R., Teupe, C., Abolmaali, N. D., Vogl, T. J., Hofmann, W. K., Martin, H., Dimmeler, S. and Zeiher, A. M. (2004). Transplantation of progenitor cells and regeneration enhancement in acute myocardial infarction: final one-year results of the TOPCARE-AMI Trial. *J Am Coll Cardiol*, 44. 8, pp.(1690-9).

Schachinger, V., Erbs, S., Elsasser, A., Haberbosch, W., Hambrecht, R., Holschermann, H., Yu, J., Corti, R., Mathey, D. G., Hamm, C. W., Suselbeck, T., Assmus, B., Tonn, T., Dimmeler, S. and Zeiher, A. M. (2006). Intracoronary bone marrow-derived progenitor cells in acute myocardial infarction. *N Engl J Med*, 355. 12, pp.(1210-21), 1533-4406 (Electronic), 0028-4793 (Linking).

Schaefer, A., Meyer, G. P., Fuchs, M., Klein, G., Kaplan, M., Wollert, K. C. and Drexler, H. (2006). Impact of intracoronary bone marrow cell transfer on diastolic function in patients after acute myocardial infarction: results from the BOOST trial. *Eur Heart J*, 27. 8, pp.(929-35), 0195-668X (Print), 0195-668X (Linking).

Scheubel, R. J., Zorn, H., Silber, R. E., Kuss, O., Morawietz, H., Holtz, J. and Simm, A. (2003). Age-dependent depression in circulating endothelial progenitor cells in patients undergoing coronary artery bypass grafting. *J Am Coll Cardiol*, 42. 12, pp.(2073-80).

Scorsin, M., Hagege, A., Vilquin, J. T., Fiszman, M., Marotte, F., Samuel, J. L., Rappaport, L., Schwartz, K. and Menasche, P. (2000). Comparison of the effects of fetal cardiomyocyte and skeletal myoblast transplantation on postinfarction left ventricular function. *J Thorac Cardiovasc Surg*, 119. 6, pp.(1169-75), 0022-5223 (Print).

Shake, J. G., Gruber, P. J., Baumgartner, W. A., Senechal, G., Meyers, J., Redmond, J. M., Pittenger, M. F. and Martin, B. J. (2002). Mesenchymal stem cell implantation in a swine myocardial infarct model: engraftment and functional effects. *Ann Thorac Surg*, 73. 6, pp.(1919-25; discussion 1926).

Shi, Q., Rafii, S., Wu, M. H., Wijelath, E. S., Yu, C., Ishida, A., Fujita, Y., Kothari, S., Mohle, R., Sauvage, L. R., Moore, M. A., Storb, R. F. and Hammond, W. P. (1998). Evidence for circulating bone marrow-derived endothelial cells. *Blood*, 92. 2, pp.(362-7).

Shintani, S., Murohara, T., Ikeda, H., Ueno, T., Honma, T., Katoh, A., Sasaki, K., Shimada, T., Oike, Y. and Imaizumi, T. (2001). Mobilization of endothelial progenitor cells in patients with acute myocardial infarction. *Circulation*, 103. 23, pp.(2776-9).

Singla, D. K., Hacker, T. A., Ma, L., Douglas, P. S., Sullivan, R., Lyons, G. E. and Kamp, T. J. (2006). Transplantation of embryonic stem cells into the infarcted mouse heart: formation of multiple cell types. *J Mol Cell Cardiol*, 40. 1, pp.(195-200), 0022-2828 (Print).

Smadja, D. M., Cornet, A., Emmerich, J., Aiach, M. and Gaussem, P. (2007). Endothelial progenitor cells: characterization, in vitro expansion, and prospects for autologous cell therapy. *Cell Biol Toxicol*, 23. 4, pp.(223-39), 0742-2091 (Print)0742-2091 (Linking).

Smits, A. M., van Vliet, P., Hassink, R. J., Goumans, M. J. and Doevendans, P. A. (2005). The role of stem cells in cardiac regeneration. *J Cell Mol Med*, 9. 1, pp.(25-36), 1582-1838 (Print).

Song, L. and Tuan, R. S. (2004). Transdifferentiation potential of human mesenchymal stem cells derived from bone marrow. *Faseb J*, 18. 9, pp.(980-2).

Soriano, P. (1999). Generalized lacZ expression with the ROSA26 Cre reporter strain. *Nat Genet*, 21. 1, pp.(70-1).

Stamm, C., Westphal, B., Kleine, H. D., Petzsch, M., Kittner, C., Klinge, H., Schumichen, C., Nienaber, C. A., Freund, M. and Steinhoff, G. (2003). Autologous bone-marrow stem-cell transplantation for myocardial regeneration. *Lancet*, 361. 9351, pp.(45-6).

Strauer, B. E., Brehm, M., Zeus, T., Kostering, M., Hernandez, A., Sorg, R. V., Kogler, G. and Wernet, P. (2002). Repair of infarcted myocardium by autologous intracoronary mononuclear bone marrow cell transplantation in humans. *Circulation*, 106. 15, pp.(1913-8).

Strauer, B. E. and Kornowski, R. (2003). Stem cell therapy in perspective. *Circulation*, 107. 7, pp.(929-34).

Strauer, B. E., Yousef, M. and Schannwell, C. M. (2010). The acute and long-term effects of intracoronary Stem cell Transplantation in 191 patients with chronic heARt failure: the STAR-heart study. *Eur J Heart Fail*, 12. 7, pp.(721-9), 1879-0844 (Electronic)1388-9842 (Linking).

Taylor, D. A., Atkins, B. Z., Hungspreugs, P., Jones, T. R., Reedy, M. C., Hutcheson, K. A., Glower, D. D. and Kraus, W. E. (1998). Regenerating functional myocardium: improved performance after skeletal myoblast transplantation. *Nat Med*, 4. 8, pp.(929-33).

Tendera, M., Wojakowski, W., Ruzyllo, W., Chojnowska, L., Kepka, C., Tracz, W., Musialek, P., Piwowarska, W., Nessler, J., Buszman, P., Grajek, S., Breborowicz, P., Majka, M. and Ratajczak, M. Z. (2009). Intracoronary infusion of bone marrow-derived selected CD34+CXCR4+ cells and non-selected mononuclear cells in patients with acute STEMI and reduced left ventricular ejection fraction: results of randomized, multicentre Myocardial Regeneration by Intracoronary Infusion of Selected Population of Stem Cells in Acute Myocardial Infarction (REGENT) Trial. *Eur Heart J*, 30. 11, pp.(1313-21), 1522-9645 (Electronic).

Teng, C. J., Luo, J., Chiu, R. C. and Shum-Tim, D. (2006). Massive mechanical loss of microspheres with direct intramyocardial injection in the beating heart: implications for cellular cardiomyoplasty. *J Thorac Cardiovasc Surg*, 132. 3, pp.(628-32), 1097-685X (Electronic).

Tepper, O. M., Galiano, R. D., Capla, J. M., Kalka, C., Gagne, P. J., Jacobowitz, G. R., Levine, J. P. and Gurtner, G. C. (2002). Human endothelial progenitor cells from type II diabetics exhibit impaired proliferation, adhesion, and incorporation into vascular structures. *Circulation*, 106. 22, pp.(2781-6).

Terada, N., Hamazaki, T., Oka, M., Hoki, M., Mastalerz, D. M., Nakano, Y., Meyer, E. M., Morel, L., Petersen, B. E. and Scott, E. W. (2002). Bone marrow cells adopt the phenotype of other cells by spontaneous cell fusion. *Nature*, 416. 6880, pp.(542-5).

Timmermans, F., De Sutter, J. and Gillebert, T. C. (2003). Stem cells for the heart, are we there yet? *Cardiology*, 100. 4, pp.(176-85).

Toma, C., Pittenger, M. F., Cahill, K. S., Byrne, B. J. and Kessler, P. D. (2002). Human mesenchymal stem cells differentiate to a cardiomyocyte phenotype in the adult murine heart. *Circulation*, 105. 1, pp.(93-8).

Tomita, S., Li, R. K., Weisel, R. D., Mickle, D. A., Kim, E. J., Sakai, T. and Jia, Z. Q. (1999). Autologous transplantation of bone marrow cells improves damaged heart function. *Circulation*, 100. 19 Suppl, pp.(II247-56).

Tomita, S., Mickle, D. A., Weisel, R. D., Jia, Z. Q., Tumiati, L. C., Allidina, Y., Liu, P. and Li, R. K. (2002). Improved heart function with myogenesis and angiogenesis after autologous porcine bone marrow stromal cell transplantation. *J Thorac Cardiovasc Surg*, 123. 6, pp.(1132-40),

Topol, E. J. (2003). Current status and future prospects for acute myocardial infarction therapy. *Circulation*, 108. 16 Suppl 1, pp.(III6-13), 1524-4539 (Electronic).

Towbin, J. A. and Bowles, N. E. (2002). The failing heart. *Nature*, 415. 6868, pp.(227-33),

Tse, H. F., Thambar, S., Kwong, Y. L., Rowlings, P., Bellamy, G., McCrohon, J., Thomas, P., Bastian, B., Chan, J. K., Lo, G., Ho, C. L., Chan, W. S., Kwong, R. Y., Parker, A., Hauser, T. H., Chan, J., Fong, D. Y. and Lau, C. P. (2007). Prospective randomized trial of direct endomyocardial implantation of bone marrow cells for treatment of severe coronary artery diseases (PROTECT-CAD trial). *Eur Heart J*, 28. 24, pp.(2998-3005), 0195-668X (Print).

Urbich, C. and Dimmeler, S. (2004). Endothelial progenitor cells functional characterization. *Trends Cardiovasc Med*, 14. 8, pp.(318-22).

van Ramshorst, J., Bax, J. J., Beeres, S. L., Dibbets-Schneider, P., Roes, S. D., Stokkel, M. P., de Roos, A., Fibbe, W. E., Zwaginga, J. J., Boersma, E., Schalij, M. J. and Atsma, D. E. (2009). Intramyocardial bone marrow cell injection for chronic myocardial ischemia: a randomized controlled trial. *JAMA*, 301. 19, pp.(1997-2004), 1538-3598 (Electronic).

Vanderheyden, M., Vercauteren, S., Mansour, S., Delrue, L., Vandekerckhove, B., Heyndrickx, G. R., Van Haute, I., De Bruyne, B., Timmermans, F., Wijns, W. and Bartunek, J. (2007). Time-dependent effects on coronary remodeling and epicardial conductance after intracoronary injection of enriched hematopoietic bone marrow stem cells in patients with previous myocardial infarction. *Cell Transplant*, 16. 9, pp.(919-25), 0963-6897 (Print).

Vasa, M., Fichtlscherer, S., Aicher, A., Adler, K., Urbich, C., Martin, H., Zeiher, A. M. and Dimmeler, S. (2001). Number and migratory activity of circulating endothelial progenitor cells inversely correlate with risk factors for coronary artery disease. *Circ Res*, 89. 1, pp.(E1-7).

Vassilopoulos, G. and Russell, D. W. (2003). Cell fusion: an alternative to stem cell plasticity and its therapeutic implications. *Curr Opin Genet Dev*, 13. 5, pp.(480-5), 0959-437X (Print).

Verfaillie, C. M., Pera, M. F. and Lansdorp, P. M. (2002). Stem cells: hype and reality. *Hematology (Am Soc Hematol Educ Program)*, pp.(369-91).

Vulliet, P. R., Greeley, M., Halloran, S. M., MacDonald, K. A. and Kittleson, M. D. (2004). Intra-coronary arterial injection of mesenchymal stromal cells and microinfarction in dogs. *Lancet*, 363. 9411, pp.(783-4).

Wang, J. S., Shum-Tim, D., Chedrawy, E. and Chiu, R. C. (2001). The coronary delivery of marrow stromal cells for myocardial regeneration: pathophysiologic and therapeutic implications. *J Thorac Cardiovasc Surg*, 122. 4, pp.(699-705).

Wang, J. S., Shum-Tim, D., Galipeau, J., Chedrawy, E., Eliopoulos, N. and Chiu, R. C. (2000). Marrow stromal cells for cellular cardiomyoplasty: feasibility and potential clinical advantages. *J Thorac Cardiovasc Surg*, 120. 5, pp.(999-1005).

Watanabe, E., Smith, D. M., Jr., Delcarpio, J. B., Sun, J., Smart, F. W., Van Meter, C. H., Jr. and Claycomb, W. C. (1998). Cardiomyocyte transplantation in a porcine myocardial infarction model. *Cell Transplant*, 7. 3, pp.(239-46).

Weisel, R. D., Fazel, S., Fedak, P. W. and Li, R. K. (2004). Cardiac restoration by cell transplantation. *Int J Cardiol*, 95 Suppl 1. pp.(S5-7).

Wohrle, J., Merkle, N., Mailander, V., Nusser, T., Schauwecker, P., von Scheidt, F., Schwarz, K., Bommer, M., Wiesneth, M., Schrezenmeier, H. and Hombach, V. (2010). Results of intracoronary stem cell therapy after acute myocardial infarction. *Am J Cardiol*, 105. 6, pp.(804-12), 1879-1913 (Electronic)0002-9149 (Linking).

Wollert, K. C. (2008). Cell therapy for acute myocardial infarction. *Curr Opin Pharmacol*, 8. 2, pp.(202-10), 1471-4892 (Print).

Wollert, K. C. and Drexler, H. (2005). Clinical applications of stem cells for the heart. *Circ Res*, 96. 2, pp.(151-63).

Wollert, K. C. and Drexler, H. (2010). Cell therapy for the treatment of coronary heart disease: a critical appraisal. *Nat Rev Cardiol*, 7. 4, pp.(204-15), 1759-5010 (Electronic) 1759-5002 (Linking).

Wollert, K. C., Meyer, G. P., Lotz, J., Ringes-Lichtenberg, S., Lippolt, P., Breidenbach, C., Fichtner, S., Korte, T., Hornig, B., Messinger, D., Arseniev, L., Hertenstein, B., Ganser, A. and Drexler, H. (2004). Intracoronary autologous bone-marrow cell transfer after myocardial infarction: the BOOST randomised controlled clinical trial. *Lancet*, 364. 9429, pp.(141-8).

Ying, Q. L., Nichols, J., Evans, E. P. and Smith, A. G. (2002). Changing potency by spontaneous fusion. *Nature*, 416. 6880, pp.(545-8).

Yoon, C. H., Hur, J., Park, K. W., Kim, J. H., Lee, C. S., Oh, I. Y., Kim, T. Y., Cho, H. J., Kang, H. J., Chae, I. H., Yang, H. K., Oh, B. H., Park, Y. B. and Kim, H. S. (2005). Synergistic neovascularization by mixed transplantation of early endothelial progenitor cells and late outgrowth endothelial cells: the role of angiogenic cytokines and matrix metalloproteinases. *Circulation*, 112. 11, pp.(1618-27), 1524-4539 (Electronic)0009-7322 (Linking).

Yoon, J., Choi, S. C., Park, C. Y., Choi, J. H., Kim, Y. I., Shim, W. J. and Lim, D. S. (2008). Bone marrow-derived side population cells are capable of functional cardiomyogenic differentiation. *Mol Cells*, 25. 2, pp.(216-23), 1016-8478 (Print).

Zhang, M., Methot, D., Poppa, V., Fujio, Y., Walsh, K. and Murry, C. E. (2001). Cardiomyocyte grafting for cardiac repair: graft cell death and anti-death strategies. *J Mol Cell Cardiol*, 33. 5, pp.(907-21).

Zhang, S., Wang, D., Estrov, Z., Raj, S., Willerson, J. T. and Yeh, E. T. (2004). Both cell fusion and transdifferentiation account for the transformation of human peripheral blood CD34-positive cells into cardiomyocytes in vivo. *Circulation*, 110. 25, pp.(3803-7).

Zhang, S. N., Sun, A. J., Ge, J. B., Yao, K., Huang, Z. Y., Wang, K. Q. and Zou, Y. Z. (2009). Intracoronary autologous bone marrow stem cells transfer for patients with acute myocardial infarction: a meta-analysis of randomised controlled trials. *Int J Cardiol*, 136. 2, pp.(178-85), 1874-1754 (Electronic) 0167-5273 (Linking).

Ziegelhoeffer, T., Fernandez, B., Kostin, S., Heil, M., Voswinckel, R., Helisch, A. and Schaper, W. (2004). Bone marrow-derived cells do not incorporate into the adult growing vasculature. *Circ Res*, 94. 2, pp.(230-8).

13

Stem Cell Transplantation for the Treatment of Peripheral Arterial Disease

Nikolaos Papanas
Outpatient Clinic of the Diabetic Foot in the Second Department of Internal Medicine, Democritus University of Thrace
Greece

1. Introduction

Peripheral Arterial Disease (PAD) is a chronic occlusion of lower-extremity arteries distal to the aortic bifurcation [1, 2]. It affects 3%-12% in the general population and its frequency increases with age [1-3]. In the vast majority of patients, PAD is ascribable to atherosclerosis [1,2]. It is a progressive disease leading to severe haemodynamic compromise of the affected extremity and may be even limb threatening in the event of critical ischaemia [1-3]. Not only does PAD have a considerable unfavourable impact on the quality of life, but it also poses a tremendous economic burden on society [1-3] and is an ominous harbinger of high morbidity and mortality due to concomitant coronary artery disease and cerebrovascular disease [4-6].
Management of PAD may be divided into medical and surgical. The former includes risk factor modification and medical treatment [3, 5, 7, 8], while the latter comprises surgical revascularisation (bypass surgery, intraluminal angioplasty, endovascular surgery) [9-11]. Revascularisation may be divided into percutaneous transluminal angioplasty (PTA) and by-pass surgery [9, 10, 12, 13]. The former can restore adequate blood flow, while recurrence is infrequent and amputation rates are rare [9-11]. This technique yields favourable results in patients with critical limb ischaemia as well [9, 10]. By-pass graft surgery has also been extensively practised [12, 13]. Depending on the location of affected arteries, it may be performed to the femoral, popliteal or even distal arteries, such as the dorsalis pedis artery [12, 13].
However, vascular atherosclerotic lesions may diffusely affect several anatomical regions. This holds especially true for elderly and diabetic patients, who may, therefore, be poor candidates for surgical intervention [1, 14]. Regrettably, it may also apply to patients with critical limb ischaemia or those who have had prior revascularisation [1, 15]. Hence, there is an undeniable need to develop alternative therapeutic modalities to restore limb blood flow [1, 5, 15, 16]. This chapter reviews the progress achieved with autologous stem cell transplantation, an important innovation involving intramuscular and/or intra-arterial stem cell administration into the affected lower extremity [15-17].

2. Stem cell therapy: Principles and cell types

Stem cell therapy is based on administration of autologous stem cells, taken either from the bone marrow, from the peripheral blood or, more rarely, from adipose tissue [17-23].

Human bone marrow contains stem cells with the potential for differentiation into a variety of tissues, including endothelium, liver, muscle, bone and skin [15-17, 22, 23]. Consequently, bone marrow-derived cells may differentiate into endothelial cells and also provide progenitor and/or stem cells to wounds during healing [15-17, 22, 23]. Importantly, the newly derived endothelial cells promote angiogenesis, i.e. new vessel formation (also called neovascularisation), by two mechanisms [15, 17, 24-26]. First, they get connected to each other and are organised into capillary tubes, thus forming the primary capillary plexus, which is further refined by remodelling [25, 26]. Secondly, they release growth factors promoting angiogenesis (also called angiogenic growth factors), mainly Vascular Endothelial Growth Factor (VEGF), Fibroblast Growth Factor (FGF), Hypoxia inducible factor-1a (HIF-1 a), Hepatocyte Growth Factor (HGF), Placental Growth Factor (PlGF), angiopoietin, Nerve Growth Factor (NGF), Developmental endothelial locus-1 (Del-1) and others [15, 24]. Similar efficacy appears to be also accomplished with peripheral blood stem cells and adipose tissue-derived stem cells [18-21, 23].

Bone marrow-derived cells include autologous bone marrow-derived bone marrow stem cells (BMCs), bone marrow mononuclear cells (BMMNCs) and endothel progenitor cells (EPCs) [16, 18-20, 22, 23]. Peripheral blood-derived cells include mononuclear cells (PBMNCs), polymorphonuclear leukocytes (PMNCs), erythroid colony-forming cells (ECFCs), circulating blood-derived progenitor cells (CPCs) and EPC, while other cells mainly include mesenchymal stem cells (MSCs) and adipose tissue-derived cells (ADSCs) [16, 18-23].

3. Pre-clinical studies

Preclinical studies have evaluated a variety of stem cells from bone marrow, peripheral blood or other sources, using the ischaemic hindlimb model in the mouse, rat or rabbit. Experimental hindlimb ischaemia is achieved by surgical ligation of the superficial femoral and/or external iliac artery [15]. This unilateral hindlimb ischaemia has been widely used as a model of PAD, although it virtually equates to acute ischaemia, not entirely representative of chronic PAD [15]. Major studies are briefly described in this section.

3.1 Bone marrow-derived stem cells

In the mouse ischaemic hindlimb, Ziegelhoeffer et al. [27] investigated whether BMCs act by incorporation into vessels. They found a threefold increase of such cells around growing collateral arteries in the ischaemic vs. control limb. The authors concluded that BMCs do not promote vascular growth by incorporating into vessel walls but may function as supporting cells [27]. In another study, intravenous BMC administration was successful in improving blood flow as well as increasing capillary density and expression of the proliferation-associated protein Ki-67 in mouse ischaemic hindlimbs [28]. Importantly, this beneficial effect was enhanced by concomitant metabolic vascular protection (1.0% vitamin E added to the chow and 0.05% vitamin C and 6% L-arginine added to the drinking water), suggesting an additive effect, attribitable in part to NO activation and reduced systemic oxidative stress [28].

A further study provided evidence for increased neovascularisation with bone marrow-derived cells. Takagi et al. [29] examined whether granulocyte-colony stimulating factor (G-CSF) can enhance neovascularisation and collateral vessel formation induced by BMNC transplantation in the hindlimb ischaemia model. Administration of G-CSF significantly increased blood perfusion on Laser Doppler imaging (LDI), number of angiographically detectable collateral vessels, and capillary density [29]. Similarly, neovascularisation was significantly increased in rats receiving BMNCs, as evidenced by angiographical findings and capillary density. The combination of G-CSF and BMNCs augmented neovascularisation in comparison to BMNCs alone, as manifested by angiographical findings and capillary density [29]. It was concluded that G-CSF and BMNCs combination could emerge as a helpful therapeutic approach to restore blood flow [29].

Sica et al. [30] explored the effect of intravenous BMC administration and concurrent metabolic vascular protection (1.0% vitamin E, 0.05% vitamin C, and 6% L-arginine) in the ischaemic hindlimb of diabetic and non-diabetic mice. In both mice, BMC treatment increased blood flow and capillary density and decreased interstitial fibrosis [30]. This effect was amplified by metabolic vasculoprotective treatment. The latter had by itself no effect on capillary density, but could reduce interstitial fibrosis in non-diabetic mice [30].

One year later, Jeon et al. [31] examined whether the angiogenic efficacy of a combination of two angiogenic strategies (EPC mobilization with G-CSF and BMMNC transplantation) is superior to either strategy alone in the mouse ischaemic hindlimb. Both treatments were efficacious in increasing microvessel density and expression of bFGF and VEGF [31]. Combination therapy accomplished more extensive expression of bFGF and VEGF than either form of single therapy [31].

More recently, a study randomised ApoE knockout mice to receive either phosphate buffered saline (PBS) or intra-arterial BMCs [32]. Efficacy was assesed by LDI of the ischaemic limbs. It was demonstrated that BMCs significantly ($p<0.05$) increased blood flow recovery in ischaemic limbs, in comparison to saline (61.8±15% vs. 41.9±13.9%, respectively) [32]. Of note, it was revealed that BMCs differentiated not only into small blood vessels, but also into skeletal myofibres and supporting membranes [32]. The latter changes were associated with increased serum levels of VEGF, FGF-2, transforming growth factor beta (TGFbeta), interleukin 4 (IL-4), and tumour necrosis factor alpha (TNF-alpha). These workers proposed that the benefit of new skeletal muscle formation might be added to that of angiogenesis in long-standing lower extremity ischaemia [32].

3.2 Stem cells from peripheral blood

Asahara et al. [33] showed increased capillary density and augmented blood flow recovery following administration of heterologous, homologous and autologous EPCs in the mouse ischaemic hindlimb. Athymic nude rats with unilateral hindlimb ischaemia, were randomly assigned to intramuscular administration of PBMNCs plus platelets or PBMNCs plus platelets plus PMNs [34]. Blood perfusion, as evaluated by LDI, improved significantly ($p<0.001$) by 44% in rats receiving PBMNC plus platelets. There was a further significant ($p<0.01$) increase with the addition of PMNs. Density of newly formed capillaries was increased by PBMNCs plus platelets (3.5-fold increase) or platelets alone (2.4-fold increase), with a significant between-group difference ($p<0.001$), whereas PMNs exerted a significant ($p<0.05$) inhibitory role (32% inhibition) [34].

Yoon et al. [35] utilised the synergistic effect of EPCs and outgrowth endothelial cells (OECs), both cultured from the peripheral blood. Each cell type alone succeeded in a significant ($p<0.05$) increase of new vessel formation in the mouse experimental model [35]. In vitro, there was a synergistic effect by mutual interaction through cytokines and matrix metalloproteinases (MMPs). In vivo, injection of both cell types resulted in significantly ($p<0.05$) more pronounced neovascularisation than either cell type alone [35].

In another study, rats were randomly allocated to MSCs, MNCs, or vehicle infusion (control group) [36]. Significant improvement of hindlimb ischemia was observed both in rats receiving MNCs and in those treated with MSCs, as compared to the control group. LDI perfusion index was highest in the MSC group (0.81±0.08), followed by the MNC (0.69±0.1) and control group (0.57±0.06) [36]. Capillary density was significantly ($p<0.01$) increased with both treatments compared to controls. MSC treatment was significantly ($p<0.01$) more efficacious in increasing capillary density than MNC treatment [36]. The number of transplanted cell-derived endothelial cells was also highest in MSC rats. Finally, MSCs were significantly ($p<0.05$) more tolerant to apoptotic stimuli (serum starvation and hypoxia) in vitro than MNCs [36].

Two further works confirmed the beneficial effect of EPCs from peripheral blood in hindlimb ischaemia of the mouse [37] and rat [38]. In the first study, EPCs improved histologically confirmed muscle healing, blood flow and vessel density [37]. These changes were associated with modulation of proangiogenic pathways: increased VEGF-A levels and sensitivity to VEGF family ligands, increased levels of monocyte chemotactic protein-1 (MCP-1) [37]. In the second work, EPC treatment was combined with targeted extracorporeal shock wave application to facilitate their tissue recruitment [38]. Combined treatment achieved a significant ($p<0.01$) increase in the number of histologically confirmed vascular endothelial growth factor-positive endothelial cells per myocyte, as well as a significant ($p<0.05$) enhancement in EPC recruitment and homing [38]. Moreover, there was a significant ($p<0.05$) increase in relative blood flow recovery assessed by LDI [38].

Using an immunodeficient hindlimb ischemia model and LDI, Sasaki et al. [39] evaluated limb salvage rate and blood perfusion after intramuscular implantation of ECFCs, as compared to PBS treatment in the controls. Salvage rate and blood perfusion were increased by 38% ($p<0.05$) and 82.8% ($p<0.01$), respectively [39]. Vascular smooth muscle cell recruitment was also increased and the capillary density was 1.6-fold higher ($p<0.05$) than in the control group [39]. Finally, ECFCs were confirmed to supply angiogenic cytokines (VEGF and FGF-2), suggesting a possible novel strategy for therapeutic angiogenesis [39].

3.3 Other stem cells

A variety of other cells have also been investigated. Niagara et al. [40] studied the effect of rabbit autologous primary skeletal myoblasts on angiographically confirmed new vessel formation. Treatment induced a significant increase in neovascularisation ($p<0.05$) and capillary density ($p<0.01$) [40].

Scientists from the Cardiovascular Research Institute in the Washington Hospital Center have experimented with MSCs in the mouse [41, 42]. They showed that these cells effectively increased VEGF and b-FGF levels, collateral blood flow and limb function [41, 42]. At the same time, auto-amputation and muscle atrophy were reduced. Their data indicated the great importance of paracrine signaling and, additionally, showed that cell incorporation into vessels was not a prerequisite for their effects [41, 42].

ADSCs represent another alternative. Cultured ADSCs from C57Bl/6 mouse inguinal adipose tissue were transplanted into the ischaemic mouse hindlimb and improved blood flow assessed by LDI and capillary density assessed by anti-CD31 immunostaining antibody [43]. The therapeutic effect appeared to be mainly achieved by their ability to secrete angiogenic growth factors: relatively high expression of HGF, VEGF, placental growth factor (PGF), and transforming growth factor beta (TGF)-β, as well as moderate expression of FGF-2 and angiopoietin 1 were noted [43]. Moon et al. [21] have also shown that hADSC can improve blood flow (evidenced by LDI), even when transplanted relatively late, i.e. 7 days post induction of ischaemia in mice. The therapeutic effect was related to the number of transplanted cells. Conditioned media from these cells increased proliferation of human aortic endothelial cells [21].

More recently, Kim and collagues [44] have compared human adipose stromal cells (hADSC) with human bone marrow stromal cells (hBMSC) in a nude mice model of hindlimb ischaemia. The former showed superior recovery of blood flow and higher expression of matrix metalloproteinases (MMP3 and MMP9) than the latter, prompting further exploration of this therapeutic alternative [44].

4. Clinical studies

Clinical studies have evaluated stem cells taken from bone marrow, peripheral blood or other sources. These works differ substantially in number of patients recruited and endpoints used. Major studies are briefly described in this section (Tables 1, 2).

4.1 Bone marrow-derived stem cells

4.1.1 BMCs

Esato et al. [45] administered BMCs in 8 selected patients with chronic lower extremity ischaemia (4 patients with PAD, 4 patients with thromboangiitis obliterans [TAO, Bürger's disease]), in whom prior treatment had failed. Symptoms improved in 7/8 patients. Moreover, complete ulcer healing was achieved in 2/3 patients, and partial healing in the 3rd patient [45]. Temperature increase was documented in 2 patients and new collateral formation in 2 out of 3 patients who underwent angiography.

Nizankowski et al. [46] delivered BMCs intramuscularly by repeated injections into the pedal and tibial regions in 10 patients suffering from chronic leg ischaemia staged Fontaine IV. Efficacy was assessed by LDI, transcutaneous oxygen pressure ($TcPO_2$), ankle-brachial index (ABI), visual analogue pain scale, analgesic therapy requirement, ulceration area, angiography and scintigraphy. Improved blood flow was documented by LDI and elevation in $TcPO_2$ [46]. Painful symptoms were relieved in the majority of patients, and only 3 amputations were finally needed at 12-month follow-up. Treatment efficacy did not depend on the number of cells injected [46].

Another centre recruited 22 patients, who were divided into those with severe and those with moderate ischaemia [47]. Both patient groups received BMCs, the number of which was higher in those with severe ischaemia. Evaluation included improvement of pain, cold sensation and numbness, ABI, $TcPO_2$, angiography, amputation rate, and foot ulcer healing. At 4 weeks, pain was alleviated in 90.0% in patients with severe and 16.7% with moderate ischaemia ($p<0.01$) [47]. The corresponding rates of relief in cold sensation were 90.5% and 5.3% ($p<0.01$), while those in improved numbness were 62.5% and 9.1% ($p<0.01$) [47]. ABI was increased by 31.8% in subjects with severe ischaemia, but not in those with moderate ischaemia ($p<0.01$). The corresponding increments in $TcPO_2$ were 94.4% and 11.1% ($p<0.01$) [47]. Angiography demonstrated new collateral vessel network in 100% of subjects with severe ischaemia but not in the presence of moderate ischaemia ($p<0.01$) [47]. Subjects with severe ischaemia exhibited 4.5% amputation rates and 75% rates improved ulcer healing. The corresponding rates in those with moderate ischaemia were 27.3% and 0% ($p<0.05$). Thus, it was concluded that efficacy was significantly superior with higher number of implanted cells [47].

In a small series of 5 patients with advanced PAD and foot ulcers, intramuscular (gastrocnemius muscle) BMC transplantation achieved substantial improvement in pain intensity and pain-free walking distance in all of them after 12 months [48]. Complete ulcer healing was achieved in 3 and partial healing in one patient [48]. There was a progressive improvement of ABI and $TcPO_2$ over 12 months: at the end of this period, the average ABI rose from 0.41 to 0.83 ($p<0.05$) and the average $TcPO_2$ from 18.8 to 37.5 mmHg [48]. Duplex ultrasonography showed improvement in one patient, while angiography detected new collaterals in 3 patients [48].

Eighteen patients with advanced PAD (staged Fontaine III/IV) were recruited in another long-term study with 18-month follow-up. The control group included 18 matched patients taking maximal drug therapy [49]. In the treatment arm, patients received BMCs in two intra-arterial doses and, concurrently, daily antioxidants and L-arginine. Among BMC-treated patients, mean walking distance started to increase at 3 months and exhibited further improvement at 18 months ($p<0.05$) [49]. Significant ($p<0.05$) improvement in ABI was seen in 10/18 patients at 3 months and in 12/18 patients at 18 months [49]. Ischaemic ulcers improved in 13/18 patients after 6-12 months. Amputation rates were 13.3% (2/18) in BMC-treated patients and 55.6% (10/18) in controls ($p=0.014$) [49].

A total of 37 patients suffering from PAD staged Fontaine IV with an ulcerated limb (including diabetic foot) receiving BMCs intramuscularly were recruited in a further trial [50]. Efficacy parameters included toe pressure, toe-pressure index (TBI), ABI, $TcPO_2$, LDI, skin perfusion pressure, wound healing and amputation rates. Limb salvage was achieved in 30 patients (81%) and amputation rate was 19% (7 patients) [50]. In the limb salvage group, significant ($p<0.05$) improvements in toe pressure, TBI, LDI and $TcPO_2$ were noted [50]. Significant ($p<0.05$) improvement in pain-free walking distance at treadmill testing was also shown in a study of 42 subjects with chronic PAD involving the femoropopliteal-tibial segment by intramuscular BMC administration [51].

To evaluate the long-term efficacy of BMC transplantation in subjects with PAD without option for revascularisation or with unsuccessful revascularisation, the BONe Marrow Outcome Trial in Critical Limb Ischemia (BONMOT-CLI) was designed [52]. This is a double-blinded, 1:1 randomised, placebo-controlled multi-centre study enrolling patients from 4 German centres. Patients will be randomised to autologous BMCs (expected 45 patients) injected at 40 sites into the ischaemic limb or sham bone marrow aspiration and 40 saline injections [52]. The composite primary endpoint of major amputation or persisting critical limb ischaemia will be evaluated at 3 months. Subjects will be then followed up for up to two years. Secondary endpoints will include death, changes in perfusion, quality of life, pain-free walking distance, minor amputations, wound healing, collateral density and cancer incidence [52].

4.1.2 BMMNCs

In 2002, the Therapeutic Angiogenesis using Cell Transplantation (TACT) Study Investigators recruited 25 patients with unilateral leg ischaemia (group A), who received BMMNCs into the gastrocnemius muscle and saline into the less ischaemic limb, as well as 22 patients with bilateral leg ischaemia (group B), who randomly received BMMNCs in one leg and peripheral blood-mononuclear cells in the other as a control [53]. Efficacy parameters included ABI, $TcPO_2$ and rest pain. In group B, the BMMNC-treated limb exhibited significant increases in ABI, as compared to the contralateral limb, by 0.09 (CI 0.06-0.11, $p<0.0001$), in $TcPO_2$ by 13 mmHg (CI 9-17, $p<0.0001$) and in pain-free walking time by

1.2 min (CI 0.7-1.7, p=0.0001) [53]. Pain intensity was also reduced by -0.85 (-1.6 to -0.12, p=0.025) in comparison to the contralateral limb [53]. Similar improvements were seen in the BMMNC-treated vs. saline-treated extremity in group A. The new therapy was well tolerated. Two patients in group A died after myocardial infarction, but this was unrelated to treatment [53].

Kirana et al. [54] described a 60-year old type 2 diabetic patient with infected gangrenous ulcers of the 3rd and 4th left toes. The patient has severe stenosis of anterior tibial, posterior tibial and dorsalis pedis arteries, leading to critical ischaemia. After amputation of the affected toes, he received intramuscular BMMNC transplantation. After 20 weeks, the stump had completely healed, and there were improvements in ABI, Laser Doppler reactive hyperaemia and $TcPO_2$ [54].

Higashi et al. [55] studied the effect of BMMNCs in 7 patients with PAD. Treatment improved ABI (from 0.33±0.21 to 0.39±0.17, p=0.06), $TcPO_2$ (from 28.4±11.5 to 36.6±5.2 mm Hg, p=0.03), and pain-free walking time (from 0.8±0.6 to 2.9±2.2 minutes, p=0.02) [55]. Moreover, it improved endothelial function, evaluated by blood flow in response to acetylcholine (from 19.3±6.8 to 29.6±7.1 mL/min per 100 mL, p=0.002) [55]. In the same year, two other small studies the efficacy of BMMNCs. The first work showed that these cells improved blood flow in 12 patients with PAD staged Fontaine IV [56]. Specifically, there was an increase in pain-free walking time (from 140±53 s to 451±74 s, p=0.034), ABI (from 0.65±0.08 to 0.73±0.07, p=0.055) and perfusion index (proximal area from 1.32±0.10 to 1.56±0.11, p=0.007) [56]. Perfusion scintigraphy could identify new collaterals after BMMNC transplantation [56]. The second work showed increased ABI (from 0.54±0.47 to 0.61±0.50, p<0.05) and $TcPO_2$ (28.4±15.4 mmHg to 37.1±24.4 mmHg) following BMNNC transplantation in 8 patients with PAD [57].

A group from the Heinrich Heine University of Düsseldorf, Germany has performed a series of small studies enquiring into the efficacy of BMMNCs. A 62-year old man with PAD staged Fontaine IIb received combined intra-arterial and intramuscular transplantation of BMMNCs [58]. After 10 weeks, there was a 7-fold improvement of walking distance, a greater than 50% increase of tissue oxygen saturation and a 24% increment of ABI on exercise [58]. Then, 10 patients received intra-arterial (into the common femoral artery) and intramuscular (into the muscles of the thigh and the lower leg) BMMNC transplantation [59]. After 2 months, pain-free walking distance increased significantly (p<0.05) in all patients. Moreover, significant (p<0.05) improvement of ABI, capillary-venous oxygen saturation and parameters of venous occlusion plethysmography were observed [59]. A 63-year-old type 2 diabetic patient with severe intermittent claudication and a recalcitrant ulcer in the right hallux was treated with intra-arterial and intramuscular BMMNC transplantation [60]. At 8 weeks, the ulcer had healed completely. Six months later, there were significant (p<0.05) increments in the claudication-free walking distance by >100% and in resting blood flow on venous occlusion plethysmography by 23% [60]. The next study in 8 patients with PAD staged Fontaine II or III demonstrated a significant (p<0.05) 3.7-fold increment in pain-free walking distance [61]. ABI on exercise (from 0.62±0.17 to 0.77±0.15, p=0.018), capillary-venous oxygen saturation (from 50±15 to 62±6%, p=0.027) and venous occlusion plethysmography (from 4.6±1.7 to 6.5±1.7 ml/100 ml tissue/min, p<0.05) exhibited significant improvements as well [61]. Finally, 13 subjects with chronic PAD staged Fontaine IIb received combined intra-arterial and intramuscular BMMNC transplantation [62]. After 2 months, the pain-free walking distance increased significantly by more than 300% (from 147±90 to 500±614 m, p=0.001). ABI rose, both at rest (from

0.66±0.18 to 0.80±0.15, p=0.003) and on exercise (from 0.64±0.19 to 0.76±0.16, p=0.006) [62]. The authors also found significant improvements in capillary-venous oxygen saturation (from 56±14 to 63±5, p=0.021) and venous occlusion plethysmography (from 2.1±0.7 to 2.5±0.7, p=0.009). Importantly, beneficial effects were sustained at 13 months [62].

A4 mixed series of 35 patients (30 with ischaemic diabetic foot, 2 with PAD and 3 with TAO) received BMMNC transplantation after bone marrow mobilisation with GcSF [63]. Pain was entirely relieved in 94.7% and improved in 97.1% of patients [63]. Numbness was alleviated in 93.3% of patients. Claudication-free walking distance was prolonged in all subjects, while 47.9% exhibited a significant ($p<0.05$) increase in ABI and 92.3% a significant ($p<0.05$) increase in $TcPO_2$ [63]. Ulcer healing was accomplished in 9.1% and ulcer area reduction in 27.3% [63]. Amputation rate was 6.3%. New collateral vessels were identified by angiography in 91.2%. Complications included transient fever and mild fatigue in one patient and acute myocardial infarction in one patient. The latter occurred 7 days following transplantation and easily recovered with treatment [63].

These authors then compared intramuscular (group A, n=16) with intra-arterial (group B, n=16) BMMNC injection [64]. Efficacy parameters included rest pain, coldness, ABI, intermittent claudication, $TcPO_2$ and angiography. Treatment was equally efficacious in both groups [64]. Rest pain was improved in 76.5% of group A and in 93.3% of group B patients. Coldness improved in 100% of patients from both groups. ABI increased in 44.4% of group A and in 41.2% of group B patients [64]. Limb salvage was achieved in 83.3% of group A and in 94.1% of group B patients [64]. $TcPO_2$ increased to ≥20 mmHg in 20 limbs, while a rich new collateral network was identified in 9/15 limbs which underwent angiography. There were 2 deaths from heart failure [64].

Longer data from this group have recently become available [65]. A total of 65 patients with PAD have received BMMNCs: 12 patients were transplanted 2-4 times and 53 patients only once [65]. Mean follow-up was 21.5 months (range 8-56) [65]. In both treatment groups, coldness improved in all patients, while there were significant ($p<0.05$) increments in ABI and $TcPO_2$. Overall efficacy was 70.8% and the recurrence rate was 10.7%. Response duration was over 12 months in 91.3% of patients, over 24 months in 52.2% and over 37 months in 26.1%. Efficacy was significantly ($p<0.001$) higher in subjects transplanted 2-4 times (100%) than in those transplanted once (64.2%) [65]. Mortality rate was 12.3%: 5 patients died of myocardial infarction and heart failure, and 3 died of cerebral infarction [65].

Twelve patients received BMMNC transplantation by two different techniques (either sorted on a blood cell separator or isolated by density gradient on Ficoll-Hypaque) [66]. Both modalities were equally effective in improving ABI at rest, oxygen saturation, pain-free walking time and rest pain intensity. Improvement was sustained at 24 weeks and limb salvage rate was 41.67% [66]. In another small series, 7 patients (3 with TAO and 4 with PAD undergoing haemodialysis) received mononuclear cell transplantation (BMMNC in 6, PBMNC in 1) into the gastrocnemius and quadriceps femoris muscles [67]. Patients with TAO exhibited improvements in painful symptoms, ABI, $TcPO_2$ and thermography, while patients with PAD did not respond [67]. It was concluded that transplantation was more effective in TAO than in PAD, but this must be interpreted with caution, given the very small number of patients [67].

Van Tongeren et al. [68] compared combined intra-arterial plus intramuscular (n=12) to exclusive intramuscular (n=15) BMMNC transplantation. Efficacy was assessed at 1, 6 and 12 months by means of limb salvage, pain-free walking distance, ABI and pain scores [68].

Both modalities were equally effective. Pain-free walking distance improved from 81±56 m to 257±126 m at 6 months (p=0.0002). Mean ABI increased by 23% after 6 months (p=0.01) and pain score was reduced up to 50% (p=0.001). Two patients in the combined treatment group vs. 7 patients in the intramuscular group (p=0.17) required amputation [68].
In the same year, the 3-year outcomes of the TACT (Therapeutic Angiogenesis using Cell Transplantation) follow-up study were published [69]. Primary endpoints comprised mortality and amputation-free interval. Median follow-up was 25.3 months (range 0.8-69.0 months) and 3-year survival rates were 80% (CI 68-91) in PAD (n=74) and 100% in TAO (n=41). Amputation-free rates were 60% (CI 46-74) and 91% (CI 82-100), respectively [69]. At 2 years, ABI and $TcPO_2$ had not changed significantly, but there was a sustainable significant improvement in pain scale, ulcer size and pain-free walking distance [69].
Ten patients with end-stage PAD underwent 2 BMMNC transplantations, while 10 matched patients served as controls [70]. In the treatment arm, there was a significant ($p<0.05$) improvement in ABI, claudication-free walking distance and capillary density, which were maintained at 12 months [70]. Eight patients with critical limb ischaemia (CLI) and no alternative treatment option received BMMNC transplantation into the gastrocnemius muscle [71]. Pain, angiography and non-invasive vascular workup were evaluated. At 4 months, pain was reduced in 5 patients. At 8 months, 5 patients could be evaluated and showed stability or insignificant improvement [71].
Twenty-four patients with CLI received intra-arterial BMMNC transplantation [72]. After 12 months, all patients survived and only 2 of them had undergone amputation. Ulcer healing rate was 78% and median Fontaine stage had improved from 3.5 to 2 ($p<0.0001$) [72]. Collateral vessel formation had improved by 1.13 and 1.3 points on a four-point semiquantitative scale in calf and foot, respectively ($p<0.0001$) [72]. Impressively, significant improvements were reported in all items of the SF-36 quality of life questionnaire [72].
De Vriese et al. [73] included 16 very old patients (mean age 78±2 years) with CLI and substantial comorbidities (hypertension, smoking, diabetes, hypercholesterolaemia and uraemia), who underwent intramuscular BMMNC transplantation. $TcPO_2$ improved from 0.51±0.11 to 0.86±0.03 mm Hg ($p<0.001$) after 12 weeks, whereas ABI showed no significant change (0.42±0.15 vs. 0.59±0.1, p=0.23) [73]. On digital subtraction angiography, the number of collateral vessels increased by 0.89±0.86 (p=0.33), but capillary surface area on gastrocnemius muscle biopsy increased from 0.61±0.07% to 2.38±0.73% ($p<0.05$) [73]. Two patients died of gangrene, 3 patients were amputated and one patient required by-pass surgery. Two further patients died of unrelated causes [73]. Of note, symptomatic relief was mainly achieved in patients with less severe ischaemia [73]. Thus, despite some objective improvement in vascular parameters, BMMNC administration was only associated with very modest overall improvement in these high-risk patients.
More recently, 51 patients with limb-threatening CLI facing risk of major amputation received BMMNCs intramuscularly [74]. This treatment was offered after unsuccessful or impossible revascularisation procedure and optimal medical therapy. Limb salvage was 59% at 6 months and 53% at the end of the study (mean follow-up 411±261 days) [74]. At 6 months, ABI increased from 0.33±0.18 to 0.46±0.15 (p=0.005) and $TcPO_2$ increased from 12±12 to 25±15 mmHg (p=0.001) in patients with limb salvage, but not in those ultimately amputated [74]. The former were also downstaged from a mean Rutherford category of 4.9 to 3.3 (p=0.0001). Wound area was reduced from 11.6±20 to 4.4±11 cm^2. Median walking distance improved from 0 to 40 m, but only in those escaping amputation. Finally, analgesic requirement dropped by 62% [74].

A further work looked at the short-term results of dual intramuscular and intra-arterial autologous BMMNC transplantation in 9 patients facing the risk of lower-extremity amputation [75]. Eight patients had rest pain, 7 had diabetes mellitus and 8 recalcitrant ulcers. Efficacy parameters comprised ABI, rest pain, ulcer healing and amputation. The primary composite endpoint was defined as improved ABI, relief of rest pain, ulcer healing and absence of major amputations [75]. Overall success rate was 33.3% (3 patients), while success in at least one of the 4 components of the primary endpoint was noted in 5 additional patients (55.6%) [75]. ABI exhibited a non-significant improvement by 0.12 (dorsalis pedis artery) and 0.08 (posterior tibial artery). Three patients (33.3%) sustained major amputations. Those remaining free from amputations showed improvement in patient severity and could be downstaged by at least one level in Rutherford and Fontaine classifications at a mean follow-up of 7.8 months [75]. Complete ulcer healing was accomplished at 3 months in all ulcerated patients not needing amputation. These short-term findings favoured BMMNC transplantation for limb salvage in patients with severe PAD, but it must be borne in mind that patient numbers were small [75].

Authors	Number of Patients	Type of Cells	Clinical outcomes
Nizankowski et al. [46]	10	BMCs	Improvement of symptoms, increased blood flow (LDI, $TcPO_2$)
Gu et al. [47]	22	BMCs	Improvement of symptoms (pain, cold sensation), increase in ABI and $TcPO_2$, new collateral vessels
Napoli et al. [49]	18	BMCs	Increase in ABI and walking distance, ulcer healing, reduction of amputation rates
Procházka et al. [50]	37	BMCs	Improvement in toe pressure, TBI, LDI and $TcPO_2$
Korymasov et al. [51]	42	BMCs	Improvement in pain-free walking distance
Tateishi-Yuyama et al. [53]	25	BMMNCs vs. PBMNCs	Increase in ABI, $TcPO_2$ and walking distance
Higashi et al. [55]	7	BMMNCs	Increase in ABI, $TcPO_2$, endothelial function and walking distance
Miyamoto et al. [56]	12	BMMNCs	Increase in ABI, perfusion index and walking distance

Authors	Number of Patients	Type of Cells	Clinical outcomes
Bartsch et al. [62]	13	BMMNCs	Long-term improvement in walking distance, ABI and oxygen saturation
Gu et al. [63]	35 (30 ischaemic diabetic foot, 2 PAD, 3 TAO)	BMMNCs	Improvement in pain, numbness and walking distance, increase in ABI and $TcPO_2$, ulcer healing, new collateral vessels
Gu et al. [64]	32	BMMNCs	Improvement in symptoms, ABI and $TcPO_2$, limb salvage, new collateral vessels
Gu et al. [65]	65	BMMNCs	Long-term improvement in coldness, ABI and $TcPO_2$, new collateral vessels
Van Tongeren et al. [68]	27	BMMNCs	Improvement in walking distance and ABI, reduced amputation rates
Matoba et al. [69]	115 (74 PAD, 41 TAO)	BMMNCs	Long-term improvement in pain scale, ulcer size and walking distance, reduced amputation rates
Chochola et al. [72]	24	BMMNCs	Ulcer healing, new collateral vessels, Fontaine downstaging, improved quality of life
De Vriese et al. [73]	16	BMMNCs	Increased capillary surface area on muscle biopsy, increase in $TcPO_2$, insignificant increase in ABI and number of collateral vessels
Amann et al. [74]	51	BMMNCs	Limb salvage, increase in ABI and $TcPO_2$,

Table 1. Major clinical studies with bone marrow-derived stem cell therapy

Finally, in a phase I clinical trial, 10 patients with advanced PAD (Fontaine stages IIb to IV) received a cell product consisting of autologous BMMNCs and MSCs [76]. ABI improved significantly (from 0.34±0.19 to 0.69±0.18, $p<0.002$) at 2 months. Claudication-free walking time increased significantly ($p<0.05$) by 3.48±1.72 folds at 6 months [76]. Significant ($p<0.05$) improvements were also seen in 99mTc-TF perfusion scintigraphy scores and in quality of life scores at 6 months [76]. $TcPO_2$ rose insignificantly ($p=0.067$) from 33±6 mm Hg to 46±10 mm Hg at 6 months, but angiographic score increased from 0.90± 0.30 to 1.89±0.78 ($p=0.002$) [76].

4.2 Stem cells from peripheral blood

4.2.1 PBMNCs

The main cell population used is PBMNCs, with or without GcSF mobilisation. Inaba et al. [77] included 7 patients with PAD, who received GcSF subcutaneously for 5 days and then intramuscular PBMNC transplantation. Pain was relieved as early as at 3 days [77]. Maximum pain-free walking distance was increased by day 7. The heel ulcer completely healed in one patient. These improvements were sustained at 12 months. Improvements in ABI and angiographic findings were minor [77].

Ishida et al. [78] examined the feasibility and safety of PBMNC administration after GcSF mobilisation in 6 patients (1 with PAD, 5 with TAO). A slight increase in ABI and ulcer area were seen in 4 and 3 patients, respectively [78]. Mean walking distance significantly increased from 203 m to 559 m ($p=0.031$) at 4 weeks and was sustainable for 24 weeks. There was also a significant improvement in physiological functioning, as evaluated by the SF-36 questionnaire on quality of life [78]. No serious adverse events were noted.

In 92 patients with CLI, intramuscular PBMNC administration after GcSF mobilisation was most effective for non-diabetic non-dialysis subjects in terms of averting amputations [79]. In those with diabetes and/or undergoing haemodialysis, treatment was efficacious for milder ischaemia (up to Fontaine stage III), but not for advanced disease (staged Fontaine IV) [79]. Indeed, most amputations (n=37) were carried out in Fontaine staged IV diabetic or dialysis patients. Characteristically, amputation rate was as high as 71% in Fontaine staged IV diabetic patients on haemodialysis [79]. Interestingly, serum VEGF levels increased in all patients (mean increase 176%), regardless of clinical outcome.

In 15 patients with CLI, intramuscular PBMNC administration after GcSF mobilisation induced significant ($p<0.05$) increases in mean ABI (from 0.3 to 0.46), mean pain-free walking distance (from 0.15 to 0.72 km) and mean maximal walking distance (from 0.96 to 2.13 km) at 12 months [80]. Moreover, 5/6 foot ulcers healed completely.

Two studies compared PBMNC to BMMNC transplantation. Gu et al. [19] carried out this comparison in 42 patients with unilateral lower extremity ischaemia (28 with ischaemic diabetic foot, 8 with TAO and 6 patients with PAD). Each therapy was administered to 21 patients. No difference was seen between the two groups [19]. All but one ulcers healed. At 4 weeks, pain was relieved in 88.2% of patients in the BMMNc and in 89.5% of patients in the PBMNC group [19]. Cold sensation was relieved in 94.4% of BMMNC and 94.7% of PBMNC patients. Numbness was improved in 69.2% of BMMNC and 66.7% of PBMNC patients. ABI increased in 38.1% of BMMNC and 33.3% of PBMNC patients. $TcPO_2$ increased in 85.7% of BMMNC and 90.5% of PBMNC patients. New collateral vessels were identified in 83.3% of BMMNC and 77.8% of PBMNC patients. Wound healing was noted in 60.0% of BMMNC and 66.7% of PBMNC patients, and amputation rate was 9.1% in each group. Follow-up was extended to a mean of 8 months (range 3-15) in 40 patients [19]. At the end of follow-up, painful symptoms improved in 75.0% of BMMNC and 70.0% of PBMNC patients, ABI increased in 60.0% of BMMNC and 65.0% of PBMNC patients. $TcPO_2$ increased in 80.0% of BMMNC and 75.0% of PBMNC patients. Finally, new collateral vessels were detected in 90.0% of BMMNC and 84.6% of PBMNC patients [19].

Similarly, 150 patients with PAD were randomised to PBMNC (n=76) or BMMNC (n=74) and followed for 12 weeks [20]. In the PBMNC group, improvements in ABI ($p<0.0001$), skin temperature ($p=0.028$), and rest pain ($p<0.0001$) were significantly more pronounced in comparison to the BMNC group [20]. However, no between-group difference was found in terms of pain-free walking distance, $TcPO_2$, ulcers, and amputation rates. Thus, while both treatment options were effective, PBMNC administration yielded higher overall efficacy [20].

Finally, Zhang et al. [23] studied 15 patients (10 with PAD, 5 with TAO) suffering from severe ischaemia affecting the popliteal and distal arterial segment who could not be surgically revascularised. Four patients were Fontaine stage II, five patients were stage III, and six patients were stage IV. Three patients had diabetes mellitus and one had chronic renal failure [23]. All patients received PBMNCs intramuscularly. After two and 12 months, ABI, $TcPO_2$, claudication-free walking distance and pain intensity improved significantly ($p<0.005$) [23]. A significant increase in new vessel formation was also evidenced by angiography at 24 weeks. No adverse vents were noted.

4.2.2 Other stem cells

Seven patients with CLI (Rutherford stages 4 or 5) were treated with an intra-arterial infusion of autologous CPCs isolated from peripheral blood following GcSF mobilisation [81]. After 12 weeks, a 30-fold increase in pain-free walking distance from 6.4±12.5 to 195±196 m ($p=0.016$) was evident. At the same time, pain intensity was significantly reduced (from 8±1 to 2±2; $p=0.001$). There were also significant increments in ABI (from 0.48±0.09 to 0.64±0.11 $p=0.001$) and $TcPO_2$ (from 15±10 to 35±9 mmHg, $p=0.001$) [81]. Moreover, a 5-fold increase in flow-dependent vasodilation (from 0.9±0.3 to 5.0±1.4% $p=0.016$) and a 140% increase in adenosine-dependent flow reserve (from 3.6±1.3 to 4.9±1.7, $p=0.004$) in the superficial femoral artery were documented. Local or systemic adverse events were not seen [81].

In 6 patients with CLI due to PAD involving the infrapopliteal segment, intramuscular injections of non-mobilised peripheral blood angiogenic cell precursors (NMPB-ACPs) were carried out [82]. Five patients showed significant improvement in blood flow (ABI, $TcPO_2$) and 4 had complete healing of ulcers or amputation stumps. However, major amputation could finally not be avoided in 2 patients [82]. Obviously, more experience with this treatment is needed.

5. Safety issues

The vast majority of studies suggests that stem cell therapy for PAD is very safe. Local, injection-related side effects are, indeed, extremely rare. As regards systemic toxicity, there have been concerns with the use of GcSF, given that this growth factor has been reported to occasionally cause coronary ischaemia [83] or acute arterial thrombosis [84]. However, safety profiles of studies using GcSF have not been worse than those without GcSF.

In some studies, deaths have been reported. These were mostly due to acute myocardial infarction [53, 65, 69] congestive heart failure [64, 65, 69] and stroke [65, 69], while perforation peritonitis [69], sepsis [69] and suicide [69] have been reported as exceptionally

Authors	Number of Patients	Type of Cells	Clinical outcomes
Inaba et al. [77]	7	PBMNCs, with or without GcSF	Increased walking distance, minor improvements in ABI and angiographic findings
Ishida et al. [78]	6 (1 PAD, 5 TAO)	PBMNCs with GcSF	Increase in walking distance, slight improvement in ABI and ulcer area
Kawamura et al. [79]	92	PBMNCs with GcSF	Reduced amputations, mostly in non-diabetic non-dialysis patients
Zhang et al. [80]	15	PBMNCs	Increased walking distance and ABI, ulcer healing
Gu et al. [19]	42 (28 ischaemic diabetic foot, 8 TAO, 6 PAD)	PBMNCs vs. BMMNCs	Equal efficacy: ulcer healing, improvement of symptoms, increase in ABI and $TcPO_2$, new collateral vessels
Huang et al. [20]	150	PBMNCs vs. BMMNCs	PBMNC administration: higher overall efficacy. Improvement in ABI, skin temperature, rest pain, walking distance, $TcPO_2$, ulcers, and amputation rates (both treatments)
Zhang et al. [23]	15 (10 PAD, 5 TAO)	PBMNCs	Long-term improvement in ABI, $TcPO_2$, walking distance and pain, new collateral vessels
Lenk et al. [81]	7	CPCs with GcSF	Reduced pain, increase in walking distance, ABI, $TcPO_2$, flow-dependent vasodilation and adenosine-dependent flow reserve

Table 2. Major clinical studies with peripheral blood-derived stem cell therapy

rare causes. However, death events were never clearly related to stem cell therapy. Conversely, some of them could be characterised as clearly unrelated [53, 65]. Indeed, patients recruited in stem cell studies had severe or even limb-threatening vascular disease with considerable comorbidity. Such patients are generally a priori anticipated to suffer further vascular events, mainly in the coronary or cerebral arteries [1, 2, 4, 5]. Of note, in the study with the longest follow-up and most information on safety [69], mortality rate was 14.9% in PAD and 0% in TAO. Thus, caution is needed before attributing this mortality to stem cell treatment, and more information is required.
Serious adverse events have included acute myocardial infarction, stroke, post-surgery restenosis of coronary arteries, sepsis and peritonitis, while mild adverse events have included transient fever and myalgia [69]. Matoba et al. [69] have provided the most detailed and long-term data. Again, these adverse events were only observed in PAD and not in TAO, casting doubt on their relation to treatment. Regrettably, there was no control group to compare event rates.
In summary, stem cell transplantation appears to be generally safe. Deaths and adverse events have never been unambiguously linked to treatment. Nonetheless, it should not escape our notice that patient series have been very small, some studies only presenting case series of up to 10 subjects [45, 46, 48, 54, 58-62, 75-78, 82]. Follow-up has also been relatively short, in most studies less than 1 year. More importantly, event rates could not be compared to a control group. Hence, additional data is eagerly awaited to confirm or refute this safety profile.

6. Clinical implications

Available evidence suggests that stem cell therapy should be considered for those patients with severe PAD who are poor candidates for revascularisation. The question then to ask would be: bone marrow-derived or peripheral blood-derived stem cells? Not to be underestimated, the latter is easier to perform and, theoretically, offers the incremental advantage that it might be repeated. Moreover, it does not appear to be inferior in efficacy, but more comparative data is required. At least as regards mononuclear cells, PBMNCs show comparable [19] or even superior [20] efficacy in comparison to BMMNCs. If comparable efficacy is generalised for all bone marrow-derived vs. peripheral blood-derived stem cells, then the individual decision may rest with the treating physician, depending on previous experience and availability issues.
BMCs, BMMNCs and PBMNCs are the main cell types used [19, 20, 23, 29, 47, 50, 55, 56, 63-65, 68, 69, 73, 74, 77, 78, 80, 81], and there is no clear superiority of one cell type over the others. Administration has been intramuscular into the ischaemic extremity in the vast majority of trials. Combined intra-arterial and intramuscular administration has also been used [58-60, 62]. Combined intra-arterial and intramuscular BMMNC transplantation has been shown to be equally effective to intramuscular transplantation [68]. Gu et al. [64] showed equal treatment efficacy for intra-arterial vs. intramuscular BMMNC transplantation. Arguably, intramuscular administration may, for the time being, constitute the route of choice, since it has been more extensively studied.
The beneficial effect of stem cell therapy has been evaluated by diverse outcomes. These include alleviation of pain [19, 23, 46, 47, 51, 63, 69], increased walking distance [19, 20, 23, 47, 51, 53, 55, 56, 63, 68, 69, 77, 78, 80], improved ABI and $TcPO_2$ [19, 20, 23, 47, 53, 55, 56, 50, 63, 68, 69, 73, 74, 77, 78, 80, 81], but also hard endpoints, notably ulcer healing [19, 20, 47, 49,

63, 69, 74, 78, 80], quality of life [72] and limb salvage [20, 49, 68, 69, 79]. Moreover, new collateral vessels have been identified on angiography [19, 23, 47, 63, 65, 72, 73]. Thus, despite the variety of efficacy parameters in the individual works, it is true that studies converge to suggest an objectively demonstrable and clinically meaningful benefit.

Importantly, however, long-term efficacy needs to be better studied as well. Indeed, follow-up in most studies has been up to one year and few works have provided long-term data [23, 62, 65, 69]. Based on the latter, treatment-induced improvements in painful symptoms, walking distance and blood flow parameters (ABI, $TcPO_2$) appear to be, at least in part, sustainable at 2-3 years [23, 62, 65, 69]. If long-term efficacy becomes finally established, the argument in favour of stem cell therapy for severe inoperable PAD will be strongly enhanced.

Two further issues deserve careful consideration. First, safety profile appears, as already discussed, very good, but more information is needed. Ideally, we need a robust analysis balancing improvement in clinical outcomes to adverse event rates. In this context, it is extremely important to define the number needed to treat (NNT) and the number needed to harm (NNH) before attempting more widespread use of this treatment option. NNT and NNH could even vary, according to the type of patient population (e.g. PAD, CLI, TAO, acute CLI, elderly patients with substantial comorbidity and so on) [69]. However, this information is vital for clinical decisions outside randomised clinical trials. Secondly, stem cell therapy is an expensive treatment and its cost-effectiveness has not been determined. Thus, a detailed cost-benefit analysis is desirable.

Last, but not least, stem cell therapy has hitherto only been explored as a salvage therapy in severe lower extremity ischaemia not responding to standardised treatment. Such approach has precluded comparison with revascularisation. Still, given the promising results in clinical benefits achieved with stem cell transplantation, it may be worthwhile to consider this therapeutic option earlier in the course of PAD, instead of reserving it for patients with no other choice. Earlier utilisation of stem cell therapy might thus yield even better results. Moreover, it might be combined with revascularisation, if both are used in the appropriate time frame. These are important aspects for future investigation.

7. Conclusions

Stem cell therapy is emerging as a promising treatment option for patients with PAD who are poor candidates for established treatment, including revascularisation and best medical therapy. There are two options, bone marrow-derived and peripheral blood-derived stem cells. Theoretically, transplantation of peripheral blood-derived stem cells offers the additional advantage that it can be repeated, but this potential has not been utilised. In the vast majority of cases, stem cell administration has been intramuscular into the ischaemic limb, but intra-arterial administration appears to be equally efficacious. The number of cells that need to be transplanted merits more precise study. Finally, while the new treatment appears generally safe, more data from larger patient series and longer follow-up are awaited for confirmation.

8. References

[1] Criqui MH. Peripheral vascular disease – epidemiological aspects. *Vasc Med* 2001; 6: 3-7.
[2] Ouriel K. Peripheral arterial disease. *Lancet* 2001; 358: 1257-64.

[3] Aronow WS. Management of peripheral arterial disease. *Cardiol Rev 2005: 13:* 61-8.
[4] Aronow WS, Ahn C. Prevalence of coexistence of coronary artery disease, peripheral arterial occlusive disease and atherothrombotic brain infarction in men and women ≥62 years of age. *Am J Cardiol 1994;* 74: 64-65.
[5] Cimminiello C. Peripheral arterial disease as a global vascular risk factor. *Haematologica* 2001; 86: 16-8.
[6] Yusuf S, Reddy S, Ounpuu S, Anand S. Global burden of cardiovascular diseases: part I: general considerations, the epidemiologic transition, risk factors, and impact of urbanization. *Circulation* 2001; 104: 2746-53.
[7] Burns P, Lima E, Bradbury AW. What constitutes best medical therapy for peripheral arterial disease? *Eur J Vasc Endovasc Surg* 2002; 24: 6-12.
[8] Hiatt WR. Medical treatment of Peripheral Arterial Disease and Claudication. *New Engl J Med* 2001; 344: 1608-21.
[9] Faglia E, Mantero M, Caminiti M et al. Extensive use of peripheral angioplasty, particularly infrapopliteal, in the treatment of ischaemic diabetic foot ulcers: clinical results of a multicentric study of 221 consecutive diabetic subjects. *J Intern Med* 2002; 252: 225-32.
[10] Faglia E, Dalla Paola L, Clerici G et al. Peripheral angioplasty as the first-choice revascularization procedure in diabetic patients with critical limb ischemia: prospective study of 993 consecutive patients hospitalized and followed between 1999 and 2003. *Eur J Vasc Endovasc Surg* 2005; 29: 620-7.
[11] Lepantalo M, Biancari F, Tukiainen E. Never amputate without consultation of a vascular surgeon. *Diabetes Metab Res Rev* 2000; 16: Suppl 1: S27-S32.
[12] Pomposelli FB, Kansal N, Hamdan AD et al. A decade of experience with dorsalis pedis artery bypass: analysis of outcome in more than 1000 cases. *J Vasc Surg* 2003; 37: 307-15.
[13] Sigala F, Georgopoulos S, Langer S et al. Outcome of infrainguinal revascularization for critical limb ischemia in diabetics with end stage renal disease. *Vasa* 2006; 35: 15-20.
[14] Faries PL, Teodorescu VJ, Morrissey NJ et al. The role of surgical revascularization in the management of diabetic foot wounds. *Am J Surg* 2004; 187: 34S-37S.
[15] Kalka C, Baumgartner I. Gene and stem cell therapy in peripheral arterial occlusive disease. *Vasc Med* 2008; 13: 157-72.
[16] Rafii S, Lyden D. Therapeutic stem and progenitor cell transplantation for organ vascularization and regeneration. *Nat Med* 2003; 9: 702-2.
[17] Lachmann N, Nikol S. Therapeutic angiogenesis for peripheral artery disease: stem cell therapy. *Vasa* 2007; 36: 241-51.
[18] Capiod JC, Tournois C, Vitry F et al. Characterization and comparison of bone marrow and peripheral blood mononuclear cells used for cellular therapy in critical leg ischaemia: towards a new cellular product. *Vox Sang* 2009; 96: 256-65.
[19] Gu Y, Zhang J, Qi L. Comparative study on autologous implantation between bone marrow stem cells and peripheral blood stem cells for treatment of lower limb ischemia. *Zhongguo Xiu Fu Chong Jian Wai Ke Za Zhi* 2007; 21: 675-8.
[20] Huang PP, Yang XF, Li SZ et al. Randomised comparison of G-CSF-mobilized peripheral blood mononuclear cells versus bone marrow-mononuclear cells for the treatment of patients with lower limb arteriosclerosis obliterans. *Thromb Haemost* 2007; 98: 1335-42.

[21] Moon MH, Kim SY, Kim YJ et al. Human adipose tissue-derived mesenchymal stem cells improve postnatal neovascularization in a mouse model of hindlimb ischemia. *Cell Physiol Biochem* 2006; 17: 279-90.
[22] Schannwell CM. Stem cell therapy for cardiovascular diseases. Experiences in Düsseldorf. *Dtsch Med Wochenschr* 2008; 133: Suppl 8: S274-S279.
[23] Zhang H, Zhang N, Li M et al. Therapeutic angiogenesis of bone marrow mononuclear cells (MNCs) and peripheral blood MNCs: transplantation for ischemic hindlimb. *Ann Vasc Surg* 2008; 22: 238-47.
[24] Felmeden DC, Blann AD, Lip GYH. Angiogenesis: basic pathophysiology and implications for disease. *Eur Heart J* 2003; 24: 586-603.
[25] Flamme I, Risau W. Induction of vasculogenesis and hematopoiesis in vitro. *Development* 1992; 116: 435-9.
[26] Risau W. Mechanisms of angiogenesis. *Nature* 1997; 386: 671-4.
[27] Ziegelhoeffer T, Fernandez B, Kostin S et al. Bone marrow-derived cells do not incorporate into the adult growing vasculature. *Circ Res* 2004; 94: 230-8.
[28] Napoli C, Williams-Ignarro S, de Nigris F et al. Beneficial effects of concurrent autologous bone marrow cell therapy and metabolic intervention in ischemia-induced angiogenesis in the mouse hindlimb. *Proc Natl Acad Sci USA* 2005; 102: 17202-6.
[29] Takagi Y, Omura T, Yoshiyama M et al. Granulocyte-colony stimulating factor augments neovascularization induced by bone marrow transplantation in rat hindlimb ischemia. *J Pharmacol Sci* 2005; 99: 45-51.
[30] Sica V, Williams-Ignarro S, de Nigris F et al. Autologous bone marrow cell therapy and metabolic intervention in ischemia-induced angiogenesis in the diabetic mouse hindlimb. *Cell Cycle* 2006; 5: 2903-8.
[31] Jeon O, Song SJ, Bhang SH et al. Additive effect of endothelial progenitor cell mobilization and bone marrow mononuclear cell transplantation on angiogenesis in mouse ischemic limbs. *J Biomed Sci* 2007; 14: 323-30.
[32] Liu Q, Chen Z, Terry T et al. Intra-arterial transplantation of adult bone marrow cells restores blood flow and regenerates skeletal muscle in ischemic limbs. *Vasc Endovasc Surg* 2009; 43: 433-43.
[33] Asahara T, Murohara T, Sullivan A et al. Isolation of putative progenitor endothelial cells for angiogenesis. *Science* 1997; 275: 964-7.
[34] Iba O, Matsubara H, Nozawa Y et al. Angiogenesis by implantation of peripheral blood mononuclear cells and platelets into ischemic limbs. *Circulation* 2002; 106: 2019-25.
[35] Yoon C.H, Hur J, Park KW et al. Synergistic neovascularization by mixed transplantation of early endothelial progenitor cells and late outgrowth endothelial cells: the role of angiogenic cytokines and matrix metalloproteinases. *Circulation* 2005; 112: 1618-27.
[36] Iwase T, Nagaya N, Fujii T et al. Comparison of angiogenic potency between mesenchymal stem cells and mononuclear cells in a rat model of hindlimb ischemia. *Cardiovasc Res* 2005; 66: 543-51.
[37] Awad O, Dedkov EI, Jiao C et al. Differential healing activities of CD34+ and CD14+ endothelial cell progenitors. *Arterioscler Thromb Vasc Biol* 2006; 26: 758-64.

[38] Aicher A, Heeschen C, Sasaki K et al. Low-energy shock wave for enhancing recruitment of endothelial progenitor cells: a new modality to increase efficacy of cell therapy in chronic hind limb ischemia. *Circulation* 2006; 114: 2823-30.

[39] Sasaki S, Inoguchi T, Muta K et al. Therapeutic angiogenesis by ex vivo expanded erythroid progenitor cells. *Am J Physiol Heart Circ Physiol* 2007; 292: pp. H657-H665.

[40] Niagara MI, Haider HK, Ye L et al. Autologous skeletal myoblasts transduced with a new adenoviral bicistronic vector for treatment of hind limb ischemia. *J Vasc Surg* 2004; 40: 774-85.

[41] Kinnaird T, Stabile E, Burnett MS et al. Marrow-derived stromal cells express genes encoding a broad spectrum of arteriogenic cytokines and promote in vitro and in vivo arteriogenesis through paracrine mechanisms. *Circ Res* 2004; 94: 678-85.

[42] Kinnaird T, Stabile E, Burnett MS et al. Local delivery of marrow-derived stromal cells augments collateral perfusion through paracrine mechanisms. *Circulation* 2004; 109: 1543-9.

[43] Nakagami H, Maeda K, Morishita R et al. Novel autologous cell therapy in ischemic limb disease through growth factor secretion by cultured adipose tissue-derived stromal cells. *Arterioscler Thromb Vasc Biol* 2005; 25: 2542-7.

[44] Kim Y, Kim H, Cho H et al. Direct comparison of human mesenchymal stem cells derived from adipose tissues and bone marrow in mediating neovascularization in response to vascular ischemia. *Cell Physiol Biochem* 2007; 20: 867-76.

[45] Esato K, Hamano K, Li TS et al. Neovascularization induced by autologous bone marrow cell implantation in peripheral arterial disease. *Cell Transplant* 2002; 11: 747-52.

[46] Nizankowski R, Petriczek T, Skotnicki A, Szczeklik A. The treatment of advanced chronic lower limb ischaemia with marrow stem cell autotransplantation. *Kardiol Polsk 2005;* 63: 351-60.

[47] Gu Y, Zhang J, Qi L. Effective autologous bone marrow stem cell dosage for treatment of severe lower limb ischemia. *Zhongguo Xiu Fu Chong Jian Wai Ke Za Zhi* 2006; 20: 504-6.

[48] Boda Z, Udvardy M, Farkas K et al. Autologous bone marrow-derived stem cell therapy in patients with severe peripheral arterial disorder. *Orv Hetil* 2008; 149: 531-40.

[49] Napoli C, Farzati B, Sica V et al. Beneficial effects of autologous bone marrow cell infusion and antioxidants/L-arginine in patients with chronic critical limb ischemia. *Eur J Cardiovasc Prev Rehab* 2008; 15: 709-18.

[50] Procházka V, Gumulec J, Chmelová J et al. Autologous bone marrow stem cell transplantation in patients with end-stage chronical critical limb ischemia and diabetic foot. *Vnitr Lék* 2009; 55: 173-8.

[51] Korymasov EA, Tiumina OV, Aiupov AM et al. Use of autologous progenitor cells of the bone marrow in treatment of patients with lower-limb atherosclerosis obliterans. *Angiol Sosud Khir* 2009; 15: 28-31.

[52] Amann B, Lüdemann C, Rückert R et al. Design and rationale of a randomized, double-blind, placebo-controlled phase III study for autologous bone marrow cell transplantation in critical limb ischemia: the BONe Marrow Outcomes Trial in Critical Limb Ischemia (BONMOT-CLI). *Vasa* 2008; 37: 319-25.

[53] Tateishi-Yuyama E, Matsubara H, Murohara T et al. Therapeutic angiogenesis for patients with limb ischaemia by autologous transplantation of bone-marrow cells: a pilot study and a randomised controlled trial. *Lancet* 2002; 360: 427-35.
[54] Kirana S, Stratmann B, Lammers D et al. Wound therapy with autologous bone marrow stem cells in diabetic patients with ischaemia-induced tissue ulcers affecting the lower limbs. *Int J Clin Pract* 2007; 61: 690-2.
[55] Higashi Y, Kimura M, Hara K et al. Autologous bone-marrow mononuclear cell implantation improves endothelium-dependent vasodilation in patients with limb ischemia. *Circulation* 2004; 109: 1215-8.
[56] Miyamoto M, Yasutake M, Takano H et al. Therapeutic angiogenesis by autologous bone marrow cell implantation for refractory chronic peripheral arterial disease using assessment of neovascularization by 99mTc-tetrofosmin (TF) perfusion scintigraphy. *Cell Transplant* 2004; 13: 429-37.
[57] Saigawa T, Kato K, Ozawa T et al. Clinical application of bone marrow implantation in patients with arteriosclerosis obliterans, and the association between efficacy and the number of implanted bone marrow cells. *Circ J* 2004; 68: 1189-93.
[58] Bartsch T, Falke T, Brehm M et al. Intra-arterial and intramuscular transplantation of adult, autologous bone marrow stem cells. Novel treatment for therapy-refractory peripheral arterial occlusive disease. *Dtsch Med Wochenschr* 2006; 131: 79-83.
[59] Bartsch T, Falke T, Brehm M et al. Transplantation of autologous adult bone marrow stem cells in patients with severe peripheral arterial occlusion disease. *Med Klin (Munich)* 2006; 101: 195-7.
[60] Bartsch T, Brehm M, Falke T et al. Rapid healing of a therapy-refractory diabetic foot after transplantation of autologous bone marrow stem cells. *Med Klin (Munich),* 2005; 100: 676-80.
[61] Bartsch T, Brehm M, Zeus T, Strauer BE. Autologous mononuclear stem cell transplantation in patients with peripheral occlusive arterial disease. *J Cardiovasc Nurs 2006;* 21: 430-2.
[62] Bartsch, T, Brehm M, Zeus T et al. Transplantation of autologous mononuclear bone marrow stem cells in patients with peripheral arterial disease (the TAM-PAD study). *Clin Res Cardiol* 2007; 96: 891-9.
[63] Gu Y, Zhang J, Qi L. A clinical study on implantation of autologous bone marrow mononuclear cells after bone marrow stimulation for treatment of lower limb ischemia. *Zhongguo Xiu Fu Chong Jian Wai Ke Za Zhi* 2006; 20: 1017-20.
[64] Gu YQ, Zhang J, Guo LR et al. Transplantation of autologous bone marrow mononuclear cells for patients with lower limb ischemia. *Chin Med J (Engl)* 2008; 121: 963-7.
[65] Gu Y, Qi L, Zhang J et al. Middle-term outcome of autologous bone marrow mononuclear cells transplantation for treatment of lower limb ischemia. *Zhongguo Xiu Fu Chong Jian Wai Ke Za Zhi* 2009; 23: 341-4.
[66] Hernández P, Cortina L, Artaza H et al. Autologous bone-marrow mononuclear cell implantation in patients with severe lower limb ischaemia: a comparison of using blood cell separator and Ficoll density gradient centrifugation. *Atherosclerosis* 2007; 194: e52-6.

[67] Kajiguchi M, Kondo T, Izawa H et al. Safety and efficacy of autologous progenitor cell transplantation for therapeutic angiogenesis in patients with critical limb ischemia. *Circ J* 2007; 71: 196-201.
[68] Van Tongeren RB, Hamming JF, Fibbe WE et al. Intramuscular or combined intramuscular/intra-arterial administration of bone marrow mononuclear cells: a clinical trial in patients with advanced limb ischemia. *J Cardiovasc Surg (Torino) 2008;* 49: 51-8.
[69] Matoba S, Tatsumi T, Murohara T et al. Long-term clinical outcome after intramuscular implantation of bone marrow mononuclear cells (Therapeutic Angiogenesis by Cell Transplantation [TACT] trial) in patients with chronic limb ischemia. *Am Heart J* 2008; 156: 1010-8.
[70] Cobellis G, Silvestroni A, Lillo S et al. Long-term effects of repeated autologous transplantation of bone marrow cells in patients affected by peripheral arterial disease. *Bone Marrow Transplant* 2008; 42: 667-72.
[71] Wester T, Jørgensen JJ, Stranden E et al. Treatment with autologous bone marrow mononuclear cells in patients with critical lower limb ischaemia. A pilot study. *Scand J Surg* 2008; 97: 56-62.
[72] Chochola M, Pytlík R, Kobylka P et al. Autologous intra-arterial infusion of bone marrow mononuclear cells in patients with critical leg ischemia. *Int Angiol* 2008; 27: 281-90.
[73] De Vriese AS, Billiet J, Van Droogenbroeck J et al. Autologous transplantation of bone marrow mononuclear cells for limb ischemia in a caucasian population with atherosclerosis obliterans. *J Intern Med* 2008; 263: 395-403.
[74] Amann B, Luedemann C, Ratei R, Schmidt-Lucke JA. Autologous bone marrow cell transplantation increases leg perfusion and reduces amputations in patients with advanced critical limb ischemia due to peripheral artery disease. *Cell Transplant 2009;* 18: 371-80.
[75] Franz RW, Parks A, Shah KJ et al. Use of autologous bone marrow mononuclear cell implantation therapy as a limb salvage procedure in patients with severe peripheral arterial disease. *J Vasc Surg* 2009; 50: 1378-90.
[76] Lasala GP, Silva JA, Gardner PA, Minguell JJ. Combination stem cell therapy for the treatment of severe limb ischemia: safety and efficacy analysis. *Angiology* 2010; 61: 551-6.
[77] Inaba S, Egashira K, Komori K. Peripheral-blood or bone-marrow mononuclear cells for therapeutic angiogenesis? *Lancet* 2002; 360: 2083.
[78] Ishida A, Ohya Y, Sakuda H et al. Autologous peripheral blood mononuclear cell implantation for patients with peripheral arterial disease improves limb ischemia. *Circ J* 2005; 69: 1260-5.
[79] Kawamura A, Horie T, Tsuda I et al. Clinical study of therapeutic angiogenesis by autologous peripheral blood stem cell (PBSC) transplantation in 92 patients with critically ischemic limbs. *Int J Artif Organs* 2006; 9: 226-33.
[80] Zhang HK, Li M, Feng H. Autologous transplantation of peripheral blood stem cell intreatment of critical limb ischemia. *Zhejiang Da Xue Xue Bao Yi Xue Ban* 2007; 36: 360-3.

[81] Lenk K, Adams V, Lurz P et al. Therapeutical potential of blood-derived progenitor cells in patients with peripheral arterial occlusive disease and critical limb ischaemia. *Eur Heart J 2005;* 26: 1903-9.
[82] Mutirangura P, Ruangsetakit C, Wongwanit C et al. Enhancing limb salvage by non-mobilized peripheral blood angiogenic cell precursors therapy in patients with critical limb ischemia. *J Med Assoc Thai* 2009; 92: 320-7.
[83] Fukumoto Y, Miyamoto T, Okamura T et al. Angina pectoris occurring during granulocyte colony-stimulating factor-combined preparatory regimen for autologous peripheral blood stem cell transplantation in a patient with acute myelogenous leukaemia. *Br J Haematol* 1997; 97: 666-8.
[84] Kawachi Y, Watanabe A, Uchida T et al. Acute arterial thrombosis due to platelet aggregation in a patient receiving granulocyte colony-stimulating factor. *Br J Haematol* 1996; 94: 413-6.

14

Myogenic Potential of Murine Embryonic Stem Cells in the Dmd^{mdx} Mouse Model for Duchenne Muscular Dystrophy

Danielle Ayub-Guerrieri[1], Poliana C. M. Martins-Machado[1], Paula C.G. Onofre-Oliveira[1], Lygia V. Pereira[2], Camila F. Almeida[1], Vanessa F. Lopes[1] and Mariz Vainzof[1]

[1]*Laboratory of Muscle Proteins and Comparative Histopathology, Human Genome Research Center, Biosciences Institute, University of São Paulo*

[2]*Dept of Genetics and Evolutionary Biology, Biosciences Institute, University of São Paulo*

Brazil

1. Introduction

The neuromuscular disorders are a heterogeneous group of genetic diseases, causing a progressive loss of the motor ability. More than 30 genetically defined forms are recognized, and in the last decade, mutations in several genes have been reported, resulting in the deficiency or loss of function of different important muscle proteins. Biochemical and imunohistological analysis have localized these proteins in several compartments of the muscle fiber. The proteins dystrophin, sarcoglycans and dysferlin are sarcolemmal or peri-sarcolemmal proteins, α2-laminin and collagen VI are extracellular matrix proteins, telethonin and actin are sarcomeric proteins, calpain 3 and FKRP are cytosolic enzymes, and emerin and lamin A/C are nuclear proteins (Vainzof et al., 2008).

Defects in components of the dystrophin-glycoprotein complex (DGC) are known to be an important cause of different forms of muscular dystrophies (Ervasti & Campbell, 1993; Yoshida & Ozawa, 1990). The DGC is an oligomeric complex which connects the subsarcolemmal cytoskeleton to the extracellular matrix. The DGC consists of dystroglycan (α- and β-DG), sarcoglycan (α, β-, γ-, δ- and ε-SG) and syntrophin/dystrobrevin subcomplexes. The intracellular link of the DGC is the protein dystrophin, that plays an important structural role in muscle fibers. Mutations in the dystrophin gene cause the most common form of X-linked Duchenne muscular dystrophy (DMD) (Hoffman et al., 1987). Dystrophin binds its amino-terminal and rod domain to actin and with its carboxy terminal to the integral membrane protein β-DG. The sarcoglycan sub-complex is also linked to β-DG and includes α-SG, β-SG, γ-SG, and δ-SG, which are tightly associated and inserted into the membrane. Mutations in the genes coding the 4 SG proteins cause severe forms of limb-girdle muscular dystrophies type LGMD2D, 2E, 2C and 2F, respectively. The peripheral membrane glycoprotein α-DG, a receptor for the heterotrimeric basement membrane protein laminin-2, binds to β-DG and so completes the connection from the inside to the outside of the cell (Straub & Campbell, 1997). Mutations in the LAMA2 gene, encoding the α2 chain of laminin-2, cause α2-laminin deficiency, and a severe form of congenital muscular dystrophy (CMD1A) linked to chromosome 6q (Tomé et al., 1994). In addition, some forms of muscular dystrophy have recently been associated with genes encoding putative or known glycosyltransferases. Muscle protein analysis in these patients show a hypoglycoslation of α-dystroglycan and a consequent reduction of numerous ligands components of the extracellular matrix, such as laminin 2 (Muntoni et al., 2004). Other milder forms of muscular dystrophy are caused by mutations in genes coding the enzyme calpain 3 (LGMD2A), the sarcolemmal protein dysferlin (LGMD2B), and the sarcomeric protein telethonin (LGMD2G) (revision in Vainzof & Zatz, 2007).

Several animal models, manifesting phenotypes observed in neuromuscular diseases have been identified in nature or generated in laboratory. These models generally present physiological alterations observed in human patients, and can be used as important tools for genetic, clinic and histopathological studies (Vainzof et al., 2008)

The *Dmd^mdx^* mouse is the most widely used animal model for Duchenne muscular dystrophy (DMD). Although it is a good genetic and biochemical model, presenting total deficiency of the protein dystrophin in the muscle, this mouse is not useful for clinical trials, because of its very mild phenotype. The canine golden retriever MD model represents a more clinically similar model of DMD due to its larger size and significant muscle weakness. Autosomal recessive limb-girdle MD forms models include the *SJL/J* mice that develop a spontaneous myopathy resulting from a mutation in the Dysferlin gene, being a model for LGMD2B. For the human sarcoglycanopahties (SG), the BIO14.6 hamster is the spontaneous animal model for δ-SG deficiency, while some canine models with deficiency of SG proteins have also been identified. More recently, using the homologous recombination technique in embryonic stem cell, several mouse models have been developed with null mutations in each one of the 4 SG genes. All sarcoglycan-null animals display a progressive muscular dystrophy of variable severity, and share the property of a significant secondary reduction in the expression of the other members of the sarcoglycan subcomplex, and other components of the Dystrophin-glycoprotein complex.

Mouse models for congenital MD include the $Lama2^{dy}/J$ (dystrophia-muscularis) mouse, and the allelic mutant $Lama2^{dy2J}/J$ mouse, both presenting significant reduction of α2-laminin in the muscle and a severe phenotype. The myodystrophy mouse ($Large^{myd}$), harbors a mutation in the glycosyltransferase Large, which leads to altered glycosylation of α-DG, and also a severe phenotype.

The study of animal models for genetic diseases, in spite of the existence of differences in some phenotypes can provide important clues to the understanding of the pathogenesis of these disorders and are also very valuable for testing strategies for therapeutic approaches.

In all forms of muscular dystrophies, in the early phase of the disease, there is a continuous process of degeneration and regeneration. The ability of adult skeletal muscle to regenerate has been attributed to the special properties of satellite cells. In mature skeletal muscle satellite cells are normally in a dormant state. Upon injury, many growth factors are secreted at the site of the lesion that recruits satellite cells, triggering the process of regeneration. In the progression of the dystrophic process, however, the endogenous satellite cell pool becomes exhausted and degenerated muscle fibers are replaced by fibrotic and adipose tissues, responsible for the observed muscle weakness.

Therefore, stem cell therapy should be an ideal treatment for all forms of muscular dystrophies, in which the deficiency of a specific muscle protein leads to muscle degeneration. Stem cell transplantation would either prevent or break the cycles of degeneration and regeneration in the dystrophic process, by replacing the deficient protein

(Meng et al., 2011). The majority of the therapeutic trial under investigation are done using Duchenne muscular dystrophy as a model for, because it is the most common and severe mendelian form of muscular dystrophy.

2. Stem cell therapy for muscular dystrophies

The use of normal stem cells to rescue the effects associated with mutant tissue is a promising avenue of research. To correct the dystrophic phenotype healthy stem cells transplanted to the diseased muscles must first be attracted to the injured area, answer to the endogenous signals that stimulate muscle differentiation and, ultimately, fuse to dystrophic myofibers, or form new myotubes. Incorporated healthy nuclei would thus contribute to restore the normal levels of functional skeletal muscle proteins such as dystrophin.

According to a recent revision by Meng et al. (2011), the ideal stem cell for treating DMD should fulfill several criteria, and should be able to: expandable in vivo, maintaining stem cell properties, be immune-competent, be systemically-delivered reaching all muscles of the body, survive, proliferate and migrate upon arrival within the host muscle, differentiate into muscle fibres, reconstitute the satellite cell pool with functional stem cells, be capable of expressing the missing protein, lead to improvement in muscle strength.

The most appropriated type of stem cell for these therapies is still under investigation. Among adult stem cells, skeletal muscle satellite cells have been considered the only source of stem cells for post-natal muscle regeneration (Seale & Rudnicki, 2000). Other stem cells within the skeletal muscle have, however been identified, including muscle-derived stem cells, muscle side population cells, myogenic endothelial cells, and mesoangioblasts (pericytes) (Meng et al, 2011). Human skeletal muscle-derived pericytes have been shown to form muscle fibers, after intra-arterial transplantation in the dystrophin deficiency host mice (Dellavalle et al., 2007). However, in contrast, muscle-derived cells (mdcs), which are very similar to the pericytes, did not contribute to muscle regeneration after systemic delivery in mdx nu/nu host (Meng et al., 2011b). These findings illustrate the opinion of several authors suggesting that further work is necessary to prepare pure cell populations from skeletal muscle that maintain their phenotype in culture and make a robust contribution to skeletal muscle regeneration after systemic delivery in dystrophic mouse models (Meng et al., 2011b)

Among the other non-muscle stem cells, previous attempts at developing stem cell therapies for the treatment of DMD were performed using bone marrow and blood-derived stem cells, mesenchimal stem cells from other origins, and pluripotent stem cells. Bone marrow-derived stem cells (BM) are a population of circulating cells with myogenic potential, present in the bone marrow, as already demonstrated as early as in the 1960s. Afterward, some authors confirmed that BM-derived cells can undergo myogenic differentiation and participate in muscle repair after injury, albeit at very low levels (Ferrari et al., 1998; Gussoni et al., 1999). Transplantation studies with BM cells injected into *Dmd*^{mdx} mouse confirmed that these cells may persist in the musculature for long periods of time, and that they are able to express exogenous dystrophin protein. However, the amount of muscle generated after a BM transplant was still not therapeutically relevant, consisting of about 0.5% of regenerating fibers containing donor cells (Ferrari et al., 2002; Gussoni et al., 2002). Therefore, none of these experiments provide consistent data with the idea that transplanted BM cells can actually correct the dystrophic phenotype.

Mesenchymal stem cells from other origins have been shown to regenerate muscle fibers, but at a very low efficiency, in mouse models. However, some therapeutic effects have been observed, such as by reducing inflammation in the dystrophin deficient muscle (Ichim et al., 2010).

Because of their pluripotency, much attention has been devoted to the potential applications of Embryonic Stem (ES) (Evans & Kaufman, 1981). However, to date, ES cells have not had a significant impact on the development of cell-based therapies to treat muscular dystrophy. While a few studies have been performed in *Dmd^mdx^* mice, mainly involving the transplantation of embryoid bodies co-cultured with dissociated skeletal muscle, no evidence of long-term regenerative capacity of transplanted cells has been observed (Bhagavati & Xu, 2005). In table 1, the published experiments to date are summarized.

ES cell type	Strategy	Animal models	Markers	Delivery	Reference
mESC	media supplementation with DMSO	only in vitro experiments	none	only in vitro experiments	Rohwedel et al., 1994
mESC	IGF-II overexpression	only in vitro experiments	none	only in vitro experiments	Prelle et al., 2000
mESC	co-culture with muscle cells culture	*mdx* mice/ Rag2-/-γc-/- mice	none	intramuscular	Bhagavati & Xu, 2005
hESC	media supplementation 5-azacytidine	*NOD-SCID* mice	none	intramuscular	Zheng et al., 2006
ZHTc6 mESC	MyoD gene-inducible system	*mdx* nude mice	Sca+/c-kit+/CD34-	intramuscular	Ozasa et al., 2007
hESC	FACS of mesenchymal precursors	*SCID/Beige* mice	CD73+/CD56+	intramuscular	Barberi et al., 2007
mESC	Pax3 induction/ FACS	*mdx* mice/ *Rag2-/-γc-/-* mice	$PDGF\alpha R^{+}/Flk1^{-}$	intramuscular/systemic	Darabi et al., 2008
mESC	FACS of paraxial mesoderm precursors	*KSN* nude mice	PDGFαR+/Flk1- or Flk+	intramuscular	Sakurai et al., 2008
mESC	media supplementation DMSO and RA	only in vitro experiments	none	only in vitro experiments	Kennedy et al., 2009
mESC	Pax3 induction/ FACS	*Frg1* mice	$PDGF\alpha R^{+}/Flk1^{-}$	intramuscular/systemic	Darabi et al., 2009
mESC	media composed of 5% HS and 10% FBS	*mdx* mice	SM/C-2.6+	intramuscular	Chang et al., 2009

Table 1. Summary of experiments for testing the myogenesis potential of ESCs in NMD.

Thus, stem cell therapies for muscle disorders are still in the early days. There are many outstanding questions such as the immunogenic capacity of stem cells, whether local or systemic injections are the best route of administration in the diseased tissue, whether there is an optimal amount of cells to be administered, or whether the regenerative growth factors present in the dystrophic muscle are sufficient to promote the survival and the fusion of these cells to the damaged muscle (Bradley et al., 2002; Mimeault et al., 2007).

In order to answer at least some of the questions summarized above, we investigated the capacity of ES and EB cells to generate muscle in vivo, and to express the deficient protein dystrophin in the *Dmd^mdx^* mouse model for DMD.

3. In vivo experiments with ES cells and potential therapies for NMD

3.1 ES cell culture

The embryonic stem cell line USP-1 derived from 129/Sv mice (Sukoyan et al., 2002) was cultivated as described (Rohwedel et al., 1994). Briefly, cells grew on irradiated mouse feeder layer on gelatin-coated flasks (Nunc) in DMEM (Gibco) supplemented with 10% fetal bovine serum (Hyclone) and additives (Rohwedel et al., 1994). During the isolation and early stages of ES cell cultivation, the medium was supplemented with human recombinant leukemia inhibitory factor (hLIF) at 1000 units/ml (Chemicon) and was replaced every day. When reached the confluence (after 4 days in culture) these cells were trypsinized, and transferred to another culture flask (75 cm^2) previously covered by a layer of mouse fibroblast feeders, plated at a density of 1.0 x 10^6 cells/cm2 for proliferation. For differentiation, aliquots of 20 µl cultivation medium containing 800 cells were placed on the lids of Petri dishes filled with PBS, based in protocols developed by Wobus et al. in 1988 (2002), which are based in previous differentiation by embryoid body formation and posterior treatment with 1% DMSO (dimethyldisulfoxide). The ES cell aggregates (Embryoid Bodies - EB) were cultivated in hanging drops for 2 days and subsequently in suspension on bacteriological Petri dishes for additional 3 days in a specific skeletal muscle differentiation medium (proliferation medium with 1% DMSO, but no LIF). At day 5, EBs were plated separately onto gelatin-coated 24 well-microwell plates for morphological analysis. EB were dissociated before injection and ES cells were marked with red dye Vybrant® Dil cell-labeling solution (Invitrogen) to facilitate their tracking in injected muscles. This dye efficiently label live cells, and is diluted and disappears with cells proliferation and cellular death.

3.2 Animals, immunosuppression and transplantation

Four to six-week-old *Dmd*^{mdx} mice recipients were obtained from our animal house, in Human Genome Research Center at University of São Paulo. The mice received routine required cares for good health, and all experiments were approved by the research ethics committee of the Biosciences Institute, University of São Paulo.

For the transplantation experiments, the *Dmd*^{mdx} mice were divided into groups according to cell treatment (Table 2). In the immunosuppressed group, the animals received a daily dose of 1 mg/kg FK506 (Tacrolimus, Sigma-Aldrich) intraperitoneal injections, from the day before stem cells injection until the time of euthanasia.

Each animal was injected with 1.0 x 10^6 ES or EB cells into the gastrocnemius muscle, or through intravenous injections in the tail vein.

Animals were euthanized using CO_2 chamber 2 days, 1 week, 2 weeks, 4 weeks and 8 weeks post-transplantation, and muscles were analyzed for dystrophin expression, and identification of the presence of the injected stem cells.

3.3 Tissue processing and dystrophin analysis

Gastrocnemius muscles were dissected and collected from all mice and additionally, several other tissues such as tail, liver and spleen were collected from the mice injected systemically. Samples of the contra lateral muscles were also collected and used as controls in the intramuscular injected mice. All tissue were immediately frozen in liquid nitrogen. For histological analysis, frozen sections of 7 µm were prepared on slides using a cryostat (Zeiss, Jena, Germany).

Cryosections were stained with hematoxilin and eosin or used for immunohistochemistry to evaluate the expression of the muscle proteins. In brief, they were incubated with 1:100 diluted rabbit antibody anti-dystrophin AB 15277 (Abcam) and mouse antibody anti-fetal myosin NCL-MHCd (Novocastra). Cy3-conjugated anti-rabbit and anti-mouse (1:100) were used as secondary antibodies (Sigma-Aldrich). Samples were analyzed under fluorescence microscopy using appropriate filters for the fluorophore.

Treatment: ES cells		
	Administration via	*Euthanasia and analysis*
Nonimmunosuppressed mice		
Mdx 19	Local	2 days
Mdx 20	Local	2 week
Mdx 21	Local	4 weeks
Mdx 22	Systemic	2 weeks
Mdx 23	Systemic	4 weeks
Immunosuppressed mice		
Mdx 10	Local	1 week
Mdx 11	Local	2 weeks
Mdx 12	Local	4 weeks
Mdx 13	Systemic	1 week
Mdx 14	Systemic	2 weeks
Treatment: EB cells		
Immunosuppressed mice	*Administration via*	*Euthanasia and analysis*
Mdx 15	Local	4 weeks
Mdx 16	Local	8 weeks
Mdx 17	Systemic	4 weeks
Mdx 18	Systemic	8 weeks

Table 2. Intramuscular and systemic injected Dmd^{mdx} mice.

Total proteins were extracted from muscle and other tissues, separated by 6% SDS-PAGE polyacrylamide gel electrophoresis and were transferred onto nitrocellulose membrane (GE Healthcare) for 60 min at 0.35A at 4°C. Membranes were then pre-stained in 0.2% Ponceau S, to ensure protein transfer and equal loading of the lanes with protein. Membranes were blocked with 5% nonfat milk in PBS, 0.1% Tween 20 (PBS-T) for 1 h and subsequently incubated with mouse antibody directed against dystrophin VP-D508 (Vector). After an overnight period of incubation with the primary antibody, membranes were washed three times with PBS-T for 10 min. The blots were then reacted with alkaline phosphatase-conjugated secondary antibody (1:1000) for 1 h at room temperature. The detection of protein was done using colorimetric reaction for the enzyme, using nitroblue tetrazolium and 5-bromo-4-chloro-3-indolyl phosphate as substrate (Vainzof et al., 1993).

Data on the injected cells (ES or EB cells), immunosupression and detection of cells both through red dye presence and DNA analysis, as well as screening for dystrophin expression in all injected animals are summarized in Table 3.

To check the ability of these cells to differentiate into new muscle fibers, and to express muscle proteins, we analyzed recipient muscles by immunohistochemistry using an anti-dystrophin antibody. The immunohistochemistry revealed no significant labeling, with only scattered fibers partially labeled, in a pattern observed also in non-injected Dmd^{mdx} mice (Fig. 1). Western blot analysis confirmed the lack of dystrophin expression in all tested animals.

	Local injections									
	ES					EB cells				
	Analysis		Detection			Analysis		Detection		
		time	red dye	DNA	Dystrophin		time	red dye	DNA	Dystrophin
	Mdx 19	2 d	+++	-	-					
Nonimmunosuppressed mice	Mdx 20	2 w	-	-	-					
	Mdx 21	4 w	-	-	-					
	Mdx 10	1 w	+++	-	-	Mdx 15	4 w	-	-	-
Immunosuppressed mice	Mdx 11	2 w	++	+	-	Mdx 16	8 w	-	-	-
	Mdx 12	4 w	+	-	-					
	Systemic injections									
	ES					EB cells				
	Analysis		Detection			Analysis		Detection		
		time	red dye	DNA	Dystrophin		time	red dye	DNA	Dystrophin
Nonimmunosuppressed mice	Mdx 22	2 w	-	-	-					
	Mdx 23	4 w	-	-	-					
Immunosuppressed mice	Mdx 13	1 w	+ (tail)	-	-	Mdx 17	4 w	-	-	-
	Mdx 14	2 w	+ (tail)	-	-	Mdx 18	8 w	-	-	-

Table 3. Results observed in the injected mice.

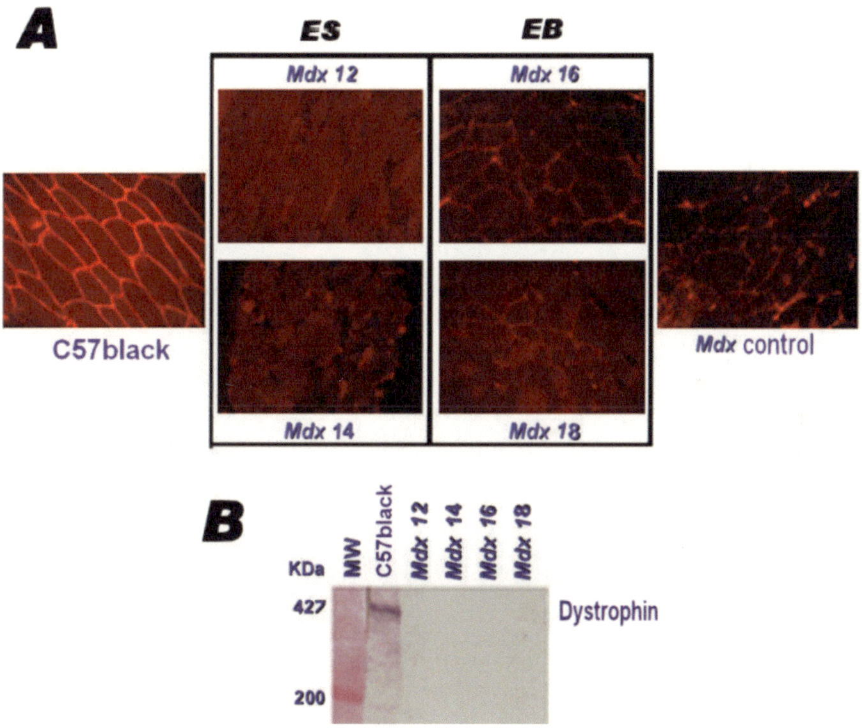

Fig. 1. Dystrophin analyses in the injected and control muscles. (A) Dystrophin immunofluorescence and (B) Western blotting analysis showing the positive pattern of labeling in normal muscle of *C57black* mice and the absence dystrophin in immunosuppresed mice injected with ES (Mdx12 and Mdx14) and EB cells (Mdx 16 and Mdx 18). Cells were injected in the muscle in Mdx 12 and Mdx 16, and intra-venous in Mdx 14 and Mdx 18.

3.4 ES and EB cells identification

Genomic DNA was obtained from cells and muscle tissue of injected mice. Fragments of injected and not injected muscles were placed in an extraction solution (100mM Tris HCl pH 8.5; 5mM EDTA, 0.2% SDS; 200mM NaCl), containing 2.5 µl proteinase K (10mg/ml). These samples were kept overnight in dry bath at 55°C. The samples were subjected to centrifugation of 1000g for 10 minutes, and added 500 µl isopropanol in the supernatant. After the new centrifugation of 13500g for 15 minutes, the DNA was precipitated in TE buffer (TRIS-EDTA). The PCR primers used were from the same region of the genome of mice, which produce products of different sizes due to the presence of polymorphic regions, referring to strains of *129/Sv* mice and *Dmd*^{mdx} mice [http://www.informatics.jax.org/searches/probe.cgi?37495] for the identification of ES cells. PCR conditions were 95°C 5 minutes, with 40 cycles of 95°C 1 minute, 55°C 1 minute and 72°C 1 minute. DNA fragments were separated by electrophoresis in 10% acrylamide gels and marked with ethidium bromide to its visualization under UV transillumination.

As almost 100% of the injected ES and EB cells were previously labeled with a red fluorescent dye, we analyzed the presence of these marked cells in histological sections of the injected mice. We identified their presence for longer periods (after 1 week) only in the muscle of the ES injected mice (Table 3 and Fig. 2). This may mean that the ES remained in the injected place.

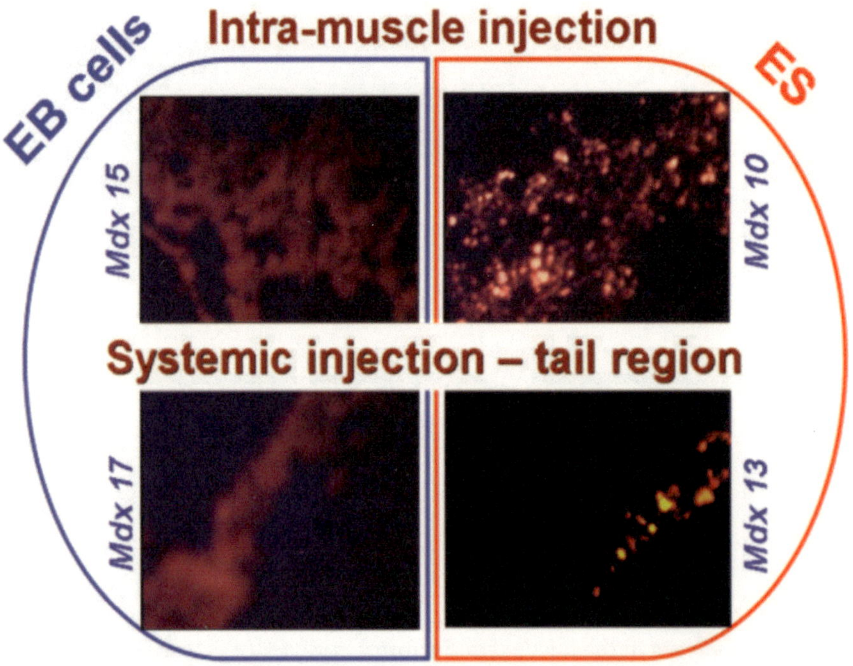

Fig. 2. Identification of dye labbeling of ES and EB cells. Screening for the red-labeled ES and EB cells injected in the different *Dmd*^{mdx} mice. Intramuscular injections (Mdx 15 and Mdx 10) and systemically-injected (Mdx 17 and Mdx 13), showing that only the ES cells were identified.

In the animals Mdx 10, 11 and 12 (local injections of ES cells, after 1, 2 and 4 weeks), we noticed an increase of the muscles in the leg, and the hematoxilin-eosin (H&E) staining showed an intense degeneration and presence of several types of undifferentiated cells (Fig. 3). A histopathological examination of frozen sections identified the proliferation of various types of immature cells, suggesting the formation of a tumor, consistent with the embryonic nature of these cells. But we did not have clear evidences that it was teratoma.

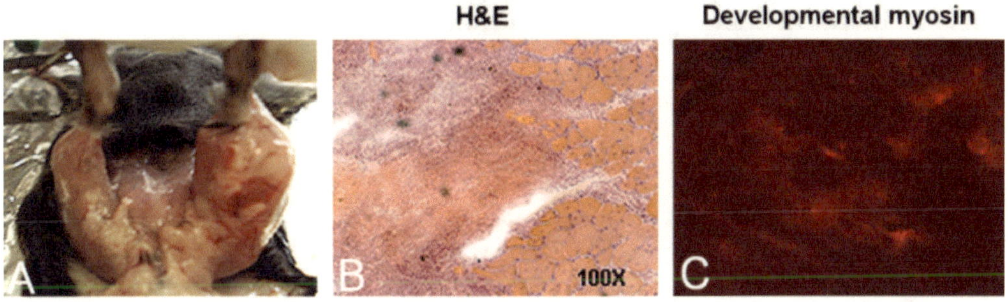

Fig. 3. Macroscopic and microscopic analyses of the muscle injected with ES cells. Analyses of the immunosuppressed Mdx 11, with local injections of ES cells, after 2 weeks post-transplantation. A- A significant increase in the muscles of the injected leg was observed. B- Hematoxilin-eosin (H&E) staining showed the presence of several types of undifferentiated cells. C- Immunohistological reaction using anti-developmental myosin showed no positive evidence of the presence of new fibers.

To check if the new generated tissue included differentiation into muscle fibers, we probed the same region described above with the antibodies against fetal myosin. However, there was no immunoreactivity for fetal myosin and no evidence of formation of new fibers (Fig. 3).

3.4.1 ES cell quantification

To investigate the presence and try to quantity of injected ES and EB cells in the muscle we used PCR analysis with specific primers to polymorphic markers of each cell strain. This method detected the presence of stem cells only in the muscle of Mdx 11 (Fig. 4A), who also showed an enlargement of the leg, with significant mass growth.

The experiment revealed a 121 bp band corresponding to the molecular weight of *Dmd*^{mdx} mice polymorphism, while a band of 142 bp is observed in the donor ES cells strain (line 4, Fig. 4A).

To estimate the relative amount of the injected cells and verify their possible multiplication in vivo, we performed a standard curve with known concentrations of DNA from the two different mice strains (Fig. 4B). The ES cells strain was only detected starting with the concentration of 30%. In Mdx 11, the number of ES cells in the mice injected muscle were in a similar concentration of the limit of detection of 30% (Fig. 4A). No 142 bp band was observed in other mice, suggesting lower concentration of the injected ES and EB cells in these muscles.

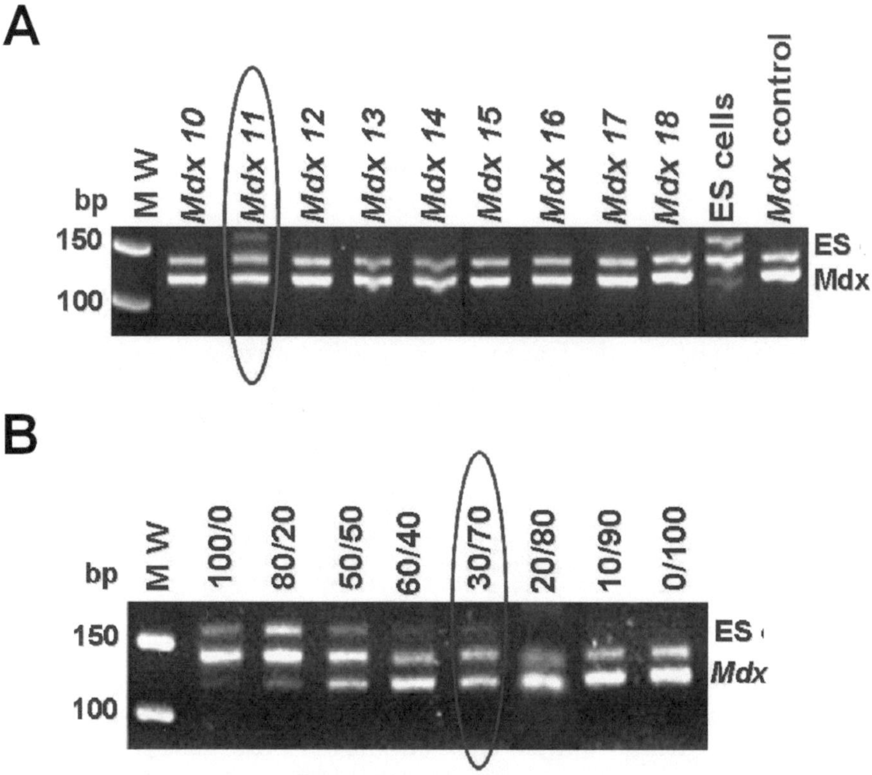

Fig. 4. DNA screening for the ES and EB presence. (A) PCR reaction for polymorphic markers of *C57Black* (*Dmd*^{mdx} strain) and *129/Sv* (ES and EB cells) strains in the muscles from all injected animals (all immunosupressed). The band of 121 bp is from the *Dmd*^{mdx} strain, while the band of 142 bp is from of the *129/Sv* strain of the ES cells. In Mdx 11 both bands can be observed, indicating the presence of the ES cells in the injected muscles. (B) The same PCR reaction, in a standard curve with known concentrations of DNA from the two different mice strains, showing the identification of the ES cells starting in the concentration of 30%.

4. Efficient stem cell therapies for dystrophic muscle

Replacement therapies using stem cells are emerging as promising avenues for treatment of genetic diseases because the transplantation of healthy stem cells into affected individuals can potentially rescue defects and injuries.

The identification of mesenchymal stem cells (MSCs) as a sub-population of fibroblast-like adherent cells in the BM opened new perspectives for therapies for the different forma of muscular dystrophies, since these cells were able to differentiate into several other lineages, producing in vitro skeletal muscle, in addition to osteoblasts, chondroblasts, and adipocytes (Friedenstein et al., 1966; Prockop, 1997). Very recently, we showed that hMSC were not rejected when transferred to *SJL/J* mice by systemic and repeated injections (Vieira et al., 2008), while mMSC intramuscularly injected cells were eliminated after 2 days in the *Dmd*^{mdx} mouse (Ayub-Guerrieri et al., 2009). The injected cells in the *SJL/J* mouse were able to fuse with the host muscle cell and to express the exogenous protein. However, in this previous study, the number of cells and injections were significantly higher and the *SJL/J* mouse model for LGMD2B presents a milder phenotype with no active muscle degeneration. This suggests that either the number of injected cells must be higher, or dystrophin deficient dystrophic muscle probably constitutes an adverse environment for the newly introduced cells. Thus, the high degree of degeneration, sclerosis and fat infiltration, as well as inflammation and activation of cascades of degeneration/regeneration genes of advanced muscular dystrophy reduce the efficacy of cell delivery, and can result in their poor survival in the injected muscles and limit the effectiveness of replacement therapies with stem cells. A very recent study (Gargioli et al., 2008) corroborates this hypothesis. Modified tendon fibroblasts expressing angiogenic factors and metalloproteinases were injected in Dmd^{mdx} mice and were able to restore the vascular network and reduce collagen deposition, allowing efficient cell therapy in aged dystrophic mice (Gargioli et al., 2008). Therefore, we suggest that efficient stem cell therapies for degenerated dystrophic muscle must include additional procedures to improve the general health of the diseased tissue before stem cells can be delivered and integrated.

Here we show that undifferentiated ES cells hold a better promise of success of retention than differentiated EB cells, since after intramuscular injections, ES cells, but not EB cells, were found in the muscle of animals after 1, 2 and 4 weeks post-injections. Additionally, systemic injection of both ES and EB cells were not directed nor retained by the dystrophic muscles. Two other reports on the use of ES cells in the treatment of DMD showed evidence for engraftment (Bhagavati & Xu, 2005; Darabi et al., 2008). In the first study, the transplantation of EB co-cultured with dissociated skeletal muscle into mdx mice was limited to qualitative detection of donor-derived cells in recipient muscle. However, in the second study, authors showed that paraxial mesoderm is not generated efficiently during ES differentiation induced by conventional protocols, which may further explain our lack of success in integrating ES cells into dystrophic muscles (Darabi et al., 2008).

A second important question relates to the extent to which stem cells are immunogenic, since the safe and successful clinical will be indirectly dependent on the strength of rejection reactions (Mimeault et al., 2007; Trounson 2006). The first studies on the immunologic properties of ES cells and their derivatives have suggested their ability to induce immunologic tolerance (Drukker et al., 2002; Bonde & Zavazava, 2006). Because of this, we investigated the myogenic potential of ES and EB cells when injected into muscle of immunosuppressed mdx mice. Our results showed that the non differentiated ES cells were retained, while pre-differentiated EB cells were all eliminated. Although this could happen because of the non-integration of the EB cells, this could also suggest that even with immunosupression, more differentiated cells may cause a stronger immunological response than undifferentiated ES cells, as also suggested by other group (Drukker et al., 2002). These data are in accordance with recent publication showing that hESCs were highly immunogenic, triggered both cellular and humoral-mediated pathways, and as a result, were rapidly rejected in xenogeneic hosts. These authors observed that combinations of immunosuppressive regimens are necessary for stem cell survival in vivo (Swijnenburg et al., 2008).

Several reports have already described that, when injected sub-cutaneousely into severe combined immunodeficient mice, undifferentiated ES cells will produce teratoma or teratocarcinoma, with derivatives of the three germ layers (Bradley et al., 2002; Mimeault et al., 2007; Reubinoff et al., 2000; Andrews et al., 2005; Wu et al., 2007). Here we observed an

increase of the leg of ES cells injected mice. Our histology findings suggest that the donor cells may have multiplied and have been embedded in the mice muscles, in a pattern similar to a teratoma. In one hand, the formation of tumor confirms the ability of these ES injected cells to multiplication in vivo. On the other hand, this new tissue had no muscle characteristics, since no dystrophin, nor developmental myosin positive fibers were identified. Ultimately, it is clear that the signaling environment of the severely dystrophic muscle was not able to induce the differentiation of stem cells into muscle cells.
The use of single injections of stem cells for future therapies in children would have the significant benefit of being less invasive, and less traumatic from a clinical and immunological stand point. But, it also leads to a limitation in the number of injected cells, which can be insufficient to promote an appropriate restoration of the degenerated muscle. However, the finding of the injected ES cells in the recipient muscle, even using only one application of the same quantity of 10^6 cells, indicates that this number was sufficient for their retention in the injected muscle. Using polymorphic markers for the ES cell line we identified a concentration of at least 30% of these cells in the injected muscle. This result indicates that muscle enlargement observed in injected mice was caused by at least this amount of cells. In the other locally injected mice with leg enlargement, probably a lower amount of these cells were sufficient to promote the tumor, but this amount was not sufficient to allow its detection by our methodology.

5. Conclusion

Our results suggest that the dystrophic muscle is an adverse ambient for the homing and maintenance of injected stem cells, since when the "still pluripotent" stem cells were injected they formed tumors, and when the muscle pre-differentiated cells were injected they were cleared/migrated out of the tissue. Injected embryonic stem cells, therefore, must receive more or different stimulation from the dystrophic environment to differentiate or to fuse into muscle. Additional studies are necessary to increase the therapeutic potential of these cells in dystrophic murine models.

6. Acknowledgments

The authors would like to thank the following researchers, for scientific and technical support: Dra. Mayana Zatz, Dra. Helga C. Silva, Lydia Yamamoto, Dinorah Zilberztajn, Marta Canovas, Patrícia Kossugue, Adriano Senkevics, Priscila Clara Calyjur and André Santos. FVB transgenic mice carrying the enhanced green fluorescent protein (eGFP) were generated and kindly donated by the Transgenic Unit of the Faculty of Medicine of the University of São Paulo. This work was supported by Fundação de Amparo a Pesquisa do Estado de São Paulo - Centro de Pesquisa, Inovação e Difusão (FAPESP-CEPID), Conselho Nacional de Desenvolvimento Científico e Tecnológico (CNPq), and Instituto Nacional de Ciencia e Tecnologia (INCT), FINEP, and Associação Brasileira de Distrofia Muscular (ABDIM).

7. References

Andrews, P.W.; Martin, M.M.; Bahrami, A.R.; Damjanov, I.; Gokhale, P. & Draper, J.S. (2005). Embryonic stem (ES) cells and embryonal carcinoma (EC) cells: opposite sides of the same coin. *Biochem Soc Trans*, 33: 1526–1530.

Ayub-Guerrieri, D.; Onofre-Oliveira, P.C.G.; Lopes, V.F.; Martins, P.C.M. & Vainzof, M. (2009). Poor maintenance of eGFP- mesenchymal stem cells in the dystrophic muscle. *Neuromuscular Disorders*, 19: 634.

Barberi, T.; Bradbury, M.; Dincer, Z.; Panagiotakus, G.; Socci, N. D. & Studer, L. (2007). Derivation of engraftable skeletal myoblasts from human embryonic stem cells. *Nature Medicine,* 13 (5): 642-648.

Bhagavati, S. & Xu, W. (2005). Generation of skeletal muscle from transplanted embryonic stem cells in dystrophic mice. *Biochem Biophys Res Commun,* 333: 644–649.

Bonde, S. & Zavazava, N. (2006). Immunogenicity and Engraftment of Mouse Embryonic Stem Cells in Allogeneic Recipients. *Stem Cells,* 24: 2192–2201.

Bradley, J.A.; Bolton, E.M. & Pedersen, R.A. (2002). Stem cell medicine encounters the immune system. *Nature reviews,* 2: 859 - 871.

Chang, H.; Yoshimoto, M.; Umeda, K.; Iwasa, T.; Mizuno, Y.; Fukada, S.; Yamamoto, H.; Motobashi, N.; Suzuki, Y. M.; Takeda, S.; Heike, T. & Nakahata, T. (2009). Generation of transplantable, functional satellite-like cells from mouse embryonic stem cells. *The FASEB Journal,* 23 (6): 1907-1919.

Dangain J., & Vrbova G. (1984). Muscle development in (mdx) mutant mice. *Muscle Nerve,* 7: 700-704.

Darabi, R.; Gehlbach, K.; Bachoo, R.M.; Kamath, S.; Osawa, M.; Kamm, K.E.; Kyba, M. & Perlingeiro, R.C.R. (2008). Functional skeletal muscle regeneration from differentiating embryonic stem cells. *Nature Med,* 14 (2): 134-143.

Darabi, R.; Baik, J.; Clee, M.; Kyba, M.; Tupler, R. & Perlingeiro, R. C. R. (2009). Engraftment of embryonic stem cell-derived myogenic progenitors in a dominant model of muscular dystrophy. *Experimental Neurology,* 220 (1): 212-216.

Drukker, M.; Katz, G.; Urbach, A.; Schuldiner, M.; Markel, G.; Itskovitz-Eldor, J.; Reubinoff, B.; Mandelboim, O. & Benvenisty, N. (2002). Characterization of the expression of MHC proteins in human embryonic stem cells. *Proc Natl Acad Sci U S A,* 99: 9864 - 9869.

Emery, A. (2002). The muscular dystrophies. Lancet , 359: 687–695.

Evans, M.J. & Kaufman, M.H. (1981). Establishment in culture of pluripotential cells from mouse embryos. *Nature,* 292: 154–156.

Ervasti, E. & Campbell, K.P. (1993). A role for the dystrophin-glycoprotein complex as a transmembrane linker between laminin and actin. *J. Cell Biol.,* 122: 809-823.

Ferrari, G. & Mavilio, F. (2002). Myogenic stem cells from the bone marrow: a therapeutic alternative for muscular dystrophy? *Neuromuscul Disord,* 12 (Suppl 1): S7–S10.

Ferrari, G.; Cusella de Angelis, G.; Coletta, M.; Paolucci, E.; Stornaiuolo, A.; Cossu, G. & Mavilio, F. (1998). Muscle regeneration by bone-marrow-derived myogenic progenitors. *Science,* 279:1528-1530.

Friedenstein, A.J.; Piatetzky-Shapiro, I.I. & Petrakova, K.V. (1966). Osteogenesis in transplants of bone marrow cells. *J Embryol Exp Morphol,* 16: 381–390.

Gargioli, C.; Coletta, M.; De Grandis, F.; Cannata, S.M. & Cossu, G. (2008). PIGF-MMP-9-expressing cells restore microcirculation and efficacy of cell therapy in aged dystrophic muscle. *Nature Med,* 14 (9):973-978.

Gussoni, E.; Bennett, R.R.; Muskiewicz, K.R.; Meyerrose, T.; Nolta, J.A.; Gilgoff, I.; Stein, J.; Chan, Y.M.; Lidov, H.G.; Bönnemann, C.G.; Von Moers, A.; Morris, G.E.; Den Dunnen, J.T.; Chamberlain, J.S.; Kunkel, L.M. & Weinberg, K. (2002). Long-term persistence of donor nuclei in a Duchenne muscular dystrophy patient receiving bone marrow transplantation. *J Clin Invest*, 110 (6): 807-814.

Gussoni, E.; Soneoka, Y.; Strickland, C.D.; Buzney, E.A.; Khan, M.K.; Flint, A.F. & et al. (1999). Dystrophin expression in the mdx mouse restored by stem cell transplantation. *Nature*, 401: 390–394.

Hoffman E.P., Morgan J.E., Watkins S.C. & Partridge T.A. (1990). Somatic reversion/suppression of the mouse mdx phenotype in vivo. *J Neurol Sci*, 99: 9-25.

Hoffman, E.P.; Brown, R.H. & Kunkel, L.H. (1987). Dystrophin: the protein product of the Duchenne Muscular Dystrophy Locus. *Cell*, 51: 919-928.

Ichim TE, Alexandrescu DT, Solano F et al. (2010). Mesenchymal stem cells as anti-inflammatories : implications for treatment of Duchenne msucualr dystrophy. *Cell Immunol*, 260 : 75-82

Kennedy, K. A. M.; Porter, T.; Mehta, V.; Ryan, S. D.; Price, F.; Peshdary, V.; Karamboulas, C.; Savage, J.; Drysdale, T. A.; Li, S.; Bennett, S. A. L. & Skerjanc, I. L. (2009). Retinoic acid enhances skeletal muscle progenitor formation and bypasses inhibition by bone morphogenetic protein 4 but not dominant negative β-catenin. *BMC Biology*, 7: 67-88.

Mimeault, M.; Hauke, R. & Batra, S.K. (2007). Stem Cells: A revolution in therapeutics - Recent advances in stem cell biology and their therapeutic applications in regenerative medicine and cancer therapies. *Nature*, 83 (3): 252-264.

Meng, J. ; Muntoni, F. & Morgan, J.E. (2011). Stem cells to treat muscular dystrophies - where are we ? *Neuromuscl Disord*, 21 : 4-12.

Meng, J. ; Adkin, C.F. ; Xu, S-w. ; Muntoni, F. ; Morgan, J.E. (2011). Contribution of Human muscle-derived cells to skeletal muscle regeneration in dystrophic host mice. *PlosOne*, 6 :e17454

Muntoni, F. ; Brockington, M. ; Torelli, S. ; Brown S.C. (2004). Defective glycosylation in congenital muscular dystrophies. *Curr Opin Neurol*, 17 : 205-209.

Ohlendieck, K. & Campbel, K.P. (1991). Dystrophin-associated proteins are greatly reduced in skeletal muscle from mdx mice. *J Cell Biol*, 115: 1685-1694.

Ozasa, S.; Kimura, S.; Ito, K.; Ueno, H.; Ikezawa, M.; Matsukura, M.; Yoshioka, K.; Araki, K.; Yamamura, K.; Abe, K.; Niwa, H. & Miike, T. (2007). Efficient conversion of ES cells into myogenic lineage using the gene-inducible system. *Biochemical and Biophysical Research Communications*, 357 (4): 957-963.

Prelle, K.; Wobus, A. M.; Krebs, O.; Blum, W. F. & Wolf, E. (2000). Overexpression of insulin-like growth factor-II in mouse embryonic stem cells promotes myogenic differentiation. *Biochemical and Biophysical Research Communications*, 277 (3): 631-638

Prockop, D.J. (1997). Marrow stromal cells as stem cells for nonhematopoietic tissues. *Science*, 276: 71-74.

Reubinoff, B.E.; Pera, M.F.; Fong, C.; Trounson, A. & Bongso, A. (2000). Embryonic stem cell lines from human blastocysts: somatic differentiation in vitro. *Nat Biotechnol*, 18: 399-404.

Rohwedel, J.; Maltsev, V.; Bober, E.; Arnold, H.H.; Hescheler, J. & Wobus, M. (1994). Muscle cell differentiation of embryonic stem cells reflects myogenesis in vivo: developmentally regulated expression of myogenic determination genes and functional expression of ionic currents. *Dev Biol*, 64: 87–101.

Sakurai, H.; Okawa, Y.; Inami, Y.; Nishio, N. & Isobe, K. (2008). Paraxial mesodermal progenitors derived from mouse embryonic stem cells contribute to muscle regeneration via differentiation into muscle satellite cells. *Stem Cells*, 26 (7): 1865-1873.

Seale, P. & Rudnicki, M.A. (2000). A new look at the origin, function, and "stem-cell" status of muscle satellite cells. *Dev Biol*, 218: 115-124.

Sicinski, P.; Geng, Y., Ryder-Cook, A.S.; Barnard, E.A.; Darlison, M.G. & Barnard, P.J. (1989). The molecular basis of muscular dystrophy in the mdx mouse. A point mutation. *Science*, 244: 1578–1580.

Straub, V. & Campbell, K.P. (1997). Muscular dystrophies and the dystrophin-glycoprotein complex. *Curr Opin Neurol*, 10 :168-175.

Sukoyan, M.A.; Kerkis, A.Y.; Mello, M.R.; Kerkis, I.E.; Visintin, J.A. & Pereira, V. (2002). Establishment of new murine embryonic stem cell lines for the generation of mouse models of human genetic diseases. *Braz J Med Biol Res*, 35 (5): 535-542.

Swijnenburg, R.J.; Schrepfer, S.; Govaert, J.A.; Cao, F.; Ransohoff, K.; Sheikh, A.Y.; Haddad, M.; Connolly, A.J.; Davis, M.M.; Robbins, R.C. & Wu, J.C. (2008). Immunosuppressive therapy mitigates immunological rejection of human embryonic stem cell xenografts. *Proc Natl Acad Sci U S A*, 105 (35): 12991- 12996.

Trounson, A. (2006). The production and directed differentiation of human embryonic stem cells. Endocr Rev, 27: 208–219.

Tome, F.M.; Evangelista, T.; Leclerc, A. ; Sunada, Y.; Manole, E.; Estournet, B.; Barois, A.; Campbell, K.P. & Fardeau, M. (1994). Congenital muscular dystrophy with merosin deficiency. C. R. Acad. Sci. III, 317 :351-357.

Vainzof, M. & Zatz, M. (2007). Muscular Dystrophies and Protein Mutations, in: Uversky VN and Fink AL. Protein Misfolding, Aggregation, and Conformational Diseases. Part B: Molecular Mechanisms of Conformational Diseases. Serie Protein Reviews, Ed. Zouhair Atassi, Volume 6, Springer Sciences and Business Media LLC, USA. Pp 391-403.

Vainzof, M.; Ayub-Guerrieri, D.; Onofre, P.C.G.; Martins, P.C.M.; Lopes, V.F.; Zilberztajn, D.; Maia, L.S.; Sell, K. & Yamamoto, L.U. (2008). Animal Models for genetic neuromuscular diseases. J Molec Neurosc, 34(3): 241-8.

Vainzof, M.; Passos-Bueno, M.R. & Zatz, M. (2003). Immunological methods for the analysis of protein expression in neuromuscular diseases. *Methods Mol Biol*, 217: 355–378.

Vieira, N.M.; Bueno, C.R.Jr.; Brandalise, V.; Moraes, L.V.; Zucconi, E.; Secco, M.; Suzuki, M.F.; Camargo, M.M.; Bartolini, P.; Brum, P.C.; Vainzof, M. & Zatz, M. (2008). SJL dystrophic mice express a significant amount of human muscle proteins following systemic delivery of human adipose-derived stromal cells without immunosuppression. *Stem Cells*, 26 (9):2391-8.

Wobus, A. M.; Guan, K.; Yang, H. T. & Boheler, K. R. (2002). Embryonic stem cells as a model to study cardiac, skeletal muscle and vascular smooth muscle cell differentiation. *Methods in Molecular Biology*, 185: 127-156.

Wu, D.C.; Boyd, A.S. & Wood, K.J. (2007). Embryonic stem cell transplantation: potential applicability in cell replacement therapy and regenerative medicine. *Front Biosci*, 12: 4525-4535.

Yoshida, M. & Ozawa, E. (1990). Glycoprotein complex anchoring dystrophin to sarcolemma. *J. Biochem.*, 108: 748-752.

Zheng, J. K.; Wang, Y.; Karandikar, A.; Wang, Q.; Gai, H.; Liu, A. L.; Peng, C. & Sheng, H. Z. (2006). Skeletal myogenesis by human embryonic stem cells. *Cell Research,* 16 (8): 713-722.

15

Stem Cell in Neurological Disorders

Nirmeen Kishk and Noha Abokrysha
Cairo University
Egypt

1. Introduction

There are various sources of stem cells which are being studied for their potential in stem cell-based therapies for CNS diseases (Yu D & Silva, 2008):

1.1 Embryonic stem cells

Embryonic stem cells are pluripotent cells with indefinite self-renewal capabilities as well as the ability to differentiate into all cell types derived from the 3 embryonic germ layers (How Embryonic, 2010). Embryonic stem cells are favorable in the research community because they are relatively easy to isolate, can grow indefinitely, and have the potential to develop into any type of adult cell.

1.2 Adult stem cells

Adult stem cells (ASCs) play a critical role in tissue maintenance and repair (Stem Cell Basics, 2010). Research on adult stem cells began in the 1950s with the discovery of multipotent hematopoietic and mesenchymal stem cells in bone marrow, which can generate a number of tissues (Stem Cell Basics, 2010). **Bone Marrow-Derived Mesenchymal Stem Cells** can be expanded and differentiated in vitro using various media formulations and culture surface conditions to direct them to different cell lineages (Ho et al., 2006). BMSCs have the ability to migrate to areas of injury, even crossing the blood-brain barrier (Akiyama et al., 2002; Tang et al., 2007). **Although the reproducibility of BMSC therapies needs to be thoroughly examined, these early experiments suggest that BMSCs can be administered intravenously to CNS targets.** (Rice & Scolding, 2008)

1.3 Neural stem cells

The adult mammalian CNS contains NSCs which were first inferred from evidence of neuronal turnover in the olfactory bulb and hippocampus in the adult. (Altman & Das, 1965, 1966). Neural stem cells are able to differentiate into neurons, astrocytes, oligodendrocytes and various forms of neural precursors (Flax et al.,1998;Gage,2000;Palmer et al.,1997;Takahashi et al.,1999;Weiss et al.,1996) .Moreover, in vivo delivery of these cells to animal models of neurodegenerative diseases was associated with varying degrees of functional recovery (Ourednik et al.,2002). Figure (1) (Lindvall & Kokaia., 2006)

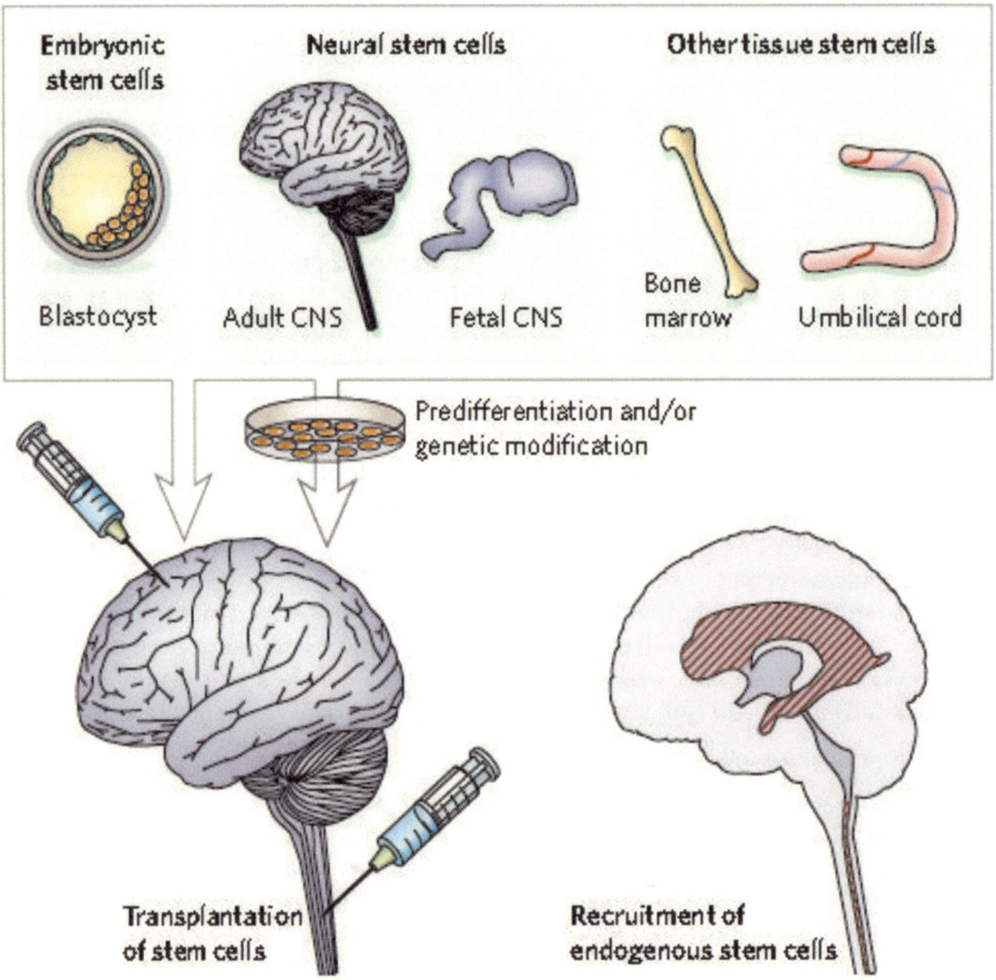

Fig. 1. Application of stem cells for neurogical disorders. Steam cells would be isolated and transplanted to the diseased brain and spinal cord, either directly or after predifferentiation/genetic modification in culture to form specific types of neuron and glial cell, or cells producing neuroprotective molecules. In strategies relying on stimulation of the patient's own repair mechanisms, endogenous stem cells would be recruited to areas of the adult brain and spinal cord affected by disease, where they would produce new neurons and glia (neurogenetic and gliogenic areas along lateral ventricle and central canal are shown in hatched red). Stem cells could provide clinical benefits by neuronal replacement, remyelination and neuroprotection.

2. Strategies for using stem cells to treat neurological disorders

These include strategies in which:

i. Stem cells are transplanted within the brain, are infused by blood circulation (Figure 2); or delivered through bone marrow transplantation (BMT; Figure 2).

ii. Stem cells are stimulated by cytokines, or trophic and growth factors, into the brain in vivo;
iii. Stem cells are engineered to correct the genetic defect (Figure 2), delivery of therapeutic agents.
iv. Stem cells are combined with biomaterials (Figure 2).
http://www.discoverymedicine.com/Antonio-Orlacchio/ n.d

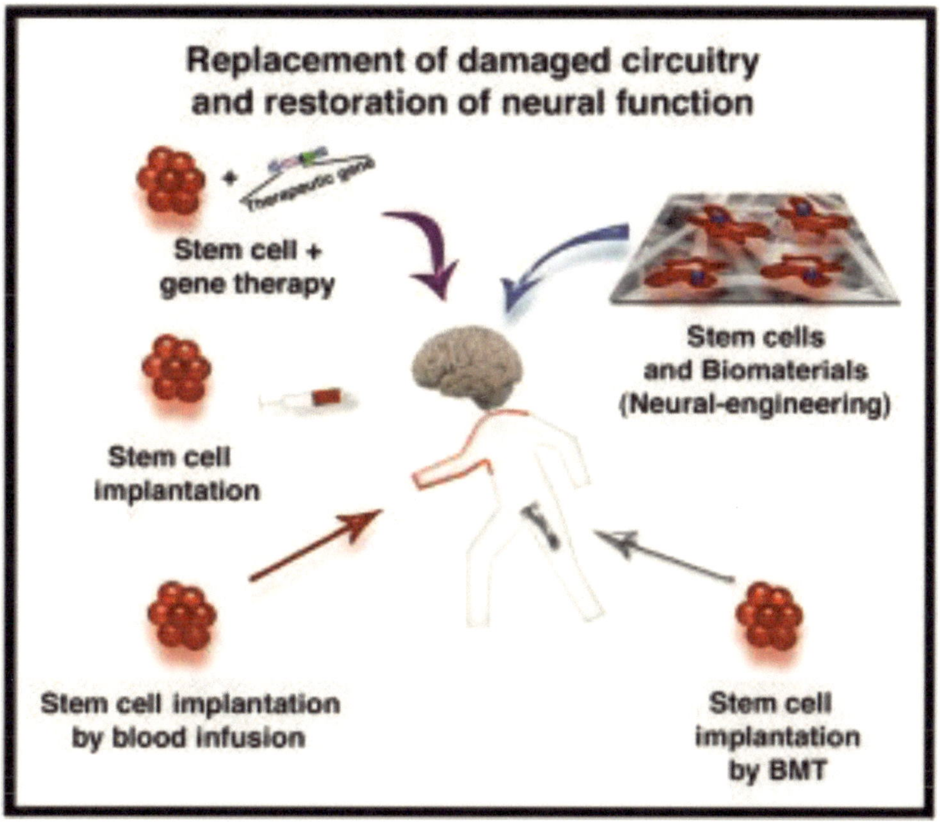

Fig. 2. Stem cells replacement therapy for neurological diseases. Cartoon schematizes the different strategies for stem cell delivery in order to repair the degenerated tissue. http://www.discoverymedicine.com/Antonio-Orlacchio/files/2010/06/discovery_medicine_orlacchio_no49_figure

2.1 Cell replacement through transplantation of exogenous cells

Transplantated donor cells into a host organism offer the chance to expand and manipulate cells in vitro. This method has proven successful in the haematopoietic system. Such method to be successful in the CNS the transplanted cells must (1) survive the transplantation; (2) migrate to the site of replacement; (3) differentiate into appropriate cell phenotype(s); and (4) make appropriate connections with the host tissue.

The condition is further complicated during neurodegenerative disease, since the same factors that cause mature neurons to die (oxidative stress, accumulation of toxic protein aggregates, etc.), may also lead to cell death in transplanted cells. There are a number of criteria for improving the opportunities for successful cell replacement using transplanted cells. Generally, selecting cells from the same region as the region to which the cells will be transplanted increases the transplantation success rate. (Tobi& Mahendra., 2005)

2.2 Mobilization of endogenous stem cells

A second therapeutic approach is the mobilization of endogenous cells to replace lost or damaged cells . The production of new neurons (neurogenesis) occurs at a very low rate in the adult brain, in two regions: the subventricular zone of the lateral ventricle, which produces new neurons of the olfactory bulb, and in the hippocampus. Neurogenesis can be stimulated in the adult brain by factors, including diet, exercise and modification of hormone levels. (Tobi& Mahendra., 2005)

Moreover, neurogenesis can be stimulated in the adult brain by growth factors, such as epidermal growth factor (EGF) or basic fibroblast growth factor (bFGF).However, delivery of these factors is complicated, since growth factors are large molecules that cannot penetrate the blood-brain barrier when it is intact. Transfer of growth factors into the adult CNS can be accomplished through direct injection into the brain and/or the ventricles within the brain or through transplantation of cells genetically engineered to secrete growth factors into the surrounding environment. Additionally, growth factors can be used to protect endogenous NPCs neural progenitor cells from dying. (Tobi& Mahendra., 2005)

2.3 Delivery of therapeutic agents

Drug Delivery to the CNS is complicated by the blood-brain barrier, which prevents many large, hydrophilic molecules from entering the brain through the bloodstream. One potential use of NPCs neural progenitor cells is as delivery agents for pharmacological compounds . A large number of studies provides evidence that delivery of supportive factors can slow the degeneration process in several neurodegenerative diseases. Although this method has many problems, including controlling the survival rate of transplanted cells and controlling the rate of drug secretion, this is one path currently being explored for therapy using stem cells. (Tobi& Mahendra., 2005)

2.4 Stem cells are combined with biomaterials

Tissue engineering approach includes the transplantation of stem cells in combination with natural or synthetic biomaterials (Dawson et al., 2008). Verification came from data showing that chemical and biological modifications of biomaterials could directly influence stem cell behavior (e.g., change of substrate properties, nanopattern design, scaffold degradation rate) (Atala, 2009; Martino et al., 2009).

3. Stem cells for different neurological disorders

The use of stem cells in neurological diseases is much more complex than in other systems. Many challenges are unique to the nervous system are as follows: (a) The need to integrate into a sophisticated array of interconnected cells that extend over great distances; (b) The absence of developmental cues in adults that guided the establishment of neural networks during development, thus making regeneration more difficult; (c) The possibility in progressive or recurrent neurologic diseases that the transplanted cells may be attacked and injured (Potter et al, 2007)

3.1 Parkinson's disease

A widespread loss of dopamine neurons (DA) in the substantia nigra pars compacta and their terminals in the striatum occurs in Parkinson's disease (PD) (Kish et al., 1988; Agid, 1991). Many issues for the dopaminergic depletion associated with the disease have been suggested, including programmed cell death, viral infection, and environmental toxins. As an efficient treatment for PD, patients have been given L-dihydroxyphenyl alanine (L-DOPA), a precursor of dopamine, but long-term administration of L-DOPA consequently produces side effects (Lang and Lozano, 1998 a, b). So, human fetal ventral mesencephalic tissues were transplanted of into the striatum of PD patients as a successful therapy for patients with advanced disease, since the late 1980s (Lindval et al., 1990; Olanow et al., 1996; Kordower et al, 1997; Dunnett and Bjorklund, 1999). This fetal tissue transplantation has serious problems associated with ethical and religious questions and logistics of acquiring fetal tissues (Hagell et al., 1999). To avoid these difficulties, utilization of neurons with a DA phenotype generated from ESCs, MSCs, or NSCs could serve as a practical and effective alternative for fetal brain tissues for transplantation. DA neurons were generated from mouse ESCs or mouse NSCs (Lee et al., 2000; Hagell & Brundin, 2002; J.H. Kim et al., 2002; T.E. Kim et al., 2003). Neural cells with a DA phenotype have been generated from monkey ESCs by coculturing with mouse bone marrow stromal cells (Takagi et al., 2005) and also from human NSCs derived from fetal brain (Redmond et al., 2007), and improvement was seen in MPTP lesioned monkeys following intrastriatal transplantation of these cells (Takagi et al., 2005; Redmond et al., 2007).

NSCs which were transplanted in the brain attenuate anatomic or functional deficits associated with injury or disease in the CNS via cell replacement, release of specific neurotransmitters, and production of neurotrophic factors that protect injured neurons and promote neuronal growth. Recently, continuously dividing immortalized cell lines of human NSC have been generated from fetal human brain cell culture via a retroviral vector (Kim, 2004; Lee et al., 2007; Kim et al., 2008), and one of the immortalized NSC lines, induced functional improvement in a rat model of PD following transplantation into the striatum (Yasuhara et al., 2006).

3.2 Alzheimer's disease

Alzheimer disease is characterized by degeneration and loss of neurons and synapses throughout the brain, particularly in the basal forebrain, amygdala, hippocampus, and cortical area. Cognitive function of patients progressively declines, and patients become demented and die prematurely (Coyle et al., 1983). No successful treatment is currently available except for acetylcholinesterase inhibitors, which augment cholinergic function, but this is not curative and is only a temporary measure.

A recent study has reported that Nerve growth factor (NGF) prevents neuronal death and improves memeory in animal models of aging, excitotoxicity, and amyloid toxicity (Tuszynski, 2002), suggesting that NGF may be used for treating neuronal degeneration and cell death in AD. However, convey of NGF into the brain is not possible via peripheral administration; because of its size and polarity, NGF does not cross the blood-brain barrier. To avoid this difficulty, gene therapy approach could be adopted. By utilizing an ex vivo gene therapy approach (genetically modify cells), NGF can be given directly to the brain and diffuse for a distance of 2-5 mm (Tuszynski et al., 1990).

Ex vivo NGF gene delivery was clinically tried in eight mild-AD patients, implanting autologous fibroblasts genetically modified to express human NGF into the forebrain. After follow-up of 22 months in six subjects, long-term adverse effects were not found. PET scans showed significant increases in cortical fluorodeoxyglucose after treatment (Tuszynski et al., 2005). Genetically modified stem cells could be used in place of fibroblasts to carry new genes for delivery of NGF to prevent degeneration of basal forebrain cholinergic neurons (Flax et al., 1998; Kim, 2004; Lee et al., 2007, Kang et al., 1993).

A blood stem cell growth factor (granulocyte-colony stimulating factor (G-CSF) is routinely administered to cancer patients whose blood stem cells and white blood cells have been depleted following chemotherapy or radiation. the bone marrow was stimulated by G-CSF to produce more white blood cells needed to fight infection; and is also used to enhance the stem cells circulating in the blood of donors before the cells are harvested for bone marrow transplants. (Pavlović et al., 2009; Lee et al., 2010)

Advanced clinical trials are now investigating the effectiveness of G-CSF to treat stroke, and the compound was safe and well tolerated in early clinical studies of ischemic stroke patients. This growth factor could potentially provide a powerful new therapy for Alzheimer's disease that may actually reverse disease, not just alleviate symptoms like currently available drugs(Ramos & Raj., 2009). The researchers showed that injections under the skin of filgrastim (Neupogen®) - one of three commercially available G-CSF compounds - mobilized blood stem cells in the bone marrow and neural stem cells within the brain and both of these actions led to improved memory and learning behavior in the Alzheimer's mice on the basis of reactive microglia derived from stem cells that are destroying deposits of amyloid plaques in brain tissue. So far, a human growth factor that stimulates blood stem cells to proliferate in the bone marrow reverses memory impairment in mice genetically altered to develop Alzheimer's disease (Ramos & Raj., 2009). The G-CSF significantly reduced levels of the brain-clogging protein beta amyloid deposited in excess in the brains of the Alzheimer's mice increased the production of new neurons and promoted nerve cell connections.

Researchers at the Kyungpook National University in Daegu, South Korea, tested the potential therapeutic effects of bone marrow-derived MSCs in mice. The 2009 study was able to successfully confirm that BM-MSC transplantation accelerated the removal of amyloid-β plaques from the brains of acute AD mice, this study also showed that the BM-MSCs can induce normally-quiescent microglia to clear out amyloid-β build-up (Lee et al., 2009). The three professors who performed the 2009 study in Korea reported in early 2010 that the decreased amyloid-β deposition was directly related to microglial activation (Lee et al., 2010). They also showed that the microglia ameliorate memory deficiencies in the AD mice. This finding was supported by a version of the Morris water maze test known as the hidden platform. Subjects which received PBS injections deteriorated as expected in their ability to learn and memorize the maze, while those treated with BM-MSCs displayed navigational patterns that resembled control subjects. Moreover, they reported that the MB-MSC transplantation was able to reduce tau hyperphosphorylation (Lee et al., 2010).

3.3 Huntington's disease

A neurodegenerative disorder (Huntington's disease (HD) is characterized by involuntary choreiformic movements, cognitive impairment, and emotional disturbances (Greenmayre and Shoulson, 1994; Harper, 1996). In spite of identification of the HD gene and associated protein, the mechanisms involved in the pathogenesis of HD remain largely unknown.

A recent research has documented improvements in motor and cognition performance in HD patients following fetal cell transplantation (Bachaud-Le´vi et al., 2000). This research

follows previous reports on experimental HD animals that positive effects of fetal striatal cell transplantation ameliorate neuronal dysfunction (Nakao and Itakura, 2000) and that striatal graft tissue could integrate and survive within the progressively degenerated striatum in a transgenic HD mouse model (Dunnett et al., 1998; Freeman et al., 2000).
A restrictive factor in the transplantation of fetal striatal cells is the difficulty in supplying sufficient amounts of embryonic striatal tissue and the concomitant ethical issues associated with the use of human embryonic tissue. A perfect source of cell transplantation in HD would be NSCs, which could participate in normal CNS development and differentiate into regionally appropriate cell types in response to environmental factors. Prior studies have shown that NSCs isolated from embryonic or adult mammalian CNS can be propagated in vitro and subsequently implanted into the brain of animal models of human neurological disorders, including HD (Brustle and McKay, 1996; Flax et al., 1998; Gage, 2000; Temple, 2001; Gottlieb, 2002; Lindvall and Kokaia, 2006).
Genetically modified NSCs producing neurotrophic factors and transplantation of NSCs to replace degenerated neurons have been used to protect striatal neurons against excitotoxic insults (Bjorklund and Lindvall, 2000).
Recently, human NSCs were injected intravenously to counteract neural degeneration in HD model rats and demonstrated functional recovery in grafted animals (Lee et al., 2005, 2006).

3.4 Amyotrophic Lateral Sclerosis

3.4.1 Introduction

Amyotrophic lateral sclerosis (ALS) is a fatal neurodegenerative disorder characterized by progressive dysfunction and degeneration of motor neurons occur not only in the spinal cord (lower motor neurons) but also in the cerebral cortex and brainstem (upper motor neurons). Muscle weakness progresses rapidly and death occurs within a few years. There are currently no effective treatments for ALS (Wolfson et al,2009).
The expectation for ALS patients is that stem cell transplantation will replace motor neurons, leading to the recovery of neuromuscular functions. Unfortunately, the expectation that stem cells will play such a regenerative role in patients with ALS is unrealistic because of the complexity of the task, a more realistic expectation for stem cells is that they play a supportive role in maintaining the viability of or extending the function of surviving motor neurons (Silani et al,2002).
Inducing stem cells to differentiate into supporting cells, glia, or interneurons that might produce factors that would support motor neurons, or perhaps the stem cells themselves might produce such factors (Svendsen et al,2004) .

3.4.2 Stem cells for treating ALS: current developments

3.4.2.1-Neural stem cells (NSCs) is a challenging therapeutic strategy for treatment of ALS. To provide insight into the potential of the intravenous delivery of NSCs, Mitrecić, and his colleagues, (2010) evaluated the delivery of NSCs marked with green fluorescent protein to the central nervous system (CNS) via intravenous in an ALS model. Highly efficient cell delivery to the CNS was found in symptomatic ALS (up to 13%), moderate in presymptomatic ALS (up to 6%), and was the lowest in wild animals (up to 0.3%). The study provides basic facts about the process occurring after NSCs leave the blood stream and enter the nervous tissue affected by inflammation or degeneration, which should help facilitate the planning of future bench-to-bedside translational projects.

3.4.2.2-Haemopiotic stem cell : Preliminary trials with autologous hematopoietic stem cells have been reported in humans. In one, peripheral blood-purified CD34+ cells were injected intrathecally into 3 patients with ALS (Appel et al,2008, Janson et al 2001)) None reported side effects after 6-12 months, but no clinical efficacy was reported. In another, Deda and collaborators(2009) reported follow-up results one year after stem cell transplantation. The post-operative status of nine patients indicate a significant improvement in comparison to the pre-operative status, as confirmed by electroneuromyography.

3.4.2.3-Mesenchymal bone marrow stem cell: are currently used as an alternative therapy in amyotrophic lateral sclerosis (ALS) patients. Cho, et al 2010 isolated BM-SCs from 11 ALS patients and characterized their potential secretory capacity of neurotrophic factors and they noticed that ALS-SCs have diminished capacity as trophic mediators and may have reduced beneficial effects in cell therapy& suggested that MSCs at early passages are more suitable for stem cell therapy in ALS patients because of their stability and more potent anti-inflammatory and neuroprotective properties (Choi, et al,2010).

Marzzini 2003 ,study the effect intraspinal transplantation of MSCs on 7 patients , Minor postoperative adverse events were transient, but muscle strength continued to decline. Three months later, however, the investigators reported a trend toward slowing of the decline in the proximal muscle groups of the lower limb in 4 patients and a mild increase in strength in 2 patients. The absence of placebo controls and longer follow-up preclude any inferences of efficacy from this study. *Karussis* and his collauge 2010, prove in phase I trial using intrathecal with or without intravenous MSCs in patients with ALS, Transplantation is clinically feasible and relatively safe procedure with stability of the mean Amyotrophic Lateral Sclerosis Functional Rating Scale [ALSFRS] score.

3.4.2.4- Induced pluripotent stem cell: The generation of pluripotent stem cells from an individual patient would enable the large-scale production of the cell types affected by the patient's disease, using induced pluripotent stem (iPS) cells overcomes the ethical problems of using embryos. (John et al,2008) can make reprogramming of human fibroblasts to induced pluripotent stem (iPS) cells from skin biopsy of an 82-year-old woman diagnosed with a familial form of amyotrophic lateral sclerosis (ALS). These patient-specific iPS cells possess properties of embryonic stem cells, they were successfully directed to differentiate into motor neurons, the cell type destroyed in ALS.

These preliminary hope needs to be tempered with caution because of the early stages of stem cell research in general, and in ALS in particular.

3.5 Spinal Muscular Atrophy

Spinal muscular atrophies (SMA) present a heterogenic group of hereditary neurological diseases and one of the types of motor neuron disease. It is a common "rare disorder", affecting approximately 1 in 6000 babies born, to date no cure exists. The primary approaches to treating or curing SMA have now focused on two strategy options, the first, genetic therapy - manipulating the genetic material responsible for producing SMA. While the second is cellular replacement therapy - replacing dead or dying motor neurons.

Stem cell strategies are presently under investigation, although significant preclinical work and methodological advances remain ahead before these approaches can become clinically relevant (Douglas et al; 2010).

The goal of transplantation is providing a pool of cells that are able to support endogenous neurons through delivery of neuroprotective factors and providing a replacement population for lost motor neurons. In Vitro Stem-cell-derived motor neurons have been shown to grow axons and successfully form neuromuscular junctions (Gao *et al;*2007, 2005; Wichterle *et al;*2002), and stem cell transplants have lead to growth of axons and some recovery in paralyzed rats (Deshpande *et al;*2006; Harper *et al;* 2004). Induced pluripotent stem cells have been derived from patients with SMA and used to generate motor neurons that show selective deficits compared with wild-type motor neurons in culture (Ebert *et al;*2009).

Spinal-cord-derived neural stem cells have been successfully transplanted into the mouse model of SMA, with a modest improvement of the clinical phenotype and generation of a viable population of motor neurons (Corti *et al;*2008).

A subsequent study using pluripotent stem cells demonstrated similar successful stem cell engraftment and differentiation with improved survival and functional improvement in treated *SMA Model* compared with controls, demonstrating the therapeutic potential of this approach (Corti *et al;* 2010).

Although these reports suggest the possibility of stem cell therapy, several challenges must be addressed before the successful implementation of stem cell therapy can be fully realized. For this strategy to be applicable, large numbers of stem cells need to be generated, to successfully populate the nervous system, properly differentiate into motor neurons and, critically, must successfully and correctly extend axons to and synapse upon muscle targets. It still remains unclear when this therapeutic strategy will become a practical approach in the treatment for SMA in the human population.

3.6 Brain tumor

Despite extensive surgical excision and radiotherapy and chemotherapy; malignant brain tumors such as glioblastoma multiforme remain virtually untreatable and lethal (Black and Loeffler, 1997). The opposition to treatment is associated with their exceptional migratory nature and ability to insinuate themselves seamlessly and extensively into normal brain tissue, often migrating great distances from the primary tumor masses(Dunn and Black, 2003; Sanai et al., 2005). Medulloblastoma is the most common among childhood brain tumors and is incurable. Available treatments including radical surgical resection followed by radiation and chemotherapy have substantially improved the survival rate in this disease; however, it remains incurable in about one-third of patients (Packer et al., 1999). As well, in the case of recurrence, frequently associated with tumor dissemination and the main cause of death, therapeutic options are rarely available (Patrice et al., 1995; Graham et al., 1997).

The capability of human NSCs as an effective delivery system to target and disseminate therapeutic agents to medulloblastomas was demonstrated for the first time (Kim et al., 2006). One of the causes for the recurrence of medulloblastoma in children after standard treatment is the inherent tendency of tumor cells to metastatize through cerebrospinal fluid, leading to leptomeningeal dissemination. Throughout the entire spinal cord, human NSC F3.CD cells were found to distribute diffusely to metastatic medulloblastoma cells after injection in the cisterna magna, and the CD gene in NSCs functioned effectively and killed tumor cells (Shimato et al., 2007).

3.7 Temporal lobe epilepsy

Current studies have shown that transplanted neurons can restore neurogenesis (Kuruba et al., 2009), and GABAergic neurons can reduce seizures (Alvarez-Dolado et al., 2006).

Moreover, one study has shown that a specific type of "stem cell," transplanted into the dentate gyrus often matures into normal GCs, which could be used in a restorative manner if neurogenesis declines (Carpentino et al., 2008). Remarkably, these stem cells can lead to abnormal growths if a normal animal is used, but in an animal that has had seizures, the stem cells become GCs and tumors do not appear to develop (Carpentino et al., 2008). Infusion of neuropeptide Y may be a particularly effective strategy after such stem cell infusion, because it stimulates precursor division (Howell et al., 2003; Scharfman and Gray, 2006; 2007) and reduces seizures (Noe et al., 2006).

There are four distinct stem cell-based approaches for treating TLE. The first approach involves development of methods for inhibiting increased proliferation of hippocampal NSCs during the first few weeks following the SE. Addressing this issue is important in light of studies suggesting that epileptic seizures such as SE not only increase dentate neurogenesis but also lead to abnormal migration of newly born granule cells into the dentate, where they exhibit spontaneous epileptiform bursts and may contribute to the development of chronic epilepsy (Dashtipour et al., 2001; Parent et al., 2006).

The second approach focuses on developing strategies that activate endogenous NSCs in the chronically epileptic hippocampus to produce a large number of new neurons including GABA-ergic interneurons. This approach has significance because studies in both TLE and animals models of TLE suggest that chronic TLE is associated with dramatically declined production of new neurons in the adult DG. Decreased neurogenesis during chronic epilepsy may contribute to the persistence of seizures possibly due to decreased addition of new GABA-ergic interneurons.

The third strategy comprises rigorous analyses of the efficacy of grafts of NSCs placed into the hippocampus after the onset of chronic epilepsy for suppressing seizures and learning and memory deficits. This is because the initial results of stem cell grafting studies in TLE models reported (Shetty & Hattiangady.,2006, Acharya et al.,2007) are promising in terms of their short term survival and their effectiveness for reducing the frequency of seizures and findings of delivery of anticonvulsant compounds such as NPY, glial-derived neurotrophic factor, and adenosine is efficacious for reducing seizures in animal models of TLE (Noe et al., 2007; Li et al., 2007).

A fourth approach would be a combination therapy comprising NSC cell transplants and cell or recombinant viral vector-based delivery of anticonvulsant compounds into the hippocampus during chronic epilepsy. This plan may be very efficient, as seizure control would likely be mediated by both GABA-ergic interneurons derived from NSC transplants and anti-convulsant compounds released by genetically engineered cells. (Shetty& Hattiangady., 2007)

3.8 Lysosomal storage diseases

Most affected babies by lysosomal storage diseases show a diffuse CNS involvement (Meikle et al., 1999). At present, no effective treatment is available for most of the lysosomal diseases, because the blood–brain barrier bars entry of enzyme preparations into the brain (Sly and Vogler, 2002). However, therapeutic levels of enzymes could be achieved in the brain of animal models of lysosomal diseases by direct inoculation of genetically engineered mouse (Snyder et al., 1995), fibroblasts (Taylor and Wolfe, 1997), or amniotic epithelial cells (Kosuga et al., 2001). In consideration of their widespread migratory ability, normal or genetically modified stem cells would allow widespread delivery of missing enzymes all over the brain. In a mouse model of mucopolysaccharidosis VII (MPS VII), a lysosomal disease caused by a genetic defect in the activity of b-glucuronidase (b-gluc), genetically engineered mouse overexpressing b-gluc were transplanted into the cerebral ventricle and resulted in reduction of lysosomal storage in the mouse brain (Snyder et al., 1995). Similarly, the transplantation of (b-gluc) overexpressing human NSCs into MPS VII mice and human NSCs migrated extensively all over the brain, produced high levels of b-gluc enzyme, and cleared lysosomal storage in the neuronal cytoplasm (Meng et al., 2003). In an earlier study, immortalized human NSCs were transplanted in a mouse model of Tay-Sachs disease in which abnormal lysosomal storage of GM2 ganglioside is found in the brain, resulting from total absence of hexosaminidase enzyme activity. After transplantation of human NSCs, there was a clearance of storage in neuronal cytoplasm in Tay-Sachs model mice (Flax et al., 1998). The results indicate that NSCs could serve as an excellent gene transfer vehicle for the treatment of diffuse CNS pathology in human lysosomal storage diseases, including Krabbe disease, Gaucher's disease, metachromatic leukodystrophy, and adrenoleucodystrophy.

3.9Multiple sclerosis

3.9.1 Introduction

MS is a chronic, demyelinating disease of the brain and spinal cord ,MS is heterogeneous disease, and so the degree of the disease can range from fairly benign to extremely debilitating and the stages of disease can range from only relapses to progressive (Weiner ;2009)

Unfortunately, the available treatments (Immunomodulatory and immunosuppressive)are not curative, they can reduce CNS inflammation and may delay progression, but control of disease is unsatisfactory in many patients ,a logical treatment approach to enhance neuroprotective mechanisms and to induce neuroregeneration through stem cell transplantation, stem cell therapy for MS can categorize to immune reconstruction or tissue reconstruction (remylination),two distinct approaches can be considered to promote myelin repair, in one the endogenous myelin repair processes are stimulated through the delivery of growth factors, and in the second the repair process are augmented through the delivery of exogenous cells with myelination potential. Also, the effective treatment of MS requires modulation of the immune system, since demyelination is associated with specific immunological activation (Karussis & Kassis; 2007)

Several types of stem cells having the capacity for promoting myelin repair, as well as modulating the immune response, are potential candidates for MS therapy(Emily et al;2007).

3.9.2 Stem Cells for treating MS: current developments

3.9.2.1 Embryonic stem cells (ESCs)

When transplanted in rodent models of induced demyelination, embryonic stem cells were shown able to differentiate into glial cells and re-ensheath demyelinated axons in vivo (Brustle etal,1999 ; McDonald et al;1999). Researchers have underlined that ESCs could be a "double-edged sword" since they may cause the formation of a non-homologous implant and teratomas within the organ of transplantation (Brustle etal;1999, Deacon etal,1998;,Yanai et al;1995).

3.9.2.2 Adult stem cells

can be detected in both fetuses, and adults. They can be harvested from different tissues: adipose tissue (Gimble& Guilak ;2003), bone-marrow (hematopoietic-HSC (Wognum et al;2003), mesenchymal-MSC (Pittenger,1999 ; Prockop etal;1997), CNS (neural stem cells-NSC, neurospheres) , olfactory bulb (Roisen et al;2001) and others. Several studies suggested the neuroregenerative, the immunomodulatory potential of these adult stem cells (puissant etal;2005 ,Yu et al;2006).

3.9.2.2.1 Adult neural stem cells (NSCs)

In a study by Pluchino et al; 2003 it was shown that adult neural stem cells cultured and injected into EAE-mice—intravenously (iv) or intracerebroventricularly (icv), could migrate into the demyelinating CNS area and differentiate into mature brain cells. It was apparent in this study, that oligodendrocyte progenitors were especially increased, in this model. Clinically, EAE symptoms were strongly down-regulated in the transplanted animals. Despite these promising results, NSC are still not considered the perfect stem cell population for cell replacement therapy and are associated with significant drawbacks. The difficulty is in culturing neurospheres from regions of the adult brain that do not normally undergo self-renewal (Shihabuddin etal;2000) neurosphere-derived cells do not necessarily behave as stem cells when transplanted back into the brain and thus form a focus for immune rejection.

3.9.2.2.2 Hematopoietic stem cells

Autologous Hematopoietic stem cells (HSCT) was largely preferred to allogeneic transplantation because of the lower risk of severe toxicity (van Gelder & van Bekkum; 1995). Briefly, patients with autoimmune diseases can be considered for HSCT if: (i) their disease is severe enough to cause an increased risk of mortality or advanced and irreversible disability; (ii) the disease has been unresponsive to conventional treatments, so, Fassas and his collaeuge; 2003 recommended practice points to AHSCT for MS patients
(a) AHSCT seems to be the best anti-inflammatory treatment as evidenced in MRI scans. Its clinical value remains to be validated in controlled trials.(b) MS types characterized by neurodegenerative pathogenic components are unlikely to benefit from ASCT.(c) Good candidates are young patients with rapidly evolving RR-MS or "malignant" MS. Also, patients with SP-MS having EDSS scores below 6.5, evidence of inflammation in the CNS, and clinical worsening during the last year.(d)Intense conditioning or extensive T depletion increase the morbidity & mortality risk

3.9.2.2.3 Mesenchymal stem cells (MSCs)

initially isolated from bone marrow but are now shown to reside in almost every type of connective tissue (Da silva etal; 2006). The use of bone marrow derived MSC provides several advantages over conventional neuronal, embryonic and hematopoietic stem cells: (1) they can be obtained from the adult bone marrow; (2) they can be easily cultured and expanded in large numbers; (3) they can be injected autologously without the need of immunosuppressive means to risk for induction of malignancies, as compared to other types of stem cells prevent rejection; and (4) they are less prone to genetic abnormalities during multiple in vitro passages these cells have been shown to have diff eren tiation capacities as well as paracrine eff ects via the secretion of growth factors, cytokines, antifi brotic or angiogenic mediators (Djouad et al ; 2009). A large body of studies also indicates that MSCs possess an immunosuppressive function both *in vitro* and *in vivo* (Karussis & Kassis; 2008). How mesenchymal stem cells affect functional recovery in the damaged adult CNS is not well understood. Fig (3) represent MSC Potential for Therapeutic Applications in autoimmune disorders

They have therefore been tested in the EAE (Ren et al ;2009) MSCs were shown to decrease the clinical signs associated with demyelination when injected before or at the onset of the disease, thus demonstrating the therapeutic efficacy of MSCs (Zappia eta;l2005). This effect was associated with immune suppression of effector T cells leading to IL-2 reversible T- cell anergy. Subsequently, it was reported that MSCs inhibited T-cell activation with reduced IL-17 and TNFα levels via the secretion of CCL2 by MSCs (Rafie et al;2009)The preclinical studies, .(Asano et al; 2006, Draper etal,2004, Lee et al;2001).

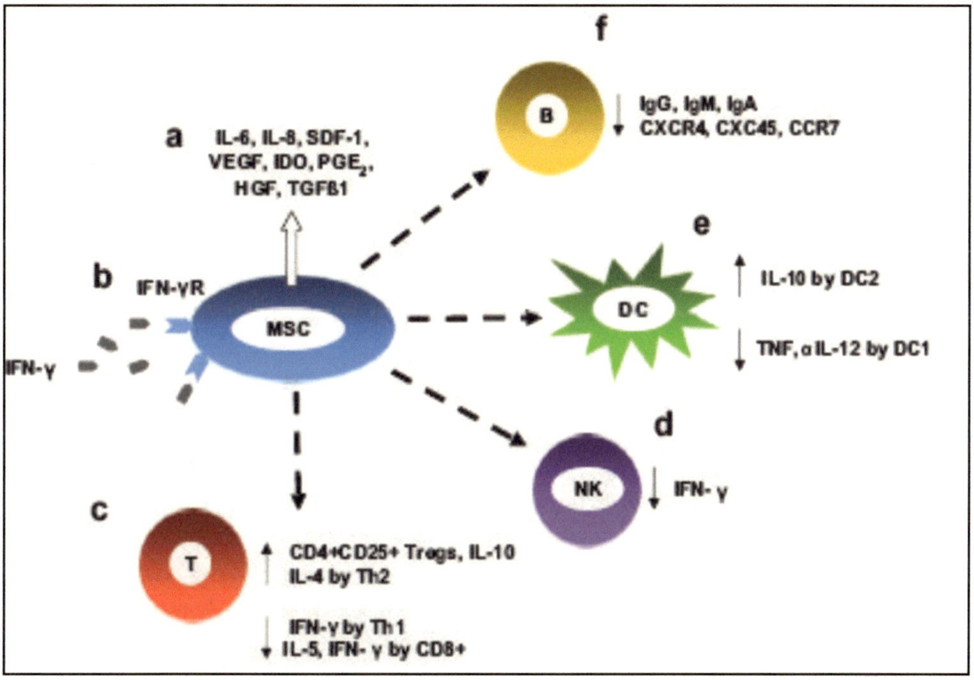

Fig. 3. The immunomodulatory effect of MSCs. (a) A number of soluble factors secreted by MSCs can suppress the activity of inflammatory immune cells. (b) This suppressive activity can be enhanced by the presence of pro-inflammatory cytokines secreted by immune cells such as IFN-g. (c) The proportion of regulatory T cells and levels of IL-10 production are increased by MSCs. MSCs induce a bias towards a Th2 response and upregulate the production of IL-4. The production of IFN-g by Th1 cells is reduced. MSCs suppress CD8 T cells prior to activation and reduce production of IL-5 and IFN-g. (d) MSCs can reduce the production of IFN-g by NK cells. (e) DC differentiation, maturation and antigen presentation is inhibited by MSCs. The production of pro-inflammatory cytokines is reduced while IL-10 production is upregulated. (f) At high concentrations MSCs inhibit proliferationof B-cells, reduce the levels of IgM, IgG and IgA and downregulate the expression of several cytokine receptors (Payne et al;2008)

Together with the cumulative data from ongoing clinical trials with MSCs in various clinical conditions (reviewed by Giordano et al2007), provided the scientific basis for many (*Karussis* et al ;2010,Mohyeddin Bonab et al ;2007, Riordan et al;2009). phase 1/2 pilot clinical trial using combined intrathecal and / or intravenous injection of bone marrow- derived autologous MSCs which preliminary prove the feasibility and safety of this type of cell therapy,in the form of early clinical stabilization or improvement in some of the patients which could be related to these immunomodulating effects and he possibility of neuroprotection and neuroregeneration through transdifferentiation of MSCs into cells of the neuronal or glial lineage, although Further controlled trials are warranted to evaluate the long term safety and the potential clinical efficacy of MSC transplantation.The current data indicate that MSCs represent a promising alternative strategy in the treatment of MS however, Many questions remain to be addressed, about a better understanding of the underlying mechanisms of immuno suppression as well as satisfying safety concerns as regards the *in vivo* survival, formation of ectopic tissue and malignant transformation.

3.10 Stroke

3.10.1 Introduction

Stoke is the third leading cause of death in the USA and can be caused by the occlusion of small vessels in the brain that resulting in subsequent neuronal death. This will trigger a cascade of events including a wide spread inflammatory response. Current therapies for ischemic insults include thrombolysis through treatment with tissue plasminogen activator (tPA). (Bliss etal;2010).Stem cell transplantation offers an exciting new therapeutic avenue for stroke not only to prevent damage, which has been the focus of conventional therapeutic strategies, but also to actually repair the injured brain in ischemic(Bliss et al., 2007) and hemorrhagic stroke (Andres et al., 2008).

3.10.2 Theraputic time window for stem cell therapy in stroke patients

The majority of pre-clinical studies transplanted the stem cells within the first 3 days after stroke and they have mostly used bone marrow- or blood-derived cells (Bliss et al., 2007; Locatelli et al., 2009). This time window is greater than the 3- to 6-h window required for t-PA therapy, the only treatment for stroke that currently exists. Cell enhanced recovery has also been reported with sub-acute (1 week post-stroke) and chronic (3 weeks post-stroke) delivery of many cell types including neural cells (Borlongan et al., 1998; Chen et al., 2001a,b; Daadi et al., 2008; Shen et al., 2007; Zhao et al., 2002). Comparison of the results to identify an optimum time for transplantation is difficult as the studies used different models of stroke, cell types, methods of cell delivery, and behavioral tests to assess efficacy. The optimum time for transplantation may be dependent on (a) the cell type used , (b) their mechanism of action. If the treatment strategy is focused on neuroprotective mechanisms, the acute delivery will be critical, however, if the cells transplanted were meant to enhance endogenous repair mechanisms(e.g. plasticity and angiogenesis), then sub-acute delivery would be essential as these events are more prevalent in the first few weeks after ischemia (Carmichael, 2006; Hayashi et al., 2003), (c) route of administration such as intravascular transplantation may require early administration as the cells use inflammatory signals to home to the injured brain (Guzman et al., 2008; Park et al., 2009; Pluchino et al., 2005) and (d) location of infarction , the majority of pre-clinical studies show cell-enhanced recovery after striatal lesions (Bliss et al., 2007; Guzman et al., 2008; Hicks and Jolkkonen, 2009; Locatelli et al., 2009) although cell-induced improvements with cortical lesions are also reported (Hicks et al., 2009; Shyu et al., 2006; Zhao et al.,2002). However, not all studies find that cell therapy is effective (Hicks et al., 2008).As shown in table (1) that the timing of transplantation affected the outcome of these trials is not clear, but they at least demonstrate that delivery of cells at different times is feasible.

3.10.3 Ishemic versus haemorhagic stroke

Ischemic and hemorrhagic strokes differ in their pathophysiology and mechanism of recovery (Xi et al., 2006). For example, there is no salvageable penumbra with intracerebral hemorrhage (ICH) unlike ischemic stroke (Qureshi et al., 1999), and patients with ICH do not suffer from reperfusion injury with its burst of free radical production (Kleinig and Vink, 2009). Toxic blood breakdown products like thrombin, hemoglobin, and iron additionally contribute to neuronal damage after ICH (Hua et al., 2007; Wang et al., 2002). Therefore, it is plausible that hemorrhagic and ischemic stroke may respond differently to cell therapy and should be tested separately in clinical trials (Andres et al., 2008; Wechsler et al., 2009).

Clinical identifier and clinical phase	Cell type	Estimated enrollment	Time of delivery*	Route of delivery	Country
NCT0047357 Phase I	Autologous bone marrow	10	3h-90 days	Intra-arterial	Brazil
NCT00859014 Phase I	Autologous mononuclear bone marrow	10	24 h-72 h	Intravenous	USA
Nct00525197 Phase I/II	Autologous CD34 + bone marrow	10	7 days	Intra-arterial	UK
NCT00950521 Phase II	Autologous CD34 + peripheral blood	30	6-60 months	Intracerebral	China
NCT00875654 Phase II	Autologous MSCs	30	< 6weeks	Intravenous	France

Clinical identifier fromclinical trial gov; time of delivery after stroke onset
Table 1. Current clinical cell transplantation trial for stroke

3.10.4 Route of administration

Many of the studies using systemic delivered cells find significant functional recovery with very few (Guzman et al., 2008; Hicks and Jolkkonen, 2009; Li et al., 2002; Vendrame et al., 2004) or sometimes no cells (Borlongan et al., 2004) entering the brain. Modo et al. (2002) found equal functional recovery when cells were grafted in the ipsi- or contralesional hemispheres, the optimum route of human stem cell delivery has not been determined but will ultimately depend on the timing of delivery, the cell type used, and their mechanism of action.human bone marrow cells (HBMC), human umbilical cord blood cells (HUCBC), peripheral blood progenitor cells, and adipose tissue mesenchymal progenitor cells have all been reported to enhance recovery after stroke with intracerebral or intravascular delivery, and with acute (1 day), subacute (1 week), or chronic (1 month) delivery after stroke (Bliss et al., 2007, Guzman et al., 2008,Hicks &Jolkkonen;2009,Shen et al., 2007).The only clinical use of intravenous injection of ex vivo-cultured autologous MSCs for the treatment of stroke patients was reported by Bango, et,al, 2005. showed improvement in Neurological outcomes as determined by the Barthel index and modified Rankin score, Although this trial described the success and safety of intravenous injection of ex vivo-cultured autologous MSCs, but only five patients were treated with MSCs and therefore the results should be interpreted with caution. Furthermore, evidence that the intravenously injected MSCs were biologically active is indirect and based on a presumed improved functional recovery (Dekeyser; 2005)

3.10.5 Stem cells types for treating stroke: Current developments

A variety of human cell types have been tested in experimental stroke (Bliss et al., 2007): (1) neural stem/progenitor cells,(2) immortalized cell lines ,(3) hematopoietic/endothelial progenitors and stromal cells isolated from bone marrow, umbilical cord blood, peripheral blood, or adipose tissue. To become a useful therapeutic option, cells must show efficacy, have a large expansion capacity in culture to meet the eventual clinical demand.Cell transplantation has shown much promise in experimental models of stroke with a diverse array of cell types which reported to enhance functional recovery after ischemic (Bliss et al., 2007) and hemorrhagic stroke (Andres et al., 2008). Such results led to early Phase I and II clinical trials using a cell line of immature neurons (hNT) derived from a human teratocarcinoma, fetal porcine cells, or autologous mesenchymal stem cells (MSCs). These studies focused on the safety and feasibility of cell transplantation therapy. No cell-related adverse effects were reported with the hNT (Kondziolka et al., 2005, 2000) and MSC transplants (Bang et al., 2005). However, 2 out of the 5 patients receiving the porcine cells developed either seizures or aggravation of motor deficits (Savitz et al., 2005); the value of the cell therapy to these adverse effects is unclear.

3.10.5 Potential mechanisms of transplanted cell-mediated stroke recovery

3.10.5.1 Induction of neurogenesis and synapses formation

Human NPCs form synapses with host circuits (Ishibashi et al., 2004; Daadi et al., 2009a), However, only very few synapses are seen, and recovery occurred too early to be attributable to newly formed neuronal connections (Englund et al., 2002, Song et al., 2002).

3.10.5.2 Neuroprotective mechanism

Through secretion of trophic factors such as vascular endothelial growth factor (VEGF), fibroblast growth factor (FGF), glial cell-derived neurotrophic factor (GDNF), and brain-derived neurotrophic factor (BDNF) that are likely to contribute for recovery (Kurozumi et al., 2005; Llado et al., 2004). Li and Chopp, 2009 suggested that MSCs regulate the levels of cell death through release of trophic factors as well as altering the gap junction coupling between astrocytes, this allows these cells to respond more effectively to control damage

3.10.5.3 Immunomodulation and anti-inflammatory mechanism

Intravenous injection of HUCB or direct injection of human MSCs into the hippocampus after global ischemia lead to down regulate many inflammatory and immune response genes and shifted the balance from a pro- to anti-inflammatory response (Ohtaki et al., 2008).

3.10.5.4 Induction of Angiogenesis

Increased vascularization in the penumbra within a few days after stroke correlates with improved neurological recovery in stroke patients (Krupinski et al., 1993; Senior, 2001) transplanted cell-induced blood vessel formation has been reported with BMSCs, NPCs,

HUCBCs and cells from human peripheral blood (Chen et al., 2003; Horie et al., 2008; Shen et al., 2006; Shyu et al., 2006; Taguchi et al., 2004).

3.10.5.5 Activation of endogenous restorative processes

Induction of host brain plasticity and increase in endogenous brain structural plasticity and motor remapping after ischemia is postulated to underlie the spontaneous recovery seen after stroke (Benowitz and Carmichael, 2010; Carmichael, 2006;2008), and cell transplantation may enhance these process. HUCBCs increased sprouting of nerve fibers from the contralateral to the ischemic hemisphere (Xiao et al., 2005), a similar phenomenon recorded with fetal-derived NPCs (Daadi et al., 2009b, Horie et al., 2009). these restorative process not well understood but may signify a natural repair mechanism of the brain that could be enhanced by transplanted cells. MSC-treated rats demonstrated elevated oligodendrocyte precursors, which increased in concert with enhanced white matter areas (Taguchi et al., 2004,Li et al., 2005, 2006; Shen et al., 2006).and also Xin et al., 2010, suggest that MSCs may also locally increase the levels of tPA in astrocytes around the stroke lesion and that this increases neuroprotection and enhances neurite outgrowth.

The pre-clinical evidence shows great promise for cell transplantation as a therapy for stroke. While we can be cautiously optimistic about the reality of such a therapy, many fundamental questions related to the optimal patient (including age, sex, etiology, anatomic location and size of infarct, and medical history), the most appropriate cell type, cell dose, the timing of surgery, the route and site of delivery, the need for immunosuppression, and mechanism of action remain to be answered.

3.11 Cerebral palsy

3.11.1 Introduction

Cerebral palsy is a group of brain diseases which produce chronic motor disability in children, that affect children from all countries and all ethnic backgrounds. The causes are quite varied and range from damage to the brain during pregnancy, labor or shortly following birth and due to the increased survival of very premature infants, the incidence of cerebral palsy may be increasing. While premature infants and term infants who have suffered neonatal hypoxic-ischaemic (HI) injury represent only a minority of the total cerebral palsy population,(Bartley & Carroll.,2003) Maximum repair and regeneration for cerebral palsy patients as listed by Filip et al. (2004) include:treatment of any infections, chemical toxicities, heavy metal poisoning, Oxygen therapies, Neuroprotective diet and therapies that include antioxidant and endogenous stem cell/stress reduction program that continues to promote repair and regeneration

The similar logistics of stem cell therapy in ischemic stroke also applies for the management of cerebral palsy,however, studies in this population are sparse (Mueller et al., 2005).

3.11.2 Stem cell therapy for cerebral palsy: Current development

3.11.2.1 Human neural stem cells (hNSCs)

hNSCs replaced lost cells in a newborn mouse model of brain damage. Mice received brain parenchymal or intraventricular injections of hNSCs derived from embryonic germ (EG) cells. The stem cells migrated away from the injection site they can survive and disseminate into the lesioned areas, differentiate into neuronal and glial cells and replace lost neurons (Mueller et al ;2005)

3.11.2.2 Human umbilical cord stem cells (hUCSC)

1.5 million CD34+/CD133 human umbilical cord stem cells had been injected subcutaneous in adipose tissue adjacent to the umbilicus. Patient with CP experienced clinically significant improvements in cognitive and motor skill function following this. What is intriguing is that many of these children began demonstrating benefit within the first day or so of receiving the injection. The children who demonstrated improvement were infants and toddlers (Singh and Roy, 2008).Clearly too little time elapsed to attribute these positive changes to hUCSC migration to the brain, engraftment and proliferation, however, these early onset clinically significant improvements become explicable when viewed as the end result of growth factor and neurotrophin activity. The hUCSC deposited in adipose tissue causes adipocytes to synthesize blood brain barrier disruptive TNF-alpha and NGF. This would be consistent with published laboratory and animal studies, and with the rapid improvements seen in the treated children (Singh and Roy, 2008, Payne, 2005).Medical College of Georgia researchers are conducting the first FDA-approved clinical trial to determine whether an infusion of stem cells from umbilical cord blood can improve the quality of life for children with cerebral palsy. The study will include 40 children age 2-12 whose parents have stored cord blood at the Cord Blood Registry in Tucson, Ariz (Medical College of Georgia, 2010)..

3.11.2.3.Mesenchymal Bone Marrow stem cells (MSCs)

Padma, 2005 use an intra-arterial infusion of autologous bone marrow stem cells to patients with static encephalopathy including cerebral palsy, it was found that this procedure was feasible, safe and caused improve in neurological functional outcome Chen et al; 2010,proved that MSC transplanted to animal model of Periventricular white matter injury (PVWMI) in preterm infants may have been neuroprotective and indirectly contributed to brain repair which proved by in vivo MRI demonstrated that labeled cells migrated away from the injection site toward lesioned areas in both hemispheres, confirmed by microscopy postmortem, but double-labeling studies found little evidence of differentiation into neural phenotypes. By expert opinion Carroll and Mays (2011), stem cells may be beneficial in acute injuries of the CNS the biology of stem cells is not well enough understood in chronic injuries or disorders such as cerebral palsy. More work is required at the basic level of stem cell biology, in the development of animal models, and finally in well-conceived clinical trials.

3.12 Spinal cord injury

3.12.1 Introduction

Spinal cord injuries result in long-term functional deficits as a result of the failure of severed adult CNS neurons to regrow long distances, connect to their original targets, and restore circuitry. Several factors are thought to contribute to the lack of regeneration of spinal cord axons. These include a reduction in the intrinsic growth capacity of adult CNS projection neurons, the presence of inhibitory cues derived from damaged CNS myelin, and the formation of a glial scar by local astrocytes in response to inflammatory stimuli (Fitch and Silver;2008).There is no cure, and the most common current treatment – high-dose methylprednisolone – is of questionable value (Lindvall & Kokaia;2006).Multiple approaches will be required to generate functional recovery. This hypothesis has recently received strong support from the use of combinatorial therapies directed at intrinsic and environmental regulators of regeneration Cell-based therapy is currently one of the promising approachs as many studies have shown improvement in sensory or motor function in the presence of various types of grafted stem cells or ex vivo pre-differentiated stem cells (Kadoya et al., 2009).

3.12.2 Rationales for therapeutic use of stem cells for SCI include (figure 4)

3.12.2.1 Replacement of damaged neurons and glial cells

One possible effect of cell therapy is "replacement," meaning that the grafted cells integrate into the host tissue and replace damaged or lost cells. Several studies have been performed using in vitro expanded neural stem/progenitor cells, which were then implanted into injured animal model for spinal cord. The cells survived and differentiated into neurons, astrocytes, and oligodendrocytes and had a positive effect on functional outcome (Ogawa et al;2002,Okada et al; 2005) Similarly, MSCs can also differentiate into neuron-like cells and glia which stained for the neural proteins (Azizi;1998, Brazelton ;2000, Prockop ;1997, Woodbury;2000, Okano ;2005 , Mezey; 2000, Hofstetter *et al*;2002)

3.12.2.2 Environmental change in the spinal cord that would encourage regeneration

Create a more favorable environment for limiting damage and promoting regeneration, via immunoregulation (Aggarwal & Pittenger, 2005; Noel et al., 2007), expression of growth factors and cytokines (Song et al., 2004), improved vascularization, providing a permissive growth substrate, and/or suppressing cavity formation (Hofstetter et al., 2002). Enhance remylination and increase the survival of oligodendrocytes Zhang et al; 2008.
3.12.2.3- cell fusion: The implanted adult stem cells may even fuse with the endogenous stem cells of the spinal cord. Some experiments have shown that MSCs have the ability to fuse with a variety of cells(Alvarez-Dolad;2003 ,Terada et al; 2002). Other studies have shown that cell fusion does not exist or if it does, it is specific to the liver (Newsome et al;2003). This concept should be tested within the framework of spinal cord injury in the future.However,the attendant risks of stem cell therapy for SCI—including tumor formation, or abnormal circuit formation leading to dysfunction—must be weighed against the potential benefits of this approach.

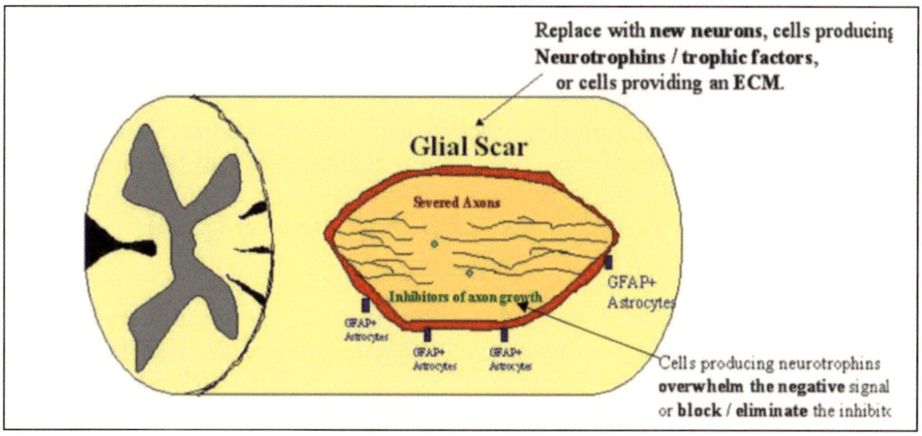

Fig. 4. Possible ways adult stem cells improve recovery in the injured spinal cord (*Sherri & Schultz*,2005)

3.12.3 Stem cell therapy of SCI: current development

Cell-based therapy is currently a promising approach as many studies have shown improvement in sensory or motor function in the presence of various types of stem cells embryonic stem cells, Blesch et al;2002 olfactory ensheathing glial cells, and Schwann cells (Lima;2006) The bone marrow also contains at least 2 discernable stem cell populations(Jiang et al;2002 , *Mazurier et al ;2003*)

3.12.3.1.Embryonic stem (ES) cells

Several studies have indicated that rodent ESC-derived neurons can survive, integrate and help restore function following transplantation into spinal cord injured rats (Finley et al., 1996; Lang et al., 2004), Deshpande et al., 2006). Human ESCs have been directed to differentiate into multipotent neural precursors (Carpenter et al., 2001; Reubinoff et al., 2001),and to high purity oligodendrocyte progenitors (Keirstead et al., 2005; Nistor et al., 2005). embryonic and fetal neural stem cells demonstrated stability, sustainability, and expandability in long-term culture systems in order for them to be considered as a possibility in human application. However, serious ethical dilemmas, also lack the ease of accessibility and practicality limit the routine clinical use (Reubinoff et al., 2001).

3.12.3.2 Adult stem cells

Unlike embryonal or fetal origin stem cells, using adult stem cells avoids ethical and moral problems as well as teratogenic and oncogenic risks, a variety of adult stem cells have been implanted in a rat model of spinal cord injury, ranging from olfactory ensheathing cells,and schwann cells *(Lima etal;2010)* cultured spinal cord stem cells, umbilical cord SC, bone marrow derived stem cells, dermis derived stem cells (Sahni & Kessler;2010)

3.12.3.2.1 Human Neuronal stem cell (NSCs)

NSCs have been preferred in SCI studies because NSCs have the definite ability to differentiate into functional neurons and glial cells after being transplanted in the injured spinal cord (Kim et al; 2007, Johnson et al; 2010. Mothe et al;2008) However, like embryonic stem cells, clinical application of adult NSCs, requires careful preclinical evaluation of their safety , efficacy, purity of the neural cultures as well as there are bioethical issues to be considered (Daar et al;2004, Henon;2003, Riaz et al; 2002)

3.12.3.2.2 Olfactory ensheathing glia cells (OEC's)

"OEC's have been shown to penetrate the inhibitory glial scar at the injury site, and then migrate to their correct targets, restoring function. *OEC's* could also provide an extracellular matrix and other types of neurotrophins to the injured neurons and neural differentiated adult stem cells, *OECs are themselves not considered stem cells(Lima etal;2010)*. AS they are the patient's own cells, there is no concern regarding rejection (Lima et al;2006).

3.12.3.2.3 Human umbilical cord stem cells (hUCB)

The hUCB cells are immune naïve and so they subsequently cause less graft rejection, GvHD and post-transplant infections (Knutsen & Wall;1999,Newcomb et al ; 2007, Tse & Laughlin; 2005,) and they are able to differentiate into neural lineage. Evidence has emerged suggesting alternative pathways of graft-mediated neural repair that involve neurotrophic effects. These effects are caused by the release of various growth factors that promote cell survival, angiogenesis and anti-inflammation, and this is all a side from a cell replacement mechanism (Park et al ; 2011), Willing and his colleague (2003) prove that, hUCB cells can be administered by an intraarterial or intravenous route as well as by the direct intralesional approach. Intravenous injection of mononuclear hUCB cells was at least as effective and/or more effective at some points than direct implantation.

3.12.3.2.4 Mesenchymal stem cells (MSCs)

Autologous bone marrow- derived stem cells are ideal candidates for treating SCI in emerging clinical studies, because there are no ethical obstacles to their use and the health risk for patients with SCI is rather small (Park et al; 2005). Numerous electrophysiological and histological preclinical studies have revealed feasibility and beneficial potential of implantation of stem cells from bone marrow in animal models of SCI which showed, neuronal and axonal regeneration, astrocyte proliferation, remyelination, neovascularization, and functional improvement (Akiyama et al ;2002a,b, Chopp et al ;2000, Hofstetter et al;2002, Inoue et al;2003, Jendelova´et al;2004, Kalyani et al;1998, Saporta etal;2003, Sykova´& Jendelova´;2006 &2005, Sykova´ et al; 2005, Urdzı´kova´et al;2006, Wu et al ;2003). These studies have also shown that the optimal therapeutic window for implantation in animal models of SCI is 7–21 days after injury. As regard mode of cell delivery preclinical experiments in rats with SCI demonstrated that intravenously implanted human bone marrow-MSCs labeled in vitro with iron oxide nanoparticles and followed in vivo by magnetic resonance imaging (MRI), migrate, survive, and home only to the lesion site (Jendelova´et al;2004, Sykova´& Jendelova´;2005). All these data encourage scientists to initiate many nonrandomized phase I/II clinical studies using autologousBM-MSCs ,delivered to the patient by many routes (Geffner et al ;2008) either implanted direct intralesional Park and colleagues 2005 or intra- arterially via a. vertebralis (i.e., close to the lesion site) (Sykova et al ;2006)or less invasive which include intravenous, or Intrathecal (Bakshi et al ;2006 , Kishk, et,al2010) into patients with subacute (Park et al ;2005, Sykovaet,al2006) or chronic SCI (Sykova etal;2006 , Kishk, et al;2010)at the cervical or thoracic level, the outcome from BMMC implantation in acute and chronic patients is promising .however, the therapeutic window will play an important role in any type of SCI treatment. There seems to be a similar therapeutic window in humans as in animals, which is up to 3–4 weeks after SCI. Sykova and his colleague ;2006 suggest that administering the cells closer to the injury site, such as through the catheterization of a. vertebralis, or into the cerebrospinal fluid (Kishk et al; 2010,Ohata et al; 2004), or even intraspinally at the lesion border (Park et al; 2005), might be important for a better outcome. The observed partial recovery might be attributable to a "rescue effect," a reduction in tissue loss from secondary injury processes, as well as to diminished glial scarring. MSCs may induce an allodynia-like response by producing intrathecal proinflammatory cytokines, especially interleukin-1, tumor necrosis factor, and interleukin-6 (. Chae etal; 2009). Neither Abrams et al;2009, Sykova et al;2006 or Geffner et al; 2008 reported central pain as a complication. The deployment of MSCs in patients with subacute or chronic traumatic SCI will need longer follow-up, more studies that explore the best timing post injury , the dose and duration of MSC interventions, Objective assessment of the bladder and bowel function with urodynamic studies and anal sphincter EMG is necessary. Imaging of the lesions by MRI using particle-labeled MSCs could determine whether the cells reach the lesion. Future in vivo markers for neuronal regeneration or remyelination could give more insight into the mechanisms of any biological effects (Dobkin 2010)

Clinical studies are necessary , However, the question of which cell type is most beneficial for SCI treatment is still unresolved as what are the mechanisms underlying the beneficial effect(s) The therapeutic window, the implantation strategy, the method of administration, the number of cells, and the possible side effects can only be tested in human clinical trials guided by Guidelines for trials of cellular therapies (Fawcett etal;2007) .

4. Conclusion

It is realistic to believe that stem cells will be used clinically, not as a cure-all but as part of a therapeutic armamentarium The key, however, will be in applying the right cell type to the right disease and conveying the right amount of expectation to the patient. Meticulous attention to the ethics and Collaboration between basic scientists, clinicians, industry partners, and funding bodies is required to translate the potential of cell therapy into a reality in a timely, but safe and effective manner .

5. References

Abrams MB, Dominguez C, Pernold K, et al. (2009)Multipotent mesenchymal stromal cells attenuate chronic inflammation and injury-induced sensitivity to mechanical stimuli in experimental spinal cord injury. *Restor Neurol Neurosci.*;27: 307-321.

Acharya MM, Rai KS, Hattiangady B *et al (*2007). Potential of hippocampal stem/progenitor cell grafts for restraining spontaneous motor seizures during chronic temporal lobe epilepsy. *Society for Neuroscience.* Abstracts; 11 3-7.

Aggarwal, S., Pittenger, M.F., (2005). Human mesenchymal stem cells modulate allogeneic immune cell responses. *Blood* 105, 1815–1822.

Agid Y (1991). Parkinson's disease: pathophysiology. *Lancet;* 337:1321–1324.

Akiyama, Y.; Radtke, C.; Honmou, O.; Kocsis, J. D. ;(2002a): Remyelination of the spinal cord following intravenous delivery of bone marrow cells. Glia 39:229–236.

Akiyama, Y.; Radtke, C.; Kocsis, J. D. (2002b) Remyelination of the rat spinal cord by transplantation of identified bone marrow stromal cells. *J. Neurosci.* 22:6623–6630.

Altman J, Das GD (1965). Autoradiographic and histological evidence of postnatal hippocampal neurogenesis in rats. *J Comp Neurol;* 124:319–335.

Altman J, Das GD (1966). Autoradiographic and histological studies of postnatal neurogenesis. I. A longitudinal investigation of the kinetics, migration and transformation of cells incorporating tritiated thymidine in neonate rats, with special reference to postnatal neurogenesis in some brain regions. *J Comp Neurol;* 126:337–389.

Alvarez-Dolado M, Calcagnotto M, Karkar K, Southwell D, Jones-Davis D, Estrada R, et al (2006). Cortical inhibition modified by embryonic neural precursors grafted into the postnatal brain. *J Neurosci.;* 26:7380–7389.

Alvarez-Dolado, M.; Pardal, R.; Garcia-Verdugo, J.M.; Fike, J.R.; etal. (2003) Fusion of bone-marrow-derived cells with Purkinje neurons, cardiomyocytes and hepatocytes *Nature,* 10, 1-6.

and iron. *Stroke* 38, 759–762.

Andres, R.H., Guzman, R., Ducray, A.D., Mordasini, P., Gera, A et al.(2008). Cell replacement therapy for intracerebral hemorrhage. *Neurosurg. Focus* 24, E16.

Appel SH, Engelhardt JI, Henkel JS, Siklos L, Beers DR, (2008): Hematopoietic stem cell transplantation in patients with sporadic amyotrophic lateral sclerosis. *Neurology* 71(17):1326-34.

Asano T, Sasaki K, Kitano Y, Terao K, Hanazono Y (2006). In vivo tumor formation from primate embryonic stem cells. *Methods Mol Biol*;329:459–67.

Atala A(2009). Engineering organs.*Curr Opin Biotechnol.*; 20(5):575-92.

Azizi, S. A.; Stokes, D.; Augelli, B. J.; Di Girolamo, C.;Prockop, D. J.(1998): Engraftment and migration of human bone marrow stromal cells implanted in the brains of albino rats—similarities to astrocyte grafts. *Proc. Natl. Acad.* Sci. USA 95:3908–3913.

Bachoud-Le´vi AC, Re´my P, Nguyen JP, Brugie`res P, Lefaucheur JP, Bourdet C,et al, (2000). Motor and cognitive improvements in patients with Huntington's disease after neural transplnatation. *Lancet*. 356:1975–1979

Bakshi A, Barshinger AL, Swanger SA, Madhavani V, et,al. (Jan 2006)Lumbar puncture delivery of bone marrow stromal cells in spinal cord contusion: a novel method for minimally invasive cell transplantation ,*Journal of Neurotrauma.*, Vol. 23, No. 1:55-65.

Bang OY, Lee JS, Lee PH, et al(2005). Autologous mesenchymal stem cell transplantation in stroke patients. Ann *Neurol*;57: 874–82.

Bartley J.; Carroll J.E,(2003) Stem cell therapy forcerebral palsy *Expert Opinion on Biological Therapy*, Volume 3, Number 4, 1 July, pp. 541-549(9).

Benowitz, L.I., Carmichael, S.T.,(2010):. Promoting axonal rewiring to improve outcome after stroke. *Neurbiol. Dis.* 37, 259–266.

Bjorklund A, Lindvall O(2000). Cell replacement therapies for central nervous system disorders. *Nat Neurosci.*; 3:537–544.

Black PM, Loeffler J. (1997). Cancer of the nervous system. Oxford: Blackwell.

Blesch A, Lu P, Tuszynski M.(2002) Neurotrophic factors, gene therapy, and neural stem cells for spinal cord repair. *Brain Res Bull.*;57:833-838.

Bliss T.M., Andres R.H., Steinberg GK (2010) Optimizing the success of cell transplantation therapy for stroke *Neurobiology of Disease* 37-275–283

Bliss, TM ., Guzman, R., Daadi, M., Steinberg, G.K., (2007). Cell transplantation therapy for stroke. *Stroke* 38, 817–826.

Borlongan, C.V., Hadman, M., Sanberg, C.D., Sanberg, P.R.,(2004). Central nervous system entry of peripherally injected umbilical cord blood cells is not required for neuroprotection in stroke. *Stroke* 35, 2385–2389.

Borlongan, C.V., Tajima, Y., Trojanowski, J.Q., Lee, V.M., Sanberg, P.R., (1998): Transplantation of cryopreserved human embryonal carcinoma-derived neurons (NT2N cells) promotes functional recovery in ischemic rats. *Exp. Neurol.* 149, 310–321.

Brazelton, T. R.; Rossi, F. M.; Keshet, G. I.; Blau, H. M.(2000) From marrow to brain: Expression of neuronal phenotypes in adult mice. Science 290:1775–1779.

Brustle O, Jones KN, Learish RD, Karram K, Choudhary K,Wiestler OD, et al (1999). Embryonic stem cell-derived glial precursors: a source of myelinating transplants. *Science*;285:754–6. 894*D*.

Brustle O, McKay RG. (1996). Neuronal progenotors as tools for cell replacement in the nervous system. *Curr Opin Neurobiol.*; 6:688–695.

Busch SA, Horn KP, Cuascut FX, Hawthorne AL, Bai L, et al.(2010): Adult NG2+ cells are permissive to neurite outgrowth and stabilize sensory axons during macrophage-induced axonal dieback after spinal cord injury. *J Neurosci* 30:255-265.

Carmichael, S.T.,(2006). Cellular and molecular mechanisms of neural repair after stroke: making waves. *Ann. Neurol.* 59, 735–742.

Carmichael, S.T.,(2008). Themes and strategies for studying the biology of stroke recovery in the post stroke epoch. *Stroke* 39, 1380–1388.

Carpenter, M.K., Inokuma, M.S., Denham, J., Mujtaba, T., Chiu, C.P., etal., (2001). Enrichment of neurons and neural precursors from human embryonic stem cells. Exp. *Neurol.* 172, 383–397.

Carpentino J, Hartman N, Grabel L, Naegele J. (2008). Region-specific differentiation of embryonic stem cellderived neural progenitor transplants into the adult mouse hippocampus following seizures. *J Neurosci Res;* 86:512–524.

Carroll JE, Mays RW (2011) Update on stem cell therapy for cerebral palsy.*Expert Opin Biol Ther.* Apr;11(4):463-71

Chae WP, Keun SK, Sohyun B, et al. (2009)Cytokine secretion profiling of human mesenchymal stem cells by antibody array. *IntJ Stem Cells.*;2:59-68.

Chen A, Siow B, Blamire AM, Lako M, Clowry GJ (2010): Transplantation of magnetically labeled mesenchymal stem cells in a model of perinatal brain injury.*Stem Cell Res.* Nov;5(3):255-66.

Chen J, Chopp M. :(2006)Neurorestorative treatment of stroke: cell and pharmacological approaches. *NeuroRx* ;3:466–73.

Chen J, Li Y, Wang L, et al. (2001b):Therapeutic benefit of intravenous administration of bone marrow stromal cells after cerebral ischemia in rats. *Stroke*;32:1005–11.

Chen, J., Sanberg, P.R., Li, Y., Wang, L., Lu, M., Willing, A.E., Sanchez-Ramos, J., Chopp, M., (2001a). Intravenous administration of human umbilical cord blood reduces behavioral deficits after stroke in rats *Stroke* 32, 2682–2688.

Chen, J., Zhang, Z.G., Li, Y., Wang, L., Xu, et al (2003). Intravenous administration of human bone marrow stromal cells induces angiogenesis in the ischemic boundary zone after stroke in rats. *Circ. Res.* 92, 692–699.

Cho GW, Noh MY, Kim HY, Koh SH, Kim KS, Kim SH (2010)Bone marrow-derived stromal cells from amyotrophic lateral sclerosis patients have diminished stem cell capacity. *Stem Cells Dev.*Jul;19(7):1035-42

Cho JS, Park HW, Park SK, Roh S, Kang SK, etal. (2009)Transplantation of mesenchymal stem cells enhances axonal outgrowth and cell survival in an organotypic spinal cord slice culture. *Neurosci Lett* 454:43-8.

Choi MR, Kim HY, Park JY, Lee TY, Baik CS, et al;(2010):Selection of optimal passage of bone marrow-derived mesenchymal stem cells for stem cell therapy in patients with amyotrophic lateral sclerosis. *Neurosci Lett.* Mar 19;472(2):94-8

Chopp, M.; Zhang, X. H.; Li, Y.; Wang, L.; Chen, J.; et al (2000): Treatment with bone marrow stromal cell transplantation Spinal cord injury in rat:. *Neuroreport* 11:3001–3005.

cord injury. *J. Neurosci.* 25, 4694–4705.

Corti, S., Nizzardo, M., Nardini, M., Donadoni, C., Salani, S., Ronchi, D. et al. (2010) Embryonic stem cell-derived neural stem cells improve spinal muscular atrophy phenotype in mice. *Brain* 133: 465_481.

Corti, S., Nizzardo, M., Nardini, M., Donadoni, C.,Salani, S., Ronchi, D. et al. (2008) Neural stem cell transplantation can ameliorate the phenotype of a mouse model of spinal muscular atrophy. *J Clin Invest* 118: 3316_3330.

Coyle JT, Price DL, DeLong MR(1983). Alzheimer's disease: a disorder of cortical cholinergic innervation. *Science.*; 219:1184–1190.

Da Silva Meirelles L, Chagastelles PC, Nardi NB(2006): Mesenchymal stem cells reside in virtually all post-natal organs and tissues. *J Cell Sci* , 119:2204-2213.

Daadi, M., Arac, A., Davis, A., Maag, A. L., Batnagar, R., et al., (2009b). Grafts of human embryonic stem cell-derived neural stem cells promote neuroanatomical rewiring and connectivity with host in hypoxia ischemia model of neonates. *Stroke.* 40(4), e169, P72.

Daadi, M.M., Li, Z., Arac, A., Grueter, B.A., Sofilos, M., Malenka, R.C., Wu, J.C., Steinberg, G.K., (2009a). Molecular and magnetic resonance imaging of human embryonic stem cell-derived neural stem cell grafts in ischemic rat brain. *Molecular Therapy* 17, 1282–1291.

Daadi, M.M., Maag, A.L., Steinberg, G.K., (2008);. Adherent self-renewable human embryonic stem cell-derived neural stem cell line: functional engraftment in experimental stroke model. *PLoS ONE* 3, e1644.

Daar, A. S., Bhatt, A., Court, E., & Singer, P. A. (2004). Stem cell research and transplantation: science leading ethics. *Transplantation Proceedings*, 36(8), 2504–2506.

Dashtipour K, Tran PH, Okazaki MM *et al* (2001). Ultrastructural features and synaptic connections of hilar ectopic granule cells in the rat dentate gyrus are different from those of granule cells in the granule cell layer. *Brain Res;* 890:261–271.

Dawson E, Mapili G, Erickson K, Taqvi S, Roy K (2008). Biomaterials for stem cell differentiation. *Adv Drug Deliv Rev;* 60(2):215-28.

De Keyser J (2005). Autologous mesenchymal stem cell transplantation in stroke patients. *Ann Neurol,* 58:653–4, author reply 4–5.

Deacon T, Dinsmore J, Costantini LC, Ratliff J, Isacson O. etal (1998):stem cells can differentiate into dopaminergic and serotonergic neurons after transplantation. *Exp Neurol*,149:28–41.

Deda H, Inci MC, Kürekçi AE, Sav A, Kayihan Ozgün E, Ustünsoy GE, Kocabay S.(2009): Treatment of amyotrophic lateral sclerosis patients by autologous bone marrow-derived hematopoietic stem cell transplantation: a 1-year follow-up. *Cytotherapy* 11(1):18-25.

Deshpande, D.M., Kim, Y.S., Martinez, T., Carmen, J., Dike, S., Shats, I. et al. (2006) Recovery fromparalysis in adult rats using embryonic stem cells. *Ann Neurol* 60: 32_44.

Djouad F, Bouffi C, Ghannam S, Noel D, Jorgensen C (2009): Mesenchymal stem cells: innovative therapeutic tools for rheumatic diseases. *Nat Rev Rheumatol* , 5:392-399.

DobkinBH (2010). What matters in cellular transplantation for spinal cord injury: the cells, the rehabilitation, or the best mix? *Neurorehabil Neural Repair.* 2010;24:7-9

Doetsch, F., Caille, I., Lim, D.A., Garcia-Verdugo, J.M., Keirstead, H.S., etal. (2005). Human embryonic stem cell-derived oligodendrocyte progenitor cell transplants remyelinate and restore locomotion after spinal cord injury. *J. Neurosci.* 25, 4694–4705.

Douglas M. Sproule and Petra Kaufmann (2010) Therapeutic developments in spinal muscular atrophy Ther *Adv Neurol Disord* 3(3) 173_185.

Draper JS, Smith K, Gokhale P, Moore HD, Maltby E, Johnson J, et al(2004). Recurrent gain of chromosomes 17q and 12 in cultured human embryonic stem cells. *Nat Biotechnol*;22:53–4.

Dunn IF, Black PM (2003). The neurosurgeon as local oncologist: cellular and molecular neurosurgery in malignant glioma therapy. *Neurosurgery;* 52:1411–1422.

Dunnett SB, Bjorklund A (1999). Prospects for new restorative and neuroprotective treatments in Parkinson's disease. *Nature;* 399:A32–A39.

Dunnett SB, Carter RJ, Watts C, Torres EM, Mahal A, Mangiarini L, Bates G, Morton AJ (1998). Striatal transplantation in a transgenic mouse model of Huntington's disease. *Exp Neurol.;* 154:31–40.

Ebert AD, Yu J, Rose FF Jr, Mattis VB, Lorson CL, Thomson JA, Svendsen CN (2009)Induced pluripotent stem cells from a spinal muscular atrophy patient.*Nature.* Jan 15;457.

Englund, U., Bjorklund, A., Wictorin, K., Lindvall, O., Kokaia, M., (2002). Grafted neuralstem cells develop into functional pyramidal neurons and integrate into host cortical circuitry. *Proc. Natl. Acad. Sci.* U. S. A. 99, 17089–17094.

Fassas A, Vassilios K. Kimiskidis (2003) : Stem cell transplantation for multiple sclerosis: What is the evidence? *Elsevier , Blood Reviews* 17, 233–240

Fawcett JW, Curt A, Steeves JD, et al.(2007) Guidelines for the conductof clinical trials for spinal cord injury as developed by the ICCP panel: spontaneous recovery after spinal cord injury and statistical power needed for therapeutic clinical trials. *Spinal Cord.;*45:190-205.

Filip S, Mokry J., Karbanova J., et al (2004) :Local environmental factor determine hematopoietic differentiation of neural stem cells . *Stem Cells Dev.*, 13:113-120

Finley, M.F., Kulkarni, N., Huettner, J.E., (1996). Synapse formation and establishment of neuronal polarity by P19 embryonic carcinoma cells and embryonic stem cells. J. *Neurosci.* 16, 1056–1065.

Fitch MT, Silver J.(2008): CNS injury, glial scars, and inflammation: Inhibitory extracellular matrices and regeneration failure. *Exp Neurol* 209:294-301, .14.

Flax JD, Aurora S, Yang C, Simonin C, Wills AM, Billinghurst LL, et al (1998). Engraftable human neural stem cells respond to developmental cues, replace neurons, and express foreign genes. *Nat Biotechnol;* 16:1033–1039.

Freeman TB, Cicchetti F, Hauser RA, Deacon TW, Li XJ, Hersch SM, Nauert G,et al, (2000). Transplanted fetal striatum in Huntington's disease: phenotypic development and lack of pathology. *Proc Natl Acad Sci USA.;* 97:13877–13882.

Gage FH (2000). Mammalian neural stem cells. *Science;* 287:1433–1438.

Gao, J., Coggeshall, R.E., Chung, J.M., Wang, J. and Wu, P. (2007) Functional motoneurons develop from human neural stem cell transplants in adult rats. *Neuroreport* 18: 565_569.

Geffner LF, Santacruz P, Izurieta M, et al. (2008)Administration of autologous bone marrow stem cells into spinal cord injury patients via multiple routes is safe and improves their quality of life: comprehensive case studies. *Cell Transplant.;* 17:1277-1293.

Gimble J, Guilak F.(2003): Adipose-derived adult stem cells: isolation, characterization, and differentiation potential. *Cytotherapy;*5:362–9.

Giordano A, Galderisi U, Marino IR.(2007) From the laboratory bench to the patient's bedside: an update on clinical trials with mesenchymal stem cells. *J Cell Physiol.*211(1):27-35.

Gottlieb D I (2002). Large scale sources of neural stem cells. *Annu Rev Neurosci .;* 25:381–407

Graham ML, Herndon JE, Casey JR, Chaffee S, Ciocci GH, Krischer JP, Kurtzberg et al,(1997). High dose chemotherapy with autologous stem-cell rescue in patients with recurrent and high-risk pediatric brain tumors. *J Clin Oncol.*; 15:1814–23.

Greenmayre JT, Shoulson I (1994). Huntington disease. In: *Calne D, editor. Neurodegenrative disease*. Philadelphia: W.B. Saunders.; p 658–704.

Guzman, R., Choi, R., Gera, A., De Los Angeles, A., Andres, R.H., Steinberg, G.K., (2008). Intravascular cell replacement therapy for stroke. *Neurosurg. Focus* 24, E15.

Hagell P, Brundin P (2002). Cell survival and clinical outcome following intrastriatal transplantation in Parkinson's disease. *J Neuropathol Exp Neurol.*; 60:741–752.

Hagell P, Schrag A, Piccini P, Jahanshahi M, Brown R, Rehncrona S, Widner H, et al,(1999). Sequential bilateral transplantation in Parkinson's disease: effects of the second graft. *Brain*; 122:1121–1132.

Harper PS (1996). Huntington's disease. Philadelphia; W.B. Saunders.

Harper, J.M., Krishnan, C., Darman, J.S., Deshpande, D.M., Peck, S., Shats, I. et al. (2004) Axonal growth of embryonic stem cell-derived motoneurons in vitro and in motoneuron-injured adult rats. *Proc Natl Acad Sci U S A* 101: 7123_7128.

Hayashi, T., Noshita, N., Sugawara, T., Chan, P.H.,(2003). Temporal profile of angiogenesis and expression of related genes in the brain after ischemia. *J. Cereb. Blood Flow Metab.* 23, 166–180.

Henon, P. R. (2003). Human embryonic or adult stem cells: an overview on ethics and perspectives for tissue engineering. *Advances in Experimental Medicine and Biology*, 534, 27–45.

Hicks, A., Jolkkonen, J.,(2009):. Challenges and possibilities of intravascular cell therapy in stroke. *Acta Neurobiol. Exp. (Wars)* 69, 1–11.

Hicks, A.U., Lappalainen, R.S., Narkilahti, S., Suuronen, R., Corbett, D.,(2009). Transplantation of human embryonic stem cellderived neural precursor cells and enriched environment after cortical stroke in rats: cell survival and functional recovery. *Eur. J. Neurosci.* 29, 562–574.

Ho M, Yu D, Davidsion MC, Silva GA(2006). Comparison of standard surface chemistries for culturing mesenchymal stem cells prior to neural differentiation. *Biomaterials;* 27:4333–4339.

Hofstetter, C.P., Schwarz, E.J., Hess, D., Widenfalk, J., El Manira, A., etal, (2002). Marrow stromal cells form guiding strands in the injured spinal cord and promote recovery. *Proc. Natl. Acad. Sci. U. S. A.* 99, 2199–2204.

Horie, N., Andres, R. H., Keren-Gill, H., Slikker, W., Sun, G. Het al,(2009) Transplanted human neural progenitor cells enhance endogenous dendritic plasticity and axonal sprouting after stroke in rat. *Stroke.* 40(4), e150, 190.

Horie, N., Bliss, T., Shichinohe, H., Palmer, T., Steinberg, G., (2008):. Ischemia-induced Angiogenesis Is Enhanced By hCNS-SCn Transplantation. *Stroke* 39(2) 658, p364..

How Embryonic Stem Cell Lines Are Made (2010)." Dolan DNA Learning Center. Cold Spring Harbor Laboratory. Web. 1 (July 2010). http://www.dnalc.org/resources/animations/stemcells.html

Howell OW, Scharfman HE, Herzog H, Sundstrom LE, Beck-Sickinger A, Gray WP(2003). Neuropeptide y is neuroproliferative for post-natal hippocampal precursor cells. *J Neurochem;* 86:646–659.

http://www.discoverymedicine.com/Antonio-Orlacchio/files/2010/06/discovery_medicine_orlacchio_no49_figure
http://www.discoverymedicine.com/Antonio-Orlacchio/nd

Hua, Y., Keep, R.F., Hoff, J.T., Xi, G., (2007). Brain injury after intracerebral hemorrhage: the role of thrombin and iron. *Stroke* 38, 759–762.

Inoue, M.; Honmou, O.; Oka, S.; Houkin, K.; Hashi, K.; etal.(2003) Comparative analysis of remyelinating potential of focal and intravenous administration of autologous bone marrow cells into the rat demyelinated spinal cord. *Glia* 44:111–118.

Ishibashi, S., Sakaguchi, M., Kuroiwa, T., Yamasaki, M., Kanemura, Y., et al, (2004). Human neural stem/ progenitor cells, expanded in long-term neurosphere culture, promote functional recovery after focal ischemia in Mongolian gerbils. *J. Neurosci. Res.* 78, 215–223.

Janson CG, Ramesh TM, During MJ, et al (2001):. Human intrathecal transplantation of peripheral blood stem cells in amyotrophic lateral scle rosis. *J Hematother Stem Cell Res*;10:913-915.

Jendelova´, P.; Herynek, V.; Urdzı´kova´, L.; Glogarova´, K.; Kroupova´, J.; et al.(2004) :MR tracking of transplanted bone marrow and embryonic stem cells labeled by iron oxide nanoparticles in rat brain and spinal cord. *J. Neurosci. Res.* 76:232–243.

Jiang Y, Balkrishna N, Jahagirdar R, et al.(2002) Pluripotency of mesenchymal stem cells derived from adult marrow. *Nature.*;418:41-49.

John T. Dimos, Kit T. Rodolfa, Kathy K. Niakan, et,al (2008): *Induced Pluripotent Stem Cells Generated from Patients with ALS Can Be Differentiated into Motor Neurons." Science 29 August : Vol. 321 no. 5893 pp. 1218-1221.*

Johnson, P. J., Tatara, A., Shiu, A., & Sakiyama-Elbert, S. E. (2010). Controlled release of neurotrophin-3 and platelet-derived growth factor from fibrin scaffolds containing neural progenitor cells enhances survival and differentiation into neurons in a sub acute model of SCI. *Cell Transplantation*, 19(1), 89–101.

Kadoya K, Tsukada S, Lu P, et al.(2009)Combined intrinsic and extrinsic neuronal mechanisms facilitate bridging axonal regeneration one year after spinal cord injury. *Neuron.* ;64:165-172.

Kadoya K, Tsukuda S, Lu P, Coppola G, Geeschwind D, etal.(2009) Combined intrinsic and extrinsic neuronal mechanisms facilitate bridging axonal regeneration one year after spinal cord injury. *Neuron* 64:165-72.

Kalyani, A. J.; Piper, D.; Mujtaba, T.; Lucero, M. T.; Rao, M. S. (1998)Spinal cord neuronal precursors generate multiple neuronal phenotypes in culture. *J. Neurosci.* 18:7856–78

Kang UJ, Fisher LJ, Joh TH, O'Malley KL, Gage FH. (1993). Regulation of dopamine production by genetically modified primary fibroblasts. *J Neurosci* 13:5203–5211.

Karussis D& KassisI (2007). Use of stem cells for the treatment of MS *Expert Review of eurotherapeutics* Sept:7(9)1189-1201 .

Karussis D; Karageorgiou C, Dembinsky,A et,al (2010): Safety and Immunological Effects of Mesenchymal Stem Cell Transplantation in Patients With Multiple Sclerosis and Amyotrophic Lateral Sclerosis *Arch Neurol. ;67(10):1187-1194*

KarussisD, Kassis I (2008): The potential use of stem cells in multiple sclerosis: An overview of the preclinical experience *Clinical Neurology and Neurosurgery* 110, 889–896

Keirstead, H.S., Nistor, G., Bernal, G., Totoiu, M., Cloutier, F., (2005):. Human embryonic stem cell-derived oligodendrocyte progenitor cell transplants remyelinate and restore locomotion after spinal

Keyoung, H.M.; Roy, N.S.; Benraiss, A.; Louissaint, A.; Suzuki, A.; etal. (2001) high yield selection and extraction of two premoter -defined phenotypes from neural stem cells from the fetal human brain *Nature Biotechnol.*, 19, 843-850.

Kim JH, Auerbach JM, Rodrı´guez-Go´mez JA, Velasco I, Gavin D, Lumelsky N, Lee SH, et al, (2002). Dopamine neurons derived from embryonic stem cells function in an animal model of Parkinson's disease. *Nature.*; 418:50–56.

Kim SU (2004). Human neural stem cells genetically modified for brain repair in neurological disorders. *Neuropathology.*; 24:159–174.

Kim SU, de Vellis J(2009). Stem cell-based cell therapy in neurological diseases: a review. *J Neurosci Res.*; 1; 87(10):2183-200.

Kim SU, Nagai A, Nakagawa E, Choi HB, Bang JH, Lee HJ, Lee MA, Lee YB, Park IH(2008). Production and characterization of immortal human neural stem cell line with multipotent differentiation property. *Methods Mol Biol.*; 438:103–121.

Kim SU, Park IH, Kim TH, Kim KS, Choi HB, Hong SH, Bang JH, et al(2006). Brain transplantation of human neural stem cells transduced with tyrosine hydroxylase and GTP cyclohydrolase 1 provides functional improvement in animal models of Parkinson disease. *Neuropathology*; 26:129–140.

Kim TE, Lee HS, Lee YB, Hong SH, Lee YS, Ichinose H, Kim SU, Lee MA (2003). Sonic hedgehog and FGF8 collaborate to induce dopaminergic phenotype in Nurr-1 over-expressing neural stem cells. *Biochem Biophys Res Commun.*; 305:1040–1048.

Kim, B. G., Hwang, D. H., Lee, S. I., Kim, E. J., & Kim, S. U. (2007). Stem cell-based cell therapy for spinal cord injury. *Cell Transplantation*, 16(4), 355–364.

Kish SJ, Shannak K, Hornykiewitcz O ,(1988). Uneven pattern of dopamine loss in the striatum of patients with idiopathic Parkinson's disease. Pathophysiologic and clinical implications. *N Engl J Med.*; 318:876–880.

Kishk NA, GabrH,HamdyS, AfifiL, Abokrysha NT, and Mahmoud H (2010):Case Control Series of Intrathecal Autologous Bone Marrow Mesenchymal Stem Cell Therapy for Chronic Spinal Cord Injury *Neurorehabil Neural Repair* published on line July 26, 2010

Kleinig, T.J., Vink, R., (2009):. Suppression of inflammation in ischemic and hemorrhagic stroke: therapeutic options. *Curr. Opin. Neurol.* 22, 294–301.

Knutsen, A. P., & Wall, D. A. (1999). Kinetics of T-cell development of umbilical cord blood transplantation in severe T-cell immunodeficiency disorders. *Journal of Allergy and Clinical Immunology*, 103(5 Pt 1), 823–832.

Kondziolka, D., Steinberg, G.K., Wechsler, L., Meltzer, C.C., Elder, E.,et al, (2005). Neurotransplantation for patients with subcortical motor stroke: a phase 2 randomized trial. *J. Neurosurg.* 103, 38–45.

Kondziolka, D., Wechsler, L., Goldstein, S., Meltzer, C., Thulborn, K.R., et al,(2000). Transplantation of cultured human neuronal cells for patients with stroke. *Neurology* 55, 565–569.

Kordower JH, Goetz CG, Freeman TB, Olanow CW(1997). Doparminergic transplants in patients with Parkinson's disease: neuroanatomical correlates of clinical recovery. *Exp Neurol.*; 144:41–46.

Kosuga M, Sasaki K, Tanabe A, Li XK, Okawa H, Ogino I, Okuda O, Arai H, Sakuragawa N, et al,(2001). Engraftment of genetically engineered amniotic epithelial cells corrects lysosomal storage in multiple areas of the brain in MPS VII mice. *Mol Ther.;* 3:39–48.

Krupinski, J., Kaluza, J., Kumar, P., Wang, M., Kumar, S.,(1993). Prognostic value of blood vessel density in ischaemic stroke. *Lancet* 342, 742.

Kurozumi, K., Nakamura, K., Tamiya, T., Kawano, Y., Ishii, K.,et al(2005). Mesenchymal stem cells that produce neurotrophic factors reduce ischemic damage in the rat middle cerebral artery occlusion model. *Mol. Ther.* 11, 96–104.

Kuruba R, Hattiangady B, Shetty AK (2009). Hippocampal neurogenesis and neural stem cells in temporal lobe epilepsy. *Epilepsy Behav;* 14(S1):65–73.

Lang AE, Lozano AM (1998a). Parkinson's disease. First of two parts. *N Engl J Med.;* 339:1044–1053.

Lang AE, Lozano AM(1998b). Parkinson's disease. Second of two parts. *N Engl J Med.;* 339:1130–1143.

Lang, K.J., Rathjen, J., Vassilieva, S., Rathjen, P.D., (2004). Differentiation of embryonic stem cells to a neural fate: a route to re-building the nervous system? *J. Neurosci.* Res. 76, 184–192.

Lee HJ, Kim KS, Kim EJ, Choi HB, Lee KH, Park IH, Ko Y, Jeong SW, Kim SU(2007). Brain transplantation of human neural stem cells promotes functional recovery in mouse intracerebral hemorrhage stroke model. *Stem Cells;* 25:211–224.

Lee K, Majumdar MK, Buyaner D, Hendricks JK, Pittenger MF, Mosca JD(2001): Human mesenchymal stem cells maintain transgene expression during expansion and differentiation. *Mol Ther;*3:857–66.

Lee SH, Lumelsky N, Studer L, Auerbach JM, McKay RD. (2000). Efficient generation of midbrain and hindbrain neurons from mouse embryonic stem cells. *Nat Biotechnol;* 18:675–679.

Lee ST, Chu K, Park JE, Lee K, Kang L, Kim SU, Kim M (2005). Intravenous administration of human neural stem cells induces functional recovery in Huntington's disease rat model. *Neurosci Res.;* 52:243–249.

Lee ST, Park JE, Lee K, Kang L, Chu K, Kim SU, Kim M, Roh JK (2006). Noninvasive method of immortalized neural stem-like cell transplantation in an experimental model of Huntington's disease. *J Neurosci Methods.;* 52:250–254.

Lee, J. K., Jin, H. K., & Bae, J.-s. (2009). Bone marrow-derived mesenchymal stem cells reduce brain amyloid-β deposition and accelerate the activation of microglia in an acutely induced Alzheimer's disease mouse model. *Neuroscience Letters; 450,* 136-141.

Lee, J. K., Jin, H. K., Endo, S., Schuchman, E. H., Carter, J. E., & Bae, J.-s. (2010).Intracerebral Transplantation of Bone Marrow-Derived Mesenchymal Stem Cells Reduces Amyloid-Beta Deposition and Rescues Memory Deficits in Alzheimer's Disease Mice by Modulation of Immune Responses. *Stem Cells; 28,* 329-343.

Li T, Steinbeck JA, Lusardi T *et al*(2007). Suppression of kindling epileptogenesis by adenosine releasing stem cell-derived brain implants. *Brain;* 130:1276–1288.

Li Y, Chen J, Chen XG, et al.(2002): Human marrow stromal cell therapy for stroke in rat: neurotrophins and functional recovery. *Neurology* ;59:514–23.

Li Y, Chopp M.(2009): Marrow stromal cell transplantation in stroke and traumatic brain injury. *Neurosci Lett* 456:120-3.

Li, Y., Chen, J., Zhang, C.L., Wang, L., Lu, D., et al,(2005).: Gliosis and brain remodeling after treatment of stroke in rats with marrow stromal cells. *Glia* 49, 407–417.

Li, Y., McIntosh, K., Chen, J., Zhang, C., Gao, Q., et al,(2006). Allogeneic bone marrow stromal cells promote glial-axonal remodeling without immunologic sensitization after stroke in rats. *Exp. Neurol.* 198, 313–325.

Lima C, Escada P, Pratas-Vital J, et al.(2010)Olfactory mucosal autografts and rehabilitation for chronic traumatic spinal cord injury. *Neurorehabil Neural Repair.*;24:10-24.

Lima C, Pratas-Vital J, Escada P, Hasse-Ferreira A, Capucho C, etal.(2006) Olfactory mucosa autografts in human spinal cord injury: a pilot clinical study. *Journal of Spinal Cord Medicine*;29(3):191-203.

Lindvall O and Kokaia Z(2006). Stem cells for the treatment of neurological disorders. *Nature*; 441, 1094-1096.

Lindvall O, Brundin P, Widner H, Rehncrona S, Gustavii B, Frackowiak R, et al(1990). Grafts of fetal dopamine neurons survive and improve motor function in Parkinson's disease. *Science*; 247:574–577.

Llado, J., Haenggeli, C., Maragakis, N.J., Snyder, E.Y., Rothstein, J.D.,(2004). Neural stem cells protect against glutamate-induced excitotoxicity and promote survival of injured motor neurons through the secretion of neurotrophic factors. *Mol. Cell. Neurosci.* 27, 322–331.

Locatelli, F., Bersano, A., Ballabio, E., Lanfranconi, S., Papadimitriou, D.,(2009). Stem cell therapy in stroke. Cell Mol. Life Sci. 66, 757–772.

Martino S, D'Angelo F, Armentano I, Tiribuzi R, Pennacchi M, et al,(2009). Hydrogenated amorphous carbon nanopatterned film designs drive human bone marrow mesenchymal stem cell cytoskeleton architecture. *Tissue Eng Part A.*; 15(10):3139-49.

Marzzini L, Fagioli F, Boccaletti R, et al (2003). Stem cell therapy in amyotrophic lateral sclerosis: a methodologic approach in humans. *Amyotroph Lateral Scleros Other Motor Neuron Disord.* ;4:158-161.

Mazurier F, Doedens M, Gan OI, et al. (2003)Rapid myeloerythroid repopulation after intrafemoral transplantation of NOD-SCID mice reveals a new class of human stem cells. *Nature Med.*;9:959-963.

McDonald JW, Liu XZ, Qu Y, Liu S, Mickey SK, Turetsky D, et al.(1999): Transplanted embryonic stem cells survive, differentiate and promote recovery in injured rat spinal cord. *Nat Med*;5:1410–2.

Medical College of Georgia, 2010:First FDA-approved stem cell trial in pediatric cerebral palsy . Retrieved Febrauary,26 ScienceDaily

Meikle PJ, Hopwood JJ, Clague AE, Carey WF (1999). Prevalence of lysosomal storage disorders. *JAMA*; 281:249–254.

Meng XL, Shen JS, Ohashi T, Maeda H, Kim SU, Eto Y(2003). Brain transplantation of genetically engineered human neural stem cells transduced with beta-glucuronidase globally corrects lysosomal storage and brain lesions in mucopolysaccharidosis VII mice. *J Neuosci Res.*; 74:266–277.

Mezey, E.; Chandross, K. J.; Harta, G.; Maki, R. A.; McKercher, S. R.(2000) Turning blood into brain: Cells bearing neuronal antigens generated in vivo from bone marrow. *Science* 290:1779–1782.

Mitrecić D, Nicaise C, Gajović S, Pochet R (2010):Distribution, differentiation, and survival of intravenously administered neural stem cells in a rat model of amyotrophic lateral sclerosis.*Cell Transplant.*;19(5):537-48

Modo, M., Stroemer, R.P., Tang, E., Patel, S., Hodges, H., 2002. Effects of implantation site of stem cell grafts on behavioral recovery from stroke damage. *Stroke* 33, 2270–2278.

Mohyeddin Bonab M. Yazdanbakhsh S.,Alimoghaddom K.,Ghavamzadeh A.,Hooshmand F et al(2007) ,Does mesenchymal stem cell therapy help multiple sclerosis patients? report of a pilot study. *Iran J Immunol.*;4(1):50-57.

Mothe, A. J., Kulbatski, I., Parr, A., Mohareb, M., & Tator, C. H. (2008). Adult spinal cord stem/progenitor cells transplanted as neurospheres preferentially differentiate into oligodendrocytes in the adult rat spinal cord. *Cell Transplant*, 17(7), 735–751.

Mueller D, Shamblott MJ, Fox HE, Gearhart JD, Martin LJ. (2005)Transplanted human embryonic germ cell-derived neural stem cells replace neurons and oligodendrocytes in the forebrain of neonatal mice with excitotoxic brain damage ,*J Neurosci. Res*;82 (5):592-608.

Nakao N, Itakura T. (2000). Fetal tissue transplants in animal models of Huntington's disease: the effects on damaged neuronal circuitry and behavioral deficits. *Prog Neurobiol.*; 61:313–338.

Newcomb, J. D., Sanberg, P. R., Klasko, S. K., & Willing, A. E. (2007). Umbilical cord blood research: current and future perspectives. *Cell Transplantation*, 16(2), 151–158.

Newsome, P.N.; Johannessen, I.; Boyle, S.; Dalakas, E.; McAulay, K.A.; etal. (2003) Human cord-blood derived cells can differentiate into hepatocytes in the mouse liver with no evidence of cellular fusion. *Gastroenterology*, 124, 1891-1900.

Nistor, G.I., Totoiu, M.O., Haque, N., Carpenter, M.K., Keirstead, H.S.,(2005). Human embryonic stem cells differentiate into oligodendrocytes in high purity and myelinate after spinal cord transplantation. *Glia* 49, 385–396.

Noe' F, Nissinen J, Pitkanen A *et al* (2007). Gene therapy in epilepsy: The focus on NPY. *Peptides*; 28:377–383.

Noe F, Nissinen J, Pitkänen A, Gobbi M, Sperk G, During M, Vezzani A(2006). Gene therapy in epilepsy: The focus on NPY. *Peptides*; 28:377–383.

Noel, D., Djouad, F., Bouffi, C., Mrugala, D., Jorgensen, C., (2007). Multipotent mesenchymal stromal cells and immune tolerance. Leuk. Lymphoma 48, 1283–1289.

Ogawa, Y.; Sawamoto, K.; Miyta, T.; Watanabe, M.; Nakamura, M.; etal.(2002) Transplantation of in vitroexpanded fetal neural progenitor cells results in neurogenesis and functional recovery after spinal cord contusion injury in adult rats. J. *Neurosci.* Res. 69:925–933.

Ohta, M.; Suzuki, Y.; Noda, T.; Ejiri, Y.; Dezawa, M.; etal.(2004). Bone marrow stromal cells infused into the cerebrospinal fluid promote functional recovery of the injured rat spinal cord with reduced cavity formation. *Exp. Neurol.* 187:266–278.

Ohtaki, H., Ylostalo, J.H., Foraker, J.E., Robinson, A.P., Reger, R.L., Shioda, S., Prockop, D.J., (2008). Stem/progenitor cells from bone marrow decrease neuronal death in global ischemia by modulation of inflammatory/immune responses. *Proc. Natl. Acad. Sci.* USA 105, 14638–14643.

Okada, S.; Ishii, K.; Yamane, J.; Iwanami, A.; Ikegami, T.; etal. (2005)In vivo imaging of engrafted neural stem cells: Its application in evaluating the optimal timing of transplantation for spinal cord injury. *FASEB J*. 19:1839–1841.

Olanow CW, Kordower J, Freeman T (1996). Fetal nigral transplantation as a therapy for Parkinson's disease.*Trends Neurosci*.; 19:102–109.

Ourednik J, Ourednik V, Lynch WP, Schachner M, Snyder EY (2002). Neural stem cells display an inherent mechanism for rescuing dysfunctional neurons. *Nat Biotechnol;* 20:1103–1110.

Packer RJ, Cogen P, Vezina G, Rorke LB(1999). Medulloblastoma: clinical and biologic aspects. *Neurooncology* .; 1:232–50.

Padma SMV (2005):Stem cell therapy in static encephalopathy including cerebral palsy. *Indian Journal of Medical and Paediatric Oncology;* 26: 16-18.

Palmer TD, Takahashi J, Gage FH (1997). The adult rat hippocampus contains primordial neural stem cells. *Mol Cell Neurosci;* 8:389–404.

Parent JM, Elliott RC, Pleasure SJ *et al (*2006). Aberrant seizure-induced neurogenesis in experimental temporal lobe epilepsy. *Ann Neurol;* 59:81–91.

Park, DH& Lee JH & Borlongan CV & Sanberg PS & Chung YG et al, (2011) : Transplantation of Umbilical Cord Blood Stem Cells for Treating Spinal Cord Injury Stem Cell Rev and Rep 7:181–194

Park HC, Shims YS, Ha Y, et al.(2005) Treatment of complete spinal cord injury patients by autologous bone marrow cell transplantation and administration of granulocytemacrophage colony stimulating factor. *Tissue Eng.;11:913-922.*

Park, D.H., Eve, D.J., Musso III, J., Klasko, S.K., et,al (2009):. Inflammation and stem cell migration to the injured brain in higher organisms. *Stem. Cells Dev.* 18, 693–702.

Park, H. C.; Shims, Y. S.; Ha, Y.; Yoon, S. H.; Park, S. R.; etal (2005)Treatment of complete spinal cord injury patients by autologous bone marrow cell transplantation and administration of granulocyte macrophage colony stimulating factor. *Tissue Eng.* 11: 913–922.

Park, *I.H., Zhao, R., West, J.A., Yabuuchi, A., et,al (2006). A conditionally immortal clonal stem cell line from human cortical neuroepithelium for the treatment of ischemic stroke. Exp. Neurol. 199, 143–155.*

Patrice SJ, Tarbell NJ, Goumnerova LC, Shrieve DC, Black PM, Loeffler JS(1995). Results of radiosurgery in the management of recurrent and residual medulloblastoma. *Pediatr Neurosurg*.; 22:197–203.

Pavlović, M., Todorović, M., Tyagi, V., Todorović, V., & Balint, B (2009). Adult stem cell research and regenerative therapy in neurological diseases: Limitations and perspectives, Part II: Neurological diseases and stem cell therapy. *Bilten za transfuziologiju*.; 55(1-2), 15-30.

Payne N, Siatskas C, Claude C. Bernard A (2008):The promise of stem cell and regenerative therapies for multiple sclerosis *Journal of Autoimmunity* 31 , 288–294

Payon AG(2005), Beneficial effects of subcutaneously injected human umbilical cord stem cellson cerebral palsyand traumatic brain injury in children and a posited mechanism. *Med Hypotheses Res*2:497-501.

Pittenger MF, Mackay AM, Beck SC, Jaiswal RK, Douglas R, Mosca JD, et al (1999). Multilineage potential of adult human mesenchymal stem cells. *Science* ;284:143–7.

Pluchino S, Quattrini A, Brambilla E, Gritti A, Salani G, et al(2003). Injection of adult neurospheres induces recovery in a chronic mode of multiple sclerosis. *Nature*;422:688–94.

Pluchino, S., Zanotti, L., Rossi, B., Brambilla, E.et,al, (2005). Neurosphere-derived multipotent precursors promote neuroprotection by an immunomodulatory mechanism. *Nature* 436, 266–271

Potter E; CardonaM; KerrPosted D.(2007): Stem Cells and Neurologic Diseases: Hope or Hype? Future Neurology. ;2(5):471-47

Prockop, D. J. (1997)Marrow stromal cells as stem cells for nonhematopoietic tissues. *Science* 276:71–74.

Puissant B, Barreau C, Bourin P, Clavel C, Corre J, Bousquet C, et al(2005):. Immunomodulatory effect of human adipose tissue-derived adult stem cells: comparison with bone marrow mesenchymal stem cells. *Br J Haematol* ;129:118–29.

Qureshi, A.I., Wilson, D.A., Hanley, D.F., Traystman, R.J.,(1999). No evidence for an ischemic penumbra in massive experimental intracerebral hemorrhage. *Neurology* 52, 266–272.

Rafei M, Campeau PM, Aguilar-Mahecha A, Buchanan M, Williams P,et al(2009).: Mesenchymal stromal cells ameliorate experimental autoimmune encephalomyelitis by inhibiting CD4 Th17 T cells in a CC chemokine ligand 2-dependent manner. *J Immunol* , 182:5994-6002.

Ramos S and Raj A.: Blood stem cell growth factor reverses memory decline in mice. (2009). The randomized, controlled trial, at: www.physorg.com/news 165684042.html

Redmond DE Jr, Bjugstad KB, Teng YD, Ourednik V, Ourednik J, Wakeman DR, et al,(2007). Behavioral improvement in a primate Parkinson's model is associated with multiple homeostatic effects of human neural stem cells. *Proc Natl Acad Sci.*; 104:12175–12180.

Ren G, Su J, Zhang L, Zhao X, Ling W, L'Huillie A, Zhang J, Lu Y, Roberts AI, Ji W et al.(2009): Species variation in the mechanisms of mesenchymal stem cell-mediated immunosuppression. *Stem Cells*, 27:1954-1962.

Reubinoff, B.E., Itsykson, P., Turetsky, T., Pera, M.F., Reinhartz, E., et al., (2001). Neural progenitors from human embryonic stem cells. Nat. *Biotechnol.* 19, 1134–1140.

Riaz, S. S., Jauniaux, E., Stern, G. M., & Bradford, H. F. (2002). The controlled conversion of human neural progenitor cells derived from foetal ventral mesencephalon into dopaminergic neurons in vitro. *Brain Research. Developmental Brain Research*, 136(1), 27–34.

Rice CM, Scolding NJ (2008). Autologous bone marrow stem cells—properties and advantages. *J Neurol Sci*; 265:59–62.

Roisen FJ, Klueber KM, Lu CL, Hatcher LM, Dozier A, Shields CB, et al.(2001): Adult human olfactory stem cells. *Brain Res*;890:11–22.

Sahni V& Kessler J.A(2010): Stem cell therapies for spinal cord injury Nature Reviews Neurology 6, 363-372 .

Sanai N, Alvarez-Buylla A, Berger MS (2005). Mechanisms of disease. neural stem cells and the origin of gliomas. *N Engl J Med.*; 353:811–822.

Saporta, S.; Kim, J. J.; Willing, A. E.; Fu, E. S.; Davis, C. D.; etal.(2003) Human umbilical cord blood stem cells infusion in spinal cord injury: Engraftment and beneficial influence on behavior. J. *Hematother. Stem Cell Res.* 12:271–278.

Sasaki, M.; Honmou, O.; Akiyama, Y.; Uede, T.; Hashi, K.; etal(2001) Transplantation of an acutely isolated bone marrow fraction repairs demyelinated adult rat spinal cord axons. *Glia* 35:26-34.

Savitz, S.I., Dinsmore, J., Wu, J., Henderson, G.V., Stieg, P., Caplan, L.R.,(2005). Neurotransplantation of fetal porcine cells in patients with basal ganglia infarcts: a preliminary safety and feasibility study. *Cerebrovasc. Dis.* 20, 101–107.

Scharfman HE, Gray WP (2006). Plasticity of neuropeptide y in the dentate gyrus after seizures, and its relevance to seizure-induced neurogenesis. *EXS*; 95:193–211.

Scharfman HE, Gray WP (2007). Relevance of seizure-induced neurogenesis to the etiology of temporal lobe epilepsy. *Epilepsia.*; 48(S2):33–41.

Senior, K., 2001. Angiogenesis and functional recovery demonstrated after minor stroke. *Lancet* 358, 817.

Shen, L.H., Li, Y., Chen, J., Zacharek, A., Gao, Q., et al (2007). Therapeutic benefit of bone marrow stromal cells administered 1 month after stroke. J. *Cereb. Blood Flow Metab.* 27, 6–13.

Shen, L.H., Li, Y., Chen, J., Zhang, J., Vanguri, P., Borneman, J., Chopp, M., 2006. Intracarotid transplantation of bone marrow stromal cells increases axon-myelin remodeling after stroke. *Neuroscience* 137, 393–399.

Sherri S. Schultz, (2005), Adult Stem Cell Application in Spinal Cord Injury *Current Drug Targets, Vol. 6, No. 1 6,* 63-73

Shetty AK and Hattiangady B (2006). Survival and differentiation of stem/progenitor cells from the postnatal hippocampus following grafting into the intact or injured young adult and aged hippocampus. *Society for Neuroscience* .Abstracts; 10- 16.

Shetty, A. K. and Hattiangady, B (2007). Concise Review: Prospects of Stem Cell Therapy for Temporal Lobe Epilepsy. *STEM CELLS;* 25: 2396–2407.

Shihabuddin LS, Horner PJ, Ray J, Gage FH.(2000) Adult spinal cord stem cells generate neurons after transplantation in the adult dentate gyrus. *J Neurosci*;20:8727–35.

Shimato S, Natsume A, Takeuchi H, Wakabayashi T, Fujii M, Ito M, Ito S, et al, (2007). Human neural stem cells target and deliver therapeutic gene to experimental leptomeningeal medulloblastoma. *Gene Ther;* 15:1132–1142.

Shyu, W.C., Lin, S.Z., Chiang, M.F., Su, C.Y., Li, H.,(2006). Intracerebral peripheral blood stem cell (CD34+) implantation induces neuroplasticity by enhancing beta1 integrin-mediated angiogenesis in chronic stroke rats. *J. Neurosci.* 26, 3444–3453.

Silani V, Fogh I, Ratti A, Sassone J, Ciammola A, Cova L (2002). Stem cells in the treatment of amyotrophic lateral sclerosis. *Amyotroph Lateral Scleros Other Motor Neuron Disord.*;3:173-181.

Singh and Roy, 2008: Stem Cells in PMR practices, *IJPMR*; 19 (1):1-5

Sly WS, Vogler C(2002). Brain directed gene therapy for lysosomal storage disease: going well beyond the blood–brain barrier. Proc Natl Acad Sci USA.; 99:5760–5762.

Snyder EY, Taylor RM, Wolfe JH(1995). Neural progenitor cell engraftment corrects lysosomal storage throughout the MPS VII mouse brain. *Nature;* 374:367–370.

Song, H.J., Stevens, C.F., Gage, F.H., (2002). Neural stem cells from adult hippocampus develop essential properties of functional CNS neurons. *Nat. Neurosci.* 5, 438–445.

Song, S., Kamath, S., Mosquera, D., Zigova, T., Sanberg, P., etal., (2004). Expression of brain natriuretic peptide by human bone marrow stromal cells. Exp. *Neurol.* 185, 191–197.

Stem Cell Basics (2010). Stem Cell Information. National Institue of Health, Aug.. Web. July 2010. http://stemcells.nih.gov/info/basics/basics4.asp

Svendsen CN, Langston JW (2004). Stem cells for parkinson disease and ALS: replacement or protection? *Nature Med.* ;10:224-225.

Sykova´, E.; Jendelova´, P. (2005)Magnetic resonance tracking of implanted adult and embryonic stem cells in injured brain and spinal cord. *Ann. NY Acad. Sci.* 1049:146–160.

Sykova´, E.; Jendelova´, P. (2006)Magnetic resonance tracking of transplanted stem cells in rat brain and spinal cord. *Neurodegen. Dis.* 3:62–67.

Sykova´, E.; Urdzı´kova´, L.; Jendelova´, P.; Burian, M.; Glogarova´, K.; etal. (2005) Bone marrow cells—a tool for spinal cord injury repair. Exp. *Neurol.* 193:261–262.

SykovaE, Homola A, Mazanec R, LachmannH, Langkramer S dova K Kobylka P,Pa´dr, R etal (2006): Autologous Bone Marrow Transplantation in Patients With Subacute and Chronic Spinal Cord Injury *Cell Transplantation*, Vol. 15, pp. 1–100.

Taguchi, A., Soma, T., Tanaka, H., Kanda, T., Nishimura, H., et al(2004).: Administration of CD34+ cells after stroke enhances neurogenesis via angiogenesis in a mouse model. *J. Clin. Invest.* 114, 330–338.

Takagi Y, Takahashi J, Saiki H, Morizane A, Hayashi T, Kishi Y, Fukuda H, Okamoto Y, et al,(2005). Dopaminergic neurons generated from monkey ES cells function in a Parkinson primate model. *J Clin Invest.*; 115:102–108.

Takahashi J, Palmer TD, Gage FH (1999). Retinoic acid and neurotrophins collaborate to regulate neurogenesis in adult-derived neural stem cell cultures. *J Neurobiol*; 38:65–81.

Tang Y, Yasuhara T, Hara K, Matsukawa N, Maki M, Yu G, et al (2007). Transplantation of bone marrow derived stem cells: a promising therapy for stroke. *Cell Transplant*; 16:159–169.

Taylor RM, Wolfe JH(1997). Decreased lysosomal storage in the adult MPS VII mouse brain in the vicinity of grafts of retroviral vector-corrected fibroblasts secreting high levels of betaglucuronidase. *Nat Med.*; 3:771–774.

Temple S(2001). The development of neural stem cells. *Nature.*; 414:112–117.

Terada, N.; Hamazaki, T.; Oka, M.; Hoki, M.; Mastalerz, D.M.; etal. (2002) Bone marrow cells adopt the phenotype of other cells by spontaneous cell fusion. *Nature*, 416(6880), 542-545.

Tobi L, Mahendra S R (2005). *Stem Cells and Treatment of Neurodegenerative Disorders* .ENCYCLOPEDIA OF LIFE SCIENCES ©, John Wiley & Sons, Ltd. www.els.net.

Tse, W. & Laughlin, M. J. (2005). Umbilical cord blood transplantation: a new alternative option: alternative option.Hematology/the Education Program of the American Society of Hematology. American Society of Hematology. *Education Program*, 377–383.

Tuszynski MH, Thal L, Pay M, Salmon DP, U HS, Bakay R, Patel P, Blesch A, et al,(2005). A phase 1 clinical trial of nerve growth factor gene therapy for Alzheimer disease. *Nat Med.*; 11:551–555.

Tuszynski MH, U HS, Amaral DG, Gage FH(1990). Nerve growth factor infusion in primate brain reduces lesion-induced cholinergic neuronal degeneration. *J Neurosci .*; 10:3604–3614.

Tuszynski MH2002. Gene therapy for neurodegenerative disorders. *Lancet Neurol.*; 1:51–57.

Urdzı´kova´, L.; Jendelova´, P.; Glogarova´, K.; Burian, M.; Ha´jek, M.; etal´,(2006). Transplantation of bone marrow stem cells as well as mobilization by granulocyte-colony stimulating factor promote recovery after spinal cord injury in rat. *J. Neurotrauma* 23:1379–1391.

van Gelder M, van Bekkum DW.(1995): Treatment of relapsing EAE in rats with allogeneic bone marrow transplantation from a resistant strain. *Bone Marrow Transplant*;16:343–51.

Vendrame, M., Cassady, J., Newcomb, J., Butler, T., et,al,(2004). Infusion of human umbilical cord blood cells in a rat model of stroke dose-dependently rescues behavioral deficits and reduces infarct volume. *Stroke* 35, 2390–2395.

Wachs F.-P., Couillard-Despres S., Engelhardt M., Wilhelm D., Ploetz S., et al, (2003). High efficacy of clonal growth and expansion of adult neural stem cells. Lab. *Invest.*, 83: 949-962.

Wang, X., Mori, T., Sumii, T., Lo, E.H., (2002). Hemoglobin-induced cytotoxicity in rat cerebral cortical neurons: caspase activation and oxidative stress. *Stroke* 33, 1882–1888.

Wechsler, L., Steindler, D., Borlongan, C., Chopp, M., Savitz, S.,et al,(2009). Stem Cell Therapies as an Emerging Paradigm in Stroke (STEPS): bridging basic and clinical science for cellular and neurogenic factor therapy in treating stroke. *Stroke* 40, 510–515.

Weiner HL (2009): The challenge of multiple sclerosis: how do we cure a chronic heterogeneous disease? *Ann Neurol* , 65:239-248.

Weiss S, Dunne C, Hewson J, Wohl C, Wheatley M, Peterson AC, et al(1996). Multipotent CNS stem cells are present in the adult mammalian spinal cord and ventricular neuroaxis. *J Neurosci*; 16:7599–7609.

Wichterle, H., Lieberam, I., Porter, J.A. and Jessell, T.M. (2002) Directed differentiation of embryonic stem cells into motor neurons. *Cell* 110: 385_397.

Willing AE, Lixian J, Milliken M, Poulos S, Zigova T, Song S,et al, (2003) Intravenous versus intrastriatal cord blood administration in a rodent model of stroke. *J Neurosci Res.*; 73(3):296-307.

Wognum AW, Eaves AC, Thomas TE(2003):. Identification and isolation of hematopoietic stem cells. Arch *Med Res*;34:461–75.

Wolfson C, Kilborn S, Oskoui M, Genge A.(2009) Incidence and prevalence of amyotrophic lateral sclerosis in Canada: a systematic review of the literature. *Neuroepidemiology* 33(2):79-88.

Woodbury, D.; Schwarz, E. J.; Prockop, D. J.; Black, I. B.(2000) Adult rat and human bone marrow stromal cells differentiate into neurons. J. *Neurosci*. Res. 61:364–370

Wu, S.; Suzuki, Y.; Ejiri, Y.; Noda, T.; Bai, H.; etal.(2003):Bone marrow stromal cells enhance differentiation of cocultured neurospheres cells and promote regeneration of injured spinal cord. J. *Neurosci*. Res. 72:343–351.

Xi, G., Keep, R.F., Hoff, J.T., (2006). Mechanisms of brain injury after intracerebral haemorrhage. *Lancet Neurol*. 5, 53–63.

Xiao, J., Nan, Z., Motooka, Y., Low, W.C., (2005). Transplantation of a novel cell line population of umbilical cord blood stem cells ameliorates neurological deficits associated with ischemic brain injury. *Stem. Cells Dev*. 14, 722–733.

Xin H, Li Y, Shen LH, Liu X, Wang X, et al. (2010) Increasing tPA Activity in Astrocytes Induced by Multipotent Mesenchymal Stromal Cells Facilitate Neurite Outgrowth after Stroke in the Mouse. *PLoS ONE* 5(2): e9027.

Yanai J, Doetchman T, Laufer N, Maslaton J, Mor-Yosef S, et al (1995). Embryonic cultures but not embryos transplanted to the mouse's brain grow rapidly without immunosuppression. *Int J Neurosci*;81:21–6.

Yasuhara T, Matsukawa N, Hara K, Yu G, Xu L, Maki M, Kim SU, Borlongan CV(2006). Transplantation of neural stem cells exerts neuroprotection in a rat model of Parkinson disease. *J Neurosci.*; 26:124497–12511.

Yu D & Silva GA(2008). Stem cell sources and therapeutic approaches for central nervous system and neural retinal disorders. *Neurosurg Focus.*; 24(3- 4)

Yu JJ, Sun X,Yuan X, Lee JW, Snyder EY,Yu JS.(2006) Immunomodulatory neural stem cells for brain tumour therapy. *Expert Opin Biol Ther* ;6:1255–62.

Zhang J, Li Y, Zheng X, et al.(2008)Bone marrow stromal cells protect oligodendrocytes from oxygen-glucose deprivation injury. *J Neurosci Res.*;86:1501-1510.

Zhao LR, Duan WM, Reyes M, et al(2002). Human bone marrow stem cells exhibit neural phenotypes and ameliorate neurological deficits after grafting into the ischemic brain of rats. *Exp Neurol* ;174: 11–20.

16

Cell Therapy for Parkinson's Disease: Failure or Success?

Magdalena Guerra-Crespo[1,2], Alberto K. De la Herrán-Arita[1], Arturo Hernández-Cruz[1,2], José Bargas[1,2] and René Drucker-Colín[1,2]
[1]*Departamento de Neuropatología Molecular, Instituto de Fisiología Celular*
[2]*Grupo Células troncales adultas, regeneración neuronal y enfermedad de Parkinson (IMPULSA-02), Universidad Nacional Autónoma de México*
México

1. Introduction

The mature central nervous system (CNS) is probably the most complex structure known in nature. This fact and the irreversibility of most forms of clinical brain damage are the basis for the long-held belief that the adult brain cannot restore itself and cannot be repaired. Parkinson's disease (PD) is a chronic and progressive neurodegenerative disorder characterized by the loss of dopaminergic neurons in the substantia nigra pars compacta (SNPc) with a concomitant loss of the catecholamine called dopamine (DA), the neurotransmitter released at the axon terminals of the SNPc neurons that project to the striatum (caudate nucleus and putamen) (Fig.1).

Historically, the therapy for PD is aimed at reinstalling proper stimulation of the dopamine receptors in striatum. The dramatic breakthrough was the introduction of L-DOPA treatment in 1969 (Cotzias *et al.*, 1969). L-DOPA, the precursor of dopamine, passes the blood-brain barrier (BBB) and is converted to dopamine, which becomes available for dopamine receptors in striatum, thereby improving the balance between excitatory and inhibitory influences in this brain region. Together with L-DOPA treatment, dopamine reuptake inhibitor, dopaminergic agonists and muscarinic antagonists also have a clinical effect. Despite this pharmacologic advance in treatment, there remains no cure for PD.

Because PD is a neurodegenerative process and long term therapy is necessary, development of severe side effects such as dyskinesias (movement disorder), limits the usefulness of L-DOPA therapy over time and progressively becomes less effective; consequently, patients become more troubled by freezing or akinesias. In addition, L-DOPA will not only reach the striatum, but the entire CNS as well as the rest of the body, where it can develop unwanted side effects.

Additionally, surgical treatment is being used to treat people with advanced PD for whom drug therapy is no longer sufficient. The more frequently employed techniques are thalamotomy, lesion of the internal globus pallidus or subthalamic nucleus and chronic implantation of electrodes for deep brain stimulation, amongst others. Even though there is a clinical recovery in PD patients after surgical therapy, as seen with pharmacological therapy, the progression of the disease cannot be avoided. Hence, the basic principle of

neural transplantation for PD is to provide DA from the graft in a stable fashion directly into the striatum where the intrinsic dopaminergic system has been degenerated (Fig.1). This procedure attracted the attention of the entire scientific community over the past decades.
In this chapter we will describe the experimental work and clinical trials that provided the basis for the development of cell therapy in PD. Afterward, this chapter will depict the suitability and the therapeutic potential of stem cells from different origin that could be employed for regeneration therapy in human clinical trials. Finally, this chapter will discuss the current strategies for the assessment of tissue integration after grafting and our proposal of an alternative method for evaluating the effectiveness of the transplants in rodent models of PD.

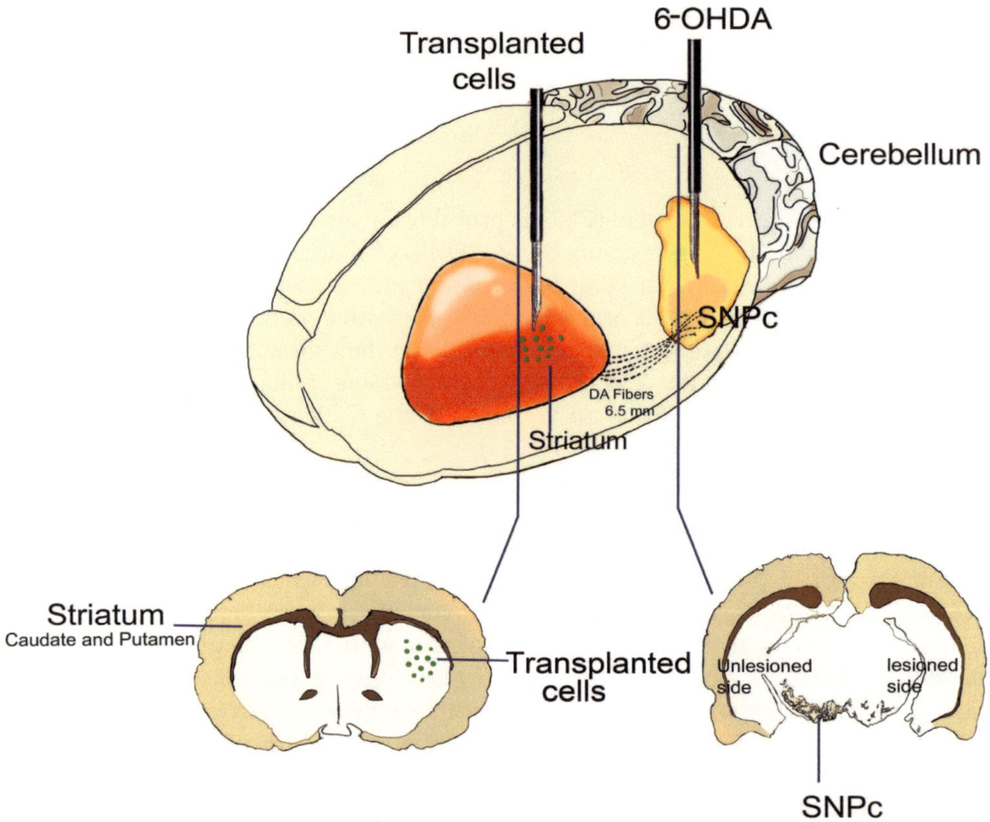

Fig. 1. Transplant therapy in a rodent model of PD. Grafted cells are located in the denervated striatum of the 6-hydroxydopamine (6-OHDA) lesioned rat.

2. The experimental approach in animal models of PD: the foundation of transplant therapy

Even though PD is mostly idiopathic, it is well known that loss of a specific and highly specialized neuronal subpopulation underlie the process of the disease. The optimal tactic to evaluate the benefits and/or drawbacks of different treatments, devoid of using human

patients, is to develop an animal model that resembles the pathology observed in PD. To this point, animal models of PD are the best method to evaluate the success or failure of transplant therapy.
Section 2 depicts the evolution of transplant research done in animal models. A summary of the trials can be found in Table 1.

2.1 Animal models of PD

The most commonly used PD model is the 6-OHDA unilateral lesioned rat. 6-OHDA is a specific neurotoxin for catecholaminergic neurons; uptake of 6-OHDA by these neurons is performed in a similar fashion to that for intrinsic catecholamines (Sachs & Jonsson, 1975). Because 6-OHDA is not able to cross the BBB, unilateral direct injection into the substantia nigra or into the medial forebrain bundle is sufficient to destroy >95% of the midbrain dopaminergic neurons. Unilateral lesioned rats will rotate contralaterally in response to DA agonists such as apomorphine, which is the result of the supersensitivity of striatal DA receptors in the lesioned side. On the opposite, unilateral lesioned rats will rotate ipsilaterally to the lesion in response to amphetamine. Circling behavior can be analyzed by gross visual observation, video recording or rotometer.
Another neurotoxin commonly employed for the development of animal models of PD is 1-methyl-4-phenyl-1,2,3,6-tetrahydropyridine (MPTP). This toxin was discovered in young drug addicts who used illicit drugs contaminated with meperidine, causing symptoms very similar to idiopathic PD (Langston et al., 1983). Sensitivity to MPTP varies considerably among animal species. Non-human primates are most sensitive in a manner similar to humans; mice, cats, and dogs are rather sensitive, whereas rats are resistant to MPTP (Kopin & Markey, 1988).

2.2 Fetal substantia nigra transplants

With the extensive variety of animal models of PD, experimental transplant therapy for PD swiftly began in the late 70's; fetal substantia nigra was the initial choice for transplants, since fetal tissue was considered to have lower possibility of rejection from the host's immune system than adult cells.
The firsts fetal substantia nigra experimental transplant performed in a PD model, reported a reduction of the motor abnormalities in a unilaterally DA-denervated rat (Björklund & Stenevi, 1979; Perlow *et al.*, 1979). Since then, the technique has been improved in several ways and established that DA-containing grafts can alleviate some of the symptoms in different experimental models of PD, including mice and rats (Aguayo *et al.*, 1984; Annett *et al.*, 1994; Brundin *et al.*, 1988; Di Porzio & Zuddas, 1992; Dunnett *et al.*, 1981; Hefti *et al.*, 1985; Herman *et al.*, 1991; Zhu *et al.*, 1992).
Once the first experiments using fetal substantia nigra grafts in rodents were concluded and giving preliminary and promising results, the next rational step was to extrapolate these experiments to a higher member of the evolutionary chain, the non-human primates. Morihisa *et al.*, were the first ones to explore the potential of this therapeutic approach, transplanting fetal substantia nigra into the denervated caudate nucleus of the rhesus monkey with unilateral lesion of the striatum secondary to 6-OHDA direct administration. For their surprise, under the specific conditions of their experiment, fetal substantia nigra graft did not survive in either of two animals tested. The complete loss of the transplanted dopaminergic graft could be due to an immunologic incompatibility; which placed the notion that the brain is an immunologically privileged site under inquiry and questioned the

idea that rejection processes would be less severe inside the brain. In order to circumvent graft rejection, they employed grafts from autolog adrenal medulla tissue; of particular interest is that adrenal medulla tissue from four animals tested did survive (Morihisa *et al.*, 1984). Even though they did not report improvement after adrenal medullary chromaffin transplant, they were the pioneers in the usage of autologous transplants for PD treatment, one that took notoriety in the years to come.

2.3 Sympathetic neuron transplants

The potential therapeutic value of autologous transplants was also explored using sympathetic neurons for cell transplant in PD. Sympathetic ganglion cells mainly release noradrenaline whereas a minute amount of DA is also synthesized and secreted by a small subpopulation of the ganglion cells (Lyon *et al.*, 1987). With this taken into consideration, autologous sympathetic neurons derived from the superior cervical ganglion were grafted in animal models of PD, monkey and rat. Tissue pieces of the sympathetic ganglion were grown in serum-free medium supplemented with nerve growth factor; short neurites were found to emanate from sympathetic ganglion tissue after 6 days in culture. These studies have revealed that sympathetic neuron autografts placed in the DA-denervated striatum survive and improve motor deficits such as drug induced circling behavior and hypokinetic disorders, seen in rodent and non-human primate models of PD (Itakura *et al.*, 1988; Nakao *et al.*, 1995).

2.4 Human retinal pigment epithelial (hRPE) cells transplant

With the arrival of new and more advanced cell culture procedures, innovative sources were available for the acquisition of alternative DA-producing cells. Amongst them, the hRPE cell was one of the most promising ones; the study of hRPE cells demonstrated that these cells express vesicular monoamine transporter and the D1 receptor. Moreover, hRPE cells secrete L-DOPA and small quantities of dopamine. This alternative graft was transplanted in a rat and monkey model of PD with significant improvement of circling behavior and motor task, respectively. Nevertheless, a thorough depiction of dopaminergic transplanted cell survival and function was not provided (Doudet *et al.*, 2004; Subramanian *et al.*, 2002).

2.5 The dim truth about experimental transplant in animal models of PD

Regardless of the improvement reported in these animal models of PD after transplant, several factors were not taken into account in order to make an adequate evaluation of the enhancement produced by the graft. Even though there is evidence of a decrease in circling behavior, the survival of DA-producing transplanted neurons is not fully described in many of the experiments, and for that reason, a correlation between neuronal graft survival and behavior improvement cannot be supported.

With the purpose of making an un-doubtful conclusion that a functional recovery is clearly undergoing, a clear correlation between the extent and location of dopaminergic re-innervation within the striatum and recovery of dopaminergic function should be done. The remission of circling behavior may relate to the fact that DA receptors on the denervated side are supersensitive and/or that striatum is normally supplied with a far greater innervation than is needed for minimal maintenance of function. It is also possible that a low number of neurons and terminals may increase their synthesis and turnover of DA to help compensate for the deficit.

Cell Source	Donor	Recipient	Result	Time period (after transplant)	TH neuron count (after transplant)	Reference
Fetal 19 days Nigra	Rat	6-OHDA unilateral lesion Rat	Decrease in 33% of ipsilateral rotation (amphetamine)	3.5 months	300 neurons	Björklund & Stenevi, 1979
Fetal Nigra	Rat	6-OHDA unilateral lesion Rat	Decrease in 70% of contralateral rotation (apomorphine)	4 weeks	NA	Perlow *et al.*, 1979
Fetal 19 days Nigra	Rat	6-OHDA bilateral lesion Rat	Decrease in 50% of spontaneous akinesia	20 days	NA	Dunnett *et al.*, 1981
Fetal 16-17 days Nigra	Rat	6-OHDA unilateral lesion Rat	Augmentation of DA fiber density	10 months	NA	Aguayo *et al.*, 1984
Adrenal medulla autografts	Rhesus macaque (non-human primate)	6-OHDA unilateral lesion Rhesus macaque	NA	8 months	150 cathecolamine neurons	Morihisa *et al.*, 1984
PC12 cells (rat adrenal medulla pheochromocytoma)	Rat PC12 cell culture	6-OHDA unilateral lesion Rat	Decrease in 27% of ipsilateral rotation (amphetamine)	2 weeks	NA	Hefti *et al.*, 1985
Superior cervical ganglia autografts	Rhesus macaque (non-human primate)	MPTP lesion Rhesus macaque	Increase of 60% of homovanillic acid content in cerebrospinal fluid.	3 months	NA	Itakura *et al.*, 1988
Fetal 6.5-8 weeks Nigra	Human	6-OHDA unilateral lesion Rat	Decrease in 94% of ipsilateral rotation (amphetamine)	19 weeks	1, 600 neurons	Brundin et al., 1988
Fetal 3 weeks Nigra	Pig	6-OHDA unilateral lesion Rat	Abolishment of circling behavior (amphetamine)	17 weeks	Up to 12, 000 neurons	Huffaker *et al.*, 1989
Fetal 14 days Nigra	Rat	6-OHDA unilateral lesion Rat	Decrease in 50% of contralateral rotation (apomorphine)	9 months	3, 400 neurons	Herman *et al.*, 1991
Fetal 12-13 days Nigra	Mouse	MPTP bilateral lesion Mouse	NA	4 months	2, 600 neurons	Di Porzio & Zuddas, 1992
Fetal 14 days Nigra expressing PKCβ1 cDNA (second messenger)	Rat	6-OHDA unilateral lesion Rat	Decrease in 64% of ipsilateral rotation (amphetamine)	8 weeks	NA	Zhu *et al.*, 1992
Fetal 13-16 days Nigra	Rat	6-OHDA unilateral lesion Rat	Decrease in 70% of contralateral rotation. (apomorphine)	3 months	NA	Chung *et al.*, 1993
Fetal 74 days Nigra	Marmoset (non-human primate)	6-OHDA unilateral lesion Marmoset	Ipsilateral/contralateral rotations per minute ratio of 0.63 vs 5.15 of control lesion	6 months	1, 800 neurons	Annett *et al.*, 1994
Superior cervical ganglia (4-week-old culture)	Rat	6-OHDA unilateral lesion Rat	Decrease in 40% of contralateral rotation (apomorphine)	4 weeks	NA	Nakao *et al.*, 1995.
Fetal 14 days Nigra mixed with cultured bFGF-producing cells	Rat	6-OHDA unilateral lesion Rat	Total reversal of ipsilateral rotation (amphetamine)	8 weeks	2, 800 neurons	Takayama *et al.*, 1995

Cell Source	Donor	Recipient	Result	Time period	TH neuron count	Reference
				(after transplant)		
Engineered grafts						
Fetal 17 weeks retinal pigment epithelial cells attached to microcarriers	Human	6-OHDA unilateral lesion Rat	Decrease in 50% of contralateral rotation (apomorphine)	12 weeks	NA	Subramanian *et al.*, 2002
Fetal retinal pigment epithelial cells attached to microcarriers	Human	MPTP lesion Rhesus macaque	Improvement of 38% in motor tasks (personalized scale)	8 weeks	NA	Doudet *et al.*, 2004

Table 1. Cell transplant therapy in animal models of PD.

In spite of the methodological scarcity, these investigations gave the stepping stone towards a clinical approach in PD transplantation and supported the idea that it could be extrapolated to human clinical trials.

3. Transplant therapy in PD patients: A leap forward to a potential successful treatment

Transplantation of dopaminergic cells into the denervated striatum is an experimental approach that was brought to practice to overcome the disadvantages of medications used in the treatment of PD in an attempt to provide the brain with an unlimited source of DA synthesized by the grafted cells. On section 3 we depict the clinical trials that have been taking place until nowadays (Table 2).

The results of the first human trials were published in 1980's. More than 30 years ago, the major difficulty for performing a transplant was the shortage of dopaminergic cells suitable for transplant and cell-based manipulation techniques were not sophisticated enough at that time. The most logical selection was to employ autologus dopaminergic tissue. Several of the technical and ethical problems, and probably all of the immunologic difficulties, would be solved if catecholamine-producing cells from the patient's own body were used; this reasoning led two independent groups (Backlund *et al.*, 1985; Drucker-Colín *et al.*, 1988) to investigate if adult adrenal medullary chromaffin cells were able to function as an alternative DA source for transplant. Because the systemic administration of L-DOPA or DA agonist has been successfully applied in the treatment of PD, the logical inference was to expect that adrenal medullary chromaffin cells might act as a chronic pump for delivering DA into the denervated striatum of PD patients.

The first attempt to use autologus medullary chromaffin cell transplant in PD patients was done in Sweden (Backlund *et al.*, 1985), though, the clinical improvement was over a period of one week and one of the patients developed a resistance to L-DOPA treatment.

The first report of significant improvement of symptoms in PD patients after autologous medullary chromaffin cell transplant was done by our (Drucker-Colín) group in México (Drucker-Colin et al., 1988; Madrazo, 1987, 1988). Our team chose a microsurgical transcortical intraventricular approach and implanted autologous adrenal medulla directly into the unilateral caudate nucleus. Although the grafts were unilateral, the improvement was bilateral and symmetric. The direct contact with the caudate nucleus (without the use of a stainless-steel tissue holder that the Swedish group used), exposure to cerebrospinal fluid, and the shorter interval between adrenal dissection and transplantation almost certainly made the difference. However, there were significant risks and disadvantages associated with autologous adrenal transplants; surgical time in a double operation is considered perilous for a PD patient (since most of them are over 60 years) and DA producing cells obtained from adrenal medulla might be not sufficient to compensate the neural DA deficit. These trials were a breakthrough result in the context of transplant therapy in neurodegenerative diseases; the notion that peripheral tissue would be able to survive and integrate within the host's CNS as well as to induce clinical improvement was not considered to be possible before these studies.

Given the promissory results obtained using adrenal chromaffin cells, which showed variability and technical difficulties; tissue from fetal substantia nigra became to be considered a better option as source of DA cells. Madrazo and Drucker-Colín performed the first transplant of fetal substantia nigra using tissue from cadaveric consent donation (Madrazo *et al.*, 1988). The two transplanted patients showed an improvement of the UPDRS score at 8 weeks post-surgery.

Some years later, in another attempt to improve autolog transplantation of adrenal medullary chromaffin cells, our group took advantage that cultured adrenal chromaffin cells differentiate into neurons in a higher rate when they are stimulated by a low frequency magnetic field (Drucker-Colín *et al.*, 1999). This study proved that adult differentiated adrenal chromaffin cells can be successfully transplanted into the caudate nucleus of a patient with PD and substantially mitigate the clinical symptoms and reduce the intake of L-DOPA medication (see reference video of pre and post-transplanted stages of the surgically treated patient). Postoperative clinical assessments revealed a significant amelioration of visuospatial deficits and a progressive fading of rigidity and akynesia, as well as an improvement in memory tasks; furthermore, a decrease in approximately 70% of L-DOPA intake to ameliorate parkinsonian symptoms was reported.

In spite of the political controversy involving this pioneer studies, the clinical improvement obtained in some patients set off the interest of transplant therapy in PD patients around the world (Deacon *et al.*, 1997; Freed 1990, 2001; Hirsch *et al.*, 1990; Itakura *et al.*, 1997; Kordower 1995, 1996; Lindvall *et al.*, 1990; Peterson *et al.*, 1989; Sawle *et al.*, 1992; Watts *et al.*, 2003) (See table 2).

However, the employment of human fetal dopaminergic neurons as a DA source for transplant is a complicated approach, due to its availability, which is imposed by ethical affairs; for this reason, they have been only employed in some human cases (Freed *et al.*, 1990; Hauser *et al.*, 1999; Kordower *et al.*, 1995; Lindvall, *et al.*, 1990; Madrazo *et al.*, 1988; Mendez *et al.*, 2008; Sawle *et al.*, 1992).

All these initial trials were performed as "open label" (no placebo group as a control), which arose the debate regarding the inclusion of a control placebo group, which is envisaged in general as unethical.

Freed's group published in 2001 the first double-blind placebo trials using embryonic tissue. Their results were considered disappointing by some researchers and promising by others (Dunnet *et al.*, 2001; Isacson *et al.*, 2001). The controversy is based in the modest recovery observed in the grafted patients compared with the sham group; moreover, the serious dyskinetic side effects observed after several years of transplantation, places this procedure under a doubtful perception of acceptability. However, the fact they found improvement in subjects younger than 60 years, leads to consider better methodologically controlled approaches.

One of the disadvantages in employing this cell source (fetal nigral tissue) is the variability of the results; improvements of parkinsonian symptoms occur in patients who received the largest amounts of fetal mesencephalic tissue (Hauser *et al.*, 1999; Sawle *et al.*, 1992), whereas in other cases, the transplanted cells did not survive for more than a few months and the patients usually return to preoperative state (Goetz *et al.*, 1991; Hirsch *et al.*, 1990). Additionally, to maintain the fetal tissue viable after transplantation, long-term immunosuppressive treatment is needed. Nonetheless, without a doubt the major limiting factor of this approach is the difficulty in obtaining a sufficient amount of human viable fetal tissue.

Dejectedly, as has been observed in those studies, there are some patients that clearly demonstrate improvement and others that do not show any recovery; however, it is still not possible to unravel this mystery of selectivity. Whitin the possible mechanisms accounting for these changes, there is patient age, transplant technique, cell graft source and as the most probable candidate the immune system; which remains the most formidable barrier to transplantation as a routine medical treatment (Freed *et al.*, 2004). The immune system has developed elaborate and effective mechanisms to combat foreign agents and we are now in pursuit of elucidating them.

The successes of human to human organ transplantation, coupled with severe limitations in the availability of human organs, propelled research into the use of non-human organs in the 1990's. The domesticated pig was the preferred donor at the time. In an audacious attempt to replace fetal tissue, surgery to transplant brain tissue from pig fetuses into the caudate-putamen of individuals diagnosed with PD was performed. They reported a clinical improvement of rigidity in a period of 7 months and that autoimmune processes did not become established in the patients (Deacon *et al.*, 1997). Unfortunately, safety and efficacy have not yet been clearly demonstrated, and information is still preliminary to warrant this therapy. Current opinion is that the risk of transfer of a significant infectious microorganism from the pig to the human recipient of a xenograft is small and ethically acceptable if the transplant will be highly beneficial to the host. However, pig specific pathogens are considered to be of potential risk, thus, limiting the use of pig xenografts. Clinical xenotransplantation is becoming feasible and attractive as a routine therapy, nonetheless, some attention should be set on this matter.

In another attempt to evade the host immune response after transplantation, human retinal pigment epithelial (hRPE) cells were taken into consideration. As mentioned previously, experimental studies suggested that RPE cells were a suitable cell type to serve as a potential enduring source of L-DOPA for implantation into the striatum of animal models of PD (Subramanian *et al.*, 2002). These implants ameliorated the motor deficits in rodent and non-human primate models of PD and immune suppression was not required even when transplants were made from one species to another (Doudet *et al.*, 2004; Subramanian *et al.*, 2002).

Those results were taken into account in a novel tissue engineering strategy; hRPE cells were attached to gelatin microcarriers (spheramine™) and transplanted into the putamen contralateral to the more symptomatic side of patients with Parkinson disease (Watts *et al.*, 2003). They reported an average improvement of 48% in motor score 12 months after implantation evaluated through the Unified Parkinsons's Disease Rating Scale (UPDRS) with the patient in the "off" state, which was sustained for 24 months. Improvement was also observed in activities of daily living, quality of life, and motor fluctuations.

However, in a more recent work, a post-mortem study of a patient enrolled in a similar clinical trial reported that only 118 cells from the transplant endured (estimated 0.036% survival) and that clinical improvement was not observed (Farag *et al.*, 2009). This engineered tissue approach is still an emerging therapy alternative for PD and the existing body of literature is still inadequate to allow conclusions regarding this procedure. Concerns related to small sample sizes and a limited number of controlled trials are joined with a wide array of methodological issues with the procedure itself. We have no doubt that better procedures will arrive alongside the advances in tissue development.

The leading inquiry raised by transplant therapy hitherto is the variability in functional recovery either in animal models and human patients. In human studies, this unevenness is not only seen between different trials, but also within groups of PD patients transplanted in the same assay. In order to attend such matters, a comprehensive clinical trial must be done, one that includes a meticulous analysis of patient selection (e.g., age, level of PD), knowledge of the number of DA neurons transplanted, type of surgery, follow up of the patient before surgery, monitoring of the clinical improvement after grafting, analysis of cell survival and neuron functionality. A successful cell therapy must settle advantages over current treatments for lessening motor symptoms in PD patients. Cell replacement should provide long-lasting, major improvements of mobility and no manifestation of dyskinesias without the need of further therapeutic interventions, an uneasy task that most likely will not be achieved by using solid fetal ventral midbrain grafts. This is suggested in the recent report from Mendez (Mendez *et al.*, 2008), who transplanted fetal ventral midbrain as cellular suspension. The analysis did not reveal development of diskynesias in the transplanted PD subjects in a period spanning from 9 to 14 years; moreover, the postmortem analysis of the transplanted patients' brains did not show any sign of degeneration (alpha synunclein, ubiquitin), and only a minimum of microglial reaction from the host was found. In contrast, in the two double blind placebo trials, where solid pieces of fetal midbrain tissue were used, severe microglial reaction was found and diskynesias were also observed (Freed *et al.*, 2001; Olanow *et al.*, 2003). Mendez's results supports a new "hope" in medical treatment for PD patients based in stem cell tissue procedures, after the setback given by the report from Kordower of transplant analysis following 14 years after the procedure (Kordower *et al.*, 2008), in which the postmortem tissue revealed that the grafted cells in the striatum were affected by a neurodegenerative mechanism, suggesting that the neurodegenerative process was not exclusive of the SNPc. Taken together, the present observations suggest that the methodological procedures and not the cell source, are directly involved in the limited success of transplant therapy that employs solid portion of fetal ventral tissue.

Currently, there are various ongoing stem cell based clinical trials registered by the National Institutes of Health (NIH): (http://www.clinicaltrials.gov/ct2/results?term=transplant +parkinson), and new assessments are being developed by the Transeuro consortium (Allan *et al.*, 2010) (http://www.transeuro.org.uk/index.html). The European group coordinated by Dr. Roger Barker is planning highly controlled studies in the near future (2012). Interestingly, the cells to be employed in these trials will be obtained from the midbrain of human fetuses, the same tissue we used the first time more than 3 decades ago. Again, as in past, the idea still creates controversy between those who concur with such methods and those who are unfavorable to it due to the limited success percentage (Holden, 2009).

We expect that the new trails should provide better motor recovery to the ones observed in previous assays. Even though behavioral improvement is seen in experimental transplant trials, going from 30% to total reversal of circling behavior (Tables 1 and 3), the motor symptoms observed in grafted human subjects when analyzed by UPDRS do not surpass 60% of clinical improvement (Table 2 and Dunnet *et al.*, 2001).

Cell Source	Donor	Implant Zone	Result	Time period	TH neuron count	Reference
				(after transplant)		
Autologous adrenal medullary tissue	Human	Striatum (Putamen)	Improvement of rigidity	1 week	NA	Backlund *et al.*, 1985
Autologous adrenal medullary tissue	Human	Lateral Ventricle	Disappearance of rigidity and akinesia	5 months	NA	Madrazo *et al.*, 1987
Autologous adrenal medullary tissue	Human	Striatum (Caudate)	Decrease in 50% of L-DOPA dose	1 year	NA	Drucker-Colín *et al.*, 1988
Autologous adrenal medullary tissue/fetal 13 weeks substantia nigra tissue	Human	Lateral Ventricle	Disappearance of rigidity and akinesia. Decrease in 70% of L-DOPA dose	2 years	NA	Madrazo *et al.*, 1988
Autologous adrenal medullary tissue	Human	Striatum (Putamen)	Improvement of rigidity	2 weeks	Necrotic tissue	Peterson *et al.*, 1989
Autologous adrenal medullary tissue	Human	Striatum (Caudate)	None	4 months	Necrotic tissue	Hirsch *et al.*, 1990
Fetal 45-55 days Nigra tissue	Human	Striatum (Caudate and Putamen)	Improvement of Hoehn –Yahr scale from 3.71 to 2.5	4 years	N.A.	Freed *et al.*, 1990
Fetal 8-9 weeks Nigra tissue	Human	Striatum (Caudate)	Increase in 130% of [18F)-DOPA intake in PET scan. Increase of L-DOPA effectiveness from 2 to 14 hours	5 months	NA	Lindvall *et al.*, 1990
Autologous adrenal medullary tissue	Human	Striatum (Caudate)	Clinical improvements of 19% in UPDRS score	2 years	NA	Goetz *et al.*, 1991
Autologous adrenal medullary tissue	Human	Striatum (Caudate)	Decrease in 60% of L-DOPA dose	18 months	NA	López-Lozano *et al.*, 1991
Fetal 8-9 months Nigra tissue	Human	Striatum (Putamen)	Increase in 200% of [18F)-DOPA intake.	1 year	NA	Sawle *et al.*, 1992
Fetal 6.5-9 weeks Nigra tissue	Human	Striatum (Putamen)	Improvement of UPDRS from 78 to 49.5	18 months	210, 000 neurons	Kordower *et al.*, 1995/1997
Superior cervical ganglia autografts (stellate ganglion)	Human	Striatum (Putamen)	Reduction of 33% in the time taken to perform motor task. No changes in UPDRS score	3 years	NA	Itakura *et al.*, 1997
Fetal 9 weeks Nigra tissue	Pig	Striatum (Caudate and Putamen)	Improvement of rigidity	7 months	NA	Deacon *et al.*, 1997
Adrenal medullary tissue differentiated to DA cells in 60Hz magnetic field	Human	Striatum (Caudate)	Decrease in 70% of L-DOPA dose	1 month	NA	Drucker-Colín *et al.*, 1999

Cell Source	Donor	Implant Zone	Result	Time period (after transplant)	TH neuron count (after transplant)	Reference
Fetal 6-9 weeks Nigra tissue	Human	Substantia Nigra Pairs compacta	Clinical improvements of 32% in UPDRS score	20 months	80, 000 neurons	Hauser *et al.*, 1999
Fetal Nigra tissue	Human	Striatum (Putamen)	Increase in 40% of [18F)-DOPA intake in PET scan. Clinical improvements of 30% in UPDRS score.	1 year	30, 000 neurons	Freed *et al.*, 2001
Fetal Nigra tissue	Human	Striatum (Putamen)	Clinical improvements of 15% in UPDRS score	2 years	NA	Olanow *et al.*, 2003
Suspended fetal ventral midbrain cells from 6-10 weeks	Human	Striatum (Putamen)	Increase in 185% of [18F)-DOPA intake in PET scan. Clinical improvements of 60% in UPDRS score	1 year	NA	Mendez *et al.*, 2008
Adult neural stem cells grown as neurospheres	Human	Striatum (Putamen)	Increase in 30% of [18F)-dopa intake in PET scan. Clinical improvements of 80% in UPDRS score	1 year	NA	Lévesque *et al.*, 2009
Autologous bone marrow-derived mesenchymal stem cells	Human	Sublateral ventricular zone	Clinical improvements of 22% in UPDRS score	3 years	NA	Venkataramana *et al.*, 2010
Engineered grafts						
Retinal Pigment Epithelial cells attached to microcarriers	Human	Striatum (Putamen)	Clinical improvements of 48% in UPDRS score	1 year	NA	Watts *et al.*, 2003
Retinal Pigment Epithelial cells attached to microcarriers	Human	Striatum (Putamen)	None	6 months	118 cells	Farag *et al.*, 2009

Table 2. Cell and stem cell transplant therapy in patients with PD.

4. Stem cell transplantation: Is it the idyllic approach?

Early research gazed at ventral mesencephalic fetal dopaminergic tissue for transplantation in animal models of PD and PD patients with partial success. Nonetheless, while fetal primary tissue showed promise, the widespread clinical application of this approach is considered limited due to the matter of rejection and low cell survival. The search for alternative sources able to circumvent at least some of the problems inherent to fetal tissue is actually undergoing.

Since the advent of stem cells, they have been proposed as potential candidates to generate dopaminergic mesencephalic neurons and replace the cell loss that takes place in PD.

Stem cells are of an undifferentiated nature and possess an extend capacity to proliferate, as well they are capable to endow cells with the same sort of undifferentiated state. These cells are sorted in accordance to their intrinsic capacity to generate different cellular types, ability that diminishes throughout embryonic development.

They are considered totipotent when they are able to originate any kind of embryonic or extra-embryonic cell (few days embryo) (Thomson & Odorico, 2000). Further in embryonic development, in the blastocyst stage, cells of the inner cell mass (ICM) have the potential to contribute to the three embryonic germinal layers, ectoderm, mesoderm and endoderm, and they are classified as pluripotent. Finally, when the embryo evolves to the gastrula stage, the ICM cells have already differentiated and compromised to a specific lineage, based on the elapsed time and location within the embryo. These cells are multipotent and they are present in the embryo as well as in the adult organism; an example is the neural stem cells (NSC) discussed below.

We should classify stem cells in three broad groups: Adult neural stem cells (neural stem cells and mesenchymal), embryonic stem (ES) cells and the recently obtained induced pluripotent stem (iPS) cells. All of them have been considered in greater or lesser degree accordingly to their specific characteristics as a potential source of dopaminergic cells. With this wide diversity of cells available for transplant, the question directly arising is, "is there an idyllic stem cell to be employed as source of dopaminergic neurons?" said it in other way, "is there a cell capable of long-term survival, steady release of DA, integration into the host brain and therefore induce functional benefits without side effects?" In this section we will describe the properties and capacity of stem cells from different origin to become dopaminergic neurons, with the purpose of elucidating that question. Additionally, we will describe the results of the first stem cell transplant experiments performed in animal models of PD and the few trials implemented in humans.

4.1 Adult neural stem cells

Several studies indicated the existence of dividing cells in the CNS (Altman, 1969) but the discovery and isolation of a subtype of multipotent cells, the NSC, in specific regions of the mice adult brain (Reynolds & Weiss, 1992) and later in the human brain (Eriksson *et al.*, 1998) were the events that clearly revealed that the adult CNS posses an inherent plastic capability. This instigated the relentless pursue to replace the cellular loss that takes place in the CNS after injury or neurodegenerative diseases.

NSC are self-renewing progenitors specified to give rise only nervous tissue-specific cell types, neurons, glia and oligodendroglia (Reynolds and Weiss, 1996). In adult rodent brain, the SVZ and the dentate gyrus of the hippocampus contain NSC population with permanent capacity of proliferation (Doetsch *et al.*, 1999; Gage, 2000). These brain areas with potential to generate new neurons are defined as neurogenic niches. The advantage of containing an adult source of stem cells is the possibility of avoiding the use of human fetal tissue and embryonic derived stem cells for replacement therapy in neurodegenerative diseases.

4.1.1 NSC from adult subventricular zone (SVZ)

Neural stem cells from the SVZ move in chain migration along the rostral migratory stream (RMS) and differentiate into periglomerular interneurons at the olfactory bulb throughout the rat lifespan (Doestch *et al.*, 1999). The functional relevance of permanent replacement of periglomerular neurons in the rodent brain is attributed to the olfactory adjustment to odor changes in the environment.

Noteworthy, stem cells in the SVZ are able to respond to adverse damage in the brain. An increase in proliferation is observed under traumatic conditions such as acute stroke (Arvidsson *et al.*, 2002; Parent *et al.*, 2002) or in chronic stroke after being stimulated by transforming growth factor alpha (TGFa) (Guerra *et al.*, 2009). Conversely, in animal models of PD, decreased proliferation of the progenitors in the SVZ takes place (Hoglinger *et al.*, 2004). Interestingly, a simultaneous increase of Paired box gene 6 (Pax6) dopaminergic interneurons in the periglomerular layer is also occurring (Winner *et al.*, 2006). The same increase of TH+ cells is observed in the olfactory bulb of postmortem tissue in PD patients (Huisman *et al.*, 2004).

The reasons to consider the adult multipotent progenitors of the SVZ as an option to regenerate the dopaminergic population in the affected striatum of PD patients are their commitment to differentiate into dopaminergic neurons in the olfactory bulb and that they remain responsive to different signals.

However, it is well known that different set of transcription factors determine the correct dopaminergic fate in the olfactory bulb and SNPc. In the olfactory bulb, transcription factors such as the ETS transcription factor Er81 (Er81), Pax6 and Distal-less homeobox 2 (Dlx2), regulate the terminal differentiation of periglomerular dopaminergic neurons (Brill *et al.*, 2008; Cave and Baker 2009; Hack *et al.*, 2005; Kohwi *et al.*, 2005). Instead, the initial specification of dopaminergic mesencephalic neurons from the SNPc is regulated by the LIM homeobox transcription factor 1 (Lmx1a) and the homeobox transcription factor 1 (Msx1) (Andersson *et al.*, 2006). This evidence suggests that the neuroblasts arising from the SVZ do not develop into A9 dopaminergic neurons, the cells mainly affected in PD (German *et al.*, 1992). Additionally, when NSC are expanded in culture as floating cellular aggregates in the neurospheres assay they mainly generate glial cells (Storch *et al.*, 2004). Together, these results on animal models suggest that adult NSC arising from the SVZ should not be an optimal selection for cell therapy in PD.

4.1.2 Neural stem cells-fetal derived transplants.

Different protocols *in vitro* have been developed to generate dopaminergic neurons from NSC (for review see Deirborg *et al.*, 2008). In spite the expansion process of NSC, it decreases their potential to differentiate into dopaminergic neurons (Ptak et al., 1995); in rats, it was demonstrated that fetal ventral mesencephalic (VM) precursor cells have the potential to proliferate and differentiate into dopaminergic neurons in culture when stimulated by the growth factor FGF-2 (Studer *et al.*, 1998). An important fact involving the use of the expanded fetal neural stem cells is the low survival observed after grafting the lesioned striatum (around 3-5%), therefore, the improvement of behavioral deficits is not great compared to the one obtained with primary fetal cells (not amplified) (Brundin *et al.*, 1988). Even thought fetal neural stem cells are not the most adequate source for transplant, they could be considered a better option than NSC from the SVZ.

4.1.3 Neural stem cells-adult derived transplant

NSC from adult tissue have been employed in a relevant clinical trial, neural stem cell-derived neurons were isolated from cortical and subcortical tissue and expanded *in vitro* for several months. Nine months after harvesting, autologous cell suspensions containing differentiated dopaminergic and GABAergic neurons were microinjected unilaterally in a patient with advanced Parkinson's disease. Over the next 3 years, the overall UPDRS improved by 80%. However, at five years post-operatively, clinical motor scores returned to baseline (Table 2) (Lévesque *et al.*, 2009).

Even though this result is not the most promising one, it laid the foundation for the development of autologous neural stem cell based therapy in PD patients and pointed out that this source represents another option for cell replacement therapy, avoiding the employment of embryonic tissue.

4.1.4 Mesenchymal stem cells (MSCs)

Efforts in employing non-fetal stem cells for transplant therapy in PD are also currently undergoing. Venkataramana *et al.* performed a unilateral transplantation of autologous bone marrow-derived mesenchymal stem cells (BM-MSCs). The BM-MSCs were transplanted into the sublateral ventricular zone by stereotaxic surgery. The transplanted patients showed a clinical improvement of 23% in their UPDRS score. Moreover, a subjective improvement was found in symptoms like facial expression, gait, and freezing episodes (Venkataramana *et al.*, 2010). These results indicate that this new protocol seems to be safe, and no serious adverse events occurred after stem-cell transplantation in PD patients, additionally, no fetal tissue was needed to obtain human stem cells. However, more efforts should be done to improve these stem cell-based techniques in experimental models of PD before using them in PD patients.

As far as we know, this is the last report of a clinical trial in PD patients. Taking this and all of the previous works done for transplant therapy in PD patients, we can state that the international quest for the best transplant therapy should not have been taken with such haste; a more detailed analysis of graft survival and long term clinical improvement are considered necessary. Relevant concerns prevail about methodology aspects, and also about the most debated complication of cell therapy in PD, the occurrence of post-operative graft-induced dyskinesias in the majority of the clinical protocols.

4.2 Embryonic stem cells: The most promising source of dopaminergic neurons

The information described above shows that even thought fetal midbrain precursors have capacity to proliferate and differentiate into dopaminergic neurons, the efficiency of this process in culture is low, therefore, restricting their potential as donor tissue. ES cells instead have the unique ability to self-renew indefinitely while maintaining the potential to give rise to all cell types in the human body. These two properties of ES cells make them gain a remarkable interest as promising tools for regenerative medicine, specially in PD transplant therapy. Another advantage that propelled the use of stem cells is that genetic manipulations has became easily practicable in the last few years.

Nevertheless, currently there are three major aspects limiting the success of stem cell therapy; one being the low number of stem cells sources with the potential to differentiate into the mesencephalic dopaminergic phenotype; the limited survival of grafted cells transplanted in both animal models and humans; and the most important, the danger of teratomas.

In this section, we summarize the different attempts to increase the number of stem cells using basically 3 different culture systems (feeder stromal cells, embryoid bodies and neural rossetes) to induce differentiation towards a dopaminergic phenotype. Evaluation of survival, integration and function of ES-dopaminergic cells after transplantation in animal models of PD is also described.

4.2.1 Mouse Embryonic stem (mES) cells

In 2000, Kawasaki *et al.* performed the first grafted study of mice ES-dopaminergic derived cells into the mouse striatum that had been previously treated with 6-OHDA. The system used to derivate TH mesencephalic-type neurons consisted in culturing ES cells onto the bone marrow stromal cell line PA6, which they identified to have the feature (by unknown mechanism) of promoting specific dopaminergic differentiation. The number of TH neurons obtained by this co-culture method is significantly higher (~16% of the total cells) to the five stages protocol reported in the same year by Lee et and collaborators (Lee *et al.*, 2000). Two weeks after transplantation, 22% of the grafted TH-positive neurons survived and even though the improvement of rotational behavior after transplant was not analyzed, they confirmed that ES-dopaminergic cells are capable of being transplanted into a parkinsonian model and survive, which laid the interest to explore further possibilities for therapeutic application (Kawasaki *et al.*, 2000).
In continuance work, other research groups have been improving culture protocols in order to increase the number of dopaminergic neurons to be grafted. In 2002, the first transplant of mES cells took place, Kim *et al.* reported that ~78% of dopaminergic neurons could be derived from mES cells *in vitro*. The highest number of TH cells reported until now was promoted by the overexpression of the nuclear receptor related-1 (Nurr1), a transcription factor relevant in the induction of mesencephalic precursors into dopamine neurons (Wallen *et al.*, 1999). Interestingly, a remarkable behavioral improvement after grafting was observed; the transplanted parkinsonian rats presented a total reversal of amphetamine-induced rotational behavior 8 weeks after transplant (Kim *et al.*, 2002). The behavioral enhancement correlates with the evaluated release of dopamine and functional synapses of the trasplanted cells. Additionally, no tumor formation was found.
That same year, Bjorklund *et al.* followed an opposite transplant approach. The group grafted ES cells on embryoid body (EB) stage. It is important to mention that EBs are aggregated of cells with spherical shape that differentiate stochastically from ES cells under specific *in vitro* conditions, they have the inherent property of recapitulate embryonic development and were developed the first time by Lee in a 5 stage protocol (Lee *et al.*, 2000). Rats transplanted with EB presented a decrease in 46% of amphetamine-induced rotation 9 weeks after transplantation (Bjorklund *et al.*, 2002). The reasoning of the improvement in the motor asymmetry is that the ES cell-derived neurons released dopamine in sufficient amount when they were stimulated by amphetamine. However, the number of TH+ neurons produced after grafting was not studied. They instead analyzed the functional activity of ES cells by PET imaging, finding high similarity to that observed in transplanted dopaminergic midbrain fetal neurons. Unfortunately, in contrast to the study realized by Kim *et al.*, 25% of the animals developed tumors, a result expected given the fact that not all of the grafted cells were in a differentiated state.
To improve the therapeutic potential of ES cells Barberi *et al.* implemented a faster and simpler culture method. Their protocol consisted in the induction of the differentiation of mice ES cells into dopaminergic neurons using the feeder stromal cell line (MS5). They found that nuclear transfer-derived ES dopaminergic neurons following this protocol and

transplantation into the striatum of parkinsonian mice provoked a decrease in 80% of circling behavior; additionally, some animals contained up to 40,000 transplanted TH+ neurons 8 weeks after transplant (Barberi *et al.*, 2003).
In another effort to obtain functional DA neurons derived from ES cells but without any genetic modification, Rodríguez-Gómez *et al.* generated CNS progenitor populations from mES cells that were later expanded and promoted to differentiate into dopaminergic neurons in the presence of mitogen and specific signaling molecules. Mitogen withdrawal from the growth medium was done after expressing Engrailed 1 (En1), Paired box gene 2 (Pax2), and Orthodenticle homeobox 2 (Otx2) specific neuronal transcription factors, hence, achieving differentiation (Rodríguez-Gómez *et al.*, 2007). They also described that in the grafted animals, PET imaging showed that the number of postsynaptic DA D2 receptors was normalized in the host striatum. Additionally, microdialysis in grafted animals displayed that dopaminergic transplanted neurons release was induced by depolarization and pharmacological stimulants. Their data suggest that ES cell-derived neurons release DA in a physiological manner and that reuptake postsynaptic responses remains after implantation.
On the other hand, to optimize the procedure for generation of mesencephalic dopaminergic neurons from ES cells, Jönsson et al. elaborated a protocol to determine the optimal stage of development of cells used for grafting. By means of fluorescence-activated cell sorting procedures, they isolated DA precursors from mouse ventral mesencephalon in two defined stages of differentiation, at 10.5 and 12.5 embryonic days, when the dopaminergic mesencephalic neurons are arising. After transplantation into the striatum of 6-OHDA-lesioned rats, the histological analysis showed that TH-expressing cells poorly survived sorting and transplantation; however, this study demonstrated that the differentiation state of the progenitors is important. The transplanted neuroblasts originate more TH+ cells when obtained from an early state of differentiation (Jönsson *et al.*, 2009). These outcomes have inferences for recent efforts to develop well-characterized stem cell-derived mesencephalic DA progenitor cell preparations for future cell therapy.
Even though the 6-OHDA lesioned rat is a great model of PD for analyzing ES cell transplant therapy, before considering preliminary transplantation trials in PD patients, experiments on non-human primates are quintessential. With this in mind, Takagi *et al.* generated neurospheres composed of neural progenitors from monkey ES cells cultured on PA6 stromal feeder cells and transplanted them into the putamen of MPTP lesioned monkeys. They reported a significant behavioral recovery and an average survival of 2,100 TH+ transplanted neurons 14 weeks after transplantation; additionally, PET imaging revealed a 50% increase in 18^{F}-fluorodopa uptake. This study demonstrated that the transplanted ES cells functioned as DA neurons and alleviated the motor symptoms in a parkinsonian non-human primate (Takagi *et al.*, 2005); furthermore, they demonstrated that ES cell transplant therapy might be an appropriate treatment alternative to human PD patients.

4.2.2 Human ES cells

After Thomson isolated stem cells from human embryo (Thomson *et al.*, 1998), the next boundary to overcome was to generate neural progenitors derived from human ES (hES) cells. In an innovating experiment, undifferentiated hES cells were plated on fresh mitotically inactivated feeders and cultured for 8 days to induce neural differentiation, these progenitors were transplanted into the striatum of parkinsonian rats. The grafts survived for at least 12 weeks, managed to differentiate *in vivo* to DA neurons and improved circling behavior with a 45% reduction of amphetamine-induced ipsilateral rotation (Ben-Hur *et al.*, 2004). Induced dopaminergic differentiation from hES was shortly addressed by other groups with different culture approaches but with similar relatively successful outcomes (Brederlau *et al.*, 2006; Chiba *et al.*, 2008; Cho *et al.* 2008; Roy *et al.*, 2006).

Roy used a combined strategy to improve dopaminergic neuron amounts. Sonic hedgehog (Shh) and fibroblast growth factor 8 (FGF-8) were co-cultured with telomerase-immortalized hES cells derived from human fetal midbrain astrocytes in order induce a dopaminergic neuronal fate. After achieving a high number of TH+ cells specific for the A9 dopaminergic lineage (colocalizing with the G protein–gated inwardly rectifying K+ channel, Girk2), they were transplanted into the striatum of 6-OHDA lesioned parkinsonian rats. The authors reported that the dopaminergic implants generated a significant, substantial and long-lasting restitution of motor function. Nonetheless, the grafts displayed increasing cores of undifferentiated mitotic neuroepithelial cells, which can turn tumorigenic (Roy *et al.*, 2006). These data dictate the need for extreme caution in developing a clinical application of hES cell-derived grafts, given their potential for undifferentiated expansion, yet they also proved that TH+ neurons can be obtained and/or induced from a diverse source of cells.

Chiba *et al.* followed a different method to improve hES cell transformation to phenotypically stable DA neurons. They blocked the effect of the neuronal differentiation promoter Noggin by means of the bone morphogenic protein (BMP) antagonist. With this strategy they found that BMP inhibitor Noggin increased production of DA neurons from hES cells differentiated on PA6 stromal cells. In addition, these DA neurons survived transplantation and led to behavioral improvement in parkinsonian rats 4 weeks after grafting (Chiba *et al.*, 2008). These neurons derived from hES cells may not be suitable for eventual transplantation into PD patients (due to a triple trisomy created by continuous passaging), but with this work, they demonstrated that Noggin has a critical role for determining midbrain dopaminergic phenotype at an early stage in ES cell differentiation, which will be of great value for future research in dopaminergic cell engineering.

The recent generation of dopaminergic mesencephalic neurons obtained from human ES cells by employing spherical neural masses (SNMs), also called neural rossete cells, is considered the protocol that allows to yield the higher number of dopaminergic neurons in cell culture (roughly 66%) (Cho *et al.*, 2008). SNMs are columnar epitelial structures that represent a novel NSC type, distinct and more primitive than those previously characterized from other NSC stages (Elkabetz *et al.*, 2008; Pankratz *et al.*, 2007). NSC in neural rossetes mimic the neural plate stage and can be directed towards a dopaminergic mesencephalic lineage by employing the signaling molecules Shh and FGF-8 (Elkabetz *et al.*, 2008; Cho et al., 2008). However, whether the fully differentiated DA cells obtained from this novel protocol have an ameliorative effect on parkinsonian rodent models or PD patients still remains to be determined. Cho and collaborators decided to transplant cells originated from neural rossetes in the initial stage of dopaminergic neuronal specification into the lesioned striatum of parkinsonian rats in order to avoid a low survival rate. In this specific time point, approximately 50% of the cells were TH+ and immunohistochemical analysis demonstrated that only 2.7% of the grafted neurons were TH+ 12 weeks after transplantation. Despite this low cell number, a significant behavioral recovery was observed in 3 different motor behavioral tests. Although this stem cell source is promissory due to the high number of yield DA cells without the need of genetic modifications or employment of feeder cells; this technique stills needs to circumvent the problem of low survival ratio, which is generally observed when terminal differentiated cells are grafted. Additionally, in order to generate an efficient long-term culture (30 to 40 days), experimented handling and dedicated efforts are required. These results also point out the necessity of better and rational procedures to direct cell specification to a desired phenotype *in vitro.*

The bulk of results suggest ES cell transplantation is approaching the point of technical practicability toward clinical therapy for PD. hES cell transplantation has the potential to be considered one of the most promising therapies for PD treatment; however, this excitement is gradually fading as a result of political and ethical concerns. Amongst the limiting methodological factors, is the difficulty to maintain normal karyotype and to avoid tumor formation. Furthermore, the differentiation rate of ES cells towards a dopaminergic lineage is not high enough to obtain sufficient dopaminergic cells for transplantation.

4.3 Induced pluripotent cells: The most recent breakthrough in stem cells field

Induced pluripotent (iPS) cells are derived from somatic differentiated cells by overexpression of specific transcription factors, which induces cell reprogramming. They are fairly similar to ES cells in terms of self-renewal and pluripotency, which provides them with the potential to differentiate into any cell type in the organism.

iPS cells were obtained for the first time in 2006 by Yamanaka´s group; they were acquired from embryonic and adult mouse fibroblasts reprogrammed to a pluripotent-like state by viral transduction of four transcription factors: Klf4, cMyc, Oct3/4 and Sox2. The obtained cells demonstrated to possess typical ES cells properties (Takahashi *et al.,* 2006); however, they did not form effective fertile chimeras, meaning that the ability to transmit genotype through the germinal line was not achieved, this being one of the most important properties that defines a pluripotent cell. This was later rein-vindicated in 2007 in a parallel work by Jaenisch´s and Yamanaka´s groups, they successfully reprogrammed fibroblasts and obtained viable chimeras (Okita *et al.,* 2007; Wernig *et al.,* 2007); nonetheless, they emphasized that the transduction of c-Myc should be avoided in iPSC reprogramming, due to its tumorigenic capacity.

At the end of 2007, and almost simultaneously, Takahashi and Yu were able to manufacture iPS cells from human dermal fibroblast. Takahashi enforced the expression of the same transcription factors employed in the previous study (Klf4, cMyc, Oct3/4, Sox2); while Yu *et al.,* demonstrated that other two transcription factors, Nanog and LIN28 (besides Oct3/4, Sox2) were also relevant in cell reprogramming (Takahashi *et al.,* 2007; Yu *et al.,* 2007). One year later, iPS cells were derived from human fetal tissue, neonatal and adult somatic cells by employing the same combination of genes used by the pioneer team, they also reported that c-Myc only increases the efficiency of reprogrammation, but it is not essential for the establishment of a pluripotency state, suggesting that c-Myc could be nonessential for reprogrammation (Park *et al.,* 2008). Taken together, these results reveal that iPS cells are an attractive resource for replacement therapy, given the fact that its use would overcome the technical and ethical difficulties of ES cells by avoiding the use of human embryos, circumventing transplant rejection and therefore avoiding the use of immune suppressants. However, concerns for the transduction of the oncogen c-Myc were still latent and attempts to remove it from the transcription factor cocktail were in progress.

Later in the same year, it was demonstrated that c-Myc is absolutely expendable to obtain viable iPS cells, either from mice and human samples (Nakagawa *et al.,* 2008). With the removal of this specific proto-oncogen, the possibility of tumor formation was reduced, and therefore, research on possible clinical applications achieved important notoriety.

iPS cell lines obtained from patients afflicted with different diseases have been developed with the purpose of investigating the therapeutic potential of autologous transplants, but also to offer an unprecedented opportunity to recapitulate pathological conditions *in vitro* (Gunaseeli *et al.,* 2010; Park *et al.,* 2008). Neurodegenerative diseases are particularly taken into consideration, since they are secondary to a relatively selective loss of neurons, for that reason, the quest for replacing the neuronal loss of a specific population employing iPS cells is currently undergoing in a similar way that occur for ES cells.

The iPS cells neuronal differentiation *in vitro* assays have been successful and were achieved by employing established ES cell differentiation protocols. The first dopaminergic differentiation was obtained from neural precursors derived from mice iPS cells. These cells were grafted into the mice fetal encephalon (chimerical brain) and revealed a high capacity for migration, differentiation and integration within different brain areas. The iPS-derived dopaminergic neurons generated in this study were grafted in the 6-OHDA rat model of PD, such transplants improved significantly circling behavior stimulated by amphetamine (Wernig *et al.,* 2008). However, it was reported that the presence of undifferentiated cells in the graft promoted tumor formation; this fact contributed to the idea that obtaining a purified population is essential for a satisfactory transplant outcome.

It has been proved that pluripotent cell production can be obtained from patients suffering from other neurodegenerative disease. *Dimos et al.* developed iPS cells from patients with amyotrophic lateral sclerosis (ALS), which was effectively differentiated *in vitro* to motoneurons, the degenerated cells that are accountable for the development of ALS (Dimos *et al.,* 2008). This corroborated that iPS cells patient-specified manufacture is a viable option and possess certain advantages over ES cells since iPS cells evade the ethical and methodological difficulties presented in ES cells production.

A new study performed in 2009 widened the possibility of inducing patient-specified neural differentiation *in vitro* into iPS cells engineering. Jaenisch´s group managed to generate fibroblast-derived iPS cells from 5 idiophatic PD patients, with the advantage that once the cellular reprogrammation employing three transcription factors (Oct4, Klf4 y Sox2) was achieved, they were able to remove all viral vectors using Cre-recombinase (iPS cells factor-free). These cells maintained all the properties of a pluripotent cell and were capable to induce neuronal differentiation towards a dopaminergic phenotype *in vitro*; additionally, they have the advantages of being a free viral transcription factor cell line and that overall gene expression patterns were more similar to ES cells than carrying-factor iPS cells (Soldner *et al.* 2009).

Different approaches have been developed in order to analyze the latent benefits of human iPS cells in animal models of PD. Cai and collaborators employed a commercially available iPS cell line, the IMR90 clone 4, to induce its differentiation towards a mesencephalic dopaminergic lineage in culture. Once they achieved the dopaminergic phenotype, they were transplanted into the striatum of 6-OHDA unilateral lesioned rats. Six weeks after the transplant, they analyzed survival and integration of the transplanted cells within the host brain. They reported that several transplanted cells survived, but some of them expressed nestin and Ki67 which are expressed in dividing cells, such as tumoral cells (Cai *et al.,* 2010). This work however, does not allow conclusions regarding physiological effects, since behavioral outcome was not analyzed.

Later in that year, Hargus *et al.* employed dopaminergic neurons derived from factor-free iPS cells and transplanted them into the striatum of 6-OHDA unilateral lesioned rats; they later analyzed development, growth, survival and degeneration in an *in vivo* longitudinal study. They observed a high level of cell survival with no apparent degeneration; they also detected axonal outgrowth in different areas, this can be translated to a soaring capacity of the adult brain to keep axonal guidance instructions that allow the transplanted cells to integrate within the adult brain. In addition, a progressive reduction of motor asymmetry (about 70%) was observed in the transplanted animals over a period of sixteen weeks, 50% of the transplanted DA neurons were identified as SNPc mesencephalic cells and presence of tumoral cells was null (Hargus *et al.,* 2010). However, synaptic function analysis and a longer examination of transplant survival still remain to be done. Besides, the improvement of motor asymmetry was similar to the one reported in other experimental trials that employed different cells, in that aspect, there is no greater advantage of iPS cells over ES cells.

In 2010, reprogramming mice fibroblast directly into a neuronal phenotype directly to a pluripotency stage *in vitro* was achieved. A combination of 3 genes specifically expressed in neural tissue, Ascl1, Brn2 y Myt1l, was transduced by means of a lentiviral vector. These kind of cells were called induced neurons (iN), since they expressed neural proteins, generate action potentials and form functional synapses; most of these cells express an excitatory phenotype that was later confirmed by cortical neural markers and only a low number of these cells belong to the GABAergic phenotype (Vierbuchen *et al.,* 2010). Yet, it still remains to be determined if iN can be derived from fibroblasts obtained from patients afflicted with a neurodegenerative disease, and which transcription factors must be overexpressed in order to induce different neuronal phenotypes.

Regardless of all of the possibilities that iPS cells might offer, the uncertainty about the latency of tumor formation by being employed in experimental trials still remains. Albeit, better methods to induce reprogrammation are currently being developed, but even with viral vector removal, this possibility cannot be totally excluded.

Additionally, recent results demonstrate that production of aberrant methylation sites are originated during the epigenomic reprogrammation of iPS cells. Methylation of human iPS cells was observed recurrently at different rates throughout the reprogrammation process. Furthermore, the variability of the methylated sites is often seen in iPS cell lines, which indicates that there are some regions more prone to insufficient or aberrant reprogrammation. These methylated sites are categorized as "hotspots" for incorrect epigenomic reprogrammation, this alteration is not a simple anomaly exclusive of the pluripotent state, it can also be transmitted through iPS cells differentiation (Lister *et al.,* 2011).

So far, iPS cells cannot be employed as a feasible resource for cellular replacement therapy and further work needs to be developed before a clinical trial could take place; however, without a doubt, they possess great potential to be employed as tools for patient-specific disease modeling.

Finally, if we were to choose the type of cell that would bestow the best results after transplantation, taking into consideration the information aforementioned; we believe that the most promising candidate today is the ES cell. In order to provide this personal opinion, we are leaving aside the ethical issues that embrace them, albeit this kind of concerns are of the outmost importance, they are out of the scope of this chapter.

Cell Source	*In vitro* protocol for TH cells production	Recipient	Result	Time period (after transplant)	TH neuron count (after transplant)	Reference
Mouse Embryonic Stem Cells						
mES cells-derived dopamine neurons	ES cultured on PA6 stromal cell	6-OHDA unilateral lesion Mouse	N/A	2 weeks	13, 000 (~22% of the original graft)	Kawasaki *et al.*, 2000
mES cells (no differentiated)	Embryoid body formation stage	6-OHDA unilateral lesion Rat	Decrease in 46% of ipsilateral rotation (amphetamine)	9 weeks	NA	Bjorklund *et al.*, 2002
mES cells-derived dopamine neurons	Genetically-engineered to over express Nurr1	6-OHDA bilateral lesion Rat	Total reversal of ipsilateral rotation (amphetamine)	8 weeks	20, 000	Kim *et al.*, 2002
ntES cell-derived dopamine neurons	Nuclear transfer-derived ES (ntES). Coculture with stromal cells	6-OHDA unilateral lesion Mouse	Decrease in 80% of ipsilateral rotation (amphetamine)	8 weeks	~22,000 (some reaching 40,000)	Barberi *et al.*, 2003
mES cells-derived dopamine neurons	Mice transfected with GFP differentiated using 5 stages protocol	6-OHDA unilateral lesion Mouse	Decrease in 50% of ipsilateral rotation (amphetamine)	2 weeks	NA	Nishimura *et al.*, 2003
mES cells-derived dopamine neurons	Differentiated using 5 stages protocol	6-OHDA unilateral lesion Rat	Total reversal of ipsilateral rotation (amphetamine) Starting at 4 weeks	32 weeks	5, 000	Rodriguez-Gomez *et al.*, 2007
ntES cell-derived from parkinsonian mice	Nuclear transfer-derived ES (ntES). Coculture with stromal cells	6-OHDA unilateral lesion mouse	Total reversal of ipsilateral rotation (amphetamine)	10 weeks	Avg 8,784 ± 4,293 cells in the group with better results	Tabar *et al.*, 2008
Mesencephalic DA neurons	Ngn2-GFP mice Pitx3-GFP mice Nestin-GFP mice Sox2-GFP mice	6-OHDA unilateral lesion Rat	Total reversal of ipsilateral rotation (amphetamine) Starting at 3 weeks	6 weeks	440	Jönsson *et al.*, 2009
Non-human primates Embryonic Stem Cells (monkey ES cells)						
Non-human primates ES cells	Neural progenitors expanded as neurospheres. FGF2 addition	MPTP bilateral lesion Cynomolgus monkeys	Behavioral improvement in neurological score	14 weeks	2, 100 per side	Takagi *et al.*, 2005
Human Embryonic Stem cells						
hES cells	Culture of hES cells enriched with neural progenitors	6-OHDA unilateral lesion Rat	Decrease in 45% of ipsilateral rotation (amphetamine)	12 weeks	389 (0.18% of the graft)	Ben-Hur *et al.*, 2004

Cell Source	*In vitro* protocol for TH cells production	Recipient	Result	Time period (after transplant)	TH neuron count (after transplant)	Reference
			Decrease in 31% of contralateral rotation (apomorphine)			
hES cells-derived dopamine neurons	Culture on PA6 cells (stromal cells)	6-OHDA unilateral lesion Rat	No improvement (amphetamine)	13 weeks	10-50	Brederlau *et al.*, 2006
hES cells-derived dopamine neurons	Cocultured with human telomerase-immortalized midbrain astrocytes	6-OHDA unilateral lesion Rat	Total reversal of ipsilateral rotation (apomorphine)	Since 6 weeks	26,000 per mm^3	Roy *et al.*, 2006
hES cells-derived dopamine neurons	Generation of spherical neural masses (SNMs) (rossetes)	6-OHDA unilateral lesion Rat	Decrease in 49.43% of contralateral rotation (apomorphine) Decrease in 58.37% of ipsilateral rotation (amphetamine)	12 weeks	10,700 (2.7% of transplanted cells)	Cho *et al.* 2008
Induced Pluripotent Stem cells						
iPS clone O9	Reprogrammed fibroblasts (mouse)	6-OHDA unilateral lesion Rat	Decrease in 60% of ipsilateral rotation (amphetamine)	8 weeks	29, 000	Wernig *et al.*, 2008
iPS cells IMR90 clone 4	Reprogrammed fibroblasts (human)	6-OHDA unilateral lesion Rat	Decline in rotation score	6 weeks	NA	Cai *et al.*, 2010
iPS cells excisable virus	Human PD (patients)	6-OHDA unilateral lesion Rat	Decrease in 70% of ipsilateral rotation (amphetamine)	16 weeks	122 neurons per mm^3	Hargus *et al.*, 2010

Table 3. Stem cell therapy in animal models of PD.

5. Evaluating the functional effectiveness of Stem Cell Therapy for PD

One of the major difficulties in restoring the motor functionality in PD through transplant therapy is the correct integration of grafted cells into the nigrostriatal pathway; they are required to improve motor symptoms and circumvent the appearance of undesirable side effects, such as dyskinesia (Politis *et al.*, 2010). Grafted cells should not only provide a DA source for the correct modulation of the basal ganglia, but also be able to integrate the nigrostriatal pathway and receive the regulatory input that allows compensatory mechanisms, such as feedback regulation. In addition, the complexity of the adult brain and the lack of a permissive environment to neuronal regeneration make it particularly complicated for even the best non-modified or engineered stem cell to integrate into the basal ganglia motor loop.

Cell grafts might be able to compensate the loss of modulatory inputs from the SNPc by forming local neural circuits within the striatum (Barker *et al.*, 1999; Kreitzer *et al.*, 2008; Lindvall *et al.*, 2004). Ideally, these circuits should be capable to modulate basal ganglia pathways through interaction with D2 and the D1 dopamine receptors, as well as to be subject to feedback regulation in order to prevent variations of DA availability.

In order to evaluate these neural circuits, the need of joined functional and behavioral tests to assess improvement is one of the main objectives. The use of positron emission tomography (PET) coupled to magnetic resonance imaging (MRI) techniques has diminished the gap between behavioral and functional data.

5.1 MRI assessments

Conventional MRI reveals brain structural changes as reductions in volume (atrophy) and alterations in water-proton relaxation T1 and T2 signals. Water normally flows along neural tracts in the brain. Diffusion-weighted or diffusion tensor MRI can be used to quantify loss of anisotropy (directionality) or increase in amplitude of water diffusion in order to demonstrate disruption of neural tracts.

Conventional T1 and T2 weighted MRI show normal nigral structure in idiopathic PD and therefore is not diagnostically helpful. Volumetric T1 weighted MRI studies have also failed to detect a reduction in nigral volume in PD, possibly because of difficulties in accurately defining the border of the SNPc (Geng *et al.*, 2006) and thus it would be the most appropriate technique for evaluating the survival of dopamine-derived transplants.

MRI inversion recovery sequences can be designed to suppress either gray or white matter signal. Segmented inversion recovery ratio imaging generates ratio images of gray matter and white matter suppressed signal at a voxel level. With the segmented inversion recovery ratio approach, PD patients were reported to show a gradient of altered nigral signal that was absent in healthy controls (Hutchinson & Raff, 2000). Although the use of gray matter and white matter suppressing inversion recovery sequences can detect changes in nigral structures in PD, it is a complicated approach to implement and currently not sensitive enough to be of diagnostic value in transplant therapy.

Regardless of the limitations previously mentioned, MRI is a non-invasive, real-time cellular imaging modality that has no radiation, and for these reasons, several attempts have been made in order to take advantage of these qualities. New MRI techniques currently employ magnetic labeling using superparamagnetic iron oxide particles (SPIOs) coupled to stem cells transplanted in the rat striatum with the purpose of assessing the survival time of transplanted tissue (Berman *et al.*, 2011; Obenaus *et al.*, 2011). Nevertheless, this innovative technique is only useful for quantifying cell survival and does not provide data to evaluate transplanted cell function.

5.2 PET assessments

Terminal dopa decarboxylase activity and dopamine turnover can be assessed using radioactive markers such as 6-[^{18}F]-fluoro-L-DOPA (^{18}F-DOPA), which could be helpful for the evaluation of PD progression *in vivo* (Brooks *et al.*, 2008) and graft survival in transplanted PD patients (Mendez *et al.*, 2002; Sawle *et al.*, 1992). Early hemiparkinsonian patients show a bilateral reduced putamen dopaminergic terminal function. Clinical parkinsonism occurs when PD patients have lost 40%–50% of posterior putamen-dopamine terminal function (Morrish et al., 1995). Levels of ^{18}F-dopa uptake in the putamen and DAT binding correlate inversely with bradykinesia and rigidity of PD patients, and could be valuable for estimation of cell therapy functional recovery.

However, this type of analysis has limitations; ^{18}F-DOPA is not specific for evaluation of DA re-uptake in small animal PET scanning, which is an essential feature for the assessment of emergent cell therapy in animal models of PD. Other radiotracers of DAT binding, such as [^{11}C]-(+)-α-dihydrotetrabenazine (^{11}C-(+)DTBZ), have better specificity and spatial resolution when evaluating animal models of PD (Collantes *et al.*, 2008) and could be used to examine the survival of DA neurons.

Still, in order to correctly evaluate DA release, it is necessary to use radioactively labeled dopamine receptor antagonists, such as [^{11}C]-raclopride, which can be displaced from DA receptors by DA released from nerve terminals located in situ or exposed to amphetamine (Piccini *et al.*, 2000). ^{11}C-raclopride PET is able to indirectly detect synaptic dopamine fluxes by monitoring changes in striatal D2 receptor availability (Laruelle, 2000). The higher the extracellular dopamine level, the lower the dopamine D2 site availability to the tracer. Animal microdialysis studies suggest that a 25% fall in putamen 11C-raclopride uptake after amphetamine equates to a 10-fold rise in synaptic dopamine levels (Breier *et al.*, 1997). In PD patients, when given L-DOPA treatment, they show a fall in striatal 11C-raclopride binding; the response of bradykinesia and rigidity to L-DOPA in PD correlates with the resulting increases instriatal dopamine levels detected with 11C-raclopride PET (Pavese *et al.*, 2006).

By means of ^{18}F-DOPA and [^{11}C]-raclopride PET scans, it has been shown that grafted (human embryonic mesencephalic and fetal dopaminergic tissue) cells remain active and form connections with neurons residing in the host striatum when transplanted in human brains (Mendez *et al.*, 2008; Piccini *et al.*, 1999; Spencer *et al.*, 1992). The major drawback is that the spatial resolution achieved with the current available PET scanners (2×10^{-3} m) is insufficient to evaluate single cell events, which can only be resolved with 100-fold higher resolutions (10–20 μm).

So far, the information already described shows that current MRI and PET analysis have some limitations for a correct cell graft evaluation of survival and integration within the host, specifically at the cellular level.

5.3 Microcircuits on brain slice preparations: the need for functional studies

It is clear that better protocols for cell replacement therapy would benefit from a better understanding of normal basal ganglia circuit functions. One of us (Bargas' group) designed a brain slice preparation in which individual cell activity can be observed and evaluated, using calcium-imaging techniques (Carrillo-Reid *et al.*, 2008). When corticostriatal slices are loaded with a calcium-sensitive fluorescent dye, it is possible to analyze single-cell activity in a widespread area of the striatum. Additionally, it is possible to examine the activity of cell clusters that have an associated firing pattern (neural ensembles) and represent functional microcircuits within the striatum (Carrillo-Reid, 2008, 2009). Different cells within the striatum belong to different ensembles, and there are some cells that belong to most of the ensembles present in a given slice. These cells serve as central pattern generators (CPGs), which are considered to represent memory traces in the nervous system (Carrillo-Reid, 2008, 2009; Grillner, 2005, 2006). CPG activity within the striatum might be related to the encoding of motor programs that are activated secondary to cortical stimulation. In this regard, it is worth of notice that unstimulated corticostriatal slices mostly show silent cells, with some uncoordinated activity.

Ensemble activity is triggered by stimulation of cortical fibers or after exposure to glutamatergic agonist NMDA. These results suggest the existence of intrinsic microcircuits in the striatum, which could be the actual modulators of motor activity in this brain area.
Thus, it would be necessary to determine if the segregated pathways represented by neurons expressing either the D1 or the D2 dopaminergic receptors, belong to distinct activity ensembles; this would be of major interest, since deregulation of the D1 receptor pathway has been proposed to be responsible for the dyskinesias observed in PD patients after prolonged exposure to L-DOPA (Dupre *et al.*, 2008; Santini *et al.*, 2008; Taylor *et al.*, 2005). Of particular interest would be to demonstrate if there is a difference between ensembles in the intact brain and those in the affected striatum of PD patients.
We propose that the use of this brain slice preparation will allow us to study if transplanted dopaminergic cells can integrate into local microcircuits within the striatum, and if formation of new modulatory connections could compensate for the loss of the dopaminergic input from the SNPc and reestablish control of motor function.

6. Conclusion

This chapter highlights that although almost three decades have passed since the Backlund-Lindvall and Drucker-Colín *et al.* work was published, transplant therapy for PD is still retained as a potential clinical approach; stem cell research is currently providing a more adequate source of cells. So far, the large body of evidence appears to validate the use of manipulated stem cells in the coming years.
Although stem cell therapy shows promise, it is still in development, and even with a wide range of methods, currently there is no ideal scenario for clinical transplantation and perhaps it will not arrive in the near future. ES cell therapy still remains elusive and is burdened with social and logistical concerns. In contrast, iPS generation circumvents some previous limitations, since it does not require embryonic material. iPS cell technology might have a significant impact on regenerative medicine in the near future.
In order for cell therapy to become a viable option of treatment, a pure source of dopaminergic neurons is essential. However, the poor survival of transplanted differentiated cells limits transplant therapy; ergo a pure population would not be the best option, since cell survival decreases after transplantation. Additionally, the combination with other cell populations could also be necessary for a successful transplant, given the fact that primary fetal cells that provide behavioral improvement are only ~20% of the total implanted cells.
Finally, it should be particularly noted, that aside of a couple of studies, all report incomplete motor improvements. The search for the basis of this result is today still in its embryonic stage. As a final point, research with better surgical procedures that provides controlled assessments consequently will lead to a more clear understanding of the phenomena involved grafting in PD.
In sum, despite several failures, we believe that cell replacement therapy has a viable future.

7. Acknowledgments

We thank to Diana Millán-Aldaco, Marcela Palomero-Rivero, Alejandra Boronat-García, José Rubén García-Montes and M. Guadalupe Maya-Espinosa for their critical input on the manuscript. This work was supported by IMPULSA of the Universidad Nacional

Autónoma de México (UNAM) and by Programa de Apoyo a Proyectos de Investigación e Innovación Tecnológica (PAPIIT), UNAM Grant No. IN225209-3.

8. References

Aguayo, A.J., Bjorklund, A., Stenevi, U. & Carlstedt, T. (1984) Fetal mesencephalic neurons survive and extend long axons across peripheral nervous system grafts inserted into the adult rat striatum. *Neuroscience Letters,* 45, 53-58.

Allan, L.E., Petit, G.H. & Brundin, P. (2010) Cell transplantation in Parkinson's disease: problems and perspectives.*Curr Opin Neurol,* 23, 426-432.

Altman, J. (1969) Autoradiographic and histological studies of postnatal neurogenesis. IV. Cell proliferation and migration in the anterior forebrain, with special reference to persisting neurogenesis in the olfactory bulb. *The Journal of Comparative Neurology,* 137, 433–457.

Andersson, E., Tryggvason, U., Deng, Q., Friling, S., Alekseenko, Z., Robert, B., Perlmann, T. & Ericson, J. (2006) Identification of intrinsic determinants of midbrain dopamine neurons. *Cell,* 124, 393–405.

Annett, L.E., Martel, F.L., Rogers, D.C., Ridley, R.M., Baker, H.F. & Dunnett, S.B. (1994) Behavioral Assessment of the Effects of Embryonic Nigral Grafts in Marmosets with Unilateral 6-OHDA Lesions of the Nigrostriatal Pathway. *Exp Neurol,* 125, 228-246

Arvidsson, A., Collin, T., Kirik, D., Kokaia, Z, & Lindvall, O. (2002) Neuronal replacement from endogenous precursors in the adult brain after stroke. *Nat Med,* 8, 963-970.

Backlund, E.O., Granberg, P.O., Hamberger, B., Knutsson, E., Mårtensson, A., Sedvall, G., Seiger, A. & Olson, L. (1985) Transplantation of adrenal medullary tissue to striatum in parkinsonism: First clinical trials. *J. Neurosurg,* 62, 169-173.

Barberi, T., Klivenyi, P., Calingasan, N.Y., Lee, H., Kawamata, H., Loonam, K., Perrier, A.L., Bruses, J., Rubio, M.E., Topf, N., Tabar, V., Harrison, N.L., Beal, M.F., Moore, M.A. & Studer, L. (2003) Neural subtype specification of fertilization and nuclear transfer embryonic stem cells and application in parkinsonian mice. *Nat Biotechnol,* 1200-1207.

Barker, R.A. & Dunnett, S.B. (1999) Functional integration of neural grafts in Parkinson's disease. *Nat Neurosci,* 2, 1047–1048.

Ben-Hur, T., Idelson, M., Khaner, H., Pera, M., Reinhartz, E., Itzik, A. & Reubinoff, B.E. (2004) Transplantation of human embryonic stem cell-derived neural progenitors improves behavioral deficit in Parkinsonian rats. *Stem Cells,* 22, 1246-1255.

Berman, S.M., Walczak, P. & Bulte, J.W. (2011) MRI of transplanted neural stem cells. *Methods Mol Biol,* 711, 435-449.

Björklund, A. & Stenevi, U. (1979) Reconstruction of the nigrostriatal dopamine pathway by intracerebral nigral transplants. *Brain Res,* 177, 555-560.

Bjorklund, L.M., Sánchez-Pernaute, R., Chung, S., Andersson, T., Chen, I.Y., McNaught, K.S., Brownell, A.L., Jenkins, B.G., Wahlestedt, C., Kim, K.S. & Isacson, O. (2002) Embryonic stem cells develop into functional dopaminergic neurons after transplantation in a Parkinson rat model. *Proc Natl Acad Sci U S A,* 99, 2344-2349.

Bongso, A., Fong, C.Y., Ng, S.C. & Ratnam, S. (1994) Human embryonic behavior in a sequential human oviductendometrial coculture system. *Fertil Steril,* 6, 976-978.

Brederlau, A., Correia, A.S., Anisimov, S.V., Elmi, M., Paul, G., Roybon, L., Morizane, A., Bergquist, F., Riebe, I., Nannmark, U., Carta, M., Hanse, E,, Takahashi, J., Sasai, Y., Funa, K., Brundin, P., Eriksson, P.S. & Li, J.Y. (2006) Transplantation of human embryonic stem cell-derived cells to a rat model of Parkinson's disease: effect of *in vitro* differentiation on graft survival and teratoma formation. *Stem Cells*, 24, 1433-1440.

Breier, A., Su, T.P., Saunders, R., Carson, R.E., Kolachana, B.S., de Bartolomeis, A., Weinberger, D.R., Weisenfeld, N., Malhotra, A.K., Eckelman, W.C. & Pickar, D. (1997) Schizophrenia is associated with elevated amphetamine-induced synaptic dopamine concentrations: evidence from a novel positron emission tomography method. *Proc Natl Acad Sci U S A*, 94, 2569-2574.

Brill, M.S., Snapyan, M., Wohlfrom, H., Ninkovic, J., Jawerka, M., Mastick, G.S., Ashery-Padan, R., Saghatelyan, A., Berninger, B. & Götz, M.A. (2008) Dlx2- and pax6-dependent transcriptional code for periglomerular neuron specification in the adult olfactory bulb. *J Neurosci*, 28, 6439-6452.

Brooks, D., Sawle, G., Schroter, G. & Ansari, A.A. (1992) Survival of implanted fetal dopamine cells and neurologic improvement 12 to 46 months after transplantation for Parkinson's disease. *N Engl J Med*, 327, 1549-55.

Brooks, DJ. (2008) The role of structural and functional imaging in parkinsonian states with a description of PET technology. *Semin Neurol*, 28, 435–445.

Brundin, P., Strecker, R.E., Widner, H., Clarke, D.J., Nilsson, O.G., Astedt, B., Lindvall, O. & Björklund, A. (1988) Human fetal dopamine neurons grafted in a rat model of Parkinson's disease: immunological aspects, spontaneous and drug-induced behaviour, and dopamine release. *Exp Brain Res*, 70, 192-208.

Cai, J., Yang, M., Poremsky, E., Kidd, S., Schneider, J.S. & Iacovitti, L. (2010) Dopaminergic neurons derived from human induced pluripotent stem cells survive and integrate into 6-OHDA-lesioned rats. *Stem cells and development*, 19, 1017-1023.

Cave, J.W. & Baker, H. (2009) Dopamine systems in the forebrain. *Adv Exp Med Biol*, 651, 15-35.

Carrillo-Reid, L., Tecuapetla, F., Tapia, D., Hernández-Cruz, A., Galarraga, E., Drucker-Colin & R., Bargas, J. (2008) Encoding network states by striatal cell assemblies. *J Neurophysiol*, 99, 1435–1450.

Carrillo-Reid, L., Tecuapetla, F., Ibáñez-Sandoval, O., Hernández-Cruz, A., Galarraga, E. & Bargas, J. (2009) Activation of the cholinergic system endows compositional properties to striatal cell assemblies. *J Neurophysiol* 101, 737–749.

Chiba, S., Lee, Y.M., Zhou, W. & Freed, C.R. (2008) Noggin enhances dopamine neuron production from human embryonic stem cells and improves behavioral outcome after transplantation into Parkinsonian rats. *Stem Cells*, 26, 2810-2820.

Chung, S.S., Kim, S.H., Yang, W.I., Choi, I.J., Lee, W.Y., Moon, J.G., Park, H.S., Shin, H.S., Kim, D.S. & Ahn, Y.M. (1993) Homogenous fetal dopaminergic cell transplantation in rat striatum by cell suspension methods. *Yonsei Med J*, 34, 145-151.

Cho, M.S., Lee, Y.E., Kim, J.Y., Chung, S., Cho, Y.H., Kim, D.S., Kang, S.M., Lee, H., Kim, M.H., Kim, J.H., Leem, J.W., Oh, S.K., Choi, Y.M., Hwang, D.Y., Chang, J.W. & Kim, D.W. (2008) Highly efficient and large-scale generation of functional dopamine neurons from human embryonic stem cells. *Proc Natl Acad Sci U S A*, 105, 3392-3397.

Collantes, M., Peñuelas, I., Alvarez-Erviti, L., Blesa, J., Martí-Climent, J.M., Quincoces, G., Delgado, M., Ecay, M., Martín, A., Arbizu, J., Rodríguez-Oroz, M.C., Obes,o J. & Richter, J.A. (2008) Use of 11C-(+)-alpha-dihydrotetrabenazine for the assessment of dopaminergic innervation in animal models of Parkinson's disease. *Rev Esp Med Nucl*, 27, 103–111.

Cotzias, G.C., Papavasiliou, P.S. & Gellene, R. (1969) Modification of parkinsonism: chronic treatment with L-dopa. *N Engl J Med*, 280, 337-345

Deacon, T., Schumacher, J., Dinsmore, J., Thomas, C., Palmer, P., Kott, S., Edge, A., Penney, D., Kassissieh, S., Dempsey, P. & Isacson, O. (1997) Histological evidence of fetal pig neural cell survival after transplantation into a patient with Parkinson's disease. *Nat Med*, 3, 350-353.

Deierborg, T., Soulet, D., Roybon, L., Hall, V. & Brundin, P. (2008) Emerging restorative treatments for Parkinson's disease. *Prog Neurobiol*, 85, 407-432.

Di Porzio, U. & Zuddas, A. (1992) Embryonic dopaminergic neuron transplants in MPTP lesioned mouse striatum. *Neurochemistry International*, 20, 309-320

Diamond, S.G., Markham, C.H., Rand, R.W., Becker, D.P. & Treciokas, L.J. (1994) Four-year follow-up of adrenal-to-brain transplants in Parkinson's disease. *Arch Neurol*, 51, 559-63.

Dimos, J.T., Rodolfa, K.T., Niakan, K.K., Weisenthal, L.M., Mitsumoto, H., Chung, W., Croft, G.F., Saphier, G., Leibel, R., Goland, R., Wichterle, H., Henderson, C.E. & Eggan, K. (2008) Induced pluripotent stem cells generated from patients with ALS can be differentiated into motor neurons. *Science*, 321, 1218-1221.

Doetsch, F., Caillé, I., Lim, D.A., García-Verdugo, J.M. & Alvarez-Buylla, A. (1999) Subventricular zone astrocytes are neural stem cells in the adult mammalian brain. *Cell*, 97, 703-716.

Doudet, D.J., Cornfeldt, M.L., Honey, C.R., Schweikert, A.W. & Allen, R.C. (2004) PET imaging of implanted human retinal pigment epithelial cells in the MPTP-induced primate model of Parkinson's disease. *Exp Neurol*, 189, 361-368.

Drucker-Colín, R., Madrazo, I., Ostrosky-Solís, F., Shkurovich, M., Franco, R. & Torres, C. (1988) Adrenal medullary tissue transplants in the caudate nucleus of Parkinson's patients. *Prog Brain Res*, 78, 567-574.

Drucker-Colín, R., Verdugo-Díaz, L., Morgado-Valle, C., Solís-Maldonado, G., Ondarza, R., Boll, C., Miranda, G., Wang, G.J. & Volkow, N. (1999) Transplant of cultured neuron-like differentiated chromaffin cells in a Parkinson's disease patient. A preliminary report. *Arch Med Res*, 30, 33-39.

Dunnett, S.B., Björklund, A. & Lindvall, O. (2001) Cell therapy in Parkinson's disease - stop or go? *Nat Rev Neurosci*, 2, 365-369.

Dunnett, S.B., Bjorklund A., Stenevi, U. & Iversen, S.D. (1981) Grafts of embryonic substantia nigra reinnervating the ventrolateral striatum ameliorate sensorimotor impairments and akinesia in rats with 6-OHDA lesions of the nigrostriatal pathway. *Brain Res*, 229, 209-217.

Dupre, K.B., Eskow, K.L., Barnum, C.J. & Bishop, C. (2008) Striatal 5-HT1A receptor stimulation reduces D1 receptor-induced dyskinesia and improves movement in the hemiparkinsonian rat. *Neuropharmacology*, 55, 1321–1328.

Elkabetz, Y., Panagiotakos, G., Al Shamy, G., Socci, N.D., Tabar, V. & Studer, L. (2008) Human ES cell-derived neural rosettes reveal a functionally distinct early neural stem cell stage. *Genes Dev,* 22, 152-65.

Eriksson, P.S., Perfilieva, E., Bjork-Eriksson, T., Alborn, A.M., Nordborg, C., Peterson, D.A. & Gage, F.H. (1998) Neurogenesis in the adult human hippocampus. *Nat Med,* 4, 1313-1317.

Farag, E.S., Vinters, H.V. & Bronstein, J. (2009) Pathologic findings in retinal pigment epithelial cell implantation for Parkinson disease. *Neurology,* 73, 1095-1102.

Fazzini, E., Dwork, A.J., Blum, C., Burke, R., Cote, L., Goodman, R.R., Jacobs, T.P., Naini, A.B., Pezzoli, G. & Pullman, S. (1991) Stereotaxic implantation of autologous adrenal medulla into caudate nucleus in four patients with parkinsonism. One-year follow-up. *Arch Neurol,* 48, 813-820.

Forno, L.S. & Langston, J.W. (1991) Unfavorable outcome of adrenal medullary transplant for Parkinson's disease. *Acta Neuropathol,* 81, 691-694.

Freed, C.R., Breeze, R.E., Rosenberg, N.L., Schneck, S.A., Kriek, E., Qi, J.X., Lone, T., Zhang, Y.B., Snyder, J.A., Wells, T.H., Ramig, L.O., Thompson, L., Mazziotta, J.C., Huang, S.C., Grafton, S.T., Gildenberg, P.L., Pettigrew, L.C., Merrell, R., Butler, I., Conklin, R., Katz, J. & DeFrance, J. (1990) Transplantation of adrenal medullary tissue to caudate nucleus using stereotactic techniques. *Stereotact Funct Neurosurg,* 54-55, 268-271.

Freed, C.R., Greene, P.E., Breeze, R.E., Tsai, W.Y., DuMouchel, W., Kao, R., Dillon, S., Winfield, H., Culver, S., Trojanowski, J.Q., Eidelberg, D. & Fahn, S. (2001) Transplantation of embryonic dopamine neurons for severe Parkinson's disease. *N Engl J Med,* 344, 710-719.

Gage, F.H. (2000) Mammalian neural stem cells. *Science,* 287, 1433-1438.

Geng, D.Y., Li, Y.X. & Zee, C.S. (2006) Magnetic resonance imaging-based volumetric analysis of basal ganglia nuclei and substantia nigra in patients with Parkinson's disease. *Neurosurgery,* 58, 256–262.

German, D.C., Manaye, K.F., Sonsalla, P.K. & Brooks, B.A. (1992) Midbrain dopaminergic cell loss in Parkinson's disease and MPTP-induced parkinsonism: sparing of calbindin D28k-containing cells. *Ann NY Acad Sci,* 648, 42–62.

Goetz, C.G., Stebbins, G.T. 3rd, Klawans, H.L., Koller ,W.C., Grossman, R.G., Bakay, R.A. & Penn, R.D. (1991) United Parkinson Foundation Neurotransplantation Registry on adrenal medullary transplants: presurgical, and 1- and 2-year follow-up. *Neurology,* 41, 1719-1722.

Grillner, S., Hellgren, J., Ménard, A., Saitoh, K. & Wikström, M.A. (2005) Mechanisms for selection of basic motor programs-roles for the striatum and pallidum. *Trends Neurosci,* 28, 364–370.

Grillner, S. (2006) Biological pattern generation: the cellular and computational logic of networks in motion. *Neuron,* 52, 751–766.

Guerra-Crespo, M., Gleason, D., Sistos, A., Toosky, T., Solaroglu, I., Zhang, J.H., Bryant, P.J. & Fallon, J.H. (2009) Transforming growth factor-alpha induces neurogenesis and behavioral improvement in a chronic stroke model. *Neuroscience,* 160, 470-483.

Gunaseeli, I., Doss, M.X., Antzelevitch, C., Hescheler, J. & Sachinidis, A. (2010) Induced pluripotent stem cells as a model for accelerated patient and disease-specific drug discovery. *Current Med Chem,* 17, 759-766.

Hack, M.A., Saghatelyan, A., de Chevigny, A., Pfeifer, A., Ashery-Padan, R., Lledo, P.M. & Götz, M. (2005) Neuronal fate determinants of adult olfactory bulb neurogenesis. *Nat Neurosci*, 8, 865-872.

Hargus, G., Cooper, O., Deleidi, M., Levy, A., Lee, K., Marlow, E., Yow, A., Soldner, F., Hockemeyer, D., Hallett, P.J., Osborn, T., Jaenisch, R. & Isacson, O. (2010) Differentiated Parkinson patient-derived induced pluripotent stem cells grow in the adult rodent brain and reduce motor asymmetry in Parkinsonian rats. *Proc Natl Acad Sci U SA*, 107, 15921-15926.

Hauser, R.A., Freeman, T.B., Snow, B.J., Nauert, M., Gauger, L., Kordower, J.H. & Olanow, C.W. (1999) Long-term Evaluation of Bilateral Fetal Nigral Transplantation in Parkinson Disease. *Arch neurol*, 56, 179-187

Hefti, F., Hartikka, J. & Schlumpf, M. (1985) Implantation of PC12 Cells into the Corpus Striatum of Rats with Lesions of the Dopaminergic Nigrostriatal Neurons. *Brain Res*, 348, 283-288.

Herman, J.P., Abrous, D.N. & Le Moal, M. (1991) Anatomical and behavioral comparison of unilateral dopamine-rich grafts implanted into the striatum of neonatal and adult rats. *Neuroscience*, 40, 465-475.

Hirsch, E.C., Duyckaerts, C., Javoy-Agid, F., Hauw, J.J. & Agid, Y. (1990) Does adrenal graft enhance recovery of dopaminergic neurons in Parkinson's disease? *Ann Neurol*, 27, 676-82.

Höglinger, G.U., Rizk, P., Muriel, M.P., Duyckaerts, C., Oertel, W.H., Caille, I. & Hirsch, E.C. (2004) Dopamine depletion impairs precursor cell proliferation in Parkinson disease. *Nat Neurosci*, 7, 726-735.

Holden, C. (2009) Fetal Cells Again? *Science*, 326, 358-359.

Huffaker, T.K., Boss, B.D., Morgan, A.S., Neff, N.T., Strecker, R.E., Spence, M.S. & Miao, R. (1989) Xenografting of fetal pig ventral mesencephalon corrects motor asymmetry in the rat model of Parkinson's disease. *Exp Brain Res*, 77, 329-336.

Huisman, E., Uylings, H.B. & Hoogland, P.V. (2004) A 100% increase of dopaminergic cells in the olfactory bulb may explain hyposmia in Parkinson's disease. *Mov Disord*, 19, 687-892.

Hutchinson, M. & Raff, U. (2000) Structural changes of the substantia nigra in Parkinson's disease as revealed by MR imaging. *AJNR Am J Neuroradiol*, 21, 697-701.

Isacson, O., Bjorklund, L. & Pernaute, R.S. (2001) Parkinson's disease: interpretations of transplantation study are erroneous. *Nat Neurosci*, 4, 553.

Itakura, T., Kamei, I., Nakai, K., Naka, Y., Nakakita, K., Imai, H. & Komai, N. (1988) Autotransplantation of the superior cervical ganglion into the brain. A possible therapy for Parkinson's disease. *J Neurosurg*, 68, 955–959.

Itakura, T., Uematsu, Y., Nakao, N., Nakai, E. & Nakai, K. (1997) Transplantation of autologous sympathetic ganglion into the brain with Parkinson's disease. Long-term follow-up of 35 cases. *Stereotact Funct Neurosurg*, 69, 112-115.

Jankovic, J., Grossman, R., Goodman, C., Pirozzolo, F., Schneider, L., Zhu, Z., Scardino, P., Garber, A.J., Jhingran, S.G. & Martin, S. (1989) Clinical, biochemical, and neuropathologic findings following transplantation of adrenal medulla to the caudate nucleus for treatment of Parkinson's disease. *Neurology*, 39, 1227-1234.

Jönsson, M.E., Ono, Y., Björklund, A. & Thompson, L.H. (2009) Identification of transplantable dopamine neuron precursors at different stages of midbrain neurogenesis. *Exp Neurol,* 219, 341-54.

Kawasaki, H., Mizuseki, K., Nishikawa, S., Kaneko, S., Kuwana, Y., Nakanishi, S., Nishikawa & S.I., Sasai, Y. (2000) Induction of midbrain dopaminergic neurons from ES cells by stromal cell-derived inducing activity. *Neuron,* 28, 31-40.

Kim, J.H., Auerbach, J.M., Rodríguez-Gómez, J.A., Velasco, I., Gavin, D., Lumelsky, N., Lee, S.H., Nguyen, J., Sánchez-Pernaute, R., Bankiewicz, K. & McKay, R. (2002) Dopamine neurons derived from embryonic stem cells function in an animal model of Parkinson's disease. *Nature,* 418, 50-56.

Kohwi, M., Osumi, N., Rubenstein, J.L. & Alvarez-Buylla, A. (2005) Pax6 is required for making specific subpopulations of granule and periglomerular neurons in the olfactory bulb. *J Neurosci,* 25, 6997-7003.

Kopin, I.J. & Markey, S.P. (1988) MPTP toxicity: Implications for research in Parkinson's disease. *Ann Rev Neurosci,* 11, 81-96.

Kordower, J.H., Chu, Y., Hauser, R.A., Freeman, T.B.& Olanow, C.W. (2008) Lewy body-like pathology in long-term embryonic nigral transplants in Parkinson's disease. *Nat Med,* 14, 504–506.

Kordower, J.H., Freeman, T.B., Snow, B.J., Vingerhoets, F.J., Mufson, E.J., Sanberg, P.R., Hauser, R.A., Smith, D.A., Nauert, G.M., Perl, D.P. & Olanow, C.W. (1995) Neuropathological evidence of graft survival and striatal reinnervation after the transplantation of fetal mesencephalic tissue in a patient with Parkinson's disease. *N Engl J Med,* 332, 1118-1124.

Kordower, J.H., Rosenstein, J.M., Collier, T.J., Burke, M.A., Chen, E.Y., Li, J.M., Martel, L., Levey, A.E., Mufson, E.J., Freeman, T.B. & Olanow, C.W. (1996) Functional fetal nigral grafts in a patient with Parkinson's disease: chemoanatomic, ultrastructural, and metabolic studies. *J Comp Neurol,* 370, 203-230.

Kreitzer, A.C. & Malenka, R.C. (2008). Striatal plasticity and basal ganglia circuit function. *Neuron,* 60, 543–554.

Laruelle, M. (2000) The role of model-based methods in the development of single scan techniques. *Nucl Med Biol,* 27, 637-642.

Langston, J.W., Ballard, P., Tetrud, J.W. & Irwin, I. (1983) Chronic parkinsonism in humans due to a product of meperidine analog synthesis. *Science,* 219, 979-980.

Lee, S.H., Lumelsky, N., Studer, L., Auerbach, J.M. & McKay, R.D. (2000) Efficient generation of midbrain and hindbrain neurons from mouse embryonic stem cells. *Nat Biotechnol,* 18, 675–679.

Lévesque, M.F., Neuman, T. & Rezak, M. (2009) Therapeutic Microinjection of Autologous Adult Human Neural Stem Cells and Differentiated Neurons for Parkinson's Disease: Five-Year Post-Operative Outcome. *The Open Stem Cell Journal,* 1, 20-29

Lindvall, O., Backlund, E.O., Farde, L., Sedvall, G., Freedman, R., Hoffer, B., Nobin, A., Seiger, A. & Olson, L. (1987) Transplantation in Parkinson's disease: two cases of adrenal medullary grafts to the putamen. *Ann Neurol,* 22, 457-468.

Lindvall, O., Brundin, P., Widner, H., Rehncrona, S., Gustavii, B., Frackowiak, R., Leenders, K.L., Sawle, G., Rothwell, J.C., Marsden, C.D. & Björklund, A. (1990) Transplantation strategies in the treatment of Parkinson's disease: experimental basis and clinical trials. *Science,* 247, 574-577.

Lindvall, O., Kokaia, Z. & Martinez-Serrano, A. (2004) Stem cell therapy for human neurodegenerative disorders-how to make it work. *Nat Med,* 10, S42–S50.

Lister, R., Pelizzola, M., Kida, Y.S., Hawkins D., Nery, J.R., Hon, G., Antosiewicz-Bourget, J., O'Malley, R., Castanon, R., Klugman, S., Downes, M., Stewart, R., Ren, B., Thomson, J.A., Evans, R.M. & Ecker, J.R. (2011) Hotspots of aberrant epigenomic reprogramming in human induced pluripotent stem cells. *Nature,* 471, 68-73.

López-Lozano, J.J., Bravo, G. & Abascal, J. (1991) Grafting of perfused adrenal medullary tissue into the caudate nucleus of patients with Parkinson's disease. Clinica Puerta de Hierro Neural Transplantation Group. *J Neurosurg,* 75, 234-243.

Lyon, R.A., Titeler, M., Bigornia, L. & Schneider, A.S. (1987) D2 dopamine receptors on bovine chromaffin cell membranes: identification and characterization by [3H] N-methylspiperone binding. *J Neurochem,* 48, 631-635.

Madrazo, I., Drucker-Colín, R., Díaz, V., Martínez-Mata, J., Torres, C. & Becerril, J.J. (1987) Open microsurgical autograft of adrenal medulla to the right caudate nucleus in two patients with intractable Parkinson's disease. *N Engl J Med,* 316, 831-834.

Madrazo, I., León, V., Torres, C., Aguilera, M.C., Varela, G., Alvarez, F., Fraga, A., Drucker-Colín, R., Ostrosky, F. & Skurovich, M. (1988) Transplantation of fetal substantia nigra and adrenal medulla to the caudate nucleus in two patients with Parkinson's disease. *N. Engl. J. Med,* 318, 51.

Mendez, I., Dagher, A., Hong, M., Gaudet, P., Weerasinghe, S., McAlister, V., King, D., Desrosiers, J., Darvesh, S., Acorn, T. & Robertson, H. (2002) Simultaneous intrastriatal and intranigral fetal dopaminergic grafts in patients with Parkinson disease: a pilot study. Report of three cases. *J Neurosurg,* 96, 589–596.

Mendez, I., Viñuela, A., Astradsson, A., Mukhida, K., Hallett, P., Robertson, H., Tierney, T., Holness, R., Dagher, A., Trojanowski, J.Q. & Isacson, O. (2008) Dopamine neurons implanted into people with Parkinson's disease survive without pathology for 14 years. *Nat Med,* 14, 507-9.

Morihisa, J.M., Nakamura, R.K., Freed, W.J., Mishkin, M. & Wyatt, R.J. (1984) Adrenal medulla grafts survive and exhibit catecholamine-specific fluorescence in the primate brain. *Exp Neurol,* 84, 643-653.

Morrish, P.K., Sawle, G.V. & Brooks, D.J. (1995) Clinical and [18F] dopa PET findings in early Parkinson's disease. *J Neurol Neurosurg Psychiatry,* 59, 597-600.

Nakagawa, M., Koyanagi, M., Tanabe, K., Takahashi, K., Ichisaka, T., Aoi, T., Okita, K., Mochiduki, Y., Takizawa, N. & Yamanaka, S. (2008) Generation of induced pluripotent stem cells without Myc from mouse and human fibroblast. *Nature Biotechnology,* 26, 101-106.

Nakao, N., Itakura, T., Uematsu, Y. & Komai, N. (1995) Transplantation of cultured sympathetic ganglionic neurons into parkinsonian rat brain: survival and function of graft. *Acta Neurochir,* 133, 61-67.

Okita, K., Ichisaka, T. & Yamanaka, S. (2007) Generation of germline-competent induced pluripotent stem cells. *Nature,* 448, 313-317.

Obenaus, A., Dilmac, N., Tone, B., Tian, H.R., Hartman, R., Digicaylioglu, M., Snyder, E.Y. & Ashwal, S. (2011) Long-term magnetic resonance imaging of stem cells in neonatal ischemic injury. *Ann Neurol,* 69, 282-291.

Olanow, C.W., Goetz, C.G., Kordower, J.H., Stoessl, A.J., Sossi, V., Brin, M.F., Shannon, K.M., Nauert, G.M., Perl, D.P., Godbold, J. & Freeman, T.B. (2003) A double-blind placebo-controlled trial of bilateral fetal nigral transplantation in Parkinson's disease. *Ann Neurol,* 54, 403–414.

Pankratz, M.T., Li, X.J., Lavaute, T.M., Lyons, E.A., Chen, X. & Zhang, S.C. (2007) Directed neural differentiation of human embryonic stem cells via an obligated primitive anterior stage. *Stem Cells,* 25, 1511-1520.

Pavese, N., Evans, A.H., Tai, Y.F., Hotton, G., Brooks ,D.J., Lees, A.J. & Piccini, P. (2006) Clinical correlates of levodopa-induced dopamine release in Parkinson disease: a PET study. *Neurology,* 67, 1612-1617.

Parent, J.M., Vexler, Z.S., Gong, C., Derugin, N. & Ferriero, D.M. (2002) Rat forebrain neurogenesis and striatal neuron replacement after focal stroke. *Ann Neurol* 52, 802-813.

Park, I.H., Arora, N., Huo, H., Maherali, N., Ahfeldt, T., Shimamura, A., Lensch, M.W., Cowan, C., Hochedlinger K. & Daley, G.Q. (2008) Disease-specific induced pluripotent stem (iPS) cells. *Cell,* 134, 877-886.

Park, I.H., Zhao, R., West, J.A., Yabuuchi, A., Huo, H., Ince, T.A., Lerou, P.H., Lensch, M.H. & Daley, G.Q. (2008) Reprogramming of human somatic cells to pluripotency with defined factors. *Nature,* 451, 141-146.

Perlow, M.J., Freed, W.J., Hoffer, B.J., Seiger, A., Olson, L. & Wyatt, R.J. (1979) Brain grafts reduce motor abnormalities produced by destruction of nigrostriatal dopamine system. *Science,* 204, 643-7.

Peterson, D.I., Price, M.L. & Small, C.S. (1989) Autopsy findings in a patient who had an adrenal-to-brain transplant for Parkinson's disease. *Neurology,* 39, 235-238.

Petruk, K.C., Wilson, A.F., Schindel, D.R., Witt, N.J., McLean, D.R., McFarland, P.A., Johnston, R.G., McPhee, M.S., Martin, W.R. & Calne, D.B. (1990) Treatment of refractory Parkinson's disease with adrenal medullary autografts utilizing two-stage surgery. *Prog Brain Res,* 82, 671-676.

Piccini, P., Brooks, D.J., Björklund, A., Gunn, R.N., Grasby, P.M., Rimoldi, O., Brundin, P., Hagell, P., Rehncrona, S., Widner, H. & Lindvall, O. (1999) Dopamine release from nigral transplants visualized in vivo in a Parkinson's patient. *Nat Neurosci,* 2, 1137–1140.

Piccini, P., Lindvall, O., Björklund, A., Brundin, P., Hagell, P., Ceravolo, R., Oertel, W., Quinn, N., Samuel, M., Rehncrona, S., Widner, H. & Brooks, D.J. (2000) Delayed recovery of movement-related cortical function in Parkinson's disease after striatal dopaminergic grafts. *Ann Neurol,* 48, 689-95.

Politis, M. & Piccini, P. (2010) Brain imaging after neural transplantation. *Prog Brain Res,* 184, 193-203.

Ptak, L.R., Hart, K.R., Lin, D. & Carvey, P.M. (1995) Isolation and manipulation of rostral mesencephalic tegmental progenitor cells from rat. *Cell Transplant,* 4, 335–342.

Rafael, H., Moromizato, P., Espinoza., M, Ayulo, V. & González-Portillo, M. (1991) Transplantation of adrenal medulla and omentum to the putamen by a transinsular pathway for Parkinson's disease. *Neurosurgery,* 28, 481.

Reynolds, B.A. & Weiss, S. (1992) Generation of neurons and astrocytes from isolated cells of the adult mammalian central nervous system. *Science,* 255, 1707-1710.

Reynolds, B.A. & Weiss, S. (1996) Clonal and population analyses demonstrate that an EGF-responsive mammalian embryonic CNS precursor is a stem cell. *Dev Biol,* 175, 1–13.

Rodríguez-Gómez, J.A., Lu, J.Q., Velasco, I., Rivera, S., Zoghbi, S.S., Liow, J.S., Musachio, J.L., Chin, F.T., Toyama, H., Seidel, J., Green, M.V., Thanos, P.K., Ichise, M., Pike, V.W., Innis, R.B. & McKay, R.D. (2007) Persistent dopamine functions of neurons derived from embryonic stem cells in a rodent model of Parkinson disease. *Stem Cells,* 25, 918-928.

Roy, N.S., Cleren, C., Singh, S.K., Yang, L., Beal, M.F. & Goldman, S.A. (2006) Functional engraftment of human ES cell-derived dopaminergic neurons enriched by coculture with telomerase-immortalized midbrain astrocytes. *Nat Med,* 12, 1259-1268.

Sachs, C. & Jonsson, G. (1975) Mechanism of action of 6-hydroxydopamine. *Biochem Phannacol,* 24, 1-8

Santini, E., Valjent, E. & Fisone, G. (2008) Parkinson's disease: levodopa-induced dyskinesia and signal transduction. *FEBS J,* 275, 1392–1399.

Sawle, G.V., Bloomfield, P.M., Björklund, A., Brooks, D.J., Brundin, P., Leenders, K.L., Lindvall, O., Marsclen, C.D., Rehncrona, S., Widner, H. & Frackowiak, R.S.J. (1992) Transplantation of fetal dopamine neurons in Parkinson's disease: PET [^{18}F]-6-L-fluurodopa studies in two patients with putaminal implants. *Ann Neurol,* 31, 166-173.

Soldner, F., Hockemeyer, D., Beard, C., Gao, Q., Bell, G.W., Cook, E.G., Hargus, G., Blak, A., Cooper, O., Mitalipova, M., Isacson, O. & Jaenisch, R. (2009) Parkinson´s disease patient-derived induced pluripotent stem cells free of viral reprogramming factors. *Cell,* 136, 964-977.

Spencer, D.D., Robbins, R.J., Naftolin, F., Marek, K.L., Vollmer, T., Leranth, C., Roth, R.H., Price, L.H., Gjedde, A., Bunney, B.S., Sass, K.J., Elsworth, J.D., Kier, E.L., Makuch, R., Hoffer, P.B., & Redmond, D.E. (1992) Unilateral transplantation of human fetal mesencephalic tissue into the caudate nucleus of patients with Parkinson's disease. *N Engl J Med,* 327, 1541-8.

Storch, A., Sabolek, M., Milosevic, J., Schwarz, S.C. & Schwarz, J. (2004) Midbrain-derived neural stem cells: from basic science to therapeutic approaches. *Cell Tissue Res,* 318, 15–22.

Studer, L., Tabar, V. & McKay, R.D. (1998) Transplantation of expanded mesencephalic precursors leads to recovery in parkinsonian rats. *Nat Neurosci,* 1, 290 -295.

Subramanian, T., Marchionini, D., Potter, E.M. & Cornfeldt, M.L. (2002) Striatal xenotransplantation of human retinal pigment epithelial cells attached to microcarriers in hemiparkinsonian rats ameliorates behavioral deficits without provoking a host immune response. *Cell Transplant,* 11, 207-214.

Takahashi, K. & Yamanaka, S. (2006) Induction of pluripotent stem cells from mouse embryonic and adult fibroblast cultures by defined factors. *Cell,* 126, 663–676.

Takahashi, S., Tanabe, K., Ohnuki, M., Narita, M., Ichisaka, T., Tomoda, K. & Yamanaka, S. (2007) Induction of pluripotent stem cells from adult human fibroblast by defined factors. *Cell,* 131, 861–872.

Takagi, Y., Takahashi, J., Saiki, H., Morizane, A., Hayashi, T., Kishi, Y., Fukuda, H., Okamoto, Y., Koyanagi, M., Ideguchi, M., Hayashi, H., Imazato, T., Kawasaki, H., Suemori, H., Omachi, S., Iida, H., Itoh, N., Nakatsuji, N., Sasai, Y. & Hashimoto, N. (2005) Dopaminergic neurons generated from monkey embryonic stem cells function in a Parkinson primate model. *J Clin Invest,* 115, 102-109.

Takayama, H., Ray, J., Raymon, H.K., Baird, A., Hogg, J., Fisher, L.J. & Gage, F.H. (1995) Basic fibroblast growth factor increases dopaminergic graft survival and function in a rat model of Parkinson's disease. *Nat Med,* 1, 53-58.

Taylor, J.L., Bishop, C. & Walker, P.D. (2005) Dopamine D1 and D2 receptor contributions to l-DOPA-induced dyskinesia in the dopamine-depleted rat. *Pharmacol Biochem Behav,* 81, 887-893.

Thomson, J.A. & Odorico, J.S. (2000) Human embryonic stem cell and embryonic germ cell lines. *Trends Biotechnol,* 18, 53-57.

Thomson, J.A., Itskovitz-Eldor, J., Shapiro, S.S., Waknitz, M.A., Swiergiel, J.J., Marshall, V.S. & Jones, J.M. (1998) Embryonic stem cell lines derived from human blastocysts. *Science,* 282, 1145–1147.

Venkataramana, N.K., Kumar, S.K., Balaraju, S., Radhakrishnan, R.C., Bansal, A., Dixit, A., Rao, D.K., Das, M., Jan, M., Gupta, P.K., & Totey, S.M. (2010) Open-labeled study of unilateral autologous bone-marrow-derived mesenchymal stem cell transplantation in Parkinson's disease. *Transl Res,* 155, 62-70.

Vidaltamayo, R., Bargas, J., Covarrubias, L., Hernández-Cruz, A., Galarraga, E., Gutiérrez-Ospina, G. & Drucker-Colin R. (2010) Stem cell therapy for Parkinson's disease: a road map for a successful future. Stem *Cells Dev,* 19, 311-320.

Vierbuchen, T., Ostermeier, A., Pang, Z.P., Kokubu, Y., Südhof, T.C. & Wernig, M. (2010) Direct conversion of fibroblasts to functional neurons by defined factors. *Nature,* 463, 1035-1041.

Wallén, A., Zetterström, R.H., Solomin, L., Arvidsson, M., Olson, L. & Perlmann, T. (1999) Fate of mesencephalic AHD2-expressing dopamine progenitor cells in NURR1 mutant mice. *Exp Cell Res,* 253, 737–746.

Waters, C.H., Apuzzo, M.L., Neal, J.H. & Weiner, L.P. (1992) Long-term follow-up of adrenal medullary transplantation for Parkinson's disease. *J Geriatr Psychiatry Neurol,* 5, 35-9.

Watts, R.L., Raiser, C.D., Stover, N.P., Cornfeldt, M.L., Schweikert, A.W., Allen, R.C., Subramanian, T., Doudet, D., Honey, C.R. & Bakay, R.A. (2003) Stereotaxic intrastriatal implantation of human retinal pigment epithelial (hRPE) cells attached to gelatin microcarriers: a potential new cell therapy for Parkinson's disease. *J Neural Transm Suppl,* 65, 215-227.

Wernig, M., Meissner, A., Foreman, R., Brambrink, T., Manching, K., Hochedlinger, K., Bernstein, B.E. & Jaenisch, R. (2007) *In vitro* reprogramming of fibroblats into pluripotent ES-cell-like state. *Nature,* 448, 318-324.

Wernig, M., Zhao, J.P., Pruszak, J., Hedlund, E., Fu, D., Soldner, F., Broccoli, V., Constantine-Paton, M., Isacson, O. & Jaenisch, R. (2008) Neurons derived from reprogrammed fibroblasts functionally integrate into the fetal brain and improve symptoms of rats with Parkinson's disease. *Proc Natl Acad Sci U SA,* 105, 5856-5861

Winner, B., Geyer, M., Couillard-Despres, S., Aigner, R., Bogdahn, U., Aigner, L., Kuhn, G. & Winkler, J. (2006) Striatal deafferentation increases dopaminergic neurogenesis in the adult olfactory bulb. *Exp Neurol,* 197, 113-121.

Yu, J., Vodyanik, M.A., Smuga-Otto, K., Antosiewicz-Bourget, J., Frane, J.L., Tian, S., Nie, J., Jonsdottir, G.A., Ruotti, V., Stewart, R., Slukvin, I.I. & Thompson, J.A. (2007) Induced pluripotent stem cell lines derived from human somatic cells. *Science,* 318, 1917-1920.

Zhu, S.M., Kujirai, K., Dollison, A., Angulo, J., Fahn, S. & Cadet, J.L. (1992) Implantation of genetically modified mesencephalic fetal cells into the rat striatum. *Brain Res Bull,* 29, 81-93.

17

Stem Cell Therapy for Neuromuscular Diseases

Mirella Meregalli, Andrea Farini and Yvan Torrente
Università degli Studi di Milano, Fondazione IRCCS Cà Granda Ospedale Maggiore Policlinico di Milano – Centro Dino Ferrari
Italy

1. Introduction

Neuromuscular disease is a very broad term that encompasses many diseases and aliments that either directly, via intrinsic muscle pathology, or indirectly, via nerve pathology, impair the functioning of the muscles. Neuromuscular diseases affect the muscles and/or their nervous control and lead to problems with movement. Many are genetic; sometimes, an immune system disorder can cause them. As they have no cure, the aim of clinical treatment is to improve symptoms, increase mobility and lengthen life. Some of them affect the anterior horn cell, and are classified as acquired (*e.g.* poliomyelitis) and hereditary (*e.g.* spinal muscular atrophy) diseases. SMA is a genetic disease that attacks nerve cells, called motor neurons, in the spinal cord. As a consequence of the lost of the neurons, muscles weakness becomes to be evident, affecting walking, crawling, breathing, swallowing and head and neck control. Neuropathies affect the peripheral nerve and are divided into demyelinating (*e.g.* leucodystrophies) and axonal (*e.g.* porphyria) diseases. Charcot-Marie-Tooth (CMT) is the most frequent hereditary form among the neuropathies and it's characterized by a wide range of symptoms so that CMT-1a is classified as demyelinating and CMT-2 as axonal (Marchesi & Pareyson, 2010). Defects in neuromuscular junctions cause infantile and non-infantile Botulism and Myasthenia Gravis (MG). MG is a antibody-mediated autoimmune disorder of the neuromuscular junction (NMJ) (Drachman, 1994; Meriggioli & Sanders, 2009). In most cases, it is caused by pathogenic autoantibodies directed towards the skeletal muscle acetylcholine receptor (AChR) (Patrick & Lindstrom, 1973) while in others, non-AChR components of the postsynaptic muscle endplate, such as the muscle-specific receptor tyrosine kinase (MUSK), might serve as targets for the autoimmune attack (Hoch et al., 2001). Although the precise origin of the autoimmune response in MG is not known, genetic predisposition and abnormalities of the thymus gland such as hyperplasia and neoplasia could have an important role in the onset of the disease (Berrih et al., 1984; Roxanis et al., 2001).

Several diseases affect muscles: they are classified as acquired (*e.g.* dermatomyositis and polymyositis) and hereditary (*e.g.* myotonic disorders and myopaties) forms. Among the myopaties, muscular dystrophies are characterized by the primary wasting of skeletal muscle, caused by mutations in the proteins that form the link between the cytoskeleton and the basal lamina (Cossu & Sampaolesi, 2007). Mutations in the dystrophin gene cause severe form of hereditary muscular diseases; the most common are Duchenne Muscular Dystrophy (DMD) and Becker Muscular Dystrophy (BMD). DMD patients suffer for complete lack of dystrophin that causes progressive degeneration, muscle wasting and death into the second/third decade of life. Beside, BMD patients show a very mild phenotype, often asymptomatic primarily due to the expression of shorter dystrophin mRNA transcripts that maintain the coding reading frame. DMD patients' muscles show absence of dystrophin and presence of endomysial fibrosis, small fibers rounded and muscle fiber degeneration/regeneration. Untreated, boys with DMD become progressively weak during their childhood and stop ambulation at a mean age of 9 years, later with corticosteroid treatment (12/13 yrs). Proximal weakness affects symmetrically the lower (such as quadriceps and gluteus) before the upper extremities, with progression to the point of wheelchair dependence. Eventually distal lower and then upper limb weakness occurs. Weakness of neck flexors is often present at the beginning, and most patients with DMD have never been able to jump. Wrist and hand muscles are involved later, allowing the patients to keep their autonomy in transfers using a joystick to guide their wheelchair. Musculoskeletal contractures (ankle, knees and hips) and learning difficulties can complicate the clinical expression of the disease. Besides this weakness distribution in the same patient, a deep variability among patients does exist. They could express a mild phenotype, between Becker and Duchenne dystrophy, or a really severe form, with the loss of deambulation at 7-8 years. Confinement to a wheelchair is followed by the development of scoliosis, respiratory failure and cardiomyopathy. In 90% of people death is directly related to chronic respiratory insufficiency (Rideau et al., 1983). The identification and characterization of dystrophin gene led to the development of potential treatments for this disorder (Bertoni, 2008). Even if only corticosteroids were proven to be effective on DMD patient (Hyser and Mendell, 1988), different therapeutic approaches were attempted, as described in detail below (see section 7).

2. Treatment for neuromuscular diseases: gene and cell therapy

The identification and characterization of the genes whose mutations caused the most common neuromuscular diseases led to the development of potential treatments for those disorders. Gene therapy for neuromuscular disorders embraced several concepts, including replacing and repairing a defective gene or modifying or enhancing cellular performance, using gene that is not directly related to the underlying defect (Shavlakadze et al., 2004). As an example, the finding that DMD pathology was caused by mutations in the dystrophin gene allowed the rising of different therapeutic approaches including growth-modulating agents that increase muscle regeneration and delay muscle fibrosis (Tinsley et al., 1998), powerful antisense oligonucleotides with exon-skipping capacity (McClorey et al., 2006), anti-inflammatory or second-messenger signal-modulating agents that affect immune responses (Biggar et al., 2006), agents designed to suppress stop codon mutations (Hamed, 2006). Viral and non-viral vectors were used to deliver the full-length - or restricted versions - of the dystrophin gene into stem cells; alternatively, specific antisense oligonucleotides were designed to mask the putative splicing sites of exons in the mutated region of the primary RNA transcript whose removal would re-establish a correct reading frame. In parallel, the biology of stem cells and their role in regeneration were the subject of intensive and extensive research in many laboratories around the world because of the promise of stem cells as therapeutic agents to regenerate tissues damaged by disease or injury (Fuchs and Segre, 2000; Weissman, 2000). This research constituted a significant part of the rapidly developing field of regenerative biology and medicine, and the combination of gene and cell therapy arose as one of the most suitable possibility to treat degenerative disorders. Several

works were published in which stem cell were genetically modified by ex vivo introduction of corrective genes and then transplanted in donor dystrophic animal models.
Stem cells received much attention because of their potential use in cell-based therapies for human disease such as leukaemia (Owonikoko et al., 2007), Parkinson's disease (Singh et al., 2007), and neuromuscular disorders (Endo, 2007; Nowak and Davies, 2004). The main advantage of stem cells rather than the other cells of the body is that they can replenish their numbers for long periods through cell division and, they can produce a progeny that can differentiate into multiple cell lineages with specific functions (Bertoni, 2008). The candidate stem cell had to be easy to extract, maintaining the capacity of myogenic conversion when transplanted into the host muscle and also the survival and the subsequent migration from the site of injection to the compromise muscles of the body (Price et al., 2007). With the advent of more sensitive markers, stem cell populations suitable for clinical experiments were found to derive from multiple region of the body at various stage of development. Numerous studies showed that the regenerative capacity of stem cells resided in the environmental microniche and its regulation. This way, it could be important to better elucidate the molecular composition - cytokines, growth factors, cell adhesion molecules and extracellular matrix molecules - and interactions of the different microniches that regulate stem cell development (Stocum, 2001).
Several groups published different works concerning adult stem cells such as muscle-derived stem cells (Qu-Petersen et al., 2002), mesoangioblasts (Cossu and Bianco, 2003), blood- (Gavina et al., 2006) and muscle (Benchaouir et al., 2007)-derived CD133+ stem cells. Although some of them are able to migrate through the vasculature (Benchaouir et al., 2007; Galvez et al., 2006; Gavina et al., 2006) and efforts were done to increase their migratory ability (Lafreniere et al., 2006; Torrente et al., 2003a), poor results were obtained.
Embryonic and adult stem cells differ significantly in regard to their differentiation potential and in vitro expansion capability. While adult stem cells constitute a reservoir for tissue regeneration throughout the adult life, they are tissue-specific and possess limited capacity to be expanded ex vivo. Embryonic Stem (ES) cells are derived from the inner cell mass of blastocyst embryos and, by definition, are capable of unlimited in vitro self-renewal and have the ability to differentiate into any cell type of the body (Darabi et al., 2008b). ES cells, together with recently identified iPS cells, are now broadly and extensively studied for their applications in clinical studies.

3 Embryonic Stem Cells (ESCs)

Embryonic stem cells are pluripotent cells derived from the early embryo that are characterized by the ability to proliferate over prolonged periods of culture remaining undifferentiated and maintaining a stable karyotype (Amit and Itskovitz-Eldor, 2002; Carpenter et al., 2003; Hoffman and Carpenter, 2005). They are capable of differentiating into cells present in all 3 embryonic germ layers, namely ectoderm, mesoderm, and endoderm, and are characterized by self-renewal, immortality, and pluripotency (Strulovici et al., 2007).

3.1 Methods for ESCs isolation

hESCs are derived by microsurgical removal of cells from the inner cell mass of a blastocyst stage embryo (Fig. 1). The ES cells can be also obtained from single blastomeres. This technique creates ES cells from a single blastomere directly removed from the embryo bypassing the ethical issue of embryo destruction (Klimanskaya et al., 2006). Although maintaining the viability of the embryo, it has to be determined whether embryonic stem cell lines derived from a single blastomere that does not compromise the embryo can be considered for clinical studies. Cell Nuclear Transfer (SCNT): Nuclear transfer, also referred to as nuclear cloning, denotes the introduction of a nucleus from an adult donor cell into an enucleated oocyte to generate a cloned embryo (Wilmut et al., 2002).

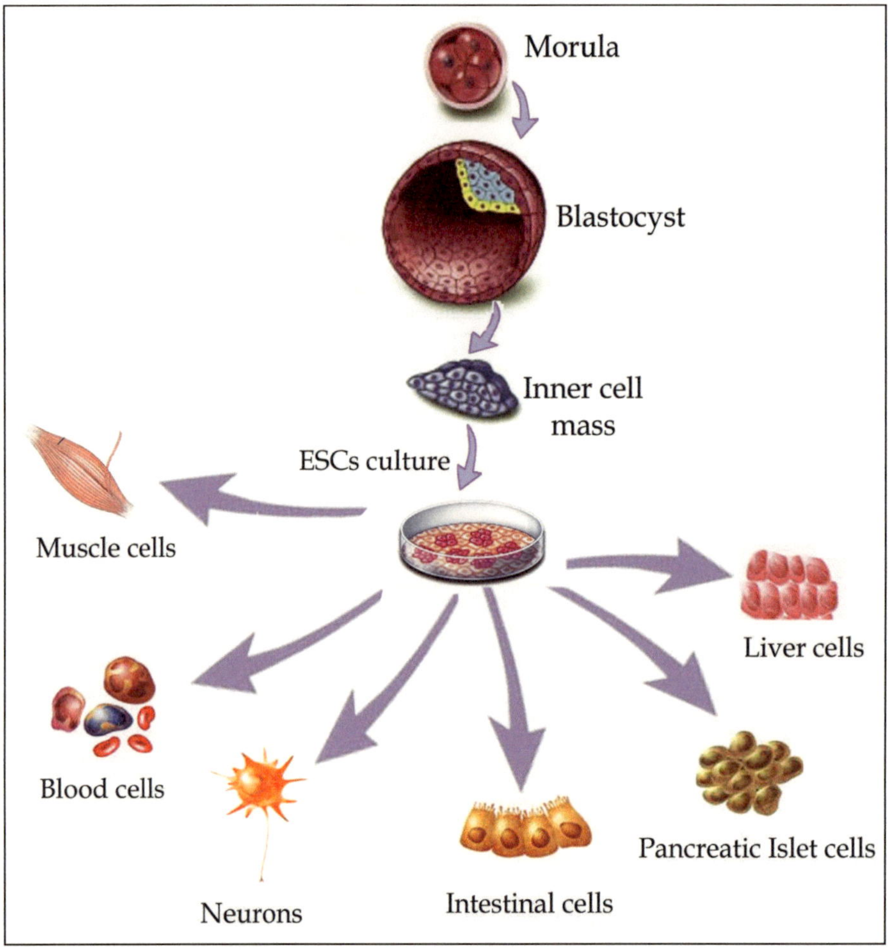

Fig. 1. ESCs differentiation. Differentiation potentiality of human embryonic stem cell lines. Human embryonic stem cell pluripotency is evaluated by the ability of the cells to differentiate into different cell types.

3.1.1 ES cells from single blastomeres

This technique creates ES cells from a single blastomere directly removed from the embryo bypassing the ethical issue of embryo destruction (Klimanskaya et al., 2006). Although maintaining the viability of the embryo, it has to be determined whether embryonic stem cell lines derived from a single blastomere that does not compromise the embryo can be considered for clinical studies.

3.1.2 Cell Nuclear Transfer (SCNT)

Nuclear transfer, also referred to as nuclear cloning, denotes the introduction of a nucleus from an adult donor cell into an enucleated oocyte to generate a cloned embryo (Wilmut et al., 2002). The first application of this technique was in 1996 the creation of Dolly the sheep (Campbell et al., 1996). Transferred to the uterus of a female recipient, this embryo has the potential to grow into a clone of the adult donor cell, a process termed "reproductive cloning." When explanted in culture, this embryo can give rise to embryonic stem cells that have the potential to become any or almost any type of cell present in the adult body. This process is also called "nuclear transplantation therapy" or "therapeutic cloning" because embryonic stem cells derived by nuclear transfer are genetically identical to the donor and thus potentially useful for therapeutic applications. It might substantially improve the treatment of neurodegenerative diseases, blood disorders, or diabetes, whose therapies are currently limited by the availability or immunocompatibility of tissue transplants (Hochedlinger and Jaenisch, 2003). Unfortunately, reproductive cloning is a largely inefficient and error-prone process that results in the failure of most clones during development due both to activation of inadequate pathways of early embryonic development (Solter, 2000) and suppression of pathway of differentiation (Rideout et al., 2001). In contrast, reprogramming errors do not appear to interfere with therapeutic cloning, because the process appears to select for functional cells. Recent advances in the field of nuclear cloning showed that most clones die early in gestation while cloned animals share abnormalities regardless of the type of donor cell or the species used, correlating with aberrant gene expression (Hochedlinger and Jaenisch, 2003). Although experiments in animals showed that by SCNT it is possible to obtain primate ES cells (Byrne et al., 2007) and nuclear cloning combined with gene and cell therapy represents a valid strategy for treating genetic disorders (Rideout et al., 2002), the low efficiency of the technique, the difficulties in obtaining human eggs and the arising ethical problems are significant challenges to the widespread use of SCNT for the production of hESC.

3.2 Characterization of ESCs

Human embryonic stem cells (hESCs) were first derived from the inner cell mass (ICM) of the blastocyst stage (100-200 cells) of embryos generated by *in vitro* fertilization (Thomson et al., 1998), but methods have been developed to derive hESCs from the late morula stage (30-40 cells) (Strelchenko et al., 2004), from arrested embryos (16-24 cells incapable of further development) (Zhang et al., 2006) and single blastomeres isolated from 8-cell embryos (Klimanskaya et al., 2006). Because hESCs have the potential to differentiate into normal tissues of all types, the ability to derive and maintain hESCs in culture gave rise to the possibility of having an unlimited supply of normal differentiated cells to engineer diseased tissues to regain normal function (Moon et al., 2006; Skottman et al., 2006).

Nowadays, several studies demonstrating hESC differentiation into specific cell lineages use feeder layers of heterologous cells to maintain hESCs in culture and to signal the hESCs to differentiate into specific cell types (Conrad et al., 2008; Takahashi and Yamanaka, 2006). After transplantation into the recipient, the hESCs and their progeny could be exogenously controlled if they differentiated into malignant cells or if they otherwise grew and/or functioned in an unwanted lineage; if hESCs are to be useful in generating normal tissues for the treatment of human disease, the tissues to be transplanted must be compatible with the host such that the cells derived from the hESCs will not be recognized as "foreign" and rejected as would any transplanted tissue from an unrelated donor.

The first human stem cell line bank was opened in 2004 in the United Kingdom (http://www.ukstemcellbank.org.uk/). The National Institutes of Health registry (http://stemcells.nih.gov/research/registry/) has also archived a number of hESC lines and established criteria for demonstration of the pluripotency of these lines. Cells should be able to give rise to any cell lineage of the body and thus to form a teratoma (a tumour containing tissues from the 3 primary germ layers) after injection in an immune-suppressed animal and should be capable of unlimited self-renewal. A number of scientific/medical issues need to be addressed before stem cells can be considered safe for clinical applications. In fact for hESCs to be useful for therapy, technologies must be developed to provide them with the specific signals required to differentiate in a controlled lineage, to regulate and/or shut down the growth of hESCs and their progeny once they have been transferred to the recipient. Pluripotency is evidenced by the ability to form teratomas when transplanted in immunodeficient mice, the concern exists that these cells could form malignant tumours in their new host. One strategy for dealing with this problem is to select pure populations of more committed cells for transfer. Demonstrating genetic and epigenetic stability will therefore be important before these cells are used clinically. Moreover attention has to be focused on circumventing the host rejection of transplanted, non-autologous hESC-derived cells (Strulovici et al., 2007).

Therefore, karyotypic abnormalities have been described in several hESC lines, although changes might be at least partially dependent on culture techniques (Mitalipova et al., 2005). In additional to biologic issues directly affecting the stem cell product, it is fundamental that controlled, standardized practices and procedures be followed to maintain the integrity, uniformity, of the human stem cell preparations. Because of stem cells are maintained and expanded in vitro before transplantation, culture conditions compatible with human administration must be used. Feeder cells and sera of animal origin have to be reduced and ideally avoided to reduce the potential risk of contamination by xenogeneic protein. Consequently, life-long immunosuppressive therapy, which can lead to infections and organ-based toxic side effects, such as nephropathy, might be required to prevent graft rejection (Brignier and Gewirtz, 2010).

3.2.1 Expression markers

Cell origins are often defined by one or more cell surface and or intracellular epitopes unique to that particular cell type. Stage-specific embryonic antigen (SSEA) markers are used to distinguish early stages of cell development, denoting pluripotency. These markers are globo-series glycolipids and are recognised by monoclonal antibodies. The SSEA-4 epitope is the globo-series glycolipid GL7. It has been demonstrated that GL7 can react with antibodies to both SSEA-3 and SSEA-4 (Kannagi et al., 1983a; Kannagi et al., 1983b). Human ES cells will express SSEA-3 and -4 during pluripotency and only SSEA-1 upon differentiation (Andrews et al., 1996; Reubinoff et al., 2001; Thomson et al., 1998; Thomson and Marshall, 1998). The TRA-1-60 epitope adheres to a particular epitope of the proteoglycan and is sialidase sensitive, whereas antibody TRA-1-81 reacts with another unknown epitope of the same core proteoglycan molecule. Nanog is a NK-2-type homeodomain gene thought to encode a transcription factor that is critically involved in the self-renewal of stem cells. Thus, it may possibly act to repress genes necessary for differentiation and activate those involved in self-renewal. Lin and colleagues (Lin et al., 2005) demonstrated that the tumour suppressor p53 binds to the promoter of Nanog.

Therefore, p53 can stimulate p53-dependent cell-cycle arrest and apoptosis when genetic integrity is not preserved. Oct-4, a POU-domain transcription factor, is highly expressed in ES cells (Niwa, 2001; Reubinoff et al., 2001; Thomson et al., 1998) and has been shown to be essential for maintaining pluripotency (Niwa, 2001). It has been reported that Oct-4 transcripts are nearly exclusively found in pluripotent cells *in vivo* and within culture. Oct-4 down-regulation is observed in differentiating cells (Rosner et al., 1990). Not only is Oct-4 necessary for the maintenance of pluripotency, but its expression level governs three cell fates once differentiation occurs (Hay et al., 2004; Niwa et al., 2000). Several candidate genes have been reported as targets of Oct-4 based on stem cell expression patterns and immunoprecipitation, but few have been conclusively verified. Target genes of Oct4 include Rex-1, Lefty-1, PDGFalfaR and Utf-1, and those cooperating with Oct4 include Sox2. In the ongoing search for the identification of pluripotent markers, Xu and colleagues have reported that the catalytic component of telomerase, telomerase reverse transcriptase or hTERT, is expressed in undifferentiated cells and down-regulated upon differentiation (Xu et al., 2001).

3.3 Potentialities of ESCs

ES-derived progenitors possess excellent self-renewal and regenerative potential, but the research on these cells is at the beginning. Recently, Jaenisch and collaborators published that the adult cells contain unipotent and multipotent stem cells such as haematopoietic stem cells even if totipotent and pluripotent cells are restricted to the early embryo (Jaenisch and Young, 2008). Although the decrease in developmental potential, the nuclei of most of adult cells maintain nuclear plasticity to reset to an embryonic state. It's possible to enhance this process by exposing the oocyte to specific factors through nuclear transfer or the cells to pluripotent cell-specific factors by driving over-expression of defined transcription factors. However before the clinical applications of these cells, it's needed to optimize the engraftment of hESCs and the development of a protocol to obtain similar populations of muscle precursors from human ES cells. ESCs are derived rather easily, and they can grow indefinitely in culture. Second, embryonic stem cells can be manipulated genetically by homologous recombination to correct a genetic defect (Rideout et al., 2001). Recently, Jiang and collaborators demonstrated pluripotency of mesenchymal stem cells derived from adult marrow, differentiating into cells of all three germ layers both in vitro and in vivo after being injected into blastocysts (Jiang et al., 2002). Unfortunately, they did not assess the ability of these cells in correcting a disease phenotype into both human or mouse animal models. Embryonic stem cells can become any type of cell through the use of specific culture conditions or genetic manipulation. To avoid the ethical and practical limitations of therapeutic cloning mentioned above, it would be useful to reprogram somatic cells directly into embryonic stem cells without the use of oocytes. For this reason, it's necessary to deeply understand the role of several molecular factors in establishing and maintaining pluripotency, such as Oct-4 (Niwa et al., 2002). Oct-4 null embryos cannot form a pluripotent inner cell mass, consequently their development is arrested. To circumvent the need for human oocytes, it could be possible to modify the expression of Oct-4 and its related genes in somatic cells to reprogram their nuclei to an embryonic state.

3.4 Drawbacks to clinic use of ESCs

3.4.1 Host response to grafted tissue

It is probable that ESCs will suffer from the same acute and chronic rejection problems that accompany other grafts and it is likely that this question will not be answered until these cells are implanted into humans for the first time. Three methods have been proposed to avoid this problem. The first requires the use of somatic cell nuclear transfer (SCNT) techniques, as used to clone animals, to personalize ESCs. The nucleus of a somatic cell from the individual to be treated would be transferred into an enucleated donor oocyte, which would then be used to derive a blastocyst and subsequently isolate ESCs lines that would be genetically identical to the patient (Yang et al., 2007). In this case, any cell generated from the personalized ESC line should not be rejected. However, despite the claims of the South Korean Dr Hwang Woo-suk in 2005 that have subsequently been shown to be fraudulent, SCNT has not been successfully performed in human oocytes. It appears that the process is more complicated in humans than in animals where this technology has been used successfully in many species. The second method makes use of the capacity of ESCs to differentiate into multiple tissues and would involve replacing the recipients immune system with haematopoietic cells generated from the same ESC line as that used for tissue replacement. This technique has been used in solid organ transplants where patients have previously received bone marrow transplantation, and these patients did not require immune suppression (Helg et al., 1994). The third method used genetic manipulation to engineer ESCs in which MHC molecules or other immune effectors have been deleted (Hyslop et al., 2005). All these methods are under development. In addition to the immune response to the cells themselves, animal products are used to isolate ESCs in every methods, and it cause expression of animal proteins on the surface of the ES cells. This will also induce an immune response and a large amount of work is currently going into deriving and maintaining ESCs in total absence of these animal products and also undertaking these processes to the good manufacturing practice (GMP) standards required for clinical use. It is in the development of GMP protocols for the derivation and manipulation of ES cells.

3.4.2 Cell proliferation and differentiation

In addition to the problems connected with immune interactions, there are other important problems in providing a suitable number of cells for transplantation. The first problem is related to have the very large numbers of cells required for tissue replacement without tumour formation. As previously discussed, ES cells can form teratomas and therefore all undifferentiated cells will need to be removed from a graft. It is possible to circumvent this problem genetically modifying the cells by a suicide gene system. The expression of the suicide gene could be driven by the promoter of a gene such as Oct4 that is only expressed in undifferentiated cells; this technology has been used in gene therapy systems. Activation of the suicide gene by drug treatment of cell cultures would give rise to the death of the undifferentiated cells not affecting differentiating cells. Other methods include the expression of a genetic label such as the green fluorescent protein or a marker that enables cell sorting by FACS (Strulovici et al., 2007). In this case, cells could be selected using FACS for the surface markers expressed either excluding undifferentiated cells or positively selecting for the required differentiated cell type. Even if contaminating undifferentiated cells are removed, remain the problem due to the necessity to have a good number of mature cells able to form a robust graft and also maintaining the cells where they have to perform their therapeutically action. Differentiation protocols are being studying to generate very large number of cells (Joannides et al., 2007) and scaffolds able to keep the cells in place (Ferreira et al., 2007). Another aspect of tissue replacement that cannot be ignored is that few tissues are formed from a single uniform cell type. In order to regenerate a functional organ, it will be necessary to develop other structures such as vasculature and lymphatic drainage systems as well as complex mixed cell populations (Caspi et al., 2007). These problems present a major challenge to tissue culture technology not only for ESCs but also for adult stem cells, and the development of new systems will be important in order to utilize ES cells to their maximum potential.

3.4.3 Ethical implications to clinical use of ESCs

The use of hESCs in medical research has focused much attention from many sectors of the public. Religious, historical, cultural, medical, and other points of view have contributed to a very vigorous and wide-ranging discourse over the use of these materials (Leist et al., 2008). Some consider research with hESCs to be inherently immoral because these individual's believe that life begins with fertilization of the ovum, and the destruction of an embryo with the potential to develop into a viable human being is thought like an infanticide. For this reason, the American federal government severely restricted access and use of hESCs in 2001. These restrictions have now been largely overturned by the Obama administration. In contrast, proponents of this line of research insist that the potential benefits to human from this research mitigate such concerns. They also argue that hESCs are made from unwanted fertilized ovum that would likely be destroyed in any event. Stem cells created by means of nuclear transfer share the same ethical concerns. Furthermore, because these cells have the potential to generate a complete embryo, they also raise the even more highly charged possibility of cloning human beings, so-called reproductive cloning (Brignier and Gewirtz, 2010). Many organizations and countries have already banned reproductive cloning of human beings. Because this procedure can be used to generate stem cells for therapeutic purposes, in countries where this type of cloning is legal, such as Australia and the United Kingdom, the created embryos must be destroyed within 14 days. A human nucleus is transferred into an animal's oocyte, creating a hybrid embryo that must be destroyed within 2 weeks and cannot be implanted. Clearly, creation of such tissues raises even more complex issues. Finally, the issue of financial compensation for embryo and gamete donors is also controversial, with guidelines for this problem being proposed by the International Society of Stem Cell Research (http://www.isscr.org/guidelines /index.htm). Everyone involved in the debate want very much to avoid the development of an underground black market in spare embryos ((Brignier and Gewirtz, 2010).

3.4.4 Clinical applications of ESCs

Among the important potential applications of gene therapy to hESCs is the correction of genetic diseases. Although many hereditary disorders can be targeted by gene therapy vectors alone, the combination of gene therapy and stem cell therapy may have added utility, where cells differentiated from hESCs would act as factories to produce therapeutic proteins or where a high proportion of corrected cells could be developed. In the case of circulating proteins (*e.g.*, factor IX, factor VIII, von Willebrand factor, a1-antitrypsin), it may be possible to establish tissue reservoirs distant from the normal site of the secreted product (Mountford, 2008). The cells differentiated from hESCs can be delivered to a site accessible and receptive to transplantation, even if that tissue is not the normal site of production of the protein of interest. When the product is not secreted, the hESCs with their regulated genetic characteristics must be differentiated to the correct cell type (*e.g.*, the cystic fibrosis transmembrane conductance regulator for cystic fibrosis or dystrophin for muscular dystrophy). Such proteins can exert their influence only at the appropriate site, and there is no known mechanism by which cells expressing the protein remote from the affected tissue could have a therapeutic effect. Additional drawbacks are to be solved in order to obtain the successful therapeutic application of gene-modified hESCs including whether the hESC themselves, or the expressed product, will be toxic or immunogenic in the recipient. If the recipient of the cells, never exposed to the protein before, as in deletion or nonsense mutants, can be showed an immune reaction against the protein, limiting the effectiveness of the therapy (Mountford, 2008).

To develop hESC-based therapies, it is obvious that strategies capable of mitigating risks related to the therapeutic use of hESCs should be pursued the development of the therapies. These strategies might include several non-mutually exclusive mechanism for ablating all genetically modified cells while sparing most endogenous cells. The introduction of a step in the development of the therapy at which a single genetically modified cell would be isolated, expanded, and characterized with respect to the location of the mutation would allow an analysis of the relative risk of the insertion site. Similar limiting dilution cloning strategies are now routinely performed during the original isolation of a stem cell line to ensure that only one karyotype is represented (Mountford, 2008). Progress in understanding how insertional mutagenesis can lead to uncontrolled growth of stem cells is an essential prerequisite for this analysis and is currently an active area of research. Genetic modification can be used to enhance our ability to conduct such an isolation step by adding a convenient ligand for cell isolation. Genetic modification is also potentially useful for solving the problem of uncontrolled cell growth. Incorporating the genes for an ablation strategy at the same time as the genes for the therapeutic strategy would give the best chance of ensuring that the safety mechanism will be present when and if needed. Initial applications of genetically modified hESCs are likely to occur where the risk/benefit ratio tilts in favour of benefit, as in fatal disorders for which there is no therapy. The risks of the hESC therapies will have to be understood and probably reduced to maintain an appropriate risk/benefit ratio before these technologies can be applied to diseases that are inherently less dangerous to the patient. Gene therapy should prove to be valuable in reducing the risks associated with making hESC therapy a reality (Mountford, 2008).

In particular, the ability to generate cells with in vivo muscle regenerative potential in culture and systemically transfer them to recipients is an important step towards the therapeutic application of ES cell-derived cells. Unfortunately, it's not still provided a reproducible method to generate ES-derived myogenic progenitors for skeletal muscle regeneration. An ES cell-based therapy would have many advantages: it could allow the transplantation of a more primitive cell with greater replicative potential and patient-specific ES cells could be induced from adult somatic cells. Moreover, the derivation of an ES cell-derived myogenic population with proliferative and regenerative potential has not been accomplished. Only two papers described some evidences for engraftment on transplantation of an ES cell-derived population but they were limited to qualitative detection of donor derived cells in recipient muscle (Bhagavati and Xu, 2005; Kamochi et al., 2006). The Pax3 ES cell-derived population exhibited good potential for skeletal muscle regeneration but several studies concerning their capacity of replenish satellite cells-niche are in progress (Darabi et al., 2008a). On the other hand there is enough optimism about the ESC-based therapies because of it may offer reliable and cost-effective therapeutic substitute for treatment of neuromuscular disorders as DMD or BMD. Moreover there are critical issues that need attention in case of ESC-based approaches. No enough knowledge there are about genetic and epigenetic stability of hESC lines over longer time periods and it's not negligible the possible uncontrolled cell proliferation and reprogramming of ESCs in vitro.

The injection of these cells could probably generate an immune rejection needed an immunosuppressive therapy. Optimization of a generic differentiation protocol and its empirical testing with a better understanding of the molecular processes governing ESC differentiation can guarantee the clinical use of these cells. In conclusion the success of the clinical application of adult or embryonic stem cells will be employed to a large-scale production of desired cell type with appropriate functionality, an optimal number of cells for transplant, a modification of less invasive delivery systems and a technique to label cells for transplant and subsequent tracking of cell fate.

4. Induced Pluripotent Stem Cells (iPS)

The major impediment to ES based therapies in humans involves the moral and ethical problems linked to the blastocyst destruction and oocyte donation necessary to generate patient-specific pluripotent stem cell lines. These limitations have encouraged researchers to understand the mechanisms regulating pluripotency and to experimentally determine its gene expression program. Recent works describe the derivation of ES-like iPS cells from adult mouse and human cells (Nakagawa et al., 2008; Takahashi et al., 2007; Yu et al., 2009) by introducing specific sets of genes encoding transcription factors expressed in undifferentiated ES cells to reprogram the adult cells. Although initial studies indicating these cells to share characteristics of "true" ES cells, more detailed work is needed to determine how closely they resemble ES cells. Like ESCs, iPS cells can differentiate into all adult cell types. Researchers now have the ability to create tissue-based models of human disease based on cells derived from individual patients. This technology has the potential to herald a new era of patient-specific, cell-based medicine; however, given the oncogenic potential of undifferentiated iPS cells due to the unsafe reintroduction of these genes (Takahashi and Yamanaka, 2006), the safety of these cells has to be tested accurately before attempting any therapies. It has been demonstrated that continuous over-expression of transcription factors, especially the c-myc oncogene, may be associated with tumorigenesis (Takahashi et al., 2007). Even if it was demonstrated that the promoters of these viral vectors can be silenced by endogenous gene expression during reprogramming, chimeric mice derived from iPS cells were showed to be more prone to tumour formation. Following ameliorations in iPS technology, Nakagawa and co-workers generated pluripotent stem cells without c-myc over-expression both from mouse and human fibroblasts, with lower efficiency (Nakagawa et al., 2008). In addition, chimeric mice created from these non-myc iPS cells do not form tumours at an elevated rate. Recently, Chuang et al. proposed the use of baculoviral systems as a new gene delivery vector for stem cell engineering, and in particular for transgenic expression in human ESCs (Chuang et al., 2007). These vectors can be used for large segments, more than 30 kb, that do not fit into adenoviral or lentiviral vectors and could limit the risk derived from the great immunogenicity of adenoviral vectors.

4.1 Methods for reprogramming human fibroblasts into iPS cells

4.1.1 Production of iPS cells using viral integration

In 2006, Yamanaka et al. identified four transcription factors—Klf4, Sox2, Oct4, and c-Myc able to transform mouse fibroblasts in pluripotent clones through retroviral transduction. The clones were selected for their ability to reactivate the non essential Oct4 downstream target gene, Fbx15 (Takahashi and Yamanaka, 2006). This first-generation of

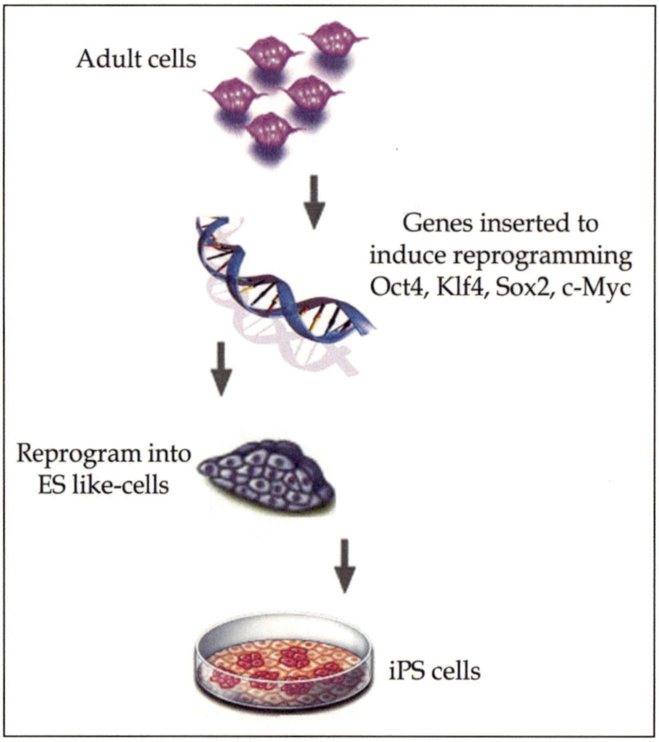

Fig. 2 Generation of iPS cells. Reprogramming of adult stem cells in iPS cells mediated by Oct-4, Klf4, Sox2 and c-Myc give raise to ES like cells with embryonic potential.

iPS cells exhibited partial demethylation and reactivation of the canonical pluripotency governing genes Oct4 and Nanog. In 2007, fully pluripotent iPS cells were generated by increasing the stringency of the selection strategy and selecting for reactivation of the pluripotency regulators Oct4 and Nanog themselves (Maherali et al., 2007; Okita et al., 2007; Wernig et al., 2007). These new second-generation iPS cells had fully demethylated Oct4 and Nanog promoters. These cells could not form viable mice by tetraploid complementation; an assay in which ES or iPS cells are injected into tetraploid blastocysts, resulting in embryos derived entirely from the injected cells, while extraembryonic tissues are derived from the host blastocyst. It is controversial if this failure is due to effects of the randomly integrated retroviral vectors used for reprogramming or represents a more fundamental defect in the developmental potency of iPS cells. Pluripotent iPS cells could be identified on the basis of their morphology, eliminating the need for genetically modified reporter genes and permitting the isolation of iPS cells derived from human fibroblasts (Park et al., 2008; Takahashi et al., 2007; Yu et al., 2009). After the work of Yamanaka, Yu et al generated iPS cells starting from a combination of Oct4, Sox2, Nanog, and LIN28. The finding that direct epigenetic reprogramming with different combinations of transcription factors can be applied to human cells represents the breaking of a species barrier that SCNT has yet to overcome. The therapeutic value of iPS cells is the presence of proviral integrations harboring known oncogenes, particularly c-Myc, as well as Oct4 and Klf4. c-Myc was dispensable for iPS generation from fibroblast target cells (although iPS formation occurred with reduced efficiency), and chimeric mice derived from three-factor iPS cells (Oct4, Sox2, and Klf4) did not exhibit tumour formation while the cells derived from the first four-factors showed tumorigenic characteristics (Nakagawa et al., 2008; Wernig et al., 2007). Despite the non essential role of c-Myc in the reprogramming process, the potential for insertional mutagenesis and the oncogenic properties of the other reprogramming factors has prompted several groups to undertake direct epigenetic reprogramming approaches using either nonviral methods, or retrospectively eliminate proviral integrations after iPS cell generation.

4.1.2 Production of iPS cells without viral integration

Different reasons have drive the researchers to evolve strategies for lentivirus-free iPS cell generation. First, the introduction of foreign DNA able to integrate in random positions in the genome represent a risk for cell physiology giving rise to an insertional mutagenesis. The foreign DNA could destroy reading frames of genes or influences gene regulation. Moreover silencing of lentiviral transgenes is incomplete in iPS cells leading to reactivation of lentiviral vectors (Wernig et al., 2007). In fact basal expression of lentiviral reprogramming factors is found even in fully reprogrammed cells (Hotta and Ellis, 2008; Pfannkuche et al.). Pfannkuche et al. demonstrated that in murine iPS cultures, there was different expression of viral Oct4 by three orders of magnitude between spontaneously differentiating iPS and stable subclones of the same origin (Pfannkuche et al.). From several studies, it is known that altered levels of pluripotency factors influence the fate of pluripotent cells. Overexpression of Oct4 seems to direct ES cells towards an endodermal fate (Niwa et al., 2000; Pesce and Scholer, 2001). More strikingly, elevated levels of Sox2 can alter the whole network of pluripotency factors and abolish maintenance of pluripotency (Kopp et al., 2008; Rizzino, 2009). The best strategy to eliminate reprogramming factors or to prevent integration of foreign transgenes for generation of iPS cells for therapeutic issues could be to employ techniques that are able to reprogram somatic cells without the use of any type of DNA. One possibility to achieve this aim is the use of cell-permeable transcription factors. The application of cell permeable proteins for direct reprogramming has been shown with human newborn fibroblasts, but, again, the efficiency of reprogramming appears low (Kim et al., 2009a). Kim and co-workers includes incubation with the cell permeable factors for 42 days with six passage steps during incubation. Even if the efficiency appears low, this study demonstrated for the first time that cell-permeable proteins can reprogram somatic cells. However, the long incubation time of 6 weeks in the presence of recombinant reprogramming factors in culture media, could be difficult. The application of cell-permeable reprogramming factors is a very promising approach that certainly justifies further investigation to achieve efficient, less time consuming reprogramming of cells from adult donors. Recent results suggest that a substantial amount of cell permeable reprogramming factors is enclosed in endosomes after cell uptake and might therefore be unavailable for reprogramming (Pan et al.). Another possibility to perform reprogramming is the introduction of RNA that can be transfected into cells. Transport of synthetic mRNAs to the nucleus is not necessary and, therefore, the transfection efficiency to deliver mRNA to the cytoplasm might be high. The stability of RNA limit the reprogramming efficiency and to obtain the right amount of protein translated it will be necessary several rounds of transfection to achieve reprogramming. A further strategy to induce pluripotency without the generation of stably transfected cells is the use of non integrating viruses. Adenoviral vectors were designed to express reprogramming factors and transport the coding sequence transiently into the target cells.

Application of the adenoviral approach on fetal human fibroblasts resulted in the generation of stable human iPS cells without viral integration. The human iPS cells form teratomas *in vivo*, which is the most stringent assay applicable to test their pluripotency (Zhou and Freed, 2009). Adenoviral vectors could enhance the production of virus- free iPS cells but the use of viruses that are DNA-based still bears a residual risk of integration into the host genome. This might be overcome with an RNA-based virus to generate safe iPS cells without application of DNA based vectors.

Another possible approach is the use of a single plasmid containing Oct4, Klf4, c-myc and Sox2 stably transfected into somatic cells for reprogramming. The plasmid expressed a single mRNA that codes for all four reprogramming factors and the red-fluorescent protein (RFP) in order to discriminate the transfected cells (Kaji et al., 2009). Different studies are also focused on the application of transposon-based systems. The application of the piggy-BAC transposon allows delivery of reprogramming factors into cells of different organisms and stable integration of the reprogramming genes into the genome by action of the transposase enzyme transiently transfected in a plasmid, and it is able to catalyse the integration of the reprogramming factors (Woltjen et al., 2009). The whole cassette is flanked by inverted terminal repeats, which mediate insertion into the genome by the host-factor independent activity of a transposase. The principal advantage of the system is not only the higher efficiency to deliver reprogramming factors but also the possibility to remove the transgene. Expression of reprogramming factors from the transposon was controlled by a doxicyclin inducible promoter, in order to guarantee a temporal control of transcription factors.

4.1.3 Generation of iPS cells with less than Yamanaka's four factors

Researchers are looking for solutions to reduce the number of factors needed for reprogramming. One of the reasons is to reprogram the cells using chemical compounds in order to standardize the procedure conditions, to control the response of each compound and to regulate the reprogramming factor expression. Another reason is that reduction of reprogramming factors to perhaps a single transgene could guarantee the understanding of mechanisms underlying reprogramming. Between the four factors used by Yamanaka, c-myc has been thought to have the highest oncogenic potential. The three other factors were not associated with tumorigenesis. The role of Klf4 in carcinogenesis is ambiguous. It can act as a tumour suppressor gene, especially in gastrointestinal tumours, or as oncogene, in the development of breast carcinoma where it is involved in the early phase of malignant transformation (Rowland and Peeper, 2006). The first reduction to three factors Oct4, Sox2 and Klf4 induced pluripotency in mouse and human fibroblasts (Nakagawa et al., 2008; Wernig et al., 2007) and other somatic cell types (Park et al., 2008). Moreover the generation of iPS cells was possible combinating Oct4, Sox2 and c-myc in the absence of Klf4. It was concluded that Oct4 and Sox2 in combination with either c-myc or Klf4 were sufficient (Park et al., 2008). Other researchers studied the use of another set of factors including Sox2, Oct4 and the nuclear receptor Essrb, which could replace Klf4 for reprogramming (Feng et al., 2009). However, the estimated reprogramming efficiency was ten times higher than in the original 4-factor approach in fibroblasts. Finally, reprogramming using only Oct4 in the absence of any other factor became possible in neural stem cells derived from adult mice (Kim et al., 2009b) and humans (Kim et al., 2009a). Another strategy is the substitution of single reprogramming factors by small molecules. For example c-myc or Klf4, could be substitute with valproic acid (VPA), thus making reprogramming of human fibroblasts possible only by retroviral transduction of Oct4 and Sox2 (Huangfu et al., 2008), with a 200-fold reduction of reprogramming efficiency compared to Oct4, Sox2 and Klf4 in combination with VPA. VPA acts as histone deacetylase inhibitor (Huangfu et al., 2008). Other authors utilized high-throughput assays to identify chemical factors which can substitute reprogramming genes. For example, Klf4 could be chemically substituted by the application of kenpaullone (Lyssiotis et al., 2009), and two other small molecules (BIX-01294 and Bayk8644) could enable reprogramming of embryonic fibroblasts with only Oct4 and Klf4.

4.1.4 Enhancement of iPS production

Due to the low efficiency of iPS production, different factors have been investigated in order to improve this procedure. Simultaneous transduction of c-myc, Oct4, Sox, Klf4, with Nanog and Lin28 increases efficiency dramatically (Liao et al., 2008). Other studies have shown that expression of p53 siRNA and UTF1 dramatically increases iPS colonies (Zhao et al., 2008). While p53 siRNA remarkably increases the number of iPS colonies, UTF1, a pluripotent marker and necessary for the maintenance of pluripotency in mES cells (Gaspar-Maia et al., 2009). By affecting different pathways, these two factors together can synergize iPS generation (Zhao et al., 2008). Other pathways which have an impact on iPS cells formation include the TGFβ and the MEK-ERK pathway. The role of these pathways in cell survival has made them beneficial in reprogramming studies. Chemical inhibitors of the MEK pathway apparently inhibit the growth of non-iPS cells while increasing the growth rate in reprogrammed iPS cells. Suppressing MEK and TGFβ pathways concurrently with chemical compounds increases iPS amounts. This study showed that combined suppression of both pathways resulted in extensive amounts of iPS generation in comparison to individual inhibition of each pathway (Lin et al., 2009). Repression of the Ink4/Arf locus which is a regulator of the p53 p21 pathway has a positive effect on reprogramming. Different pluripotency genes are controlled by the expression of different micro RNAs expressed in the cell. For example, let-7 is one of the miRNAs which targets the 3'UTR and ORF's of several pluripotency expressed genes including c-myc, Oct4, Sox2 and Nanog. Suppressing let-7 by antisense inhibitors improves iPS generation several fold in mouse embryonic fibroblasts (MEFs) (Melton et al. , 2010). Reduced oxygen levels favour the growth of haematopoietic stem cells and maintain ES cells in a pluripotent state. Induction of pluripotency in hypoxic conditions increases the number of iPS colonies and vitamin C elevates iPS reprogramming efficiency (Esteban et al. , 2010). The increase in iPS formation is not due to its antioxidant activity but vitamin C inhibits p53, thus facilitating the induction of iPS cells (Shi et al. , 2010). Vitamin C extends the life span of both iPS and MEF cells pointing at roles for vitamin C in anti-ageing. Composition of the iPS culture medium also defines the efficiency of iPS formation. The use of transcription factors together with different chemical agents can enhance iPS induction. This can result in the derivation of more pluripotent cells in a shorter time, making its application more convenient for clinical purposes in a not too far future.

4.2 Clinical applications of iPS

Following the success of haematopoietic stem cell therapy in the treatment of haematological diseases, the potential application of cell based therapy has been extended to the treatment of other human diseases. In particular, different types of adult stem cells, including bone marrow, peripheral haematopoietic, and mesenchymal stem cells (MSC) have been evaluated in the treatment of different diseases (Charwat et al., 2010; Siu et al.,

2010). ESCs have been explored for tissue regeneration because of their ability to differentiate into various therapeutic relevant cell types *in vitro* (Murry and Keller, 2008). Despite this, there is limited progress in the use of ESCs for tissue regeneration in humans due to various technical, social and religious issues (Kiskinis and Eggan). The generation of patient-specific iPS cells has the advantage of avoiding many of the ethical concerns associated with the use of embryonic or foetal material, and have no risk of immune rejection. Currently, several therapeutic relevant cell types, including motor neuron (Dimos et al., 2008), hepatocytes (Song et al., 2009), pancreatic insulin producing cells (Zhang et al., 2006), haematopoietic cells (Hanna et al., 2007), retinal cells (Carr et al., 2009), cardiomyocytes (Zwi et al., 2009) and MSCs (Lian et al., 2010), have been successfully derived from human iPS cells, and some of them have been tested to treat diseases in animal models. The use of iPS cells has thus been proposed as diagnostic and therapeutic tools for different haematological disorders (Ye et al., 2009). Nelson et al. (Nelson et al., 2009) reported the use of iPS cells for myocardial repair in animal models of acute myocardial infarction. There are several major challenges to overcome before iPS cell technology is applied in clinical practice. Now, current iPS cells are not "clinical grade". Genome-integrating viral vectors used for reprogramming are known oncogenes, particularly c-Myc, Oct4 and Klf4, such that iPS cells thus generated are unlikely to be safe for clinical application. Nonetheless recent technological advances, including reprogramming without viral integration such as plasmids or direct reprogramming protein delivery assays can solve this problem (Kiskinis and Eggan, 2010; Saha and Jaenisch, 2009). Despite the challenges in the therapeutic use of iPS cells, preclinical studies have provided the proof-of-concept that patient-specific iPS cells can provide an unlimited cell source to produce massive therapeutic cell types, such as cardiomyocytes and MSCs, and can be prepared in an "off the- shelf " format for cell transplantation.

Given the many potential risks of applying autologous iPS cell treatment to human subjects, iPS cell therapies may encounter strict regulatory restrictions. For instance, it took Geron Corporation more than 6 years to receive approval from the Food and Drug Administration (FDA) for its human ES cell-derived neuronal cell (GRNOPC1) therapies in terms of cell product safety and reliability. Recently, a second company has presented an investigational new drug for a phase I/II trial using human ES cell-derived retinal pigment epithelial (RPE) cells to treat patients with Stargardt's Macular Dystrophy (SMD), one of the most common causes of juvenile blindness. The sponsoring company, Advanced Cell Technology (ACT), has performed years of testing to show that differentiated RPE cells can improve the visual performance of rats without adverse effects (e.g., teratomas) in hundreds of treated animals. Another issue that may hinder the clinical translation of iPS cell therapies is the economic feasibility of producing individualized iPS cell therapeutic products. The viability of a business model for patient-specific iPS treatment is still unknown. It may well be the case that few if any pharmaceutical companies will be able to produce cost-effective individualized iPS cell products tailored for a single patient at a time. To be commercially feasible, these cells will need to be made in standardized, large-scale production, and the individual needs or profiles of patients will need to be easily assessed to allow matching and wide distribution (Sun et al., 2010). Several groups have also begun the generation of patient-specific human iPS cell lines. Park et al. generated a library of patient-derived iPS lines from numerous disorders including Huntington's disease, juvenile diabetes mellitus, Down syndrome, muscular dystrophy, and several others (Park et al., 2008). Of particular interest are iPS cells derived from neurodegenerative diseases (Dimos et al., 2008; Ebert et al., 2009). These iPS lines can be differentiated *in vitro* into the affected neuronal cell type, generating for the first time a model for idiopathic neurodegenerative disorders which can be screened in culture for the onset, cell autonomy, and contribution of environmental factors to the phenotype. Ultimately, if human neurodegenerative phenotypes can be recapitulated in iPS cell-derived culture models, these cells could be screened using chemical libraries to identify molecules that can arrest or even reverse the progression of these disorders.

5. Similarities and differences between ES and iPS cells

The comparison of iPS and ES cells revealed that these cells are very similar. The differentiation capacity of iPS cells seems to resemble that of ES cells; iPS derived somatic cells are comparable to those derived from ES cells. Several studies describe the derivation of a variety of cell types from murine and human iPS cells, among them cardiomyocytes, smooth muscle cells, hepatic cells and neurons with similar differentiation behaviour of iPS and ES cells. The transcriptome of iPS and ES cells was analyzed by Gene chip analysis; the results showed that these cells are very similar but they are not identical. Chin and co-workers compared the expression pattern of different human ES and iPS cells; they analysed histone modifications and the expression of non-coding miRNAs in both type of cells and they constructed a fingerprint of iPS cells that distinguishes them from ES pluripotent cells (Pfannkuche et al., 2010). The comparison of the transcriptome of early and late passage iPS cells with ES cells revealed two datasets of differentially regulated genes. They identified a subset of 318 genes differentially expressed between human ES and iPS cells at any stage. The genes that are higher expressed in iPS cells were also found higher expressed in fibroblasts than in ES cells. The same conclusion is valid for genes that were expressed at a lower level in iPS than ES cells are usually also lower expressed in fibroblasts than ES cells. Together, these findings point at an imperfect reprogramming of a small set of genes. It is not known if there are implications of iPS fingerprint for the physiology of these cells. It is important determine the genes that are differentially regulated and that could influence the cellular physiology. It is likely that some of these gene functions are redundant with others that are not affected by incomplete reprogramming and, therefore, do not influence cell physiology (Pfannkuche et al., 2010).

Beside genes that constitute a potential iPS cell fingerprint, there are varying gene sets that are differentially expressed in individual iPS cell lines. It will be interesting to see if properties of the ancestral cell types are transmitted to the iPS cell line generated. In this regard, partial reprogramming plays a role and one fascinating aspect to address is if partial reprogramming alters the differentiation capacity of a cell in a way that it potentially influences the fate decision of the partially reprogrammed iPS cells. The differences between iPS and ES cells could be an assays to measure the quality of iPS cells. Although it has been shown that the overall gene expression of iPS cells differs from normal ES cells, this comparison has never been made between cells from the same individual. Usually iPS cells are compared with those ES cells either derived from another species or from a different individual; raising concerns about whether these are informative approaches. It is clear, therefore, that iPS cells derived from the trophoblast of an embryo compared with ES cells derived from the inner cell mass of the same embryo would give a more explicit view of how distinct or similar these cells really are (Pfannkuche et al., 2010).

6. Adult stem cells versus ESC&iPS

Although ESCs and iPS cells are now the most studied cells for clinical applications in neuromuscular diseases, different stem cells isolated from adult tissues was extensively used - and are still used - unfortunately with poor results. For several years after they were discovered, the satellite cell were considered as the only cells responsible for the growth and maintenance of skeletal muscle. With the improvements of cell-isolation technology, a number of markers were described to identify muscular and non-muscular subpopulations able to actively participate in myogenesis (Meregalli et al., 2010). In the skeletal muscle, adult multi-lineage progenitor cell populations were showed to have myogenic potential, such as muscle-derived stem cells (MDSCs) and muscle-derived CD133+ progenitors. Moreover, it was also shown that non-muscular resident stem cells could participate in myogenesis (Krause et al., 2001; Mezey et al., 2000; Pittenger et al., 1999; Prockop, 1997). In particular, a subpopulation of CD133+ cells was isolated form the blood, playing an important role in myogenic development (Torrente et al., 2003b) while mesoangioblasts were identified in the dorsal aorta of avian and mammalian species (Cossu and Bianco, 2003).

6.1 Satellite cells

Satellite cells are small progenitor cells that lie between the basement membrane and sarcolemma of individual muscle fibers: normally they are quiescent, consequently they cannot differentiate nor undergo cell division. Oxidative stress and specific stimuli form the environment can activate them, so that they differentiate, proliferate as skeletal myoblasts and activate myogenic differentiation to form new myofibers. Recently, Montarras and colleagues were able to directly isolate a pure population of satellite cells from diaphragm muscle of a Pax3-GFP knock-in mouse (Montarras et al., 2005). After FACS and gene expression analysis, they purified a predominantly quiescent population of satellite cells expressing Pax3, CD34 and Pax7. These cells were firstly injected into dystrophic dogs and restored dystrophin expression 3 weeks post-transplantation. Transplanted into irradiated dystrophic mice, they also formed a small amount of the satellite cell pool that expressed both Pax7 and Pax3 (Montarras et al., 2005). If compared with the results obtained after the injection of human cells isolated from adult muscle, these cells showed an incredibly efficient level of muscular regeneration (Morgan et al., 1996). Since they were doubtless highly myogenic, satellite cells were not considered in a clinical point of view, because it was difficult to isolate them and above all to proliferate and expand them to obtain the right number for transplantation experiments. Moreover, the growth of freshly isolated satellite cells in vitro significantly reduced their in vivo myogenic potential.

6.2 Muscle-derived stem cells (MDSCs)

Muscle-derived stem cells (MDSCs) are a recently-identified subpopulation of cells that resides within skeletal muscle and possess the ability to self renew and to differentiate into other mesodermal cell types (Sarig et al., 2006; Tamaki et al., 2007). Furthermore, it's known that these cells are distinct from satellite cells (Asakura and Rudnicki, 2002; Qu-Petersen et al., 2002) and that, when appropriately stimulated, they could preserve their myogenic potential in vitro even after differentiation into other lineages (Negroni et al., 2006). In the last years, different works assessed the capacity of MDSCs to differentiate and regenerate skeletal muscle when transplanted into animal models. Sca-1+CD34+ stem cells were purified from the muscle tissues of newborn mice, showing multipotency in vitro.

Moreover, after intra-arterial injection, these cells were able to interact and firmly adhere to endothelium in mdx muscles microcirculation and then participated in muscle regeneration (Torrente et al., 2003a). Qu-Petersen and collaborators isolated a MDSCs population Sca-1+/-CD34+/-c-kit-CD45- and demonstrated that they displayed a better transplantation efficiency than satellite cells (Qu-Petersen et al., 2002). MDSCs were also identified in human muscle, expressing in proliferation the CD133 antigen and also desmin and α-SMA when cultured in myogenic conditions (Miraglia et al., 1997). Moreover, among human MDSCs, it was identified a subpopulation of progenitor stem cells with neurogenic properties (Alessandri et al., 2004). According to these evidences, MDSCs are suitable for clinical perspectives as they are easy to proliferate, migrate through the vasculature, and are multipotent, although it could be necessary to better investigate their physiological location (Deasy et al., 2005; Deasy et al., 2001).

6.3 Mesoangioblasts

Mesoangioblasts are multipotent progenitors of mesodermal tissues, physically associated with the embryonic dorsal aorta in avian and mammalian species, expressing Flk-1, stem cell antigen 1, CD34 and various leukocyte molecules (Cossu and Bianco, 2003; Tagliafico et al., 2004). It was shown that mesoangioblasts treated with a lentiviral vector expressing human microdystrophin were able to produce dystrophin-positive myofibers after injection in animal model of DMD and ameliorated muscle function and mobility (Cossu and Sampaolesi, 2007; Sampaolesi et al., 2006). Furthermore, to improve their efficiency of muscle repair, mesoangioblasts were treated to increase their migration to skeletal muscle and to reduce unspecific trapping in the capillary filters of the body, such as liver and lung (Galvez et al., 2006).

6.4 CD133+ stem cells

A role for CD133 as a marker of stem cells with the capacity to engraft and differentiate to form functional non-haematopoietic adult lineages and contribute to disease amelioration via tissue regeneration emerged in the last years. Human CD133+ cells, isolated from peripheral blood and manipulated *in vitro* to undergo myogenesis, were shown to ameliorate disease via a direct contribution to muscular regeneration when transplanted into dystrophic mice (Torrente et al., 2004). In particular, they restored dystrophin expression and eventually regenerate the murine satellite cells pool after intramuscular and intra-arterial delivery. Human CD133+ cells colonized the mouse muscle and formed hybrid regenerated fibers expressing human dystrophin. Moreover, they were detected in several vessels near areas of regeneration, where they expressed human ve-cadherin and CD31 (Torrente et al., 2004). A CD133+ stem cell subpopulation was also identified in normal and dystrophic muscles. They were positive for CD45 antigen, indicating their hematopoietic commitment while the expression of Pax-7, Myf-5, MyoD, m-cadherin, MRF-4, and myogenin after 24 days of culture in the proliferation medium and their ability to differentiate into multinucleated myotubes expressing MyHCs suggested a myogenic commitment (Torrente et al., 2007). According to these data, CD133+ stem cells were considered as a possible tool in the treatment of degenerating diseases. Stamm and collaborators showed that transplanted BM-derived CD133+ cells improved function of infracted myocardium probably as a result of the amelioration in blood vessel formation (Stamm et al., 2003) while Torrente and co-workers demonstrated that intramuscular transplantation of muscle-derived CD133+ cells in DMD patients was a safe procedure and feasible. DMD patients showed an increased number of capillaries per muscle fiber and expressed a change in the ratio of slow-to-fast myosin myofibers (Torrente et al., 2007). Human dystrophic blood- and muscle-derived CD133+ expressed an exon-skipped version of human dystrophin after transduction with a lentivirus carrying a construct designed to skip exon 51 and participated in vivo in muscle regeneration (Benchaouir et al., 2007). This combination of cell- and gene-based approaches via the *ex vivo* introduction of corrective genes into dystrophic CD133+ cells permitted - in a clinical point of view- the use of patient's own cells: autologous transplantation would reduce the risk of implant rejection.

7. Therapeutic potential of adult and embryonic stem cells

Several questions remain to be answered before any of the previously described cell preparations can be moved into clinical trials even if there has been great advance in the generation of cell populations showed *in vivo* myogenic potential. Based on their unique characteristics and in vivo skeletal muscle regeneration potential, adult stem cell populations discussed in this review are excellent clinical candidates. As shown before, mesoangioblasts (Cossu and Bianco, 2003) and blood derived CD133+ (Gavina et al., 2006) have the ability to migrate through the vasculature, most do not. Potential future methods to increase the migratory ability of stem cell population include the identification of cell surface markers like adhesion molecules (Torrente et al., 2003b) and appropriate growth factors (Horsley et al., 2003; Torrente et al., 2003b). Mesoangioblasts serve as a paradigm for widespread distribution, and after treatment with growth factors are able to correct efficiently the dystrophic phenotype. For now the intra-arterial injection of mesoangioblasts represent a hope for patients suffering from various muscular dystrophies. Satellite cells was one of the first cell types used in cell-based therapy of muscular dystrophy. Expanded satellite cells or myoblasts were isolated from wild-type mice and intramuscularly injected in dystrophic mdx mice (Conway et al., 1997; Tremblay et al., 1998); unfortunately it was demonstrated that myoblast transplantation is an inefficient technique because of the low efficiency of the dystrophin production in muscle fibers of DMD patients and no functional or clinical improvement in the children (Peault et al., 2007). In possible future clinical trials, adult stem cells purified from patients suffering for neuromuscular disorders could be ex vivo engineered and re-injected in the initial donor intra-arterially. The intra-arterial injections of the patient's own stem cells transduced allow the distribution of the cells to the whole body musculature so that it could be possible to take care of severe-affected patients that have reduced mass body, as in DMD and BDM pathology (Brignier and Gewirtz, 2010). One of the most important problem to solve for future clinical application is the amelioration in safety procedures of gene's modifications. One of the most reliable methods for gene therapy, fully utilized in DMD clinical approaches, seems to be the exon skipping mediated by AONs or molecules like PTC124. Ongoing phase I/II studies try to assess the efficacy and the safety of intramuscular administered morpholino oligomer directed against exon 51 (AVI-4658 PMO). Morpholinos can interfere with mRNA splicing processes by preventing the formation of the snRNP complex or by interfering with the binding sites for other regulatory proteins (Vetrini et al., 2006). They mediate the exclusion of exons from the mature mRNA as AONs. PTC124 was shown to partially restore dystrophin production in animals with DMD due to a nonsense mutation. The main purpose of a phase II study completed on May 2007 was to understand whether PTC124 can safely increase functional

dystrophin protein in the muscles of patients with DMD due to a nonsense mutations. This study demonstrated the safety and the efficacy of the PTC124 treatment; now three ongoing phase 2a and 2b studies are started in DMD and BMD patients (www.clinicaltrials.gov).
A decade of studies in human ESCs has yielded remarkable progress and understanding in stem cell biology. The technical challenge of creating patient-specific ESCs, the ethical issues arising from the foetal origin of human ESCs and the potential risk of immune rejection make broad clinical application of this cell type difficult. Recent advances in human iPS cell technology can potentially circumvent these disadvantages: iPS cells thus provide an invaluable resource of cell types for modelling diseases, drug or toxicology screening, and patient-specific cell therapy. Significant challenges remain to be overcome before the full potential of human iPS cell technology can be realised. The utilization and practical application of ESC in cell replacement therapy are still in a preliminary stage and need more investigation and clinical trials before they can be accepted as ideal for the treatment of neuromuscular diseases. Nevertheless, the daily increase in experimental findings is reinforcing the hope that ESC will be a versatile source of renewable cells for application in cell replacement therapy (Brignier & Gewirtz, 2010). Therefore, there is enough optimism among the scientists that ESC-based therapies may offer reliable and cost-effective therapeutic substitute for treatment of severe degenerative disorders in the near future. Major objection to hES cell research is focused one ethical reasons. The core reason for objection to hES cell research is that it destroys human blastocysts or embryos, which means it destroys human lives and eventually violates human dignity because of the blastosysts have the same moral value as that of human beings or at least that blastocysts have the potential to develop into human beings (Jung, 2009). Accordingly, research with hESCs is increasing exponentially worldwide, particularly in the United States, where important limitations on research with such cells were overturned in 2009. Furthermore, the US Food and Drug Administration trial using hESC-based therapy in patients with spinal cord injury is now on-going. Nonetheless, a number of substantive scientific and ethical issues remain to be resolved before hESCs can enter the therapeutic mainstream. In the meantime, recent breakthroughs in generating iPSCs would obviate the need to solve the most vexing of these problems. In fact, it seems reasonable to hope that in the next few years many of the enabling issues relevant to iPSCs will be solved, allowing the field of regenerative medicine to deliver on its vast potential promise. Although it is difficult to predict the ultimate utility of stem cell-based therapy at this time, it is not difficult to conclude that this is an extremely important area of scientific research. Open discussions between political bodies and the various interest groups in the scientific, medical, and religious communities need to take place to address the concerns of each and to provide an ultimate solution that is clearly in the interest of humanity.

8. Conclusions

European Medicines Agency (EMEA) issued the guideline to replace the Points to Consider on the Manufacture and Quality Control of Human Somatic Cell Therapy Medicinal Products (CPMP/BWP/41450/98). In general, when a cell-based medicinal product (CBMP) enters the clinical development phase, the same requirements as for other medicinal products apply. The clinical development plan should include pharmacodynamic studies, pharmacokinetic studies, mechanism of action studies, dose finding studies and randomised clinical trials in accordance to the Directive 2001/20/EC and to the existing general guidances and specific guidances for the condition evaluated. It takes into account the current legislation and the heterogeneity of human cell-based products, including combination products. A risk analysis approach can be used by the applicants to justify the development and evaluation plans and can be a basis for the preparation of a risk management plan. Special problems might be associated with the clinical development of human cell-based medicinal products. Guidance is therefore provided on the conduct of pharmacodynamic/pharmacokinetic studies, dose finding and clinical efficacy and safety studies. The guideline describes the special consideration that should be given to pharmacovigilance aspects and the risk management plan for these products. The active substance of a CBMP is composed of the engineered (manipulated) cells and/or tissues. When the cells in the active substance are genetically modified, the "Note for Guidance on the quality, preclinical and clinical aspects of gene transfer medicinal products" should be followed, which gives details on the quality control, characterisation and preclinical testing of gene transfer vectors. Cell populations which are transformed should be assayed for appropriate and reproducible expression of the newly acquired characteristics. Special attention should be paid to the level and length of expression and quality of the gene product(s) produced by the cells. As far as applicable and practicable, the new characteristics of the cells should be quantified and controlled. During in vitro cell culture, consideration should be given to ensure acceptable growth and manipulation of the isolated cells. The processing steps should be properly designed to preserve the integrity and control the function of the cells and their manipulation should be documented in detail and closely monitored according to specific process controls. Moreover, the duration of cell culture and maximum number of cell passages should be clearly specified and validated. The relevant genotypic and phenotypic characteristics of the primary cell cultures, of the established cell lines and the derived cell clones should be defined and their stability with respect to culture longevity determined. Consistency/repeatability of the cell culture process should be demonstrated and the culture conditions including the media and the duration should be optimised with respect to the intended clinical function of the cells. If genetically modified cells are used in the product, any additional proteins expressed from the vector, such as antibiotic resistance factors, selection markers, should be analysed to determine their presence in the product. Microassay for gene expression profile, flow cytometry and other different techniques allowed these expression studies. CBMP might require administration through specific surgical procedures, method of administration or the presence of concomitant treatments to obtain the intended therapeutic effect. The biological effects of CBMP are highly dependent on the *in vivo* environment, and may be influenced by the replacement process or the immune reaction either from the patient or from the cell-based product. These requirements coming from the clinical development should be taken into account for the final use of these products. Their standardisation and optimisation should be an integral part of the clinical development studies. Ahead of these considerations it's not still provided a reproducible method to isolate ESCs even if an ES cell–based therapy would have many advantages: it could allow the transplantation of a more primitive cell with greater replicative potential and patient-specific ES cells could be induced from adult somatic cells. The development of several ESC-based technologies, such as genetic manipulation tools and their potential applications, could accelerate the use of these cells into clinical therapy, even if ethical, logistics and economics concerns need attention in case of ESC-based techniques. There are several major challenges to overcome before iPS cell technology is applied in clinical practice. First, current iPS cells are not "clinical grade".

Genome-integrating viral vectors used for reprogramming are known oncogenes, particularly c-Myc, Oct4 and Klf4, such that iPS cells thus generated are unlikely to be safe for clinical application. Nonetheless recent technological advances, including reprogramming without viral integration such as plasmids or direct reprogramming protein delivery assays could solve this problem (Kiskinis and Eggan, 2010; Saha and Jaenisch, 2009). Despite the challenges in the therapeutic use of iPS cells, preclinical studies provided the proof-of-concept that patient-specific iPS cells can provide an unlimited cell source to produce massive therapeutic cell types, such as cardiomyocytes and MSCs, and can be prepared in an "off the- shelf " format for cell transplantation. However given the many potential risks of applying autologous iPS cell treatment to human subjects, iPS cell therapies may encounter strict regulatory restrictions.
At now in our opinion the most promising results in the treatment of neuromuscular disorders were obtained using adult stem cells because of many questions are needed to be answer regarding the ES and iPS cells. According to the results described, the most promising possibility for the therapy of muscular dystrophies is a combination of different approaches to obtain the beneficial impact of multiple strategies combined into a single approach, such as cellular therapy associated with gene therapy or pharmacological treatments. One of the most used approach is called autologous transfer in which patient's own cells are genetically corrected in vitro with lentiviral vectors and then re-implanted to allow the re-expression of functional dystrophin protein. The 'exon skipping' approach is an alternative strategy for gene therapy and it is done through AONs that hybridize with the donor and/or acceptor sites of the mutated exon, causing its exclusion from the intact transcript. On the other side, the allogenic transfer implies that the cells isolated from an individual with functional dystrophin will be injected into the patient, allowing problems due to immunorejections or administration of specific immunosuppressive drugs. Several problems arose quickly, such as the low efficiency with which stem cells enter muscle via vasculature, the potential to enhance proliferation of stem cells in culture, the time required in culture for autologous cells prior to implantation back into the patient, the longevity of the transplanted muscle nuclei in vivo, the development of tumours as a consequence of hazardous integration of the provirus. In conclusion the success of the clinical application of adult or embryonic stem cells will be employed to a large-scale production of desired cell type with appropriate functionality, an optimal number of cells for transplant, a modification of less invasive delivery systems and a techniques to label cells for transplant and subsequent tracking of cell fate.

9. Acknowledgments

This work has been supported by the Association Monégasque contre les Myopathies (AMM), Telethon grant E36840, Optistem European Project n39'00i8, the Duchenne Parent Project de France (DPP France), the Associazione La Nostra Famiglia Fondo DMD Gli Amici di Emanuele, Fondazione Cariplo, Fondazione Telethon and the Associazione Amici del Centro Dino Ferrari.

10. References

Alessandri, G.; S. Pagano, A. Bez, A. Benetti, S. Pozzi, G. Iannolo, M. Baronio, G. Invernici, A. Caruso, C. Muneretto, G. Bisleri, & Parati, E. (2004). Isolation and culture of human muscle-derived stem cells able to differentiate into myogenic and neurogenic cell lineages. *Lancet*. 364:1872-83.

Amit, M. & Itskovitz-Eldor, J. (2002). Derivation and spontaneous differentiation of human embryonic stem cells. *J Anat*. 200:225-32.

Andrews, P.W.; J. Casper, I. Damjanov, M. Duggan-Keen, A. Giwercman, J. Hata, A. von Keitz, L.H. Looijenga, J.L. Millan, J.W. Oosterhuis, M. Pera, M. Sawada, H.J. Schmoll, N.E. Skakkebaek, W. van Putten, & Stern P. (1996). Comparative analysis of cell surface antigens expressed by cell lines derived from human germ cell tumours. 1*Int J Cancer*. 66:806-16.

Asakura, A. & M.A. Rudnicki M.A. (2002). Side population cells from diverse adult tissues are capable of in vitro hematopoietic differentiation. *Exp Hematol*. 30:1339-45.

Benchaouir, R.; M. Meregalli, A. Farini, G. D'Antona, M. Belicchi, A. Goyenvalle, M. Battistelli, N. Bresolin, R. Bottinelli, L. Garcia, & Torrente Y. (2007). Restoration of human dystrophin following transplantation of exon-skipping-engineered DMD patient stem cells into dystrophic mice. *Cell Stem Cell*. 1:646-57.

Berrih, S.; E. Morel, C. Gaud, F. Raimond, H. Le Brigand, & Bach J.F. (1984). Anti-AChR antibodies, thymic histology, and T cell subsets in myasthenia gravis. *Neurology*. 34:66-71.

Bertoni, C. (2008). Clinical approaches in the treatment of Duchenne muscular dystrophy (DMD) using oligonucleotides. *Front Biosci*. 13:517-27.

Bhagavati, S. & Xu W. (2005). Generation of skeletal muscle from transplanted embryonic stem cells in dystrophic mice. *Biochem Biophys Res Commun*. 333:644-9.

Biggar, W.D.; V.A. Harris, L. Eliasoph, & Alman B. (2006). Long-term benefits of deflazacort treatment for boys with Duchenne muscular dystrophy in their second decade. *Neuromuscul Disord*. 16:249-55.

Brignier, A.C. & Gewirtz A.M. (2010). Embryonic and adult stem cell therapy. *J Allergy Clin Immunol*. 125:S336-44.

Byrne, J.A.; D.A. Pedersen, L.L. Clepper, M. Nelson, W.G. Sanger, S. Gokhale, D.P. Wolf, and Mitalipov SM. (2007). Producing primate embryonic stem cells by somatic cell nuclear transfer. *Nature*. 450:497-502.

Campbell, K.H.; J. McWhir, W.A. Ritchie, & Wilmut I. (1996). Sheep cloned by nuclear transfer from a cultured cell line. *Nature*. 380:64-6.

Carpenter, M.K.; E. Rosler, & Rao M.S. (2003). Characterization and differentiation of human embryonic stem cells. *Cloning Stem Cells*. 5:79-88.

Carr, A.J.; A.A. Vugler, S.T. Hikita, J.M. Lawrence, C. Gias, L.L. Chen, D.E. Buchholz, A. Ahmado, M. Semo, M.J. Smart, S. Hasan, L. da Cruz, L.V. Johnson, D.O. Clegg, & Coffey P.J. (2009). Protective effects of human iPS-derived retinal pigment epithelium cell transplantation in the retinal dystrophic rat. *PLoS One*. 4:e8152.

Caspi, O.; A. Lesman, Y. Basevitch, A. Gepstein, G. Arbel, I.H. Habib, L. Gepstein, & Levenberg S. (2007). Tissue engineering of vascularized cardiac muscle from human embryonic stem cells. *Circ Res*. 100:263-72.

Charwat, S.; I. Lang, M. Dettke, S. Graf, N. Nyolczas, R. Hemetsberger, S. Zamini, A. Khorsand, H. Sochor, G. Maurer, D. Glogar, & Gyongyosi M. (2010). Effect of intramyocardial delivery of autologous bone marrow mononuclear stem cells on the regional myocardial perfusion. NOGA-guided subanalysis of the MYSTAR prospective randomised study. *Thromb Haemost*. 103:564-71.

Chuang, C.K.; L.Y. Sung, S.M. Hwang, W.H. Lo, H.C. Chen, & Hu Y.C. (2007). Baculovirus as a new gene delivery vector for stem cell engineering and bone tissue engineering. *Gene Ther*. 14:1417-24.

Conrad, S.; M. Renninger, J. Hennenlotter, T. Wiesner, L. Just, M. Bonin, W. Aicher, H.J. Buhring, U. Mattheus, A. Mack, H.J. Wagner, S. Minger, M. Matzkies, M. Reppel, J. Hescheler, K.D. Sievert, A. Stenzl, & Skutella T. (2008). Generation of pluripotent stem cells from adult human testis. *Nature*. 456:344-9.

Conway, S.J.; D.J. Henderson, & Copp A.J. (1997). Pax3 is required for cardiac neural crest migration in the mouse: evidence from the splotch (Sp2H) mutant. *Development*. 124:505-14.

Cossu, G. & Bianco P. (2003). Mesoangioblasts--vascular progenitors for extravascular mesodermal tissues. *Curr Opin Genet Dev*. 13:537-42.

Cossu, G. & Sampaolesi M. (2007). New therapies for Duchenne muscular dystrophy: challenges, prospects and clinical trials. *Trends Mol Med*. 13:520-6.

Darabi, R.; K. Gehlbach, R.M. Bachoo, S. Kamath, M. Osawa, K.E. Kamm, M. Kyba, & Perlingeiro R.C. (2008a). Functional skeletal muscle regeneration from differentiating embryonic stem cells. *Nat Med*. 14:134-43.

Darabi, R.; F.N. Santos, & Perlingeiro R.C. (2008b). The therapeutic potential of embryonic and adult stem cells for skeletal muscle regeneration. *Stem Cell Rev*. 4:217-25.

Deasy, B.M.; B.M. Gharaibeh, J.B. Pollett, M.M. Jones, M.A. Lucas, Y. Kanda, & Huard J. (2005). Long-term self-renewal of postnatal muscle-derived stem cells. *Mol Biol Cell*. 16:3323-33.

Deasy, B.M.; R.J. Jankowski, & Huard J. (2001). Muscle-derived stem cells: characterization and potential for cell-mediated therapy. *Blood Cells Mol Dis*. 27:924-33.

Dimos, J.T.; K.T. Rodolfa, K.K. Niakan, L.M. Weisenthal, H. Mitsumoto, W. Chung, G.F. Croft, G. Saphier, R. Leibel, R. Goland, H. Wichterle, C.E. Henderson, & Eggan K. (2008). Induced pluripotent stem cells generated from patients with ALS can be differentiated into motor neurons. *Science*. 321:1218-21.

Drachman, D.B. (1994). Myasthenia gravis. *N Engl J Med*. 330:1797-810.

Ebert, A.D.; J. Yu, F.F. Rose, Jr., V.B. Mattis, C.L. Lorson, J.A. Thomson, & Svendsen C.N. (2009). Induced pluripotent stem cells from a spinal muscular atrophy patient. *Nature*. 457:277-80.

Endo, T. (2007). Stem cells and plasticity of skeletal muscle cell differentiation: potential application to cell therapy for degenerative muscular diseases. *Regen Med*. 2:243-56.

Esteban, M.A.; T. Wang, B. Qin, J. Yang, D. Qin, J. Cai, W. Li, Z. Weng, J. Chen, S. Ni, K. Chen, Y. Li, X. Liu, J. Xu, S. Zhang, F. Li, W. He, K. Labuda, Y. Song, A. Peterbauer, S. Wolbank, H. Redl, M. Zhong, D. Cai, L. Zeng, & Pei J. (2010). Vitamin C enhances the generation of mouse and human induced pluripotent stem cells. *Cell Stem Cell*. 6:71-9.

Feng, B.; J. Jiang, P. Kraus, J.H. Ng, J.C. Heng, Y.S. Chan, L.P. Yaw, W. Zhang, Y.H. Loh, J. Han, V.B. Vega, V. Cacheux-Rataboul, B. Lim, T. Lufkin, & Ng H.H. (2009). Reprogramming of fibroblasts into induced pluripotent stem cells with orphan nuclear receptor Esrrb. *Nat Cell Biol*. 11:197-203.

Ferreira, L.S.; S. Gerecht, J. Fuller, H.F. Shieh, G. Vunjak-Novakovic, & Langer R. (2007). Bioactive hydrogel scaffolds for controllable vascular differentiation of human embryonic stem cells. *Biomaterials*. 28:2706-17.

Fuchs, E. & Segre J.A. (2000). Stem cells: a new lease on life. *Cell*. 100:143-55.

Galvez, B.G.; M. Sampaolesi, S. Brunelli, D. Covarello, M. Gavina, B. Rossi, G. Constantin, Y. Torrente, & Cossu G. (2006). Complete repair of dystrophic skeletal muscle by mesoangioblasts with enhanced migration ability. *J Cell Biol*. 174:231-43.

Gaspar-Maia, A.; A. Alajem, F. Polesso, R. Sridharan, M.J. Mason, A. Heidersbach, J. Ramalho-Santos, M.T. McManus, K. Plath, E. Meshorer, & Ramalho-Santos M.

(2009). Chd1 regulates open chromatin and pluripotency of embryonic stem cells. *Nature*. 460:863-8.

Gavina, M.; M. Belicchi, B. Rossi, L. Ottoboni, F. Colombo, M. Meregalli, M. Battistelli, L. Forzenigo, P. Biondetti, F. Pisati, D. Parolini, A. Farini, A.C. Issekutz, N. Bresolin, F. Rustichelli, G. Constantin, & Torrente Y. (2006). VCAM-1 expression on dystrophic muscle vessels has a critical role in the recruitment of human blood-derived CD133+ stem cells after intra-arterial transplantation. *Blood*. 108:2857-66.

Hamed, S.A. (2006). Drug evaluation: PTC-124--a potential treatment of cystic fibrosis and Duchenne muscular dystrophy. *IDrugs*. 9:783-9.

Hanna, J.; M. Wernig, S. Markoulaki, C.W. Sun, A. Meissner, J.P. Cassady, C. Beard, T. Brambrink, L.C. Wu, T.M. Townes, & Jaenisch R. (2007). Treatment of sickle cell anemia mouse model with iPS cells generated from autologous skin. *Science*. 318:1920-3.

Hay, D.C.; L. Sutherland, J. Clark, & Burdon T. (2004). Oct-4 knockdown induces similar patterns of endoderm and trophoblast differentiation markers in human and mouse embryonic stem cells. *Stem Cells*. 22:225-35.

Helg, C.; B. Chapuis, J.F. Bolle, P. Morel, D. Salomon, E. Roux, V. Antonioli, M. Jeannet, & Leski M. (1994). Renal transplantation without immunosuppression in a host with tolerance induced by allogeneic bone marrow transplantation. *Transplantation*. 58:1420-2.

Hoch, W.; J. McConville, S. Helms, J. Newsom-Davis, A. Melms, & Vincent A. (2001). Auto-antibodies to the receptor tyrosine kinase MuSK in patients with myasthenia gravis without acetylcholine receptor antibodies. *Nat Med*. 7:365-8.

Hochedlinger, K.; & Jaenisch R. (2003). Nuclear transplantation, embryonic stem cells, and the potential for cell therapy. *N Engl J Med*. 349:275-86.

Hoffman, L.M. & Carpenter M.K. (2005). Characterization and culture of human embryonic stem cells. *Nat Biotechnol*. 23:699-708.

Horsley, V.; K.M. Jansen, S.T. Mills, & Pavlath G.K. (2003). IL-4 acts as a myoblast recruitment factor during mammalian muscle growth. *Cell*. 113:483-94.

Hotta, A. & Ellis J. (2008). Retroviral vector silencing during iPS cell induction: an epigenetic beacon that signals distinct pluripotent states. *J Cell Biochem*. 105:940-8.

Huangfu, D.; K. Osafune, R. Maehr, W. Guo, A. Eijkelenboom, S. Chen, W. Muhlestein, & Melton D.A. (2008). Induction of pluripotent stem cells from primary human fibroblasts with only Oct4 and Sox2. *Nat Biotechnol*. 26:1269-75.

Hyser, C.L. & Mendell J.R. (1988). Recent advances in Duchenne and Becker muscular dystrophy. *Neurol Clin*. 6:429-53.

Hyslop, L.A.; L. Armstrong, M. Stojkovic, & Lako M. (2005). Human embryonic stem cells: biology and clinical implications. *Expert Rev Mol Med*. 7:1-21.

Jaenisch, R. & Young R. (2008). Stem cells, the molecular circuitry of pluripotency and nuclear reprogramming. *Cell*. 132:567-82.

Jiang, Y.; B.N. Jahagirdar, R.L. Reinhardt, R.E. Schwartz, C.D. Keene, X.R. Ortiz-Gonzalez, M. Reyes, T. Lenvik, T. Lund, M. Blackstad, J. Du, S. Aldrich, A. Lisberg, W.C. Low, D.A. Largaespada, & Verfaillie C.M. (2002). Pluripotency of mesenchymal stem cells derived from adult marrow. *Nature*. 418:41-9.

Joannides, A.J.; C. Fiore-Heriche, A.A. Battersby, P. Athauda-Arachchi, I.A. Bouhon, L. Williams, K. Westmore, P.J. Kemp, A. Compston, N.D. Allen, & Chandran S. (2007). A scaleable and defined system for generating neural stem cells from human embryonic stem cells. *Stem Cells*. 25:731-7.

Jung, K.W. (2009). Perspectives on human stem cell research. *J Cell Physiol*. 220:535-7.

Kaji, K.; K. Norrby, A. Paca, M. Mileikovsky, P. Mohseni, & Woltjen K. (2009). Virus-free induction of pluripotency and subsequent excision of reprogramming factors. *Nature*. 458:771-5.

Kamochi, H.; M.S. Kurokawa, H. Yoshikawa, Y. Ueda, C. Masuda, E. Takada, K. Watanabe, M. Sakakibara, Y. Natuki, K. Kimura, M. Beppu, H. Aoki, & Suzuki N. (2006). Transplantation of myocyte precursors derived from embryonic stem cells transfected with IGFII gene in a mouse model of muscle injury. *Transplantation*. 82:516-26.

Kannagi, R.; N.A. Cochran, F. Ishigami, S. Hakomori, P.W. Andrews, B.B. Knowles, & Solter D. (1983a). Stage-specific embryonic antigens (SSEA-3 and -4) are epitopes of a unique globo-series ganglioside isolated from human teratocarcinoma cells. *Embo J*. 2:2355-61.

Kannagi, R.; S.B. Levery, F. Ishigami, S. Hakomori, L.H. Shevinsky, B.B. Knowles, & Solter D. (1983b). New globoseries glycosphingolipids in human teratocarcinoma reactive with the monoclonal antibody directed to a developmentally regulated antigen, stage-specific embryonic antigen 3. *J Biol Chem*. 258:8934-42.

Kim, D.; C.H. Kim, J.I. Moon, Y.G. Chung, M.Y. Chang, B.S. Han, S. Ko, E. Yang, K.Y. Cha, R. Lanza, & Kim K.S. (2009a). Generation of human induced pluripotent stem cells by direct delivery of reprogramming proteins. *Cell Stem Cell*. 4:472-6.

Kim, J.B.; V. Sebastiano, G. Wu, M.J. Arauzo-Bravo, P. Sasse, L. Gentile, K. Ko, D. Ruau, M. Ehrich, D. van den Boom, J. Meyer, K. Hubner, C. Bernemann, C. Ortmeier, M. Zenke, B.K. Fleischmann, H. Zaehres, & Scholer H.R. (2009b). Oct4-induced pluripotency in adult neural stem cells. *Cell*. 136:411-9.

Kiskinis, E. & Eggan K. (2010). Progress toward the clinical application of patient-specific pluripotent stem cells. *J Clin Invest*. 120:51-9.

Klimanskaya, I.; Y. Chung, S. Becker, S.J. Lu, & Lanza R. (2006). Human embryonic stem cell lines derived from single blastomeres. *Nature*. 444:481-5.

Kopp, J.L.; B.D. Ormsbee, M. Desler, & Rizzino A. (2008). Small increases in the level of Sox2 trigger the differentiation of mouse embryonic stem cells. *Stem Cells*. 26:903-11.

Krause, D.S.; N.D. Theise, M.I. Collector, O. Henegariu, S. Hwang, R. Gardner, S. Neutzel, Sharkis S.J. (2001). Multi-organ, multi-lineage engraftment by a single bone marrow-derived stem cell. *Cell*. 105:369-77.

Lafreniere, J.F.; P. Mills, M. Bouchentouf, & Tremblay J.P. (2006). Interleukin-4 improves the migration of human myogenic precursor cells in vitro and in vivo. *Exp Cell Res*. 312:1127-41.

Leist, M.; S. Bremer, P. Brundin, J. Hescheler, A. Kirkeby, K.H. Krause, P. Poerzgen, M. Puceat, M. Schmidt, A. Schrattenholz, N.B. Zak, & Hentze H. (2008). The biological and ethical basis of the use of human embryonic stem cells for in vitro test systems or cell therapy. *Altex*. 25:163-90.

Lian, Q.; Y. Zhang, J. Zhang, H.K. Zhang, X. Wu, Y. Zhang, F.F. Lam, S. Kang, J.C. Xia, W.H. Lai, K.W. Au, Y.Y. Chow, C.W. Siu, C.N. Lee, & Tse H.F. (2010). Functional mesenchymal stem cells derived from human induced pluripotent stem cells attenuate limb ischemia in mice. *Circulation*. 121:1113-23.

Liao, J.; Z. Wu, Y. Wang, L. Cheng, C. Cui, Y. Gao, T. Chen, L. Rao, S. Chen, N. Jia, H. Dai, S. Xin, J. Kang, G. Pei, & Xiao L. (2008). Enhanced efficiency of generating induced pluripotent stem (iPS) cells from human somatic cells by a combination of six transcription factors. *Cell Res*. 18:600-3.

Lin, T.; R. Ambasudhan, X. Yuan, W. Li, S. Hilcove, R. Abujarour, X. Lin, H.S. Hahm, E. Hao, A. Hayek, & Ding S. (2009). A chemical platform for improved induction of human iPSCs. *Nat Methods*. 6:805-8.

Lin, T.; C. Chao, S. Saito, S.J. Mazur, M.E. Murphy, E. Appella, & Xu Y. (2005). p53 induces differentiation of mouse embryonic stem cells by suppressing Nanog expression. *Nat Cell Biol*. 7:165-71.

Lyssiotis, C.A.; R.K. Foreman, J. Staerk, M. Garcia, D. Mathur, S. Markoulaki, J. Hanna, L.L. Lairson, B.D. Charette, L.C. Bouchez, M. Bollong, C. Kunick, A. Brinker, C.Y. Cho, P.G. Schultz, & Jaenisch R. (2009). Reprogramming of murine fibroblasts to induced pluripotent stem cells with chemical complementation of Klf4. *Proc Natl Acad Sci U S A*. 106:8912-7.

Maherali, N.; R. Sridharan, W. Xie, J. Utikal, S. Eminli, K. Arnold, M. Stadtfeld, R. Yachechko, J. Tchieu, R. Jaenisch, K. Plath, & Hochedlinger K. (2007). Directly reprogrammed fibroblasts show global epigenetic remodeling and widespread tissue contribution. *Cell Stem Cell*. 1:55-70.

Marchesi, C & Pareyson, D. (2009). Diagnosis, natural history, and management of Charcot-Marie-Tooth disease. *Lancet Neurol*. 8(7):654-67

McClorey, G.; A.M. Fall, H.M. Moulton, P.L. Iversen, J.E. Rasko, M. Ryan, S. Fletcher, & Wilton S.D. (2006). Induced dystrophin exon skipping in human muscle explants. *Neuromuscul Disord*. 16:583-90.

Melton, C.; R.L. Judson, & Blelloch R. (2010). Opposing microRNA families regulate self-renewal in mouse embryonic stem cells. *Nature*. 463:621-6.

Meregalli, M.; A. Farini, D. Parolini, S. Maciotta, & Torrente Y. (2010). Stem cell therapies to treat muscular dystrophy: progress to date. *BioDrugs*. 24:237-47.

Meriggioli, M.N.; & Sanders D.B. (2009). Autoimmune myasthenia gravis: emerging clinical and biological heterogeneity. *Lancet Neurol*. 8:475-90.

Mezey, E.; K.J. Chandross, G. Harta, R.A. Maki, & McKercher S.R. (2000). Turning blood into brain: cells bearing neuronal antigens generated in vivo from bone marrow. *Science*. 290:1779-82.

Miraglia, S.; W. Godfrey, A.H. Yin, K. Atkins, R. Warnke, J.T. Holden, R.A. Bray, E.K. Waller, & Buck D.W. (1997). A novel five-transmembrane hematopoietic stem cell antigen: isolation, characterization, and molecular cloning. *Blood*. 90:5013-21.

Mitalipova, M.M.; R.R. Rao, D.M. Hoyer, J.A. Johnson, L.F. Meisner, K.L. Jones, S. Dalton, & Stice S.L. (2005). Preserving the genetic integrity of human embryonic stem cells. *Nat Biotechnol*. 23:19-20.

Montarras, D.; J. Morgan, C. Collins, F. Relaix, S. Zaffran, A. Cumano, T. Partridge, & Buckingham M. (2005). Direct isolation of satellite cells for skeletal muscle regeneration. *Science*. 309:2064-7.

Moon, S.Y.; Y.B. Park, D.S. Kim, S.K. Oh, & Kim D.W. (2006). Generation, culture, and differentiation of human embryonic stem cells for therapeutic applications. *Mol Ther*. 13:5-14.

Morgan, J.E.; R.M. Fletcher, & Partridge T.A. (1996). Yields of muscle from myogenic cells implanted into young and old mdx hosts. *Muscle Nerve*. 19:132-9.

Mountford, J.C. (2008). Human embryonic stem cells: origins, characteristics and potential for regenerative therapy. *Transfus Med*. 18:1-12.

Murry, C.E. & Keller G. (2008). Differentiation of embryonic stem cells to clinically relevant populations: lessons from embryonic development. *Cell*. 132:661-80.

Nakagawa, M.; M. Koyanagi, K. Tanabe, K. Takahashi, T. Ichisaka, T. Aoi, K. Okita, Y. Mochiduki, N. Takizawa, & Yamanaka S. (2008). Generation of induced pluripotent stem cells without Myc from mouse and human fibroblasts. *Nat Biotechnol.* 26:101-6.

Negroni, E.; G.S. Butler-Browne, & Mouly V. (2006). Myogenic stem cells: regeneration and cell therapy in human skeletal muscle. *Pathol Biol (Paris).* 54:100-8.

Nelson, T.J.; A. Martinez-Fernandez, S. Yamada, C. Perez-Terzic, Y. Ikeda, & Terzic A. (2009). Repair of acute myocardial infarction by human stemness factors induced pluripotent stem cells. *Circulation.* 120:408-16.

Niwa, H. (2001). Molecular mechanism to maintain stem cell renewal of ES cells. *Cell Struct Funct.* 26:137-48.

Niwa, H.; S. Masui, I. Chambers, A.G. Smith, & Miyazaki J. (2002). Phenotypic complementation establishes requirements for specific POU domain and generic transactivation function of Oct-3/4 in embryonic stem cells. *Mol Cell Biol.* 22:1526-36.

Niwa, H.; J. Miyazaki, & Smith A.G. (2000). Quantitative expression of Oct-3/4 defines differentiation, dedifferentiation or self-renewal of ES cells. *Nat Genet.* 24:372-6.

Nowak, K.J. & Davies K.E. (2004). Duchenne muscular dystrophy and dystrophin: pathogenesis and opportunities for treatment. *EMBO Rep.* 5:872-6.

Okita, K.; T. Ichisaka, & Yamanaka S. (2007). Generation of germline-competent induced pluripotent stem cells. *Nature.* 448:313-7.

Owonikoko, T; M. Agha, R. Balassanian, R. Smith, & Raptis A. (2007). Gemtuzumab therapy for isolated extramedullary AML relapse following allogeneic stem-cell transplant. *Nat Clin Pract Oncol.* 4:491-5.

Pan, C.; B. Lu, H. Chen, & Bishop C.E. (2010). Reprogramming human fibroblasts using HIV-1 TAT recombinant proteins OCT4, SOX2, KLF4 and c-MYC. *Mol Biol Rep.* 37:2117-24.

Park, I.H.; R. Zhao, J.A. West, A. Yabuuchi, H. Huo, T.A. Ince, P.H. Lerou, M.W. Lensch, & Daley G.Q. (2008). Reprogramming of human somatic cells to pluripotency with defined factors. *Nature.* 451:141-6.

Patrick, J. & Lindstrom J. (1973). Autoimmune response to acetylcholine receptor. *Science.* 180:871-2.

Peault, B.; M. Rudnicki, Y. Torrente, G. Cossu, J.P. Tremblay, T. Partridge, E. Gussoni, L.M. Kunkel, & Huard J. (2007). Stem and progenitor cells in skeletal muscle development, maintenance, and therapy. *Mol Ther.* 15:867-77.

Pesce, M.; & Scholer H.R. (2001). Oct-4: gatekeeper in the beginnings of mammalian development. *Stem Cells.* 19:271-8.

Pfannkuche, K.; A. Fatima, M.K. Gupta, R. Dieterich, & Hescheler J. (2010). Initial colony morphology-based selection for iPS cells derived from adult fibroblasts is substantially improved by temporary UTF1-based selection. *PLoS One.* 5:e9580.

Pittenger, M.F.; A.M. Mackay, S.C. Beck, R.K. Jaiswal, R. Douglas, J.D. Mosca, M.A. Moorman, D.W. Simonetti, S. Craig, & Marshak D.R. (1999). Multilineage potential of adult human mesenchymal stem cells. *Science.* 284:143-7.

Price, F.D.; K. Kuroda, & Rudnicki M.A. (2007). Stem cell based therapies to treat muscular dystrophy. *Biochim Biophys Acta.* 1772:272-83.

Prockop, D.J. (1997). Marrow stromal cells as stem cells for nonhematopoietic tissues. *Science.* 276:71-4.

Qu-Petersen, Z.; B. Deasy, R. Jankowski, M. Ikezawa, J. Cummins, R. Pruchnic, J. Mytinger, B. Cao, C. Gates, A. Wernig, & Huard J. (2002). Identification of a novel population of muscle stem cells in mice: potential for muscle regeneration. *J Cell Biol*. 157:851-64.

Reubinoff, B.E.; P. Itsykson, T. Turetsky, M.F. Pera, E. Reinhartz, A. Itzik, & Ben-Hur T. (2001). Neural progenitors from human embryonic stem cells. *Nat Biotechnol*. 19:1134-40.

Rideau, Y. ; B. Glorion, & Duport G. (1983). Prolongation of ambulation in the muscular dystrophies. *Acta Neurol (Napoli)*. 5:390-7.

Rideout, W.M. 3rd; K. Eggan, & Jaenisch R. (2001). Nuclear cloning and epigenetic reprogramming of the genome. *Science*. 293:1093-8.

Rideout, W.M. 3rd; K. Hochedlinger, M. Kyba, G.Q. Daley, & Jaenisch R. (2002). Correction of a genetic defect by nuclear transplantation and combined cell and gene therapy. *Cell*. 109:17-27.

Rizzino, A. (2009). Sox2 and Oct-3/4: a versatile pair of master regulators that orchestrate the self-renewal and pluripotency of embryonic stem cells. *Wiley Interdiscip Rev Syst Biol Med*. 1:228-36.

Rosner, M.H.; M.A. Vigano, K. Ozato, P.M. Timmons, F. Poirier, P.W. Rigby, & Staudt L.M. (1990). A POU-domain transcription factor in early stem cells and germ cells of the mammalian embryo. *Nature*. 345:686-92.

Rowland, B.D. & Peeper D.S. (2006). KLF4, p21 and context-dependent opposing forces in cancer. *Nat Rev Cancer*. 6:11-23.

Roxanis, I.; K. Micklem, & Willcox N. (2001). True epithelial hyperplasia in the thymus of early-onset myasthenia gravis patients: implications for immunopathogenesis. *J Neuroimmunol*. 112:163-73.

Saha, K. & Jaenisch R. (2009). Technical challenges in using human induced pluripotent stem cells to model disease. *Cell Stem Cell*. 5:584-95.

Sampaolesi, M.; S. Blot, G. D'Antona, N. Granger, R. Tonlorenzi, A. Innocenzi, P. Mognol, J.L. Thibaud, B.G. Galvez, I. Barthelemy, L. Perani, S. Mantero, M. Guttinger, O. Pansarasa, C. Rinaldi, M.G. Cusella De Angelis, Y. Torrente, C. Bordignon, R. Bottinelli, & Cossu G. (2006). Mesoangioblast stem cells ameliorate muscle function in dystrophic dogs. *Nature*. 444:574-9.

Sarig, R.; Z. Baruchi, O. Fuchs, U. Nudel, & Yaffe D. (2006). Regeneration and transdifferentiation potential of muscle-derived stem cells propagated as myospheres. *Stem Cells*. 24:1769-78.

Shavlakadze, T.; J. White, J.F. Hoh, N. Rosenthal, & Grounds M.D. (2004). Targeted expression of insulin-like growth factor-I reduces early myofiber necrosis in dystrophic mdx mice. *Mol Ther*. 10:829-43.

Shi, Y.; Y. Zhao, & Deng H. (2010). Powering reprogramming with vitamin C. *Cell Stem Cell*. 6:1-2.

Singh, N.; V. Pillay, & Choonara Y.E. (2007). Advances in the treatment of Parkinson's disease. *Prog Neurobiol*. 81:29-44.

Siu, C.W.; S.Y. Liao, Y. Liu, Q. Lian, & Tse H.F. (2010). Stem cells for myocardial repair. *Thromb Haemost*. 104:6-12.

Skottman, H.; M.S. Dilber, & Hovatta O. (2006). The derivation of clinical-grade human embryonic stem cell lines. *FEBS Lett*. 580:2875-8.

Solter, D. (2000). Mammalian cloning: advances and limitations. *Nat Rev Genet*. 1:199-207.

Song, Z.; J. Cai, Y. Liu, D. Zhao, J. Yong, S. Duo, X. Song, Y. Guo, Y. Zhao, H. Qin, X. Yin, C. Wu, J. Che, S. Lu, M. Ding, & Deng H. (2009). Efficient generation of hepatocyte-like cells from human induced pluripotent stem cells. *Cell Res*. 19:1233-42.

Stamm, C.; I. Friehs, Y.H. Choi, D. Zurakowski, F.X. McGowan, & del Nido P.J. (2003). Cytosolic calcium in the ischemic rabbit heart: assessment by pH- and temperature-adjusted rhod-2 spectrofluorometry. *Cardiovasc Res.* 59:695-704.

Stocum, D.L. (2001). Stem cells in regenerative biology and medicine. *Wound Repair Regen.* 9:429-42.

Strelchenko, N.; O. Verlinsky, V. Kukharenko, & Verlinsky Y. (2004). Morula-derived human embryonic stem cells. *Reprod Biomed Online.* 9:623-9.

Strulovici, Y.; P.L. Leopold, T.P. O'Connor, R.G. Pergolizzi, & Crystal R.G. (2007). Human embryonic stem cells and gene therapy. *Mol Ther.* 15:850-66.

Sun, N.; M.T. Longaker, & Wu J.C. (2010). Human iPS cell-based therapy: considerations before clinical applications. *Cell Cycle.* 9:880-5.

Tagliafico, E.; S. Brunelli, A. Bergamaschi, L. De Angelis, R. Scardigli, D. Galli, R. Battini, P. Bianco, S. Ferrari, G. Cossu, & Ferrari S. (2004). TGFbeta/BMP activate the smooth muscle/bone differentiation programs in mesoangioblasts. *J Cell Sci.* 117:4377-88.

Takahashi, K.; K. Tanabe, M. Ohnuki, M. Narita, T. Ichisaka, K. Tomoda, & Yamanaka S. (2007). Induction of pluripotent stem cells from adult human fibroblasts by defined factors. *Cell.* 131:861-72.

Takahashi, K. & Yamanaka S. (2006). Induction of pluripotent stem cells from mouse embryonic and adult fibroblast cultures by defined factors. *Cell.* 126:663-76.

Tamaki, T.; Y. Okada, Y. Uchiyama, K. Tono, M. Masuda, M. Wada, A. Hoshi, T. Ishikawa, & Akatsuka A. (2007). Clonal multipotency of skeletal muscle-derived stem cells between mesodermal and ectodermal lineage. *Stem Cells.* 25:2283-90.

Thomson, J.A.; J. Itskovitz-Eldor, S.S. Shapiro, M.A. Waknitz, J.J. Swiergiel, V.S. Marshall, & Jones J.M. (1998). Embryonic stem cell lines derived from human blastocysts. *Science.* 282:1145-7.

Thomson, J.A. & Marshall V.S. (1998). Primate embryonic stem cells. *Curr Top Dev Biol.* 38:133-65.

Tinsley, J.; N. Deconinck, R. Fisher, D. Kahn, S. Phelps, J.M. Gillis, & Davies K. (1998). Expression of full-length utrophin prevents muscular dystrophy in mdx mice. *Nat Med.* 4:1441-4.

Torrente, Y.; M. Belicchi, C. Marchesi, G. Dantona, F. Cogiamanian, F. Pisati, M. Gavina, R. Giordano, R. Tonlorenzi, G. Fagiolari, C. Lamperti, L. Porretti, R. Lopa, M. Sampaolesi, L. Vicentini, N. Grimoldi, F. Tiberio, V. Songa, P. Baratta, A. Prelle, L. Forzenigo, M. Guglieri, O. Pansarasa, C. Rinaldi, V. Mouly, G.S. Butler-Browne, G.P. Comi, P. Biondetti, M. Moggio, S.M. Gaini, N. Stocchetti, A. Priori, M.G. D'Angelo, A. Turconi, R. Bottinelli, G. Cossu, P. Rebulla, & Bresolin N. (2007). Autologous transplantation of muscle-derived CD133+ stem cells in Duchenne muscle patients. *Cell Transplant.* 16:563-77.

Torrente, Y.; M. Belicchi, M. Sampaolesi, F. Pisati, M. Meregalli, G. D'Antona, R. Tonlorenzi, L. Porretti, M. Gavina, K. Mamchaoui, M.A. Pellegrino, D. Furling, V. Mouly, G.S. Butler-Browne, R. Bottinelli, G. Cossu, & Bresolin N. (2004). Human circulating AC133(+) stem cells restore dystrophin expression and ameliorate function in dystrophic skeletal muscle. *J Clin Invest.* 114:182-95.

Torrente, Y.; G. Camirand, F. Pisati, M. Belicchi, B. Rossi, F. Colombo, M. El Fahime, N.J. Caron, A.C. Issekutz, G. Constantin, J.P. Tremblay, & Bresolin N. (2003a). Identification of a putative pathway for the muscle homing of stem cells in a muscular dystrophy model. *J Cell Biol.* 162:511-20.

Torrente, Y.; E. El Fahime, N.J. Caron, R. Del Bo, M. Belicchi, F. Pisati, J.P. Tremblay, & Bresolin N. (2003b). Tumor necrosis factor-alpha (TNF-alpha) stimulates chemotactic response in mouse myogenic cells. *Cell Transplant.* 12:91-100.

Tremblay, P.; S. Dietrich, M. Mericskay, F.R. Schubert, Z. Li, & Paulin D. (1998). A crucial role for Pax3 in the development of the hypaxial musculature and the long-range migration of muscle precursors. *Dev Biol.* 203:49-61.

Vetrini, F.; R. Tammaro, S. Bondanza, E.M. Surace, A. Auricchio, M. De Luca, A. Ballabio, & Marigo V. (2006). Aberrant splicing in the ocular albinism type 1 gene (OA1/GPR143) is corrected in vitro by morpholino antisense oligonucleotides. *Hum Mutat.* 27:420-6.

Weissman, I.L. 2000. Stem cells: units of development, units of regeneration, and units in evolution. *Cell.* 100:157-68.

Wernig, M.; A. Meissner, R. Foreman, T. Brambrink, M. Ku, K. Hochedlinger, B.E. Bernstein, & Jaenisch R. (2007). In vitro reprogramming of fibroblasts into a pluripotent ES-cell-like state. *Nature.* 448:318-24.

Wilmut, I.; N. Beaujean, P.A. de Sousa, A. Dinnyes, T.J. King, L.A. Paterson, D.N. Wells, & Young L.E. (2002). Somatic cell nuclear transfer. *Nature.* 419:583-6.

Woltjen, K.; I.P. Michael, P. Mohseni, R. Desai, M. Mileikovsky, R. Hamalainen, R. Cowling, W. Wang, P. Liu, M. Gertsenstein, K. Kaji, H.K. Sung, & Nagy A. (2009). piggyBac transposition reprograms fibroblasts to induced pluripotent stem cells. *Nature.* 458:766-70.

Xu, C.; M.S. Inokuma, J. Denham, K. Golds, P. Kundu, J.D. Gold, & Carpenter M.K. (2001). Feeder-free growth of undifferentiated human embryonic stem cells. *Nat Biotechnol.* 19:971-4.

Yang, X.; S.L. Smith, X.C. Tian, H.A. Lewin, J.P. Renard, & Wakayama T. (2007). Nuclear reprogramming of cloned embryos and its implications for therapeutic cloning. *Nat Genet.* 39:295-302.

Ye, L.; J.C. Chang, C. Lin, X. Sun, J. Yu, & Kan Y.K. (2009). Induced pluripotent stem cells offer new approach to therapy in thalassemia and sickle cell anemia and option in prenatal diagnosis in genetic diseases. *Proc Natl Acad Sci U S A.* 106:9826-30.

Yu, J.; K. Hu, K. Smuga-Otto, S. Tian, R. Stewart, Slukvin, II, & Thomson J.A. (2009). Human induced pluripotent stem cells free of vector and transgene sequences. *Science.* 324:797-801.

Zhang, X.; P. Stojkovic, S. Przyborski, M. Cooke, L. Armstrong, M. Lako, & Stojkovic M. (2006). Derivation of human embryonic stem cells from developing and arrested embryos. *Stem Cells.* 24:2669-76.

Zhao, Y.; X. Yin, H. Qin, F. Zhu, H. Liu, W. Yang, Q. Zhang, C. Xiang, P. Hou, Z. Song, Y. Liu, J. Yong, P. Zhang, J. Cai, M. Liu, H. Li, Y. Li, X. Qu, K. Cui, W. Zhang, T. Xiang, Y. Wu, Y. Zhao, C. Liu, C. Yu, K. Yuan, J. Lou, M. Ding, & Deng H. (2008). Two supporting factors greatly improve the efficiency of human iPSC generation. *Cell Stem Cell.* 3:475-9.

Zhou, W.; & Freed C.R. (2009). Adenoviral gene delivery can reprogram human fibroblasts to induced pluripotent stem cells. *Stem Cells.* 27:2667-74.

Zwi, L.; O. Caspi, G. Arbel, I. Huber, A. Gepstein, I.H. Park, & Gepstein L. (2009). Cardiomyocyte differentiation of human induced pluripotent stem cells. *Circulation.* 120:1513-23.

18

Treatment of the end Stage Liver Cirrhosis by Human Umbilical Cord Blood Stem Cells: Preliminary Results

Jong Yoon Bahk[1*], Zhengfu Piao[2*], Jae Hun Jung[3] and Hoon Han[4]
[1]*ChungAng University Medical School*
[2]*Beijing You-an Hospital, affiliated Capital Medical University & Beijing Institute of Liver Disease*
[3]*Gyeongsang National University Medical School*
[4]*Histostem & Huenna Ltd*
[1]*Korea*
[2]*China*

1. Introduction

The liver is a vital organ essential for life of vertebrates and other animals and plays a variety of metabolic functions, including glycogen storage, detoxification, plasma protein synthesis, the production of biochemicals for digestion and other roles. In the normal liver, hepatocytes physiologically renewed at very slow tempo. However, when injured by acute damage or drug intoxications, dormant hepatocytes re-enter the cell cycle and hepatic progenitor cells (HPCs) or oval cells are also thought to differentiate into hepatocytes, resulting in restoration of the structure and functions of the liver parenchyma **(1-2):** thus the liver has regenerative capacity. In severely overwhelmed cases, mature hepatocytes seemed to be blocked their proliferation activity, but HPCs are thought to be profoundly activated and play an important role in compensation of liver function (3, **4)**. When liver injury and inflammation occur chronically, normal liver tissue is replaced by fibrosis, scare tissue, diffuse necrosis and regenerative nodules **(5)**, thus resulting in an irreversible chronic inflammation loss of liver function at the final stage: liver cirrhosis (LC) **(6, 7, 8, 9)**. Regardless of its underlying causes, the morphologic figures and complications caused by LC are similar and as disease progresses, complications develop. They include portal hypertension, varix, ascities, hepatic encephalopathy, idiopathic peritonitis and hepatic coma. The classification of LC is based on the etiology, such as alcoholics, post-hepatitic, biliary, cardiac, metabolic, inherited and drug induced cirrhosis. Incidence of the specific types of liver cirrhosis is different from area to areas. Alcoholic cirrhosis is the most common type in North America, Western Europe and South America. Abstinence from alcohol would prevent the complicating liver cirrhosis **(10)**. One-fourth of the patients with repeated liver injuries and three fourth of post-hepatitic cirrhosis has the history of viral hepatitis (hepatitis B or hepatitis C). Post-hepatitic cirrhosis induced by those viruses is the

---

* Dr. Bahk and Dr. Piao are co-corresponding authors

most common cirrhosis found in South Eastern Asia and China including Korea, and second most common in America and Western Europe. In America, over 20% of patients infected chronically with HCV for more than 20 years were reported to develop post-hepatitic cirrhosis **(11)**. Biliary cirrhosis also occurs in association with injury of the biliary system or its prolonged impairment, which destroys liver parenchyme with progressive fibrosis. Billary, cardiac, metabolic, inherited and drug induced liver cirrhosis are much less common than alcoholic or posthepatitic cirrhosis (12). Avoidance of causative drugs, excessive protein intake and anti-infectious medicine were usually used as a first choice of treatment. Overall, about 75% of patients have progression and die within 5 years after the findings of the disease. Immune system is also deeply involved in the magnitude of cirrhosis. For example, in some instances cirrhosis can cause immune system dysfunction, leading to high susceptibility to infections or in other cases, autoimmune responses caused by the immunologic damage to the liver causing inflammation and eventually scaring and cirrhosis.

Until now, no simple hematologic diagnostic test for the LC has been developed **(13).** LC is diagnosed by clinical findings, such as chronic liver disease, reduced platelet count, esophageal varix, ascitis, portal hypertension symptoms, splenomegaly and changes in hematologic parameters **(14).** For the definite diagnosis, the liver biopsy is used, although invasive.

LC is the common end feature of an excessive and persistent scaring resulted from secondary fibrotic tissue remodeling followed by liver injuries and is regarded as irreversible disease.

In the rodents, both findings that carbon tetrachloride (CCl_4) treatment and bile duct ligation injury were reported to induce fibrogenesis **(15)**, and spontaneous recovery from once established liver fibrosis **(16-17)**, suggest that the hepatic fibrosis is a dynamic bi-directional event, and highlighted that the liver fibrosis is potentially reversible by antifibrotic therapy **(18)**.

After hepatic injury, the hepatic stellate cells (HSCs), positioned in the center of the ongoing fibrogenesis of liver **(19)**, are activated to proliferate and to produce contractile elements such as α smooth muscle actin (α-sma), a major determinant of sinusoidal portal hypertension **(27)**, and collagens (mainly type I and III) **(21)**. This process is most likely mediated through the stimulation of cytokines, such as transforming growth factor-β (TGF-β), platelet-derived growth factor (PDGF), and endothelin (ET-1) **(20)**, that are produced in a synergistic fashion of paracrine and autocrine from injured hepatic cells and inflammatory cells. The positive feedback loops between extracellular matrix and produced cytokines may also play an important roles in the acceleration of fibrosis (**22**).

Among them, TGF-β1 is thought to be the most significant factor for HSCs, because it upregulates the expression of its receptor **(23, 32)**, leading to raising the susceptibility to other mitogens such as PDGF, thrombin and angiotensin-II, etc. and eventually inducing proliferation of HSCs. It also accelerated the accumulation of HSCs in response to migration stimuli such as PDGF, vasoactive substance ET-1, and monocyte chemotatic protein **(24)**. Extracellular matrix (ECM) modulates HSC functions by interacting it with PDGF, fibroblast growth factor (FGF), TGF-β, and matrix metalloproteinases (MMPs) that present in the space of Disse **(25)**. Integrins expressed on HSCs play important roles in cell-matrix interactions for activation of latent TGF-β1 that modifies fibrogenesis.

In contrast, basement membrane tends to suppress the proliferation activity of HSCs and their collagen synthesis **(26)**. The cirrhotic liver overexpressed the ET-1 **(28),** a stimulant of nitric oxide production from HSCs and autocrine, that plays important roles for their own proliferation. HSCs also secrete neutrophil and monocyte chemoattractants, such as colony stimulating factors (CSF), MCP-1, and IL-8 that amplify the hepatic inflammatory response (**31**). Matrix synthesis is highly dependent on TGF β-1 and liver cirrhosis is closely related with ECM alteration in quantity and quality. Hepatic fibrosis is enhanced by synthesis of the neomatrix with type-1 collagen and the neomatrix degradation is inhibited by TIMPs (tissue inhibitor of metalloproteinases, TIMPs) (**33**). Thus activated HSCs are responsible for hepatic fibrosis. TIMP-1 inhibits apoptosis of HSCs (**33**) and induced promotion of perpetuation in hepatic fibrosis (**34**).

Prevention of the unnecessary medication is important for maintaining physiological conditions of the liver because it is very sensitive to medicines as well as microbial infection. The principle of treatment in LC is to remove such potential risk factors, and to correct the underlying invasive causes. Currently, treatments are focused on preventing complications such as ascites, esophageal varix and hepatic encephalopathy etc. Detection of serological changes at the early stage and regular follow up studies of those features are important for reducing the risky liver diseases such as LC. Usually, diet, bed rest, fluid restriction and diuretics, paracentesis and the liver transplantation are included in the treatment of the LC (**35-36**). For the treatment of decompensated LC, orthotopic liver transplantation is regarded as the only definitive therapeutic option (**36**). Several factors such as lack of available donors, combined with operative risks, complications associated with rejection, usage of immunosuppressive agents and cost-intensiveness make this strategy available to only a few people (**37, 41**). Those problems inherent in the liver transplantation have prompted the search for alternative therapeutic methods for intractable liver diseases **(42).** Due to these problems, many LC patients die from life-threatening complications at relatively early age. Recently, along with the development of regenerative medicine, the use of pluripotent stem cells are proposed for recovering from the disease states as an alternative therapy (**38**). Fetal liver derived stem cells transplanted showed some improvement for conservative management in the end stage liver diseases (**39**). In clinical cell therapy, highly enriched cell numbers and the high repopulation potential are essential (**40**).

Liver cells obtained from the post-mortem have been also used for transplantation as the promising alternative (**43-46**). The liver cell transplantation is regarded as a less invasive, less expensive, and can alleviate the problem of organ shortage. Although there are some advantages in the liver cell transplantation over the orthotopic liver transplantation, long-term observation seems to be important for the evaluation.

Reversal of hepatic fibrosis and cirrhosis may be achieved by resolution of liver fibrosis, an increase of apoptosis of activated HSCs and collagenolysis **(47)**. In the CCl_4 rat model, the removal of CCl_4 increased the apoptosis of HSCs: 50% decrease in cell numbers at 72 hours post-removal. And antifibrotic therapy with NO donor decreased the HSCs number to 50% at 18 hours after the treatment **(48).** Induction of HSCs apoptosis degraded the matrix and recovered from the liver fibrogenetic conditions by removing the source of fibrotic neomatrix and TIMPs. **(49-50)** Biliary fibrosis was reversed after decompression of the bile duct ligation (**51**). Apoptosis of HSCs induces pro-MMP-2 activation **(52)** suggesting matrix remodeling, because MMPs specifically degrade collagens and non-collagenous substrates in matrix. The matrix degradation seems a pivotal initial step in the processes of liver fibrosis resolution. The most important event is the action of the interstitial collagenases, MMP-1, which cleaves the collagen-1 molecule **(53).** Collagenase activity resulted from the gap between activated MMPs to TIMPs. Increase of TIMP-1 expression occurs in parallel with progressive fibrosis. Sustained TIMP-1 expression means the failure in degradation of fibrosis and there is intrinsic link between HSC apoptosis and matrix degradation **(47).** This suggests that the proteolysis facilitates the resolution and repair of the injured liver **(54)**. The failure of degradation of fibrosis by collagenase impairs HSC apoptosis to induce the delaying or blocking the hepatocyte regenerative response **(54)**.

Recovery from the hepatic fibrosis has been reported in animal models after removal of CCl_4, which is used for induction of acute hepatitis **(55-61)** and clinical patients **(62-66).** Hepatic fibrosis, induced by several toxins **(57-59)** or ligation of bile duct **(60),** was reversed after the removal of the causes. In human, reversal of the hepatic fibrosis was reported in patients with alcoholic liver disease, hemochromatosis and other liver diseases **(62-64).** Reversal of the posthepatitic cirrhosis was also noted after improvement from the hepatitis B infection **(65).** Moreover, posthepatitic cirrhosis due to hepatitis C infection responded to the interferon treatment. **(66)**

Due to the limitation of the donor livers, the stem cell-based therapy has been suggested as a possible alternative therapy **(67).** Plasticity (trans-differentiation) and fusion activities inherent in stem cells are important for the development, regeneration and repair of liver organs **(68-70, 72)**. The stem cells produce various humoral substances (cytokines, growth factors) and factors for homing or migration, also. Those characteristics of stem cells are also important for the therapy **(71)**. **R**ecently, mesenchymal stem cell (MSC) derived from umbilical cord (UC) is regarded as a promising form for cell therapy because of their easy accessibility, MSC is much primitive than other tissue sources and do not express the major histocompatibility complex (MHC) class II (HLA-DR) antigens **(74)**. Previous studies have shown that

UCMSC are still viable and not rejected at 4 months after xenografts, without the need for immune suppression, suggesting that they are a favorable cell source for transplantation **(75-77)**. UCMSC are able to differentiate into adipocytes, osteocytes, chondrocytes **(78,79)**, neurons **(80,81)**, cardiomyocytes **(78,82)** and renal tubule epithelial cells **(83)** upon cultured in induction media.

Initially, stem cell was isolated from bone marrow and characterized as plastic adherent fibroblastoid cell which has the capacity to generate some tissues **(84).** As study on stem cell has progressed, the stem cells were also found in the many other adult organs. The discovery of trans-differentiation potential **(85)** led us to use them in currently incurable or intractable diseases. Stem cells are used for two purposes: one is the trans-differentiation of stem cells into the specific cell types to replace the damaged or destroyed cells or tissues, and the other is the stimulation of the pre-existing native cellular repair mechanism in damaged organs, which may contain resident dormant stem cells. Currently, stem cell-based cell therapy is increasingly applied to the variety of diseases including cardiovascular issues **(86)**, diabetes **(87)**, musculoskeletal disorders **(88)**, renal problems **(89)**, impotence **(90)** and hepatic cirrhosis **(91)**. Stem cells can also be used for cytokine producer. Stem and other cells, which are genetically engineered for the production of cytokines, are also used as a vehicle of cytokines for targeting injury or disease sites, such as cancer sites **(92).** Stem cells, in particular committed to hepatocytes, can also be used for screening of drug toxicity **(93).**

Stem cells are classified into three, according to the differences in sources; embryonal stem cell (ESC), adult stem cell (ASC) and induced pluripotent stem (iPS) cell. ESC was developed by Thomson in Wisconsin **(94).** Fetal stem cells and amniotic stem cells may belong to this category. The differentiation potential of ESCs is great but the clinical use has been strictly limited due to the ethical problem and the tumorigenesis. The iPS cells were generated from mouse fibroblasts in 2006 by Yamanaka **(95)**, with introduction of four genes; Oct 3/4, Sox2, c-Myc, and Klf 4. The potentiality of iPS cells seems appears to be unlimited for clinical applications, but the bio-characterization of iPS cells has not established yet, although they are under intensive study. The origin of ASCs first discovered is bone marrows (BMSC), and then, the ASCs were discovered from the many mature organs **(96-100).** Among these, bone marrow, adipose tissues (ADSC) and umbilical cord blood (UCBSC) derived stem cells are currently available for cell-based clinical therapies. BMSCs were actively applied for the treatment of hematologic diseases, already. The spectrum of clinical application of BMSCs continues to expand with the time. UCBSCs are collected from venous blood of the cord, but the cell number collected is not great. The UCBSC was applied for Fanconi's anemia **(101)** in 1986 between siblings. UCBSC seems to contain the most immature form of ASCs that have few chances to contact with environmental antigens, immunologically immature **(102)**. The UCBSCs do not raises any ethical concern at umbilical cord blood collection issues and collection can be processed without invasiveness for donors. The regenerative capabilities of UCBSC are similar to other types of ASC **(103).** Adipose tissue derived stem cell (ADSC) was first described by plastic surgeons **(104)**. The procedure harvesting adipose tissue is a common work among plastic surgeons for cosmetic or regenerative purposes, although it is somewhat invasive. The processed lipoaspirated tissues harbor abundant multipotent stem cells, which have similar potential with BMSC or UCBSC **(103)**. ADSCs are permitted for autologous cell-based therapy in many countries.

The angiogenic, neurogenic and trans-differentiation potential of stem cells are the principal targets in stem cell based treatment **(105-110)**. Vascular endothelial growth factor (VEGF), basic fibroblast growth factor (bFGF), platelet derived growth factor AB (PDGF-AB), transforming growth factor-β (TGF-β), and integrin β are stem cell angiogenic factors **(111,112).** Brain-derived neurotrophic factors, and neurotrophin-4/5 (NT4/5) **(113,114)** are stem cell neurotrophic factors. Reports on the effects of stem cells for improvement of vascular insufficiencies **(115-118)** and neuropathies **(119-121)** are accumulating these days.

Liver cirrhosis has been thought to be an irreversible disease. However, recent studies on cirrhosis of animals and humans suggest that liver cirrhosis is a potentially reversible disease. Stem cells, which are of anti-fibrotic and trans-differential potential, raised the possibility of their??? use for the treatment of liver fibrosis and cirrhosis as an alternative treatment replacing for liver transplantation.

In the present study we report our pre-clinical and clinical experiences of umbilical cord blood derived stem cell treatment for end stage liver cirrhosis, and discuss on the stem cell therapy in liver cirrhosis.

2. Materials and methods

2.1 Preclinical study

For the evaluation of the stem cell effect, we prepared the rat liver cirrhosis model with carbon tetrachloride (CCl_4). Male Sprague-Dawley rats (6weeks old, 180--200g) were used. The rats were grouped into 3, one control group (A) and two experimental groups, CCl_4 treated group (B) and CCl_4 treatment plus stem cell treated group (C). Cirrhosis was induced by intraperitoneal administration of CCl_4 (4:1 olive oil) at a dose of 0.1 mL/100 g body weights, three times a week. The same volume of olive oil only was intraperitoneally infused for control. For cirrhosis induction, CCl_4 was infused for 8 weeks. Human umbilical cord blood stem cells were infused at a dose of $1x10^6$ cells in 0.2 mL through the tail vein, and saline of same volume for control and CCl_4 only group. Rats were sacrificed at one, two and four weeks and one pathologist evaluated the pathologic changes of liver. For pathologic evaluation, sections of approximately 4 μm thick were made and stained with haematoxylin and eosin (H&E) staining for routine histology, and Masson's trichrome (MT) staining for collagen.

2.2 Clinical study

Total 51 patients were participated in this treatment. The exclusion criteria for patient recruitment in stem cell treatment with liver cirrhosis were age limitation, over 70, and cancer. Among them, 46 patients (male 27 and 19 female) were classified as the Child-Pugh (CP) class C liver cirrhosis and 5 (four male and one female) were classified as the CP class B, finally, although they were classified as the class C from the other hospital, at the time of initiation. All participants are randomly involved in this treatment after reviewed the medical records of patients whose life expectancy were less than 6 months, evaluated by doctors in charge before involved in these treatments. 43 patients graduated from the college or upper rank school. The cause of cirrhosis was alcholic in 18 and posthepatitic in 36. Although 17 had the heart problem, it was not definite whether cardiac problem was the solitary cause of the cirrhosis or not, except two. The common clinical complications for portal hypertension were varix (41) and ascitis (34). For all patients, the explanation was made on the treatments rationale and material, and informed consents were obtained. All patients understood the treatment and no body has the opinion on baseline studies (Table 1). All participating patients were negative on cancer especially at baseline studies. They had specific conditions that could be related with liver cirrhosis and had the several complications induced from portal hypertension (Table 2).

Imaging	Ultrasound Exam. on Heart, Liver, Kidney CT on Abdomen
Endoscopic	Gastroscopy, Esophagoscopy
Blood & Serology	CBC, Serologic series 12
Urine	UA, Microscopic Examination
Cancer Marker	AFP, CEA
Others	VDRL, AIDS, Hepatitis B Ag/Ab, Hepatitis C AG/Ab., Electrolyte, C-reactive protein, ASO titer, EKG, ESR, RA factor
Coagulation	PT (INR)

Table 1. Base line studies.

The human umbilical cord blood stem cells (hUCBSC) were supplied by Histostem Co. Ltd. (Seoul, Korea) that was permitted for clinical use from the Korean government (KFDA).
The supplied stem cells were ABO, HLA-ABC, DR and sex matched for each patient, and the donor of each stem cell unit does not have any specific familial medical history. The stem cell markers of hUCBSCs were studies by flow cytometry. Total cell numbers that infused for each patient were around 1.5×10^7. hUCBSCs were infused percutaneous directly into liver parenchyma using needle under the ultrosonographic guide. All patients

were informed for usual life after one-day bed rest. Patients were followed every month from treatment at least for 6 times. At follow up check, patients were evaluated for prothrombin time, albumin, ascities and encephalopathy before and after each month from the stem cell therapy. Patients were evaluated for serological results, ascities and encephalopathy before and after each month from the stem cell therapy. The final survival was checked at November 30, 2010, 7 to 75 months from the initiation of the stem cell therapy.

Conditions manifested at initial visit	**Positive Patients Number**	**Negative Patients Number**
Alcohol Intoxication	18	33
Hepatitis History	36	15
Heart Problem	17	34
GB Problem	0	51
Varix	41	10
Ascitis	34	17
Hepatic encephalopathy	19	32
Peritonitis	1	50

Table 2. Patient condition at presentation.

Study Items	Normal Value	Normal Patients	Range of Measure value / Patients Number		
Total Protein	6.3 - 8.2 g/dL	12	39		
Albumin	3.5 - 5.1 g/dL	5	1.0 - 2.0	2.1 - 3.0	>3.1
			17	26	3
ALP	38 - 126	13	38		
sGOT	5 - 40 IU/L	2	49		
sGPT	5 - 35 IU/L	1	50		
Total Bilirubin	0.2 - 1.3 mg/dL	14	<2mg/dL	2-3mg/dL	>3mg/dL
			18	16	3
Direct Bilirubin	0 - 0.3 mg/dL	17	34		
PT (INR)	<1.7	18	<1.7	1.7 - 2.3	>2.3
			18	22	11

Table 3. The results of serologic test before stem cell treatment.

3. Results

3.1 Stem cells

The flow cytometric results (Figure 1) of hUCBSCs were CD13(+), CD14(-), CD29(+), CD31(-), CD34(-), CD44(+), CD45(-), CD49e(+), CD54(+), CD90(+), CD106(-), AMSA(+), SH2(+), SH3(+), HLA-ABC(+) and HLA-DR(-).

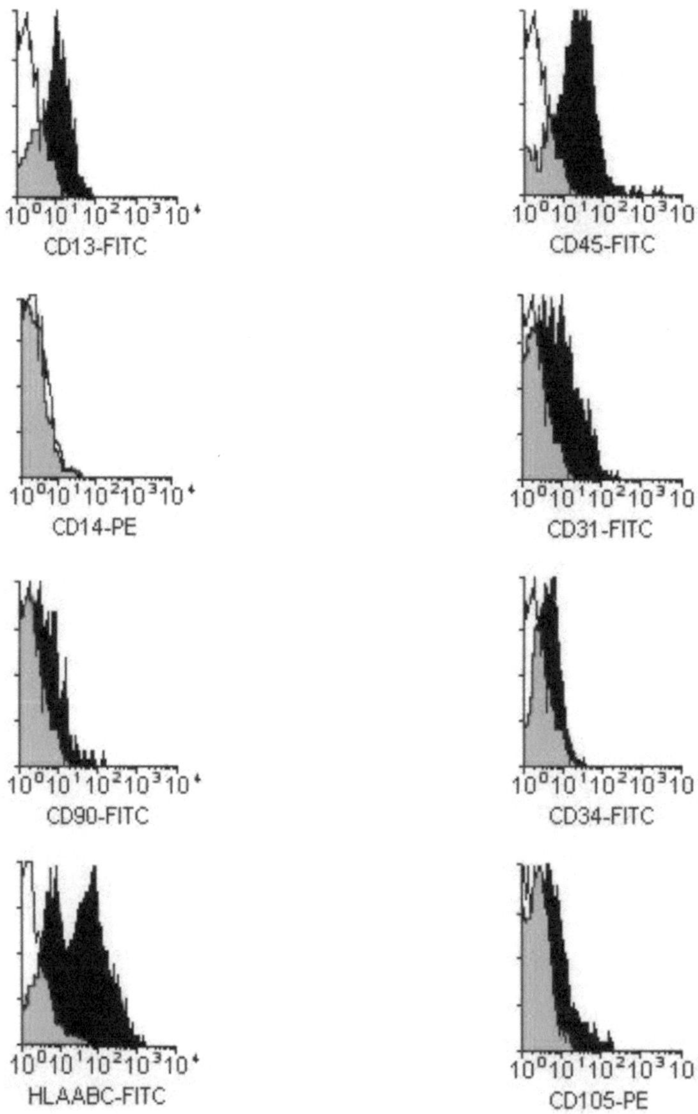

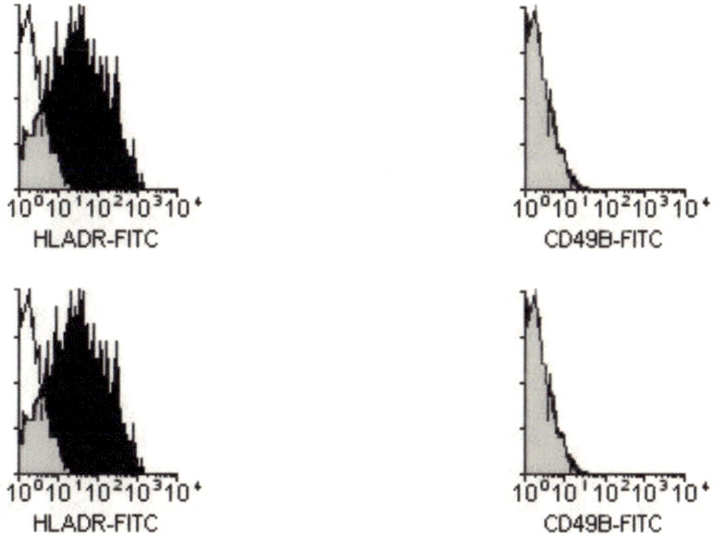

Fig. 1. Flow cytometric findings of the surface markers in hUCBSCs.

3.2 Preclinical study

The gross finding of the liver was examined (Figure 2) at 12 weeks after the starting the infusion of CCl_4. Compared to the control which have normal hepatic configuration (A), the CCl_4 infused rats (B) without stem cell treatment showed nodular surfaced distorted liver but CCl_4 infusion with stem cell treatment rat (C) showed recovered from the distorted nodular liver but much distorted compared with control (A).

The microscopic study (Figure 3), with haematoxylin and eosin (H&E) staining for routine histology and Masson's trichrome (MT) staining for collagen, were done on liver section at 1, 2 and 4 weeks after CCl_4 injection for 8 weeks. The control group showed normal

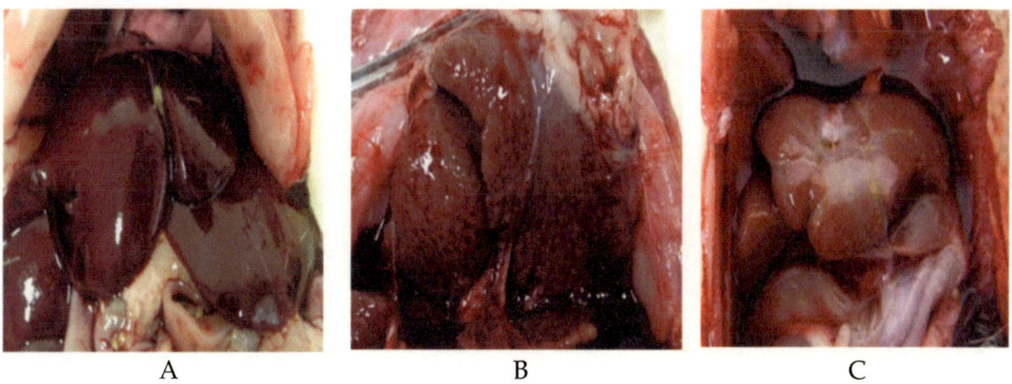

Fig. 2. Gross findings of the liver at 12 weeks after initiation of the CCl_4 intraperitoneal injections. A is control, B is CCl_4 intraperitoneal injection and C is CCl_4 intraperitoneal injection plus hUCBSC treated rat.

architecture at all livers (A). At 1 week after CCl_4 injection for 8 weeks, the experimental group showed the cirrhosis was induced in both without (B1) and with stem cell treatment group (C1), the fibrosis and deposition of collagen, separating liver parenchyma into large lobules. At 2 week after CCl_4 injection for 8 weeks, there was definite difference between without stem cell treatment group and with stem cell group. The without stem cell treatment group (B2) showed increased septa composed of fibrosis and collagen. Compared to without stem cell group, with stem cell group (C2) showed reduced septa than first week after CCl_4 injection for 8 weeks. At 4 week after CCl_4 injection for 8 weeks, the without stem cell treatment group (B3) showed much increased septa composed of fibrosis and collagen and there is new fine septa were appeared within large lobules. In with stem cell treatment group (C3), the septa composed of fibrosis and collagen were reduced and the lobules are closely approximated each other. In these pathologic findings, experimental group without stem cell therapy induced the extensive fibrosis / cirrhosis and the fibrosis was progressed as the time passing (B). These CCl_4 induced fibrosis / cirrhosis was proved by disruption of liver parenchyma architecture, extension of fibers, large fibrous septa formation, pseudolobe separation and collagen accumulation. These alterations were progressed and increased in fibrosis and collagen deposition with time passing. The histopathological findings confirmed that the cirrhosis was significantly reduced by stem cell infusion.

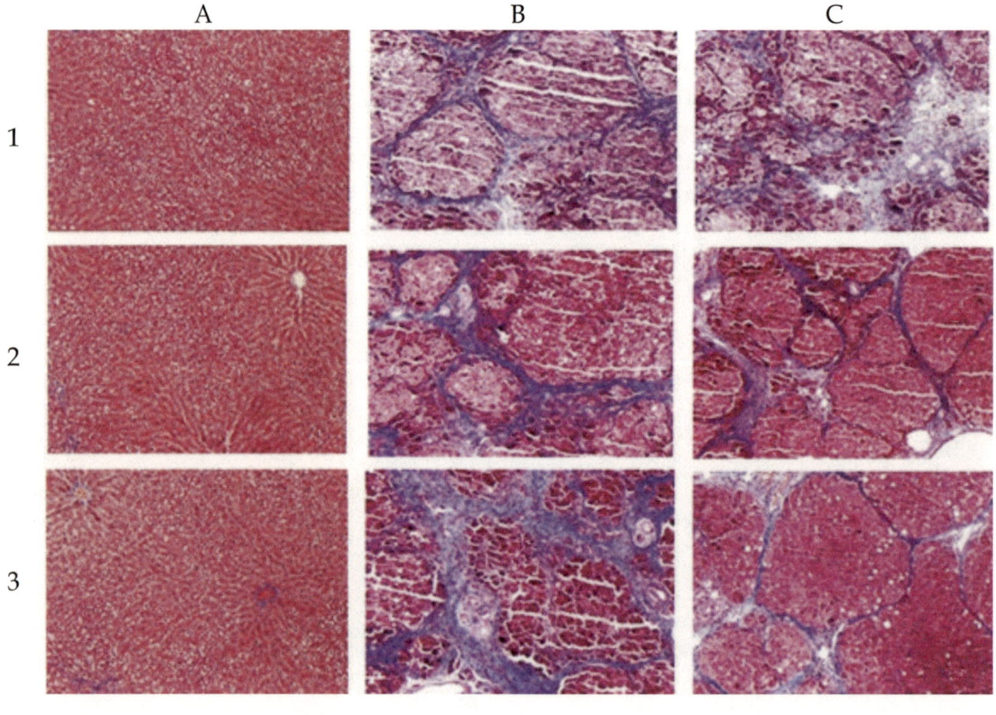

Fig. 3. Microscopic findings of the rat liver in control and experimental rats. A from control rat, B from CCl_4 intraperitoneal injection only rat and C from CCl_4 intraperitoneal injection with hUCBSC therapy rat. 1 at 9 week from CCl_4 intraperitoneal injection, 2 at 10 weeks and 3 at 12 weeks.

3.3 Clinical study

The participants' age ranged from 49 to 68 years old (mean 54.7 years). The mean age of the CP class C were 56.2 and class B were 48.9. After the stem cell therapy, there was no death within 6 months from the initiation of the stem cell therapy. There were 9 deaths from 7 months within 12 months, 19 deaths from 13 months within 2 years, 13 deaths over 2 year within 3 years, 3 deaths over 3 years within 5 years. 7 are living now, from 11 to 75 months from stem cell therapy, and among them, two are living over 5 year. During the procedure (Figure 4), the bleeding from the liver was not remarkable. Clinically, ascitis was improved in 21 and 5 did not show any ascitis within 6 months. Among 5, 2 are alive more than 5 years (Figure 5). In hepatic encephalopathy, all patients showed improvement in symptom. But there is no patient who lived more than 30 months. Serologic follow up check was done in all patients (Table 3 and 4). In serologic test, although there were some changes in the

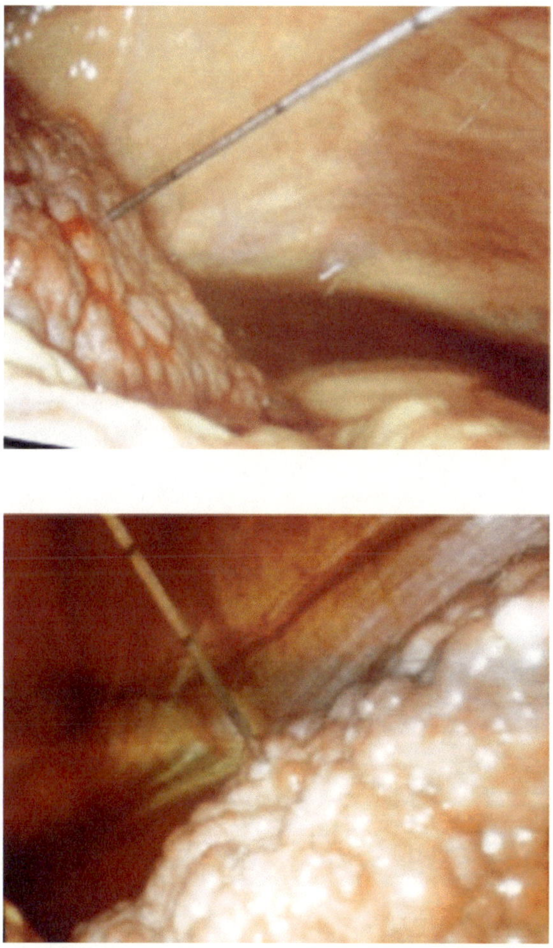

Fig. 4. Laparascopic view of percutaneous injection of the stem cells into liver. The needle is inserted into the nodular surfaced liver.

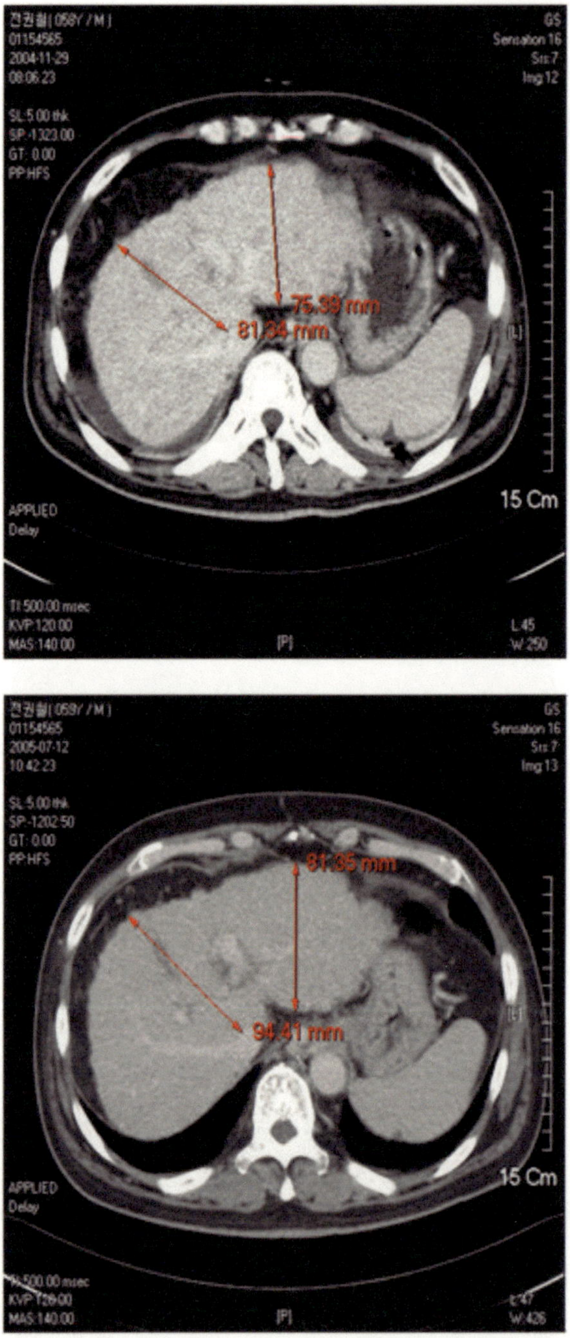

Fig. 5. CT findings from the patient who is living more than 5 years. A is CT taken just before (Feb. 21, 2005) stem cell therapy and B is CT taken was after 5 months (July 23, 2005). The measured liver size was increased from 689.98 CC to 915.36 CC and hepatic density is lowered suggesting release of fibrosis.

results, it was impossible to get the uniform information in trends of change. Among 39, those who had abnormal protein level, total protein level was normalized at 6, but among 39, only 2 were normalized in albumin. In these 2, one had normalization of total protein level but one did not, meaning some difference in improvements among sub-group.

Study Items	Normal Value	Normal Patients	Range of Measure value Patients Number		
Total Protein	6.3 - 8.2	18(+6)	33(-6)		
Albumin	3.5 - 5.1	7(+2)	1.0 - 2.0	2.1 - 3.0	>3.1
			12(-5)	25(-1)	7(+4)
ALP	38 - 126	17(+4)	34(-4)		
sGOT	M: 17 - 59 F: 14 - 36	9(+7)	42(-7)		
sGPT	M: 21 - 72 F: 9 - 52	12(+11)	39(-11)		
Total Bilirubin	0.2 - 1.3	16((+2)	<2mg/dL	2-3mg/dL	>3mg/dL
			21(+3)	14(-2)	0(-3)
Direct Bilirubin	0 - 0.3	18(+1)	33(-1)		
PT (INR)	<1.7	23(+5)	<1.7	1.7 - 2.3	>2.3
			23	22	6

Table 4. Serologic test results at post stem cell treatment 6 months.

4. Discussion

Most of the chronic liver injuries, regardless of their causes, progress to liver fibrosis and eventually result in cirrhosis that is thought to be irreversible, and the liver cirrhosis (LC) results in impairment of the hepatic function becoming a massive health care burden worldwide. LC is induced by many different causes, such as chronic viral hepatitis, toxic damage including alcohol, parasitic disease, inborn errors of metabolism, and non-alcoholic fatty liver disease etc. The cause of LC has a wide geographic variation. Alcohol is the most common cause in western countries (122) and liver disease is the 5th most common cause of death in the United Kingdom (123.) Recent reports on pre-clinical and clinical studies suggested that LC could be reversible. HSCs activated upon liver injury are thought to be responsible for collagenogenesis and fibrosis in extracelluar matrix (124, 125). Modulation of HSC activity (126.), promotion of HSC apoptosis (127), blocking of matrix formation (128) or anti-proliferation measures on matrix fibrogenic and contractile response to HSC and degradation of established matrix (125,129) could be taken as potent strategies for reversal of LC. In addition to these strategies, stem cell therapy with multiple cytokine production and trans-differentiation potential would be a new option (130) for reversal of established LC.

HSCs are responsible for the production of extracellular matrix at the center of LC. Therefore, the inhibition of the activation of HSCs (131-133) or promotion of HSC apoptosis

(134-135) would be an eradicating measure for fibrogenesis. The platelet-derived growth factor (PDGF) is the most potent mitogen for HSCs, thereby inhibitors against PDGF have been tried (136-138). For the patients whose fibrosis is progressing, the matrix formation blockers (139) or anti-proliferation measures to matrix fibrogenesis (140) would be another measure for the reduction of LC. HSCs contribute to portal hypertension through multiple mechanisms including collagen deposition, vasoconstriction, and regulation of sinusoidal structure. So, the reduction of contractile response to HSC (141) would be a measure reducing critical complications induced by portal hypertension. Increasing the degradation of established matrix (125,142,142) would be an ideal strategy for reversal of LC.

Recently developed stem cell therapy (143-144) may be a promising strategy for LC, because we can expect them to trans-differentiate into hepatocytes and to stimulate the differentiation of hepatocytes from the dormant stem cells remaining in injured host liver and stem cells in bone marrow through cytokine production (145). BMSC, ADSC, and UCBSC are currently used for stem cells therapy. Unseparated bone marrow cells and purified BMSCs have been used for the treatment of the hematopoietic diseases for more than 50 years. Their application fields are expanding day by day. For advanced liver diseases such as cirrhosis, stem cell engraftment can be most promising strategy after the organ transplantation (146). In animals treated with CCl_4 intraperitoneal infusion, the percentage of lin- Sca-1+ cells in the bone marrow and peripheral blood increases twofold at twenty-four hours later (147) and this increase peaks at day one in the bone marrow and day two in the peripheral blood (147). This finding suggests that bone stem cells started to proliferate and migrate to the periphery (147). Requirement of engraftment of stem cells into the patients with the severe liver diseases are increasing. For BMSC therapy, bone marrow aspiration is inevitable and bone marrow aspiration is a invasive procedure. Differentiation potential, and maximal life span of BMSC decrease with increasing age (148-149). In post-hepatitic LC patients, the autologous BMSCs are not suitable for the treatment, because their proliferation capacity is restricted significantly in the tissue environment of end stage of the disease (150)

ADSCs were relatively recently found. Their easy isolation (151) and high differentiation potential (152) may be attractive as alternative promising stem cells in place of BMSCs. In the meantime, UCBSCs has been used for several diseases. UCBSCs have advantages over other types of stem cells, because umbilical cord blood (UCB) can be obtained without invasiveness or harm to donor. Cells from UCB have many advantages because of the immaturity of stem cells compared to other types. Moreover, UCBSCs provide no ethical barriers for basic and clinical applications (153-154). Recently, UCB banking for transplantation of haematopoietic stem cells is increasingly in many institutions (155) due to their easy availability and the ready on shelf system. All clinical papers are of autologous or allogenic stem cell therapy. Among the adult stem cells, UCBSCs are youngest stem cells and they had only few chances to contact with environmental antigens. UCBSCs have higher proliferative potential than BMSC and higher expression of the endothelial-specific markers following endothelial differentiation (156).

When we decided to attempt to treat with UCBSCs for some diseases, we carefully consider suitable cell (origin) for target organs, cell numbers, route for administration, supplementation of growth factors and post-treatment cares. In spite of increasing requirement of stem cell treatment, the protocol for stem cell treatment has not established until now. Because the preparation of the stem cells for enough quantity that has the identical or similar biology is not easy, we should consider the measure to secure the enough number of stem cells, culture or mixing of different donors. Although many trials with stem cell were done, most of them are autologous rather than allogenic. Basically, many diseases caused by different pathogenesis, it needs different treatment. In case of LC, bone marrow stem cells from liver cirrhotic caused by chronic hepatitis B infection showed significantly lower S-phase fractions and growth factor (IGF-1, PDGF α, PDGF β) receptor expression than normal people (157). It indicates that allogenic stem cell therapy is better than autologous in LC. We suppose that allogenic stem cell therapy would be reasonable in case of the hereditary or genetic disease. The number of stem cells in specific treatment has not decided yet, too. Moriya et al. (158) used $1x10^5$ embryonal stem cells in animal study model (mice). Liu et al.(159) used $5x10^5$ endothelial precusor cells in 150 g animal study model(rat). Yan et al. (160) infuse $3x10^6$ hunman UCBSCs in mice. Although these reports are animal study, when we calculate the cell number in weight to weight (animal to human) base, the cell number for human treatment should be over $1x10^8$ cells. But Mohamadnejad et al. (161) infused 31.73×10^6 autologous bone marrow mesenchymal stem cells for four liver cirrhotic patients via peripheral vein. So, the cell number for treatment of disease should be standardized according to the disease, although there should be many trials for standardization.

In introduction part of this chapter, we mentioned on the characteristics of stem cells, homing or migration (someone uses 'targeting'). Homing in stem cell biology means stem cell moving toward injury site along the some chemotactic signal (162). So, many animal experiments localized the targeted stem cells, which were infused via tail vein (260-164) or intraperitoneal infusion(165) and sometimes via specific route such as portal vein(166). And in case of cellular cardioplasty for myocardiac infarction model, several different infusion has been used, direct intracoronary (167), intravenous with homing(168) and mobilization with homing from bone marrow or peripheral blood(169). In cellular cardiomyoplasty, according to the cell delivery system, the results are different. Least effective method is mobilization from peripheral blood or bone marrow(170-171) and most effective method is intracoronary infusion (172-173). And in case of intravenous infusion, according to the blood flow, all blood in body has to pass the lung and some proportion of infused stem cells are trapped at the lung (174-175). Some are trapped at other organs(176).

There are many growth factors which support or enhancing stem cell activities. The supplement of growth factors into stem cell culture media or coupled treatment of stem cell with growth factors was reported in vitro and in vivo, already. Supplement of growth factors in culture system has been done from long ago in culturing technique and sometimes essential job for culture, but the coupled clinical treatment of stem cell with growth factors (177-178) are in different conditions (177-178). Although we use the various growth factors for in vitro culture system, there are a few growth factors which are permitted to use for human from the health concerning governmental bureau. So, it is not easy to search the simultaneous application of the growth factor as an auxiliary measure for stem cell therapy (178-180).

Post-treatment care are difficult to mention at present time because there is no report for large sized world wide data related on the stem cell treatment on specific disease until now. Although the data on the post-treatment care for better maintenance of end result of stem cell therapy are needed, it is not proposed yet. As in the material and method, we had the clinical treatment for end-stage LC whose survival was expected to be less than 6 months. The result showed that the all patients had improved survival although their initial life expectancy was based on the doctor's experiences. Other laboratory data had some differences among patient to patient, but most of the patients who had the hepatic encephalopathy and more than half of the ascitic patient had the improved symptoms.

In author's stem cell treatment for LC, we used umbilical cord blood stem cells (UCBSCs), only. UCBSCs produce and secrete various humoral factors (cytokine or growth factors)(261-262). They are stem cell factor (SCF), macrophage colony stimulating factor (M-CSF), granulocyte macrophage colony stimulating factor (GM-CSF), vascular endothelial growth factor (VEGF), interleukin 1β (IL-1β), IL-6, IL-8, IL-11, IL-12, IL-15, stromal cell-derived factor 1α (SDF-1α), hepatocyte growth factor (HGF), epitehlial cell-derived protein 78 (ENA-78), Growth-related oncogene (GRO), oncostain M (OSM), monocyte chemoattractant protein-1 (MCP-1), fibroblast growth factor 4 (FGF-4), FGF-7, FGF-9, granulocyte chemotactic protein 2 (GCP-2), Insulin like growth factor (IGF), Insulin-like growth factor binding protein 1 (IGFBP-1), IGFBP-2, IGFBP-3, IGFBP-4, Interferon-gamma-inducible protein 10 (IP-10), Leukemia inhibitory factor (LIF), migration inhibitory factor (MIF), macrophage inflammatory protein 3α (MIP-3alpha), osteoprotegerin, (OPG), pulmonary and activation-regulated chemokine (PARC), Placenta growth factor (PIGF), transforming growth factor β1(TGF-β1), TGF-β2, TGF-β3, tissue inhibitors of matrix metalloproteinases 1 (TIMP-1) and TIMP-2.

Among these cytokines, some of them are favorable for antifibrosis and some are prone to enhance the fibrosis in liver. The results of balance between favorable and unfavorable effects among the various cytokines secreted from the UCBSC would predict treatment or aggravation of LC. Among above cytokines that UCBSCs secrete, HGF, IGF-1, IGFBP and interferon would be closely related with fibrolysis of liver but TGF-β1 and TIMP will promote hepatic fibrosis. Although the results of UCBSC treatment on LC patient showed much improvement on symptoms and liver function related laboratory data, we do not know the interaction between those cytokines, which has some antagonistic effects.

We injected stem cells into liver parenchyme directly under the ultrasound guide. Total cell numbers were around 1.5×10^7 hUCBSCs in each patient. Until now, there are many trials with animal model of LC but no author has presented the proper cell number for stem cell treatment on LC. Around 1×10^6 cells were infused intravascularly in rat or mouse liver cirrhosis model and they showed improvement in hepatic functions. Although the most of patients in our treatment group showed favorable effects on LC after stem cell therapy, the role of cell number in LC treatment was not definite. We infused the stem cells directly into the target organ, liver, the cell number that authors used were much less than animal experiments. In spite of our small numbered stem cells, there was some improvement in data for hepatic function.

As mentioned before, we infused the stem cells directly into the liver parenchyma. The infusion route would be closely related with results of the stem cell treatment. In systemic intravenous administration would be the easiest method in stem cell treatment but the trapping of stem cell by lung or some reticuloendothelial system, during circulation before homing, would lessen the treatment effects. Although there are several data related with homing after intravenous infusion of stem cells, it is unclear that how many portions of infused cells are homed. To increase the homed stem cells in target organ, some therapies try to infuse the stem cells via selective artery, although it is invasive. More invasive therapy would be the orthotopic infusion/transplantation of stem cells under direct of indirect visualization. Surgically exposed target lesion would be treated with high accuracy and possibly permit reduced numbered stem cell implantation with same effects. Modern developed imaging modalities enabled us target the lesions with very high accuracy with less damage. The choice of the route for best effective targeting the lesion would be depend on the imaging modality and technique of operator.

Average life expectancy of the LC patients chose randomly by authors is expected to no more than 6 months. After stem cell therapy, no body had gone within 6 month and two of 51 patients survived more than 5 years. There would be many causes contributing to the improvement of patients' survival. Among them, the role of stem cell for improvement of survival and laboratory data should get attention. Although Kögler and Liu reported the various cytokines, which they detected from UCBSCs, there would be many unknown cytokines (humoral factors), which is secreted by the UCBSCs. We do not know the mechanisms that improved survival and laboratory data in liver function, but the HGF, INF and IGF and IGFBP would be closely related with fibrolysis and collagenolysis.

Originally HGF has mitogenic, motogenic, morphogenic and anti-cell death activities (181), and identified and cloned as a mitogen protein for hepatocyte (182-183). HGF stimulates expression and activity of proteases involved in breakdown of ECM proteins, including urokinase-type plasminogen activator and matrix metalloproteinases. In LC, HGF suppress the proliferation while promoting apoptosis of α-SMA-positive cells in the liver, that histological resolution from liver cirrhosis.(181) Growth inhibition and promotion of apoptosis in portal myofibroblasts by HGF would be an ancillary resolution for liver fibrosis/cirrhosis.(184) Mesenchymal stem cells (MSCs) can prevent the development of liver fibrosis, and hepatocyte growth factor (HGF) can also attenuate liver cirrhosis. (185) HGF/MSCs significantly inhibit the formation of liver fibrosis in rats, while MSCs and HGF had synergistic effects in the process. The antifibrosis effect of HGF/MSCs may have contributed in modulating the activation and apoptosis of HSCs, elevating the rHGF expression level, and decreasing the TGF-beta 1 secretion of activated HSCs.(185). In CCl_4-induced liver cirrhosis rat model, hBMSCs treatment results were induced by two mechanisms that work together: the differentiation of transplanted hBMSCs into liver cells that are able to restore normal liver functions, and expression by production of MMP by hBMSCs which is involved in the repair of liver fibrosis. (186).

In interferon, there are many subtypes but only a few of them are reported to be produced from the UCBSCs. Among them, interferon alpha (IFN-α) suppress the progression of hepatic fibrosis (187), and lowers fibrosis scores, tissue hydroxyproline levels, liver TIMP-1 and elevate MMP-13 levels. IFN-α has role to increases HSC apoptosis, too(188). TIMP is the representative fibrosis favoring molecule and MMP is the representative molecule for fibrolysis. IFNγ have the hepatic protective effect by inhibition of adenosine A2A receptor function in hepatic stellate cells (191). Among those cytokines, M-CSF promotes the interferone production.

Insulin-like growth factor-1 (IGF-1) and major portion of the circulating IGF-1 synthesized in the liver are hepatic origin in normal state, but in LC the plasma levels are diminished (190-191). When the deficient IGF-1 in LC was replaced by daily administration of recombinant IGF-1, it induces a significant improvement of liver function (192). In liver, the expression of IGF-1 receptor is poor (193-194) and it seems that IGF-I acts on nonparenchymal cells. IGF-1 improves the liver structures and function through the activation of tissue-repair mechanism at non-parenchyma. It seems that there is an amplification loop which are favoring the efficacy of the therapy because the IGF-1 upregulate the IGF-1R in hepatic septa (195). In liver cirrhosis, the supplementation of IGF-1 induces the antifibrogenic and hepatoprotective effects. (196) Aside from the antifibrogenic effects, patients treated with stem cell showed improved serum albumin and enzyme levels indicating liver functions and these results are supposed due to IGF effects.

This discussion reviewed the mechanism that stem cell have favoring results related with LC. But it would not be the total mechanisms that explain the favoring effect of stem cells on LC. Among the above mentioned cytokines produced by UCBSCs, there are reports for characterization of cytokins, such as SDF-1(197) IL-8 (198) M-CSF (199), RANTES (200), MIP-1α (201), IP-10 (202) and EGF (203), but the most of them are not directly related with improvement of LC.

Until now, there are several clinical papers, around 10, on the results of autologous stem cell treatments, either autologous bone marrow derived stem cells or mobilized stem cells and fetal stem cell for the end stage liver cirrhotic patients. All of them are backing further clinical application of stem cell on liver cirrhosis. Most of the reports were related with the treatment with small number of patients, less than 10 patients, except for India (204). And the routes of stem cell infusion, aside from the mobilized hepatopoietic stem cells, are portal vein or hepatic artery with a few intravenous infusions. With these data, it is practically difficult to compare the superiority of the route of stem cell infusion due to small patient number in each case. Although there are common findings are improvement of hepatic conditions.

In the clinical data that we introduced in this chapter, all patients were supposed to be gone within 6 months before they were involved in stem cell therapy, meaning terminal state LC patients. Two third in 51 had the ascitis and more than one third had the hepatic encephalopathy. After single stem cell therapy using UCBSC, most of the hepatic encephalopathy patients and more than two third of ascitic patients were improved in symptom or responded to medical therapy who were refractory to previous therapy. But other data for liver function, albumin, aminotransferases, bilirubin or prothrombin time has irregular responses. About half of patients of ascites who had the improved symptoms showed improved albumin values but difficult to pull out any reproducible protocol, requiring further study. Because we choose the patients randomly in loose exclusion criteria, it would be difficult to get the objectively reproducible data from our experience for base of any protocol in stem cell therapy. But we can get some idea of trends of stem cell therapy for LC, although we had a single treatment. The stem cell therapy improves the ascitis and hepatic encephalopathy definitely in some group of patient, although we do not characterized this group, yet. In case of hepatic encephalopathy, the most of patients showed objectively improved symptom but we do not presume the mechanism, too. At the point of patient survival, all patients survived more than 6 months, although it would be difficult to believe the patients' initial life expectancy as the objective data. The patients' survival would not be related with liver function only but would be the summation of the various indexes of life signs. The most common side reaction that related with the stem cell treatment was pain, experiencing during stem cell infusion procedures. As mentioned before, much of the stem cell biology including producing cytokines has been proven already but it would be a tip of iceberg. We treated the patients who were regarded as the hopeless condition in point of survival improvement based on the reasonable solid animal study results of stem cell therapy, but they lack objective clinical efficacy aside from trend of stem cell therapy. So, to get the further acquisition of objective clinical effects and base on proper protocol, large multicenter trial on stem cell therapy for liver cirrhosis would be needed.

5. Conclusion

The conclusion of our review and experience is that there are lots of beneficial effects of stem cells on end stage liver cirrhosis and stem cell therapy serves for prolongation of the life and improvement of quality of life. The analysis of author's experience lacks objectivity. And if there is the more systemic multicenter large numbered study, we can make the proper guideline for stem cell therapy on liver cirrhosis.

6. References

[1] Heim D, Wege H. (2009) Hepatic stem and progenitor cells in liver diseases and hepatocarcinogenesis. Minerva Gastroenterol Dietol. 55(2):111-210, ISSN1827-1642.

[2] Kung JW, Forbes SJ. (2009) Stem cells and liver repair. Curr Opin Biotechnol. 20(5):568-74, ISSN 0958-1669.

[3] Bird TG, Lorenzini S, Forbes SJ. (2008) Activation of stem cells in hepatic diseases. Cell Tissue Res. 331(1):283-300, ISSN 0302-766X.

[4] Heim D, Wege H. (2009) Hepatic stem and progenitor cells in liver diseases and hepatocarcinogenesis. Minerva Gastroenterol Dietol. 55(2):111-121, ISSN 1121-421X.

[5] Manuelpillai U, Tchongue J, Lourensz D, Vaghjiani V, Samuel CS, Liu A, Williams ED, Sievert W. (2010) Transplantation of Human Amnion Epithelial Cells Reduces Hepatic Fibrosis in Immunocompetent CCl Treated Mice. Cell Transplant. 19(9):1157-68, ISSN 0963-6897.

[6] Friedman, SL. (2003) Liver fibrosis - from bench to bedside. *J. Hepatol.* 38(Suppl. 1):S38 S53, ISSN 0168-8278.

[7] Tasci I, Mas MR, Vural SA, Comert B, Alcigir G, Serdar M, Mas N, Isik AT, Ates Y. (2006) Rat liver fibrosis regresses better with pegylated interferon alpha2b and ursodeoxycholic acid treatments than spontaneous recovery. Liver Int. 26(2):261-268, ISSN 1478-3223.

[8] Lefton HB, Rosa A, Cohen M. (2009) Diagnosis and epidemiology of cirrhosis. Med Clin North Am. Jul;93(4):787-99, vii, ISSN 0025-7125.

[9] Fernández M, Semela D, Bruix J, Colle I, Pinzani M, Bosch J. (2009) Angiogenesis in liver disease. J Hepatol. 50(3):604-20, ISSN 0168-8278.

[10] Sandahl TD, Jepsen P, Thomsen KL, Vilstrup H. Incidence and mortality of alcoholic hepatitis in Denmark 1999-2008: A nationwide population based cohort study. J Hepatol. 2011 Apr;54(4):760-764, ISSN 0168-8278.

[11] Seeff LB. (2002) Natural history of chronic hepatitis C. Hepatology 36(5 suppl 1):S35-S46, ISSN 0270-9139.

[12] Wallaert B, Bonniere P, Prin L, Cortot A, Tonnel AB, Voisin C. (1986) Primary biliary cirrhosis. Subclinical inflammatory alveolitis in patients with normal chest roentgenograms. Chest. 90(6):842-8 ISSN 0012-3692.

[13] Oh S, Afdhal NH. (2001) Hepatic fibrosis: are any of the serum markers useful? Curr Gastroenterol Rep. 3(1):12-18. ISSN 1522-8037.

[14] Lu LG, Zeng MD, Mao YM, Li JQ, Qiu DK, Fang JY, Cao AP, Wan MB, Li CZ, Ye J, Cai X, Chen CW, Wang JY, Wu SM, Zhu JS, Zhou XQ. (2003) Relationship between clinical and pathologic findings in patients with chronic liver diseases. World J Gastroenterol. 9(12):2796-2800, ISSN 1007-9327.

[15] Chang ML, Yeh CT, Chang PY, Chen JC. (2005) Comparison of murine cirrhosis models induced by hepatotoxin administration and common bile duct ligation. World J Gastroenterol. 21;11(27):4167-72. ISSN 1007-9327.

[16] Issa R, Williams E, Trim N, Kendall T, Arthur MJ, Reichen J, Benyon RC, Iredale JP. (2001) Apoptosis of hepatic stellate cells: involvement in resolution of biliary fibrosis and regulation by soluble growth factors. Gut. 48(4):548-57. ISSN 0017-5749.

[17] Issa R, Zhou X, Constandinou CM, Fallowfield J, Millward-Sadler H, Gaca MD, Sands E, Suliman I, Trim N, Knorr A, Arthur MJ, Benyon RC, Iredale JP. (2004) Spontaneous recovery from micronodular cirrhosis: evidence for incomplete resolution associated with matrix cross-linking. Gastroenterology. 126(7):1795-1808. ISSN 0016-5085.

[18] Tasci I, Mas MR, Vural SA, Deveci S, Comert B, Alcigir G, Mas N, Akay C, Bozdayi M, Yurdaydin C, Bozkaya H, Uzunalimoglu O, Isik AT, Said HM. (2007) Pegylated interferon-alpha plus taurine in treatment of rat liver fibrosis. World J Gastroenterol. 21;13(23):3237-44. ISSN 1007-9327.

[19] Friedman, S.L. (1993) Seminar in medicine of the Beth Islael Hospital, Boston. The cellular basis of hepatic fibrosis. Mechanisms and treatment strategies. N Engl J Med 328, 1828-1835. ISSN 0028-4793.

[20] Geremias AT, Carvalho MA, Borojevic R, Monteiro AN. (2004) TGF beta1 and PDGF AA override collagen type I inhibition of proliferation in human liver connective tissue cells. BMC Gastroenterol. Dec 3;4:30. ISSN 1471-230X.

[21] Kharbanda KK, Rogers DD 2nd, Wyatt TA, Sorrell MF, Tuma DJ. (2004) Transforming growth factor-beta induces contraction of activated hepatic stellate cells. J Hepatol. 41(1):60-6. ISSN 0168-8278

[22] Parsons CJ, Takashima M, Rippe RA. (2007) Molecular mechanisms of hepatic fibrogenesis. J Gastroenterol Hepatol. 22 Suppl 1:S79-84. ISSN 0815-9319.

[23] Dooley S, Delvoux B, Lahme B, Mangasser-Stephan K, Gressner AM. (2000) Modulation of transforming growth factor beta response and signaling during transdifferentiation of rat hepatic stellate cells to myofibroblasts. Hepatology. 31(5):1094-106. ISSN 0270-9139.

[24] Chang YZ, Yang L, Yang CQ. (2008) Migration of hepatic stellate cells in fibrotic microenvironment of diseased liver model. Hepatobiliary Pancreat Dis Int. 7(4):401-5. ISSN 1499-3872.

[25] Friedman SL. (2003) Liver fibrosis – from bench to bedside. J Hepatol 38:S38–53, ISSN 0168-8278.

[26] Jesnowski R, Fürst D, Ringel J, Chen Y, Schrödel A, Kleeff J, Kolb A, Schareck WD, Löhr M. (2005) Immortalization of pancreatic stellate cells as an in vitro model of pancreatic fibrosis: deactivation is induced by matrigel and N-acetylcysteine. Lab Invest. 85(10):1276-91, ISSN 0023-6837.

[27] Rockey DC. (2001) Hepatic blood flow regulation by stellate cells in normal and injured liver. Semin Liver Dis. 21(3):337-49. ISSN 0272-8087

[28] Pinzani M, Milani S, De Franco R, Grappone C, Caligiuri A, Gentilini A, Tosti-Guerra C, Maggi M, Failli P, Ruocco C, Gentilini P. (1996) Endothelin 1 is overexpressed in human cirrhotic liver and exerts multiple effects on activated hepatic stellate cells. Gastroenterology. 110(2):534-48. ISSN 0016-5085

[29] Rockey DC, Chung JJ. (1998) Reduced nitric oxide production by endothelial cells in cirrhotic rat liver: endothelial dysfunction in portal hypertension. Gastroenterology. 114(2):344-51. ISSN 0016-5085

[30] Pinzani M, Marra F. (2001) Cytokine receptors and signaling in hepatic stellate cells. Semin Liver Dis. 21(3):397-416. ISSN 0272-8087.

[31] Marra F. (1999) Hepatic stellate cells and the regulation of liver inflammation. J Hepatol. 31(6):1120-30. ISSN 0168-8278.

[32] Friedman SL. (2000) Molecular regulation of hepatic fibrosis, an integrated cellular response to tissue injury. J Biol Chem. 275(4):2247-50. ISSN 0021-9258.

[33] Iredale JP, Benyon RC, Arthur MJ, Ferris WF, Alcolado R, Winwood PJ, Clark N, Murphy G. (1996) Tissue inhibitor of metalloproteinase-1 messenger RNA expression is enhanced relative to interstitial collagenase messenger RNA in experimental liver injury and fibrosis. Hepatology. 24(1):176-84. ISSN 0270-9139

[34] Murphy FR, Issa R, Zhou X, Ratnarajah S, Nagase H, Arthur MJ, Benyon C, Iredale JP. (2002) Inhibition of apoptosis of activated hepatic stellate cells by tissue inhibitor of metalloproteinase-1 is mediated via effects on matrix metalloproteinase inhibition: implications for reversibility of liver fibrosis. J Biol Chem. 29;277(13):11069-76, ISSN 0021-9258.

[35] Schouten J, Michielsen PP. (2007) Treatment of cirrhotic ascites. Acta Gastroenterol Belg. 70(2):217-22. ISSN 0001-5644

[36] Heidelbaugh JJ, Sherbondy M. (2006) Cirrhosis and chronic liver failure: part II. Complications and treatment. Am Fam Physician. 74(5):767-76, 0002-838X.

[37] Zhao Q, Ren H, Zhu D, Han Z. (2009) Stem/progenitor cells in liver injury repair and regeneration. Biol Cell. 101(10):557-71, ISSN 0248-4900.

[38] Kharaziha P, Hellström PM, Noorinayer B, Farzaneh F, Aghajani K, Jafari F, Telkabadi M, Atashi A, Honardoost M, Zali MR, Soleimani M. (2009) Improvement of liver function in liver cirrhosis patients after autologous mesenchymal stem cell injection: a phase I-II clinical trial. Eur J Gastroenterol Hepatol. 21(10):1199 205. ISSN 0954-691X

[39] Khan AA, Shaik MV, Parveen N, Rajendraprasad A, Aleem MA, Habeeb MA, Srinivas G, Raj TA, Tiwari SK, Kumaresan K, Venkateswarlu J, Pande G, Habibullah CM. (2010) Human fetal liver-derived stem cell transplantation as supportive modality in the management of end-stage decompensated liver cirrhosis. Cell Transplant. 19(4):409-18. ISSN 0963-6897

[40] Oertel M, Menthena A, Chen YQ, Teisner B, Jensen CH, Shafritz DA. (2008) Purification of fetal liver stem/progenitor cells containing all the repopulation potential for normal adult rat liver. Gastroenterology. 134(3):823-32 ISSN 0016-5085

[41] Abbasoglu O. (2008) Liver transplantation: yesterday, today and tomorrow. World J Gastroenterol. 14(20):3117-22. ISSN 1007-9327

[42] Gaudio E, Carpino G, Cardinale V, Franchitto A, Onori P, Alvaro D. (2009) New insights into liver stem cells. Dig Liver Dis. 41(7):455-62. ISSN 1590-8658

[43] Strom SC, Roy-Chowdhury J, Fox IJ. (1999) Hepatocyte transplantation for the treatment of human disease. Sem Liver Dis 19: 39–.48. ISSN 0272-8087.

[44] Allen KJ, Soriano H. (2001) Liver cell transplantation: The road to clinical application. J Lab Clin Med 138: 298–312. ISSN 0038-2299.

[45] Najimi M, Sokal E. (2004) Update on liver cell transplantation. J Pediatr Gastroenterol Nutr 39: 311–319. ISSN 0277-2116.

[46] Dhawan A, Mitry RR, Hughes RD. (2006) Hepatocyte transplantation for liver-based metabolic disorders. J Inherit Metab Dis 29: 431–.435 ISSN 0141-8955.

[47] Iredale JP, Benyon RC, Pickering J, McCullen M, Northrop M, Pawley S, Hovell C, Arthur MJ. (1998) Mechanisms of spontaneous resolution of rat liver fibrosis. Hepatic stellate cell apoptosis and reduced hepatic expression of metalloproteinase inhibitors. J Clin Invest. 102(3):538-549. ISSN 0021-9738

[48] Dai L, Ji H, Kong XW, Zhang YH. (2010) Antifibrotic effects of ZK14, a novel nitric oxide-donating biphenyldicarboxylate derivative, on rat HSC-T6 cells and CCl4-induced hepatic fibrosis. Acta Pharmacol Sin. 31(1):27-34. ISSN 1671-4083

[49] Iredale JP, Benyon RC, Pickering J, McCullen M, Northrop M, Pawley S, Hovell C, Arthur MJ. (1998) Mechanisms of spontaneous resolution of rat liver fibrosis.

Hepatic stellate cell apoptosis and reduced hepatic expression of metalloproteinase inhibitors. J Clin Invest. 102(3):538-549. ISSN 0021-9738
[50] Wright MC, Issa R, Smart DE, Trim N, Murray GI, Primrose JN, Arthur MJ, Iredale JP, Mann DA. (2001) Gliotoxin stimulates the apoptosis of human and rat hepatic stellate cells and enhances the resolution of liver fibrosis in rats. Gastroenterology. 121(3):685-698. ISSN 0016-5085
[51] Issa R, Williams E, Trim N, Kendall T, Arthur MJ, Reichen J, Benyon RC, Iredale JP. (2001) Apoptosis of hepatic stellate cells: involvement in resolution of biliary fibrosis and regulation by soluble growth factors. Gut. 48(4):548-557. ISSN 0017-5749
[52] Hartland SN, Murphy F, Aucott RL, Abergel A, Zhou X, Waung J, Patel N, Bradshaw C, Collins J, Mann D, Benyon RC, Iredale JP. (2009) Active matrix metalloproteinase-2 promotes apoptosis of hepatic stellate cells via the cleavage of cellular N-cadherin. Liver Int. 29(7):966-78. ISSN 1478-3223
[53] Leroy V, Monier F, Bottari S, Trocme C, Sturm N, Hilleret MN, Morel F, Zarski JP. (2004) Circulating matrix metalloproteinases 1, 2, 9 and their inhibitors TIMP-1 and TIMP-2 as serum markers of liver fibrosis in patients with chronic hepatitis C: comparison with PIIINP and hyaluronic acid. Am J Gastroenterol. 99(2):271-9. ISSN 0002-9270
[54] Issa R, Zhou X, Trim N, Millward-Sadler H, Krane S, Benyon C, Iredale J. (2003) Mutation in collagen-1 that confers resistance to the action of collagenase results in failure of recovery from CCl4-induced liver fibrosis, persistence of activated hepatic stellate cells, and diminished hepatocyte regeneration. FASEB J. 17(1):47-9. ISSN 0892-6638
[55] CameronGR, Karunaratne WAE. (1936) Carbon tetrachloride cirrhosis in relation to liver regeneration. J Path Bact 42: 1-21, 0368-3494.
[56] Quinn PS, Higginson J. (1965) Reversible and irreversible changes in experimental cirrhosis. Am J Pathol 47: 353-369. ISSN 0002-9440
[57] Morrione TG, Levine J: Collagenolytic activity and collagen resorption in experimental cirrhosis. Arch Pathol 1967; 84: 59-63, 0363-0153.
[58] Hutterer F, RubinE, Popper H. (1964) Mechanism of collagen resorption in reversible hepatic fibrosis. Exp Mol Pathol 86: 215-223 ISSN 0014-4800
[59] Takada A, Porta EA, Hartroft WS. (1967) The recovery of experimental dietary cirrhosis. Am J Pathol 51: 929-957, 0002-9440.
[60] Jacques WE, McAdams AI. (1957) Reversible biliary cirrhosis in rat after partial ligation of common bile duct. AMA Arch Path 63:149-153, ISSN 0096-6711.
[61] Okazaki I, Oda M, Maruyama K, Funatsu eversible biliary cirrhosis in rat after partial ligation of common bile duct. AMA Arch Path 63: 149-153 K, Matsuzaki S, Kamegaya K, Tsuchiya M. (1974) Mechanism of collagen resorption in experimental hepatic fibrosis with special reference to the activity of lysosomal enzymes. Biochem Exp Biol 11: 15-28, 0366-0060.
[62] RubinE, Hutterer F: Hepatic fibrosis. (1967) Studies in the formation and resorption. In: Wagner BM, Smith DE, eds, The Connective Tissue, Baltimore, Williams and Wilkins, 142-160
[63] Perez-Tamayo R. (1965) Some aspects of connective tissue of the liver. In: Popper H, Schaffner F, eds, Progress in Liver Diseases 2, New York, Grune & Stratton, 192-210
[64] Rojkind M, Dunn MA. (1979) Hepatic fibrosis. Gastroenterology 76: 849-863, ISSN 0016-5085.

[65] Maruyama K, Okazaki I, Kashiwazaki K, Oda M, Ishii H, Tsuchiya M. (1981) A case of subacute hepatitis with reversible liver fibrosis. Gastroenterol Jpn 16: 611-615, ISSN 0435-1339.

[66] Arai M, Niioka M, Maruyama K, Wada N, Fujimoto N, Nomiyama T, Tanaka S, Okazaki I. (1996) Changes in serum levels of metalloproteinases and their inhibitors by treatment of chronic hepatitis C with interferon. Dig Dis Sci 41: 995-1000, ISSN 0163-2116.

[67] Guettier C. (2005) Which stem cells for adult liver? Ann Pathol. 25(1):33-44, ISSM 0242-6498.

[68] Shackel N, Rockey D. (2005) In pursuit of the ''Holy Grail''—stem cells, hepatic injury, fibrogenesis and repair. *Hepatology* 41: 16–18, ISSN 0270-9139.

[69] Vieyra DS, Jackson KA, Goodell MA. (2005) Plasticity and tissue regenerative potential of bone marrow-derived cells. *Stem Cell Rev.* 1: 65–69, ISSN 1550-8943.

[70] Filip S, English D, Mokry J. (2004) Issues in stem cell plasticity. *J. Cell. Mol. Med.* 8: 572–7, ISSN 1582-1838.

[71] Smart N, Riley PR. (2008) The stem cell movement. Circ Res. 102(10):1155-68, ISSN 0009-7330.

[72] Eckersley-Maslin MA, Warner FJ, Grzelak CA, McCaughan GW, Shackel NA. (2009) Bone marrow stem cells and the liver: are they relevant? J Gastroenterol Hepatol. 24(10):1608-16, ISSN 0815-9316.

[73] Oertel M, Menthena A, Chen YQ, Teisner B, Jensen CH, Shafritz DA. (2008) Purification of fetal liver stem/progenitor cells containing all the repopulation potential for normal adult rat liver. Gastroenterology. 134(3):823-832, ISSN 0016-5085.

[74] Zhao Q, Ren H, Li X, Chen Z, Zhang X, Gong W, Liu Y, Pang T and Han ZC. (2009) Differentiation of human umbilical cord mesenchymal stromal cells into low immunogenic hepatocyte-like cells. Cytotherapy 11(4):414-426, ISSN 1465-3249.

[75] Weiss ML, Medicetty S, Bledsoe AR, Rachakatla RS, Choi M, Merchav S, Luo Y, Rao MS, Velagaleti G, Troyer D. (2006) Human umbilical cord matrix stem cells: preliminary characterization and effect of transplantation in a rodent model of Parkinson's disease. Stem Cells 24: 781 - 92. ISSN 0250-6793

[76] Medicetty S, Bledsoe AR, Fahrenholtz CB, Troyer D, Weiss ML. (2004) Transplantation of pig stem cells into rat brain: proliferation during the first 8 weeks. Exp Neurol 190: 32 - 41, ISSN 0014-4886.

[77] Fu YS, Cheng YC, Lin MY, Cheng H, Chu PM, Chou SC, Shih YH, Ko MH, Sung MS. (2006) Conversion of human umbilical cord mesenchymal stem cells in Wharton's jelly to dopaminergic neurons in vitro : potential therapeutic application for parkinsonism. Stem Cells 24: 115 - 124. ISSN 0250-6793

[78] Kögler G, Sensken S, Airey JA, Trapp T, Müschen M, Feldhahn N, Liedtke S, Sorg RV, Fischer J, Rosenbaum C, Greschat S, Knipper A, Bender J, Degistirici O, Gao J, Caplan AI, Colletti EJ, Almeida-Porada G, Müller HW, Zanjani E, Wernet P. (2004) A new human somatic stem cell from placental cord blood with intrinsic pluripotent differentiation potential. J Exp Med 200:123-135, ISSN 0022-1007.

[79] Strauer BE, Brehm M, Zeus T, Bartsch T, Schannwell C, Antke C, Sorg RV, Kögler G, Wernet P, Müller HW, Köstering M. (2005) Regeneration of human infarcted heart muscle by intracoronary autologous bone marrow cell transplantation in chronic coronary artery disease: IACT study. J Am Coll Cardiol 46:1651-1658, ISSN 0735-1097.

[80] Yoo KJ, Li RK, Weisel RD, Mickle DA, Jia ZQ, Kim EJ, Tomita S, Yau TM. (2000) Heart cell transplantation improves heart function in dilated cardiomyopathic hamsters. Circulation 102:III204-III209, ISSN 0009-7322.
[81] Kim SW, Han H, Chae GT, Lee SH, Bo S, Yoon JH, Lee YS, Lee KS, Park HK, Kang KS. (2006) Successful stem cell therapy using umbilical cord blood-derived multipotent stem cells for Bueger's disease and ischemic limb disease animal model. Stem Cells 24:1620-1626, ISSN 0250-6793.
[82] Lim JH, Byeon YE, Ryu HH, Jeong YH, Lee YW, Kim WH, Kang KS, Kweon OK. (2007) Transplantation of canine umbilical cord blood-derived mesenchymal stem cells in experimentally induced spinal cord injured dogs. J Vet Sci 8:275-282, ISSN 1229-845X
[83] Min JY, Yang Y, Sullivan MF, Ke Q, Converso KL, Chen Y, Morgan JP, Xiao YF. (2003) Long-term improvement of cardiac function in rats after infarction by transplantation of embryonic stem cells. Thorac Cardiovasc Surg 125:361-369. ISSN 0022-5223.
[84] Gordon MY, Gordon-Smith EC. (1981) Bone marrow fibroblastoid colony-forming cells (F-CFC) in aplastic anaemia: colony growth and stimulation of granulocyte-macrophage colony-forming cells (GM-CFC). Br J Haematol. 49(3):465-77, ISSN 0006-0291
[85] Kadivar M, Khatami S, Mortazavi Y, Shokrgozar MA, Taghikhani M, Soleimani M. (2006) In vitro cardiomyogenic potential of human umbilical vein-derived mesenchymal stem cells. Biochem Biophys Res Commun. 340(2):639-47, ISSN 0006-291X
[86] Reinecke H, Minami E, Zhu WZ, Laflamme MA. (2008) Cardiogenic differentiation and transdifferentiation of progenitor cells. Circ Res. 103(10):1058-71, ISSN 0009-7330.
[87] Mishra PK, Singh SR, Joshua IG, Tyagi SC. (2010) Stem cells as a therapeutic target for diabetes. Front Biosci. 15:461-77, ISSN 1945-0494.
[88] Toyoda M, Cui Ch, Umezawa A. (2007) Myogenic transdifferentiation of menstrual blood-derived cells. Acta Myol. 26(3):176-178, ISSN 1128-2460.
[89] Bussolati B, Camussi G. (2007) Stem cells in acute kidney injury. Contrib Nephrol. 156:250-258, ISSN 0302-5114.
[90] Bahk JY, Jung JH, Han H, Min SK, Lee YS. (2010) Treatment of diabetic impotence with umbilical cord blood stem cell intracavernosal transplant: preliminary report of 7 cases. Exp Clin Transplant. 8(2):150-60 ISSN 1304-0855.
[91] Nikeghbalian S, Pournasr B, Aghdami N, Rasekhi A, Geramizadeh B, Hosseini Asl SM, Ramzi M, Kakaei F, Namiri M, Malekzadeh R, Vosough Dizaj A, Malek-Hosseini SA, Baharvand H. (2011) Autologous transplantation of bone marrow-derived mononuclear and CD133(+) cells in patients with decompensated cirrhosis. Arch Iran Med. 14(1):12-7, ISSN 1029-2977.
[92] Ren C, Kumar S, Chanda D, Kallman L, Chen J, Mountz JD, Ponnazhagan S. (2008) Cancer gene therapy using mesenchymal stem cells expressing interferon-beta in a mouse prostate cancer lung metastasis model. Gene Ther. 15(21):1446-53, ISSN 0969-7128.
[93] Dan YY, Yeoh GC. (2008) Liver stem cells: a scientific and clinical perspective. J Gastroenterol Hepatol. 23(5):687-698, ISSN 0815-9319.
[94] Thomson JA, Kalishman J, Golos TG, Durning M, Harris CP, Becker RA, Hearn JP. (1995) Isolation of a primate embryonic stem cell line. Proc Natl Acad Sci U S A. 92(17):7844-8, ISSN 0027-8424.

[95] Takahashi K, Yamanaka S. (2006) Induction of pluripotent stem cells from mouse embryonic and adult fibroblast cultures by defined factors. Cell. 126(4):663-76, ISSN 0062-8674.
[96] Yoshimura K, Sato K, Aoi N, Kurita M, Inoue K, Suga H, Eto H, Kato H, Hirohi T, Harii K. (2008) Cell-assisted lipotransfer for facial lipoatrophy: efficacy of clinical use of adipose-derived stem cells. Dermatol Surg. 34(9):1178-85, ISSN 1076-0512.
[97] Alvarez-Buylla A, Lois C. (1995) Neuronal stem cells in the brain of adult vertebrates. Stem Cells. 13(3):263-72, ISSN 0250-6793.
[98] Ohyama M. (2007) Hair follicle bulge: A fascinating reservoir of epithelial stem cells. J Dermatol Sci 46, 81-89, ISSN 0923-1811.
[99] Crigler L, Kazhanie A, Yoon TJ, Zakhari J, Anders J, Taylor B, and Virador VM. (2007) Isolation of a mesenchymal cell population from murine dermis that contains progenitors of multiple cell lineages. FASEB J. 21(9): 2050--2063, ISSN 0892-6638.
[100] van der Bogt KEA, Sheikh AY, Schrepfer S, Hoyt G, Cao F, Ransohoff KJ, Swijnenburg J, Pearl J, Lee A, Fischbein M, Contag CH, Robbins RC, Wu JC. (2008) Comparison of Different Adult Stem Cell Types for Treatment of Myocardial Ischemia. Circulation 118;S121-S129, ISSN 0009-7322.
[101] Gluckman E, Broxmeyer HA, Auerbach AD, Friedman HS, Douglas GW, Devergie A, Esperou H, Thierry D, Socie G, Lehn, P. (1989) Hematopoietic reconstitutions in a patient with fanconi''s anemia by means of umbilical-cord blood fro an HLA identical sibling. N. Engl. J. Med., 321(17), 1174-1178, ISSN 0028-4793.
[102] Bradley MB, Cairo MS. (2005) Cord blood immunology and stem cell transplantation. Hum Immunol. 66(5):431-46, ISSN 0198-8859.
[103] Bieback K, Kern S, Kocaömer A, Ferlik K, Bugert P. (2008) Comparing mesenchymal stromal cells from different human tissues: bone marrow, adipose tissue and umbilical cord blood. Biomed Mater Eng. 18(1 Suppl):S71-6, ISSN 0959-2989.
[104] Tholpady SS, Katz AJ, Ogle RC. (2003) Mesenchymal stem cells from rat visceral fat exhibit multipotential differentiation in vitro. Anat Rec A Discov Mol Cell Evol Biol. May;272(1):398-402, ISSN 1552-4884.
[105] Escolar ML, Poe MD, Provenzale JM, Richards KC, Allison J, Wood S, Wenger DA, Pietryga D, Wall D, Champagne M, Morse R, Krivit W, Kurtzberg J. (2005) Transplantation of umbilical-cord blood in babies with infantile Krabbe's disease. N Engl J Med 352: 2069, ISSN 0028-4793.
[106] Reffelmann T, Koenemann S, Kloner RA. (2009) Promise of blood-and bone marrow-derived stem cell transplantation for functional cardiac repair: putting it in perspective with existing therapy. J Am coll Cardiool 53: 305, ISSN 0735-1097.
[107] Lim JH, Byeon E, Ryu HH, Jeong YH, Lee YW, Kim WH, *et al.* (2007) Transplantation of canine umbilical cord blood-derived mesenchymal stem cells in experimentally induced spinal cord injured dogs. J Vet Sci 8: 275, ISSN 1229-845X.
[108] Zhang QH, She MP. (2007) Biological behaviour and role of endothelial progenitor cells in vascular diseases. Chin Med J (Engl). 120:2297, ISSN 0366-6999.
[109] Kim Kim SW, Han H, Chae GT, Lee SH, Bo S, Yoon JH, Lee YS, Lee KS, Park HK, Kang KS. (2006) Successful stem cell therapy using umbilical cord blood-derived multipotent stem cells for Bueger's disease and ischemic limb disease animal model. Stem Cells 24: 1620. ISSN 0250-6793
[110] Savitz SI, Rosenbaum DM, Dinsmore JH, Wechsler LR, Caplan LR. (2002) Cell Transplantation for Stroke. Ann Neurol 52: 266, ISSN 0364-5134.
[111] Gomes ME, Bossano CM, Johnston CM, Reis RL, Mikos AG. (2006) In vitro localization of bone growth factors in constructs of biodegradable scaffolds seeded with

marrow stromal cells and cultured in a flow perfusion bioreactor. Tissue Eng. 12(1):177-88, ISSN 1076-3279.
[112] Lee MY, Huang JP, Chen YY, Aplin JD, Wu YH, Chen CY, Chen PC, Chen CP. (2009) Angiogenesis in differentiated placental multipotent mesenchymal stromal cells is dependent on integrin alpha5beta1. PLoS One. 22;4(10):e6913, ISSN 1392-6203.
[113] Fan CG, Zhang QJ, Tang FW, Han ZB, Wang GS, Han ZC. (2005) Human umbilical cord blood cells express neurotrophic factors. Neurosci Lett. 3;380(3):322-5, ISSN 0304-3940.
[114] Youssoufian M, Walmsley B. (2007) Brain-derived neurotrophic factor modulates cell excitability in the mouse medial nucleus of the trapezoid body. Eur J Neurosci. 25(6):1647-52, ISSN 0953-816X.
[115] Eton D, Yu H. (2010) Enhanced cell therapy strategy to treat chronic limb-threatening ischemia. J Vasc Surg. 52(1):199-204, ISSN 0741-5214.
[116] Chang SA, Kang HJ, Lee HY, Kim KH, Hur J, Han KS, Park YB, Kim HS. (2009) Peripheral blood stem cell mobilisation by granulocyte-colony stimulating factor in patients with acute and old myocardial infarction for intracoronary cell infusion. Heart. 95(16):1326-30, ISSN 1355-6037.
[117] Kang HJ, Kim HS. (2008) G-CSF- and erythropoietin-based cell therapy: a promising strategy for angiomyogenesis in myocardial infarction. Expert Rev Cardiovasc Ther. 6(5):703-13, ISSN 1477-9072.
[118] Savitz SI, Chopp M, Deans R, Carmichael ST, Phinney D, Wechsler L. (2011) Stem Cell Therapy as an Emerging Paradigm for Stroke (STEPS) II. Stroke. 42(3):825-9, ISSN 0039-2499.
[119] Urbaniak Hunter K, Yarbrough C, Ciacci J. (2010) Gene- and cell-based approaches for neurodegenerative disease. Adv Exp Med Biol. 671:117-30, ISSN 0065-2598.
[120] Lee HJ, Lee JK, Lee H, Shin JW, Carter JE, Sakamoto T, Jin HK, Bae JS. (2010) The therapeutic potential of human umbilical cord blood-derived mesenchymal stem cells in Alzheimer's disease. Neurosci Lett. 30;481(1):30-5, ISSN 0304-3940.
[121] Fornai F, Meininger V, Silani V. (2011) Future therapeutical strategies dictated by pre-clinical evidence in ALS. Arch Ital Biol. 149(1):169-74, ISSN 0003-9829.
[122] Ramachandran P, Iredale JP. (2009) Reversibility of liver fibrosis. Ann Hepatol. 8(4):283-91, ISSN 1665-2681.
[123] Durante AJ, St Louis T, Meek JI, Navarro VJ, Sofair AN. (2008) The mortality burden of chronic liver disease may be substantially underestimated in the United States. Conn Med. 72(7):389-92, ISSN 0011-6178.
[124] Hui AY, Friedman SL. (2003) Molecular basis of hepatic fibrosis. Expert Rev Mol Med. 5(5):1-23, ISSN 1462-3994.
[125] Li JT, Liao ZX, Ping J, Xu D, Wang H. (2008) Molecular mechanism of hepatic stellate cell activation and antifibrotic therapeutic strategies. J Gastroenterol. 43(6):419-28, ISSN 0944-1774.
[126] Friedman SL. (1999) Cytokines and fibrogenesis. Semin Liver Dis. 19(2):129-40, ISSN 0272-8087.
[127] Han YP. (2006) Matrix metalloproteinases, the pros and cons, in liver fibrosis. J Gastroenterol Hepatol. 21 Suppl 3:S88-91, ISSN 0815-9319.
[128] Hartland SN, Murphy F, Aucott RL, Abergel A, Zhou X, Waung J, Patel N, Bradshaw C, Collins J, Mann D, Benyon RC, Iredale JP. (2009) Active matrix metalloproteinase-2 promotes apoptosis of hepatic stellate cells via the cleavage of cellular N-cadherin. Liver Int. 29(7):966-78, ISSN 1478-3223.

[129] Povero D, Busletta C, Novo E, di Bonzo LV, Cannito S, Paternostro C, Parola M. (2010) Liver fibrosis: a dynamic and potentially reversible process. Histol Histopathol. 25(8):1075-91, ISSN 0213-3911.

[130] Cantz T, Sharma AD, Jochheim-Richter A, Arseniev L, Klein C, Manns MP, Ott M. (2004) Reevaluation of bone marrow-derived cells as a source for hepatocyte regeneration. Cell Transplant. 13(6):659-66, ISSN 0963-6897.

[131] Friedman SL. (1999) Cytokines and fibrogenesis. Semin Liver Dis. 19(2):129-40, ISSN 0272-8087.

[132] Jiang JX, Venugopal S, Serizawa N, Chen X, Scott F, Li Y, Adamson R, Devaraj S, Shah V, Gershwin ME, Friedman SL, Török NJ. (2010) Reduced nicotinamide adenine dinucleotide phosphate oxidase 2 plays a key role in stellate cell activation and liver fibrogenesis in vivo. Gastroenterology. 139(4):1375-84. 0016-5085.

[133] Myung SJ, Yoon JH, Kim BH, Lee JH, Jung EU, Lee HS. (2009) Heat shock protein 90 inhibitor induces apoptosis and attenuates activation of hepatic stellate cells. J Pharmacol Exp Ther. 330(1):276-282, ISSN 0022-3565.

[134] Han YP. (2006) Matrix metalloproteinases, the pros and cons, in liver fibrosis. J Gastroenterol Hepatol. 21 Suppl 3:S88-91, ISSN 0815-9319.

[135] Langer DA, Das A, Semela D, Kang-Decker N, Hendrickson H, Bronk SF, Katusic ZS, Gores GJ, Shah VH. (2008) Nitric oxide promotes caspase-independent hepatic stellate cell apoptosis through the generation of reactive oxygen species. Hepatology. 47(6):1983-93, ISSN 0270-9139.

[136] Liu Y, Wen XM, Lui EL, Friedman SL, Cui W, Ho NP, Li L, Ye T, Fan ST, Zhang H. (2009) Therapeutic targeting of the PDGF and TGF-beta-signaling pathways in hepatic stellate cells by PTK787/ZK22258. Lab Invest. 89(10):1152-60, ISSN 0023-6837.

[137] Gonzalo T, Beljaars L, van de Bovenkamp M, Temming K, van Loenen AM, Reker-Smit C, Meijer DK, Lacombe M, Opdam F, Kéri G, Orfi L, Poelstra K, Kok RJ. (2007) Local inhibition of liver fibrosis by specific delivery of a platelet-derived growth factor kinase inhibitor to hepatic stellate cells. J Pharmacol Exp Ther. 321(3):856-65, ISSN 0022-3565.

[138] Yoshiji H, Noguchi R, Kuriyama S, Ikenaka Y, Yoshii J, Yanase K, Namisaki T, Kitade M, Masaki T, Fukui H. (2005) Imatinib mesylate (STI-571) attenuates liver fibrosis development in rats. Am J Physiol Gastrointest Liver Physiol. 288(5):G907-13, ISSN 0193-1857.

[139] Hartland SN, Murphy F, Aucott RL, Abergel A, Zhou X, Waung J, Patel N, Bradshaw C, Collins J, Mann D, Benyon RC, Iredale JP. (2009) Active matrix metalloproteinase-2 promotes apoptosis of hepatic stellate cells via the cleavage of cellular N-cadherin. Liver Int. 29(7):966-78, ISSN 1478-3223.

[140] Wang Y, Gao J, Zhang D, Zhang J, Ma J, Jiang H. (2010) New insights into the antifibrotic effects of sorafenib on hepatic stellate cells and liver fibrosis. J Hepatol. 53(1):132-144, ISSN 0168-8278.

[141] Tao J, Mallat A, Gallois C, Belmadani S, Méry PF, Nhieu JT, Pavoine C, Lotersztajn S. (1999) Biological effects of C-type natriuretic peptide in human myofibroblastic hepatic stellate cells. J Biol Chem. 20;274(34):23761-9, ISSN 0021-9258.

[142] Povero D, Busletta C, Novo E, di Bonzo LV, Cannito S, Paternostro C, Parola M. (2010) Liver fibrosis: a dynamic and potentially reversible process. Histol Histopathol. 25(8):1075-91, ISSN 0213-3911.

[143] Cantz T, Sharma AD, Jochheim-Richter A, Arseniev L, Klein C, Manns MP, Ott M. (2004) Reevaluation of bone marrow-derived cells as a source for hepatocyte regeneration. Cell Transplant. 13(6):659-66, ISSN 0963-6897.

[144] Kharaziha P, Hellstroem PM, Noorinayerb B, Farzanehd F, Aghajanib K, Jafarib F, Telkabadib M, Atashid A, Honardoostd M, Zalic MR and Soleimanie M. (2009) Improvement of liver function in liver cirrhosis patients after autologous mesenchymal stem cell injection: a phase I--II clinical trial. Eur J Gastroenterol Hepatol 21:1199--1205, ISSN 0954-691X

[145] Shackel N, Rockey D. (2005) In pursuit of the ''Holy Grail''--stem cells, hepatic injury, fibrogenesis and repair. *Hepatology* 41: 16-18, ISSN 0279-9139.

[146] Jensen CH, Jauho EI, Santoni-Rugiu E, Holmskov U, Teisner B, Tygstrup N, Bisgaard HC. (2004) Transit-amplifying ductular (oval) cells and their hepatocytic progeny are characterized by a novel and distinctive expression of delta-like protein/ preadipocyte factor 1/fetal antigen 1. *Am. J. Pathol.* 164: 1347--1359. ISSN 0002-9440.

[147] Khurana S, Mukhopadhyay A. (2007) Characterization of the potential subpopulation of bone marrow cells involved in the repair of injured liver tissue. Stem Cells 25: 1439--47, ISSN 0250-6793.

[148] Stenderup K, Justesen J, Clausen C, Kassem M. (2003) Aging is associated with decreased maximal life span and accelerated senescence of bone marrow stromal cells. Bone. 33:919–26. ISSN 8756-3282.

[149] Mueller SM, Glowacki J. (2001) Age-related decline in the osteogenic potential of human bone marrow cells cultured in three-dimensional collagen sponges. J Cell Biochem. 82:583–90, ISSN 0730-2312.

[150] Zhong YS, Lin N, Deng MH, Zhang FC, Tang ZF, Xu RY. (2010) Deficient proliferation of bone marrow-derived mesenchymal stem cells in patients with chronic hepatitis B viral infections and cirrhosis of the liver. Dig Dis Sci. 55(2):438-45. ISSN 0163-2116.

[151] Zuk PA, Zhu M, Mizuno H, Huang J, Futrell JW, Katz AJ, Benhaim P, Lorenz HP, Hedrick MH. (2001) Multilineage cells from human adipose tissue: Implications for cell-based therapies. Tissue Eng 7:211–228. ISSN 1937-3341.

[152] Lee RH, Kim B, Choi I, Kim H, Choi HS, Suh K, Bae YC, Jung JS. (2004) Characterization and expression analysis of mesenchymal stem cells from human bone marrow and adipose tissue. Cell Physiol Biochem 14:311–324. ISSN 1015-8987

[153] Gluckman E, Rocha V, Boyer-Chammard A, Locatelli F, Arcese W, Pasquini R, Ortega J, Souillet G, Ferreira E, Laporte JP, Fernandez M, Chastang C. (1997) Outcome of cord-blood transplantation from related and unrelated donors. Eurocord Transplant Group and the European Blood and Marrow Transplantation Group. N Engl J Med. 337:373–81, ISSN 0028-4793.

[154] Grewal SS, Barker JN, Davies SM, Wagner JE. (2003) Unrelated donor hematopoietic cell transplantation: marrow or umbilical cord blood. Blood. 101:4233–44, ISSN 0006-4971.

[155] Warkentin PI. (2003) Foundation for the Accreditation of Cellular Therapy Voluntary accreditation of cellular therapies: Foundation for the Accreditation of Cellular Therapy (FACT) Cytotherapy. 5:299–305, ISSN 1465-3249.

[156] Chen MY, Lie PC, Li ZL, Wei X. (2009) Endothelial differentiation of Wharton's jelly-derived mesenchymal stem cells in comparison with bone marrow-derived mesenchymal stem cells. Exp Hematol. 37(5):629-40, ISSN 0301-482X.

[157] Zhong YS, Lin N, Deng MH, Zhang FC, Tang ZF, Xu RY. (2010) Deficient proliferation of bone marrow-derived mesenchymal stem cells in patients with chronic hepatitis B viral infections and cirrhosis of the liver. Dig Dis Sci. 55(2):438-45, ISSN 0163-2116.

[158] Moriya K, Yoshikawa M, Ouji Y, Saito K, Nishiofuku M, Matsuda R, Ishizaka S, Fukui H. (2008) Embryonic stem cells reduce liver fibrosis in CCl4-treated mice. Int J Exp Pathol. 89(6):401-9, ISSN 0959-9673.

[159] Liu P, Liu ZD, Wu N, cong X, Fei R, Chen HS, Wei L. (2009) Transplanted Endothelial Progenitor Cells Ameliorate Carbon Tetrachloride–.Induced Liver Cirrhosis in Rats. Liver Transpl 15:1092-1100, ISSN 1527-6465.

[160] Yan Y, Xu W, Qian H, Si Y, Zhu W, Cao H, Zhou H, Mao F. (2009) Mesenchymal stem cells from human umbilical cords ameliorate mouse hepatic injury in vivo. Liver Int. 29(3):356-65, ISSN 1478-4223.

[161] Mohamadnejad M, Alimoghaddam K, Mohyeddin-Bonab M, Bagheri M, Bashtar M, Ghanaati H, Baharvand H, Ghavamzadeh A, Malekzadeh R. (2007) Phase 1 trial of autologous bone marrow mesenchymal stem cell transplantation in patients with decompensated liver cirrhosis. Arch Iran Med. 10(4):459-66, ISSN 1029-2977.

[162] Li M, Yu J, Li Y, Li D, Yan D, Qu Z, Ruan Q. (2010) CXCR4 positive bone mesenchymal stem cells migrate to human endothelial cell stimulated by ox-LDL via SDF-1alpha/CXCR4 signaling axis. Exp Mol Pathol. 88(2):250-5, ISSN 0014-4800.

[163] Magnasco A, Corselli M, Bertelli R, Ibatici A, Peresi M, Gaggero G, Cappiello V, Chiavarina B, Mattioli G, Gusmano R, Ravetti JL, Frassoni F, Ghiggeri GM. (2008) Mesenchymal stem cells protective effect in adriamycin model of nephropathy. Cell Transplant. 17(10-11):1157-67, ISSN 0963-6897.

[164] Zhao X, Huang L, Yin Y, Fang Y, Zhou Y. (2007) Autologous endothelial progenitor cells transplantation promoting endothelial recovery in mice. Transpl Int. 20(8):712-21, ISSN 0943-0874.

[165] Di Campli C, Piscaglia AC, Rutella S, Bonanno G, Vecchio FM, Zocco MA, Monego G, Michetti F, Mancuso S, Pola P, Leone G, Gasbarrini G, Gasbarrini A. (2005) Improvement of mortality rate and decrease in histologic hepatic injury after human cord blood stem cell infusion in a murine model of hepatotoxicity. Transplant Proc. 37(6):2707-10, ISSN 0041-1345.

[166] Okura H, Saga A, Fumimoto Y, Soeda M, Moriyama M, Moriyama H, Nagai K, Lee CM, Yamashita S, Ichinose A, Hayakawa T, Matsuyama A. (2011) Transplantation of human adipose tissue-derived multilineage progenitor cells reduces serum cholesterol in hyperlipidemic watanabe rabbits. Tissue Eng Part C Methods. 17(2):145-54, ISSN 1937-3384.

[167] Kang HJ, Kim MK, Kim MG, Choi DJ, Yoon JH, Park YB, Kim HS. (2011) A multicenter, prospective, randomized, controlled trial evaluating the safety and efficacy of intracoronary cell infusion mobilized with granulocyte colony-stimulating factor and darbepoetin after acute myocardial infarction: study design and rationale of the 'MAGIC cell-5-combination cytokine trial'. Trials. 7;12(1):33, ISSN 1745-6215.

[168] Hare JM, Traverse JH, Henry TD, Dib N, Strumpf RK, Schulman SP, Gerstenblith G, DeMaria AN, Denktas AE, Gammon RS, Hermiller JB Jr, Reisman MA, Schaer GL, Sherman W. (2009) A randomized, double-blind, placebo-controlled, dose-escalation study of intravenous adult human mesenchymal stem cells (prochymal) after acute myocardial infarction. J Am Coll Cardiol. 54(24):2277-86, ISSN 0735-1097.

[169] Chang SA, Kang HJ, Lee HY, Kim KH, Hur J, Han KS, Park YB, Kim HS. (2009) Peripheral blood stem cell mobilisation by granulocyte-colony stimulating factor in patients with acute and old myocardial infarction for intracoronary cell infusion. Heart. 95(16):1326-30, ISSN 1335-6037.
[170] Zohlnhöfer D, Dibra A, Koppara T, de Waha A, Ripa RS, Kastrup J, Valgimigli M, Schömig A, Kastrati A. (2008) Stem cell mobilization by granulocyte colony-stimulating factor for myocardial recovery after acute myocardial infarction: a meta-analysis. J Am Coll Cardiol. 51(15):1429-37, ISSN 0735-1097.
[171] Fan L, Chen L, Chen X, Fu F. (2008) A meta-analysis of stem cell mobilization by granulocyte colony-stimulating factor in the treatment of acute myocardial infarction. Cardiovasc Drugs Ther. 22(1):45-54, ISSN 0920-3206.
[172] Zhang S, Sun A, Xu D, Yao K, Huang Z, Jin H, Wang K, Zou Y, Ge J. (2009) Impact of timing on efficacy and safetyof intracoronary autologous bone marrow stem cells transplantation in acute myocardial infarction: a pooled subgroup analysis of randomized controlled trials. Clin Cardiol. 32(8):458-66, ISSN 0160-9289.
[173] Sun L, Zhang T, Lan X, Du G. (2010) Effects of stem cell therapy on left ventricular remodeling after acute myocardial infarction: a meta-analysis. Clin Cardiol. 33(5):296-302, ISSN 0160-9289.
[174] Pendharkar AV, Chua JY, Andres RH, Wang N, Gaeta X, Wang H, De A, Choi R, Chen S, Rutt BK, Gambhir SS, Guzman R. (2010) Biodistribution of neural stem cells after intravascular therapy for hypoxic-ischemia. Stroke. 41(9):2064-70, ISSN 0093-2499.
[175] Harting MT, Jimenez F, Xue H, Fischer UM, Baumgartner J, Dash PK, Cox CS. (2009) Intravenous mesenchymal stem cell therapy for traumatic brain injury. J Neurosurg. 110(6):1189-97, ISSN 0022-3085.
[176] Feng Q, Chow PK, Frassoni F, Phua CM, Tan PK, Prasath A, Khee Hwang WY. (2008) Nonhuman primate allogeneic hematopoietic stem cell transplantation by intraosseus vs intravenous injection: Engraftment, donor cell distribution, and mechanistic basis. Exp Hematol. 36(11):1556-66, ISSN 0301-472X.
[177] Hohenstein B, Kuo MC, Addabbo F, Yasuda K, Ratliff B, Schwarzenberger C, Eckardt KU, Hugo CP, Goligorsky MS. (2010) Enhanced progenitor cell recruitment and endothelial repair after selective endothelial injury of the mouse kidney. Am J Physiol Renal Physiol. 298(6):F1504-14, ISSN 1931-857X.
[178] Herrmann JL, Abarbanell AM, Weil BR, Wang Y, Poynter JA, Manukyan MC, Meldrum DR. (2010) Postinfarct intramyocardial injection of mesenchymal stem cells pretreated with TGF-alpha improves acute myocardial function. Am J Physiol Regul Integr Comp Physiol. 299(1):R371-8, ISSN 0363-6119.
[179] Kögler G, Radke TF, Lefort A, Sensken S, Fischer J, Sorg RV, Wernet P. (2005) Cytokine production and hematopoiesis supporting activity of cord blood-derived unrestricted somatic stem cells. Exp Hematol. 33(5):573-583, ISSN 0301-472X.
[180] Liu CH, Hwang SM. (2005) Cytokine interactions in mesenchymal stem cells from cord blood. Cytokine. 32(6):270-279, ISSN 1014-4666.
[181] Gak E, Taylor WG, Chan AM, Rubin JS. (1992) Processing of hepatocyte growth factor to the heterodimeric form is required for biological activity. FEBS Lett. 311(1):17-21, ISSN 0014-5793.
[182] Nakamura T, Nawa K, Ichihara A. (1984) Partial purification and characterization of hepatocyte growth factor from serum of hepatectomized rats. Biochem Biophys Res Commun 122:1450-1459, ISSN 0006-291X.

[183] Nakamura T, Nishizawa T, Hagiya M, Seki T, Shimonishi M, Sugimura A, Tashiro K, Shimizu S. (1989) Molecular cloning and expression of human hepatocyte growth factor. Nature 342:440-443, ISSN 0028-0836.

[184] Kim WH, Matsumoto K, Bessho K, Nakamura T. (2005) Growth inhibition and apoptosis in liver myofibrolasts promoted by hepatocyte growth factor leads to resolution from liver cirrhosis. Am J Pathol 16:1017-1028, ISSN 0002-9440

[185] Yu Y, Lu L, Qian X, Chen N, Yao A, Pu L, Zhang F, Li X, Kong L, Sun B, Wang X. (2010) Antifibrotic effect of hepatocyte growth factor-expressing mesenchymal stem cells in small-for-size liver transplant rats. Stem Cells Dev. 19(6):903-14, ISSN 1547-3287.

[186] Chang YJ, Liu JW, Lin PC, Sun LY, Peng CW, Luo GH, Chen TM, Lee RP, Lin SZ, Harn HJ, Chiou TW. (2009) Mesenchymal stem cells facilitate recovery from chemically induced liver damage and decrease liver fibrosis. Life Sci. 85(13-14):517-25, ISSN 0024-3205.

[187] Tanabe J, Izawa A, Takemi N, Miyauchi Y, Torii Y, Tsuchiyama H, Suzuki T, Sone S, Ando K. (2007) Interferon-beta reduces the mouse liver fibrosis induced by repeated administration of concanavalin A via the direct and indirect effects. Immunology. 122(4):562-70, ISSN 0019-2805.

[188] Tasci I, Mas MR, Vural SA, Comert B, Alcigir G, Serdar M, Mas N, Isik AT, Ates Y. (2006) Rat liver fibrosis regresses better with pegylated interferon alpha2b and ursodeoxycholic acid treatments than spontaneous recovery. Liver Int 26: 261–268, ISSN 1478-3223.

[189] Block ET, Cronstein BN. (2010) Interferon-gamma inhibits adenosine A2A receptor function in hepatic stellate cells by STAT1-mediated repression of adenylyl cyclase. Int J Infereron Cytokine Mediator Res. 2010(2):113-126, ISSN 1179-139X.

[190] Jones JI, Clemmons DR. (1995) Insulin-like growth factors and their binding proteins: biological actions. Endocr Rev. 16:3–32, ISSN 0163-769X

[191] Clemmons DR, Van Wyk JJ. (1984) Factors controlling blood concentration of somatomedin C. Clin Endocrinol Metab 13:113-43, ISSN 0300-595X.

[192] Conchillo M, de Knegt RJ, Payeras M, Quiroga J, Sangro B, Herrero JI, Castilla-Cortazar I, Frystyk J, Flyvbjerg A, Yoshizawa C, Jansen PL, Scharschmidt B, Prieto J. (2005) Insulin-like growth factor I(IGF-1) replacement therapy increases albumin concentration in liver cirrhosis: results of a pilot randomized controlled clinical trial. J Hepatol 43:630-636. ISSN 0168-8278.

[193] Caro JF, Poulos J, Ittoop O, Pories WJ, Flickinger EG, Sinha MK. (1988) Insulin like growth factor I binding in hepatocytes from human liver, heman hepatoma, and normal, regenerating, and fetal rat liver. J Clin Invest 81;976-981. ISSN 0021-9738.

[194] Scharf JG, Knittel T, Dombrowski F, Müller L, Saile B, Braulke T, Hartmann H, Ramadori G. (1998) Characterization of the IGF axis components in isolate rat hepatic stellate cells. Hepatology 27:1275-1284. ISSN 0270-9139.

[195] Sobrevals L, Rodriguez C, Romero-Trevejo JL, Gondi G, Monreal I, Pañeda A, Juanarena N, Arcelus S, Razquin N, Guembe L, González-Aseguinolaza G, Prieto J, Fortes P. (2010) Insulin-like growth factor I gene transfer to cirrhotic liver induces fibrolysis and reduces fibrogenesis leading to cirrhosis reversion in rats. Hepatology. 51(3):912-21. ISSN 0270-9139.

[196] Tutau F, Rodríguez-Ortigosa C, Puche JE, Juanarena N, Monreal I, García Fernández M, Clavijo E, Castilla A, Castilla-Cortázar I. (2009) Enhanced actions of insulin-like growth factor-I and interferon-alpha co-administration in experimental cirrhosis. Liver Int. 29(1):37-46. ISSN 1478-3223.

[197] Gilchrist ES, Plevris JN. (2010) Bone Marrow-Derived Stem Cells in Liver Repair: 10 Years Down the Line. Liver Transpl 16:118-129. ISSN 1527-6465.

[198] Kögler G, Radke TF, Lefort A, Sensken S, Fischer J, Sorg RV, Wernet P. (2005) Cytokine production and hematopoiesis supporting activity of cord blood-derived unrestricted somatic stem cells. Exp Hematol. 33(5):573-83. ISSN 0301-472X

[199] Schibler KR, Liechty KW, White WL, Christensen RD. (1993) Production of granulocyte colony-stimulating factor in vitro by monocytes from preterm and term neonates. Blood. 82(8):2478-84. ISSN 0006-4971.

[200] Jenhani F, Durand V, Ben Azouna N, Thallet S, Ben Othmen T, Bejaoui M, Domenech J. (2011) Human cytokine expression profile in various conditioned media for in vitro expansion bone marrow and umbilical cord blood immunophenotyped mesenchymal stem cells. Transplant Proc. 43(2):639-43. ISSN 0041-1345.

[201] Suehiro Y, Muta K, Umemura T, Abe Y, Nishimura J, Nawata H. (1999) Macrophage inflammatory protein 1alpha enhances in a different manner adhesion of hematopoietic progenitor cells from bone marrow, cord blood, and mobilized peripheral blood. Exp Hematol. 27(11):1637-45. ISSN 0301-472X.

[202] Liu CH, Hwang SM. (2005) Cytokine interactions in mesenchymal stem cells from cord blood. Cytokine. 32(6):270-9. ISSN 1043-4666.

[203] Gang EJ, Jeong JA, Han S, Yan Q, Jeon CJ, Kim H. (2006) In vitro endothelial potential of human UC blood-derived mesenchymal stem cells. Cytotherapy. 8(3):215-27. ISSN 1465-3249.

[204] Khan AA, Shaik MV, Parveen N, Rajendraprasad A. Aleem MA, Habeeb MA, Srinivas G, Raj TA, Tiwari SK, Kumaresan K, Venkateswarlu J, Pande G, Habibullah CM. (2010) Human fetal liver derived stem cell transplantation as supportive modalitity in the management of end-stage decompensated liver cirrhosis. Cell Transplant 19(4) 409-418. ISSN 0963-6897.

19

Stem Cell Therapy for Patients with Chronic Liver Disease

Reebye, V; Levičar, N and Habib, N.
Dept of Surgery and Cancer, Imperial College London
UK

1. Introduction

Adult stem cells are a population of immature cells that have the capability of self renewal to provide the human body with a constant source of cells for maintaining healthy tissues or replacing those that are damaged. Our present day scientific developments alongside clinical experiences have collectively consolidated our understanding of the signals that mediate stem cell lineage commitment and differentiation. From this, the concept of using adult stem cells to repopulate chronically diseased organs is now a feasible option for regenerative medicine.

Whilst the use of human embryonic stem cells in the clinical setting has been overwhelmingly marred by its oncogenic potential as well as ethical concerns, the rate of progress in understanding the complex developmental plasticity of adult stem cells, in particular those derived from the bone marrow, has successfully allowed their use for translational research with unattached ethical issues. The pioneering use of bone marrow stem cells in the 1960s as a viable treatment option for leukaemia, myeloma and lymphoma has come a long way since then. Today we are able to use these stem cells to give rise to bone and cartilage (1) and to repair both cardiac and liver function (2, 3). Since a detailed review for each of these exciting developments is too broad for this chapter, we will mainly focus on the use of bone marrow derived stem cell for the treatment of chronic liver disease.

2. Chronic liver disease

Liver cirrhosis is the end stage of chronic liver disease and is associated with many serious systemic complications resulting from both liver failure and portal hypertension. This condition has a poor prognosis and is difficult to treat. In the UK alone as many as one in ten people have some form of liver disease where many die prematurely as a result (4). Liver disease is currently the fifth most common cause of mortality in the UK for both men and women and whilst the mortality rates for coronary heart disease, cancer, respiratory and cerebro-vascular diseases are falling, the death rates from all types of liver disease in England and Wales over the last 25 years showed over 150% increase in men and 100% in women (5). This significant upward trend clearly indicates a need to plan ahead for the health service. Worldwide, the common causes of liver fibrosis and cirrhosis include hepatitis B, hepatitis C and alcohol consumption. Other causes include immune mediated damage, genetic abnormalities, and non-alcoholic steatohepatitis which are mostly associated with diabetes and metabolic syndrome (6-8).

3. Cirrhosis

Liver cirrhosis or fibrosis is a common progressive pathological lesion of chronic liver disease that occurs in response to various liver-damaging factors. Cirrhosis is defined as the histological development of regenerative nodules surrounded by fibrous bands which leads to portal hypertension; the development of hepatocellular carcinoma and end-stage liver disease. Over the past two decades the main cellular and molecular mechanisms of cirrhotic initiation and progression have been clearly defined. Various causative factors including hepatitis virus infection; ischemia; parasite infection; abnormal copper or iron load, all result in chronic inflammation which initially leads towards an excessive synthesis of the extracellular matrix (ECM). Activated hepatic stellate cells; injured or regenerated hepatocytes; Kupffer cells; sinusoidal cells and natural killer (NK) cells all produce cytokines and immunoreactive factors which exert various biological effects in an autocrine and paracrine manner to initiate fibrotic growth (9-12).
The detailed understanding of the natural history and pathophysiology of cirrhosis has resulted in better management of its complications to improve the quality of life of patients. Whilst pharmacological treatments can halt progression to decompensated cirrhosis, orthotopic liver transplantation (OLT) still remains the only highly successful and curative option for end stage cirrhosis. Survival data from the United Network of Organ Sharing (UNOS) study in 2004 indicated a survival rate of only 61% at 8 year post transplantation (13). OLT cannot be sustained as the only option for advanced liver failure. The limited availability of organs for a constantly growing list of patients requiring transplant in addition to issues of compatibility and comorbid factors means that not everyone is eligible for transplantation. There is therefore is an unmet need to find a suitable alternative.

4. Hepatocyte transplantation

Hepatocyte transplantation has been identified as a promising alternative to OLT for certain liver-based metabolic disorders and acute liver failure. With the advantage of preserving the native liver, hepatocyte transplantation is less invasive and can be performed repeatedly to allow a higher chance for improved recovery in patients with acute liver failure (14-16). The mechanisms by which intraportally injected donor hepatocytes engraft into the host liver parenchyma involves cell migration within the liver sinusoids (17). These cells become trapped causing portal hypertension and ischemic–reperfusion (18, 19). This consequently initiates clearance by the innate immune system where the associated release of cytokines induces vascular permeability thus allowing surviving hepatocytes to translocate through the sinusoidal fenestrations and integrate into the liver parenchyma (20). Although the initial engraftment success of hepatocytes after transplantation is low, repeated hepatocyte transplantation in animal models have shown a sufficient increase in the number of engrafted cells to allow recovery from some metabolic defects (21). Isolation of hepatocytes is carried out from donor liver that are unused or not suitable for whole organ transplantation. Cells harvested from a single donor have the advantage of being used for multiple recipients in addition to being cryopreserved for use in emergency treatment of acute liver failure or for repeat treatment of liver-based metabolic disorders (22-24). To date hepatocyte transplantation has only been used to address single gene defects and some metabolic disorders where the risks associated with OLT are not justified or are aimed at avoiding or postponing liver transplantation. The most successful clinical outcomes of OLT have been in patients with metabolic disorders including those with urea cycle defects (23,

24); Crigler-Najjar syndrome type 1 (CN1) (25, 26); glycogen storage disease type 1 (27) and Factor VII deficiency (28). Although it was initially thought that transplantation of hepatocytes would be less immunogenic than whole organ transplantation, data from animal models demonstrated that the innate immune system clears a significant amount of hepatocyte irrespective of whether syngeneic or allogeneic cells are used (18). As a result most transplant facilities use the same immunosupressive treatment as for liver transplantation (29). Since the clinical outcome following hepatocyte transplantation is still highly variable due to a multitude of factors including fibrotic damage to the liver which limits the proper engraftment of cells; the availability of the cells required to provided sufficient liver function over repeated infusion per patient, and often the very marginal quality of the cells isolated from donor liver, a new and readily available cell source must be identified. This strong demand has fuelled the interest in haematopoietic stem cells for treating acute liver failure.

5. The developmental plasticity of adult stem cells

It had previously been assumed that adult stem cells were lineage restricted, however the culmination of research over the past decade has now fully confirmed the remarkable developmental plasticity of human adult stem cells. The ability of adult bone marrow derived myogenic progenitors to participate in the regeneration of damaged skeletal muscle (30); ischemic myocardium (31-33); neurogenesis (34, 35) and the conversion of adult neural stem cells back into hematopoietic cells (36), redefined the biology of development. It was now accepted that the original three germs layers were not necessarily vital for transdifferentiation to occur (37). This implied that adult stem cells could exhibit similar pluripotency as embryonic stem cells.

6. Generation of hepatocytes by haematopoietic stem cells

The demonstration that adult bone marrow stem cells could differentiate towards a hepatic lineage was made almost a decade ago. Two independent research groups showed that the adult rat bone marrow contains a subpopulation of cells (about 3%) co-expressing the haematopoietic stem cell markers ($CD34^{+}$, c-Kit, Thy-1); α-fetoprotein (AFP) and c-met. When they cultured this crude extract of bone marrow with hepatocyte growth factor (HGF) and epidermal growth factor (EGF), they were able to detect the expression of albumin, a marker of fully differentiated hepatocyte (38, 39). Over the years further refinement of the *in vitro* culture conditions led to the differentiation of haematopoietic cells that expressed more mature liver specific transcription factors including human Hepatocyte nuclear factor-1α (HNF-1α); cytokeratin (CK8); CK19 and AFP (40, 41). Studies on human adult bone marrow similarly confirmed their capability of differentiating into liver like cells (42, 43) where numerous cytokines and growth factors were shown to be important for driving this differentiation under *in vitro* conditions (44-46). More interestingly was the discovery that exposing rat haematopoietic stem cells to injured liver tissues induced the expression of functional hepatocyte factors (CK18, albumin and transferrin) (47). This suggested that the liver micro-environment alone was sufficient in providing the appropriate cues for inducing stem cell conversion towards the hepatic lineage. Several human post-mortem studies confirmed this observation where the presence of bone marrow derived cells were found in the liver. Theise *et al.*, (2000) investigated archival autopsy and biopsy liver specimens from recipients of sex-mismatched therapeutic bone marrow transplantation and OLT (48). By immunohistochemical staining for CK8, CK18 and CK19, they identified hepatocytes and cholangiocytes of bone marrow origin as well as the detection of the Y-chromosome by fluorescent *in situ* hybridization (FISH). Using double staining analysis, they found a large number of engrafted hepatocytes (4%-43%) and cholangiocytes (4%-38%) - that were bone-marrow derived-replenishing the hepatic parenchyma. Despite a general variability of this observation amongst parallel studies (49, 50), the ultimate conclusion was that bone marrow derived haematopoietic stem cells were able to successfully engraft into the liver and gradually adopt a phenotype very similar to the hepatic lineage.

7. Human studies

Today, the most widely studied adult stem cells are derived from the hematopoietic lineage. Hematopoietic stem cells (HSCs) form part of the bone marrow compartment which constitutes a heterogenous population of mesenchymal stem cells; committed progenitor cells and non-circulating stromal cells. The identification of HSCs from the bone marrow compartment is largely based on the expression of cell surface markers. Detecting the cluster of differentiation, CD34 is generally used in human studies as a surrogate marker for progenitor stem cells although there also exists a subpopulation of more primitive HSCs which do not express CD34 (51, 52). Despite the numerous studies that have shown bone marrow stem cells can give rise to hepatocytes, their use as a therapeutic agent is still in its infancy. Many studies are still at a pilot stage requiring randomisation and controls, however those reported show interesting results that require conformation. The first clinical study was performed by a German group, am Esch *et al.*, (2005), where three patients were infused with autologous $CD133^+$ cells subsequent to portal vein embolisation of right liver segments (3). A computerised tomography (CT) scan for volume analysis of the left lateral segments showed a 2.5 fold increase in growth rate of the three patients when compared to the control group without bone marrow stem cell administration. Despite the small number of patients and the lack of adequately sized randomised control group, this data showcased the promising potential of bone marrow derived stem cells in enhancing liver regeneration. A year later, a parallel study was reported by Gordon *et al.*, (2006) which included a phase I clinical trial on five patients with liver insufficiency (53). These patients were given granulocyte colony-stimulating factor (G-CSF) to mobilise their stem cells for collection by leukapheresis followed by purification of $CD34^+$ expressing cells which were then injected into either the portal vein or hepatic artery. Three of the five patients showed improvement in serum bilirubin whereas four out of the five patients displayed significant increase in serum albumin. Clinically, the procedure was well tolerated with no observed procedure-related complications and the data concluded that there was a marked contribution of the stem cells towards regeneration of the damaged liver (53). In another independent study, Terai *et al.*, (2006) enrolled nine patients with liver cirrhosis where they were injected with autologously derived bone marrow stem cells enriched for the expression of $CD34^+$, $CD45^+$ and c-kit^+. These patients were followed for 24 weeks where they showed a significant improvement in serum albumin levels in addition to significantly improved Child-Pugh scores (54). A year later, Rajkumar *et al.*, (2007) reported another small scale trial where 22 patients with chronic liver disease (Child-Pugh scores B-C) were enrolled (55). 200ml to 300ml of bone marrow aspirate were subjected to density gradient fractionation for isolation of $CD34^+$ cells. These were then administered intravenously through the median cubital vein. Liver function tests; ultrasound and CT scans were performed before, 4 weeks and 8 weeks following cell infusion. 32% of the patients showed a drop in bilirubin levels; 67% of the patients showed an increase in serum albumin levels and 73% showed reduction in ascites. No patients from this study showed severe adverse effects following transfusion and the overall quality of life index was significantly improved in the majority (82%) (55). In 2007, Mohamadnejad *et al.*, (56, 57) performed two clinical studies in patients with decompensated liver cirrhosis. In the first study they treated four patients each with 31.73 x 10^6 of cultured autologous mesenchymal stem cells, infused through a peripheral vein. The Phase I study demonstrated no side effects and the quality of life of all four patients improved by the end of follow-up. Furthermore, the model for end-stage liver disease scores of two patients (patient 1 & patient 4) showed marked amelioration in their Child Pugh scores by the end of follow-up. In the second study, they treated four patients with autologous $CD34^+$ cells isolated from the bone marrow, which was slowly infused through the hepatic artery of the patients. In two patients they observed mild albumin improvement, however, the health of two patients further deteriorated where one died of liver failure a few days after the transplantation. This trial was prematurely stopped due to the severity of the side effects where it was concluded that infusion of $CD34^+$ stem cells through the hepatic artery was not safe in decompensated cirrhosis, but suggested that it may be beneficial to use alternative routes to transplant $CD34^+$ cells. Having corroborated the use of bone marrow stem cells as a regenerative therapy and demonstrated their safety in patients with liver insufficiency, Pai *et al.*,(2008) conducted a prospective clinical efficacy study of expanded adult $CD34^+$ stem cells infused into the hepatic artery in nine patients with alcoholic liver cirrhosis (ALC) to determine whether clinical benefit was conferred (58). The primary end point was to assess the safety of infusing autologous stem cells into the hepatic artery of these patients; the secondary end point was to assess the improvement in liver function through serological and biochemical analysis; and to determine whether there were any symptomatic improvements. Following $CD34^+$ stem cell infusion, CT scans showed normal enhancement in the liver parenchyma, with no evidence of focal liver lesions; additionally, duplex Doppler ultrasound scans showed normal flow in the portal veins and hepatic artery. To corroborate the safety and efficacy of improving liver function in patients with liver cirrhosis, Levicar *et al.*, (2008) then reported the results from a long-term follow-up (12-18 months) of patients with chronic liver disease injected with $CD34^+$ enriched stem cells (59). During this time the patients were monitored for side effects, toxicity and changes in clinical, haematological and biochemical parameters. All the patients tolerated the treatment protocol without any complications or side effects related to the procedure. Four patients showed an initial improvement in serum bilirubin level, which was maintained for up to 6 months; whilst only a marginal increase in serum bilirubin was observed in three of the patients at 12 months. Only one patient showed an increase in serum bilirubin over slightly longer period of 18 months post-infusion. CT scans and serum AFP monitoring did not show any lesions or tumor formation in the patients. This successful study provided the basis for subsequent trials where more recently Li Nan *et al.*, (2010) reported the clinical outcome of autologous $CD34^+$ infusion in patients with hepatitis induced liver failure (60). Twenty seven patients with Child- Pugh C cirrhosis were enrolled of which 22 were positive for Hepatitis B and 5 were positive for Hepatitis C. 50ml to 120ml of bone marrow aspirate were subjected to a Percoll gradient for the isolation of myeloid stem cells. 20 of the patients received cell infusion via the hepatic portal vein whilst 7 patients were infused via the hepatic artery using the Seldinger percutaneous technique. The patients were closely monitored for 3 months following transplantation where it was initially noticed that liver function tests decreased for 3 days. The authors attributed this with the contrast media used to perform arteriography. Since all the patients had Child- Pugh C cirrhosis, the contrast media may have aggravated liver injury. Despite this, the patients did subsequently show an improvement in liver function by 1 week of transplantation. The most significant recovery was seen at 3 months post procedure where total bilirubin and albumin levels were highest. Overall, an improved clinical outcome was observed in the majority of the patients where jaundice was resolved and ascites improved 3 months after therapy. From this cohort of patients, two died, one due to peritoneal cavity infection and liver failure, and the second from a massive hemorrhage of the upper alimentary tract. The latest clinical study on patients with end stage liver disease has been reported by Nikeghbalian *et al.*, (2011) (61). Six patients were intraportally injected with autologous bone marrow-derived $CD133^{+}$ cells. This arm of patients were compared with patients subjected short term infusion (6months) and long term infusion (12 months) of mononuclear cells. Liver function test was positive at 24 months of follow up in all the patients enrolled in this trial. Since there were no differences between the groups receiving MNC or $CD133^{+}$ cell, this recent study has highlighted the versatility of bone marrow-derived progenitor mononuclear stem cells irrespective of the subpopulation of their cluster of differentiation markers. Furthermore this study also confirms their safe use to circumvent the need for liver transplantation in end stage liver disease.

The culmination of clinical trials thus far mentioned have shown a trend towards decreased serum fibrosis markers and improvements in bilirubin, albumin and Child-Pugh scores following stem cell infusion. The past decade has marked an important progress for regenerative liver therapy; however the accumulated data is still in its infancy. It is still not clear whether the route of infusion is important or whether mobilization of myeloid progenitor stem cells with G-CSF without leukapheresis is sufficient. Until more is learnt about the mechanisms by which HSCs contribute to hepatocyte regeneration, and the mechanism of clinical benefit in the recipient patients; controversy will inevitably shadow this form of treatment. Some authors have proposed that the observed conversion of infused stem cell to hepatocytes is simply a byproduct of cell fusion with no true induction of transdifferentiation (59, 60). Others have suggested that instead of adopting a new lineage, the stem cells indirectly serve a regenerative capability by stimulating activation of tissue specific stem cells and by inducing the release of vascular endothelial growth factor (VEGF) thereby increasing the blood supply to the cells and aiding in repair of the damaged tissue (61, 62). There are reports that also suggest HSCs may act in a regenerative capacity simply by inducing the expression of the B-Cell leukemia/lymphoma-2 gene (Bcl-2) and interleuking-6 (IL-6) in a paracrine manner thus suppressing apoptosis and inflammation of the surrounding tissue (62-64). In addition to these uncertainties, the homing mechanism of infused adult stem cells to the liver also remains unclear. There are suggestions that chemokines similar to stroma derived factor 1 may be involved, however, this remains to be validated (65). Although the application of stem cell therapy will not be prevented despite an incomplete understanding of the mechanisms involved, a clearer picture will ultimately enable better tailoring of stem cell therapy to the disease in question.

8. Current limitations

For cell replacement therapy is still too early to demonstrate its long-term effectiveness or its improvement in survival rate and in the quality of life. This will undoubtedly have to be evaluated by more randomised control studies in the future. A standardised protocol will need to be established for the most efficient route of delivering infused HSCs; for the efficient long term culture of HSCs; for the optimal cell density and repeated transplantation required to re-establish liver function.

9. Future prospects

Autologous bone marrow infusion for patients with chronic liver disease has been shown to improve liver function parameters in contrast to observations accompanied by abstinence from alcohol. The degree of effectiveness of this therapy will likely to vary among different patients. Although the mechanisms underlining adult stem cell plasticity is still far from being fully characterised, the general enthusiasm regarding its potential clinical implication is heightened by the numerous clinical trials currently underway. In hepatology, the data presented here provides hope that human adult stem cells could eventually be used in tissue replacement protocols for the treatment of inherited and acquired end-stage liver diseases.

10. References

[1] Alhadlaq A & Mao JJ (2005) Tissue-engineered osteochondral constructs in the shape of an articular condyle J Bone Joint Surg Am 87, 936-944.

[2] Saito T, Kuang JQ, Lin CC, & Chiu RC (2003) Transcoronary implantation of bone marrow stromal cells ameliorates cardiac function after myocardial infarction J Thorac Cardiovasc Surg 126, 114-123.

[3] am Esch JS, 2nd, Knoefel WT, Klein M, Ghodsizad A, Fuerst G, Poll LW, Piechaczek C, Burchardt ER, Feifel N, Stoldt V, et al. (2005) Portal application of autologous CD133+ bone marrow cells to the liver: a novel concept to support hepatic regeneration Stem Cells 23, 463-470.

[4] Oberdorfer A, Oberdorfer AL, & Tran DT (2002) in Cochrane Database of Systematic Reviews.

[5] Health Statistics Quarterly (2007). Office for National Statistics. Deaths: Selected causes (International Classification) and sex, 1971 onwards in England and Wales.

[6] Department of Health, Annual Report of the Chief Medical Officer (2001). 'Liver cirrhosis-starting to strike at younger ages. www.doh.gov.uk/cmo/annualreport2001/livercirrhosis.htm

[7] Day CP (2002) Non-alcoholic steatohepatitis (NASH): where are we now and where are we going? Gut 50, 585-588.

[8] National Center For Chronic Disease Prevention and Health Promotion. US Obesity Trends from 1985 to 2000. www.cdc.gov/nccdphp/dnpa/obesity/trends/maps/index.htm

[9] Friedman SL (1993) Seminars in medicine of the Beth Israel Hospital, Boston. The cellular basis of hepatic fibrosis. Mechanisms and treatment strategies N Engl J Med 328, 1828-1835.

[10] Friedman SL (2000) Molecular regulation of hepatic fibrosis, an integrated cellular response to tissue injury J Biol Chem 275, 2247-2250.

[11] Bissell DM (1998) Hepatic fibrosis as wound repair: a progress report J Gastroenterol 33, 295-302.

[12] Schuppan D & Afdhal NH (2008) Liver cirrhosis Lancet 371, 838-851.

[13] Roberts MS, Angus DC, Bryce CL, Valenta Z, & Weissfeld L (2004) Survival after liver transplantation in the United States: a disease-specific analysis of the UNOS database Liver Transpl 10, 886-897.

[14] Dhawan A, Puppi J, Hughes RD, & Mitry RR Human hepatocyte transplantation: current experience and future challenges Nat Rev Gastroenterol Hepatol 7, 288-298.

[15] Dhawan A, Mitry RR, & Hughes RD (2006) Hepatocyte transplantation for liver-based metabolic disorders J Inherit Metab Dis 29, 431-435.

[16] Fisher RA & Strom SC (2006) Human hepatocyte transplantation: worldwide results Transplantation 82, 441-449.

[17] Fisher RA, Bu D, Thompson M, Wolfe L, & Ritter JK (2004) Optimization of conditions for clinical human hepatocyte infusion Cell Transplant 13, 677-689.

[18] Krohn N, Kapoor S, Enami Y, Follenzi A, Bandi S, Joseph B, & Gupta S (2009) Hepatocyte transplantation-induced liver inflammation is driven by cytokines-chemokines associated with neutrophils and Kupffer cells Gastroenterology 136, 1806-1817.

[19] Joseph B, Malhi H, Bhargava KK, Palestro CJ, McCuskey RS, & Gupta S (2002) Kupffer cells participate in early clearance of syngeneic hepatocytes transplanted in the rat liver Gastroenterology 123, 1677-1685.

[20] Gupta S, Rajvanshi P, Sokhi R, Slehria S, Yam A, Kerr A, & Novikoff PM (1999) Entry and integration of transplanted hepatocytes in rat liver plates occur by disruption of hepatic sinusoidal endothelium Hepatology 29, 509-519.

[21] Rajvanshi P, Kerr A, Bhargava KK, Burk RD, & Gupta S (1996) Efficacy and safety of repeated hepatocyte transplantation for significant liver repopulation in rodents Gastroenterology 111, 1092-1102.

[22] Meyburg J, Das AM, Hoerster F, Lindner M, Kriegbaum H, Engelmann G, Schmidt J, Ott M, Pettenazzo A, Luecke T, et al. (2009) One liver for four children: first clinical series of liver cell transplantation for severe neonatal urea cycle defects Transplantation 87, 636-641.

[23] Puppi J, Tan N, Mitry RR, Hughes RD, Lehec S, Mieli-Vergani G, Karani J, Champion MP, Heaton N, Mohamed R, et al. (2008) Hepatocyte transplantation followed by auxiliary liver transplantation--a novel treatment for ornithine transcarbamylase deficiency Am J Transplant 8, 452-457.

[24] Stephenne X, Najimi M, Smets F, Reding R, de Ville de Goyet J, & Sokal EM (2005) Cryopreserved liver cell transplantation controls ornithine transcarbamylase deficient patient while awaiting liver transplantation Am J Transplant 5, 2058-2061.

[25] Lysy PA, Najimi M, Stephenne X, Bourgois A, Smets F, & Sokal EM (2008) Liver cell transplantation for Crigler-Najjar syndrome type I: update and perspectives World J Gastroenterol 14, 3464-3470.

[26] Ambrosino G, Varotto S, Strom SC, Guariso G, Franchin E, Miotto D, Caenazzo L, Basso S, Carraro P, Valente ML, et al. (2005) Isolated hepatocyte transplantation for Crigler-Najjar syndrome type 1 Cell Transplant 14, 151-157.

[27] Lee KW, Lee JH, Shin SW, Kim SJ, Joh JW, Lee DH, Kim JW, Park HY, Lee SY, Lee HH, et al. (2007) Hepatocyte transplantation for glycogen storage disease type Ib Cell Transplant 16, 629-637.
[28] Dhawan A, Mitry RR, Hughes RD, Lehec S, Terry C, Bansal S, Arya R, Wade JJ, Verma A, Heaton ND, et al. (2004) Hepatocyte transplantation for inherited factor VII deficiency Transplantation 78, 1812-1814.
[29] Sokal EM, Smets F, Bourgois A, Van Maldergem L, Buts JP, Reding R, Bernard Otte J, Evrard V, Latinne D, Vincent MF, et al. (2003) Hepatocyte transplantation in a 4-year-old girl with peroxisomal biogenesis disease: technique, safety, and metabolic follow-up Transplantation 76, 735-738.
[30] Ferrari G, Cusella-De Angelis G, Coletta M, Paolucci E, Stornaiuolo A, Cossu G, & Mavilio F (1998) Muscle regeneration by bone marrow-derived myogenic progenitors Science 279, 1528-1530.
[31] Orlic D, Kajstura J, Chimenti S, Jakoniuk I, Anderson SM, Li B, Pickel J, McKay R, Nadal-Ginard B, Bodine DM, et al. (2001) Bone marrow cells regenerate infarcted myocardium Nature 410, 701-705.
[32] Jackson KA, Majka SM, Wang H, Pocius J, Hartley CJ, Majesky MW, Entman ML, Michael LH, Hirschi KK, & Goodell MA (2001) Regeneration of ischemic cardiac muscle and vascular endothelium by adult stem cells J Clin Invest 107, 1395-1402.
[33] Orlic D, Kajstura J, Chimenti S, Limana F, Jakoniuk I, Quaini F, Nadal-Ginard B, Bodine DM, Leri A, & Anversa P (2001) Mobilized bone marrow cells repair the infarcted heart, improving function and survival Proc Natl Acad Sci U S A 98, 10344-10349.
[34] Mezey E, Chandross KJ, Harta G, Maki RA, & McKercher SR (2000) Turning blood into brain: cells bearing neuronal antigens generated in vivo from bone marrow Science 290, 1779-1782.
[35] Brazelton TR, Rossi FM, Keshet GI, & Blau HM (2000) From marrow to brain: expression of neuronal phenotypes in adult mice Science 290, 1775-1779.
[36] Bjornson CR, Rietze RL, Reynolds BA, Magli MC, & Vescovi AL (1999) Turning brain into blood: a hematopoietic fate adopted by adult neural stem cells in vivo Science 283, 534-537.
[37] Clarke DL, Johansson CB, Wilbertz J, Veress B, Nilsson E, Karlstrom H, Lendahl U, & Frisen J (2000) Generalized potential of adult neural stem cells Science 288, 1660-1663.
[38] Oh SH, Miyazaki M, Kouchi H, Inoue Y, Sakaguchi M, Tsuji T, Shima N, Higashio K, & Namba M (2000) Hepatocyte growth factor induces differentiation of adult rat bone marrow cells into a hepatocyte lineage in vitro Biochem Biophys Res Commun 279, 500-504.
[39] Miyazaki M, Akiyama I, Sakaguchi M, Nakashima E, Okada M, Kataoka K, & Huh NH (2002) Improved conditions to induce hepatocytes from rat bone marrow cells in culture Biochem Biophys Res Commun 298, 24-30.
[40] Okumoto K, Saito T, Hattori E, Ito JI, Adachi T, Takeda T, Sugahara K, Watanabe H, Saito K, Togashi H, et al. (2003) Differentiation of bone marrow cells into cells that

express liver-specific genes in vitro: implication of the Notch signals in differentiation Biochem Biophys Res Commun 304, 691-695.
[41] Ratajczak MZ, Kucia M, Reca R, Majka M, Janowska-Wieczorek A, & Ratajczak J (2004) Stem cell plasticity revisited: CXCR4-positive cells expressing mRNA for early muscle, liver and neural cells 'hide out' in the bone marrow Leukemia 18, 29-40.
[42] Fiegel HC, Lioznov MV, Cortes-Dericks L, Lange C, Kluth D, Fehse B, & Zander AR (2003) Liver-specific gene expression in cultured human hematopoietic stem cells Stem Cells 21, 98-104.
[43] Kucia M, Ratajczak J, Reca R, Janowska-Wieczorek A, & Ratajczak MZ (2004) Tissue-specific muscle, neural and liver stem/progenitor cells reside in the bone marrow, respond to an SDF-1 gradient and are mobilized into peripheral blood during stress and tissue injury Blood Cells Mol Dis 32, 52-57.
[44] Heng BC, Yu H, Yin Y, Lim SG, & Cao T (2005) Factors influencing stem cell differentiation into the hepatic lineage in vitro J Gastroenterol Hepatol 20, 975-987.
[45] Block GD, Locker J, Bowen WC, Petersen BE, Katyal S, Strom SC, Riley T, Howard TA, & Michalopoulos GK (1996) Population expansion, clonal growth, and specific differentiation patterns in primary cultures of hepatocytes induced by HGF/SF, EGF and TGF alpha in a chemically defined (HGM) medium J Cell Biol 132, 1133-1149.
[46] Michalopoulos GK, Bowen WC, Mule K, & Luo J (2003) HGF-, EGF-, and dexamethasone-induced gene expression patterns during formation of tissue in hepatic organoid cultures Gene Expr 11, 55-75.
[47] Jang YY, Collector MI, Baylin SB, Diehl AM, & Sharkis SJ (2004) Hematopoietic stem cells convert into liver cells within days without fusion Nat Cell Biol 6, 532-539.
[48] Theise ND, Badve S, Saxena R, Henegariu O, Sell S, Crawford JM, & Krause DS (2000) Derivation of hepatocytes from bone marrow cells in mice after radiation-induced myeloablation Hepatology 31, 235-240.
[49] Korbling M, Katz RL, Khanna A, Ruifrok AC, Rondon G, Albitar M, Champlin RE, & Estrov Z (2002) Hepatocytes and epithelial cells of donor origin in recipients of peripheral-blood stem cells N Engl J Med 346, 738-746.
[50] Ng IO, Chan KL, Shek WH, Lee JM, Fong DY, Lo CM, & Fan ST (2003) High frequency of chimerism in transplanted livers Hepatology 38, 989-998.
[51] Osawa M, Hanada K, Hamada H, & Nakauchi H (1996) Long-term lymphohematopoietic reconstitution by a single CD34-low/negative hematopoietic stem cell Science 273, 242-245.
[52] Goodell MA, Rosenzweig M, Kim H, Marks DF, DeMaria M, Paradis G, Grupp SA, Sieff CA, Mulligan RC, & Johnson RP (1997) Dye efflux studies suggest that hematopoietic stem cells expressing low or undetectable levels of CD34 antigen exist in multiple species Nat Med 3, 1337-1345.
[53] Gordon MY, Levicar N, Pai M, Bachellier P, Dimarakis I, Al-Allaf F, M'Hamdi H, Thalji T, Welsh JP, Marley SB, et al. (2006) Characterization and clinical application of

human CD34+ stem/progenitor cell populations mobilized into the blood by granulocyte colony-stimulating factor Stem Cells 24, 1822-1830.

[54] Terai S, Ishikawa T, Omori K, Aoyama K, Marumoto Y, Urata Y, Yokoyama Y, Uchida K, Yamasaki T, Fujii Y, et al. (2006) Improved liver function in patients with liver cirrhosis after autologous bone marrow cell infusion therapy Stem Cells 24, 2292-2298.

[55] Rajkumar JS, Baskar S, Senthil Nagarajan R, Murugan P, Terai S, Sakaida I, & Abraham S (2007) Autologous Bone Marrow stem cell Infusion (AMBI) therapy for Chronic Liver Diseases. Journal of Stem Cells & Regenerative Medicine 3, 26-27.

[56] Mohamadnejad M, Alimoghaddam K, Mohyeddin-Bonab M, Bagheri M, Bashtar M, Ghanaati H, Baharvand H, Ghavamzadeh A, Malekzadeh R (2007) Phase 1 trial of autologous bone marrow mesenchymal stem cell transplantation in patients with decompensated liver cirrhosis. Arch Iran Med 10, 459-466.

[57] Mohamadnejad M, Namiri M, Bagheri M, Hashemi SM, Ghanaati H, Zare Mehrjardi N, Kazemi Ashtiani S, Malekzadeh R, Baharvand H (2007) Phase 1 human trial of autologous bone marrow-hematopoietic stem cell transplantation in patients with decompensated cirrhosis. World J Gastroenterol. 13, 3359-3363.

[58] Pai M, Zacharoulis D, Milicevic MN, Helmy S, Jiao LR, Levicar N, Tait P, Scott M, Marley SB, Jestice K, et al. (2008) Autologous infusion of expanded mobilized adult bone marrow-derived CD34+ cells into patients with alcoholic liver cirrhosis Am J Gastroenterol 103, 1952-1958.

[59] Levicar N, Pai M, Habib NA, Tait P, Jiao LR, Marley SB, Davis J, Dazzi F, Smadja C, Jensen SL, et al. (2008) Long-term clinical results of autologous infusion of mobilized adult bone marrow derived CD34+ cells in patients with chronic liver disease Cell Prolif 41 Suppl 1, 115-125.

[60] Li N, Wang Y-M, Feng K, Wang Y-W, Sha L-N, Ao G-K, Zhai J-S, Zhang L, & Shi Y-L (2010) Intrahepatic transplantation of bone marrow stem cells via hepatic artery or portal vein in patients with liver cirrhosis Journal of Chinese Clinical Medicine 5, 3.

[61] Nikeghbalian S, Pournasr B, Aghdami N, Rasekhi A, Geramizadeh B, Hosseini Asl SM, Ramzi M, Kakaei F, Namiri M, Malekzadeh R, Vosough Dizaj A, Malek-Hosseini SA, Baharvand H (2011) Autologous transplantation of bone marrow-derived mononuclear and CD133(+) cells in patients with decompensated cirrhosis. Arch Iran Med14, 12-17.

[62] Wang X, Willenbring H, Akkari Y, Torimaru Y, Foster M, Al-Dhalimy M, Lagasse E, Finegold M, Olson S, & Grompe M (2003) Cell fusion is the principal source of bone-marrow-derived hepatocytes Nature 422, 897-901.

[63] Vassilopoulos G, Wang PR, & Russell DW (2003) Transplanted bone marrow regenerates liver by cell fusion Nature 422, 901-904.

[64] Tang J, Xie Q, Pan G, Wang J, & Wang M (2006) Mesenchymal stem cells participate in angiogenesis and improve heart function in rat model of myocardial ischemia with reperfusion Eur J Cardiothorac Surg 30, 353-361.

[65] Rehman J, Traktuev D, Li J, Merfeld-Clauss S, Temm-Grove CJ, Bovenkerk JE, Pell CL, Johnstone BH, Considine RV, & March KL (2004) Secretion of angiogenic

and antiapoptotic factors by human adipose stromal cells Circulation 109, 1292-1298.
[66] Chen Z, Chua CC, Ho YS, Hamdy RC, & Chua BH (2001) Overexpression of Bcl-2 attenuates apoptosis and protects against myocardial I/R injury in transgenic mice Am J Physiol Heart Circ Physiol 280, H2313-2320.
[67] Wang M, Tsai BM, Crisostomo PR, & Meldrum DR (2006) Pretreatment with adult progenitor cells improves recovery and decreases native myocardial proinflammatory signaling after ischemia Shock 25, 454-459.
[68] Hatch HM, Zheng D, Jorgensen ML, & Petersen BE (2002) SDF-1alpha/CXCR4: a mechanism for hepatic oval cell activation and bone marrow stem cell recruitment to the injured liver of rats Cloning Stem Cells 4, 339-351.

20

Pancreatic Stem Cells: Unresolved Business

Fang-Xu Jiang and Grant Morahan
Centre for Diabetes Research, Western Australian Institute for Medical Research, and Centre for Medical Research, The University of Western Australia, Perth 6000 Australia

1. Introduction

Diabetes mellitus has recently manifested a global trend in increased prevalence and is now a major public health problem around the world including in developing countries, such as China and India. It currently affects approximately 200 million people, and this number is likely to increase to 400 million by 2030 (Lock and Tzanakakis, 2007). Approximately 10% of these cases will have type 1 diabetes mellitus (T1D), caused by absolute deficiency of insulin-producing β cells. Therefore, a cellular therapy is the best prospect for a cure of T1D, provided autoimmunity to β cells can be controlled and there is a sufficient supply of insulin-secreting β cells for transplantation. However, the poor availability of donor islets has severely restricted the broad clinical use of islet transplantation. The lack of sufficient donor islets is why much attention has recently been paid to stem cells as a renewable source of β cells.

The term 'stem cell' was initially used in embryology in the late 19th century in the context of the origin of the blood system and gametes (Ramalho-Santos and Willenbring, 2007). Stem cells are undifferentiated cells that are capable of both self-renewal and giving rise to specialized functional cells. Depending on the developmental stages of their origin, stem cells can be divided into embryonic stem cells (derived from the inner cell mass of pre-implanted embryos) (Evans and Kaufman, 1981; Martin, 1981); epiblast stem cells (derived from post-implanted epiblast-stage embryos) (Brons et al, 2007; Tesar et al, 2007); germline-derived stem cells (derived from embryonic gonadial ridges or postnatal testes) (Shamblott et al, 1998; Kanatsu-Shinohara et al, 2004; Guan et al, 2006); induced pluripotent stem cells (from foetal or adult cells) (Takahashi and Yamanaka, 2006; Aoi et al, 2008; Hanna et al, 2008; Park et al, 2008) or adult stem cells (derived from postnatal tissues). Adult stem cells are a rare population in specific tissues but show powerful potential for regeneration. They can be further divided based on their tissue origin into a number of categories such as haematopoietic stem cells, neuronal stem cells, skin stem cells as well as mesenchymal stem cells. Unlike other tissue-specific stem cells, pancreatic stem cells (PSC) were proposed only relatively recently (Ramiya et al, 2000). However, despite intense research, the presence and origin of PSC are hotly debated. In order to understand the role and potential of PSC, better knowledge of pancreas development and function is required. In this review, we will particularly discuss several types of β-cell regeneration in physiological and pathophysiological conditions, and explore the mechanisms of regeneration of β cells. We hope that this will give readers a taste of this controversial but important area of research.

2. Pancreas development and physiology

2.1 Embryology

The pancreas is an organ derived from endoderm. The endoderm is one of the three primitive germ layers formed during the early embryonic stage known as gastrulation. Taking the mouse as an example, the pancreas originates from the thickened endodermal epithelium along the dorsal and ventral surfaces of the posterior foregut. These thickenings can be identified histologically at embryonic day (E) 9.0-9.5 (Pictet et al, 1972). Subsequently, these epithelia evaginate into the surrounding mesoderm-derived mesenchymal tissue and form dorsal and ventral pancreatic buds. These buds continue to expand, branch and fuse as a result of gut rotation that brings the buds together. The fused developing pancreas continues to proliferate, differentiate and, ultimately, develop into the mature pancreas. The adult pancreas consists of digestive enzyme-secreting exocrine tissue, digestive enzyme-transporting ductal tissue and hormone-producing tissue located in the islets of Langerhans. In humans, the dorsal bud can be detected as early as 26 days postcoitum (dpc, an equivalent stage to E9.5 mouse embryos), but insulin-positive cells are not visible until 52 dpc, approximately 2 weeks later than the equivalent stage seen in mice. The appearance of human insulin-positive cells precedes that of glucagon-positive cells at 8-10 weeks of development (Piper et al, 2004). All islet cells are detectable at the end of the first trimester in humans (Piper et al, 2004), but at later stages in mice (Herrera et al, 1991). These data indicate a human-mouse temporal difference in lineage development (Richardson et al, 1997), and this is supported by differences in gene expression patterns during developmental and disease processes in these two species (Fougerousse et al, 2000). More reviews of human pancreas development can be found elsewhere (De Krijger et al, 1992; Lukinius et al, 1992; Polak et al, 2000).

2.2 Pancreatic progenitors

In the thickened DE epithelium along the dorsal and ventral surfaces of the posterior foregut at E9.0-9.5, there are a group of cells express a parahox homeobox transcription factor (TF) gene termed *Pdx1* (pancreas and duodenum TF 1). This is essential for pancreas expansion but not for pancreas initiation in early development and β-cell function in adults (Ohneda et al, 2000; Gu et al, 2002). The specification and differentiation of the various lineages in the pancreas along with transcriptional regulation has been extensively studied and reviewed (Wilson et al, 2003; Jensen, 2004; Servitja and Ferrer, 2004; Habener et al, 2005; Murtaugh, 2007; Best et al, 2008) and is briefly summarised in Figure 1.

Genetic lineage tracing experiment demonstrated that Pdx1-expressing ($Pdx1^+$) cells are multipotent pancreatic progenitors because they give rise to exocrine, endocrine and duct tissues in the pancreas (Gu et al, 2002). These cells are located at the tip of the branching pancreatic tree marked by $Pdx1^+Ptf1a^+$(pancreas transcription factor 1a) $Cpa1^+$(carboxypeptidase 1) (Zhou et al, 2007). To provide cues whether these cells are proliferative, transcriptome profiling analysis was performed showing that when DE cells commit to $Pdx1^+$ pancreatic progenitors there are at least 28 out of 69 (40.6%) cell-cycle and cell-proliferation genes up-regulated (Figure 2) (Jiang et al, 2010).

Surprisingly, direct evidence of proliferation and self-renewal of these $Pdx1^+Ptf1a^+Cpa1^+$ cells has not been produced. However, indirect evidence shows that these $Pdx1^+$ cells may take up bromodeoxyuridine (BrdU), a thymidine analogue that may be incorporated into DNA during S-phase of the cell cycle, indicative of proliferation (Seymour et al, 2007). Unfortunately, due to the lack of a specific marker, this ability of purified $Pdx1^+$ cells has not been examined *in vitro* either. Development of technologies that allow the demonstration of Pdx1⁺ cells to self-renew *in vitro* and to specify all pancreatic lineages will be not only important for pancreas developmental biology, but also crucial for a future ESC-derived cellular therapy for T1D. To serve this purpose, there are two ways to obtain purified Pdx1⁺ cells: from embryos *in vivo* or from directed ESC culture *in vitro*.

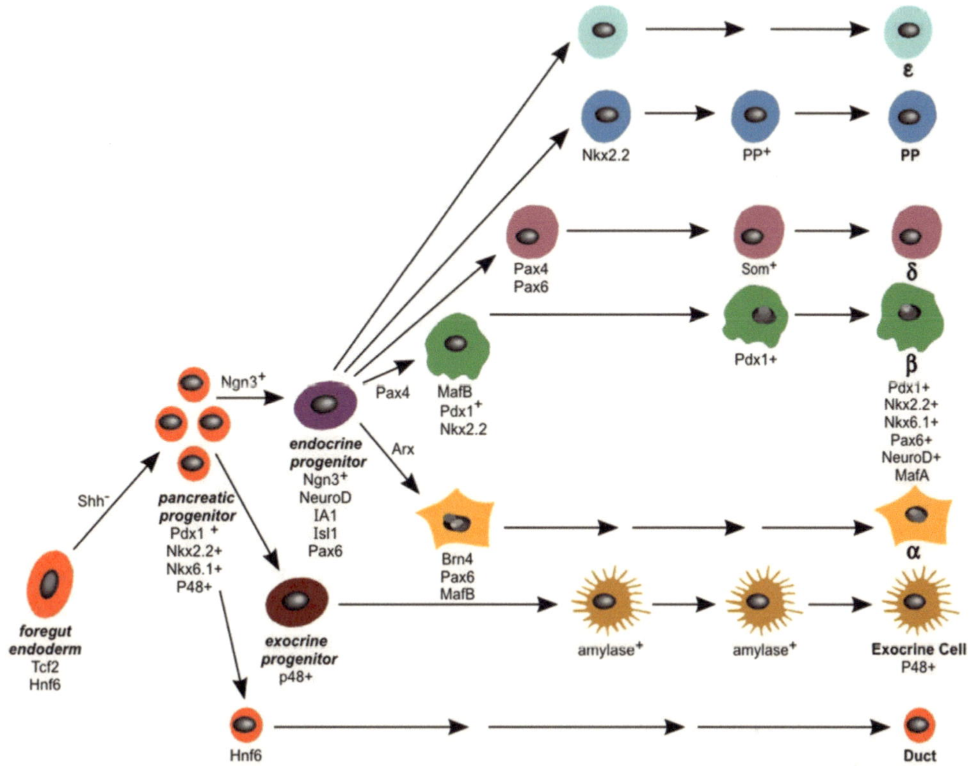

Fig. 1. Lineage development in the pancreas showing role of defined transcription factors. Progenitors within a defined domain of foregut endoderm express *Tcf2* (*T cell factor 2*, also known as *hepatocyte nuclear factor 1b, Hnf1b)* and *Hnf6* (also known as *Onecut 1*). Suppression of *sonic hedgehog* (*Shh*) signalling leads to the development of pancreatic progenitor cells, which are marked by expression of a number of transcription factors, especially Pdx1 (pancreas and duodenum transcription factor 1, also known as Ipf1), pancreas transcription factor 1a (Ptf1a), Nkx2.2 (Nk family homeobox factor 2.2), Nkx61 and Hb9 [also known as motor neuron and pancreas homeobox 1 (Mnx1)]. The Sry-related HMG box transcription factor 9 (Sox9) and suppression of the Notch signalling activates an *neurogenin-3* (*Ngn3*) which allows these pancreatic progenitors to commit to precursors of the endocrine islet lineages; these endocrine progenitors also express *NeuroD* (*neural differentiation* 1), *IA1* (*insulinoma associated* 1), *Isl1* (*Islet* 1) and *Pax6* (*paired box factor* 6). The endocrine progenitors then may differentiate into five types of islet cells [α, β, δ (somatostatin), PP (pancreatic polypeptide) and ε (ghrelin)]. For example, a group of *MafB*- (musculoaponeurotic fibrosarcoma oncogene family protein B), *Pdx1-*, *Pax4-* and *Nkx2.2*-expressing cells will give rise to mature insulin-secreting β cells, whereas cells that express *Brn4* (the brain-specific POU-box factor) and *Pax6* are destined to become glucagon-secreting α cells.

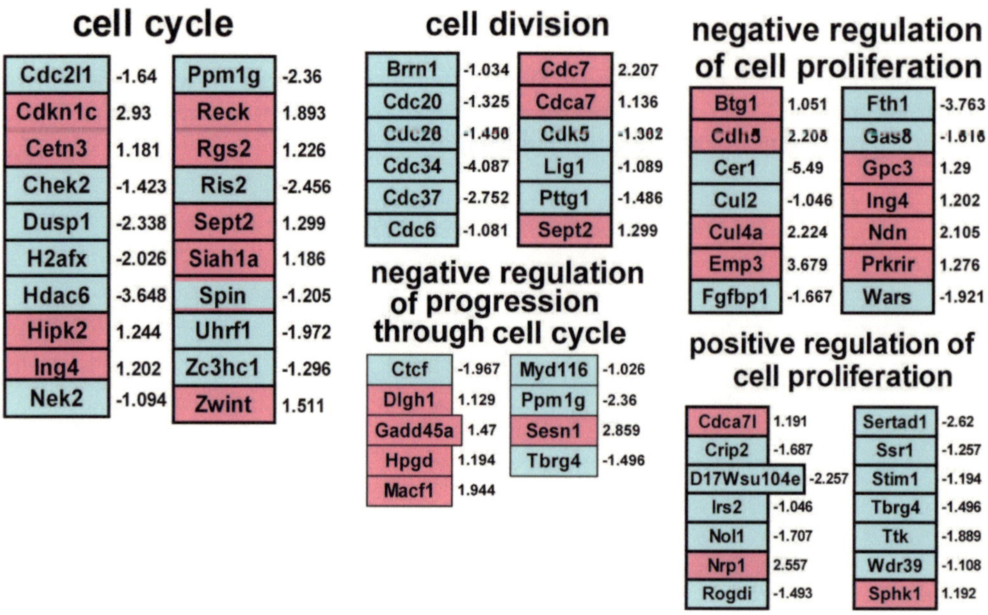

Fig. 2. Dynamic patterns of genes in Pdx1+ cells that regulate cell cycle and cell proliferation. Microarray datasets were generated from dissected definitive endodermal (DE) cells and purified Pdx1+ pancreas progenitors (Gu et al, 2004), followed by GenMapp analysis. Each box identifies a gene. The colour of each box in red or blue represents up- or down-regulation of the gene compared to its expression in DE. The number on the right-hand side is up or down (-) regulation of gene expression in the Log_2 scale. [more detail see (Jiang et al, 2010)].

In humans, numerous PDX1+ cells can be easily detected in the pancreas between 8 and 21 weeks of age (Lyttle et al, 2008; Jeon et al, 2009). The number of PDX1+ cells colocalized with insulin or somatostain is progressively increasing during this period of development (Lyttle et al, 2008). Unfortunately again, these studies did not provide data to show whether the PDX1+ cells are generated by self-renewal themselves or commitment from their progenitors.

2.3 Islet progenitors

At around E9.5 in mice, a small group of cells in the thickened DE epithelium begin to express the basic helix-loop-helix TF neurogenin 3 (Ngn3) (Gradwohl et al, 2000; Gu et al, 2002; Xu et al, 2008). Accumulating evidence indicates that Ngn3-expressing (Ngn3+) cells in the pancreas are islet progenitors that give rise to all islet lineage cells because: (1) in *Ngn3* knockout mice, islet cells do not develop (Gradwohl et al, 2000); (2) gene lineage tracing shows that Ngn3+ cells give rise to all pancreatic endocrine cells (Gu et al, 2002); (3) in adult pancreas, Ngn3+ cells can be activated by partial duct ligation and (4) after injection into a foetal pancreas *in vitro*, purified Ngn3+ cells differentiate into all islet cell types (Xu et al, 2008). Whereas mouse *Ngn3* mRNA expression peaks around E15.5 (equivalent to weeks 7-8 in humans), human *NGN3* expression was low prior to 9 weeks, but from 9 weeks onward, its expression increased sharply and remained high untill 17 weeks (Jeon et al, 2009). Although a few studies showed that Ngn3+ cells were proliferative (Jensen et al, 2000;

Oliver-Krasinski et al, 2009), the recent genetic clonal analyses of mosaic analysis with double marker (MADM) demonstrated that Ngn3⁺ cells are quiescent and these cells give rise to a single islet cell type (Desgraz and Herrera, 2009). Consistent with this, a recent seminal study shows that the expression of Ngn3 inhibits proliferation by inducing *cyclin-dependent kinase inhibitor 1a* (*Cdkn1a*) (Miyatsuka et al, 2011). Our recent study demonstrate that in addition to induction of *Cdkn1a*, numerous genes involved in cell-cycle and cell-proliferation are down-regulated (95/127 or 74.8%) (Figure 3) (Jiang et al, 2010). Even though there is no literature reporting whether Ngn3⁺ cells proliferate *in vitro*, it is not likely that Ngn3⁺ cells are the so-called PSC.

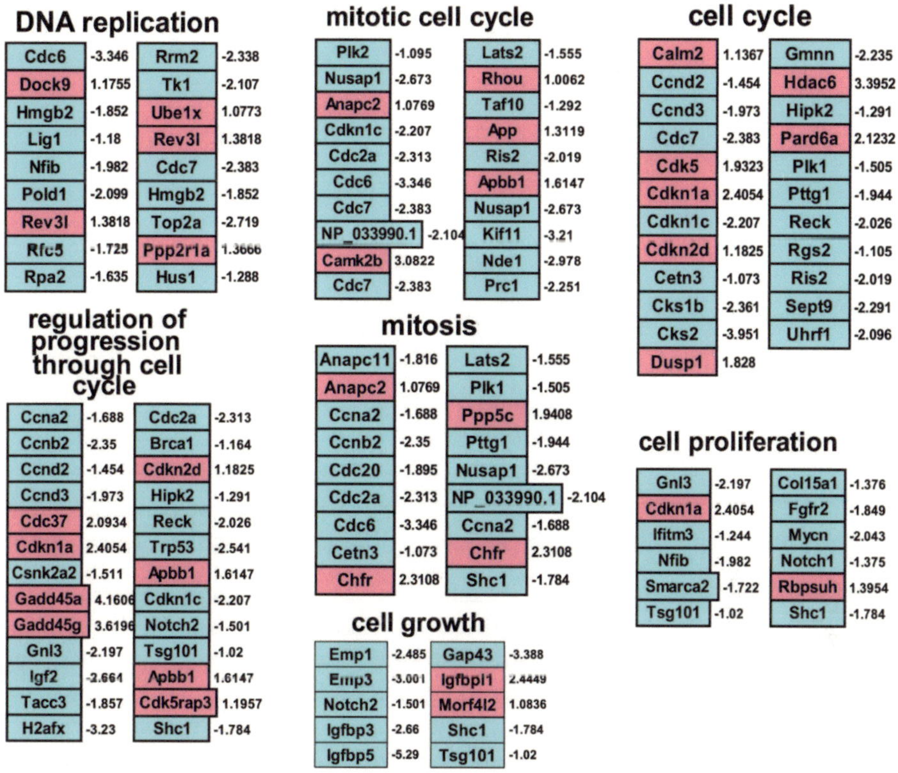

Fig. 3. Dynamic patterns of genes in Ngn3⁺ cells that regulate cell cycle and cell proliferation. Microarray datasets were generated from purified Pdx1⁺ pancreas progenitors and Ngn3⁺ islet progenitors (Gu et al, 2004), followed by GenMapp analysis. Each box identifies a gene. The colour of each box in red or blue represents up- or down-regulation of the gene compared to Pdx1⁺ cells. The number on the right-hand side is up or down (-) regulation of gene expression in the Log_2 scale. [more detail see (Jiang et al, 2010)].

2.4 Physiology

The islets are composed mainly of α, β, δ, ε and PP cells (Figure 1) that secrete glucagon, insulin, somatostatin, ghrelin and pancreatic polypeptide respectively (Jorgensen et al, 2007). These hormones are generally responsible for the regulation of glucose homeostasis. For this function, there is a set of fine tuned paracrine interactions among these endocrine cells which is summarized in Figure 4. In adult humans, there are 2,000–3,000 β cells/islet of Langerhans, with approximately 1 million islets scattered throughout the pancreas (Stefan et al, 1982). The β cells sense the fluctuation of blood glucose levels and secrete insulin in a manner dependent on the glucose concentration. Insulin regulates circulating blood glucose concentrations through its actions on peripheral tissues, such as to inhibit hepatic glucose release and stimulate glucose uptake and storage by skeletal muscle and adipocyte tissue.

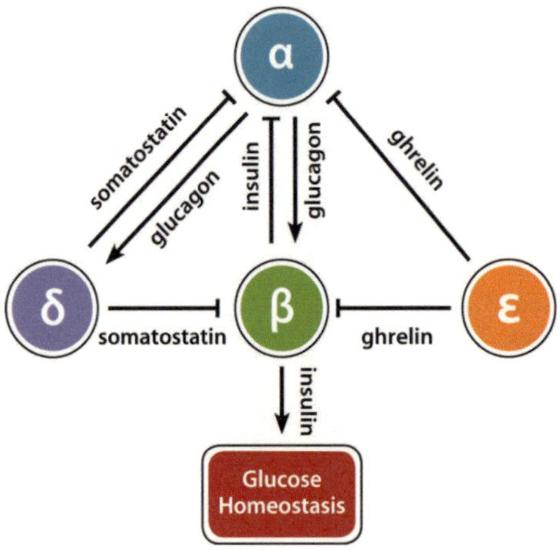

Fig. 4. Paracrine interactions between islet cells.
Insulin secreted by β cells acts as a prime hormone of glucose homeostasis and inhibits glucagon secretion by α cells. Whereas glucagon activates insulin and somatostatin secretion, somatostatin secreted by δ cells and ghrelin by ε cells inhibit insulin secretion.

3. Regeneration of β cells occurs physiologically and pathophysiologically

β cells are indeed observed to be regenerated at least during pregnancy, partial pancreatectomy and obesity. These observations lead to the birth of the PSC concept (Bonner-Weir and Sharma, 2002). The existence of PSC is also inferred from the continued function of islets after transplantation (Ryan et al, 2002; Ryan et al, 2005). Since there is no convincing evidence of contribution of haematopoietic stem cells to islet cells (Wagers et al, 2002) nor that β cells are long-lived (Bonner-Weir, 2000), the continued function of transplanted islets suggests that PSC reside inside the islets and/or the functional β cells are capable of self-renewal.

3.1 Regeneration of β cells during pregnancy

To cope with physiological demand, pancreatic β cells do regenerate during pregnancy in humans and experimental animals. For example, the uptake of BrdU increases 3-fold at E10 and 10-fold at E14 in islets of pregnant rats (Parsons et al, 1992), providing indirect evidence that proliferation of islet cells contributes significantly to the increase of islet volume (Hellman, 1960; Van Assche, 1974). However, there only is a 2-fold increase in islet volume and a 3-fold increase in BrdU labelling at E15.5 maternal mouse islets (Karnik et al, 2007). This discrepancy may reflect a species difference in regeneration capacity or a difference in sensitivity of detection methods. During human pregnancy, both an increase in volume of maternal islets and hyperplasia of 'β' cells have also been observed (Van Assche et al, 1978), but direct evidence of proliferation in these islets is still lacking. In rodents and humans, the proliferation of β cells during pregnancy may be stimulated by prolactin and placental lactogens (Nielsen et al, 1999).

Recently, genetic studies provided molecular insights into how β-cell proliferation occurs during pregnancy. The pregnancy hormone prolactin suppresses the transcriptional co-activator menin, encoded by the gene *multiple endocrine neoplasia type 1 (MEN1)*, resulting in β-cell proliferation. This is demonstrated by experiments in which a short infusion of prolactin is sufficient to reduce menin expression and stimulate proliferation of mouse islet cells (Karnik et al, 2007). In addition, the pregnancy hormones prolactin and placental lactogen also induce expression of *tryptophan hydroxylase 1* (*Tph1*), encoding the enzyme Tph1 essential for serotonin production, which induces BrdU incorporation into isolated islet cells (Kim et al, 2010). However, it is still unknown whether proliferation comes from functional β cells and/or from PSC within islets. Future research should be directed to recapitulate this effect *in vitro* with purified subpopulations of islet cells, from which novel stimuli and molecular pathways may be identified. Defining these pathways may establish a platform on which novel strategies can be developed for a cure of T1D.

3.2 Regeneration of β cells during obesity

β cells regenerate in response to pathological processes such as obesity. For example, up to 10-fold increase in β-cell mass has been observed in obese rodents, responding to their insulin resistance (Butler et al, 2003a). Double staining of pancreas sections from obese mice and humans can detect insulin-producing cells that express Ki-67, a marker strictly associated with cell proliferation (Butler et al, 2003a; Butler et al, 2003b), indicating that regeneration may occur in the islets. Again, the regeneration capacity seems to be significantly greater in mice than in humans, although the underlying mechanism is unclear yet. Studies of one obese mutant mouse line, termed A^y, showed that reduction of menin contributes to adaptive β-cell proliferation (Karnik et al, 2007). Taken together, these data suggest that a similar mechanism for β-cell regeneration may operate in physiological pregnancy and pathophysiological obesity in mice. It would be very interesting to determine whether this mechanism is also at work during human pregnancy or obesity.

3.3 Regeneration of β cells after partial pancreatectomy

Like many other organs in the body, islets do regenerate in response to injury, in this case, pancreatectomy. In rats 4 weeks after 90% pancreatectomy, for example, there is a regeneration to 27% and 45% of sham-operated pancreas and islet mass, respectively (Bonner-Weir and Sharma, 2002). However, there are species differences in regeneration capacity. Even a 50% pancreatectomy in adult dogs would cause impaired fasting glucose in the short term (Matveyenko et al, 2006) and diabetes mellitus in the longer term (Stagner and Samols, 1991). Likewise, a 50% pancreatectomy in adult humans also leads to subsequent obesity and diabetes mellitus (Robertson et al, 2002). These studies again indicate that there is a difference between species in their capacity to mount islet regeneration: this is much more powerful in rodents than in larger mammals. Additional studies are needed to confirm this capacity difference and understand its underlying mechanism. Furthermore, it is less clear to what extent islet regeneration contributes to maintain β-cell mass in adult humans from existing islet cells and how much is from cells of any other origins. This knowledge is critical for a viable strategy to promote β-cell regeneration both *in vivo* and *in vitro*. Additionally, it is unknown whether the PSC also contribute to this type of regeneration. In the following sections, we will examine several possible mechanisms that may lead to β-cell generation and regeneration.

4. Evidence that β cells are generated from PSC

In addition to the observations of long-term survival of transplanted islets mentioned above, substantial *in vitro* evidence has indicated that "pluripotent PSC" may be present in all three major pancreas compartments, i.e. the ductal epithelium (Cornelius et al, 1997; Ramiya et al, 2000; Suzuki et al, 2002), islets, and acinar tissue (Zulewski et al, 2001; Seaberg et al, 2004). This evidence comes from studies of both rodent and human pancreas. For example, a potential PSC candidate has been purified by flow cytometry in the developing and adult mouse pancreas. These cells are identified by expression of the receptor for hepatocyte growth factor, c-Met, and absence of blood cell surface markers such as CD45, TER119, c-Kit, and Flk-1. These cells can differentiate into multiple pancreatic lineage cells from individual cells *in vitro* and give rise to pancreatic endocrine and acinar cells *in vivo* following transplantation (Suzuki et al, 2004). However, the *in vivo* localization and the molecular characteristics of these c-met expressing cells are largely unknown and clonogenesis at the single cell level has not been established.

4.1 PSC may house in ductal epithelium

Over the last several years, we investigated the differentiation and proliferation of foetal mouse pancreatic cells, believed to be a rich source for potential PSC. We first demonstrated *in vitro* that bone morphogenetic proteins-2, -4, -5 and -6, members of the transforming growth factor β superfamily, can promote the proliferation of pancreatic precursors and the development of pancreatic cystic epithelial colonies containing β cells (Jiang et al, 2002; Jiang and Harrison, 2005a), a process partially recapitulating an *in vivo* developmental stage. In addition, we also found that various isoforms of epidermal growth factors can stimulate colony formation (Jiang and Harrison, 2005b). These data indicate that extracellular signalling molecules, including various families of growth factors, modulate fate changes of pancreas precursor/stem cells. However, currently these colony-forming cells may at most be considered precursors because their self-renewal has not been demonstrated *in vitro* (F.X.J. unpublished observations).

Recently, *in vitro* and *in vivo* experiments have indicated that PSC are localized to the ductal epithelium. Bonner-Weir and colleagues were the first to report that adult human pancreatic ductal epithelial cells can form islet-like clusters and differentiate into insulin-secreting β cells (Bonner-Weir et al, 2000). Ramiya and colleagues reported that transplantation of *in vitro* generated islet-like structures from mouse PSC in the ductal epithelium can reverse diabetes mellitus (Ramiya et al, 2000). Another set of experiments used cultures of "pancreatic ductal cell aggregates" that were left over from pancreas digests after purification of human islets for transplantation. From these cultures, fibroblast-like cells grew out; these have been termed "pancreatic mesenchymal stem cells (pMSC)". These cells can undergo at least 12 passages and express a range of bone marrow-derived MSC markers including CD13, CD29, CD44, CD54, CD105, α_6 integrin subunit (also known as CD49f) and Thy1 (also known as CD90). These pMSC can give rise to cells of at least two germ layer origins including endoderm-derived cells, but not convincingly pancreatic lineage cells (Seeberger et al, 2006). *In vivo,* the large, small, and centrolobular ducts of the rat pancreas contain foci of cells that express the neural stem cell–specific marker, nestin, but do not express the ductal marker cytokeratin 19 (CK19) (Zulweski et al. 2001), suggestive of "islet progenitor" cells.

However, all the above studies have used mixed cell populations and have failed to demonstrate clonogenesis. Using culture conditions suitable for generating neurospheres ex vivo, mouse pancreatic ductal cells gave rise to neurosphere-like structures that can subsequently be differentiated into several types of islet cells including β cells (Seaberg et al, 2004). The molecular phenotype of progenitor cells for these islet cells remains unknown. On the other hand, after pancreatic duct ligation, numerous $CK19^+$ ductal cells are regenerated and then $Ngn3^+$ cells are observed, transplantation of the latter has further resulted in their differentiation into functional β cells (Xu et al, 2008), suggesting that the regeneration process may resemble that of embryonic pancreas development.

These studies did not explore the immediate origin of $Ngn3^+$ cells in adult pancreas and whether the cells that give rise to these $Ngn3^+$ cells possess PSC features. Recently, the use of the *in vivo* genetic tracing Cre-loxP system has generated further knowledge in this regard. In these experiments, Cre expression was directed by the promoter of carbonic anhydrase II, a marker of mature ductal cells, resulting in the excision of the stop cassette (Rosa-loxP-stop-loxP-lacZ) in the transgenic Rosa26 (R26R) mice. Therefore β-galactosidase activity is expressed in the cells that express Cre (ductal cells). At 4 weeks, β-galactosidase is detected in many ducts, patched acinar cells and 35-40% of islet cells (Bonner-Weir et al, 2008). These data provide direct evidence that adult ductal epithelial cells can give rise to islet cells, at least in mice. It is still unknown whether the carbonic anhydrase II-expressing cell population is homogenous or heterogenous, and whether this is a differentiation process from PSC or a transdifferentiation process from mature ductal cells. Anyway, these results should be repeated by independent laboratories and even in other experimental animals.

4.2 PSC may reside in the islets

Accumulated *in vitro* evidence suggests that a subpopulation of islet cells has PSC potential. A well-planned *in vitro* study found that a subset of unspecified cells in mouse islets and ducts named as pancreas-derived multipotent precursor cells give rise to neurosphere-like structures and subsequently could differentiate into islet β-like cells. This was a minor property, as most of these cells developed into neural lineage cells (Seaberg et al, 2004). The phenotype of these potential PSC cells was not defined.

In rat and human islets, a distinct population of $nestin^+$ cells that do not express the hormones insulin, glucagon, somatostatin or pancreatic polypeptide has been identified. When cultured *in vitro,* these cells proliferate extensively (~8 months), give rise to cells that express liver and exocrine pancreas markers, such as α-fetoprotein and pancreatic amylase, and display a ductal/endocrine phenotype with expression of CK19, neural-specific cell adhesion molecule, insulin, glucagon, and PDX1. These $nestin^+$ putative progenitor cells may therefore participate in the neogenesis of islet endocrine cells (Zulewski et al, 2001), mediated at least partially by glucagon-like peptide-1, an incretin hormone derived from processing proglucagon (Abraham et al, 2002). However, recent studies *in vivo* indicate that $Nestin^+$ cells are mostly restricted in non-endodermal-derived cells (Lardon et al, 2002; Selander and Edlund, 2002). In addition, from donated adult human islets, outgrown fibroblast-like cells do not express hormones, but proliferate readily and give rise *in vitro* to hormone-expressing cell aggregates. However, these aggregates cannot be induced to become typical functional islet cells (Gershengorn et al, 2004). Furthermore, a rigorous genetic-based lineage tracing in mice under the control of *Pdx1* or *rat insulin promoter* (*RIP*) demonstrated that neither PDX1- nor RIP-expressing cells contribute significantly to these fibroblast-like cells *in vitro* (Chase et al, 2007). In contrast, human islet RIP-expressing cells could be dedifferentiated to fibroblast-like cells in which insulin expression was not detectable and proliferate *in vitro* up to 16 population doublings (Russ et al, 2008). Whether this discrepancy is due to species differences remains to be clarified. Because of its critical importance, such tracing experiments should be reproduced by independent research groups. Further investigations are required to resolve the inconsistent results from the current studies.

4.3 PSC may locate in the exocrine tissue

In the clinic, a large population of nonendocrine pancreatic acinar cells would be discarded after purification of islets from donated pancreas for transplantation. The possibility of making use of these cells has attracted significant interest in recent years. After co-transplantation with foetal pancreatic cells under the kidney capsule of immunodeficient mice, these nonendocrine pancreatic epithelial cells have been shown to be capable of endocrine differentiation though without evidence of β-cell replication or cell fusion. These experiments suggest the existence of PSC or progenitor cells within the acinar compartment of the adult human pancreas (Hao et al, 2006). More recently, analysis using the Cre/loxP-based tracing system demonstrated that amylase/elastase-expressing acinar cells can give rise to insulin-positive cells in a suspension culture (Minami et al, 2005). However, because clonal assay of these amylase/elastase-expressing cells and their intermediate steps have not been investigated, this study may simply reveal that mouse and rat pancreatic acinar cells are able to transdifferentiate into surrogate insulin-expressing cells (Baeyens et al, 2005; Minami et al, 2008). This possibility was further supported by a recent study which shows that mouse acinar cells can be directly re-programmed *in vivo* to β-like cells with just three transcription factor genes, namely, *Pdx1*, *Ngn3* and *MafA* (Zhou et al, 2008). On the other hand, similar lineage tracing experiments *in vivo* demonstrate that after 70-80% pancreatectomy, pre-existing mouse pancreatic acinar cells do not contribute to regeneration of islet β cells (Desai et al, 2007). The contradiction of *in vitro* and *in vivo* findings requires further reconciliation.

5. Evidence that islet β cells are capable of self-replication

There are several pieces of strong evidence demonstrating that islet β cells act as functional "stem" cells to reproduce themselves. Using RIP-driven reporter genes to genetically trace the fate of functional insulin-secreting cells, Dor and colleagues (Dor et al, 2004) first revealed that adult mouse pancreatic β cells are duplicated by RIP-expressing cells within the islets, either physiologically or after partial pancreatectomy. This study assumed that all RIP-expressing cells in adult islets are functional β cells and did not exclude the presence of PSC. Similarly, by using a transgenic model, in which the expression of diphtheria toxin was directed by RIP to β cells, diphtheria expression results in apoptosis of 70%-80% of β cells, destruction of islet architecture and, finally, diabetes mellitus. Withdrawal of diphtheria expression led to a significant regeneration of β-cell mass and a spontaneous normalization of blood glucose levels and islet architecture. Simultaneously, RIP-based lineage tracing analysis indicated that the proliferation of 20-30% surviving 'β' cells played a major role in this regeneration and in recovery of euglycemia (Nir et al, 2007).

Using the more sophisticated MADM system in mice known as RIP-CreER; Rosa26^{GR}/Rosa26^{RG}, each RIP-expressing clone has been demonstrated to consist of 5.1±5.4 or 8.2±6.9 cells after one or two months of chase (Brennand et al, 2007). These RIP-expressing clones have been interpreted as further evidence of regeneration of functional β cells. An additional loss-of-function study following knockout of the Hnf4α (hepatocyte nuclear factor 4α) gene suggested that the β-cell regeneration may involve the Ras/Erk signalling cascade (Gupta et al, 2007) and ultimately be regulated by cycling modulators including cyclin D2 (Georgia and Bhushan, 2004). Taken together, further identification and characterization of the so-called self-replicative or dedifferentiative RIP-expressing cells both *in vivo* and *in vitro* will be urgently needed because they may hold the key for a regenerative therapy for type 1 diabetes mellitus.

Again using thymidine-based lineage tracing, β cells were demonstrated to be produced within an islet by rare self-renewing cells with a slow replication-refractory period (Teta et al, 2007), although the identity of these unique cells and the length of their replication-refractory period remain to be determined. The frequency of these self-renewal cells can be significantly increased after partial pancreatectomy or during pregnancy. Further studies should determine the molecular signature and biological potential of these replicating self-renewal cells. Because of ethical issues, similar studies cannot be performed in human islet tissues, but such investigation should at least be repeated in larger mammals.

Nevertheless, the β-cell population in the adult islets is in fact functionally heterogeneous (Heimberg et al, 1993; Pipeleers et al, 1994; Szabat et al, 2009). By using a dual fluorescence reporter mouse line, a few Ngn3$^+$ cells in the developing pancreas have been observed to coexpress insulin (Hara et al, 2006). In humans a few NGN3$^+$ cells in the foetal pancreas are also observed to coexpress insulin from 10 to 21 weeks of age (Lyttle et al, 2008). Consistent with these studies, the insulin gene expression has been detected from Pdx1$^+$ progenitors through Ngn3$^+$ cells during development to mature islet cells (Jiang et al, 2010) (Figure 5).

An early study indicated that the RIP-expressing cells in the developing pancreas gave rise to other islet cell types in addition to β cells (Alpert et al, 1988). The CD105$^+$CD37$^+$CD90$^+$ mesenchymal stromal cells present in adult human islets have been shown to express a low level of insulin mRNA (Davani et al, 2007). After considering all these studies, therefore, lineage tracing studies under the control of other mature β-cell specific transcription factor gene promoters such as Pdx1 or MafA should be performed. Just like the β-cell line Min6 cells (Miyazaki et al, 1990), the glucose-responsive β cells in the islets may indeed duplicate themselves if the Pdx1- or MafA-expressing cells can be shown to be proliferative similar to the RIP-expressing cells.

6. The identity of PSC is inconclusive

Whereas investigation of β-cell duplication as a mechanism of islet regeneration has attracted great attention in recent years, much progress has been made to identify PSC. However, their identity is still not known. This is due, in addition to the knowledge gap that the signals required for late stage differentiation of functional β cells are largely unknown, at least partially to lack of the following factors: specific cell surface markers to characterize and purify cells that may have PSC potential; a simple, effective and reproducible *in vitro* assay to examine self-renewal and differentiation potential of a purified cell population; *in vivo* functional assays to determine biological function for both experimental animals and humans. Finally, well-agreed general criteria for defining a PSC are required.

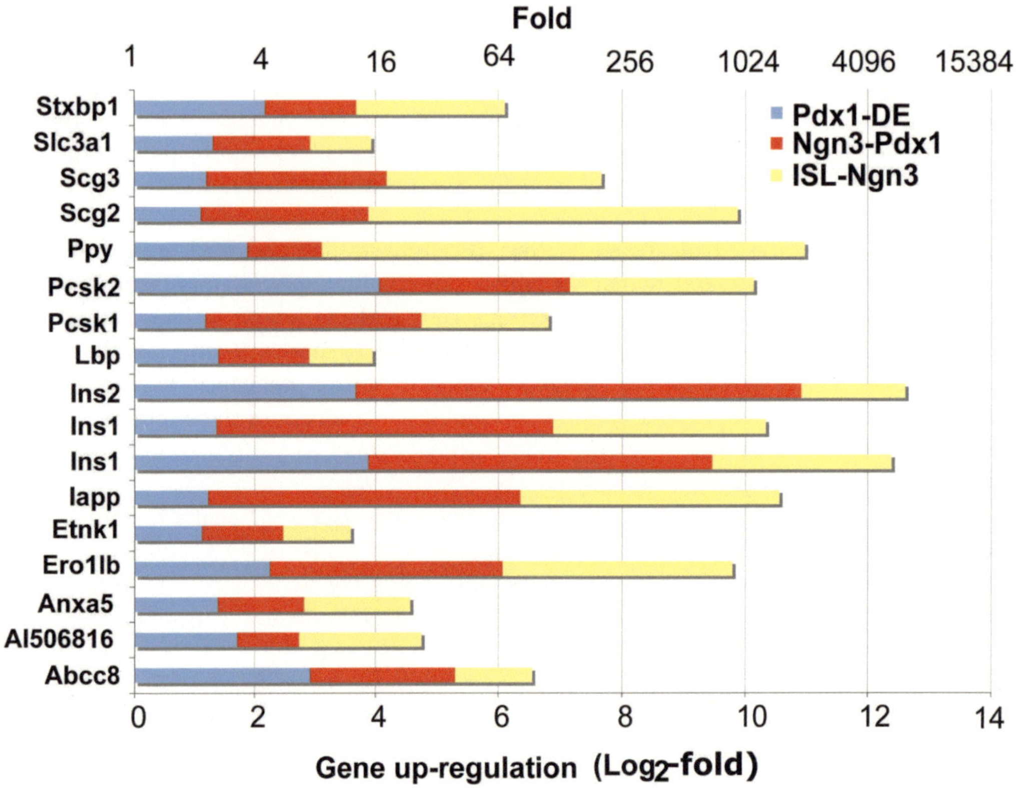

Fig. 5. Insulin genes being continuously up-regulated from pancreas progenitors to mature islet cells.
Microarray datasets were generated from dissected definitive endodermal (DE) cells, purified Pdx1+ pancreas progenitors, Ngn3+ islet progenitors and mature islet cells (ISL) (Gu et al, 2004). Bioinformatics algorithms for time-series datasets were used to calculate expression levels of insulin genes and other genes involved in post-translational modifications of insulin during the stages from Pdx1+ through Ngn3+ islet progenitors to ISL. [more detail see (Jiang et al, 2010)].

7. Future directions for PSC research

We believe arguably that the successful establishment of *in vitro* culture and *in vivo* functional assay systems (Daniel et al, 1989; Dexter, 1989) facilitates the investigation of haematopoietic stem cells and subsequently leads to the application in clinics as a stem cell therapy. Therefore our future work should reach a consensus that PSC at least satisfy the following criteria: (1) clonogenesis should be demonstrated at the level of single cells sorted by flow cytometry; (2) sorted single cells should self-renew *in vitro*; (3) clonogenic cells should give rise *in vitro* to more than one specialized cell lineage and (4) these cells should differentiate *in vivo* into several functional cell types after transplantation. In the future, the investigation of PSC should be particularly encouraged based on three considerations. First, differentiation of embryonic stem cells into insulin-expressing cells was initially exciting and promising, but safety consideration aside, the generation of functional β cells proves extremely difficult and progress has been far slower than originally expected. PSC and dedifferentiation or transdifferentiation of other cell types in the pancreas may therefore provide an alternative renewable source of surrogate β cells. Second, β-cell duplication theory was mostly established from rodent models. It cannot completely exclude the existence of PSC, and it assumed that all insulin-expressing cells in the adult islets are uniformly functional β cells, but in fact these cells are heterogeneous. Last, as there is a significant difference in regeneration capacity between rodent and human islets, it may be wise not to directly extrapolate regeneration data from rodents to humans. Unlike previous studies that were only performed on *in vivo* or *in vitro* experiments without targeting specific cell types, future PSC work may need to employ integrated approaches, for example, applying cell surface markers to target particular cell populations; examining *in vitro* their potential for self-renewal and clonogenesis, using genetic approaches to dissect molecular mechanisms of their phenotypic changes *in vivo* and finally to examine their lineage contribution and biological function. Performing such integrated research is likely beyond the capacity of most individual laboratories, and therefore requires multi-discipline, multi-laboratory and even multi-national collaborations including the participation of the pharmaceutical sector to generate a therapeutic grade of surrogate β cells for a replacement therapy to diabetes mellitus.

8. Note to add

On the eve of submission, an important study from Dr Derek van der Kooy's laboratory sheds a shiny ray of new light on this inconclusive topic. By using state-of-art genetic lineage tracing techniques, Smukler and colleagues now provide conclusive evidence that the PSC cells (originally called as pancreas-derived multipotent precursor cells) in the adult pancreas were derived from the embryonic pancreatic lineage, but not from the neural crest, as previously supposed. The PSC cells express insulin, along with an array of markers typical of islet progenitors; are distinct from mature functional ells; and give rise to non-beta cells in vivo. In addition to mouse PSC cells, human PSC cells have also been identified. After transplantation to diabetic mice, both mouse and human PSC cells could ameliorate their diabetes mellitus (Smukler et al, 2011).

9. Acknowledgement

The authors are supported by grants from Juvenile Diabetes Research Foundational International (4-2006-1025), the Diabetes Research Foundation of Western Australia, the University of Western Australia, and the Medical Research Foundation of Royal Perth Hospital.

10. References

Abraham, E. J., Leech, C. A., Lin, J. C., Zulewski, H. and Habener, J. F. (2002) Insulinotropic hormone glucagon-like peptide-1 differentiation of human pancreatic islet-derived progenitor cells into insulin-producing cells, *Endocrinology* 143(8): 3152-61.

Alpert, S., Hanahan, D. and Teitelman, G. (1988) Hybrid insulin genes reveal a developmental lineage for pancreatic endocrine cells and imply a relationship with neurons, *Cell* 53(2): 295-308.

Aoi, T., Yae, K., Nakagawa, M., Ichisaka, T., Okita, K., Takahashi, K., Chiba, T. and Yamanaka, S. (2008) Generation of pluripotent stem cells from adult mouse liver and stomach cells, *Science* 321(5889): 699-702.

Baeyens, L., De Breuck, S., Lardon, J., Mfopou, J. K., Rooman, I. and Bouwens, L. (2005) In vitro generation of insulin-producing beta cells from adult exocrine pancreatic cells, *Diabetologia* 48(1): 49-57.

Best, M., Carroll, M., Hanley, N. A. and Piper Hanley, K. (2008) Embryonic stem cells to beta-cells by understanding pancreas development, *Mol Cell Endocrinol* 288(1-2): 86-94.

Bonner-Weir, S. (2000) Life and death of the pancreatic beta cells, *Trends Endocrinol Metab* 11(9): 375-8.

Bonner-Weir, S., Inada, A., Yatoh, S., Li, W. C., Aye, T., Toschi, E. and Sharma, A. (2008) Transdifferentiation of pancreatic ductal cells to endocrine beta-cells, *Biochem Soc Trans* 36(Pt 3): 353-6.

Bonner-Weir, S. and Sharma, A. (2002) Pancreatic stem cells, *J Pathol* 197(4): 519-26.

Bonner-Weir, S., Taneja, M., Weir, G. C., Tatarkiewicz, K., Song, K. H., Sharma, A. and O'Neil, J. J. (2000) In vitro cultivation of human islets from expanded ductal tissue, *Proc Natl Acad Sci U S A* 97(14): 7999-8004.

Brennand, K., Huangfu, D. and Melton, D. (2007) All beta Cells Contribute Equally to Islet Growth and Maintenance, *PLoS Biol* 5(7): e163.

Brons, I. G., Smithers, L. E., Trotter, M. W., Rugg-Gunn, P., Sun, B., Chuva de Sousa Lopes, S. M., Howlett, S. K., Clarkson, A., Ahrlund-Richter, L., Pedersen, R. A. et al. (2007) Derivation of pluripotent epiblast stem cells from mammalian embryos, *Nature* 448(7150): 191-5.

Butler, A. E., Janson, J., Bonner-Weir, S., Ritzel, R., Rizza, R. A. and Butler, P. C. (2003a) Beta-cell deficit and increased beta-cell apoptosis in humans with type 2 diabetes, *Diabetes* 52(1): 102-10.

Butler, A. E., Janson, J., Soeller, W. C. and Butler, P. C. (2003b) Increased beta-cell apoptosis prevents adaptive increase in beta-cell mass in mouse model of type 2 diabetes: evidence for role of islet amyloid formation rather than direct action of amyloid, *Diabetes* 52(9): 2304-14.

Chase, L. G., Ulloa-Montoya, F., Kidder, B. L. and Verfaillie, C. M. (2007) Islet-derived fibroblast-like cells are not derived via epithelial-mesenchymal transition from Pdx-1 or insulin-positive cells, *Diabetes* 56(1): 3-7.

Cornelius, J. G., Tchernev, V., Kao, K. J. and Peck, A. B. (1997) In vitro-generation of islets in long-term cultures of pluripotent stem cells from adult mouse pancreas, *Horm Metab Res* 29(6): 271-7.

Daniel, C. P., Ponting, I. L. and Dexter, T. M. (1989) Growth and development of haemopoietic cells: a deterministic process?, *Hamatol Bluttransfus* 32: 172-7.

Davani, B., Ikonomou, L., Raaka, B. M., Geras-Raaka, E., Morton, R. A., Marcus-Samuels, B. and Gershengorn, M. C. (2007) Human islet-derived precursor cells are mesenchymal stromal cells that differentiate and mature to hormone-expressing cells in vivo, *Stem Cells* 25(12): 3215-22.

De Krijger, R. R., Aanstoot, H. J., Kranenburg, G., Reinhard, M., Visser, W. J. and Bruining, G. J. (1992) The midgestational human fetal pancreas contains cells coexpressing islet hormones, *Dev Biol* 153(2): 368 75.

Desai, B. M., Oliver-Krasinski, J., De Leon, D. D., Farzad, C., Hong, N., Leach, S. D. and Stoffers, D. A. (2007) Preexisting pancreatic acinar cells contribute to acinar cell, but not islet beta cell, regeneration, *J Clin Invest* 117(4): 971-7.

Desgraz, R. and Herrera, P. L. (2009) Pancreatic neurogenin 3-expressing cells are unipotent islet precursors, *Development* 136(21): 3567-74.

Dexter, T. M. (1989) Regulation of hemopoietic cell growth and development: experimental and clinical studies, *Leukemia* 3(7): 469-74.

Dor, Y., Brown, J., Martinez, O. I. and Melton, D. A. (2004) Adult pancreatic beta-cells are formed by self-duplication rather than stem-cell differentiation, *Nature* 429(6987): 41-6.

Evans, M. J. and Kaufman, M. H. (1981) Establishment in culture of pluripotential cells from mouse embryos, *Nature* 292(5819): 154-6.

Fougerousse, F., Bullen, P., Herasse, M., Lindsay, S., Richard, I., Wilson, D., Suel, L., Durand, M., Robson, S., Abitbol, M. et al. (2000) Human-mouse differences in the embryonic expression patterns of developmental control genes and disease genes, *Hum Mol Genet* 9(2): 165-73.

Georgia, S. and Bhushan, A. (2004) Beta cell replication is the primary mechanism for maintaining postnatal beta cell mass, *J Clin Invest* 114(7): 963-8.

Gershengorn, M. C., Hardikar, A. A., Wei, C., Geras-Raaka, E., Marcus-Samuels, B. and Raaka, B. M. (2004) Epithelial-to-mesenchymal transition generates proliferative human islet precursor cells, *Science* 306(5705): 2261-4.

Gradwohl, G., Dierich, A., LeMeur, M. and Guillemot, F. (2000) neurogenin3 is required for the development of the four endocrine cell lineages of the pancreas, *Proc Natl Acad Sci U S A* 97(4): 1607-11.

Gu, G., Dubauskaite, J. and Melton, D. A. (2002) Direct evidence for the pancreatic lineage: NGN3+ cells are islet progenitors and are distinct from duct progenitors, *Development* 129(10): 2447-57.

Gu, G., Wells, J. M., Dombkowski, D., Preffer, F., Aronow, B. and Melton, D. A. (2004) Global expression analysis of gene regulatory pathways during endocrine pancreatic development, *Development* 131(1): 165-79.

Guan, K., Nayernia, K., Maier, L. S., Wagner, S., Dressel, R., Lee, J. H., Nolte, J., Wolf, F., Li, M., Engel, W. et al. (2006) Pluripotency of spermatogonial stem cells from adult mouse testis, *Nature* 440(7088): 1199-203.

Gupta, R. K., Gao, N., Gorski, R. K., White, P., Hardy, O. T., Rafiq, K., Brestelli, J. E., Chen, G., Stoeckert, C. J., Jr. and Kaestner, K. H. (2007) Expansion of adult beta-cell mass in response to increased metabolic demand is dependent on HNF-4alpha, *Genes Dev* 21(7): 756-69.

Habener, J. F., Kemp, D. M. and Thomas, M. K. (2005) Minireview: transcriptional regulation in pancreatic development, *Endocrinology* 146(3): 1025-34.

Hanna, J., Markoulaki, S., Schorderet, P., Carey, B. W., Beard, C., Wernig, M., Creyghton, M. P., Steine, E. J., Cassady, J. P., Foreman, R. et al. (2008) Direct reprogramming of terminally differentiated mature B lymphocytes to pluripotency, *Cell* 133(2): 250-64.

Hao, E., Tyrberg, B., Itkin-Ansari, P., Lakey, J. R., Geron, I., Monosov, E. Z., Barcova, M., Mercola, M. and Levine, F. (2006) Beta-cell differentiation from nonendocrine epithelial cells of the adult human pancreas, *Nat Med* 12(3): 310-6.

Hara, M., Dizon, R. F., Glick, B. S., Lee, C. S., Kaestner, K. H., Piston, D. W. and Bindokas, V. P. (2006) Imaging pancreatic beta-cells in the intact pancreas, *Am J Physiol Endocrinol Metab* 290(5): E1041-7.

Heimberg, H., De Vos, A., Vandercammen, A., Van Schaftingen, E., Pipeleers, D. and Schuit, F. (1993) Heterogeneity in glucose sensitivity among pancreatic beta-cells is correlated to differences in glucose phosphorylation rather than glucose transport, *Embo J* 12(7): 2873-9.

Hellman, B. (1960) The islets of Langerhans in the rat during pregnancy and lactation, with special reference to the changes in the B/A cell ratio, *Acta Obstet Gynecol Scand* 39: 331-42.

Herrera, P. L., Huarte, J., Sanvito, F., Meda, P., Orci, L. and Vassalli, J. D. (1991) Embryogenesis of the murine endocrine pancreas; early expression of pancreatic polypeptide gene, *Development* 113(4): 1257-65.

Jensen, J. (2004) Gene regulatory factors in pancreatic development, *Dev Dyn* 229(1): 176-200.

Jensen, J., Heller, R. S., Funder-Nielsen, T., Pedersen, E. E., Lindsell, C., Weinmaster, G., Madsen, O. D. and Serup, P. (2000) Independent development of pancreatic alpha- and beta-cells from neurogenin3-expressing precursors: a role for the notch pathway in repression of premature differentiation, *Diabetes* 49(2): 163-76.

Jeon, J., Correa-Medina, M., Ricordi, C., Edlund, H. and Diez, J. A. (2009) Endocrine cell clustering during human pancreas development, *J Histochem Cytochem* 57(9): 811-24.

Jiang, F. X. and Harrison, L. C. (2005a) Convergence of bone morphogenetic protein and laminin-1 signaling pathways promotes proliferation and colony formation by fetal mouse pancreatic cells, *Exp Cell Res* 308(1): 114-22.

Jiang, F. X. and Harrison, L. C. (2005b) Laminin-1 and epidermal growth factor family members co-stimulate fetal pancreas cell proliferation and colony formation, *Differentiation* 73(1): 45-9.

Jiang, F. X., Mehta, M. and Morahan, G. (2010) Quantification of Insulin Gene Expression During Development of Pancreatic Islet Cells, *Pancreas* 39: 201-208.

Jiang, F. X., Stanley, E. G., Gonez, L. J. and Harrison, L. C. (2002) Bone morphogenetic proteins promote development of fetal pancreas epithelial colonies containing insulin-positive cells, *J Cell Sci* 115(Pt 4): 753-760.

Jorgensen, M. C., Ahnfelt-Ronne, J., Hald, J., Madsen, O. D., Serup, P. and Hecksher-Sorensen, J. (2007) An illustrated review of early pancreas development in the mouse, *Endocr Rev* 28(6): 685-705.

Kanatsu-Shinohara, M., Inoue, K., Lee, J., Yoshimoto, M., Ogonuki, N., Miki, H., Baba, S., Kato, T., Kazuki, Y., Toyokuni, S. et al. (2004) Generation of pluripotent stem cells from neonatal mouse testis, *Cell* 119(7): 1001-12.

Karnik, S. K., Chen, H., McLean, G. W., Heit, J. J., Gu, X., Zhang, A. Y., Fontaine, M., Yen, M. H. and Kim, S. K. (2007) Menin controls growth of pancreatic beta-cells in pregnant mice and promotes gestational diabetes mellitus, *Science* 318(5851): 806-9.

Kim, H., Toyofuku, Y., Lynn, F. C., Chak, E., Uchida, T., Mizukami, H., Fujitani, Y., Kawamori, R., Miyatsuka, T., Kosaka, Y. et al. (2010) Serotonin regulates pancreatic beta cell mass during pregnancy, *Nat Med* 16(7): 804-8.

Lardon, J., Rooman, I. and Bouwens, L. (2002) Nestin expression in pancreatic stellate cells and angiogenic endothelial cells, *Histochem Cell Biol* 117(6): 535-40.

Lock, L. T. and Tzanakakis, E. S. (2007) Stem/Progenitor cell sources of insulin-producing cells for the treatment of diabetes, *Tissue Eng* 13(7): 1399-412.

Lukinius, A., Ericsson, J. L., Grimelius, L. and Korsgren, O. (1992) Ultrastructural studies of the ontogeny of fetal human and porcine endocrine pancreas, with special reference to colocalization of the four major islet hormones, *Dev Biol* 153(2): 376-85.

Lyttle, B. M., Li, J., Krishnamurthy, M., Fellows, F., Wheeler, M. B., Goodyer, C. G. and Wang, R. (2008) Transcription factor expression in the developing human fetal endocrine pancreas, *Diabetologia* 51(7): 1169-80.

Martin, G. R. (1981) Isolation of a pluripotent cell line from early mouse embryos cultured in medium conditioned by teratocarcinoma stem cells, *Proc Natl Acad Sci U S A* 78(12): 7634-8.

Matveyenko, A. V., Veldhuis, J. D. and Butler, P. C. (2006) Mechanisms of impaired fasting glucose and glucose intolerance induced by an approximate 50% pancreatectomy, *Diabetes* 55(8): 2347-56.

Minami, K., Okano, H., Okumachi, A. and Seino, S. (2008) Role of cadherin-mediated cell-cell adhesion in pancreatic exocrine-to-endocrine transdifferentiation, *J Biol Chem* 283(20): 13753-61.

Minami, K., Okuno, M., Miyawaki, K., Okumachi, A., Ishizaki, K., Oyama, K., Kawaguchi, M., Ishizuka, N., Iwanaga, T. and Seino, S. (2005) Lineage tracing and characterization of insulin-secreting cells generated from adult pancreatic acinar cells, *Proc Natl Acad Sci U S A* 102(42): 15116-21.

Miyatsuka, T., Kosaka, Y., Kim, H. and German, M. S. (2011) Neurogenin3 inhibits proliferation in endocrine progenitors by inducing Cdkn1a, *Proc Natl Acad Sci U S A* 108(1): 185-90.

Miyazaki, J., Araki, K., Yamato, E., Ikegami, H., Asano, T., Shibasaki, Y., Oka, Y. and Yamamura, K. (1990) Establishment of a pancreatic beta cell line that retains glucose-inducible insulin secretion: special reference to expression of glucose transporter isoforms, *Endocrinology* 127(1): 126-32.

Murtaugh, L. C. (2007) Pancreas and beta-cell development: from the actual to the possible, *Development* 134(3): 427-38.

Nielsen, J. H., Svensson, C., Galsgaard, E. D., Moldrup, A. and Billestrup, N. (1999) Beta cell proliferation and growth factors, *J Mol Med* 77(1): 62-6.

Nir, T., Melton, D. A. and Dor, Y. (2007) Recovery from diabetes in mice by beta cell regeneration, *J Clin Invest* 117(9): 2553-61.

Ohneda, K., Mirmira, R. G., Wang, J., Johnson, J. D. and German, M. S. (2000) The homeodomain of PDX-1 mediates multiple protein-protein interactions in the formation of a transcriptional activation complex on the insulin promoter, *Mol Cell Biol* 20(3): 900-11.

Oliver-Krasinski, J. M., Kasner, M. T., Yang, J., Crutchlow, M. F., Rustgi, A. K., Kaestner, K. H. and Stoffers, D. A. (2009) The diabetes gene Pdx1 regulates the transcriptional network of pancreatic endocrine progenitor cells in mice, *J Clin Invest* 119(7): 1888-98.

Park, I. H., Zhao, R., West, J. A., Yabuuchi, A., Huo, H., Ince, T. A., Lerou, P. H., Lensch, M. W. and Daley, G. Q. (2008) Reprogramming of human somatic cells to pluripotency with defined factors, *Nature* 451(7175): 141-6.

Parsons, J. A., Brelje, T. C. and Sorenson, R. L. (1992) Adaptation of islets of Langerhans to pregnancy: increased islet cell proliferation and insulin secretion correlates with the onset of placental lactogen secretion, *Endocrinology* 130(3): 1459-66.

Pictet, R. L., Clark, W. R., Williams, R. H. and Rutter, W. J. (1972) An ultrastructural analysis of the developing embryonic pancreas, *Dev Biol* 29(4): 436-67.

Pipeleers, D., Kiekens, R., Ling, Z., Wilikens, A. and Schuit, F. (1994) Physiologic relevance of heterogeneity in the pancreatic beta-cell population, *Diabetologia* 37 Suppl 2: S57-64.

Piper, K., Brickwood, S., Turnpenny, L. W., Cameron, I. T., Ball, S. G., Wilson, D. I. and Hanley, N. A. (2004) Beta cell differentiation during early human pancreas development, *J Endocrinol* 181(1): 11-23.

Polak, M., Bouchareb-Banaei, L., Scharfmann, R. and Czernichow, P. (2000) Early pattern of differentiation in the human pancreas, *Diabetes* 49(2): 225-32.

Ramalho-Santos, M. and Willenbring, H. (2007) On the origin of the term "stem cell", *Cell Stem Cell* 1(1): 35-8.

Ramiya, V. K., Maraist, M., Arfors, K. E., Schatz, D. A., Peck, A. B. and Cornelius, J. G. (2000) Reversal of insulin-dependent diabetes using islets generated in vitro from pancreatic stem cells, *Nat Med* 6(3): 278-282.

Richardson, M. K., Hanken, J., Gooneratne, M. L., Pieau, C., Raynaud, A., Selwood, L. and Wright, G. M. (1997) There is no highly conserved embryonic stage in the vertebrates: implications for current theories of evolution and development, *Anat Embryol (Berl)* 196(2): 91-106.

Robertson, R. P., Lanz, K. J., Sutherland, D. E. and Seaquist, E. R. (2002) Relationship between diabetes and obesity 9 to 18 years after hemipancreatectomy and transplantation in donors and recipients, *Transplantation* 73(5): 736-41.

Russ, H. A., Bar, Y., Ravassard, P. and Efrat, S. (2008) In vitro proliferation of cells derived from adult human beta-cells revealed by cell-lineage tracing, *Diabetes* 57(6): 1575-83.

Ryan, E. A., Lakey, J. R., Paty, B. W., Imes, S., Korbutt, G. S., Kneteman, N. M., Bigam, D., Rajotte, R. V. and Shapiro, A. M. (2002) Successful islet transplantation: continued insulin reserve provides long-term glycemic control, *Diabetes* 51(7): 2148-57.

Ryan, E. A., Paty, B. W., Senior, P. A., Bigam, D., Alfadhli, E., Kneteman, N. M., Lakey, J. R. and Shapiro, A. M. (2005) Five-year follow-up after clinical islet transplantation, *Diabetes* 54(7): 2060-9.

Seaberg, R. M., Smukler, S. R., Kieffer, T. J., Enikolopov, G., Asghar, Z., Wheeler, M. B., Korbutt, G. and van der Kooy, D. (2004) Clonal identification of multipotent precursors from adult mouse pancreas that generate neural and pancreatic lineages, *Nat Biotechnol* 22(9): 1115-24.

Seeberger, K. L., Dufour, J. M., Shapiro, A. M., Lakey, J. R., Rajotte, R. V. and Korbutt, G. S. (2006) Expansion of mesenchymal stem cells from human pancreatic ductal epithelium, *Lab Invest* 86(2): 141-53.

Selander, L. and Edlund, H. (2002) Nestin is expressed in mesenchymal and not epithelial cells of the developing mouse pancreas, *Mech Dev* 113(2): 189-92.

Servitja, J. M. and Ferrer, J. (2004) Transcriptional networks controlling pancreatic development and beta cell function, *Diabetologia* 47(4): 597-613.

Seymour, P. A., Freude, K. K., Tran, M. N., Mayes, E. E., Jensen, J., Kist, R., Scherer, G. and Sander, M. (2007) SOX9 is required for maintenance of the pancreatic progenitor cell pool, *Proc Natl Acad Sci U S A* 104(6): 1865-70.

Shamblott, M. J., Axelman, J., Wang, S., Bugg, E. M., Littlefield, J. W., Donovan, P. J., Blumenthal, P. D., Huggins, G. R. and Gearhart, J. D. (1998) Derivation of pluripotent stem cells from cultured human primordial germ cells, *Proc Natl Acad Sci U S A* 95(23): 13726-31.

Smukler, S. R., Arntfield, M. E., Razavi, R., Bikopoulos, G., Karpowicz, P., Seaberg, R., Dai, F., Lee, S., Ahrens, R., Fraser, P. E. et al. (2011) The Adult Mouse and Human Pancreas Contain Rare Multipotent Stem Cells that Express Insulin, *Cell Stem Cell* 8(3): 281-93.

Stagner, J. I. and Samols, E. (1991) Deterioration of islet beta-cell function after hemipancreatectomy in dogs, *Diabetes* 40(11): 1472-9.

Stefan, Y., Orci, L., Malaisse-Lagae, F., Perrelet, A., Patel, Y. and Unger, R. H. (1982) Quantitation of endocrine cell content in the pancreas of nondiabetic and diabetic humans, *Diabetes* 31(8 Pt 1): 694-700.

Suzuki, A., Nakauchi, H. and Taniguchi, H. (2004) Prospective isolation of multipotent pancreatic progenitors using flow-cytometric cell sorting, *Diabetes* 53(8): 2143-52.

Suzuki, A., Oyama, K., Fukao, K., Nakauchi, H. and Taniguchi, H. (2002) Establishment of clonal colony-forming assay system for pancreatic stem/progenitor cells, *Cell Transplant* 11(5): 451-3.

Szabat, M., Luciani, D. S., Piret, J. M. and Johnson, J. D. (2009) Maturation of adult beta-cells revealed using a Pdx1/insulin dual-reporter lentivirus, *Endocrinology* 150(4): 1627-35.

Takahashi, K. and Yamanaka, S. (2006) Induction of pluripotent stem cells from mouse embryonic and adult fibroblast cultures by defined factors, *Cell* 126(4): 663-76.

Tesar, P. J., Chenoweth, J. G., Brook, F. A., Davies, T. J., Evans, E. P., Mack, D. L., Gardner, R. L. and McKay, R. D. (2007) New cell lines from mouse epiblast share defining features with human embryonic stem cells, *Nature* 448(7150): 196-9.

Teta, M., Rankin, M. M., Long, S. Y., Stein, G. M. and Kushner, J. A. (2007) Growth and regeneration of adult beta cells does not involve specialized progenitors, *Dev Cell* 12(5): 817-26.

Van Assche, F. A. (1974) Quantitative morphologic and histoenzymatic study of the endocrine pancreas in nonpregnant and pregnant rats, *Am J Obstet Gynecol* 118(1): 39-41.

Van Assche, F. A., Aerts, L. and De Prins, F. (1978) A morphological study of the endocrine pancreas in human pregnancy, *Br J Obstet Gynaecol* 85(11): 818-20.

Wagers, A. J., Sherwood, R. I., Christensen, J. L. and Weissman, I. L. (2002) Little evidence for developmental plasticity of adult hematopoietic stem cells, *Science* 297(5590): 2256-9.

Wilson, M. E., Scheel, D. and German, M. S. (2003) Gene expression cascades in pancreatic development, *Mech Dev* 120(1): 65-80.

Xu, X., D'Hoker, J., Stange, G., Bonne, S., De Leu, N., Xiao, X., Van De Casteele, M., Mellitzer, G., Ling, Z., Pipeleers, D. et al. (2008) Beta cells can be generated from endogenous progenitors in injured adult mouse pancreas, *Cell* 132(2): 197-207.

Zhou, Q., Brown, J., Kanarek, A., Rajagopal, J. and Melton, D. A. (2008) In vivo reprogramming of adult pancreatic exocrine cells to beta-cells, *Nature* 455(7213): 627-32.

Zhou, Q., Law, A. C., Rajagopal, J., Anderson, W. J., Gray, P. A. and Melton, D. A. (2007) A multipotent progenitor domain guides pancreatic organogenesis, *Dev Cell* 13(1): 103-14.

Zulewski, H., Abraham, E. J., Gerlach, M. J., Daniel, P. B., Moritz, W., Muller, B., Vallejo, M., Thomas, M. K. and Habener, J. F. (2001) Multipotential nestin-positive stem cells isolated from adult pancreatic islets differentiate ex vivo into pancreatic endocrine, exocrine, and hepatic phenotypes, *Diabetes* 50(3): 521-33.

21

Cell Sources for Treating Diabetes

Subhshri Sahu[2], Mugdha V. Joglekar[2], Sundy N. Y. Yang[1] and Anandwardhan A. Hardikar[1,2,3]
[1]*Diabetes &Pancreas Biology Group, O'Brien Institute, University of Melbourne,*
[2]*Stem Cells and Diabetes Section, National Centre for Cell Science, Pune,*
[3]*Australian Catholic University, Melbourne,*
[1,3]*Australia*
[2]*India*

1. Introduction

Diabetes is one of the most prevailing metabolic disorders worldwide. The number of individuals suffering from diabetes is increasing significantly. There are two main types of diabetes; Type 1 Diabetes mellitus (T1D) and Type 2 Diabetes mellitus (T2D). T1D results from autoimmune destruction of insulin-producing pancreatic beta-cells. This results in a complete loss of insulin, a hormone that is essential for loading any excess circulating glucose into target organs. Individuals with T1D therefore need daily injections of insulin so as to regulate their circulating blood glucose concentrations and survive with the disease. Although such lifelong insulin therapy supports glucose metabolism, patients often develop complications related to renal function, endothelial damage and retinopathy. T1D usually develops in children or young adults. T2D on the other hand results from insulin resistance, a condition in which insulin is not properly recognized by receptors on cell surface. It is sometimes combined with an absolute insulin deficiency as eventually these patients loose their β cells and are then classified as type 2 diabetic individuals requiring insulin. It usually occurs in people over 40 years of age, but is also being observed at younger age. A cure for diabetes came in with the demonstration to support insulin action and regain the patient's quality of life by transplantation of cadaveric insulin-producing cells (Tzakis et al., 1990). Such a therapy, which in practice includes whole pancreas transplantation or transplantation of the endocrine portion; the islet of Langerhans, has been carried out in several patients suffering from Diabetes. In year 2000, the Edmonton group successfully demonstrated transplantation of cadaveric islets using a steroid-free immunosuppressive regimen, which was a major advance in the field. The Edmonton Protocol involves isolating islets from a cadaveric donor pancreas using enzymatic digestion. Islets from one to as many as three donors were transplanted in to each recipient. Cadaveric islets are infused into the patient's portal vein, and are then kept from being destroyed by the recipient's immune system through the use of two immunosuppressants, sirolimus and tacrolimus (Shapiro et al., 2000). Several hundred people have received islet transplants since 2000. By a year after transplantation, about 50 - 68% of patients were observed off insulin, but by five years after the procedure, less than 10% of total patients were seen to be free of daily insulin injections. One of the major factors in success of these transplants has been the number of transplants / islet cell grafts received by these individuals. For such a therapy to be available for several millions of people suffering from T1D it is necessary to identify other human cell sources that can be used for replacement therapy in diabetes.

Several possible cell sources have been considered for obtaining insulin-producing cells. Our previous studies demonstrate that lineage committed stem cells such as pancreatic islet-derived are better cell sources for treatment of diabetes. We and others showed that progenitor cells generated from insulin-producing cells retain the active chromatin conformation of pancreatic or islet-specific lineage and hence can be differentiated into insulin-producing cells with better efficiency (Gershengorn et al., 2005; Gershengorn et al., 2004; Joglekar and Hardikar; Joglekar et al., 2009a; Mutskov et al., 2007; Russ et al., 2008; Russ et al., 2009). Apart from beta cells themselves, other pancreatic cell types (endocrine, acinar or ductal) are also demonstrated as potential candidate cell types (Baeyens et al., 2005; Bonner-Weir et al., 2004; Lardon et al., 2004a; Lardon et al., 2004b). Differentiation from human embryonic stem cells is also shown to generate insulin-producing cells (Blyszczuk et al., 2004). D'Amour et al illustrated this phenomenon (D'Amour et al., 2005; D'Amour et al., 2006) by recapitulating the normal sequence of embryonic developmental events that lead to beta-cell differentiation (Van Hoof et al., 2009). It was demonstrated (Aoi et al., 2008; Nakagawa et al., 2008; Okita et al., 2007; Takahashi et al., 2007; Takahashi and Yamanaka, 2006) that forced expression of "reprogramming factors" (also known as the "Yamanaka factors") allowed conversion of fibroblasts or somatic cells to an embryonic stem cell-like state. These cells, referred to as induced pluripotent stem (iPS) cells are popular since they can be derived from adult somatic cells, including those from diabetic patients (Huangfu et al., 2008; Maehr et al., 2009), and could be potentially used for autologous transplantation in diabetes, if efficient differentiation is achieved. Another study (Zhou et al., 2008; Zhou and Melton, 2008) also demonstrated that forced expression of pancreatic transcription factors (Pdx1, Ngn3 and MafA) induced insulin expression in exocrine pancreatic (acinar) cells. We and others have also demonstrated (Bar et al., 2008; Joglekar et al., 2009a; Joglekar et al., 2009b; Russ et al., 2009) the role of several regulators in generation of islet-derived progenitors (via EMT) and their conversion to insulin-producing cells (via reverse-EMT or MET). Mesenchymal stem cells obtained from non-endodermal tissues such as bone marrow, umbilical cord blood placenta, Wharton's jelly, adipose tissue are also observed to differentiate to beta-like cells in vitro or in vivo (Chandra et al., 2009; Kadam et al.; Parekh et al., 2009; Phadnis et al.). Although non-endodermal cells can be differentiated to insulin-producing cells in vitro, they are limited in their ability to efficiently transcribe insulin gene.

Transdifferentiation, is a phenomenon that refers to differentiation of an already differentiated tissue to another differentiated cell type, possibly via de-differentiation and re-differentiation. An example is the conversion of liver cells to pancreas (Ferber et al., 2000). More recently, we demonstrated evidence that human gallbladder epithelial cells contain pancreatic hormone-producing cells (Sahu et al., 2009a). These observations are in line with similar observations made in mice (Coad et al., 2009). Although the mechanisms that contribute to generation of hormone-producing cells in human gallbladder remain to be identified, these observations reveal that gallbladder epithelial cells can be a potentially important source of islet progenitors for cell replacement therapy in diabetes. It is important to note that gallbladder and pancreas are next door neighbours during embryonic development (Figure 1). In fact, the gallbladder develops from the ventral pancreatic bud.

Thus understanding the development of gallbladder-derived hormone-producing immature islet cells will help us in generating an alternate source of pancreatic stem cells for replacement therapy in diabetes. In this chapter, we further emphasis on gallbladder and liver development and cells derived from these organs as potential candidates in cell replacement therapy for diabetes.

2. Liver and gallbladder as cell sources to generate insulin producing cells

Low yield and availability of transplantable islets from cadaveric pancreatic samples is a major concern in the field of cell replacement therapy for diabetes. There is a wider search for cell sources with the potential to produce insulin at physiologically significant levels. Liver and gallbladder are next door neighbors to pancreas during embryonic development and share common expression territories for several transcription factors during embryonic development (Sahu et al., 2009b). In this regard, the role of liver and especially gallbladder has come into picture recently. During embryonic development, the development of pancreas begins with dorsal and ventral out-pouching of the gut tube. Interestingly, the ventral pancreatic bud also gives rise to the gallbladder. In this section we will discuss the developmental plans of liver and gallbladder in context of pancreas development.

2.1 Liver function

The liver is a large glandular organ that is well known due to its' extensive regenerative capabilities as well as ability to trans-differentiate into pancreatic fate. The liver is made up of hepatocytes with their basal surface facing the capillaries invading sinusoidal spaces. The liver secretes bile through its apical surface into the canaliculi that join to form the biliary duct. Biliary duct carries bile to gallbladder via hepatic duct. The function of gallbladder is to concentrate and store bile, which is required for digestion of fats. The adult rat liver consists of 60% hepatocytes while remaining mass consist of cholangiocytes, kuppfer cells, stellate cells and endothelial cells even those existing in sinusoidal spaces. Although majority of liver parenchyma consists of hepatocytes, the interaction of hepatocytes with the non-parenchymal cells is important for normal liver function (Kmiec, 2001). However, rats do not have a gallbladder while mice do. Recently considerable amount of advances have been made in understanding the mechanisms that direct the development of bile duct. Although mechanisms involving gallbladder development have been less explored, studies from liver development have provided us with clues to understand the developmental biology of biliary duct and gallbladder.

2.2 Embryonic development of the liver

Liver development is a complex process involving fate decision, specification, proliferation and differentiation. Development of liver begins with foregut development (involving Wnt/β-catenin and FGF4), hepatic induction (by FGFs and BMPs), hepatic morphogenesis (involving transition from a simple cuboidal to pseudostratified columnar epithelium) and hepatic bud growth and maturation (influenced by signals from developing blood vesels). With the help of molecular genetic tools, tissue explant cultures and molecular biological techniques, liver development has been well understood in mice (Zaret, 1998; Zaret, 2000; Zorn and Wells, 2007). The liver primordium appears as a diverticulum in the ventral region of the anterior intestinal portal at 9.0, 10.5, and 21days of gestation in mouse, rat, and human embryos, respectively (Du Bois, 1963; Shiojiri et al., 1991). This part of the ventral foregut endoderm lies adjacent to heart and hence receives signals from cardiac mesoderm (Douarin, 1975; Fukuda-Taira, 1981; Fukuda, 1979). As the development proceeds, liver primordium diverges into two portions in mouse and human embryos: the cranial and caudal lobe. The cranial diverticulum develops into the liver parenchyma and hepatic ducts while the caudal portion gives rise to gallbladder and the ventral pancreas (Shiojiri, 1997). At the early somite stage (14-20) or early stage of liver development i.e at embryonic day (e)9.5, the liver primordium is separated from septum transversum by basement membrane (Medlock and Haar, 1983). As the liver bud grows this basement membrane is disrupted gradually and the pre-hepatic cells separate as layers from the foregut and migrate as cords into the surrounding septum transversum (Douarin, 1975; Medlock and Haar, 1983). Fibroblasts and stellate cells of liver are generated form septum transversum mesenchyme (STM). Liver bud undergoes a period of accelerated growth between e10–15 to become the major hematopoietic organ, which is further colonized and vascularized by hematopoietic cells.

The hepatoblasts are bi-potential cells that give rise to different types of cells including the biliary epithelial cells (BECs), intrahepatic bile ducts and hepatocytes depending on their distribution. For example BECs are generated from hepatoblasts residing near portal vein while hepatocytes mainly originate from the hepatoblasts residing in the parenchyma. BECs are also known as cholangiocytes and line the lumen of intrahepatic bile ducts (IHBD) (Lemaigre, 2003; Shiojiri, 1984). The formation of biliary network, which connects to extrahepatic bile duct (EHBD) and maturation of hepatocytes is a gradual process that begins at e13 and continues after birth to achieve the characteristic tissue architecture.

2.3 Endoderm formation

The definitive endoderm emerges as a sheet of cells starting from the anterior end of the primitive streak during gastrulation. The glandular and epithelial cells of lung, thyroid, pancreas, gastrointestinal and respiratory epithelium and of course the hepatocytes and the biliary epithelium originate from multipotent endoderm. Signaling by Nodal, a transforming growth factor beta (TGFβ) family growth factor initiates both endoderm and mesoderm formation in a concentration dependent manner. Low nodal doses are seen to induce mesoderm while high dose induce endoderm (Shen, 2007; Zorn and Wells, 2007). Nodal signaling stimulates core group of endoderm specific transcription factors including Sox17 (HMG domain DNA binding factor) and the fork head domain proteins Foxa1-3 (hepatocyte nuclear factor or HNF 3α/β/γ) (Zorn and Wells, 2007), which in turn regulates the downstream genes committing cells to endodermal lineage.

2.4 Endoderm patterning

The cells that first emerge from the primitive streak give rise to the anterior part while those that arise later give rise to the posterior part (Lawson and Pedersen, 1987). Anterior endoderm development requires higher nodal signaling than the posterior endoderm. It expresses Foxa2, as it is preferentially required for the formation of anterior definitive endoderm (Dufort et al., 1998; Zorn and Wells, 2007). During the early somite stages of development, morphogenetic movement transforms the endoderm into an epithelial gut tube. This epithelial gut tube is surrounded by mesoderm. The mesoderm secretes several factors that regulate pattern formation along the epithelial gut tube along anterior- posterior (A-P) axis to form the foregut, midgut and hindgut regions. These regions of endoderm are distinguishable by the expression of Hhex in foregut, Pancreatic and duodenal homeobox 1

(Pdx1) in midgut and Cdx in posterior endoderm or hindgut (Grapin-Botton, 2005; Moore-Scott et al., 2007). The foregut endoderm contains common precursors for liver, gallbladder and pancreas. Initially the regional identity is "plastic" as recombination of the early foregut endoderm with the posterior endoderm results into intestinal fate instead of liver and pancreas development (Gualdi et al., 1996; Horb and Slack, 2001; Kumar et al., 2003; Wells and Melton, 2000).

2.5 Repression of Wnt/β-catenin and FGF4 is required for foregut development

Overlapping spatiotemporal gradients of Wnt, fibroblast growth factor (FGF), bone morphogenetic proteins (BMP) and retinoic acid secreted from the adjacent mesoderm, patterns the endoderm to express regionally specific genes (Chen et al., 2004; Dessimoz et al., 2006; Kumar et al., 2003; Martin et al., 2005; McLin et al., 2007; Roberts et al., 1995; Stafford et al., 2004; Tiso et al., 2002). For the development of foregut endoderm, inhibition of Wnt and FGF4 signaling by Wnt antagonist like Sfrp5 (secreted by the foregut endoderm) is essential (Dessimoz et al., 2006; McLin et al., 2007; Wells and Melton, 2000). Obviously the posterior endoderm depends on the secretion of Wnt and FGF4 by the posterior mesoderm. Experimentally it has been demonstrated that blocking the canonical Wnt effector β-catenin in the posterior endoderm of Xenopus embryos activates Hex expression and results in ectopic liver bud formation in intestine (McLin et al., 2007).

2.6 Hepatic competence

The foregut endoderm is the only competent candidate in vivo to generate liver and its establishment is the first step towards hepatogenesis (Douarin, 1975; Fukuda-Taira, 1981). This intrinsic property to develop into liver is due to expression of transcription factors including Foxa2, Gata4-6 and Hhex (or Hex). These transcription factors have important role in early organogenesis. Foxa and Gata have been shown to bind to enhancer element of albumin gene and enhances chromatin accessibility thereby increasing the competency of such liver specific genes to be transcribed (Bossard and Zaret, 1998; Bossard and Zaret, 2000; Gualdi et al., 1996). Evidently, deletion of Foxa1 and Foxa 2 prevents induction of liver development in early foregut.

2.7 Hepatic induction by cardiac mesoderm and STM

At an early somite stage of 4-7, FGF signals from the cardiac mesoderm of developing heart and BMPs from the STM induces hepatic fate in the ventral foregut endoderm (Douarin, 1975; Fukuda-Taira, 1981; Gualdi et al., 1996; Jung et al., 1999; Rossi et al., 2001). This has been illustrated in mouse models where the mouse foregut explants isolated at 2-4 somite stage when cultured in the presence of cardiac mesoderm start expressing albumin after 1-2 days. However removal of cardiac mesoderm or blocking FGF signals inhibited liver induction (Calmont et al., 2006; Gualdi et al., 1996; Rossi et al., 2001). It has also been illustrated that exogenous FGF1 and FGF2 can replace cardiac mesoderm and induce albumin expression in foregut endoderm explants. However BMP signaling may act by maintaining Gata4/6 expression and is not sufficient for hepatic induction in the explants (Rossi et al., 2001). Analysis of downstream signal transduction pathway of FGF signaling suggests that P13 kinase pathway promotes hepatic growth while MAP kinase pathway regulates hepatic gene expression (Calmont et al., 2006). Although the target genes of FGF and BMP signal have not been revealed due to early embryonic lethality, it is predicted that they act cooperatively to regulate liver development. Additionally, wnt2 and wnt2b are also expressed in the mouse lateral plate mesoderm at the time of hepatic induction but their role is yet to be determined. Mouse explant studies have suggested that high, low and medium concentrations of FGFs are critical for segregation of different organs lung, liver and gallbladder respectively from a common progenitor cell. It is not clear whether it is the proximity or the duration that the endoderm is in contact with the cardiac mesoderm that controls FGF doses. *Hhex,* a homeobox protein regulates this process. It is responsible for the proliferation of foregut cells and the movement of epithelium in relation to the FGF-secreting cardiac mesoderm during development (Bort et al., 2004).

2.8 Endothelial cells signals in hepatogenesis

At around e9.5 endothelial cells intersperse and form simple continuous vessels into the future parenchymal cells as it expands into septum transversum mesenchyme. As development proceeds, the morphology of the endothelial cells changes, becoming fenestrated and forming endothelial sheets that line the hepatic sinusoids (Bankston and Pino, 1980; Enzan et al., 1997). The role of endothelial cells in hepatogenesis was illustrated by utilizing Flk-1$^{-/-}$ mouse which lack mature endothelial cells or blood vessels (Shalaby et al., 1997; Shalaby et al., 1995). In control mice, at an early stage, when liver starts developing, the endothelial cells are found between the hepatic bud and the septum transversum however Flk-1$^{-/-}$ mice lack endothelial cells as expected. Comparison of control and Flk-1$^{-/-}$ embryos liver revealed that although hepatic specification occurred hepatogenesis failed and there were no signs of expansion or morphogenetic movement into the septum transversum. Flk expression is restricted to endothelial cells and not in STM and endoderm. Hence it was confirmed that endothelial cell signaling is a crucial aspect for normal hepatogenesis.

2.9 Transcription factors and their roles in controlling hepatogenesis

During embryonic development, hepatogenesis is defined by the precise timing and the location of tissue interactions and the signaling molecules involved therein. However the intracellular response to these signaling events explains the mechanisms through which the cells of endoderm actually adopt a hepatic fate. Large changes in gene expression patters accompany the differentiation of ventral endoderm towards hepatic phenotype. It is thus quite obvious that a lot of transcription factors are involved as gene regulation is tightly regulated by transcription factors (Cereghini, 1996; Duncan, 2000). Molecular biology and molecular genetics have been helpful in identifying a number of transcription factors that are crucial candidates to regulate various aspects of hepatogenesis.

2.9.1 Transcription factors controlling hepatic competency

Competency is described as the innate ability of a cell type to follow a developmental fate given appropriate inductive signals (Gilbert and MA., 2000; Waddington, 1940). The work of Gauldi et al suggests that both ventral and dorsal endoderm have the potential to express albumin gene (Gualdi et al., 1996). However the laboratory of Ken Zaret have suggested a mechanism whereby Hnf3 (FoxA) and Gata4 could modulate competency intracellularly (Bossard and Zaret, 1998; Chaya et al., 2001; Cirillo et al., 2002; Gualdi et al., 1996; Shim et al., 1998). Extracts from e9.5 liver buds when used for in vivo foot printing analysis of *Albumin* enhancer revealed that several binding sites including Hnf3, Gata and Nf-1 sites were occupied. However uncommitted cells from endoderm showed occupancy at Hnf3 and Gata sites only. In addition, in vitro chromatin reconstitution assays elucidated that the binding of either Hnf3 or Gata4 resulted in the opening of compacted chromatin (Cirillo et al., 2002). Thus Hnf3 and Gata were found responsible for remodeling of chromatin structure and hence were able to generate available receptive sites on them for the occupancy of transcriptional activators. Supporting examples of this model is from the finding that Hnf3 can relieve the transcriptional repression of α-fetoprotein gene in transcription assays *in vitro* (Crowe et al., 1999). A similar situation exists for the Gata factor as Gata 4 has been shown to bind the Albumin enhancer. The expressions of these transcriptional activators are in turn controlled by inductive events.

2.9.2 Role of Hex and Prox in early stages of liver development

Hex (or Prh) encodes for divergent homeobox transcription factor known as hematopoietically-expressed homeobox protein. At an early head fold stage (pre-somite, e8.0), prior to the initiation of hepatogenesis, expression analysis of the Hex mRNA revealed that it is highly expressed throughout the ventral endoderm. Expression within the ventral endoderm is restricted to two discrete areas that are future sites for liver and thyroid development by e8.5 or 10 somite stage (e8.5) (Keng et al., 1998; Thomas et al., 1998). Hence the early expression of Hex mRNA in ventral endoderm defines Hex as one of the earliest markers of liver development. Role of Hex was discerned using Hex $^{-/-}$ embryos at e9.5 to 10.5. These embryos show thickening of the ventral foregut endoderm but the expansion and morphogenetic outgrowth of liver was undetectable in contrast to the control embryos. In situ hybridization studies on Hex $^{-/-}$ embryos revealed the absence of α-fetoprotein mRNA expression at e9.5 and e10.5. In another study using RT-PCR, absence of albumin was reported in these mutants (Keng et al., 2000; Martinez Barbera et al., 2000). Knockout studies have also revealed the role of Prox 1 in early stages of liver development. Analysis of Prox1-/- embryo found that the liver of these mutant embryos were smaller than the wild types and although these mutants formed distinct liver lobes the hepatic parenchymal cells

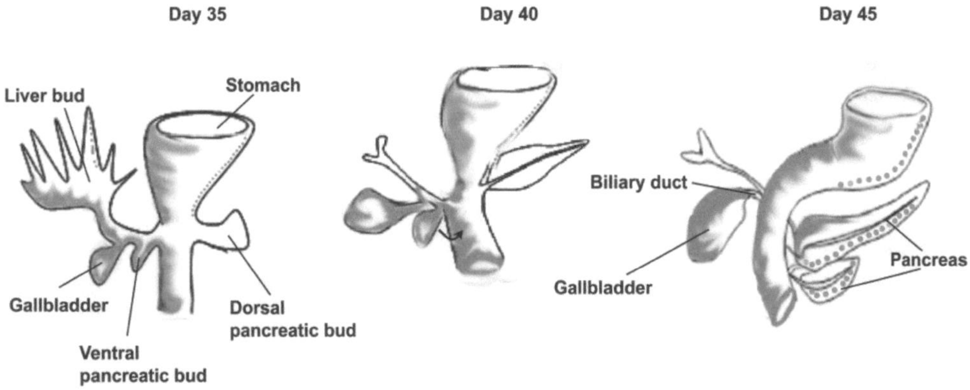

Fig. 1. The human liver, gallbladder and pancreas are next door neighbors during embryonic development. The gallbladder actually develops from the ventral pancreatic bud and is in close proximity to the developing pancreas until the ventral pancreas rotates to fuse with the dorsal pancreatic bud. We therefore believe that this close proximity and overlapping expression pattern of pancreatic transcription factors in the domain of developing gallbladder induces expression of endocrine pancreatic hormones in the gallbladder.

were restricted to central rudiment (Sosa-Pineda et al., 2000). These mutant embryos showed reduced number of hepatoblasts than control embryos. These two characteristic of the mutant phenotype explains why these embryos failed to inhibit the mesenchymal portion of the liver. At e10 Prox1$^{-/-}$ hepatoblasts mice failed to exhibit the delamination of hepatic chords into the septum transversum in contrast to control hepatoblasts which normally migrate as chords into the septum transversum. Although these mutants failed to expand and migrate, they were found positive for albumin and α-fetoprotein implying their specification toward hepatic fate (Sosa-Pineda et al., 2000). In short, Prox 1 is dispensable for hepatic specification but it is required for morphogenic expansion of the liver primordium.

2.9.3 Transcription factors in regulation of hepatocyte differentiation

During 1980s enormous amount of work from several laboratories allowed the identification of the transcription factors that predominantly regulates the liver specific genes (Lai and Darnell, 1991; Sladek and Darnell, 1992). These transcription factors included Hnf1α and β , the c/EBP family, Hnf3α, β, and γ, Hnf4α, and more recently Hnf6 (Lai and Darnell, 1991; Lemaigre et al., 1996; Samadani and Costa, 1996; Sladek and Darnell, 1992). Knockout models for these factors have helped in studying the role of these factors in hepatic differentiation. It was found that loss of a single factor generally has negligible effect. For example disruption of either Hnf3α, Hnf3β or Hnf3γ in the liver does not affect hepatocyte differentiation, although moderate changes in expression of a subsets of hepatocyte genes have been described (Kaestner et al., 1998; Shih et al., 1999; Sund et al., 2000). Functional redundancy could be the reason for this insignificant effect of disruption of single gene but its note worthy that disruption of Hnf3β inhibits β-cell differentiation in pancreas although they express Hnf3α (Kaestner et al., 1998; Sund et al., 2000). Noticeably disruption of some genes has serious effects in liver development. For example the early liver development in c/EBPα$^{-/-}$ embryos is normal although the neonates die from hypoglycemia due to failure of liver to accumulate glycogen. In these mice, expression of glycogen synthase is disrupted and the bile canaliculi form abnormally (Wang et al., 1995). In addition hepatocyte proliferation was increased in the newborns c/EBPα$^{-/-}$ liver. This was supported by the fact that c/EBPα inhibits cell proliferation by regulating S-phase-specific E2F–p107 complexes and by stabilisizing the cyclin dependent kinase inhibitor p21 (Harris et al., 2001; Timchenko et al., 1997; Timchenko et al., 1999). Hnf1α mutants do not show significant effect on liver development, however mice lacking Hnf1α suffer from various different abnormalities depending on the targeted allele. In one case, mice suffer from hepatic dysfunction, Phenylketonuria and renal Faconi's syndrome, while a different gene targeting strategy generated mice that suffered from non-insulin-dependent diabetes mellitus (NIDDM) and Laron dwarfism. Loss of function of phenylalanine hydroxylase (Pah) gene due to the inability of Hnf1α to facilitate modification of chromatin around the Pah promoter results Phenylketonuria (Pontoglio et al., 1996; Pontoglio et al., 1997). Hnf4α mRNA was first detected in the primitive endoderm of the blastocyst at the time of implantation. Hnf4α expression was first detected in the liver bud at around 10 somite stage. Hnf4α expression was restricted to the ventral extraembryonic visceral endoderm which forms the endodermal portion of the yolk sac just before this developmental stage. Hnf4α acts on the important effector in the hepatocyte development. Hnf4α positively regulates Hnf1α expression as evident by the presence of Hnf4α binding sites on the transcriptional regulatory element of Hnf1α. Hnf4α slightly precedes the expression of Hnf1α in number of tissues in which both factors are expressed. Although hepatoblasts during liver development are relatively immature, they still initiate expression of genes that define mature hepatocyte phenotype. When the mRNA levels of such genes were compared between control and Hnf4α$^{-/-}$ livers it was found that the expression of 14 genes that have important roles in hepatocyte function were either undetectable in some cases or down-regulated in others in the absence of Hnf4α (Li et al., 2000). These results have also been consistent with the adult hepatocytes that lack Hnf4α (Hayhurst et al., 2001). All of these studies strongly support the notion that Hnf4α is a central regulator of hepatocyte differentiation.

2.10 Transcription factors in gallbladder and intrahepatic bile duct development

The gallbladder and liver share several common transcription factors during embryonic development. In developing mouse liver, HNF6 is expressed in hepatocytes and in the epithelial cells of both the intrahepatic and extrahepatic bile ducts (Landry et al., 1997; Rausa et al., 1997). Recent studies illustrate that Hnf6$^{-/-}$ mouse embryos fail to develop a gallbladder and exhibit severe abnormalities in both extrahepatic and intrahepatic bile ducts (IHBD), which is associated with diminished expression of the Hnf1β transcription factor (Clotman et al., 2002; Jacquemin et al., 2000). Thus Hnf6 is essential for regulating the expression of Hnf1β, which in turn plays important role in intrahepatic bile duct and gall bladder development. Foxf1 is another transcription factor involved in gallbladder development. It is expressed in the embryonic gut mesoderm and septum transversum. It participates in the mesenchymal-epithelial cell inductive signaling of the internal organs. Foxf1$^{+/-}$ mice show several developmental abnormalities in gallbladder, esophagus, lung and trachea which supports this hypothesis (Kalinichenko et al., 2001; Kalinichenko et al., 2002; Mahlapuu et al., 2001). Foxf1$^{+/-}$ gallbladder fails to express vascular cell adhesion molecule-1, α5 integrin, and platelet-derived growth factor receptor α and hepatocyte growth factor genes required for mesenchymal epithelial cell induction (Kalinichenko et al., 2002). Foxf1$^{+/-}$ gallbladder shows malformations in external smooth muscle cell layer, reduction in mesenchymal cell number and in some cases, lack of a discernible biliary epithelial cell layer (Kalinichenko et al., 2002). There is little information available regarding the development of gallbladder. Present knowledge is gained based on a few knockout studies that are summarized in the table below. It suggests a need to generate more similar information for better understanding of molecular pathways regulating the development of hormone-producing cells in gallbladder.

No.	Gene name	Phenotype	Ref
1	forkhead box F1a (Foxf1)	Heterozygotes of Foxf1 knockout animals (haplosufficient) have severely abnormal and small gall bladder and they lack epithelial cell layer	Kalinichenko et al, Journal of Biological Chemistry, 2002, 227:12369-12374
2	One cut domain family member 1 (Onecut1/ Hnf6)	Hnf6 null mice lack gall bladder. Extrahepatic as well as intrahepatic bile ducts are abnormally developed	Clotman et al, Development, 2002, 129:1819-1828

No.	Gene name	Phenotype	Ref
3	Hairy and enhancer of split 1 (Hes 1)	Hes1 knockout animals do not develop gall bladder and have severely reduced extrahepatic bile ducts. Such mutant bile ducts express Ngn3 and ectopically differentiate into pancreatic exocrine and endocrine tissue	Sumazaki et al, Nature Genetics, 2004, 36:83-87
4	Hematopoietically expressed homeobox (Hhex)	Conditional deletion of Hhex in hepatic diverticulum is embryonic lethal and results in small, cystic liver. Gall bladder and extrahepatic bile duct are absent. Conditional deletion of Hhex in embryonic liver is not lethal, however it precipitates into irregular development leading to polycystic liver disease in adulthood	Hunter et al, Developmental Biology, 2007, 308:355-367
5	Leucine-rich repeat-containing G protein-coupled receptor 4 (LGR4)	LGR4 hypomorphic embryos develop normal gall bladder bud at e10.25, however it does not elongate in later stages of development. Adult mouse completely lacks gall bladder and cystic duct. Liver and pancreas are unaffected.	Yamashita et al, Developmental Dynamics, 2009, 238:993-1000
6	Hepatocyte nuclear factor beta Hnf1β	HNF1b deletion results into severe jaundice caused by abnormalities in intrahepatic bile ducts and gallbladder. It also leads to reduced expression of Organic Anion Transporting Polypeptide 1 or Oatp1, bile acid transporter and fatty acid dehydrogenase or Very long-chain acyl-CoA dehydrogenase VLCAD.	Coffinier C et al, Development 2002; 129:1829–1838.
7	Hlx -/-	Expression of Hlx is seen in visceral mesenchyme of the developing liver, gallbladder, and gut. Hlx -/- embryos exhibit normal formation of the liver diverticulum and differentiation of hepatocytes , but the cells fail to proliferate and therefore develop into under-sized liver (3% of normal size).	Hentsch B et al, Genes Dev 1996;10:70-79.

Table 1. Phenotypes of the knockout mouse models for the genes involved in gallbladder development.

3. Concluding remarks

Liver, gallbladder and pancreas originate from nearby regions of endoderm and develop in close association with each other (Figure 1). It is quite evident and exciting to note that these organs share several transcription factors (Sahu et al., 2009b). The liver shares expression of genes such as Glut2 and glucokinase and as discussed earlier has the same embryological origin as ventral pancreas. It is hence encouraging that the cells derived from organs outside pancreas can be useful in generating insulin producing cells for cell replacement therapy in diabetes. It has been demonstrated that ectopic expression of Pdx1 in liver ameliorates type 1 diabetes through conversion of hepatocytes in to insulin-producing cells (Ferber et al., 2000). Precisely diabetic CAD-NOD mice treated with Ad-CMV-PDX-1 became normoglycemic and maintained a stable body weight. Ectopic PDX-1 expression induced pancreatic gene expression and insulin production in the mice livers (Ferber et al., 2000). Retrovirally immortalized human fetal liver progenitor cells when transduced using lentiviral vector containing PDX1 underwent successful differentiation into insulin-expressing cells. These cells showed responsiveness to glucose, exhibit a β-cell-like phenotype and are able to restore euglycemia upon transplantation in diabetic mice (Zalzman et al., 2005; Zalzman et al., 2003). Forced expression of Pdx1 is now known to induce conversion of human fetal liver cells to insulin-producing cells. These modified cells express multiple beta cell specific genes. However, they also activated genes expressed in exocrine pancreas and continued expressing hepatic genes as well. Similarly expression of another beta cell transcription factor; NeuroD, in mouse liver cells in vivo resulted in reversal of hyperglycemia (Kojima et al., 2003). This phenomenon was promoted by the presence of betacellulin. Manipulation of the culture conditions (by using activin A in serum free medium for differentiation) of the Pdx1-expressing human fetal liver cells was shown to further promote the differentiation of these cells towards the beta-cell phenotype, as judged by insulin content and gene expression (Zalzman et al., 2005; Zalzman et al., 2003). All of the above studies involve genetic manipulation of stem / propgenitor cells. Such in vitro manipulated cells cannot be used for treatment of diabetes in humans. It is therefore necessary to identify a source of human insulin-producing or islet progenitor cells that can efficiently transcribe, translate and secrete insulin.

At present, human cadaveric pancreas is the ultimate source of such insulin-producing cells for replacement therapy in diabetes. Although this therapy has potential to cure diabetes, there is a significant scarcity in number of cadaveric pancreas available for transplantation (~30 "good" pancreas/year in Australia). In contrast to this, there are several hundreds of gallbladders available every year from just one surgical team in Melbourne, Australia. In other speciality GI centres, this number is over a thousand every year. The level of insulin transcript seen in gallbladder is 1000-fold less (Ct value 22-26) than that observed in pancreas (Ct value 12-16). This level of insulin gene expression is already significantly higher than any of the differentiated stem cells reported until now. Gallbladder cells naturally exhibit insulin expression higher that the differentiated stem cells and hence there is scope for enhancement of expression levels of insulin in gallbladder derived cells upon differentiation/maturation. Further research in understanding the maturation / expansion and differentiation of gallbladder-derived islet progenitor cells will help in providing alternative source for cell replacement therapy in diabetes.

4. Acknowledgements

The work discussed in this chapter was funded by the UK-India Educational Research Initiative grant through the British Council and Department of Science and Technology, India. The authors acknowledge research fellowship support from CSIR to SS and MVJ and from O'Brien Institute to SY. The O'Brien Institute acknowledges the Victorian State Government's Department of Innovation, Industry and Regional Development's Operational Infrastructure Support Program. Authors apologize to researchers whose work could not be cited due to space limitations.

5. References

Aoi, T., Yae, K., Nakagawa, M., Ichisaka, T., Okita, K., Takahashi, K., Chiba, T. and Yamanaka, S. (2008). Generation of pluripotent stem cells from adult mouse liver and stomach cells. Science 321, 699-702.

Baeyens, L., De Breuck, S., Lardon, J., Mfopou, J. K., Rooman, I. and Bouwens, L. (2005). In vitro generation of insulin-producing beta cells from adult exocrine pancreatic cells. Diabetologia 48, 49-57.

Bankston, P. W. and Pino, R. M. (1980). The development of the sinusoids of fetal rat liver: morphology of endothelial cells, Kupffer cells, and the transmural migration of blood cells into the sinusoids. Am J Anat 159, 1-15.

Bar, Y., Russ, H. A., Knoller, S., Ouziel-Yahalom, L. and Efrat, S. (2008). HES-1 is involved in adaptation of adult human beta-cells to proliferation in vitro. Diabetes 57, 2413-20.

Blyszczuk, P., Asbrand, C., Rozzo, A., Kania, G., St-Onge, L., Rupnik, M. and Wobus, A. M. (2004). Embryonic stem cells differentiate into insulin-producing cells without selection of nestin-expressing cells. Int J Dev Biol 48, 1095-104.

Bonner-Weir, S., Toschi, E., Inada, A., Reitz, P., Fonseca, S. Y., Aye, T. and Sharma, A. (2004). The pancreatic ductal epithelium serves as a potential pool of progenitor cells. Pediatr Diabetes 5 Suppl 2, 16-22.

Bort, R., Martinez-Barbera, J. P., Beddington, R. S. and Zaret, K. S. (2004). Hex homeobox gene-dependent tissue positioning is required for organogenesis of the ventral pancreas. Development 131, 797-806.

Bossard, P. and Zaret, K. S. (1998). GATA transcription factors as potentiators of gut endoderm differentiation. Development 125, 4909-17.

Bossard, P. and Zaret, K. S. (2000). Repressive and restrictive mesodermal interactions with gut endoderm: possible relation to Meckel's Diverticulum. Development 127, 4915-23.

Calmont, A., Wandzioch, E., Tremblay, K. D., Minowada, G., Kaestner, K. H., Martin, G. R. and Zaret, K. S. (2006). An FGF response pathway that mediates hepatic gene induction in embryonic endoderm cells. Dev Cell 11, 339-48.

Cereghini, S. (1996). Liver-enriched transcription factors and hepatocyte differentiation. FASEB J 10, 267-82.

Chandra, V., G, S., Phadnis, S., Nair, P. D. and Bhonde, R. R. (2009). Generation of pancreatic hormone-expressing islet-like cell aggregates from murine adipose tissue-derived stem cells. Stem Cells 27, 1941-53.

Chaya, D., Hayamizu, T., Bustin, M. and Zaret, K. S. (2001). Transcription factor FoxA (HNF3) on a nucleosome at an enhancer complex in liver chromatin. J Biol Chem 276, 44385-9.

Chen, Y., Pan, F. C., Brandes, N., Afelik, S., Solter, M. and Pieler, T. (2004). Retinoic acid signaling is essential for pancreas development and promotes endocrine at the expense of exocrine cell differentiation in Xenopus. Dev Biol 271, 144-60.

Cirillo, L. A., Lin, F. R., Cuesta, I., Friedman, D., Jarnik, M. and Zaret, K. S. (2002). Opening of compacted chromatin by early developmental transcription factors HNF3 (FoxA) and GATA-4. Mol Cell 9, 279-89.

Clotman, F., Lannoy, V. J., Reber, M., Cereghini, S., Cassiman, D., Jacquemin, P., Roskams, T., Rousseau, G. G. and Lemaigre, F. P. (2002). The onecut transcription factor HNF6 is required for normal development of the biliary tract. Development 129, 1819-28.

Coad, R. A., Dutton, J. R., Tosh, D. and Slack, J. M. (2009). Inhibition of Hes1 activity in gall bladder epithelial cells promotes insulin expression and glucose responsiveness. Biochem Cell Biol 87, 975-87.

Crowe, A. J., Sang, L., Li, K. K., Lee, K. C., Spear, B. T. and Barton, M. C. (1999). Hepatocyte nuclear factor 3 relieves chromatin-mediated repression of the alpha-fetoprotein gene. J Biol Chem 274, 25113-20.

D'Amour, K. A., Agulnick, A. D., Eliazer, S., Kelly, O. G., Kroon, E. and Baetge, E. E. (2005). Efficient differentiation of human embryonic stem cells to definitive endoderm. Nat Biotechnol 23, 1534-41.

D'Amour, K. A., Bang, A. G., Eliazer, S., Kelly, O. G., Agulnick, A. D., Smart, N. G., Moorman, M. A., Kroon, E., Carpenter, M. K. and Baetge, E. E. (2006). Production of pancreatic hormone-expressing endocrine cells from human embryonic stem cells. Nat Biotechnol 24, 1392-401.

Dalvi, M. P., Umrani, M. R., Joglekar, M. V. and Hardikar, A. A. (2009). Human pancreatic islet progenitor cells demonstrate phenotypic plasticity in vitro. J Biosci 34, 523-8.

Dessimoz, J., Opoka, R., Kordich, J. J., Grapin-Botton, A. and Wells, J. M. (2006). FGF signaling is necessary for establishing gut tube domains along the anterior-posterior axis in vivo. Mech Dev 123, 42-55.

Douarin, N. M. (1975). An experimental analysis of liver development. Med Biol 53, 427-55.

Du Bois, A. M. (1963). The embryonic liver. In: The Liver, Morphology, Biochemistry, Physiology. C. Rouiller, ed. Academic Press, NewYork, pp. 1–39. In In: The Liver, Morphology, Biochemistry, Physiology., (ed., pp. pp. 1–39.

Dufort, D., Schwartz, L., Harpal, K. and Rossant, J. (1998). The transcription factor HNF3beta is required in visceral endoderm for normal primitive streak morphogenesis. Development 125, 3015-25.

Duncan, S. A. (2000). Transcriptional regulation of liver development. Dev Dyn 219, 131-42.

Efrat, S. (2008). In vitro expansion of human beta cells. Diabetologia 51, 1338-9.

Enzan, H., Himeno, H., Hiroi, M., Kiyoku, H., Saibara, T. and Onishi, S. (1997). Development of hepatic sinusoidal structure with special reference to the Ito cells. Microsc Res Tech 39, 336-49.

Ferber, S., Halkin, A., Cohen, H., Ber, I., Einav, Y., Goldberg, I., Barshack, I., Seijffers, R., Kopolovic, J., Kaiser, N. et al. (2000). Pancreatic and duodenal homeobox gene 1 induces expression of insulin genes in liver and ameliorates streptozotocin-induced hyperglycemia. Nat Med 6, 568-72.

Fukuda-Taira, S. (1981). Hepatic induction in the avian embryo: specificity of reactive endoderm and inductive mesoderm. J Embryol Exp Morphol 63, 111-25.

Fukuda, S. (1979). The development of hepatogenic potency in the endoderm of quail embryos. J Embryol Exp Morphol 52, 49-62.

Gershengorn, M. C., Geras-Raaka, E., Hardikar, A. A. and Raaka, B. M. (2005). Are better islet cell precursors generated by epithelial-to-mesenchymal transition? Cell Cycle 4, 380-2.
Gershengorn, M. C., Hardikar, A. A., Wei, C., Geras-Raaka, E., Marcus-Samuels, B. and Raaka, B. M. (2004). Epithelial-to-mesenchymal transition generates proliferative human islet precursor cells. Science 306, 2261-4.
Gilbert, S. F. and MA. (2000). Developmental Biology, Sinauer Associates, Sunderland,.
Grapin-Botton, A. (2005). Antero-posterior patterning of the vertebrate digestive tract: 40 years after Nicole Le Douarin's PhD thesis. Int J Dev Biol 49, 335-47.
Gualdi, R., Bossard, P., Zheng, M., Hamada, Y., Coleman, J. R. and Zaret, K. S. (1996). Hepatic specification of the gut endoderm in vitro: cell signaling and transcriptional control. Genes Dev 10, 1670-82.
Hardikar, A. A., Marcus-Samuels, B., Geras-Raaka, E., Raaka, B. M. and Gershengorn, M. C. (2003). Human pancreatic precursor cells secrete FGF2 to stimulate clustering into hormone-expressing islet-like cell aggregates. Proc Natl Acad Sci U S A 100, 7117-22.
Hardikar AA , T. D. (2009). β-cells in the lab. Nature India.
Harris, T. E., Albrecht, J. H., Nakanishi, M. and Darlington, G. J. (2001). CCAAT/enhancer-binding protein-alpha cooperates with p21 to inhibit cyclin-dependent kinase-2 activity and induces growth arrest independent of DNA binding. J Biol Chem 276, 29200-9.
Hayhurst, G. P., Lee, Y. H., Lambert, G., Ward, J. M. and Gonzalez, F. J. (2001). Hepatocyte nuclear factor 4alpha (nuclear receptor 2A1) is essential for maintenance of hepatic gene expression and lipid homeostasis. Mol Cell Biol 21, 1393-403.
Horb, M. E. and Slack, J. M. (2001). Endoderm specification and differentiation in Xenopus embryos. Dev Biol 236, 330-43.
Huangfu, D., Osafune, K., Maehr, R., Guo, W., Eijkelenboom, A., Chen, S., Muhlestein, W. and Melton, D. A. (2008). Induction of pluripotent stem cells from primary human fibroblasts with only Oct4 and Sox2. Nat Biotechnol 26, 1269-75.
Jacquemin, P., Durviaux, S. M., Jensen, J., Godfraind, C., Gradwohl, G., Guillemot, F., Madsen, O. D., Carmeliet, P., Dewerchin, M., Collen, D. et al. (2000). Transcription factor hepatocyte nuclear factor 6 regulates pancreatic endocrine cell differentiation and controls expression of the proendocrine gene ngn3. Mol Cell Biol 20, 4445-54.
Joglekar, M. V. and Hardikar, A. A. Epithelial-to-mesenchymal transition in pancreatic islet beta cells. Cell Cycle 9, 4077-9.
Joglekar, M. V. and Hardikar, A. A. (2010). Epithelial-to-mesenchymal transition in pancreatic islet beta cells. Cell Cycle 9, 4077-9.
Joglekar, M. V., Joglekar, V. M., Joglekar, S. V. and Hardikar, A. A. (2009a). Human fetal pancreatic insulin-producing cells proliferate in vitro. J Endocrinol 201, 27-36.
Joglekar, M. V., Patil, D., Joglekar, V. M., Rao, G. V., Reddy, D. N., Mitnala, S., Shouche, Y. and Hardikar, A. A. (2009b). The miR-30 family microRNAs confer epithelial phenotype to human pancreatic cells. Islets 1, 137-47.
Jung, J., Zheng, M., Goldfarb, M. and Zaret, K. S. (1999). Initiation of mammalian liver development from endoderm by fibroblast growth factors. Science 284, 1998-2003.
Kadam, S., Muthyala, S., Nair, P. and Bhonde, R. Human placenta-derived mesenchymal stem cells and islet-like cell clusters generated from these cells as a novel source for stem cell therapy in diabetes. Rev Diabet Stud 7, 168-82.

Kaestner, K. H., Hiemisch, H. and Schutz, G. (1998). Targeted disruption of the gene encoding hepatocyte nuclear factor 3gamma results in reduced transcription of hepatocyte-specific genes. Mol Cell Biol 18, 4245-51.

Kalinichenko, V. V., Lim, L., Stolz, D. B., Shin, B., Rausa, F. M., Clark, J., Whitsett, J. A., Watkins, S. C. and Costa, R. H. (2001). Defects in pulmonary vasculature and perinatal lung hemorrhage in mice heterozygous null for the Forkhead Box f1 transcription factor. Dev Biol 235, 489-506.

Kalinichenko, V. V., Zhou, Y., Bhattacharyya, D., Kim, W., Shin, B., Bambal, K. and Costa, R. H. (2002). Haploinsufficiency of the mouse Forkhead Box f1 gene causes defects in gall bladder development. J Biol Chem 277, 12369-74.

Keng, V. W., Fujimori, K. E., Myint, Z., Tamamaki, N., Nojyo, Y. and Noguchi, T. (1998). Expression of Hex mRNA in early murine postimplantation embryo development. FEBS Lett 426, 183-6.

Keng, V. W., Yagi, H., Ikawa, M., Nagano, T., Myint, Z., Yamada, K., Tanaka, T., Sato, A., Muramatsu, I., Okabe, M. et al. (2000). Homeobox gene Hex is essential for onset of mouse embryonic liver development and differentiation of the monocyte lineage. Biochem Biophys Res Commun 276, 1155-61.

Kmiec, Z. (2001). Cooperation of liver cells in health and disease. Adv Anat Embryol Cell Biol 161, III-XIII, 1-151.

Kojima, H., Fujimiya, M., Matsumura, K., Younan, P., Imaeda, H., Maeda, M. and Chan, L. (2003). NeuroD-betacellulin gene therapy induces islet neogenesis in the liver and reverses diabetes in mice. Nat Med 9, 596-603.

Kumar, M., Jordan, N., Melton, D. and Grapin-Botton, A. (2003). Signals from lateral plate mesoderm instruct endoderm toward a pancreatic fate. Dev Biol 259, 109-22.

Lai, E. and Darnell, J. E., Jr. (1991). Transcriptional control in hepatocytes: a window on development. Trends Biochem Sci 16, 427-30.

Landry, C., Clotman, F., Hioki, T., Oda, H., Picard, J. J., Lemaigre, F. P. and Rousseau, G. G. (1997). HNF-6 is expressed in endoderm derivatives and nervous system of the mouse embryo and participates to the cross-regulatory network of liver-enriched transcription factors. Dev Biol 192, 247-57.

Lardon, J., De Breuck, S., Rooman, I., Van Lommel, L., Kruhoffer, M., Orntoft, T., Schuit, F. and Bouwens, L. (2004a). Plasticity in the adult rat pancreas: transdifferentiation of exocrine to hepatocyte-like cells in primary culture. Hepatology 39, 1499-507.

Lardon, J., Huyens, N., Rooman, I. and Bouwens, L. (2004b). Exocrine cell transdifferentiation in dexamethasone-treated rat pancreas. Virchows Arch 444, 61-5.

Lawson, K. A. and Pedersen, R. A. (1987). Cell fate, morphogenetic movement and population kinetics of embryonic endoderm at the time of germ layer formation in the mouse. Development 101, 627-52.

Lemaigre, F. P. (2003). Development of the biliary tract. Mech Dev 120, 81-7.

Lemaigre, F. P., Durviaux, S. M., Truong, O., Lannoy, V. J., Hsuan, J. J. and Rousseau, G. G. (1996). Hepatocyte nuclear factor 6, a transcription factor that contains a novel type of homeodomain and a single cut domain. Proc Natl Acad Sci U S A 93, 9460-4.

Li, J., Ning, G. and Duncan, S. A. (2000). Mammalian hepatocyte differentiation requires the transcription factor HNF-4alpha. Genes Dev 14, 464-74.

Maehr, R., Chen, S., Snitow, M., Ludwig, T., Yagasaki, L., Goland, R., Leibel, R. L. and Melton, D. A. (2009). Generation of pluripotent stem cells from patients with type 1 diabetes. Proc Natl Acad Sci U S A 106, 15768-73.

Mahlapuu, M., Enerback, S. and Carlsson, P. (2001). Haploinsufficiency of the forkhead gene Foxf1, a target for sonic hedgehog signaling, causes lung and foregut malformations. Development 128, 2397-406.
Martin, M., Gallego-Llamas, J., Ribes, V., Kedinger, M., Niederreither, K., Chambon, P., Dolle, P. and Gradwohl, G. (2005). Dorsal pancreas agenesis in retinoic acid-deficient Raldh2 mutant mice. Dev Biol 284, 399-411.
Martinez Barbera, J. P., Clements, M., Thomas, P., Rodriguez, T., Meloy, D., Kioussis, D. and Beddington, R. S. (2000). The homeobox gene Hex is required in definitive endodermal tissues for normal forebrain, liver and thyroid formation. Development 127, 2433-45.
McLin, V. A., Rankin, S. A. and Zorn, A. M. (2007). Repression of Wnt/beta-catenin signaling in the anterior endoderm is essential for liver and pancreas development. Development 134, 2207-17.
Medlock, E. S. and Haar, J. L. (1983). The liver hemopoietic environment: I. Developing hepatocytes and their role in fetal hemopoiesis. Anat Rec 207, 31-41.
Moore-Scott, B. A., Opoka, R., Lin, S. C., Kordich, J. J. and Wells, J. M. (2007). Identification of molecular markers that are expressed in discrete anterior-posterior domains of the endoderm from the gastrula stage to mid-gestation. Dev Dyn 236, 1997-2003.
Mutskov, V., Raaka, B. M., Felsenfeld, G. and Gershengorn, M. C. (2007). The human insulin gene displays transcriptionally active epigenetic marks in islet-derived mesenchymal precursor cells in the absence of insulin expression. Stem Cells 25, 3223-33.
Nakagawa, M., Koyanagi, M., Tanabe, K., Takahashi, K., Ichisaka, T., Aoi, T., Okita, K., Mochiduki, Y., Takizawa, N. and Yamanaka, S. (2008). Generation of induced pluripotent stem cells without Myc from mouse and human fibroblasts. Nat Biotechnol 26, 101-6.
Okita, K., Ichisaka, T. and Yamanaka, S. (2007). Generation of germline-competent induced pluripotent stem cells. Nature 448, 313-7.
Parekh, V. S., Joglekar, M. V. and Hardikar, A. A. (2009). Differentiation of human umbilical cord blood-derived mononuclear cells to endocrine pancreatic lineage. Differentiation 78, 232-40.
Phadnis, S. M., Joglekar, M. V., Dalvi, M. P., Muthyala, S., Nair, P. D., Ghaskadbi, S. M., Bhonde, R. R. and Hardikar, A. A. Human bone marrow-derived mesenchymal cells differentiate and mature into endocrine pancreatic lineage in vivo. Cytotherapy 13, 279-93.
Pontoglio, M., Barra, J., Hadchouel, M., Doyen, A., Kress, C., Bach, J. P., Babinet, C. and Yaniv, M. (1996). Hepatocyte nuclear factor 1 inactivation results in hepatic dysfunction, phenylketonuria, and renal Fanconi syndrome. Cell 84, 575-85.
Pontoglio, M., Faust, D. M., Doyen, A., Yaniv, M. and Weiss, M. C. (1997). Hepatocyte nuclear factor 1alpha gene inactivation impairs chromatin remodeling and demethylation of the phenylalanine hydroxylase gene. Mol Cell Biol 17, 4948-56.
Rausa, F., Samadani, U., Ye, H., Lim, L., Fletcher, C. F., Jenkins, N. A., Copeland, N. G. and Costa, R. H. (1997). The cut-homeodomain transcriptional activator HNF-6 is coexpressed with its target gene HNF-3 beta in the developing murine liver and pancreas. Dev Biol 192, 228-46.
Roberts, D. J., Johnson, R. L., Burke, A. C., Nelson, C. E., Morgan, B. A. and Tabin, C. (1995). Sonic hedgehog is an endodermal signal inducing Bmp-4 and Hox genes during induction and regionalization of the chick hindgut. Development 121, 3163-74.

Rossi, J. M., Dunn, N. R., Hogan, B. L. and Zaret, K. S. (2001). Distinct mesodermal signals, including BMPs from the septum transversum mesenchyme, are required in combination for hepatogenesis from the endoderm. Genes Dev 15, 1998-2009.

Russ, H. A., Bar, Y., Ravassard, P. and Efrat, S. (2008). In vitro proliferation of cells derived from adult human beta-cells revealed by cell-lineage tracing. Diabetes 57, 1575-83.

Russ, H. A., Ravassard, P., Kerr-Conte, J., Pattou, F. and Efrat, S. (2009). Epithelial-mesenchymal transition in cells expanded in vitro from lineage-traced adult human pancreatic beta cells. PLoS One 4, e6417.

Sahu, S., Joglekar, M. V., Dumbre, R., Phadnis, S. M., Tosh, D. and Hardikar, A. A. (2009a). Islet-like cell clusters occur naturally in human gall bladder and are retained in diabetic conditions. J Cell Mol Med 13, 999-1000.

Sahu, S., Tosh, D. and Hardikar, A. A. (2009b). New sources of beta-cells for treating diabetes. J Endocrinol 202, 13-6.

Samadani, U. and Costa, R. H. (1996). The transcriptional activator hepatocyte nuclear factor 6 regulates liver gene expression. Mol Cell Biol 16, 6273-84.

Shalaby, F., Ho, J., Stanford, W. L., Fischer, K. D., Schuh, A. C., Schwartz, L., Bernstein, A. and Rossant, J. (1997). A requirement for Flk1 in primitive and definitive hematopoiesis and vasculogenesis. Cell 89, 981-90.

Shalaby, F., Rossant, J., Yamaguchi, T. P., Gertsenstein, M., Wu, X. F., Breitman, M. L. and Schuh, A. C. (1995). Failure of blood-island formation and vasculogenesis in Flk-1-deficient mice. Nature 376, 62-6.

Shapiro, A. M., Lakey, J. R., Ryan, E. A., Korbutt, G. S., Toth, E., Warnock, G. L., Kneteman, N. M. and Rajotte, R. V. (2000). Islet transplantation in seven patients with type 1 diabetes mellitus using a glucocorticoid-free immunosuppressive regimen. N Engl J Med 343, 230-8.

Shen, M. M. (2007). Nodal signaling: developmental roles and regulation. Development 134, 1023-34.

Shih, D. Q., Navas, M. A., Kuwajima, S., Duncan, S. A. and Stoffel, M. (1999). Impaired glucose homeostasis and neonatal mortality in hepatocyte nuclear factor 3alpha-deficient mice. Proc Natl Acad Sci U S A 96, 10152-7.

Shim, E. Y., Woodcock, C. and Zaret, K. S. (1998). Nucleosome positioning by the winged helix transcription factor HNF3. Genes Dev 12, 5-10.

Shiojiri, N. (1984). The origin of intrahepatic bile duct cells in the mouse. J Embryol Exp Morphol 79, 25-39.

Shiojiri, N. (1997). Development and differentiation of bile ducts in the mammalian liver. Microsc Res Tech 39, 328-35.

Shiojiri, N., Lemire, J. M. and Fausto, N. (1991). Cell lineages and oval cell progenitors in rat liver development. Cancer Res 51, 2611-20.

Sladek, F. M. and Darnell, J. E. (1992). Mechanisms of liver-specific gene expression. Curr Opin Genet Dev 2, 256-9.

Sosa-Pineda, B., Wigle, J. T. and Oliver, G. (2000). Hepatocyte migration during liver development requires Prox1. Nat Genet 25, 254-5.

Stafford, D., Hornbruch, A., Mueller, P. R. and Prince, V. E. (2004). A conserved role for retinoid signaling in vertebrate pancreas development. Dev Genes Evol 214, 432-41.

Sund, N. J., Ang, S. L., Sackett, S. D., Shen, W., Daigle, N., Magnuson, M. A. and Kaestner, K. H. (2000). Hepatocyte nuclear factor 3beta (Foxa2) is dispensable for maintaining the differentiated state of the adult hepatocyte. Mol Cell Biol 20, 5175-83.

Takahashi, K., Tanabe, K., Ohnuki, M., Narita, M., Ichisaka, T., Tomoda, K. and Yamanaka, S. (2007). Induction of pluripotent stem cells from adult human fibroblasts by defined factors. Cell 131, 861-72.

Takahashi, K. and Yamanaka, S. (2006). Induction of pluripotent stem cells from mouse embryonic and adult fibroblast cultures by defined factors. Cell 126, 663-76.

Thomas, P. Q., Brown, A. and Beddington, R. S. (1998). Hex: a homeobox gene revealing peri-implantation asymmetry in the mouse embryo and an early transient marker of endothelial cell precursors. Development 125, 85-94.

Timchenko, N. A., Harris, T. E., Wilde, M., Bilyeu, T. A., Burgess-Beusse, B. L., Finegold, M. J. and Darlington, G. J. (1997). CCAAT/enhancer binding protein alpha regulates p21 protein and hepatocyte proliferation in newborn mice. Mol Cell Biol 17, 7353-61.

Timchenko, N. A., Wilde, M. and Darlington, G. J. (1999). C/EBPalpha regulates formation of S-phase-specific E2F-p107 complexes in livers of newborn mice. Mol Cell Biol 19, 2936-45.

Tiso, N., Filippi, A., Pauls, S., Bortolussi, M. and Argenton, F. (2002). BMP signalling regulates anteroposterior endoderm patterning in zebrafish. Mech Dev 118, 29-37.

Tzakis, A. G., Ricordi, C., Alejandro, R., Zeng, Y., Fung, J. J., Todo, S., Demetris, A. J., Mintz, D. H. and Starzl, T. E. (1990). Pancreatic islet transplantation after upper abdominal exenteration and liver replacement. Lancet 336, 402-5.

Van Hoof, D., D'Amour, K. A. and German, M. S. (2009). Derivation of insulin-producing cells from human embryonic stem cells. Stem Cell Res 3, 73-87.

Waddington, C. H. (1940). Organisers and Genes, Cambridge University Press, Cambridge, MA.

Wang, N. D., Finegold, M. J., Bradley, A., Ou, C. N., Abdelsayed, S. V., Wilde, M. D., Taylor, L. R., Wilson, D. R. and Darlington, G. J. (1995). Impaired energy homeostasis in C/EBP alpha knockout mice. Science 269, 1108-12.

Wells, J. M. and Melton, D. A. (2000). Early mouse endoderm is patterned by soluble factors from adjacent germ layers. Development 127, 1563-72.

Zalzman, M., Anker-Kitai, L. and Efrat, S. (2005). Differentiation of human liver-derived, insulin-producing cells toward the beta-cell phenotype. Diabetes 54, 2568-75.

Zalzman, M., Gupta, S., Giri, R. K., Berkovich, I., Sappal, B. S., Karnieli, O., Zern, M. A., Fleischer, N. and Efrat, S. (2003). Reversal of hyperglycemia in mice by using human expandable insulin-producing cells differentiated from fetal liver progenitor cells. Proc Natl Acad Sci U S A 100, 7253-8.

Zaret, K. (1998). Early liver differentiation: genetic potentiation and multilevel growth control. Curr Opin Genet Dev 8, 526-31.

Zaret, K. S. (2000). Liver specification and early morphogenesis. Mech Dev 92, 83-8.

Zhou, Q., Brown, J., Kanarek, A., Rajagopal, J. and Melton, D. A. (2008). In vivo reprogramming of adult pancreatic exocrine cells to beta-cells. Nature 455, 627-32.

Zhou, Q. and Melton, D. A. (2008). Extreme makeover: converting one cell into another. Cell Stem Cell 3, 382-8.

Zorn, A. M. and Wells, J. M. (2007). Molecular basis of vertebrate endoderm development. Int Rev Cytol 259, 49-111

22

Stem Cell Therapy for Islet Regeneration

Phuc Pham Van
University of Science,
Vietnam National University HCM city
Vietnam

1. Introduction

Diabetes mellitus is an endocrine disorder characterised by inadequate production or use of insulin, resulting in abnormally high blood glucose levels. High blood glucose leads to the formation of reactive advanced glycation end-products (Feldman et al., 1997), which are responsible for complications such as blindness, kidney failure, cardiovascular disease, stroke, neuropathy and vascular dysfunction. Diabetes mellitus is classified as either type 1 or type 2. Type 1 diabetes mellitus (insulin-dependent diabetes mellitus) results from the autoimmune destruction of the pancreatic beta cells, whereas type 2 diabetes mellitus (non-insulin-dependent diabetes mellitus) results from insulin resistance and impaired glucose tolerance.

Approximately 7.8% (23.6 million people) of the US population has been diagnosed with diabetes mellitus, and another 57 million people are likely to develop diabetes mellitus in the coming years (American Diabetes Association, 2007). The number of people with diabetes mellitus is set to continue to rapidly increase between now and 2030, especially in developing countries.

Over the last decade, a new form of treatment called islet transplantation therapy was thought to provide good patient outcomes; however, few islets are available for transplantation. Typically, the pooled islets isolated from two pancreases are enough to treat a single patient. Since the enormous potential of stem cells was discovered, it was hoped that they would provide the most effective treatment for diabetes mellitus. Over the past two decades, hundreds of studies have looked at the potential of stem cell therapy for treating diabetes mellitus. Successful stem cell therapy would eliminate the cause of the disease and lead to stable, long-term results; hence, the term "pancreatic regeneration" was coined. The hypothesis was that stem cells could regenerate the damaged pancreas. After careful consideration of the aetiology of diabetes mellitus, scientists have put forward two general treatment strategies: stem cell therapy to treat the autoimmune aspect of the disease, and stem cell therapy to treat the degenerative aspect of the disease. In this review, we focus on stem cell-based therapies aimed at islet regeneration through stem cell or insulin-producing cell (IPC) transplantation. We will also discuss the latest strategies for treating both type 1 and type 2 diabetes mellitus using stem cell therapy, along with the (initially promising) results.

2. Islet regeneration by cell replacement

2.1 Stem cell sources

Many different types of stem cells have been used in the research, testing and treatment of diabetes mellitus, including stem cells that can be used to regenerate pancreatic islets, e.g. embryonic stem cells, adult stem cells and infant stem cells (umbilical cord stem cells isolated from umbilical cord blood).

2.1.1 Embryonic stem cells

Human embryonic stem cells (ESCs) were first isolated at the University of Wisconsin-Madison in 1998 by James Thomson (Thomson et al., 1998). These cells were established as immortal pluripotent cell lines that are still in existence today. The ESCs were derived from blastocysts donated by couples undergoing treatment for infertility using methodology developed 17 years earlier to obtain mouse ESCs. Briefly, the trophectoderm is first removed from the blastocyst by immunosurgery and the inner cell mass is plated onto a feeder layer of mouse embryonic fibroblasts (Trounson et al., 2001; 2002). However, cells can also be derived from early human embryos at the morula stage (Strelchenko et al. 2004) after the removal of the zona pellucida using an acidified solution, or by enzymatic digestion by pronase (Verlinsky et al., 2005). Nowadays, ESCs can be isolated from many different sources (Fig. 1).

ESCs are pluripotent, which means that they can differentiate into any of the functional cells derived from the three germ layers, including beta cells or insulin-producing cells (IPCs). The differentiation of ESCs into IPCs is prerequisite for their use as a diabetes mellitus treatment, and may occur either *in vivo* (after transplantation) or *in vitro* (before transplantation). *In vivo* differentiation is based on micro environmental conditions at the graft site, whereas *in vitro* differentiation requires various external factors that induce the phenotypic changes required to produce IPCs. This means that diabetes mellitus can be treated either by direct transplantation of ESCs, or by indirect transplantation of IPCs that have been differentiated from ESCs. However, Naujok et al. (2009) showed that ESCs could modify gene expression and exhibit a phenotype similar to that of islet cells when transplanted into the pancreas only if they are first differentiated *in vitro,* and that *in vitro* differentiation is a prerequisite for successful *in vivo* differentiation (Naujok et al., 2009). Moreover, using ESCs for pancreatic regeneration carries with it the risk of tumour formation after transplantation.

Therefore, the *in vitro* differentiation of ESCs into IPCs is necessary before they can be used to treat diabetes mellitus. Studies looking at the *in vitro* differentiation of ESCs into IPCs were first performed in 2001 using mouse cells (Lumelsky et al., 2001). However, the results could not be repeated in subsequent studies (Rajagopal et al., 2003; Hansson et al., 2004; Sipione et al., 2004). Researchers then developed a strategy for selecting ESCs expressing genes related to pancreatic cells (e.g. nestin), and successfully generated IPCs from these ESCs (Soria et al., 2000; Leon-Quinto et al., 2004). Other workers succeeded in creating IPCs from ESCs using gene transfer (Blyszczuk et al., 2003; Schroeder et al., 2006), or phosphoinositol-3 kinase inhibitors (Hori et al., 2002). The differentiation of ESCs into IPCs usually involves differentiation into embryoid bodies. This relatively long process comprises two phases: the embryoid body stage (4–5 days) and the differentiation stage (30–40 days). In 2005, Shi et al. decreased the time taken for this differentiation process to 15 days (Shi et al., 2005).

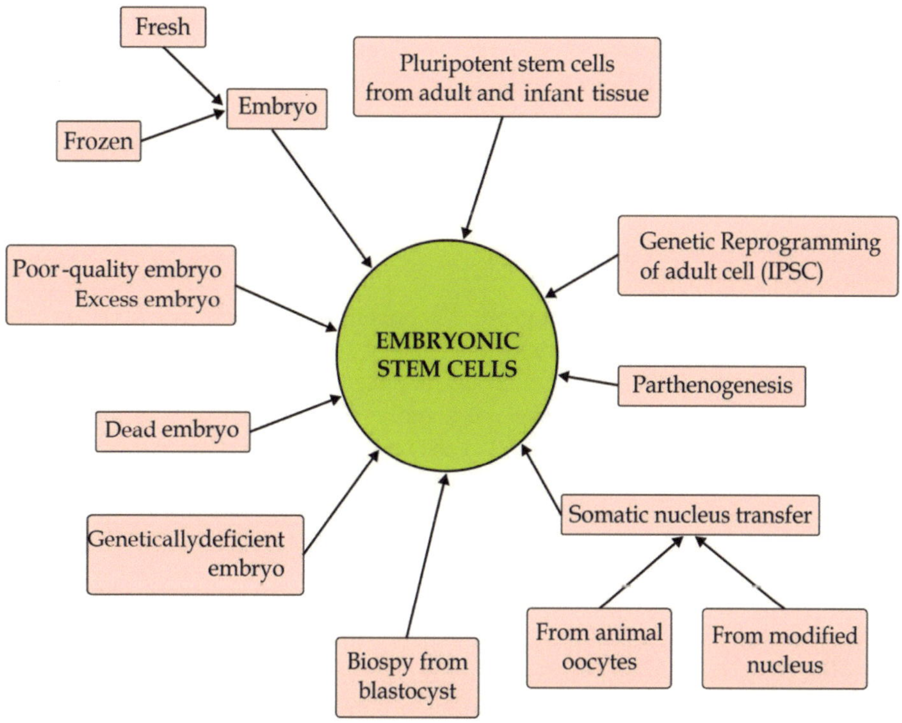

Fig. 1. ESC sources. ESCs can be isolated from fresh, frozen, dead, excess and genetically deficient embryos, by parthenogenesis and somatic nucleus transfer, from biopsies, and from pluripotent stem cells obtained from adult tissues.

In 2001, Assady et al. reported that IPCs could be generated by spontaneous differentiation of human ESCs. Although the IPC number and insulin content of these cells was low, this was the first proof-of-principle experiment indicating that human ESCs were a potential source of β-like cells. Recent reports from D'Amour et al. and Kroon et al. described the differentiation of pancreatic lineage cells from human ESCs *in vitro*. To date, many groups have reported the *in vitro* generation of IPCs from human ESCs (**D'Amour** et al., 2006; Jiang et al., 2007; Jiang et al., 2007).

2.1.2 Induced pluripotent stem cells

First created by Takahashi et al. (2007) and Yu et al. (2007), induced pluripotent stem cells (IPSCs) are a new source of embryonic-like stem cells, and are considered a technical breakthrough in stem cell research. IPSCs have several advantages over ESCs. One major advantage is that IPSCs can be created from any cell-type; thus, creating patient-specific stem cells (Park et al., 2008; Dimos et al., 2008). Similar to ESCs, IPSCs can differentiate into many different cell types, including neurons (Dimos et al., 2008; Chambers et al., 2008), heart muscle cells (Zhang et al., 2009) and insulin-secreting cells (Tateishi et al., 2008; Zhang et al., 2009).

IPSCs can be created from many different cell types via a simple process. First-generation IPSCs are obtained by transferring four genes (*Oct-3/4, Sox-2, c-Myc* and *Klf4*; Shinya Yamanaka et al., 2006) or *Oct-3 / 4, Sox-2, Nanog* and *LIN28* into mice. Second-generation

IPSCs are derived using only *Oct-3/4, Sox-2* and *Klf4,* because *c-Myc* is an oncogene (Nakagawa et al., 2008). Third-generation IPSCs are generated using only two genes, *Oct-3/4* and *Sox-2*, and the histone deacetylase inhibitor, valproic acid (VPA) (Danwei Huangfu et al., 2008).
A recent study shows that IPSCs can be successfully created from adult fibroblasts derived from type 1 diabetic patients (Rene'Maehr et al., 2009). These cells were differentiated into IPCs and used to successfully treat diabetic rats (Alipio et al., 2010).

2.1.3 Pancreatic stem cells

A recent report by Harry Heimberg's group (Heimberg et al., 2008) describes the existence of pancreatic stem cells in mice. In their most recent study, Heimberg's group ligated the ducts that secrete pancreatic enzymes in adult mice. The result was a doubling in the number of beta cells within two weeks. Also, the pancreases of the experimental animals began to produce more insulin; evidence that the newly generated beta cells were functional (Xu et al., 2008). Another research team showed that the production of new beta cells was dependent on the gene *neurogenin 3 (Ngn3),* which plays a role in the pancreas during embryonic development, and successfully isolated and established a murine pancreatic stem cell line (Noguchi et al., 2008; 2009).
Human pancreatic stem cells have also been successfully differentiated into IPCs (Noguchi et al., 2010). Islet cells were isolated from the pancreases of human donors using the Ricordi technique modified by the Edmonton protocol. The isolated cells were then cultured in media specifically designed for mouse or human pancreatic embryonic stem cell culture. The cells were differentiated for 2 weeks in induction media containing exendin-4, nicotinamide, keratinocyte growth factor, PDX-1 protein, or protein BETA2/NeuroD. However, according to Davani et al. (2007), human islet precursor cells derived from human pancreases exhibit the properties of mesenchymal stem cells (MSCs) in that they adhere to plastic, express CD73, CD90 and CD105, and differentiate *in vitro* into adipocytes, chondrocytes, and osteocytes. Davani et al. also identified a rare population of CD105+/CD73+/CD90+ cells in adult human islets that express low levels of insulin mRNA (Davani et al., 2007).

2.1.4 Mesenchymal stem cells

MSCs are multipotent stem cells that can differentiate into a variety of cell types, such as osteoblasts (bone cells), chondrocytes (cartilage cells) and adipocytes (fat cells) (Anna et al., 2008). This cell type was first discovered in 1924 by the cell morphologist Alexander A. Maximo, who described a type of cell within the mesenchyme that develops into various types of blood cell. Ernest A. McCulloch and James E. Till first revealed the clonal nature of marrow cells in 1963 (Becker et al., 1963; Siminovitch et al., 1963). Subsequently, *ex vivo* clonogenic assays were used to examine the potential of multipotent marrow cells (Friedenstein et al., 1974, 1976). In these assays, stromal cells or MSCs were used as colony-forming unit-fibroblasts (CFU-f). The characteristics of MSCs are as follows: they adhere to culture vessels; they have a fibroblast-like shape; they express Stro-1, CD133, CD29, CD44, CD90, CD105 (SH2), SH3, SH4 (CD73), c-kit, CD71, and CD106; and they can differentiate into specialised cells, e.g. bone, cartilage and fat.
MSCs have been isolated from many different tissues, including bone marrow (Oyajobi et al.. 1999; Majumdar et al., 2000; Prockop et al., 2001; Smith et al., 2004; Titorencu et al., 2007;

Wolfe et al., 2008; Gronthos and Zannettino et al., 2008; Phadnis et al., 2011; Bao et al., 2011), adipose tissue (Katz et al., 2005; Baptista et al., 2009; Caviggioli et al., 2009; Baer et al., 2010; Bruyn et al., 2010; Estes et al., 2010; Tucker, Bunnell, 2011), peripheral blood (Kassis et al., 2006), umbilical cord blood (Erices et al., 2000; Rosada et al., 2003; Hutson et al., 2005; Reinisch et al., 2007; Bieback and Klüter et al., 2007; Perdikogianni et al., 2008; Zhang et al., 2011), banked umbilical cord blood (Phuc et al., 2011), umbilical cords (Cutler et al., 2010; Farias et al., 2011), umbilical cord membranes (Deuse et al., 2010; Kita et al., 2010), umbilical cord veins (Santos et al., 2010), Wharton's jelly from the umbilical cord (Zeddou et al., 2010; Peng et al., 2011), placenta (Miao et al., 2006; Battula et al., 2007; Huang et al., 2009; Semenov et al., 2010; Pilz et al., 2011), decidua basalis (Macias et al., 2010; Lu et al., 2011), the ligamentum flavum (Chen et al., 2011), amniotic fluid (Feng et al., 2009; Choi et al., 2011, Shuang-Zhi et al., 2010), amniotic membrane (Chang et al., 2010; Marongiu et al., 2010), dental pulp (Agha-Hosseini et al., 2010; Karaöz et al., 2010; Yalvac et al., 2010; Spath et al., 2010), chorionic villi from human placenta (Poloni et al., 2008), foetal membranes (Soncini et al., 2007), menstrual blood (Meng et al., 2007; Hida et al., 2008; Musina et al., 2008; Kyurkchiev et al., 2010), and breast milk (Patki et al., 2010) (Fig. 2).

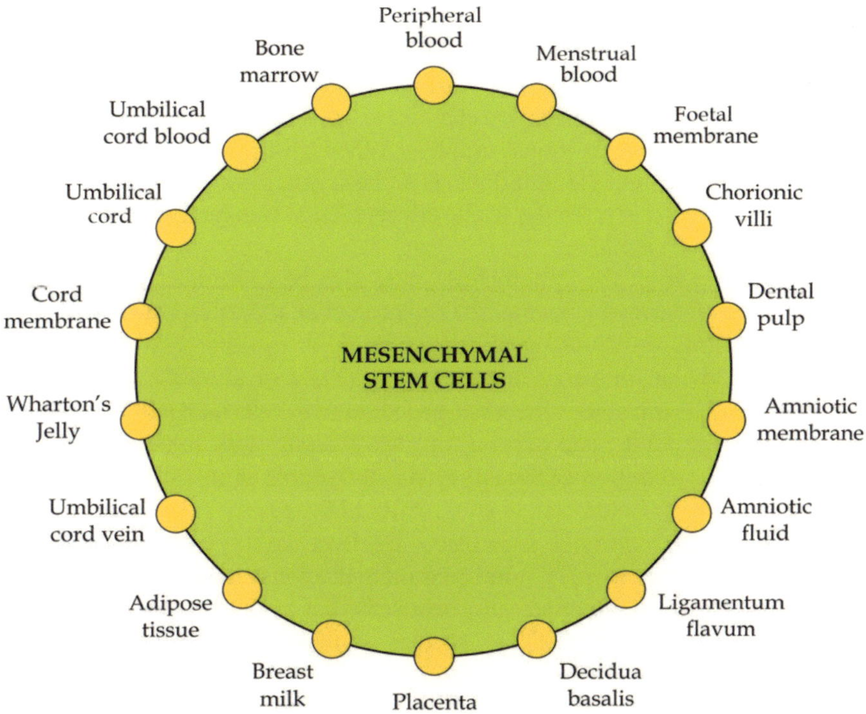

Fig. 2. Sources of MSCs. MSCs can be derived from several adult or infant tissues.

MSCs have been successfully differentiated into IPCs *in vitro* and can reduce blood glucose levels in both animals and humans after transplantation. The *in vitro* differentiation of MSCs into IPCs requires certain substances combined with medium stress. Most successful protocols for the differentiation of MSCs into IPCs used nicotinamide and/or exendin-4 inducers. Changes in the glucose concentration within the culture medium are necessary to trigger this process. MSCs are commonly cultured in low glucose medium to initiate differentiation before they can be induced to differentiate into IPCs by nicotinamide. In some studies, epidermal growth factor (EGF) was added to the culture medium during the IPC maturation phase in addition to nicotinamide. Currently, IPCs can be generated from MSCs obtained from human umbilical cord blood (Gao et al., 2008; Parekh et al., 2009; Wang et al., 2010), banked human umbilical cord blood (Phuc et al., 2011), placenta (Kadam et al., 2010), bone marrow (Sun et al., 2007; Xie et al., 2009; Phadnis et al., 2011), menstrual blood (Li et al., 2010), amniotic fluid (Trovato et al., 2009), Wharton's jelly (Chao et al., 2008; Wu et al., 2009), amnion (Kadam et al., 2010), and adipose tissue (Chandra et al., 2009). Other studies report the successful use of transgenesis to differentiate MSCs into IPCs, or up-regulation of genes (mainly *PDX-1* or *betacellulin*) related to signalling pathways that trigger this process (Karnieli et al., 2007; Li et al., 2007; Li et al., 2008; Hisanaga et al., 2008; Limbert and Seufert., 2009; Yuan et al., 2010; Paz et al., 2011). Moreover, coating the tissue culture flasks with substrates such as fibronectin or laminin can also induce MSCs to differentiate into IPCS (Moriscot et al., 2005; Chang et al., 2008; Gao et al., 2008; Lin et al., 2010; Lin et al., 2011).

2.1.5 Other sources

Recent reports suggest that pancreatic duct cells, liver cells, spleen cells, and other cell types have the ability to differentiate into islet cells. Although it is difficult to differentiate adult cells into insulin-producing pancreatic cells, some researchers have shown evidence of pancreatic duct regeneration in mouse models. When gastrin was injected into mice to induce acinar cells to differentiate into duct cells, these cells became a cellular substrate for the formation of new beta cells, similar to the effects seen in rats receiving glucose injections (Weir and Bonner-Weir et al., 2004).

Liver cells originating from the endothelium may also be candidates for this specialised insulin-secreting role (Meivar-Levy et al., 2006). Yang et al. (2002) reported that exposure to high glucose concentrations caused oval cells in the liver to differentiate into cells with a phenotype similar to that of pancreatic islet cells (Yang et al., 2002). Another strategy involves the *in vivo* gene transfer of the *pdx-1* gene into liver cells using an adenovirus vector to induce endogenous *pdx-1* gene expression. *Pdx-1*, along with other beta cell genes, is associated with insulin secretion (Zalzman et al., 2005; Sapir et al., 2005; Shternhall-Ron et al., 2007; Aviv et al., 2009; Gefen-Halevi et al., 2010; Meivar-Levy and Ferber, 2010). Similar to *pdx-1*, *betacellulin* and *neuro-D* expression by liver cells yielded sufficient insulin-producing cells in a streptozocin (STZ)-induced diabetic mouse model. These techniques not only induce liver cells to differentiate into beta cells, but also create new islets within the liver itself (Kojima et al., 2003). Other studies showed that human foetal liver cells transfected with *telomerase* and *pdx-1* can produce insulin and release it into the body. These cells cured diabetes mellitus when transplanted into immuno-deficient diabetic mice.

Fibroblasts are a relatively new source of islets and are easily isolated from skin. In a recent study, 61 single-cell-derived dermal fibroblast clones were established from human foreskin using a limiting dilution technique. These cells were able to differentiate into islet-like clusters when induced using pancreatic-inducing medium and several hormones, including insulin, glucagon and somatostatin, were detectable at both the mRNA and protein levels after induction. Moreover, transplantation of these islet-like clusters resulted in the release insulin in response to glucose *in vitro* (Bi et al. 2010).

2.2 Stem cell transplantation strategies

2.2.1 Transplantation methods

Transplantation of stem cells/IPCs to treat diabetes mellitus has been investigated in both animal models and humans. Many different types of stem cells have been tested using different methods. Cells can be grafted underneath the kidney capsule (Rackham et al., 2011; Figliuzzi et al., 2009; Ito et al., 2010; Lin et al., 2009; Kodama et al., 2009; Kodama et al., 2008; Zhang et al., 2010; Ohmura et al., 2010; Xiao et al., 2008; Berman et al., 2010), delivered via intra-peritoneal injection (Boroujeni et al., 2011; Chandra et al., 2009; Koya et al., 2008; Shao et al., 2011; Kadam et al., 2010; Phuc et al., 2011; Lin et al., 2009) or intra-portally (Shyu et al., 2011; Trivedi et al., 2008; Li et al., 2010; Wu et al., 2007; Longoni et al., 2010; Itakura et al., 2007), grafted into the liver (Chao et al., 2008; Zhu et al., 2009; Xu et al., 2007; Chen et al., 2009; Wang et al., 2010) or injected into the tail vein (Dinarvand et al., 2010; Koblas et al., 2009; Kajiyama et al., 2010; Jurewicz et al., 2010) (Fig. 3). However, there is little research comparing the efficiency of these methods. Chen et al. (2009) showed that transplantation of stem cells into the liver produces better results than transplantation into the renal capsule. Although diabetes mellitus is caused by destruction of the beta cells within the pancreatic islets, no studies have attempted transplantation directly into the pancreas. This is because the pancreas is very sensitive organ and is vulnerable to mechanical intervention.

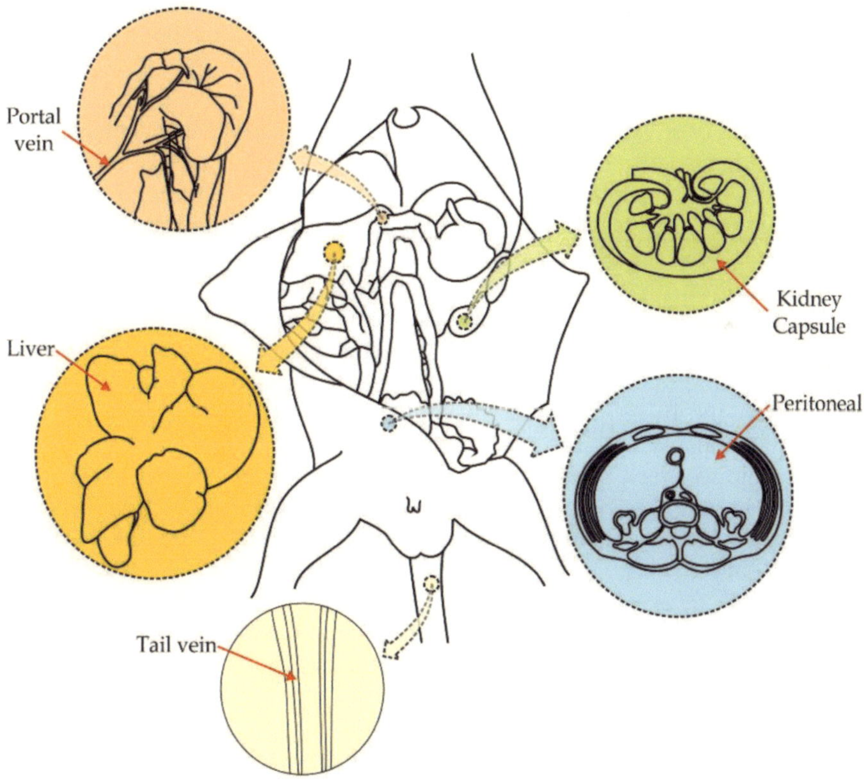

Fig. 3. Methods of stem cell/IPC transplantation. Stem cells or IPCs can be transplanted via the tail vein, intraperitoneally, under the kidney capsule, into the liver, or via the portal vein.

2.2.2 Stem cell transplantation

Unlike IPC transplantation, the mechanisms underlying islet regeneration and the reductions in blood glucose levels seen in diabetic patients require further study. The main questions that need to be answered are: 1) what role do grafted stem cells play in the regeneration of pancreatic islets? 2) How will stem cells behave when grafted into the body rather than the pancreas?

One type of stem cell that has been used to treat diabetes mellitus and investigated extensively in animal models is MSCs. Almost all research on MSC transplantation shows that *in vitro* or *in vivo* transplantation of MSCs results in a reduction of blood glucose levels, weight gain and increased longevity. However, MSCs can play multiple roles. Grafted stem cells can move into the pancreatic islets and differentiate into IPCs (Sorvi et al., 2005; Sordi, 2009). In an *in vitro* model using MSCs derived from human bone marrow and pancreatic islets, Sorvi *et al.* (2005) demonstrated crosstalk between MSCs and pancreatic cells mediated by various chemokines and their receptors. A minority of BM-MSCs (2–25%) express chemokine receptors (CXC receptor 4 [CXCR4], CX3C receptor 1 [CX3CR1], CXCR6, CC chemokine receptor 1 [CCR1], and CCR7) and, accordingly, show chemotactic migration in response to chemokine CXC ligand 12 (CXCL12), CX3CL1, CXCL16, CC chemokine ligand 3 (CCL3), and CCL19. These factors, released from the islets, were then able to attract MSCs. Moreover, MSCs were detected within the pancreatic islets of mice injected with green fluorescent protein (GFP)-positive MSCs (Sorvi et al., 2005). This result was subsequently confirmed in 2009 by Sordi, who hypothesised that the crosstalk between MSCs and pancreatic islets was driven by the CXCR4-CXCL12 and CX3CR1-CX3CL1 axes (Sordi, 2009). Movement of MSCs into the pancreas after transplantation was also confirmed by Lin et al. (2009) and Phadnis et al. (2011). Using bone marrow-derived MSC transplantation coupled with down-regulation of neurogenin 3 (Ngn3) induced by a recombinant lentivirus encoding two different small hairpin RNAs (shRNAs) for specific interference, they showed the successful engraftment of MSCs. In addition, they found that the endogenous pancreatic stem cells differentiated into IPCs and played a major role in reversing hyperglycaemia (Lin et al., 2009). However, there are cases in which stem cells derived from human umbilical cord blood also move into the pancreas and differentiate into IPCs in immunocompromised diabetic animals without improving hyperglycaemia (Koblas et al., 2009). Hasegawa et al. (2007) used Nos3 (-/-) mice as a model of impaired bone marrow-derived cell mobilisation and showed that the hyperglycaemia-improving effects of bone marrow transplantation were inversely correlated with the severity of myelo-suppression and delays in peripheral white blood cell recovery. Thus, stem cell mobilisation is critical for bone marrow transplantation-induced beta cell regeneration after injury. Therefore, they suggested that, during bone marrow transplantation, grafted cells first move into the recipient's bone marrow and, subsequently, into the injured periphery to regenerate the recipient's pancreatic beta cells (Hasegawa et al., 2007).

Another study showed that MSCs display immunomodulatory functions. MSCs prevented beta-cell destruction and development of diabetes mellitus by inducing regulatory T cells (Madec et al., 2009). Thus, MSC transplantation may prevent islet cell destruction by the immune system seen in type 1 diabetes mellitus and the pancreatic islets can be gradually restored. The result was a decrease in blood sugar levels and weight gain. While in a more recent study, it is said that MSCs protected islets from hypoxia/reoxygenation (H/R)-induced injury by decreasing apoptosis and increasing the expression of HIF-1α, HO-1, and COX-2 mRNA. The MSCs induced the expression of anti-apoptotic genes, thereby enhancing resistance to H/R-induced apoptosis and dysfunction (Lu et al., 2010).

The use of ESCs for treating diabetes mellitus is limited because of high levels of tumour formation. So there were a few researches using the ESCs for treating diabetes mellitus. In one study, pancreatic cell ontogeny within ESCs transplanted into the renal capsule of STZ-induced mice resulted in pancreatogenesis *in situ* or beta cell neogenesis. Immunohistochemistry was performed on excised pancreatic tissues using antibodies against stage- and lineage-specific pancreatic markers. Twenty-one days post-transplantation, PDX-1+ pancreatic foci appeared in the renal capsule, which expressed exocrine enzymes (amylase) and endocrine hormones (insulin, glucagon, and somatostatin). These multi-hormonal endocrine cells, a characteristic of beta cell regeneration, suggested possible divergence from embryonic islet cell development (Kodama et al., 2008). In another study, Kodama et al. (2009) showed that transplanted ESCs could migrate into the injured pancreas. Cell tracing analysis showed that significant beta cell neogenesis occurred 2 to 3 weeks after injury. Importantly, whereas pancreas-localised ESC or their derivatives were found adjacent to the sites of regeneration, neogenic pancreatic epithelia, including Ngn3+ cells, were endogenous. Transplantation efficiency was confirmed by enhanced endogenous regeneration and increased beta cell differentiation from endogenous progenitor cells (Kodama et al., 2009).

2.2.3 Transplantation of differentiated stem cells

Based on the successful transplantation of beta cells, or pancreatic islets, for the treatment of diabetes mellitus (Ris et al., 2011; Wahoff et al., 1995; 1996), transplantation of IPCs differentiated from stem cells is seen as a promising therapy for diabetic patients, particularly in light of the lack of tissue donors and the many side effects of insulin injections. Unlike stem cells, transplanted IPCs produce insulin directly. IPC transplantation using different grafting methods has been studied in mouse models. Routes of administration include the portal vein, intra-peritoneal injection, the liver, the tail vein, and the kidney capsule. IPCs, differentiated from bone marrow-derived MSCs, were successfully allografted into the portal vein in a rat model of diabetes mellitus. After transplantation, the IPCs migrated into the liver where they expressed islet hormones, resulting in reduced glucose levels between Days 6 and 20 post-injection (Wu et al., 2007). Xenotransplantation of IPCs derived from fresh or banked human umbilical cord blood into diabetic mice also showed positive results. These IPCs, transplanted via the portal vein (Wang et al., 2010) or intraperitoneally (Phuc et al., 2011), reduced the blood glucose levels in diabetic mice. When IPCs were grafted into the portal vein, human C-peptides were detected in the mouse livers by immunohistochemistry (Wang et al., 2010). Similar to these results, xenotransplantation of IPCs differentiated from the Wharton's jelly from human umbilical cords restored normoglycaemia, body weight and a normal glucose tolerance test, indicating that the cells are functional when grafted via the portal vein (Kadam et al., 2010) or liver (Chao et al., 2008).

Zang et al. (2010) injected IPCs differentiated from human islet-derived progenitor cells under the renal capsule of immunodeficient mice. One month later, 19/28 mice transplanted with progenitor cells and 4/14 mice transplanted with IPCs produced human C-peptide that was detectable in the blood. This indicates that the *in vivo* environment further facilitates the maturation of progenitor cells. Moreover, 9/19 mice transplanted with progenitor cells, and 2/4 mice transplanted with IPCs, secreted C-peptide in response to glucose (Zang et al., 2010). Allotransplantation of IPCs differentiated from islet progenitor cells produced similar results (Shyu et al. 2011). In this study, progenitor cells cultured in matrigel differentiated into IPCs following transplantation into diabetic mice.

ESCs were also differentiated into IPCs and used to treat diabetes mellitus in animal models. After transplantation, these cells did not induce teratoma formation in STZ-induced mice and treatment reduced blood glucose levels to almost normal levels (Kim et al., 2003). Another study indicated that ESCs could differentiate into IPCs; however, transplantation of these pancreatic progenitor clusters into STZ-induced mice failed to reverse the hyperglycaemic state. This indicates that ESCs can differentiate into pancreatic progenitor cells and commit to a pancreatic islet cell fate, but are unable to perform the normal functions of beta cells (Chen et al., 2008). While most studies have focused on experimental treatments using IPC transplantation, another study used liver cells (rather than IPCs) to treat diabetes mellitus. Hepatic cells were differentiated from bone marrow-derived MSCs. Transplantation of syngeneic hepatic cells into STZ-induced mice cured their diabetes mellitus. Treatment of mice with hyperglycaemia and islet cell destruction resulted in repair of the pancreatic islets. Blood glucose levels, intra-peritoneal glucose tolerance tests, and serum insulin levels recovered significantly in the treated group. In addition, both body weight and the number of islets were significantly increased (Dinarvand et al., 2010).

2.2.4 Stem cell gene therapy

Due to their properties of self-renewal and capacity for multipotent differentiation, stem cells are thought to be the best vector for delivering genes and therapeutic gene-coded proteins into the body. Gene transfer experiments that cause stem cells to differentiate into beta cells, or that transfer specific genes coding for insulin, have also been conducted in recent years. There are several possible reasons why the use of stem cell gene therapies can be used to treat diabetes mellitus. However, no study has compared the difference between IPCs produced by chemical induction and those derived from gene transfer. Some researchers hypothesise that the key is the genetic transfer of the signalling pathways related to differentiation from stem cells into IPCs, which will create IPCs more similar to stem cells *in vivo*. Others argue that genetic modifications, e.g. *PDX-1, betacellulin*, or *Neuro-D* transfer, induce cells to differentiate into beta cells, while yet others suggest that the efficiency of IPC transplantation is low because IPCs are mature, specialised cells. For long-term effectiveness, a source of insulin with a long-term regenerative capacity is needed. Early studies by Xu et al. (2007), looking at transferring insulin into MSCs, showed that the resulting MSCs did express human insulin. The body weight of diabetic mice treated with these MSCs increased by 6% within 6 weeks of treatment, and average blood glucose levels were 10.40 +/- 2.80 mmol/l (Day 7) and 6.50 +/- 0.89 mmol/l (Day 42), compared with 26.80 +/- 2.49 mmol/l (Day 7) and 25.40 +/- 4.10 mmol/l (Day 42) in untreated animals ($p < 0.05$). Experimental diabetes mellitus was effectively relieved for up to 6 weeks after intrahepatic transplantation of murine MSCs expressing human insulin (Xu et al., 2007). In other studies, STZ-treated mice transplanted with *Pdx1*-transduced adipose tissue-derived MSCs (*Pdx1*-MSCs) showed significant decreases in blood glucose levels and increased survival compared with control mice (Lin et al., 2009; Kajiyama et al., 2010).

2.2.5 Transplantation of immuno-isolated IPCs

Transplantation IPCs offers a potential cell replacement therapy for patients with type 1 diabetes mellitus. However, because of the inadequate number of cells obtained from donors, other stem cell sources have drawn significant attention from many research groups. The efficacy of these approaches is limited because they typically necessitate the administration of immunosuppressive agents to prevent rejection of transplanted cells. The use immunosuppressive drugs can have deleterious side effects, such as increased susceptibility to infection, liver and kidney damage, and an increased risk of cancer. In addition, immunosuppressive drugs may have unexpected effects on the transplanted tissues. For example, some reports have shown that cyclosporine can inhibit insulin secretion by pancreatic cells.

Immuno-isolation is a promising technique that protects the implanted tissues from rejection. One of the most common immuno-isolation techniques is to encapsulate cells within a semi-permeable membrane, such as alginate, that physically protects the grafts from the host's immune cells while simultaneously allowing nutrients and metabolic products to diffuse into or out of the capsule. To achieve this, the cells are encapsulated within a hydrogel or alginate membrane using gravity, electrostatic forces, or coaxial airflow to form the capsule. Allogeneic and xenogeneic transplantation of encapsulated islets of Langerhans restores normal blood glucose levels in mice (Dufrane et al., 2006; Fan et al., 1990; Omer et al., 2003), dogs (Soon-Shiong et al., 1992a,b; 1993) and non-human primates (Sun et al., 1996) with diabetes mellitus induced by autoimmune diseases or chemical injury, without on the need for immunosuppressive agents. In most of these studies, transplantation was via intraperitoneal injection of islet cells. However, Dufrane et al. (2006) recently reported the generation of encapsulated porcine islets using a Ca-alginate material. These capsules were implanted under the kidney capsule of nondiabetic *Cynomolgus* monkeys. The implanted porcine islets survived for up to 6 months after implantation without immunosuppression, even in animals injected with porcine IgG. Moreover, C-peptide was detected in 71% of the animals. After 135 and 180 days, the explanted capsules still synthesised insulin and responded to glucose stimulation (Dufrane et al., 2006).

In another study, transplantation of alginate-encapsulated IPCs from an embryo-derived mouse embryo progenitor-derived insulin-producing-1 (MEPI-1) cell line lowered hyperglycaemia in immuno-competent, allogeneic diabetic mice. After transplantation, hyperglycaemia was reversed and was followed by a 2.5-month period of normal to moderate hypoglycaemia before relapse. Relapse occurred within 2 weeks in mice transplanted with non-encapsulated MEPI-1 cells. Blood glucose levels, insulin levels, and the results of an oral glucose tolerance test all correlated directly with the number of viable cells remaining in the capsules in the transplanted animals (Shao et al., 2011). Moreover, encapsulation of IPCs differentiated from amnion-derived MSCs, or adipose tissue-derived MSCs in polyurethane-polyvinyl pyrrolidone macrocapsules, or IPCs in calcium alginate, resulted in the restoration of normoglycaemia without immunorejection (Chandra et al., 2009; Kadam et al., 2010) in diabetic rats

2.2.6 Co-transplantation of stem cells and IPCs

Allogeneic islet/IPC transplantation is an efficient method for maintaining normal glucose levels and for the treatment of diabetes mellitus. However, limited sources of islets/IPCs, high rates of islet/IPC graft failure and the need for long-term immunosuppression are major obstacles to the widespread application of these therapies. To overcome these problems, co-transplantation of pancreatic islets/IPCs and adult stem cells is considered as a potential target for the near future. In fact, new results suggest that co-transplantation of stem/precursor cells, particularly MSCs, and islets/IPCs promotes tissue engraftment and beta cell/IPC survival. This theory proposes that stem cells also act as "feeder" cells for the islets, supporting graft protection, tissue revascularisation, and immune acceptance (Sordi et al., 2010).

Overcoming the loss of islet mass is important for successful islet transplantation. Adipose tissue-derived stem cells (ADSCs; referred to as MSCs by some authors) have angiogenic and anti-inflammatory properties. Co-transplantation of ADSCs and islets into mice promotes survival, improves insulin secretion by the graft, and reduces the islet mass required for treatment (Ohmura et al., 2010). In another study, MSCs derived from adipose tissue were differentiated into IPCs and co-transplanted with cultured bone marrow cells into 11 diabetic patients (7 male, 4 female; disease duration, 1–24 years; age range, 13–43 years). Their mean exogenous insulin requirements were 1.14 units/kg BW/day, the mean glycosylated haemoglobin (Hb1Ac) level was 8.47%, and the mean c-peptide level was 0.1 ng/mL. All the patients received successful transplants and the mean follow-up period was 23 months. The results showed a decreased mean exogenous insulin requirement of 0.63 units/kgBW/day, a reduced Hb1Ac of 7.39%, and raised serum c-peptide levels (0.38 ng/mL). The patients reported no diabetic ketoacidosis events and a mean weight gain of 2.5 kg on a normal vegetarian diet and physical activity (Vanikar et al., 2010). However, a previous report indicated that similar results were obtained with undifferentiated MSC-derived adipose tissue co-transplanted with cultured bone marrow. In this study, human adipose tissue-derived MSCs were transfused along with unfractionated cultured bone marrow into five insulinopenic diabetic patients (2 male, 3 female; age range, 14–28 years; disease duration, 0.6 to 10 years) being treated with human insulin (14–70 U/d). The patients had postprandial blood sugar levels between 156 and 470 mg%, Hb1Ac levels of 6.8% to 9.9%, and c-peptide levels of 0.02 to 0.2 ng/mL. After successful transplantation, all patients showed a 30% to 50% reduction in their insulin requirements along with a 4–26-fold increase in serum c-peptide levels during a mean follow-up period of 2.9 months (Trivedi et al., 2008).

After transplantation, MSCs appear to play an immunomodulatory role, thereby promoting graft acceptance. In a cynomolgus monkey model, allogeneic MSCs were co-transplanted with islets intra-portally on postoperative Day 0 and intravenously with donor marrow on postoperative Days 5 and 11. Increased co-transplantation efficiency was associated with increased numbers of regulatory T-cells in the peripheral blood, indicating that co-transplantation of MSCs and islets may be an important method of enhancing islet engraftment and, thereby, decreasing the number of islets required (Berman et al., 2010). Co-transplantation may also downregulate the production of pro-inflammatory cytokines. These results also suggest that MSCs may prevent acute rejection and improve graft function after portal vein pancreatic islet transplantation (Longoni et al., 2010), or that they may induce haematopoietic chimerism and subsequent immune tolerance without causing graft-versus-host disease (Itakura et al., 2007). Moreover, MSC-stimulated graft vascularisation and improved islet graft function are both associated with co-transplanted islets (Figliuzzi et al. 2009; Ito et al. 2010). In addition, interleukin (IL)-6, IL-8, vascular endothelial growth factor-A, hepatocyte growth factor, and transforming growth factor-beta were detected at significant levels in MSC culture medium. These are trophic factors secreted by human MSCs that enhance the survival and function of the islets after transplantation (Park et al., 2010).

3. Islet regeneration by immune correction

There is increasing evidence suggesting that both autoimmune and autoinflammatory mechanisms are involved in the development of type 1 and type-2 diabetes mellitus. Type 1 diabetes mellitus is currently treated with anti-inflammatory drugs and immunosuppressive and immunomodulatory agents. However, despite their profound effects on immune responses, these drugs do not induce clinically significant remission in certain patients. In recent years, stem cells have come to be regarded as the best treatment for autoimmune disorders, including type 1 diabetes mellitus.

In a phase 1/2 study of autologous non-myeloablative haematopoietic stem cell (HSC) transplantation, C-peptide levels were detected in 23 type 1 diabetes mellitus patients (age range, 13–31 years). During a 7–58 month follow-up (mean, 29.8 months; median, 30 months), 20/23 patients with no previous history of ketoacidosis and not receiving corticosteroids were found to be insulin free. Twelve patients maintained normal blood glucose levels for up to 31 months (range, 14–52 months). Eight patients suffered a relapse and resumed insulin injections at a lower dose (0.1–0.3 IU/kg). No mortality was reported. Thus, C-peptide levels increased significantly and the majority of patients achieved insulin independence with good glycemic control (Couri et al., 2009).

In another study, bone marrow from *gfp* transgenic mice was isolated and transplanted into diabetic mice. Repair of diabetic islets was evidenced by a reduction in hyperglycaemia, an increase in the number of islets, and altered pancreatic histology. Transplanted cells in the recipient pancreases expressed CD34 (an HSC marker), but not *insulin, PDX-1, Ngn3, Nkx2.2, Nkx6.1, Pax4, Pax6,* or CD45. It was concluded that BM-derived cells, especially HSCs, were able to repair islets by stimulating the proliferation of beta cells and the differentiation of pancreatic stem cells; however, they could not differentiate into beta cells or IPCs (Gao et al., 2008). Chamson-Reig et al. (2010) used a *Vav-iCre* double transgenic mouse model to investigate the use of HSC transplantation for the treatment of type 1 diabetes mellitus. Only haemopoietic lineage cells expressed the *Vav1* gene promoter, evidenced by expression of a *R26R*-enhanced yellow fluorescent protein (YFP) reporter gene. Between postnatal Days 2 and 4, mice were injected with either STZ or vehicle (control). Mice were sacrificed between Days 10 and 130 and the pancreases examined by immunofluorescence microscopy. The results showed that approximately 30% of YFP-positive cells within the islets co-stained for the endothelial cell marker, CD31. The number of haemopoietic-derived cells and the proportion of CD31-positive cells significantly increased 21 and 40 days after STZ treatment respectively. This indicates that haemopoietic lineage cells promote intra-islet angiogenesis following beta cell loss due to STZ treatment, which supports a partial recovery of the islets (Chamson-Reig et al., 2010). Huang et al. (2010) demonstrated that the role of bone marrow transplantation was to supply a source of very small embryonic-like cells that exist in the bone marrow without the need for HSCs. The authors concluded that these very small embryonic-like cells mobilise to the injured pancreatic tissue and contribute to beta cell regeneration after bone marrow transplantation (Huang et al., 2010).

The initial results of some studies investigating the treatment of type 2 diabetes mellitus show that transplantation of stem cells produces good results. Intra-bone marrow-bone marrow transplantation plus thymus transplantation can be used to treat type 2 diabetes mellitus by normalising the T cell imbalance. Recipient *db/db* mice showed increases in body weight, reduced blood glucose levels, and a reduction in plasma IL-6 and IL-1β levels 7 weeks after transplantation. More importantly, treatment resulted in the restoration of normal CD4/CD8 ratios, increased plasma adiponectin levels, improved insulin sensitivity, and an increase in the number of insulin-producing cells. Furthermore, expression of pancreatic pAKT, pLKB1, pAMPK and HO-1 increased after transplantation. In short, this treatment normalises T cell subsets, and restores cytokine balance and insulin sensitivity in the *db/db* mouse model (Li et al. 2010).

The results from some preclinical or clinical trials to treat type 1 and type 2 diabetes were summarized in Table 1.

Species	Diabetic	Kind of Stem cell/IPCs	Transplantation methods	Auto/Allo/ Xenograft	Immunoco-mpression	Effects	Refferences
Mouse	1;STZ	MSC (Human endometrium)	N/A	Xeno	Yes	Restore blood glucose levels	Li et al., 2010
Rat	1;STZ	Proliferating islet Cells (Rat)	Portal vein	Allo	None	Restore euglycemia, weight gain for up to 130 days	Li et al., 2010
Mouse	NOD	MSCs (BM)	Tail vein	Allo	None	Long-term reversal of hyperglycemia (reduce blood glucose from week 2-12)	Jurewicz et al., 2010
Rat	1;STZ	Pancreatic stem cell (Human fetal pancreatic tissues)	Kidney capsule	Xeno	None	Reduce blood glucose levels; prolong the life time	Xiao et al., 2008
Rat	1;STZ	MSCs (BM)	Tail vein	Allo	None	Reduce blood glucose level up to 45 days	Dong et al., 2008
Dog	1;STZ	Insulin expressing-MSCs (dog bone)	Liver	Auto	None	Reversal of hyperglycemia for up to 16 week	Zhu et al., 2009
Mouse	1;STZ	MSCs + Islets	Kidney capsule	Syngenic	None	92% of mice reverting to normoglycaemia, (42% islets alone) after 1 month	Rackham et al., 2011
Rat		MSCs + Islets	Kidney capsule	Syngenic	None	Normoglycemia maintained until killing	Figliuzzi et al., 2009
Mouse	NOD SCID	MSCs + Islets	Renal capsule	Allo	None	Reversed diabetes in 8/8; (3/10 in control)	Ito et al., 2010
Rat	1;STZ	BM-MSC	Kidney capsule	Allo	None	Increasing beta-cell mass	Lin et al., 2009
Mouse	1;STZ	ESCs	Kidney capsule	Allo	None	Beta cell neogenesis occurred between 2 to 3 weeks	Kodama et al., 2009

Table 1. The results of some preclinical and clinical experiment about treating diabetes type 1 & 2 by stem cell/IPCs transplantation.

4. Conclusion

Taking into account all the currently available results (Table 1), we can expect that diabetes mellitus will be successfully treated using stem cell therapy in the near future. However, questions regarding the survival of the cells after grafting and improvements in the vitality and maintenance of cellular function after transplantation remain to be answered. On the basis of evidence supporting the many advantages of bone marrow transplantation, umbilical cord blood transplantation, and HSC therapy for blood-related diseases, the strategy of HSC/BM/UCB may produce several positive results in the coming years and become the treatment of choice for both type 1 and type 2 diabetes mellitus. Although more difficult, ESCs or adult stem cell-derived IPC transplantation are also important treatments for diabetes mellitus, especially when HSCs are in short supply.

5. References

Agha-Hosseini, F.; Jahani, M.A.; Jahani, M.; Mirzaii-Dizgah, I. & Ali-Moghaddam, K. (2010). In vitro isolation of stem cells derived from human dental pulp, *Clin Transplant*, Vol.24, No.2, (Mar 2010), pp.E23-8.

Alipio, Z.; Liao, W.; Roemer, E.J.; Waner, M.; Fink, L.M.; Ward, D.C. & Ma, Y. (2010). Reversal of hyperglycemia in diabetic mouse models using induced-pluripotent stem (iPS)-derived pancreatic beta-like cells, *Proc Natl Acad Sci U S A*, Vol.107, No.30, (Jul 2010), pp.13426-31.

American Diabetes Association (2007). Economic costs of diabetes in the U.S. *Diabetes Care*, Vol.31, (2008), pp.596-615.

Anna, M.; Wobus (2008). *Stem Cells*, Springer, pp.248, ISBN 9783540778547.

Assady, S.; Maor, G.; Amit, M.; Itskovitz-Eldor, J.; Skorecki, K.L. & Tzukerman, M. (2001). Insulin production by human embryonic stem cells. *Diabetes*, (2001), Vol.50, pp.1691–1697.

Aviv, V.; Meivar-Levy, I.; Rachmut, I.H.; Rubinek, T.; Mor, E. & Ferber, S. (2009) Exendin-4 promotes liver cell proliferation and enhances the PDX-1-induced liver to pancreas transdifferentiation process, *J Biol Chem*, (Nov 2009), Vol.284, No.48, pp.33509-20.

Baer, P.C.; Griesche, N.; Luttmann, W.; Schubert, R.; Luttmann, A. & Geiger, H. (2010). Human adipose-derived mesenchymal stem cells in vitro: evaluation of an optimal expansion medium preserving stemness, *Cytotherapy*, (2010), Vol.12, No.1, pp:96-106.

Bao, X.; Wei, J.; Feng, M.; Lu, S.; Li, G.; Dou, W.; Ma, W.; Ma, S.; An, Y. & Qin, C. (2011). Zhao RC, Wang R. Transplantation of human bone marrow-derived mesenchymal stem cells promotes behavioral recovery and endogenous neurogenesis after cerebral ischemia in rats, *Brain Res*, (Jan 2011), Vol.7, No.1367, pp.103-113.

Baptista, L.S.; do Amaral, R.J.; Carias, R.B.; Aniceto, M.; Claudio-da-Silva, C. & Borojevic, R. (2009). An alternative method for the isolation of mesenchymal stromal cells derived from lipoaspirate samples, *Cytotherapy, (2009), Vol.*11, No.6, pp.706-15.

Battula, V.L.; Treml, S.; Abele, H. & Bühring, H.J. (2008) Prospective isolation and characterization of mesenchymal stem cells from human placenta using a frizzled-9-specific monoclonal antibody, *Differentiation*, (Apr 2008), Vol.76, No.4, pp.326-36.

Becker, A.J.; McCulloch, E.A. & Till, J.E. (1963). Cytological demonstration of the clonal nature of spleen colonies derived from transplanted mouse marrow cells, *Nature*, Vol.197, pp.452–4.

Berman, D.M.; Willman, M.A.; Han, D.; Kleiner, G.; Kenyon, N.M.; Cabrera, O.; Karl J.A.; Wiseman, R.W.; O'Connor, D.H.; Bartholomew, A.M. & Kenyon, N.S. (2010) Mesenchymal stem cells enhance allogeneic islet engraftment in nonhuman primates, *Diabetes*, (Oct 2010), Vol.59, No.10, pp.2558-68.

Bi, D.; Chen, F.G.; Zhang, W.J.; Zhou, G.D.; Cui, L.; Liu, W. & Cao, Y. (2010) Differentiation of human multipotent dermal fibroblasts into islet-like cell clusters, *BMC Cell Biol*, (Jun 2010), Vol.25, pp.11:46.

Bieback, K. & Klüter, H. (2007) Mesenchymal stromal cells from umbilical cord blood, *Curr Stem Cell Res Ther*, (Dec 2007), Vol.2, No.4, pp.310-23.

Blyszczuk, P.; Czyz, J. & Kania, G. (2003). Expression of Pax4 in embryonic stem cells promotes differentiation of nestin-positive progenitor and insulin-producing cells, *Proc Natl Acad Sci USA*, Vol.100, pp.998–1003

Boroujeni, N.B.; Hashemi, S.M.; Khaki, Z. & Soleimani, M. (2011). The reversal of hyperglycemia after transplantation of mouse embryonic stem cells induced into early hepatocyte-like cells in streptozotocin-induced diabetic mice, *Tissue Cell*, (Jan 2011). [Epub ahead of print]

Caviggioli, F.; Vinci, V.; Salval, A. & Klinger, M. (2009). Human adipose-derived stem cells: isolation, characterization and applications in surgery, *ANZ J Surg*, (Nov 2009), Vol.79, No.11, pp.856.

Chamson-Reig, A.; Arany, E.J. & Hill, D.J. (2010) Lineage tracing and resulting phenotype of haemopoietic-derived cells in the pancreas during beta cell regeneration, *Diabetologia*, (Oct 2010), Vol. 53, No.10, pp.2188-97.

Chandra, V.G.S.; Phadnis, S.; Nair, P.D. & Bhonde, R.R. (2009). Generation of pancreatic hormone-expressing islet-like cell aggregates from murine adipose tissue-derived stem cells, *Stem Cells*, (Aug 2009), Vol.27, No.8, pp.1941-53.

Chang, C.F.; Hsu, K.H.; Chiou, S.H.; Ho, L.L.; Fu, Y.S. & Hung, S.C. (2008). Fibronectin and pellet suspension culture promote differentiation of human mesenchymal stem cells into insulin producing cells, *J Biomed Mater Res A*, (Sep 2008), Vol.86, No.4, pp.1097-105.

Chang, Y.J.; Hwang, S.M.; Tseng, C.P.; Cheng, F.C.; Huang, S.H.; Hsu, L.F.; Hsu, L.W. & Tsai, M.S. (2010). Isolation of mesenchymal stem cells with neurogenic potential from the mesoderm of the amniotic membrane, *Cells Tissues Organs*, (2010), Vol.192, No.2, pp.93-105.

Chao, K.C.; Chao, K.F.; Fu, Y.S. & Liu, S.H. (2008). Islet-like clusters derived from mesenchymal stem cells in Wharton's Jelly of the human umbilical cord for transplantation to control type 1 diabetes, *PLoS One*, (Jan 2008), Vol.3, No.1, pp.e1451.

Chen, B.; Zhou, L.; Wang, L.; Hu, S. & Wang, R. (2009). Better induction and differentiation strategy for rat pancreatic stem cells: transplant in liver niche, *Transplant Proc*, (Nov 2009), Vol.41, No.9, pp.3898-904.

Chen, C.; Zhang, Y.; Sheng, X.; Huang, C. & Zang, Y.Q. (2008). Differentiation of embryonic stem cells towards pancreatic progenitor cells and their transplantation into streptozotocin-induced diabetic mice, *Cell Biol Int*, (Apr 2008), Vol.32, No.4, pp.456-61.

Chen, Y.T.; Wei, J.D.; Wang, J.P.; Lee, H.H.; Chiang, E.R.; Lai, H.C.; Chen, L.L.; Lee, Y.T.; Tsai, C.C.; Liu, C.L. & Hung, S.C. (2011). Isolation of mesenchymal stem cells from human ligamentum flavum: implicating etiology of ligamentum flavum hypertrophy, *Spine (Phila Pa 1976)*, (Feb 2011).

Choi, S.A.; Lee, J.H.; Kim, K.J.; Kim, E.Y.; Park, K.S.; Park, Y.B.; Li, X.; Ha, Y.N.; Park, J.Y. & Kim, M.K. (2011). Isoaltion and characterization of mesenchymal stem cells derived from human amniotic fluid, *Reprod Fertil Dev,* (Jan 2011), Vol.23, No.1, pp.243.

Couri, C.E.; Oliveira, M.C.; Stracieri, A.B.; Moraes, D.A.; Pieroni, F.; Barros, G.M.; Madeira, M.I.; Malmegrim, K.C.; Foss-Freitas, M.C.; Simões, B.P.; Martinez, E.Z.; Foss, M.C.; Burt, R.K. & Voltarelli, J.C. (2009) C-peptide levels and insulin independence following autologous nonmyeloablative hematopoietic stem cell transplantation in newly diagnosed type 1 diabetes mellitus, *JAMA,* (Apr 2009), Vol.301, No.15, pp.1573-9.

Cutler, A.J.; Limbani, V.; Girdlestone, J. & Navarrete, C.V. (2010). Umbilical cord-derived mesenchymal stromal cells modulate monocyte function to suppress T cell proliferation, *J Immunol,* (Dec 2010), Vol.185, No.11, pp.6617-23.

D'Amour, K.A.; Bang, A.G.; Eliazer, S.; Kelly, O.G.; Agulnick, A.D.; Smart, N.G.; Moorman, M.A.; Kroon, E.; Carpenter, M.K. & Baetge, E.E. (2006). Production of pancreatic hormone-expressing endocrine cells from human embryonic stem cells, *Nat Biotechnol* 24:1392–1401.

Danwei Huangfu, Kenji Osafune, René Maehr, Wenjun Guo, Astrid Eijkelenboom, Shuibing Chen, Whitney Muhlestein & Douglas A Melton. Induction of pluripotent stem cells from primary human fibroblasts with only Oct4 and Sox2, *Nature Biotechnology,* (2008), Vol.26, pp.1269 - 1275.

Davani, B.; Ikonomou, L.; Raaka, B.M.; Geras-Raaka, E.; Morton, R.A.; Marcus-Samuels, B. & Gershengorn, M.C. (2007). Human islet-derived precursor cells are mesenchymal stromal cells that differentiate and mature to hormone-expressing cells in vivo, *Stem Cells,* (Dec 2007), Vol.25, No.12, pp.3215-22.

Deuse, T.; Stubbendorff, M.; Tang-Quan, K.; Phillips, N.; Kay, M.A.; Eiermann, T.; Phan, T.T.; Volk, H.D.; Reichenspurner, H.; Robbins, R.C. & Schrepfer, S. (2010). Immunogenicity and immunomodulatory properties of umbilical cord lining mesenchymal stem cells, *Cell Transplant,* (Nov 2010). [Epub ahead of print]

Dimos, J.T.; Rodolfa, K.T. & Niakan, K.K. (2008). Induced pluripotent stem cells generated from patients with ALS can be differentiated into motor neurons, *Science,* (2008), Vol.321, pp.1218—1221

Dinarvand, P.; Hashemi, S.M. & Soleimani, M. (2010). Effect of transplantation of mesenchymal stem cells induced into early hepatic cells in streptozotocin-induced diabetic mice, *Biol Pharm Bull,* (2010), Vol.33, No.7, pp.1212-7

Dong, Q.Y.; Chen, L.; Gao, G.Q.; Wang, L.; Song, J.; Chen, B.; Xu, Y.X. & Sun, L. (2008). Allogeneic diabetic mesenchymal stem cells transplantation in streptozotocin-induced diabetic rat, *Clin Invest Med,* (Dec 2008), Vol.31, No.6, pp.E328-37.

Dufrane, D.; Steenberghe, M.; Goebbels, R.M.; Saliez, A.; Guiot, Y. & Gianello, P. (2006). The influence of implantation site on the biocompatibility and survival of alginate encapsulated pig islets in rats, *Biomaterials* (2006), Vol.17, pp.3201-8.

Erices, A.; Conget, P. & Minguell, J.J. (2000). Mesenchymal progenitor cells in human umbilical cord blood, *Br J Haematol,* (Apr 2000), Vol.109, No.1, pp.235-42.

Estes, B.T.; Diekman, B.O.; Gimble, J.M. & Guilak, F. (2010). Isolation of adipose-derived stem cells and their induction to a chondrogenic phenotype, *Nat Protoc,* (Jul 2010), Vol.5, No.7, pp.1294-311.

Evans, M.J. & Kaufman, M.H. (1981). Establishment in culture of pluripotent cells from mouse embryos. *Nature,* (1981), Vol.292, pp.154- 156.

Fan, M.Y.; Lum, Z.P.; Fu, X.W.; Levesque, L.; Tai, I.T. & Sun, A.M. (1990). Reversal of diabetes in BB rats by transplantation of encapsulated pancreatic islets, *Diabetes,* (1990), Vol4, pp.519-22.

Farias, V.A.; Linares-Fernández, J.L.; Peñalver, J.L.; Payá Colmenero, J.A.; Ferrón, G.O.; Duran, E.L.; Fernández, R.M.; Olivares, E.G.; O'Valle, F.; Puertas, A.; Oliver, F.J. & Ruiz de Almodóvar, J.M. (2011). Human umbilical cord stromal stem cell express CD10 and exert contractile properties, *Placenta,* (Jan 2011), Vol.32, No.1, pp.86-95

Feldman, E.L.; Stevens, M.J. & Greene, D.A. *(1997).* Pathogenesis of diabetic neuropathy, *Clin Neurosci,* (1997), Vol.4, No.6, pp.365-70

Feng, J.X.; La, X.L.; Ma, Y.; Bi, X.J. & Wen, H. (2009). Isolation of human pluripotent mesenchymal stem cells from second-trimester amniotic fluid using two kinds of culture protocol and their differentiation into neuron-like cells, *Zhongguo Wei Zhong Bing Ji Jiu Yi Xue,* (Dec 2009), Vol.21, No.12, pp.729-33

Figliuzzi, M.; Cornolti, R.; Perico, N.; Rota, C.; Morigi, M.; Remuzzi, G.; Remuzzi, A. & Benigni, A. (2009). Bone marrow-derived mesenchymal stem cells improve islet graft function in diabetic rats, *Transplant Proc,* (Jun 2009), Vol.41, No.5, pp.1797-800.

Friedenstein, A.J.; Deriglasova, U.F.; Kulagina, N.N.; Panasuk, A.F.; Rudakowa, S.F.; Luria, E.A & Ruadkow, I.A. (1974). Precursors for fibroblasts in different populations of hematopoietic cells as detected by the in vitro colony assay method, *Exp Hematol,* Vol.2, No.2, pp.83–92

Friedenstein, A.J.; Gorskaja, J.F. & Kulagina, N.N. (1976). Fibroblast precursors in normal and irradiated mouse hematopoietic organs, *Exp Hematol,* Vol.4, No.5, pp.267–74.

Gao, F.; Wu, D.Q.; Hu, Y.H. & Jin, G.X. (2008). Extracellular matrix gel is necessary for in vitro cultivation of insulin producing cells from human umbilical cord blood derived mesenchymal stem cells, *Chin Med J (Engl),* (May 2008), Vol.121, No.9, pp.811-8

Gao, F.; Wu, D.Q.; Hu, Y.H.; Jin, G.X.; Li, G.D.; Sun, T.W. & Li, F.J. (2008). In vitro cultivation of islet-like cell clusters from human umbilical cord blood-derived mesenchymal stem cells, *Transl Res,* (Jun 2008), Vol.151, No.6, pp.293-302

Gao, X.; Song, L.; Shen, K.; Wang, H.; Niu, W. & Qin, X. (2008). Transplantation of bone marrow derived cells promotes pancreatic islet repair in diabetic mice, *Biochem Biophys Res Commun,* (Jun 2008), Vol.371, No.1, pp.132-7.

Gefen-Halevi, S.; Rachmut, I.H.; Molakandov, K.; Berneman, D.; Mor, E.; Meivar-Levy, I. & Ferber, S. (2010). NKX6.1 promotes PDX-1-induced liver to pancreatic β-cells reprogramming, *Cell Reprogram,* (Dec 2010), Vol.12, No.6, pp.655-64

Gronthos, S. & Zannettino, A.C. (2008). A method to isolate and purify human bone marrow stromal stem cells, *Methods Mol Biol,* (2008), Vol.449, pp.45-57.

Hansson, M.; Tonning, A. & Frandsen, U. (2004). Artifactual insulin release from differentiated embryonic stem cells. *Diabetes,* (2004), Vol.53, pp.2603–2609

Hasegawa, Y.; Ogihara, T.; Yamada, T.; Ishigaki, Y.; Imai, J.; Uno, K.; Gao, J.; Kaneko, K.; Ishihara, H.; Sasano, H.; Nakauchi, H.; Oka, Y. & Katagiri, H. (2007). Bone marrow (BM) transplantation promotes beta-cell regeneration after acute injury through BM cell mobilization, *Endocrinology,* (May 2007), Vol.148, No.5, pp.2006-15

Hida, N.; Nishiyama, N.; Miyoshi, S.; Kira, S.; Segawa, K.; Uyama, T.; Mori, T.; Miyado, K.; Ikegami, Y.; Cui, C.; Kiyono, T.; Kyo, S.; Shimizu, T.; Okano, T.; Sakamoto, M.; Ogawa, S. & Umezawa, A. (2008). Novel cardiac precursor-like cells from human menstrual blood-derived mesenchymal cells, *Stem Cells, (Jul* 2008), Vol.26, No.7, pp.1695-704

Hisanaga, E.; Park, K.Y.; Yamada, S.; Hashimoto, H.; Takeuchi, T.; Mori, M.; Seno, M.; Umezawa, K.; Takei, I. & Kojima, I. (2008). A simple method to induce differentiation of murine bone marrow mesenchymal cells to insulin-producing cells using conophylline and betacellulin-delta4, *Endocr J,* (Jul 2008), Vol.55, No.3, pp.535-43

Hori, Y.; Rulifson, I.C.; Tsai, B.C.; Heit, J.J.; Cahoy, J.D. & Kim, S.K. (2002). Growth inhibitors promote differentiation of insulin-producing tissue from embryonic stem cells, *Proc Natl Acad Sci USA,* (2002), Vol.99, pp.16105–16110

Huang, Y.; Kucia, M.; Hussain, L.R.; Wen, Y.; Xu, H.; Yan, J.; Ratajczak, M.Z. & Ildstad, S.T. (2010). Bone marrow transplantation temporarily improves pancreatic function in streptozotocin-induced diabetes: potential involvement of very small embryonic-like cells, *Transplantation,* (Mar 2010), Vol.89, No.6, pp.677-85

Huang, Y.C.; Yang, Z.M.; Chen, X.H.; Tan, M.Y.; Wang, J.; Li, X.Q.; Xie, H.Q. & Deng, L. (2009). Isolation of mesenchymal stem cells from human placental decidua basalis and resistance to hypoxia and serum deprivation. *Stem Cell Rev,* (Sep 2009), Vol.5, No.3, pp.247-55

Hutson, E.L.; Boyer, S. & Genever, P.G. (2005). Rapid isolation, expansion, and differentiation of osteoprogenitors from full-term umbilical cord blood, *Tissue Eng* (Sep 2005), Vol.11, No.9-10, pp.1407-20

Itakura, S.; Asari, S.; Rawson, J.; Ito, T.; Todorov, I.; Liu, C.P.; Sasaki, N.; Kandeel, F. & Mullen, Y. (2007). Mesenchymal stem cells facilitate the induction of mixed hematopoietic chimerism and islet allograft tolerance without GVHD in the rat, *Am J Transplant,* (Feb 2007), Vol.7, No.2, pp.336-46

Ito, T.; Itakura, S.; Todorov, I.; Rawson, J.; Asari, S.; Shintaku, J.; Nair, I.; Ferreri, K.; Kandeel, F.; Mullen, Y. (2010). Mesenchymal stem cell and islet co-transplantation promotes graft revascularization and function, *Transplantation,* (Jun 2010), Vol.89, No.12, pp.1438-45

Jiang, J.; Au, M.; Lu, K.; Eshpeter, A.; Korbutt, G.; Fisk, G. & Majumdar, A.S. (2007). Generation of insulin-producing islet-like clusters from human embryonic stem cells, *Stem Cells,* Vol.25, pp.1940–1953

Jiang, W.; Shi, Y.; Zhao, D.; Chen, S.; Yong, J.; Zhang, J.; Qing, T.; Sun, X.; Zhang, P.; Ding, M.; Li, D. & Deng, H. (2007). In vitro derivation of functional insulin-producing cells from human embryonic stem cells, *Cell Res,* (2007), Vol.17, pp.333–344

Jurewicz, M.; Yang, S.; Augello, A.; Godwin, J.G.; Moore, R.F.; Azzi, J.; Fiorina, P.; Atkinson, M.; Sayegh, M.H. & Abdi, R. (2010). Congenic mesenchymal stem cell therapy reverses hyperglycemia in experimental type 1 diabetes, *Diabetes,* (Dec 2010), Vol.59, No.12, pp.3139-47

Kadam, S.; Muthyala, S.; Nair, P. & Bhonde, R. (2010). Human placenta-derived mesenchymal stem cells and islet-like cell clusters generated from these cells as a novel source for stem cell therapy in diabetes, *Rev Diabet Stud,* (2010), Vol.7, No.2, pp.168-82

Kadam, S.S. & Bhonde, R.R. (2010). Islet neogenesis from the constitutively nestin expressing human umbilical cord matrix derived mesenchymal stem cells, *Islets,* (Mar 2010), Vol.2, No.2, pp.112-20

Kadam, S.S.; Sudhakar, M.; Nair, P.D. & Bhonde, R.R. (2010). Reversal of experimental diabetes in mice by transplantation of neo-islets generated from human amnion-derived mesenchymal stromal cells using immuno-isolatory macrocapsules, *Cytotherapy,* (Dec 2010), Vol.12, No.8, pp.982-91

Kajiyama, H.; Hamazaki, T.S.; Tokuhara, M.; Masui, S.; Okabayashi, K.; Ohnuma, K.; Yabe, S.; Yasuda, K.; Ishiura, S.; Okochi, H. & Asashima, M. (2010). Pdx1-transfected adipose tissue-derived stem cells differentiate into insulin-producing cells in vivo and reduce hyperglycemia in diabetic mice, *Int J Dev Biol,* (2010), Vol.54, No.4, pp.699-705

Karaöz, E.; Doğan, B.N.; Aksoy, A.; Gacar, G.; Akyüz, S.; Ayhan, S.; Genç, Z.S.; Yürüker, S.; Duruksu, G.; Demircan, P.C. & Sariboyaci, A.E. (2010). Isolation and in vitro characterisation of dental pulp stem cells from natal teeth, *Histochem Cell Biol,* (Jan 2010), Vol.133, No.1, pp.95-112

Karnieli, O.; Izhar-Prato, Y.; Bulvik, S.; Efrat, S. (2007). Generation of insulin-producing cells from human bone marrow mesenchymal stem cells by genetic manipulation, *Stem Cells,* (Nov 2007), Vol.25, No.11, pp.2837-44

Kassis, I.; Zangi, L.; Rivkin, R.; Levdansky, L.; Samuel, S.; Marx, G. & Gorodetsky, R. (2006). Isolation of mesenchymal stem cells from G-CSF-mobilized human peripheral blood using fibrin microbeads, *Bone Marrow Transplant,* (May 2006), Vol.37, No.10, pp.967-76

Katz, A.J.; Tholpady, A.; Tholpady, S.S.; Shang, H. & Ogle, R.C. (2005). Cell surface and transcriptional characterization of human adipose-derived adherent stromal (hADAS) cells, *Stem Cells,* (Mar 2005), Vol.23, No.3, pp.412-23

Kim, D.; Gu, Y.; Ishii, M.; Fujimiya, M.; Qi, M.; Nakamura, N.; Yoshikawa, T.; Sumi, S. & Inoue, K. (2003). In vivo functioning and transplantable mature pancreatic islet-like cell clusters differentiated from embryonic stem cell, *Pancreas,* (Aug 2003), Vol.27, No.2, pp.e34-41

Kita, K.; Gauglitz, G.G.; Phan, T.T.; Herndon, D.N.; Jeschke, M.G. (2010). Isolation and characterization of mesenchymal stem cells from the sub-amniotic human umbilical cord lining membrane, *Stem Cells Dev,* (Apr 2010), Vol.19, No.4, pp.491-502

Koblas, T.; Zacharovová, K.; Berková, Z.; Leontovic, I.; Dovolilová, E.; Zámecník, L. & Saudek, F. (2009). In vivo differentiation of human umbilical cord blood-derived cells into insulin-producing beta cells, *Folia Biol (Praha),* (2009), Vol.55, No.6, pp.224-32

Kodama, M.; Takeshita, F.; Kanegasaki, S.; Ochiya, T. & Quinn, G. (2008). Pancreatic endocrine and exocrine cell ontogeny from renal capsule transplanted embryonic stem cells in streptozocin-injured mice, *J Histochem Cytochem,* (Jan 2008), Vol.56, No.1, pp.33-44

Kodama, M.; Tsukamoto, K.; Yoshida, K.; Aoki, K.; Kanegasaki, S. & Quinn, G. (2009). Embryonic stem cell transplantation correlates with endogenous neurogenin 3 expression and pancreas regeneration in streptozotocin-injured mice, *J Histochem Cytochem,* (Dec 2009), Vol.57, No.12, pp.1149-58

Kojima, H.; Fujimiya, M.; Matsumura, K.; Younan, P.; Imaeda, H.; Maeda, M. & Chan, L. (2003). NeuroD-betacellulin gene therapy induces islet neogenesis in the liver and reverses diabetes in mice, *Nat Med,* (May 2003), Vol.9, No.5, pp.596-603

Koya, V.; Lu, S.; Sun, Y.P.; Purich, D.L.; Atkinson, M.A.; Li, S.W. & Yang, L.J. (2008). Reversal of streptozotocin-induced diabetes in mice by cellular transduction with recombinant pancreatic transcription factor pancreatic duodenal homeobox-1: a novel protein transduction domain-based therapy, *Diabetes,* (Mar 2008), Vol.57, No.3, pp.757-69

Kroon, E.; Martinson, L.A.; Kadoya, K.; Bang, A.G.; Kelly, O.G.; Eliazer, S.; Young, H.; Richardson, M.; Smart, N.G.; Cunningham, J.; Agulnick, A.D.; D'Amour, K.A.; Carpenter, M.K. & Baetge, E.E. (2008). Pancreatic endoderm derived from human embryonic stem cells generates glucose-responsive insulin-secreting cells in vivo, *Nat Biotechnol,* Vol.26, pp.443–452

Kyurkchiev, S.; Shterev, A. & Dimitrov, R. (2010). Assessment of presence and characteristics of multipotent stromal cells in human endometrium and decidua, *Reprod Biomed Online,* (Mar 2010), Vol.20, No.3, pp.305-13

Leon-Quinto, T.; Jones, J.; Skoudy, A.; Burcin, M. & Soria, B. (2004). In vitro directed differentiation of mouse embryonic stem cells into insulin-producing cells, *Diabetologia,* Vol.47, pp.1442–1451

Li, H.Y.; Chen, Y.J.; Chen, S.J.; Kao, C.L.; Tseng, L.M.; Lo, W.L.; Chang, C.M.; Yang, D.M.; Ku, H.H.; Twu, N.F.; Liao, C.Y.; Chiou, S.H. & Chang, Y.L. (2010). Induction of insulin-producing cells derived from endometrial mesenchymal stem-like cells, *J Pharmacol Exp Ther,* (Dec 2010), Vol.335, No.3, pp.817-29

Li, L.; Li, F.; Qi, H.; Feng, G.; Yuanm, K.; Deng, H. & Zhou, H. (2008). Coexpression of Pdx1 and betacellulin in mesenchymal stem cells could promote the differentiation of nestin-positive epithelium-like progenitors and pancreatic islet-like spheroids, *Stem Cells Dev,* (Aug 2008), Vol.17, No.4, pp.815-23

Li, M.; Abraham, N.G.; Vanella, L.; Zhang, Y.; Inaba, M.; Hosaka, N.; Hoshino, S.; Shi, M.; Ambrosini, Y.M.; Gershwin, M.E. & Ikehara, S. (2010). Successful modulation of type 2 diabetes in db/db mice with intra-bone marrow--bone marrow transplantation plus concurrent thymic transplantation, *J Autoimmun,* (Dec 2010), Vol.35, No.4, pp.414-23

Li, Y.; Zhang, R.; Qiao, H.; Zhang, H.; Wang, Y.; Yuan, H.; Liu, Q.; Liu, D.; Chen, L. & Pei, X. (2007). Generation of insulin-producing cells from PDX-1 gene-modified human mesenchymal stem cells, *J Cell Physiol,* (Apr 2007), Vol.211, No.1, pp.36-44

Limbert, C. & Seufert, J. (2009). In vitro (re)programming of human bone marrow stromal cells toward insulin-producing phenotypes, *Pediatr Diabetes,* (Sep 2009), Vol.10, No.6, pp.413-9

Lin, G.; Wang, G.; Liu, G.; Yang, L.J.; Chang, L.J.; Luc, T.F. & Lin, C.S. (2009). Treatment of type 1 diabetes with adipose tissue-derived stem cells expressing pancreatic duodenal homeobox 1, *Stem Cells Dev,* (Dec 2009), Vol.18, No.10, pp.1399-406

Lin, H.Y.; Tsai, C.C.; Chen, L.L.; Chiou, S.H.; Wang, Y.J. & Hung, S.C. (2010). Fibronectin and laminin promote differentiation of human mesenchymal stem cells into insulin producing cells through activating Akt and ERK, *J Biomed Sci,* (Jul 2010), Vol.12, No.17, pp.56.

Lin, P.; Chen, L.; Yang, N.; Sun, Y.; Xu, Y.X. (2009). Evaluation of stem cell differentiation in diabetic rats transplanted with bone marrow mesenchymal stem cells, *Transplant Proc,* (Jun 2009), Vol.41, No.5, pp.1891-3

Longoni, B.; Szilagyi, E.; Quaranta, P.; Paoli, G.T.; Tripodi, S.; Urbani, S.; Mazzanti, B.; Rossi, B.; Fanci, R.; Demontis, G.C.; Marzola, P.; Saccardi, R.; Cintorino, M. & Mosca, F. (2010). Mesenchymal stem cells prevent acute rejection and prolong graft function in pancreatic islet transplantation, *Diabetes Technol Ther,* (Jun 2010), Vol.12, No.6, pp.435-46

Lu, D.R. (2010). Implantation of bFGF-treated islet progenitor cells ameliorates streptozotocin-induced diabetes in rats, *Acta Pharmacol Sin,* (Nov 2010), Vol.31, No.11, pp.1454-63

Lu, G.H.; Zhang, S.Z.; Chen, Q.; Wang, X.F.; Lu, F.F.; Liu, J.; Li, M.; Li, Z.Y. (2011). Isolation and multipotent differentiation of human decidua basalis-derived mesenchymal stem cells, *Nan Fang Yi Ke Da Xue Xue Bao,* (Feb 2011), Vol.31, No.2, pp.262-265

Lu, Y.; Jin, X.; Chen, Y.; Li, S.; Yuan, Y.; Mai, G.; Tian, B.; Long, D.; Zhang, J.; Zeng, L.; Li, Y. & Cheng, J. (2010). Mesenchymal stem cells protect islets from hypoxia/reoxygenation-induced injury, *Cell Biochem Funct*, (Dec 2010), Vol.28, No.8, pp.637-43

Lumelsky, N.; Blondel, O.; Laeng, P.; Velasco, I.; Ravin, R. & McKay, R. (2001). Differentiation of embryonic stem cells to insulin-secreting structures similar to pancreatic islets, *Science*, (2001), Vol.292, pp.1389–1394

Macias, M.I.; Grande, J.; Moreno, A.; Domínguez, I.; Bornstein, R. & Flores, A.I. (2010). Isolation and characterization of true mesenchymal stem cells derived from human term decidua capable of multilineage differentiation into all 3 embryonic layers, *Am J Obstet Gynecol*, (Nov 2010), Vol.203, No.5, pp.495.e9-495.e23

Madec, A.M.; Mallone, R.; Afonso, G.; Abou Mrad, E.; Mesnier, A.; Eljaafari, A. & Thivolet, C. (2009). Mesenchymal stem cells protect NOD mice from diabetes by inducing regulatory T cells, *Diabetologia*, (Jul 2009), Vol.52, No.7, pp.1391-9.

Majumdar, M.K.; Banks, V.; Peluso, D.P. & Morris, E.A. (2000). Isolation, characterization, and chondrogenic potential of human bone marrow-derived multipotential stromal cells, *J Cell Physiol*, (Oct 2000), Vol.185, No1, pp.98-106

Mao, G.H.; Chen, G.A.; Bai, H.Y.; Song, T.R.; Wang, Y.X. (2009). The reversal of hyperglycaemia in diabetic mice using PLGA scaffolds seeded with islet-like cells derived from human embryonic stem cells, *Biomaterials*, (Mar 2009), Vol. 30, No.9, pp.1706-14.

Marongiu, F.; Gramignoli, R.; Sun, Q.; Tahan, V.; Miki, T.; Dorko, K.; Ellis, E.; & Strom, S.C. (2010) Isolation of amniotic mesenchymal stem cells, *Curr Protoc Stem Cell Biol*, (Mar 2010), Chapter 1:Unit 1E.5.

Meivar-Levy, I. & Ferber, S. (2006). Regenerative medicine: using liver to generate pancreas for treating diabetes, *Isr Med Assoc J*, (Jun 2006), Vol.8, No.6, pp.430-4.

Meivar-Levy, I. & Ferber, S. (2010). Adult cell fate reprogramming: converting liver to pancreas, *Methods Mol Biol*, (2010), Vol.636, pp.251-83

Meng, X.; Ichim, T.E.; Zhong, J.; Rogers, A.; Yin, Z.; Jackson, J.; Wang, H.; Ge, W.; Bogin, V.; Chan, K.W.; Thébaud, B. & Riordan, N.H. (2007). Endometrial regenerative cells: a novel stem cell population, *J Transl Med*, (Nov 2007), Vol.5, pp.57.

Miao, Z.; Jin, J.; Chen, L.; Zhu, J.; Huang, W.; Zhao, J.; Qian, H. & Zhang, X. (2006). Isolation of mesenchymal stem cells from human placenta: comparison with human bone marrow mesenchymal stem cells, *Cell Biol Int*, (Sep 2006), Vol.30, No.9, pp.681-7.

Moriscot, C.; de Fraipont, F.; Richard, M.J.; Marchand, M.; Savatier, P.; Bosco, D.; Favrot, M.; Benhamou, P.Y. (2005). Human bone marrow mesenchymal stem cells can express insulin and key transcription factors of the endocrine pancreas developmental pathway upon genetic and/or microenvironmental manipulation in vitro, *Stem Cells*, (Apr 2005), Vol.23, No.4, pp.594-603

Musina, R.A.; Belyavski, A.V.; Tarusova, O.V.; Solovyova, E.V. & Sukhikh, G.T. (2008). Endometrial mesenchymal stem cells isolated from the menstrual blood, *Bull Exp Biol Med*, (Apr 2008), Vol.145, No.4, pp.539-43

Nakagawa, M.; Koyanagi, M.; Tanabe, K.; Takahashi, K.; Ichisaka, T.; Aoi, T.; Okita, K.; Mochiduki, Y.; Takizawa, N.; & Yamanaka, S. (2007). Generation of induced pluripotent stem cells without Myc from mouse and human fibroblasts, *Nat Biotechnol*, (Jan 2008), Vol.26, No.1, pp.101-6

Naujok, O.; Francini, F.; Picton, S.; Bailey, CJ.; Lenzen, S.; Jörns, A. (2009). Changes in gene expression and morphology of mouse embryonic stem cells on differentiation into insulin-producing cells in vitro and in vivo, *Diabetes Metab Res Rev,* (Jul 2009), Vol.25, No.5, pp.464-76

Noguchi, H.; Matsumoto, S.; Ueda, M.; Hayashi, S.; Kobayashi, N.; Jackson, A.; Naziruddin, B. & Levy, M.F. (2008). Method for isolation of mouse pancreatic stem cells, *Transplant Proc,* (Mar 2008), Vol.40, No.2, pp.422-3

Noguchi, H.; Naziruddin, B.; Shimoda, M.; Fujita, Y.; Chujo, D.; Takita, M.; Peng, H.; Sugimoto, K.; Itoh, T.; Tamura, Y.; Olsen, G.S.; Kobayashi, N.; Onaca, N.; Hayashi, S.; Levy, M.F. & Matsumoto S. (2010). Induction of insulin-producing cells from human pancreatic progenitor cells, *Transplant Proc,* (Jul 2010), Vol.42, No.6, pp.2081-3

Noguchi, H.; Oishi, K.; Ueda, M.; Yukawa, H.; Hayashi, S.; Kobayashi, N.; Levy, M.F. & Matusmoto, S. (2009). Establishment of mouse pancreatic stem cell line, *Cell Transplant,* (2009), Vol.18, No.5, pp.563-71

Ohmura, Y.; Tanemura, M.; Kawaguchi, N.; Machida, T.; Tanida, T.; Deguchi, T.; Wada, H.; Kobayashi, S.; Marubashi, S.; Eguchi, H.; Takeda, Y.; Matsuura, N.; Ito, T.; Nagano, H.; Doki, Y. & Mori, M. (2010). Combined transplantation of pancreatic islets and adipose tissue-derived stem cells enhances the survival and insulin function of islet grafts in diabetic mice, *Transplantation,* (Dec 2010), Vol.90, No.12, pp.1366-73

Omer, A.; Keegan, M.; Czismadia, E.; De Vos, P.; Van Rooijen, N.; Bonner-Weir, S. & Weir, G.C. (2003). Macrophage depletion improves survival of porcine neonatal pancreatic cell clusters contained in alginate macrocapsules transplanted into rats, *Xenotransplantation,* (2003), Vol.3, pp.240-51

Oyajobi, B.O.; Lomri, A.; Hott, M. & Marie, P.J. (1999). Isolation and characterization of human clonogenic osteoblast progenitors immunoselected from fetal bone marrow stroma using STRO-1 monoclonal antibody, *J Bone Miner Res,* (Mar 1999), Vol.14, No.3, pp.351-61

Parekh, V.S.; Joglekar, M.V. & Hardikar, A.A. (2009). Differentiation of human umbilical cord blood-derived mononuclear cells to endocrine pancreatic lineage, *Differentiation,* (Nov 2009), Vol.78, No.4, pp.232-40

Park, I.H.; Arora, N. & Huo, H. (2008). Disease-specific induced pluripotent stem cells, *Cell,* (2008), Vol.134, pp.877–886

Park, K.S.; Kim, Y.S.; Kim, J.H.; Choi, B.; Kim, S.H.; Tan, A.H.; Lee, M.S.; Lee, M.K.; Kwon, C.H.; Joh, J.W.; Kim, S.J. & Kim, K.W. (2010). Trophic molecules derived from human mesenchymal stem cells enhance survival, function, and angiogenesis of isolated islets after transplantation, *Transplantation,* (Mar 2010), Vol.89, No.5, pp.509-17

Patki, S.; Kadam, S.; Chandra, V. & Bhonde, R. (2010). Human breast milk is a rich source of multipotent mesenchymal stem cells, *Hum Cell,* (May 2010), Vol.23, No.2, pp.35-40

Paz, A.H.; Salton, G.D.; Ayala-Lugo, A.; Gomes, C.; Terraciano, P.; Scalco, R.; Laurino, C.C.; Passos, E.P.; Schneider, M.R.; Meurer, L. & Cirne-Lima, E. (2011). Betacellulin overexpression in mesenchymal stem cells induces insulin secretion in vitro and ameliorates streptozotocin-induced hyperglycemia in rats, *Stem Cells Dev,* (Feb 2011), Vol.20. No.2, pp.223-32

Peng, J.; Wang, Y.; Zhang, L.; Zhao, B.; Zhao, Z.; Chen, J.; Guo, Q.; Liu, S.; Sui, X.; Xu, W. & Lu, S. (2011). Humanumbilical cord Wharton's jelly-derived mesenchymal stem cells differentiate into a Schwann-cell phenotype and promote neurite outgrowth in vitro, *Brain Res Bull,* (Feb 2011), Vol.84, No.3, pp.235-43

Perdikogianni, C.; Dimitriou, H.; Stiakaki, E.; Martimianaki, G. & Kalmanti, M. (2008). Could cord blood be a source of mesenchymal stromal cells for clinical use? *Cytotherapy,* (2008), Vol.10, No.5, pp.452-9

Phadnis, S.M.; Joglekar, M.V.; Dalvi, M.P.; Muthyala, S.; Nair, P.D.; Ghaskadbi, S.M.; Bhonde, R.R. & Hardikar, A.A. (2011) Human bone marrow-derived mesenchymal cells differentiate and mature into endocrine pancreatic lineage in vivo, *Cytotherapy,* (Mar 2011), Vol.13, No.3, pp.279-93

Phuc, P.V.; Nhung, T.H.; Loan, D.T.; Chung, D.C. & Ngoc, P.K. (2010). Differentiating of banked human umbilical cord blood-derived mesenchymal stem cells into insulin-secreting cells, *In Vitro Cell Dev Biol Anim,* (Jan 2011), Vol.47, No.1, pp.54-63.

Pilz, G.A.; Ulrich, C.; Ruh, M.; Abele, H.; Schäfer, R.; Kluba, T.; Bühring, H.J.; Rolauffs, B. & Aicher, W.K. (2010). Human Term Placenta-Derived Mesenchymal Stromal Cells Are Less Prone to Osteogenic Differentiation Than Bone Marrow-Derived Mesenchymal Stromal Cells, *Stem Cells Dev,* (Jan 2011). [Epub ahead of print]

Poloni, A.; Rosini, V.; Mondini, E.; Maurizi, G.; Mancini, S.; Discepoli, G.; Biasio, S.; Battaglini, G.; Berardinelli, E.; Serrani, F. & Leoni, P. (2008). Characterization and expansion of mesenchymal progenitor cells from first-trimester chorionic villi of human placenta, *Cytotherapy,* (2008), Vol.10, No.7, pp.690-7

Prockop, D.J.; Sekiya, I.; Colter, D.C. (2001). Isolation and characterization of rapidly self-renewing stem cells from cultures of human marrow stromal cells, *Cytotherapy,* (2001), Vol.3, No.5, pp.393-6

Rackham, C.L.; Chagastelles, P.C.; Nardi, N.B.; Hauge-Evans, A.C.; Jones, P.M. & King, A.J. (2011). Co-transplantation of mesenchymal stem cells maintains islet organisation and morphology in mice, *Diabetologia,* (Jan 2011) [Epub ahead of print]

Rajagopal, J.; Anderson, W.J.; Kume, S.; Martinez, O.I.; Melton, D.A. (2003). Insulin staining of ES cell progeny from insulin uptake, *Science,* (2003), Vol.299, pp.363

Reinisch, A.; Bartmann, C.; Rohde, E.; Schallmoser, K.; Bjelic-Radisic, V.; Lanzer, G.; Linkesch, W. & Strunk, D. (2007). Humanized system to propagate cord blood-derived multipotent mesenchymal stromal cells for clinical application, *Regen Med,* (Jul 2007), Vol.2, No.4, pp.371-82

Rene ´ Maehra, Shuibing Chena, Melinda Snitowa, Thomas Ludwigb, Lisa Yagasaki a, Robin Golandc, Rudolph L. Leibel & Douglas A. Meltona. (2009). Generation of pluripotent stem cells from patients with type 1 diabetes, *Proc Natl Acad Sci U S A,* (Sep 2009), Vol.106, No.37, pp.15768-73

Ris, F.; Niclauss, N.; Morel, P.; Demuylder-Mischler, S.; Muller, Y.; Meier, R.; Genevay, M.; Bosco, D. & Berney, T. (2011). Islet Autotransplantation After Extended Pancreatectomy for Focal Benign Disease of the Pancreas, *Transplantation,* (Mar 2011) [Epub ahead of print]

Rosada, C.; Justesen, J.; Melsvik, D.; Ebbesen, P. & Kassem, M. (2003). The human umbilical cord blood: a potential source for osteoblast progenitor cells, *Calcif Tissue Int,* (Feb 2003), Vol.72, No.2, pp.135-42

Santos, T.M.; Percegona, L.S.; González, P.; Calil, A.; Corradi Perini, C.; Faucz, F.R.; Câmara, N.O.; Aita, C.A. (2010). Expression of pancreatic endocrine markers by mesenchymal stem cells from human umbilical cord vein, *Transplant Proc,* (Mar 2010), Vol.42, No.2, pp.563-5

Sapir, T.; Shternhall, K.; Meivar-Levy, I.; Blumenfeld, T.; Cohen, H.; Skutelsky, E.; Eventov-Friedman, S.; Barshack, I.; Goldberg, I.; Pri-Chen, S.; Ben-Dor, L.; Polak-Charcon, S.; Karasik, A.; Shimon, I.; Mor, E. & Ferber, S. (2005). Cell-replacement therapy for diabetes: Generating functional insulin-producing tissue from adult human liver cells, *Proc Natl Acad Sci U S A,* (May 2005), Vol.102, No.22, pp.7964-9

Schroeder, I.S.; Rolletschek, A.; Blyszczuk, P.; Kania, G. & Wobus, A.M. (2006). Differentiation of mouse embryonic stem cells to insulin-producing cells, *Nature Protoc,* Vol.1, pp.495–507

Semenov, O.V.; Koestenbauer, S.; Riegel, M.; Zech, N.; Zimmermann, R.; Zisch, A.H. & Malek, A. (2010). Multipotent mesenchymal stem cells from human placenta: critical parameters for isolation and maintenance of stemness after isolation, *Am J Obstet Gynecol,* (Feb 2010), Vol.202, No.2, pp.193.e1-193.e13

Shao, S.; Gao, Y.; Xie, B.; Xie, F.; Lim, S.K.; Li, G. (2011). Correction of hyperglycemia in type 1 diabetic models by transplantation of encapsulated insulin-producing cells derived from mouse embryo progenitor, *J Endocrinol,* (Mar 2011), Vol.208, No.3, pp.245-55

Shi, Y.; Hou, L. & Tang, F. (2005). Inducing embryonic stem cells to differentiate into pancreatic β cells by a novel three-step approach with activin a and all-trans retinoic acid, *Stem Cells,* Vol.23, pp.656–662

Shternhall-Ron, K.; Quintana, F.J.; Perl, S.; Meivar-Levy, I.; Barshack, I.; Cohen, I.R. & Ferber, S. (2007). Ectopic PDX-1 expression in liver ameliorates type 1 diabetes, *J Autoimmun,* (Mar 2007), Vol.28, No.2-3, pp.134-42

Shuang-Zhi, H.; Ping, S. & Xi-Ning, P. (2010). Culture and identification of human amniotic mesenchymal stem cells, *Chin Med Sci J,* (Dec 2010), Vol.25, No.4, pp.211-4.

Shyu, J.F.; Wang, H.S.; Shyr, Y.M.; Wang, S.E.; Chen, C.H.; Tan, J.S.; Lin, M.F.; Hsieh, P.S.; Sytwu, H.K. & Chen, T.H. (2011). Alleviation of hyperglycemia in diabetic rats by intraportal injection of insulin-producing cells generated from surgically resected human pancreatic tissue, *J Endocrinol,* (Mar 2011), Vol.208, No.3, pp.233-244.

Siminovitch, L.; McCulloch, E.A. & Till, J.E. (1963). The distribution of colony-forming cells among spleen colonies, *Journal of Cellular and Comparative Physiology,* Vol.62, pp.327–36

Sipione, S.; Eshpeter, A.; Lyon, J.G.; Korbutt, G.S.; Bleackley, R.C. (2004). Insulin expressing cells from differentiated embryonic stem cells are not beta cells, *Diabetologia*, Vol.47, (2004), pp.499–508

Smith ,J.R.; Pochampally, R.; Perry, A.; Hsu, S.C. & Prockop, D.J. (2004). Isolation of a highly clonogenic and multipotential subfraction of adult stem cells from bone marrow stroma, *Stem Cells,* (2004), Vol.22, No.5, pp.823-31

Soncini, M.; Vertua, E.; Gibelli, L.; Zorzi, F.; Denegri, M.; Albertini, A.; Wengler, G.S. & Parolini, O. (2007). Isolation and characterization of mesenchymal cells from human fetal membranes, *J Tissue Eng Regen Med,* (Jun 2007), Vol.1, No.4, pp.296-305

Soon-Shiong, P.; Feldman, E.; Nelson, R.; Heintz, R.; Merideth, N.; Sandford, P.; Zheng, T. & Komtebedde, J. (1992). Long-term reversal of diabetes in the large animal model by encapsulated islet transplantation, *Transplant Proc,* (1992), Vol.6, pp.2946-7.

Soon-Shiong, P.; Feldman, E.; Nelson, R.; Komtebedde, J.; Smidsrod, O.; Skjak-Braek, G.; Espevik, T.; Heintz, R. & Lee, M. (1992). Successful reversal of spontaneous diabetes in dogs by intraperitoneal microencapsulated islets, *Transplantation* (1992), Vol.5, pp.769-74

Sordi V, Malosio ML, Marchesi F, Mercalli A, Melzi R, Giordano T, Belmonte N, Ferrari G, Leone BE, Bertuzzi F, Zerbini G, Allavena P, Bonifacio E, Piemonti L. Bone marrow mesenchymal stem cells express a restricted set of functionally active chemokine receptors capable of promoting migration to pancreatic islets. *Blood.* 2005 Jul 15;106(2):419-27. Epub 2005 Mar 22.

Sordi, V. & Piemonti, L. (2010). Mesenchymal stem cells as feeder cells for pancreatic islet transplants, *Rev Diabet Stud*, (2010), Vol.7, No.2, pp.132-43

Sordi, V. (2009). Mesenchymal stem cell homing capacity, *Transplantation*, (2009), Vol.87, 9 Suppl, pp.S42-5

Soria, B.; Roche, E.; Berna, G.; Leon-Quinto, T.; Reig, J.A.; Martin, F. Insulin-secreting cells derived from embryonic stem cells normalize glycemia in streptozotocin-induced diabetic mice, *Diabetes*, Vol.49, (2000), pp.157–162

Spath, L.; Rotilio, V.; Alessandrini, M.; Gambara, G.; De Angelis, L.; Mancini, M.; Mitsiadis, T.A.; Vivarelli, E.; Naro, F.; Filippini, A.; Papaccio, G. (2010). Explant-derived human dental pulp stem cells enhance differentiation and proliferation potentials, *J Cell Mol Med*, (Jun 2010), Vol.14, No.6B, pp.1635-44

Strelchenko, N.; Verlinsky, O.; Kukharenko, V.; Verlinsky, Y. (2004). Morula-derived human embryonic stem cells, *Reprod Biomed Online*, Vol.9, (2004), pp.623–629

Sun, Y.; Chen, L.; Hou, X.G.; Hou, W.K.; Dong, J.J.; Sun, L.; Tang, K.X.; Wang, B.; Song, J.; Li, H. & Wang, K.X. (2007). Differentiation of bone marrow-derived mesenchymal stem cells from diabetic patients into insulin-producing cells in vitro, *Chin Med J (Engl)*, (May 2007), Vol.120, No.9, pp.771-6

Sun, Y.; Ma, X.; Zhou, D.; Vacek, I. & Sun, A.M. (1996). Normalization of diabetes in spontaneously diabetic cynomologus monkeys by xenografts of microencapsulated porcine islets without immunosuppression, *J Clin Invest*, (1996), Vol.6, pp.1417-22

Takahashi, K. & Yamanaka, S. (2006). Induction of pluripotent stem cells from mouse embryonic and adult fibroblast cultures by defined factors, *Cell*, (Aug 2006), Vol.126, No.4, pp.663-76

Takahashi, K.; Tanabe, K. & Ohnuki, M. (2007). Induction of pluripotent stem cells from adult human fibroblasts by defined factors, *Cell*, (2007), 131, pp.861—872

Titorencu, I.; Jinga, V.V.; Constantinescu, E.; Gafencu, A.V.; Ciohodaru, C.; Manolescu, I.; Zaharia, C. & Simionescu, M. (2007). Proliferation, differentiation and characterization of osteoblasts from human BM mesenchymal cells, *Cytotherapy*, (2007), Vol.9, No.7, pp.682-96

Trivedi, H,L.; Vanikar, A.V.; Thakker, U.; Firoze, A.; Dave, S.D.; Patel, C.N.; Patel, J.V.; Bhargava, A.B. & Shankar, V. (2008). Human adipose tissue-derived mesenchymal stem cells combined with hematopoietic stem cell transplantation synthesize insulin, *Transplant Proc*, (May 2008), Vol.40, No.4, pp.1135-9

Trounson, A. (2002). Human embryonic stem cells: mother of all cell and tissues, *Reprod Biomed Online*, (2002), Vol.4, Suppl.1, pp.58–63

Trovato, L.; De Fazio, R.; Annunziata, M.; Sdei, S.; Favaro, E.; Ponti, R.; Marozio, L.; Ghigo, E.; Benedetto, C. & Granata, R. (2009). Pluripotent stem cells isolated from human amniotic fluid and differentiation into pancreatic beta-cells, *J Endocrinol Invest*, (Dec 2009), Vol.32, No.11, pp.873-6

Tucker, H.A. & Bunnell, B.A. (2011). Characterization of human adipose-derived stem cells using flow cytometry, *Methods Mol Biol*, (2011), Vol.702, pp.121-31

Vanikar, A.V.; Dave, S.D.; Thakkar, U.G. & Trivedi, H.L. (2010). Cotransplantation of adipose tissue-derived insulin-secreting mesenchymal stem cells and hematopoietic stem cells: a novel therapy for insulin-dependent diabetes mellitus, *Stem Cells Int*, (Dec 2010), Vol.2010, pp.582382

Verlinsky, Y.; Strelchenko, N.; Kukharenko, V.; Rechitsky, S.; Verlinsky, O.; Galat, V.; Kuliev, A. Human embryonic stem cell lines with genetic disorders, *Reprod Biomed Online*, (2005), Vol.10, pp.105–110

Wahoff, D.C.; Papalois, B.E.; Najarian, J.S.; Kendall, D.M.; Farney, A.C.; Leone, J.P.; Jessurun, J.; Dunn, D.L.; Robertson, R.P. & Sutherland, D.E. (1995). Autologous islet transplantation to prevent diabetes after pancreatic resection, *Ann Surg*, (Oct 1995), Vol.222, No.4, pp.562-75

Wahoff, D.C.; Paplois, B.E.; Najarian, J.S.; Farney, A.C.; Leonard, A.S.; Kendall, D.M.; Roberston, R.R. & Sutherland, D.E. (1996). Islet Autotransplantation after total pancreatectomy in a child, *J Pediatr Surg*, (Jan 1996), Vol.31, No.1, pp.132-5

Wang, H.S.; Shyu, J.F.; Shen, W.S.; Hsu, H.C.; Chi, T.C.; Chen, C.P.; Huang, S.W.; Shyr, Y.M.; Tang, K.T. & Chen, T.H. (2010). Transplantation of insulin producing cells derived from umbilical cord stromal mesenchymal stem cells to treat NOD mice, *Cell Transplant*, (Aug 2010) [Epub ahead of print]

Weir, G.C. & Bonner-Weir, S. (2004). Beta-cell precursors--a work in progress, *Nat Biotechnol*, (Sep 2004), Vol.22, No.9, pp.1095-6

Wolfe, M.; Pochampally, R.; Swaney, W.; Reger, R.L. (2008). Isolation and culture of bone marrow-derived human multipotent stromal cells (hMSCs), *Methods Mol Biol*, (2008), Vol.449, pp.3-25

Wu, L.F.; Wang, N.N.; Liu, Y.S. & Wei, X. (2009). Differentiation of Wharton's jelly primitive stromal cells into insulin-producing cells in comparison with bone marrow mesenchymal stem cells, *Tissue Eng Part A*, (Oct 2009), Vol.15, No.10, pp.2865-73

Wu, X.H.; Liu, C.P.; Xu, K.F.; Mao, X.D.; Zhu, J.; Jiang, J.J.; Cui, D.; Zhang, M.; Xu, Y.; Liu, C. (2007). Reversal of hyperglycemia in diabetic rats by portal vein transplantation of islet-like cells generated from bone marrow mesenchymal stem cells, *World J Gastroenterol*, (Jun 2007), Vol.13, No.24, pp.3342-9

Xiao, M.; An, L.; Yang, X.; Ge, X.; Qiao, H.; Zhao, T.; Ma, X.; Fan, J.; Zhu, M. & Dou, Z. (2008). Establishing a human pancreatic stem cell line and transplanting induced pancreatic islets to reverse experimental diabetes in rats, *Sci China C Life Sci*, (Sep 2008), Vol.51, No.9, pp.779-88

Xie, Q.P.; Huang, H.; Xu, B.; Dong, X.; Gao, S.L.; Zhang, B.; Wu, Y.L. (2009). Human bone marrow mesenchymal stem cells differentiate into insulin-producing cells upon microenvironmental manipulation in vitro, *Differentiation*, (Jun 2009), Vol.77, No.5, pp.483-91

Xu, J.; Lu, Y.; Ding, F.; Zhan, X.; Zhu, M. & Wang, Z. (2007). Reversal of diabetes in mice by intrahepatic injection of bone-derived GFP-murine mesenchymal stem cells infected with the recombinant retrovirus-carrying human insulin gene, *World J Surg*, (Sep 2007), Vol.31, No.9, pp.1872-82

Xu, X.; D'Hoker, J.; Stangé, G.; Bonné, S.; De Leu, N.; Xiao, X.; Van de Casteele, M.; Mellitzer, G.; Ling, Z.; Pipeleers, D.; Bouwens, L.; Scharfmann, R.; Gradwohl, G. & Heimberg, H. (2008). Beta cells can be generated from endogenous progenitors in injured adult mouse pancreas, *Cell*, (Jan 2008), Vol.132, No.2, pp.197-207

Yalvac, M.E.; Ramazanoglu, M.; Rizvanov, A.A.; Sahin, F.; Bayrak, O.F.; Salli, U.; Palotás, A. & Kose, G.T. (2010). Isolation and characterization of stem cells derived from human third molar tooth germs of young adults: implications in neo-vascularization, osteo-, adipo- and neurogenesis, *Pharmacogenomics J*, (Apr 2010), Vol.10, No.2, pp.105-13

Yang, L.; Li, S.; Hatch, H.; Ahrens, K.; Cornelius, J.G.; Petersen, B.E. & Peck, A.B. (2002). In vitro trans-differentiation of adult hepatic stem cells into pancreatic endocrine hormone-producing cells, *Proc Natl Acad Sci U S A*, (Jun 2002), Vol.99, No.12, pp.8078-83

Yu, J.; Vodyanik, M.A.; Smuga-Otto, K. (2007). Induced pluripotent stem cell lines derived from human somatic cells, *Science*, (2007), Vol.318, pp.1917–1920

Yuan, H.; Li, J.; Xin, N.; Zhao, Z. & Qin, G. (2010). Expression of Pdx1 mediates differentiation from mesenchymal stem cells into insulin-producing cells, *Mol Biol Rep*, (Dec 2010), Vol.37, No.8, pp.4023-31

Zalzman, M.; Anker-Kitai, L.; Efrat, S. (2010). Differentiation of human liver-derived, insulin-producing cells toward the beta-cell phenotype, *Diabetes*, (Sep 2005), Vol.54, No.9, pp.2568-75

Zeddou, M.; Briquet, A.; Relic, B.; Josse, C.; Malaise, M.G.; Gothot, A.; Lechanteur, C. & Beguin, Y. (2010). The umbilical cord matrix is a better source of mesenchymal stem cells (MSC) than the umbilical cord blood, *Cell Biol Int*, (Jul 2010), Vol.34, No.7, pp.693-701

Zhang, X.; Hirai, M.; Cantero, S.; Ciubotariu, R.; Dobrila, L.; Hirsh, A.; Igura, K.; Satoh, H.; Yokomi, I.; Nishimura, T. & Takahashi, T.A. (2011). Isolation and characterization of mesenchymal stem cells from human umbilical cord blood: Reevaluation of critical factors for successful isolation and high ability to proliferate and differentiate to chondrocytes as compared to mesenchymal stem cells from bone marrow and adipose tissue, *J Cell Biochem*, (Feb 2011), doi: 10.1002/jcb.23042

Zhang, Y.; Ren, Z.; Zou, C.; Wang, S.; Luo, B.; Li, F.; Liu, S. & Zhang, Y.A. (2010). Insulin-producing cells from human pancreatic islet-derived progenitor cells following transplantation in mice, *Cell Biol Int*, (Nov 2010), [Epub ahead of print]

Zhu, S.; Lu, Y.; Zhu, J.; Xu, J.; Huang, H.; Zhu, M.; Chen, Y.; Zhou, Y.; Fan, X. & Wang, Z. (2009). Effects of Intrahepatic Bone-Derived Mesenchymal Stem Cells Autotransplantation on the Diabetic Beagle Dogs, *J Surg Res*, (Oct 2009) [Epub ahead of print]

23

Topical Stem and Progenitor Cell Therapy for Diabetic Foot Ulcers

Aonghus O'Loughlin and Timothy O'Brien
Regenerative Medicine Institute, National University of Ireland, Galway
Ireland

1. Introduction

The prevalence of diabetes mellitus is increasing to epidemic proportions worldwide. Diabetic foot ulceration can affect up to 25 percent of people with diabetes mellitus throughout their lives. The most significant complication of foot ulceration is lower limb amputation, which arises from pre-existing ulcers in the majority of cases. Despite current clinical care protocols for ulcer treatment, there exists a high amputation rate. This presents a major burden for individual patients' health and well-being in addition to significant financial cost for health care systems. There is an urgent need for new medicinal products to treat diabetic ulcers. Cell-based therapies offer a novel treatment strategy to augment diabetic wound healing, increase ulcer healing rate and prevent amputation. The field of tissue engineering has developed commercially available skin substitutes for diabetic cutaneous wound repair. These products have incorporated somatic cells delivered in a bioengineered scaffold. However, having been available for the last decade, the majority have demonstrated only moderate clinical benefit in small clinical trials. In comparison, stem and progenitor cell therapy offer the potential for accelerated wound repair in addition to structural skin regeneration with functional recovery.

Stem cells have the ability to self-renew and differentiate into other cell types and are classified into adult stem and progenitor cells, embryonic stem cells and induced pluripotent stem cells. The mechanisms of action of stem and progenitor cells are not fully elucidated but include 1) differentiation to specialised cells e.g. skin cells of the dermis and epidermis 2) acting by paracrine or autocrine effects through the secretion of trophic factors e.g. the production of soluble mediators for neo-angiogenesis and 3) immuno-modulatory functions. Much research endeavour is determining the benefit of stem cell treatment on diabetic cutaneous wound healing with encouraging results in animal models. Regenerative medicine and tissue engineering specialties are rapidly elucidating the mechanisms of action of stem cells and translating the results of in-vitro and in-vivo experiments to human clinical trials. The requirements for success will be patient safety, clinical efficacy and convenience of use.

The focus of this chapter is to review the area of topical stem and progenitor cell therapy as a treatment for non-healing diabetic foot ulcers. It will focus on adult stem cells as these are nearer to use in human trials and do not pose the ethical constraints associated with the use of embryonic stem cells. Topical treatment with endothelial progenitor cell (EPC) and mesenchymal stem cell (MSC) therapy is presented in this review, and more specifically the delivery of these cells using biomaterial scaffolds. The currently available cell therapy products for wound repair will be presented. The case for adopting stem and progenitor cell therapy in research and treatment of diabetic foot ulcers will be discussed. The benefits of biomaterials and functionalised scaffolds for mediating cell therapy to a wound will be described. For both endothelial progenitor cells and mesenchymal stem cells, the potential mechanisms of action will be discussed with reference to key pre-clinical and clinical studies. The chapter will also describe strategies to enhance the therapeutic potential of stem and progenitor cells for wound healing. These will include the employment of matri-cellular proteins i.e. proteins associated with the extracellular matrix that mediate diverse biological functions, gene therapy, conditioned media experiments and the delivery of several cell types. A section of the chapter will focus on translational of these advanced biological medicines to clinical trials. This includes issues regarding pre-clinical animal models, optimal cell source, safety and regulatory approval. Finally the chapter will highlight the potential of cell based therapies in other conditions causing cutaneous wounding, i.e burns, decubitus ulcers and other rare blistering conditions e.g. epidermolysis bullosa.

2. The biology of cutaneous wounds

The repair of cutaneous wounds is a highly complex biological process. After injury, multiple biological pathways immediately become activated and are synchronised to respond.(Gurtner et al., 2008) Adult wound healing occurs by tissue repair with consequent scarring. The goal of adult wound healing is to repair a skin defect, to ensure the restoration of a barrier and to regain tensile strength. There is involvement of several cell types, cytokines and extra-cellular matrix components. The physiological overlapping pathways that are required for optimal wound healing include haemostasis (which occurs immediately on wounding), inflammation with cell migration and proliferation (neutrophils initially and subsequently macrophages). The proliferation of fibroblasts results in extra-cellular matrix deposition. Remodeling and wound contraction occur once closure of the wound takes place. Angiogenesis (growth of new blood vessels from pre-existing blood vessels) and re-epithelialisation are central processes in wound healing. This is a superficial description of wound healing and conveys the complexity of the process, but highlights the potential for disruption in a difficult to heal wound. (Breen et al., 2008 ;Harding er al., 2002) The physiological response to acute cutaneous wounds usually takes 3-14 days to complete. (Liu et al., 2008) Wound healing involves activation of keratinocytes, fibroblasts, endothelial cells, macrophages and platelets.(Brem et al., 2007) Figure I details the stages of normal cutaneous wound healing.

2.1 Diabetic wound healing

Delayed wound healing as occurs in diabetes mellitus results from dysregulation of the normal healings pathways. The diabetic wound is complex with contribution from infection, neuropathy and impaired vascular supply. There are many physiological defects in diabetic wounds. These include decreased or impaired growth factor production, angiogenic response, macrophage function, collagen accumulation, epidermal barrier function, quantity of granulation tissue, keratinocyte, fibroblast migration and proliferation and bone healing. There is an imbalance between the accumulation of extra-cellular matrix components and their re-modeling by matrix metallo-proteinases.(Brem et al. 2007) In addition fibroblasts from diabetic wounds become senescent and show a decreased proliferative response to growth factors.(Falanga et al., 2005) There is a chronic inflammatory environment associated with diabetic wounds. This is associated with a persistent increase in pro-inflammatory cytokines by various immune and non-immune cells and it is hypothesized that this blunts the acute, focused cytokine response needed to progress through the normal phases of wound healing.(Pradhan et al., 2009)

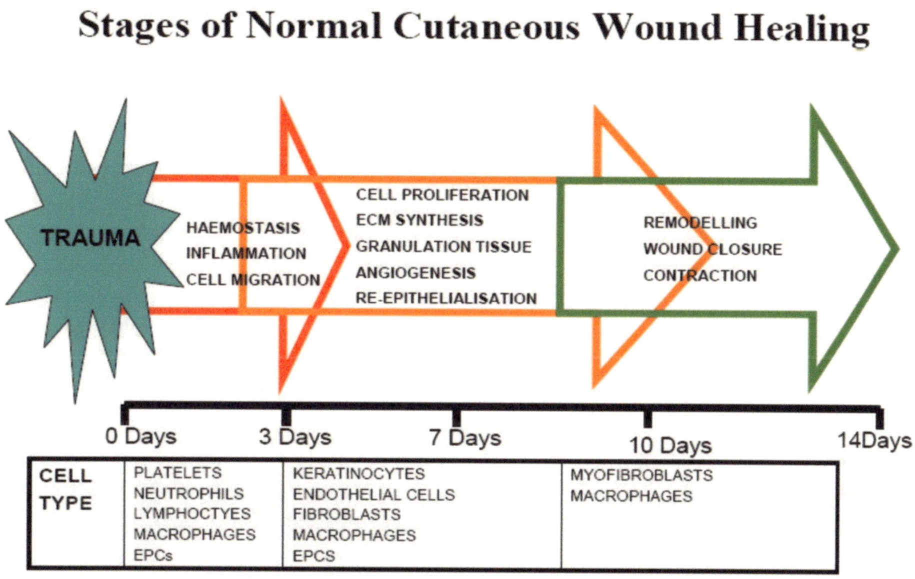

Fig. I. Stages of normal wound healing with predominant cell types involved at each stage of process. The wound healing spectrum is a continuum with overlapping phases.

2.2 Angiogenesis and wound Healing

The impaired vascular supply associated with diabetes leads to poor blood flow at the wound site impeding the optimal endogenous reparative response (Jeffcoate & Harding 2003) Impaired angiogenesis is a feature of diabetic wounds. In addition neovascularisation, or the de novo formation of new blood vessels is critical for granulation tissue formation and tissue regeneration in wound healing. (Gurtner et al., 2008) The impaired angiogenic response that occurs in diabetes mellitus leads to hypoxia at the wound site. Temporary hypoxia is requisite for normal wound healing. In the non-diabetic situation, hypoxia leads to activation of the transcription factor complex HIF-1α (Hypoxia inducible factor-1α), which leads to transcription of multiple genes required for successful wound healing. With diabetes, hyperglycaemia affects the stability and activation of HIF-1α. This suppresses platelet-derived growth factor, vascular endothelial growth factor and transforming growth factor-β, which are required for angiogenesis, in vitro and in vivo wound healing.(Botusan et al., 2008)

2.3 Wound repair versus regeneration

Adult wound healing occurs by repair. Wound repair leads to scarring and results in decreased tensile strength of wounds. Skin regeneration is the regeneration of wounds with restoration of the normal function and anatomy of skin. In biology, foetal wound repair is a regenerative process, and some vertebrate species demonstrate successful tissue regeneration where the initial phase of wound repair is followed by perfect structural and functional regeneration of the organ. An example of this is Xenopus limb regeneration. The challenge for scientists is to produce tissue engineered products that exhibit extra-cellular matrix re-modeling characteristics seen in embryonic wound repair to produce functional and durable skin. (Metcalfe & Ferguson 2007)

3. The case for novel topically applied stem and progenitor cell therapies

3.1 Burden of diabetic ulceration

There exists a growing global epidemic of diabetes mellitus. It is predicted that the prevalence of diabetes mellitus will be 4.4% of the global population or 366 million people by the year 2030.(Wild et al., 2004) In 2010, the prevalence of diabetes in China was reported as 9.7%.(Yang et al., 2010) This will likely continue to increase based on the prevalence of obesity in populations. Foot ulcers can affect 12 to 25 percent of persons with diabetes mellitus throughout their lives.(Brem et al., 2006) Lower limb disease is the most common source of complications and hospitalisation in the diabetic population. (Boyko et al., 2006) Major lower limb amputations in patients with diabetes arise from preceding ulcers in 85% of cases. (Frykberg et al., 2006) The cost of treating diabetic foot ulcers creates a burden on healthcare resources. Boulton et al. reviewed the epidemiology and cost of treating foot ulceration globally and one report estimated the cost of diabetic foot ulceration treatment including amputation at €10.9 billion in the United States of America for the year 2001.(Boulton et al., 2005) In addition to the cost to healthcare system budgets, for individual patients, the parameters of pain, social isolation, physical morbidity, restrictions in work capacity, and psychological well-being are negatively affected by leg ulceration.(Herber et al., 2007)

3.2 Classification of diabetic ulcers

Diabetic foot ulcers can be classified as ischaemic, neuropathic or neuro-ischaemic. The ability to heal ulcers is predicated on the restoration of an adequate blood supply. The typical angiographic pattern of ischaemic diabetic vasculopathy is occluded distal blood vessels. The optimal treatment of ischemic lower extremity ulcers is the restoration of blood flow. This review paper focuses on treatment of neuropathic ulcers. Neuropathic ulcers develop due to distal sensory loss and consequent foot deformity. Ulceration develops at sites of excessive pressure predominantly under the first metatarsalphalangeal joint, in the majority due to unperceived trauma. Neuroischaemic ulcers are a combination of ischaemic and neuropathic ulcers.

3.3 Current treatment strategies

The management of the diabetic foot is complex requiring a multidisciplinary approach. A non-healing ulcer is an ulcer which has been present for > 8 weeks. Our group has reviewed the current standards of care required to investigate, treat and prevent diabetic foot ulceration and consequent amputation.(O'Loughlin et al.,2010) This manuscript highlights the benefit of routine examination and evaluation of the diabetic foot with identification of risk factors for ulceration. There are published risk stratification guidelines for diabetic foot ulceration based on the presence or absence of sensory loss, foot deformity and vascular insufficiency.(Boulton et al.,2008) The current standard care involves removal of pressure from the ulcer, restoration of blood flow if peripheral vascular disease is present, debridement of the ulcer and institution of antibiotic therapy to control infection. Topical dressings, patient education, podiatry review, and orthotics are beneficial. A systematic review of the control arms of trials investigating novel treatments reported that for standard treatment of neuropathic diabetic ulcers, where blood supply had been adequate (as defined by a transcutaneous oxygen pressure of > 30 mmHg or an ankle-brachial index > 0.7), after 20 weeks 31% of diabetic neuropathic ulcers were healed and at 12 weeks, 24% of neuropathic ulcers were completely healed.(Margolis et al., 1999)A protocol for the management of diabetic foot ulcers suggested treatment with growth factors and/or cellular therapy if wound healing is not is not observed after 2 weeks of standard therapy and a new epithelial layer has not formed.(Brem et al., 2004)

3.4 Benefit of a cell-based therapy for non-healing diabetic ulcers

It is evident that there is a critical clinical need to develop novel therapies for treatment of non-healing diabetic ulcers in order to prevent amputation and reduce the significant financial drain on healthcare budgets and burden on individuals health. The understanding of the patho-physiology of diabetic wound healing is important in the development of advanced wound healing treatments. It allows therapeutic targeting of the different phases of wound healing. Cell therapy may reverse the biological defects in diabetic wounds by acting as reservoirs for cell and growth factor production. Gurtner *et al.* states that the ultimate solution to both under-healing and over-healing is likely to be administration of cells that retain the ability to elaborate the full complexity of biological signaling, together with the environmental cues that are needed to regulate the differentiation and proliferation of these cells (Gurtner et al., 2008)

3.5 Limitations with current cell-based therapy

To date clinical trials of topical cell based therapy for non-healing diabetic foot ulcers have yielded limited results. There are several reasons for this. One reason is methodological flaws in the clinical trials which have raised concerns over the validity of the results. Systematic reviews on skin replacement therapy have reported statistical benefit in wound healing endpoints. However there was a lack of information reported on safety, method of recruitment, randomization methods and blinding strategy for outcome assessments. There is a lack of power size calculations in some of the trials and little mention of dropouts in trial. The interventions did appear as safe as standard treatments. (Barber et al., 2008) It is felt that the deficiencies in clinical trials investigating skin replacement therapies for diabetic foot ulcers affect the conclusions of systematic reviews.(Blozik et al., 2008; Barber et al., 2008;Teng et al., 2010) Further larger scale trials are required.

However the lack of clinical success with these advanced medicinal products is most likely not solely due to the aforementioned flaws in trial design. The current somatic cell therapies do not address the underlying pathology in the diabetic wound i.e. chronic inflammation and impaired angiogenesis. An efficient blood supply is central to normal wound healing, and delayed or inefficient angiogenesis will prolong ulceration and increase the probability of amputation. The current cell treatments do not target angiogenesis (blood vessel formation from pre-existing blood vessels) or neo-vasculogenesis (de novo blood vessel formation). Somatic cells do not differentiate into other cell types of the dermis and epidermis. The most frequently studied somatic cells include fibroblasts and keratinocytes. The employment of these cell treatments result in wound healing by repair and not by regeneration.

3.6 Potential superiority of treatment with stem and progenitor cells

Endothelial progenitor cells are a newly described cell type involved in angiogenesis. They can migrate to a site of injury/ischaemia and play a central role in vascular maintenance, angiogenesis and neo-vascularisation. (Marrotte et al., 2010). Adult mesenchymal stem cell treatment holds promise as this cell type addresses the key wound impairments seen in non-healing diabetic ulcers. They are immuno-modulatory and may create a more favourable inflammatory environment of the diabetic wound. They also promote angiogenesis by paracrine effects. Adult mesenchymal stem cells in diabetic wounds may in addition to beneficial paracrine activity, differentiate into other cell types e.g. epidermal keratincocytes, endothelial cells and pericytes in vivo.(Wu et al., 2007) In fact there is a growing body of evidence that the use of stem cells in wound healing in addition to augmenting wound repair, also promote skin regeneration and scarless wound healing.(Fu et al., 2009)

4. Endothelial Progenitor Cells (EPCs)

4.1 Background

The discovery of putative EPCs by Ashara et al in 1997 (Asahara, et al., 1997) has illuminated the fields of vascular biology and diabetes related vascular dysfunction. For the first time, vasculogenesis or de novo blood vessel formation was determined to occur post-natally, as previously it was assumed to occur only during embryogenesis. The delivery of EPCs to ischaemic sites in the body offers the possibility of successful treatment of diabetic vascular disease. Worldwide, research groups are testing the hypothesis that EPC therapy may treat peripheral vascular disease and prevent the progression of non-healing diabetic foot ulcers to amputation. These cells are suitable for autologous therapy without immunological rejection but this approach may be hindered due to disease associated cell dysfunction.

EPC research is complicated by several issues. These include a lack of a standardised definition of the cell-type. The reports in the literature describe different identities, sources of isolation, culture methodologies and function. The cells maybe isolated from the peripheral blood, umbilical cord blood or bone marrow. They are referred to as progenitor cells or stem cells. In a comprehensive review, Hirschi *et al.* describe three different EPC types isolated from mononuclear cells.(Hirschi et al., 2008) This classification reflects the different cell types reported as EPCs.

All three cell types are cultured in endothelial based media. The first cell type is named colony forming unit-Hill cells which arise from peripheral blood mononuclear cells which are non-adherent and give rise to a colony after 5 days in culture. The second cell type is a heterogenous collection of cells termed circulating angiogenic cells or early EPCs. These arise from mononuclear cells which are adherent to fibronectin or other matrix adhesion proteins after 4-7 days. They do not form colonies and have a low proliferative potential. They retain monocytic properties, secrete angiogenic factors and die after approximately 4 weeks in culture.(Liew et al.,2008) The third cell type is the endothelial colony forming cell or late EPC. These cells are derived from mononuclear cells that adhere to fibronectin and appear after 6-21 days. They display cobblestone morphology and from blood vessels in vitro. They are highly proliferative. (Hirschi et al.,2008) The cells maybe further characterised by their ability to ingest acetylated low density lipoprotein and bind *Ulex europaeusagglutinin 1* plant lectin. The different cell types may also be characterized by flow cytometry for surface immunophenotype. Late EPCs display markers CD 34, CD 133, VEGFR2, CD 31 and are negative for CD 45.

4.2 Benefit in wound healing

Topical and systemic EPC therapy is beneficial in wound healing. The predominant mechanism is the augmentation of angiogenesis and neo-vascularisation. Suh *et al.* reported that EPC therapy increased recruitment of monocytes and macrophages in addition to augmenting angiogenesis. (Suh et al., 2005) This highlights the benefit in early stages of wound healing. It is known that EPCs in wounds result in increased granulation tissue and wound closure. (Asai et al., 2006)It is intuitive that this is the case as a multitude of in vitro studies have shown the production of growth factors and cytokines from EPCs which are closely involved in wound healing. Table 1 presents the in vivo studies of EPC treatment for diabetic ulcers. These studies support the benefit of topical EPC therapy in diabetic wound healing. The mechanism is reported as via paracrine effect, direct incorporation in blood vessels and differentiation into endothelial cells. The field of topical EPC therapy is in the early stages with benefit demonstrated in these studies. Intramuscular EPC therapy has shown benefit in critical limb ischaemia. (Huang et al., 2005) Further research is required to determine the benefit of EPCs delivered in a biomaterial. In addition the standardisation of cell dose, definition of cell type and animal model is required. The use of human cells in immunocompromised animals are required to further elucidate therapeutic efficacy

4.3 Mechanisms of actions

4.3.1 Paracrine effect

Early EPCs and Late EPCs may contribute to post-natal neovascularisation by secretion of angiogenic cytokines and growth factors. The secretome of EPCs contains cytokines and growth factors which stimulate wound healing by increasing proliferation, migration and cell survival of the different cell types required for wound healing i.e. keratinocytes, endothelial cells and fibroblasts. The conditioned media from EPC cultures revealed production of interleukin-8, Stromal-derived factor-1α, vascular endothelial growth factor, platelet-derived growth factor and monocyte chemo-attractant protein-1(Di Santo et al.,2009;Barcelos et al., 2009;Zhang et al.,2009) These cytokines are central to cutaneous wound healing. Extensive secretome analysis can be undertaken using mass spectrometry to determine novel factors involved in EPC biology.(Pula et al., 2009)

4.3.2 Direct incorporation in blood vessels

The second mechanism of action is the direct incorporation of EPCs into the growing blood vessel wall or the differentiation of these cells into mature endothelial cells. This mechanism is associated with late EPCs This mechanism has been shown in animal models and may not be as significant as the paracrine effect of cell therapy. (Di Santo et al., 2009) The comparison of EPC conditioned media as compared to EPC therapy alone for wound healing is important. The transplantation of conditioned media or identified therapeutic factors would allow for protein-based therapy. One study compared conditioned media from EPCs to EPC treatment alone in an animal model of cutaneous wound healing. Injection of EPC conditioned media alone into the same diabetic wound in mice promoted wound healing and increased neovascularization to a similar extent as achieved with EPC transplantation alone.(Kim et al., 2010) However Marrote et al. did not find similar therapeutic efficacy with less wound healing effect from EPC conditioned media.(Marrotte et al., 2010)

4.4 Impaired angiogenesis in diabetes due to EPC dysfunction

It is known that EPCs are decreased in number and dysfunctional in people suffering from diabetes mellitus. The decrease in number of circulating EPCs in people with diabetes is still under investigation but defects in the SDF-1α/CXCR-4 pathway are becoming evident. (Tepper et al., 2010)There are defects in EPC recruitment to wound sites. This is due to decreased mobilisation from the bone marrow and decreased homing to cutaneous wounds. (Brem et al., 2007) With diabetes there is decreased EPC participation in neoangiogenesis and neovsacularisation. Studies show that there are defects in cell migration, adhesion and tube formation. (Tepper et al., 2002) There is also an increase in reactive oxygen species in EPCs isolated from diabetes patients leading to cellular dysfunction. There is a body of evidence indicating that diabetes mellitus related EPC cell dysfunction represents a mechanism for impaired angiogenesis and impaired wound healing seen in diabetic patients.(Marrotte et al.,2010) The obstacle with autologus EPC therapy for diabetic complications is that there is a decreased number of cells available for transplantation. In addition, these autologous cells are dysfunctional.

4.5 Strategies to increase EPC efficacy

4.5.1 Topical delivery

In normal healing EPCs are released into the circulation from the bone marrow in response to ischaemia and travel to sites of tissue injury and participate in angiogenesis. (Takahashi et al., 1999) Diabetes-related vascular dysfunction arises from impairments in EPC mobilisation and homing to sites of ischaemia and cutaneous wounds. This has been shown in animal models of diabetic wound healing. In mice with cutaneous wounds and 4 weeks of streptozocin induced hyperglycaemia, the levels of circulating EPCs were unchanged but the levels of bone marrow derived EPCs within the wound granulation tissue were decreased as compared to non-diabetic controls. The bone marrow derived EPCs from diabetic mice showed increased apoptosis and decreased proliferation in diabetic wound tissue as compared to non-diabetic controls. (Albiero et al., 2011) The topical delivery of cells to a wound would overcome this homing defect and in addition would allow for ex-vivo manipulation during the cell isolation process. This ex-vivo manipulation may restore the EPC functional defect and succeed in restoring diabetic wound healing to the non-diabetic phenotype. Systemic delivery of stem cell results in cells being taken from the circulation in the lungs, spleen and liver and not reaching the wound. (Sorrell & Caplan 2010) The high prevalence of peripheral vascular disease in people with disease also inhibits the intravascular delivery of cell to the affect foot ulcer. The topical delivery of cells allows for concentrated doses of cells to be delivered to a skin wound and not become trapped in other sites in the body.

4.5.2 Matricellular proteins: Osteopontin

Osteopontin (OPN) is a matricellular protein and is involved in tissue repair and angiogenesis. These proteins modulate cell function by interacting with cell-surface receptors, proteases, hormones, and other bioeffector molecules, as well as with structural matrix proteins such as collagens (Bornstein, 2009) Decreased OPN is found in EPCs in diabetes mellitus. Dysfunction is reversed by exposure of EPCs to Osteopontin. (Vaughan EE, Liew A et al. 2011 In Press) Osteopontin is involved in angiogenesis. Osteopontin knockout mice have decreased myocardial angiogenesis in response to ischaemia and delayed recovery after hindlimb ischaemia. OPN is involved in wound healing. Wound healing studies in osteopontin knockout mice show more residual debris and less matrix organisation than wildtype mice. (Scatena et al., 2007) OPN expression is associated with enhanced angiogenesis and collagenisation of the wound bed. Delay in diabetic wound healing may arise in part because of the low expression of OPN early in the wound bed after wounding, resulting in the reduced migration of immune cells to the site of injury leading to the accumulation of cell debris, decreased recruitment of endothelial cells, delayed angiogenesis and poor matrix organization. (Sharma et al., 2006)

4.5.3 Biomaterials and encapsulated cells

Adhesion to a substrate allows transplanted cell survival over even short time frames, and manipulation of major cellular processes (e.g., migration, proliferation, and differentiation) over longer time scales.(Mooney & Vanderburg 2008) Sufficient numbers of cells do not remain in place when applied to the wound surface.(Falanga, 2007). The use of biomaterials allows for more control in mediating delivery of cells to a wound. Current delivery options include injection of cells, delivery in extra-cellular matrix, delivery on a scaffold and delivery as part of a tissue engineering skin equivalents. (Sorrell & Caplan 2010) Silva *et al.* reported that delivery of EPCs using an alginate scaffold created a depot of endothelial progenitor cells which ensured sustained viability and function of cells in a mouse model of hind-limb ischaemia. This method was more successful than direct injection of cells alone. The vascular progenitor cells exit the biomaterial over time and repopulate damaged tissue and participate in the vascular network. (Silva et al., 2008) Cell encapsulation using biomaterials holds promise for both autologous and allogeneic cell therapy. The potential benefit of cell encapsulation with biomaterials includes sustained viability , the ability of the cell to avoid immune rejection, secrete therapeutic proteins and protect against mechanical stress(Orive et al., 2003;Freimark et al., 2010) Encapsulation of adult mesenchymal stem cells permits cell survival, proliferation and differentiation.(Anderson et al., 2011)

4.5.4 Co-culture, gene therapy and hyperoxia

It is hypothesised that endothelial progenitor cells act as angiogenic support cells by their paracrine activity. Co-administration of EPCs with smooth muscle progenitor cells increased vessel density in a mouse model of hind-limb ischaemia to a greater degree than administration of either cell alone. (Foubert et al., 2008) Endothelial cells increase mesenchymal stem cell proliferation. (Saleh et al. 2010) Gene therapy may rescue diabetic EPC dysfunction. Using an ex vivo gene transfer strategy, EPC cell cultures can serve as gene carriers and function as a temporal local production unit of de novo synthesized growth factors within the wound or skin replacement. (Dickens et al., 2010)Increased reactive oxygen species and oxidative stress has been shown to give rise to the dysfunction of diabetic EPCs, leading to inhibition of cell proliferation, nitric oxide production, matrix metalloproteinase-9 activity and migration. Manganese superoxide dismutase gene therapy reverses this dysfunction restoring the cells ability to mediate angiogenesis and wound repair.(Marrotte et al., 2010) Hyperoxia increases nitric oxide mediated EPC activity. (Gallagher et al., 2007) The diabetes related dysfunction in hypoxia inducible factor -1α which reduces vascular endothelial growth factor production (required for EPC activity) can be reversed by topical wound administration of the iron chelating agent desferoxamine. (Thangarajah et al., 2010)

4.5.5 Increase number of EPCs

Increasing EPC number for topical treatment increases the wound healing benefit of EPCs. (Marrotte et al., 2010) Granulocyte macrophage-colony stimulating factor (GM-CSF) increases monocyte derived peripheral blood EPCs. In-vitro animal studies reveal that proliferation of EPCs derived from the bone marrow can be accelerated by GM-CSF. (Wang et al., 2009) GM-CSF is routinely used in the patients receiving chemotherapy. It has been used in human clinical trials for investigation of autologous therapy in critical limb ischaemia. (Huang et al., 2005) In diabetic patients medications such as statins and angiotensin-converting enzyme inhibitor therapy can increase EPC number. (Liew et al., 2008)

Wound Model	EPC type	Delivery	Results	Mechanism	Ref.
Diabetic immuno-deficient mouse Ischemic ulcer	Human fetal $CD133^+$ progenitor cells	Topical type 1 collagen seeded with EPCs	↑ wound closure ↑ angiogenesis	Paracrine signalling	{Barcelos, et al. 2009}
Diabetic Mouse Full thickness ulcer	$CD34^+$ EPCs	Intradermal injection	↑ wound closure. ↑epithelial coverage ↑ vascularisation	Not addressed	{Sivan-Loukianova et al., 2003 }
Diabetic Mouse full thickness ulcer	bone marrow derived $CD34^+$ EPCs	Intradermal injection	↑vascularisation ↑wound closure	Paracrine signalling	(Stepanovic et al., 2003)
Diabetic immuno-deficient mouse Full thickness ulcer	Human umbilical cord blood EPCs	Intradermal injection of EPCs and Topical EPC-CM	↑angiogenesis ↑ wound closure. Conditioned media showed therapeutically equivalent effect	Paracrine signalling	{Kim et al., 2010}
Genetically Diabetic mouse full thickness ulcer	Early EPCs	Topical delivery of genetically modified EPCs	↑wound closure ↑angiogenesis ↑ benefit with gene therapy and ↑ cell dose	Paracrine signaling EPCs present in capillaries	(Marrotte, et al.)
Diabetic mice full thickness cutaneous wounds	Lineage Negative progenitor cells (EPCs)	topically applied in a collagen scaffold	↑Wound Closure ↑ Vascular density	Differentiate into endothelial cells	(Lin et al., 2008)
Human diabetic critical limb ischaemia and foot ulceration	Autologous GM-CSF mobilized EPCs	Intra-muscular injections	Ulcer healing,	↑vessel density	(Huang et al., 2005)

Table I. Animal and human trials of EPC therapy for diabetic wounds

5. Mesenchymal Stem Cells (MSCs)

MSCs are adult fibroblast-like cells that differentiate along multiple mesenchymal pathways when exposed to appropriate stimuli. They adhere to tissue culture plastic and express cell surface markers for CD 105, CD 73, CD 90, and fail to express cell surface markers for CD 45, CD 34, CD 14, CD 11b, CD 79a and CD 19.(Sorrell & Caplan 2010) MSCs were originally isolated from bone marrow by Friedenstein *et al.* in 1968.(Friedenstein et al., 1968) They may also be known as fibroblast colony forming units, marrow stromal cells, multipotent adult progenitor cells, connective tissue progenitor cells or multipotent mesenchymal stromal cells. MSCs may be found in almost all postnatal organs and tissues, including adipose, periosteum, synovial membrane, synovial fluid, muscle, dermis, deciduous teeth, pericytes, trabecular bone, infrapatellar fat pad, articular cartilage and umbilical cord blood. (Si, et al., 2011) Stem cells located outside of the bone marrow are generally referred to as "tissue stem cells". Tissue stem cells are located in sites called niches, which differ among various tissues e.g. a stem cell niche in the bulge area of hair follicles. (Cha & Falanga 2007)

5.1 MSC treatment and wound healing

The complex pathology of diabetic foot ulceration requires that novel treatments are developed. The factors which are central to ongoing ulceration include poor blood supply, inflammation and decreased functioning of resident wound healing cells. MSC treatment has been shown to augment angiogenesis, suppress inflammation and augment wound healing cell functions. The focus of this review is the topical application of MSCs directly to the wound. There have been animal and human studies showing benefit of MSC therapy in the treatment of cutaneous wounds. (Fu & Li 2009) Table 2 details the animal and human trials investigating topical MSC therapy in diabetic wounds. Topical MSC therapy is further advanced than EPC therapy. The in vivo studies in table 2 demonstrate that topical delivery of MSCs result in benefit in diabetic animal cutaneous wounds. It is clear that augmented wound repair occurs by differentiation of MSCs to cells with keratinocyte markers and paracrine mediated increases in angiogenesis and vessel density. Human studies although with a small number of patients have shown benefit with several treatments. Further evidence is required from human cells in immunocompromised animal models to assess wound healing response. Standardisation in wound healing endpoints in both human and animal studies will allow comparison of effect between MSCs and modified MSCs. More research is required on the benefit of cells delivered using biomaterials.

Previous reports have investigated the benefit of topically applied fresh autologous bone marrow to wounds and have not been included in the table. In response to wounding and ischaemic conditions there is a mobilisation and homing of bone marrow MSCs to the wound. MSCs can undergo differentiation and act in a paracrine manner to reduce inflammation, stimulate angiogenesis and cause proliferation and migration of other cell types involved in wound healing. The MSC secretome is of central importance in realising the beneficial paracrine effects of the cells.

5.2 MSC: Mechanisms of action

5.2.1 Differentiation

MSCs may differentiate into mesodermal tissue including osteocytes, chondrocytes and adipocytes. They can differentiate into several cell types including cardiomyocytes, vascular endothelial cells, neurons, hepatocytes and epithelial cells, making them a potential cell based treatment for human disease.(Volarevic et al. 2011) Allogeneic green fluorescent protein labelled bone marrow-derived MSCs have been applied directly to and injected around a cutaneous wound. MSC treatment accelerated wound closure, with increased re-epithelialisation, cellularity and angiogenesis. In the wound the MSCs expressed keratinocyte-specific protein keratin and formed glandular structures suggesting MSCs contribute to tissue regeneration by differentiating into keratinocytes.(Wu et al., 2007) MSCs differentiate into epidermal keratinocytes in vivo and in-vitro and also into skin appendages{Sasaki et al., 2008;Li et al., 2006).

5.2.2 Migration/Homing of MSCs

Bone marrow-derived MSCs contribute to cutaneous wound healing. The homing mechanisms are complex. Potential mechanisms include specific receptors or ligands undergoing up-regulation in response to injury. This not only facilitates trafficking, adhesion and infiltration of MScs but also provide MSCs with a specialised niche to support self-renewal and maintain pluripotency.(Si et al., 2011) MSCs become arrested in blood vessels of injured or ischaemic tissues and secrete a variety of growth factors and cytokines beneficial for wound healing.(Karp & Leng Teo 2009)

5.2.3 Paracrine effects of MSCs

MSCs act in a paracrine fashion to exert their beneficial effects. MSC-conditioned media medium augments wound repair with accelerated epithelialisation. (Wu et al, 2007) The analysis of MSC conditioned media revealed cytokines and growth factors required for wound healing. Vascular endothelial growth factor-a, Insulin like growth factor-1, epidermal growth factor, keratinocyte growth factor, angiopoietin-1, stromal derived factor-1, macrophage inflammatory protein-1, alpha and beta erythropoietin were increased in MSC conditioned media when compared to dermal fibroblast conditioned media. Bone marrow-derived MSC conditioned medium attracts macrophages and endothelial progenitor cells to wounds.(Chen et al., 2008) MSC paracrine signaling has potential beneficial effects on angiogenesis, epithelialisation and fibro-proliferation during wound repair (Hocking & Gibran 2010) Wu *et al.* reported that BM-MSC treated diabetic wounds had increased capillary density, but the bone marrow-derived MSCs were not found in the new capillary structures. This paracrine effect was supported by analysis of the conditioned media which revealed high levels of VEGF-α and angiopoeitin-1 with increased endothelial tube formation. (Wu et al., 2007)

5.2.4 Immunomodulation

An important characteristic of MSCs is that they express low levels of major histocompatibility complex-I (MHC-I) molecules and do not express MHC-II molecules, CD 80, CD 40 or CD 86 on their cell surface.(Zhang et al., 2010) This allows for allogeneic transplantation as MSCs. Human clinical trials have been conducted using allogeneic MSCs for the treatment of many conditions including graft-versus-host disease, type-1 diabetes, ischaemic heart disease, and neurological disorders e.g. stroke. MSCs possess immunosuppressive and anti-inflammatory properties in vitro and in vivo. They may suppress the proliferation and function of the innate and adaptive immune response and the immunomodulatory functions may occur by direct cell-cell contact or by paracrine means.(Zhang et al., 2010) Macrophages are a fundamental cell type in wound healing and immunity. They can be classified as having a pro-inflammatory M1 phenotype or polarisation and an anti-inflammatory M2 or wound healing phenotype. MSCs are capable of eliciting M2 polarisation of macrophages which contributes to marked acceleration of wound healing (Zhang et al., 2010)

5.3 Optimising MSC therapeutic effect

The high proliferation capacity of MSCs mean that there is less dose limiting obstacles with MSC therapy. The allogeneic treatment allows for an "off-the-shelf" product. This is possible as the cells maybe cryopreserved for use in the future. MSCs are amenable to ex-vivo manipulation by gene therapy to provide cellular protection in an ischaemic environment. (McGinley et al., 2011). Highly concentrated cell doses can be directly applied to the wound surface or adjacent to the wound and delivery can be mediated using biomaterials. (Sorrell & Caplan 2010) As is the case with EPCs, biomaterials ensure sustained viability of cells and cell encapsulation technology may protect cells from mechanical stress common in diabetic foot ulceration. (Anderson et al., 2011;Orive, et al. 2003) Table 2 summarises the published research, and includes studies showing the benefit of MSCs on wound healing. There is also a need to better understand the stem cell niche involved in diabetic cutaneous wounds. This is required as this niche is the necessary microenvironment for controlling stem cell fate. Tissue engineering should provide both cells and adequately functionalised biomaterials in order to restore the elements of the stem cell niche. (Becerra et al., 2010)

Wound	MSC type	Delivery	Results	Mechanism	Ref.
Diabetic Mouse ulcers Human chronic ulcers DFU, n=1	Autologous Bone Marrow-Derived MSCs (BM-MSCs)	Topical Fibrin spray	↑ Wound Closure in mice and humans. No adverse events	↑ elastin fibres in MSC treated wound	(Falanga et al 2007)
Human chronic ulcers DFU, n=2	Autologous BM-MSCs	Collagen sponge with silicone film	Healing of wounds in 18 of 20 patients	↑ fibrous, fat and vascular tissue	(Yoshikawa et al.,2008)
Human DFU, n=1	Autologous BM-MSC	Fresh Bone marrow isolate applied to wound then covered with collagen seeded with MSCs	↓wound size with closing and healing of ulcer.	N/A	(Vojtassak et al., 2006)
Human chronic wounds DFU, n=6	Autologous BM-MSCs + standard wound dressing	MSCs injected in and around ulcer, and ulcer covered by dressing	↓ ulcer size at 12 weeks	Increased inflammatory cells and capillary proliferation	(Dash et al., 2009)

Wound	MSC type	Delivery	Results	Mechanism	Ref.
Diabetic rats Full thickness wounds	BM-MSCs transfected with hepatocyte growth factor	Direct injection to wound dermis	Decreased wound healing time with adHGF MSCs	↑blood vessels ↓ collagen formation, ↓ AGEs with AdHGF MSCs	(Ha et al., 2010)
Diabetic mouse with full thickness ulcer	Allogeneic BM-MSCs	Topical application and injection around wound edge	↑wound closure ↑epithelia ↑cellularity ↑angiogenesis	Differentiate MSCs to keratinocytes Paracrine ↑angiogenesis	(Wu et al., 2007)
Diabetic mouse Full thickness ulcer	ATSC over-expressing SDF-1	Topical cell application to wound	↑ % wound closure ↓epithelial gap, ↑ cellularity	Differentiation and paracrine effect on wound cells	(Di Rocco, et al. 2010)
Diabetic Mouse Full thickness ulcer	Diabetic MSCs co-applied with14S,21R - diHDHA	Topical MSCs applied to wound bed and injected intra-dermally	↑ epithelium ↑ GT	↑angiogenesis due to paracrine effect	(Tian et al., 2011)
Diabetic mouse Full thickness Ulcer	Umbilical cord-MSCs	Topical MSCs or systemic MSCs injection	↓ wound size with topically applied MSCs	TGF-β Paracrine effect	(Tark et al., 2010)
Diabetic mouse Full thickness ulcer	Autologous ATSC	Topical delivery using collagen scaffold	↑GT ↑epithelium ↑ no, capillary	Paracrine	(Nambu et al., 2009)
Diabetic Mouse ulcer	Allogeneic BM-MSCs	Topical Delivery	↑epithelium ↑ GT ↑ blood vessels	Paracrine	(Javazon et al., 2007)
DFU = Diabetic Foot Ulcer, BM = Bone Marrow, AGE = Advanced Glycation Endproducts ATSC = Adispose Tissue-derived stromal cells, GT = Granulation Tissue					

Table II. Animal and human trial on Topical MSC treatment of diabetic wounds

6. Biomaterial scaffolds for cell therapy in diabetic wound healing

6.1 Benefit of cell delivery using scaffolds for cell therapy

As explained above, a limitation of systemic delivery of stem cells is the poor engraftment efficiency to the target site, specifically to the wound. It is known that cell infusions e.g. into ischaemic muscle, typically result in > 90% of cells rapidly dying. (Silva et al., 2008) Therefore some of the failures experienced in clinical cell transplantation may directly arise from the manner of administration of the cells rather than a lack of intrinsic bioactivity of the cells. (Silva et al., 2008) The use of a matrix is vital to the integrity of cell maintenance and growth because cells are anchorage dependent and require an appropriate milieu of mechanical strength, material support, controlled porosity and interconnected channelling (Yang et al., 2002)

6.2 Determining the optimal biomaterial for topical treatment of diabetic wounds

The goal of developing novel wound healing treatments is to reduce the time to complete wound closure and restore the barrier function of the skin. The ideal qualities of a skin substitute for diabetic ulcer wound repair is that it will be clinically effective, safe to the patient, inexpensive, easy to use, readily available, durable and encourage cell-matrix interactions. The ideal biomaterial should support reconstruction of new tissues without inflammation. (Huang & Fu 2010)

There is a multitude of biomaterials for wound treatments commercially available and undergoing research. They may have different physicochemical profiles with differing mechanical and degradation properties. They may be synthetic or natural. Natural biomaterials are generally considered more biocompatible and similar to the host extra-cellular matrix. The drawback of synthetic biomaterials is their lack of cellular recognition signals. (Huang & Fu 2010) Skin substitutes can be classified based on **1.** anatomical structure (dermal, epidermal, dermo-epidermal), **2.** duration of cover (permanent, semi-permenant, temporary), **3.** type of biomaterial (biological: autologous, allogeneic, xenogeneic or synthetic: biodegradable, non-biodegradable), **4.** skin substitute composition (cellular, acellluar) and **5.** Where primary biomaterial loading with cellular components occurs(in vitro, in-vivo).(Shevchenko et al., 2010) There are techniques used for development of tissue engineered ulcer healing products. These include 1. Transplantation of cells without matrix or scaffold, 2. Transplantation of biomaterials alone or with the addition of proteins e.g. cytokines and 3. Transplantation of cells in a 3-D scaffold. (Jimenez and Jimenez 2004)

6.3 Currently available cell-based biomaterial dressings for wound healing

The focus of this chapter is on cell-based treatments using a 3-D scaffold. There are several terms that encompass such skin substitutes i.e. tissue-engineered skin, tissue engineered skin constructs, skin substitute bioconstructs, bioengineered skin, living skin replacements and living skin equivilants. (Shevchenko et al., 2010) The gold standard skin replacement treatment for many conditions has been full-thickness skin grafting. There are inherent risks associated with autologous grafts e.g. donor site pain, scarring and infection or delayed healing and failure of graft at recipient site. The risks with non-autologous skin grafts include immune rejection and infection transmission. (Wu et al., 2010) A disadvantage of the currently available cell-based topical therapies is that they do not address the lack of angiogenic properties of the skin substitute. This is important as the successful ability of a skin graft to take to an ulcer is an adequate vascular supply. Table III summarises some of the commercially available skin substitutes and the clinical indications for their use. Apligraf and Dermagraf are temporary treatments for non-healing diabetic ulcers. These skin substitutes are biomaterials seeded with keratinocytes and/or fibroblasts. They are indicated as a topical treatment for non-healing diabetic ulcers in the USA.

6.4 Collagen as a biomaterial

Collagen is the major extra-cellular matrix protein of the dermal layer of the skin. It forms an intrinsic part of blood vessels and supports angiogenesis. It is a commonly used biomaterial for topical cell based wound dressings e.g Apligraf (Organogenesis). It displays low antigenicity with purification techniques available to eliminate the immunogenic telo-peptides.(Huang & Fu 2010) Collagen is appropriate for temporary dressings as it is mechanically weak and undergoes degradation on implantation.(Huang & Fu 2010) It is possible to manipulate collagen by cross-linking and enhance its physico-chemical properties. There are widely used commercial collagen based dressings for diabetic foot ulcers (e.g Promogram, which contains oxidised regenerated cellulose by Johnson & Johnson). (Zhong et al., 2010) Integra (LifeSciences) is a wound healing product consisting of bovine type 1 collagen cross-linked with chondroitin-6-sulphate which is bonded to a silicone membrane. It acts as a template for fibroblast migration and capillary growth in vivo. (Zhong et al., 2010) We have successfully seeded stem and progenitor cells in a collagen scaffold. Figure 2 is a scanning electron microscope image of EPCs and MSCs seeded in a collagen scaffold for 24 hours.

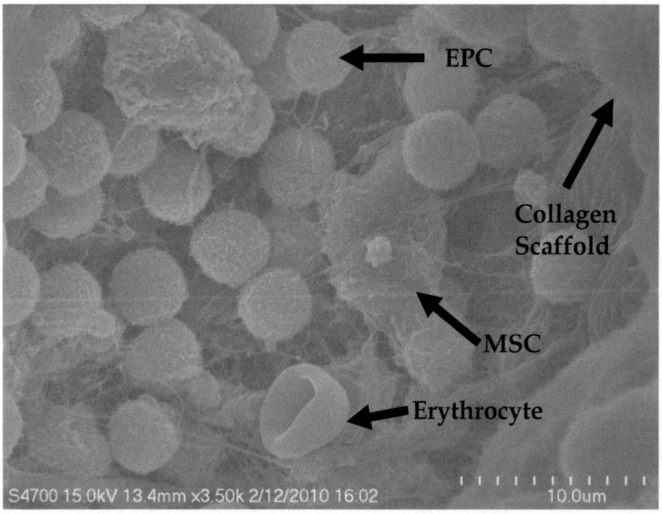

Fig. II. Scanning electron microscope of co-culture of mesenchymal stem cells and early endothelial progenitor cells in a type 1 bovine collagen scaffold.

Product	Description	Indication
Apligraft /Graftskin Organogenesis Canton, MA, USA	Allogeneic neonatal foreskin keratinocytes and fibroblasts seeded in a type 1 bovine collagen	Diabetic foot ulcers venous leg ulcers Partial thickness burns Epidermolysis Bullosa
Dermagraft Advanced Biohealing Inc Lojalla, Ca, USA.	Allogeneic neonatal fibroblasts seeded in a polyglycolic acid (Dexon) or polyglactin-9-10Vicryl scaffold.	Full thickness DFU Epidermolysis Bullosa
TissueTech Autograft system. Laserskin and Hyalograft Fidia Farmaceutical Abano Terme Italy	Autologous fibroblasts and keratinocytes cultured on a hyaluronic acid laser perforated membrane	DFU and Chronic wounds
Epicel Genzyme Biosurgery Cambridge, MA, USA	Autologous keratinocytes and xenogenic proliferation-arrested mouse fibroblasts in petroleum gauze dressing	Full thickness burns burns taking >30% of body area
Transcyte Advanced Biohealing Inc, Lojolla California	Human allogeneic fibroblasts cultured on a nylon mesh pre-coated with collagen	Burns Transparent dressing
Orcel Ortec International New York ,NY USA	Type 1 Bovine collagen seeded with allogeneic neonatal fibroblasts and keratinocytes	donor sites for autografting, DFU Epidermolysis Bullosa
Epidex Modex Therapeutics Luzanne Switzerland	Cultured epidermal skin equivalent derived from keratinocyte precursors of human hair follicles	Chronic Leg uclers
Myskin Altrika Sheffield UK	Autologous keratinocytes grown on a silicone layer with irradiated murine fibroblasts	Non-healing wounds DFU, Burns, Pressure ulcers
Bioseed-S BioTissue Technologies Freiburg, Germany	Autologous keratinocytes resuspended in a fibrin sealant	Venous leg ulcers
Permaderm Regenicin www.regenicin.com	Autologous keratinocytes and fibroblasts seeded on collagen biomaterial	Burns Chronic Wounds
DFU = Diabetic foot ulcers		

Table III. Sample of currently available Cell-Scaffold skin replacement therapies and their indications

7. Translation to human therapy

7.1 Safety and regulatory approval

With any new cell-based therapy, it is mandatory to ensure safety for the patient. Any negative toxic side-effect of cell-based therapies would be a set back for the field of tissue engineering and regenerative medicine. In Europe, the European Medicines Agency (EMA) controls regulation and clinical trials of new cell based products. In North America, this process is under the remit of the Food and Drugs Administration (FDA). The EMA also advises on the development of stem cell products which are an example of an advanced therapy medicinal product (ATMP). In February 2011, the EMA published a document entitled "Reflection paper on stem cell-based medicinal products", highlighting the current situation in the field of stem cell therapy. (EMA 2011) Safety and clinical efficacy is first proven by scientifically robust methodology in pre-clinical studies. It is required that the product is produced and clinical trials carried out according to international standards. These standards include GLP (good lab practice), GMP (good manufacturing practice), and GCP (Good clinical Practice). There is a requirement for quality checks in the manufacturing process. This includes analysis of cell treatment batches to ensure cell quality, identity, viability and traceability of cells. The goal is a robust, stringently controlled production and manufacturing process.

7.2 Preclinical animal models: choice of model and regulatory issues

It is necessary to prove treatment efficacy in an animal model. An in vitro wound healing model is not sufficient to confirm treatment efficacy. The complexity of diabetic foot ulceration with its multi-factorial pathology cannot be realised in an animal model. There are over 10 different animal models of diabetic ulceration in the reported literature. There are inherent differences between animals and humans. These include cutaneous anatomy, vascular supply, duration of diabetes and the presence of other cardiovascular risk factors e.g. smoking.

In addition there are a myriad of endpoints reported in animal wound healing studies. The most robust clinically relevant wound healing endpoints are percentage wound closure and time to complete healing. The myriad of new treatment modalities under investigation have effects on different phases of the wound healing spectrum. The pig has skin felt to be the most close to humans but these are large expensive animals. The genetically modified, leptin receptor deficient diabetic mouse is widely used as a model of type 2 diabetes, but wound healing occurs by contraction in this model and does not reflect the human situation. The rabbit ear dermal ulcer model is a powerful model for examining re-epithelialisation and granulation tissue formation in an excisional wound. (Breen et al., 2008) A comprehensive review by Lammers et al. recommends a more systematic evaluation of tissue-engineered constructs in animal models to enhance the comparison of different constructs, accelerating the trajectory to application in human patients. (Lammers et al., 2010)

The EMA provides advice on the animal models to use for translation of cell-based therapy to humans. The choice of the most relevant animal model should be determined by the specific safety aspect to be evaluated. It advises the use of human cells to be tested in proof of concept and safety studies. This methodology requires the use of immuno-compromised models either genetically immuno-suppressed or treated with immuno-suppressants.(EMA 2011) The persistence of cells and the functionality of the cells should be assessed. The potential of undifferentiated pluripotent stem cells to form tumours and

be genetically unstable due to ex-vivo manipulation requires this to be assessed in animal models. This is more likely with embryonic stem cells and pluripotent stem cells. Bio-distribution of cells to other organs and ectopic tissue formation need to be investigated. Prior to first-in-man studies, there are guidelines published by the EMA to identify and mitigate risks. Dose finding studies, immunological, pharmacokinetic, pharmco-dynamic and long term pharmaco-vigilant studies should be undertaken and planned. (EMA 2011)
The use of biomaterials in conjunction with stem and progenitor cells is defined by the EMA as a 'tissue-engineered product' and falls under the term ATMP.(EU 2007) The experience with the development of allogeneic bi-layered skin has provided valuable information on the development of skin replacement therapy. Apligraf (Organogenesis), a living bi-layered skin substitute has received approval from FDA. It is described as a Class III medical device via premarket approval and meets requirements for a human cell, tissue, cellular and tissue-based product. As the product is made from viable human skin cells, it cannot be terminally sterilized, but safety concerns have been addressed. These include risk of transmission of infection, immunogenicity, immunological graft rejection and tumour formation. As cells are derived from neonatal foreskin, maternal blood of the neonatal donor and the cell banks are thoroughly screened for infectious agents, pathogens and other contaminants.(EU 2007;Wu et al., 2010)

7.3 Structured diabetic foot care

Stem and progenitor cell-based topical treatments will not be used in isolation to treat diabetic foot ulceration. Ideally, these advanced biological treatments will be part of a treatment algorithm, which would see the implementation of standard care prior to use of cell therapy. If the restoration of vascular supply, removal of pressure, control of infection and debridement of the wound does not succeed in ulcer healing, then the indication for cell based therapy would apply. There are analyses of factors associated with lack of healing with fibroblast dermal substitutes. An episode of infection during 12 weeks of treatment was associated with a 3.4 times increased risk of non-closure of a wound. (Wu et al., 2010) High bacterial load in the wound negatively affects wound healing with Dermagraft and Browne et al. recommend reducing the bacterial load with combination antibiotics prior to the application of skin substitutes. (Browne et al., 2001) New treatment modalities are under investigation which may augment wound healing and reduce bacterial load. Plasma therapy may reduce bacterial burden and enhance wound healing. (Heinlin et al., 2010)

7.4 Cost: Benefit analysis

To ensure development of a successful topical cell based therapy, the product must have potential widespread use in the clinical arena. It must demonstrate clinical efficacy in clinical trials. In randomised controlled clinical trials the new product must show superiority both in comparison to standard care and to other market leaders in the field. It is expensive to conduct human clinical trials, therefore the product must demonstrate favourable health economics so as to be attractive to health care providers and industrial partners. To gain market access, manufacturers have to establish not only the efficacy of the product but also whether the product provides a cure at an acceptable cost per unit of health gain. (Langer et al., 2009) Several studies have investigated the cost-effectiveness of these products. The results feature favourable cost-effectiveness ratios in selected patient groups with chronic wounds. The cost of the product and product development should be offset against the total cost of care of the patient with a non-healing diabetic foot ulcer. (Langer et al., 2009) There is a need for high quality clinical trials in this area.

8. Cell-based therapies in other dermatological conditions

MSCs and EPCs have the potential to treat other dermatological conditions apart from diabetic foot ulceration. As seen in table III there are several conditions which may be suitable for these therapies including chronic venous and pressure ulcers, burns and epidermolysis bullosa. The economic burden of chronic wounds is potentially the largest burden on healthcare systems. Stem and progenitor cells may be used as orphan medications for life-threatening or extremely rare debilitating conditions. These drugs are not developed by large pharmaceuticals and are not subject to the same regulatory process. An example of this is the blistering disorder epidermolysis bullosa. In addition research into basic stem cell biology will elucidate mechanisms of action of stem cells which may guide the development of future therapies. The development of successful skin regeneration and elucidation of key molecules and biological systems will allow for scar free repair and increased strength of healed wounds. There are further exciting developments in the field of stem and progenitor cell therapy for tissue regeneration. Hair follicle biology is important for skin biology and epidermal haemostasis. There are resident stem cells in the bulge area of the hair follicle which are required for re-epithelialisation during wound healing. (Wu et al., 2011) They are a readily isolatable source of adult stem cell suitable for autologous therapy.(Amoh et al., 2010)

9. Conclusions

This book chapter has reviewed the current state of Stem and Progenitor cell therapy for non-healing diabetic foot ulceration. The urgent clinical need for developing improved novel cell treatments is stressed. The scientific basis for potential success with topical stem and progenitor therapy is reviewed. The advantage of using biomaterials to mediate cell delivery is discussed. Further developments in tissue engineering will provide more intelligent biomaterials which ensure better viability and control of stem cell fate and function. The logistical hurdles to translation of bench-side discoveries are reviewed and information provided on accelerated development of these advanced medicinal products.
The importance of translational science is being recognised as a key driver to the realisation of basic science discoveries for humans. There are strategic efforts to translate basic science to clinical benefit. This bench-to-bedside approach is the focus of government policies throughout the world with collaborations developing between pharmaceutical and biotechnology industries, academia and clinicians. The success of treatments will rely on clinical efficacy, safety, ease of use and cost-effectiveness. The potential to translate this technology to a variety of clinical dermatological disorders increases the attractiveness for industrial investment for further research and development of these products. A central component to the successful translation of this treatment will be the performance of robust randomised controlled trials. Stem cell therapy is a new field encompassing both tissue engineering and regenerative medicine science and holds promise for the improved treatment of diseases which are suboptimally managed with current therapies.

10. Acknowledgement

This work was supported from Molecular Medicine Ireland who funded Aonghus O'Loughlin for a clinical scientist research program.

11. References

Albiero, M., L. Menegazzo, et al. (2011) "Defective recruitment, survival and proliferation of bone marrow-derived progenitor cells at sites of delayed diabetic wound healing in mice." *Diabetologia* 54(4): 945-53.

Amoh, Y., K. Katsuoka, et al. (2010) "The advantages of hair follicle pluripotent stem cells over embryonic stem cells and induced pluripotent stem cells for regenerative medicine." *J Dermatol Sci* 60(3): 131-7.

Anderson, S. B., C. C. Lin, et al. (2011) "The performance of human mesenchymal stem cells encapsulated in cell-degradable polymer-peptide hydrogels." *Biomaterials* 32(14): 3564-74.

Asahara, T., T. Murohara, et al. (1997). "Isolation of putative progenitor endothelial cells for angiogenesis." *Science* 275(5302): 964-7.

Asai, J., H. Takenaka, et al. (2006). "Dibutyryl cAMP influences endothelial progenitor cell recruitment during wound neovascularization." *J Invest Dermatol* 126(5): 1159-67.

Barber, C., A. Watt, et al. (2008). "Influence of bioengineered skin substitutes on diabetic foot ulcer and venous leg ulcer outcomes." *J Wound Care* 17(12): 517-27.

Barcelos, L. S., C. Duplaa, et al. (2009). "Human CD133+ progenitor cells promote the healing of diabetic ischemic ulcers by paracrine stimulation of angiogenesis and activation of Wnt signaling." *Circ Res* 104(9): 1095-102.

Becerra, J., L. Santos-Ruiz, et al. (2010) "The Stem Cell Niche Should be a Key Issue for Cell Therapy in Regenerative Medicine." *Stem Cell Rev.*

Blozik, E. and M. Scherer (2008). "Skin replacement therapies for diabetic foot ulcers: systematic review and meta-analysis." *Diabetes Care* 31(4): 693-4.

Bornstein, P. (2009). "Matricellular proteins: an overview." *J Cell Commun Signal* 3(3-4): 163-5.

Botusan, I. R., V. G. Sunkari, et al. (2008). "Stabilization of HIF-1alpha is critical to improve wound healing in diabetic mice." *Proc Natl Acad Sci USA* 105(49): 19426-31.

Boulton, A. J., D. G. Armstrong, et al. (2008). "Comprehensive foot examination and risk assessment: a report of the task force of the foot care interest group of the American Diabetes Association, with endorsement by the American Association of Clinical Endocrinologists." *Diabetes Care* 31(8): 1679-85.

Boulton, A. J., L. Vileikyte, et al. (2005). "The global burden of diabetic foot disease." *Lancet* 366(9498): 1719-24.

Boyko, E. J., J. H. Ahroni, et al. (2006). "Prediction of diabetic foot ulcer occurrence using commonly available clinical information: the Seattle Diabetic Foot Study." *Diabetes Care* 29(6): 1202-7.

Breen, A., G. Mc Redmond, et al. (2008). "Assessment of wound healing in the alloxan-induced diabetic rabbit ear model." *J Invest Surg* 21(5): 261-9.

Brem, H., P. Sheehan, et al. (2004). "Protocol for treatment of diabetic foot ulcers." *Am J Surg* 187(5A): 1S-10S.

Brem, H., P. Sheehan, et al. (2006). "Evidence-based protocol for diabetic foot ulcers." *Plast Reconstr Surg* 117(7 Suppl): 193S-209S; discussion 210S-211S.

Brem, H. and M. Tomic-Canic (2007). "Cellular and molecular basis of wound healing in diabetes." *J Clin Invest* 117(5): 1219-22.

Browne, A. C., M. Vearncombe, et al. (2001). "High bacterial load in asymptomatic diabetic patients with neurotrophic ulcers retards wound healing after application of Dermagraft." *Ostomy Wound Manage* 47(10): 44-9.

Cha, J. and V. Falanga (2007). "Stem cells in cutaneous wound healing." *Clin Dermatol* 25(1): 73-8.

Chen, L., E. E. Tredget, et al. (2008). "Paracrine factors of mesenchymal stem cells recruit macrophages and endothelial lineage cells and enhance wound healing." *PLoS One* 3(4): e1886.

Dash, N. R., S. N. Dash, et al. (2009). "Targeting nonhealing ulcers of lower extremity in human through autologous bone marrow-derived mesenchymal stem cells." *Rejuvenation Res* 12(5): 359-66.

Di Rocco, G., A. Gentile, et al. (2010) "Enhanced healing of diabetic wounds by topical administration of adipose tissue-derived stromal cells overexpressing stromal-derived factor-1: biodistribution and engraftment analysis by bioluminescent imaging." *Stem Cells Int* 2011: 304562.

Di Santo, S., Z. Yang, et al. (2009). "Novel cell-free strategy for therapeutic angiogenesis: in vitro generated conditioned medium can replace progenitor cell transplantation." *PLoS One* 4(5): e5643.

Dickens, S., S. Van den Berge, et al. (2010) "Nonviral transfection strategies for keratinocytes, fibroblasts, and endothelial progenitor cells for ex vivo gene transfer to skin wounds." *Tissue Eng Part C Methods* 16(6): 1601-8.

EMA. (2011). "Reflection paper on stem cell-based medicinal products." Retrieved 15 March 2011,fromhttp://www.ema.europa.eu/docs/en_GB/document_library/Scientific_guideline/2011/02/WC500101692.pdf.

EU. (2007). "Regulation (EC) No 1394/2007 of the European Parliament and of the Council on advanced therapy medicinal products and amending Directive 2001/83/ECand Regulation (EC) No 726/2004." Retrieved 10 March 2011, from http://ec.europa.eu/health/files/eudralex/vol1/reg_2007_1394/reg_2007_1394_en.pdf.

Falanga, V. (2005). "Wound healing and its impairment in the diabetic foot." *Lancet* 366(9498): 1736-43.

Falanga, V., S. Iwamoto, et al. (2007). "Autologous bone marrow-derived cultured mesenchymal stem cells delivered in a fibrin spray accelerate healing in murine and human cutaneous wounds." *Tissue Eng* 13(6): 1299-312.

Foubert, P., G. Matrone, et al. (2008). "Coadministration of endothelial and smooth muscle progenitor cells enhances the efficiency of proangiogenic cell-based therapy." *Circ Res* 103(7): 751-60.

Freimark, D., P. Pino-Grace, et al. (2010) "Use of Encapsulated Stem Cells to Overcome the Bottleneck of Cell Availability for Cell Therapy Approaches." *Transfus Med Hemother* 37(2): 66-73.

Friedenstein, A. J., K. V. Petrakova, et al. (1968). "Heterotopic of bone marrow. Analysis of precursor cells for osteogenic and hematopoietic tissues." *Transplantation* 6(2): 230-47.

Frykberg, R. G., T. Zgonis, et al. (2006). "Diabetic foot disorders. A clinical practice guideline (2006 revision)." *J Foot Ankle Surg* 45(5 Suppl): S1-66.
Fu, X. and H. Li (2009). "Mesenchymal stem cells and skin wound repair and regeneration: possibilities and questions." *Cell Tissue Res* 335(2): 317-21.
Gallagher, K. A., Z. J. Liu, et al. (2007). "Diabetic impairments in NO-mediated endothelial progenitor cell mobilization and homing are reversed by hyperoxia and SDF-1 alpha." *J Clin Invest* 117(5): 1249-59.
Gurtner, G. C., S. Werner, et al. (2008). "Wound repair and regeneration." *Nature* 453(7193): 314-21.
Ha, X., Q. Yin, et al. (2010) "[Study on bone marrow mesenchymal stem cells transfected with adenovirus hepatocyte growth factor gene promoting wounds repair in diabetic rats]." *Zhongguo Xiu Fu Chong Jian Wai Ke Za Zhi* 24(12): 1520-4.
Harding, K. G., H. L. Morris, et al. (2002). "Science, medicine and the future: healing chronic wounds." *BMJ* 324(7330): 160-3.
Heinlin, J., G. Morfill, et al. (2010)F "Plasma medicine: possible applications in dermatology." *J Dtsch Dermatol Ges* 8(12): 968-76.
Herber, O. R., W. Schnepp, et al. (2007). "A systematic review on the impact of leg ulceration on patients' quality of life." *Health Qual Life Outcomes* 5: 44.
Hirschi, K. K., D. A. Ingram, et al. (2008). "Assessing identity, phenotype, and fate of endothelial progenitor cells." *Arterioscler Thromb Vasc Biol* 28(9): 1584-95.
Hocking, A. M. and N. S. Gibran (2010) "Mesenchymal stem cells: paracrine signaling and differentiation during cutaneous wound repair." *Exp Cell Res* 316(14): 2213-9.
Huang, P., S. Li, et al. (2005). "Autologous transplantation of granulocyte colony-stimulating factor-mobilized peripheral blood mononuclear cells improves critical limb ischemia in diabetes." *Diabetes Care* 28(9): 2155-60.
Huang, S. and X. Fu (2010) "Naturally derived materials-based cell and drug delivery systems in skin regeneration." *J Control Release* 142(2): 149-59.
Javazon, E. H., S. G. Keswani, et al. (2007). "Enhanced epithelial gap closure and increased angiogenesis in wounds of diabetic mice treated with adult murine bone marrow stromal progenitor cells." *Wound Repair Regen* 15(3): 350-9.
Jeffcoate, W. J. and K. G. Harding (2003). "Diabetic foot ulcers." *Lancet* 361(9368): 1545-51.
Jimenez, P. A. and S. E. Jimenez (2004). "Tissue and cellular approaches to wound repair." *Am J Surg* 187(5A): 56S-64S.
Kim, J. Y., S. H. Song, et al. (2010) "Human cord blood-derived endothelial progenitor cells and their conditioned media exhibit therapeutic equivalence for diabetic wound healing." *Cell Transplant* 19(12): 1635-44.
Lammers, G., P. D. Verhaegen, et al. (2010) "An overview of methods for the in vivo evaluation of tissue-engineered skin constructs." *Tissue Eng Part B Rev* 17(1): 33-55.
Langer, A. and W. Rogowski (2009). "Systematic review of economic evaluations of human cell-derived wound care products for the treatment of venous leg and diabetic foot ulcers." *BMC Health Serv Res* 9: 115.
Li, H., X. Fu, et al. (2006). "Adult bone-marrow-derived mesenchymal stem cells contribute to wound healing of skin appendages." *Cell Tissue Res* 326(3): 725-36.
Liew, A., J. H. McDermott, et al. (2008). "Endothelial progenitor cells for the treatment of diabetic vasculopathy: panacea or Pandora's box?" *Diabetes Obes Metab* 10(5): 353-66.

Lin, C. D., A. C. Allori, et al. (2008). "Topical lineage-negative progenitor-cell therapy for diabetic wounds." *Plast Reconstr Surg* 122(5): 1341-51.

Liu, Z. J. and O. C. Velazquez (2008). "Hyperoxia, endothelial progenitor cell mobilization, and diabetic wound healing." *Antioxid Redox Signal* 10(11): 1869-82.

Margolis, D. J., J. Kantor, et al. (1999). "Healing of diabetic neuropathic foot ulcers receiving standard treatment. A meta-analysis." *Diabetes Care* 22(5): 692-5.

Marrotte, E. J., D. D. Chen, et al. (2010) "Manganese superoxide dismutase expression in endothelial progenitor cells accelerates wound healing in diabetic mice." *J Clin Invest* 120(12): 4207-19.

McGinley, L., J. McMahon, et al. (2011) "Lentiviral vector mediated modification of mesenchymal stem cells & enhanced survival in an in vitro model of ischaemia." *Stem Cell Res Ther* 2(2): 12.

Metcalfe, A. D. and M. W. Ferguson (2007). "Bioengineering skin using mechanisms of regeneration and repair." *Biomaterials* 28(34): 5100-13.

Mooney, D. J. and H. Vandenburgh (2008). "Cell delivery mechanisms for tissue repair." Cell *Stem Cell* 2(3): 205-13.

Nambu, M., S. Kishimoto, et al. (2009). "Accelerated wound healing in healing-impaired db/db mice by autologous adipose tissue-derived stromal cells combined with atelocollagen matrix." *Ann Plast Surg* 62(3): 317-21.

O'Loughlin, A., C. McIntosh, et al. (2010) "Review paper: basic concepts to novel therapies: a review of the diabetic foot." *Int J Low Extrem Wounds* 9(2): 90-102.

Orive, G., R. M. Hernandez, et al. (2003). "Cell encapsulation: promise and progress." *Nat Med* 9(1): 104-7.

Pradhan, L., C. Nabzdyk, et al. (2009). "Inflammation and neuropeptides: the connection in diabetic wound healing." *Expert Rev Mol Med* 11: e2.

Pula, G., U. Mayr, et al. (2009). "Proteomics identifies thymidine phosphorylase as a key regulator of the angiogenic potential of colony-forming units and endothelial progenitor cell cultures." *Circ Res* 104(1): 32-40.

Saleh, F. A., M. Whyte, et al. (2010) "Regulation of Mesenchymal Stem Cell Activity by Endothelial Cells." *Stem Cells Dev.*

Sasaki, M, R. Abe et al. (2008) "Mesenchymal stem cells are recruited into wounded skin and contribute to wound repair by transdifferentiation into multiple skin cell type. *J Immunol* 15;180(4):2581-7.

Scatena, M., L. Liaw, et al. (2007). "Osteopontin: a multifunctional molecule regulating chronic inflammation and vascular disease." *Arterioscler Thromb Vasc Biol* 27(11): 2302-9.

Sharma, A., A. K. Singh, et al. (2006). "Differential regulation of angiogenic genes in diabetic wound healing." *J Invest Dermatol* 126(10): 2323-31.

Shevchenko, R. V., S. L. James, et al. (2010) "A review of tissue-engineered skin bioconstructs available for skin reconstruction." *J R Soc Interface* 7(43): 229-58.

Si, Y. L., Y. L. Zhao, et al. (2011) "MSCs: Biological characteristics, clinical applications and their outstanding concerns." *Ageing Res Rev* 10(1): 93-103.

Silva, E. A., E. S. Kim, et al. (2008). "Material-based deployment enhances efficacy of endothelial progenitor cells." *Proc Natl Acad Sci USA* 105(38): 14347-52.

Sorrell, J. M. and A. I. Caplan (2010) "Topical delivery of mesenchymal stem cells and their function in wounds." *Stem Cell Res Ther* 1(4): 30.

Stepanovic, V., O. Awad, et al. (2003). "Leprdb diabetic mouse bone marrow cells inhibit skin wound vascularization but promote wound healing." *Circ Res* 92(11): 1247-53.

Suh, W., K. L. Kim, et al. (2005). "Transplantation of endothelial progenitor cells accelerates dermal wound healing with increased recruitment of monocytes/macrophages and neovascularization." *Stem Cells* 23(10): 1571-8.

Takahashi, T., C. Kalka, et al. (1999). "Ischemia- and cytokine-induced mobilization of bone marrow-derived endothelial progenitor cells for neovascularization." *Nat Med* 5(4): 434-8.

Tark, K. C., J. W. Hong, et al. (2010) "Effects of human cord blood mesenchymal stem cells on cutaneous wound healing in leprdb mice." *Ann Plast Surg* 65(6): 565-72.

Teng, Y. J., Y. P. Li, et al. (2010) "Bioengineered skin in diabetic foot ulcers." *Diabetes Obes Metab* 12(4): 307-15.

Tepper, O. M., J. Carr, et al. (2010) "Decreased circulating progenitor cell number and failed mechanisms of stromal cell-derived factor-1alpha mediated bone marrow mobilization impair diabetic tissue repair." *Diabetes* 59(8): 1974-83.

Tepper, O. M., R. D. Galiano, et al. (2002). "Human endothelial progenitor cells from type II diabetics exhibit impaired proliferation, adhesion, and incorporation into vascular structures." *Circulation* 106(22): 2781-6.

Thangarajah, H., I. N. Vial, et al. (2010) "HIF-1alpha dysfunction in diabetes." *Cell Cycle* 9(1): 75-9.

Tian, H., Y. Lu, et al. (2011) "14S,21R-dihydroxydocosahexaenoic acid remedies impaired healing and mesenchymal stem cell functions in diabetic wounds." *J Biol Chem* 286(6): 4443-53.

Vaughan EE., Liew A, et al (2011 In press) "Osteopontin mediates endothelial progenitor cell secretion of factors to enhance angiogenesis" *Cell Transplantation*.

Vojtassak, J., L. Danisovic, et al. (2006). "Autologous biograft and mesenchymal stem cells in treatment of the diabetic foot." *Neuro Endocrinol Lett* 27 Suppl 2: 134-7.

Volarevic, V., N. Arsenijevic, et al. (2011) "Concise review: Mesenchymal stem cell treatment of the complications of diabetes mellitus." *Stem Cells* 29(1): 5-10.

Wang, Q. R., F. Wang, et al. (2009). "GM-CSF accelerates proliferation of endothelial progenitor cells from murine bone marrow mononuclear cells in vitro." *Cytokine* 45(3): 174-8.

Wild, S., G. Roglic, et al. (2004). "Global prevalence of diabetes: estimates for the year 2000 and projections for 2030." *Diabetes Care* 27(5): 1047-53.

Wu, S. C., W. Marston, et al. (2010) "Wound care: the role of advanced wound healing technologies." *J Vasc Surg* 52(3 Suppl): 59S-66S.

Wu, X., Q. T. Shen, et al. (2011) "Skin stem cells orchestrate directional migration by regulating microtubule-ACF7 connections through GSK3beta." *Cell* 144(3): 341-52.

Wu, Y., L. Chen, et al. (2007). "Mesenchymal stem cells enhance wound healing through differentiation and angiogenesis." *Stem Cells* 25(10): 2648-59.

Yang, S., K. F. Leong, et al. (2002). "The design of scaffolds for use in tissue engineering. Part II. Rapid prototyping techniques." *Tissue Eng* 8(1): 1-11.

Yang, W., J. Lu, et al. (2010) "Prevalence of diabetes among men and women in China." *N Engl J Med* 362(12): 1090-101.

Yoshikawa, T., H. Mitsuno, et al. (2008). "Wound therapy by marrow mesenchymal cell transplantation." *Plast Reconstr Surg* 121(3): 860-77.

Zhang, Q. Z., W. R. Su, et al. (2010) "Human gingiva-derived mesenchymal stem cells elicit polarization of m2 macrophages and enhance cutaneous wound healing." *Stem Cells* 28(10): 1856-68.

Zhang, Y., D. A. Ingram, et al. (2009). "Release of proinflammatory mediators and expression of proinflammatory adhesion molecules by endothelial progenitor cells." *Am J Physiol Heart Circ Physiol* 296(5): H1675-82.

Zhong, S. P., Y. Z. Zhang, et al. 2010 "Tissue scaffolds for skin wound healing and dermal reconstruction." *Wiley Interdiscip Rev Nanomed Nanobiotechnol* 2(5): 510-25.

24

Regeneration of the Kidney - Viewed from ES Cell

Hirotsugu Iwatani[1] and Enyu Imai[2]
[1]*Department of Geriatric Medicine and Nephrology, Osaka University Graduate School of Medicine, Suita, Osaka*
[2]*Department of Nephrology, Nagoya University Graduate School of Medicine, Nagoya, Aichi*
Japan

1. Introduction

Embryonic stem (ES) cells are derived from inner cell mass of blastocyst stage of fertilized eggs and remain undifferentiated in the presence of leukemia inhibitory factor (LIF) (Evans M.J & Kaufman M.H. (1981), Martin G.R. (1981).). Depletion of LIF leads to ES cell to form embryoid body (EB) that can differentiate into several different lineage cells such as endodermal, mesodermal and ectodermal lineage cells. Until now, ES cells have been induced to differentiate in vitro into several cell types such as hematopoietic cell (Nakano T.,et al. (1994), Nishikawa S.I.,et al. (1998), Palacios R.,et al. (1995).), pancreatic cell (Soria B.,et al. (2000).), neuronal cell (Brustle O.,et al. (1999), Scholer H.R.,et al. (1991).) and cardiomyocyte (Boheler K.R.,et al. (2002), Doetschman T.C.,et al. (1985), Klug M.G.,et al. (1996).).
To be sure that ES cells are pluripotent, but mere transplantation of non-differentiated ES cells leads to teratomas as Yamamoto reported the teratoma formation after the transplantation of undifferentiated ES cells into retroperitoneum of nude mice (Yamamoto M.,et al. (2006).). This is one of the main obstacles that limit the clinical application of ES cell transplantation therapy. Therefore, it is essential to differentiate ES cells to some degree before transplantation. But the information as to the directed differentiation of ES cells to kidney lineages are limited although spontaneous differentiation of ES cells to various cells is observed. We have gathered pieces of information of the developmental biology and the differentiation conditions of several stem/progenitor cells to matured kidney. In this review, we focus on a dream of the generation of kidney from ES cells or stem/progenitor cells.

2. Developmental biology of the kidney

Mammalian kidney traces three successive stages during the development: pronephros, mesonephros and metanephros. In mammals, pronephros, a rudimentary and temporary structure consisting of one nephron, retrogrades at the first stage of nephrogenesis. Next, mesonephros arises as Wolffian duct from the intermediate mesoderm, elongates along the oral-caudal axis, and forms immature glomerular and tubular structures consisting of several dozens of nephrons. However, both of the structures at this stage are not functional and then retrograde.

Metanephros develops from the caudal part of Wolffian duct and finally produces functional glomeruli and tubules as follows: A single ureteric bud emerges from the caudal part of the retrograding mesonephros and starts to branch. Immature mesenchymal cells condense around the ureteric buds to form the metanephric blastema. Cells around an each tip of the branching ureteric buds and another type of cells surrounding such cells are termed the metanephric mesenchyme and stromal cells, respectively. The former metanephric mesenchymal cells differentiate through mesenchymal-to-epithelial transformation (MET) into various types of epithelial cells including visceral glomerular epithelial cells (podocytes), parietal glomerular epithelial cells, proximal tubular epithelial cells, and distal tubular epithelial cells, and develop into a cylindrical epithelial structure at an each tip of the ureteric buds. At an end of the differentiating epithelial structures, podocytes and parietal glomerular epithelial cells interact with mesangial cells and fenestrated endothelial cells, and form globular networks of capillaries (glomeruli). At the other end, the epithelial structures connect in a one-to-one manner to collecting ducts each of which is derived from one branch of the ureteric buds. This whole structure is called the nephron, a unit to produce urine. The final number of nephrons depends both on the number of branching of the ureteric buds and on the differentiation potentials of the metanephric mesenchyme independently. In human, one kidney has one million nephrons. The interaction in a reciprocal and inductive manner between the metanephric mesenchyme and the ureteric bud plays an important role in the nephrogenesis. A mouse or rat embryonic kidney further develops when placed in culture, but is not functional because vascular networks are not integrated. In contrast, rat kidney rudiments at early stage of the development, which contain the metanephric mesenchyme and the ureteric bud, can form functional nephrons when it is transplanted in the omentum of other living animals (Hammerman M.R. (2004).). These facts indicate that a set of the metanephric mesenchyme and the ureteric bud has the perfect intrinsic program to regulate immature cells, to recruit and integrate vascular components, and to form functional nephrons.

3. Toward the regeneration of the kidney from ES cells

3.1 Differentiation of rodent ES cells into renal lineage

The kidney is a highly terminated and complex organ. It needs to be formed so that the adjacent placement in three dimensions of several components of kidney cells are critical and only work when the placement is perfect unlike hematopoietic cells or insulin-secreting cells working as a cell itself. In this section, we focus on the differentiation trials of murine ES cells into renal lineage. Steenhard B.M. et al showed an integration of ES cells in metanephric kidney organ culture (Steenhard B.M.,et al. (2005).). The authors microinjected murine ROSA26 ES cells that express ubiquitously β-galactosidase when cultured in the presence of LIF to suppress differentiation, into E12 or E13 metanephroi and kept them on organ culture system. ES cell-derived, β-galactosidase-positive cells were found in epithelial structure resembling tubules and, on rare occasions, in structures resembling glomerular tufts. By the analysis of electron microscopy, the ES cell-derived tubules were surrounded by basement membrane and had epithelial microvilli and junctional complexes. Subset of these epithelial tubules expressed markers of *Lotus tetragonolobus* (LTA) and α_1 Na^+/K^+ ATPase. By infecting cytomegalovirus-promoter green fluorescent protein (GFP) adenovirus to ES cells before injection, injected ES cells were traced by GFP from 18h to 48h in cultured kidney, indicating that injected ES cells were definitely alive and functioning in the cultured kidney.

Kobayashi T. et al reported that Wnt4-transformed mouse ES cells differentiated into renal tubular cell in vitro and in vivo. The authors stably transformed Wnt4 cDNA to mouse ES cells (Wnt4-ES cells) and incubated them by hanging drop culture method (Kobayashi T.,et al. (2005).). Hepatocyte growth factor (HGF) and activin-A enhanced the expression of aquaporin (AQP)2 in Wnt4-ES cells and induced tubular-like structure formation in three dimensional culture. The authors confirmed that the addition of anti-Wnt4 antibody in the culture system downregulated the frequency of tubular-like structure formation. When Wnt4-ES cells were transplanted in the renal cortex of four-week-old mouse, they also formed tubular-like structure in teratoma. The authors showed that embryoid bodies from the Wnt4-ES cells had an ability to differentiate into renal tubular cells. The most significant point is that Wnt4 expressing ES cells can be integrated into tubular cells in the adolescent (4 week old) kidney microenvironment which would be less suitable for ES cells to differentiate than in the microenvironment of E12 or E13 of kidney adopted by Steenhard. Wnt4, HGF and activin-A may be promoting factors for the differentiation of ES cells to renal tubular cells.

Kim D.et al reported that mouse ES cells can differentiate into renal epithelial cells with the combination of retinoic acid, activin-A and Bone morphogenetic protein (Bmp)-7 (Kim D. & Dressler G.R. (2005).). The authors showed that cultured ES cells can be induced to express markers specific for the intermediate mesoderm with the combination of these three nephrigenic growth factors. They also showed that treated ES cells can contribute to tubular epithelial cells with almost 100% efficacy by injecting into developmental kidney rudiments. The injected Rosa 26 ES cells were traced by lac-Z staining and the tubular cells were confirmed morphologically by immunohistochemical staining with LTA and laminin. The transplantation or injection of cells entails a contentious problem of cell fusion although the authors insist that the occurrence of cell fusion in their experiment is very rare. Investigation of karyotype would be ideal for the detailed analysis.

Vigneau C. et al used mosue ES cell line with GFP knocked into the functional brachyury locus as well as lacZ in the ROSA26 locus (LacZ/brachyury/GFP) in selection and lineage tracing (Vigneau C.,et al. (2007).). The authors firstly optimized the culture conditions using activin-A to give rise to maximal numbers of renal progenitors identified by expression of the combination of renal markers, cadherin11, WT1, Pax2 and Wnt4. Then, they next enriched LacZ/brachyury/GFP$^+$ cells by FACS and injected these cells into embryonic kidney explants in organ culture. Five days after the injection, β-galactosidase positive cells were incorporated into blastema of the nephrogenic zone. Even after the single injection into developing live newborn mouse kidneys, the LacZ/brachyury/GFP$^+$ cells were stably integrated into proximal tubules with normal morphology and normal polarization of alkaline phosphatase and AQP1 for 7 months without teratoma formation. These results suggest that differentiation of ES cells with activin-A and selection for brachyury expression lead to the isolation and purification of renal proximal tubular progenitor cells.

Bruce S.J. et al reported that ES cells can be differentiated into renal lineage using BMP4 in vitro (Bruce S.J.,et al. (2007).). With their induction culture system, the authors confirmed that gene expressions of WT1, Wnt4, GDNF, Nephrin, mineral corticoid receptor, podocalyxin, Wnt11, Pod1 or AQP1 were upregulated at later stage.

Ross E.A. et al reported a study of ES cells showing an importance of scaffolds (Ross E.A.,et al. (2009).). The authors firstly decellularized intact rat kidney in a fashion that retained the matrix and secondly seeded them with pluriopotent rat ES cells either through the artery or through the ureter. The transplanted cells populated and proliferated within the glomerular, vascular and tubular structures. The cells lost their original embryonic appearance and expressed differentiated immunohistochemical markers. But the cells that lost the contact with basement membrane matrix became apoptotic. This study suggests the importance of extracellular matrix for the kidney regeneration to occur as a scaffold.

Morizane R. et al reported that differentiation of murine ES ells and iPS cells into renal lineage *in vitro* (Morizane R.,et al. (2009).). By hanging drop of murine ES cells, renal inducing factors were screened. GDNF and BMP7 enhanced the expression of Pax2 and WT1, markers of metanephric mesenchyme. Activin enhanced the expression of Pax-2 and KSP, suggesting activin enhanced the differentiation of ES cells into not only to mesoderm but also to tubular cells. According to their analysis, it is difficult to promote ES cells to differentiate into podocyte, because no inducer was found to enhance the expression of nephrin. To be sure, their analysis is based on *in vitro* marker analysis only and morphogenic or functional analyses are lacking, but several differentiation markers are investigated to compare the differentiation of ES cells and iPS cells. The information is useful to understand the effects of cytokines or growth factors for the designated differentiation of ES cells.

Sall1, a multi-zinc finger transcription factor which is expressed at metanephric mesenchyme, is involved in mouse kidney organogenesis because Sall1-deficient mouse show kidney agenesis or severe dysgenesis (Nishinakamura R.,et al. (2001).). It is highly expressed in ES cells and, most recently, it was reported to regulate ES cell differentiation in accordance with Nanog (Karantzali E.,et al. (2010).). Sall1 inhibits the ectodermal and mesodermal differentiation. More specifically, the induction of the gastrulation markers T brachyury, Goosecoid, and Dkk1 and the neuroectodermal markers Otx2 and Hand1 was inhibited by Sall1 overexpression during embryoid differentiation. In consideration of the fact that kidney is derived from intermediate mesoderm, switching off at an appropriate timing the signal Sall1 once positive at an early stage of development would be important for the kidney organogenesis.

Luis M. et al reported a method to isolate renal stem cells from embryonic kidney not from ES cells (Lusis M.,et al. (2010).). Metanephric mesenchyme of the developing kidney gives rise to various types of epithelial cells including visceral glomerular epithelial cells (podocytes), parietal glomerular epithelial cells, proximal tubular epithelial cells, and distal tubular epithelial cells. It is regarded as progenitor population pool of the developing kidney. But isolated metanephric mesenchyme does not self renew and requires immortalization for survival in culture. The authors developed a method of isolation and sustained culture of long-term repopulating, clonal progenitors from mouse embryonic kidney (E12.5) as free floating nephrospheres just as neurosphere often observed in neural stem cell culture system. These cells showed clonal self renewal for more than twenty passages when cultured with bFGF and thrombin, showed broad mesodermal multipotency, but held the expression of renal transcription factors such as Wt1, Sall1, Eya1, Six1, Six2, Osr1 and Hoxa11. These cells' capacity to contribute to developing embryonic kidney was limited, and nephrospheres did not display in vitro renal epithelial capacity. Holding mesodermal multipotency but renal transcription markers means that nephrosphere would be a developmentally transient state between pluripotent ES cells and renal progenitors. The nephrospheres can be cultured from Sall1-positive and –negative fraction of embryonic kidney, which suggests that they were derived from metanephric mesenchyme as a whole. This renal stem cell population could be cultured from only embryonic kidney, not from postnatal kidney.

3.2 Differentiation of human ES cells into renal lineage

As to the human cells, there are only a few reports of differentiation of ES cells to renal lineage. The concomitant use of retinoic acid, activin-A, and BMP-4 or BMP-7 induced the directed differentiation of human ES cells to renal precursors when analyzed by differentiation marker (Batchelder C.A.,et al. (2009).). As to the culture substrate, gelatin was reported to most closely recapitulate the directed developmental pattern of renal gene expression. Human ES cells also have an ability to differentiate into WT-1 and renin-expressing cells with the combination of several growth factors including either nerve growth factors (NGF) or HGF (Schuldiner M.,et al. (2000).). Tarantal A.F. et al reported an imaging study of human ES cells (Tarantal A.F.,et al. (2011).). The authors radiolabeled human ES cells expressing firefly luciferase, transplanted them to fetal monkey kidney at early second trimester under ultrasound guidance, and traced them by non-invasive imaging (positron emission tomography (PET) and bioluminescence imaging (BLI)). Transplanted cells were imaged in vivo and identified at the site of injection. Lin et al reported an isolation method by flow cytometer of putative renal progenitor fraction of human ES cells (Lin S.A.,et al. (2010).). The authors reduced serum concentration and feeder cell density to differentiate human ES cells in vitro for 14 days. Then they fractioned the differentiated ES cells based on the expression marker CD24, podocalyxin and GCTM2. These cells upregulated the renal transcription factor, PAX2, LHX1 and WT1 compared to unfractionated ES cells. Immunohistochemical analysis also confirmed these cells coexpressed nuclear factor WT1 and PAX2 protein. Microarray analysis showed that the most differentially up-regulated genes in the fraction were associated with kidney development compared to other fractions.

3.3 Lessons from animal cap study

The differentiation conditions of murine or human ES cells to renal lineage are similar to the developmental biology in amphibians. In amphibian eggs, the ectodermal cell mass of mid-blastula embryos, what is celled "animal cap", is similar to ES and iPS cells in mammals. When animal caps are treated with the combination of retinoic acid and activin-A, they differentiated into pronephric tubules (Osafune K.,et al. (2002).). Using this in vitro system to generate pronephros, the marker gene expression for pronephric tubules were observed (Brennan H.C.,et al. (1999).) and the molecular mechanism underlying pronephros development was investigated. These differentiation-inducing factors, retinoic acid and activin-A have been applied to directed differentiation trials of ES cells to renal lineage.

3.4 ES cells and kidney stem/progenitor cells

In consideration of these previous reports as to the differentiation of ES cells to renal lineage, the information is limited and we have to learn the microenvironment from signaling during the embryonic development as an early stage of differentiation and from transdifferentiation or conversion condition from other precursor and/or stem cells to the renal lineage as a late stage of development. In general, stem/progenitor cells and the proper microenvironment are essential for the repair or regeneration of damaged tissue. The same is true to the ES cell for the directed differentiation. A suitable microenvironment includes the presence of local cytokines as well as of extracellular matrix which can function as a scaffold and give rise to cytokines or growth factors around the stem/progenitor cell. Micro RNA coated with microvesicle which is discussed later would be added.

3.5 Lessons from the repair of kidney by non-ES cells

In thinking of the natural development of ES cells into cells of the kidney, ES cells would develop into component cells of the kidney via renal stem/progenitor cells (pathway II in fig 1). Or, ES cells may develop into component cells of the kidney without going through renal stem/progenitor cells (pathway I in fig 1). But renal stem/progenitor cells can give rise to all component cells of the organ, developing ES cells into renal stem/progenitor cells would be the most efficient way to create component cells of the kidney. We have focused on the renal stem/progenitor cell for the regeneration or repair, and ES cell is one of the promising candidates apart from the ethical problem. Let us first overview the repair of kidney by the stem/progenitor cell other than ES cells, and then try to find out the clue to manipulate the ES cells to the renal lineage.

As an extra-renal source of the kidney stem/progenitor cell, bone marrow cells or other cells (Wang Y.,et al. (2004).) have been investigated. In the early 2000s, pluripotent bone marrow-derived stem cells were thought to contribute to kidney repair. Bone marrow comprises several cell types such as endothelial progenitor cells, hematopoietic stem cells (HSCs), and mesenchymal stromal cells (MSCs) which is often referred to as mesenchymal stem cells. Bone marrow-derived stem cells appeared to be able to replace damaged renal tissue by replacing mesangial cells (Imasawa T., et al. (2001), Ito T.,et al. (2001).), tubular epithelia cells (Poulsom R.,et al. (2001).), endothelial cells (Rookmaaker M.B.,et al. (2003).) and even podocytes (Prodromidi E.I., et al. (2006), Sugimoto H.,et al. (2006).). But these researches were based on the transplantation method and the kidney specific marker analysis. Therefore, an event of cell fusion appears to represent a part of the engraftment of the recipient cells by the donor cells which was interpreted as transdifferentiation in the publication of that era. In a strict meaning, the transdifferentiation is conversion from one cell lineage to another different lineage with conversion of cellular functions and markers and maintenance of a normal karyotype. In studies of these periods, the concept of transdifferentiation was loosely defined and clear transdifferentiation was not demonstrated in many of the studies. In thinking of the ES cell differentiation via extrarenal cells into renal lineage (pathway III or IV in Fig 1), these *in vivo* transplantation studies seem helpful, however, they should be interpreted with caution. *In vitro* differentiation condition from extrarenal source into renal component cells (Suzuki A.,et al. (2004).) may be helpful in manipulating ES cells into renal lineage. As to the cell fusion, Bonde S. et al recently reported an interesting research. When bone marrow cells are cocultured with ES cells, the phenomenon of cell fusion occurs. Almost all the hybrid cells had lost CD45 expression. In short, after cell fusion, the ES cells can reprogram the cellular contents of somatic cells (bone marrow cells in this research) to adopt an ES cell phenotype.

Bone marrow-derived pluripotent MSCs, but not HSCs have an ability to engraft to tubular epithelial cells in an experimental kidney injury model and to help to recover the lost renal function (Morigi M.,et al. (2004).). It has recently been reported that kidney-derived MSC contribute to vasculogenesis, angiogenesis, and endothelial repair (Chen J.,et al. (2008).). This finding supports the possibility that MSC residing in kidney can participate in kidney repair or regeneration.

But recent studies argue against the direct differentiation of bone marrow-derived stem cells or MSC into kidney. The relatively small frequency of engraftment of bone marrow-derived cells (3-22%) means that most of the kidney component cells are composed of kidney resident cells. Moreover the functional improvement in the recipient transplanted with stem/progenitor cells is not due to the existence of transplanted cell itself in the kidney. For example, after ischemia-reperfusion injury, recipients transplanted with MSC keep improved renal function in spite of the small percentage of engraftment by the donor cells. This may be due to the various cytokines secreted from transplanted MSC that organize an appropriate microenvironment for the kidney repair in autocrine and paracrine fashion. To be sure, MSC secretes various cytokines such as VEGF, HGF and IGF-1 (Togel F.,et al. (2005).). Injection of bone marrow-derived MSC into peritoneal cavity attenuated the cisplatin-induced renal failure without the engraftment of MSC in the kidney. The conditioned media from culture of these stromal cells induced migration and proliferation of kidney-derived epithelial cells and significantly diminished cisplatin-induced proximal tubule cell death (Bi B.,et al. (2007).). In considering the cell administration pathway, the finding from these two experiments indicates that it is the humoral factors, not cellular factors that contribute to the kidney repair or regeneration by MSC.

3.6 Humoral factors important for the kidney repair or regeneration

Then the next question is "what are the humoral factors important for the kidney repair or regeneration?" Many factors have been proposed as candidates. One candidate is various cytokines secreted by MSC. There is a report indicating that administration of MSC to a rat model of ischemia-reperfusion-induced acute renal failure improved renal function, whereas administration of syngeneic fibroblasts did not. MSC expressed more growth factors VEGF, HGF and IGF-1 than fibroblasts (Togel F., et al (2005)). Some of these factors are known to modulate kidney function or repair. For example, VEGF attenuates glomerular inflammation and accelerates glomerular capillary repair (Shimizu A.,et al. (2004).). HGF, an angiogenic growth factor, prevents epithelial cell death and enhances regeneration and remodeling of injured or fibrotic renal tissue (Matsumoto K. & Nakamura T. (2001).). The effect of these growth factors or cytokines is evaluated as individual effect. If these factors are mixed and administered at a time, or if the administration timing is sequential, the effect might be different and complicated. Understanding the microenvironment in which the stem/progenitor cells is put in time and space in terms of cytokine expression would be firstly vital. This is the point from protein expression level.

Another humoral candidate factor would be RNA in microvesicle. This is a newly proved communication system. Microvesicles derived from MSC activate the proliferation of surviving renal tubular cells after injury by transferring mRNA (Bruno S.,et al. (2009).). Thus, MCS can repair the damaged tubuli by secreting microvesicles which function in a paracrine fashion. This is a communication tool in nucleotide level whose regulation system is not well known. Moreover, a microRNA are reported to mediate transforming growth factor-β1 autoregulation in renal glomerular mesangial cells (Kato M.,et al. (in press).). Ho J. et al reported that the loss of microRNAs in nephron progenitors resulted in a premature depletion of the population during kidney development, and as a consequence, a marked decrease in nephron number, which was accompanied by the increased apoptosis and expression of the pro-apoptotic protein Bim (Ho J.,et al. (in press).). Expression profiling in the embryonic kidney revealed several microRNAs (mmu-miR-10a, mmu-miR-17-5p and mmu-miR-106b) that are expressed in nephron progenitors and that are thought to target Bim transcript. MicroRNA-mediated regulation of Bim expression affects nephron survival. The importance of RNA is not restricted to only nephrogenesis. Recently, non-coding RNA has been emerging as key regulators of embryogenesis in general. Transcription factors OCT4 and NR2F2 and microRNA miR-302 are linked in the regulatory circuitry that critically regulate both pluripotency and differentiation in human ES cells (Rosa A. &

Brivanlou A.H. (2011).). The understanding of the microenvironment in terms of nucleotide level, microRNA during embryogenesis including nephrogenensis is important and the research has just begun.
Then, the last question would be "what are the cellular targets of these humoral factors?" The answer would be definitely the renal stem/progenitor cells including ES cells. Many researchers have undertaken this challenging theme to get the clearly defined renal stem/progenitor cells. Based on the characteristics of slow cycling property of stem/progenitor cells, the renal stem cell has been reported to exist as a label-retaining cell in renal papilla (Oliver J.A.,et al. (2004).) or tubular epithelial cells (Maeshima A.,et al. (2003).). Bowman's capsule (Sagrinati C.,et al. (2006).) and the S3 segment of the proximal tubules (Kitamura S.,et al. (2005).) have also been reported to contain renal stem/progenitor cells. A fraction of $CD133^{+}$ $CD24^{+}$ cells within the population of parietal epithelial cells are reported to engraft to tubular cells (Sagrinati C. (2006).). As markers of renal stem/progenitor cells are not clear, side population (SP) phenotype was adopted as a tool to investigate stem/progenitor cells in the kidney. Under our experimental conditions, kidney-derived side population cells did not engraft to kidney component cells (Iwatani H.,et al. (2004).), while other researchers reported the cells differentiated into multiple lineages *in vitro* and the transplantation of these cells *in vivo* ameliorated the impaired renal function without apparent integration into kidney component cells (Challen G.A.,et al. (2006).). Induction of ES cells into the above mentioned candidates of kidney stem/progenitor cells might be a target of the future study (pathway II-a in Fig 1).

3.7 Epigenetic factors important for the kidney repair or regeneration

A proliferative potential of the kidney is much lower than that of epithelial cells of the intestine. The characteristics of kidney are also borne out by the incidence of malignant tumors, as the kidney is less likely to be the origin of a tumor than is the gastrointestinal tract. The less proliferative potential of the kidney may be due to gene programming of its constituent cells. More precisely, partial and programmed gene inactivation of genomic DNA that is controlled in time and space, may be involved in the regulation of kidney fate. Different gene expressions in different organs for all the same genome: this mysterious but important fact cannot be solved without the concept of organ specific gene activation or inactivation. The system of epigenetic regulation could explain the concept. Actually, the epigenetic regulations are thought to be important for developmental biology. In terms of the renal progenitor cells, de Groh E.D. et al reported an interesting investigation (de Groh E.D.,et al. (2010).). The authors found that the treatment with 4-(phenylthio)butanoic acid (PTBA), which functions as histone deacetylase (HDAC) inhibitor expanded the renal progenitor cells in zebrafish. The PTBA-mediated expansion of renal progenitor cells was involved in retinoic acid signaling. Finally, the authors suggest a mechanistic link between renal progenitor cells, HDAC and the retinoic acid signaling pathway, although the precise target of HDAC is not clarified.
Aiden A.P. et al reported a comparison of genome-wide chromatin profiles of Wilms tumors, ES cells and normal kidney (Aiden A.P.,et al. (2010).). Wilms tumor is a pediatric cancer, characterized by a triphasic histology that includes undifferentiated blastemal components and varying amounts of epithelial and stromal elements (Rivera M.N. & Haber D.A. (2005).). The tumor can be associated with developmental abnormalities of nephrogenic rest and thought to be closely related to kidney organogenesis. Wilms tumor exhibited active chromatin domains previously observed in ES cells in terms of epigenetic

state. In the tumor cells, the domains often correspond to the genes that are critical for renal development and are expressed in renal stem cells. Wilms tumor cells also expressed "embryonic" chromatin regulators and maintained stem cell-like p16 silencing. It is suggested that Wilms tumor cells share transcriptional and epigenetic characteristics with normal renal stem cells As the developmental biology of the kidney is closely related to epigenetic state, the investigation of the mechanism of the epigenetic regulation of pluripotent stem cells would be highlighted.
In terms of the current difficulty in making many homogenous, high quality iPS cells available due to the lack of quality checking system, ES cell would be the best target for the development of the induction or differentiation conditions for pluripotent stem cells to the renal lineage. ES cell study will strongly lead iPS cell study and make the iPS cell study more matured.

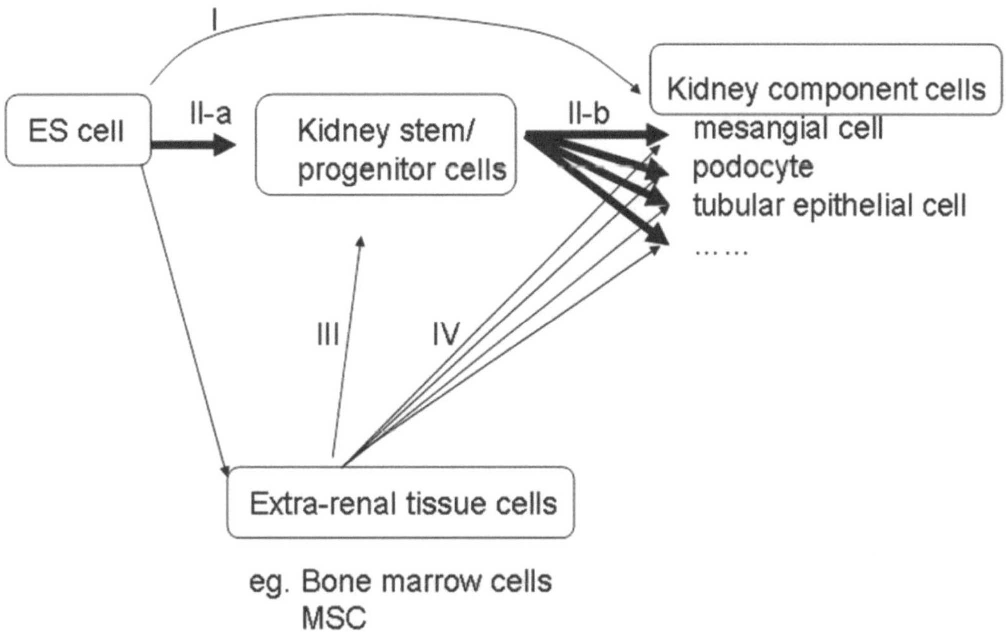

Fig. 1. Possible pathways for ES cells to differentiate into renal lineage
There may be several pathways for ES cells to differentiate into renal lineage.
ES cells would develop into component cells of the kidney via renal stem/progenitor cells (pathway II-a and II-b). Or, ES cells may develop into component cells of the kidney without going through renal stem/progenitor cells (pathway I), although the differentiation condition into kidney from ES cell is not clarified yet. Thick arrow (pathway II-a and II-b) would be the putative mainstream of ES cell differentiation. Previous reports of pathway IV may be helpful in understanding pathway II-b.

4. References

Aiden A. P., Rivera M. N., Rheinbay E., Ku M., Coffman E. J., Truong T. T., Vargas S. O., Lander E. S., Haber D. A. &Bernstein B. E. (2010). Wilms tumor chromatin profiles highlight stem cell properties and a renal developmental network. *Cell Stem Cell*, Vol.6, No.6, pp. 591-602

Batchelder C. A., Lee C. C., Matsell D. G., Yoder M. C. &Tarantal A. F. (2009). Renal ontogeny in the rhesus monkey (Macaca mulatta) and directed differentiation of human embryonic stem cells towards kidney precursors. *Differentiation*, Vol.78, No.1, pp. 45-56

Bi B., Schmitt R., Israilova M., Nishio H. &Cantley L. G. (2007). Stromal cells protect against acute tubular injury via an endocrine effect. *J Am Soc Nephrol*, Vol.18, No.9, pp. 2486-96

Boheler K. R., Czyz J., Tweedie D., Yang H. T., Anisimov S. V. &Wobus A. M. (2002). Differentiation of pluripotent embryonic stem cells into cardiomyocytes. *Circ Res*, Vol.91, No.3, pp. 189-201

Brennan H. C., Nijjar S. &Jones E. A. (1999). The specification and growth factor inducibility of the pronephric glomus in Xenopus laevis. *Development*, Vol.126, No.24, pp. 5847-56

Bruce S. J., Rea R. W., Steptoe A. L., Busslinger M., Bertram J. F. &Perkins A. C. (2007). In vitro differentiation of murine embryonic stem cells toward a renal lineage. *Differentiation*, Vol.75, No.5, pp. 337-49

Bruno S., Grange C., Deregibus M. C., Calogero R. A., Saviozzi S., Collino F., Morando L., Busca A., Falda M., Bussolati B., Tetta C. &Camussi G. (2009). Mesenchymal stem cell-derived microvesicles protect against acute tubular injury. *J Am Soc Nephrol*, Vol.20, No.5, pp. 1053-67

Brustle O., Jones K. N., Learish R. D., Karram K., Choudhary K., Wiestler O. D., Duncan I. D. &McKay R. D. (1999). Embryonic stem cell-derived glial precursors: a source of myelinating transplants. *Science*, Vol.285, No.5428, pp. 754-6

Challen G. A., Bertoncello I., Deane J. A., Ricardo S. D. &Little M. H. (2006). Kidney side population reveals multilineage potential and renal functional capacity but also cellular heterogeneity. *J Am Soc Nephrol*, Vol.17, No.7, pp. 1896-912

Chen J., Park H. C., Addabbo F., Ni J., Pelger E., Li H., Plotkin M. &Goligorsky M. S. (2008). Kidney-derived mesenchymal stem cells contribute to vasculogenesis, angiogenesis and endothelial repair. *Kidney Int*, Vol.74, No.7, pp. 879-89

de Groh E. D., Swanhart L. M., Cosentino C. C., Jackson R. L., Dai W., Kitchens C. A., Day B. W., Smithgall T. E. &Hukriede N. A. (2010). Inhibition of histone deacetylase expands the renal progenitor cell population. *J Am Soc Nephrol*, Vol.21, No.5, pp. 794-802

Doetschman T. C., Eistetter H., Katz M., Schmidt W. &Kemler R. (1985). The in vitro development of blastocyst-derived embryonic stem cell lines: formation of visceral yolk sac, blood islands and myocardium. *J Embryol Exp Morphol*, Vol.87, 27-45

Evans M.J. & Kaufman M.H. (1981). Establishment in culture of pluripotential cells from mouse embryos. *Nature*, Vol.292, No.5819, pp. 154-6

Hammerman M. R. (2004). Renal organogenesis from transplanted metanephric primordia. *J Am Soc Nephrol*, Vol.15, No.5, pp. 1126-32

Ho J., Pandey P., Schatton T., Sims-Lucas S., Khalid M., Frank M. H., Hartwig S. &Kreidberg J. A. The Pro-Apoptotic Protein Bim Is a MicroRNA Target in Kidney Progenitors. *J Am Soc Nephrol*, (in press)

Imasawa T., Utsunomiya Y., Kawamura T., Zhong Y., Nagasawa R., Okabe M., Maruyama N., Hosoya T. & Ohno T. (2001). The potential of bone marrow-derived cells to differentiate to glomerular mesangial cells. *J Am Soc Nephrol*, Vol.12, No.7, pp. 1401-9

Ito T., Suzuki A., Imai E., Okabe M. &Hori M. (2001). Bone marrow is a reservoir of repopulating mesangial cells during glomerular remodeling. *J Am Soc Nephrol*, Vol.12, No.12, pp. 2625-35

Iwatani H., Ito T., Imai E., Matsuzaki Y., Suzuki A., Yamato M., Okabe M. &Hori M. (2004). Hematopoietic and nonhematopoietic potentials of Hoechst^{low}/side population cells isolated from adult rat kidney. *Kidney Int*, Vol.65, No.5, pp. 1604-14

Karantzali E., Lekakis V., Ioannou M., Hadjimichael C., Papamatheakis J. &Kretsovali A. (2010). Sall1 regulates embryonic stem cell differentiation in association with nanog. *J Biol Chem*, Vol.286, No.2, pp. 1037-45

Kato M., Arce L., Wang M., Putta S., Lanting L. &Natarajan R. A microRNA circuit mediates transforming growth factor-β1 autoregulation in renal glomerular mesangial cells. *Kidney Int*, (in press)

Kim D. & Dressler G.R. (2005). Nephrogenic factors promote differentiation of mouse embryonic stem cells into renal epithelia. *J Am Soc Nephrol*, Vol.16, No.12, pp. 3527-34

Kitamura S., Yamasaki Y., Kinomura M., Sugaya T., Sugiyama H., Maeshima Y. &Makino H. (2005). Establishment and characterization of renal progenitor like cells from S3 segment of nephron in rat adult kidney. *FASEB J*, Vol.19, No.13, pp. 1789-97

Klug M. G., Soonpaa M. H., Koh G. Y. &Field L. J. (1996). Genetically selected cardiomyocytes from differentiating embronic stem cells form stable intracardiac grafts. *J Clin Invest*, Vol.98, No.1, pp. 216-24

Kobayashi T., Tanaka H., Kuwana H., Inoshita S., Teraoka H., Sasaki S. &Terada Y. (2005). Wnt4-transformed mouse embryonic stem cells differentiate into renal tubular cells. *Biochem Biophys Res Commun*, Vol.336, No.2, pp. 585-95

Lin S. A., Kolle G., Grimmond S. M., Zhou Q., Doust E., Little M. H., Aronow B., Ricardo S. D., Pera M. F., Bertram J. F. &Laslett A. L. (2010). Subfractionation of differentiating human embryonic stem cell populations allows the isolation of a mesodermal population enriched for intermediate mesoderm and putative renal progenitors. *Stem Cells Dev*, Vol.19, No.10, pp. 1637-48

Lusis M., Li J., Ineson J., Christensen M. E., Rice A. &Little M. H. (2010). Isolation of clonogenic, long-term self renewing embryonic renal stem cells. *Stem Cell Res*, Vol.5, No.1, pp. 23-39

Maeshima A., Yamashita S. &Nojima Y. (2003). Identification of renal progenitor-like tubular cells that participate in the regeneration processes of the kidney. *J Am Soc Nephrol*, Vol.14, No.12, pp. 3138-46

Martin G. R. (1981). Isolation of a pluripotent cell line from early mouse embryos cultured in medium conditioned by teratocarcinoma stem cells. *Proc Natl Acad Sci U S A*, Vol.78, No.12, pp. 7634-8

Matsumoto K. & Nakamura T. (2001). Hepatocyte growth factor: renotropic role and potential therapeutics for renal diseases. *Kidney Int*, Vol.59, No.6, pp. 2023-38

Morigi M., Imberti B., Zoja C., Corna D., Tomasoni S., Abbate M., Rottoli D., Angioletti S., Benigni A., Perico N., Alison M. &Remuzzi G. (2004). Mesenchymal stem cells are renotropic, helping to repair the kidney and improve function in acute renal failure. *J Am Soc Nephrol*, Vol.15, No.7, pp. 1794-804

Morizane R., Monkawa T. &Itoh H. (2009). Differentiation of murine embryonic stem and induced pluripotent stem cells to renal lineage in vitro. *Biochem Biophys Res Commun*, Vol.390, No.4, pp. 1334-9

Nakano T., Kodama H. &Honjo T. (1994). Generation of lymphohematopoietic cells from embryonic stem cells in culture. *Science*, Vol.265, No.5175, pp. 1098-101

Nishikawa S. I., Nishikawa S., Hirashima M., Matsuyoshi N. &Kodama H. (1998). Progressive lineage analysis by cell sorting and culture identifies FLK1+VE-cadherin+ cells at a diverging point of endothelial and hemopoietic lineages. *Development*, Vol.125, No.9, pp. 1747-57

Nishinakamura R., Matsumoto Y., Nakao K., Nakamura K., Sato A., Copeland N. G., Gilbert D. J., Jenkins N. A., Scully S., Lacey D. L., Katsuki M., Asashima M. &Yokota T. (2001). Murine homolog of SALL1 is essential for ureteric bud invasion in kidney development. *Development*, Vol.128, No.16, pp. 3105-15

Oliver J. A., Maarouf O., Cheema F. H., Martens T. P. &Al-Awqati Q. (2004). The renal papilla is a niche for adult kidney stem cells. *J Clin Invest*, Vol.114, No.6, pp. 795-804

Osafune K., Nishinakamura R., Komazaki S. &Asashima M. (2002). In vitro induction of the pronephric duct in Xenopus explants. *Dev Growth Differ*, Vol.44, No.2, pp. 161-7

Palacios R., Golunski E. &Samaridis J. (1995). In vitro generation of hematopoietic stem cells from an embryonic stem cell line. *Proc Natl Acad Sci U S A*, Vol.92, No.16, pp. 7530-4

Poulsom R., Forbes S. J., Hodivala-Dilke K., Ryan E., Wyles S., Navaratnarasah S., Jeffery R., Hunt T., Alison M., Cook T., Pusey C. &Wright N. A. (2001). Bone marrow contributes to renal parenchymal turnover and regeneration. *J Pathol*, Vol.195, No.2, pp. 229-35

Prodromidi E. I., Poulsom R., Jeffery R., Roufosse C. A., Pollard P. J., Pusey C. D. &Cook H. T. (2006). Bone marrow-derived cells contribute to podocyte regeneration and amelioration of renal disease in a mouse model of Alport syndrome. *Stem Cells*, Vol.24, No.11, pp. 2448-55

Rivera M.N. & Haber D.A. (2005). Wilms' tumour: connecting tumorigenesis and organ development in the kidney. *Nat Rev Cancer*, Vol.5, No.9, pp. 699-712

Rookmaaker M. B., Smits A. M., Tolboom H., Van 't Wout K., Martens A. C., Goldschmeding R., Joles J. A., Van Zonneveld A. J., Grone H. J., Rabelink T. J. &Verhaar M. C. (2003). Bone-marrow-derived cells contribute to glomerular endothelial repair in experimental glomerulonephritis. *Am J Pathol*, Vol.163, No.2, pp. 553-62

Rosa A. & Brivanlou A.H. (2011). A regulatory circuitry comprised of miR-302 and the transcription factors OCT4 and NR2F2 regulates human embryonic stem cell differentiation. *EMBO J*, Vol.30, No.2, pp. 237-48

Ross E. A., Williams M. J., Hamazaki T., Terada N., Clapp W. L., Adin C., Ellison G. W., Jorgensen M. & Batich C. D. (2009). Embryonic stem cells proliferate and differentiate when seeded into kidney scaffolds. *J Am Soc Nephrol*, Vol.20, No.11, pp. 2338-47

Sagrinati C., Netti G. S., Mazzinghi B., Lazzeri E., Liotta F., Frosali F., Ronconi E., Meini C., Gacci M., Squecco R., Carini M., Gesualdo L., Francini F., Maggi E., Annunziato F., Lasagni L., Serio M., Romagnani S. &Romagnani P. (2006). Isolation and characterization of multipotent progenitor cells from the Bowman's capsule of adult human kidneys. *J Am Soc Nephrol*, Vol.17, No.9, pp. 2443-56

Scholer H. R., Ciesiolka T. &Gruss P. (1991). A nexus between Oct-4 and E1A: implications for gene regulation in embryonic stem cells. *Cell*, Vol.66, No.2, pp. 291-304

Schuldiner M., Yanuka O., Itskovitz-Eldor J., Melton D. A. &Benvenisty N. (2000). Effects of eight growth factors on the differentiation of cells derived from human embryonic stem cells. *Proc Natl Acad Sci U S A*, Vol.97, No.21, pp. 11307-12

Shimizu A., Masuda Y., Mori T., Kitamura H., Ishizaki M., Sugisaki Y. &Fukuda Y. (2004). Vascular endothelial growth factor165 resolves glomerular inflammation and accelerates glomerular capillary repair in rat anti-glomerular basement membrane glomerulonephritis. *J Am Soc Nephrol*, Vol.15, No.10, pp. 2655-65

Soria B., Roche E., Berna G., Leon-Quinto T., Reig J. A. &Martin F. (2000). Insulin-secreting cells derived from embryonic stem cells normalize glycemia in streptozotocin-induced diabetic mice. *Diabetes*, Vol.49, No.2, pp. 157-62

Steenhard B. M., Isom K. S., Cazcarro P., Dunmore J. H., Godwin A. R., St John P. L. &Abrahamson D. R. (2005). Integration of embryonic stem cells in metanephric kidney organ culture. *J Am Soc Nephrol*, Vol.16, No.6, pp. 1623-31

Sugimoto H., Mundel T. M., Sund M., Xie L., Cosgrove D. &Kalluri R. (2006). Bone-marrow-derived stem cells repair basement membrane collagen defects and reverse genetic kidney disease. *Proc Natl Acad Sci U S A*, Vol.103, No.19, pp. 7321-6

Suzuki A., Iwatani H., Ito T., Imai E., Okabe M., Nakamura H., Isaka Y., Yamato M. &Hori M. (2004). Platelet-derived growth factor plays a critical role to convert bone marrow cells into glomerular mesangial-like cells. *Kidney Int*, Vol.65, No.1, pp. 15-24

Tarantal A. F., Lee C. C., Batchelder C. A., Christensen J. E., Prater D. &Cherry S. R. (2011). Radiolabeling and In Vivo Imaging of Transplanted Renal Lineages Differentiated from Human Embryonic Stem Cells in Fetal Rhesus Monkeys. *Mol Imaging Biol*,

Togel F., Hu Z., Weiss K., Isaac J., Lange C. &Westenfelder C. (2005). Administered mesenchymal stem cells protect against ischemic acute renal failure through differentiation-independent mechanisms. *Am J Physiol Renal Physiol*, Vol.289, No.1, pp. F31-42

Vigneau C., Polgar K., Striker G., Elliott J., Hyink D., Weber O., Fehling H. J., Keller G., Burrow C. &Wilson P. (2007). Mouse embryonic stem cell-derived embryoid bodies generate progenitors that integrate long term into renal proximal tubules in vivo. *J Am Soc Nephrol*, Vol.18, No.6, pp. 1709-20

Wang Y., Iwatani H., Ito T., Horimoto N., Yamato M., Matsui I., Imai E. &Hori M. (2004). Fetal cells in mother rats contribute to the remodeling of liver and kidney after injury. *Biochem Biophys Res Commun*, Vol.325, No.3, pp. 961-7

Yamamoto M., Cui L., Johkura K., Asanuma K., Okouchi Y., Ogiwara N. &Sasaki K. (2006). Branching ducts similar to mesonephric ducts or ureteric buds in teratomas originating from mouse embryonic stem cells. *Am J Physiol Renal Physiol*, Vol.290, No.1, pp. F52-60

25

Periodontal Ligament Stem Cells

Hidefumi Maeda, Naohisa Wada, Shinsuke Fujii,
Atsushi Tomokiyo and Akifumi Akamine
Department of Endodontology & Operative Dentistry, Kyushu University
Japan

1. Introduction

Millions of people experience consequential tooth loss related to irretrievable damage of the periodontium caused by deep caries, severe periodontal diseases or irreversible trauma, resulting in a decreased quality of life. Hence, dental scientists have focused much attention on tissue engineering techniques in an effort to address this condition. The periodontium is composed mainly of two soft tissues and two hard tissues; the former includes the periodontal ligament (PDL) tissue and gingival tissue, and the latter includes alveolar bone and cementum covering the tooth root. In particular, PDL is a dynamic connective tissue that is subjected to continual adaptation to maintain tissue size and width, as well as structural integrity, including fibers and bone modeling. The main role of PDL is to anchor the tooth root to the alveolar bone socket tightly, cushioning mechanical load that proceeds from mastication. Thus, PDL tissue constitutes the bedrock of periodontium to determine the life-span of tooth. In this chapter, we describe the phenotypes of multipotent clonal human PDL progenitor/stem cell lines that we have recently established, and present a comparison with PDL stem cells (PDLSCs).

PDLSCs represent typical properties of bone marrow-derived mesenchymal stem cells (BMMSCs). PDLSCs exhibit a self-renewing capacity and express cell surface markers similar to BMMSCs. PDLSCs also possess the multipotential to differentiate into various types of cells, such as osteoblast-like cells, adipocytes, chondrocytes and neurocytes, in vitro. Additionally, PDLSCs represent a distinctive potential to form cementum- and PDL-like tissues in vivo, suggesting that PDLSCs might belong to a unique population of somatic stem cells.

Since a very small number of stem cells are included in PDL tissue, researchers have tried to establish immortalized PDL stem cell lines for the convenience and consistency of analyses. Recently, we succeeded in establishing two clonal human PDL cell lines with multipotential by transduction with Simian Virus (SV) 40 large T-antigen and human telomerase reverse transcriptase. These clonal cell lines exhibited unique characteristics: cell line 1-11 demonstrated osteoblastic and adipocytic differentiation capabilities, and cell line 1-17 exhibited the potential to differentiate into osteoblasts, adipocytes, chondrocytes and neurocytes. Flow cytometric analysis demonstrated that the percentage of cells expressing BMMSC markers was very high in both cell lines. On the other hand, cell line 1-17 expressed embryonic stem (ES) cell marker genes at higher levels, compared with cell line 1-11. Furthermore, both cell lines expressed PDL phenotype-related molecules at equivalent levels while BMMSCs showed little expression. When transplanted into PDL defects formed in immunodeficient rats, cell line 1-11 homed on the surface of bone and cementum, and between bone and cementum, while cell line 1-17 was localized only between bone and cementum. Thus, we believe that both cell lines met the prerequisite for PDL stem cells, and furthermore, that the differentiation stage of cell line 1-17 may be more immature than that of cell line 1-11.

Differential analyses between these two clones will allow us to further reveal the mechanism of differentiation of PDL stem cells, as well as to elucidate key signals to direct stem cells, including other MSCs, induced pluripotent stem (iPS) cells or ES cells, toward the PDL-lineage.

2. What is the origin of PDL?

PDL tissue originates from cranial neural crest (CNC)-derived ectomesenchymal cells (Chai et al., 2000). Neural crest cells are multipotent cell populations derived from the dorsal surface of the neural tube and migrate extensively during early stages of embryogenesis (Bronner-Fraser, 1993). The neural crest is divided into four specific regions that include the CNC, trunk neural crest, vagal and sacral neural crest, and cardiac neural crest, depending on the crest cell potentials to develop into particular tissues. During craniofacial development, CNC cells migrate ventrolaterally and contribute significantly to the formation of mesenchymal structures in the head and neck (Le et al., 1993). When tooth development begins, the primitive oral epithelium thickens and proliferates into the underlying CNC-derived ectomesenchyme, promoting mesenchymal cellular condensation. The dental follicle (DF) derived from CNC (Chai et al., 2000) is a sac including these structures, namely the odontogenic organ during tooth germ development (Cho et al., 2000). DF cells differentiate into PDL cells and are also believed to contain the progenitor cells that differentiate into cementoblasts and osteoblasts (Ten Cate, 2008). Thus, the DF plays a crucial role in forming the PDL tissue, namely the fabrication of the periodontium.

3. What are PDL tissues and what are their roles?

PDL tissue is a major constituent of the periodontium, along with other components, such as gingival soft tissue, the cementum covering the tooth root and alveolar bone (Melcher, 1984). PDL tissue is a dynamic soft connective tissue that is situated between the tooth root and the bone socket, and the width of PDL tissue ranges from 0.15 to 0.38 μm (Ten Cate, 2008). PDL tissue is composed of heterogeneous cell populations, various fibers, non-collagenous extracellular matrix (ECM) proteins, nerve fibers and blood vessels. PDL cell populations include fibroblasts, which are principal cells in PDL tissue (Berkovitz et al., 1995; Beertsen et al., 1997), undifferentiated mesenchymal cells, epithelial cell rests of Malassez and endothelial cells.

PDL cells not only produce collagen fibers and other ECM proteins, but have osteoblastic/cementoblastic properties (Roberts et al., 1982; McCulloch et al., 1991), such as high alkaline phosphatase (ALP) activity (Lekic et al., 1996) and the expression of osteoblast-related genes, including Runt-related transcription factor-2 (Fujii et al., 2006), Osterix (Kato et al., 2005), bone sialoprotein (Lekic et al., 2001), osteopontin (Chutivongse et al., 2005), osteocalcin (Li et al., 2001) and periostin (Yamada et al., 2001). Additionally, PDL cells synthesize an osteoclastogenesis activator, receptor activator of NF-kappa B ligand (Kanzaki et al., 2001), as well as its inhibitor, osteoprotegerin (Wada et al., 2001; Hasegawa et al., 2002), suggesting that these cells participate in alveolar bone metabolism.

The fibers in PDL tissues mainly comprise collagens and elastic system fibers. The predominant collagen in PDL tissue is type I collagen, and other collagens (types III, VI and XII) are also deposited in PDL tissue (Lukinmaa et al., 1992; MacNeil et al., 1998). The collagen fibers in PDL tissue form in definite and distinct bundles; the terminus of the principal collagen bundles insert into the cementum and alveolar bone, as known as Sharpey's fibers. Oxytalan fibers, a member of the elastic system fibers, are present within PDL tissue (Beertsen et al., 1997). The fibers run vertically along the cementum surface of the apical root, forming a three-dimensional branching meshwork that surrounds the roots (Ten Cate, 2008). PDL tissue plays a crucial role in fixing the tooth to the alveolar bone socket through these three-dimensional structures, which involve cushioning mechanical loads such as compressive and tensile forces that proceed from mastication, as well as adjusting occlusal forces depending on food hardness via sensory nerve perception in the PDL tissue. Thus, PDL tissue constitutes the bedrock of periodontium, which determines the life-span of tooth.

Proteoglycans (PGs), including biglycan, CD44, decorin, fibromodulin, perlecan and versican are also present in PDL tissue as non-collagenous ECM proteins (Häkkinen et al., 1993; Cheng et al., 1999). PGs are extracellular and cell surface-associated macromolecules that regulate cell adhesion, cell growth, matrix formation and bind growth factors.

4. What causes PDL tissue loss?

PDL tissue loss is caused mainly by the inflammatory or physical tissue destruction stimulated by severe periodontitis, deep caries reaching infrabony levels or irreversible trauma. Bacterial infection is essential for the initiation of periodontitis. One of the components of the cell wall of gram-negative anaerobic bacteria, lipopolysaccharide, which is a complex glycolipid (Lamont et al., 1998), promotes the expression of pro-inflammatory cytokines, such as interleukin-1 beta and tumor necrosis factor-alpha, in macrophages and PDL cells (Lindemann et al., 1988; Wada et al., 2004). These cytokines are reported not only to disturb the function of PDL cells (Quintero et al., 1995; Agarwal et al., 1998), but also to stimulate the production of matrix metalloproteases (MMPs), such as MMP-1, in PDL cells (Oyama et al., 2007; Xiang et al., 2009). MMP-1 is the major proteolytic enzyme that can cleave native type I and type III collagens, suggesting its aggressive contribution to the destruction of the PDL structure (Birkedal-Hansen, 1993).

5. What is needed for the regeneration of PDL tissues?

To create effective and practical therapies for the defects of PDL tissue, new approaches based on tissue engineering concepts have been developing. Tissue engineering, first described in the late 1980s, is an emerging interdisciplinary field in biomedical engineering. This concept aims to create or regenerate functional tissues through the use of an appropriate combination of three fundamental tools, namely signaling molecules, engineering scaffolds and cells, which together are also known as the tissue engineering triad (Langer & Vacanti, 1993). In addition to these triads, we think that blood supply and the innervation of peripheral nerves are requisites in PDL tissue regeneration.

5.1 Signaling molecules

Several growth factors have been reported to promote the regeneration of PDL tissue in vitro and in vivo. Growth factors are natural biological molecules that bind to cell-surface receptors and regulate key cellular activities during tissue repair, including cell proliferation, chemotaxis, differentiation and matrix synthesis (Chen et al., 2010a).

5.1.1 Basic Fibroblast Growth Factor

Basic fibroblast growth factor (bFGF) is a single-chain polypeptide mainly associated with mesoderm- and neuroectoderm-derived cells, and belongs to a member of the heparin-binding growth factor family (Gospodarowicz et al., 1986). bFGF is involved in controlling cell proliferation, migration, differentiation and survival, and also has a role in angiogenesis (Savona et al., 1997). In PDL cells, bFGF has been shown to increase proliferation, the production of hyaluronan, heparan sulfate and osteopontin, and the expressions of MMP-1, MMP-3 and MMP-9 mRNA, while inhibiting the induction of ALP activity, mineralized nodule formation and the expressions of ALP, type I collagen, MMP-2 and tropoelastin mRNA (Takayama et al., 1997; Palmon et al., 2001; Shimabukuro et al., 2008; Terashima et al., 2008; Hakki et al., 2009). Additionally, the topical application of bFGF was shown to significantly induce PDL regeneration without ankylosis, root resorption and epithelial down-growth in an experimental alveolar bone defect model in dogs (Murakami et al., 2003). Moreover, the regenerative effect of bFGF on PDL tissue in patients with periodontitis was confirmed in a randomized controlled phase II clinical trial and a multi-center randomized clinical trial (Kitamura et al., 2011).

5.1.2 Transforming Growth Factor

Transforming growth factor-β (TGF-β) consists of three 25-kDa homodimeric mammalian isoforms (β1, β2 and β3). TGF-β1 plays an important role in wound healing and repair through the induction of angiogenesis, the increase of ECM formation and the inhibition of inflammation and matrix degradation in mesenchymal cells (Ling & Robinson, 2002). In PDL cells, TGF-β1 has been reported to increase cell-surface proteoglycan genes such as syndecan-2 and betaglycan (Worapamorn et al., 2001), and has been reported to promote the synthesis of DNA, fibronectin, secreted protein acidic and rich in cytokine/osteonectin (SPARC), and connective tissue growth factor (CTGF) (Fujita et al., 2004; Takeuchi et al., 2009). Previous reports have demonstrated the opposite effects of TGF-β1 on PDL cells: Si & Liu (2001) have shown a significant increase of ALP activity, while Brady et al. (1998) reported ALP suppression. Importantly, Fujii et al. (2010) reported the contrastive effects of TGF-β1 on primary PDL cells and a PDL stem/progenitor cell line: TGF-β1 promoted proliferation and showed no alternation of alpha-smooth muscle actin, type I collagen and fibrillin 1 mRNA expressions in primary PDL cells, whereas TGF-β1 suppressed proliferation and up-regulated these mRNA expressions in a PDL stem/progenitor cell line. These results suggest the dual effects of TGF-β1 on PDL cells depending on their differentiation stage.

5.1.3 Bone Morphogenetic Protein

Bone morphogenetic proteins (BMPs) and growth and differentiation factors (GDFs) together form a single family of cystine-knot cytokines in the TGF-β superfamily. A feature of the BMPs is their ability to induce bone formation and regulate morphogenetic activities in the development of various tissues. Many reports have demonstrated the promotion of alveolar bone and cementum regeneration by BMP-2 in several experimental periodontal defects (Kinoshita et al., 1997). Unfortunately, BMP-2 treatment did not induce the formation of functional PDL structures, such as Sharpey's fibers, and sometimes elicited ankylosis (Selvig et al., 2002). Osteogenic protein-1 (OP-1/BMP-7) promoted ALP activity and osteocalcin expression in PDL cells (Dereka et al., 2009). Moreover, OP-1-treated PDL precursor cells increased proliferation (Rajshankar et al., 1998). In vivo assays with OP-1 treatment have demonstrated the stimulation of osteogenesis, cementogenesis and new attachment formation in Class III furcation defects (Giannobile et al., 1998). BMP-12 has been reported to act as a signaling molecule during embryonic tendon, ligament and joint formation, and is reportedly involved in tendon and ligament healing (Wolfman et al., 1997). Wikesjö et al. (2004) demonstrated the promotion of new bone and cementum formation with BMP-12 treatment in supra-alveolar periodontal defects. Interestingly, they also showed that BMP-12 treatment induced to form functionally oriented PDL bridging the gaps between bone and cementum, whereas this was a rare observation in BMP-2 treatment. GDF-5, known as BMP-14, was reported to significantly induce formation of new bone, cementum and fibers along the root surface in one-wall intrabony defects without root resorption and ankylosis (Kim et al., 2009).

5.1.4 Brain-Derived Neurotrophic Factor

Brain-derived neurotrophic factor (BDNF) belongs to the second member of the neurotrophin family and is involved in differentiation, proliferation, survival, homeostasis and regeneration of central and peripheral neurons by binding to cell surface receptors TrkB and p75NTR (Ebendal, 1992). PDL cells have been reported to increase proliferation and the expressions of ALP and osteocalcin mRNA and osteopontin, BMP-2 and type I collagen synthesis through BDNF exposure (Tsuboi et al., 2001). Moreover, BDNF or BDNF/high-molecular-weight-hyaluronic acid reportedly increased the formation of new alveolar bone, cementum, Sharpey's fiber-like connective tissues and blood capillary in Class III furcation defects in a dog model (Takeda et al., 2005; Takeda et al., 2011).

5.1.5 Connective Tissue Growth Factor

Connective tissue growth factor (CTGF) is a cysteine-rich, matrix-associated, and heparin-binding protein that was originally discovered as a chemotactic and mitogenic factor for fibroblast-like cells in vitro (Bradham et al., 1991). After CTGF treatment, hMSCs revealed a typical fibroblastic phenotype and failed to show osteogenic or chondrogenic differentiation (Lee et al., 2006). CTGF stimulated proliferation, ALP, type I collagen, and periostin mRNA expression, ALP activity and collagen synthesis in PDL cells, while CTGF induced little change of osteopontin and osteocalcin mRNA expressions (Asano et al., 2005). Transplantation of CTGF-treated PDL progenitors exhibited an increase of type III collagen, biglycan and periostin mRNA expressions and enhanced the formation of PDL-like tissue, namely dense collagen fiber bundles with a high expression level of periostin in a mouse model (Dangaria et al., 2009).

5.1.6 Wnt

Wnts are a family of 19 secreted cysteine-rich glycoproteins that participate in embryonic development and adult homeostasis by regulating cellular proliferation, differentiation and apoptosis (Logan & Nusse, 2004). Wnt proteins transduce multiple signaling cascades including the canonical Wnt/β-catenin pathway, the Wnt/Ca2+ pathway and the Wnt/polarity pathway. Among those, canonical Wnt/β-catenin has been reported to increase bone formation through the stimulation of osteoblast development and the inhibition of osteoblast and osteocyte apoptosis (Kato et al., 2002; Bodine et al., 2004). The canonical Wnt/β-catenin pathway has been shown to promote mineralized nodule formation, ALP activity and osteogenic transcription factor expression in PDL cells (Heo et al., 2010), while the pathway reportedly suppressed ALP activity and cementum function-associated gene expression in cementoblasts (Sodek & Limeback, 1979).

5.1.7 Platelet-Derived Growth Factor

Platelet-derived growth factor (PDGF) is composed of two disulfide bounded polypeptide chains that are encoded by two different genes, PDGF-A and PDGF-B. In nature, PDGF can exist as a homodimer: PDGF-AA and PDGF-BB. PDGF-BB was reported to promote the proliferation, adhesion, type I collagen synthesis and migration of PDL cells (Belal et al., 2006), as well as to enhance periodontal regeneration of Class III furcation defects in beagle dogs without significant ankylosis or root resorption (Cho et al., 1995). Moreover, PDGF-BB has been shown to stimulate a significant increase in the rate of clinical attachment level gain and bone defect fill, as well as to reduce gingival recession (Nevins et al., 2005).

5.1.8 Enamel Matrix Derivative

Porcine enamel matrix protein was reported to induce new bone and cementum formation in periodontal defects in monkeys (Hammarstrom, 1997). Since then, many in vivo and in vitro studies, as well as clinical trials, have revealed the progression of PDL regeneration in the use of enamel matrix derivative (EMD). EMD has been shown to promote the growth, proliferation, migration, adhesion and protein synthesis of PDL fibroblasts, osteoblasts and cementoblasts, as well as to regulate their bone-related gene expressions (Carinci et al., 2006). Additionally, EMD has been shown to induce the production of several growth factors in PDL cells and osteoblasts (Mizutani et al., 2003; Heng et al., 2007).

5.2 Scaffolds

Scaffolds that have the capabilities of growth factor delivery, easy cellular infiltration and subsequent tissue integration become powerful tools for tissue engineering. PDL cells are subjected to a multitude of factors and microenvironments presented by local substrates and their surface conditions. There are two classes of scaffold biomaterials: natural materials and synthetic materials.

Of the natural materials, collagens are ubiquitous interstitial proteins responsible for maintaining structural integrity and are rapidly synthesized and degraded in PDL tissue (Sodek & Limeback, 1979). Several reports have revealed the promotion of PDL regeneration using collagen gels incorporating BMP-2, as well as gelatin hydrogels containing bFGF (King et al., 1998; Nakahara et al., 2003).

Chitosan is a deacetylated derivative of chitin, the second most abundant natural polymer found in the shells of crustaceans (Kim et al., 2008). Chitosan has been shown to increase type I collagen and ALP mRNA expressions in PDL cells, and chitosan scaffolds have been reported to induce the formation of new bone and cementum in one-wall intrabony defects (Yeo et al., 2005; Pang et al., 2005). Additionally, chitosan scaffolds that release growth factors such as PDGF or bFGF have been shown to accelerate PDL regeneration (Inanç et al., 2007; Akman et al., 2010).

Calcium hydroxyapatite is a main component of root cementum and alveolar bone. PDL cells cultured with nanoparticles of hydroxyapatite have been reported to significantly increase proliferation and ALP activity (Sun et al., 2007), and porous hydroxyapatite scaffolds have induced pocket depth reduction and attachment gain in PDL tissue (Carraro, 1988). Moreover, numerous hybrid calcium hydroxyapatite scaffolds, including nanohydroxyapatite/chitosan (Zhang et al., 2007), nanohydroxyapatite/collagen (Fukui et al., 2008) and hydroxyapatite/gelatin components (Rungsiyanont et al., 2011) have shown promise for PDL tissue regeneration. GEM 21S®, which consists of a combination of PDGF-BB and β-tricalcium phosphate (TCP) matrix ranging in size from 250 to 1000 μm ("small β-TCP"), has been used widely in the treatment of periodontitis.

Synthetic polymers have also been used as scaffold materials for tissue engineering. In particular, polyesters such as polyglycolic acid (PGA), polylactic acid (PLA) and polylacticcoglycolic acid (PLGA) have been frequently used for PDL regeneration (Chen et al., 2010b). Moreover, several reports have demonstrated the utility of the combination of growth factors and synthetic polymers such as GDF-5/PLGA (Kwon et al., 2010). Interestingly, Chen et al. revealed that BMP-2-loaded pellets incorporated into glycidyl methacrylated dextran/gelatin scaffolds stimulated not only bone and cementum formation, but also PDL-like connective tissue regeneration (Chen et al., 2007). Implantation of TGF-β3 in lyophilized Matrigel® scaffolds in baboon Class III furcation defects with minced fragments of rectus abdominis muscle reportedly induced responding stem cells to stimulate the formation of large amounts of alveolar bone with associated PDL fibers and newly formed cementum (Ripamonti et al., 2009).

5.3 Cells

The use of stem cells as primary cell sources is of enormous interest to clinicians and researchers for their great potential in tissue engineering applications. Sources of stem cells from human (Feng et al., 2010) and swine (Ding et al., 2010) PDL tissues are reported to be the most potent for PDL tissue regeneration, as well as canine (Kawaguchi et al., 2004) and human (Yamada et al., 2006) bone marrow and rat adipose-tissue (Tobita et al., 2008). Recent studies have attempted topical applications of stem cells combined with growth factors and/or scaffolds for PDL tissue regeneration. Indeed, the combination of platelet-rich plasma with growth factors, fibrin scaffolds and BMMSCs successfully regenerated PDL tissue in a clinical setting (Yamada et al., 2006). Adipose-derived stem cells mixed with platelet-rich plasma promoted PDL regeneration in rat PDL tissue defects (Tobita et al., 2008). Moreover, the PDL cell population has been described to include stem/progenitor cells that have the possibility to repair destructed PDL tissues by recruiting their latent regenerative potential to differentiate into fibroblasts, osteoblasts or cementoblasts in response to their microenvironment (Melcher, 1976; McCulloch et al., 1987). Bovine derived bone substitute scaffolds mixed with fibrin were shown to represent high biological affinity and support for the growth, migration and differentiation of PDL stem cells (Trubiani et al., 2008). Collagen scaffolds were shown to stimulate PDL stem/progenitor cells to attach, proliferate, exhibit a PDL spindle-like morphology and subsequently form PDL-like tissue (Luan et al., 2009). BMP-2 stimulation was shown to induce mineralized nodule formation of multipotent PDL cells, while bFGF stimulation elicited capillary-like structure formations (Shirai et al., 2009). Hydroxyapatite/tricalcium phosphate blocks, into which stem cells from root apical papilla were loaded, when furthermore coated with Gelfoam containing PDL stem cells and implanted into extracted sockets, developed PDL-like tissue around the blocks (Sonoyama et al., 2006). These results suggest the efficacy of therapeutic approaches based on tissue engineering triads in PDL tissue regeneration.

5.4 Blood supply

PDL tissues are nourished from supraperiosteal arterioles along the surface of alveolar bone, vessels of the PDL region and arterioles from the interdental septum extending into the gingival and sulcus area (Egelberg, 1967). Therefore, the lack of sufficient blood supply may restrict PDL regeneration. bFGF is a potent mediator for angiogenesis and plays an essential role in the formation of new blood vessels, including the differentiation of angioblasts and the subsequent formation of vascular tubules (Savona et al., 1997). Moreover, a recent report showed that tissue engineered PDL constructs incorporating human umbilical vein endothelial cells in vitro induced the formation of capillary-like structures, ALP activity and collagen deposition within the constructs (Nagai et al., 2009)

5.5 Innervation of peripheral nerves

Diverse functions of the nervous system have been reported in peripheral tissues. Peripheral nerves regulate DNA synthesis and proliferation of abdominal cells (Kiba et al., 1996). Denervation has been shown to suppress osteoclast activities in vertebral and mandibular bone (Imai & Matsusue, 2002). Furthermore, alveolar nerve denervation leads to a reduction in the distribution of epithelial cell rests of Malassez (Yamashiro et al., 2000), resulting in a narrowing of the PDL width and dento-alveolar ankylosis (Fujiyama et al., 2004). Neuropeptides, such as substance P and calcitonin gene-related peptides that are released from axons of sensory neurons in PDL tissue (Jacobsen et al., 1998), have been shown to stimulate angiogenesis, bone remodeling (Hukkanen et al., 1993), fibroblast proliferation and collagen organization (Burssens et al., 2005). Therefore, the peripheral nervous system also plays important roles in homeostasis and the regeneration of PDL tissue.

6. Characteristics of PDL stem cells

A "stem cell" is capable of self-renewal and multi-lineage differentiation. In other words, stem cells possess the potential to propagate and generate additional stem cells, and can differentiate and commit to maturation along multiple lineages that produce a range of specialized cell types. Human mesenchymal stem cells (MSCs) were originally isolated in aspirates of adult bone marrow by their capacity to form clonogenic clusters of adherent fibroblast colony-forming units (Friedenstein et al., 1974). Additionally, they displayed the potential to undergo extensive proliferation and to differentiate into mesodermal lineage cells, such as adipocytes, chondrocytes and osteoblasts, both in vitro and after transfer in vivo (Pittenger et al., 1999). Human MSCs are also reported to exhibit endodermic and neuroectodermic differentiation potentials (Castro-Malaspina et al., 1980; Petersen et al., 1999; Kopen et al., 1999; Gronthos & Simmons, 1995). In recent years, MSCs have been identified in various tissues throughout the human adult body including bone marrow, adipose tissue, placenta and muscle (Baroffio et al., 1996; Fukuchi et al., 2004; Pittenger et al., 1999; Zannettino et al., 2008).

It is considered that PDL tissue also harbors a small number of MSCs in niches that may be derived from DF cells, responsible for maintaining and regenerating periodontal tissue structure and function. The MSC population in PDL tissue has the potential to differentiate into fibroblasts, osteoblasts, and cementoblasts, which are critical cell populations to maintain and remodel periodontal tissue. In 2004, a multipotent PDL stem cell population, termed as PDLSC, was first isolated from the PDL tissue of extracted human third molar teeth (Seo et al., 2004). PDLSCs represent typical properties of MSCs, as well as unique properties. PDLSCs exhibit the self-renewing capacity to generate clonogenic adherent cell colonies to tissue culture plastic and resemble fibroblasts in their spindle-like shapes in vitro. Several studies have reported that PDLSCs express cell surface markers, such as STRO-1 (putative stem cell marker), CD146 (perivascular cell marker), STRO-3 (tissue nonspecific ALP), CD13, CD29 (integrin β-1), CD44, CD90 (Thy-1), CD105 (endoglin), CD106 (VCAM-1), and CD166 (ALCAM) (Nagatomo et al., 2006; Seo et al., 2004; Trubiani et al., 2005; Wada et al., 2009), similar to BMMSCs (Shi & Gronthos 2003; Gronthos et al., 2007), as well as embryonic markers, including OCT-4, SSEA-1 and SSEA-4 (Trubiani et al., 2010). PDLSCs have so far been isolated from PDL heterogeneous cell populations using STRO-1 and CD146 antibodies (Seo et al., 2004), although the percentage of STRO-1-positive cells in the PDL tissue is reportedly not to be very high (STRO-1 positive cells: 1.2-7.68%) (Nagatomo et al., 2006; Xu et al., 2009). Additionally, PDLSCs express a PDL marker, scleraxis, suggesting that PDLSCs may belong to a unique population of MSCs.

PDLSCs also possess the multipotential to differentiate in vitro into various types of cells, such as osteoblasts, adipocytes and chondrocytes, similar to BMMSCs (Gay et al., 2007). In addition to a common multipotency with BMMSCs, PDLSCs represent a unique potential to form cementum and PDL-like tissues including Sharpey's fiber-like structure in vivo, when subcutaneously transplanted into the dorsal surfaces of immunocompromised mice with a hydroxyapatite or β-TCP carrier (Seo et al., 2004). Furthermore, human PDLSCs that were transplanted into surgically created defects at the periodontal area of mandibular molars in immunocompromised rats were integrated into the PDL tissue, resulting in homing on the surfaces of alveolar bone and tooth root (Seo, et al., 2004). Swine or canine PDLSCs have also been reported to regenerate PDL tissues when transplanted into surgically created periodontal defect areas in swine or canine models (Liu et al., 2008; Park et al., 2010).

Immunomodulatory property is also known to be one of the characteristics of MSCs (Bartholomew et al., 2002). MSCs have been found to inhibit mixed lymphocyte reactions (MLR) and T-lymphocyte proliferation following mitogenic stimulation in vitro, despite expression of leucocyte antigen (HLA) major histocompatibility complex (MHC) (Bartholomew et al., 2002; Di Nicola et al., 2002), suggesting allogeneic BMMSC administration may be an appropriate stem cell therapy technique. Similarly, human and canine PDLSCs have been reported to possess the capacity to suppress immune responses and inflammatory reactions (Kim et al., 2010; Wada et al., 2009). It has been demonstrated that human PDLSCs inhibited peripheral blood mononuclear cell proliferation stimulated with mitogen or in MLR, and this inhibition was mediated by soluble factors including TGF-β1, hepatocyte growth factor (HGF) and indoleamine 2, 3-dioxygenase (IDO), known to be induced by interferon-gamma (IFN-γ) (Wada et al., 2009). In a swine model with experimental periodontal bone defects, the transplantation of a sheet of allogenic PDLSCs reversed the periodontitis through the low immunogenicity and immunosuppressive function possessed by PDLSCs (Ding et al., 2010).

In the first human trial report using ex-vivo-expanded primary PDL cells including PDLSCs, the cells were mixed with bone grafting material and transplanted into the intrabony defects of deep periodontal pockets of three patients. The results of the transplantations showed a significant improvement of periodontal diseases, suggesting that PDLSC transplantation could be a safe and promising treatment for clinical periodontitis (Feng et al., 2010). Furthermore, it has been reported that human PDLSCs recovered from cryopreserved human PDL tissue maintained normal PDLSC characteristics, including expression of

STRO-1, single-colony-strain generation, multipotency and cementum/PDL-like tissue generation, providing a possible and practical clinical approach for the utilization of PDLSCs isolated from frozen PDL tissues (Seo et al., 2005).

These findings demonstrate that PDLSCs not only possess stem cell characteristics, including self-renewal, multipotency and immunomodulatory properties similar to BMMSCs, but also hold a unique potential to fabricate three-dimensional PDL tissue.

7. PDL stem cell lines

Because there are a small number of stem cells in PDL tissues, researchers have tried to establish immortalized PDL stem cell lines for the convenience and consistency of analyses. Several genes, such as SV 40 large T-antigen, human telomerase transcriptase, human papillomavirus 16-related E6E7, Bmi-1 and BMP4, have been transduced into mammalian PDL cells. As a consequence, clonal PDL cell lines derived from mice (Saito et al., 2002), swine (Ibi et al., 2007) and humans (Kamata et al., 2004; Fujita et al., 2005; Pi et al., 2007; Mi et al., 2011) have been developed. In addition, clonal cell lines from mouse dental follicle (DF) cells as a source of PDLSCs have also been reported (Saito et al., 2005; Luan et al., 2006). However, the multipotency in these cell lines has not been analyzed in detail. One group reported that the clonal swine PDL cell lines exhibited osteoblastic differentiation and the formation of tube-like structures (Shirai et al., 2009). Recently, our group succeeded in establishing two clonal human PDL cell lines with multipotency by transducing primary PDL cells with both SV 40 large T-antigen and human telomerase reverse transcriptase genes (Fujii et al., 2006; Fujii et al., 2008; Tomokiyo et al., 2008). These clonal cell lines had unique characteristics: Cell line 1-11 demonstrated osteoblastic and adipocytic differentiation (Fig. 1) (Fujii et al., 2008), and cell line 1-17 could differentiate into osteoblasts, chondrocytes, adipocytes and neurocytes (Fig. 2) (Tomokiyo et al., 2008).

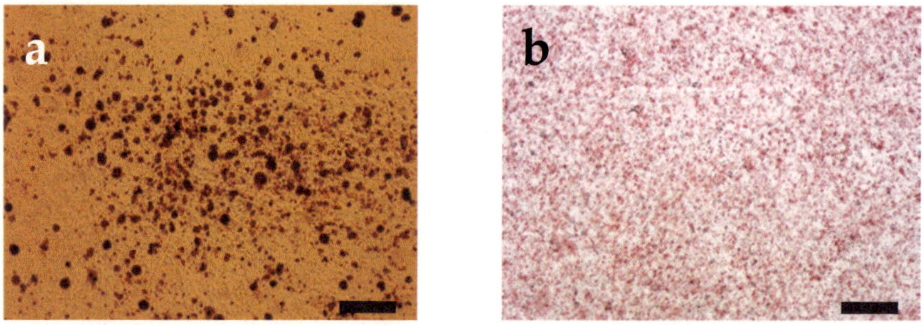

Fig. 1. Osteoblastic (a) and adipocytic (b) differentiation of cell line 1-11. Cell line 1–11 was cultured for 4 weeks in osteoblastic induction medium. von Kossa-positive deposits (dark brown dots) were observed in (a). (b): Cell line 1–11 was cultured for 18 days in adipocytic induction medium. Oil red O-positive clusters (red) were observed. Bars: 150 μm.

Flow cytometric analysis demonstrated that the percentage of cells expressing CD13, CD29, CD44, CD71, CD90, CD105, and CD166 was very high in both cell lines, while the number of cells expressing an endothelial marker, CD18 and a hematopoietic marker, CD34 was low (data shown partially in Fig. 3). Lindroos et al. (2008) also reported the expressions of CD13, CD 29, CD44, CD90, CD105, and CD166 in human PDLSCs. Cell line 1-11 expressed other MSC markers, STRO-1 and CD146/MUC18, whereas cell line 1-17 contained STRO-1 only (data not shown), although these markers are also believed to be expressed in human and mouse PDLSCs (Seo et al., 2004; Xu et al., 2009).

On the other hand, cell line 1-17 intensely expressed ES cell marker genes, OCT-4 and Nanog, whereas cell line 1-11 expressed these markers weakly (Fig. 4). However, these expressions were reduced when the cells were cultured in osteoblast induction medium (data not shown). Furthermore, both cell lines expressed neural crest (NC)-related genes, Nestin, p75NTR and Slug at almost equivalent levels. Based on these results, these cell lines were exactly derived from NC, and cell line 1-17 may be more immature than cell line 1-11.

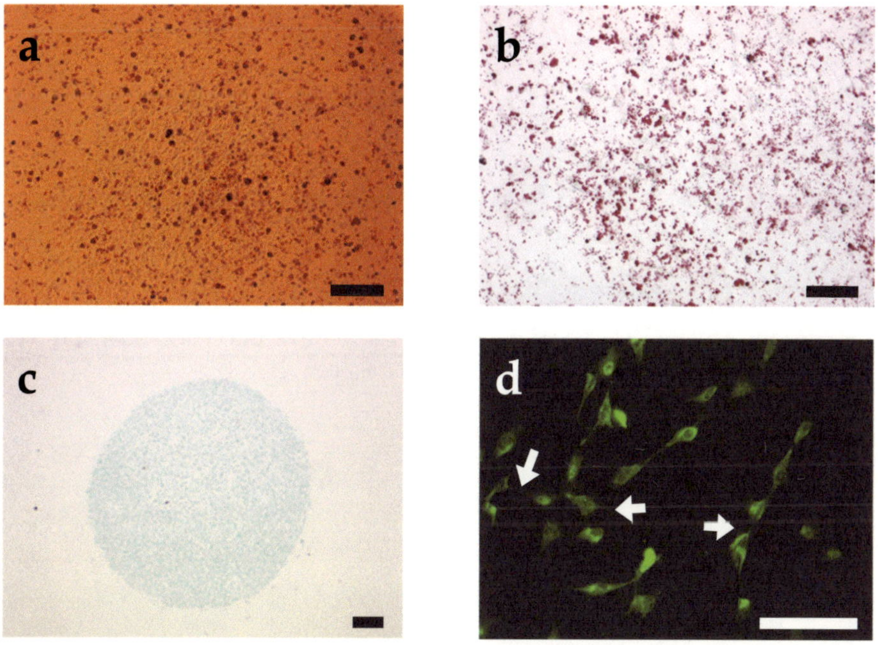

Fig. 2. Osteoblastic (a), adipocytic (b), chondrocytic (c), and neurocytic (d) differentiation of cell line 1–17. (a): Cell line 1–17 was cultured for 4 weeks in osteoblastic induction medium and then subjected to von Kossa staining. The von Kossa-positive calcified deposits (dark brown dots) were detected. (b): Cell line 1–17 was cultured for 18 days in adipocytic induction medium. Oil red O-positive clusters (red) were observed. (c): Cell line 1-17 was cultured for 4 weeks in chondrocytic induction medium and then subjected to Alcian blue staining. (d): Cell line 1-17 was cultured for 4 weeks in neurocytic induction medium and then subjected to immunocytochemical staining of human neurofilament (green). The extension of dendrite-like processes (arrows) was observed. Bars: 150 μm.

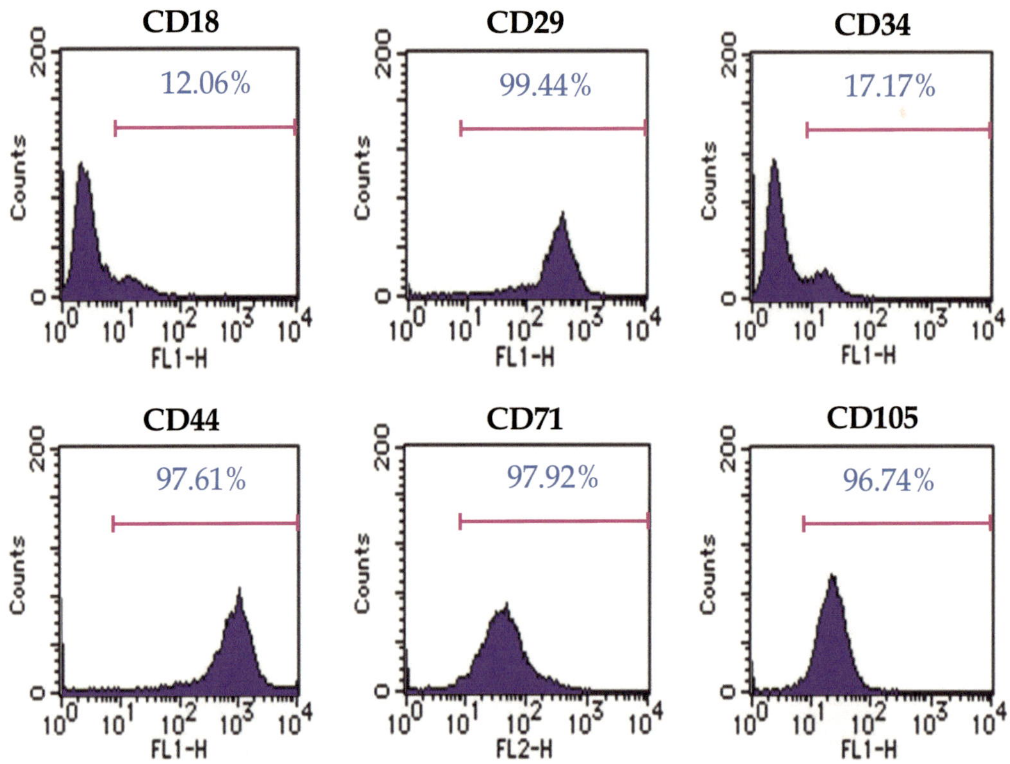

Fig. 3. Flow cytometric analysis of cell line 1-11 with antibodies reactive to the stem cell-related surface molecules CD29, CD44, CD71 and CD105, and the endothelial and hematopoietic surface molecules, CD18 and CD34, respectively.

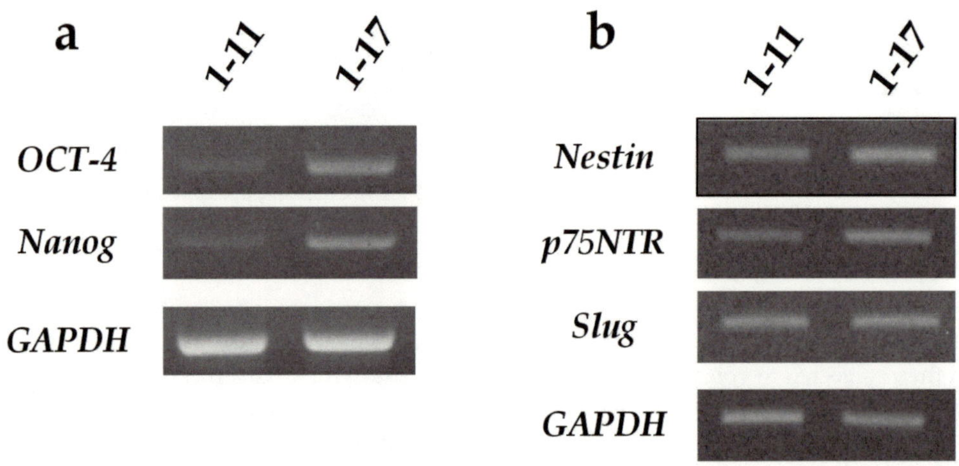

Fig. 4. Semi-quantitative RT-PCR analysis shows transcription levels of the ES cell-related pluripotency genes, OCT-4 and Nanog (a), and CNC-related genes, Nestin, p75NTR and Slug (b), in cell lines 1-11 and 1–17.

In addition, both cell lines expressed PDL phenotype-related genes, periostin and scleraxis (Yokoi et al., 2007), at almost equivalent levels while human MSCs showed little expression of the respective genes (Fig. 5a). Immunofluorescent data demonstrated the higher expression of periostin in cell line 1-17 compared with hMSCs (Fig. 5b). These data indicate that the cell lines are obviously PDL-committed cells, exhibiting the phenotypes of stem cells.

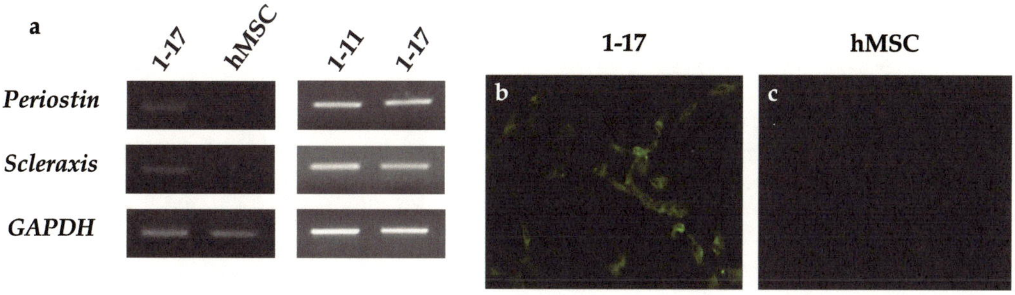

Fig. 5. Semi-quantitative RT-PCR analysis concerning the transcription of the PDL-related genes, periostin and scleraxis in cell lines 1-11 and 1-17 and in hMSCs (a). Immunocytochemical analysis of the expression of periostin in cell line 1-17 and MSCs (b).

When cell line 1-11 was subcutaneously transplanted with β-TCP into the dorsal side of immunodeficient mice for 8 weeks, the cells fabricated PDL-like tissue including Sharpey's fiber-like structures (Fig. 6a), while cell line 1-17 did not exhibit such potential (data not shown). Furthermore, when transplanted into periodontal defects formed in immunodeficient rats for 3 weeks, cell line 1-11 homed on the surfaces of bone and root and between bone and root, namely within the ligament tissue, while cell line 1-17 was localized only within the ligament tissue (Fig. 6b, c). These observations were very similar to a previous report by Seo, et al. (2004).

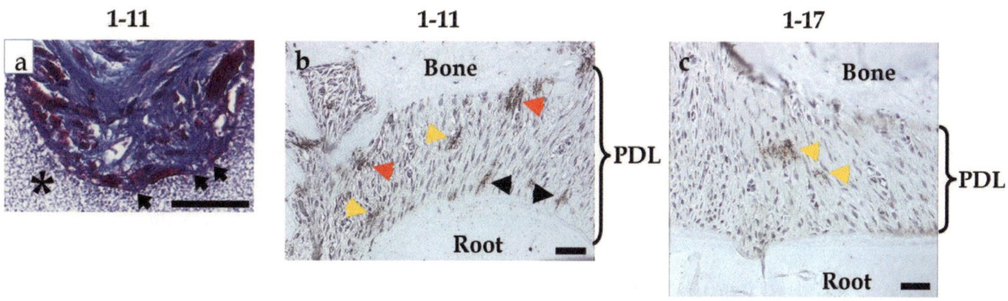

Fig.6. Trichrome staining of cell line 1-11 transplanted with β-TCP (*) into immunodeficient mice for 8 weeks (a). Sharpey's fiber-like structures were observed (arrows). Immunolocalization of anti-human mitochondria-positive cells after transplantation of cell lines 1-11 (b) or 1-17 (c) into periodontal defects in immunodeficient rats. Immuno-positive cells were detected on the surface of bone (red arrows) and tooth root (black arrows) and within ligament tissue (yellow arrows). Bars: 50 μm.

The treatment of two cell lines by bFGF provided intriguing results concerning osteoblastic differentiation. Although both cell lines cultured in osteoblastic differentiation medium produced calcified deposits at almost equivalent volumes, bFGF provoked these productions in cell line 1-11, while inhibited the productions in cell line 1-17 almost completely (Fig. 7).

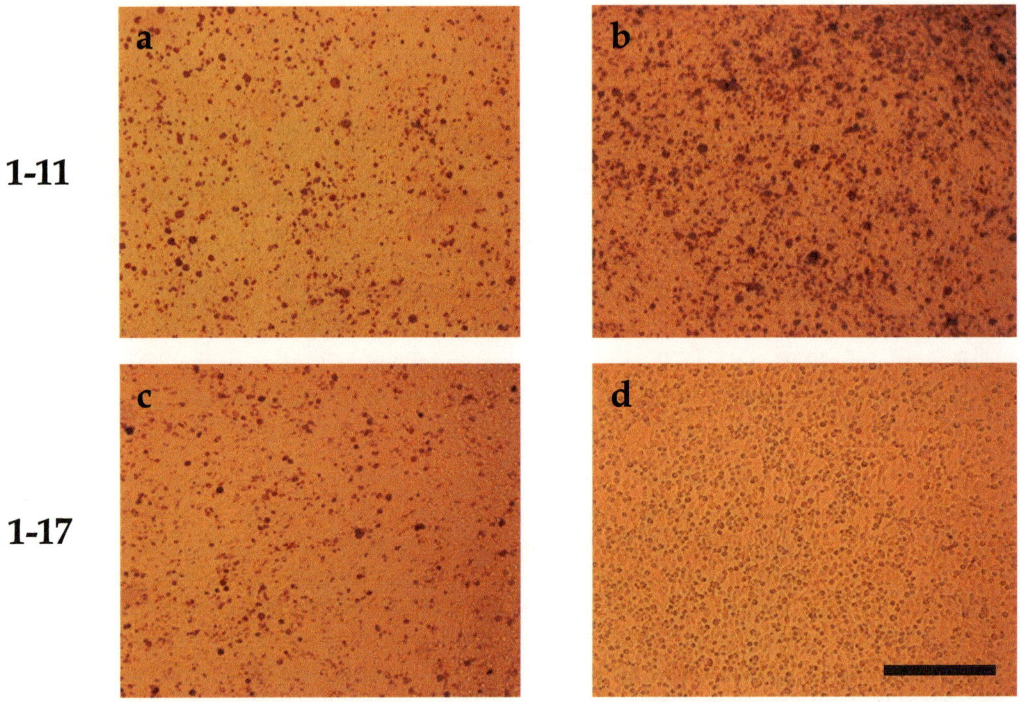

Fig. 7. von Kossa staining of cell lines 1-11 (a, b) and 1-17 (c, d) treated with (b, d) or without (a, c) bFGF in osteoblastic differentiation medium. Dark brown dots show von Kossa-positive calcified deposits. Bar: 200 μm.

Thus, our results indicate that both cell lines met the prerequisite for PDL stem cells, and furthermore, that the differentiation stage of cell line 1-17 may be lower than that of cell line 1-11, based on the differences of multipotency and phenotypes between the lines and the transplantation analyses. These differences may be related to the expression of CD146 or the responsiveness to bFGF. Therefore, we believe that there may be distinguishing factors in these cell lines that influence the direction of stem cell differentiation within the PDL tissues.

8. Conclusions and future directions

Both cell lines exhibited characteristics of PDLSCs and permitted the routine analyses associated with PDLSCs. Their phenotypes, however, were somewhat different from each other. We believe that PDLSCs originate from DF cells, which stay in a niche in the PDL tissue during and after its development, and that our established cell lines would signify the existence of stem cells at various differentiation stages in the PDL tissue (Fig. 8). A differential analysis between these two clones will allow us to further reveal the mechanism of PDLSC differentiation. These results will enable elucidation of signals that direct stem cells including MSCs, iPS cells or ES cells toward PDL-lineage cells, as well as to identify the optimum signaling molecules and scaffolds for PDL regeneration. With the integration of these elements, the potential development of an efficient therapy for PDL regeneration is likely.

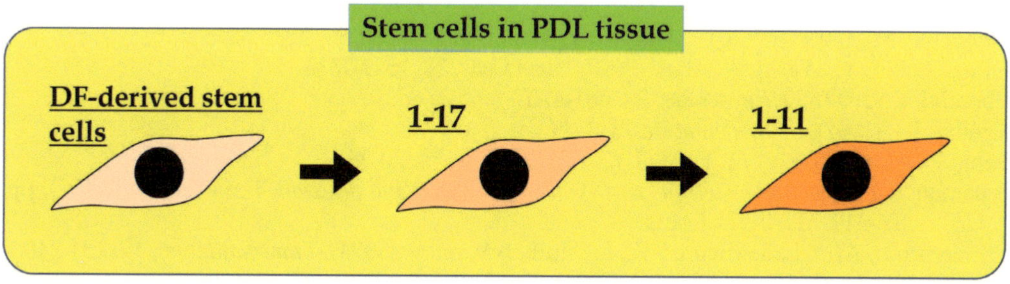

Fig. 8. Schema explaining the differentiation stages of undifferentiated cells included in the PDL tissue.

9. Acknowledgments

We thank Drs. Monnouchi, Kono, Koori, Yamamoto and Teramatsu for their great support in preparation of this chapter. This work was financially supported by grants-in-aid (Project Nos. 21390510, 21791942, 22390359 , 22592123, and 23659890) for scientific research from the Ministry of Education, Culture, Sports, Science and Technology (Japan).

10. References

Agarwal, S., Chandra, CS., Piesco, NP., et al. (1998). *Infect Immun,* 66: 932-937
Akman, AC., Tiğl,i RS., Gümüşderelioğlu, M., et al. (2010). *J Biomed Mater Res A* 92:,953-962
Asano, M., Kubota, S., Nakanishi, T., et al. (2005). *Cell Commun Signal,* 5: 3-11
Baroffio, A., Hamann, M., Bernheim, L., et al. (1996). *Differentiation,* 60: 47-57
Bartholomew, A., Sturgeon, C., Siatskas, M., et al. (2002). *Exp Hematol,* 30: 42-48
Beertsen, W., McCulloch, CAG. & Sodek, J. (1997). *Periodontology,* 13: 20-40
Belal, MH., Watanabe, H., Ichinose, S., et al. (2006). *Oral Dis,* 12: 543-552
Berkovitz, BKB. & Shore, RC. (1995). *Cells of periodontal ligament,* London: Mosby-Wolfe: p 9-33
Birkedal-Hansen, H. (1993). *J Periodontol,* 64 : 474-484
Bodine, PV., Zhao, W., Kharode, YP., et al. (2004). *Mol Endocrinol,* 18: 1222-1237
Bradham, DM., Igarashi, A., Potter, RL., et al. (1991). *J Cell Biol,* 114: 1285-1294
Brady, TA., Piesco, NP., Buckley, MJ., et al. (1998). *J Dent Res,* 77: 1779-1790
Bronner-Fraser M. (1993). *BioEssays,* 15, 221-230
Burssens, P., Steyaert, A., Forsyth, R., et al. (2005). *Foot Ankle Int,* 26: 832-839
Carinci, F., Piattelli, A., Guida, L., et al. (2006). *Oral Dis,* 12: 329-342
Carraro, JJ. (1988). *Int Dent J,* 38:,170-176.
Castro-Malaspina, H., Gay, RE., Resnick, G., et al. (1980). *Blood,* 56: 289-301
Chai, Y., Jiang, X., Ito, Y., et al. (2000). *Development,* 127: 1671–1679
Chen, FM., Zhao, YM., Zhang, R., & et al. (2007). *J Control Release,* 121: 81-90
Chen, FM., Zhang, M. & Wu, ZF. (2010a). *Biomaterials,* 31: 6279-6308
Chen, FM., Zhang, J., Zhang, M., et al. (2010b). *Biomaterials,* 31: 7892-7927
Cheng, H., Caterson, B. & Yamauchi, M. (1999). *Connect Tissue Res,* 40: 37-47

Cho, MI. & Garant, PR. (2000). *Periodontol 2000,* 24: 9–27

Cho, MI., Lin, WL. & Genco, RJ. (1995). *J Periodontol,* 66: 522-530

Chutivongse, N., Sumrejkanchanaki, P., Yongchaitrakul, T., et al. (2005). *Biochem Biophys Res Commun,* 336: 836–841

Dangaria, SJ., Ito, Y., Walker, C., et al. (2009). *Differentiation,* 78: 79-90

Dereka, XE., Markopoulou, CE., Mamalis, A., et al. (2009). *Growth Factors,* 27: 274-279

Di Nicola, M., Carlo-Stella, C., Magni, M., et al. (2002). *Blood,* 99: 3838-3843

Ding, G., Liu, Y., Wang, W., et al. (2010). *Stem Cells,* 28: 1829-1838

Ebendal, T. (1992). *J Neurosci Res,* 32: 461-470

Egelberg, J. (1967). *J Periodontal Res,* 1: 1-39

Feng, F., Akiyama, K., Liu, Y., et al. (2010). *Oral Dis,* 16: 20-28

Freeman, E. (1994) *Oral histology: development, structure, and function,* Ten Cate, A.R. (Ed.), pp. 276-312, Mosby, St Louis.

Friedenstein, AJ., Chailakhyan, RK., Latsinik, NV., et al. (1974). *Transplantation,* 17: 331-340

Fujii, S., Maeda, H., Wada, N., et al. (2006). *Cell Tissue Res,* 324: 117-125

Fujii, S., Maeda, H., Wada, N., et al. (2008). *J Cell Physiol,* 215: 743-749

Fujii, S., Maeda, H., Tomokiyo, A., et al. (2010). *Cell Tissue Res,* 342: 233-242

Fujita, T., Otsuka-Tanaka, Y., Tahara, H., et al. (2005). *J Oral Sci,* 47: 177-184

Fujita, T., Shiba, H., Van Dyke, TE., et al. (2004). *Cell Biol Int,* 28: 281-286

Fujiyama, K., Yamashiro, T., Fukunaga, T., et al. (2004). *J Dent Res,* 83: 625-629

Fukuchi, Y., Nakajima, H., Sugiyama, D., et al. (2004). *Stem Cells,* 22: 649-658

Fukui, N., Sato, T., Kuboki, Y., et al. (2008). *Biomed Mater Eng* 18:25-33

Gay, IC., Chen, S. & MacDougall, M. (2007). *Orthod Craniofac Res,* 10: 149-160

Giannobile, WV., Ryan, S., Shih, MS., et al. (1998). *J Periodontol,* 69: 129-137

Gospodarowicz, D., Neufeld, G. & Schweigerer, L. (1986). *Mol Cell Endocrinol,* 46: 187-204

Gronthos, S. & Simmons, PJ. (1995). *Blood,* 85: 929-940

Gronthos, S., Mankani, M., Brahim, J., et al. (2000). *Proc Natl Acad Sci U S A,* 97: 13625-13630

Gronthos, S., Fitter, S., Diamond, P., et al. (2007). *Stem Cells Dev,* 16: 953-963

Hakki, SS., Hakki, EE. & Nohutcu, RM. (2009). *J Periodontal Res,* 44: 794-802

Häkkinen, L., Oksala, O., Salo, T., et al. (1993). *J Histochem Cytochem,* 41: 1689-1699

Hammarstrom, L. (1997). *Ciba Found Symp,* 205: 246-255

Hasegawa, T., Yoshimura, Y., Kikuiri, T., et al. (2002). *J Periodontal Res,* 37: 405-411

Heng, NH., N'Guessan, PD., Kleber, BM., et al. (2007). *J Periodontol,* 78: 2369-2379

Heo, JS., Lee, SY. & Lee, JC. (2010). *Mol Cells,* 30: 449-454

Hukkanen, M., Konttinen, YT., Santavirta, S., et al. (1993). *Neuroscience,* 54: 969-979

Ibi, M., Ishisaki, A., Yamamoto, M., et al. (2007). *Arch Oral Biol,* 52: 1002-1008

Imai, S. & Matsusue, Y. (2002). *Microsc Res Tech,* 58: 61-69

Inanç, B., Eser Elçin, A., Koç, A., et al. (2007). *J Biomed Mater Res A,* 82: 917-926

Jacobsen, EB., Fristad, I. & Heyeraas, KJ. (1998). *Acta Odontol Scand,* 56: 220-228

Kamata, N., Fujimoto, R., Tomonari, M., et al. (2004). *J Oral Pathol Med,* 33: 417-423

Kanzaki, H., Chiba, M., Shimizu, Y., et al. (2001). *J Dent Res,* 80: 887-891

Kato, C., Kojima, T., Komaki, M., et al. (2005). *Biochem Biophys Res Commun,* 326: 147-153

Kato, M., Patel, MS., Levasseur, R., et al. (2002). *J Cell Biol,* 157: 303-314

Kawaguchi, H., Hirachi, A., Hasegawa, N., et al. (2004). *J Periodontol,* 75: 1281-1287

Kiba, T., Tanaka, K., Numata, K., et al. (1996). *Gastroenterology,* 110: 885-893

Kim, HS., Kim, KH., Kim, SH., et al. (2010). *J Periodontal Implant Sci,* 40: 265-270

Kim, IY., Seo, SJ., Moon, HS., et al. (2008). *Biotechnol Adv,* 26: 1-21

Kim, TG., Wikesjö, UM., Cho, KS., et al. (2009). *J Clin Periodontol,* 36: 589-597

Kinoshita, A., Oda, S., Takahashi, K., et al. (1997). *J Periodontol,* 68: 103-109

King, GN., King, N. & Hughes, FJ. (1998). *J Periodontol,* 69: 561-570
Kitamura, M., Akamatsu, M., Machigashira, M., et al. (2011). *J Dent Res,* 90: 35-40
Kopen, GC., Prockop, DJ. & Phinney, DG. (1999). *Proc Natl Acad Sci U S A,* 96: 10711-10716
Kwon, DH., Bennett, W., Herberg, S., et al. (2010). *J Clin Periodontol,* 37: 390-397
Lamont, RJ. & Jenkinson, HF. (1998). *Microbiol Mol Biol Rev,* 62: 1244-1263
Langer, R. & Vacanti, JP. (1993). *Science,* 260: 920-926
Le Douarin. NM., Ziller, C. & Coul, G. (1993). *Dev Biol,* 159: 24-49
Lee, CH., Moioli, EK. & Mao, JJ. (2006). *Conf Proc IEEE Eng Med Biol,* 1: 775-778
Lekic, P. & McCulloch, CA. (1996). *Anat Rec,* 245: 327-341
Lekic, P., Rojas, J., Birek, C., et al. (2001). *J Periodontal Res,* 36: 71-79
Li, H., Bartold, PM., Young, WG., et al. (2001). *J Bone Miner Res,* 16: 1068-1076
Lindemann, RA., Economou, JS. & Rothermel, H. (1988). *J Dent Res,* 67: 1131-1135
Lindroos, B., Mäenpää, K., Ylikomi, T., et al. (2008). *Biochem Biophys Res Commun,* 368: 329-335
Ling, E. & Robinson, DS. (2002). *Clin Exp Allergy,* 32: 175-178
Liu, Y., Zheng, Y., Ding, G., et al. (2008). *Stem Cells,* 26: 1065-1073
Logan, CY. & Nusse, R. (2004). *Annu. Rev. Cell Dev. Biol,* 20: 781-810
Luan, X., Ito, Y., Dangaria, S., et al. (2006). *Stem Cells Dev,* 15: 595-608
Luan, X., Dangaria, S., Ito, Y., et al. (2009). *J Dent Res,* 88: 781-791
Lukinmaa, PL. & Waltimo, J. (1992). *J Dent Res,* 71: 391-397
MacNeil, RL., Berry, JE., Strayhorn, CL., et al. (1998). *Arch Oral Biol,* 43: 779-787
McCulloch, CA., Nemeth, E., Lowenberg, B., et al. (1987). *Anat Rec,* 219: 233-242
McCulloch, CA, & Bordin, S. (1991). *J Periodontal Res,* 26: 144-154
Melcher, AH. (1976). *J Periodontol,* 47: 256-260
Melcher, AH. (1984). *Michigan: University of Michigan Press,* pp. 1-15
Mi, HW., Lee, MC., Fu, E., et al. (2011). *Gene Ther,* in press
Miura, M., Gronthos, S., Zhao, M., et al. (2003). *Proc Natl Acad Sci U S A,* 100: 5807-5812
Mizutani, S., Tsuboi, T., Tazoe, M., et al. (2003). *Oral Dis,* 9: 210-217
Murakami, S., Takayama, S., Kitamura, M., et al. (2003). *J Periodontal Res,* 38: 97-103
Nagai, N., Hirakawa, A., Otani, N., t al. (2009). *Cells Tissues Organs,* 190: 303-312
Nagatomo, K., Komaki, M., Sekiya, I., et al. (2006). *J Periodontal Res,* 41: 303-310
Nakahara, T., Nakamura, T., Kobayashi, E., et al. (2003). *Tissue Eng,* 9: 153-162
Nemoto, E., Koshikawa, Y., Kanaya, S., et al. (2009). *Bone,* 44: 805-812
Nevins, M., Giannobile, WV., McGuire, MK., et al. (2005). *J Periodontol,* 76: 2205-2215
Oyama, T., Matsushita, K., Sakuta, T., et al. (2007). *J Periodont Res,* 42: 53-61
Palmon, A., Roos, H., Reichenberg, E., et al. (2001). *J Periodontal Res,* 36: 65-70
Pang, EK., Paik, JW., Kim, SK., et al. (2005). *J Periodontol,* 76: 1526-1533
Park, JY., Jeon, SH. & Choung, PH. (2010). *Cell Transplant,* in press.
Petersen, BE., Bowen, WC., Patrene, KD., et al. (1999). *Science,* 284: 1168-1170
Pi, SH., Lee, SK., Hwang, YS., et al. (2007). *J Periodontal Res,* 42: 104-113
Pittenger, MF., Mackay, AM., Beck, SC., et al. (1999). *Science,* 284: 143-147
Quintero, JC., Piesco, NP., Langkamp, HH., et al. (1995). *J Dent Res,* 74: 1802-181
Rajshankar, D., McCulloch, CA., Tenenbaum, HC., et al. (1998). *Cell Tissue Res,* 294: 475-483
Ripamonti, U., Parak, R. & Petit, JC. (2009). *J Periodontal Res,* 44: 81-87
Roberts, WE., Mozsary, PG. & Klingler, E. (1982). *Am J Anat,* 165: 373-384
Rungsiyanont, S., Dhanesuan, N., Swasdison, S., et al. (2011). *J Biomater Appl,* in press.
Saito, M., Handa, K., Kiyono, T., et al. (2005). *J Bone Miner Res,* 20: 50-57
Saito, Y., Yoshizawa, T., Takizawa, F., et al. (2002). *J Cell Sci,* 115: 4191-4200
Savona, C., Javerzat, S., Perollet, C., et al. (1997). *Rev Prat,* 47: 2239-2243
Selvig, KA., Sorensen, RG., Wozney, JM., et al. (2002). *J Periodontol,* 73: 1020-1029

Seo, BM., Miura, M., Gronthos, S., et al. (2004). *Lancet,* 364: 149-155
Seo, BM., Miura, M., Sonoyama, W., et al. (2005). *J Dent Res,* 84: 907-912
Shi, S. & Gronthos, S. (2003). *J Bone Miner Res,* 18: 696-704
Shimabukuro, Y., Ichikawa, T., Terashima, Y., et al. (2008). *Matrix Biol,* 27: 232-241
Shirai, K., Ishisaki, A., Kaku, T., et al. (2009). *J Periodontal Res,* 44: 238-247
Si, X. & Liu, Z. (2001). *Zhonghua Kou Qiang Yi Xue Za Zhi,* 36: 23-26
Sodek, J. & Limeback, HF. (1979). *J Biol Chem,* 254: 10496-10502
Sonoyama, W., Liu, Y., Fang, D., et al. (2006). *PLoS One,* 1:e79
Sun, W., Chu, C., Wang, J., et al. (2007). *J Mater Sci Mater Med,* 18: 677-683
Takayama, S., Murakami, S., Miki, Y., et al. (1997). *J Periodontal Res,* 32: 667-675
Takeda, K., Shiba, H., Mizuno, N., et al. (2005). *Tissue Eng,* 11: 1618-1629
Takeda, K., Sakai, N., Shiba, H., et al. (2011). *Tissue Eng Part A,* in press
Takeuchi, H., Kubota, S., Murakashi, E., et al. (2009). *J Periodontal Res,* 44: 161-169
Ten Cate, AR. (2008). *Oral histology, development, structure, and function,* St. Louis: Mosby. pp. 239-267
Terashima, Y., Shimabukuro, Y., Terashima, H., et al. (2008). *J Cell Physiol,* 216: 640-650
Tobita, M., Uysal, AC., Ogawa, R., et al. (2008). *Tissue Eng Part A,* 14: 945-953
Tomokiyo, A., Maeda, H., Fujii, S., et al. (2008). *Differentiation,* 76: 337-347
Trubiani, O., Di Primio, R., Traini, T., et al. (2005). *Int J Immunopathol Pharmacol,* 18: 213-221
Trubiani, O., Orsini, G., Zini, N., et al. (2008). *J Biomed Mater Res A,* 87: 986-993
Trubiani, O., Zalzal, SF., Paganelli, R., et al. (2010). *J Cell Physiol,* 225: 123-131
Tsuboi, Y., Nakanishi, T., Takano-Yamamoto, T., et al. (2001). *J Dent Res,* 80: 881-886
Wada, N., Maeda, H., Tanabe, K., et al. (2001). *J Periodontal Res,* 36: 56-63
Wada, N., Maeda, H., Yoshimine, Y., et al. (2004). *Bone,* 35: 629-635
Wada, N., Menicanin, D., Shi, S., et al. (2009). *J Cell Physiol,* 219: 667-676
Wikesjö, UM., Sorensen, RG., Kinoshita, A., et al. (2004). *J Clin Periodontol,* 31: 662-670
Wolfman, NM., Hattersley, G., Cox, K., et al. (1997). *J Clin Invest,* 100: 321-330
Worapamorn, W., Haase, HR., Li, H., et al. (2001). *J Cell Physiol,* 186: 448-456
Xiang, J., Li, C., Dong, W., et al. (2009). *J Periodont Res,* 44: 784-793
Xu, J., Wang, W., Kapila, Y., et al. (2009). *Stem Cells Dev,* 18: 487-496
Yamada, S., Murakami, S., Matoba, R., et al. (2001). *Gene,* 275: 279-286
Yamada, Y., Ueda, M., Hibi, H., et al. (2006). *Int J Periodontics Restorative Dent,* 26: 363-369
Yamashiro, T., Fujiyama, K., Fujiyoshi, Y., et al. (2000). *Bone,* 26: 663-669
Yeo, YJ., Jeon, DW., Kim, CS., et al. (2005). *J Biomed Mater Res B Appl Biomater,* 72: 86-93
Yokoi, T., Saito, M., Kiyono, T., et al. (2007). *Cell Tissue Res,* 330: 133-145
Zannettino, AC., Paton, S., Arthur, A., et al. (2008). *J Cell Physiol,* 214: 413-421
Zhang, YF., Cheng, XR., Chen, Y., et al. (2007). *J Biomater Appl,* 21: 333-349

26

Clinical Stem Cell Imaging and *In vivo* Tracking

Sahar Mirpour and Ali Gholamrezanezhad
The Russell H. Morgan Department of Radiology and Radiological Science,
Johns Hopkins Medical Institutions
USA

1. Introduction

The field of stem cell biology and regenerative medicine is rapidly moving toward translation to clinical application (Gholamrezanezhad, Bagheri et al. 2007; Gholamrezanezhad, Mirpour et al. 2011). Stem cells are able to perform restorative functions in patients with different disorders and hence stem cell therapy may be a potential alternative to some of the current modalities of treatment, such as organ transplantation. In fact, Cell transplantation therapy offers a means to stimulate tissue repair either by direct (graft-induced) or indirect (host-induced) tissue regeneration or angiogenesis (Kraitchman and Bulte 2009). During the past years, there has been enormous progress in understanding multipotent and pluripotent stem cells and their possible applications in clinical settings (Jang, Ye et al. 2011). After preliminary observations of the benefits of these cells in preclinical studies of different human disorders, the first clinical studies using bone-marrow-derived stem cells were initiated (Kraitchman and Bulte 2008). At this instant, stem cell-based therapies have been explored to potentially regenerate healthy tissues and it is clear that they can be the source of regenerate cells for different disorders (Jang, Ye et al. 2011). Although to-date numerous trials have been reported, yet there are remarkable debates on the efficacy of stem cell and the optimal treatment regime.

Many aspects of stem cell biology and their behavior inside of the human body are still beyond our understanding. One of the main obstacles and barriers in front of clear perception of the fate of stem cells in human body is lack of a perfect and ideal tool of cell tracking in patients receiving stem cells of different origin. In fact, it has been challenging to study developmental potentials of stem cells because they reside in complex cellular environments and aspects of their distribution, migration, engraftment, survival, proliferation, and differentiation often could not be sufficiently elucidated based on limited snapshot images of location or environment or molecular markers. Therefore, reliable imaging methods to follow or track the outcome of stem cells are highly required (Kraitchman and Bulte 2008; Jang, Ye et al. 2011; McColgan, Sharma et al. 2011), in order to confirm adequate homing of the injected cells to the target organ. The cell fate can be verified in animal models using histopathologic examination of tissue. However, in clinical settings and in human studies, noninvasive methods for assessment of cell survival and engraftment will be needed to assess therapeutic efficacy. Like detecting cells microscopically, cell labeling for noninvasive *in vivo* imaging relies on targeting contrast agents to stem cells to enhance their conspicuity relative to local tissue (Kraitchman and

Bulte 2009). Both short- and long-term monitoring of stem cells in cultures and in living organisms have benefited from recently developed imaging techniques which are applied to assess cell behavior and function. Confocal and multiphoton microscopy, time-lapse imaging technology, transfection with reporter genes and also noninvasive clinical imaging modalities have provided the opportunity to monitor cell behavior in the context of a living organism. In turn, the knowledge gained through these data has brought the understanding of stem cell behavior to a new stage (Jang, Ye et al. 2011).

"Successful *in vivo* imaging requires that a contrast agent associated with a stem cell exert an "effect size" sufficient for detection by imaging hardware. Although the most attractive contrast agents for tracking are endogenous ones (ie, normal components of the stem cell), their effect size is extremely small" (Frangioni and Hajjar 2004). Frangioni et al described the 8 characteristics of an ideal imaging technology for stem cell tracking, among them biocompatibility and safety of the exogenous contrast agent was the first, and foremost factor. No need for genetic modification or perturbation to the stem cells, single-cell detection at any anatomic location, quantification of cell number, minimal or no dilution with cell division, minimal or no transfer of contrast agent to nonstem cells, noninvasive imaging in the living subject over months to years, and no requirement for injectable contrast agent were the other characteristics (Frangioni and Hajjar 2004). Unfortunately most of these features are not fulfilled by the currently applied methods of clinical *in vivo* stem cell tracking for human trials or daily clinical practice.

A. Techniques of Stem Cell Imaging

Among the different methods examined in clinical settings, radioactive labeling with radiopharmaceuticals and magnetic resonance imaging (MRI) are the most widely used methods for tracking infused stem cells, and therefore will be described in more details. However, the current noninvasive imaging approaches for tracking stem cells *in vivo* include also imaging with quantum dots (QDs), and reporter genes. We will review all the current imaging modalities and evaluate their relative advantages versus disadvantages. Because no single contrast agent/detector pair will satisfy all essential requirements of stem cell tracking in clinical settings, dual- and multimodality techniques, which combine the best features of each technology, have been recommended (Josephson, Kircher et al. 2002; Hill, Dick et al. 2003; Wickline and Lanza 2003; Doubrovin, Serganova et al. 2004; Frangioni and Hajjar 2004; Wu 2007; Zhang and Wu 2007; Higuchi, Anton et al. 2009; Gera, Steinberg et al. 2010).

2. Plain radiography and computed tomography

Plain X ray radiography and computed tomography (CT) are the most readily available clinical imaging modalities with low cost. In order to obtain images with optimum quality on plain radiographs or CT and to render stem cells visible, contrast generation requires extremely high concentrations of high-density/high–atomic number materials (e.g. iodine, gadolinium, or metals). Such contrast is difficult to achieve in clinical settings, rendering these methods unlikely to play a direct role for stem cell tracking purposes (Frangioni and Hajjar 2004).

3. Nanoparticle labeling

Semiconductor quantum dots (QDs), as new class of fluorescent probes, has been applied in noninvasive imaging, using fluorescent semiconductor nanocrystals to detect membrane molecules of interest (Jaiswal, Mattoussi et al. 2003; Lin, Xie et al. 2007; Zhang and Wu 2007; Lu, Xu et al. 2010; Pi, Zhang et al. 2010; Ramot, Steiner et al. 2010). Depending on the size and composition of these probes, they can be designed to emit different wavelengths of light, ranging from ultraviolet to nearly infrared. Although there are minimal experiences and published reports with their application for the purpose of *in vivo* stem cell tracking (Lu, Xu et al. 2010; Yukawa, Kagami et al. 2010), the photostability of QDs, as seen in their resistance to photobleaching and long-lasting fluorescence, suggest them as attractive tools for tracking stem cells *in vivo* (Lin, Xie et al. 2007; Zhang and Wu 2007). However, some concerns about the effects of QDs on stem cell proliferation and differentiation (Hsieh, Wang et al. 2006) and safety of this approach have been raised (Ramot, Steiner et al. 2010). Fortunately, recent evidences are negative regarding any adverse effect of QDs on viability, proliferation, and pluripotency of stem cells as compared with non-labeled control cells (Ruan, Shen et al. 2010; Rutten, Janes et al. 2010). Also some other obstacles, including the tendency for aggregation of QDs in the cytosol, impenetrability of QDs into cells, nonspecific binding to multiple molecules, and degradation or excretion of QDs must be overcome before QDs can realize their full potential in bioimaging and clinical settings (Zhang and Wu 2007; Pi, Zhang et al. 2010). Among different QDs suggested for the purpose of stem cell tracking, MNPs@SiO2(RITC) seems to be promising, as they offer advantages of photostability against ultraviolet light exposure, nontoxicity to human stem cells, do not affect the surface phenotype or morphology of human stem cells, and also have stable retention properties in stem cells *in vitro*. These results demonstrate that MNPs@SiO2(RITC) are biocompatible and useful tools for human MSC labeling and bioimaging. Moreover, MNPs@SiO2(RITC) have multimodal fluorescent and magnetic characteristics. Furthermore, Park et al using optical and magnetic resonance imaging, successfully detected a visible signal from labeled stem cells that were transplanted into mice (Park, Tae et al. 2010).

4. Reporter gene labeling

Reporter gene imaging has introduced as an exceptional tool to localize specific molecules of interest within living subjects (Higuchi, Anton et al. 2009; Kraitchman and Bulte 2009). The reporter gene of interest usually encodes a specific protein, which encompass the potential to interact with an imaging probe and generating signals captured and quantified by conventional imaging modalities, such as SPECT, PET, MRI, or an optical charge-coupled device (Zhang and Wu 2007). Although it has been noted that "these imaging techniques can be used to assess cell trafficking with methods that are easily translatable to humans" (Acton and Zhou 2005), several shortcoming (which are discussed later) should be overcome before reaching to clinical applications.

5. Optical imaging

Bioluminescence and fluorescence are two complementary optical imaging techniques which are used for stem cell tracking. Bioluminescence utilizes light generated by the enzyme luciferase to detect cells *in vivo*, the effectiveness of which has been confirmed in small animal models (Tang, Shah et al. 2003; Wang, Rosol et al. 2003; Wu, Chen et al. 2003; Cao, Wagers et al. 2004; Frangioni and Hajjar 2004; Min, Ahn et al. 2006; Bradbury, Panagiotakos et al. 2007; Min, Le et al. 2008; Reumers, Deroose et al. 2008; Zhang, Qiao et al. 2011). Unfortunately, luciferase genes and other candidate substrates described to date generate only visible light (spectrum: 400 to 700 nm), which has very high absorption and scatter in living tissue, precluding its application in animals larger than rats [6]. The need to stable expression of nonhuman genes and the injection of high concentrations of potentially immunogenic, nonhuman substrates, (e.g. luciferin and coelenterazine) are the other limitations of the technique. It is therefore unlikely that bioluminescence reaches clinical application [6].

Fluorescence imaging, another widely used technique of optical imaging, uses a fluorescent protein for *in vivo* imaging (eg, green fluorescent protein, small-molecule polymethines) or organic/inorganic hybrids (eg, quantum dots) (Frangioni and Hajjar 2004; Zhang and Wu 2007). Near-infrared fluorescence imaging has decreased absorption and scatter of photons at 700 to 1000 nm, which is advantage of the technique over bioluminescence. However, similar to bioluminescence, near-infrared fluorescence imaging is limited to a shallow tissue depth (4-10 cm), even with tomographic imaging methods. Hence, clinical application of fluorescence imaging likely will be restricted to near-surface imaging, such as intraoperative procedures (Reynolds, Troy et al. 1999; Tepper-Wessels, Loeschinger et al. 2001; Nakayama, del Monte et al. 2002; Bonde, Maxwell et al. 2005; Sung, Hong et al. 2009; Polzer, Haasters et al. 2010). Similar to scintigraphic and MRI techniques of stem cell tracking, major disadvantages of fluorescence imaging are the dilution of the agent with each cell division and the possibility of uptake by non-stem cells following stem cell death (Frangioni and Hajjar 2004; Zhang and Wu 2007).

Several studies have confirmed the effectiveness and practicability of reporter gene labeling with bioluminescence and fluorescence markers to track the distribution and engraftment of stem cells in animal experiments using optical imaging (Cao, Lin et al. 2006; Sheikh, Lin et al. 2007; Zhang and Wu 2007; Li, Suzuki et al. 2008). Although the signals generated by these techniques are more sensitive than those of other imaging modalities in small animals, their emissions can easily be scattered within deep tissues. Hence, this current inability to accurately determine cell depth is a major limitation (Zhang and Wu 2007).

6. Ultrasound

Regarding the wide availability and easy applicability of ultrasound and echocardiography, tracking by these techniques would be extremely convenient. In these techniques, contrast for echocardiography is generated by acoustic interfaces such as water/gas (eg, microbubbles, perfluorocarbons) (Frangioni and Hajjar 2004; Rodriguez-Porcel, Gheysens et al. 2005). “Although a single unit of contrast is on the order of 0.25 to 1 μm in diameter, the generated acoustic perturbation appears much larger” (Frangioni and Hajjar 2004). Therefore, theoretically ultrasound-based techniques have the potential to detect a single cell loaded with a single unit of contrast (Frangioni and Hajjar 2004; Klibanov, Rasche et al. 2004). However, techniques to intracellular ultrasound contrast labeling are not yet robust, and more importantly, their possible effects on cell function are not yet identified. Exactly similar to MRI and scintigraphic techniques of stem cell tracking, contrast agents are subject to dilution during cell division and transfer to non-stem cells following leakage out of the cells or cell death. Another shortcoming of echogenic contrasts is the possibility of casting acoustic "shadow" below the first unit of contrast detected, which in conjunction with low spatial resolution of ultrasound and its inaccessibility in many anatomic sites, theoretically precludes precise quantification of cell number (Frangioni and Hajjar 2004).

7. Scintigraphic tracking

Scintigraphic techniques represents the commonest method of human therapeutic cell tracking and are better suited for whole-body biodistribution studies, as compared with iron oxide labelling for MRI (Detante, Moisan et al. 2009; McColgan, Sharma et al. 2011).

- SPECT (Single Photon Emission Computed Tomography)

Two approaches for *in vivo* stem cell tracking by nuclear medicine technology (SPECT or PET) have been described: 1) direct loading with a radiotracer, and 2) enzymatic conversion and retention of a radioactive substrate or receptor-mediated binding (Frangioni and Hajjar 2004).

a. **Direct loading:** For direct loading, three well-established cell labeling radiopharmaceuticals are ^{18}F-FDG (discussed in the section of PET), ^{99m}Tc-HMPAO and ^{111}In-Oxine, with short, intermediate and long half-lives, respectively. The time period during which radiolabeled cell distribution can be tracked scitigraphically is limited by the half life and decay of their label (Steindler 2007).

   i. **^{99m}Tc-HMPAO:** Technetium (^{99m}Tc) exametazime is a radiopharmaceutical routinely used for the labeling of leucocytes to localize infections and inflammatory bowel diseases (Weldon, Joseph et al. 1995). Because of its short half life (almost 6 hours) and its gamma ray energy well adapted to conventional gamma-cameras (thus ensuring high image quality), ^{99m}Tc-HMPAO is well suited for short tracking of stem cells *in vivo* (even better than ^{111}In) (Meyerwittkopf, Bockisch et al. 1993; Detante, Moisan et al. 2009). As a lipophilic complex that is reduced into a hydrophilic complex by a glutathione-dependent mechanism following entrapment in cells, ^{99m}Tc-HMPAO provides a stable labeling marker for *in vivo* cell tracking (Neirinckx, Burke et al. 1988), and therefore it has been already used on different cell types to study cell biodistribution over periods of up to 24 hours (Detante, Moisan et al. 2009).

   ii. **^{111}In-oxine:** Indium-111 is an Auger electron emitter, which internalizes nonspecifically into both normal and malignant cells (Bindslev, Haack-Sorensen et al. 2006; Gholamrezanezhad, Mirpour et al. 2009). Thanks to the long half life of Indium-111, serial tracking of cells over 10 days using single photo electron computed tomography (SPECT) imaging is possible (Gholamrezanezhad, Mirpour et al. 2011). Given that Indium-111 oxine was approved for clinical application for labeling white blood cells to track sites of inflammation since more than 20 years ago (Lavender, Goldman et al. 1977; Kraitchman and Bulte 2009), it was an acceptable extension to label stem cells for noninvasive tracking of cells in human biodistribution studies (Lavender, Goldman et al. 1977; Gao, Dennis et al. 2001; Aicher, Brenner et al. 2003; Barkholt, Danielsson et al. 2003; Chin, Nakamoto et al. 2003; Barkholt, Danielsson et al. 2004; Brenner, Aicher et al. 2004; Maskali, Tran et al. 2004; Maskali, Tran et al. 2005; Bindslev, Haack-Sorensen et al. 2006; Legrand, Cougnenc et al. 2007). Based on the Andersson et al. study, the decline of intracellular ^{111}In concentration is most prominent during the first 6 hours post-labeling and appears to remain stable thereafter (Andersson, ForssellAronsson et al. 1996). Hence, the radiotracer seems to be suitable for clinical imaging, regarding its *in vivo* stability. In our experience in patients with liver cirrhosis, we found no unexpected adverse events or complications during labeled stem cell infusion and within one month follow up period (Gholamrezanezhad, Mirpour et al. 2011). The cells can be visualized *in vivo* up to at least 10 days after initial infusion and it is possible to quantify and follow them using radio-labeling with ^{111}Indium derivatives (Gholamrezanezhad, Mirpour et al. 2011).

The main problem with application of ^{111}Indium-labed chemicals for stem cell tracking is the concern about its cytotoxic effects on stem cells which will adversely affect their possible therapeutic potential for patients. As the first experiment on human stem cells, we designed a study to continuously pursue the effect of Indium-111 labeling on stem cell viability during the 2-week period of post-labeling. After culturing mesenchymal stem cells (MSCs), we divided the cells into six samples, each of which contained $1x10^6$ MSCs. The first sample was considered as the control. The remaining five samples (samples 2-6) were labeled with the following doses of ^{111}In-oxine, respectively: 0.76, 1.64, 3.48, 5.33, and 7.16 MBq/10^6 MSCs. To evaluate the effects of ^{111}In-oxine labeling on cellular viability and count, all samples were examined immediately after labeling (2 h) as well as 24, 48 h, and 5, 7, and 14 days post-labeling (Gholamrezanezhad, Mirpour et al. 2009). Fortunately, no statistically significant relationship was found between labeling efficiency and administered dosage of radiopharmaceutical. In fact, the labeling efficiency showed no reduction with lower doses of the radiotracer applied to the incubation media. However, a significant inverse relationship was noted between radiotracer dose and viability of stem cells during the 2-week period of follow-up. For example, with the specific activity of 4.98 MBq/10^6 MSCs, more than 80% of the cells lost their viability within 2 weeks of follow-up. The cytotoxicity feature with the lower doses was much less (between 20%–60%), but never was equal to zero. Hence, in our opinion, one of the main disadvantages inherent to ^{111}In-oxine labeling of stem cells for the purpose of *in vivo* tracking is its time and dose dependent side effects. Although it has been shown that radio-labeling is one of the most sensitive methods for stem cell tracking, we recommended that the application of this tracking technique should be treated with great caution, and if necessary, as little of ^{111}In-oxine as possible should be added to the cells (or only a limited fraction of the cells should be labeled) to minimize cell loss. Such a conclusion is also confirmed by other similar reported experiments (Correa, Mesquita et al. 2007). Hence, in our second experience to track stem cells after intravenous infusion to patients with liver cirrhosis [3], we just labeled 50% of the stem cells. Labeled cells were infused from one arm of patients and the non-labeled cells were infused from the opposite arm to reduce any possible adverse effects. Also the labeling dose was kept to the lowest possible, providing good-quality images [Fig. 1]. Such a concern also persists when using technetium derivatives, although with much less severity. In Detante et al experiment, ^{99m}Tc labeling (even at low doses of 5 Bq/cell) induced a loss of stem cells' ability to form colonies. However, they found no deleterious effect on cell proliferation with HMPAO alone and therefore they concluded that loss of stem cells' proliferation seems to be due to the ^{99m}Technetium. Although their results of flow cytometry with 7-AAD demonstrated a good cell viability over the 2 days of experiment, the authors emphasized that the alteration of the ability of radiolabeled stem cells to proliferate must be carefully considered, as it may alter their therapeutic benefit (Detante, Moisan et al. 2009).

It also should be stated that although nuclear medicine techniques to track stem cells following injection or infusion to patients are highly helpful and lack some of the remarkable disadvantages inherent to other modalities (such as low sensitivity of MRI), they suffer from technique-specific shortcomings. These shortcomings seem to be more prominent when the technique is applied in clinical and *in vivo* settings. It is possible that stem cells labeled with radiotracers are phagocytosed by the reticuloendothelial cells. We

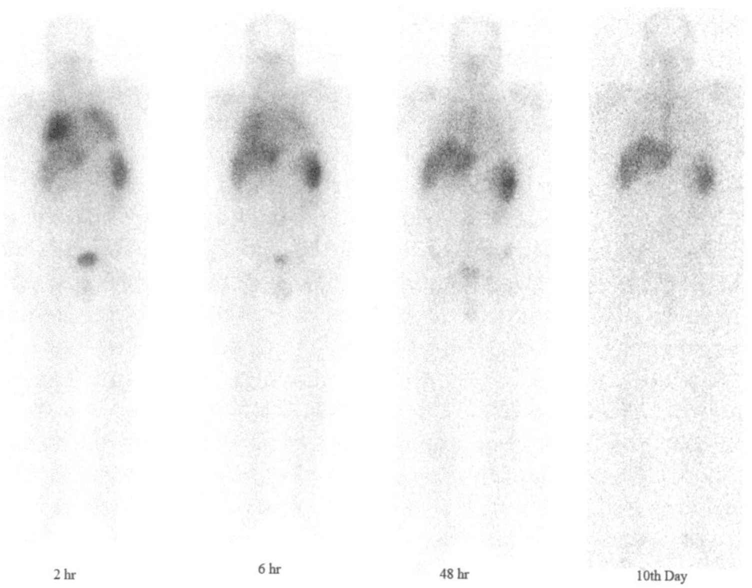

Fig 1. After I.V. infusion, the radioactivity was first observed to accumulate in the lungs, and gradually shifted to the liver and spleen. In this patient the initial hepatic accumulation of radioactivity was visualized at the second-hour images. Also, the early images (2h) demonstrated faint 111In activity in the spleen. During the following hours to days, the radioactivity became more prominent in the liver and spleen.

cannot exclude the possibility that radionuclide signals are detected from the reticuloendothelial cells, which theoretically are able to phagocytize nonviable stem cells (for example, due to time-dependent cytotoxic effect of the 111In-oxine, after infusion, a limited portion of stem cells will lose viability (Gholamrezanezhad, Mirpour et al. 2009). In fact, no one can be sure that the observed radioactivity *in vivo* reveals the presence of viable stem cells in the organs having radioactivity. However, it should be emphasized that this limitation is not limited to scintigraphic tracking methods: MR tracking of stem or other cells has a critical flaw that the ferromagnetic particles do not represent viable cells *in vivo*. We simply do not know whether MR signals mean magnetic particles in the "live" cells or "already dead" cells or even particles in the engulfed/digested cells. While label detachment from the cell is a primary concern for direct labeling techniques, similar issues exist with reporter gene imaging, where the reporter gene product may still be present after the cell has died (Gilad, Winnard et al. 2007). In other words, direct loading is problematic given the tradeoff between half-life and long-term exposure to inherently cytotoxic ionizing radiation and given the possibility of transfer of the radiolabel from stem cells to nonstem cells (Frangioni and Hajjar 2004).

Technical consideration: Region of Interest (ROI) analysis is a simple tool to measure the percentage of cells homing to the target organ. Decay and background corrections and geometric mean calculation are techniques to improve the reliability and accuracy of the data (Gholamrezanezhad, Mirpour et al. 2011). In animal studies, biodistribution analysis is possible, in which, organ, blood, and urine activities of labeled stem cells are expressed as count per minute (cpm), and then decay-correction to the time of injection, and normalization to the tissue weight (gram) and to the injected dose is done. Simply, these results can be expressed as percent of the injected dose per gram of sample weight (%ID/g) (Detante, Moisan et al. 2009).

b. Enzymatic conversion/retention and Receptor-mediated targeting methods:

As fully explained by Zhang et al, these methods have been applied with both scintigraphic [SPECT and PET studies] and MRI substrates (Zhang and Wu 2007). Hence, the following features can be applied to both MRI and scintigraphic techniques: Receptor-mediated targeting needs stable expression of a receptor not found elsewhere throughout the body and intravenous injection of a labeling agent (Acton and Zhou 2005; Serganova and Blasberg 2005; Shah 2005; Belmar, So et al. 2007; Zhang and Wu 2007; Roelants, Labar et al. 2008; Higuchi, Anton et al. 2009; Kraitchman and Bulte 2009; Nyolczas, Charwat et al. 2009; Pomper, Hammond et al. 2009). For example, inducible promoters can be employed to modulate the level of reporter and therapeutic gene expression. Sodium iodide symporter gene for reporter gene imaging by I-124-PET has been also suggested (Higuchi, Anton et al. 2009). Tissue-specific promoters may also be employed as sensors of stem cell differentiation (Klug, Soonpaa et al. 1996; Zhang and Wu 2007). The noteworthy advantages of this technique are the ability to follow stem cells indefinitely after stable integration of the transgene, no marker dilution by cell division, and the ability to destroy stem cells by administration of a suicide drug specific for the enzyme. The disadvantages are the necessity to genetically manipulate stem cells ex vivo and intravenous administration of another dosage of labeling agent for each imaging session (Zhang and Wu 2007). Before reporter gene imaging can fully and safely be applied clinically, several problems must be solved, the main example of which are transfection stability, minimizing immunologic response and minimizing the potential interference of stem cell function and differentiation by vector transfection or transduction (Chung, Andersson et al. 2002; Li, Fehse et al. 2002; Jakobsson, Rosenqvist et al. 2004; Zhang and Wu 2007; Olmer, Haase et al. 2009; Tran, Lair et al. 2010).

Receptor-mediated targeting, which is briefly described previously, requires stable expression of a receptor not found elsewhere throughout the body and intravenous injection of a labeling ligand of the receptor.

- Positron Emission Tomography (PET)

Due to its high spatial resolution and high sensitivity, PET can be considered a high-quality imaging tool for stem cell tracking (Kang, Kang et al. 2006). ^{18}F-FDG has been considered a candidate radiotracer for labeling stem cells. As glucose uptake is a physiologic process, uptake by stem cells does not need any cellular modification and occurs *in vitro* in the same manner as *in vivo* without specific manipulation (Tamura, Unno et al. 2004). Few studies have indicated that stem cells labeled with ^{18}F-FDG could be used to monitor cell homing into the target tissues and to track the biodistribution of the injected cells (Hertenstein, Wollert et al. 2004; Tamura, Unno et al. 2004; Wollert, Hofmann et al. 2004; Hofmann, Wollert et al. 2005).

^{18}F-FDG [2-deoxy-2-(^{18}F)fluoro-D-glucose], a glucose analog with the positron-emitting radioactive isotope fluorine-18 substituted for the normal hydroxyl group at the 2' position in the glucose molecule, is the most widely used label for the purpose of stem cell tracking. The radioactivity elimination half-life of 110 minutes.

Overall, PET has the advantage of the higher sensitivity than SPECT, which allows more accurate quantification of cell number (Frangioni and Hajjar 2004). Although both strategies mentioned above for SPECT stem cell tracking can be used for stem cell tracking with PET, the most advanced by far has been stated to be the stable integration of a mutant herpes simplex type 1 thymidine kinase (TK) into stem cells and periodic intravenous injection of the TK substrate 18FHBG (Wu, Chen et al. 2003; Frangioni and Hajjar 2004). The main advantage of the technique is its ability to track and quantify stem cells over the course of months; however disadvantages such as necessity of genetic manipulation of stem cells, an infrastructure for 18F chemistry, a PET scanner, and concerns about toxic effects of radiation exposure to the stem cells and subject (albeit it intermittent) should be kept in mind (Frangioni and Hajjar 2004).

Also other shortcomings inherent to both SPECT and PET-based tracking of stem cells, which are comprehensively discussed by Frangioni and Hajjar, are: nonspecific uptake of the radiotracer by normal tissue, relatively low efficiency of collimated SPECT cameras, and photon attenuation by tissue (Frangioni and Hajjar 2004). Although tissue photon attenuation correction is currently employed very easily and brings the advantage of coregistering anatomic (CT) and physiologic (SPECT or PET) data, it can theoretically reduce sensitivity, and prevents accurate quantification of stem cell number (Chin, Nakamoto et al. 2003; Frangioni and Hajjar 2004) and also increases the radiation burden to the subject and also the cost.

8. MRI

MRI and magnetic resonance spectroscopy are common imaging modalities for *in vivo* stem cell tracking in both preclinical and clinical settings (Hoehn, Kustermann et al. 2002; Modo, Cash et al. 2002; van den Bos, Wagner et al. 2003; Kustermann, Roell et al. 2005; Crowder, Brant et al. 2006; Jansen, Shamblott et al. 2006; Au, Lam et al. 2007; Hsiao, Tai et al. 2007; Magnitsky, Walton et al. 2007; Partlow, Chen et al. 2007; Politi, Bacigaluppi et al. 2007; Suzuki, Cunningham et al. 2007; Dash, Chan et al. 2008; Himmelreich and Hoehn 2008; Kim and Song 2008; Terrovitis, Stuber et al. 2008; Wang, Wang et al. 2008; Wu, Hu et al. 2008; Kedzioreik, Ouwerkerk et al. 2009; Kim, Huh et al. 2009; Lythgoe 2009; Reekmans, Tambuyzer et al. 2009; Solanky, Chung et al. 2010; Tseng, Shih et al. 2010; Chien, Hsiao et al. 2011). Extraordinary three dimensional and anatomic resolution and high safety profile are the main advantages. To be tracked in human body and the target organ, stem cells should be enriched with a contrast agent producing a sufficient positive or negative signal to distinguish them from the non-stem cell background (Zhang and Wu 2007). The three major techniques to label stem cells for the purpose of MRI tracking are: (1) non-specific direct cell labeling, (2) indirect specific (i.e., receptor-mediated) cell labeling, which is accomplished by complexing an MR-visible contrast with a ligand that binds to a stem cell-specific receptor; and (3) reporter gene probe labeling, which is transfection with a reporter gene that expresses either an enzyme or receptor that can be detected using MRI (Gilad, Winnard et al. 2007; Kraitchman and Bulte 2008). For MRI cell labeling, direct and receptor-based labeling have been most extensively used. MRI imaging techniques are also can be divided into two groups: those generating primarily T1 contrast and T2/T2* contrast (Frangioni and Hajjar 2004).

T1-weighted contrast agents are those that utilize the lanthanide gadolinium (Gd^{3+}), loaded via pino/endocytosis into stem cells. The main problem is that Gd^{3+}-based contrast requires 50- to 500-μmol/L concentrations of low-molecular-weight Gd^{3+}-containing molecules or attachment to bulky scaffolds (e.g. dendrimers and dextrans) to augment the T1 effect (Frangioni and Hajjar 2004). This methods permit stem cell tracking for up to 6 weeks (Bulte, Douglas et al. 2001; Bulte, Douglas et al. 2002; Modo, Cash et al. 2002; Frangioni and Hajjar 2004).

T2/T2* contrast is by far the most commonly applied technique of stem cell imaging using MRI. Since early 1990s, different superparamagnetic iron oxide nanoparticles and more recent formulations to improve cell loading (Tat peptides) have been used (Lewin, Carlesso et al. 2000; Daldrup-Link, Rudelius et al. 2003; Frank, Miller et al. 2003; Frangioni and Hajjar 2004). Most of them are biocompatible, safe, and nontoxic, and some already have been approved by the FDA (Frangioni and Hajjar 2004). Because of the magnification effect at high field strengths, superparamagnetic nanoparticles can be used to track extremely small numbers of cells for up to several weeks.

Previous studies have confirmed that stem cell traffic throughout the body can be noninvasively monitored in MRI by labeling cells with micrometer-sized iron oxide particles or superparamagnetic iron oxides (Mancardi, Saccardi et al. 2001; Saiz, Carreras et al. 2001; Daldrup-Link, Rudelius et al. 2003; Saiz, Blanco et al. 2004; Magnitsky, Watson et al. 2005; Cheung, Chow et al. 2006; Brekke, Williams et al. 2007; Roberts, Price et al. 2007; Suzuki, Yeung et al. 2007; Arbab and Frank 2008; Kraitchman and Bulte 2008; Kraitchman, Gilson et al. 2008; Kustermann, Himmelreich et al. 2008; Mani, Adler et al. 2008; Wang, Wang et al. 2008; Chapon, Jackson et al. 2009; Sung, Hong et al. 2009; Yamada, Gurney et al. 2009; Yang, Liu et al. 2009; Drey, Ewert et al. 2010; Nohroudi, Arnhold et al. 2010; Flexman, Cross et al. 2011; Jang, Ye et al. 2011; Liu, Wang et al. 2011). Yang et al have successfully tracked stem cells following administration to animals with myocardial infarction (MI) and showed that signal attenuation is observed at MI sites for the group of cases with MI (suggesting labeled cells accumulation), while no remarkable signal attenuation is observed for the sham group (Yang, Schumacher et al. 2011). Based on such observations, it seems that the ferromagnetically labeled stem cells could provide a tool to determine whether the stem cells reach the target organ and, if so, how long they remain there. Furthermore, tracking of labeled stem cells with MRI can be used to ensure appropriate cell delivery and homing and provide insight into the preferred sites of engraftment. From this information, application of individualized stem cell therapy will be possible, as the most appropriate dosage and timing of stem cell infusion can be determined to optimize the cell-based treatment based on the clinical characteristics of each individual patient (Kraitchman and Bulte 2008).

Drawbacks of MRI

An important limitation of MRI is that it may not distinguish iron-labeled cells from free iron particles. Transfer of iron oxide particles from originally labeled cells to macrophages can lead to false interpretation of MRI data, which imposes additional disadvantage for *in vivo* tracking of stem cells using MRI (Jang, Ye et al. 2011). Also, some authors claim that MRI tracking of stem cells suffers from low sensitivity (Boersma, Tromp et al. 2005; Kraitchman, Tatsumi et al. 2005). In fact, initial cell-labeling techniques were hampered by limited concentration of internalized contrast agent, which lead to the limited sensitivity of MRI to detect the labeled cells (Kraitchman, Tatsumi et al. 2005). To compensate for this limited sensitivity, experimental cell-tracking studies were performed using MR imagers with very high magnetic field strengths of up to 14 Tesla (Weissleder, Cheng et al. 1997), which are not routinely available for clinical purposes. Schoepf et al. asserted that with clinical 1.5 Tesla MR and clinically applicable contrast agents, it is not feasible to trace the *in vivo* biodistribution of intravenously infused cells to more than one final target organ, or to depict the migration of the transplanted cells to several subsequent target organs over time (Schoepf, Marecos et al. 1998). Other disadvantages of MRI tracking of stem cells is the limited number of probes available (Boersma, Tromp et al. 2005) and concerns about toxic effects of MRI contrast agents on stem cells (Crabbe, Vandeputte et al. 2010; Nohroudi, Arnhold et al. 2010). It has been demonstrated that cell viability and migratory potential of stem cells are negatively correlated with incorporated magnetic particles, presumably due to interference with the actin cytoskeleton of the cells (Nohroudi, Arnhold et al. 2010). A significant limitation of MRI is that presence of some implantable devices, such as pacemakers and defibrillators, should be considered as contraindications to scanning. Although some authors suggest that patients with pacemakers can be scanned safely at 1.5 T (Martin, Coman et al. 2004), the role of MRI in clinical settings of stem cell tracking remains questionable, as patients with cardiac devices are undoubtedly major candidates of stem cell therapies, and cardiac MRI is not readily available at all institutions (Frangioni and Hajjar 2004).

B. *In vivo* experiences

Regarding the difficulties and serious complications inherent to invasive techniques of administration or delivery of stem cells to the patients' target organs, such as hemorrhage or contrast nephropathy following angiography, stem cell delivery and administration via intravenous infusion has been suggested (Detante, Moisan et al. 2009). However, there is a remarkable concern that following peripheral vein infusion of stem cells, many of cells may entrap in tissues other than the target organ, with no clinical benefit for the patient. Thus, there is a need to track stem cells after peripheral vein infusion, and to provide information about the bio-distribution of the cells delivered by such a route, in order to confirm adequate homing of the injected cells to the target organ (Gholamrezanezhad, Mirpour et al. 2011). Also these data are necessary to assess the risks of a systemic cell therapy (Detante, Moisan et al. 2009). Both animal (Gao, Dennis et al. 2001; Kraitchman, Tatsumi et al. 2005; Detante, Moisan et al. 2009) and human (Gholamrezanezhad, Mirpour et al. 2011) experiences have shown that after intravenous infusion, the labeled stem cells first accumulate in the lungs. Delivery by left ventricular cavity infusion results in drastically lower lung uptake and better uptake in the target tissue (Barbash, Chouraqui et al. 2003). During the following hours to days, the radioactivity gradually moves toward and increases in the liver and spleen. In fact, over the next 48 hours, significant proportions of cells escape from the lungs' capillaries and migrate to systemic circulation. Such an increase in spleen and liver is in contrast with the overall activity reduction which is observed in all other organs and suggests that stem cells could be sequestered in the spleen and liver (Detante, Moisan et al. 2009; Gholamrezanezhad, Mirpour et al. 2011).

Future outlook and conclusion

Although remarkable advances have been made in the development of clinically applicable stem cell tracking techniques, given the inherent limitations of currently available modalities, future investigation should focus on improving sensitivity and spatial resolution, while decreasing patient exclusion and study complexity and cost.

At this time, no imaging modality seems to be perfect in all aspects and each modality presents its own set of advantages and disadvantages. X-ray techniques (plain radiography and CT) suffer from inadequate contrast sensitivity for stem cell tracking in the clinical setting (Frangioni and Hajjar 2004). Ultrasound/echocardiography shortcomings are mainly due to their limited anatomic accessibility, resolution, and quantification, although they have the potential for single-cell detection (Frangioni and Hajjar 2004). MRI, although has the advantages of a safe profile and 3-dimensional capacity, has the lowest sensitivity (10^{-3} to 10^{-5} mol/L), has limited availability at some institutions and is contraindicated in patients with implantable devices (Wu 2007). Scintigraphic techniques have a fair sensitivity (10^{-8} to 10^{-9} mol/L), but are not suitable for long-term cell tracking in the case of labeling with short half life tracers (limited by the radioisotope decay) and concerns about cytotoxic effects limits the application of those radiopharmaceuticals with longer half life. Moreover, long-term tracking using scintigraphic techniques requires genetic manipulation of the stem cell, stable expression of a transgene, and multiple exposures to potentially hazardous ionizing radiation (Frangioni and Hajjar 2004). Solid-state nanotechnology modalities and reporter gene labeling, although far-flung at present, seem to be promising, as they could potentially provide noninvasive, real-time monitoring of anatomic location of single stem cells. However, concerns such as possible dispersals of QDs in the cytosol need to be resolved. Near-infrared fluorescence (10^{-9} to 10^{-12} mol/L) and bioluminescence (10^{-15} to 10^{-17} mol/L) imaging techniques are the most sensitive among the available modalities, but both are constrained to a relatively shallow tissue depth, which limits their efficacy in clinical settings. In fact, they are most suitable for small animal studies and near-surface or histological applications (Frangioni and Hajjar 2004). Reporter gene imaging, though able to evaluate cell fate by assessing the viability and proliferation of the cells more accurately, will need to concerns about its safety (because of the genetic modification) should be resolved (Zhang and Wu 2007).

Therefore, one should ask what biologic questions need to be addressed before selecting an appropriate imaging technique (Zhang and Wu 2007). As it is well summarized by Zhang, if the exact location of cell homing must be identified, then MRI is the preferred method. If only the short-term biodistribution of transplanted cells is to be determined, then scintigraphic modalities are suitable. If knowledge of the long-term fate of transplanted stem cells is sought, then reporter gene imaging is best available technique (Zhang and Wu 2007). As an alternative solution, multimodality imaging approaches with a tailored combination of two or more imaging techniques have been suggested, which may minimize the potential drawbacks of using each imaging modality alone and may provide the most ideal information profile for clinical applications (Zhang and Wu 2007; Higuchi, Anton et al. 2009).

9. References

Acton, P. D. and R. Zhou (2005). "Imaging reporter genes for cell tracking with PET and SPECT." Quarterly Journal of Nuclear Medicine and Molecular Imaging 49(4): 349-360.

Aicher, A., W. Brenner, et al. (2003). "Assessment of the tissue distribution of transplanted human endothelial progenitor cells by radioactive labeling." Circulation 107(16): 2134-2139.

Andersson, P., E. ForssellAronsson, et al. (1996). "Internalization of indium-111 into human neuroendocrine tumor cells after incubation with indium-111-DTPA-D-Phe(1)-octreotide." Journal of Nuclear Medicine 37(12): 2002-2006.

Arbab, A. S. and J. A. Frank (2008). "Cellular MRI and its role in stem cell therapy." Regen Med 3(2): 199-215.

Au, W. Y., W. M. Lam, et al. (2007). "A magnetic resonance imaging study of iron overload in hemopoietic stem cell transplant recipients with increased ferritin levels." Transplantation Proceedings 39(10): 3369-3374.

Barbash, I. M., P. Chouraqui, et al. (2003). "Systemic delivery of bone marrow-derived mesenchymal stem cells to the infarcted myocardium - Feasibility, cell migration, and body distribution." Circulation 108(7): 863-868.

Barkholt, L., R. Danielsson, et al. (2003). "Indium-111 labeled donor lymphocyte infusion against liver metastases after allogeneic hematopoietic stem cell transplantation in patients with renal and colon carcinoma." Bone Marrow Transplantation 31: S148-S149.

Barkholt, L., R. Danielsson, et al. (2004). "Indium-III labelled donor lymphocyte infusion via hepatic artery or IV against liver metastases of renal and colon carcinoma after allogeneic haematopoietic stem cell transplantation." Biology of Blood and Marrow Transplantation 10(2): 80-80.

Belmar, C., P. W. So, et al. (2007). "Non-invasive genetic imaging for molecular and cell therapies of cancer." Clinical & Translational Oncology 9(11): 703-714.

Bindslev, L., M. Haack-Sorensen, et al. (2006). "Labelling of human mesenchymal stem cells with indium-111 for SPECT imaging: effect on cell proliferation and differentiation." European Journal of Nuclear Medicine and Molecular Imaging 33(10): 1171-1177.

Bindslev, L., M. Haack-Sorensen, et al. (2006). "Labelling of human mesenchymal stem cells with indium-111 for SPECT imaging: effect on cell proliferation and differentiation." Eur J Nucl Med Mol Imaging 33(10): 1171-1177.

Boersma, H. H., S. C. Tromp, et al. (2005). "Stem cell tracking: Reversing the silence of the lambs..." Journal of Nuclear Medicine 46(2): 200-203.

Bonde, J., D. J. Maxwell, et al. (2005). "*In vivo* imaging and tracking of defined, repopulating human stem cell subsets using fluorescence conjugated nanoparticles in a clinically applicable ex vivo protocol." Blood 106(11): 849a-849a.

Bradbury, M. S., G. Panagiotakos, et al. (2007). "Optical bioluminescence imaging of human ES cell progeny in the rodent CNS." Journal of Neurochemistry 102(6): 2029-2039.

Brekke, C., S. C. Williams, et al. (2007). "Cellular multiparametric MRI of neural stem cell therapy in a rat glioma model." Neuroimage 37(3): 769-782.

Brenner, W., A. Aicher, et al. (2004). "111In-labeled CD34+ hematopoietic progenitor cells in a rat myocardial infarction model." J Nucl Med 45(3): 512-518.

Bulte, J. W., T. Douglas, et al. (2001). "Magnetodendrimers allow endosomal magnetic labeling and *in vivo* tracking of stem cells." Nature Biotechnology 19(12): 1141-1147.

Bulte, J. W. M., T. Douglas, et al. (2002). "Monitoring stem cell therapy *in vivo* using magnetodendrimers as a new class of cellular MR contrast agents." Academic Radiology 9: S332-S335.

Cao, F., S. Lin, et al. (2006). "*In vivo* visualization of embryonic stem cell survival, proliferation, and migration after cardiac delivery." Circulation 113(7): 1005-1014.

Cao, Y. A., A. J. Wagers, et al. (2004). "Shifting foci of hematopoiesis during reconstitution from single stem cells." Proceedings of the National Academy of Sciences of the United States of America 101(1): 221-226.

Chapon, C., J. S. Jackson, et al. (2009). "An *in vivo* multimodal imaging study using MRI and PET of stem cell transplantation after myocardial infarction in rats." Mol Imaging Biol 11(1): 31-38.

Cheung, J. S., A. M. Chow, et al. (2006). "Cell number quantification of USPIO-labeled stem cells by MRI: an *in vitro* study." Conf Proc IEEE Eng Med Biol Soc 1: 476-479.

Chien, L. Y., J. K. Hsiao, et al. (2011). "*In vivo* magnetic resonance imaging of cell tropsim, trafficking mechanism, and therapeutic impact of human mesenchymal stem cells in a murine glioma model." Biomaterials 32(12): 3275-3284.

Chin, B. B., Y. Nakamoto, et al. (2003). "111In oxine labelled mesenchymal stem cell SPECT after intravenous administration in myocardial infarction." Nucl Med Commun 24(11): 1149-1154.

Chung, S. M., T. Andersson, et al. (2002). "Analysis of different promoter systems for efficient transgene expression in mouse embryonic stem cell lines." Stem Cells 20(2): 139-145.

Correa, P. L., C. T. Mesquita, et al. (2007). "Assessment of intra-arterial injected autologous bone marrow mononuclear cell distribution by radioactive labeling in acute ischemic stroke." Clinical Nuclear Medicine 32(11): 839-841.

Crabbe, A., C. Vandeputte, et al. (2010). "Effects of MRI contrast agents on the stem cell phenotype." Cell Transplant 19(8): 919-936.

Crowder, K., J. Brant, et al. (2006). "Unique perfluorocarbon nanobeacons improve stem/progenitor cell tracking with MRI." Faseb Journal 20(4): A633-A633.

Daldrup-Link, H. E., M. Rudelius, et al. (2003). "Targeting of hematopoietic progenitor cells with MR contrast agents." Radiology 228(3): 760-767.

Dash, R., T. Chan, et al. (2008). "Magnetic Resonance Imaging with Iran-Oxide Labeling Detects Differential Cell Survival after Doxorubicin Exposure in Cardiac Myocytes, Fibroblasts, and Stem Cells." Circulation 118(18): S996-S996.

Detante, O., A. Moisan, et al. (2009). "Intravenous administration of 99mTc-HMPAO-labeled human mesenchymal stem cells after stroke: *in vivo* imaging and biodistribution." Cell Transplant 18(12): 1369-1379.

Doubrovin, M., I. Serganova, et al. (2004). "Multimodality *in vivo* molecular-genetic imaging." Bioconjugate Chemistry 15(6): 1376-1388.

Drey, F., B. Ewert, et al. (2010). "Oninvasive *in vivo* Tracking of Mesenchymal Stem Cells by Mri and Evaluation of Cell Therapeutic Effects." Transplant International 23: 22-22.

Flexman, J. A., D. J. Cross, et al. (2011). "Quantitative analysis of neural stem cell migration and tracer clearance in the rat brain by MRI." Mol Imaging Biol 13(1): 104-111.

Frangioni, J. V. and R. J. Hajjar (2004). "*In vivo* tracking of stem cells for clinical trials in cardiovascular disease." Circulation 110(21): 3378-3383.

Frangioni, J. V. and R. J. Hajjar (2004). "*In vivo* tracking of stem cells for clinical trials in cardiovascular disease." Circulation 110(21): 3378-3383.

Frank, J. A., B. R. Miller, et al. (2003). "Clinically applicable labeling of mammalian and stem cells by combining superparamagnetic iron oxides and transfection agents." Radiology 228(2): 480-487.

Gao, J., J. E. Dennis, et al. (2001). "The dynamic *in vivo* distribution of bone marrow-derived mesenchymal stem cells after infusion." Cells Tissues Organs 169(1): 12-20.

Gera, A., G. K. Steinberg, et al. (2010). "*In vivo* neural stem cell imaging: current modalities and future directions." Regenerative Medicine 5(1): 73-86.

Gholamrezanezhad, A., M. Bagheri, et al. (2007). "The First Experience of Stem Cell Labeling in Iran Using 111In- Oxine." Iranian Journal of Nuclear Medicine 15(2): 25-27.

Gholamrezanezhad, A., S. Mirpour, et al. (2009). "Cytotoxicity of 111In-oxine on mesenchymal stem cells: a time-dependent adverse effect." Nucl Med Commun 30(3): 210-216.

Gholamrezanezhad, A., S. Mirpour, et al. (2011). "*In vivo* tracking of 111In-oxine labeled mesenchymal stem cells following infusion in patients with advanced cirrhosis." Nuclear Medicine and Biology. 2011 Jun 22. [Epub ahead of print]

Gilad, A. A., P. T. Winnard, Jr., et al. (2007). "Developing MR reporter genes: promises and pitfalls." NMR Biomed 20(3): 275-290.

Hertenstein, B., K. C. Wollert, et al. (2004). "Monitoring of bone marrow cell homing in the infarcted human myocardium by PET." Blood 104(11): 736a-736a.

Higuchi, T., M. Anton, et al. (2009). "Combined Reporter Gene PET and Iron Oxide MRI for Monitoring Survival and Localization of Transplanted Cells in the Rat Heart." Journal of Nuclear Medicine 50(7): 1088-1094.

Hill, J. M., A. J. Dick, et al. (2003). "Serial cardiac magnetic resonance imaging of injected mesenchymal stem cells." Circulation 108(8): 1009-1014.

Himmelreich, U. and M. Hoehn (2008). "Stem cell labeling for magnetic resonance imaging." Minimally Invasive Therapy & Allied Technologies 17(2): 132-142.

Hoehn, M., E. Kustermann, et al. (2002). "Monitoring of implanted stem cell migration *in vivo*: A highly resolved *in vivo* magnetic resonance imaging investigation of experimental stroke in rat." Proceedings of the National Academy of Sciences of the United States of America 99(25): 16267-16272.

Hofmann, M., K. C. Wollert, et al. (2005). "Monitoring of bone marrow cell homing into the infarcted human myocardium." Circulation 111(17): 2198-2202.

Hsiao, J. K., M. F. Tai, et al. (2007). "Magnetic nanoparticle labeling of mesenchymal stem cells without transfection agent: Cellular behavior and capability of detection with clinical 1.5 T magnetic resonance at the single cell level." Magnetic Resonance in Medicine 58(4): 717-724.

Hsieh, S. C., F. F. Wang, et al. (2006). "The inhibition of osteogenesis with human bone marrow mesenchymal stem cells by CdSe/ZnS quantum dot labels." Biomaterials 27(8): 1656-1664.

Jaiswal, J. K., H. Mattoussi, et al. (2003). "Long-term multiple color imaging of live cells using quantum dot bioconjugates." Nature Biotechnology 21(1): 47-51.

Jakobsson, J., N. Rosenqvist, et al. (2004). "Dynamics of transgene expression in a neural stem cell line transduced with lentiviral vectors incorporating the cHS4 insulator." Experimental Cell Research 298(2): 611-623.

Jang, Y. Y., Z. Ye, et al. (2011). "Molecular imaging and stem cell research." Mol Imaging 10(2): 111-122.

Jansen, J. F. A., M. J. Shamblott, et al. (2006). "Stem cell profiling by nuclear magnetic resonance spectroscopy." Magnetic Resonance in Medicine 56(3): 666-670.

Josephson, L., M. F. Kircher, et al. (2002). "Near-infrared fluorescent nanoparticles as combined MR/optical imaging probes." Bioconjugate Chemistry 13(3): 554-560.

Kang, W. J., H. J. Kang, et al. (2006). "Tissue distribution of 18F-FDG-labeled peripheral hematopoietic stem cells after intracoronary administration in patients with myocardial infarction." J Nucl Med 47(8): 1295-1301.

Kedziorek, D., R. Ouwerkerk, et al. (2009). "X-Ray- and MRI-Visible Microencapsulated Mesenchymal Stem Cell for Cell Delivery and Tracking on Clinical Scanners." Journal of Nuclear Medicine 50(4): 657-657.

Kim, D. and J. Song (2008). "Stem cell tracking using magnetic resonance imaging." Tissue Engineering and Regenerative Medicine 5(3): 420-424.

Kim, Y. J., Y. M. Huh, et al. (2009). "*In vivo* magnetic resonance imaging of injected mesenchymal stem cells in rat myocardial infarction; simultaneous cell tracking and left ventricular function measurement." International Journal of Cardiovascular Imaging 25: 99-109.

Klibanov, A. L., P. T. Rasche, et al. (2004). "Detection of individual microbubbles of ultrasound contrast agents: imaging of free-floating and targeted bubbles." Invest Radiol 39(3): 187-195.

Klug, M. G., M. H. Soonpaa, et al. (1996). "Genetically selected cardiomyocytes from differentiating embryonic stem cells form stable intracardiac grafts." Journal of Clinical Investigation 98(1): 216-224.

Kraitchman, D. L. and J. W. Bulte (2009). "*In vivo* imaging of stem cells and Beta cells using direct cell labeling and reporter gene methods." Arterioscler Thromb Vasc Biol 29(7): 1025-1030.

Kraitchman, D. L. and J. W. M. Bulte (2008). "Imaging of stem cells using MRI." Basic Research in Cardiology 103(2): 105-113.

Kraitchman, D. L. and J. W. M. Bulte (2009). "*In vivo* Imaging of Stem Cells and Beta Cells Using Direct Cell Labeling and Reporter Gene Methods." Arteriosclerosis Thrombosis and Vascular Biology 29(7): 1025-U1103.

Kraitchman, D. L., W. D. Gilson, et al. (2008). "Stem cell therapy: MRI guidance and monitoring." J Magn Reson Imaging 27(2): 299-310.

Kraitchman, D. L., M. Tatsumi, et al. (2005). "Dynamic imaging of allogeneic mesenchymal stem cells trafficking to myocardial infarction." Circulation 112(10): 1451-1461.

Kraitchman, D. L., M. Tatsumi, et al. (2005). "Dynamic imaging of allogeneic mesenchymal stem cells trafficking to myocardial infarction." Circulation 112(10): 1451-1461.

Kustermann, E., U. Himmelreich, et al. (2008). "Efficient stem cell labeling for MRI studies." Contrast Media Mol Imaging 3(1): 27-37.

Kustermann, E., W. Roell, et al. (2005). "Stem cell implantation in ischemic mouse heart: a high-resolution magnetic resonance imaging investigation." Nmr in Biomedicine 18(6): 362-370.

Lavender, J. P., J. M. Goldman, et al. (1977). "Kinetics of indium-III labelled lymphocytes in normal subjects and patients with Hodgkin's disease." Br Med J 2(6090): 797-799.

Legrand, J. F., O. Cougnenc, et al. (2007). "Evaluation of a method of stem cell labelling with Indium-111 oxine: effects on viability and proliferation." European Journal of Nuclear Medicine and Molecular Imaging 34: S284-S284.

Lewin, M., N. Carlesso, et al. (2000). "Tat peptide-derivatized magnetic nanoparticles allow *in vivo* tracking and recovery of progenitor cells." Nature Biotechnology 18(4): 410-414.

Li, Z., B. Fehse, et al. (2002). "Persisting multilineage transgene expression in the clonal progeny of a hematopoietic stem cell." Leukemia 16(9): 1655-1663.

Li, Z. J., Y. Suzuki, et al. (2008). "Comparison of reporter gene and iron particle labeling for tracking fate of human embryonic stem cells and differentiated endothelial cells in living subjects." Stem Cells 26(4): 864-873.

Lin, S., X. Xie, et al. (2007). "Quantum dot imaging for embryonic stem cells." BMC Biotechnol 7: 67.

Liu, G., Z. Wang, et al. (2011). "Low molecular weight alkyl-polycation wrapped magnetite nanoparticle clusters as MRI probes for stem cell labeling and *in vivo* imaging." Biomaterials 32(2): 528-537.

Lu, S., X. Xu, et al. (2010). "Targeting of Embryonic Stem Cells by Peptide-Conjugated Quantum Dots." Plos One 5(8): -.

Lythgoe, M. F. (2009). "High-field magnetic resonance imaging: stem cell tracking and phenotyping of transgenic mice." Neurogastroenterology and Motility 21(2): V-V.

Magnitsky, S., R. M. Walton, et al. (2007). "Magnetic resonance imaging as a tool for monitoring stem cell migration." Neurodegenerative Diseases 4(4): 314-321.

Magnitsky, S., D. J. Watson, et al. (2005). "*In vivo* and ex vivo MRI detection of localized and disseminated neural stem cell grafts in the mouse brain." Neuroimage 26(3): 744-754.

Mancardi, G. L., R. Saccardi, et al. (2001). "Autologous hematopoietic stem cell transplantation suppresses Gd-enhanced MRI activity in MS." Neurology 57(1): 62-68.

Mani, V., E. Adler, et al. (2008). "Serial *in vivo* positive contrast MRI of iron oxide-labeled embryonic stem cell-derived cardiac precursor cells in a mouse model of myocardial infarction." Magn Reson Med 60(1): 73-81.

Martin, E. T., J. A. Coman, et al. (2004). "Magnetic resonance imaging and cardiac pacemaker safety at 1.5-Tesla." Journal of the American College of Cardiology 43(7): 1315-1324.

Maskali, F., N. Tran, et al. (2004). "Indium-111 oxine labeling of rat bone marrow-derived mesenchymal stem cells for *in vivo* imaging during autologous cell therapy of myocardial infarction." European Journal of Nuclear Medicine and Molecular Imaging 31: S204-S204.

Maskali, F., N. Tran, et al. (2005). "Indium-111 oxine labeling of rat bone marrow-derived mesenchymal stem cells for *in vivo* imaging during autologous cell therapy of myocardial infarction." Journal of the American College of Cardiology 45(3): 227a-227a.

McColgan, P., P. Sharma, et al. (2011). "Stem Cell Tracking in Human Trials: A Meta-Regression." Stem Cell Rev.

Meyerwittkopf, M., A. Bockisch, et al. (1993). "Examination in-Vivo of Cell-Migration of Tc-99m Hmpao Labeled Transplanted Liver Stem-Cells in the Rabbit." Archives of Gynecology and Obstetrics 254(1-4): 1059-1061.

Min, J. J., Y. Ahn, et al. (2006). "*In vivo* bioluminescence imaging of cord blood derived mesenchymal stem cell transplantation into rat myocardium." Annals of Nuclear Medicine 20(3): 165-170.

Min, J. J., U. C. N. Le, et al. (2008). "Visualization of Embryonic Stem Cell Survival *In vivo* by Adenoviral-driven Bioluminescence Reporter in Rat Corpus Cavernosum." Tissue Engineering and Regenerative Medicine 5(4-6): 743-749.

Modo, M., D. Cash, et al. (2002). "Tracking transplanted stem cell migration using bifunctional, contrast agent-enhanced, magnetic resonance imaging." Neuroimage 17(2): 803-811.

Nakayama, A., F. del Monte, et al. (2002). "Functional near-infrared fluorescence imaging for cardiac surgery and targeted gene therapy." Mol Imaging 1(4): 365-377.

Neirinckx, R. D., J. F. Burke, et al. (1988). "The retention mechanism of technetium-99m-HM-PAO: intracellular reaction with glutathione." J Cereb Blood Flow Metab 8(6): S4-12.

Nohroudi, K., S. Arnhold, et al. (2010). "*In vivo* MRI stem cell tracking requires balancing of detection limit and cell viability." Cell Transplant 19(4): 431-441.

Nohroudi, K., S. Arnhold, et al. (2010). "*In vivo* MRI Stem Cell Tracking Requires Balancing of Detection Limit and Cell Viability." Cell Transplantation 19(4): 431-441.

Nyolczas, N., S. Charwat, et al. (2009). "Tracking the migration of cardially delivered therapeutic stem cells *in vivo*: state of the art." Regenerative Medicine 4(3): 407-422.

Olmer, R., A. Haase, et al. (2009). "Human Oct3/4 Promoter Dependent Lentiviral Transgene Expression for Simplified Monitoring of Primate Embryonic Stem Cell Culture." Tissue Engineering Part A 15(3): 727-727.

Park, K. S., J. Tae, et al. (2010). "Characterization, *in vitro* cytotoxicity assessment, and *in vivo* visualization of multimodal, RITC-labeled, silica-coated magnetic nanoparticles for labeling human cord blood-derived mesenchymal stem cells." Nanomedicine 6(2): 263-276.

Partlow, K. C., J. J. Chen, et al. (2007). "F-19 magnetic resonance imaging for stem/progenitor cell tracking with multiple unique perfluorocarbon nanobeacons." Faseb Journal 21(8): 1647-1654.

Pi, Q. M., W. J. Zhang, et al. (2010). "Degradation or excretion of quantum dots in mouse embryonic stem cells." Bmc Biotechnology 10: -.

Politi, L. S., M. Bacigaluppi, et al. (2007). "Magnetic resonance-based tracking and quantification of intravenously injected neural stem cell accumulation in the brains of mice with experimental multiple sclerosis." Stem Cells 25(10): 2583-2592.

Polzer, H., F. Haasters, et al. (2010). "Quantification of Fluorescence Intensity of Labeled Human Mesenchymal Stem Cells and Cell Counting of Unlabeled Cells in Phase-Contrast Imaging: An Open-Source-Based Algorithm." Tissue Engineering Part C-Methods 16(6): 1277-1285.

Pomper, M. G., H. Hammond, et al. (2009). "Serial imaging of human embryonic stem-cell engraftment and teratoma formation in live mouse models." Cell Research 19(3): 370-379.

Ramot, Y., M. Steiner, et al. (2010). "Pulmonary thrombosis in the mouse following intravenous administration of quantum dot-labeled mesenchymal cells." Nanotoxicology 4(1): 98-105.

Reekmans, K., B. Tambuyzer, et al. (2009). "Combined *in vivo* Bioluminescence and Magnetic Resonance Imaging of Adherently Cultured Neural Stem/Progenitor Cell Implants in Brain of Mice." Glia 57(13): S76-S76.

Reumers, V., C. M. Deroose, et al. (2008). "Noninvasive and quantitative monitoring of adult neuronal stem cell migration in mouse brain using bioluminescence imaging." Stem Cells 26(9): 2382-2390.

Reynolds, J. S., T. L. Troy, et al. (1999). "Imaging of spontaneous canine mammary tumors using fluorescent contrast agents." Photochemistry and Photobiology 70(1): 87-94.

Roberts, T. J., J. Price, et al. (2007). "Pharmacological MRI of stem cell transplants in the 3-nitropropionic acid-damaged striatum." Neuroscience 144(1): 100-109.

Rodriguez-Porcel, M., O. Gheysens, et al. (2005). "Image-guided cardiac cell delivery using high-resolution small-animal ultrasound." Molecular Therapy 12(6): 1142-1147.

Roelants, V., D. Labar, et al. (2008). "Comparison Between Adenoviral and Retroviral Vectors for the Transduction of the Thymidine Kinase PET Reporter Gene in Rat Mesenchymal Stem Cells." Journal of Nuclear Medicine 49(11): 1836-1844.

Ruan, J., J. Shen, et al. (2010). "Viability and pluripotency studying of human embryo stem cells labeled with quantum dots." Nano Biomedicine and Engineering 2(4): 245-251.

Rutten, M., M. Janes, et al. (2010). "Comparison of Quantum Dots and CM-DiI for Labeling Porcine Autologous Bone Marrow Mononuclear Progenitor Cells." The Open Stem Cell Journal 2: 25-36.

Saiz, A., Y. Blanco, et al. (2004). "Clinical and MRI outcome after autologous hematopoietic stem cell transplantation in MS." Neurology 62(2): 282-284.

Saiz, A., E. Carreras, et al. (2001). "MRI and CSF oligoclonal bands after autologous hematopoietic stem cell transplantation in MS." Neurology 56(8): 1084-1089.

Schoepf, U., E. M. Marecos, et al. (1998). "Intracellular magnetic labeling of lymphocytes for *in vivo* trafficking studies." Biotechniques 24(4): 642-+.

Serganova, I. and R. Blasberg (2005). "Reporter gene imaging: potential impact on therapy." Nuclear Medicine and Biology 32(7): 763-780.

Shah, K. (2005). "Current advances in molecular imaging of gene and cell therapy for cancer." Cancer Biology & Therapy 4(5): 518-523.

Sheikh, A. Y., S. A. Lin, et al. (2007). "Molecular imaging of bone marrow mononuclear cell homing and engraftment in ischemic myocardium." Stem Cells 25(10): 2677-2684.

Solanky, B., Y. L. Chung, et al. (2010). "Characterization of a Human Neural Stem Cell Biomarker Using Magnetic Resonance Spectroscopy." Cell Transplantation 19(3): 362-363.

Steindler, D. A. (2007). "Stem cells, regenerative medicine, and animal models of disease." Ilar Journal 48(4): 323-338.

Sung, C. K., K. A. Hong, et al. (2009). "Dual-Modal Nanoprobes for Imaging of Mesenchymal Stem Cell Transplant by MRI and Fluorescence Imaging." Korean Journal of Radiology 10(6): 613-622.

Sung, C. K., K. A. Hong, et al. (2009). "Dual-modal nanoprobes for imaging of mesenchymal stem cell transplant by MRI and fluorescence imaging." Korean J Radiol 10(6): 613-622.

Suzuki, Y., C. H. Cunningham, et al. (2007). "*In vivo* magnetic resonance imaging of stem cell viability." Journal of the American College of Cardiology 49(9): 158a-158a.

Suzuki, Y., A. C. Yeung, et al. (2007). "Cardiovascular MRI for stem cell therapy." Curr Cardiol Rep 9(1): 45-50.

Tamura, M., K. Unno, et al. (2004). "*In vivo* trafficking of endothelial progenitor cells and their possible involvement in tumor neovascularization." Life Sciences 75(5): 575-584.

Tang, Y., K. Shah, et al. (2003). "*In vivo* tracking of neural progenitor cell migration to glioblastomas." Hum Gene Ther 14(13): 1247-1254.

Tepper-Wessels, K., M. Loeschinger, et al. (2001). "Detection of rare stem cell populations from umbilical cord blood by multi fluorescence imaging." Molecular Biology of the Cell 12: 134a-134a.

Terrovitis, J., M. Stuber, et al. (2008). "Magnetic resonance imaging overestimates ferumoxide-labeled stem cell survival after transplantation in the heart." Circulation 117(12): 1555-1562.

Tran, N. D., D. Lair, et al. (2010). "Transgene Independent Germ Cell Differentiation From Human Embryonic Stem Cells (hESCs)." Fertility and Sterility 93(5): S22-S22.

Tseng, C. L., I. L. Shih, et al. (2010). "Gadolinium hexanedione nanoparticles for stem cell labeling and tracking via magnetic resonance imaging." Biomaterials 31(20): 5427-5435.

van den Bos, E. J., A. Wagner, et al. (2003). "Improved efficacy of stem cell labeling for magnetic resonance imaging studies by the use of cationic liposomes." Cell Transplantation 12(7): 743-756.

Wang, X. L., M. Rosol, et al. (2003). "Dynamic tracking of human hematopoietic stem cell engraftment using *in vivo* bioluminescence imaging." Blood 102(10): 3478-3482.

Wang, Y. X., H. H. Wang, et al. (2008). "Pitfalls in employing superparamagnetic iron oxide particles for stem cell labelling and *in vivo* MRI tracking." Br J Radiol 81(972): 987-988.

Wang, Y. X. J., H. H. Wang, et al. (2008). "Pitfalls in employing superparamagnetic iron oxide particles for stem cell labelling and *in vivo* MRI tracking." British Journal of Radiology 81(972): 987-988.

Weissleder, R., H. C. Cheng, et al. (1997). "Magnetically labeled cells can be detected by MR imaging." Jmri-Journal of Magnetic Resonance Imaging 7(1): 258-263.

Weldon, M. J., A. E. Joseph, et al. (1995). "Comparison of 99m technetium hexamethylpropylene-amine oxime labelled leucocyte with 111-indium tropolonate labelled granulocyte scanning and ultrasound in the diagnosis of intra-abdominal abscess." Gut 37(4): 557-564.

Wickline, S. A. and G. M. Lanza (2003). "Nanotechnology for molecular imaging and targeted therapy." Circulation 107(8): 1092-1095.

Wollert, K. C., M. Hofmann, et al. (2004). "Monitoring of bone marrow cell homing to the infarcted human myocardium." Circulation 110(17): 436-436.

Wu, J. C. (2007). "Comparison of imaging techniques for tracking cardiac stem cell therapy." Journal of Nuclear Medicine 48(12): 1916-1919.

Wu, J. C., I. Y. Chen, et al. (2003). "Molecular imaging of cardiac cell transplantation in living animals using optical bioluminescence and positron emission tomography." Circulation 108(11): 1302-1305.

Wu, X., J. Hu, et al. (2008). "*In vivo* tracking of superparamagnetic iron oxide nanoparticle-labeled mesenchymal stem cell tropism to malignant gliomas using magnetic resonance imaging." Journal of Neurosurgery 108(2): 320-329.

Yamada, M., P. T. Gurney, et al. (2009). "Manganese-guided cellular MRI of human embryonic stem cell and human bone marrow stromal cell viability." Magn Reson Med 62(4): 1047-1054.

Yang, J., J. Liu, et al. (2009). "*In vivo* MRI of endogenous stem/progenitor cell migration from subventricular zone in normal and injured developing brains." Neuroimage 48(2): 319-328.

Yang, Y., A. Schumacher, et al. (2011). "Monitoring bone marrow-originated mesenchymal stem cell traffic to myocardial infarction sites using magnetic resonance imaging." Magn Reson Med.

Yukawa, H., Y. Kagami, et al. (2010). "Quantum dots labeling using octa-arginine peptides for imaging of adipose tissue-derived stem cells." Biomaterials 31(14): 4094-4103.

Zhang, H. L., H. Qiao, et al. (2011). "Utility of Dual-modality Bioluminescence and MRI in Monitoring Stem Cell Survival and Impact on Post Myocardial Infarct Remodeling." Academic Radiology 18(1): 3-12.

Zhang, S. J. and J. C. Wu (2007). "Comparison of imaging techniques for tracking cardiac stem cell therapy." J Nucl Med 48(12): 1916-1919.

27

Invasive Fungal Infections in Patients with Acute Leukemia and Hematopoietic Stem Cell Transplant Recipients

Maxim Norkin and John R. Wingard
University of Florida College of Medicine, Division of Hematology/Oncology, Gainesville USA

1. Introduction

Patients with acute leukemia (AL) and hematologic stem cell transplant (HCT) recipients are at increased risk for the development of invasive fungal infections (IFI) due to prolonged neutropenia and severe immunosupression. The incidence of IFI in these patients has increased in recent years [1], which likely reflects changes in clinical practice including escalation of treatment intensity; more frequent use of unrelated, mismatched or alternative donors as stem cell source; rising number of multiple transplantations; and more frequent use of T-cell depleted grafts [2-3]. IFI are associated with high morbidity and mortality in the patients with AL and represent a leading cause of infectious mortality in allogeneic HCT (allo-HCT) recipients. Therefore, effective preventive strategies, accurate diagnostic techniques and optimal treatment approaches are required for an optimal management of high risk patients with AL and HCT recipients.

2. Clinical syndromes

Clinical symptoms of IFI are often non-specific, therefore a high index of suspicion in high risk patients is necessary for timely diagnosis and prompt treatment. Diagnosis of the IFI could be challenging, particularly in severely immunocompromised patients as a result of blunted inflammatory response due to profound neutropenia and altered T-cell function; hence, in such patients clinical and radiologic findings may be very subtle. In the case of clinical suspicion of IFI aggressive pursuit of a specific microbiologic diagnosis with early recognition and prompt initiation of antifungal therapy are the key components of optimizing treatment outcomes. The most common clinical signs are the symptoms of local or generalized inflammatory response, primarily persistent fever despite the use of broad spectrum antibiotics. Other symptoms significantly depend upon the involved anatomic site and causative pathogen. For example, for mold pathogens, lungs and less commonly sinuses are involved; accordingly, the most common symptomatology include dyspnea, cough, atypical chest pain, hemoptysis, rash, facial pain, nasal congestion and visual symptoms such as periorbital pain, blurred vision and proptosis.
Among numerous potential pathogens *Candida* and *Aspergillus spp.* are the major causative fungi leading to life-threatening infectious complications in AL and HCT recipients [4].

Invasive Candidasis

Candida fungemia and invasive Candidasis (IC) often occur during breakdown of mucosal barriers because *Candida spp.* are ubiquitous colonizers of mucosal surfaces [5-6]. Skin was previously considered as major source for dissemination of *Candida spp.,* however as it was recently demonstrated endogenous Candida spp. could frequently disseminate from a patient's gastrointestinal tract [6]. IC occurs more frequently in patients with GVHD and is often associated with the presence of indwelling venous catheters, particularly if they are used for the administration of parental nutritional supplementation.

The most common manifestation of IC is persistent and unexplained fever. In contrast to other fungal pathogens, *Candida spp.* are often identified in the blood by fungal blood cultures, but 50% of patients with autopsy-proven IC had no positive blood cultures isolated prior to death [7]. Hematogenous dissemination may lead to endophthalmitis, which is often associated with visual changes and pain [8]. The lesions of endophthalmitis are due to the inflammatory response to the *Candida* organisms, and thus cannot be seen in neutropenia; it most commonly occurs after neutrophil engraftment and frequently is associated with only subtle retinal lesions, which can be detected by indirect ophthalmoscopy. During hematogenous dissemination *Candida spp.* can invade the skin, joint, muscles and renal tubules, which clinically present as maculopapular skin lesions with a tendency to ulcerate, polyarthralgias, polymyalgias and azotemia [9]. Right upper abdominal discomfort, elevated transaminases, particularly alkaline phosphatase in the persistently febrile patents shortly after neutrophil recovery may indicate the presence of hepatosplenic candidiasis [10]. Disseminated *Candida* can cause central nervous system (CNS) syndromes, including meningitis and encephalitis. In HCT recipients who developed brain abscesses, *Candida spp.* accounted for 33% of all cases [11]. Very rarely *Candida spp.* can cause fungal endocarditis, which is the most serious manifestation of IC with symptoms similar to those of bacterial endocarditis.

Mold infections

Despite being ubiquitous, molds are unable to effectively invade tissues in immunocompetent individuals; thus, invasive mold infections (IMI) such as Aspergillosis, Zygomycosis, and Fusariosis almost exclusively develop in high-risk patients with prolonged neutropenia or in severely immunosuppressed patients [12-13]. Nosocomial IMI among patients with AL or HCT recipients primarily develop as a result of spore transmission via inhalation or direct contact [14]. Opportunistic molds, particularly *Aspergillus and Fusarium spp.* can possibly disseminate by a contaminated water, therefore, water might be a potential source of IMI in the treatment facilities [15]. *Aspergillus spp.* account for the majority of mold fungal infections, primarily mold pneumonias. *Zygomycetes* are indentified in 10% to 20% of mold pneumonias, *Scedosporium spp., Fusarium spp.,* and other molds are responsible for a very small percent of cases [16].

In the case of invasive Aspergillosis (IA), when infection spreads beyond the respiratory tract, patients often develop signs of systemic inflammation and multiorgan failure. In such cases, hematogenous dissemination can lead to involvement of the eye, sinuses and abdominal organs; however *Aspergillus spp.* may involve virtually any organ including CNS. When involvement of the sinuses occurs, IA often clinically resembles Zygomycosis (mucormycosis) of the sinuses. Profoundly immunocompromized patients can occasionally develop gastrointestinal IA from swallowed organisms that subsequently invade the gut mucosa, which can present with gastrointestinal bleeding or nonspecific signs of an acute abdomen and fever and might be confused with GVHD in allo-HCT recipients or neutropenic eneterocolitis in severely neutropenic patients [17-18]. A very serious complications occur when fungal invasion directly spreads to central nervous system involving various anatomic structures of the brain [19] or to ethmoid sinus with a farther advance to cavernous sinus [20]. Hematogenous dissemination to the brain can present as a clinical presentation of a cerebral infarction. In HCT recipients with brain abscesses *Aspergillus spp.* were causative pathogens in 58% of cases and in the majority of cases brain abscess in these patients were associated with concomitant pulmonary disease [11].

Endocarditits is another serious complication of IA and primarily happens in patients with prosthetic heart valves or prolonged fungemia due to fungal colonolization of the central venous catheters [21]. Symptoms are similar to those of bacterial endocarditis: persistent fever and thromboembolic complication due to septic emboli. Blood cultures are usually negative, in contrast to classical bacterial endocarditis and diagnosis is usually made by demonstration of hyphae in the embolus with a subsequent culture of the organism. Prognosis of these patients is very poor despite an intensive medical and surgical treatment with a mortality approaching 100% [22]. Endophthalmitis and cutaneous septic lesions by *Aspergillus spp.* are similar in appearance to those caused by disseminated Candidasis.

Zygomycetes typically enter through the sinuses, lung parenchyma, skin, and gastrointestinal tract with a tendency to disseminate in immunocompromised individuals and clinically resemble IA. A distinctive feature of *Zygomycetes* is their propensity to invade blood vessels and cause thrombosis leading to subsequent necrosis of involved tissues.

Paronychia is often caused by bacteria or yeasts; however, in severely immunocompromised recipients it can be caused by *Fusarium spp.* or other molds and lead to life-threatening IFI [16].

2. Epidemiology

Knowledge of the epidemiology of IFI is crucial for development of optimal prophylactic approaches and effective therapeutic strategies for patients with AL and HCT recipients. The epidemiology of IFI has changed significantly over the last two decades. *Candida spp.* were by far the most common causative fungi more than twenty years ago, however the incidence of IMI, particularly IA is increasing. The apparent change may, in part, be due to control of *Candida* by fluconazole prophylaxis in high risk patients.

After its peak in 1980's when *Candida spp.* became the fourth-commonest cause of bloodstream infection in the USA [23], the epidemiology of IC dramatically changed. Before the widespread use of antifungal prophylaxis the incidence of IC in HCT recipients was as high as 25%, but after the introduction of fluconazole prophylaxis in high risk patients the incidence of IC has been steadily decreasing [24]. This phenomenon is mostly caused by a reduction in incidence of *C. albicans* [25], whereas the incidence of azole resistant *Candida spp.* such as *C. glabrata* [26] and *C. krusei* [27] has risen. Analysis of multi-institutional surveys from Europe showed that between September 1997 and December 1999 in patients with hematological malignancies *C. albicans* was responsible for only 35% of cases of candidemia. The majority of cases were caused by non-albicans *Candida spp*, predominantly *C. tropicalis* and *C. krusei* in 24% and 12% of cases respectively [28]. Another retrospective study from a single center in Brazil showed that during a 9-year period (from 1995 to 2003) non-albicans *Candida spp.* were responsible for as high as 79% of all candidemia episodes in patients with hematological malignancies [29]. Therefore, despite the fact that azole prophylaxis led to a significant reduction in the incidence of IC in high-risk patients, azole-resistant species such as *C. krusei* or *C. glabrata* could be responsible for breakthrough infections [30-31].

The epidemiology of *Aspergillus spp.* in the US is also evolving. Similar to the changing incidence of *Candida* infections, it is likely that the prophylactic use of agents with anti-mold activity is responsible for a change in epidemiology of IA. Over the last 15 years the incidence of IA has tripled in allo-HCT recipients in one large transplant center [32]. If earlier epidemiologic data showed that around 90% of all IA cases were caused by *A. fumigatus* [32], the recent multicenter report demonstrated that in HCT recipients the proportion of IA caused by *A. fumigatus* has decreased to 56% with a relative increase in the incidence of other *Aspergillus spp.* including *A. flavus, A. terreus, A. niger* and *A. versicolor* [33].

In recent years previously rare molds have been emerging as important pathogens in high risk patients. There is a rising incidence of IFI caused by *Fusarium spp., Rhizopus spp., Scedosporium spp.,* although the absolute numbers of these infections are still relatively low [32]. In allo-HCT recipients the incidence of both *Fusarium spp.* and *Rhizopus spp.* infections doubled in the period 1985–1999 and was accounting for 18 % of all IMI [32]. It was noted that in some centers an increase in incidence of Zygomycosis infections occurred after the introduction of voriconazole for prophylaxis and treatment of IFI [34]. Although, data from 25 US transplant centers over a 3-year period showed very low incidence of *Zygomycetes* (less than five cases per 1,000 transplants) with no increased incidence after introduction of voriconazole [35].

3. Risk factors

Risk factors for IFIs in patients with leukemia and HCT recipients depend on both causative pathogens and host factors. Since the most common fungal pathogens are opportunitistic, they capable of causing life-threatening infections almost exclusively in immunosuppressed host and the IFI risk directly depends on the duration of neutropenia and severity of immunosuppression. Patients with AL are considered high risk for IFI if prolonged (>7 days duration) neutropenia (absolute neutrophil count <100 cells/mm3) following cytotoxic chemotherapy is anticipated and/or significant medical co-morbid conditions such as hypotension, pneumonia, new-onset abdominal pain, or neurologic changes are present. The same criteria apply for HCT recipients alongside with additional risk factors, mainly graft versus host disease (GVHD) and prolonged immunosuppressive therapy. In ASCT recipients significant alterations of humoral and cellular immunity typically resolve within 3 months, therefore, for such patients the risk for IFI beyond that time is minimal. Immune reconstitution happens slower after allo-HCT and usually approaches normality by 1 year if GVHD does not develop. The occurrence of chronic GVHD significantly impairs immune reconstitution by requiring prolonged and often intensive immunosuppressive therapy. These patients remain at a high risk for IFI if chronic GVHD persists and requires continuing immunosuppressive therapy.

Neutropenic patients with AL and HCT recipients are at a risk for IC, particularly in the absence of antifungal prophylaxis. The risk of IC in both AL patients and HCT recipients also depends on the duration of neutropenia, severity of mucosal injury, and the presence of a central venous catheter [36]. Patients with neutropenia are also at risk for IMI, particularly IA. The underlying disease itself and treatment intensity directly influence the risk of AI; patients with AML have the highest risk for IA with an incidence of approximately 20 times greater than that among patients with lymphoma and multiple myeloma [37]. The precise risk for IA in AML patients varies in different published series, however in most studies it ranges between 5 and 10%, depending on the disease status, duration of neutropenia, and the types of anti-neoplastic agents used for a leukemia treatment [37-42], where patients undergoing chemotherapy for relapsed or refractory leukemia are at greatest risk, whereas patients undergoing an induction chemotherapy for a newly diagnosed leukemia are at a lower risk and those, who are receiving consolidation therapy are at lowest risk. In HCT recipients the risk for IA significantly depends on the type of transplant (ASCT versus allo-HCT); conditioning regimen (myeloablative, non-myeloablative, reduced intensity); stem cell graft source (related, unrelated, haploidentical, umbilical cord, HLA matched or mismatched, T-cell depleted); post-transplant interventions (salvage chemotherapy, prolonged treatment with glucocorticoids and anti-rejection medications); and the development of post-transplant complications, particularly GVHD [43-45]. Overall risk for IA is low in ASCT represents (1–2%) because there is only a brief period of immunosupression and neutropenia in these patients [46]. In allo-HCT recipients, the risk for IA is substantially higher with a trimodal incidence distribution [32, 47-48]. The first peak occurs during the pre-engraftment period where the main risk factor is prolonged neutropenia which is similar to neutropenic patients with leukemia who did not receive transplantation. The second peak occurs between 2 to 3 months after the allo-HCT in patients with acute GVHD being treated with corticosteroids. The third peak occurs after one year post-transplant in patients who continue to require systemic immunosupression for extensive chronic GVHD. Late onset (41–180 days) occurrences of IA were more common in recipient of mismatched or unrelated donor HCT; in patients who received T-cell depleted or CD34-selected stem cell products; in patients receiving corticosteroids; in patients with neutropenia, lymphopenia, GVHD, CMV disease; and in patients with respiratory virus infections. Other factors such as iron overload and a toll-like receptor 4 polymorphism are recognized as independent risk factors for IA in HCT recipients [49-50]. Among less frequent IMI, Zygomycosis tends to develop relatively late, usually after 90 days post HSCT and is more commonly seen in patients with underlying MDS, chronic GVHD and patients receiving treatment of GVHD [32]. Alternatively, severe, but rare infections with *Scedosporium spp,* occurring early, usually in neutropenic AL patients and during the pre-engraftment period in HCT recipients [51].

4. Diagnostic approaches

To improve outcomes of high risk patients with AL and HCT transplant recipients it is critical to establish the diagnosis of IFI early, but currently there is no single diagnostic method that has a sufficient sensitivity and specificity to determine IFI. Therefore, timely diagnosis of IFI should be made on the basis of a constellation of clinical signs, confirmatory imaging studies and laboratory findings.

Imaging studies.

Even minor abnormalities on chest radiographs (CXR) in high risk patients should prompt further investigation, which often includes computed tomography (CT) of the chest. When suspicion of invasive pulmonary Aspergillosis is high, it is very important to pursue CT scanning of the chest, because CXR may be negative in up to 10% cases of invasive pulmonary Aspergillosis, whereas only 3% CT scans are falsely negative [52]. If possible, images need to be compared with prior ones to exclude over interpreting persistent abnormalities, which are not related to acute infection, such as scarring or other tissue changes caused by previous radiation or administration of chemotherapy. The pulmonary infiltrates due to IMI are generally nodular. CT imaging findings that are more specific radiologic signs of IA are early on a "halo sign" when the central nodular area of fungal invasion is surrounded by a ground-glass appearing hemorrhage and a "crescent sign" may occur later as a result of necrosis and cavitation of lung tissue. In a large multicenter study, most patients with documented IA had one or more dense nodules and nearly two thirds had a halo sign [53]. The chest CT defines the extent of the disease process with a greater accuracy than CXR and can also be used repeatedly to monitor the response to therapy. Better visualization of lung tissue on CT scan also helps to define optimal sampling sites and to select the most appropriate invasive diagnostic procedure, such as with bronchoalveolar lavage with or without transbronchial biopsy, image guided needle biopsy, video-assisted thoracoscopic or open lung biopsy. Importantly, during therapy lung infiltrates and clinical symptoms may appear or get worse with immune reconstitution and resolution of severe neutropenia when an inflammatory response develops at the site of tissue damage by an invasive fungal pathogen. It was demonstrated that initiation of therapy directed against Aspergillus when a halo sign has been identified resulted in significantly improved survival [54-55].

Laboratory approaches.

Candida spp. are often detected in peripheral blood of patients with IC, however, blood cultures have a low sensitivity for the diagnosis, being negative in up to 50% of cases with autopsy-proven disseminated Candidiasis [56]. Blood cultures can also be positive for a variety of other fungal pathogens, but blood cultures are practically never positive in the case of disseminated Aspergillosis or Zygomycosis [7].

Aspergillus was isolated from sputum in only 8% to 34%, and from BAL in 45% to 62% of patients with invasive Aspergillosis [57]. In case of pulmonary Zygomycosis, it is poorly isolated from sputum, with positive cultures <25% and the yield of BAL is not higher [58]. Fungal element identification in tissue of the presumed site of infection can be considered as a "gold standard" however, possibility of colonization of the respiratory secretions by environmental fungi may lead to over diagnosis. Tissue biopsies in the case of visceral fungal infections are frequently difficult to obtain due to critical condition of the patient and concomitant severe thrombocytopenia. If there is cutaneous involvement, punch biopsy of skin lesions are often required to establish the diagnosis. For a suspected fungal infection special stains for fungal organisms, such as the Gomori methenamine silver stain, are necessary because routine histologic stains may often not be sufficient to visualize the tissue invasion by fungi.

Detection of yeasts or fungal elements (hyphae) in the tissue by microscopic examination should be followed by a subsequent culture to identify a specific organism, since morphology is not sufficiently specific. *Aspergillus* is characterized by narrow, septated hyphae with acute angle branching and, in comparison with other fungi, it grows relatively rapidly with a visible growth within 3 days if inoculum size is adequate. The hyphae of *Fusarium* in tissue resemble those of *Aspergillus*. *Zygomycetes* characteristically are pauciseptate, wider, ribbon like in contrast to *Aspergillus* in tissue. Positive cultures of tissue of most molds, but especially Zygomycosis is challenging given the difficulty of extracting fungal elements from infected tissues [59]

Because of numerous limitations related to culture techniques, adjunctive non-cultural methods such as galactomannan (GM) antigen, b-(1-3)-D glucan (BG) and PCR-based methods are now being implemented for a timely diagnosis of IFI, primarily IA, particularly when tissue sample is not obtained [60-63]. If PCR-based techniques are still considered investigational [63], the clinical value of serum GM and BG assays was confirmed in prospective trials. Both tests now are accepted as supplementary diagnostic tests for an early detection of common IFI in high risk patients. A positive GM test preceded the development of clinical symptoms of IFI in the majority of patients in studies [60]. The BG has high sensitivity and specificity with a capability to detect many clinically relevant fungal pathogens including *Candida spp*, *Aspergillus spp*, *Pneumocystis spp*, and *Fusarium spp*, but not *Zygomycetes spp* or *Cryptococcus spp.* [62, 64]. Its strength is the ability to detect a variety of the most common relevant fungal pathogens in AL and HCT patients; its weakness is its inability to determine which pathogen is present. In patients with AML and MDS receiving chemotherapy the negative predictive value of BG approached 100%; a single positive test and two or more sequentially positive results had the specificity of 90% and > 95% respectively [64]. The role of BG assay in HCT recipients has not been well studied [65] and requires further investigation.

The serum GM assay detects *Aspergillus* and *Penicillium spp.* (a rare pathogen in AL and HCT patients). GM is a major component of *Aspergillus* cell walls and released during a rapid growth of hyphae. The serum GM assay can identify the presence of fungal growth in tissues during invasive infection and often quite early even before the development of any clinical signs, radiologic abnormalities or fungal growth in culture. Detection of GM antigen by ELISA in the serum has been widely retrospectively and prospectively studied with a wide range of reported sensitivities depending on the study population and definitions of the test positivity [60, 66-74] . The GM assay had only moderate accuracy in the patients with hematologic malignancies with a pooled specificity of 58-70% and sensitivity of 92-95%, and in HCT recipients a pooled specificity of 65-86%- and sensitivity 65-82% [75]. The predictive accuracy of GM is negatively affected by a concomitant use of piperacillin-tazobactam [76] and amoxicillin-clavulanate [77] due to the presence of GM in the antibiotic formulations, and a false positive GM signal may be detected up to 5 days after discontinuation of these medications. The use of antifungal prophylaxis with agents having anti-mold activity could be associated with falsely negative GM tests [67]. False positivity of GM assay was also described in case of infection with *Penicillium spp.* [78] and *Histoplasma capsulatum* [79] because these microorganisms share cross reactive antigens.

GM can also be detected in other body fluids during IA infection involving those fluids or adjacent tissues, including urine, CSF and bronchoalveolar lavage (BAL) [80-82]; however, clinical utility of the assay obtained from other sources than serum is under clinical investigation. Preliminary data has suggested that detection of GM in BAL fluid [83] could be an reliable adjunct test with a specificity and sensitivity exceeding of those in BAL fungal culture [84-85] and some studies suggest BAL GM to be more sensitive than serum GM [86]

PCR assays to detect and identify fungal pathogens are being developed and ongoing clinical testing, but not yet available for a commercial use [74]. Although development and clinical implementation of PCR assays for fungal species such as *Aspergillus spp.* holds an enormous promise, multiple obstacles, particularly widespread environmental fungal contamination and difficulties in DNA extraction, must be overcome before PCR-based techniques will be clinically useful [87-89].

5. Treatment and prevention

There are four common antifungal strategies used in current clinical practice to combat IFI. These include antifungal prophylaxis, empirical therapy, pre-emptive (or diagnostics-driven) therapy and treatment of proven and probable IFI.

Antifungal prophylaxis

The goal of antifungal prophylaxis is to prevent IFI with an administration of the antifungal agent(s) in high risk patients. Antifungal prophylaxis is a rapidly evolving field. Prophylaxis is considered clinically beneficial when the risk of a life-threatening IFI outweighs the risks of toxic effects and drug interactions, and the risk for emergence of drug resistance associated with the antifungal agent used. Optimally, antifungal prophylaxis should also be cost-effective. Therefore, the choice of empirical antifungal treatment is considered based on the prevalence of the most likely fungal pathogens, along with considerations about toxicities, resistance, and cost.

The most widely used agent for IFI prophylaxis in patients with AL and HCT recipients is fluconazole, an agent with activity against most (but not all) *Candida spp.*. Prospective studies demonstrated reductions in IFI caused by yeast organisms when fluconazole was given prophylactically at the dose of 400 mg daily in allo- HCT and ASCT recipients [90-91]. In one trial prophylaxis with fluconazole resulted in a statistically significant reduction of IFI to 3% as compared to 16% in the placebo group [90]. Administration of fluconazole up to engraftment was able to effectively prevent IFI caused by *Candida spp.* with the exception of *C. krusei.* Although there was no difference in overall mortality in patients who received prophylaxis with fluconazole as compared to those who received placebo, infection related mortality was significantly reduced [90]. In another trial [91], patients who received fluconazole prophylaxis during the first 75 post-transplant demonstrated a statistically significant clinical benefit with a reduction of IFI from 18% to 7%, a decrease in infection-related mortality from 22% to 12%, and improvement of overall mortality from 35% to 20% as compared to placebo group [91]. The enduring benefit of fluconazole prophylaxis in HCT recipients even beyond the time the drug was given was noted in a follow-up analysis of the Slavin study [92]. Based on such benefits fluconazole at a daily dose of 400 mg was widely accepted as a standard of care for IFI prophylaxis in high risk HCT recipients, endorsed by consensus guidelines [93]; however controversy remains regarding the optimal duration of prophylaxis and whether to stop it at the time of engraftment or continue until GVHD resolves and its treatment has ceased. It is also not completely clear if patient groups who at a lesser risk for a development of IC than recipients of myeloablative HCT, such as patients with AL, ASCT and non-ablative transplant recipients derive similar clinical benefit from fluconazole prophylaxis.

Another widely studied agent for antifungal prophylaxis in high risk patients is itraconazole, which has activity against *Aspergillus spp.* in addition to *Candida spp.* [94]. Comparison of itraconazole to fluconazole as a long term prophylactic agent (until day 100 or 180 post-transplant) in HCT recipients showed no difference in overall survival [95-96]; in one of the trials a reduction in the incidence of IFI in itraconazole arm was noted [95] but not in the other [96]. Despite its attractive anti-mold activity, the lack of proven clinical advantage over fluconazole [96-98], suboptimal bioavailability [95], higher mortality rate [95-96] and a higher incidence of adverse events, particularly more renal and hepatic toxicities [96] associated with itraconazole administration make this agent suboptimal for routine prophylaxis of IFI in high risk patients. In a meta-analysis of itraconazole trials, an *Aspergillus* protective effect was noted in patients given a bioavailable daily dose of at least 200 mg twice daily (typically seen when itraconazole is given as oral solution at a twice daily dose of 200mg per dose [94] .

Voriconazole is a well-tolerated triazole with an extended-spectrum of activity including activity against *Aspergillus spp.*. A large multicenter, randomized trial compared voriconazole with fluconazole for prevention of IFI after myeloablative allo-HCT [99]. In that trial patients were randomized to receive antifungal prophylaxis with study drugs for 100 days or for 180 days if patients were considered high risk. All patients were intensively monitored with twice weekly serum GM assays for initial 60 days, then at least weekly until day 100. In the case of positive GM or clinical signs suggestive of IFI patients were followed with radiographic studies and invasive diagnostic procedures for confirmation of IFI. Despite statistically non-significant trends to fewer IFI, particularly IA, and less frequent use of empiric antifungal therapy in voriconazole arm, the primary endpoint- fungal-free survival (alive and free from proven, probable, or presumptive IFI patients) at 6 months was similar in both groups. In addition, there was no difference in incidence of severe adverse events, relapse-free and overall survival at 6 months. Thus, prophylaxis with voriconazole did not improve 6-month fungal-free and overall survival as compared to fluconazole if intensive monitoring and a structured approach for empiric antifungal therapy were used [99]. Voriconazole is available in both intravenous and oral formulations, which allows its extended use to prolonged periods of high risks for the development of IFI. One of the major drawbacks of voriconazole is the lack of activity against *Zygomycetes*. Although multicenter studies with voriconazole [35, 99-100] showed no increased incidence of IFI caused by *Zygomycetes* in patients received voriconazole prophylaxis, there are single center retrospective case series suggesting an association of voriconazole use with more Zygomycoses. Thus, this issue bears further scrutiny over time. This is quite difficult because of the difficulty in culturing *Zygomycetes* in tissue. Echinocandins have a potential advantage over fluconazole given the broader spectrum of activity, in particular, against fluconazole resistant *C. krusei* or *C. glabrata*, and the anti-*Aspergillus* activity, although the latter has not yet been proved in clinical trials. A prospective comparison of micafungin to fluconazole for antifungal prophylaxis in neutropenic patients undergoing HCT showed compatible efficacy, safety profile, overall survival and reduced the need for empiric antifungal therapy in patients receiving micafungin [101]. The major disadvantage of micafungin use for a fungal prophylaxis is the need for intravenous administration and substantially greater cost; however this agent might be preferred over fluconazole in the centers with high prevalence of infections caused by non-albicans *Candida spp.*

Posaconazole is a triazole with a broad spectrum of activity against both yeasts and molds. It is active against pathogens such as the *Zygomycetes* which are resistant to many widely used antifungal agents. Such a broad spectrum of activity makes this agent an attractive alternative to fluconazole for prophylaxis of IFI. Posaconazole is approved for prophylaxis in neutropenic patients with AML or MDS based on the results of randomized multicenter studies where it was shown to be superior to fluconazole or itraconazole in the prevention of IFI [102]. In that study prophylactic administration of posaconazole resulted in improved overall survival, although at the cost of higher incidence of serious adverse events attributable to study drug [102]. Since significant numbers of the patients in the posaconazole group were receiving chemotherapy there was a possibility that greater toxicity of the posaconazole may have resulted from a harmful interactions with chemotherapy agents. This has not been adequately studied. In patients with GVHD posaconazole was similar to fluconazole in prevention of all types of IFI and, not surprisingly given its anti-mold activity, superior in reducing the incidence of IA and mortality related to fungal infections as compared to fluconazole [30]. Based on these results posaconazole was approved in the US for antifungal prophylaxis in high risks patients, however this agent has not been studied in the pre-engraftment phase of HCT.

Development of extensive chronic GVHD significantly increases the risk of late IFI (more than 100 days post-transplant), predominantly IA, with a reported incidence as high as almost 40% in one series [103]. Therefore, anti-mold prophylaxis with posaconazole in patients with extensive chronic GVHD could be beneficial in these patients.

Amphotericin B (AmpB) is not currently used for prophylaxis of IFI due to its excessive toxicity and infusion-related events. However, recently aerosolized AmpB was prospectively tested in patients with AL and allo-HCT recipients and led to reduction in the incidence of invasive pulmonary Aspergillosis; however administration of the drug was interrupted in a substantial proportion of patients (45%) due to cough during inhalation, weakness preventing the use the aerosol delivery system and technical problem with the aerosol delivery system. Moreover, aerosol dose and delivery device have not yet been determined, therefore further study is necessary before the clinical utility of this agent can be determined [104].

The risk for IFI is substantially lower in the majority of ASCT recipients as compared to allo-HCT recipients. These patients usually do not require routine anti-yeast prophylaxis. However, for high risk sub-populations of ASCT recipients such as patients with underlying hematological malignancies, history of prolonged neutropenia, significant mucosal damage, and treatment within fludarabine or 2-CDA within 6 months prior ASCT ant-yeast prophylaxis is recommended [15].

Empirical treatment

Empirical treatment is an initiation or modification of an existing antifungal treatment in high risk patients with persistent fever of unknown source unresponsive to antibacterial agents. Although, approximately one-third of neutropenic patients with cancer receive an antifungal drug due to persistent fever, only less than 5% of these patients subsequently demonstrate the presence of IFI [102, 105-107]. Therefore, initiation of empirical antifungal therapy solely on the basis of persistent fever can be legitimately questioned.

The most common fungal pathogen in neutropenic patients is *Candida*. Anti-yeast prophylaxis dramatically reduces the risk for IC. However, in persistently febrile neutropenic patients not given antifungal prophylaxis IMI or fluconazole-resistant *Candida* infections represent the most common threats. Oral, non-absorbable antifungal drugs, such as oral AmpB, nystatin and clotrimazole troches are capable of reducing superficial colonization and control local mucosal candidiasis, but, surprisingly, do not reduce the risk for IC.

There is not enough data for recommendations regarding a particular empirical antifungal agent for patients who are already receiving anti-mold prophylactic coverage, however changing treatment to a different class of intravenous anti-mold agent should be considered, given the possibility that breakthroughs of fungal infection could be caused by resistant organisms. Another concern is that breakthrough IFI may be due to inadequately low serum levels of voriconazole or posaconazole when these agents administered orally because of the variability of blood levels after HCT [108-109].

For more than 30 years AmpB has been used for an empirical antifungal treatment, but with development of newer agents, its therapeutic role in the current management of fungal infections is minimal, due to excessive toxicity, particularly nephrotoxicity and infusion-related events. Alternative formulations of AmpB such as liposomal AmpB, AmpB colloidal dispersion, AmpB lipid complex; azoles with anti-mold activity such as itraconazole or voriconazole, and caspofungin are widely used because of better tolerability and reduced systemic toxicity, however there are no data that have demonstrated their superior efficacy to AmpB [105, 107, 110-112].

Voriconazole is recognized by many experts as a suitable alternative to a liposomal AmpB as an empirical antifungal therapy in patients with neutropenia and persistent fever [106] and currently widely used for an empirical therapy for high risk patients, particularly for an empirical treatment of probable IMI.

Pre-emptive therapy

The need for empirical antifungal therapy in febrile high risk patients have been questioned because of the very low percentage of such fevers that are actually due to IFI [42]. The advances in our ability for early detection of fungal infections have ushered in a new strategy, pre-emptive therapy. Pre-emptive treatment refers to therapy of highly suspected IFI supported by the presence of clinical symptoms, radiological findings and/or adjunctive laboratory tests [54-55, 113].

The feasibility of preemptive antifungal therapy in high risk patients with the use of serial GM testing and early CT scans was prospectively evaluated in 131 neutropenic episodes. All patients were given fluconazole prophylaxis to eliminate *Candida* as a possible cause of persistent fever. Forty one of 117 episodes (35%) met the criteria for empirical therapy persistent or recurrent fevers [55]. Use of preemptive approach in such patients was associated with a 78% reduction (from 35% to 8%) of antifungal therapy without a compromise in identifying (and treating) IFI. Indeed, early detection allowed for initiation of antifungal therapy earlier with the pre-emptive strategy. The clinical utility of preemptive antifungal therapy was also evaluated in a prospective study, which included patients with AML receiving induction or consolidation therapy, ASCT recipients and in other patients with prolonged neutropenia, however, allo-HCT recipients were excluded [114]. In that study, clinical symptoms, chest CT findings indicative of IFI, documented *Aspergillus* colonization or a positive GM were used to initiate a preemptive therapy with either AmpB or liposomal AmpB, depending on renal function. Although there was no difference in overall mortality, probable or proven IFI were more commonly detected in preemptive treatment arm than in the empirical treatment one (13/143 vs. 4/150). An increase in IFI was seen only in the subset of patients who did not receive antifungal prophylaxis and the increase was due to an increased incidence of IC [114]. This emphasizes the need for anti-*Candida* prophylaxis, if the screening testing is designed primarily for detecting *Aspergillus*.

In another study PCR-based preemptive antifungal therapy was compared with an empirical treatment with liposomal amphotericin B in patients after allo-HCT [113]. PCR-based approach led to an increased use of anti-fungal therapy and reduced 30-day mortality, but there was no difference in the incidence of IFI or 100-day survival.

These observations suggest that some of high risk neutropenic patients with persistent fever may not need an automatic empirical antifungal therapy if they receiving anti-yeast prophylaxis and closely monitored under certain specific conditions [114-117]. However, if

laboratory, clinical or radiologic abnormality suggestive of probable fungal infection is identified, prompt initiation with of antifungal therapy with an agent that includes anti-mold activity is needed.
Despite being an attractive alternative, the preemptive approach is currently regarded as investigational by many and not widely accepted although its use is embraced and its promise is discussed in consensus guidelines [118]

Treatment of established IFI

Early recognition and treatment of invasive fungal infections in AL and HCT recipients are important clinical challenges. Early initiation is a key to optimizing treatment outcomes for both *Candida* and *Aspergillus* [119-121]. The European Organization for Research and Treatment of Cancer/Mycoses Study Group (EORTC-MSG) has proposed criteria to define proven or probable IFI [122]. New antifungal agents offer choices that in some cases are less toxic than older drugs and in other cases are more efficacious. Novel strategies that combine the new diagnostic tools with new drugs, are being evaluated to change our approaches to these deadly infections [123]

Treatment of IC.

Candidemia is associated with a high mortality, especially in patients with delayed treatment, therefore antifungal therapy in patients with IC should be started as soon as possible [119]. If a central venous catheter(s) is in place and suspected to be causative for IC, it should be removed. Although AmpB was historically the preferred therapy with excellent activity against most *Candida spp.,* lipid amphotericin formulations were just as effective with less toxicity. Other agents are also effective treatment options. Multiple studies, conducted mostly in non-neutropenic patients and patients with solid malignances, demonstrated a high efficacy of fluconazole in patients with candidemia [119]. In patients with IC fluconazole was found to be non-inferior and significantly better tolerated then AmpB [119] IC, but less effective than anidulafungin IC [124]. Although well tolerated, there are gaps in its spectrum of activity with *C. krusei* being intrinsically resistant and many strains of *C. glabrata* requiring higher fluconazaole concentrations for inhibition and some strains being outright resistant. Despite the fact that anidulafungin was mainly studied in non-neutropenic patients, better efficacy [124] and emergence of fluconazole-resistant *Candida spp.* such as *C. krusei* or *C. glabrata* [125-126] make the echinocandins preferable choices for the initial treatment of IC in patients with AL and HCT recipients. After the confirmation of susceptibility of *Candida spp.* to fluconazole clinically stable patients without prior azole exposure can be stitched to fluconazole. Initial preferred dose of fluconazole is 800 mg a day with a subsequent reduction to 400 mg a day in clinically stable patients with resolved neutropenia [127]. In a large randomized study caspofungin showed compatible efficacy and significantly better side effects profile in comparison to AmpB in neutropenic patients with candidemia [128]. Given its good safety profile and efficacy against non-albicans *Candida spp.* caspofungin is indicated in severely ill, clinically unstable neutropenic patients or patients with hepatosplenic Candidasis [129]

Treatment of IMI

Voriconazole is the preferred initial agent for the treatment of IA [130] . It is effective and well tolerated both as primary and salvage treatment [131]. A large randomized study showed a higher response and survival rate with fewer *Aspergillus*-related deaths and side effects with voriconazole compared to AmpB for primary therapy of IA [115].

Although there is a very limited data regarding posaconazole activity as a first line agent in the treatment of IA as an initial agent of choice, substantial evidence supports its activity in the salvage setting after a failure of or intolerance to other antifungal treatments [132]. There are few data regarding clinical role of caspofungin as a first-line treatment for IA, but it is effective with a response up to 50% as a salvage agent in patients with IA who failed or intolerant to standard antifungal therapy [133]. The role of micafungin or anidulafungin for the treatment of IA has not been well studied but are active as salvage therapy.

Efficacy and safety of AmpB lipid complex (ABLC) as first-line or second-line therapy was demonstrated in large numbers of patients with hematologic malignancy or HCT recipients [134]. Another study showed that one third of allo-HCT recipients with IA responded to ABLC, with a response of 41% and 21% when ABCL used as a first line treatment and in patients with GVHD respectively [135]. In both studies administration of ABLC had only minimal effects on renal function in the majority of patients. Liposomal AmpB also has been shown to be effective as salvage therapy with fewer infusion related and renal toxicities. Two doses of liposomal AmpB were evaluated (10mg/day versus 3 mg/day) to determine the optimal dose for initial therapy. Both doses had comparable response rates but the 10 mg/day dose was associated with more renal toxicity [136]

The treatment options for Zygomycosis are more limited and less well studied. Patients with Zygomycosis showed a good response to high doses of liposomal AmpB (dose at least 5 mg/kg) [59] or ABLC [137]. In the case of CNS or sinus involvement additional surgical resection of necrotic tissue in patients with Zygomycosis significantly appears to be associated with improved survival as compared to antifungal therapy alone [137-138]. Posaconazole is an effective agent with response rates of about 50–80% when uses as a salvage treatment in patients with Zygomycosis [139-140].

Therapeutic Drug Monitoring

Therapeutic drug monitoring has not been found to be useful for the polyene and echinocandin classes of antifungal drugs. However, the oral formulations of the extended spectrum triazoles have been found to have variability in blood concentrations and multiple drug interactions. Multiple studies have explored whether these variations have clinical consequences in terms of efficacy. Itraconazole oral formulations have been seen to have variable bioavailability in multiple studies [141-145]. Higher doses appear to have greater antrifungal effects in both animals and humans [141, 145-148] . Voriconazole , when taken orally also has been shown to havevariable bioavailability, especially in HCT recipients [149-151].Voriconazole is metabolized in the liver via the cytochrome P450 dependent mechanism, predominantly by CYP2C19 isoenzyme which exhibits genetic polymorphism resulting in reduced drug metabolism in 15–20% patients of Asian descent, and 3–5% of Caucasians and Blacks [150]. Low voriconazole levels have been have been reported in patients with documented breakthrough infections, whereas super therapeutic levels may lead to hepatotoxicity and encephalopathy [150-151]. Less information is known about posaconazole, but it too has been shown to have variable bioavailability [152-156]. There is some suggestion that low levels may be associated with lower effectiveness [156-157].

Controversy remains regarding the use of routine monitoring of azole drug levels, however most experts agree that measurement of levels should be performed in patients with documented breakthrough infection or infections not responding to oral triazole therapy. A change to, or addition of, different class antifungal agent is recommended until the determination of the drug blood level. If drug level is found to be sub-therapeutic, resuming the drug at a higher dose should be considered.

Combination antifungal therapy

Combination antifungal therapy may theoretically lead to better efficacy with shorter courses of therapy, reduced toxicity and emergence of resistance. However potential drawbacks of combination therapy might include potential antagonistic effects of the antifungal drugs used in combination, potential increase in toxicity, and greater cost may offset the potential value of this approach [158]. A randomized trial demonstrated that in non-neutropenic patients with candidemia due to species other than C. *krusei* combination therapy with fluconazole and AmpB resulted in a faster clearance of fungi from the bloodstream as compared to fluconazole alone [159]. Based on this result combination of fluconazole and AmpB could be recommended as an option for the treatment of IC [160], however the data regarding combination antifungal therapy for candidemia in neutropenic patients with leukemia and HCT recipients is lacking. Small retrospective studies showed better responses to the combination of caspofungin plus liposomal AmpB for primary IA infection [161] or for refractory pneumonia [162]. Another small retrospective study compared the voriconazole plus caspofungin combination to a single-agent voriconazole. In this study an improved 3-month survival in HCT recipients or patients receiving chemotherapy with IA who had failed primary therapy with the combination, but there was no difference in survival at 1 year [163]. Thus currently there is no data supporting a benefit of combination therapy in patients with IA.

6. Conclusion

Over the past two decades, significant progress has been made in the prophylaxis, diagnosis and treatment of IFI in patients with AL and HCT recipients. Introduction of newer antifungal agents, better supportive care and more effective diagnostic tools resulted to a considerable improvement of clinical outcomes for such high risk patients. However, despite advances in prevention and management of systemic fungal infections, IFI remain a significant clinical problem for AL patients and HCT recipients. Further progress should be made towards improvements in diagnostic techniques, development of novel antifungal drugs and introduction of immunotherapy approaches for high risk patients. Future efforts should also be focused on better understanding of fungal immunobiology and implementation of personalized therapy based on immunologic, metabolic and genetic profiles of high risk patients.

7. References

[1] Mahfouz, T. and E. Anaissie, *Prevention of fungal infections in the immunocompromised host.* Curr Opin Investig Drugs, 2003. 4(8): p. 974-90.

[2] Marr, K.A., et al., *Invasive aspergillosis in allogeneic stem cell transplant recipients: changes in epidemiology and risk factors.* Blood, 2002. 100(13): p. 4358-66.

[3] Nucci, M., *Emerging moulds: Fusarium, Scedosporium and Zygomycetes in transplant recipients.* Curr Opin Infect Dis, 2003. 16(6): p. 607-12.

[4] Wingard, J.R., *The changing face of invasive fungal infections in hematopoietic cell transplant recipients.* Curr Opin Oncol, 2005. 17(2): p. 89-92.

[5] Blijlevens, N.M., J.P. Donnelly, and B.E. de Pauw, *Impaired gut function as risk factor for invasive candidiasis in neutropenic patients.* Br J Haematol, 2002. 117(2): p. 259-64.

[6] Nucci, M. and E. Anaissie, *Revisiting the source of candidemia: skin or gut?* Clin Infect Dis, 2001. 33(12): p. 1959-67.
[7] Reimer, L.G., M.L. Wilson, and M.P. Weinstein, *Update on detection of bacteremia and fungemia.* Clin Microbiol Rev, 1997. 10(3): p. 444-65.
[8] Coskuncan, N.M., et al., *The eye in bone marrow transplantation. VI. Retinal complications.* Arch Ophthalmol, 1994. 112(3): p. 372-9.
[9] Wingard, J.R., *Fungal infections after bone marrow transplant.* Biol Blood Marrow Transplant, 1999. 5(2): p. 55-68.
[10] Haron, E., et al., *Hepatic candidiasis: an increasing problem in immunocompromised patients.* Am J Med, 1987. 83(1): p. 17-26.
[11] Hagensee, M.E., et al., *Brain abscess following marrow transplantation: experience at the Fred Hutchinson Cancer Research Center, 1984-1992.* Clin Infect Dis, 1994. 19(3): p. 402-8.
[12] Gerson, S.L., et al., *Prolonged granulocytopenia: the major risk factor for invasive pulmonary aspergillosis in patients with acute leukemia.* Ann Intern Med, 1984. 100(3): p. 345-51.
[13] Portugal, R.D., M. Garnica, and M. Nucci, *Index to predict invasive mold infection in high-risk neutropenic patients based on the area over the neutrophil curve.* J Clin Oncol, 2009. 27(23): p. 3849-54.
[14] Denning, D.W., *Invasive aspergillosis.* Clin Infect Dis, 1998. 26(4): p. 781-803; quiz 804-5.
[15] Marr, K.A., et al., *Fungal infection prevention after hematopoietic cell transplantation.* Bone Marrow Transplant, 2009. 44(8): p. 483-7.
[16] Wingard, J.R., J. Hsu, and J.W. Hiemenz, *Hematopoietic stem cell transplantation: an overview of infection risks and epidemiology.* Hematol Oncol Clin North Am, 2011. 25(1): p. 101-16.
[17] Prescott, R.J., M. Harris, and S.S. Banerjee, *Fungal infections of the small and large intestine.* J Clin Pathol, 1992. 45(9): p. 806-11.
[18] Yong, S., H. Attal, and G. Chejfec, *Pseudomembranous gastritis: a novel complication of Aspergillus infection in a patient with a bone marrow transplant and graft versus host disease.* Arch Pathol Lab Med, 2000. 124(4): p. 619-24.
[19] Artico, M., et al., *Intracerebral Aspergillus abscess: case report and review of the literature.* Neurosurg Rev, 1997. 20(2): p. 135-8.
[20] Segal, B.H. and T.J. Walsh, *Current approaches to diagnosis and treatment of invasive aspergillosis.* Am J Respir Crit Care Med, 2006. 173(7): p. 707-17.
[21] Ellis, M.E., et al., *Fungal endocarditis: evidence in the world literature, 1965-1995.* Clin Infect Dis, 2001. 32(1): p. 50-62.
[22] El-Hamamsy, I., et al., *Aspergillus endocarditis after cardiac surgery.* Ann Thorac Surg, 2005. 80(1): p. 359-64.
[23] Edmond, M.B., et al., *Nosocomial bloodstream infections in United States hospitals: a three-year analysis.* Clin Infect Dis, 1999. 29(2): p. 239-44.
[24] Wingard, J.R. and H. Leather, *A new era of antifungal therapy.* Biol Blood Marrow Transplant, 2004. 10(2): p. 73-90.
[25] Trick, W.E., et al., *Secular trend of hospital-acquired candidemia among intensive care unit patients in the United States during 1989-1999.* Clin Infect Dis, 2002. 35(5): p. 627-30.
[26] Tortorano, A.M., et al., *Candidosis in the intensive care unit: a 20-year survey.* J Hosp Infect, 2004. 57(1): p. 8-13.

[27] Wingard, J.R., et al., *Increase in Candida krusei infection among patients with bone marrow transplantation and neutropenia treated prophylactically with fluconazole.* N Engl J Med, 1991. 325(18): p. 1274-7.

[28] Tortorano, A.M., et al., *Candidaemia in Europe: epidemiology and resistance.* Int J Antimicrob Agents, 2006. 27(5): p. 359-66.

[29] Pasqualotto, A.C., et al., *Candidaemia and cancer: patients are not all the same.* BMC Infect Dis, 2006. 6: p. 50.

[30] Ullmann, A.J., et al., *Posaconazole or fluconazole for prophylaxis in severe graft-versus-host disease.* N Engl J Med, 2007. 356(4): p. 335-47.

[31] Kanda, Y., et al., *Prophylactic action of oral fluconazole against fungal infection in neutropenic patients. A meta-analysis of 16 randomized, controlled trials.* Cancer, 2000. 89(7): p. 1611-25.

[32] Marr, K.A., et al., *Epidemiology and outcome of mould infections in hematopoietic stem cell transplant recipients.* Clin Infect Dis, 2002. 34(7): p. 909-17.

[33] Morgan, J., et al., *Incidence of invasive aspergillosis following hematopoietic stem cell and solid organ transplantation: interim results of a prospective multicenter surveillance program.* Med Mycol, 2005. 43 Suppl 1: p. S49-58.

[34] Marty, F.M., L.A. Cosimi, and L.R. Baden, *Breakthrough zygomycosis after voriconazole treatment in recipients of hematopoietic stem-cell transplants.* N Engl J Med, 2004. 350(9): p. 950-2.

[35] Park BJ, K.D., Pappas PG, et al, *Comparison of zygomycosis and fusariosis to invasive aspergillosis (IA) among transplant recipients reporting to TRANSNET.* Program and Proceedings of the 44th Interscience Conference on Antimicrobial Agents and Chemotherapy, 2004. October 30–November 2, 2004, Washington, DC(Abstract M-666.).

[36] Meyers, J.D., *Fungal infections in bone marrow transplant patients.* Semin Oncol, 1990. 17(3 Suppl 6): p. 10-3.

[37] Pagano, L., et al., *The epidemiology of fungal infections in patients with hematologic malignancies: the SEIFEM-2004 study.* Haematologica, 2006. 91(8): p. 1068-75.

[38] Gerson, S.L., et al., *Invasive pulmonary aspergillosis in adult acute leukemia: clinical clues to its diagnosis.* J Clin Oncol, 1985. 3(8): p. 1109-16.

[39] Winston, D.J., et al., *Fluconazole prophylaxis of fungal infections in patients with acute leukemia. Results of a randomized placebo-controlled, double-blind, multicenter trial.* Ann Intern Med, 1993. 118(7): p. 495-503.

[40] Rotstein, C., et al., *Randomized placebo-controlled trial of fluconazole prophylaxis for neutropenic cancer patients: benefit based on purpose and intensity of cytotoxic therapy. The Canadian Fluconazole Prophylaxis Study Group.* Clin Infect Dis, 1999. 28(2): p. 331-40.

[41] Borlenghi, E., et al., *Usefulness of the MSG/IFICG/EORTC diagnostic criteria of invasive pulmonary aspergillosis in the clinical management of patients with acute leukaemia developing pulmonary infiltrates.* Ann Hematol, 2007. 86(3): p. 205-10.

[42] De Pauw, B.E. and J.P. Donnelly, *Prophylaxis and aspergillosis--has the principle been proven?* N Engl J Med, 2007. 356(4): p. 409-11.

[43] Wingard, J.R., *Advances in the management of infectious complications after bone marrow transplantation.* Bone Marrow Transplant, 1990. 6(6): p. 371-83.

[44] Wingard, J.R., *Infections in allogeneic bone marrow transplant recipients.* Semin Oncol, 1993. 20(5 Suppl 6): p. 80-7.

[45] Wingard, J.R., et al., *Association of Torulopsis glabrata infections with fluconazole prophylaxis in neutropenic bone marrow transplant patients.* Antimicrob Agents Chemother, 1993. 37(9): p. 1847-9.

[46] Gil, L., J. Styczynski, and M. Komarnicki, *Infectious complication in 314 patients after high-dose therapy and autologous hematopoietic stem cell transplantation: risk factors analysis and outcome.* Infection, 2007. 35(6): p. 421-7.

[47] Baddley, J.W., et al., *Invasive mold infections in allogeneic bone marrow transplant recipients.* Clin Infect Dis, 2001. 32(9): p. 1319-24.

[48] Nucci, M., et al., *Fusarium infection in hematopoietic stem cell transplant recipients.* Clin Infect Dis, 2004. 38(9): p. 1237-42.

[49] Kontoyiannis, D.P., et al., *Increased bone marrow iron stores is an independent risk factor for invasive aspergillosis in patients with high-risk hematologic malignancies and recipients of allogeneic hematopoietic stem cell transplantation.* Cancer, 2007. 110(6): p. 1303-6.

[50] Bochud, P.Y., et al., *Toll-like receptor 4 polymorphisms and aspergillosis in stem-cell transplantation.* N Engl J Med, 2008. 359(17): p. 1766-77.

[51] Cortez, K.J., et al., *Infections caused by Scedosporium spp.* Clin Microbiol Rev, 2008. 21(1): p. 157-97.

[52] Patterson, T.F., et al., *Invasive aspergillosis. Disease spectrum, treatment practices, and outcomes. I3 Aspergillus Study Group.* Medicine (Baltimore), 2000. 79(4): p. 250-60.

[53] Greene RE, S.H., Oestmann JW, Stark P, Durand C, Lortholary O, Wingard JR, Herbrecht R, Ribaud P, Patterson TF, Troke PF, Denning DW, Bennett JE, de Pauw BE, Rubin RH., *Reply to Verweij et al.* Clin Infect Dis 2007. 44(12): p. 1667-1668.

[54] Caillot, D., et al., *Improved management of invasive pulmonary aspergillosis in neutropenic patients using early thoracic computed tomographic scan and surgery.* J Clin Oncol, 1997. 15(1): p. 139-47.

[55] Maertens, J., et al., *Galactomannan and computed tomography-based preemptive antifungal therapy in neutropenic patients at high risk for invasive fungal infection: a prospective feasibility study.* Clin Infect Dis, 2005. 41(9): p. 1242-50.

[56] Ellepola, A.N. and C.J. Morrison, *Laboratory diagnosis of invasive candidiasis.* J Microbiol, 2005. 43 Spec No: p. 65-84.

[57] Paterson, D.L. and N. Singh, *Invasive aspergillosis in transplant recipients.* Medicine (Baltimore), 1999. 78(2): p. 123-38.

[58] Kontoyiannis, D.P., et al., *Zygomycosis in the 1990s in a tertiary-care cancer center.* Clin Infect Dis, 2000. 30(6): p. 851-6.

[59] Pagano, L., et al., *Mucormycosis in hematologic patients.* Haematologica, 2004. 89(2): p. 207-14.

[60] Maertens, J., et al., *Use of circulating galactomannan screening for early diagnosis of invasive aspergillosis in allogeneic stem cell transplant recipients.* J Infect Dis, 2002. 186(9): p. 1297-306.

[61] Florent, M., et al., *Prospective evaluation of a polymerase chain reaction-ELISA targeted to Aspergillus fumigatus and Aspergillus flavus for the early diagnosis of invasive aspergillosis in patients with hematological malignancies.* J Infect Dis, 2006. 193(5): p. 741-7.

[62] Ostrosky-Zeichner, L., et al., *Multicenter clinical evaluation of the (1-->3) beta-D-glucan assay as an aid to diagnosis of fungal infections in humans.* Clin Infect Dis, 2005. 41(5): p. 654-9.

[63] De Pauw, B., et al., *Revised definitions of invasive fungal disease from the European Organization for Research and Treatment of Cancer/Invasive Fungal Infections Cooperative Group and the National Institute of Allergy and Infectious Diseases Mycoses Study Group (EORTC/MSG) Consensus Group.* Clin Infect Dis, 2008. 46(12): p. 1813-21.

[64] Odabasi, Z., et al., *Beta-D-glucan as a diagnostic adjunct for invasive fungal infections: validation, cutoff development, and performance in patients with acute myelogenous leukemia and myelodysplastic syndrome.* Clin Infect Dis, 2004. 39(2): p. 199-205.

[65] Segal, B.H., et al., *Prevention and early treatment of invasive fungal infection in patients with cancer and neutropenia and in stem cell transplant recipients in the era of newer broad-spectrum antifungal agents and diagnostic adjuncts.* Clin Infect Dis, 2007. 44(3): p. 402-9.

[66] Wheat, L.J., *Rapid diagnosis of invasive aspergillosis by antigen detection.* Transpl Infect Dis, 2003. 5(4): p. 158-66.

[67] Mennink-Kersten, M.A., J.P. Donnelly, and P.E. Verweij, *Detection of circulating galactomannan for the diagnosis and management of invasive aspergillosis.* Lancet Infect Dis, 2004. 4(6): p. 349-57.

[68] Herbrecht, R., et al., *Aspergillus galactomannan detection in the diagnosis of invasive aspergillosis in cancer patients.* J Clin Oncol, 2002. 20(7): p. 1898-906.

[69] Maertens, J., et al., *Screening for circulating galactomannan as a noninvasive diagnostic tool for invasive aspergillosis in prolonged neutropenic patients and stem cell transplantation recipients: a prospective validation.* Blood, 2001. 97(6): p. 1604-10.

[70] Sulahian, A., et al., *Value of antigen detection using an enzyme immunoassay in the diagnosis and prediction of invasive aspergillosis in two adult and pediatric hematology units during a 4-year prospective study.* Cancer, 2001. 91(2): p. 311-8.

[71] Pinel, C., et al., *Detection of circulating Aspergillus fumigatus galactomannan: value and limits of the Platelia test for diagnosing invasive aspergillosis.* J Clin Microbiol, 2003. 41(5): p. 2184-6.

[72] Marr, K.A., et al., *Antifungal therapy decreases sensitivity of the Aspergillus galactomannan enzyme immunoassay.* Clin Infect Dis, 2005. 40(12): p. 1762-9.

[73] Marr, K.A., et al., *Detection of galactomannan antigenemia by enzyme immunoassay for the diagnosis of invasive aspergillosis: variables that affect performance.* J Infect Dis, 2004. 190(3): p. 641-9.

[74] Hope, W.W., T.J. Walsh, and D.W. Denning, *Laboratory diagnosis of invasive aspergillosis.* Lancet Infect Dis, 2005. 5(10): p. 609-22.

[75] Pfeiffer, C.D., J.P. Fine, and N. Safdar, *Diagnosis of invasive aspergillosis using a galactomannan assay: a meta-analysis.* Clin Infect Dis, 2006. 42(10): p. 1417-27.

[76] Walsh, T.J., et al., *Detection of galactomannan antigenemia in patients receiving piperacillin-tazobactam and correlations between in vitro, in vivo, and clinical properties of the drug-antigen interaction.* J Clin Microbiol, 2004. 42(10): p. 4744-8.

[77] Mattei, D., et al., *False-positive Aspergillus galactomannan enzyme-linked immunosorbent assay results in vivo during amoxicillin-clavulanic acid treatment.* J Clin Microbiol, 2004. 42(11): p. 5362-3.

[78] Tamma, P., *The galactomannan antigen assay.* Pediatr Infect Dis J, 2007. 26(7): p. 641-2.

[79] Wheat, L.J., et al., *Histoplasmosis-associated cross-reactivity in the BioRad Platelia Aspergillus enzyme immunoassay.* Clin Vaccine Immunol, 2007. 14(5): p. 638-40.

[80] Klont, R.R., M.A. Mennink-Kersten, and P.E. Verweij, *Utility of Aspergillus antigen detection in specimens other than serum specimens.* Clin Infect Dis, 2004. 39(10): p. 1467-74.

[81] Bergeron, A., et al., *Contribution of galactomannan antigen detection in BAL to the diagnosis of invasive pulmonary aspergillosis in patients with hematologic malignancies.* Chest, 2010. 137(2): p. 410-5.

[82] Maertens, J., et al., *Bronchoalveolar lavage fluid galactomannan for the diagnosis of invasive pulmonary aspergillosis in patients with hematologic diseases.* Clin Infect Dis, 2009. 49(11): p. 1688-93.

[83] Guo, Y.L., et al., *Accuracy of BAL galactomannan in diagnosing invasive aspergillosis: a bivariate metaanalysis and systematic review.* Chest. 138(4): p. 817-24.

[84] Clancy, C.J., et al., *Bronchoalveolar lavage galactomannan in diagnosis of invasive pulmonary aspergillosis among solid-organ transplant recipients.* J Clin Microbiol, 2007. 45(6): p. 1759-65.

[85] Meersseman, W., et al., *Galactomannan in bronchoalveolar lavage fluid: a tool for diagnosing aspergillosis in intensive care unit patients.* Am J Respir Crit Care Med, 2008. 177(1): p. 27-34.

[86] Nguyen, M.H., et al., *Galactomannan Testing in Bronchoalveolar Lavage Fluid Facilitates the Diagnosis of Invasive Pulmonary Aspergillosis in Patients with Hematologic Malignancies and Stem Cell Transplant Recipients.* Biol Blood Marrow Transplant, 2010.

[87] Harrison, E., et al., *Aspergillus DNA contamination in blood collection tubes.* Diagn Microbiol Infect Dis, 2010. 67(4): p. 392-4.

[88] Khot, P.D. and D.N. Fredricks, *PCR-based diagnosis of human fungal infections.* Expert Rev Anti Infect Ther, 2009. 7(10): p. 1201-21.

[89] Wingard, J.R., *Learning from our failures: the antifungal treatment conundrum.* Clin Infect Dis, 2008. 46(9): p. 1434-5.

[90] Goodman, J.L., et al., *A controlled trial of fluconazole to prevent fungal infections in patients undergoing bone marrow transplantation.* N Engl J Med, 1992. 326(13): p. 845-51.

[91] Slavin, M.A., et al., *Efficacy and safety of fluconazole prophylaxis for fungal infections after marrow transplantation--a prospective, randomized, double-blind study.* J Infect Dis, 1995. 171(6): p. 1545-52.

[92] Marr, K.A., et al., *Prolonged fluconazole prophylaxis is associated with persistent protection against candidiasis-related death in allogeneic marrow transplant recipients: long-term follow-up of a randomized, placebo-controlled trial.* Blood, 2000. 96(6): p. 2055-61.

[93] Tomblyn, M., et al., *Guidelines for preventing infectious complications among hematopoietic cell transplant recipients: a global perspective. Preface.* Bone Marrow Transplant, 2009. 44(8): p. 453-5.

[94] Glasmacher, A., et al., *Itraconazole prevents invasive fungal infections in neutropenic patients treated for hematologic malignancies: evidence from a meta-analysis of 3,597 patients.* J Clin Oncol, 2003. 21(24): p. 4615-26.

[95] Winston, D.J., et al., *Intravenous and oral itraconazole versus intravenous and oral fluconazole for long-term antifungal prophylaxis in allogeneic hematopoietic stem-cell transplant recipients. A multicenter, randomized trial.* Ann Intern Med, 2003. 138(9): p. 705-13.

[96] Marr, K.A., et al., *Itraconazole versus fluconazole for prevention of fungal infections in patients receiving allogeneic stem cell transplants.* Blood, 2004. 103(4): p. 1527-33.

[97] Vardakas, K.Z., A. Michalopoulos, and M.E. Falagas, *Fluconazole versus itraconazole for antifungal prophylaxis in neutropenic patients with haematological malignancies: a meta-analysis of randomised-controlled trials.* Br J Haematol, 2005. 131(1): p. 22-8.

[98] Oren, I., et al., *A prospective randomized trial of itraconazole vs fluconazole for the prevention of fungal infections in patients with acute leukemia and hematopoietic stem cell transplant recipients.* Bone Marrow Transplant, 2006. 38(2): p. 127-34.

[99] Wingard, J.R., et al., *Randomized, double-blind trial of fluconazole versus voriconazole for prevention of invasive fungal infection after allogeneic hematopoietic cell transplantation.* Blood. 116(24): p. 5111-8.

[100] Kontoyiannis, D.P., et al., *Prospective surveillance for invasive fungal infections in hematopoietic stem cell transplant recipients, 2001-2006: overview of the Transplant-Associated Infection Surveillance Network (TRANSNET) Database.* Clin Infect Dis, 2010. 50(8): p. 1091-100.

[101] Mattiuzzi, G.N., et al., *Open-label, randomized comparison of itraconazole versus caspofungin for prophylaxis in patients with hematologic malignancies.* Antimicrob Agents Chemother, 2006. 50(1): p. 143-7.

[102] Cornely, O.A., et al., *Posaconazole vs. fluconazole or itraconazole prophylaxis in patients with neutropenia.* N Engl J Med, 2007. 356(4): p. 348-59.

[103] Jantunen, E., et al., *Incidence and risk factors for invasive fungal infections in allogeneic BMT recipients.* Bone Marrow Transplant, 1997. 19(8): p. 801-8.

[104] Rijnders, B.J., et al., *Aerosolized liposomal amphotericin B for the prevention of invasive pulmonary aspergillosis during prolonged neutropenia: a randomized, placebo-controlled trial.* Clin Infect Dis, 2008. 46(9): p. 1401-8.

[105] Walsh, T.J., et al., *Liposomal amphotericin B for empirical therapy in patients with persistent fever and neutropenia. National Institute of Allergy and Infectious Diseases Mycoses Study Group.* N Engl J Med, 1999. 340(10): p. 764-71.

[106] Walsh, T.J., et al., *Voriconazole compared with liposomal amphotericin B for empirical antifungal therapy in patients with neutropenia and persistent fever.* N Engl J Med, 2002. 346(4): p. 225-34.

[107] Walsh, T.J., et al., *Caspofungin versus liposomal amphotericin B for empirical antifungal therapy in patients with persistent fever and neutropenia.* N Engl J Med, 2004. 351(14): p. 1391-402.

[108] Trifilio, S., et al., *Breakthrough fungal infections after allogeneic hematopoietic stem cell transplantation in patients on prophylactic voriconazole.* Bone Marrow Transplant, 2007. 40(5): p. 451-6.

[109] Krishna, G., et al., *Pharmacokinetics of oral posaconazole in allogeneic hematopoietic stem cell transplant recipients with graft-versus-host disease.* Pharmacotherapy, 2007. 27(12): p. 1627-36.

[110] Fleming, R.V., et al., *Comparison of amphotericin B lipid complex (ABLC) vs. ambisome in the treatment of suspected or documented fungal infections in patients with leukemia.* Leuk Lymphoma, 2001. 40(5-6): p. 511-20.

[111] Wingard, J.R., et al., *A randomized, double-blind comparative trial evaluating the safety of liposomal amphotericin B versus amphotericin B lipid complex in the empirical treatment of*

febrile neutropenia. L Amph/ABLC Collaborative Study Group. Clin Infect Dis, 2000. 31(5): p. 1155-63.

[112] Boogaerts, M., et al., *Intravenous and oral itraconazole versus intravenous amphotericin B deoxycholate as empirical antifungal therapy for persistent fever in neutropenic patients with cancer who are receiving broad-spectrum antibacterial therapy. A randomized, controlled trial.* Ann Intern Med, 2001. 135(6): p. 412-22.

[113] Hebart, H., et al., *A prospective randomized controlled trial comparing PCR-based and empirical treatment with liposomal amphotericin B in patients after allo-SCT.* Bone Marrow Transplant, 2009. 43(7): p. 553-61.

[114] Cordonnier, C., et al., *Empirical versus preemptive antifungal therapy for high-risk, febrile, neutropenic patients: a randomized, controlled trial.* Clin Infect Dis, 2009. 48(8): p. 1042-51.

[115] Herbrecht, R., et al., *Voriconazole versus amphotericin B for primary therapy of invasive aspergillosis.* N Engl J Med, 2002. 347(6): p. 408-15.

[116] Weisser, M., et al., *Galactomannan does not precede major signs on a pulmonary computerized tomographic scan suggestive of invasive aspergillosis in patients with hematological malignancies.* Clin Infect Dis, 2005. 41(8): p. 1143-9.

[117] Aguilar-Guisado, M., et al., *Empirical antifungal therapy in selected patients with persistent febrile neutropenia.* Bone Marrow Transplant. 45(1): p. 159-64.

[118] Freifeld, A.G., et al., *Clinical practice guideline for the use of antimicrobial agents in neutropenic patients with cancer: 2010 update by the infectious diseases society of america.* Clin Infect Dis, 2011. 52(4): p. e56-93.

[119] Garey, K.W., et al., *Time to initiation of fluconazole therapy impacts mortality in patients with candidemia: a multi-institutional study.* Clin Infect Dis, 2006. 43(1): p. 25-31.

[120] Morrell, M., V.J. Fraser, and M.H. Kollef, *Delaying the empiric treatment of candida bloodstream infection until positive blood culture results are obtained: a potential risk factor for hospital mortality.* Antimicrob Agents Chemother, 2005. 49(9): p. 3640-5.

[121] Greene, R.E., et al., *Comparative cost-effectiveness of voriconazole and amphotericin B in treatment of invasive pulmonary aspergillosis.* Am J Health Syst Pharm, 2007. 64(24): p. 2561-8.

[122] Ascioglu, S., et al., *Defining opportunistic invasive fungal infections in immunocompromised patients with cancer and hematopoietic stem cell transplants: an international consensus.* Clin Infect Dis, 2002. 34(1): p. 7-14.

[123] Wingard, J.R., *New approaches to invasive fungal infections in acute leukemia and hematopoietic stem cell transplant patients.* Best Pract Res Clin Haematol, 2007. 20(1): p. 99-107.

[124] Reboli, A.C., et al., *Anidulafungin versus fluconazole for invasive candidiasis.* N Engl J Med, 2007. 356(24): p. 2472-82.

[125] Hachem, R., et al., *The changing epidemiology of invasive candidiasis: Candida glabrata and Candida krusei as the leading causes of candidemia in hematologic malignancy.* Cancer, 2008. 112(11): p. 2493-9.

[126] Viscoli, C., et al., *Candidemia in cancer patients: a prospective, multicenter surveillance study by the Invasive Fungal Infection Group (IFIG) of the European Organization for Research and Treatment of Cancer (EORTC).* Clin Infect Dis, 1999. 28(5): p. 1071-9.

[127] Pappas, P.G., et al., *Clinical practice guidelines for the management of candidiasis: 2009 update by the Infectious Diseases Society of America.* Clin Infect Dis, 2009. 48(5): p. 503-35.

[128] Mora-Duarte, J., et al., *Comparison of caspofungin and amphotericin B for invasive candidiasis.* N Engl J Med, 2002. 347(25): p. 2020-9.

[129] Spellberg, B.J., S.G. Filler, and J.E. Edwards, Jr., *Current treatment strategies for disseminated candidiasis.* Clin Infect Dis, 2006. 42(2): p. 244-51.

[130] Walsh, T.J., et al., *Treatment of aspergillosis: clinical practice guidelines of the Infectious Diseases Society of America.* Clin Infect Dis, 2008. 46(3): p. 327-60.

[131] Denning, D.W., et al., *Efficacy and safety of voriconazole in the treatment of acute invasive aspergillosis.* Clin Infect Dis, 2002. 34(5): p. 563-71.

[132] Mehta, A.K. and A.A. Langston, *Use of posaconazole in the treatment of invasive fungal infections.* Expert Rev Hematol, 2009. 2(6): p. 619-30.

[133] Maertens, J., et al., *Efficacy and safety of caspofungin for treatment of invasive aspergillosis in patients refractory to or intolerant of conventional antifungal therapy.* Clin Infect Dis, 2004. 39(11): p. 1563-71.

[134] Chandrasekar, P.H. and J.I. Ito, *Amphotericin B lipid complex in the management of invasive aspergillosis in immunocompromised patients.* Clin Infect Dis, 2005. 40 Suppl 6: p. S392-400.

[135] Ito, J.I., P.H. Chandrasekar, and R. Hooshmand-Rad, *Effectiveness of amphotericin B lipid complex (ABLC) treatment in allogeneic hematopoietic cell transplant (HCT) recipients with invasive aspergillosis (IA).* Bone Marrow Transplant, 2005. 36(10): p. 873-7.

[136] Cornely, O.A., et al., *Liposomal amphotericin B as initial therapy for invasive mold infection: a randomized trial comparing a high-loading dose regimen with standard dosing (AmBiLoad trial).* Clin Infect Dis, 2007. 44(10): p. 1289-97.

[137] Perfect, J.R., *Treatment of non-Aspergillus moulds in immunocompromised patients, with amphotericin B lipid complex.* Clin Infect Dis, 2005. 40 Suppl 6: p. S401-8.

[138] Tedder, M., et al., *Pulmonary mucormycosis: results of medical and surgical therapy.* Ann Thorac Surg, 1994. 57(4): p. 1044-50.

[139] van Burik, J.A., et al., *Posaconazole is effective as salvage therapy in zygomycosis: a retrospective summary of 91 cases.* Clin Infect Dis, 2006. 42(7): p. e61-5.

[140] Greenberg, R.N., et al., *Posaconazole as salvage therapy for zygomycosis.* Antimicrob Agents Chemother, 2006. 50(1): p. 126-33.

[141] Glasmacher, A., et al., *Itraconazole trough concentrations in antifungal prophylaxis with six different dosing regimens using hydroxypropyl-beta-cyclodextrin oral solution or coated-pellet capsules.* Mycoses, 1999. 42(11-12): p. 591-600.

[142] Bradford, C.R., et al., *Comparison of the multiple dose pharmacokinetics of two formulations of itraconazole during remission induction for acute myeloblastic leukaemia.* J Antimicrob Chemother, 1991. 28(4): p. 555-60.

[143] Prentice, A.G., et al., *Multiple dose pharmacokinetics of an oral solution of itraconazole in autologous bone marrow transplant recipients.* J Antimicrob Chemother, 1994. 34(2): p. 247-52.

[144] Prentice, A.G., et al., *Multiple dose pharmacokinetics of an oral solution of itraconazole in patients receiving chemotherapy for acute myeloid leukaemia.* J Antimicrob Chemother, 1995. 36(4): p. 657-63.

[145] Boogaerts, M.A., et al., *Antifungal prophylaxis with itraconazole in prolonged neutropenia: correlation with plasma levels.* Mycoses, 1989. 32 Suppl 1: p. 103-8.

[146] Berenguer, J., et al., *Itraconazole for experimental pulmonary aspergillosis: comparison with amphotericin B, interaction with cyclosporin A, and correlation between therapeutic response and itraconazole concentrations in plasma.* Antimicrob Agents Chemother, 1994. 38(6): p. 1303-8.

[147] Odds, F.C., et al., *Evaluation of possible correlations between antifungal susceptibilities of filamentous fungi in vitro and antifungal treatment outcomes in animal infection models.* Antimicrob Agents Chemother, 1998. 42(2): p. 282-8.

[148] Glasmacher A, H.C., Molitor E, et al, *Minimal effective trough concentrations for antifungal prophylaxis with itraconazole: a case-control study.* Proc Intersc Conf Antimicrob Agents Chemother, 2002 40:M-890, (abstr M-890)

[149] Brown, J. and B.B. Freeman, *Rethinking the use of voriconazole therapeutic drug monitoring in allogeneic hematopoietic stem cell transplant recipients.* Bone Marrow Transplant, 2005. 36(2): p. 177.

[150] Trifilio, S., et al., *Voriconazole therapeutic drug monitoring in allogeneic hematopoietic stem cell transplant recipients.* Bone Marrow Transplant, 2005. 35(5): p. 509-13.

[151] Pascual, A., et al., *Variability of voriconazole plasma levels measured by new high-performance liquid chromatography and bioassay methods.* Antimicrob Agents Chemother, 2007. 51(1): p. 137-43.

[152] Thompson, G.R., 3rd, et al., *Posaconazole therapeutic drug monitoring: a reference laboratory experience.* Antimicrob Agents Chemother, 2009. 53(5): p. 2223-4.

[153] Bryant, A.M., et al., *A post-marketing evaluation of posaconazole plasma concentrations in neutropenic patients with haematological malignancy receiving posaconazole prophylaxis.* Int J Antimicrob Agents, 2011. 37(3): p. 266-9.

[154] Dodds-Ashley, E., *Management of drug and food interactions with azole antifungal agents in transplant recipients.* Pharmacotherapy, 2010. 30(8): p. 842-54.

[155] Kohl, V., et al., *Factors influencing pharmacokinetics of prophylactic posaconazole in patients undergoing allogeneic stem cell transplantation.* Antimicrob Agents Chemother, 2010. 54(1): p. 207-12.

[156] Lebeaux, D., et al., *Therapeutic drug monitoring of posaconazole: a monocentric study with 54 adults.* Antimicrob Agents Chemother, 2009. 53(12): p. 5224-9.

[157] Walsh, T.J., et al., *Treatment of invasive aspergillosis with posaconazole in patients who are refractory to or intolerant of conventional therapy: an externally controlled trial.* Clin Infect Dis, 2007. 44(1): p. 2-12.

[158] Wirk, B. and J.R. Wingard, *Combination antifungal therapy: from bench to bedside.* Curr Infect Dis Rep, 2008. 10(6): p. 466-72.

[159] Rex, J.H., et al., *A randomized and blinded multicenter trial of high-dose fluconazole plus placebo versus fluconazole plus amphotericin B as therapy for candidemia and its consequences in nonneutropenic subjects.* Clin Infect Dis, 2003. 36(10): p. 1221-8.

[160] Rex, J.H., et al., *Practice guidelines for the treatment of candidiasis. Infectious Diseases Society of America.* Clin Infect Dis, 2000. 30(4): p. 662-78.

[161] Kontoyiannis, D.P., et al., *Efficacy and toxicity of caspofungin in combination with liposomal amphotericin B as primary or salvage treatment of invasive aspergillosis in patients with hematologic malignancies.* Cancer, 2003. 98(2): p. 292-9.

[162] Aliff, T.B., et al., *Refractory Aspergillus pneumonia in patients with acute leukemia: successful therapy with combination caspofungin and liposomal amphotericin.* Cancer, 2003. 97(4): p. 1025-32.
[163] Marr, K.A., et al., *Combination antifungal therapy for invasive aspergillosis.* Clin Infect Dis, 2004. 39(6): p. 797-802.

Part 3

Perspectives in Stem Cell Technology

28

Synchrotron Radiation and Nanotechnology for Stem Cell Research

Alessandra Giuliani, Fabrizio Fiori, Adrian Manescu, Vladimir S. Komlev, Chiara Renghini and Franco Rustichelli
Polytechnic University of Marche, Department SAIFET – Physical Sciences Section, Ancona
Italy

1. Introduction

Stem cell based tissue engineering therapies involve the administration of manipulated stem cell populations with the purpose of repairing and regenerating damaged or diseased tissue. Currently available methods of monitoring transplanted cells are quite limited. The monitoring of stem cell therapy outcomes requires the development of non-destructive strategies capable to identify the location, magnitude, and duration of cellular survival and fate. The recent development of imaging techniques offers great potential to address these critical issues by non-invasively tracking the fate of the transplanted cells. This chapter offers a focused presentation of some examples of the use of imaging techniques connected to the nanotechnological world in research areas related to stem cells. In particular, investigations concerning human stem cell treatment of Duchenne muscular dystrophy in animal models, cellular therapy to generate new myocardium in infarcted rat heart, bioscaffolds for cell proliferation driving to form bone tissue will be discussed.

Tissue engineering and regenerative medicine represent an emerging research area that promises new therapeutic techniques for the repair and replacement of tissues and organs that have lost functions due to ageing, disease, damage, and congenital defects (Langer & Vacanti, 1999; Atala, 2005; Jones & Hench, 2003). Clinical applications have already begun to repair a wide variety of tissues, such as blood, skin, cornea, cartilage, and bone.

Imaging techniques are playing an increasingly important role in the rigorous characterization of biomaterial properties and function. Sophisticated 2D imaging technologies have been developed to complement histological evaluation and probe complex biological events occurring at the interface between tissues and biomaterials (Boskey & Pleshko Camacho, 2007; Campbell & Kim, 2007). However, there is a clear need for high resolution 3D imaging technologies that reveal the spatial distribution of regenerated tissues forming in vitro and in vivo conditions.

Moreover, for regeneration of vascularized tissues such as bone or muscle, the ability to quantify 3D vascular in-growth would be tremendously valuable, particularly for studies exploring the potential to enhance regeneration via therapeutic angiogenesis strategies (Silva & Mooney, 2007). The imaging modality that has been most extensively applied for this purpose, particularly for bone tissue engineering studies (Mastrogiacomo et al., 2004;

Komlev et al., 2006; Papadimitropoulos et al., 2007; Eniwumide et al., 2007), is high resolution X-ray computed tomography (CT). CT provides rapid reconstruction of 3D images and quantitative volumetric analysis of X-ray attenuating materials or tissues. In the perspective of clinical translation of stem cell research, it would be advantageous to develop new techniques to detect donor cells after transplantation to track their fate and thus better understand their role in regeneration of damaged and diseased tissues.

Several groups have reported on successful labelling of mesenchymal pig (Hill et al., 2003) and mouse (Hoehn et al., 2002) embryonic stem cells with nanoparticles of iron oxide (SPIO). These particles are used as contrast agents for magnetic resonance imaging (MRI) (Arbab et al., 2003; Wang et al., 2001). It appears that cells that are able to incorporate SPIO intracellularly are readily detectable with MRI allowing in vivo tracking of such "tagged" cells (Bulte et al., 2002). MRI can provide a non-invasive and repeated three-dimensional visualization of transplanted "tagged" stem cells in organs, making it particularly attractive for imaging studies (Nuzzo et al., 2002).

The aim of this Chapter is to present recent progress obtained by using innovative and non-invasive imaging techniques, involving nanotechnologies in research areas related to stem cells. In particular, the authors will provide some examples of studies concerning human stem cell treatment of Duchenne muscular dystrophy in animal models, cellular therapy to generate new myocardium in infarcted rat heart, bioscaffolds for cell proliferation driving to form bone tissues.

The interest will be focused on X-ray computed microtomography (micro-CT), that is an imaging technique similar to conventional CT systems usually employed in medical diagnostics but with the main difference that in micro-CT a spatial resolution in the order of a few hundred nanometers can be achieved (against about 0.5 mm in CT). On the other hand, of course, such a high spatial resolution can be obtained only for reduced size samples (a few mm^3). In particular, by using synchrotron radiation, available at some European Large Scale Facilities like ESRF - Grenoble, PSI/SLS - Zurich, BESSY HZB - Berlin, HASYLAB - Hamburg and ELETTRA - Trieste, it is possible to couple high spatial resolution to high signal-to-noise ratio (Nuzzo et al., 2002; Salomé et al., 1999). Furthermore, with respect to conventional laboratory sources, the advantages of X-rays produced at synchrotron radiation sources also include a very high photon flux and a tunable-energy monochromatic beam with high coherency, in parallel beam geometry.

In conclusion, the authors will evidence that non-destructive 3D imaging techniques, such as micro-CT, are increasingly providing a powerful set of quantitative tools to aid in the development and evaluation of porous biomaterials and new approaches to engineering tissues and organs. Emphasis will be given to the fact that the key advantage of micro-CT imaging is that this method, as well as MRI, may be applicable to monitoring the stem cell homing, after cell labelling with iron oxide nanoparticles. On the other hand, when working on biopsies of small sizes (few millimeters) or small animals, micro-CT has an appreciably higher spatial resolution as compared to magnetic resonance imaging, which, in turn, has the advantage to be applicable to human body.

Micro-CT will be also shown to be a good technique for 3D studies of bioscaffolds for tissue engineering, also allowing the 3D visualization, as well as quantitative evaluations, of features very difficult to be detected by other imaging methods, such as the vascularization network in engineered bone tissue.

2. X-ray Microtomography: Outline of the technique

A revolutionary discovery in the field of medical imaging occurred at the beginning of the 1970s when the first equipment for X-rays Computed Tomography (CT) was developed. This method of imaging avoids several important limitations of conventional X-ray radiology. The impact of the CT technique has been revolutionary, enabling to view internal sample details with unprecedented precision and in a non-destructive way. Furthermore it achieves a contrast discrimination up to one thousand times better than conventional radiography (Claesson, 2001).

Tomography refers to the cross-sectional imaging of an object from either transmission or reflection data collected by illuminating the object from many different directions (Kak & Slaney, 2001). Although the first application was in diagnostic medicine, there are indeed numerous nonmedical imaging applications. This methodology is applied to the mapping of underground resources via cross borehole imaging, to some specialized cases of cross-sectional imaging for nondestructive testing, to the determination of the brightness distribution over a celestial sphere, to three-dimensional imaging with electron microscopy, etc. (Kak & Slaney, 2001).

Fundamentally, tomography employs X-rays to form images of objects based on their attenuation coefficient. Tomographic imaging deals with reconstructing an image from its projections.

The solution to the problem of how to reconstruct a function from such projections dates back to the paper published by Radon in 1917 and it was exploited with Hounsfield's invention of the x-ray computed tomographic scanner for which the same Hounsfield received a Nobel prize in 1972.

Given the enormous success of X-ray computed tomography, it is not surprising that in recent years much attention has been focused on extending this image formation technique to nuclear medicine and magnetic resonance, on one hand, and ultrasound and microwaves on the other (Kak & Slaney, 2001).

Computed X-ray microtomography (micro-CT) is similar to conventional CT systems usually employed in medical diagnoses and industrial applied research. Unlike these systems, which typically have a maximum spatial resolution of about 0.5 mm, advanced micro-CT is capable of achieving a spatial resolution up to 0.1 microns (Weitkamp et al. 2010), i.e. about three orders of magnitude lower. Such a high spatial resolution can be obtained only for samples of reduced size i.e. for dimensions in the range of a few cubic millimeters. Synchrotron Radiation allows achieving high spatial resolution images to be generated with high signal-to-noise ratio (Nuzzo et al., 2002; Salomé et al., 1999). Use of X-rays delivered by Synchrotron Facilities has several advantages compared to X-rays produced by Laboratory or Industrial sources. These include: (i) a high photon flux which permits measurements at high spatial resolution; (ii) the X-ray source is tunable, thus allowing measurements at different energies; (iii) the X-ray radiation is monochromatic, which eliminates beam hardening effects; and (iv) parallel beam acquisition allows the use of exact tomographic reconstruction algorithms. In addition, Synchrotron Radiation allows acquisition of volumes at different energies and volume subtraction to enhance contrast. Because of this variability, it is possible to enables digital image processing of micro-CT data in order to maximize contrast between neighboring voxels in the image and to increase the range of attenuation values of the entire volume. This variability in materials and acquisition methods leads to a wide range of grayscale values (corresponding to different X-ray absorption coefficients) within and among data sets (Ashbridge et al., 2003).

The spatial resolution of the CT image is dependent on the number of parallel beam projections and the number of data points in each projection. A larger data set means a more detailed description of the depicted object and hence more pixels and of smaller dimensions, i.e. better spatial resolution.
An important issue is the choice of spatial resolution versus overall sample size. Ideally, the specimen should absorb about 90 % of the incident radiation along the most radio-opaque path to obtain the best signal to noise ratio in the reconstructed image. In a homogeneous sample, absorbing 90 % of the incident radiation, the quantity $\mu(\lambda)x$ (where λ is the X-ray wavelength, $\mu(\lambda)$ is the linear attenuation coefficient of the sample for this wavelength, and x is the sample thickness), should be approximately 2. To satisfy this condition the sample thickness and/or the X-ray energy should be optimized.
In both three-dimensional (3D) conventional CT and micro-CT, hundreds of two-dimensional (2D) projection radiographs of the specimen are taken at several different angles. The information contained in each radiograph is a projection of the absorption density distribution in the sample along the direction of X-ray beam onto the plane perpendicular to the direction of the X-ray beam propagation. If the sample is then imaged several times in different orientations, 3D (volume) information on the sample structure can be obtained by using computer algorithms. This process, referred to as "tomographic image reconstruction", consists in solving an inverse problem to estimate an image from its line integrals on different directions, in 2D, and the problem is theoretically equivalent to the inversion of the Radon Transform of the image.
In practice, there are two major classes of reconstruction algorithms that use fundamentally different approaches to accomplish this conversion (Paulus et al., 2000): (i) transform-based methods using analytic inversion formulae, and (ii) series expansion methods based on linear algebra. Fourier-transform-based algorithm is commonly used in micro-CT.
An alternative approach to image reconstruction involves the use of iterative reconstruction algorithms. These algorithms start with an initial estimate of the 2D matrix of attenuation coefficients (Webb, 2003). By comparing the projections predicted from this initial estimate with those that are actually acquired, changes are made to the estimated matrix. This process is repeated for each projection in a first step, and, in a second step, also for the whole dataset until the residual error between the measured data and estimated matrix falls below a predesignated value. Iterative schemes are used relatively sparingly in standard CT scanning.
3D renderings of the data obtained after the reconstruction may be made by electronically stacking up the slices. These 3D renderings may be also sectioned in arbitrary ways and could be planed, zoomed and rotated to better locate individual details. While the slice image and 3D renderings are very useful for making qualitative observations of an internal concrete structure, the real benefit is the quantitative information that can be extracted from the 3D data sets (Ohgughi et al., 1989).
Different methods may be applied to extract quantitative architectural parameters from the tomographic images. In the field of bone research, different ways have been proposed to quantify bone micro-architecture. The 3D Mean Intercept Length (MIL) method may provide estimation of trabecular thickness and spacing, based on structural geometry assumptions, e.g. parallel plate model (Hildebrand & Ruegsegger, 1997a). However, using 3D images, such assumptions can be avoided, allowing the achievement of new model-independent quantitative parameters (Hildebrand & Ruegsegger, 1997b). Other methods, e.g. star volume distribution and star length distribution and applications of synchrotron

and conventional CT have been reviewed in (Wellington & Vinegar, 1987; Bonse & Busch, 1999; Ketcham & Carlson, 2001).

3. Synchrotron radiation microtomography for the *ex-vivo* and *in-vivo* evaluation of nanoparticle-labelled stem cell homing in muscular tissue

In recent works (Torrente et al., 2004; Gavina et al., 2006) it was shown that, after intra-arterial delivery to murine dystrophic muscle, human blood-derived CD133+ cells localize under the basal lamina and express the satellite cells markers M-cadherin and Myf5, differentiating into human muscle fibers and causing a significant amelioration of skeletal muscle structure.

The elucidation of the mechanisms involved in muscle homing of stem cells can aid in improving a potential therapy for muscular dystrophy based on the systemic delivery of such stem cells.

Combining nanoparticle cell labelling (A.K.Gupta & M.Gupta, 2005; Reimer & Weissleder, 1996) and X-ray synchrotron radiation micro-CT it is possible to provide detailed information on the stem cell migration in 3D, which are not attainable by traditional methods based on 2D techniques. In particular, micro-CT can give a relative local snapshot of the nanoparticle distribution (Brunke et al., 2005), with high spatial resolution images (from 10 μm to 1 μm) and high signal-to-noise ratio (Peyrin et al., 1998; Salomé et al., 1999).

Synchrotron radiation microtomography can be used for investigating the capacity of human stem cells to repair muscle damage in Duchenne Muscular Dystrophy.

In a first step, human blood-derived CD133+ cells were isolated from mononucleated cells collected by centrifugation (Ficoll-Hypaque; Pharmacia Biotech, Uppsala, Sweden) of several buffy coats, diluted 1:2 in RPMI 1640 medium (GIBCO, Invitrogen Life Technologies), incubated with CD133-phycoerythrin (CD133PE Miltenyi Biotech, Bergisch-Gladbach, Germany), and sorted to obtain purified CD133+ cells.

Stem cells were labelled with 250 μg/ml iron-oxide nanoparticles (Endorem) (Villa et al., 2010). Endorem is a magnetic contrast agent, based on dextran-coated Fe_3O_4 nanoparticles, with an average size of 150 nm. Labelling was performed in RPMI 1640 medium enriched with 20 ng/ml epidermal growth factor (EGF) and 10 ng/ml basic fibroblastic growth factor (bFGF) for 24 h. The average iron oxide content was 177 pg /cell.

The labelled CD133+ cells were injected into the femoral artery of scid/mdx mice. Different stem cell numbers ($5x10^4$, $1x10^5$ and $5x10^5$) were considered, at different times (0, 2, 12 and 24 h) after injection.

Ex-vivo and *in-vivo* measurements were carried out at the BM05 and ID19 beamlines, respectively, of the European Synchrotron Radiation Facility (ESRF) in Grenoble - France.

For the *ex-vivo* experiment, at different times (up to 24 hours) after cell transplantation *Tibialis Anterior* (TA) biopsies (2x2x2 mm^3) were isolated from injected legs to be studied by μ-CT (Torrente et al., 2006).

The optimal conditions for the X-ray absorption contrast among the different phases contained in the samples under investigation was obtained by varying the X-ray energy values between 18 and 27 keV. 1000 projections were recorded from each sample over 180°, with an exposure time of 1 s per projection. The detection system (2048x2048 FreLoN CCD camera) and the associated optics gave a pixel size of 1.65 μm, giving a field of view of about 3.3 x 3.3 x 3.3 mm^3. TA biopsies (2x2x2 mm^3) were isolated from injected legs and analyzed, for different numbers of initially injected cells ($5x10^5$, $1x10^5$ and $5x10^4$), as well as different times after the injection (2, 12 and 24 h). In the *in-vivo* experiment, the living animals were anesthetized and the thigh region (about 7.5 mm height in total) of their left leg was exposed to the X-ray beam. X-ray beam energy was set to 24 keV; the pixel size was fixed at 7.5 μm, thus resulting in a field-of-view of 14.4 x 14.4 x 7.2 mm^3 (2048 x 2048 x 1024 voxels); 700 projections were collected with an acquisition time of 0.2 s/projection, giving a total radiation dose on the mouse of 40 - 45 Gy. Tomograms of the thigh region taken at 0 (actually approx. 10 min.), 2, 13 and 24 hours after injection were taken, in 3 different consecutive regions-of-interest (ROI) along the direction "parallel" to the femur, for a total thickness of 5.4 mm.

3.1 *Ex-vivo* experiments

The labelled cells are visualized as red spots in the reconstructed 3D volumes (Figure 1). A more accurate observation of the spatial distribution of the particles can be obtained by "deleting" the other phases by software (Figure 2).

The signal of labelled cells was clear at all concentrations greater than $5x10^4$ cells. No difference in the location of stem cells was observed at different times after injection, and stem cells appear to be distributed along the vessels.

The volume fraction of migrated labelled stem cells was calculated by counting their corresponding pixels (Figure 3).

In order to make sure that the Endorem signal is actually associated to the labelled CD133+ stem cells, muscle biopsies from mice injected with unlabelled cells and with "naked" Endorem nanoparticles were analyzed. No detectable signal was found in muscles injected with unlabelled CD133+ cells, as well as no Endorem nanoparticles were observed within the skeletal muscle after intra-arterial injection.

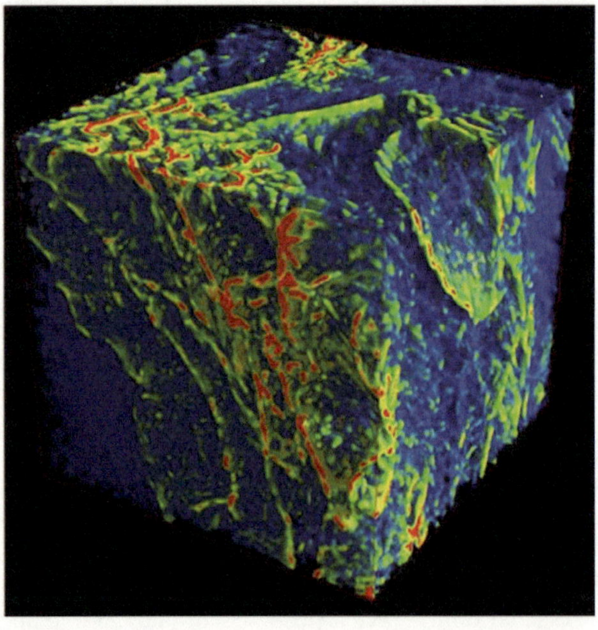

Fig. 1. 3D display showing the distribution of labelled stem cells ($5x10^5$ injected cells, 24 h after injection; red: labelled cells, green: vessels, blue: muscular tissue).

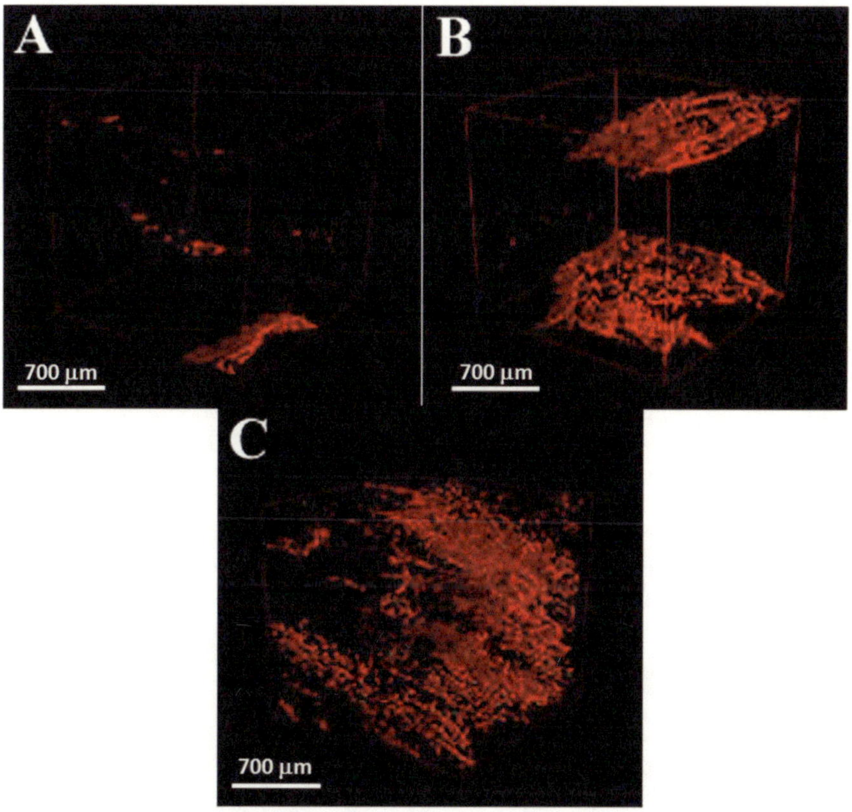

Fig. 2. 3D distribution of labelled stem cells (in red) within the muscle biopsies, 12 h after injection; A) $5x10^4$, B) $1x10^5$, C) $5x10^5$ injected cells.

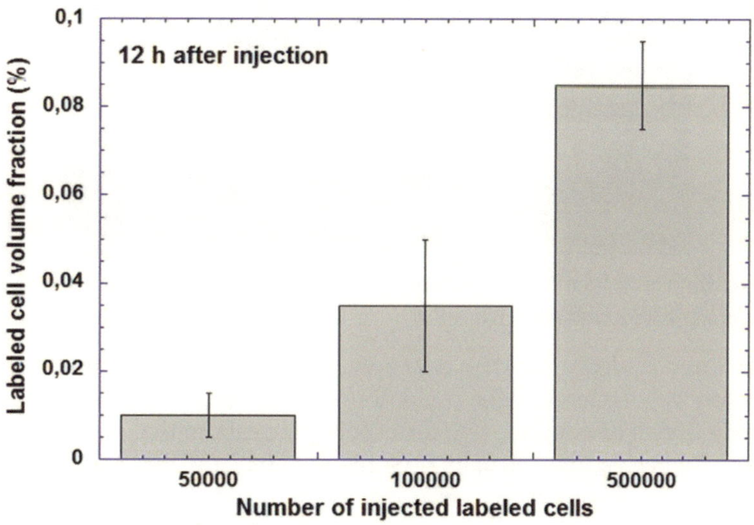

Fig. 3. Volume fraction of labelled cells in the muscle biopsies, 12 h after injection.

3.2 *In-vivo* experiments

The kinetics of the migration of stem cells into the muscular tissue was followed in the living mice, after injection of $5x10^5$ cells. To this end, one should notice that: i) μ-CT can reconstruct only objects that stay in place during the data acquisition, so that the signal coming from Endorem-labelled cells can only be due to cells that have actually migrated into the tissue, and not to those remained in the blood stream inside the vessels; ii) the small movements of the anesthetized animals, essentially due to their heartbeat (~ 10 Hz), induces two oscillations of the tissue during the acquisition time of each projection (0.2 s). Therefore the beam "sees" an apparent size of the cell as big as twice the oscillation amplitude; on this basis, assuming an oscillation amplitude of 200-300 μm, a factor ~ 50 in the apparent (measured) volume fraction with respect to the real one can be estimated.
Anyway, even though on a relative scale, the evaluation of the time evolution of this parameter is very important for the understanding of the migration kinetics.
Figure 4 shows the 3D distribution of the labelled stem cells in the investigated ROI (thigh region), for different times after injection.

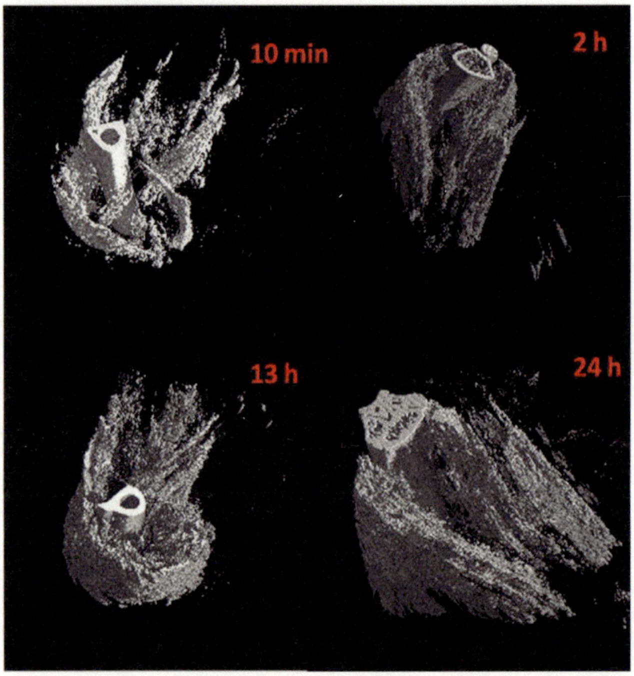

Fig. 4. 3D distribution of labelled stem cells in the femur region of the living mice, at different times after injection of $5x10^5$ cells.

In order to put into evidence the stem cells only, other tissues were eliminated by software in the image; anyway the femur bone could not be cancelled, as its absorption coefficient is similar to the Endorem one and the two corresponding peaks in the grey level histogram are superimposed. Therefore, the calculation of the apparent (in the sense described above) volume fraction of the stem cells was performed excluding the thigh region where the femur is present. The result for the calculated apparent volume fraction are shown in Figure 5, in which the most important feature evidenced is that the volume fraction is saturated after 2h.

In other words, the migration of the stem cells from the blood vessels to the muscular tissue happens within the first two hours from the injection. Synchrotron radiation micro-CT proved to be a powerful tool for the investigation of the labelled stem cells migration within the muscular tissue.

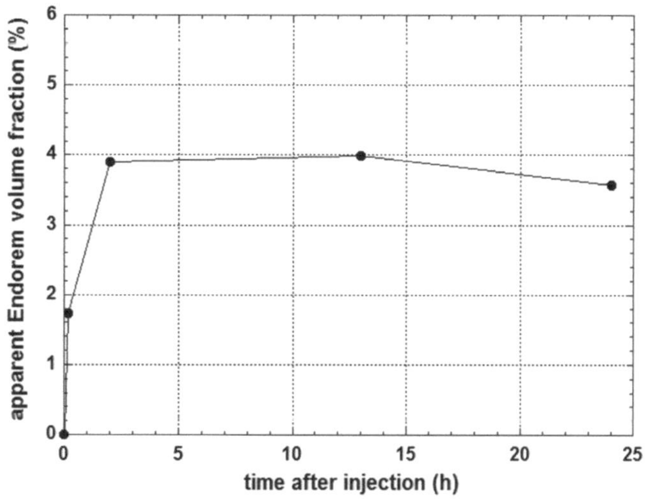

Fig. 5. Time evolution of the apparent volume fraction of labelled stem cells migrated into the muscular tissue, after injection of $5x10^5$ cells.

4. High-resolution X-ray microtomography for 3D imaging of cardiac progenitor cell homing in infarcted rat hearts

Experimental and clinical observations on the plasticity of adult stem cells has provided new tools in understanding the pathophysiology of cardiac diseases opening new strategies for the treatment of heart failure. Recent published reports (Leri et al., 2008; Kajstura et al., 2008; et al., 2006; Wollert et al., 2004; Balsam et al., 2004) have contributed to identify the possible approaches of cellular therapy to generate new myocardium, involving systemic and local mobilization of progenitor cells. Moreover, different laboratories (Beltrami et al., 2003; Oyama et al., 2007; Barile et al., 2007; Matsuura et al., 2004; Bearzi et al. 2007) have recently made available the unequivocal documentation of the existence, in the adult murine and human heart, of primitive cells able to generate all the different component structures of the myocardium. The possibility to rebuild muscle, arteries and capillaries is the necessary requirement to obtain successful approaches in cardiac regeneration, especially when taking into account the evidence that formation or implantation of a single cellular component will inevitably fail to repair the damaged organ (Menasche et al., 2008).

At tissue level of organization, microscopy techniques attempting to visualize the tissue rebuilding process, such as light, fluorescence, scanning and transmission electron microscopy are limited to two-dimensional (2D) local information or, otherwise, require laborious three-dimensional (3D) reconstruction of serial sections.

In-vivo imaging methods, including MRI, PET and micro-CT, have the potential to play a major role in the setting attempting to allow quantification of the rebuilding process, including longitudinal cell tracking (Badea et al., 2006; De Vries et al., 2005; Dhodapkar et al., 2001; Schambach et al., 2010). On the other hand, current radiologic 3D-methods, possess intrinsic limitations to identify the localization and fate of the injected cells in both clinical (Frangioni & Hajjar , 2004) and experimental (Li et al.,2009; Toyama et al., 2004; Kudo et al., 2002; Schelbert et al., 2003) settings. All these contentions related to cell tracking methodologies for myocardial regeneration have been elegantly described (Terrovitis et al., 2010). Furthermore, a recent published report (Badea et al., 2006) comparing two imaging methods - namely MRM and micro-CT - for in-vivo preclinical studies on rodents, argued that both techniques require scan times that are much longer than a single respiratory or cardiac cycle. Real time imaging is not possible with the current state-of-the-art, unless active control of ventilation – that requires complex intubation of the animal – is performed. These serious limitations related to the observation of a beating heart (existing also in the other mentioned in-vivo experiments involving other imaging techniques), do not exist in the present work, which allows the visualization in 3D and at high-resolution (10-50 times higher than MRI) of the injected cells, with the possibility to quantify them and observe their fate within the myocardium one week after the injection.

In a recent study (Giuliani et al., 2010) it is explored the use of micro-CT as an experimental technique with high spatial resolution for detection of rat Cardiac Progenitor Cells (CPCs), previously labelled with iron oxide nanoparticles, inside the infarcted rat heart, one week after injection and in ex-vivo conditions. This work contributes to understand how and to which extent the injected cells are able to migrate and regenerate the damaged myocardium. This technique was demonstrated to offer the possibility of obtaining a 3D visualization of the cell spatial distribution and a quantification of the number of cells that are able to migrate from the site of injection to different areas of the rat heart tissue, with special reference to the infarcted myocardium. Importantly, to assess the validity of this approach, two additional independent methodologies of cell tracking, Quantum Dots labelling and genetically engineered EGFP cells, were applied.

The study population consisted of male Wistar rats (Rattus norvegicus) breed in the animal facility of the University of Parma - Dept. of Pathology, age 12-14 wk, weighing 350 - 400 g. An additional group of enhanced-Green Fluorescent Protein (GFP) transgenic rats were employed and kindly provided by Dr Okabe (Okabe et al., 1997).

Myocardial infarction was produced in rats and three weeks later implantation of 5×10^5 rat clonogenic CPCs (MI-FeCell group) was performed. Cells were labelled with Feridex Nanoparticles and supplemented with Hepatocyte Growth Factor (HGF, PeproTech EC, London, UK) and Insulin-Like Growth Factor-1 (IGF-1, PeproTech EC). Two animals were treated with implantation of equal number of Quantum Dots labelled (MI-QDot group) CPCs and two additional wild type animals with green fluorescence positive (GFP^{pos})-CPCs isolated from EGFP transgenic rats, both supplemented with HGF and IGF-1. One week later animals were sacrificed and the hearts were perfusion-fixed for analysis.

CPCs and GFP^{pos}-CPCs were obtained from 3 month old Wistar rats or enhanced-Green Fluorescent Protein (GFP) rat hearts, respectively, by Langhendorff perfusion apparatus as described in (Beltrami et al., 2003) with minor modifications. The solution containing all cells was washed several times, centrifuged at 300 rpm to remove cardiomyocytes and then submitted to Percoll (Sigma, Italy) gradient to further enrich the fraction of small cells. The cell layer visualized at the interface of the desired gradient was centrifuged at 1000 rpm and cells re-suspended in 10 ml of culture medium containing Iscove Modified Dulbecco's Medium (IMDM, Sigma, Italy) supplemented with 1% Penicillin-Streptomycin (P/S, Sigma, Italy), 1% Insulin-Transferrin-Sodium Selenite (I/T/S, Sigma, Italy), 10% Fetal Bovine Serum

(FBS, Sigma, Italy) and 10 ng/ml Basic-Fibroblast Growth Factor (b-FGF, Sigma, Italy) and seeded in Petri dishes (Corning, USA) placed at 37°C-5% CO_2 for their amplification.
Microscopic observation of cultures showed the growth of two different adherent cell populations, one with mesenchymal-like and one with monomorphic blast-like characteristics. This latter population constituted the so-called Cardiac Progenitor Cells (CPCs) provided by clonogenic growth and multipotency (Beltrami et al., 2003). These cells were amplified for several passages and cryopreserved in aliquots in a medium composed by FBS supplemented with 1% Dimethylsulphoxide (DMSO, Sigma, Italy) when needed for our experimental plan.
To detect homing and engraftment of the injected cells into murine hearts, CPCs were processed using the following protocols before their injection.
Before injection CPCs were loaded with super paramagnetic iron oxide nanoparticles (Feridex -Poly-L-Lysine (PLL) complex composed of 25μgFe/ml + PLL 375 ng/ml) for 24 hours. Then, medium was removed, CPCs washed with phosphate buffered saline (PBS) and trypsinized to be injected. Furthermore, Qtracker® 585 Cell Labelling Kit (QDots, Invitrogen, Italia) were employed, possessing an emission at 585 nm wavelength able to show QDots in yellow fluorescence after UV lamp excitation.
CPCs were stained with Qtracker® 585 Cell Labelling Kit which uses a custom targeting peptide to deliver yellow-fluorescent QDot® 585 nanocrystals into the cytoplasm of living cells. The protocol was performed following manufacture's suggestions.
In order to verify the efficiency of cell labelling, immediately before injection an aliquot of CPC suspensions was analysed under a fluorescence microscope equipped with UV lamp.
A micro-CT system (Beamline BM05 at the European Synchrotron Radiation Facility - ESRF Grenoble, France) was used to non-destructively image and quantify the 3Dmicrostructural morphology of each rat heart. The micro-CT experiment was performed in two modes: with a 15keV monochromatic X-ray beam and a sample-to-detector distance of 25 mm for the absorption-contrast and 500 mm for the phase-contrast, respectively. The acquisition setup was based on previously described (Torrente et al., 2006; Salome et al., 1999) 3D parallel tomography. 1500 projections and a step of 0.12 degrees were considered for each sample, with an exposure time of 1 s per projection.
3D reconstructions of the samples were obtained from the two series of 2D projections (for each sample one in absorption and the other one in phase-contrast configuration) using a 3D filtered back projection algorithm implemented at ESRF. The different phases found in the histogram referring to the reconstruction of the in-absorption acquisitions were coloured using a 3D display software in order to make them more easily recognizable. Furthermore, as the acquisitions were performed simultaneously, the combining absorption and phase contrast images did not required their geometric warping so that the corresponding image is structured correctly. Therefore, synchronization may be interactive or partially or fully automatic with help of a fusion algorithm which is described elsewhere (Stokking et al., 2003). The full fusion automatic algorithm was used. This exact synchronization of data sets provides capacity for image fusion with superimposition of both sets of imaging data in one image data set for further 3D visualization.

4.1 Fluorescence analysis

Once inside the cells, Qtracker® label provide intense, stable fluorescence that can be traced through several generations. QDots are not transferred to adjacent cells in a population of growing cells, enabling long-term studies of live cells and tissues.

CPCs pre-loaded with 585 QDots, were implanted by 3 injections in the peri-infarct region of the left ventricular wall of chronically infarcted heart. After heart excision, tissue sections of LV, RV and S were analyzed under UV-light and the fluorescence intensity of the QDots positive areas were analysed in photomicrographs covering the entire heart (Figure 6, A-B).

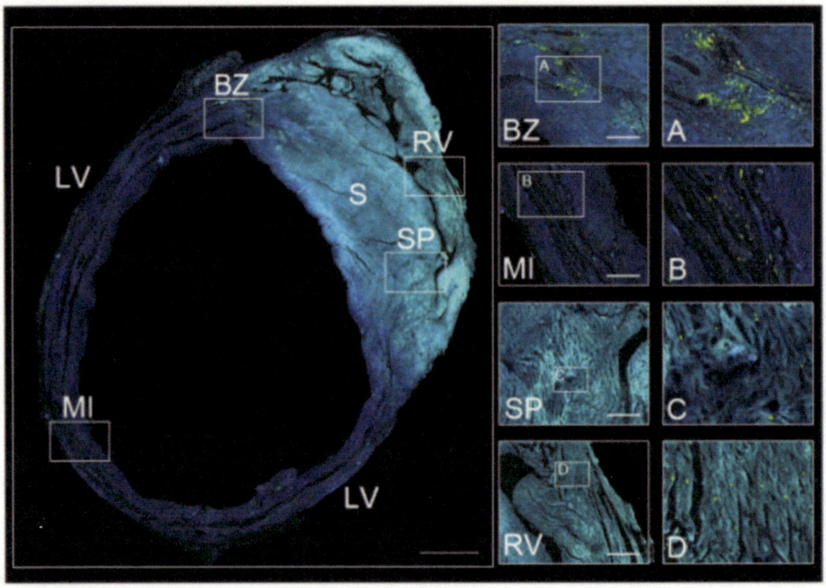

Fig. 6. Transverse section of a paraffin-embedded infarcted heart injected with CPCs and analyzed under UV-light to document the distribution of QDot labelled CPCs. Scale bars correspond to 1 mm in the left panels and to 200 μm in the right ones.

The results showed that CPC-related QDot signal was found in all regions of the heart although with different distribution and intensity. QDot-fluorescence was high and uniformly distributed in the infarcted (MI) and peri-infarcted (BZ) regions, from the site of injection following the total extension of the scar.
QDots were also occasionally found at very low occurrence in the spared myocardium (SP) and in the right ventricle (RV), far from the infarcted area and from the injection site.
To ascertain whether QDots signals were associated to living cells and not to nanoparticles released from death cells and uptaken by macrophages, the presence of GFP^{pos} cells in hearts injected with CPCs from EGFP transgenic rats was assessed. Immunofluorescence analysis showed that the distribution of GFP^{pos} cells present in the infarcted and spared myocardium one week after injection of GFP^{pos} CPCs, was similar to that observed in MI-QDots group. Interestingly, $GFP^{pos}/\alpha\text{-Sarc}^{pos}$ myocytes constituted ~15% of these cells and reached an average volume of 980±125 μm^3. The average cell volume of the remaining ~85% of GFP^{pos} cells found in the myocardium was 455 ± 76 μm^3.

4.2 Micro-CT analysis

The micro-CT analysis was used to non-destructively image and characterize, in absorption and in phase-contrast configuration, the 3D distribution of the rat CPCs one week after their injection into an infarcted rat heart. Two biopsies of infarcted rat hearts, both injected with

$5 \cdot 10^5$ stem cells labelled by iron-oxide nanoparticles, 1 week after injection were analyzed. The harvested heart samples were scanned and projection data were obtained and reconstructed into 2D images. A portion of a reconstructed 2D slice is shown in Figure 7 (A). The basic physical parameter quantified in each pixel of an absorption micro-CT image and exploited to obtain the contrast is the linear X-ray attenuation coefficient μ. The X-ray attenuation produced by the Feridex-positive stem cells is higher than attenuation referred to the other tissues of the injected hearts, allowing their visualization as bright spots in the 2D images (magenta coloured spots in Figure 7 (A)). The differences in the X-rays absorption rate within the samples translate into different peaks in the grey level scale corresponding to the different phases. The histogram of the grey level scale is shown in Figure 7 (B) and 7 (C). The volume of each phase is obtained by multiplying the volume of a voxel (~125 μm^3) by the number of voxels underlying the peak associated with the specific phase. The volume percentage of the CPCs respect to the total investigated volumes is very low (as shown in Table 1), as expected by considering the amount of the injected CPCs: for this reason the peak corresponding to the labelled CPC phase is not visible in Figure 7 (B) but can be easily detected in the inset (Figure 7 (C)), magnifying the respective portion of the histogram.

The slice images were compiled and analysed to render 3D images and to obtain a better visualization of the stem cell distribution after their injection in the infarcted rat hearts. Regions of interest were selected in each slice image and thresholded to eliminate background noise. The thresholded 3D volumes were converted into coloured images using a 3D display software, by means it is possible to correlate the bright spots of labelled cells present in the 2D images (Figure 7 (A)) with the spots distributed in the 3D reconstruction (Figure 8 (C) and (D)).

3D display of fusion images is represented on Figure 8. In combined imaging, morphological information about heart tissue from phase contrast is complemented and extended by the functional information on labelled-CPCs supplied from absorption mode.

Furthermore, the simple segmentation process provides information on the internal structure of the heart (Figure 8 (B)) and the distribution of the CPCs within it (Figure 8 (C) and (D)). Subsequently, different 3D sub-regions of interest with detailed information are illustrated in Figure 8 (E) and (F).

No dramatic difference in the location of stem cells was observed in the two hearts: the percentage of migrated labelled stem cells from the injection area to the infarcted area or towards other areas was calculated by counting their corresponding pixels, using the algorithm that automatically separates them from the other tissues. The obtained data were expressed as percentage of the analysed sample volume and are listed in Table 1.

The micro-CT was used to image and characterize the 3D spatial distribution of injected rat clonogenic cells (CPCs) inside the heart tissue of infarcted mice after cell labelling by iron oxide nanoparticles.

The 3D visualization of the spatial distribution of the grafted cells with respect to myocardium and vascular system was obtained. In particular, the X-ray absorption of the labelled cells was higher than that of host tissues, allowing their visualization as bright spots in the 2D images (Figure 7).

These slice images were compiled and analyzed to render 3D images and to obtain a better visualization of cell distribution within the samples (Figure 8). One week after injection, labelled cells were distributed mostly in proximity and towards the damaged infarcted area

(Figure 8 (F)), demonstrating migration of CPCs from the injection site made around the coronary binding. It was also possible to identify finger-like cell structures in the inner part of the left ventricular wall (Figure 8 (D) and Figure 8 (F)).
It was also observed single smaller units, in all areas of the heart, as in the atria, in large vessels (Figure 8 (E)) and in the right ventricle. These are very important and new data: in particular they constitute a confirmation that these cells can migrate through the myocardium by a biological mechanism which is still unknown.

	Total Volume of the investigated rat heart [μm³]	**Total Volume V_{tot} of CPCs* [μm³]**	**Infarcted Area (%)**	**Perinfarcted Area (%)**	**Remaining Area (%)**	**CPC Volume Fraction (%)**	**Total CPC number *N***
Sample I	**$321676 \cdot 10^6$**	**$595 \cdot 10^6$**	**77.9**	**4.3**	**17.8**	**0.18**	**~ $1.11X10^6$**
Sample II	**$331265 \cdot 10^6$**	**$995 \cdot 10^6$**	**97.8**	**1.6**	**0.6**	**0.30**	**~ $1.86X10^6$**

* Total Volume filled by the CPCs [μm³]

Table 1. Quantitative Parameters obtained in absorption configuration.

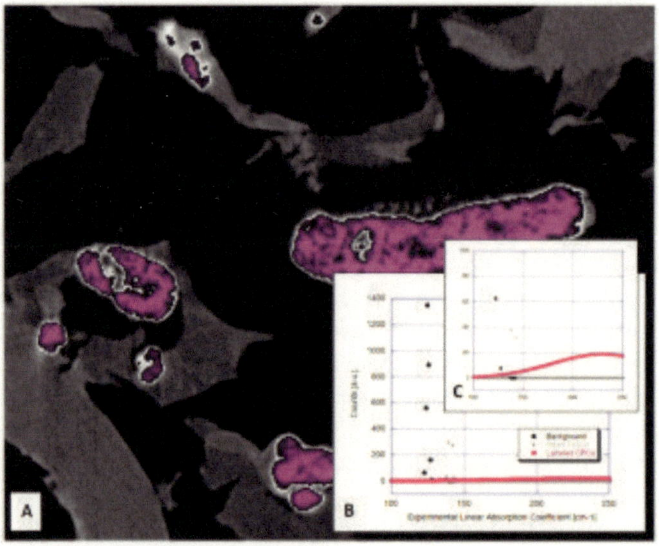

Fig. 7. (A) Portion of a reconstructed 2D slice. The X-ray attenuation produced by the labelled cells is higher than attenuation referred to the other tissues of the injected hearts, allowing their visualization as bright spots in the 2D images (magenta coloured spots). (B) The histogram of the grey level scale corresponding to the different detected phases. (C) Magnification of the portion of the histogram shown in B referred to the peak corresponding to the labelled cells grey level.

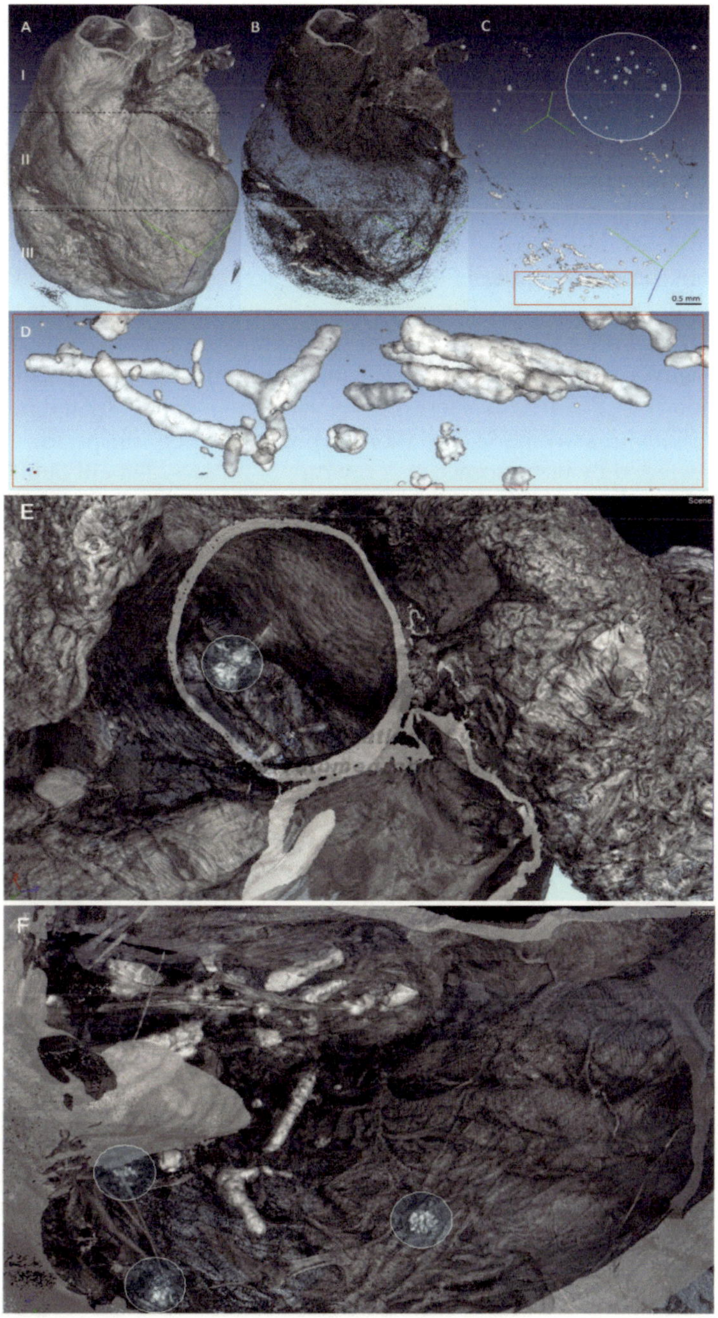

Fig. 8. (A) 3D display of fusion images. The simple segmentation process provides information on the internal structure of the heart (B) and the distribution of labelled cells (C) and (D). Fusion images of two different 3D sub regions of interest: (E) - atria and large vessels; (F) - infarcted area. Individual cells and small cell clusters are pointed out in the lighted circles.

Importantly, results obtained by micro-CT were in close agreement with imaging of the distribution of QDots, strongly suggesting the validity of these methodologies to detect cell engraftment within the damaged heart.
The issue whether either iron particles, as detected here by micro-CT, or QDots nanoparticles, as detetcted by UV excitation, could only represent unreliable signals of macrophages uptaking tracers released by dying cells, remains an unresolved problem especially on clinical ground. However, an effort was made by the present investigation to apply a third independent genetic marker to visualize the progeny of the injected cells. Thus, CPCs from rats carrying the EGFP transgene were employed and GFP labelled cells counted by immunofluorescence. This approach allowed us to ensure that the progeny of the injected cells was present in the infarcted heart and with a similar distribution to that observed with QDots labelling and micro-CT imaging. Importantly, not only survival but also differentiation of CPCs into cardiomyocytes was demonstrated by the presence of GFP positive cells with α-sarcomeric striation. Thus, whether unspecific signals were concomitantly detected by fluorescence or X ray cannot be excluded although with minimal impact on the overall results.
Indeed, it was estimated the total number N of the cells found in the hearts one week after CPC injection (as reported in Table 1). It was found out that in both hearts the number N of CPC-derived cells is more than doubled with respect to the number of injected CPCs, demonstrating that, at least for labelled CPCs and at a week from injection, the presence of Feridex does not dramatically affect the CPC life span.
In conclusion, the obtained 3D images represent a very innovative progress, as compared to the usual 2D histological images, which do not provide the overall 3D distribution of rat clonogenic cells and their progeny within the heart, providing biological insights into the early processes of cell migration.

5. Application of X-ray synchrotron radiation techniques to bone tissue engineering

Tissue engineering is a promising approach to create artificial constructs for repairing or replacing partially or entirely diseased tissues (Williams, 2004; Langer & Vacanti, 1993).
A key component in tissue engineering for bone regeneration is the scaffold, which acts as a template for cell interactions and for the growth of bone extracellular matrix to provide structural support for the newly formed tissue (Karageorgiou & Kaplan, 2005). Many researchers have tried to define which properties are required for the optimal synthetic scaffold, in particular for bone tissue replacement (Hutmacher, 2000).
Ideally, scaffolds should possess adequate mechanical strength and permeability (van Lenthe et al., 2007), high porosity (Shi et al., 2007), large surface area to volume ratio, as well as biocompatibility and non- toxicity (Murugan & Ramakrishna, 2006). Moreover, pore interconnectivity is an important scaffold criterion (Moore et al., 2004) which influences nutrient supply (Hui et al., 1996), circulation of extracellular material (Gross & Rodriguez-Lorenzo, 2004), the contact between adjacent cells (Darling & Sun, 2004), and the promotion of blood vessel and bone tissue ingrowth (Kuboki et al., 1998).
Commonly, two-dimensional (2D) investigation methods, such as scanning electron microscopy (SEM) are used for the characterization and comparison of the scaffolds. Unfortunately, these methods do not allow a complete investigation of the three dimensional (3D) spatial structure of the scaffold. More recently, micro-CT has been

proposed for the characterization of scaffolds and 3D bone ingrowth (Mastrogiacomo et al., 2004).

In the recent years, a lot of progress was obtained in the field of bone tissue engineering by using innovative analytical and diagnostic imaging techniques, such as micro-CT and pseudo-holotomography.

X-ray synchrotron microtomography was applied (Komlev et al., 2006) to investigate highly porous hydroxyapatite scaffolds, previously seeded with bone marrow stromal cells (BMSC), implanted in an immunodeficient murine model. It was proved that it was possible to obtain in a nondestructive way a quantitative analysis of tissue engineering constructs, determining the total volume and thickness distribution of newly formed bone into implants in a small animal model, by using the micro-CT technique. This methodology offers major advantages, including the possibility of investigating the influence of scaffold parameters, such as porosity and spatial distribution of the structure thickness, on the bone growth within the implant.

Very recently, different ceramic scaffolds with high porosity were characterized (Komlev et al., 2010) and it was evaluated the bone growth into the tissue engineering constructs *ex vivo* at different implantation times, by using synchrotron radiation micro-CT. In this study, three types of ceramic scaffolds with different composition and structure (namely synthetic 100% hydroxyapatite (HA; Engipore), synthetic calcium phosphate multiphase biomaterial containing 67% silicon stabilized tricalcium phosphate (Si-TCP; Skelite™) and natural bone mineral derived scaffolds (Bio-oss®), were seeded with mesenchymal stem cells (BMSC) and ectopically implanted for 8 and 16 weeks in immunodeficient mice. X-ray synchrotron radiation microtomography was used to derive 3D structural information on the same scaffolds, both before and after implantation. The images of the three scaffolds before implantation revealed an appreciable difference among their morphologies (Figure 9). In particular, the Bio-Oss®, contained elongated ellipsoidal pores, whereas the HA scaffold contained roughly spherical pores.

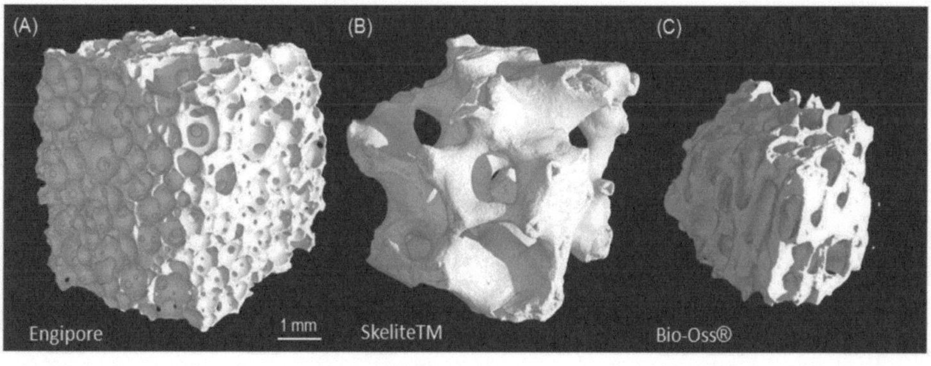

Fig. 9. 3D display of different scaffolds before implantation: **(A)** Engipore (hydroxyapatite) produced by FinCeramica, Faenza, Italy; **(B)** SkeliteTM (silicon-stabilized tricalcium phosphate) produced by Millenium Biologics Kingston, Ontario, Canada; **(C)** Bio-Oss® (natural bone mineral) produced by Geistlich Pharma AG, Wolhusen, Switzerland (Komlev et al., 2010).

Figure 10 (Panels A1-C1), obtained by an innovative imaging procedure, gives an instantaneous pictorial view of the wall thickness distribution in the scaffold and confirms the results previously obtained in (Papadimitropoulos et al., 2007), namely biodegradation for the Si-TCP scaffold and lack of it for the HA scaffold. The newly investigated Bio-Oss® showed a very little decrease of the scaffold wall thickness; the decrease was at the limit of detectability, and needs to be confirmed by additional experiments.

Based on these findings, the scaffold biodegradation in the tissue engineered implanted constructs was investigated after 16 weeks implantation only for the Skelite™. 3D displays of registered images of pre- and post implantation Skelite™ samples implanted for 8 (A) and 16 (B) weeks are presented in Figure 11 (panels A-B). As it appears in the images in Figure 10 (panel B2), blue and yellow correspond to totally or partially resorbed scaffold. The volume percentage distribution of the different phases is presented in panels A1 and B1. An increase in the percentage of the resorbed scaffold was observed in the case of a longer implantation time.

The analysis proposed in the manuscript (Komlev et al., 2010) is, therefore, a major improvement as compared to the imaging procedure adopted in our previous work (Papadimitropoulos et al., 2007), where only a comparison between different subvolumes of the implants before and after implantation was made.

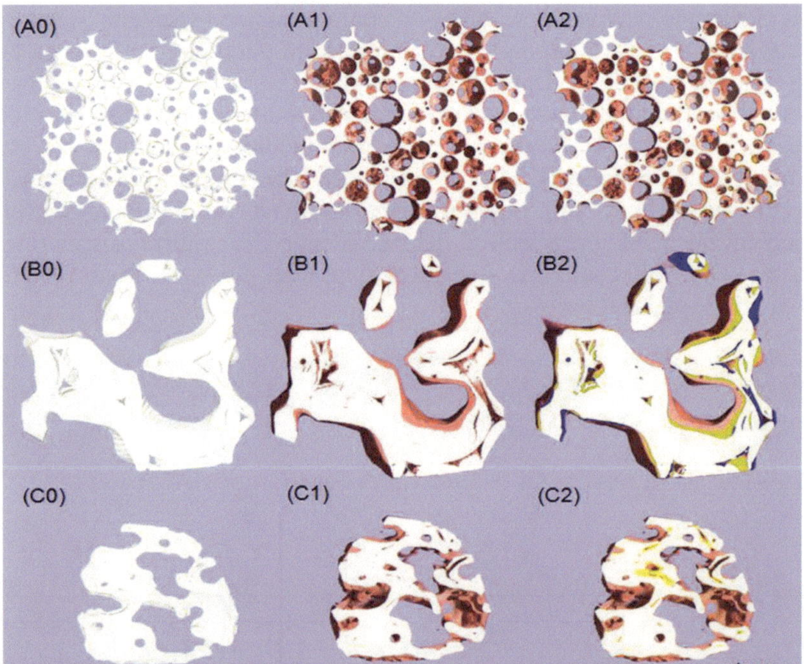

Fig. 10. 3D display of subvolumes of scaffolds before and after implantation: (A) Engipore; (B) SkeliteTM; (C) Bio-Oss®; (A1-C1). New bone (pink) on the surface of the scaffolds (white). In the panels A2-C2 are presented the images obtained by combining (registering) the data of panels A0-C0 with those of panels A1- C1. Blue volumes indicate portions of scaffolds present in panels A0-C0 (pre-implant) and absent in panels A1-C1 (after implantation) and correspond to completely resorbed scaffold. Yellow volumes indicate virgin scaffold volume in which after implantation a reduction of the sample density is observed (Komlev et al., 2010).

Finally, a high content of innovation is associated to the detailed kinetics studies on the Skelite™ scaffolds implanted for different times, not only due to the large number of the implantation times investigated, but also to the recording in the X-ray absorption histograms of separate peaks associated to HA and TCP in the same scaffold (Figure 12).

It is therefore possible to observe that the progressive biodegradation of Skelite™ scaffold is eventually due to the TCP component. It should be noted that when investigated by microdiffraction studies, the local structural study at the interface between the newly formed bone and the Skelite™ scaffold indicated that scaffold biodegradation was mainly due to TCP depletion (Papadimitropoulos et al., 2007).

Moreover, saturation in the TCP biodegradation occurred at an implantation time of about 10 weeks, whereas saturation in the tissue engineered bone occurred at an implantation time of about 22 weeks.

This could indicate that the bone growth did not occur only in the scaffold volume that was resorbed, but also in the inward direction with respect to the pore surface. This finding is in agreement with the results presented in Figure 5 of reference (Mastrogiacomo et al., 2007), and in Figure 4 of reference (Papadimitropoulos et al., 2007).

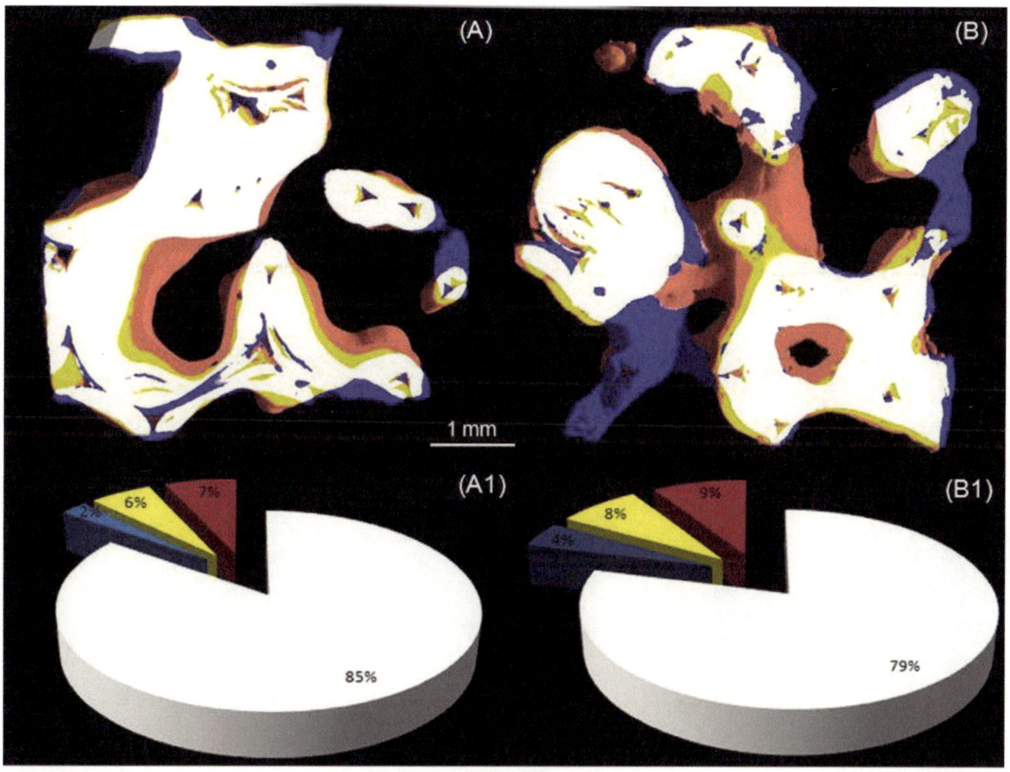

Fig. 11. Display based on a combination of the 3D structure of pre- and post implanted SkeliteTM samples for 8 (A) and 16 (B) weeks, respectively (white - scaffold; pink - new bone; blue - total resorption; yellow - partial resorption (see caption of Figure 10). (A1-B1) volume percentage distribution of the different phases (Komlev et al., 2010).

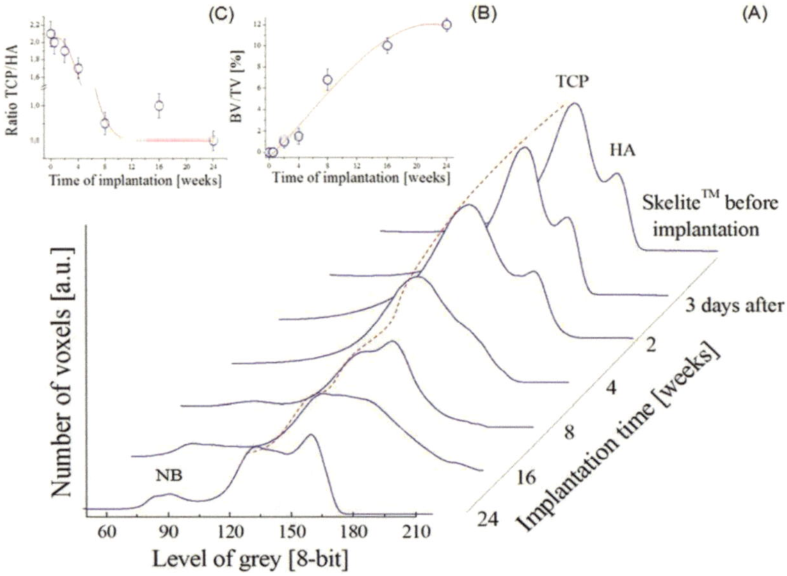

Fig. 12. Volume histograms of SkeliteTM scaffolds implanted from 3 days to 24 weeks (NB - new bone, TCP - tricalcium phosphate, HA - hydroxyapatite). (A) New bone deposition kinetics; (B) Percentage of bone volume/ total volume; (C) TCP/HA mean ratio as a function of the implantation time (Komlev et al., 2010).

Figure 13 (A) and (B) presents a 3D reconstruction of a bone tissue- engineered construct, 24 weeks after the implantation. Three phases are clearly distinguishable: the scaffold (white), the engineered bone (light brown), and the vessel networks within the pores (green). In Figure 13 (B), the engineered bone was removed by digital processing to obtain a clearer evidence of vessel network structure.

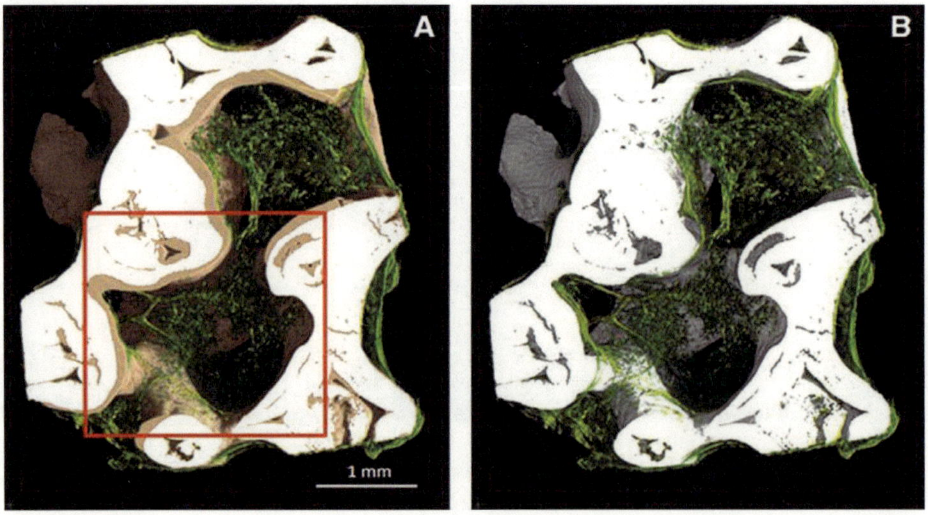

Fig. 13. 3D enhanced contrast micro-CT images of the tissue-engineered construct after 24 weeks of implantation in an immunocompromised mouse. (Komlev et al., 2009).

In absorption micro-CT, the image contrast results from the difference among different attenuation coefficients of X-ray for the different materials. Unfortunately, X-ray attenuation coefficients of soft tissues are low, and, thus, the contrast difference of these structures is low, unless a contrast agent is adopted. Therefore, in a recent study (Komlev et al., 2009), it was proposed to use X-ray synchrotron radiation pseudo-holotomography to visualize microvascular networks, at a three-dimensional (3D) level, for the first time with no need for contrast agents, and to extract quantitative structural data in a Bioceramic/MSC composite implanted for 24 weeks in a mouse.
The enhanced contrast micro-CT technique is a new imaging method based on classical micro-CT and recently developed technique of holotomography. By using a coherent X-ray beam, phase contrast may be simply obtained by free space propagation (i.e., by positioning the detector at some distance from the sample), while a 2D projection of the phase map can be obtained from three or four series of images, each series being recorded at different distance from the object for each rotation angles considered. Then, the 3D phase map is reconstructed with the same algorithm as in classical tomography (Cloetens, 2009). Anyway, through the weighted superposition of both attenuation and phase maps (enhance contrast micro-CT), it is possible to generate better images.
Vessels are easy to see also in 2D micro-CT images (Figure 14 (A)). Figure 14 (B) shows a histogram of the 3D vessel diameter distribution measured within the full volume of the sample implanted for 24 weeks. The mean vessel diameter measured from pseudo-holotomography data was 49±25 μm. This value was comparable to the 47±18 μm measured in control histology sections.

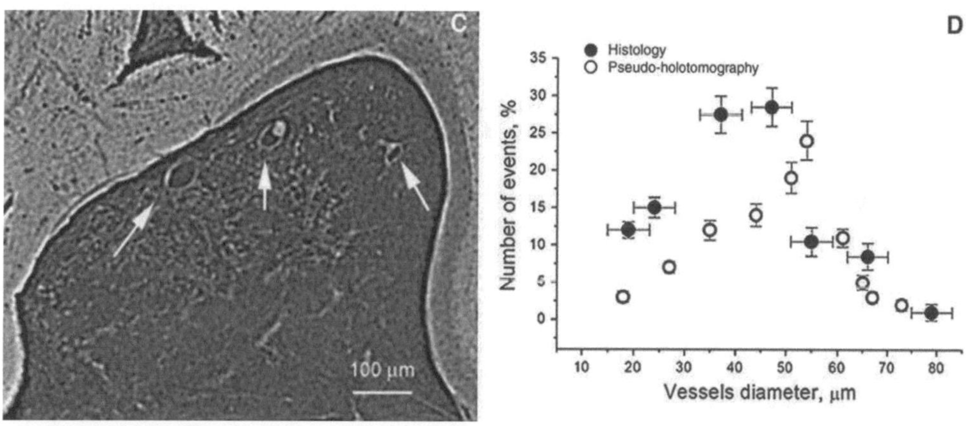

Fig. 14. 2D micro-CT image (A) and (B) histogram of the vessel diameter distribution measured for pseudo-holotomography data (open circle) and histology (solid circle) (Komlev et al., 2009).

This progress could be extrapolated to different biomedical research areas where angiogenesis and microvasculogenesis play an important role, as for the development of tissues such as bone, in regenerative medicine, or in pathologies characterized by inflammation and tissue damage such as diabetes, osteoarthritis, and muscular dystrophy. Of great interest could also be the application of the pseudo-holotomography to investigations of therapeutical roadmaps for tumor treatment involving the suppression of vascularization.

6. References

Arbab, A.S., Bashaw, L.A., Miller, B.R., Jordan, E.K., Bulte, J.W. & Frank, J.A., Intracytoplasmic tagging of cells with ferumoxides and transfection agent for cellular magnetic resonance imaging after cell transplantation: methods andtechniques. Transplantation 2003, 76: 1123.

Ashbridge, D.A., Thorne, M.S., Rivers, M.L., Muccino, J.C. & O'Day, P.A., Image optimization and analysis of synchrotron X-ray computed microtomography (CT) data, Computers & Geosciences 2003, 29: 823–36.

Atala, A. Tissue engineering, stem cells and cloning: current concepts and changing trends, Expert Opin Biol Ther. 2005, 5(7): 879-92.

Badea, C.T., Hedlund, L.W. & Johnson, G.A., Micro-CT with respiratory and cardiac gating, Med Phys., 2004, 31: 3324–3329.

Balsam, L.B., Wagers, A.J., Christensen, J.L., Kofidis, T., Weissman, I.L. & Robbins, R.C., Haematopoietic stem cells adopt mature haematopoietic fates in ischaemic myocardium. Nature,2004, 428:668 –73.

Barile, L., Chimenti, I., Gaetani, R., Forte, E., Miraldi, F. & Frati G., Cardiac stem cells: isolation, expansion and experimental use for myocardial regeneration. Nat Clin Pract Cardiovasc Med, 2007, 4 (Suppl 1):S9 –14.

Bearzi, C., Rota, M., Hosoda, T., Tillmanns, J., Nascimbene, A. & De Angelis, A., Human cardiac stem cells. Proc Natl Acad Sci USA, 2007, 104:14068 –73.

Beltrami, A.P., Barlucchi, L., Torella, D., Baker, M., Limana, F. & Chimenti, S., Adult cardiac stem cells are multipotent and support myocardial regeneration. Cell, 2007, 114:763–76.

Bonse, U. & Busch, F., X-ray computed microtomography (μCT) using synchrotron radiation (SR). Progr in Bioph and Molec Biol 1999; 65(1/2): 133–69.

Boskey, A. & Pleshko Camacho, N. FT-IR imaging of native and tissue-engineered bone and cartilage. Biomaterials 2007, 28: 2465.

Brunke, O., Odenbach, S., Fritsche, C., Hilger, I. & Kaiser, W.A., Determination of magnetic particle distribution in biomedical applications by X-ray microtomography, J. Magnet. Mag. Mater. 289 (2005) 428–430.

Bulte, J.W., Duncan, I.D. & Frank, J.A. In vivo magnetic resonance tracking of magnetically labelled cells after transplantation.J. Cereb. Blood Flow Metab. 2002, 22: 899.

Campbell, C.T. & Kim, G., SPR microscopy and its applications to high-throughput analyses of biomolecular binding events and their kinetics. Biomaterials 2007, 28: 2380.

Claesson, T., A medical imaging demonstrator of computed tomography and bone mineral densitometry. Stockholm: Universitetsservice US AB; 2001.

Cloetens, P., Holotomography: quantitative phase tomography with micrometer resolution using hard synchrotron radiation X rays. Appl Phys Lett, 2009, 75: 2912.

Darling, A.L. & Sun, W., 3D microtomographic characterization of precision extruded poly-epsilon-caprolactone scaffolds. Journal of Biomedical Materials Research Part B - Applied Biomaterials 2004, 70B (2):311–317.

De Vries, I.J.M., Lesterhuis, W.J., Barentsz, J.O., Verdijk, P., Van Krieken, J.H. & Boerman, O.C., Magnetic resonance tracking of dendritic cells in melanoma patients for monitoring of cellular therapy. Nature Biotechnology, 2005, 23: 1407 - 1413.

Dhodapkar, M.V., Steinman, R.M, Krasovsky, J., Munz, C. & Bhardwaj, N., Antigen-specific inhibition of effector T cell function in humans after injection of immature dendritic cells. J Exp Med, 2001, 193:233-8.

Eniwumide, J.O., Yuan, H., Cartmell, S.H., Meijer, C.J. & de Bruijn, J.D., Ectopic bone formation in bone marrow stem cell seeded calcium phosphate scaffolds as compared to autograft and (cell seeded) allograft. Eur. Cells Mater. 2007, 14: 30.

Frangioni, J.V. & Hajjar, R.J., In Vivo Tracking of Stem Cells for Clinical Trials in Cardiovascular Diseases. Circulation, 2004, 110:3378-3384.

Gavina, M., Belicchi, M., Rossi, B., Ottoboni, L., Colombo, F., Meregalli, M., Battistelli, M., Forzenigo, L., Biondetti, P., Pisati, F., Parolini, D., Farini, A., Issekutz, A.C., Bresolin, N., Rustichelli, F., Constantin, G. & Torrente, Y., VCAM-1 expression on dystrophic muscle vessels has a critical role in the recruitment of human blood-derived CD133+ stem cells after intra-arterial transplantation, Blood 108(8) (2006) 2857-2866.

Giuliani, A., Frati, C., Rossini, A., Komlev, V. S., Lagrasta, C., Savi, M., Cavalli, S., Gaetano, C., Quaini, F., Manescu, A. & Rustichelli, F. , High-resolution X-ray microtomography for three-dimensional imaging of cardiac progenitor cell homing in infarcted rat hearts. Journal of Tissue Engineering and Regenerative Medicine, n/a. doi: 10.1002/term.409

Gross, K.A. & Rodriguez-Lorenzo, L.M., Biodegradable composite scaffolds with an interconnected spherical network for bone tissue engineering. Biomaterials 2004, 25 (20):4955–4962.

Gupta, A.K. & Gupta, M., Synthesis and surface engineering of iron oxide nanoparticles for biomedical applications, Biomaterials 2005, 26: 3995–4021.

Hildebrand, T. & Ruegsegger, P., A new method for the model independent assessment of thickness in three-dimensional images. J Microsc, 1997; 185: 67–5.

Hildebrand, T. & Ruegsegger, P., Quantification of bone microarchitecture with the Structure Model Index. Comput Meth Biomech Biomed Engng 1997; 1: 15–23.

Hill, J.M., Dick, A.J., Raman, V.K., Thompson, R.B., Yu, Z.X. & Hinds, K.A. Serial Cardiac Magnetic Resonance Imaging of Injected Mesenchymal Stem Cells. Circulation 2003, 108: 1009.

Hoehn, M., Kustermann, E., Blunk, J., Wiedermann, D., Trapp, T., Wecker, S., Focking, M., Arnold, H., Hescheler, J., Fleischmann, B.K., Schwindt, W. & Buhrle, C. Monitoring of implanted stem cell migration in vivo: a highly resolved in vivo magnetic resonance imaging investigation of experimental stroke in rat. Proc Natl Acad Sci U S A. 2002; 99: 16267–16272.

Hui, P.W., Leung, P.C. & Sher, A. Fluid conductance of cancellous bone graft as a predictor for graft–host interface healing. Journal of Biomechanics 1996, 29 (1):123–132.

Hutmacher, D.W., Scaffolds in tissue engineering bone and cartilage. Biomaterials 2000,21:2529–43.

Jones, J.R. & Hench, L.L. Regeneration of Trabecular Bone using Porous Ceramics. Curr. Opin. Solid State Mater. Sci. 2003, 7: 301.

Kajstura, J., Urbanek, K., Rota, M., Bearzi, C., Hosoda, T., Bolli, R., Anversa, P. & Leri, A. Cardiac stem cells and myocardial disease. J Mol Cell Cardiol 2008, 45(4):505-13.

Kak, A.C. & Slaney, M., Principles of Computerized Tomographic Imaging, Society of Industrial and Applied Mathematics, 2001.

Karageorgiou, V. & Kaplan D. Porosity of 3-D biomaterial scaffolds and osteogenesis. Biomaterials 2005, 26:5474–91.

Ketcham, R.A. & Carlson, W.D. Acquisition, optimization and interpretation of X-ray computed tomographic imagery: applications to the geosciences. Comput Geosci 2001, 27(4): 381–400.

Komlev, V., Mastrogiacomo, M., Pereira, R.C., Peyrin, F., Rustichelli, F., & Cancedda, R. Biodegradation of porous calcium phosphate scaffolds in ectopic bone formation model studied by X-ray computed microtomography. European Cells and Materials 2010, 19: 136-146.

Komlev, V., Mastrogiacomo, M., Peyrin, F., Cancedda, R., & Rustichelli, F. X-Ray Synchrotron Radiation Pseudo-Holotomography as a New Imaging Technique to Investigate Angio- and Microvasculogenesis with No Usage of Contrast Agents. Tissue Engineering: Part C 2009, 15(3): 425-430.

Komlev, V.S., Peyrin, F., Mastrogiacomo, M. , Cedola, A., Papadimitropoulos, A., Rustichelli, F., & Cancedda, R. 3D analysis by X-ray computed microtomography of in vivo bone growth into porous calcium phosphate scaffolds. Tissue Eng. 2006, 12: 3449-3458.

Kuboki, Y., Takita, H., Kobayashi, D., Tsuruga, E., Inoue, M., Murata, M., Nagai, N., Dohi, Y., & Ohgushi, H. BMP-induced osteogenesis on the surface of hydroxyapatite with geometrically feasible and nonfeasible structures: topology of osteogenesis. Journal of Biomedical Materials Research 1998, 39 (2):190–199.

Kudo, T., Fukuchi, K., Annala, A.J., Chatziioannou, A.F., Allada, V., Dahlbom, M., Tai, Y.C., Inubushi, M., Huang, S.C., Cherry, S.R., Phelps, M.E. & Schelbert, H.R. Noninvasive measurement of myocardial activity concentrations and perfusion defect sizes in rats with a new small-animal positron emission tomograph. Circulation. 2002,106:118-23.

Langer, R.S. & Vacanti, J.P. Tissue engineering: the challenges ahead. Sci. Am. 1999, 280:86.

Leri, A., Kajstura, J., Anversa P. & Frishman W.H. Myocardial regeneration and stem cell repair. Curr Probl Cardiol 2008, 33:91-153.

Li, Z., Lee, A., Huang, M., Chun, H., Chung, J., Chu, P., Hoyt, G., Yang, P., Rosenberg, J., Robbins, R.C. & Wu, J.C. Imaging Survival and Function of Transplanted Cardiac Resident Stem Cells. J Am Coll Cardiol 2009, 53:1229–40.

Mastrogiacomo, M., Komlev, V.S., Hausard, M., Peyrin, F., Turquier, F., Casari, S., Cedola, A., Rustichelli, F. & Cancedda, R.Synchrotron Radiation Microtomography of Bone Engineered from Bone Marrow Stromal Cells. Tissue Eng. 2004; 10: 1767.

Mastrogiacomo, M., Papadimitropoulos, A., Cedola, A., Peyrin, F., Giannoni, P., Pearce, S.G., Alini, M., Giannini, C., Guagliardi, A., & Cancedda R. Engineering of bone using bone marrow stromal cells and a silicon-stabilized tricalcium phosphate bioceramic. Evidence for a coupling between bone formation and scaffold resorption. Biomaterials 2007, 28: 1376.

Matsuura, K., Nagai, T., Nishigaki, N., Oyama, T., Nishi, J., Wada, H., Sano, M., Toko, H., Akazawa, H., Sato, T., Nakaya, H., Kasanuki, H. & Komuro, I. Adult cardiac Sca-1-positive cells differentiate into beating cardiomyocytes. J Biol Chem 2004; 279:11384-11391.

Menasche, P., Alfieri, O., Janssens, S., McKenna, W., Reichenspurrer, H., Trinquart, L., Vilquin, J., Marolleau, J.P., Seymour, B., Larghero, J., Lake, S., Chtellier, G., Solomon, S., Desnos, M. & Hagege, A. The Myoblast Autologous Grafting in Ischemic Cardiomyopathy (MAGIC) Trial.First Randomized Placebo-Controlled Study of Myoblast Transplantation. Circulation 2008;117:1189-1200.

Moore, M.J., Jabbari, E., Ritman, E.L.,. Lu, L.C, Currier, B.L., Windebank, A.J. & Yaszemski M.J. Quantitative analysis of interconnectivity of porous biodegradable scaffolds with micro-computed tomography. Journal of Biomedical Materials Research Part A 2004, 71A (2):258–267.

Murugan, R. & Ramakrishna, S. Nano-featured scaffolds for tissue engineering: a review of spinning methodologies. Tissue Engineering 2006, 12 (3):435–447.

Nuzzo, S., Peyrin, F., Cloetens, P., Baruchel, J. & Boivin, G. Quantification of the degree of mineralization of bone in three dimension using Synchrotron Radiation Microtomography. Med Phys 2002; 19: 2672-81.

Ohgushi, H., Goldberg, V.M., Caplan, A.I. Repair of bone defects with marrow cells and porous ceramic. Experiments in rats. Acta Orthop Scand 1989, 60(3): 334-9.

Okabe, M., Ikawa, M., Kominami, K., Nakanishi, T., Nishimune, Y. 'Green mice' as a source of ubiquitous green cells. FEBS Lett.1997, 407(3):313-9.

Oyama, T., Nagai, T., Wada, H., Naito, A.T., Matsuura, K., Iwanaga, K., Takahashi, T., Goto, M., Mikami, Y., Yasuda, N., Akazawa, H., Uezumi, A., Takeda, S. & Komuro, I. Cardiac side population cells have a potential to migrate and differentiate into cardiomyocytes in vitroand in vivo. J Cell Biol 2007; 176:329–41.

Papadimitropoulos, A., Mastrogiacomo, M. , Peyrin, F., Molinari, E., Komlev, V.S., Rustichelli, F., & Cancedda, R. Kinetics of In Vivo Bone Deposition by Bone Marrow Stromal Cells Within a Resorbable Porous Calcium Phosphate Scaffold:An X-Ray Computed Microtomography Study. Biotechnol. Bioeng. 2007, 98: 271.

Paulus, M.J., Gleason, S.S., Kennel, S.J., Hunsicker, P.R. & Johnson, D.K. High resolution x-ray computed tomography: an emerging tool for small animal cancer research. Neoplasia 2000; 2: 62-70.

Peyrin, F., Salomé, M., Cloetens, P., Laval-Jeantet, A.M., Ritman, E. & Ruegsegger, P., Micro-CT examinations of trabecular bone samples at different resolutions: 14, 7 and 2 micron level. Technol. Health Care 1998, 6: 391-401.

Reimer, P. & Weissleder, R. Development and experimental application of receptor - specific MR contrast media, Radiology 1996; 36: 153–163.

Salomé, M., Peyrin, F., Cloetens, P., Odet, C., Laval-Jeantet, A.M., Baruchel, J. & Spanne, P. A synchrotron radiation microtomography system for the analysis of trabecular bone samples. Med Phys 1999; 26: 2194-2204.

Schambach, S.J., Bag, S., Groden, C., Schilling, L., Brockmann, M.A., Vascular imaging in small rodents using micro-CT, Methods 2010, Volume 50, Issue 1, 26-35.

Schelbert, H.R., Inubushi, M. & Ross, R.S., PET imaging in small animals. J Nucl Cardiol, 2003, 10:513–20.

Shi, X.F., Sitharaman, B., Pham, Q.P., Liang, F., Wu, K., Billups , W.E., Wilson, L.J. & Mikos A.G. Fabrication of porous ultra-short single-walled carbon nanotube nanocomposite scaffolds for bone tissue engineering, Biomaterials 2007,28:4078–4090.

Silva, E.A. & Mooney, D.J. Spatiotemporal control of vascular endothelial growth factor delivery from injectable hydrogels enhances angiogenesis. J Thromb. Haemost. 2007, 5: 590.

Stokking, R., Zubal, I.G. & Viergever, M.A., Display of fused images: methods, interpretation, and diagnostic improvements. Semin Nucl Med, 2003, 33:219-227.

Terrovitis, J.V., Ruckdeschel Smith, R. & Marbán, E., Assessment and Optimization of Cell Engraftment After Transplantation Into the Heart. Circ. Res., 2010, 106:479-494.

Torrente, Y., Belicchi, M., Sampaolesi, M., Pisati, F., Meregalli, M., D'Antona, G., Tonlorenzi, R., Porretti, L., Gavina, M., Mamchaoui, K., Pellegrino, M.A., Furling, D., Mouly, V., Butler-Browne, G.S., Bottinelli, R., Cossu, G. & Bresolin, N. Human circulating AC133(+) stem cells restore dystrophin expression and ameliorate function in dystrophic skeletal muscle. J. Clin. Invest. 2004; 114:182– 195.

Torrente, Y., Gavina, M., Belicchi, M., Fiori, F., Komlev, V.S., Bresolin, N. & Rustichelli, F. High-resolution X-ray microtomography for three-dimensional visualition of human stem cell Muscle homing. Febs Letter 2006; 580: 5759-64.

Toyama, H., Ichise, M., Liow, J.S., Vines, D.C., Seneca, N.M., Modell, K.J. & Seidel, J. Green MV, Innis RB.Evaluation of anesthesia effects on [18F]FDG uptake in mouse brain and heart using small animal PET. Nucl Med Biol 2004, 31:251- 6.

Van Lenthe, G.H., Hagenmüller, H., Bohner, M., Hollister, S.J., Meinel, L. & Müller, R. Nondestructive micro-computed tomography for biological imaging and quantification of scaffold-bone interaction in vivo. Biomaterials 2007, 28 (15):2479-2490.

Villa, C., Erratico, S., Razini, P., Fiori, F., Rustichelli, F., Torrente, Y. & Belicchi, M., Stem Cell Tracking By Nanotechnologies, Int. J. Mol. Sci. 11 (2010) 1070-1081.

Wang, Y.X., Hussain, S.M. & Krestin, G.P. Superparamagnetic iron oxide contrast agents: application in MR imaging. Eur. Radiol. 2001, 11: 2319.

Webb, A. G., Introduction to Biomedical Imaging. IEEE Press Series on Biomedical Engineering Sponsored by the IEEE Engineering in Medicine and Biology Society Wiley-IEEE Press, 2003.

Weitkamp, T., Tafforeau, P., Boller, E., Cloetens, P., Valade, J.-P., Bernard, P., Peyrin, F., Ludwig, W., Helfen, L. & Baruchel, J., Status and evolution of the ESRF beamline ID19, AIP Conf. Proc., April 2010, Volume 1221, pp. 33-38.

Wellington, S.L. & Vinegar, H.J., X-ray computerized tomography. J Petr Techn 1987; 39(8): 885-98.

Williams, D., Benefit and risk in tissue engineering. Mater Today 2004; 7:24e9.

Wollert, K.C., Meyer, G.P., Lotz, J., Ringes-Lichtenberg, S., Lippolt, P., Breidenbach, C., Fichtner, S., Korte, T., Hornig, B., Arseniev, L., Hertenstein, B.,Ganser, A. & Drexler, H., Intracoronary autologous bone-marrow cell transfer after myocardial infarction: the BOOST randomised controlled clinical trial. Lancet 2004; 364:141- 8.

29

Spontaneous Generation of Neurospheres from Mouse Embryonic Stem Cells

Roberto Di Maio[1], Kommu Naga Mohan[2], Brianna Heath[3] and Leonardo D'Aiuto[4]

[1] *Department of Neurology, University of Pittsburgh, & Ri.Med Foundation, Palermo*
[2] *Department of Molecular and Human Genetics, Baylor College of Medicine*
[3] *Department of Human Genetics, Graduate School of Public Health, University of Pittsburgh,*
[4] *Department of Psychiatry, University of Pittsburgh School of Medicine, Western Psychiatric Institute and Clinic*
[1]*Italy*
[2,3,4]*United States*

1. Introduction

Neural stem cells (NSCs) are self-renewing pluripotent cells that can produce different parts of the nervous system. NSCs were initially identified in the subventricular zone of the mouse brain (Temple, 1989) and subsequently in various regions of adult brains from human and mouse (Taupin & Gage, 2002). NSCs can be derived *in vitro* from embryonic stem (ES) cells and induced pluripotent stem (iPS) cells by employing specific culture conditions. NSCs and their progeny can be expanded for several passages as tridimensional floating aggregates named neurospheres or as monolayer cultures. This in turn allows derivation of neurons of different parts of the nervous system (Gaspard & Vanderhaeghen, 2011). NSCs derived from ES/iPS cells of various genetic backgrounds represent invaluable tools for the investigation of neurogenesis, development of neurologic diseases models, and screening of new drugs to treat neurological diseases.

Derived from ES cells, NSCs and their progeny, neural progenitors, are routinely studied *in vitro* by a method called neurosphere culture system (Reynolds & Weiss, 1992). Neurospheres can measure 100-300 mm. The zonal distribution of different cell types that compose human and murine neurospheres resembles an outside-in brain structure with nestin-positive progenitor cells in the periphery and $GFAP^+$ and β-tubulin III-positive cells in the centre (Moors et al., 2009; Campos et al., 2004).

Currently, there are two methods that allow derivation of neurospheres from mouse ES cells. Both of these methods utilize fibroblast growth factor 2 (FGF-2), epidermal growth factor (EGF), retinoic acid, and other supplements such as B27 and N2. FGF-2 is a critical component in neural differentiation protocols as it promotes proliferation of neuroprogenitor cells (Yoshimura et al., 2001). For selecting and expanding neural progenitor cells, culture media are supplemented with B-27 (containing retinoic acid) and N2 (which has a subset of component of B-27 that include insulin and transferrin).

In one method of derivation, ES cells are cultured in a specific medium supplemented with EGF, FGF-2 and N2. After 4-5 days, 50-80% of the cells undergo neural lineage specification. On day 7, these cells are dissociated and plated in suspension in uncoated plates in a differentiation medium supplemented with N2, B27, mouse EGF and human FGF-2. After a few days of culturing, neurospheres can be obtained (Conti et al., 2005).
Alternatively, in a second method, mouse ES cells are co-cultured for approximately one week with the PA6 stromal cell line to achieve efficient neuronal differentiation because of the stromal cell-derived inducing activity (SDIA) (Kawasaki et al., 2000). After 8 days of SDIA treatment, mouse ES cells are separated from PA6 stromal cells and cultured in suspension in serum-free medium supplemented with FGF-2. Under these conditions neurospheres appear within 4 days (Morizane et al., 2006).
Exemplifying stages of the neurosphere generation process are shown in figure 1.
The ability of mouse ES cells to spontaneously generate neurospheres and mature neurons in short time and in few steps, represents the strength for a powerful and reliable *in vitro* model to apply research strategies oriented to study the physiopathology of the biochemical and epigenetic mechanisms leading to the neurogenesis.
We recently observed that mouse ES cells when grown in simple culture conditions spontaneously form neurospheres. In this chapter, we describe a simple method for derivation of neurospheres from mouse ES cells without using FGF-2 and EGF and without co-culturing ES cells with stromal cells.

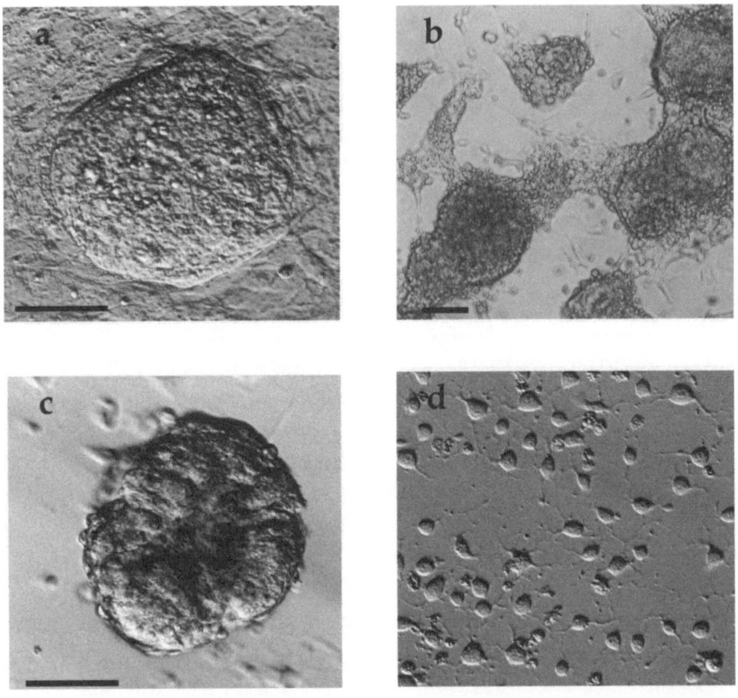

Fig. 1. Generation of neurospheres from mES cells. a) Microphotograph of a mES colony; b) typical clusters of cells generated after culturing R1 mES for three passages in a DMEM/FBS medium; c) bright field of a neurosphere; d) progenitors migrating from neurospheres. [Images taken by authors.] Scale bar 1s 100 mm.

2. Reagents and equipment

The reagents and equipment needed for the methods described in this chapter can be found listed in the tables below.

Equipment	*Company*	*Catalog #*
0.22-µm 500 ml bottle top filter	Corning	430513
0.22-µm 250 ml filter system	Corning	430767
10 cm cell culture dish	Corning	439167
24-well plates (Transwell-Clear permeable support 0.4 µm	Fischer	07-200-154
6-well ultra low attachment plate	Costar	3471
15-ml polypropylene conical tubes	BD Falcon	352096
50-ml polypropylene conical tubes	BD Falcon	352070
Transfer pipettes	Samco	262-1S
0.22-µm bottle top filters	TPP	99505
0.45-µm bottle top filters	Nalgene	165-0045
12 mm circular cover slips	Fisher	01-472A
Hemocytometer	Hycor Kova	87144
1000-ml media bottles		
Kimwipes	Kimberly-Clarks	34155
Humidified tissue culture incubator (37°C, 5% CO_2)	-	-
Centrifuge	-	-
Water bath (37°C)	-	-
Scanning confocal microscope	Olympus	FluoView 100
DM IRM inverted microscope (fitted with Leica HS N Plan BD 50X oil immersion objective)	Leica	-
75W xenon lamp-based monochromator	Ushio, Japan	-
CCD camera	Orca (Hamamatsu, Shizouka, Japan)	-
535 ± 40nm bandpass filter	Omega Optical	-
500 nm longpass dichromatic mirror	-	-
Simple PCI	Compix, Inc (Cranberry, PA)	-

Table 1. Equipment needed for the methods described in this chapter.

Reagent	*Company*	*Catalog #*
R1 ES cells	Mutant Mouse Regional Resource Centers	011979-MU
Non-essential amino acid solution (100X)	Millipore	TMS-001
Nucleosides (100X)	Millipore	ES-008-E
L-glutamine (100X)	Millipore	TMS-002-C
Penicillin/Streptomycin (100X)	Millipore	TMS-001
2-Mercaptoethanol (100X)	Millipore	ES-007-E
Sodium pyruvate (100mM)	Cellgro	25-00-CI
Trypsin-EDTA (100X)	Gibco	15400
Dulbecco's phosphate buffered saline (10X) (PBS)	Sigma	P-3813
Gelatin from porcine skin (Type A)	Sigma	G1890
Dulbecco's modified Eagle's medium (DMEM)	Millipore	SLM-220-B
Fetal bovine serum (ES cell grade) (FBS)	Gibco	16141-079
Knockout Serum Replacement (KSR)	Invitrogen	10828028
Neurobasal medium	Invitrogen	21103-049
Phosphate buffered saline (PBS)	Gibco	14040141
Poly-D-lysine (PDL) hydrobromide	Sigma	P6407
Leukemia inhibitory factor (LIF) (10^7)	Millipore	ESG1107
B27	Invitrogen	17504-044
N2 supplement	Invitrogen	17502-048
FGF-basic Recombinant Mouse	Invitrogen	PMG0035
Laminin	Sigma	L2020
BisBenzimide H 33342 trihydrochloride	Sigma	B-2261
Mounting solution	Gelvatol	-
Paraformaldehyde (16% in premixed PBS buffer at pH7 [below])	Electron Microscopy Sciences	15710-6
PBS buffer at pH7 (1X)	Roche	11666789001
10% Normal Donkey Serum (NDS)	Jackson ImmunoResearch	711-165-152
0.01% TritonX-100	Fluka	BP151-100
Tubulin III	R&D Systems	MAB1195
Rabbit Anti-NMDAR1	Abcam	Ab68144
Bovine albumin	Sigma	A-6003
Fura-2, AM cell-permeant	Invitrogen	F-1221
Cy3 AffiniPure Donkey Anti-Rabbit	Jackson ImmunoResearch	711-165-152
AffiniPure Donkey Anti-Mouse	Jackson ImmunoResearch	715-175-151

Table 2. Reagents needed for the methods described in this chapter.

3. Reagent and equipment setup

The setup for the reagents and equipment listed is as follows.

Reagent	*Setup*
ES medium	For 500 ml of ES medium, add the following components: 400 ml DMEM, 75 ml FBS, 5 ml Pen/Strep, 5 ml 2ME (100 mM), 5 ml L-glutamine, 5 ml NEAA, 5 ml nucleosides, 5 ml 2ME, and 50 µl LIF.
EB medium	For 500 ml of EB medium, add the following components: 450 ml DMEM, 25 ml KSR, 5 ml sodium pyruvate, 5 ml Pen/Strep, 5 ml 2ME (100 mM), 5 ml L-glutamine, and 5 ml NEAA.
DMEM/FBS medium	For 500 ml of feeder cells medium, add the following components: 440 ml DMEM, 50 ml FBS, 5 ml NEAA, and 5 ml Pen/Strep.
N2 medium	For 100 ml of N2 medium, add 96 ml DMEM/F12, 1 ml N-2, 1 ml Pen/Strep, 1 ml L-glutamine, 1 ml NEAA, 4 µl FGF-2, and 100 µl laminin.
B27/Neurobasal medium	For 500 ml of B27/Neurobasal medium, add the following components: 480 ml Neurobasal medium, 10 ml B27, 5 ml L-glutamine 100 X, and 5 ml Pen/Strep.
Gelatin 0.1%	Dissolve 100 mg of gelatin in 90 ml warm (~ 60 °C) distilled water. Sterilize by autoclaving and store at room temperature.
25x FGF stock solution (100 µg ml^{-1})	Dissolve 25 mg recombinant mouse FGF in 250 ml of sterile PBS. Store at ~20 °C.
5x Poly-D-Lysine	Dissolve 50 mg poly-D-lysine in 333.3 ml distilled water and rinse the package with the water. Filter the solution with a 0.22-µm bottle top filter. Aliquot 5 ml into 15-ml conical tubes and store at -20 °C.
Hepes-buffered salt solution (HBSS)	Prepare stocks of the composition described in table 4 below. To obtain the HBSS basic, combine the solutions in the following order: 10.0 ml Solution #1, 1.0 ml Solution #2, 1.0 ml Solution #3, 1.0 ml Solution #5, 86.0 ml distilled H_2O, 1.0 ml Solution #4. Prior to use, prepare the following solution: 9.8 ml HBSS basic, 0.1 ml Stock #5, 0.1 ml Stock #6. The final HBSS composition is 137 mM NaCl, 5.4 mM KCl, 0.6 mM Na_2HPO_4, 0.6 mM KH_2PO_4, 20 mM KH_2PO_4, 1.3 mM $CaCl_2$, 1.0 mM $MgSO_4$, 10 mM $NaHCO_3$, 5.5 mM glucose.

Table 3. Setup of reagents needed for the methods described.

Stock	*Preparation*
1	In 90ml of distilled H_2O, dissolve 8.0 g NaCl. 0.4 g KCl, and fill to 100 ml with distilled H_2O.
2	In 90ml of distilled H_2O, dissolve 0.41 g Na_2HPO_4 (anhydrous), 0.80 g KH_2PO_4, and fill to 100 ml with distilled H_2O
3	Add 0.72 g $CaCl_2$ to 50ml of distilled H_2O
4	Add 1.23 g $MgSO_4x7H_2O$ to 50ml of distilled H_2O
5	Add 4.76 g Hepes to 10ml of distilled H_2O
6	Add 0.9 g Glucose to 10ml of distilled H_2O
7	Add 0.84 g $NaHCO_3$ to 10ml of distilled H_2O

Table 4. Preparation of the stock solutions needed to prepare HBSS. The combination of these stock solutions in described in table 3.

Equipment	*Setup*
Gelatin-coated plate	Add 5 ml of 0.1% (wt/vol) gelatin solution into a plate so that it covers the entire bottom of the plate. Incubate the plate for 20 min at room temperature.
Poly-D-lysine (PDL) cover slips	Autoclave cover slips. Place sterilized cover slips into the wells of a 24-well plate. Dilute the 5 x poly-D-lysine to 1 x with sterile distilled water plate except for the outermost wells. Add 0.5–1 ml of the 1 x poly-D-lysine solution to cover the surface of the cover slips. After an overnight incubation, remove the PDL and wash the plates with sterile deionized water and dry in a cell culture hood.

Table 5. Setup for the equipment needed for the methods described.

4. Methods

The generation of neurospheres and subsequent neuronal differentiation from ES cells is achieved through the stages described in the protocols below.

4.1 Maintenance of ES cells

Place R1 ES cells (Nagy et al., 1993) on 10 cm gelatin-coated dishes with 12 ml of ES medium in a humidified chamber in a 5% CO_2/air mixture at 37°C. Change the medium every day.
When the cells reach 80-90% confluence split them to two dishes as follows: wash the cells with 1x PBS twice, then add 2 ml of 1x trypsin-EDTA. Rock the dish to ensure that the solution covers all the cells. Incubate the dish for 5 min at 37°C. Pipette up and down to break up clumps and obtain a good dissociation of cells. Pipette 8 ml ES medium to inhibit further tryptic activity and transfer to 15-ml conical tubes. Spin at 1000 rpm for 3 min. Aspirate the medium from cells and resuspend the pellet in 6 ml of fresh ES medium.
Transfer 3 ml of cell suspension into a new 10 cm gelatin-coated dish containing 9 ml of fresh ES medium.

4.2 Generation of embryoid bodies

Aspirate the medium, add 2 ml of 1x trypsin-EDTA and incubate for 5 min at 37 °C. Dissociate the cells by thoroughly pipetting several times with a P1000 Pipetman. Monitor

the trypsinization under an inverted microscope. If the separation of the cells is not complete, continue the treatment with trypsin for additional 2-3 min.
Transfer the cell suspension into a 15-ml sterile conical tube. Pellet the cells centrifugation for 5 min at 1200 rpm; then resuspend the cell pellet in 10 ml EB medium.
Count the cells with a cell counter and plate $5x10^5$ ES cells per well in 6-well ultra low attachment plates in EB medium. Embryoid bodies should appear by overnight incubation.
Culture plates for 7 days. Change the medium every other day as follows: Collect the EBs in suspension from the plate and transfer into a 15 ml conical tube. Leave the conical tube for 15 minutes to allow the EBs to settle to the bottom of the tube. Aspirate the medium and add 12 ml of fresh EB medium. Transfer the EBs suspension into a new 6-well ultra low attachment plates.

4.3 Neurosphere generation and neuronal differentiation

Collect EBs and remove the medium as described in the previous step.
Count the number of EBs and prepare an EBs suspension in Feeder medium at a density of 40-50 EBs/ml.
Transfer 40-50 intact EBs to non-coated 10 cm cell culture dish containing 12 ml DMEM/FBS medium.
Culture until cells reach confluence, changing medium every 2-3 days.
Aspirate the medium, add 2 ml of trypsin-EDTA solution and incubate for 5 min at 37°C. Dissociate cells thoroughly by pipetting with a P1000 Pipetman. Pipette 8 ml Feeder medium and transfer the cell suspension to a 15-ml conical tube.
Centrifuge for 5 min at 1200 rpm. Aspirate the medium and resuspend the cells in 12 ml DMEM/FBS. Culture the cells until confluence.
Trypsinize as described in section 4.2 and culture cells for other two passages.
Transfer the neurospheres floating in the culture medium into 50-ml conical tubes and allow them to settle for 10 minutes. Collect the supernatant containing neurospheres. Trypsinize cells attached to the bottom of the plate as described in section 6.2 and culture to generate new neurospheres.

4.4 Neuronal differentiation

Resuspend the neurospheres in the appropriate volume of N2 medium and culture for differentiation in a 24-well plate (15-20 neurospheres/well) containing poly-D-lysine-coated cover slips. Replace a half volume medium every other day.
At day 7 exchange N2 medium to B27/Neurobasal medium and let the cells differentiate for 10 more days with a medium change every other day.

5. Characterization and functional analysis of neurons

Selective immunocytochemistry can be performed according to standard procedures as described in the following sections.

5.1 Cell cultures

Plate the neurospheres on PDL-treated cover slips, and culture according to experimental protocols with the same conditions described above.

5.2 Immunocytochemistry procedure

Fix mouse neurospheres-derived neurons for 20 minutes at room temperature in a 4% paraformaldehyde solution (prepared from stock 16% Electron Microscopy Sciences, 15710-6 in 1X Premixed PBS Buffer at pH 7: Roche, 11666789001)

After three 10-minutes washes in PBS at room temperature, perform a blocking step for 1 hour in 10% of Normal Donkey Serum (NDS, Jackson ImmunoResearch, 017-000-121) supplemented with 0.01% TritonX-100 (Fluka, BP151-100).

Incubate the samples overnight with primary antibodies diluted in 10% of NDS in order to detect the expression of Tubulin III (R&D Systems, MAB1195) and of the ubiquitous NMDA receptor subunit NR1. (Rabbit Anti-NMDAR1, Abcam ab68144; 1/250).

After three 10-minutes washes in PBS at room temperature, use fluorescently labeled secondary antibodies for detection (Cy3 AffiniPure Donkey Anti-Rabbit [Jackson ImmunoResearch, 711-165-152] or Cy5 AffiniPure Donkey Anti- Mouse [Jackson ImmunoResearch, 715-175-151]).

After three 10-minutes washes in PBS at room temperature, treat the cells with bisBenzimide H 33342 trihydrochloride (Sigma B-2261) for 2 minutes, to stain nuclei with 405nm fluorescence.

After two 10-minutes washes, in PBS remove the cover slip from the plates and mount on glass slide with mounting solution (Gelvatol).

Images can be acquired using a laser scanning confocal microscope (FluoView 1000, Olympus). The microscope is equipped with spectral detector technology that provides precise wavelength separation of the emitted light.

5.3 Intracellular Ca^{2+} measurement

For intracellular Ca^{2+} concentration ($[Ca^{2+}]_i$) measurements, incubate cells for 45 min at 37°C with 5 μM fura-2-AM (cell permeant; Invitrogen, F-1221) and 5 μg/ml bovine serum albumin (Sigma A-6003) in Hepes-buffered salt solution (standard HBSS).

Then, wash cells 3 times for 5 min each with standard HBSS and place the cover slips onto a perfusion chamber on a DM IRM inverted microscope (Leica) fitted with a Leica HC N PLAN BD 50X oil immersion objective. $[Ca2^+]_i$ can be monitored in single cells using excitation light provided by a 75W xenon lamp-based monochromator (Ushio, Japan). Emitted light can be detected using a CCD camera (Orca; Hamamatsu, Shizouka, Japan). Alternatively, illuminate cells with 340 and 380 nm light for fura-2. Emitted fluorescence can be passed through a 500 nm longpass dichromatic mirror and a 535 ± 40 nm bandpass filter (Omega Optical).

Analyze acquired data using software such as Simple PCI (Compix, Inc., Cranberry, PA) as the 340/380 ratio. Measure fluorescence in 15–25 individual neurons for each cover slip. Subtract background fluorescence, determined from three or four cell-free regions of the cover slips, from all signals prior to calculating the ratios. Choose the excitation light exposure time and a neutral density filter to avoid saturation of the fluorescence signal.

6. Conclusion

Neural induction from pluripotent stem cells often yield heterogeneous cell populations that can hamper quantitative and comparative analyses. There is a need for improved differentiation and enrichment procedures that generate highly pure populations of neural stem cells (NSC), glia and neurons. The method described here allows an efficient generation of neurospheres from mES cells. On day-in vitro 16 after neural induction about 50% of the cells derived from neurospheres show a neural morphology, a significant expression of neuronal markers such as Tuj1 and the ubiquitous NMDA receptor subunit NR1 (figure 2), and a functional NMDAR-dependent calcium influx mediated by exogenous glutamate administration (figure 3).

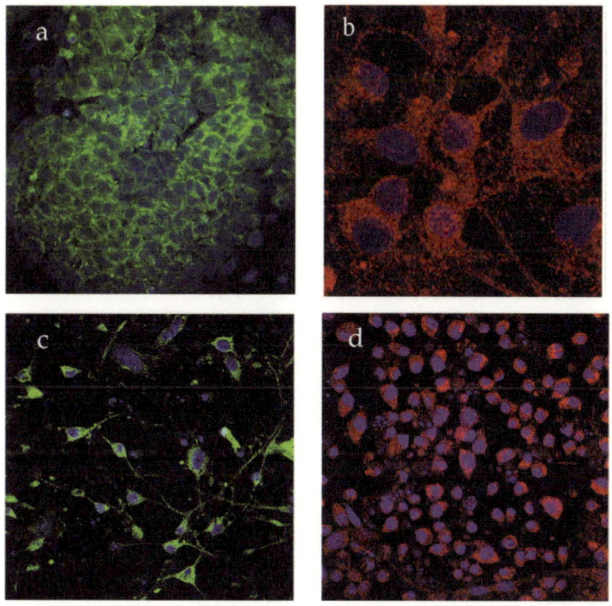

Fig. 2. Neurosphere differentiation. a) Immunostaining of a neurospheres with antibody specific to NMDA receptor (NR1 sub-unit, green). b-c) Staining of neurosphere-derived neurons with (b) NMDA receptor (NR1 sub-unit, red) and (c) β-tubulin III (green). d) staining of astrocytes differentiated from neurospheres with GFAP (red). Nuclei were counterstained with DAPI (blue). [Images taken by authors.]

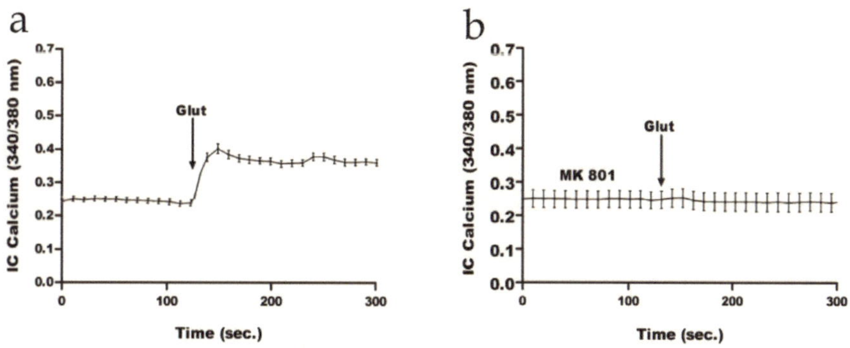

Fig. 3. Analysis of calcium influx in neurons differentiated from neurospheres indicated by glutamate. a) Administration of 10μM glutamate (glu) in neurons causes an increase in $[Ca^{2+}]_i$ (340/380 = 0.32 ± 0.05a). b) Glu-induced increase in $[Ca^{2+}]_i$ in R1 cells is blocked by treatment with MK801, an NMDAR-specific antagonist, suggesting that NMDARs are the predominant glutamate-sensitive receptors in neurospheres-derived neurons. [Data generated by authors.]

Furthermore, the rest of the cellular population show typical astroglial morphology. This chapter illustrates the utility mouse ES cells as a simple system able to develop populations of viable NSC and neurons. The method described here will enable downstream studies that require consistent and defined cell populations to study and characterize the physiopathology of the neuronal differentiation.

This procedure allows generation of a cell line derived from mouse ES cells which continually produces neuropheres in the absence of FGF-2 and EGF and coculturing. By generating neurospheres via this method, one can use them for a variety of purposes for further experiments with neurons.

9. References

Campos LS., Leone DP., Relvas JB., Brakebusch C., Fassler R., Suter U., et al. (2004). Beta1 integrins activate a MAPK signalling pathway in neural stem cells that contributes to their maintenance. *Development,* Vol. 131, No. 14 (July 2004), pp. 3433–3444. ISSN 1011-6370.

Conti L., Pollard SM., Gorba T., Reitano E., Toselli M., Biella G., Sun Y., Sanzone S., Ying QL., Cattaneo E. & Smith A. (2005). Niche-independent symmetrical self-renewal of a mammalian tissue stem cell. *PLoS Biol,* Vol. 3, No. 9 (Sep 2005)pp. e283 . ISSN 1545-7885.

Gaspard N. & Vanderhaeghen P. (2011) From stem cells to neural networks: recent advances and perspectives for neurodevelopmental disorders. *Dev Med Child Neurol,* Vol. 53, No. 1 (Jan 2011), pp. 13-7. ISSN 0012-1622.

Kawasaki H., Mizuseki K., Nishikawa S., Kaneko S., Kuwana Y., Nakanishi S., Nishikawa SI. & Sasai Y. (2000). Induction of midbrain dopaminergic neurons from ES cells by stromal cell-derived inducing activity. *Neuron,* Vol. 28, No. 1 (Oct 2000), pp. 31-40. ISSN 0986-6273.

Moors M., Rockel TD., Abel J., Cline JE., Gassmann K., Schreiber T., Schuwald J., Weinmann N. & Fritsche E. (2009). Human neurospheres as three-dimensional cellular systems for developmental neurotoxicity testing. *Environmental Health Perspectectives,* Vol. 117, No. 7 (July 2009), pp. 1131-8. ISSN 0091-6765.

Morizane A., Takahashi J., Shinoyama M., Ideguchi M., Takagi Y., Fukuda H., Koyanagi M., Sasai Y. & Hashimoto N. (2006). Generation of graftable dopaminergic neuron progenitors from mouse ES cells by a combination of coculture and neurosphere methods. *J Neurosci Res,* Vol. 83, No. 6 (May 2006), pp. 1015-27. ISSN 0306-4522.

Nagy A., Rossant J., Nagy R., Abramow-Newerly W. & Roder JC. (1993). Derivation of completely cell culture-derived mice from early-passage embryonic stem cells. *Proc Natl Acad Sci USA,* Vol. 90, No. 18, (Sep 1993), pp. 8424-8428. ISSN 0027-8424

Reynolds BA & Weiss S. (1992). Generation of neurons and astrocytes from isolated cells of the adult mammalian central nervous system. *Science,* Vol. 255, No. 5052 (Mar 1992), pp. 1707-10. ISSN 0193-4511

Taupin P. & Gage FH. (2002). Adult neurogenesis and neural stem cells of the central nervous system in mammals. *J Neurosci Res,* Vol. 89, No. 6 (Sep 2002), pp. 745-9. ISSN 0360-4012.

Temple S. (1989). Division and differentiation of isolated CNS blast cells in microculture. *Nature,* Vol. 340, No. 6233 (Aug 1989), pp. 471-3. ISSN 0028-0836.

Yoshimura S., Takagi Y., Harada J., Teramoto T., Thomas SS., Waeber C., Bakowska JC., Breakefield XO. & Moskowitz MA. (2001). FGF-2 regulation of neurogenesis in adult hippocampus after brain injury. *Proc Natl Acad Sci U S A,* Vol. 98, No. 10, (May 2001), pp. 5874-9. ISSN 0027-8424.

30

Multidrug Resistance Transporters – Roles in maintaining Cancer Stem-Like Cells

To, Kenneth K.W.[1] and Fu, L.W.[2]
[1]School of Pharmacy, The Chinese University of Hong Kong, Hong Kong
[2]State Key Laboratory of Oncology in South China, Cancer Center, Sun Yat-Sen University, Guangzhou
China

1. Introduction

Cancer has become the leading cause of death worldwide in the year 2010, according to a new edition of the *World Cancer Report* from the International Agency for Research on Cancer (World Health Organization, 2010). Despite the advance in the development of novel chemotherapeutic drugs, the dismal prognosis facing most cancer patients may result from the ability of cancer to withstand drug treatment, recur and metastasize after initial therapies. There is accumulating evidence in support of a central role for cancer stem cells (CSCs) in the initiation, propagation and recurrence of human cancers. Therefore, targeting CSCs has become an attractive research topic for the improvement of treatment outcome and prolongation of patient survival. However, CSCs are endowed with the ability to survive against chemo- and radiation therapy. A better understanding of the mechanisms underlying CSC resistance is badly needed. This chapter provides a review about evidence supporting a fundamental role for CSCs in cancer progression, and summarizes potential mechanisms of CSC resistance to treatment, with emphasis on the involvement of multidrug resistance transporters and their regulation in CSCs.

2. The Cancer Stem Cell (CSC) hypothesis

For many years, tumors have been described as the proliferation of cell clones in which multiple genetic alterations had occurred over time (Nowell, 1976). This "clonal evolution" model is a non-hierarchical model that proposes all cells within a tumor would have an equal chance of acquiring genetic mutations necessary for driving the tumor growth. Subsequently, under selective pressures, the more aggressive cells would drive the tumor progression and lead to therapy resistance. Distinct from this notion, an emerging "cancer stem cell model" is a hierarchical model, which proposes that only a subset of cells called "cancer stem cells" (CSCs) or "tumor-initiating cells" can initiate and propagate a tumor. The CSCs can self-renew, propagate the tumor, and differentiate into the diverse types of cells that are found in the original tumor, thereby mimicking stem cells.
The emergence of the CSC model can be dated back to the mid-19th century when a German pathologist Rudolf Virchow proposed that cancers arise from the activation of dormant, embryonic-like cells present in mature tissue (Virchow, 1855). His speculation was based on

the histological similarities between the developing fetus and certain types of cancer such as teratocarcinomas. Later, the term "cancer stem cells" was probably first coined by Hamburger and Salmon who postulated that a subpopulation of cells in a tumor capable of growing in soft agar are the cell renewal source of a neoplasm and also serve as the seeds of metastatic spread of cancer (Hamburger & Salmon, 1977). More recently, the first conclusive evidence for CSCs was reported in 1994 when John Dick and colleagues isolated the cancerous stem cells from acute myelogenous leukemia and documented their self-renewing capacity (Lapidot et al., 1994).

Since then, the CSC hypothesis has shifted the paradigm in our understanding of cancer tumorigenesis and has important implications in cancer chemotherapy. With respect to tumor development and progression, it could explain the well-known heterogeneous nature of cells in a tumor (Park et al., 1971). While CSCs represent the only cells with self-renewal capability driving the tumor growth, the remaining actively proliferating cells making up the bulk of the tumor could still differentiate and are therefore destined to die. Therefore, the goal of cancer chemotherapy should be to target these CSCs for complete eradication of the tumor.

It should be noted that the CSC hypothesis indeed does not specify the origin of the cancer initiating cells. They could originate from a stem, progenitor or differentiated cell. Therefore, the term "tumor-initiating cell" is often used instead of cancer stem cells to avoid the confusion. The prevailing thought is that CSCs are derived through an activation process involving one of three possible pathways (Figure 1): (1) from normal stem cells losing growth regulation; (2) from progenitor (Jamieson et al., 2004; Krivtsov et al., 2006) or differentiated cells acquiring the self-renewal capacity; or (3) by the fusion of normal stem cells with cancer cells (Pawelek & Chakraborty, 2008; Dittmar et al., 2009).

2.1 Detection and identification of CSCs

The general consensus nowadays is that CSCs can only be ultimately defined experimentally by their ability to recapitulate the generation of a continuously growing tumor (Clarke et al., 2006). However, due to technical difficulty of tumor repopulation *in vivo*, three other popular molecular or phenotypic characteristics of CSCs are being exploited for their identification and prospective isolation from tumor specimens and cancer cell lines. These include (1) the "side population (SP)" phenotype manifested by the exclusion of Hoechst 33342 dye in flow cytometric assays; (2) cell surface markers; and (3) anchorage-independent sphere formation ability. The putative CSC population thus identified will usually be further validated by their ability to initiate a tumor and subsequently recapitulate the heterogeneity of the primary tumor.

2.1.1 The "side population (SP)" phenotype

CSCs and the normal stem cells alike express high levels of the ATP-binding cassette (ABC) transporters, which help protect them from cytotoxic insult throughout their long lifespan. By using the energy of ATP hydrolysis, ABC transporters actively efflux drugs out of the cells, thereby protecting them from toxic xenobiotics (Gottesman et al., 2002). Importantly, this drug-efflux capability conferred by ABC transporters (including ABCG2 and P-gp/MDR1) has been used as a marker in the isolation and analysis of haematopoietic stem cells (HSCs). Goodell and colleagues were the first to report that when mouse bone marrow-derived cells are incubated with the dye Hoechst 33342 and then analyzed by dual-wavelength flow cytometry, a small population of cells does not accumulate an appreciable amount of the fluorescent dye and is thus identified as a Hoechst-dim side population (SP) (Goodell et al., 1997). Remarkably, the side population is highly enriched in HSCs (Goodell et al., 1996). When isolated from mice and transplanted into irradiated mice, small numbers of these SP cells can reconstitute the bone marrow, demonstrating that these cells are pluripotent. Later, it was demonstrated that the transporter Abcg2, but not P-gp/Mdr1, was responsible for the SP in mouse bone marrow (Zhou et al., 2001). Human ABCG2 was subsequently also found to be responsible for the SP phenotype in human bone marrow (Scharenberg et al., 2002).

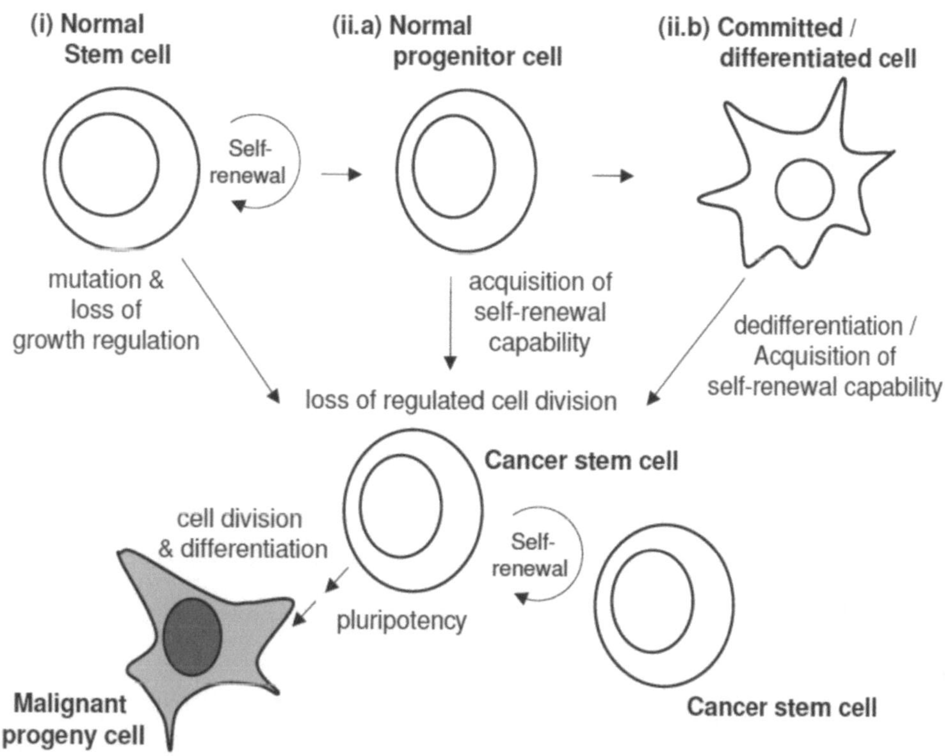

Fig. 1. Origin of CSCs (tumor-initiating cells). A CSC may arise from one of the following molecular pathways: (i) a stem cell undergoing a mutation; (iia & iib) a progenitor/differentiated cell undergoing several mutations, thus allowing them to acquire the self-renewal ability; (iii) fusion of a cancer cell with a normal stem cell, thereby equipping the former with self-renewal capability (not shown in the figure). Like normal stem cells, CSCs are capable of long-term self-renewal and dividing asymmetrically to recapitulate the generation of a continuously growing tumor (pluripotency). In all scenarios, the resulting CSC has lost normal growth regulation and progress into malignancy.

Since its initial application in bone marrow HSCs, the side population technique based on Hoechst 33342 efflux has been adapted to identify putative stem cells and progenitors in many normal tissues (Zhou et al., 2001; Asakura & Rudnucki, 2002; Leckner et al., 2002; Alvi et al.,

2003; Summer et al., 2003; Budak et al., 2005; Du et al., 2005, Hussain, et al., 2005). SP cells have also been found in a number of established cancer cell lines as well as tumor samples and have been shown to have stem cell-like properties, overexpress ABCG2, and possess inherent drug-resistance (Kondo et al., 2004; Hirschmann-Jax et al., 2004; Haraguchi et al., 2005; Seigel et al., 2005, Chiba et al., 2006). Figure 2 shows the existence of such a SP in a ABCG2-overexpressing mitoxantrone-selected resistant pancreatic cell line. The nearly complete elimination of all SP cells after treatment with the specific ABCG2 inhibitor, fumitremorgin C (FTC), suggests that ABCG2 is a major molecular determinant for the SP phenotype.

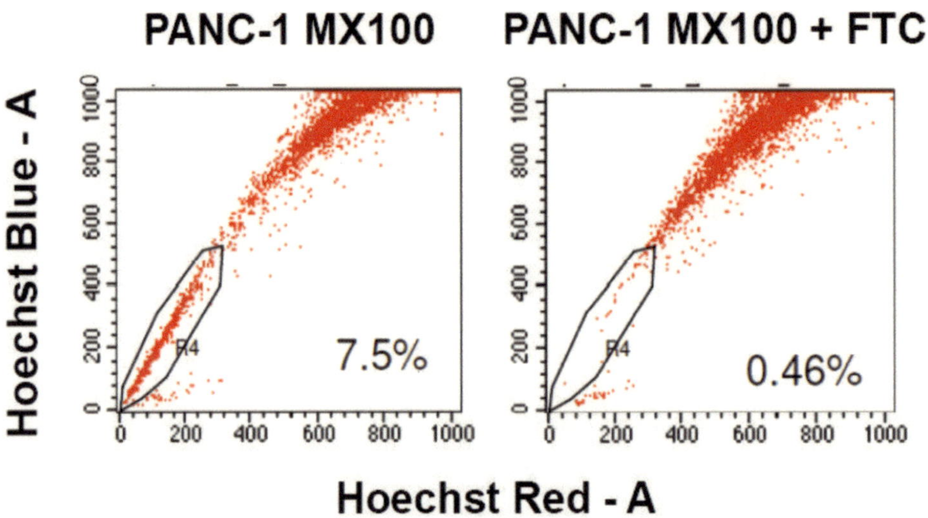

Fig. 2. Identification and isolation of SP cells for the study of putative CSCs. FACS analysis was performed for a mitoxantrone-selected drug-resistant and ABCG2-overexpressing pancreatic cancer cell line after incubation with Hoechst 33342 dye for 1 h. The gated R4 region represents a Hoechst staining-resistant cell population (i.e. SP cells); their abundance are indicated by the number in the figure.

Despite the initial excitment about using SP to identify CSCs, the ABCG2-highly expressing SP and ABCG2-negative non-SP tumor cells have been reported to be similarly tumorigenic (Patrawala et al., 2005). It is believed that the SP fraction obtained is not a pure stem-cell population, which is greatly affected by the isolation method (Montanaro et al., 2004). There were also report demonstrating that SP cells do not identify stem cell (Triel et al., 2004). Moreover, ABCG2, the molecular determinant for Hoechst exclusion, is not an absolute requirement for stem cells. Abcg2-deficient mice are viable and demonstrate no defect in steady state hematopoiesis, though the bone marrow of Abcg2-deficient mice does lack a SP (Zhou et al., 2002). Nevertheless, since CSCs lack distinct molecular markers, Hoechst 33342-dependent cell sorting remains the most widely employed technique for the identification and purification of putative CSCs.

It is also noteworthy that expression of drug transporters (especially MDR1/Pgp) can be part of the differentiated phenotype of cells in normal tissue (Triel et al., 2004). Histopathological and molecular biological studies have reported increased expression of ABCB1 in more differentiated tumors (Mizoguchi et al., 1990; Nishiyama et al., 1993; Bates et al., 1989; Mickley et al., 1989). We have also reported the cell type-specific upregulation of ABCG2 by romidepsin, a differentiating agent and anticancer drug, in cancer cell lines (To et al, 2008, 2011). Furthermore, P-gp and/or ABCG2 are usually overexpressed with the onset of multidrug resistance in cancer cell populations. In these situations, the SP phenotype will not definitively identify CSCs, but because the overexpression of these transporters allows the cells to effectively exclude the Hoechst dye.

2.1.2 CSC-specific cell surface markers

Another common way to identify putative CSCs from patient-derived tissues or cancer cell lines is by labeling the isolated cells with antibodies against various cell surface markers already known in normal stem cells. Cells bearing these cell-surface markers can be isolated by fluorescence activated cell sorting (FACS) (Woodward et al., 2005) or magnetic bead columns (Dou et al., 2007). These enriched cell populations are then tested for their ability to initiate tumorigenesis in immune deficient mice.

Several cell surface markers have been used to detect CSCs (Table 1). Among them, the cell surface protein CD133 (Prominin 1, a transmembrane glycoprotein) is probably the most extensively used marker, which was also known to define stem and progenitor cells in varuous tissues (Shmelkov et al., 2005). A cautionary technical note is worth mentioning. CD133 expression is in fact found to be indifferent to the differentiation status of most cells. On the other hand, its posttranslational glycoslyated form was found to be downregulated upon cell differentiation (Florek et al., 2005). Therefore, upon dedifferentation of the committed cells to generate CSCs as observed in oncogenesis (Figure 1), the glycosylation of CD133 (AC133, the glycosylated epitope of CD133) is expected to increase and therefore serve as a marker for the tumorigenic potential of putative CSCs. In other words, antibody against AC133, but not CD133, should be used for the prospective identification of putative CSCs. It is also noteworthy that, since tumor initiating CSCs are heterogeneous, a specific marker or set of markers has not been found to unequivocally identify CSCs in solid tumors (Welte et al., 2010). CSCs identified from solid tumors may also express other organ-specific markers.

Tumor type	Putative CSC cell surface markers	Reference
Breast	$CD44^{+}$ $CD24^{-/low}$	Al-Hajj et al. (2003); Wang et al. (2010)
Colon	$CD133^{+}$	Fang et al. (2010)
Colon	$CD44^{+}$ $EpCam^{high}$ $CD166^{+}$	Kanwar et al. (2010)
CNS	$CD133^{+}$	Pallini et al. (2011)
HNSCC	$CD44^{+}$ $ALDH^{+}$	Chen et al. (2010)
Liver	$CD13^{+}$	Haraquchi et al. (2010)
Melanoma	$CD20^{+}$	Schmidt et al. (2011)
Multiple myeloma	$CD138^{-}$	Singh et al. (2004)
NSCLC	$CD133^{+}$	Salnikov et al. (2010)
Pancreas	$CD44^{+}$ $CD24^{+}$ ESA^{+}	Hong et al. (2009)
Prostate	$CD44^{+}$ $\alpha 2\beta 1^{+}$ $CD133^{+}$	Collins et al. (2005)

(HNSCC = head and neck squamous cell carcinoma; NSCLC = non small cell lung cancer)

Table 1. Commonly employed CSC cell surface markers in various tissues

2.1.3 Anchorage-independent sphere formation assay

Putative CSCs have also been identified based on their ability to form colonies *in vitro*. Typically, putative CSCs fractions are seeded onto culture dishes coated with serum-free media containing epidermal growth factor and basic fibroblast growth factor. The growth of spherical colonies after a few weeks is considered indicative of self-renewal ability, and would be consistent with a CSC phenotype. Sphere-forming ability as a measure of stem cells was first developed for central nervous system (CNS) cells, where it has been shown that a subset of cells isolated from human fetal brain, and subsequently from human CNS tumors, can form spheres when cultured under the appropriate conditions (Tamaki et al., 2002). These spheres can self renew *in vitro*, and differentiate into all of the neuronal lineages, both *in vitro* and *in vivo*. More importantly, it was subsequently demonstrated for brain tumors that the neurosphere-forming cells could be prospectively isolated from fresh tissue using the cell surface marker CD133. These CD133+ cells did indeed initiate brain tumors *in vivo*, without any *in vitro* manipulation, indicating that they do in fact represent CSCs (Singh et al., 2003).

2.2 CSCs in hematopoietic malignancies

The hematopoietic system is the best characterized somatic tissue with respect to stem cell biology. Many of the physical, biologic, and developmental features of normal hematopoietic stem cells have been defined and useful methods for studying stem cells have been established. It is therefore not surprising that CSCs were first identified in human acute myelogenous leukemia (AML), an aggressive malignancy of immature hematopoietic cells in the bone marrow. The leukemia-initiating activity of primary human AML cells in immunodeficient mice was first demonstrated by John Dick & colleagues, where they found that the "leukemic stem cells (LSCs)" were exclusively found in the CD34+CD38- subpopulation (Bonnet & Dick, 1997). As normal hematopoietic stem cells (HSC) share the CD34+CD38- immunophenotype, it was proposed that AML stem cells arise from HSC.

2.3 CSCs in solid tumors and cancer cell lines

CSCs have been more difficult to identify in non-haematopoietic cancers because fewer well-developed phenotypic markers and definitive assay systems are available. Al-Hajj et al. were the first to identify and prospectively isolate a minority subpopulation of cells from a human solid breast cancer cell line, based on the expression of surface markers and their potential to form tumor after injection into the mammary fat pad of NOD/SCID mice (Al-Hajj et al., 2003). Cells with the phenotype of epithelial-specific antigen (ESA)+ Lineage marker (Lin)-CD24-/lowCD44+ were found to generate tumor that were histologically similar to those of primary breast tumors in mice when as few as 100 cells were transplanted. Similar findings were also published for human brain tumors (GBMs and medulloblastomas) (Singh et al., 2003; Hemmati et al., 2003). These CSCs can differentiate into cells that have characteristics of both neurons and glial cels, self-renew *in vitro* at higher levels than normal neuronal stem cells, and grow and differentiate in neonatal rat brains. Interestingly, the putative CSCs isolated from these brain tumors overexpressing CD133 were found to regenerate identical brain tumors in NOD/SCID mice. Furthermore, these tumors could also be serially transplanted (Singh et al., 2004). It is likely that, as suitable markers and assay systems become available, more solid tumor CSCs will be described.

2.4 Signaling pathways supporting the self-renewal of CSCs

There are several signaling pathways including Notch (McGovern et al., 2009), Wnt/β-catenin (Reya & Clevers, 2005), Hedgehog (Medina et al., 2009), and PI3K/Akt (Hu et al., 2005), which have known roles in maintenance and/or control of normal and cancer stem cell compartments, as well as being implicated in cancer. Since they are playing an important functional role in CSC self-renewal and survival, they also represent attractive novel therapeutic targets for complete eradication of tumor. A short list of candidate chemotherapeutic drugs designed to target these signaling pathways, currently under preclinical development or in clinical trials, is compiled in Table 2 and Table 3, respectively. Selected signaling cascades are also discussed in more detail as follows.

Target	Novel agent/combination	CSCs from specific tumor type tested	Reference
Bcl-2	TRAIL + ABT-737 (Abbott)	Brain	Tagscherer et al., 2008
CD44	CD44 antibodies	AML	Jin et al., 2006
Death receptors	TRAIL + chemotherapy	AML progenitors	Plasilova et al., 2002
Fatty acid synthase (F A S)	Resveratrol	Breast	Pandey et al., 2010
Hedgehog	Cyclopamine	Medulloblastoma	Berman et al., 2002
IL-4	IL-4 blocking antibodies + chemotherapy	Colon	Francipane et al., 2008
Notch	γ-secretase inhibitor (GSI-18)	Medulloblastoma	Fan et al., 2006
PI3K/Akt	A-443654 (Abbott)	Brain	Gallia et al., 2009
SHH/mTOR	Cyclopamine + rapamycin + chemotherapy	Pancreas	Mueller et al., 2009
TGF-β	TGF-β + imatinib	CML	Naka et al., 2010
Wnt/β-catenin/lef-1	CGP049090, PKF115-584 (Novartis)	CLL	Gandhirajan et al., 2010
Wnt (Canonical)	ZTM000990, PKF118-310, anti-WNT1 & anti-WNT2 antibodies	--	Barker & Clevers, 2006
XIAP	Small molecular XIAP inhibitors (Pfizer) + γ-irradiation	Brain	Vellanki et al., 2009

Table 2. Preclinical studies of novel drug candidates targeting various signaling pathways associated with CSCs.

Target	Drug	Cancer type (Phase)	Sponsor	Clinical trial identifier
Notch	MK0752	Breast (I)	Merck	NCT00106145
		Pancreatic (I, II)	Cancer Research, UK	NCT01098344
	PF-03084014	Leukemia (I)	Pfizer	NCT00878189
	RO4929097	Renal cell (II)	U Health Network, Toronto	NCT01141569
Sonic Hedgehog	BMS-833923	Basal cell (I)	Bristol-Myers Squibb	NCT00670189
	GDC-0449	Solid tumors (I)	Genentech	NCT00968981
		Colorectal (II)	Genentech	NCT00636610
	LDE225	Medulloblastoma (I)	Novartis	NCT00880308
	PF-04449913	Hematologic (I)	Pfizer	NCT00953758
Wnt	Resveratrol	Colon (I, II)	U California, Irvine	NCT00256334

Table 3. Clinical trials on new drugs specifically targeting CSC signaling pathways.

2.4.1 Notch

The Notch/γ-secretase/Jagged signaling pathway is involved in cellular response to intrinsic or extrinsic developmental cues to execute specific developmental programs (Artavanis-Tsakonas et al., 1999). It is an important regulator of differentiation and helps control cell fate. It is also involved in vasculogenesis and angiogenesis. Extensive crosstalk has been shown to exist between Notch and other developmental signaling pathways (Hedgehog and Wnt, see below). Notch signaling is activated by ligand binding. The Notch ligands (Jagged 1 & 2, and Delta 1-3) induce the release of the Notch intracellular (Notch-IC) domain via enzymatic cleavage by α- and γ-secretases. The released Notch-IC will then translocate to the nucleus where it turns on the transcription of Notch responsive genes (Artavanis-Tsakonas et al., 1999; Lehar et al., 2005). Notch signaling pathways are activated in both breast CSCs (Phillips et al., 2006) and in endothelial cells (Scharpfenecker et al., 2009) in response to radiation. γ-secretase inhibitors have been developed to inhibit Notch signaling to block CSC self-renewal and were shown to inhibit the engraftment of medulloblastoma in animal model (Fan et al., 2006).

2.4.2 Wnt/β-catenin

The Wnt signaling pathway is an ancient system that has been highly conserved during evolution. It has been implicated in a wide range of biological processes from maintaining stem cells in their pluripotent state to the induction of specific tissues and organs during development. Canonical Wnt signals are transduced through Frizzled family receptors and LRP5/LRP6 coreceptor to the β-catenin signaling cascade (comprehensively reviewed by Wend et al., 2010). This Wnt/β-catenin signaling pathway is important for self-renewal in stem cells and has been found to be dysregulated in solid and haematopoietic cancers (Zhao et al., 2007; Katoh & Katoh, 2007). The pathway has also been shown to promote genomic instability, thereby enhancing the DNA damage tolerance in CSCs (Eyler & Rich, 2008).

Conditional knockout of the key Wnt mediator protein β-catenin in hematopoietic progenitor cells have been shown to delay the development of CML in a bone marrow transplantation model in mice (Zhao et al., 2007).

2.4.3 Sonic hedgehog (SHH)

The SHH pathway is regulated by the binding of Shh (ligand) on the transmembrane receptor patched (Ptch). In the absence of Shh, Ptch inhibits the activity of another transmembrane protein, smoothened (Smo), resulting in inactivation of the SHH pathway (Pasca di Magliano et al., 2003). Binding of Shh to Ptch abrogates the inhibitory effect of Ptch. Smo is derepressed and the transcription factor Gli (Gli1-3) is activated. Gli1 is a potent activator of a number of downstream target genes, including *HNF-3β, cyclins D1, IGFBP-6, Snail, CXCR4,* and *Bcl-2,* to regulate neural development, cell proliferation, oncogenesis, survival, epithelial-mesenchymal transition, migration, invasion and metastasis, respectively (Katoh & Katoh, 2007). As positive and negative feedbacks, GLI1 protein respectively activates its own expression and that of PTCH1 (Agren et al., 2004). Therefore, Gli1 is considered a marker of abnormal activation of SHH pathway. While both SHH and Wnt pathways are commonly hyperactivated in tumors to sustain tumor growth, crosstalk between the two pathways has been reported (He et al., 2006), which adds to the complexity of regulation of CSCs. With the development of specific SHH inhibitors such as cyclopamine, the SHH signaling pathway has been proposed to be a druggable target in CSCs (Medina et al., 2009).

2.4.4 PI3K/PTEN/Akt

The PI3K/PTEN/Akt pathway is one of the most extensively studied signal transduction axes that control cell growth, survival, and proliferation (Sarker et al., 2009). The loss of PTEN and the consequent enhancement of Akt pathway activity has been found to constitute the major molecular events accompanying the increased stem cell character and chemoresistance of gliomas (Hu et al., 2005). Activate Akt pathway is also associated with the occurrence of a population of radiation resistant cancer stem-like cells in medulloblastomas, where Akt inhibition appears to sensitize the cells for radiation-induced apoptosis (Hambardzumyan et al., 2008).

With a better appreciation of the CSC-specific signaling pathways, it becomes logical in an attempt to eradicate the tumor by combining these CSC-targeted therapies with standard chemotherapies. Since the aforementioned pathways also govern normal stem cell development and maintenance, it will be critical to establish a dose and schedule where the tumor is suppressed or eliminated without undue toxicity of normal stem cells. Recent data on mouse leukemia models suggest that the PTEN-dependence of CSCs may be exploited for their specific targeting, while sparing the normal haematopoietic stem cells (Yilmaz et al., 2006). Rapamycin was found to selectively kill the leukemia initiating cells in mice harboring a conditional deletion of PTEN, illustrating that novel therapies may be devised specifically for CSCs.

3. Multidrug resistance and cancer stem cells

3.1 Working models of cancer drug resistance

Clinical drug resistance to anticancer therapy is well-known to be multifactorial, involving alteration in drug targets, inactivation of drug, decreased drug uptake, increased drug efflux, and dysregulation of apoptosis pathways (Gottesman, 2002). Usually, cancers that recur after an initial response to chemotherapeutic drugs become resistant to multiple drugs, giving rise to the phenomenon of multidrug resistance (MDR). The traditional belief is that a few cells in the tumor have acquired genetic alteration(s) to confer drug resistance (i.e. "clonal evolution" model, Figure 3A). These resistant clones have a selective advantage that enables them to outgrow the rest of the tumor following chemotherapy.

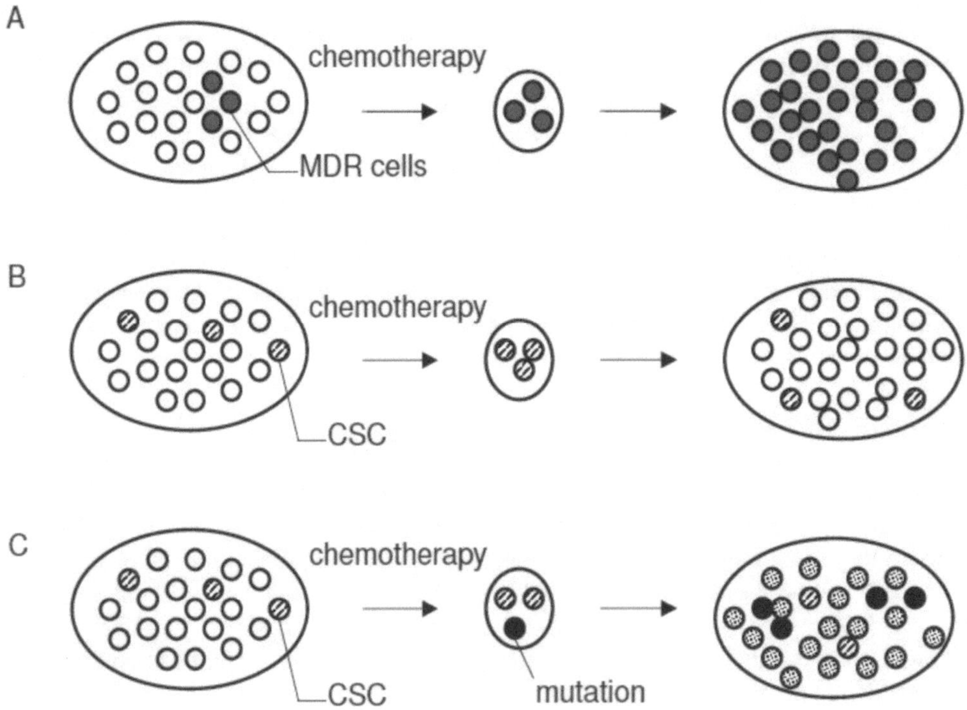

A. Conventional model of cancer drug resistance: A few resistant clones (MDR cells) have acquired drug resistance through genetic alterations. Following chemotherapy, these drug resistant clones survive and give rise to a tumor made up of their progeny cells.
B. CSC model of cancer drug resistance: The original tumor contains a small population of CSCs and their more differentiated progeny. Following chemotherapy, only the CSCs survive by their innate protective mechanisms. Thereafter, they can repopulate the tumor by asymmetrical cell division (i.e. giving rise to another CSC and a differentiated progeny originated from the CSC).
C. "Acquired resistance" CSC model of cancer drug resistance: The original tumor contains a small population of CSCs and their more differentiated progeny. Following chemotherapy, while only the CSCs survive, some of them acquire mutations that confer a high level of drug resistance.

Fig. 3. Models of cancer drug resistance

With better appreciation of the role of CSCs in cancer biology, MDR is generally considered to be ultimately caused by CSCs. As discussed above, CSCs share many properties of the normal stem cells, which help protect them from cytotoxic insult throughout their long lifespan. These properties include quiescence, resistance to xenobiotics through the expression of several ATP-binding cassette (ABC) transporters, active DNA repair capacity, and resistance to apoptosis, which collectively make CSCs naturally resistant to chemotherapy. Therefore, after exposure to conventional chemotherapeutic drugs, CSCs survive and are able to repopulate the tumor both with another CSC and with differentiated cells originated from the CSCs (Figure 3B).

This working model where the intrinsic protective mechanisms of CSCs alone provide the basis for drug resistance might be considered too simplistic. A modified "acquired resistance" stem-cell model was thus proposed to more closely resemble the real clinical situation (Figure 3C). This can be exemplified in the recent studies of imatinib resistance in leukemia patients. Imatinib, a tyrosine kinase inhibitor, is a promising molecularly targeted chemotherapeutic agent. It has been shown to be both a substrate and inhibitor of ABCG2, thus allowing its efflux by a stem cell that express this ABC transporter (Houghton et al., 2004; Burger et al., 2004). In-depth mechanistic studies in imatinib-resistant leukemia cells revealed several "acquired" mutations in the kinase domain of ABL in patient with CML or with AML associated with t(9;22)(q34;q11). These findings therefore suggest that CSCs expressing the drug transporter could facilitate, but not be solely responsible for, the acquisition of acquired mechanisms of drug resistance. As for imatinib, the acquired mutation in ABL, the ultimate drug target, could confer higher levels of drug resistance.

Different tumor types may respond differently to chemotherapy. Cancers that respond to initial chemotherapy may appear to acquire drug resistance during the course of treatment. Other cancers may appear to be intrinsically resistant. In either case, the CSC model of drug resistance applies. It is the quiescent CSCs with the innate drug resistance that survive the chemotherapy; and more importantly, they are capable of repopulating the tumor following chemotherapy. On the other hand, acquired drug resistance in more differentiated cancer cells, through gene mutation, amplification or rearrangement, may contribute to an aggressive phenotype, but it is not the primary reason for cancer recurrence or spread after therapy. Therapeutic strategies that specifically target the CSCs should eradicate tumors more effectively than current treatments and reduce the risk of relapse and metastasis.

3.2 ABC Transporters and normal stem cells/CSCs

Among the several protective mechanisms for CSCs, the overexpression of the ATP-binding cassette (ABC) efflux transporters is probably the most important. The ABC transporters belong to the largest superfamily of transport proteins (Gottesman & Ambudkar, 2001). A total of 49 ABC transporter genes have been identified in the human genome and they were grouped into seven subfamilies (designated A to G) according to their structural and sequence homologues (Vasiliou et al., 2009). By using the energy of ATP hydrolysis, these transporters actively efflux drugs from cells, serving to protect them from cytotoxic substances. The two ABC transporter-encoding genes that have been studied most extensively in stem cells are *ABCB1* (*MDR1*), which encodes P-glycoprotein, and *ABCG2*. Together with *ABCC1* (*MRP1*), they represent the three major multidrug resistance genes that have been identified in cancer cells. Table 4 summarizes the different ABC transporters that have been found to contribute to cancer drug resistance.

Gene	Protein / alias	Location on chromosome	Chemotherapeutic drugs effluxed	Other important substrates
ABCA2	ABCA2	9q34.3	estramustine	-
ABCA3	ABCA3	16p13.3	daunorubicin	surfactant?
ABCB1	P-gp/ MDR1	7q21.12	colchicine, doxorubicin, etoposide, vinblastine, paclitaxel	digoxin, saquinivir, rhodamine
ABCB4	MDR3	7q21.12	paclitaxel, vinblastine	bile salt
ABCB5	ABC19	7p15.3	doxorubicin	-
ABCB11	BSEP/SPGP	2q21.3	paclitaxel, vinblastine	bile salt, pravastatin
ABCC1	MRP1	16p13.12	doxorubicin, daunorubicin, vincristine, etoposide, colchicine, camptothecins, methotrexate	rhodamine
ABCC2	MRP2	10q24.2	vinblastine, cisplatin, doxorubicin, methotrexate	sulfinpyrazone, bilirubin
ABCC3	MRP3	17q21.33	methotrexate, etoposide	-
ABCC4	MRP4	13q32.1	6-mercaptopurine, 6-thioguanine, methotrexate	cAMP, cGMP
ABCC5	MRP5	3q27.1	6-mercaptopurine, 6-thioguanine,	cAMP, cGMP
ABCC6	MRP6	16p13.12	etoposide	-
ABCC10	MRP7	6p21.1	paciltaxel	$E_217\beta G$
ABCC11	MRP8	16q12.1	5-fluorouracil	cAMP, cGMP
ABCG2	ABCG2/BCRP	4q22	mitoxantrone, topotecan, doxorubicin, daunorubicin, irinotecan, imatinib, methotrexate	pheophorbide A, Hoechst 33342, rhodamine

Table 4. ABC transporters involved in drug resistance

3.2.1 ABCB1

ABCB1, also commonly known as P-glycoprotein (P-gp), is the most extensively studied multidrug resistance transporter, which was discovered more than 30 years ago (Jiliano & Ling, 1976). It has been found to be expressed in > 50% of all drug-resistant tumors. Human ABCB1 is the product of the *MDR1* gene and acts as an ATP-dependent pump for a multitude of structurally unrelated hydrophobic compounds, including numerous anticancer and antimicrobial drugs (Gottesman & Ambudkar, 2001).

In Hoechst dye exclusion assay using human cancer cell lines, the expression of ABCB1 (usually together with ABCG2) has been found to be higher in the isolated SP cells. As described above, the SP population has an enhanced capacity for the efflux of Hoechst dye, presumably due to ABCB1 and/or ABCG2 expression. Although Zhou et al. reported that ABCB1 may not contribute to the SP phenotype because bone marrow cells from *Mdr1a/1b-/-* mice are completely lacking in the SP population (Zhou et al., 2001), ABCB1 is still generally considered to be important in protecting CSCs from toxic insult. Result by Zhou et al. may just represent a tumor type-specific observation. Moreover, SP indeed does not define CSCs. The SP fraction is composed of both stem and non-stem cells, and some stem cells are not located in the SP compartment (Zhou et al., 2002).

3.2.2 ABCG2

ABCG2 is a more recently discovered ABC transporter responsible for cancer drug resistance. It was discovered almost simultaneously by three research groups, giving it three different names (BCRP/ABCP/MXR) in the 1990s (Doyle et al., 1998, Allikmets et al., 1998; Miyake et al., 1999). Subsequently, the sequences for these genes turned out to be nearly identical, thereafter the gene was assigned an official name *ABCG2* by the Human Gene Nomenclature Committee, which falls into the "G" subfamily of ABC transporters comprising only of half-transporters.

The list of ABCG2 substrates has been expanding rapidly, which highlights the important role of this transporter in drug disposition and treatment outcomes (Polgar et al., 2008). Numerous cancer chemotherapeutic drugs have been identified as ABCG2 substrates, such as mitoxantrone, flavopiridol, topotecan, and some of the newly developed tyrosine kinase inhibitors. There is considerable overlap in substrate drug specificity of ABCG2 and other multidrug resistance transporters, including ABCB1, ABCC1, ABCC2, and some solute carrier transporters. Besides anticancer drugs, several other therapeutic classes have also been described as ABCG2 substrates, including antibiotics, antivirals, flavonoids, and antihyperlipidemic drugs.

Numerous studies have indicated that ABCG2 overexpression plays a possible role in cancer drug resistance, particularly in leukemia (Ross et al., 2010). For example, higher expression of ABCG2 was found to be associated with AML cases (Ross et al., 2000) or with a poor response to remission induction therapy in AML cases (Steinbach et al., 2002). Of note, ABCG2 is often found to be expressed together with P-gp in AML cases, resulting in poor prognosis (Galimberti et al., 2004; van den Heuvel-Eibrink et al., 2007). Interestingly, ABCG2 and ABCB1 mRNA level was found to be higher in non-responding AML cases, only when the primitive subset of CD34+/CD38- leukaemia stem cells were analyzed (Ho et al., 2008). Although the self renewal capability was not evaluated for the CD34+/CD38- cell population in these studies, it appears that they are the tumor-initiating CSCs protected by the increased expression of the transporters.

3.2.3 Physiological role of ABCB1 and ABCG2 in CSCs

Although high expression of ABCB1 and ABCG2 is generally believed to be a marker for normal and/or cancer stem cells, their physiological role in still not clear. Mice deficient in either Abcb1, Abcc1, or Abcg2 are viable, fertile and have normal stem cell compartments (Schinkel et al., 1994; Zhou et al., 2002; Jonker al., 2002). This indicates that none of these transporter genes are necessary for stem cell growth or maintenance. On the other hand, these knockout mice are more sensitive to the effects of drugs such as vinblastine, ivermectin, topotecan and mitoxantrone, consistent with a role for these ABC transporters in protecting cells from toxins.

As discussed above, both ABCB1 and ABCG2 have been proposed to be survival factors for normal stem cells or CSCs by excluding various xenotoxins out of the cells. However and interestingly, ABCB1/P-gp expression generally tracks with the cell differentiation status, where more differentiated tumors tend to have higher expression of the transporter (Mizoguchi et al., 1990; Nishiyama et al, 1993). Moreover, ABCB1/P-gp expression is almost universally found to be upregulated, accompanied by increased expression of markers of maturation, in cancer cell lines treated with differentiating agents (Bates et al., 1989; Mickley et al., 1989). Given that CSCs need to maintain their pluripotent state for self-renewal and repopulating the rest of the tumor, they should be minimally differentiated. It follows that ABCB1/P-gp may not be an important CSC survival factor *per se*. In contrast, high level functional expression of ABCG2 has been reported in undifferentiated human embryonic stem cells (hESCs) (Apati et al., 2008). The therapeutic implication of these observations is that the undifferentiated and ABCG2-overexpressing cancer cells within a tumor may represent the most chemoresistant putative CSCs that need to be targeted for complete eradication of the tumor.

3.2.4 Overcoming drug resistance by transporter inhibition

3.2.4.1 Early generations of transporter inhibitors

An obvious strategy to restore drug sensitivity in MDR cancer cells caused by ABC drug transporters is to block transporter-mediated drug efflux. Over the past decade, tremendous efforts have been made to discover and synthesize such inhibitors/modulators. Of note, efforts to combat drug resistance caused by the MDR transporters have focused mostly on the use of functional modulators or reversal agents, rather than modulation of the transporter gene regulation. The most well-known inhibitors that have been tested for targeting ABCB1/P-gp (verapamil, cyclosporine A, and valspodar (PSC833)) and ABCG2 (fumitremorgin C (FTC) and Ko143) are also useful research tools for studying modulation of these transporters. A few of these inhibitors, including tariquidar (XR9576) (Kuhnle et al., 2009), can interact with both ABCB1/P-gp and ABCG2. They represent promising lead compounds for further development because drug-resistant tumors usually have overexpression of more than one MDR transporters.

Numerous clinical trials have been performed to evaluate the combination of P-gp modulators with standard chemotherapy regimens in enhancing anticancer efficacy (Sandor et al., 1998). However, they were mostly disappointing and failed to prove the MDR reversal hypothesis, partly because of the lack of specific and potent inhibitors against the MDR transporters. On the other hand, unpredictable pharmacokinetic drug interactions, simultaneous involvement of several drug transporters in tumor tissues, as well as the variability in drug transporter expression levels among individuals, remain obstacles to using modulators to restore drug sensitivity in the clinic.

3.2.4.2 Novel transporter inhibitors may hold promise to target CSCs

The abrupt termination of a phase III clinical trial of a second generation ABCB1/P-gp inhibitor, valspodar (also known as PSC833), due to unexpected toxicity to the patients probably have an enormous negative impact in the field. It was just until recently when the discovery of potent and specific inhibition of P-gp and/or ABCG2 by tyrosine kinase inhibitors (TKIs) has renewed the research interest in developing drug transporter inhibitors for the circumvention of MDR. TKIs are an important new class of molecularly targeted

chemotherapeutic agents that specifically inhibit several oncogenic tyrosine kinases, thereby regulating cancer proliferation, invasion, metastasis and angiogenesis. The first TKI that was approved for CML, imatinib, has been shown to reverse MDR by inhibiting ABCB1(P-gp) (Hegedus et al., 2002) and ABCG2 (Houghton et al., 2004). A few other TKIs were also demonstrated to reverse drug resistance mediated by MDR transporters in various *in vitro* and *in vivo* models (reviewed in Wang & Fu, 2010). However, it is still controversial as to whether TKIs are substrates or inhibitors of ABCB1 (P-gp) and/or ABCG2, which may depend on the concentration used. Moreover, since these TKIs are acting on oncogenic tyrosine kinases, which may have interactions/crosstalk with the other CSC-specific signaling pathways described before, the novel TKIs may prove to be good drug candidates targeting CSCs.
Given the central role played by ABCB1 and/or ABCG2 in protecting CSCs, specific transporter inhibitors theoretically could be employed as "cancer stem cell sensitizing agents" that allow the most crucial and drug resistant cells in a tumor to be destroyed. These therapies would be predicted to have toxic effects on the patients' normal stem cells. Since both ABCG2 and ABCB1 are also known to constitute the blood-brain barrier, this approach has to be carefully titrated to avoid excessive toxicity.

4. Regulation of MDR transporters and its relevance to CSCs

As mentioned above, the MDR transporter ABCG2 may be the *bona fide* CSC survival factor. Therefore, our discussion in this section will focus on ABCG2. Recently, an increasing number of studies have focused on unravelling the molecular regulation of ABCG2 because ABCG2 expression is highly sensitive to various developmental and environmental stimuli.

4.1 Transcriptional regulation of *ABCG2* at the promoter level

Early studies examining the regulation of ABCG2 have focused at the transcriptional level. A few functional cis-elements have been identified at the ABCG2 promoter, including hypoxia (Krishnamurthy et al., 2004), estrogen (Ee et al., 2004), progesterone (Wang et al., 2008), and the xenobiotic (aryl hydrocarbon receptor) response elements (Tan et al., 2010; To et al., 2011), which tightly control ABCG2 expression and serve as cellular defense mechanisms against various stimuli. A PPAR-γ response element upstream of the *ABCG2* gene has also been shown to facilitate the upregulation of ABCG2 for protecting dendritic cells (Szatmari et al., 2006). Cytokines and growth factors have also been reported to affect ABCG2 levels, though the exact mechanism is not clear.
Other studies on ABCG2 regulation are mostly related to its overexpression in drug-resistant cancer cell lines. The overexpression of ABCG2 has been found to correlate with increased binding of a set of permissive histone modification marks, RNA polymerase II and a chromatin remodelling factor Brg-1, but decreased association of a repressive histone mark, HDAC-1 and Sp1 with the proximal *ABCG2* promoter (To et al., 2008a). It has been demonstrated that chromatin dynamics and structure contribute significantly to the maintenance of pluripotency and regulation of differentiation in embryonic stem cells (Shafa et al., 2010). To this end, prolonged drug selection has been found to enrich the resulting subline with CSC characteristics (Calcagno et al., 2010). Therefore, we speculate that the chromatin remodelling observed at the *ABCG2* promoter may coincide with the enrichment of the pluripotent CSCs in the drug-selected resistant cells (Figure 4).

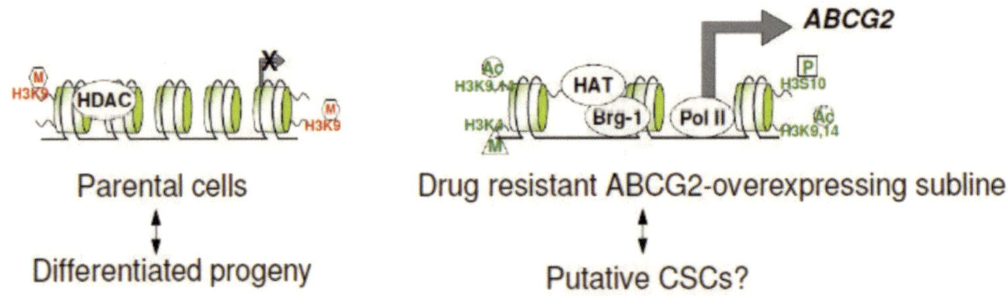

Fig. 4. Chromatin remodelling at the ABCG2 promoter allows overexpression of the transporter in drug-selected resistant cells (To et al., 2008a).

A closer look at the *ABCG2* promoter also reveal that there are a few putative binding sites for the stem cell transcription factors Oct4 and Nanog (Figure 5), which are known to promote self-renewal and pluripotency (Boyer et al., 2005). To this end, ABCG2 and Oct4/POU5F1 were found to be highly coexpressed in the resistant subline selected from the parental K562 leukemia cells (Marques et al., 2010). These observations are therefore consistent with the notion that ABCG2 is a survival factor for the pluripotent CSCs.

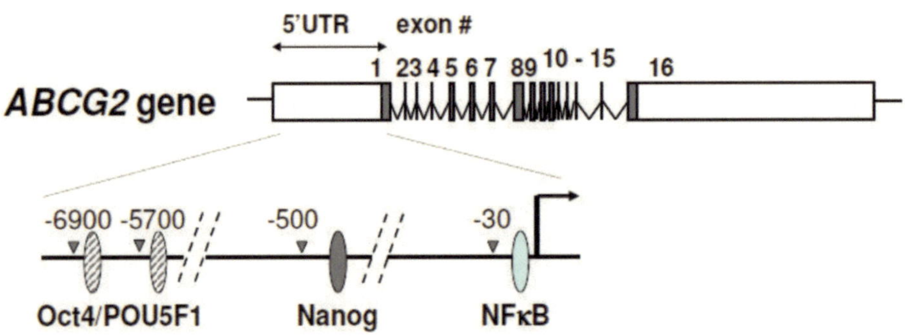

Fig. 5. Putative binding sites for the stem cell transcription factors Oct4 and Nanog at the ABCG2 promoter.

4.2 MicroRNA-mediated regulation of ABCG2

MicroRNAs (miRNAs) are small noncoding RNAs that repress gene expression in a variety of eukaryotic organisms. They play important roles in several cellular processes, such as proliferation, differentiation, apoptosis, and development, by simultaneously controlling the expression levels of hundred of genes. In human cancer, recent studies have shown that miRNA expression profiles differ between normal tissues and derived cancers and between cancer types (Lu et al., 2005). MiRNAs can also act as oncogenes or tumor suppressors, exerting a key function in tumorigenesis (Esquela-Kerscher et al., 2006; Hammond, 2007). Gene regulation by miRNAs is mediated by the formation of imperfect hybrids with the 3'untranslated region (3'UTR) sequences of the target mRNAs, leading to mRNA degradation and/or translational inhibition.

Evidence pointing to the role of miRNAs in determining drug sensitivity and MDR is emerging. First, miRNA expression is largely dysregulated in drug-resistant cancer cells

(Zhu et al., 2008; Pan et al., 2009). Second, the miRNA expression patterns in the NCI-60 drug screen cell lines are significantly correlated to the sensitivity patterns of the cancer cells for a large set of anticancer agents (Blower et al., 2008). Third, numerous miRNAs have been found to regulate drug resistance genes such as DHFR (Mishra et al., 2007) and BCL2 (Xia et al., 2008).

We and others have independently identified three miRNAs (miR-519c, -520h, and -328) regulating ABCG2 expression and determining the sensitivity of cancer cells (To et al., 2008b & 2009; Liao et al., 2008; Wang et al., 2010; Pan et al., 2009, respectively). We reported that ABCG2 mRNA is more stable in drug-selected and ABCG2-overexpressing resistant cell lines than in their parental counterparts (To et al., 2008b & 2009). This increase in mRNA stability was tied to a missing miR-519c binding site (also miR-328) in the truncated 3'UTR of ABCG2 mRNA in drug resistant cells (Figure 6). Intriguingly, the truncation of the ABCG2 3'UTR has also been reported in an undifferentiated human embryonic stem (HuES) cell line where its high ABCG2 expression was associated with the short 3'UTR variant forms (Apati et al., 2008). In contrast, another differentiated HuES cell line with lower ABCG2 levels possesses a longer 3'UTR variant (Apati et al., 2008). Sandberg et al. also found that rapidly proliferating cells express ABCG2 mRNA with shorter 3'UTRs, presumably to escape miRNA regulation (Sandberg et al., 2008). Therefore, in the resistant cells, miR-328 and miR-519c (though a proximal miR-519c binding site present also in the truncated 3'UTR has been recently discovered (Li et al., 2011)) cannot bind to ABCG2 mRNA because of the shorter 3'UTR, and thus miRNA-mediated mRNA degradation and/or protein translation block are relieved, contributing to ABCG2 overexpression (Figure 6). In a human retinoblastoma cell line model, it has been further demonstrated that low expressions of all three miRNAs (miR-328, -519c, & -520h) correlate very well with high ABCG2 expression, with concomitant expression of other stem cell markers including CD133 and ALDH1A1 (Li et al., 2011). On the other hand, hsa-miR-520h has been reported to promote differentiation of hematopoietic stem cells by inhibiting ABCG2 expression (Liao et al., 2008). These findings collectively support an important role played by miRNAs in maintaining high ABCG2 level in CSCs. It will be interesting to verify if the same phenomenon is also observed in patient tumor samples.

The regulation of the other two major multidrug resistance transporters, P-glycoprotein/MDR1 and MRP1, by miR-451, -27a and -326, respectively, have also been reported (Kovalchuk et al., 2008; Li et al., 2010; Liang et al., 2010). More importantly, modulation of miRNA expression or function can alter sensitivity of cancer cells to anticancer drugs (Zhu et al., 2008; Pan et al., 2009; Blower et al., 2008). This could be achieved by inhibiting the function of up-regulated miRNAs or restoring the expression of down-regulated miRNAs. Together, miRNAs may represent important players in intrinsic or acquired MDR in cancer cells.

With the general appreciation of the importance of miRNAs in gene regulation, an emerging role of miRNAs in regulating stem cell self-renewal and differentiation has been revealed (Kashyap et al., 2009), which are important for proper stem cell function and maintenance. Recently, the coordinated regulation of miRNAs and various stem cell transcription factors including OCT4, SOX2 and Nanog have emerged as the master regulatory mechanism for stem cells pluripotency and differentiation. Given that ABCG2 could be downstream target of these stem cell transcription factors, it remains to be seen if the miRNA/stem cell transcription factors network could intercept with the regulation of the MDR transporters in contributing to the pluripotent state and chemoresistance of the CSCs.

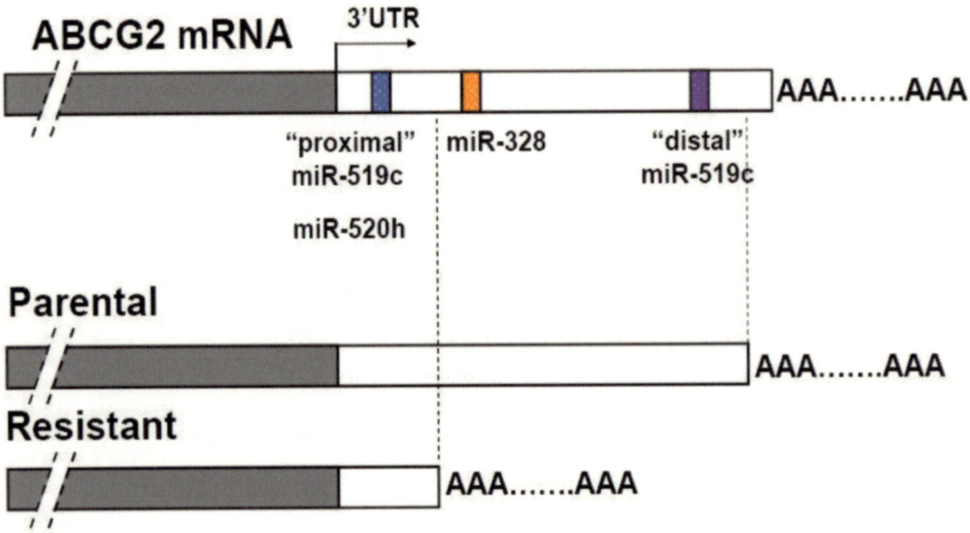

Fig. 6. A proposed model for ABCG2 upregulation in drug-resistant cells by escaping miRNA repression (To et al., 2009; Li et al., 2011)

5. Conclusion

The CSC model of drug resistance offers an appealing explanation as to why cancers that show an apparent complete clinical response to chemotherapy can relapse months or even years later. Numerous novel strategies to circumvent multidrug resistance have been designed to target the putative CSCs by exploiting pathways involved in MDR transporters-mediated drug resistance, or forcing these cells to proliferate and differentiate thus converting them into a target of conventional therapies. Given the complicated microRNA/pluripotency transcription factor/MDR transporters/CSCs network described above, a better understanding of the various molecular mechanisms regulating pluripotency is pivotal to realizing the therapeutic potential of the novel treatment modalities.

6. Acknowledgment

We acknowledge the researchers who have contributed to the advancements in cancer stem cells and ABC transporter research and whose works have not been cited here because of space limitations. The work described in this chapter was supported in part by a grant from the NSFC/RGC Joint Research Scheme sponsored by the Research Grants Council of Hong Kong and the National Natural Science Foundation of China (Project No. N_CUHK443/10) and the Seed Research Funding provided by the School of Pharmacy (CUHK) to Kenneth To.

7. References

Agren, M.; Kogerman, P.; Lleman, M.I.; Wessling, M. & Toftgard, R. (2004). Expression of the PTCH1 tumor suppressor gene is regulated by alternative promoters and a single functional Gli-binding site. *Gene*, Vol.330, (April 2004), pp.101-114.

Al-Hajj, M.; Wicha, M.S.; Benito-Hernandez, A.; Morrison, S.J. & Clarke, M.F. (2003). Prospective identification of tumorigenic breast cancer cells. *Proc Natl Acad Sci USA*, Vol.100, No.7, (April 2003), pp.3983-3988.

Allikmets, R.; Schriml, L.M.; Hutchinson, A.; Romano-Spica, V. & Dean, M. (1998). A human placenta-specific ATP-binding cassette gene (ABCP) on chromosome 4q22 that is involved in multidrug resistance. *Cancer Res*, Vol.58, No.23, (December 1998), pp.5337-5339.

Alvi, A.J.; Clayton, H.; Joshi, C.; Enver, T.; Ashworth, A.; Vivanco, M.M., Dale, T.C. & Smalley, M.J. (2003). Functional and molecular characterisation of mammary side population cells. *Breast Cancer Res*, Vol.5, No.1, (October 2002), pp.R1-R8.

Apati, A.; Orban, T.I.; Varga, N.; Nemeth, A.; Schamberger, A.; Krizsik, V.; Erdelyi-Belle, B.; Homolya, L.; Varady, G.; Padanyi, R.; Karaszi, E.; Kemna, E.; Nemet, K. & Sarkadi, B. (2008). High level functional expression of the ABCG2 multidrug transporter in undifferentiated human embryonic stem cells. *Biochim Biophy Acta*, Vol.1778, No.12, (December 2008), pp.2700-2709.

Artavanis-Tsakonas, S.; Rand, M.D. & Lake, R.J. (1999). Notch signaling: cell fate control and signal integration in development. *Science*, Vol.284, No.5415, (April 1999), pp.770-776.

Asakura, A. & Rudnicki, M.A. (2002). Side population cells from diverse adult tissues are capable of in vitro hematopoietic differentiation. *Exp Hematol*, Vol.30 No.11, (November 2002), pp.1339-1345.

Bates, S.E.; Mickley, L.A.; Chen, Y.N.; Richert, N.; Rudick, J.; Biedler, J.L. & Fojo, A.T. (1989). Expression of a drug resistance gene in human neuroblastoma cell lines: modulation by retinoic acid-induced differentiation. *Mol Cell Biol*, Vol.9, No.10, (October 1989), pp. 4337-4344.

Berman, D.M;, Karhadkar, S.S.; Hallahan, A.R.; Pritchard, J.I.; Eberhart, C.G.; Watkins, D.N.; Chen, J.K.; Cooper, M.K.; Taipale, J.; Olson, J.M. & Beachy, P.A. (2002). Medulloblastoma growth inhibition by hedgehog pathway blockade. *Science*, Vol.297, no.5586, (August 2002), pp.1559-1561.

Blower, P.E.; Chung, J.H., Verducci, J.S.; Lin, S.; Park, J.K.; Dai, Z.; Liu, C.G.; Schmmittgen, T.D.; Reinhold, W.C.; Croce, C.M.; Weinstein, J.N. & Sadee, W. (2008). MicroRNAs modulate the chemosensitivity of tumor cells. *Mol Cancer Ther*, Vol.7, No.1, (January 2008), pp.1-9.

Bonnet, D. & Dick, J.E. (1997). Human acute myeloid leukemia is organized as a hierarchy that originates from a primitive hematopoietic cell. *Nat Med*, Vol.3, No.7, (July 1997), pp.730-737.

Boyer, L.A.; Lee, T.I.; Cole, M.F.; Johnstone, S.E.; Levine, S.S.; Zucker., J.P.; Guenther, M.G.; Kumar, R.M.; Murray, H.L.; Jenner, R.G.; Gifford, D.K.;. Melton, DA, Jaenisch, R, & Young, RA. (2005). Core transcriptional regulatory circuitry in human embryonic stem cells. *Cell*, Vol.122, No.6, (September 2005), pp.947-956.

Budak, MT, Alpdogan, OS, Zhou, M, Lavker, RM, Akinci, MA, Wolosin, JM. (2005). Ocular surface epithelia contain ABCG2-dependent side population cells exhibiting features associated with stem cells. *J Cell Sci*, Vol.118, No.Pt 8, (April 2005), pp.1715-1724.

Burger, H.; van Tol, H.; Boersma, A.W.; Brok, M.; Wiemer, E.A.; Stoter, G. & Nooter, K. Imatinib mesylate (STI571) is a substrate for the breast cancer resistance protein (BCRP)/ABCG2 drug pump. *Blood*, Vol.104, No.9, (November 2004), pp.2940-2942.

Calcagno, A.M.; Salcido, C.D.; Gillet, J.P.; Wu, C.P.; Fostel, J.M.; Mumau, M.D.; Gottesman, M.M.; Varticovski, L. & Ambudkar, S.V. (2010). Prolonged drug selection of breast cancer cells and enrichment of cancer stem cell characteristics. *J Natl Cancer Inst*, Vol.102, No.21, (November 2010), pp.1637-1652.

Chen, Y.W.; Chen, K.H.; Huang, P.I.; Chen, Y.C.; Chiou, G.Y.; Lo, W.L.; Tseng, L.; Hsu, H.S.; Chang, K.W. & Chiou, S.H. (2010). Cucurbitacin I suppressed stem-like property and enhanced radiation-induced apoptosis in head and neck squamous carcinoma-derived CD44(+) ALDH(+) cells. *Mol Cancer Ther*, Vol.9, No.11, (Novermber 2010), pp.2879-2892.

Chiba, T.; Kita, K.; Zheng, Y.W.; Yokosuka, O.; Saisho, H.; Iwama, A.; Nakauchi, H. & Taniguchi, H. (2006). Side population purified from hepatocellular carcinoma cells harbors cancer stem cell-like properties. *Hepatology*, Vol.44, No.1, (July 2006), pp.240-251.

Clarke, M.F.; Dick, J.F.; Dirks, P.B.; Eaves, C.J.; Jamieson, C.H.; Jones, D.L., Visvader, J.; Weissman, I.L. & Wahl, G.M. (2006). Cancer stem cells - perspectives on current status and future directions: AACR Workshop on cancer stem cells. *Cancer Res*, Vol.66, No.19, (October 2006), pp.9339-9344.

Collins, A.T.; Berry, P.A.; Hyde, C.; Stower, M.J. & Maitland, N.J. (2005). Prospective identification of tumorigenic prostate cancer stem cells. *Cancer Res*, Vol.65, No.23, (December 2005), pp.10946-10951.

Dittmar, T.; Nagler, C.; Schwitalla, S.; Reith, G.; Niggemann, B. & Zanker, K.S. (2009) Recurrence cancer stem cells - Made by cell fusion? *Med Hypotheses*, Vol.73, No.4, (October 2009), pp.542-547.

Dou, J.; Pan, M.; Wen, P.; Li, Y.; Tang, Q.; Chu, L.; Zhao, F.; Jiang, C.; Hu, W.; Hu, K. & Gu, N. (2007) Isolation and identification of cancer stem-like cells from murine melanoma cell lines. *Cell Mol Immunol*, Vol.4, No.6, (December 2007), pp.467-472.

Doyle, L.A.; Yang, W.; Abruzzo, L.V.; Krogmann, T.; Gao, Y.; Rishi, A.K. & Ross, D.D. (1998). A multidrug resistance transporter from human MCF-7 breast cancer cells. *Proc Natl Acad Sci USA*, Vol.95, No.26, (December 1998), pp.15665-15670.

Du, Y.; Funderburgh, M.L.; Mann, M.M.; SundarRaji, N. & Funderburgh, J.L. (2005). Multipotent stem cells in human corneal stroma. *Stem Cells*, Vol.23, No.9, (October 2005), pp.1266-1275.

Esquela-Kerscher, A. & Slack, F.J. (2006). OncomiRs-microRNAs with a role in cancer. *Nat Rev Cancer*, Vol.6, No.4, (April 2006), pp.259-269.

Eyler, C.E. & Rich, J.N. (2008). Survival of the fittest: cancer stem cells in therapeutic resistance and angiogenesis. *J Clin Oncol*, Vol.26, No.17, (June 2008), pp.2839-2845.

Fan, X.; Matsui, W.; Khaki, L.; Stearns, D.; Chun, J.; Li, Y.M.; Eberhart, C.G. (2006). Notch pathway inhibition depletes stem-like cells and blocks engraftment in embryonal brain tumors. *Cancer Res*, Vol.66, No.15, (August 2006), pp.7445-52.

Fang, D.D.; Kim, Y.J.; Lee, C.N.; Aggarwal, S.; McKinnon, K.; Mesmer, D.; Norton, J.; Birse, C.E.; He, T.; Ruben, S.M. & Moore, P.A. (2010). Expansion of CD133(+) colon cancer cultures retaining stem cell properties to enable cancer stem cell target discovery. *Br J Cancer*, Vol.102, No.8, (April 2010), pp.1265-1275.

Florek, M.; Haase, M.; Marzesco, A.M.; Freund, D.; Ehninger, G.; Huttner, W.B. & Corbeil, D. (2005). Prominin-1/CD133, a neural and hematopoietic stem cell marker, is expressed in adult human differentiated cells and certain typrs of kidney cancer. *Cell Tissue Res*, Vol.319, No.1, (January 2005), pp.15-26.

Francipane, M.G.; Alea, M.P.; Lombardo, Y.; Todaro, M.; Medema, J.P. & Stassi, G. (2008). Crucial role of interleukin-4 in the survival of colon cancer stem cells. *Cancer Res*, Vol.68, No.11, (Jun 2008), pp.4022-4025.

Gallia, G.L.; Tyler, B.M.; Hann, C.L.; Siu, I.M.; Giranda, V.L.; Vescovi, A.L.; Brem, H. & Riggins, G.J. (2009). Inhibition of Akt inhibits growth of glioblastoma and glioblastoma stem-like cells. *Mol Cancer Ther*, Vol.8, No.2, (February 2009), pp.386-393.

Galimberti, S.; Guerrino, F.; Palumbo, G.A.; Consoli, U.; Fazzi, R.; Morabito, F.; Santini, V. & Petrini, M. (2004). Evaluation of BCRP and MDR-1 co-expression by quantitative molecular assessment in AML patients. *Leuk Res*, Vol.28, No.4, (April 2004), pp.367-372.

Gandhirajan, R.K.; Staib, P.A.; Minke, K.; Gehrke, I.; Plickert, G.; Schlosser, A.; Schmitt, E.K.; Hallek, M. & Kreuzer, K.A. (2010). Small molecule inhibitors of Wnt/beta-catenin/lef-1 signaling induces apoptosis in chronic lymphocytic leukaemia cells in vitro and in vivo. *Neoplasia*, Vol.12, No.4, (April 2010), pp.326-335.

Goodell, M.A.; Brose, K.; Paradis, G.; Conner, A.S. & Mulligan R.C. (1996). Isolation and functional properties of murine hematopoietic stem cells that are replicating in vivo. *J Exp Med*, Vol.183, No.4, (April 1996), pp.1797-1806.

Goodell, M.A.; Rosenzweig, M.; Kim, H.; Marks, D.F.; DeMaria, M.; Paradis, G.; Grupp, S.A.; Sieff, C.A.; Mulligan, R.C. & Johnson, R.P. (1997). Dye efflux studies suggest that hematopoietic stem cells expressing low or undetectable levels of CD34 antigen exist in multiple species. *Nat Med*, Vol.3, No.12, (December 1997), pp.1337-1345.

Gottesman, M.M. (2002). Mechanisms of cancer drug resistance. *Annu Rev Med*, Vol.53, pp.615-627.

Gottesman, M.M. & Ambudkar, S.V. (2001). Overview: ABC transporters and human disease. *J Bioenerg Biomembr*, Vol.33, No.6, (December 2001), pp.453-458.

Gottesman, M.M.; Fojo, T.; Bates, S.E. (2002). Multidrug resistance in cancer: role of ATP-dependent transporters. *Nat Rev Cancer*, Vol.2, No.1, (January 2002), pp.48-58.

Hambardzumyan, D.; Becher, O.J.; Rosenblum, M.K.; Pandolfi, P.P.; Manova-Todorova, K. & Holland, EC. (2008). PI3K pathway regulates survival of cancer stem cells residing in the pervascular niche following radiation in medulloblastoma in vivo. *Genes Dev*, Vol.22, No.4, (February 2008), pp.436-448.

Hamburger, A.W. & Salmon, S.E. (1977). Primary bioassay of human tumor stem cells. *Science*, Vol.197, No.4302, (July 1977), pp.461-463.

Hammond, S.M. (2007). MicroRNAs as tumor suppressors. *Nat Genet*, Vol.39, No.5, (May 2007), pp.582-583.

Haraquchi, N.; Ishii, H.; Mimori, K.; Tanaka, F.; Ohkuma, M.; Kim, H.M.; Akita, H.; Takiuchi, D.; Hatano, H.; Naqano, H.; Barnard, G.F.; Doki, Y. & Mori, M. (2010). CD13 is a therapeutic target in human liver cancer stem cells. *J Clin Invest*, Vol.120, No.9, (September 2010), pp.3326-3339.

He, J.; Sheng, T.; Stelter, A.A.; Li, C.; Zhang, X.; Sinha, M.; Luxon, B.A. & Xie, J. (2006). Suppressing Wnt signalling by the hedgehog pathway through sFRP-1. *J Biol Chem*, Vol.281, No.47, (November 2006), pp.35598-35602.

Hegedus, T.; Orfi, L.; Seprodi, A.; Varadi, A.; Sarkadi, B.; Keri, G. (2002). Interaction of tyrosine kinase inhibitors with the human multidrug transporter proteins, MDR1 and MRP1. *Biochim Biophys Acta*, Vol.1587, No.2-3, (July 2002), pp.318-325.

Hemmati, H.D.; Nakano, I.; Lazareff, J.A.; Masterman-Smith, M.; Geschwind, D.H.; Bronner-Fraser, M. & Kornblum, H.I. (2003). Cancerous stem cells can arise from pediatric brain tumors. *Proc Natl Acad Sci USA*, Vol.100, No.25, (December 2003), pp.15178-15183.

Hirchmann-Jax, C.; Foster, A.E.; Wulf, G.G.; Nuchtern, J.G.; Jax, T.W.; Gobel, U.; Goodell, M.A. & Brenner, M.K. (2004). A distinct "side population" of cells with high drug efflux capacity in human tumor cells. *Proc Natl Acad Sci USA*, Vol.101, No.39, (September 2004), pp.14228-14233.

Ho, M.M.; Hogge, D.E. & Ling, V. (2008). MDR1 and BCRP1 expression in leukemic progenitors correlates with chemotherapy response in acute myeloid leukemia. *Exp Hematol*, Vol.36, No.4, (April 2008), pp.433-442.

Hong, S.P.; Wen, J.; Bang, S.; Park, S. & Song, S.Y. (2009). CD44-positive cells are responsible for gemcitabine resistance in pancreatic cancer cells. *Int J Cancer*, Vol.125, No.10, (November 2009), pp.2323-2331.

Houghton, P.J.; Germain, G.S.; Harwood, F.C.; Schuetz, J.D.; Stewart, C.F.; Buchdunger, E. & Traxler, P. (2004). Imatinib mesylate is a potent inhibitor of the ABCG2 (BCRP) transporter and reverses resistance to topotecan and SN-38 in vitro. *Cancer Res*, Vol.64, No.7, (April 2004), pp.2333-2337.

Hu, X.; Pandolfi, P.; Li, Y.; Koutcher, J.A.; Rosenblum, M. & Holland, EC. (2005). mTOR promotes survival and astrocytic characteristics induced by Pten/AKT signalling in glioblastoma. *Neoplasia*, Vol.7, No.4, (April 2005), pp.356-368.

Hussain, S.Z.; Strom, S.C.; Kirby, M.R.; Burns, S.; Langemeiger, S.; Ueda, T.; Hsieh, M. & Tisdale, J.F. (2005). Side population cells derived from adult human liver generate hepatocyte-like cells in vitro. *Dig Dis Sci*, Vol.50, No.10, (October 2005), pp.1755-1763.

Jamieson, C.H.; Ailles, L.E.; Dylla, SJ, Muijtjens, M.; Jones, C.; Zehnder, J.L.; Gotlib, J.; Li, K.; Manz, M.G.; Keating, A.; Sawyers, C.L. & Weissman, I.L. (2004). Granulocyte-macrophage progenitors as candidate leukemia stem cells in blast-crisis CML. *N Engl J Med*, Vol.351, No.7, (August 2004), pp.657-667.

Jonker, J.W,; Buitelaar, M.; Wagenaar, E.; van der Valk, M.A.; Scheffer, G.L.; Scheper, R.J.; Plosch, T.; Kuipers, F.; Elferink, R.P.; Rosing, H.; Beijnen, J.H. & Schinkel, A.H. (2002). The breast cancer resistance protein protects against a major chlorophyll-derived dietary phototoxin and protoporphyria. *Proc Natl Acad Sci USA*, Vol.99, No.24, (November 2002), pp.15649-15654.

Juliano, R.L & Ling, V. (1976). A surface glycoprotein modulating drug permeability in Chinese hamster ovary cell mutants. *Biochim Biophys Acta*, Vol.455, No.1, (November 1976), pp.152-162.

Kashyap, V.; Rezende, N.C.; Scotland, K.B.; Shaffer, S.M.; Persson, J.L.; Gudas, L.J. & Mongan, NP. (2009). Regulation of stem cell pluriopotency and differentiation involves a mutual regulatory circuit of the NANOG, OCT4, and SOX2 pluriopotency transcription factors with polycomb repressive complexes and stem cell microRNAs. *Stem Cells Dev*, Vol.18, No.7, (September 2009), pp.1093-1108.

Katoh, M. & Katoh, M. (2007). Wnt signaling pathway and stem cell signaling network. *Clin Cancer Res*, Vol.13, No.14, (July 2007), pp.4042-4045.

Katoh, Y. & Katoh, M. (2009). Hedgehog target genes: mechanisms of carcinogenesis induced by aberrant hedgehog signalling activation. *Curr Mol Med*, Vol.9, No.7, (September 2009), pp.873-886.

Kondo, T.; Setoguchi, T. & Taga, T. (2004) Persistence of a small population of cancer stem-like cells in the C6 glioma cell line. *Proc Natl Acad Sci USA*, Vol.101, No.3, (January 2004), pp.781-786.

Kovalchuk, O.; Filkowski, J.; Meservy, J.; Ilnytskyy, Y.; Tryndyak, V.P.; Chekhun, V.F. & Pogribny, I.P. (2008). Involvement of microRNA-451 in resistance of the MCF-7 breast cancer cells to chemotherapeutic drug doxorubicin. *Mol Cancer Ther*, Vol.7, No.7, (July 2008), pp.2152-2159.

Krivtsov, A.V.; Twomey, D. & Feng, Z. et al. (2006). Transformation from committed progenitor to leukemia stem cell initiated by MLL-AF9. *Nature*, Vol.442, No.7104, (August 2006), pp.818-822.

Kuhnle, M.; Egger, M.; Muller, C.; Mahringer, A.; Bernhardt, G.; Fricker, G.; Konig, B. & Buschauer, A. (2009). Potent and selective inhibitors of breast cancer resistance protein (ABCG2) derived from the p-glycoprotein (ABCB1) modulator tariquidar. *J Med Chem*, Vol.52, No.4, (February 2009), pp.1190-1197.

Lapidot, T.; Sirard, C.; Vormoor, J.; Murdoch, B.; Hoang, T.; Caceres-Cortes, J.; Minden, M.; Paterson, B.; Caligiuri, M.A. & Dick, J.E. (1994). A cell initiating human acute myeloid leukaemia after transplantation into SCID mice. *Nature*, Vol.367, No.6464, (Februaru 1994), pp.645-648.

Lechner, A.; Leech, C.A.; Abraham, E.J.; Nolan, A.L. & Habener, J.F. (2002). Nestin-positive progenitor cells derived from adult human pancreatic islets of Langerhans contain side population (SP) cells defined by expression of the ABCG2 (BCRP1) ATP-binding cassette transporter. *Biochem Biophys Res Comm*, Vol..293, No.2, (May 2002), pp.670-674.

Lehar, S.M.; Dooley, J.; Farr, A.G. & Bevan M.J. (2005). Notch ligands delta1 and jagged1 transmit distinct signals to T-cell precursors. *Blood*, Vol.105, No.4, (February 2005), pp.1440-1447.

Li, Z.; Hu, S.; Wang, J.; Cai, J.; Xiao, L.; Yu, L. & Wang, Z. (2010). MiR-27a modulates MDR1/P-glycoprotein expression by targeting HIPK2 in human ovarian cancer cells. *Gynecol Oncol*, Vol.119, No.1, (October 2010), pp.125-130.

Liang, Z.; Wu, H.; Xia, J.; Li, Y.; Zhang, Y.; Huang, K.; Wagar, N.; Yoon, Y.; Cho, H.T.; Scala, S. & Shim, H. (2010). Involvement of miR-326 in chemotherapy resistance of breast cancer through modulating expression of multidrug resistance-associated protein 1. *Biochem Pharmacol*, Vol.79, No.6, (March 2010), pp.817-824.

Liao, R.; Sun, J.; Zhang, L.; Lou, G.; Chen, M.; Zhou, D.; Chen, Z. & Zhang, S. (2008). MicroRNAs play a role in the development of human hematopoietic stem cells. *J Cell Biochem*, Vol.104, No.3, (June 2008), pp.805-817.

Li, X.; Pan, Y.Z.; Seigel, G.M.; Hu, Z.H.; Huang, M. & Yu, A.M. (2011). Breast cancer resistance protein BCRP/ABCG2 regulatory microRNAs (hsa-miR-328, -519c, -

520h) and their differential expression in stem-like ABCG2+ cancer cells. *Biochem Pharmacol*, Vol.81, No.6, (March 2011), pp.783-792.

Lu, J.; Getz, G.; Miska, E.A.; Alvarez-Saavedra, E.; Lamb, J.; Peck, D.; Sweet-Cordero, A.; Ebert, B.L.; Mak, R.H.; Ferradno, A.A.; Downing J.R.; Jacks, T.; Horvitz, H.R. & Golub, T.R. (2005). MicroRNA expression profiles classify human cancers. *Nature*, Vol.435, No.7043, (June 2005), pp.834-838.

Marques, D.S.; Sandrini, J.Z.; Boyle, R.T.; Marins, L.F. & Trindade, G.S. (2010). Relationships between multidrug resistance (MDR) and stem cell markers in human chronic myeloid leukemia cell lines. *Leuk Res*, Vol.34, No.6, (June 2010), pp.757-762.

Mickley, L.A.; Bates, S.E.; Richert, N.D.; Currier, S.; Tanaka, S.; Foss, F.; Rosen, N.; Fojo, A.T. (1989). Modulation of the expression of a multidrug resistance gene (mdr-1/P-glycoprotein) by differentiating agents. *J Biol Chem*, Vol.264, No.30, (October 1989), pp.18031-18040.

McGovern, M.; Voutev, R.; Maciejowski, J.; Corsi, A.K. & Hubbard, E.J. (2009). A "latent niche" mechanism for tumor initiation. *Proc Natl Acad Sci USA*, Vol.106, No.28, (July 2009), pp.11617-11622.

Medina, V.; Calvo, M.B.; Diaz-Prado, S. & Espada, J. (2009). Hedgehog signalling as a target in cancer stem cells. *Clin Transl Oncol*, Vol.11, No.4, (April 2009), pp.199-207.

Mishra, P.J.; Humeniuk, R. & Mishra, P.J. (2007). A miR-24 microRNA binding site polymorphism in dihydrofolate reductase gene leads to methotrexate resistance. *Proc Natl Acad Sci USA*, Vol.104, No.33, (August 2007), pp.13513-13518.

Miyake, K.; Mickley, L.; Litman, T.; Zhan, Z.; Robey, R.; Cristensen, B.; Brangi, M.; Greenberger, K.; Dean, M.; Fojo, T. & Bates, S.E. (1999). Molecular cloning of cDNAs which are highly overexpressed in mitoxantrone-resistant cells: demonstration of homology to ABC transport genes. *Cancer Res*, Vol.59, No.1, (January 1999), pp.8-13.

Mizoguchi, T.; Yamada, K.; Furukawa, T.; Hidaka, K.; Hisatsugu, T.; Shimazu, H.; Tsuruo, T.; Sumizawa, T. & Akiyama, S. (1990) Expression of the MDR1 gene in human gastric and colorectal carcinomas. *J Natl Cancer Inst*, Vol.82, No.21, (November 1990), pp.1679-1683.

Montanaro, F.; Liadaki, K.; Schienda, J.; Flint, A.; Gussoni, E. & Kunkel, L.M. (2004) Demystifying SP cell purification: Viability, yield, and phenotype are defined by isolation parameters. *Exp Cell Res*, Vol.298, No.1, (August 2004), pp.144-154.

Mueller, M.T.; Hermann, P.C.; Witthauer, J.; Rubio-Viqueira, B.; Leicht, S.F.; Huber, S.; Ellwart, J.W.; Mustafa, M.; Bartenstein, P.; D'Haese, J.G.; Schoenberg, M.H.; Berger, F.; Jauch, KW, Hidalgo, M, Heeschen, C. (2009). Combined targeted treatment to eliminate tumorigenic cancer stem cells in human pancreatic cancer. *Gastroenterology*, Vol.137, No.3, (September 2009), pp.1102-1113.

Naka, K.; Hoshii, T.; Muraguchi, T.; Tadokoro, Y.; Ooshio, T.; Kondo, Y.; Nakao, S.; Motoyama, N. & Hirao, A. (2010). TGF-beta-FOXO signalling maintains leukaemia-initiating cells in chronic myeloid leukaemia. *Nature*, Vol.463, No.7281, (February 2010), pp.676-680.

Nishiyama, K.; Shirahama, T.; Yoshimura, A.; Sumizawa, T.; Furukawa, T.; Ichikawa-Haraguchi, M.; Akiyama, S. & Ohi, Y. (1993). Expression of the multidrug transporter, P-glycoprotein, in renal and transitional cell carcinomas. *Cancer*, Vol.71, No.11, (June 1993), pp.3611-3619.

Nowell, P.C. (1976). The clonal evolution of tumor cell populations. *Science*, Vol.194, No.4260, pp.23-28.

Pallini, R.; Ricci-Vitiani, L.; Montano, N.; Mollinari, C.; Biffoni, M.; Cenci, T.; Pierconti, F.; Martini, M.; De Maria, R. & Larocca, L.M. (2011). Expression of the stem cell marker CD133 in recurrent glioblastoma and its value for prognosis. *Cancer*, Vol.117, No.1, (January 2011), pp.162-174.

Pan, Y.Z.; Morris, M.E. & Yu, Z.M. (2009). MicroRNA-328 negatively regulates the expression of breast cancer resistance protein (BCRP/ABCG2) in human cancer cells. *Mol Pharmacol*, Vol.75, No.6, (June 2009), pp.1374-1379.

Park, C.H.; Bergsagel, D.E. & McCulloch, E.A. (1971). Mouse myeloma tumor stem cells: a primary cell culture assay. *J Natl Cancer Inst*, Vol.46, No.2, (February 1971), pp.411-422.

Pandey, P.R.; Okuda, H.; Watabe, M.; Pai, S.K.; Liu, W.; Kobayashi, A.; Xing, F.; Fukuda, K.; Hirota, S.; Sugai, T.; Wakabayashi, G.; Koeda, K.; Kashiwaba, M.; Suzuki, K.; Chiba, T.; Endo, M.; Fujioka, T.; Tanji, S.; Mo, Y.Y.; Cao, D.; Wilber, A.C. & Watabe, K. (2010). Resveratrol suppresses growth of cancer stem-like cells by inhibiting fatty acid synthase. *Breast Cancer Res Treat*, (December 2010), DOI 10.1007/s10549-010-1300-6

Pasca di Magliano, M. & Hebrok, M. (2003). Hedgehog signalling in cancer formation and maintenance. *Nat Rev Cancer*, Vol.3, No.12, (December 2003), pp.903-911.

Pawelek, J.M. & Chakraborty, A.K. (2008). Fusion of tumor cells with bone marrow-derived cells: a unifying explanation for metastasis. *Nat Rev Cancer*, Vol.8, No.5, (May 2008), pp.377-386.

Patrawala, L.; Calhoun, T.; Schneider-Broussard, R.; Zhou, J.; Claypool, K. & Tang, D.G. (2005). Side population is enriched in tumorigenic, stem-like cancer cells, whereas ABCG2+ and ABCG2- cancer cells are similarly tumorigenic. *Cancer Res*, Vol.65, No.14, (July 2005), pp.6207-6219.

Phillips, T.M.; McBride, W.H. & Pajonk, F. (2006). The response of CD24-/low/CD44+ breast cancer-initiaiting cells to radiation. *J Natl Cancer Inst*, Vol.98, No.24, (December 2006), pp.1777-1785.

Plasilova, M.; Zivny, J.; Jelinel, J.; Neuwirtova, R.; Cermak, J.; Encas, E.; Andera, L. & Stopka, T. (2002) TRAIL (Apo2L) suppresses growth of primary human leukaemia and myelodysplasia progenitors. *Leukemia*, Vol.16, No.1, (January 2002), pp.67-73.

Polgar, O.; Robey, R.W. & Bates, S.E. (2008). ABCG2: structure, function and role in drug response. *Expert Opin Drug Metab Toxicol*, Vol.4, No.1, (January 2008), pp.1-15.

Reya, T. & Clevers, H. (2005). Wnt signalling in stem cells and cancer. *Nature*, Vol.434, No.7035, (April 2005), pp.843-850.

Ross, D.D. & Hakanishi, T. (2010). Impact of breast cancer resistance protein on cancer treatment outcomes, In: *Multidrug resistance in cancer*, Humana Press.

Salnikov, A.V.; Gladkich, J.; Moldenhauer, G.; Volm, M.; Mattern, J. & Herr, I. (2010). CD133 is indicative for a resistance phenotype but does not represent a prognostic marker for survival of non-small cell lung cancer patients. *Int J Cancer*, Vol.126, No.4, (February 2010), pp.950-958.

Sandor, V.; Fojo, T. & Bates, S.E. (1998). Future perspectives for the development of P-glycoprotein modulators. *Drug Resist Updat*, Vol.1, No.3, pp.190-200.

Sarker, D.; Reid, A.H.; Yap, T.A. & de Bono, J.S. (2009). Targeting the PI3K/AKT pathway for the treatment of prostate cancer. *Clin Cancer Res*, Vol.15, No.15, (August 2009), pp.4799-4805.

Scharenberg, C.W.; Harkey, M.A.& Torok-Storb, B. (2002). The ABCG2 transporter is an efficient Hoechst 33342 efflux pump and is preferentially expressed by immature human hematopoietic progenitors. *Blood*, Vol.99, No.2, (January 2002), pp.507-512.

Scharpfenecker, M.; Kruse J.J.; Sprong, D.; Russell, N.S.; ten Dijke, P.; Stewart, F.A. (2009). Ionizing radiation shifts the PAI-1/ID-1 balance and activates notch signaling in endothelial cells. *Int J Rad Oncol Biol Phys*, Vol.73, No.2, (February 2009), pp.506-513.

Schinkel, A.H.; Smit, J.J.; van Tellingen, O.; Beijnen, J.H.; Wagenaar, E.; van Deemter, L.; Mol, C.A.; van der Valk, M.A.; Robanus-Maandag, E.C. & te Riele, H.P. (1994). Disruption of the mouse mdr1a P-glycoprotein gene leads to a deficiency in the blood-brain barrier and to increased sensitivity to drugs. *Cell*, Vol.77, No.4, (May 1994), pp.491-502.

Schmidt, P.; Kopecky, C.; Hombach, A.; Zigrino, P.; Mauch, C. & Abken, H. (2011). Eradication of melanomas by targeted elimination of a minor subset of tumor cells. *Proc Natl Acad Sci USA*, Vol.108, No.6, (February 2011), pp.2474-2479.

Seigel, G.M.; Campbell, L.M.; Narayan, M. & Gonzalez-Fernandez, F. (2005). Cancer stem cell characteristics in retinoblastoma. *Mol Vis*, Vol.11, (September 2005), pp.729-737.

Shafa, M.; Krawetz, R. & Rancourt, D.E. (2010). Returning to the stem state: epigenetics of recapitulating pre-differentiation chromatin structure. *Bioessays*, Vol.32, No.9, (September 2010), pp.791-799.

Shmelkov, S.V.; St Clair, R.; Lyden, D.; Rafii, S. (2005). AC133/CD133/prominin-1. *Int J Biochem Cel Biol*, Vol.37, No.4, (April 2005), pp.715-719.

Singh, S.K.; Clarke, I.D.; Terasaki, M.; Bonn, V.E.; Hawkins, C.; Squire, J. & Dirks, P.B. (2003) Identification of a cancer stem cell in human brain tumors. *Cancer Res*, Vol.63, No.18, (September 2003), pp.5821-5828.

Singh, S.K.; Hawkins, C.; Clarke, I.D.; Squire, J.A.; Bayani, J.; Hide, T.; Henkelman, R.M.; Cusimano, M.D. & Dirks, P.B. (2004) Identification of human brain tumor initiating cells. *Nature*, Vol.432, No.7015, (November 2004), pp.396-401.

Stadler, B.M. & Ruohola-Baker, H. (2008). Small RNAs: keeping stem cells in line. *Cell*, Vol.132, No.4, (February 2008), pp.563-566.

Steinbach, D.; Sell, W.; Voigt, A.; Hermann, J.; Zintl, F. & Sauerbrey, A. (2002). BCRP gene expression is associated with a poor response to remission induction therapy in childhood acute myeloid leukemia. *Leukemia*, Vol.16, No.8, (August 2002), pp.1443-1447.

Summer, R.; Kotton, D.N.; Sun, X.; Ma, B.; Fitzsimmons, K. & Fine, A. (2003). Side population cells and Bcrp1 expression in lung. *Am J Physiol Lung Cell Mol Physiol*, Vol.285, No.1, (July 2003), pp.L97-L104.

Tagscherer, K.E.; Fassl, A.; Campos, B.; Farhadi, M.; Kraemer, A.; Bock, B.C.; Macher-Goeppinger, S.; Radlwimmer, B.; Wiestler, O.D.; Herold-Mende, C. & Roth, W. (2008). Apoptosis-based treatment of glioblastomas with ABT-737, a novel small molecule inhibitor of Bcl-2 family proteins. *Oncogene*, Vol.27, No.52, (November 2008), pp.6646-6656.

Tamaki, S.; Eckert, K.; He, D.; Sutton, R.; Doshe, M.; Jain, G.; Tushinski, R.; Reitsma, M.; Harris, B.; Tsukamoto, A.; Gage, F.; Weissman, I. & Uchida, N. (2002). Engraftment of sorted/expanded human central nervous system stem cells from fetal brain. *J Neurosci Res*, Vol.69, No.6, (September 2002), pp.976-986.

To, K.K.; Polgar, O.; Huff, L.M.; Morisaki, K.; Bates, S.E. (2008a). Histone modifications at the ABCG2 promoter following treatment with histone deacetylase inhibitor mirror those in multidrug resistant cells. *Mol Cancer Res*, Vol.6, No.1, (January 2008), pp.151-164.

To, K.K.; Zhan, Z.; Litman, T. & Bates, S.E. (2008b). Regulation of ABCG2 expression at the 3'untranslated region of its mRNA through modulation of transcript stability and protein translation by a putative microRNA in the S1 colon cancer cell line. *Mol Cell Biol*, Vol.28, No.17, (September 2008), pp.5147-5161.

To, K.K.; Robey, R.W.; Knutsen, T.; Zhan, Z.; Ried, T.; Bates, S.E. (2009). Escape from hsa-miR-519c enables drug-resistant cells to maintain high expression of ABCG2. *Mol Cancer Ther*, Vol.8, No.10, (October 2009), pp.2959-2968.

To, K.K.; Robey, R.W.; Zhan, Z.; Bangiolo, L.; Bates, S.E. (2011). Upregulation of ABCG2 by romidepsin via the aryl hydrocarbon receptor pathway. *Mol Cancer Res*, (Feb 2011) [Epub ahead of print].

Triel, C.; Vestergaard, M.E.; Bolund, L.; Jensen, T.G. & Jensen, U.B. (2004). Side population cells in human can mouse epidermis lack stem cell characteristic. *Exp Cell Res*, Vol.295, No.1., (April 2004), pp.79-90.

Van den Heuvel-Eibrink, M.M.; van der Holt, B.; Burnett, A.K.; Knauf, W.U.; Fey, M.F.; Verhoef, G.E.; Vellenga, E.; Ossenkoppele, G.J.; Lowenberg, B. & Sonneveld, P. (2007). CD34-related coexpression of MDR1 and BCRP indicates a clinically resistant phenotype in patients with acute myeloid leukemia (AML) of older age. *Ann Hematol*, Vol.86, No.5., pp.329-337.

Vasiliou, V.; Vasiliou, K. & Kebert, D.W. (2009). Human ATP-binding cassette (ABC) transporter family. *Hum Genomics*, Vol.3, No.3, (April 2009), pp.281-290.

Virchow, R. (1855). Editorial. *Virchows Arch Pathol Anat Physiol Klin Med*, Vol.3, pp.23.

Wang, F.; Xue, X.; Wei, J.; An, Y.; Yao, J.; Cai, H.; Wu, J.; Dai, C.; Qian, Z.; Xu, Z. & Maio, Y. (2010). hsa-miR-520h downregulates ABCG2 in pancreatic cancer cells to inhibit migration, invasion, and side populations. *Br J Cancer*, Vol.103, No.4, (August 2010), pp.567-574.

Wang, K.H.; Kao, A.P.; Chang, C.C.; Lee, J.N.; Hou, M.F.; Long, C.Y.; Chen, H.S. & Tsai E.M. (2010). Increasing CD44+/CD24(-) tumor stem cells, and upregulation of COX-2 and HDAC6, as major functions of HER2 in breast tumorigenesis. *Mol Cancer*, Vol.9, (November 2010), pp.288.

Wang, X.K. & Fu, L.W. (2010). Interaction of tyrosine kinase inhibitors with the MDR-related ABC transporter proteins. *Curr Drug Metab*, Vol.11, No.7, (September 2010), pp.618-628.

Welte, Y.; Adjaye, J.; Lehrach, H.R. & Regenbrecht, C.R. (2010) Cancer stem cells in solid tumors: elusive or illusive? *Cell Commun Signal*, Vol.8, No.1, (May 2010), pp.6.

Wend, P.; Holland, J.D.; Ziebold, U. & Birchmeier, W. (2010). Wnt signaling in stem and cancer stem cells. *Semin Cell Dev Biol*, Vol.21, No.8, (October 2010), pp.855-863.

Woodward, W.A.; Chen, M.S.; Behbod, F. & Rosen, J.M. (2005). On mammary stem cells. *J Cell Sci*, Vol.118, No.Pt16, (August 2005), pp.3585-3594.

World Health Organization (2008) *World Cancer Report 2008*. ISBN 978-92-832-0423-7, WHO, Geneva.

Xia, L.; Zhang, D.; Du, R.; Pan, Y.; Zhao, L.; Sun, S.; Hong, L.; Liu, J. & Dan, D. (2008). miR-15b and miR-16 modulate multidrug resistance by targeting BCL2 in human gastric cancer cells. *Int J Cancer*, Vol.123, No.2, (July 2008), pp.372-379.

Yilmaz, O.H.; Valdez, R.; Theisen, B.K.; Guo, W.; Ferguson, D.O.; Wu, H. & Morrison, S.J. (2006). Pten dependence distinguishes haematopoietic stem cells from leukemia-initiating cells. *Nature*, Vol.441, No.7092, (May 2006), pp.475-482.

Zhao, C.; Blum, J.; Chen, A.; Kwon, H.Y.; Jung, S.H.; Cook, J.M.; Lagoo, A. & Rey, T. (2007) Loss of β-catenin impairs the renewal of normal and CML stem cells in vivo. *Cancer Cell*, Vol.12, No.6, (December 2007), pp.528-541.

Zhou, S.; Schuetz, J.D.; Bunting, K.D.; Colapietro, A.M.; Sampath, J.; Morris, J.J.; Lagutina, I.; Grosveld, G.C.; Osawa, M.; Nakauchi, H. & Sorrentino, B.P. (2001). The ABC transporter Bcrp1/ABCG2 is expressed in a wide variety of stem cells and is a molecular determinant of the side-population phenotype. *Nat Med*, Vol.7, No.9, (September 2001), pp.1028-1034.

Zhou, S.; Morris, J.J.; Barnes, Y.; Lan, L.; Schuetz, J.D.; Sorrentino, B.P. (2002). Bcrp1 gene expression is required for normal numbers of side population stem cells in mice, and confers relative protection to mitoxantrone in hematopoietic cells in vivo. *Proc Natl Acad Sci USA*, Vol.99, No.19, (September 2002), pp.12339-11344.

Zhu, H.; Wu, H.; Liu, X.; Evans, B.R.; Medina, D.J.; Liu, C.G. & Yang, J.M. (2008). Role of microRNA miR-27a and miR-451 in the regulation of MDR1/P-glycoprotein expression in human cancer cells. *Biochem Pharmacol*, Vol.76, No.5, (September 2008), pp.582-588.

31

Associated Adeno Virus Vector for Producing Induced Pluripotent Stem Cells (IPS) for Human Somatic Cells

Edilamar Menezes Oliveira[1] and M Ian Phillips[2]

[1]*University of São Paulo, São Paulo*

[2]*Keck Graduate Institute, Claremont, California*

[1]*Brazil*

[2]*United States*

1. Introduction

The discovery that adult somatic cells could be induced to become pluripotent stem cells with apparently all the properties of human embryonic stem cells (hESCs), independently by S .Yamanaka et al (Takahashi et al l 2007) and by J. Thompson et al (Yu et al 2007) was greeted with amazement and in some quarters with relief. The breakthrough, first reported in mice by S Yamanaka in Japan did not receive much notice, but the revelation that adult human cells could be reversed into an embryonic like state was astounding because it was so simple. It was greeted with relief where the progress of hECS was blocked at the federal level because producing hECS raised ethical concerns and political consequences. Yet four years later, although we have discovered much about iPS cells, they are far from being the ideal solution that they seemed to promise.

When human iPS cells were first published they appeared to have so many advantages over hECs. Foremost was the lack of an embryo being involved which meant they could not be logically banned or targeted as unethical (although some tried). Second the science behind making them was astonishingly simple. It only required delivery of four transcription factors found in embryos to reverse years of life as an adult cell back to an embryonic like cell. The record for number of years as an adult cell was set by Eggan et al (Dimos et al 2009) , who induced the skin cells of an 82 year old lady back to iPScells .

iPS cells offer the advantages of avoiding the religious or ethical considerations that plague the use of embryonic stem cells. They also could provide autologous transplantation, for repair and regeneration of tissue without rejection. If the donor cells retain the mutation or mutations that caused the patient's disease state, it is possible to correct those mutations before implantation by homologous recombination. Even if they can not be corrected, those mutations are open to study in lineages derived from the iPS cells. A further advantage of studying iPS cells is access to testing new drugs in those diseased human cells. At present the barriers to adopting iPS in the clinic are the limitations of methods to produce the cells.

iPS technology is expected to move health sciences forward in unique ways for diagnosis, drug screening, toxicity , repair of mutations and treatment of human diseases. iPS cells produced from an individual are embryonic-like stem cells and they can be regrown into

any of the 200 somatic cell types. iPS cell have many similarities with ESC cells including the cell morphology, surface antigens, gene expression, telomerase activity, and the epigenetic status. iPS cells have been produced by delivering transcription factors by different types of viral vectors including retroviruses (Takahasi et al 2007), lentiviruses (Chang C-W et al 2009), adenoviruses (Stadtfield et al 2008), plasmid transfections (Jia et al 2010), transposons (Woltjen et al 2010), mRNA or recombinant proteins (Warren et al 2010).
It appears that any cell type in the body can be reprogrammed. It was first achieved in skin cells for easy availability. This meant that autologous stem cells could be grown from iPS cells. Autologous transplantation using human iPS cells would not be rejected by graft versus host problems. Even if the donor cells a contain mutation that cause the disease reprogramming can be carried out in mutated cells to study diseases and to correct the mutations ex vivo by homologous recombination. IPS cells are produced to be as close to human ESCs as possible to have the advantages of pluripotency that hESCs have. However hESCs could only be transplanted allogenically into adults and rejection would always be a problem.
Despite the euphoria and literally thousands of studies, there are nagging problems with making iPS cells work in the way that was hoped. One of the reasons is that iPS cells, just like hEScs, go through a stage of producing teratomas. Several studies have found that mice produced from iPS cells are more prone to cancer. The original method for producing iPScells was by a retrovirus to deliver the four transcritption factors. Retroviruses are notorious for random insertion in the genome and being oncogenic. In Yamanka's study the method included c-Myc as a transcription factor that incrcreased cell growth. However the same property is the property that makes c-Myc an oncogene and therefore another prime suspect in causing cancer. Thompson's group avoided using c-Myc but also used a retrovirus and instead of c-Myc used LIN 28 as a transcription factor (Yu et al 2007). Eventually Yamanaka's group dispensed with c-Myc but lost efficiency. Only Oct4 and Sox2 seemed to be absolutely necessary. The other factors could be varied, with Nanog substituting for c-Myc. To avoid the dangers of retroviruses various alternatives for non-integrating delivery have been tried. Adeno-virus was used successfully. But Adenovirus while not being carcinogenic has other problems. In 1999 it was prematurely used in a Phase I gene therapy trial and caused the death of one of the participants, Jesse Gelsinger. Adenovirus produces many proteins and these induce immune reactions. It was an immune reaction to adenovirus in Jessie Gelsinger's body that made adenovirus totally unacceptable for human use (Marshall 1999). There is an alternative however, Adeno-Associated Virus (AAV). Despite its name, AAV is unrelated to Adeno-virus, and is proving to be a very safe and reliable vector for gene delivery. Gene therapy with rAAV for restoring sight to patients blind since birth or early childhood (Leber congenital amaurosis) has proven that in humans rAAV is very safe (Maguire et all 2008).
Therefore we tested if AAV could be used to induce IPS in adult cells. Eventually there will come a time when iPS will be tested for therapeutic use in treating humans. Although there are now virus free methods of making iPS we do not know which method will have the greatest efficiency, safety, reproducibility or efficacy. Therefore having several different ways of producing iPS is still a viable quest, and here we show how AAV can be used to produce iPS cells.

2. Methods

We generated iPS from with four transcription factors: Oct3/4, Sox2, Klf4 and c-Myc by recombinant adeno-associated viral (rAAV) transduction.

The parvovirus adeno-associated virus (AAV) has single-stranded genome of approximately 4.7 kb carrying capacity. Recombinant AAV (rAAV) in which the two open reading frames of AAV, designated *rep* and *cap*, have been replaced by a gene of interest has become an important tool for gene delivery (Phillips et al 2010, Maguire et al 2008, Zolotukhin et al 1999).

2.1 Construction of four rAAV plasmids

2.1.1 rAAV plasmid with transcription factor: Oct3/4

Vector: pTR-UF3 (an adeno-associated virus (AAV)-based plasmid vector) pTR-UF3 (a kind gift of Nicholas Muzyczka) contains the humanized green fluorescent protein (hGFP) gene under control of cytomegalovirus (CMV) promoter through the polio virus type 1 internal ribosomal entry site (IRES). A 1.4 kb fragment, which contains Oct3/4 (1134 bp), was extracted from pMXs-Oct3/4 (a kind gift from Shinya Yamanaka's Lab) by BstxI restriction enzyme digestion. The 1.4 kb fragment was inserted into the unique *Hin*d III site of the AAV-derived plasmid pTR-UF3, forming the AAV-derived plasmid pTR-Oct3/4-hGFP.

2.1.2 rAAV plasmid with transcription factor: Sox2

Vector: pTR-UF11 (an adeno-associated virus (AAV)-based plasmid vector) pTR-UF11 (a kind gift of Nicholas Muzyczka) is derived from pSM620 [2] in which the internal sequences have been replaced by a green fluorescent protein (GFP) gene under the control of a chicken β-actin -cytomegalovirus promoter and a *neo* gene under the control of a thymidine kinase promoter. A 1.2 kb fragment, which contains Sox2 (960 bp), was extracted from pMXs-Sox2 (a kind gift from Shinya Yamanaka's Lab) by BstxI restriction enzyme digestion. The AAV-derived plasmid, which consists of Sox2 gene, was made by digesting pTR-UF11 with NotI (cutting hGFP), blunting with large klenow fragment and ligating together with the above 1.2 kb fragment, forming pTR-Sox2-Neo.

2.1.3 rAAV plasmid with transcription factor: Klf4

Vector: pTR-UF11 (an adeno-associated virus (AAV)-based plasmid vector). The transcription factor of Klf4 was amplified by PCR (Forward Primer: GTG GTA CGG GAA ATC ACA AG and Reverse Primer: TTA AAA GTG CCT CTT CAT GTG) from the template of pMXs-Kfl4 (a kind gift from Shinya Yamanaka's Lab). The 1.5 kb PCR product consists of Klf4 factor (1425 bp). The AAV-derived plasmid, which consists of Klf4 gene, was made by digesting pTR-UF11 with XbaI and SacI, blunting with large klenow fragment and ligating together with the above 1.5 kb PCR products, forming pTR-Klf4.

2.1.4 rAAV plasmid with transcription factor: c-Myc

Vector: pTR-UF11 (an adeno-associated virus (AAV)-based plasmid vector). A 1.56 kb fragment, which contains c-Myc (1320 bp), was extracted from pMXs-c-Myc (a kind gift from Shinya Yamanaka's Lab) by BstxI restriction enzyme digestion. The AAV-derived plasmid, which consists of c-Myc gene, was made by digesting pTR-UF11 with XbaI and SacI, blunting with large klenow fragment and ligating together with the above 1.56 kb fragment, forming pTR-c-Myc. The maps of rAAV vectors (pTR-UF3 and pTR-UF11) and the rAAV plasmids with defined factor (pTR-Oct3/4-GFP, pTR-Sox2-Neo, pTR-Klf4 and pTR-c-Myc, respectively) are summarized in Figure 1.

The procedures for rAAV production, including harvesting the cells (HEK 293), extraction by freezing and thawing purification by iodixanol heparin affinity chromatography as a second step in purification following the iodixanol gradient, are used regularly in our lab and published by Zolhutkin et al (1999) Phillips et al (2010). The maps for constructing AAV with Oct3/4, Sox2, and Klf-4 and c-myc are shown in Figure 1.

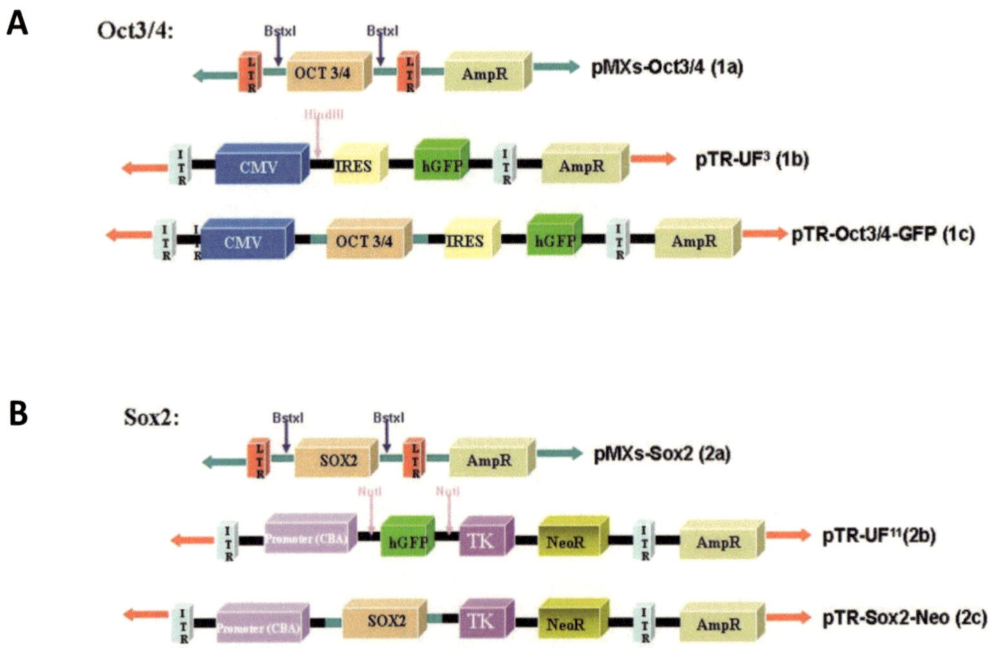

Fig. 1. **1A.** AAV-based plasmid with transcription factor of **Oct3/4**: A 1.4 kb fragment, which contains Oct3/4 (1134 bp), was extracted from pMXs-Oct3/4 (a gift from Shinya Yamanaka's Lab) by BstxI restriction enzyme digestion (1a). The 1.4 kb fragment was inserted into the unique *Hind* III site of the AAV-derived plasmid pTR-UF3 (1b), forming the AAV-derived plasmid pTR-Oct3/4-GFP (1c). pTR-UF3 (a gift of Nicholas Muzyczka), which is an adeno-associated virus (AAV)-based plasmid vector, contains the humanized green fluorescent protein (hGFP) gene under control of cytomegalovirus (CMV) promoter through the polio virus type 1 internal ribosomal entry site (IRES) (1b).
1B. AAV-based plasmid with transcription factor of **Sox2**: A 1.2 kb fragment, which contains Sox2 (960 bp), was extracted from pMXs-Sox2 by BstxI restriction enzyme digestion (2a). The AAV-derived plasmid, which consists of Sox2 gene, was made by digesting pTR-UF11 with NotI (cutting hGFP) (2b), blunting with large klenow fragment and ligating together with the above 1.2 kb fragment, forming pTR-Sox2-Neo (2c). pTR-UF11 is derived from pSM620 **[Zolhutkin et al 1999]** in which the internal sequences have been replaced by a green fluorescent protein (GFP) gene under the control of a chicken β-actin - cytomegalovirus promoter (CBA) and a *neo* gene under the control of a thymidine kinase promoter (2b). The AAV-based plasmid with transcription factors of Klf-4 and c-Myc were in the same way as to Sox2. (Phillips et al 2010).

2.2 PCR primers

The specific primers were designed to confirm the purified virus (rAAV) that contains the inserted defined factor (Myc, Klf4, Oct 4 and Sox2, respectively) as follows:

The "Universal" Forward Primer:
5′-GTG GTA CGG GAA ATC ACA AG-3′ (The primer was designed according the sequences from the backbone of rAAV plasmid vector beyond the full length of the defined factors)
The Reverse Primers:
Oct3/4
Reverse primer: 5′-AGATGGTGGTCTGGCTGAAC-3′
(Accession: M34381 Oct3/4: 581-562)
Sox2
Reverse primer: 5′-CTCCGGGAAGCGTGTACTTA-3′
(Accession: NM_011443 oct3/4: 805-786)
KLF4
Reverse primer: 5′- GGAAGACGAGGATGAAGCTG-3′
(Accession: BC010301 Klf4: 862-843)
c-Myc
Reverse primer: 5′- ATCGCAGATGAAGCTCTGGT-3′
(Accession: NM_010849 Myc: 983-964)

2.3 PCR amplification

One μl of viral DNA from the purified virus was amplified by Polymerase Chain Reaction (PCR) in 25 μl reaction volume by above universal forward primer paired with the specific reverse primer of Myc, Klf4, Oct 3/4 and Sox2, respectively. The amplification was performed in the following conditions: 4 minutes at 94°C; 15 cycles of 30 seconds at 94°C, 30 seconds at 58°C (annealing), 1 minute at 72°C and a final extension period of 10 minutes at 72°C in iCycler Thermal Cycler (Bio-Bad).

One μl of viral DNA from the purified virus of UF^3-Oct3/4-GFP was amplified as above by the forward primer 5′-CAG CGG AGA GGG TGA AGG TG-3′ (Accession: U50963 GFP: 87-106) and the reverse primer 5′-CAG GGC AGA CTG GGT GGA CA-3′ (Accession: U50963 GFP: 621-602)

2.4 PCR results

Amplification products (10 μl) were analyzed on 1% agarose stained with ethidium bromide. The results were shown in the Figure 2.

2.5 rAAV-GFP transduction

Two human somatic cell types were used: IMR-90 a fetal skin fibroblast and HEP G2 liver cells were used to induce iPS. Before the transduction we tested a normal and a starving protocol, as described below.

2.5.1 IMR-90 a fetal skin fibroblast

The IMR-90 cells were purchased from American Type Culture Collection (ATCC, Catalog No. CCL-186) and they are diploid human cells that are being extensively characterized by the ENCODE Consortium and have published DNA fingerprints that allow confirmation of the origin of reprogrammed clones. IMR-90 cells also proliferate robustly for more than 20 passages before undergoing senescence, but grow slowly in human ES cell culture conditions (Yu, J 2007).

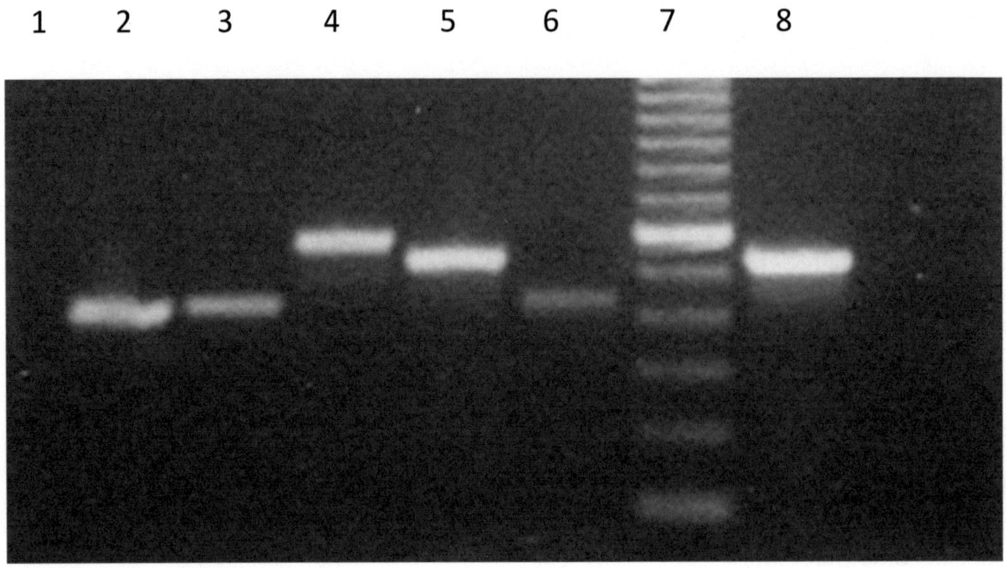

Lane 1: pUF^{11}-C-Myc (check C-Myc) (417bp)
Lane 2: pUF^{11}-KLF4 (check KLF4) (429bp)
Lane 3: pUF^{3}-Oct3/4-IRES-GFP (check Oct3/4) (583bp)
Lane 4: pUF^{3}-Oct3/4-IRES-GFP (check GFP) (535bp)
Lane 5: pUF^{11}-Sox2 (check Sox2) (433bp)
Lane 6: 100 bp DNA Ladder
Lane 7: Positive control: pUF^{11}consists of GFP (535bp)
Lane 8: Negative control

Fig. 2. PCR results with AAV delivery of defined factors for reprogramming

2.5.1.1 Normal protocol

A vector constructed with adeno-associated virus (AAV)-based plasmid vector (pUF11) containing green fluorescent protein (GFP) with 535bp (Phillips et al., 2010) was transduced in the cells. Cells were plated in 6 wells plates (5x 10^4 cells/well) and were cultured in Minimum Essential Medium (MEM Engle) supplemented with 10% heat-inactivated fetal bovine serum (10% FBS), 0.1 mM non-essential amino acids, and 1.0 mM sodium pyruvate. The cells were cultured by 24hs at 37°C, 5% CO_2. One day later, when cells were approximately 70-80% confluent, the cells were washed twice with PBS and transduced with adeno-associated virus (AAV)-GFP vector with medium (MEM Engle) without FBS and antibiotic with MOI 50, 100, 250, 500 and 1000. Cells were incubated at 37°C, 5% CO_2 for 24hs. The medium was change to MEM Engle 10% FBS. The medium was change every day. Ten days after the transduction the number fluorescent cells per well were quantified with confocal microscopy. Figure 3, the figure shows two green fluorescent cells isolated and analyzed by Confocal Microscopy (Leica Microsystems-TCS SP5) indicating successful AAV-GFP transduction. The GFP positive cells/field is shown Figure 4.

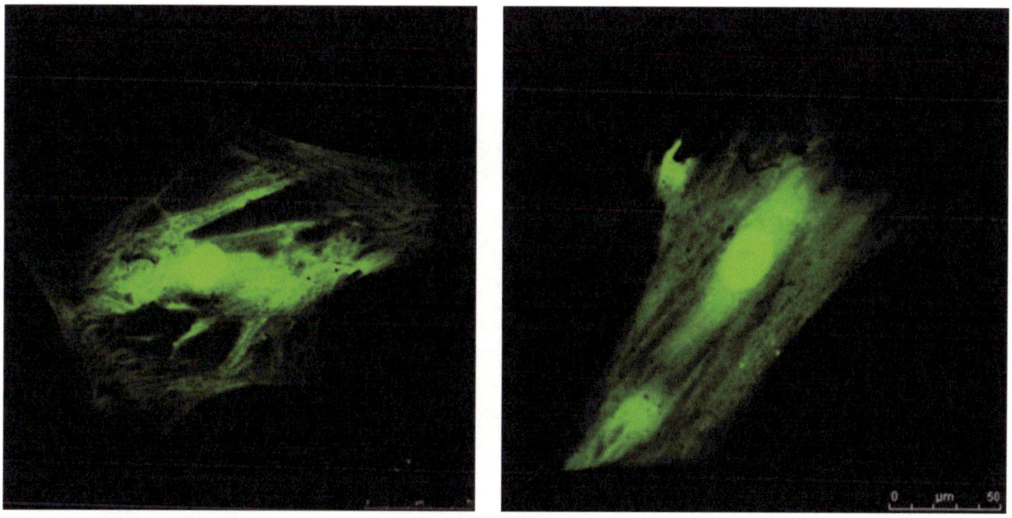

Fig. 3. AAV-GFP transduction. The figure shows two green fluorescent cells isolated and analyzed by Confocal Microscopy (Leica Microsystems-TCS SP5) indicating successful AAV-GFP transduction.

2.5.1.2 Starving protocol

To starving protocol the cells were plated in 6 wells plates (5x 10^4 cells/well) and were cultured in Minimum Essential Medium (MEM Engle) supplemented with 10% heat-inactivated fetal bovine serum (10% FBS), 0.1 mM non-essential amino acids, and 1.0 mM sodium pyruvate. The cells were cultured by 24hs at 37°C, 5% CO_2. One day later, when cells were approximately 70-80% confluent, the cells were washed twice with PBS and incubated with medium without serum for 48 hs. After, the cells were transduced with adeno-associated virus (AAV)-GFP vector with medium (MEM Engle) without FBS and antibiotic with MOI 100, 1000 and 5000 and compared to normal protocol. Cells were incubated at 37°C, 5% CO_2 for 24hs and the medium (MEM Engle) with FBS 20% for 24 hs. The medium was changed to MEM Engle 10% FBS. The medium was change every day. Ten days after the transduction the number fluorescent cells per well were quantified with Confocal microscopy. The Figure 5 shows that the transduction efficiency was most efficient with the normal protocol than the starving protocol.

2.5.1.3 Four factor transduction efficiency

To improve the transduction efficiency we tested the 4 factors together as well separately.

Using the normal protocol the cells were cultured by 24hs at 37°C, 5% CO_2. One day later, the cells were transduced with adeno-associated virus (AAV)-Oct 3/4-GFP vector for 24hs. Two days later, the cells were transduced with c-myc. Three days later, the cells were transduced with Sox2 plus Klf-4 with medium (MEM Engle) without FBS with MOI 10, 100 and MOI 1000.

Also, using the normal protocol the cells were transduced with adeno-associated virus (AAV)-Oct 3/4-GFP vector plus Sox2, Klf-4 and c-myc with medium (MEM Engle) without FBS with MOI 10, 100 and 1000. After 24 hs the medium was changed to MEM Engle 10% FBS. The medium was change every day. Ten days after the transduction the number fluorescent cells per well were quantified with confocal microscopy. The Table 1 shows that the positive GFP cells were not different when the transduction was done with the 4 factors together as separately as well.

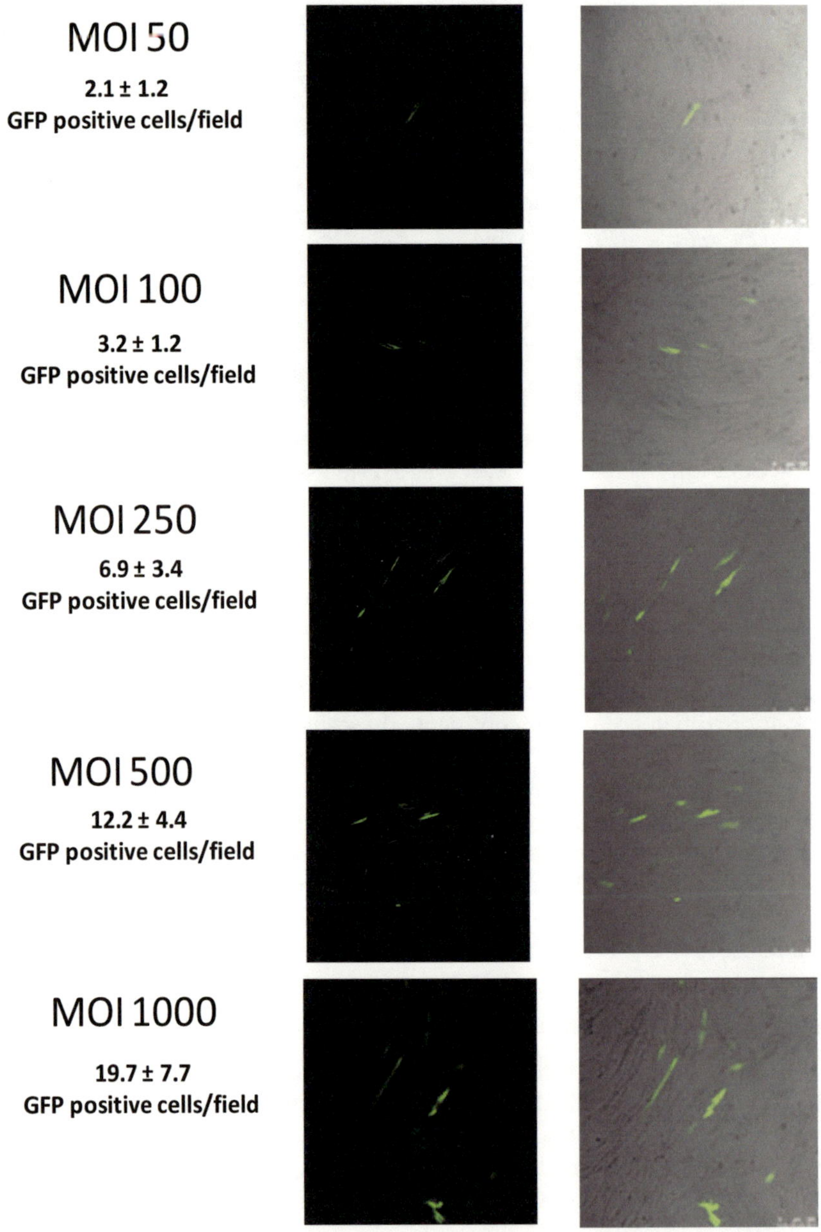

Fig. 4. AAV-GFP vector transduction. The green fluorescent cells indicate the GFP expression analyzed by Confocal Microscopy (x250) inIMR-90 cells. The figure shows on the left the GFP positive cells/ field and dark field, and on the right the merge cells. A vector **pUF**[11] (an adeno-associated virus (AAV)-based plasmid vector) consisting GFP (535bp) was transduced in human fetal fibroblasts with MOI 50, 100, 250, 500 and 1000.

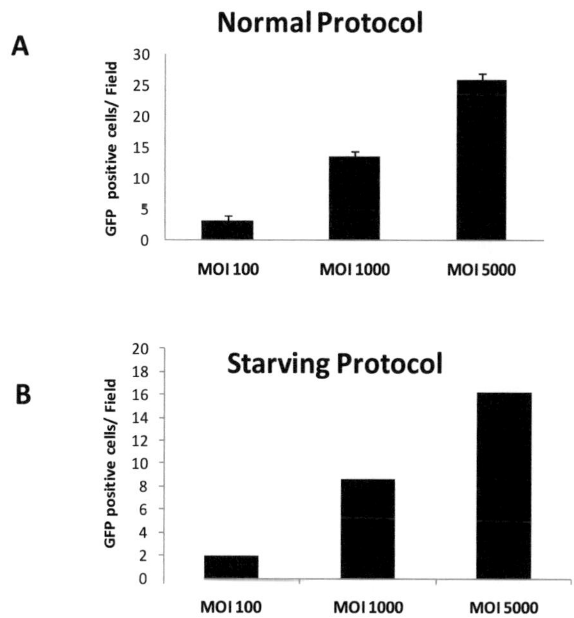

Fig. 5. The cells were cultured by 24hs at 37°C, 5% CO_2. When cells were approximately 70-80% confluent, the cells were washed PBS and transduced with adeno-associated virus (AAV)-GFP vector. A vector **pUF11** (an adeno-associated virus (AAV)-based plasmid vector) consisting GFP (535bp) was transduced in human fetal fibroblasts with MOI 100, 1000 and 5000. Ten days after the transduction the number fluorescent cells per well were quantified with confocal microscopy. **A.** Show that the transduction efficiency with the normal protocol **B.** Show the transduction efficiency with the starving protocol. The figure shows that the transduction efficiency was most efficient with the normal protocol than starving protocol.

	MOI 10	MOI 100	MOI 1000
Together	1	7	84
Separately	1	9	85

Table 1. Transduction of 4 factors together or separately as well, 10 days after the IMR-90 cells transduction. The results are expressed by GFP positive cells per well.

2.5.1.4 Four factor delivery 20 and 40 days after transduction

To test the delivery of the 4 factors transcription genes during hole the pluripotency induction time, the cells were transducted using the normal protocol with adeno-associated virus (AAV)-Oct 3/4-GFP vector plus Sox2, Klf-4 with and without c-myc with medium (MEM Engle) without FBS and antibiotic with MOI 1000 (Experiment 1 and Experiment 2) (Figure 6). After 24hs was added 1ml/well medium 30% serum for 6hs. The medium was changed MEMEngle 10% + P/S and changed each 2 or 3 days. Twenty and 40 days after transduction the RNA was extracted with Trizol to quantify the GFP, Oct3/4, Sox-2, Klf-4 and c-Myc by RT-PCR, using the primers described above. Several different groups were tested to confirm effectiveness samples as showed by RT-PCR of agarose gel samples in the Figure 6.

Experiment 1: 20 days after transduction:

a. Control cells,
b. Oct 3/4-GFP,
c. Oct 3/4-GFP + Sox-2 + Klf-4 with c-Myc,
d. Oct 3/4-GFP + Sox-2 + Klf-4 without c-Myc.

Experiment 2: 40 days after transduction

a. Control cells,
b. Oct 3/4-GFP,
c. Oct 3/4-GFP + Sox-2 + Klf-4 with c-Myc,
d. Oct 3/4-GFP + Sox-2 + Klf-4 without c-Myc.

The figure 6 shows that the 4 factors are expressed for up 40 days.

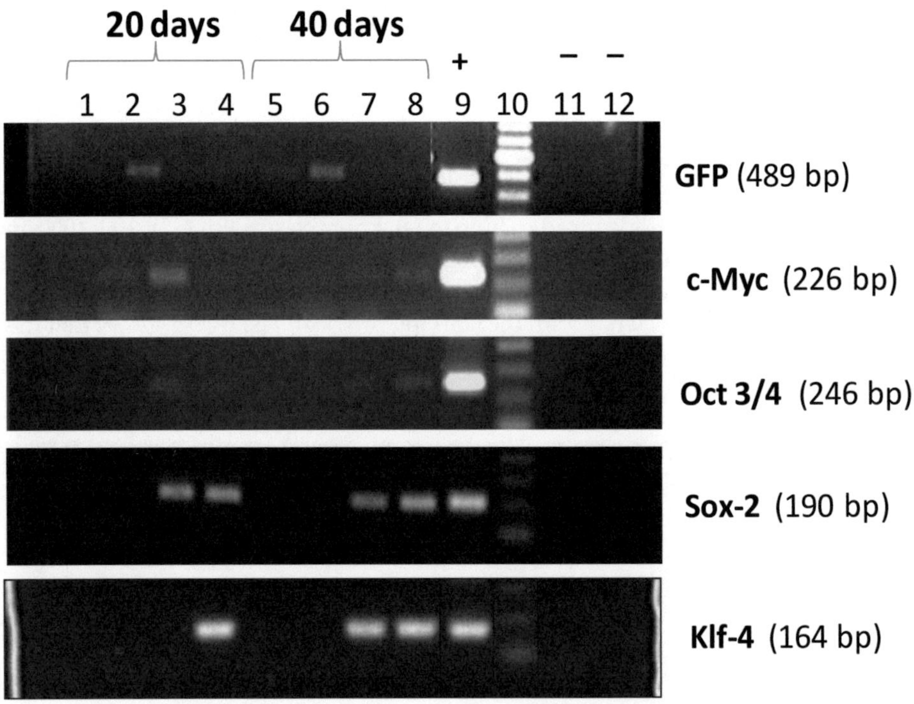

Fig. 6. RT- PCR results showing transcription factor gene expression after 20 days and 40 days transduction with AAV. All of the 4 markers are visible although Oct3/4 was weakly expressed. Sequences by RT-PCR of agarose gel samples from each experimental test group. Experiment 1: 20 days after transduction: a) Control: GFP b) With c-Myc c) Without c-Myc. Experiment 2: 40 days after transduction a) Control: GFP, b) With c-Myc c) Without c-Myc . Positive control. Ladder 100, Negative Control 1st reaction, Negative Control 2nd reaction: The cells were transduced with and without c-Myc. 1.3X10^6 cells; 70% confluence; 4 P; MOI 1000.

2.5.2 HEP G2 liver cells

Heptocyte Cells: HepG2 Cell line was derived from the hepatocellular carcinoma cell line (ATCC - Hep G2/2.2.1); adherent; epithelial; from a 15 year old male Caucasian.
For human hepatocytes cells, 6 days after rAAV-GFP (Figura 7 A and B) and rAAV-Oct 3/4-GFP (Figure 7 C and D) (MOI-100) transduction, the GFP positive cells were observed by Confocal microscopy and total RNA was extracted with Trizol and the GFP and Oct 3/4 gene expression were determined by RT-PCR (Figure 7 E and F). Microscopically the rAAV-GFP could be clearly seen expressed in the cells, however rAAV-Oct 3/4-GFP was not observed.

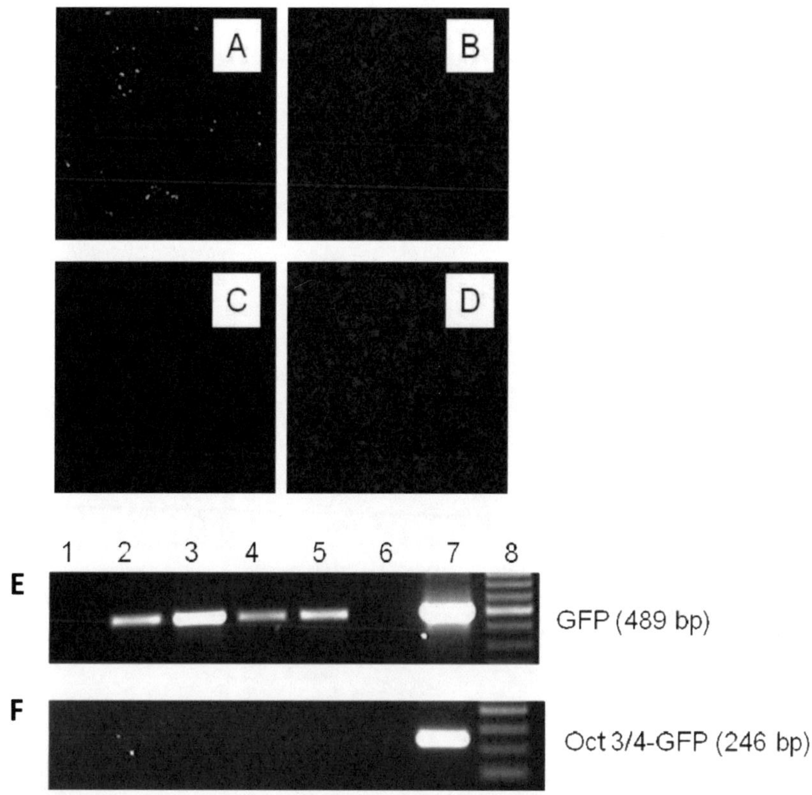

Fig. 7. Human Hepatocytes Cells, 6 days after transduction. Upper Panel: A. Confocal microscopy rAAV-GFP (MOI-100) and B. DAPI. C. rAAV-Oct3/4-GFP (MOI-100) and D. DAPI. Lower Panel: E. RT-PCR for GFP and F. Oct3/4-GFP. 1- Control; 2- GFP MOI 10; 3- GFP MOI 100; 4- Oct 3/4 - GFP MOI 10; 5-Oct 3/4 - GFP MOI 100; 6- Negative Control; 7- Positive Control; and 8- Ladder.

2.5.3 iPS Cells generation with rAAV

Since the question was: can we make iPScells with rAAV-GFP instead of retroviruses, lentivirus or adeno-virus we started with the same four transcription genes used by Yamanka who kindly provided them to us.
The IMR-90 a fetal skin fibroblast and HEP G2 liver cells were plated $1.3X10^6$ cells with 70% confluence the cells transduced with Oct3/4, Sox-2 and Klf-4 with c-Myc and without c-

Myc, MOI 1000. Before transduction the medium was changed to medium without serum and antibiotic for 24hs.The transduction was done with 4 factors together. After 24hs 30% serum was added 1ml/well for 6hs. The medium was changed to MEMEngle 10% + P/S for 3 days. The medium was changed to hESC medium 4 days after the transduction (Knockout DMEM supplemented with 15% knockout serum replacement, L-glutamine, nonessential amino acids, β-mercaptoethanol and penicillin/streptomycin). For the generation of iPS with rAAV transduction the cells were not cultured in MEFs or with chemicals treatment. The medium was changed every day. iPS colonies were picked between 20-30 days post transduction based on colony morphology. To HEP G2 liver cells 15 days after the transduction the cells were passage and re-transduced.

The medium was changed every day until the colonies become big enough to be picked out. Colonies should first become visible approximately a week after the transduction. They become large enough to be picked out around day 20. The reprogrammed clones were selected by morphological criteria (compact colonies, high nucleous to cytoplasm ratio and proeminent nucleoli) as showed Figure 8.

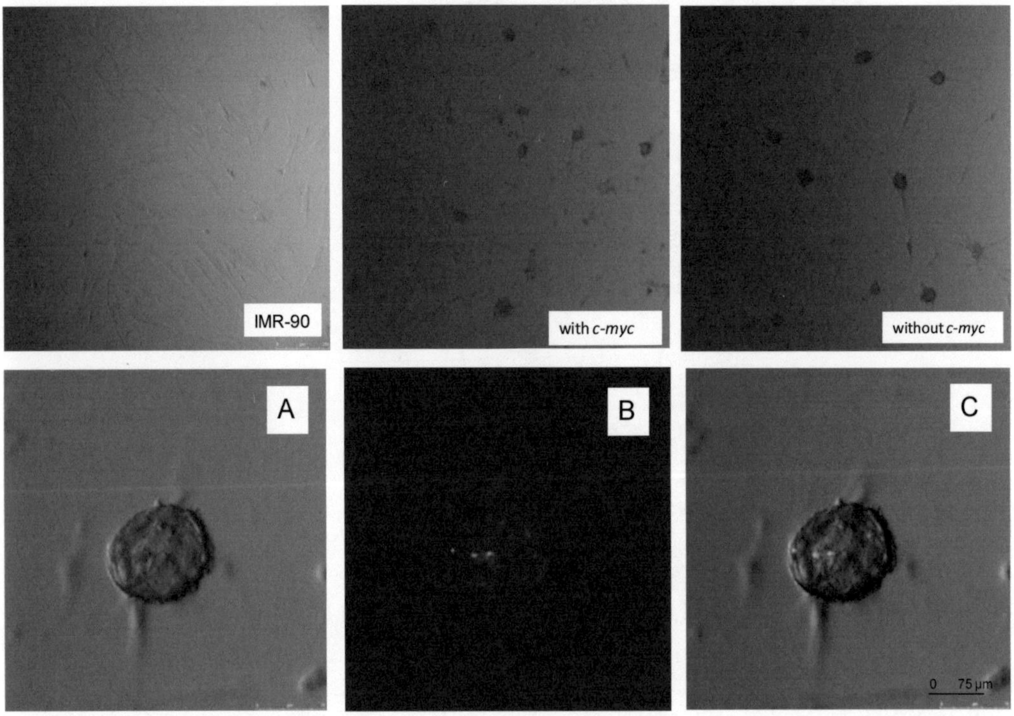

Fig. 8. iPS cells from IMR-90 fibroblast cells. Upper Panel: iPS cells confocal microscopy image. Before transduction (IMR-90), iPS cells with 4 factors (with c-Myc), and iPS cells with 3 factors (without c-Myc). Lower Panel: iPS cells without c-Myc. Confocal image with bright field (A), fluorescence (B), merge (C).

Picking out the colonies: aliquot 20 µl of 0.25% trypsin/1 mM EDTA per well of 96-well plate. The medium was removed from the dish, and added 10 ml of PBS. Aspirate PBS, and added 5 ml of PBS. Pick colonies from the dish using a Pipetman set at 2 µl, and transfer it into the 96-well trypsin plate. We picked out as many colonies as we can within 15 min.

Cells was incubated another 15 min in trypsin at 37ºC to dissociate cells in the colonies. 180 µl of ES medium was added to each well, and pipetted up and down to break up the colony to single cells. The cell suspension was transferred into the well of 24-well plates with medium without feeder cells, and 300 µl ES medium added. The cells were incubated in 37ºC, 5% CO2 until the cells reach 80-90% confluence. At this point they were passaged into 6-well plates.

2.5.3.1 Alkaline phosphatase (AP) and immunofluorescence staining

To demonstrate these cells possess characterists of embryonic stem cells, we stained them for AP activity using a kit (CHEMICON®s Alkaline Phosphatase Detection) and the expression of membrane markers, Cell surface Stage-Specific Embrionic Antigens (SSEA-4) and Keratan sulphate-associated antigens (TRA-1-60 and TRA-1-81) using a kit (CHEMICON®s ES Cell Marker Samples) and the transcription factor Oct-4.
Alkaline Phosphatase staining was used as a marker for embryonic-like stem cells .The AP was quantified 20 days after treatment with hESC medium. Figure 8 shows the AP to HEP G2 liver cells and Figure 9 show the iPS cells membrane markers from Human hepatocyte cells (HEP G2).

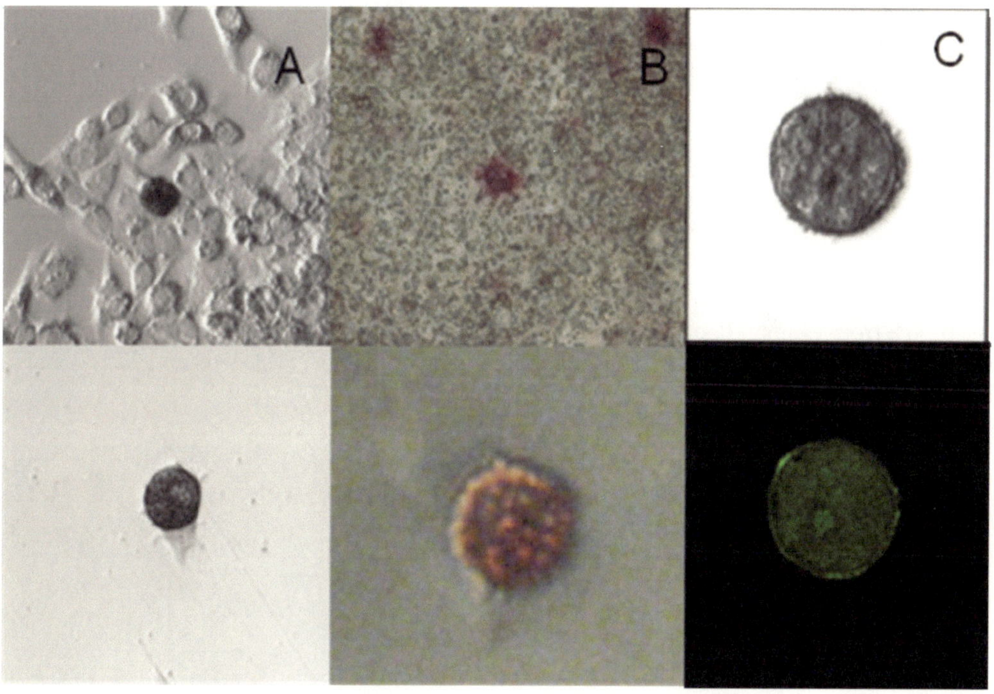

Fig. 9. iPS from Human hepatocyte cells (HepG2).Upper and down panels (A) Confocal bright-field, (B) alkaline phosphatase positive and (C) Hepatocytes cells transduced with 4 factors, 30 days after transduction. Upper, bright-field and down, GFP positive iPS (Oct-3/4 positive).

Figure 10 show the AP to IMR-90 fibroblast cells and Figure 11 show the iPS cells membrane markers from Human fibroblast cells (IMR-90). The iPS cells shows morphology such as compact colonies, high nucleus to cytoplasm ratio and prominent nucleoli. The testing for markers of human embryonic stem cells in the AAV induced iPS cells was positive.

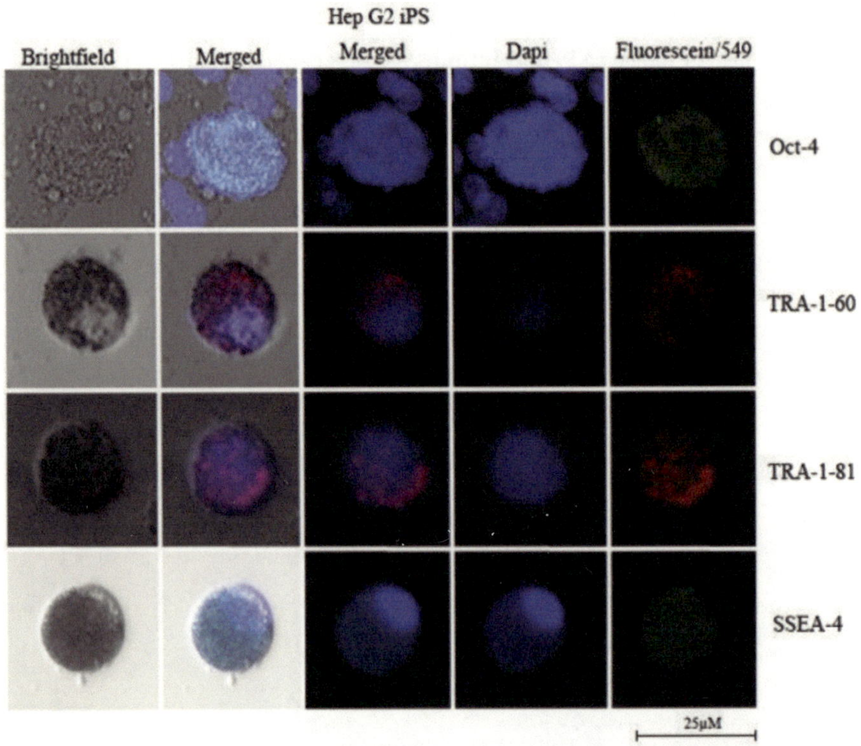

Fig. 10. iPS from Human hepatocyte cells (HepG2). iPS cells membrane markers: the transcription factor Oct-4; Keratan sulphate-associated antigens (TRA-1-60 and TRA-1-81) and Cell surface Stage-Specific Embrionic Antigens (SSEA-4).

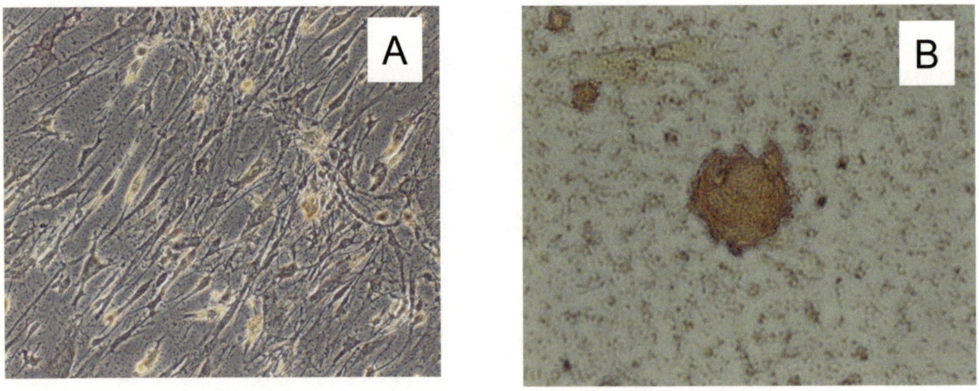

Fig 11. iPS from Fibroblast Human Cells (IMR-90). A. Control cells, B. Alkaline phosphatase positive.

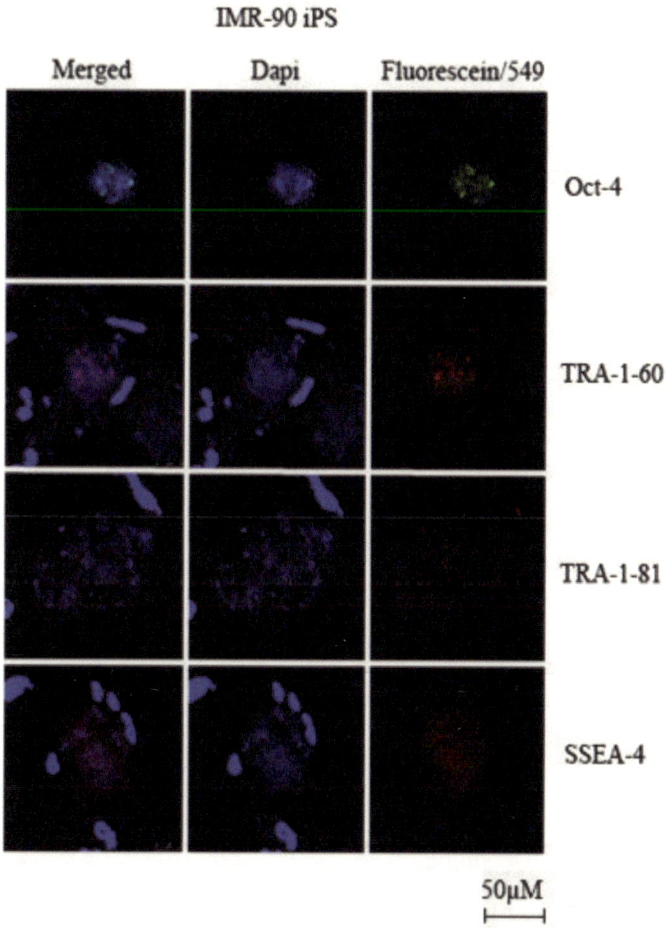

Fig. 12. iPS from Fibroblast Human Cells (IMR-90). iPS cells membrane markers: the transcription factor Oct-4; Keratan sulphate-associated antigens (TRA-1-60 and TRA-1-81) and Cell surface Stage-Specific Embrionic Antigens (SSEA-4).

3. Discussion

In the present paper we report the application of AAV serotype 2 as a vector for producing human iPS cells from adult somatic liver and cells. We delivered the transcription factors: Oct 4, Sox2, c-Myc and Klf4 that were originally used by S. Yamanaka et al (Takahashi et al 2007). We are well aware that this set of factors is not the optimal choice except for Oct4 and Sox2. Many other factors can replace cMyc and Klf4. However for the demonstration that AAV is an appropriate vector for reprogramming adult cells to iPS cells, we chose to work with the factors that have consistently been used for efficient conversion of differentiated somatic cells. In future studies with AAV we plan to test replacement factors. While more in vivo tests are necessary, this report establishes a feasible new method for producing iPS cells with AAV

rAAV has not previously been used as the viral vector for delivery of transgenes to induce pluripotent cells. AAV has several advantages over other viral vectors. Unlike retroviruses (and wild type AAV), rAAV does not integrate into the chromosomes. Therefore it does not raise concerns about insertional mutagenesis. This is confirmed by the safe and stable experience in several human trials with rAAV(Maguire et al , 2008) . Originally retroviruses were the vectors of choice for delivery of transcription factors (Takahashi et al 2007). However, they are unsuitable for human use because of oncogenic risks from insertional mutagenesis. Lentivirus has a large carrying capacity, but it is a retrovirus. As the virus type for HIV the public might be concerned about its use in iPS cells for autologous transplantation. The adenovirus has been tested as a non-integrating virus. The results were positive but not efficient (Stadtfield et al, 2008). Using adenovirus as a vector is appealing for its ease of production and use, although such studies overlook that the adenovirus is not a viable vector for humans. The adenovirus caused fatal effects in the failed gene therapy trial at University of Pennsylvania in 1999 (Marshall, 2009). Thus even though these vectors might be better developed and even though they may serve the purpose of making iPS cells, they carry a negative public image for human use. The leading viral vector in several successful clinical trials recently has been the recombinant adenoassociated virus vector (rAAV). Experience with rAAV in humans is safe, stable, producible in high titers (10^{12} infectious particles per ml) and non-immunogenic (Maguire et al 2008). In this paper we propose that rAAV is an advantageous vector for producing human iPS cells for therapeutic transplantion.

However only a few papers in the literature have explored AAV and iPS and those reports used AAV for specialized purposes. Khan et al (2010) showed that AAV can selectively target mutations and correct mutations in iPS cells by homologous recombination. They used AAV in specifically generated human iPS cells with two different gene mutations (HPRT1 and HMBA1) responsible for the Lesch-Nyan syndrome. They studied fibroblasts and mesenchymal cells as the adult cell sources that had been reprogrammed with Oct4, Sox2 Lin28 NANOG as the transgenes. Reprogramming was conventional with a lentivirus. The successful correction of a mutation in the iPS cells extends the usefulness of AAV beyond making iPS cells to making iPS cells better. If autologous adult somatic cells have a mutation that can be corrected by gene targeting with AAV, they become available for therapeutic purposes. Targeting with AAV was achieved without cytotoxicity (Kahn et al 2010) . A similar paper report from Mitsui et al (2009), who received iPS cells directly from S Yamanaka, tested AAV targeting of HPRT1.

Very recently (as we go to press) a new development in iPS production has indicated that a single microRNA may be able to effect the whole reprogramming process without resorting to transcription factors for delivery. (Anoke-Danso et al April 7, 2011). This new concept would be another use for AAV because AAV is excellent for delivery of miRNA (Qiao et al 2011). It is too soon to tell which method will ultimately make iPScell a successful alternative to embryonic stem cells or not.

4. Conclusion

In conclusion, several different types of vectors and delivery systems have been used to reprogram adult somatic cells to iPS cells but none have become predominantly useful. Here we show that rAAV is successful in delivering transcription factors to adult fibroblasts and hepatic cells and reprogramming them to iPS cells. rAAV is efficient and effective. We used low MOIs. Each of the 4 transcription factor was delivered in a separate rAAV plasmid so that in future applications the transgenes can be easily varied. There was no evidence of cytotoxicity and the iPS cells were stable. The iPS cell showed the accepted in vitro characteristics of being inducible Pluripotent Stem Cells, including cell morphology, alkaline phoshphatase positive staining, and SSEA1 positive staining. Doxocyclin was not required, making it simpler to use than the doxocyclin lentivirus method. While more in vivo tests are necessary, this report establishes a feasible new method for producing iPS cells with AAV. Other studies indicate AAV will be having the added advantage of being used for accurate targeting of mutations in iPS so that the mutations can be corrected by homologous recombination. With these advantages the use of AAV can move the technology of iPS closer to therapeutic use in human patients.

5. Aknowledgment

Oliveira EM was the recipient of a CNPq-PDE Fellowship (No.200994/2007-7) and from Dr MI Phillips Grant in USA. E. M. Oliveira holds scholarships from CNPq, Brazil. Dr. MI Phillips was supported by grant from NIH 1 R01 HL 077602.

6. References

Anokye-Danso, F., Trivedi, C.M. Juhr D., Gupta, M., Cui, Z., Tian, Y., Zhang, Y., Yang, W., Gruber, P.J., Epstein, J.A., & Morrisey, E.E. (2011). Highly efficient miRNA-mediated reprogramming of mouse and human somatic cells to pluripotency *Cell Stem Cell* 8: 376- 388.

Chang, C-W., Lai, Y-S., Pawlik, K.M., Liu, K., Sun, C.-W., Li, C., Schoeb, T.R. & Townes, T.M. (2009). Polycistronic lentiviral vector for "hit and run" reprogramming of adult skin fibroblasts to induced pluripotent stem cells, *Stem Cells* 27, 1042-1049

Dimos, J.T., Rodolfa, K.T., Niakan, K.K., Weisenthal, L.M., Mitsumoto, H., Chung, W., Croft, G.F., Saphier, G., Leibel, R., Goland, R., Wichterle, H., Henderson, C.E. & Eggan, K. 2008. Induced pluripotent stem cells generated from patients with ALS can be differentiated into motor neurons. *Science*. Aug 29;321(5893):1218-21.

Jia, F., Wilson, K.D., Sun, N., Gupta, D.M., Huang, M., Li, Z., Panetta, N.J., Chen, Z.Y., Robbins, R.C., Kay, M.A., et al. (2010). A nonviral minicircle vector for deriving human iPS cells. *Nat. Methods* 7, 197-199.

Khan, I.F., Kirata, R.K., Wang, P., Kho, J., Nelson, A., Huo, Y., Zavaljeski, M., Ware, C. & Russell, D.W. (2010). Engineering of Human Pluripotent Stem Cells by AAV-mediated Gene Technology. *Molecular Therapy* 18, 1192-1199.

Kaji, K., Norrby, K., Paca, A., Meileikovsky, M., Mohseni, P. & Woltjen, K. (2009). Virus-free induction of pluripotency and subsequent excision of reprogramming factors. *Nature* 458, 771-775.

Maguire, A.M., Bennett, J.B. et al. (2008). "Safety and efficacy of gene transfer for Leber's congenital amaurosis. *New England Journal of Medicine* 358 (21): 2240–8.

Marshall, E. "Gene Therapy Death Prompts Review of Adenovirus Vector." (1999). *Science* 286: 2244-245.

Mitsui, K., Suzuki, K., Aizawa, E., Kawase, E., Suemori, H., Nakatsuji, N. & Mitani, K. (2009). Gene targeting in human pluripotent stem cells with adeno-associated virus vectors. *BBRC 388*, 711-717.

Okita, K., Nakagawa, M., Hyenjong, H., Ichisaka, T. & Yamanaka, S. (2008). Generation of mouse induced pluripotent stem cells without viral vectors. *Science 322*, 949-953.

Phillips, M.I., Qian, K., Shen, L., Tang, Y.L. & Oliveira, E.M. (2010). Gene Therapy Strategies: Constructing an AAV Trojan Horse. In: *Genomics Essential Methods* (Ed.), pp 283-306, Mike Starkey and Ramnath Elaswarapu.(Eds). John Wiley & Sons Ltd, Hoboken, New Jersey, USA.

Qiao, C., Yuan, Z., Ji, L., He, B., Zheng, H., Mayer, C., Li, J. & Xiao, X. (2011). Liver-specific microRNA-122 target sequences incorporated in AAV vectors efficiently inhibits transgene expression in the liver *Gene Therapy* 18, 403-410

Stadtfeld, M., Nagaya, M., Utikal, J., Weir, G. & Hochedlinger, K. (2008). Induced pluripotent stems cells generated without viral integration. *Science 322*, 945-949.

Takahashi, K., Tanabe, K., Ohnuki, M., Narita, M., Ichisaka, T., Tomoda, K., & Yamanaka, S. (2007). Induction of pluripotent stem cells from adult human fibroblasts by defined factors. *Cell 131*, 861-872.

Warren, L., Manos, P.D., Ahfeldt, T., Loh, Y., Li, H., Lau, F., Ebina, W., Mandal, P.K., Smith, Z.D., Meissner, A., Daley, G.Q., Brack, A.S., Collins, J.J., Cowan, C., Schlaeger, T.M. & Rossi, D.J. (2010). Highly Efficient Reprogramming to Pluripotency and Directed Differentiation of Human Cells with Synthetic Modified mRNA. *Cell Stem Cell* 7, 1-13.

Woltjen, K., Michael, I.P., Mohseni, P., Desai, R., Mileikovsky, M., Hämäläinen, R., Cowling, R., Wang, W., Liu, P., Gertsenstein, M., et al. (2009). piggyback transportation reprograms fibroblasts to induced pluripotent stem cells. *Nature 458*, 766-770.

Yu, J., Hu, K., Smuga-Otto, K., Tian, S., Stewart, R., Slukvin, I.I. &Thomson, J.A. (2009). Human induced pluripotent stem cells free of vector and transgene sequences. *Science 324*, 797-801.

Zhou, H., Wu, S., Joo, J.Y., Zhu, S., Han, D.W., Lin, T., Trauger, S., Bein, G., Yao, S., Zhu, Y., et al. (2009) Generation of induced pluripotent stem cells using recombinant proteins. Cell Stem Cell *4*, 381-384.

Zolotukhin, S., Byrne, B.J., Mason, E., Zolotukhin, I., Potter, M., Chesnut, K., Summerford, C., Samulski, R.J. & Muzyczka, N. (1999). Gene Ther. Recombinant adeno-associated virus purification using novel methods improves infectious titer and yield. Jun;6(6):973-85.

32

Raman Spectroscopy for Noninvasive Monitoring of Umbilical Cord Mesenchymal Stem Cells Viability Transitions

P. Chen[1], F. Zhang[1], L. Lin[1], H. Bai[1], L. Zhang[1,2], G. Q. Tang[1], H. Fang[1], G. G. Mu[1], W. Gong[3], Z. P. Liu[3], Z. B. Han[3], H. Zhao[3,4] and Z. C. Han[3,5]

1. Introduction

As a burgeoning research area, stem cells' unique self-renewal and multilineage differentiation capability have quickly attracted the attention of scientists and researchers worldwide. Promising applications of stems cells include repairing and regenerating diseased or aged tissues and organs, restoring impaired body functions, and providing new treatments to various diseases.

One member in the stem cell family that deserves special mention is human umbilical cord derived mesenchymal stem cells (hUC-MSC). Great clinical application value and broad industrial development prospect of hUC-MSC have been recognized due to its large differentiation potential, strong proliferation capability, low immunogenicity, source variety, convenient availability, free of ethical restriction, and easy production in industry. Obviously, these unique features of stem cells rely on various factors including their biological composition, structure, and cell viability. The viability decrease, apoptosis, or necrosis of stem cells will seriously lead to the variation or even loss of their functions. Therefore, a key step in the production and quality control of stem cells is to accurately monitor the status of stem cells and their viability in real time. Research in this field has shown great importance in not only fundamental study of stem cell biology, but also research, development, and clinical use of stem cell drugs.

Although contemporary cell biology has provided a number of ways to acquire the information on the status of the cells, these methods are often invasive. Relevant biochemical reactions, dyeing, marking, fixation, cell lysis, and other treatments could remarkably induce the change of the environment necessary for cell growth and the normal physiological function of the cells. These biological techniques could even destroy the structures of the cells, leading to irreversible cell damage or necrosis. As a result, biology-based methods can hardly meet the requirements for the production and quality control of stem cells.

[1] *Key Laboratory of Optoelectronic Information Science and Technology, Ministry of Education of China, Institute of Modern Optics, Nankai University, Tianjin 300071, China*

[2] *Department of Chemistry, University of Pennsylvania, Philadelphia, PA 19104-6323, U.S.A*

[3] *National Engineering Research Centre of Cell Products, Tianjin Economic and Development Area, China*

[4] *Tianjin Key Laboratory of Food Biotechnology, Tianjin University of Commerce, Tianjin 300134, China*

[5] *The State Key Laboratory of Experimental Hematology, Institute of Hematology and Hospital of Blood Diseases, Chinese Academy of Medical Sciences and Peking Union of Medical College, Tianjin 300020, China*

Raman spectroscopy has prominent advantages over biological methods due to its high sensitivity to changes inside biological systems and influences from external environment. Also, it plays non-destructively on the cells. These distinguished features make Raman spectroscopy a superior choice in the study of high spatial resolution and real time monitoring of stem cells.

In this chapter, we will first introduce the basic concepts and theories of light absorption and scattering, and Raman scattering. Then, the change of the stem cell viability and its related mechanisms studied by Raman spectroscopy will be described in detail. Finally, a summary of our overall work will be given.

2. Basic concept and elementary theory

2.1 Scattering and light scattering

Scattering is a phenomenon widely seen in nature. When the incident particles collide with target particles, the incident particles interact with the target particle making the incident particles deviating from the incident direction and even changing its energy. This phenomenon is known as scattering. Analyzing the interactions of matters and their internal structures and movements by scattering has become an important tool in both macro and micro worlds.

Scattering of light is often observed in our daily lives. For example, when the light passes through a homogeneous and transparent matter such as glass and water, we cannot notice the light path from the side. However, if the media is inhomogeneous or contains impurities (like liquid with suspended particles or gels), the light beam can be clearly observed in the medium. This results from the scattering of light.

In the 19th century, the study of light scattering was focused on liquid and gas, which are widespread in nature. Based on the causes, the scatterings are classified as Tyndall scattering and molecular scattering. British physicist Lord Rayleigh studied the intensities of scattered molecular light. In 1871, he proposed that the intensity of scattered light is inversely proportional to the fourth powers of the wavelength (λ^4). He proposed that the intensity of scattered light is inversely proportional to the fourth powers of the wavelength, which is known as the famous Rayleigh scattering law. In 1908, C. Mie found that the intensity of scattered light in Tyndall scattering is not inversely proportional to the fourth powers of the wavelength, which is in contrary to molecular scattering. Therefore, Tyndall scattering is also called Mie scattering.

Since the 20th century, the study of light scattering reached deeply into the levels of atoms, electrons, and quasi-particles. Researchers began to pay attention to the energy change of the scattered light relative to the incident light (i.e., the wavelength change). It was found that the wavelength variation of the scattered light relative to the incident light corresponds to different scattering mechanisms. The types of light scattering are thus classified. In the study of light scattering, the unit of energy is commonly the wave numbers (cm^{-1}) .When the wave number change is less than 10^{-5} cm^{-1}, the scattering is called Rayleigh scattering. Wave number change of about 0.1 cm^{-1} is called Brillouin scattering. More than 1 cm^{-1} wave number change is known as Raman scattering. If the wavelength / energy of the scattered light does not change relative to the incident light, it is elastic scattering, otherwise is non-elastic scattering. Obviously, the latter two types of scattering are non-elastic scattering. The wave number change in Rayleigh scattering is caused by the rebound of the target particles, and therefore Rayleigh scattering is elastic scattering.

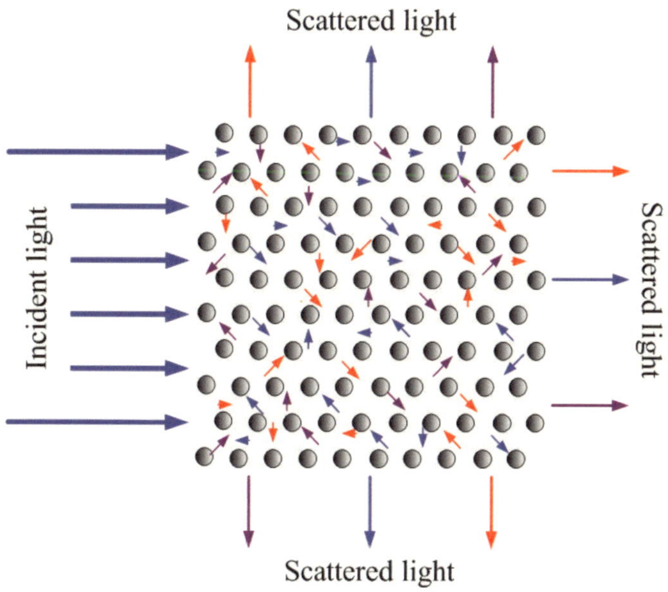

Fig.1. Scattering of light

2.2 Raman scattering

Raman spectroscopy is a scattering spectroscopy that concerns the vibrational and rotational energy levels of the molecules. It is a useful tool for the analysis of molecular structure, with advantages of being non-destructive and without the need of special ways to prepare the samples. Compared with the infrared absorption spectroscopy, water shows extremely weak Raman signals. Thus for most samples in water, the noise from water can be greatly reduced. Raman spectroscopy can also record signals from different samples in the same time, and these signals do not interfere with each other. With the developments of laser technology and weak signal detection technology, Raman spectroscopy as an effective method to study the structure of matter, will find wide and practical applications in various fields including chemistry, physics, biology, and medicine.

From the basic perspective of quantum theory, Raman scattering can be considered as an inelastic process as a result of the collisions between photons and molecules. When the incident photons collide with molecules, molecules interact with the photons and then absorb the photon energies. Molecules are thus excited into a virtual state. Since this virtual state is not stable, molecules immediately radiate photons and release energy. If the energy of the photon radiation is smaller than that of the incident photon, Stokes shift will be observed (Fig.2a). If the energy of photon radiation is larger than that of the incident photon, anti-Stokes shift will be recorded (Fig.2b). If the energy of photon radiation is equal to that of the incident photon, Rayleigh scattering will be produced (Fig.3). Since Raman scattering concerns the energy difference between the incident and scattered photons, this energy difference is associated with the corresponding vibration levels of the molecules.

Since each substance has rich vibration levels, Raman spectra carry a lot of information about the materials. Known as the fingerprint spectrum, Raman spectroscopy is of great significance to the analysis and identification of the components of materials.

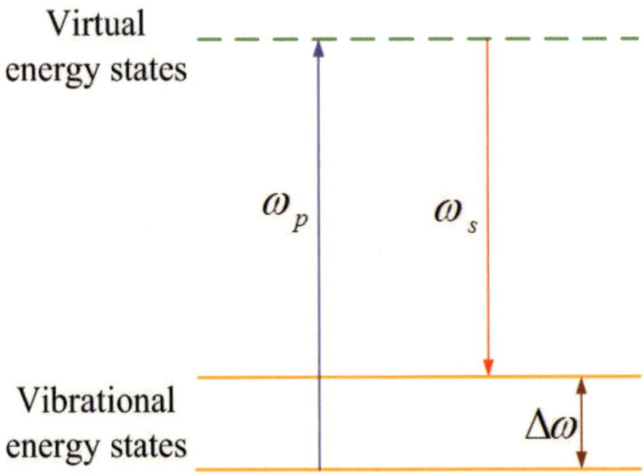

Fig. 2a. Raman scattering (Stokes effect)

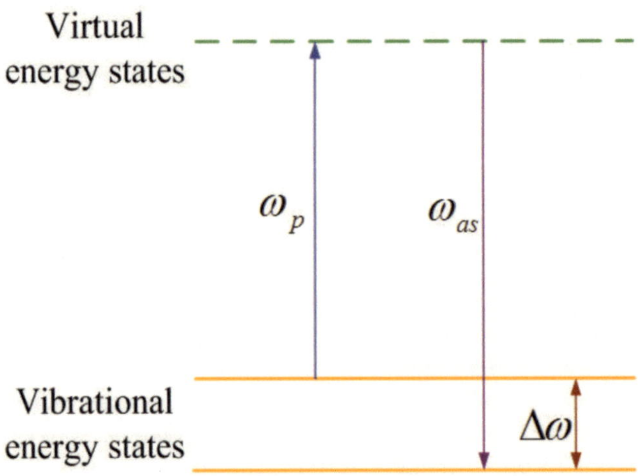

Fig. 2b. Raman scattering (Anti-Stokes effect)

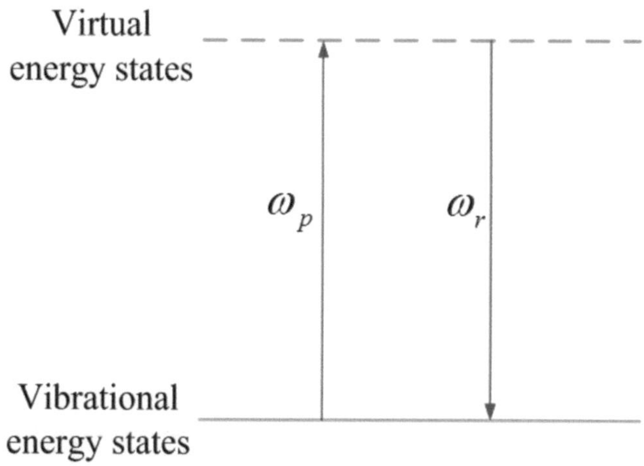

Fig. 3. Rayleigh scattering

3. Detecting viability transitions of umbilical cord mesenchymal stem cells by Raman micro-spectroscopy

Mesenchymal stem cells (MSCs) are pluripotent progenitor cells derived from the mesoderm. They are particularly attractive in cell/gene therapy for tissue regeneration and cancer therapy [1] for their own specific properties, including multi-differentiation potential, hematopoietic-supporting activity, immunomodulatory potency, and self-renewal. The major source of MSCs is bone marrow. However, the clinical use of bone marrow-derived MSCs is limited due to the relatively invasive sample collection procedure, and a significant reduction in cell number, proliferation, and differentiation capacity with age or age-related diseases. Recent advances suggest that human umbilical cord-derived mesenchymal stem cells (hUC-MSCs) could be an alternative source ideal for MSCs due to the relative ease of collection, safety, and abundance of human umbilical cord [2-5].

Our group has developed a standard procedure to acquire hUC-MSCs and established the first hUC-MSCs bank in China. Our intensive studies on the possible clinical application of this type of cells suggest that they can work as a promising candidate for cell-based therapy because of their young age and low infection rate compared to MSCs in adult tissues [6-9]. A large population of hUC-MSCs sustained in high viability condition is fundamental to successful cell-based therapy. Therefore, monitoring the cellular state transitions is an important step for the population expansion and quality assurance of these cells.

Currently, a number of methods including biological assays and optical techniques are available to handle this task. For example, laser scanning confocal fluorescence microscopy has been used to image living cells. However, the introduction of exogenous fluorescence labels is always a problem in the cell viability maintenance. On the contrary, Raman spectroscopy is a label-free, vibration spectroscopy-based technique that has been used to illustrate the intrinsic molecular structure and composition of biological samples. It has the potential to provide new and unique optical markers for cytological analysis in recent years [10-18]. At the present time, this technique has been increasingly used in the studies of stem cells. Notingher et al. monitored the differentiation of embryonic stem cells (ESCs) by recording the intensity of Raman peaks of nucleic acids as differentiation markers [19]. Chan et al. demonstrated that cell identification and classification were achieved using Raman scattering of the intrinsic biochemical signature for individual human embryonic stem cells (hESCs) [20]. More recently, Chiang and co-workers successfully applied Raman spectroscopy to monitor and quantitatively evaluate the differentiation of hMSCs into osteoblasts [21].

Here, the Raman micro-spectroscopy is used to find the molecular vibrations that are sensitive to the viability change of hUC-MSCs. The Raman spectra of hUC-MSCs with different cellular states were investigated, and the spectral differences between the cells with high viability (> 90%) and low viability (< 20%) were quantified using peak fitting and statistic t-test. It is found that the Raman peaks at 744, 877, and 1342 cm^{-1} show the most obvious changes as the cell viability decreases: the former two peaks, which correspond respectively to the vibration of C=O out-of-the-plane bending in Thymine and the symmetric stretching of C-C in lipids, are both roughly doubled; however, the latter peak, which corresponds to C-H deformation in proteins, is reduced by ca. 30%.

4. Experimental

Isolation and culturing of hUC-MSCs

hUC tissues were collected from voluntary donors, the healthy parturient women with well-developed fetus, who signed the consent form informed beforehand. The collection followed proper procedures according to the regulations of the Research Ethics Committees of the Chinese Academy of Medical Sciences and Beijing Union Medical College. The hUC tissues were then minced into $1mm^3$ small fragments with sharp scissors. The fragments were washed thoroughly with PBS to remove contamination, and then treated with 0.075% collagenaseII and 0.25% trypsin (Sigma) at 37 ^{o}C for 30 min. The digested mixture was passed through a 200μm filter to obtain cell suspensions. The dissociated cells were washed twice with PBS, planted on uncoated $25cm^2$ culture flasks, and then cultured in Dulbecco's modified Eagle's medium with low gluocose (DMEM-LG/F-12, DF12; Gibco) and 10% Fetal bovine serum (FBS, USA). The culture medium was replaced 2 days first and then after every 3 or 4 days. Daily progress of the cell culturing was monitored by phase-contrast microscopy. The cells were serially passaged and expanded in a humidified incubator at 37 ^{o}C with 5% CO_2.

hUC-MSCs with different viability were used for experiments. Decrease of the cell viability was induced by placing the cells under a condition of starvation, i.e. suspending in PBS at a low temperature of 4 ^{o}C for several days. The cell viability was accessed using the Trypan blue method in parallel with the Raman spectral measurements. The viability is defined as the ratio of the number of viable cells to the total number of all cells, and we call > 90% viability as "high viability" while < 20% viability as low viability.

Raman micro-spectroscopy of hUC-MSCs with different viability

The Raman spectra of hUC-MSCs with different viability were recorded using a Raman micro-spectrometer (LabRAM HR800, HORIBA Jobin Yvon, France), which is equipped with a 632.8 nm helium-neon laser for excitation. By means of the microscopic imaging provided by a TV camera, the Raman spectrum of each cell was measured by focusing the laser beam in the center of the cell. Under the laser power of 17 mW with a 50× objective, the signal was integrated for 100 s. Each Raman spectrum was recorded in the range of 600-1800 cm^{-1}.

Spectral processing and statistical analysis

For spectral preprocessing, the background signal was subtracted from each spectrum, and a multipoint baseline was corrected in the whole spectral region. Spectra were then normalized to the peak intensity at 1448 cm^{-1} according to ref. [11] and [20].

In order to quantitatively characterize the Raman peaks, peak fitting was carried out in the spectral regions that show obvious differences between the cells with high and low viability by using the PeakFit software (v4.12, SeaSolve Software Inc., USA). The peak shape was taken to be Gaussian.

To statistically compare the intensity difference peak by peak, the Student's t-test was performed on each fitted Raman peak for two cell-groups (one group of 22 hUCMSCs with high viability and the other of 20 with low viability). The p values obtained from the t-test are used as the indicators of the differences. Since the p value is defined as the probability of making a wrong judgment about a difference between the two sets of data, the smaller p value implies the existence of the difference with higher possibility. As a rule of thumb, when $p<0.05$ the difference is meant to be statistical significance.

5. Experimental results

Fig. 4 shows the average Raman spectra of hUC-MSCs, where curve "a" is for high viable cells and curve "b" for low viable cells. The Raman spectra show small standard deviations among the cells with the same viability (as presented as the gray shadows in curve "a" and "b"). As can be seen, there are several Raman peaks that are already discernable, i.e. the 1448 cm^{-1} peak corresponding to CH deformation, the 1659 cm^{-1} peak to the protein amide I, the 1003 cm^{-1} peak to the ring breathing of phenylalanine and the 1301 cm^{-1} peak to lipid CH_2 twist. Curve "c" plots the subtraction of "b" and "a", where visible differences can be observed, especially in the regions of 617 - 690, 703 - 770, 840 - 945, 950 - 1020, and 1250 - 1360 cm^{-1}, and at the peak of 1659 cm^{-1}.

A Raman peak, which is uniquely defined by its spectral parameters including Raman shift, intensity, and full width at half maximum (FWHM), reveals the information about a specific molecular vibration. In order to extract the Raman peaks in the spectral regions where the high viable cells and low viable ones are distinguishing, the peak fitting analysis was performed. Fig. 5 shows the fitting results of curve "a" and "b" in Fig.4, and Table 1 provides the extended peak assignments.

It can be seen from Fig. 5 that when the cell viability decreases, no obvious changes in Raman shift and FWHM are observed. However, the variations in peak intensity can be observed clearly. Apparently, as the cell viability decreases, the intensities of peaks at lower Raman shift (<1000 cm^{-1}) increase while those of peaks at 1003 and 1342 cm^{-1} weaken. These results indicate that we can choose the peak intensity as a signature for detecting the cell viability transitions. As there are several peaks whose intensity change with the cell viability decrease, the Student's t-test is used for picking out the peaks with significant difference between the high viable and low viable cells, thus to find the molecular vibrations sensitive to the cell viability changes. The result of the t-test is described by the p value, which actually reflects a reliability of the difference. Generally, $p<0.05$ represents a significant difference, and the lower the p value is, the more reliable the difference will be. As listed in Table 2, there are about half of the peaks having significant differences in peak intensity as indicated by $p<0.05$. The intensity changes of these peaks with $p<0.05$ are illustrated in Fig.

6. When the cell viability decreases, the intensity of peaks corresponding to nucleic acids at 621, 643, 667, 744, 900, and 1010 cm^{-1} and to lipids at 719, 877, and 980 cm^{-1} increase, while those assigned to proteins at 1003, 1342, and 1659 cm^{-1} decrease. Especially, the most reliable spectral differences ($p<0.001$) between the hUC-MSCs with high and low viability can be found in the C=O out of plane bending in Thymine at 744 cm^{-1}, C-C symmetric stretching in lipids at 877 cm^{-1}, and C-H deformation in proteins at 1342 cm^{-1}. In comparison of the case of high viability, the peak intensity of 744 and 877 cm^{-1} are both about doubled while that of 1342 cm^{-1} reduces by about 30% in the spectra of those with low viability. Herein, we suggest that the viability of hUC-MSCs can be characterized by intensities of these three peaks.

As mentioned in this paper, the peak intensity could be a signature for indicating the cell viability transitions. And the correlation with the peak intensity and the cell viability has been proved by our studies. To understand it further, the possible causes of the peak intensity changes are discussed briefly here. As been shown in Table 1, the Raman spectra of hUC-MSCs are contributed from all major biological macromolecules including nucleic acids, proteins, lipids and carbohydrates. The intensity changes of the Raman peaks, especially for those highly sensitive to the cell viability transitions, can be understood from the interaction model between the intracellular reactive oxygen species (ROS, such as ·OH, O_2^-, and H_2O_2, which are the products of cell metabolism) and the biological macromolecules.

In healthy cells, ROS are typically generated at controlled rates. In contrast, the production of ROS will be elevated under stress conditions (e.g. cell apoptosis). This elevation will result in changes of many cellular components, such as DNA, proteins, and lipids, which will lead to further influence on the cell viability, metabolism, and growth. In addition to the findings shown in this paper, new research results from our lab on hUC-MSCs show that about half of the cells with low viability appear to be apoptotic (these results will be published elsewhere). As mentioned above, a large amount of ROS will stack in this situation. There were results suggesting that the excessive ROS will introduce attacks on intracellular biomolecules such as nucleic acids, proteins, and lipids [26]. So we believe that the peak intensity changes can be explained by the effect of excessive ROS, as follows. Because excessive ROS destroy the base-base stacking interactions, the hyperchromicity occurs resulting in the increasing intensity of Raman peaks, which correspond to nucleic acid bases at 744, 667, 643, and 621 cm^{-1}. Moreover, the alterations in spatial structures of liposomes will inevitably change the micro-environment and spatial arrangement of chemical bonds within the molecular chains, which are probably responsible for the intensity variations of the Raman peaks assigned to lipids at 980, 877, and 719 cm^{-1}. For proteins, the conformational changes in the secondary structures of proteins, such as the reduction of α-helix as indicated by the lower intensity at 1659 cm^{-1}, were observed with as the cell viability decreases. These may be attributed to the attack of ROS that can lead to impairing or even breaking of hydrogen bonds, disulfide bonds and carbon-sulfur bonds, which play important roles in maintaining the structures of proteins. The structure variations of proteins are most likely to cause the micro-environmental changes of chemical bonds such as C-H bonds, resulting in a change of their scattering cross sections. This could be a reason responsible for the peak intensity decrease of 1342 cm^{-1}. Furthermore, ROS is able to induce protein degradation, which may also be a cause for the intensity decline of the peaks at 1003, 1342, and 1659 cm^{-1}. It should be pointed out that given the diversity of cellular components and the complexity of molecular interactions, the above discussion is only preliminary and the further investigation is needed.

6. Conclusions

In this study, we have shown that the relative intensity of Raman peaks at 1342, 877, and 744 cm^{-1}, which correspond respectively to C-H deformation in proteins, C–C symmetric stretching in lipids, and C=O out-of-the-plane bending in Thymine, are highly sensitive to the changes of hUC-MSCs viability. The intensity decrease of the peak at 1342 cm^{-1} and the intensity enhancement of the peaks at 877 and 744 cm^{-1} in the cells with low viability are related to the alterations of the secondary structures of proteins, the degradation of proteins, the structural modification of lipids, and the destruction of base-base stacking interactions in DNA. These changes could be understood by the interactions between the excessive intracellular ROS and the biomolecules.

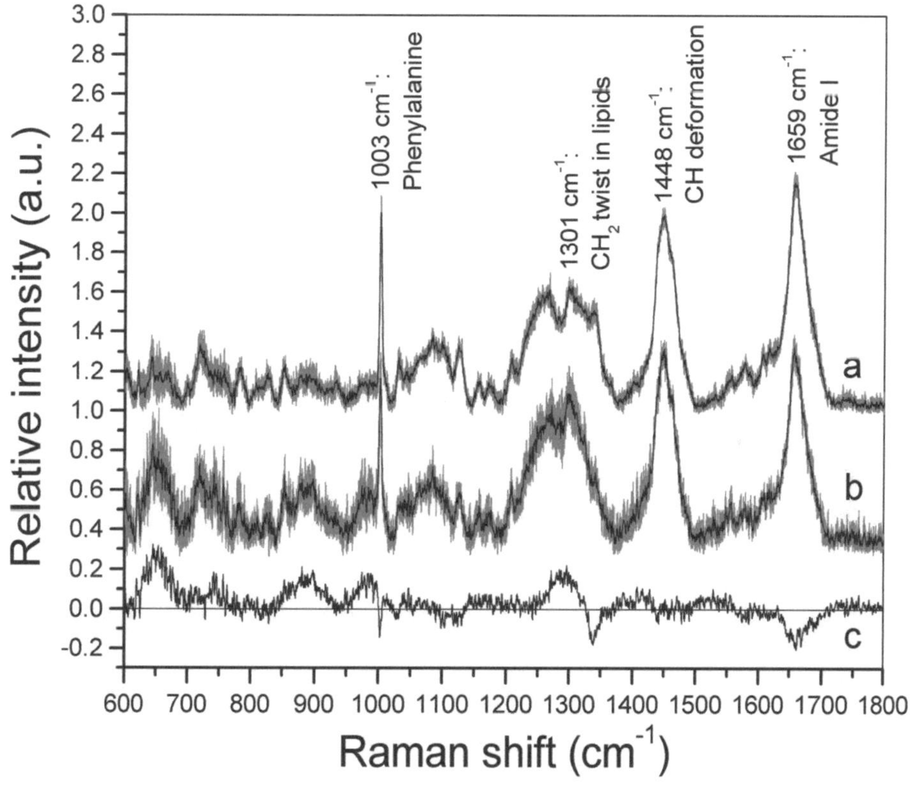

Fig. 4. Averaged Raman spectra from 22 hUC-MSCs with high viability (curve "a") and from 20 cells with low viability (curve "b"). Shaded areas represent the standard deviations. The spectra were normalized to the intensity at 1448 cm^{-1}. Difference spectrum (curve "c") was obtained by subtracting curve "a" from curve "b". Spectra are shifted vertically for clarity.

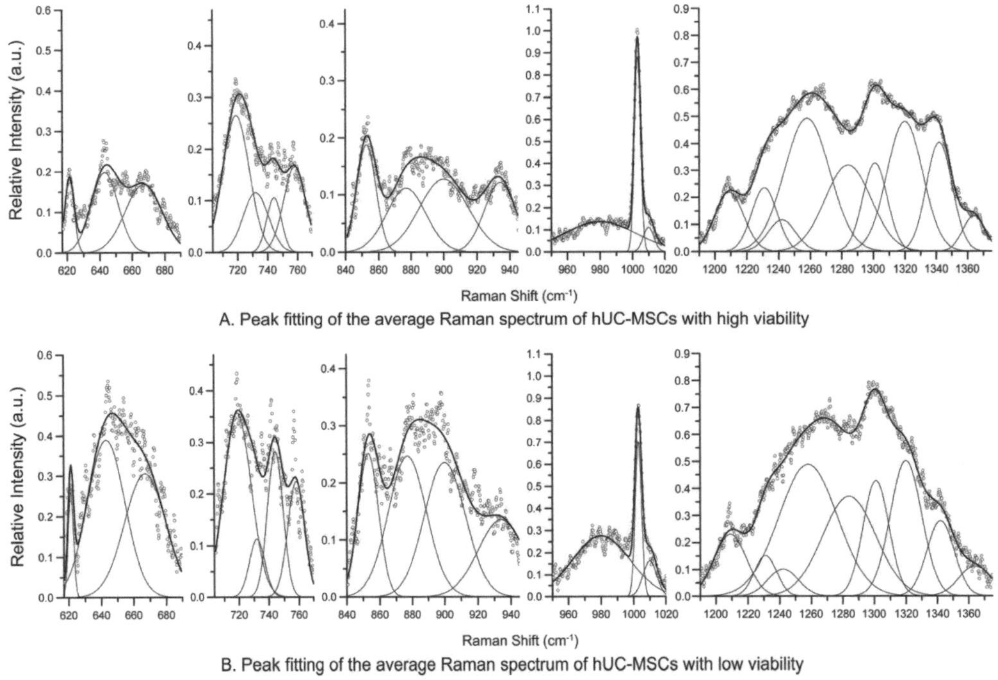

Fig. 5. Peak fitting of the averaged Raman spectra of hUC-MSCs with high viability (a) and low viability (b) in the spectral regions: 617 – 690, 703 – 770, 840 – 945, 950 – 1020, and 1190 – 1375 cm^{-1}. Empty circles – experimental data, thin lines – fitted peaks, and thick lines – fitted curves.

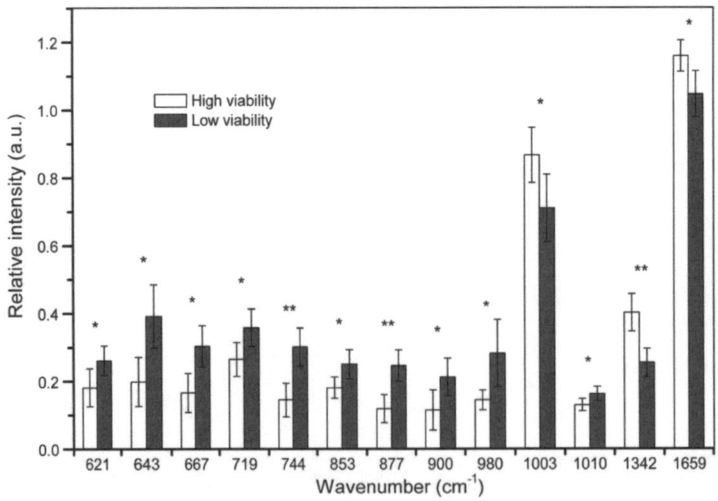

Fig. 6. Comparison of relative intensities of Raman peaks between hUC-MSCs with high and low viability. Block – averaged relative intensity, Bar – standard deviation. * means $p<0.05$, ** means $p<0.001$ obtained by the Student's t-test analysis.

Raman shift (cm^{-1})	Assignment [11,19,20,22-25]
621	A 5-member ring deformation
643	T ring angle bend
667	G ring breath
719	$CN^+(CH_3)_3$ str in lipids
732	A ring breath
744	T C=O out of plane bend
758	Trp ring breath
784	O-P-O str of DNA
811	O-P-O str of RNA
827	Tyr out of plane ring breath
853	Tyr ring breath
877	C-C symm str in lipids
900	deoxyribose
934	C-C bk str in proteins
980	=CH bend in lipids
1003	Phe symm ring breath
1010	C-O str in deoxyribose
1031	Phe
1064	Chain C-C str in lipids
1083	C-N str in proteins; Chain C-C str in lipids; C-O str in carbohydrates
1097	$O-P-O^-$ str
1127	C-N,C-C str in proteins
1157	C-C,C-N str in proteins
1174	Tyr,Phe
1209	$C-C_6H_5$ str of Phe
1231	Amide III random coil
1242	Amide III β-sheet
1258	Amide III β-sheet
1284	Amide III α-helix
1301	CH_2 twist in lipids
1320	G; CH def in proteins
1342	CH def in proteins
1364	CH_3 symm str in lipids
1448	CH def in DNA/RNA, proteins, lipids and carbohydrates
1659	Amide I α-helix

Abbreviations: asym, asymmetric; symm, symmetric; str, stretching; def, deformation; bk, backbone; A, Adenine; G, Guanine; C, Cytosine; T, Thymine; U, Uracil; Phe: phenylalanine; Tyr: tyrosine; Trp: tryptophan.

Table 1. Peak assignments for Raman spectra of hUC-MSCs

Region (cm^{-1})	Wavenumber (cm^{-1})	t-test p value
617-690	621	0.0215 *
	643	0.0026 *
	667	0.0024 *
703-770	719	0.0122 *
	732	0.2833
	744	0.0004 **
	758	0.1163
840-945	853	0.0101 *
	877	0.0005 **
	900	0.0145 *
	934	0.4063
950-1020	980	0.0173 *
	1003	0.0137 *
	1010	0.0166 *
1190-1375	1209	0.9462
	1231	0.1886
	1242	0.3941
	1258	0.2832
	1284	0.3884
	1301	0.0722
	1320	0.3496
	1342	0.0004 **
	1364	0.6455
	1659	0.0073 *

*: $p<0.05$ means significant differences; **: $p<0.001$ means very significant differences.

Table 2. p values obtained by the statistical intensity comparison of the peaks in the region 617-690, 703-770, 840-945, 950-1020, 1190-1375 cm^{-1}, and the peak at 1659 cm^{-1} between hUC-MSCs with high and low viability.

7. Relations between Reactive Oxygen Species and Raman Spectral Variations of Human Umbilical Cord Mesenchymal Stem Cells with Different Viability

Mesenchymal stem cells (MSCs) are particularly attractive in cell-based therapy because of their distinct immunosuppressive properties and multi-differentiation capacity [27,28]. MSCs are traditionally found in the bone marrow. However, they can also be isolated from other

tissues such as human umbilical cord. Human umbilical cord-derived MSCs (hUC-MSCs) have shown similar or even superior properties compared to bone marrow derived MSCs (BM-MSCs) including plastic adherence, specific surface markers, regenerative capability, collection convenience, safety, and potential of in vitro differential in certain lineages [29,30]. As a result of these advantages, hUC-MSCs have gained more attention in recent years.
Currently, hUC-MSCs are being explored as a promising candidate for therapeutic applications. Wu and Zhou [31,32] discussed the therapeutic potential of hUC-MSCs in cardiomyocyte regeneration, and their findings demonstrated that transplanted hUC-MSCs provide benefit in the restoration of cardiac function, suggesting that hUC-MSCs represent a source of stem cells for cell therapy and myocardial tissue engineering. Liao et al. reported that hUC-MSCs treatment could be an ideal strategy for cell-based therapy for central nervous system injury and disease [7]. Moreover, the impact of hUC-MSCs on liver fibrosis was also observed by Han and co-workers [33]. Recently, a series of experimental results indicate that hUC-MSCs are exceptional candidates for muscle repair [30], bone tissue engineering [34] and the treatment of rheumatoid arthritis (RA) [35].
The above data indicate that the hUC-MSCs hold great promise as therapeutic agents and as a tool for understanding the development of regeneration medicine. It is known that a large population of cells with high viability is fundamental to a successful cell-based therapy. Therefore, the detection of cell viability changes represents a key step for the population expansion and quality assurance of hUC-MSCs. As a label-free method, Raman spectroscopy is increasingly used in the study of stem cells [10, 19-21]. In our recently work [36], this technology was used to detect the viability transitions of hUC-MSCs. An encouraging finding was that the most significant Raman spectral variations associated with the cell viability decrease were at 1342, 877, and 744 cm^{-1}, which implied a great potential for characterizing the cell viability changes. Considering the biochemical effects involved in the reactions of reactive oxygen species (ROS) [26], we concluded that these Raman spectral changes are probably due to the interaction between the intracellular ROS and biological macromolecules. However, given the diversity of cellular components and the complexity of molecular interactions, the reasons and the mechanisms of the Raman spectral variations remain unclear. In this context, several studies in our group are being carried out with the purpose of providing insights into the role of ROS involved in the cell viability transitions.
As an important aspect, the dependence of the Raman spectral variations on the yield of ROS in the cells has been studied in this paper. Here, we employ a ROS probe (DCFH-DA) to monitor the change of intracellular ROS level in hUC-MSCs as their viability declines. In addition, the Raman spectra of the cells with different viability are also recorded by Raman micro-spectroscopy. The results show that as the cell viability declines, the relative intensity variations of Raman peaks at 1342, 877, and 744 cm^{-1}, and the relative yield variation of ROS in cells can be described by a same model(with different fitting parameters). Moreover, the correlations of these relative changes were evaluated by correlation coefficient method. It is found that they have perfect linear correlation coefficients respectively. This study further demonstrates that the relative yield variation of the ROS in hUC-MSCs with different viability can directly cause the relative intensity variation of Raman spectra in the cells.

7.1 Intracellular ROS measurement

The relative levels of ROS in hUC-MSCs with different viability were accessed using 2',7'-dichlorofluorescin diacetate (DCFH-DA, Sigma), a fluorescent probe for intracellular ROS measurement. The DCFH-DA itself is non-fluorescent. It can be easily diffused through the cell membrane and be hydrolyzed by intracellular esterases to produce non-fluorescent 2',7'-dichlorofluorescin (DCFH), which can no longer pass through the membrane so that the probe can easily be loaded into the cells. DCFH is then oxidized by ROS to generate fluorescent 2',7'-dichlorofluorescein (DCF), whose fluorescence intensity is proportional to the amount of intracellular ROS [37, 38]. The cells were suspended in PBS at a density of 5.5 $\times 10^5$ cells/mL. After DCFH-DA was suspended at a final concentration of 20μmol/L, the cells were incubated for 40 min at 37°C for probe loading. -DCF fluorescence was then measured by a spectrometer (SpectraPro 300i, Acton, USA) equipped with a CCD detector (Spec-10, PI, USA). Ar+ laser (488 nm) was used for excitation, and the fluorescence was recorded in the range of 510–600 nm. In addition, same measurements of the DCFH-DA and PBS mixture (final concentration of the probe: 20 μmol/L) without cells for fluorescent background acquisition were also performed.

7.2 Experimental results

ROS consist of a series of oxygen free radicals and their derivatives, such as O^{2-}, H_2O_2, and •OH, possess active chemical properties and high oxidation capability. They are generated from the sequential univalent reduction of oxygen involved in the cell metabolism [39, 40]. Since the cell itself has a highly efficient antioxidant system for eliminating ROS, the production and scavenging of ROS maintain a dynamic equilibrium under normal conditions. An appropriate amount of ROS is associated with many important physiological functions, such as biological messengers involved in cellular signal transduction [41, 42]. However, the production of intracellular ROS will greatly exceed the cellular antioxidant capacity under oxidative stress [43]. And ROS can act as the toxicants that easily react with biomolecules to damage their structures and functions and change their physical and chemical properties as Raman scattering [43–45].

Thus, to explore whether ROS play a role in the Raman spectral variations associated with the hUC-MSCs' viability transitions, we first monitored the change of intracellular ROS level as the cell viability decreased. As shown in Fig. 7, under the condition of cell starvation, the cell viability exhibits a linear decline with time. Therefore, the changes of the ROS level and the Raman spectra in cells as the cell viability decreases can be measured and discussed in the time dimension. By using ROS probe (DCFH-DA), the intracellular ROS levels were observed by the fluorescence spectra of the probe at different times. The results are shown in Fig. 8. The fluorescence spectra are from DCF, which is produced by the reactions of the probes with ROS. No fluorescence signals can be detected in the mixture solution of the probe molecules and PBS without the cells ("Background" in Fig. 8A), indicating that the fluorescence is originated only from the interaction of the probes with the intracellular ROS but not self-oxidation of the probes. Therefore, the relative intensities (F) of the fluorescence in Fig. 8A will be directly proportional to the ROS level (ROS) in cells [37, 38] under the experimental conditions we designed. The relationship can be expressed as follows

$$F = \varepsilon ROS \quad (1)$$

where ε is the proportional constant in this case. Fig. 8B shows the column diagram of the relative fluorescence intensity at different times based on Fig. 8A. It is apparent that as the viability decreases, the level of ROS increases initially. It reaches a maximum value about 1.5 times higher than the initial time (t = 0) as the viability drops to about 36%. As the viability continues to decrease, the ROS level experiences a decline.

As mentioned above, a low ROS level under normal conditions is beneficial to the cells. Few oxidative damages by ROS should be taken into consideration in this case. However, when the production of ROS exceeds the cellular antioxidant capacity, oxidative damages will occur [39]. In other words, the excessive intracellular ROS lead to the oxidative damages. Considering this, we calculate the relative variation of ROS level (ΔROS(t)) in the cells at each measurement time t based on the data in Fig. 8B, and take the first relative intensity at initial time (t = 0) as a normal value in the cells with normal viability. Therefore, the ΔROS(t) (t is the measurement time in Fig. 8B) can be calculated by the following expression.

$$\Delta ROS(t) = \frac{F(t) - F(0)}{F(0)} \times 100\% \tag{2}$$

where F(t) and F(0) represent the relative fluorescence intensity at time t and the initial time (t = 0), respectively. The change of ΔROS with time is shown by the black squares in Fig. 9. Based on the data shown in Fig. 9, we seek to find a mathematical model to describe their change with time. It is found that a rational function (shown in Eq. (3)) can make a good fitting with the fitting degree $R^2 = 0.983$. Thus, the ΔROS(t) can generally be expressed by the rational function as follows.

$$\Delta ROS(t) = \frac{a + bt}{1 + ct + dt^2} \tag{3}$$

The data in Fig. 9 show that the ROS level is always higher than the initial (t = 0) in our observation period, implying that an accumulation of the excessive ROS in cells may occur as the cell viability decreases.

It should be noted that according to our experimental methods, the DCF fluorescence intensity is proportional to the amount of ROS generated in a short time period (40 min), which can be considered as a time point when compared to the entire experimental period (7 days). Thus, the data presented in Fig. 9 can only show the variation of "transient" ROS level, but not the yield of ROS during the cell viability decline. Since the oxidative damages resulting from excessive ROS can be irreversible [42], they thus rely on the cumulative effect of excessive ROS reacting with biomolecules. This means that the oxidative damages depend in part on the yield variation of ROS as the cell viability decreases. Having this in mind, we estimate the relative yield variation of ROS by integrating Eq. (3).

$$\Delta ROS_{yield}(t_c) = \int_0^{t_c} \frac{a + bt}{1 + ct + dt^2} dt \tag{4}$$

And thus

$$= \frac{2ad - bc}{d\sqrt{4d - c^2}} \arctan\left(\frac{2dt + c}{\sqrt{4d - c^2}}\right) + \frac{b}{2d} \ln\left(t^2 + \frac{c}{d}t + \frac{1}{d}\right) \tag{5}$$

Calling the values of the parameters (a, b, c, d) from the best fitting result in Fig. 9 and considering the integral upper and lower limits, we can estimate the relative yield variation of ROS at any time point. The estimated $\Delta ROS_{yield}(t)$ at t = 0d, 1d, 2d, …, 6d is shown in Fig. 10, which exhibits a nonlinear increase with time. In particular, it is found that the change of $\Delta ROS_{yield}(t)$ can be exactly described by the logistic model (Eq. (6)) as indicated by the fitting degree R^2 = 0.999. This model describes an S-type growth, i.e., the growth gradually slows down and eventually stops when the time goes to infinity.

$$f(t) = \frac{\alpha}{1+\beta e^{-\gamma t}} \tag{6}$$

The above results show that the relative yield variation of intracellular ROS increases continuously with the cell viability drop, suggesting that excessive ROS accumulate in cells and thus may attack the biomolecules as expected.

The excessive ROS are capable of attacking biomolecules, such as proteins, nucleic acids, and lipids, by a variety of reaction mechanisms including hydrogen abstraction, addition, etc. [40, 46, 47]. For example, ROS can react directly with the amino acid residues, causing the inactivation and denaturation of proteins [46]. Their attacks on DNA can also lead to the DNA bases modification and breaking of chemical bonds and certain strands [39, 47]. In addition, ROS induce the peroxidation of unsaturated fatty acids and cause oxidative damage to lipids [39, 40]. This reaction is also an important mechanism that occurs with the oxidative damage to other biomolecules, such as malondialdehyde, an intermediate of peroxidation, that can cause the cross-linking of proteins [40, 46]. The oxidative damages from these reactions will undoubtedly lead to the structural and functional alterations of the biomolecules, and thus may cause changes in their Raman scattering.

Here we will pay attention to the dependence of the Raman spectral changes on ROS, but not the types of oxidative reactions that lead to the changes. That is, the relationship between Raman spectral variations and the relative yield variation of ROS in the cells need to be examined further in this paper. Raman spectra of hUC-MSCs with different viability (or at different time points) are shown in Fig. 11A. Raman spectral variations as the cell viability declines can be observed clearly. Thus we choose three vibration peaks that have been proven to be sensitive to the hUC-MSCs viability transitions [35]. They are protein C-H deformation at 1342 cm^{-1}, lipid C–C symmetric stretching at 877 cm^{-1}, and thymine C=O out-of-the-plane bending at 744 cm^{-1}. Their relative intensity changes as a function of time are shown in Fig. 11B. It can be seen that the relative intensity of 1342 cm^{-1} decreases with time, while those of 877 and 744 cm^{-1} exhibit the reverse orientation. Further, we will focus on the variation of the relative intensity, because any deviations of these characteristic peak intensities from their normal values will indicate the drop of cell viability [35].

Based on the data shown in Fig. 11b and the above method used to discuss ROS (Eq. (2)), the relative peak intensity variations ($\Delta I_w(t)$) of the three peaks can be calculated by the following expression.

$$\Delta I_w(t) = \frac{|I_w(t) - I_w(0)|}{I_w(0)} \times 100\% \tag{7}$$

where $I_w(t)$ and $I_w(0)$ represent the relative intensity of the peak w at the time t and the initial time (t = 0), respectively. The calculated results are shown in Fig. 12.

To investigate the change tracks of $\Delta I_{1342}(t)$, $\Delta I_{877}(t)$, $\Delta I_{744}(t)$ and their relationship to $\Delta ROS_{yield}(t)$, we also try to use the logistic model (Eq. (6)) to fit the data points in Fig.12. As presented in Fig. 12, it is interesting to find that the logistic model is also suitable for describing the changes of $\Delta I_{1342}(t)$, $\Delta I_{877}(t)$, and $\Delta I_{744}(t)$ indicated by the high fitting degrees of 0.993, 0.986, and 0.996, respectively. However, the parameters (α, β, γ) obtained by fitting

$\Delta I_{1342}(t)$, $\Delta I_{877}(t)$, and $\Delta I_{744}(t)$ differ from each other, and also show large differences from those of fitting $\Delta ROS_{yield}(t)$. This seems reasonable and understandable from the complexity and diversity of ROS reactions. There are many mechanisms showing that ROS interact with biomolecules directly or indirectly [40, 46, 47]. The attack of ROS seems impossible to cause the synchronous and homogeneous alterations in all the biomolecules. The differences in the molecular response (rate, scale, sensitivity) to the ROS attack and the types of reactions will inevitably affect the reaction rates (of the reactant depletion and the product formation) in the cells. And also, the formation of the excessive ROS themselves will lead to the component changes in the cells. All complications make the fitting parameters or the fitting curves of $\Delta I_w(t)$ of some individual Raman peaks w show larger deviation from those of $\Delta ROS_{yield}(t)$. That is, it is almost impossible for all $\Delta I_w(t)$ to correlate with $\Delta ROS_{yield}(t)$ in exact linearity. Furthermore, it should be mentioned that the parameter α is primarily determined by the maximum of the relative intensity (or yield) variation and is basically independent on the fitting curves that is defined by the shape parameters β and γ. The differences between $\Delta I_{744}(t)$ and $\Delta ROS_{yield}(t)$ in the two shape parameters are the largest (Fig. 12C) in all the $\Delta I_w(t)$, thus the linear correlation between their fitting curves is the lowest, suggesting that the reaction mechanism relating to this peak increase would be very complicated. In fact, the ROS in the cells are consumed through many channels. Therefore, we may have a better result if using the sum of the relative intensity variations of all characteristic Raman peaks (here we have three peaks) measured at the same time. The relative intensity total variation as a function of time ($\Delta I_{total}(t)$) is expressed as follows.

$$\Delta I_{total}(t) = \sum_{w} \Delta I_w(t) \tag{8}$$

where $\Delta I_w(t)$ is the relative intensity variation of the peak w at time t. Here we have w = 1342, 877, and 744 cm^{-1}. The calculated results are shown in Fig. 12d, and the fitting curve (with Eq. (6)) shows a perfect fitting degree ($R^2 = 0.999$). All the fitting constants for $\Delta I_{total}(t)$ are much better than those for any other $\Delta I_w(t)$ compared with those for $\Delta ROS_{yield}(t)$. Thus, the fitting curves of $\Delta I_{total}(t)$ and $\Delta ROS_{yield}(t)$ are basically similar, and the linear relationship between them should be better than others.

To further quantitatively evaluate the relationship, the correlation coefficient r is employed to describe the correlations of $\Delta I_{1342}(t)$, $\Delta I_{877}(t)$, $\Delta I_{744}(t)$, and $\Delta I_{total}(t)$ with $\Delta ROS_{yield}(t)$, respectively. The r is calculated as follows [48].

$$r = \frac{\sum_{i=1}^{n}(X_i - \overline{X})(Y_i - \overline{Y})}{\sqrt{\sum_{i=1}^{n}(X_i - \overline{X})^2 \times \sum_{i=1}^{n}(Y_i - \overline{Y})^2}} \tag{9}$$

where, X_i, Y_i (i = 1, 2, ..., n) express independent and dependent variables respectively; r describes the linear correlation of the two variables and takes the value in $-1 \le r \le 1$. The larger $|r|$ is, the higher the correlation will be. The calculation results show that the correlation coefficients for correlated $\Delta I_{1342}(t)$, $\Delta I_{877}(t)$, $\Delta I_{744}(t)$, and $\Delta I_{total}(t)$ to $\Delta ROS_{yield}(t)$ are 0.980, 0.985, 0.973, and 0.987, respectively. For clarity, these correlations are shown in Fig. 13. The high $|r_w|$ values provide a further evidence to confirm that the changes of the three relative peak intensities linearly correlate to the relative yield variation of ROS, especially at the beginning of the cell viability decrease. Taken together, the results demonstrate that the oxidative damages induced by excessive intracellular ROS lead to the Raman spectral variations associated with the cell viability decline. In addition, the $\Delta I_{total}(t)$ shows the best linear relationship to the $\Delta ROS_{yield}(t)$, while that of individual Raman peak intensity appears lower linear correlation coefficient, indicating that the corresponding biochemical reactions of intracellular ROS have much more complicate mechanisms and needs further study.

8. Conclusions

Herein, we reveal that the relative yield variation of ROS in hUC-MSCs with different viability can directly lead to the relative intensity variations of the three characteristic Raman peaks at 1342, 877, and 744 cm^{-1} in the cells, according to their perfect linear correlation coefficients of 0.980, 0.985, and 0.973, respectively. Our findings demonstrate that the excessive intracellular ROS can result in the almost linear variation of the relative Raman peak intensity in hUC-MSCs as the cell viability declines. These results would be very useful for understanding the mechanisms of Raman spectral changes in the cells.

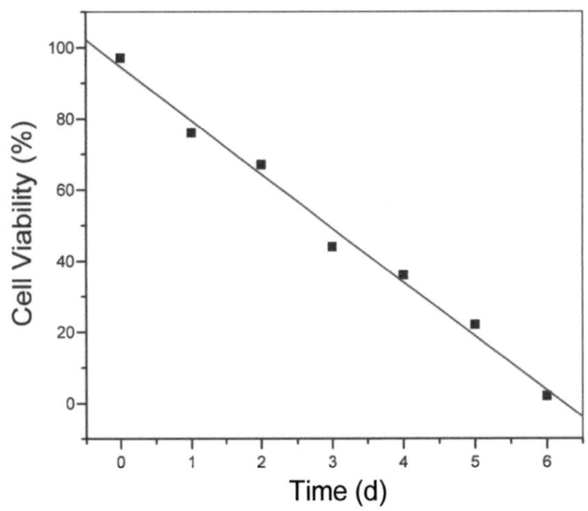

Fig. 7. Change of cell viability as a function of time (cells are placed under starvation condition). The line represents the linear fitting.

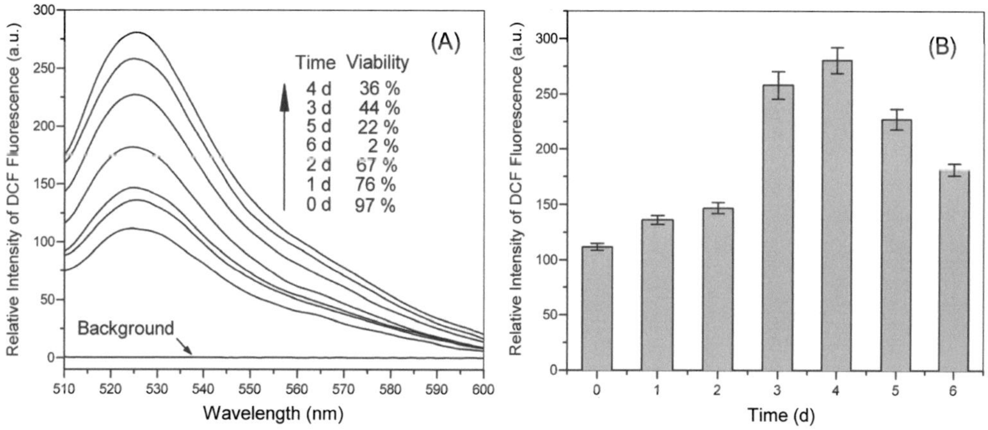

Fig. 8. Change of intracellular ROS level with time. (A) Fluorescence spectra of DCF at different times corresponding to different cell viability. (B) Histogram showing averaged relative intensities of DCF fluorescence and standard deviation.

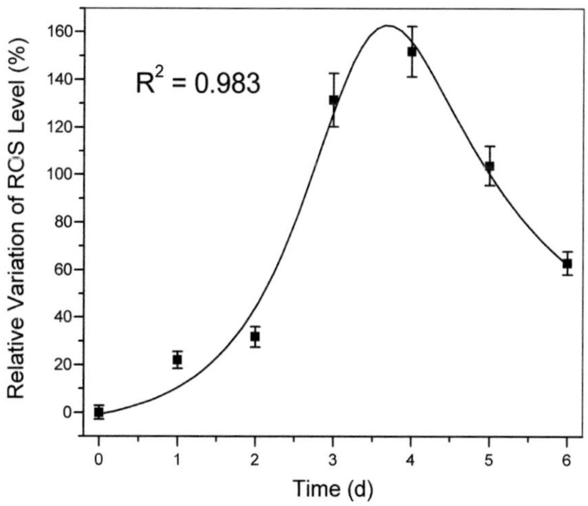

Fig. 9. Relative variation of ROS level as a function of time. The data points are from Eq. (2). The line represents the best fitting using a rational function (Eq. (3)), with the fitting degree $R^2 = 0.983$ and the parameters $a = -0.80747$, $b = 6.77836$, $c = -0.50330$, $d = 0.07387$.

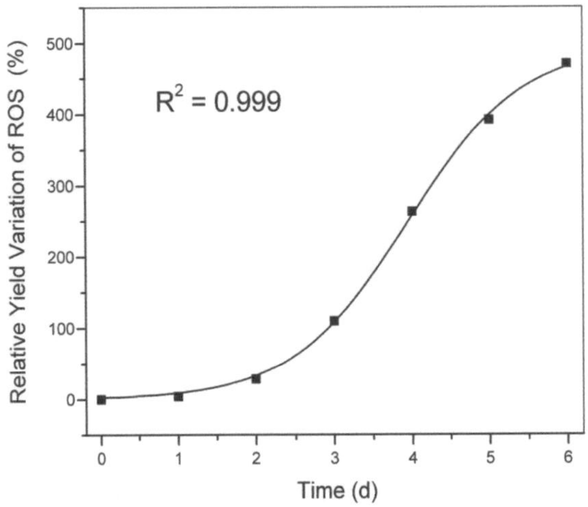

Fig. 10. Change of relative yield variation of ROS as a function of time. The line represents the best fitting using logistic model (Eq. (6)), with the fitting degree $R^2 = 0.999$ and the parameters $\alpha = 494.247$, $\beta = 208.473$, $\gamma = 1.35624$.

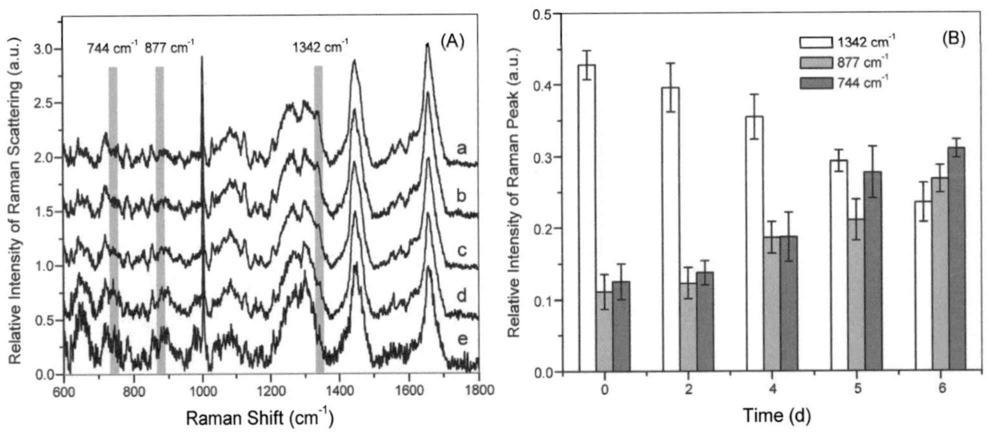

Fig. 11. (A) Raman spectra of hUC-MSCs with different cell viability: (a) cell viability 97%, $t = 0$ (b) 67%, $t = 2d$ (c) 36%, $t = 4d$ (d) 22%, $t = 5d$ (e) 2%, $t = 6d$. All spectra are normalized to the peak intensity at 1448 cm^{-1}. Spectra are shifted vertically for clarity. (B) Changes of the relative peak intensities at 1342, 877, and 744 cm^{-1} as a function of time.

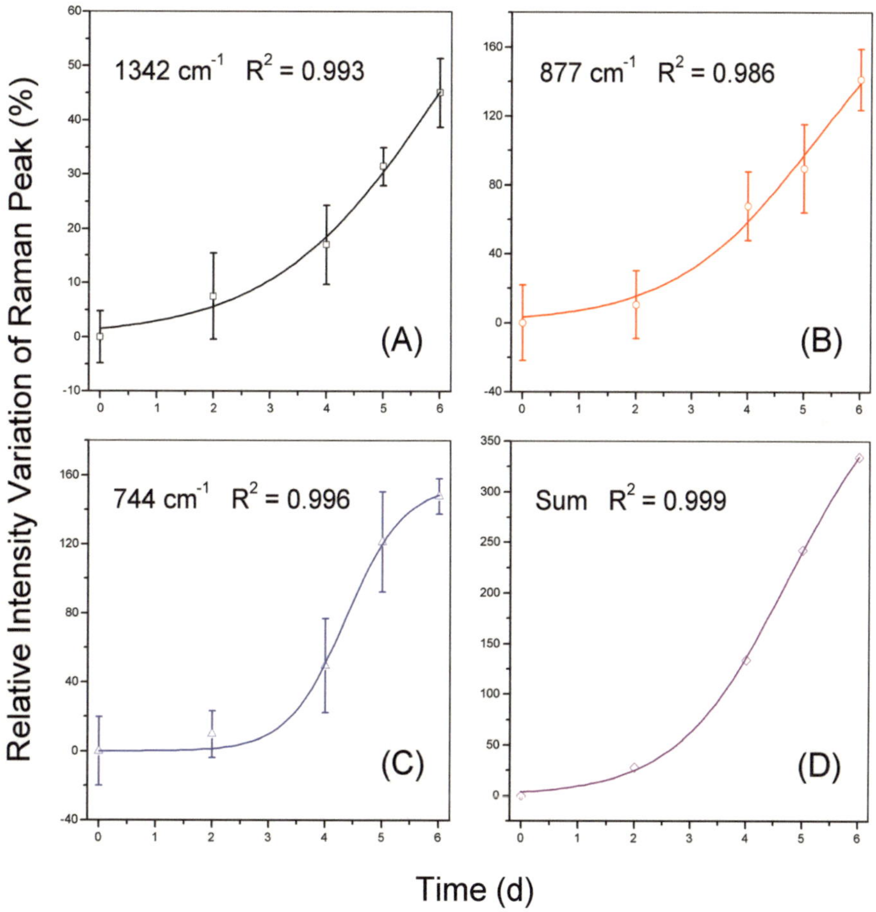

Fig. 12. Relative changes of Raman peak intensity variations at (A) 1342, (B) 877, (C) 744 cm^{-1}, and their sum (D) as a function of time. The lines represent the best fitting using logistic model (Eq. (6)). (A) Fitting degree $R^2 = 0.993$, the parameters $\alpha = 91.3853$, $\beta = 58.5466$, $\gamma = 0.67544$. (B) $R^2 = 0.986$, $\alpha = 216.993$, $\beta = 62.1816$, $\gamma = 0.78485$. (C) $R^2 = 0.996$, $\alpha = 154.484$, $\beta = 5191.54$, $\gamma = 1.95969$. (D) $R^2 = 0.999$, $\alpha = 430.957$, $\beta = 122.328$, $\gamma = 1.00860$.

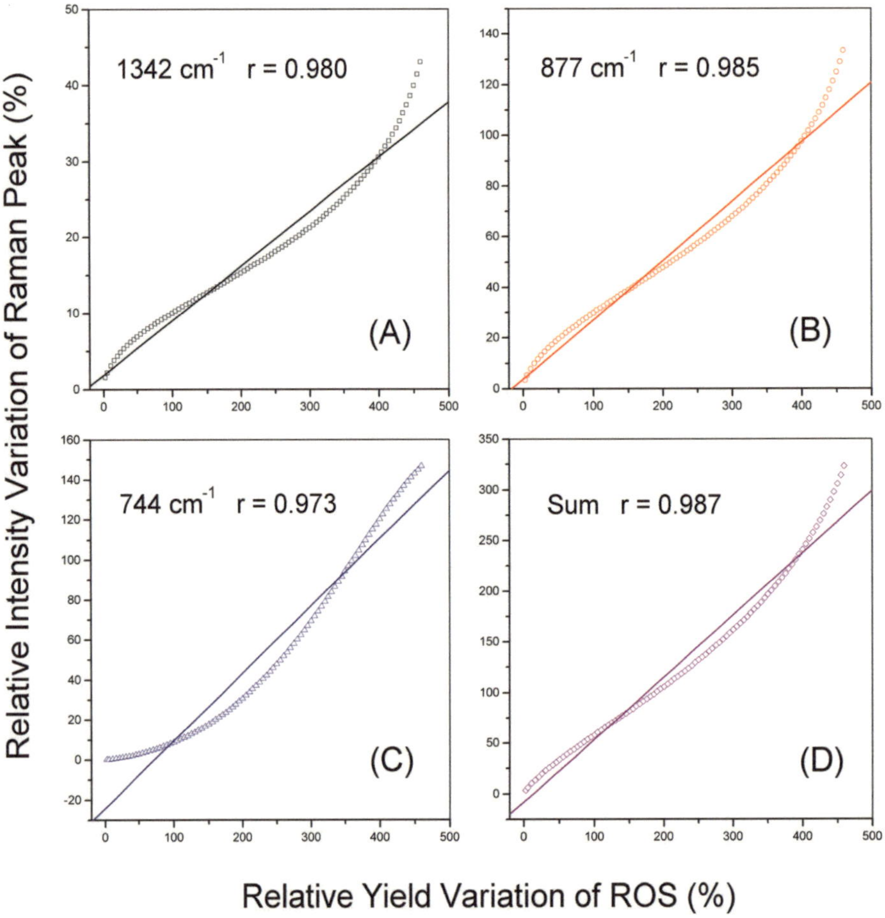

Fig. 13. Correlations between the relative yield variation of ROS and the relative intensity variations of Raman peaks at (A) 1342, (B) 877, (C) 744 cm^{-1} and their sum (D) $\Delta I_{total}(t)$. The lines represent the linear fittings for the changes of $\Delta I_{1342}(t)$, $\Delta I_{877}(t)$, $\Delta I_{744}(t)$, and $\Delta I_{total}(t)$ as a function of $\Delta ROS_{yield}(t)$, with the correlation coefficients of 0.980, 0.985, 0.973, and 0.987, respectively.

9. Summary

Mesenchymal stem cells (MSCs) are pluripotent progenitor cells derived from the mesoderm. They are particularly suitable for cell-based therapy because of their different immunosuppressive properties and multi-differentiation capacity. The main source of MSCs is bone marrow. In addition, recent progress revealed that human umbilical cord derived mesenchymal stem cells (hUC-MSCs) could be an alternative source ideal for MSCs. Comparing to bone marrow-derived MSCs (BM-MSCs), hUC-MSCs have exhibited similar or even superior properties such as plastic adherence, specific surface markers, regenerative capability, collection convenience, safety, and potential for in vitro differential in certain lineages. These advantages make hUC-MSCs a promising candidate for therapeutic applications in recent years.

As is known, a large population of cells in sustained high profitability is fundamental to successful cell-based therapy. Therefore, monitoring the variation in cell viability is an important step for the population expansion and quality assurance of hUC-MSCs. Right now there are a number of methods including biological assays and optical techniques available to handle this task. Among these techniques, Raman spectroscopy is a label-free method that allows for continued analysis. Moreover, it is proven that Raman spectroscopy has the potential to offer new and unique optical markers for cytological analysis. Currently this technology is increasingly used in the studies of stem cells.

In this work, Raman micro-spectroscopy is used to monitor the variation of hUC-MSCs viability. The Raman spectra of hUC-MSCs with different cellular states are investigated. Using peak fitting and statistical t-test, the Raman peaks are extracted with obvious differences between cells of different viability. Most significant changes ($p < 0.001$) have been found for the out-of-plane bending of C=O in thymine at 744cm^{-1}, symmetric stretching of C-C in lipids at 877cm^{-1}, and deformation of C-H in proteins at 1342cm^{-1}. As the cell viability decreases, the intensity of the peak at 1342cm^{-1} is reduced, whereas those of the peaks at 877 and 744 cm^{-1} are enhanced. Based on these results, we propose that the viability of hUC-MSCs can be characterized by these three peaks.

Given the biochemical effects involved in the reactions of reactive oxygen species (ROS), we conclude that the changes of Raman peak intensity could be explained by the interactions between excessive intracellular ROS and biological macromolecules. However, considering the diversity of cellular components and the complexity of molecular interactions, the exact reasons and mechanisms remain unclear. In this context, we further investigate the dependence of the Raman spectral variations on the yield of ROS in the cells for understanding the mechanism. Here we employ a ROS probe (DCFH-DA) to monitor the change of intracellular ROS level in hUC-MSCs as their viability declines. The results we obtained show that as the cell viability decreases, the relative intensity variations of the Raman peaks at 1342 cm^{-1}, 877 cm^{-1}, and 744 cm^{-1}, and the relative yield variations of ROS in cells can be described by the same model function (besides different fitting parameters used). In addition, the correlations of these relative changes were assessed by correlation coefficient method. Perfect linear correlation coefficients of 0.969, 0.982, and 0.975 were found. Our results further demonstrate that the relative yield variations of ROS in hUC-MSCs with different viability directly cause the relative intensity variations of Raman spectra in the cells. The results would be very useful for understanding the mechanism of the Raman spectral changes in the cells and the investigation of the basis for the application of Raman scattering in the measurement of stem cell viability.

10. Acknowledgements

We would like to thank Prof. Y. Zhao and Mrs. L.W. Duan for assistance in Raman spectral measurements. This work is supported by Tianjin Municipal Science and Technology Commission (No. 08ZCDFGX09400), and National Natural Science Fund Committee of China (No. 60508004 and No. 60778043).

11. References

[1] Motaln H, C Schichor and TT Lah. Human mesenchymal stem cells and their use in cell-based therapies. Cancer 116:2519–2530 (2010)

[2] Mitchell KE, ML Weiss, BM Mitchell, P Martin, D Davis, L Morales, B Helwig, M Beerenstrauch, K Abou-Easa, T Hildreth, D Troyer and S Medicetty. Matrix cells from Wharton's jelly form neurons and glia. Stem Cells 21:50–60 (2003).
[3] Wang HS, SC Hung, ST Peng, CC Huang, HM Wei, YJ Guo, YS Fu, MC Lai and CC Chen. Mesenchymal stem cells in the Wharton's jelly of the human umbilical cord. Stem Cells 22:1330–1337 (2004).
[4] Sarugaser R, D Lickorish, D Baksh, MM Hosseini and JE Davies. Human umbilical cord perivascular (HUCPV) cells: A source of mesenchymal progenitors. Stem Cells 23:220–229 (2005).
[5] Baksh D, R Yao and RS Tuan. Comparison of proliferative and multilineage differentiation potential of human mesenchymal stem cells derived from umbilical cord and bone marrow. Stem Cells 25:1384–1392 (2007).
[6] Li G, XA Zhang, H Wang, X Wang, CL Meng, CY Chan, DTW Yew, KS Tsang, K Li, SN Tsai, SM Ngai, ZC Han, MC Lin, ML He and HF Kung. Comparative proteomic analysis of mesenchymal stem cells derived from human bone marrow, umbilical cord, and placenta: Implication in the migration. Proteomics 9:20–30 (2009).
[7] Liao WB, J Zhong, JX Yu, J Xie, YJ Liu, L Du, SG Yang, PX Liu, J Xu, JM Wang, ZB Han and ZC Han. Therapeutic benefit of human umbilical cord derived mesenchymal stromal cells in intracerebral hemorrhage rat: Implications of anti-inflammation and angiogenesis. Cell Physiol Biochem 24:307–316 (2009).
[8] Wu KH, SG Yang, B Zhou, WT Du, DS Gu, PX Liu, WB Liao, ZC Han and YL Liu. Human umbilical cord derived stem cells for the injured heart. Med Hypotheses 68:94–97 (2007).
[9] Lu LL, YJ Liu, SG Yang, QJ Zhao, X Wang, W Gong, ZB Han, ZS Xu, YX Lu, D Liu, ZZ Chen and ZC Han. Isolation and characterization of human umbilical cord mesenchymal stem cells with hematopoiesis-supportive function and other potentials. Haematologica 91:1017–1026 (2006).
[10] Chan JW and DK Lieu. Label-free biochemical characterization of stem cells using vibrational spectroscopy. J Biophotonics 2:656–668 (2009).
[11] Notingher I, S Verrier, S Haque, JM Polak and LL Hench. Spectroscopic study of human lung epithelial cells (A549) in culture: Living cells versus dead cells. Biopolymers 72:230–240 (2003).
[12] Notingher I, G Jell, U Lohbauer, V Salih and LL Hench. In situ non-invasive spectral discrimination between bone cell phenotypes used in tissue engineering. J Cell Biochem 92:1180–1192 (2004).
[13] Verrier S, I Notingher, JM Polak and LL Hench. In situ monitoring of cell death using Raman microspectroscopy. Biopolymers 74:157–162 (2004).
[14] Short KW, S Carpenter, JP Freyer and JR Mourant. Raman spectroscopy detects biochemical changes due to proliferation in mammalian cell cultures. Biophys J 88:4274–4288 (2005).
[15] Owen CA, J Selvakumaran, I Notingher, G Jell, LL Hench and MM Stevens. In vitro toxicology evaluation of pharmaceuticals using Raman micro-spectroscopy. J Cell Biochem 99:178–186 (2006).
[16] Swain RJ and MM Stevens. Raman microspectroscopy for non-invasive biochemical analysis of single cells. Biochemical Society Transactions 35:544–549 (2007).
[17] Swain RJ, G Jell and MM Stevens. Non-invasive analysis of cell cycle dynamics in single living cells with Raman micro-spectroscopy. J Cell Biochem 104:1427–1438 (2008).
[18] Chan J, S Fore, S Wachsmann-Hogiu and T Huser. Raman spectroscopy and microscopy of individual cells and cellular components. Laser & Photon Rev 2:325–349 (2008).

[19] Notingher I, I Bisson, JM Polak and LL Hench. In situ spectroscopic study of nucleic acids in differentiating embryonic stem cells. Vib Spectrosc 35:199–203 (2004).
[20] Chan JW, DK Lieu, T Huser and RA Li. Label-free separation of human embryonic stem cells and their cardiac derivatives using Raman spectroscopy. Anal Chem 81:1324–1331 (2009).
[21] Chiang HK, FY Peng, SC Hung and YC Feng. In situ Raman spectroscopic monitoring of hydroxyapatite as human mesenchymal stem cells differentiate into osteoblasts. J Raman Spectrosc 40:546–549 (2009).
[22] Azrad E, D Zahor, R Vago, Z Nevo, R Doron, D Robinson, LA Gheber, S Rosenwaks and I Bar. Probing the effect of an extract of elk velvet antler powder on mesenchymal stem cells using Raman microspectroscopy: Enhanced differentiation toward osteogenic fate. J Raman Spectrosc 37:480–486(2006).
[23] Mohamed TA, IA Shabaan, WM Zoghaib, J Husband, RS Farag and Abd El-Nasser MA Alajhaz. Tautomerism, normal coordinate analysis, vibrational assignments, calculated IR, Raman and NMR spectra of adenine. J Mol Struct 938:263–276 (2009).
[24] Singh JS. FTIR and Raman spectra and fundamental frequencies of biomolecule: 5-Methyluracil (thymine). J Mol Struct 876:127–133 (2008).
[25] Nishimura Y, M Tsuboi, T Nakano, S Higuchi, T Sato, T Shida, S Uesugi, E Ohtsuka and M Ikehara. Raman diagnosis of nucleic acid structure: Sugar-puckering and glycosidic conformation in the guanosine moiety. Nucleic Acids Res 11:1579–1588 (1983).
[26] Beckman KB and BN Ames. The free radical theory of aging matures. Physiol Rev 78:547–581 (1998).
[27] A. J. Nauta and W. E. Fibbe, Immunomodulatory properties of mesenchymal stromal cells, Blood 110, 3499-3506 (2007).
[28] A. Uccelli, L. Moretta, and V. Pistoia, Immunoregulatory function of mesenchymal stem cells, Eur. J. Immunol. 36, 2566-2573 (2006).
[29] C. Knocaefe, D. Balci, B. B. Hayta, and A. Can, Reprogramming of human umbilical cord stromal mesenchymal stem cells for myogenic differentiation and muscle repair, Stem Cells Rev. 6, 512-522 (2010).
[30] K. Chen, D. Wang, W. T. Du, Z. B. Han, H. Ren, Y.Chi, S. G. Yang, D. L. Zhu, F. Bayard, and Z. C. Han, Human umbilical cord mesenchymal stem cells hUC-MSCs exert immunosuppressive activities through a PGE2-dependent mechanism, Clin. Immunol. 135, 448-458 (2010).
[31] K. H. Wu, B. Zhou, C. T. Yu, B. Cui, S. H. Lu, Z. C. Han, and Y. L. Liu, Therapeutic potential of human umbilical cord derived stem cells in a rat myocardial infarction model, Ann. Thorac. Surg. 83, 1491-1498 (2007).
[32] K. H. Wu, X. M. Mo, B. Zhou, S. H. Lu, S. G. Yang, Y. L. Liu, and Z. C. Han, Cardiac potential of stem cells from whole human umbilical cord tissue, J. Cell. Biochem. 107, 926-932 (2009).
[33] H. Y. Ren, Q. J. Zhao, T. Cheng, S. H. Lu, Z. Chen, L. Meng, X. F. Zhu, S. G. Yang, W. Xing, Y. D. Xiao, Q. Ren, Y. Chi, D. S. Gu, R. C. Yang, and Z. C. Han, No contribution of umbilical cord mesenchymal stromal cells to capillarization and venularization of hepatic sinusoids accompanied by hepatic differentiation in carbon tetrachloride-induced mouse liver fibrosis, Cytotherapy 12, 371-383 (2010).
[34] L. Zhao, E. F. Burguera, H. H. K. Xu, N. Amin, H. Ryou, and D. D. Arola, Fatigue and human umbilical cord stem cell seeding characteristics of calcium phosphate-chitosan-biodegradable fiber scaffolds, Biomaterials 31, 840-847 (2010).

[35] Y. Y. Liu, R. Mu, S. Y. Wang, L. Long, X. Liu, R. Li, J. Sun, J. P. Guo, X. P. Zhang, J. Guo, P. Yu, C. L. Li, X. Y. Liu, Z. Y. Huang, D. P. Wang, H. Li, Z. F. Gu, B. Liu, and Z. G. Li, Therapeutic potential of human umbilical cord mesenchymal stem cells in the treatment of rheumatoid arthritis, Arthritis Res. Ther. 12, R210 (2010).

[36] H. Bai, P. Chen, H. Fang, L. Lin, G. Q. Tang, G. G. Mu, W. Gong, Z. P. Liu, H. Wu, H. Zhao, and Z. C. Han, Detecting viability transitions of umbilical cord mesenchymal stem cells by Raman micro-spectroscopy, Laser Phys. Lett. 8, 78-84 (2011).

[37] S. I. Choi, C. S. Jeong, S. Y. Cho, and Y. S. Lee, Mechanism of apoptosis induced by apigenin in HepG2 human hepatoma cells: involvement of reactive oxygen species generated by NADPH oxidase, Arch. Pharm. Res. 30, 1328-1335 (2007).

[38] H. M. Shen, C. Y. Shi, Y. Shen, and C. N. Ong, Detection of elevated reactive oxygen species level in cultured rat hepatocytes treated with aflatoxin B1, Free Radic. Biol. Med. 21, 139-146 (1996).

[39] U. Bandyopadhyay, D. Das, and R. K. Banerjee, Reactive oxygen species: Oxidative damage and pathogenesis, Curr. Sci. 77, 658-666 (1999).

[40] D. Yin and K. Chen, The essential mechanisms of aging: Irreparable damage accumulation of biochemical side-reactions, Exp. Gerontol. 40, 455-465 (2005).

[41] B. D'Autréaux and M. B. Toledano, ROS as signalling molecules: mechanisms that generate specificity in ROS homeostasis, Nat. Rev. Mol. Cell Biol. 8, 813-824 (2007).

[42] A. Gomes, E. Fernandes, and J. L. F. C. Lima, Fluorescence probes used for detection of reactive oxygen species, J. Biochem. Biophys. Methods 65, 45-80 (2005).

[43] W. T. Chang, H. L. Lin, H. C. Chen, Y. M. Wu, W. J. Chen, Y. T. Lee, and L. Liau, Real-time molecular assessment on oxidative injury of single cells using Raman spectroscopy, J. Raman Spectrosc. 40, 1194-1199 (2009).

[44] D. M. Zhang, D. P. Jiang, M. Yanney, S. Zou, and A. Sygula, Ratiometric Raman spectroscopy for quantification of protein oxidative damage, Anal. Biochem. 391, 121-126 (2009).

[45] Y. M. Xu, Z. X. Zhou, H. Y. Yang, Y. Xu, and Z. Y. Zhang, Raman spectroscopic study of microcosmic photodamage of the space structure of DNA sensitized by Yangzhou haematoporphyrin derivative and Photofrin II, J. Photochem. Photobiol. B: Biol. 52, 30-34 (1999).

[46] G. L. Li and D. Z. Yin, Chem. Life 27, 516-520 (2007). (in Chinese)

[47] M. Dizdaroglu, P. Jaruga, M. Birincioglu, and H. Rodriguez, Free radical-induced damage to DNA: mechanisms and measurement, Free Radic. Biol. Med. 32, 1102-1115 (2002).

[48] Y. Liu, P. Chen, F. Zhang, L. Lin. G.-Q. Tang, and G.-G. Mu, Laser Phys. Lett. 6, 465 (2009).

33

Novel Platforms to Create Labeled Stem Cells

Andrew Fontes, Rene Quintanilla and Uma Lakshmipathy
Life Technologies, Primary and Stem Cell Systems
USA

1. Introduction

Stem cell research offers unlimited potential for applications in regenerative medicine. Much of the excitement surrounding this area of research comes from the cells inherent qualities; the ability to self renew and generate a variety of cells within the body. In the past two decades, research on stem cells has progressed exponentially providing a vast array of tools for a large breadth of areas. As research persists, there is a growing need for faster and more effective tools to analyze the many unknown processes that will unlock the power within these cells.

Delivery of genes to create labeled stem cells vastly enhances their utility as tools for basic research and drug screening but also as potential therapeutic agents. Gene transfer has been a routine method in stem cell research since scientists first began culturing stem cells *in vitro*. Genetic manipulation has accelerated research enabling the creation of in vitro models for drug discovery, use as a tool for dissecting basic stem cell biology and potential development of *in vivo* cell-based delivery strategies. Efficient gene transfer into stem cells is a critical step in the creation of engineered stem cells. Several viral and non-viral methods exist for the modification of stem cells, albeit with varying efficiencies (Colosimo et al., 2000). Each platform has its advantages and disadvantages that can be effectively utilized for specific applications.

Here, we review gene delivery, cloning, and modification methods and describe in detail three platforms that enable the user to create modified stem cells that transiently or stably express one or multiple genes of interest. The three platforms described here each offer a unique feature and advantage that may be ideal for a particular cell type or application. The first is a non-integrating insect virus that is ideal for the transient short term expression of genes of interest. Second is an episomal EBV based vector method that allows for populations of transgene expressing cells to be stably maintained without genomic insertion. The third platform is a site specific integrating platform that can be clonally selected and is not subject to silencing.

2. Gene delivery

Genes carried on plasmid DNA can be introduced into cells with one of the methods listed below. The choice of the method largely depends on the size of the DNA plasmid and target cell type. Working with stem cells provides a new set of challenges to gene delivery requiring substantial optimization. Human embryonic stem cells (hESC) are typically grown in tight compact colonies to maintain pluripotency thus requiring transfection methods that do not disrupt this. Apart from cell conditions, applications of the gene delivery can direct the method of choice for gene delivery. For cells with downstream clinical applications it is essential to use platforms that are "footprint-free" of transgene material. For overexpression of specific proteins or knockdown of pathways integrating virus may be ideal and required for high copy number insertion. When working with hard to transfect cell types, it is crucial to consider all options available.

2.1 Chemical methods

Chemical methods utilize lipids, polymers or proteins that form a complex with DNA. This condensed complex fuses with the cell membrane thus enabling the entry of the DNA into the cells.
Liposomes remain a popular chemical vehicle for gene delivery in many cell types. In stem cells, lipid based methods have been reported with varying efficiencies depending on the gene type and cell line (Strulovici et al., 2007). Products like Lipofectamine 2000 (Life Technologies) and FuGENE (Roche) are liposomes that utilize the characteristics of chemical compounds to form complexes with the DNA/RNA and fuse with the negatively charged cell membrane. Lipid methods show reduced toxicity as compared to $CaPO_4$ precipitation and earlier transfection protocols. Precipitation methods vary depending on pH and uniformity of the precipitate making them an unfavorable mechanism for stem cells. Synthetic polymers face similar hurdles. Cell toxicity of the reagent in the target cell type remains a limiting factor for chemical methods requiring extensive adaptations and optimization.

2.2 Physical methods

These methods involve delivery of plasmid across cell membrane using electroporation, sonoporation or particle bombardment. Electroporation utilizes electric pulses to transiently disrupt the cell membrane to create pores that allows delivery of charged RNA and DNA molecules into the targeted cells. This method is highly efficient and primary choice for delivery of large constructs into cells. In the recent years specialized electroporation based units have emerged. The Nucleofector™ from Lonza utilizes a combination of set programs with specific electric parameters and cell-type specific solutions to achieve high levels of transfection in several cell types including embryonic stem cells (Lakshmipathy et al., 2004). The Neon™ Transfection System from Life Technologies is an open system that allows optimization of electrical parameters and the flexibility of using cells from $2X10^4$ to $6X10^6$ per reaction. This technology avoids the use of standard electroporation cuvette and uses an electronic pipette tip as a transfection chamber (Kim et al., 2008). This method has been used to create labeled cells with diverse vector platforms in human embryonic stem cells (Thyagarajan et al., 2009; Liu et al., 2009). Sound waves or sonoporation has also been used to transiently create pores in the cell membrane to deliver nucleic acids into cells (Bao et al., 1997). Other mechanical methods include manual microinjection of gene into cells (King et al., 1994) or particle bombardment using gene guns (Guo et al., 1996) that are effective for certain cell types but harsh with the risk of damaging cells.

2.3 Viral methods

The use of viral vehicles for gene delivery in mammalian cells has become common practice due to the ability of virus to bypass cell membrane with ease. While viral systems retain their ability to infect the cells they have been modified to be replication incompetent. This

allows for the generation of a recombinant virus that cannot proliferate within the target cells. Viral systems infect by a variety of pathways reaching a variety of targets allowing the user to choose a virus that best suits the end need.
Unlike chemical gene transfer, some viral vectors provide a means to directly deliver the DNA to the nucleus and in some cases integrate to the genome. While this enhances the ability to create a stable cell line, it is also highly variable in copy number and genomic site of integration which may cause detrimental effects. Retrovirus and Lentivirus are integrating viral systems with the ability to infect a variety of cell types providing stable expression of the gene of interest. However, these systems have the ability to recombine with genetic material creating replication competent virus. Similarly, Adeno associated virus (AAV) provides a mechanism for genomic insertion with the added benefit of site specific integration. Conversely, adenovirus has the capability to provide a means of transient expression without genomic integration. Adenovirus and AAV are limited by genetic capacity of transgene size compared to other viral systems. (Giordano et al., 1999) Regardless of adaptations, viral systems remain unfavorable in cells for downstream clinical applications. This factor is due to the nature of virus to infect mammalian cells, thus limiting their use in cell therapy and other clinical models for stem cells.

3. Cloning strategies

Traditional cloning using restriction endonuclease and ligase procedures can be cumbersome for large and complex expression vectors. With advances in developmental biology there is a need for more intricate cloning systems that can handle customizable and high throughput screening through multiple platforms. The cloning procedures described below provide a faster and more efficient workflow for directional cloning for a variety of uses.

3.1 TOPO® TA cloning

The TOPO® TA technology is an efficient method for directionally recombining PCR (polymerase chain reaction) products into platform specific vectors. Rather than using ligase, the system utilizes vaccinia virus topoisomerase 1, an enzyme involved in the digestion and ligation of DNA during replication. TOPO® TA requires a modified plasmid equipped with the enzyme covalently bonded to a phosphate on a thymadine at the three prime end of the linearized vector. PCR products produced using Taq polymerases have three prime and five prime adenine overhangs. This overhang is utilized for the recombination with the linearized vector allowing for efficient directional recombination within five minutes at room temperature. TOPO® TA is limited by the necessity to create a TA adapted linearized vector stock for the TOPO® reactions to occur. TOPO® is not ideal for larger inserts showing decreasing efficiency with inserts larger than 5kb. The TOPO ® reaction is irreversible and the inserted gene cannot be cloned out by the enzyme. This limits the ability to use this gene for other platforms without repeating the PCR (Katzen et al., 2007).

3.2 MultiSite Gateway® cloning

Each of the cell modification platforms below has been outfitted with MultiSite Gateway® Technology. Otherwise known as Gateway®, this technology is an efficient and relatively easy way to clone multiple gene configurations into a variety of expression vectors. This improves the speed of the cloning process eliminating the need for restriction endonuclease cloning procedures. This system harnesses the power of the lambda integrase enzyme to recombine multiple fragments in an orientation specific manner with high accuracy.

The lambda integrase recognizes sequence specific recombination sites. Each gene must be amplified using Gateway® specific primers containing the proper recombination sites. As shown in Figure 1, the building process begins by cloning components of your cassette of interest into intermediate vectors. These vectors are recombined with the destination vector belonging to your platform of choice. The cloning steps are simple incubation reactions speeding up the overall process.

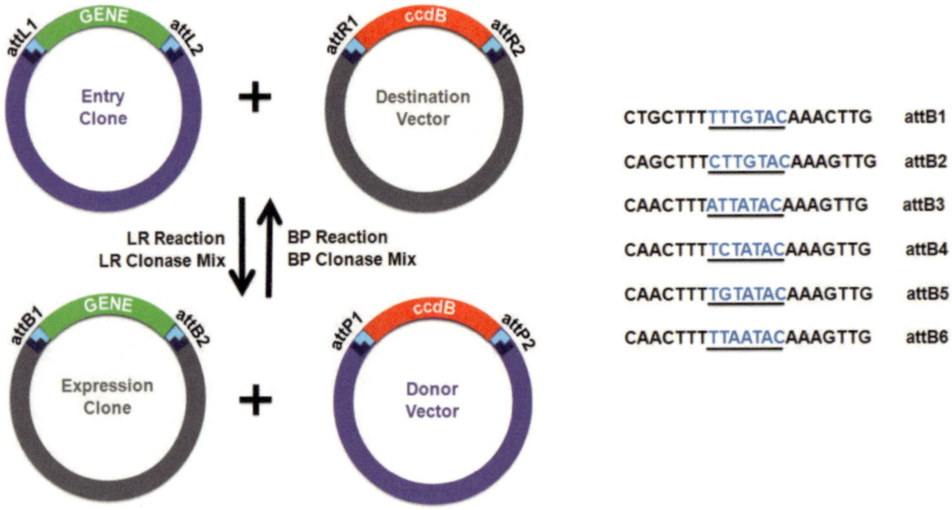

Fig. 1. Multisite Gateway® Cloning

MultiSite Gateway® Technology is a site specific recombinational cloning system where the different att specificities allows assembly of 2-4 gene fragments in a defined order and orientation. The difference in the att site specificity is determined at the single base level (underlined sequence) with virtually no cross recombination between the different sites. Here entry clones carrying the gene elements of choice are constructed into a donor vector via BP reaction to result in entry clones. Donor vectors with appropriate flanking att sites for single or multiple fragment cloning are remised with the right Destination vector and LR reaction performed to yield final expression vectors.

Outfitting a variety of platforms with Multisite Gateway® technology makes the transition between technologies a simple cloning reaction. Each expression cassette can be cloned into a variety of destination vectors expanding your ability to modify cells based on specific needs. Thus, the user can create a library of gene cassettes that can be readily used for a variety of applications. The dual arrows in Figure 1 indicate that Multisite Gateway® provides a reversible reaction (under different conditions) that retains the rapid cloning features of the TOPO® TA system. Using Multisite Gateway® we traverse a variety of cell modifying systems, based on need, with ease and speed.

4. Cell modification platforms

Besides efficient gene transfer into stem cells, a critical step is the choice of platform used to create labeled cells. Traditionally, naked plasmid DNA is used to create stable clones, but is less favored in stem cells due to varied expression levels depending on the context of its genomic integration. (Eiges et al., 2001) To overcome this issue, lentiviral methods that offer higher transduction rates and multiple integrations have been used to generate stable transgene expressing cells (Jang et al., 2006). A major limitation of lentiviral systems is the size or payload of DNA that can be packaged. Recently, methods that result in efficient random integration via transposases and site-specific genomic integration mediated via adeno-associated viruses (AAV), and site-specific integrases, have been reported as a means to modify stem cells (Surosky et al., 1997, Kowabata et al, 2006).

Site-specific integration offers an elegant method to generate clones with one copy of the gene at a specific genomic location that can be further screened to identify ideal sites that support sustained gene expression. This method require the isolation of clonal population of cells followed by rigorous characterization of the genomic integration site to ensure sustained transgene expression and is best suited for cell types that can be cultured for long periods of time. Embryonic stem cells and other cell lines, with unlimited proliferative potential, can be subjected to such methods to create engineered cells expressing the gene of interest. JumpIn™ platform utilizes phiC31 and R4 integrases to target pseudo sites in the mammalian genome that are known to support higher and longer-term expression of inserted transgenes. This method requires prolonged culture and is manipulation and ideal to create labeled pluripotent stem cells for sustained gene expression both in pluripotent state as well as differentiation.

EBV-based vectors support episomal maintenance of large genomic fragments and hence are an appealing alternative to rapidly generate labeled cells. Since the plasmid is not integrated into the host genome, expression of the transgene is relatively free from chromosomal effects associated with genomic integration. This method requires drug selection but not clonal isolation. Pooled clones can therefore be rapidly generated and suited for expression of genes or knocked down in the undifferentiated state. Since there is a risk of loss of transgene/plasmid in the absence of selective pressure, labeled pluripotent stem cells will result in attenuated transgene expression with long term differentiation.

Both the methods highlighted above require transfection of plasmid into cells and prolonged manipulation and culture, a feature not amenable to most primary and adult stem cell types due to their limited proliferation. BacMam offers an easy and fast method to deliver transgenes into a wide spectrum of stem cells with the least toxicity. The introduced transgene is diluted with passage and therefore ideal for transient expression to create assay-ready cells. Figure 2 displays a comparison of transient (BacMam), stable non-integrating (EBV-Vector), and integrating platforms(Jump-In™).

The three platforms discussed here are; Episomal vectors that results in stable retention of the plasmid (green circles) without integration into the genome; JumpIn™ platform that allows site-specific insertion of the gene (shown in green) into the host genome for stable expression and; BacMam, a viral method where the transgene does not integrate into the host genome (Green circles) but dilutes out as cell divides and hence transient.

Fig. 2. Labeling Platforms

4.1 BacMam

Transient expression systems are ideal for rapidly creating labeled cells for immediate utilization in downstream applications such as cell tracking, identification, enrichment, drug screening and other types of high throughput assays. Traditionally lipid-based methods have been used to introduce DNA carrying the gene of interest. This method, however, is not amenable to certain cell types, especially primary and stem cells that are generally hard to transfect. Baculoviruses are known to enable high efficiency labeling of hard to label cells with minimal toxicity (Zeng et al., 2009, Ho et al.,2005). These viruses are non-replicating in human cells and are stable at 4°C and thus easy to generate and use.

The BacMam platform, named for the ability of Baculovirus to transfect Mammalian cell types, is a double stranded DNA virus capable of infecting over 500 insect species. The DNA is packaged in a rod-shaped particle 40-50 nm in diameter and 200-400 nm in length. Gp64, a major glycoprotein on the envelope, is proposed to play a key role in virus attachment and entry into mammalian cells. The viral particles are endocytoced and are released into the cytoplasm before migrating to the nucleus.

BacMam is non-replicating in mammalian cells without any additional modifications. This is contrast to other viral methods such as lenti, adeno, or retro that integrate into the host genome and requires inactivation of key attributes of the virus to prevent unwanted replication within cells. In addition, integrating viral constructs also have the chance of recombining with endogenous sequences in the human genome resulting in instability. The ease of use, high efficiency of labeling and relatively low toxicity renders BacMam an ideal choice for primary, stem and progenitor cells. In addition, their ability to carry high load capacity reaching upwards of 30kb sequences adds to its appeal for simultaneous delivery of single or multiple genes of interest.

The most commonly used method to generate BacMam virus is the Bac-to-Bac® method where the gene of interest is first cloned into a transfer vector. The cloning and production process are shown in Figure 3A. This transfer vector could be created by either by Restriction endonuclease mediated cloning (pFastBac™) or MultiSite Gateway® adapted (pDest). Once the expression vector is constructed, it is transformed into a modified E coli that contains a baculovirus genome. Recombination of the transfer vector with the baculovirus genome results in BacMid which is ready for transfection into insect cells for virus production. BacMid is transfected into Sf21 (Spodoptera frugiperda) insect cells. The production is limited by variability of the virus which results in testing a variety of clones. Despite the selection of multiples clones, there is a necessity for plaque purification of the virus to ensure a clone is selected with the expression cassette.

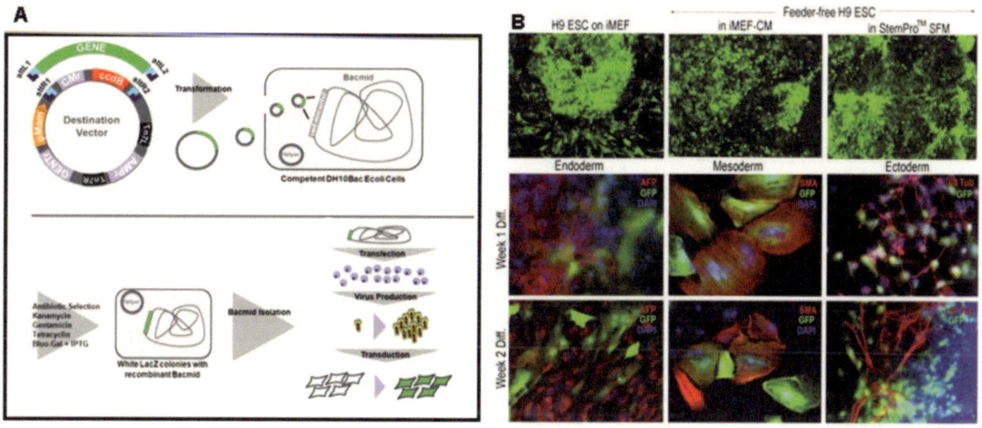

Fig. 3. BacMam

(A) Gene of interest is cloned into the baculovirus transfer vector using MultiSite Gateway® cloning. The resulting expression vector is transformed into DH10Bac™ to generate BacMid via site-specific transposition of the expression elements on the transfer vector to the baculovirus genome present in DH10Bac cells. Transfection of BacMid DNA into Sf9 or Sf21 insect cells results in BacMam virus carrying the expression cassette that can be further amplified in insect cells to generate high titer stocks. (B) BacMam carrying the expression cassette CMV-GFP can be transduced into undifferentiated H9 hESC either on feeders or feeder-free conditions. Cells that are induced to differentiate for 1-2 weeks can also be transduced by BacMam and GFP labeled cells colocalize with lineage specific markers.

The ability to transduce a variety of cells types makes BacMam an ideal system for expression in partially or terminally differentiated cells derived from embryonic stem cells (Figure 3B). These cells have a limited proliferation preventing their use with selectable markers that require clone establishment. Transduction of differentiating ESC with GFP BacMam and overlap with differentiation markers has been demonstrated.

BacMam provides an easy to clone, easy to use, non-integrating transfection system. However, this system may not be ideal for long term gene expression in dividing cell types. The inability to replicate may be a limitation of the technology since the signal will dilute out in rapidly dividing cells. Additionally, for downstream clinical applications it may undesirable to use a viral system despite its inability to infect mammalian cells.

4.2 Episomal vectors

With the ability of stem cells to divide indefinitely it is essential to have a platform that can express long term. Generating labeled stem cell lines provides a valuable tool for research. However the creation of stable lines is complicated by stability and copy number of the loci. Technologies for this purpose have been reviewed using viral, homologous recombination and integrase specific integrations (Yates et al., 2006). Plasmid DNA alone has been utilized to randomly integrate for a fast way to create a labeled line. Lentivirus works efficiently due to multiple copy insertion into the genome. However, this can have negative effects. Thus, there is a need for a stable line that does not have any genomic alterations or silencing issues. To overcome these problems, an episomally maintained DNA vector for stably expressing stem cell lines is used.

The Epstein Bar Virus (EBV), a member of the herpes virus family, is known to infect multiple cells types and remains one the of the most common virus in humans. EBV based episomal vectors have been successfully used to stably express gene of interest in multiple types of cells both in vitro and in vivo since 1985 (Yates et al., 1985). The crucial components of the viral genome that have been added to vector systems are the latent origin of replication (OriP) and the EBV Nuclear Antigen-1 region (EBNA). These elements are required in combination for stable maintenance of episomal plasmid DNA. The trans acting element, EBNA, cannot function without the cis acting element, OriP. Additionally, the EBNA-OriP combination is only functional in human, primate and canine cells. Reproduction of extrachromosomal replication using the EBNA1 has also been reported in murine models, however, this requires the additional expression of transgenes (Habel et al., 2004).

In hESC these vectors have had relative levels of success using a two step system first creating a stable EBNA1 expressing cell line and transfecting a OriP vector (Ren et al., 2006). This however does not mitigate the problem of silencing from genomic integration. The EBNA oriP destination vector (Figure 4A) is a one step vector system that includes both elements eliminating the need for an EBNA cell line step. The ability to express high levels of transgenes without genomic integration provides a useful system for downstream clinical applications. Recently, episomal vector systems using the EBNA OriP elements have been reported for the generation of induced pluripotent stem cells (Yu et al., 2009). Reprogramming by the forced expression of multiple transcription factors is an ideal example of the power of the episomal vector outfitted with MultiSite Gateway® technology. Multiple configurations can be created in a high throughput manner allowing for the adaptations of expression, stoichemetry and exchange of alternate genes.

Episomal vectors replicate once per cell cycle with activation of replication by binding of multiple EBNA1 homodimers to OriP. Similar to BacMam, this system offers an appealing alternative to integrating technologies since they are relatively free from chromosomal effects associated with genomic integration. These vectors, being rather larger in size (over 10 kb) can be transfected into embryonic and adult stem cells such as MSC albeit with lower efficiencies. Since the distribution of episomal vectors with cell division is unequal between the daughter cells, it is best to keep cells under selective pressure for sustained gene expression. It is also not necessary to generate individual clones; drug resistant colonies obtained can be pooled to rapidly create labeled cells for downstream applications. Episomal vectors have been used to generate stable hESC pooled clones that maintain transgene expression for long periods of time. Genes driven by constitutive or lineage specific promoters have been shown to express in a context specific matter (Figure 4B). In addition, majority of the cells retain transgene expression when induced to differentiate in the absence of selective pressure for 3 weeks (Thyagarajan et al., 2009).

Episomal vectors provide an easy platform to rapidly create stable cells expressing transgene of interest. A major limitation of this method is the requirement for efficient transfection method and pooled clones that have varying copy numbers per cell thereby creating a heterogeneity in expression levels that may be undesirable for certain applications.

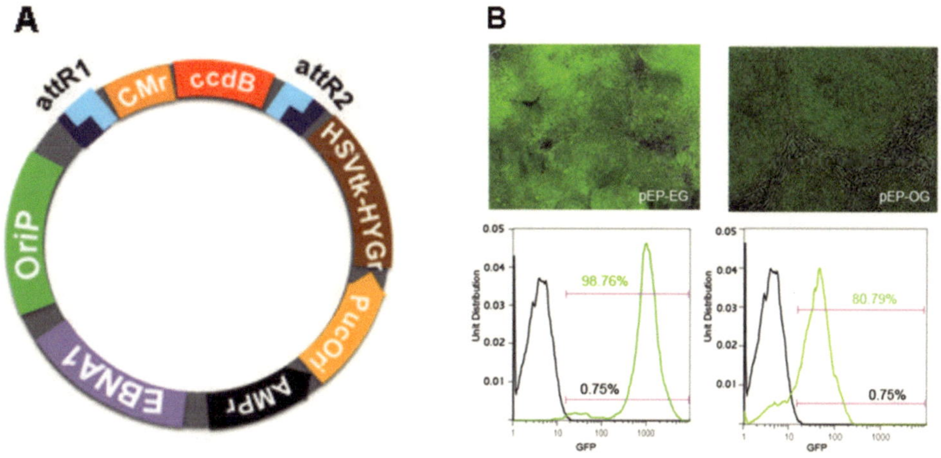

Fig. 4. Episomal Vectors.

(A) Epstein Barr virus based vectors contain Epstein-Barr virus nuclear antigen (EBNA1) and the latent origin of replication OriP that help in episomal maintenance of the vectors by facilitating the replication of the vectors once per cell cycle by binding of the multiple EBNA1 homodimers to OriP. Mutlisite Gateway® adaption of these vectors allow rapid assembly of single or multiple gene elements and a Hygromycin resistance gene provides a means to maintain selective pressure on cells carrying these vectors. (B) Human ESC can be transfected with episomal vectors to create stable pooled clones expressing GFP or gene of interest driven by constitutive promoters such as Ef1a or lineage specific promoter such as the pluripotent cell specific Oct4.

4.3 Jump-In™ technology

As research uncovers the potential of stem cell research there are many applications that require a clonal population expressing a custom cassette. Transgene cassettes can vary from lineage reporter lines, to multicistronic elements or even systems useful for higher content analysis. These complex tools will require single copy integration of the expression cassette with resistance to silencing. Single copy cell lines have been reported in human embryonic stem cells using homologous recombination, cre/lox mediated integration and zinc finger site directed integration (Costa et al., 2007, Lombardo et al., 2007, Soldner et al., 2009). Homologous recombination can lead to random integration as well as silencing depending on the locus. Unlike homologous recombination, non-viral integrases have the potential to recombine two non-identical sites. Zinc finger has been proven as an accurate method, however, require designing sequence specific zinc finger for recombination (Lombardo et al., 2007).

The Jump In system uses non-viral integrase-mediated site-specific integration to develop a platform stem cell line. The two integrases phiC31 and R4 are both members of the Streptomyces family with the potential to integrate transgenes to native attP (attachment P site) sites that are present in the bacterial genome or pseudo attP sites that are present in other species such as human, rat and mouse. Although there are a variety of pseudo attP sites within the human genome, certain hot spots have high affinity for recombination. This results in a very low background of non site specific targeting. A pseudo attP hot spot on chromosome 13 has provided an ideal location due to the known intronic areas of the chromosome. Many other gene delivery methods are affected by chromatin remodeling during differentiation. The chromosome 13 loci has been shown to remain active throughout differentiation to multiple lineages remaining resistant to silencing. The system has a much higher efficiency for large complex fragments.

Delivered via electroporation, the vectors introduce large complex gene cassettes to a specific genomic locus validated for efficiency, stability, transcriptional activity and resistance to silencing (Liu et al., 2009). The phiC31 integrase is employed to insert a native R4 attP site into the genome that can be accessed for future recombination. R4 integrase mediated integration occurs at the new R4 attP site activating the expression of a selection marker. Figure 5 shows stepwise creation of the target line and subsequent retargeting that results in activation of the Zeomycin gene. Stable clones can be selected and expanded for further validation.

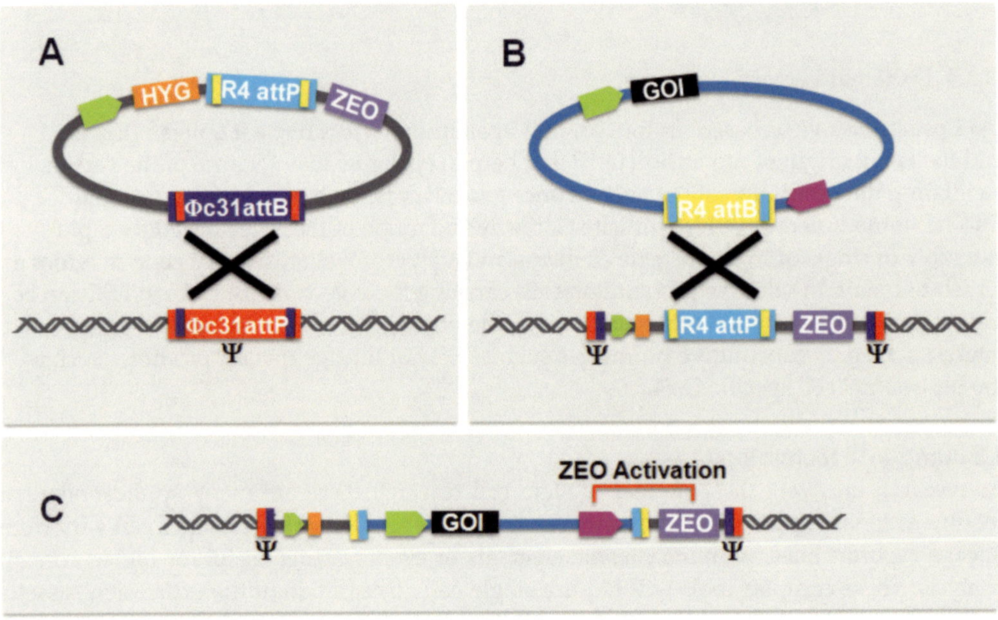

Fig. 5. Site Specific Integration

(A) A Target plasmid containing the wild type R4 integrase recognition (attP) site carrying a constitutive expressed antibiotic gene with a PhiC31 integrase attB site is transfected into cells along with a plasmid expressing phiC31 integrase. The enzyme catalyzes the integration of the target plasmid into "Pseudo attP sites" within the host cellular genome. Also contained on the Target plasmid is a promoterless antibiotic gene. The constitutive expressed antibiotic is used to select stable drug resistant clones and screened for single copy and genomic site of integration. (B) Clones carrying a single copy of the target plasmid in a desired genomic locus are retargeted. A retargeting plasmid carrying the promoter-gene expression cassette, a promoter adjacent to a R4 attB site and a plasmid expressing R4 integrase is transfected. (c) R4 integrase mediated retargeting will result in positioning of the promoter adjacent to the drug resistance gene placed in the original Target cells to activate the antibiotic gene which can then be utilized to generate drug resistant clones.

With single copy integration there is less variability between copy number and expression patterns if trying to express multiple genes. The technology has been reported for potential lineage tracking by expressing GFP from an Ef1alpha promoter throughout differentiation without silencing.

Although this platform has proven to be an accurate and efficient way to modify stem cells there are still improvements that can be made. The technology still shows a tendency to randomly integrate plasmid sequences leaving the need to verify copy number and integration sites via molecular techniques.

5. Conclusion

Gene modification provides researchers with tools to accurately dissect cellular function. The techniques described offer a toolbox with interchangeable components to create specific assays for specific applications. Utility of the various available platforms depends on several factors. First is to determine the desired length of expression needed for the desired application. Second, it is important that the cell type to be modified is amenable to gene delivery methods either via transfection or via specific viral methods. Third, the proliferative capability of cells that are being modified is important to determine if these cells can sustain long periods of manipulation and drug selection and if the resulting modified cells will retain their original stem cell characteristics.

The three platforms summarized in this chapter each have their advantages and disadvantages. The most notable difference between the three is the length of time and amount of manipulation required for generation of the labeled cells (Table 1). Starting from a vial of frozen pluripotent stem cells, 2-3 week is needed to recover and expand enough cells, typically atleast 6-10 million cells for the first step. For transient expression using BacMam, cells are ready for use 24-48 h post transduction. With EBV-episomal vector, cells after transfection are subjected to drug selection and resulting pooled clones to be characterized. This entire process of generating stable clones is approximately 8 weeks long. The Jump-In™ platform that requires two rounds of transfection, drug selection and subsequent characterization requires approximately 15 weeks to obtain modified stem cells.

Choice of platform to use is not merely based on ease or length of creating labeled cells. It is important to understand the advantages and disadvantages of each method so that features ideal for the application of interest is used to create modified stem cells (Table 2). BacMam enables efficient delivery into diverse cell types and is the fastest method to create labeled cells. However the expression is transient and since workflow does not involve drug selection resulting cells will have variable expression depending on the number of copies they carry. This method is therefore ideal for delivery of target to create assay ready cells for experiments that can be completed within 24-48 h after creation of the labeled cells. This method however is not suitable for expressing genes and monitoring its long term effect on the cells. Episomal vectors offer the advantage of creating stable pooled clones in a relatively short period of time. Disadvantages of this method are that transfection into hard to transfect cells remains an issue and since copy number is variable expression can be heterogeneous and expression level may be attenuated with long term differentiation. Finally Jump-In™ system is ideal for creation of stable pluripotent stem cells for sustained expression both in its undifferentiated state and with differentiation. The ability of this system to create clones with a single copy of the gene at a known genomic locus makes this an ideal choice for use of this line as a Platform line into which diverse targets can be inserted for comparison all from the same genomic context.

Recently efforts are focused on creating hybrid systems to generate novel engineering method which overcome some of the limitation for the system. EBV-elements EBNA and OriP cloned into BacMam transfer vectors to generate EBNA BacMam has easy delivery but also sustained gene expression. The ultimate goal is to create labeled cells that can be used for the end application.

BACMAM	EBV-vector	JUMP-IN™	STEP	Duration #Days	Total Duration #Weeks
			Thaw and Culture cells	14	2
			Feeder-free expansion	7	3
			Transfect and recover	2	3.2
			Drug select	14	5.2
			Expansion	14	7.2
			Characterize	7	8.2
			Feeder-free expansion	7	9.2
			Retarget and recover	3	10
			Drug select	14	12
			Clonal Expansion	14	14
			Characterize	7	15

Table 1. Steps and Duration of the Cell Engineering Methods
The length in time to generate labeled cells varies between the three methods (highlighted in the color bars)

	PROS	CONS	POTENTIAL APPLICATION
BACMAM	1) Ease of use 2) Efficient infection 3) Broad host range	1) Transient expression 2) heterogeneous expression 3) long viral generation process	Assay ready cells
EPISOMAL VECTORS	1) Rapid engineering 2) Stable expression 3) no genomic integration	1) need for optimal transfection 2) Heterogeneous expression 3) Variable silencing	Expression of genes and RNAi
JUMP-IN™	1) Stable integration 2) Homogeneous expression 3) defined locus, sustained expression	1) need for optimal transfection 2) clonal selection and screening 3) integrational method	Developmental reporters

Table 2. Summary of cell engineering methods
Advantages and disadvantages of each method and potential application of each method

6. Acknowledgement

We thank Stem Cell team at Life Technologies for discussions and suggestions.

7. References

Bao, S., Thrall, BD., & Miller, DL. (1997) Transfection of a reporter plamsid into culture cells by sonoporation in vitro. *Ultrasound Med Biol*, Vol. 23 No. 6, pp. 953-959

Colosimo, A., Goncz, KK., Holmes, AR., Kunzelmann, K., Novelli, G., Malone, RW., Bennett, MJ., & Gruenert, DC. (2000) Transfer and expression of foreign genes in mammalian cells. *Biotechnqiues,* Vol. 29 No. 2, pp. 314-318

Costa, M., Dottori, M., Sourris, K., Jamshidi, P., Hatzistavrou, T., Davis, R., Azzola, L., Jackson, S., Lim, SM., Pera, M., Elefanty, A., & Stanley, EG. (2007) A method for genetic modification of human embryonic stem cells using electroporation. *Nature Protocols* Vol 2, pp. 792-796

Eiges, R., Shuldiner, M., Drukker, M., Yanuka, M., Itskovitz-Elder, J., & Benvenisty, N. (2001) Establishment of a human embryonic stem cell-transfected clones carrying a marker for undifferentiated cells. *Current Biology* Vol. 11 No. 7, pp. 514-518

Guo, Z., Yang, NS., Jiao, S., Sun, J., Cheng, L., Wolff, JA., Duncan, ID. (1996) Efficient and sustained transgene expression in mature rat oligodendrocytes in primary culture. *J Neurosci Res.* Vol 43, No. 1, pp. 32-41

Habel, M., Drouin, M., & Jung., Daniel. (2004) Maintenance of Epstein-Barr virus-derived episomal vectors in the murine Sp2/0 myeloma cell line is dependant upon exogenous expression of human EBP2. *Biochemistry and Cell Biology*, Vol. 82, pp. 375-380

Ho, YC., Chung, YC., Hwang, SM., Wang, KC. & Hu, YC. (2005). Transgene expression and differentiation of baculovirus-transduced human mesenchymal stem cells. *J Gene Med* Vol. 7, pp. 860–868

Jang, JE., Shaw, K., Yu, XJ., Petersen, D., Pepper, K., Lutzko, C., & Kohn, DB. (2006) Specific and stable gene transfer to human embryonic stem cells using pseudotyped lentiviral vectors. *Stem Cells Dev*, Vol 15, No 1., Feb 2006, pp. 109-117

Kawabata, K., Sakurai F., Koizumi, N., Hayakawa, T., & Mizuguchi, H. (2006) Adenovirus vector-mediated gene transfer into stem cells. *Mol Pharm*, Vol. 3, No. 2, pp. 95-103

Kim, JA., Cho, K., Shin, & MS. (2008) A novel electroporation method using a capillary and wire-type electrode. *Biosens Bioelectron* Vol. 23, No. 9, pp. 1353–1360

King, R. (1994) Gene Delivery to Mammalian cells by microinjection, In: *Methods in Molecular Biol.* WC Heiser, Vol. 1, pp. 245, Human Press Inc., Totowa, NJ

Lakshmipathy, U., Pelacho, B., Sudo, K., Linehan, JL., Coucouvanis, E., Kaufman DS., & Verfaillie, CM. (2004) Efficient transfection of embryonic and adult stem cells. *Stem Cells* Vol. 22, pp. 531-543

Liu, Y., Thyagarajan, B., Lakshmipathy, U., Xue, H., Lieu, P., Fontes, A., MacArthur, CC., Scheyhing, K., Rao, MS., Chesnut, JD. (2009) Generation of a platform human embryonic stem cell line that allows efficient targeting at predetermined genomic location. *Stem Cells Dev.* Vol 10, Dec 2009, pp. 1459-1472

Ren, C., Zhao, M., Yang, X., Li, D., Jiang X., Wang L., Shan, W., Yang, H., Zhou, L., Zhou, W., & Zhang, H. (2006) Establishment and Applications of Epstein-Barr Virus-Based Episomal Vectors in Human Embryonic Stem Cells. *Stem Cells*, Vol. 24, pp. 1338-1347

Romano, G., Pacilio, C., & Giordano, A. (1999) Gene Transfer Technology in Therapy: Current Applications and Future Goals. *Stem Cells*, Vol. 17, pp. 191-202

Strulovici, Y., Leopold, P., O'Connor, T., Pergolizzi, R., & Crysal, R. (2007) Human Embryonic Stem Cells and Gene Therapy. *Molecular Therapy*, Vol. 15, No. 5, pp. 850-866

Surosky, RT., Urabe, M., Godwin, SG., McQuison, SA., Kurtzman, GJ., Ozawa, K., & Natsoulis, G. (1997) Adeno- Associated Virus Rep proteins Target DNA Sequences to a Unique Locus in the Human Genome. Journal of Virology, Vol 71, No 10, pp. 7951-7959

Thyagarajan, B., Olivares, EC., & Hollis, RP. et al. (2001) Site-specific genomic integration in mammalian cells mediated by phage phiC31 integrase. *Mol Cell Biol*, Vol.21, pp. 3926–3934

Yates, JL., Warren, N., & Sugden, B. (1985) Stable replication of plasmids derived from Epstein-Barr virus in various mammalian cells. *Nature*, Vol. 313, pp. 812-815

Yu, J., Hu, K., Smuga-Otto, K., Tian, S., Stewart, R., Slukvin, II., & Thomson, JA. (2009) Human Induced Pluripotent Stem Cells Free of Vector and Transgene Sequences. *Science*, Vol. 324, No. 5928, May 2009, pp. 797-801.

Zeng, J., Du, J., Lin, J., Bak, XY., Wu, C., & Wang, S. (2009) High-efficiency Transient Transduction of Human Embryonic Stem Cell-derived Neurons With Baculoviral Vectors. *Mol Therapy* Vol. 17